W9-AYE-454

AutoCAD 2009 Tutor for Engineering Graphics

AutoCAD 2009 Tutor for Engineering Graphics

ALAN J. KALAMEJA

Autodesk

Alumni Library
Wentworth Institute of Technology

DELMAR
CENGAGE Learning™

Australia • Brazil • Japan • Korea • Mexico • Singapore • Spain • United Kingdom • United States

620.004202855369
.A98 Kal
2009

905173

DELMAR
CENGAGE Learning™

AutoCAD 2009 Tutor for Engineering Graphics
Alan J. Kalameja

Vice President, Career and Professional Editorial: David Garza

Director of Learning Solutions: Sandy Clark

Acquisitions Editor: John Fedor

Managing Editor: Larry Main

Senior Product Manager: John Fisher

Senior Editorial Assistant: Dawn Daugherty

Vice President, Career and Professional Marketing: Jennifer McAvey

Marketing Director: Deborah S. Yarnell

Marketing Manager: Jimmy Stephens

Marketing Specialist: Mark Pierro

Production Director: Wendy Troeger

Production Manager: Stacy Masucci

Content Project Manager: Angela Iula

Art Director: David Arsenault

Technology Project Manager: Christopher Catalina

Production Technology Analyst: Thomas Stover

Alumni Library
Wentworth Institute of Technology

Printed in the United States of America

1 2 3 4 5 XX 10 09 08

© 2009 Delmar, Cengage Learning

ALL RIGHTS RESERVED. No part of this work covered by the copyright herein may be reproduced, transmitted, stored or used in any form or by any means graphic, electronic, or mechanical, including but not limited to photocopying, recording, scanning, digitizing, taping, Web distribution, information networks, or information storage and retrieval systems, except as permitted under Section 107 or 108 of the 1976 United States Copyright Act, without the prior written permission of the publisher.

For product information and technology assistance, contact us at
Professional & Career Group Customer Support, 1-800-658-7450

For permission to use material from this text or product,submit all requests online at **cengage.com/permissions.**
Further permission questions can be emailed to
permissionrequest@cengage.com.

Library of Congress Control Number: 2008905992

ISBN-13: 978-1-4354-0256-0

ISBN-10: 1-4354-0256-1

Delmar Cengage Learning
5 Maxwell Drive
Clifton Park, NY 12065-2919
USA

Cengage Learning is a leading provider of customized learning solutions with office locations around the globe, including Singapore, the United Kingdom, Australia, Mexico,Brazil, and Japan. Locate your local office at:
international.cengage.com/region

Cengage Learning products are represented in Canada by Nelson Education, Ltd.

For your lifelong learning solutions, visit **delmar.cengage.com**

Visit our corporate website at **cengage.com**.

NOTICE TO THE READER

Publisher does not warrant or guarantee any of the products described herein or perform any independent analysis in connection with any of the product information contained herein. Publisher does not assume, and expressly disclaims, any obligation to obtain and include information other than that provided to it by the manufacturer. The reader is expressly warned to consider and adopt all safety precautions that might be indicated by the activities herein and to avoid all potential hazards. By following the instructions contain herein, the reader willingly assumes all risks in connection with such instructions. The publisher makes no representation or warranties of any kind, including but not limited to, the warranties of fitness for particular purpose or merchantability, nor are any such representations implied with respect to the material set forth herein, and the publisher takes no responsibility with respect to such material. The publisher shall not be liable for any special, consequential, or exemplary damages resulting, in whole or part, from the readers' use of, or reliance upon, this material.

CONTENTS

INTRODUCTION

Engineering graphics is the process of defining an object graphically before it is constructed and used by consumers. Previously, this process for producing a drawing involved the use of drawing aids such as pencils, ink pens, triangles, T-squares, and so forth to place an idea on paper before making changes and producing blue-line prints for distribution. The basic principles and concepts of producing engineering drawings have not changed, even when the computer is used as a tool.

This text uses the basics of engineering graphics to produce 2D drawings and 3D computer models using AutoCAD and a series of tutorial exercises that follow each chapter. Following the tutorials in most chapters, problems are provided to enhance your skills in producing engineering drawings. A brief description of each chapter follows:

CHAPTER 1 – GETTING STARTED WITH AUTOCAD

This first chapter introduces you to the following fundamental AutoCAD concepts: Screen elements and workspaces; use of function keys; opening an existing drawing file; using Dynamic Input for feedback when accessing AutoCAD commands; basic drawing techniques using the LINE, CIRCLE, and PLINE commands; understanding absolute, relative, and polar coordinates; using the Direct Distance mode for drawing lines; using all Object snap modes, and polar and object tracking techniques; using the ERASE command; and saving a drawing. Drawing tutorials follow at the end of this chapter.

CHAPTER 2 – DRAWING SETUP AND ORGANIZATION

This chapter introduces the concept of drawing in real-world units through the setting of drawing units and limits. The importance of organizing a drawing through layers is also discussed through the use of the Layer Properties Manager palette Color, linetype, and lineweight are assigned to layers and applied to drawing objects. Advanced Layer tools such as isolating, filtering, and states, and how to create template files are also discussed in this chapter.

CHAPTER 3 – AUTOCAD DISPLAY AND BASIC SELECTION OPERATIONS

This chapter discusses the ability to magnify a drawing using numerous options of the ZOOM command. The PAN command is also discussed as a means of staying in a zoomed view and moving the display to a new location. Productive uses of real-time zooms and pans along with the effects a wheel mouse has on ZOOM and

PAN are included. All object selection set modes are discussed, such as Window, Crossing, Fence, and All, to name a few. Finally, this chapter discusses the ability to save the image of your display and retrieve the saved image later through the View Manager dialog box.

CHAPTER 4 – MODIFYING YOUR DRAWINGS

This chapter is organized into two parts. The first part covers basic modification commands and includes the following: MOVE, COPY, SCALE, ROTATE, OFFSET, FILLET, CHAMFER, TRIM, EXTEND, and BREAK. The second part covers advanced methods of modifying drawings and includes ARRAY, MIRROR, STRETCH, PEDIT, EXPLODE, LENGTHEN, JOIN, UNDO, and REDO. Tutorial exercises follow at the end of this chapter as a means of reinforcing these important tools used in AutoCAD.

CHAPTER 5 – PERFORMING GEOMETRIC CONSTRUCTIONS

This chapter discusses how AutoCAD commands are used for constructing geometric shapes. The following drawing-related commands are included in this chapter: ARC, DONUT, ELLIPSE, POINT, POLYGON, RAY, RECTANG, SPLINE, and XLINE. Tutorial exercises are provided at the end of this chapter.

CHAPTER 6 – WORKING WITH TEXT, FIELDS, AND TABLES

Use this chapter for placing text in your drawing. Various techniques for accomplishing this task include the use of the MTEXT and DTEXT commands. The creation of text styles and the ability to edit text once it is placed in a drawing are also included. A method of creating intelligent text, called Fields, is discussed in this chapter. Creating tables, table styles, and performing summations on tables are also covered here. Tutorial exercises are included at the end of this chapter.

CHAPTER 7 – OBJECT GRIPS AND CHANGING THE PROPERTIES OF OBJECTS

The topic of grips and how they are used to enhance the modification of a drawing is presented. The ability to modify objects through Quick Properties and the Properties Palette is discussed in great detail. A tutorial exercise is included at the end of this chapter to reinforce the importance of changing the properties of objects.

CHAPTER 8 – SHAPE DESCRIPTION/MULTIVIEW PROJECTION

Describing shapes and producing multiview drawings using AutoCAD are the focus of this chapter. The basics of shape description are discussed, along with proper use of linetypes, fillets, rounds, chamfers, and runouts. Tutorial exercises are available at the end of this chapter and outline the steps used for creating a multiview drawing using AutoCAD.

CHAPTER 9 – CREATING SECTION VIEWS

The basics of section views are described in this chapter, including full, half, assembly, aligned, offset, broken, revolved, removed, and isometric sections. Hatching techniques through the use of the Hatch and Gradient dialog box are also discussed. The ability to apply a gradient hatch pattern is also discussed. Tutorial exercises that deal with the topic of section views follow at the end of the chapter.

CHAPTER 10 – CREATING AUXILIARY VIEWS

Producing auxiliary views using AutoCAD is discussed in this chapter. Items discussed include rotating the snap at an angle to project lines of sight perpendicular to a surface to be used in the preparation of the auxiliary view. A tutorial exercise is provided in this chapter.

CHAPTER 11 – ADDING DIMENSIONS TO YOUR DRAWING

Basic dimensioning rules are introduced in the beginning of this chapter before all AutoCAD dimensioning commands are discussed. Placing linear, diameter, and radius dimensions is then demonstrated. The powerful QDIM command allows you to place baseline, continuous, and other dimension groups in a single operation. A tutorial exercise is provided at the end of this chapter.

CHAPTER 12 – MANAGING DIMENSION STYLES

A thorough discussion of the use of the Dimension Styles Manager dialog box is included in this chapter. The ability to create, modify, and override dimension styles is discussed. A detailed tutorial exercise is provided at the end of this chapter.

CHAPTER 13 – ANALYZING 2D DRAWINGS

This chapter provides information on analyzing a drawing for accuracy purposes. The AREA, ID, LIST, and DIST commands are discussed in detail, in addition to how they are used to determine the accuracy of various objects in a drawing. A tutorial exercise follows that allows users to test their drawing accuracy.

CHAPTER 14 – WORKING WITH DRAWING LAYOUTS

This chapter deals with the creation of layouts before a drawing is plotted out. A layout takes the form of a sheet of paper and is referred to as Paper Space. A wizard to assist in the creation of layouts is also discussed. Once a layout of an object is created, scaling through the Viewports toolbar is discussed. The creation of numerous layouts for the same drawing is also introduced, including a means of freezing layers only in certain layouts. Quick View Layouts and Quick View Drawings are also shown in this chapter as a means of better managing layouts and opened drawing. Various exercises are provided throughout this chapter to reinforce the importance of layouts.

CHAPTER 15 – PLOTTING YOUR DRAWINGS

Printing or plotting your drawings out is discussed in this chapter through a series of tutorial exercises. One tutorial demonstrates the use of the Add-A-Plotter wizard to configure a new plotter. Plotting from a layout is discussed through a tutorial. This includes the assignment of a sheet size. Tutorial exercises are also provided to create a color-dependent plot style. Plot styles allow you to control the appearance of your plot. Other tutorial exercises available in this chapter include plotting to the web.

CHAPTER 16 – WORKING WITH BLOCKS

This chapter covers the topic of creating blocks in AutoCAD. Creating local and global blocks such as doors, windows, and pipe symbols will be demonstrated. The Insert dialog box is discussed as a means of inserting blocks into drawings. The chapter continues by explaining the many uses of the DesignCenter. This feature allows the user to display a browser containing blocks, layers, and other named objects that can be dragged into the current drawing file. The use of tool palettes is also discussed as a means of dragging and dropping blocks and hatch patterns into your drawing. This chapter also discusses the ability to open numerous drawings through the Multiple Document Environment and transfer objects and properties between drawings. The creation of dynamic blocks, an advanced form of manipulating blocks, is also discussed, with numerous examples to try out. A tutorial exercise can be found at the end of this chapter.

CHAPTER 17 – WORKING WITH ATTRIBUTES

This chapter introduces the purpose for creating attributes in a drawing. A series of four commands step the user to a better understanding of attributes. The first command is ATTDEF and is used to define attributes. The ATTDISP command is used to control the display of attributes in a drawing. Once attributes are created and assigned to a block, they can be edited through the ATTEDIT command Block Attribute Manager dialog box. Global editing of attribute values can also be performed through the Find dialog box. Finally, attribute information can be extracted using the Attribute Extraction wizard. Extracted attributes can then be imported into such applications as Microsoft Excel and Access. Various tutorial exercises are provided throughout this chapter to help the user become better acquainted with this powerful feature of AutoCAD.

CHAPTER 18 – WORKING WITH EXTERNAL REFERENCES, RASTER IMAGES AND DWF FILES

The chapter begins by discussing the use of External References in drawings. An external reference is a drawing file that can be attached to another drawing file. Once the referenced drawing file is edited or changed, these changes are automatically seen once the drawing containing the external reference is opened again. Performing in-place editing of external references is also demonstrated. Importing image files is also discussed and demonstrated in this chapter. Working with DWF overlay files is also discussed in this chapter. A tutorial exercise follows at the end of this chapter to let the user practice using external references.

CHAPTER 19 – ADVANCED LAYOUT TECHNIQUES

This very important chapter is designed to utilize advanced techniques used in laying out a drawing before it is plotted. The ability to lay out a drawing consisting of various images at different scales is also discussed. The ability to create user-defined rectangular viewports through the Viewports Toolbar is demonstrated. The creation of non-rectangular viewports will also be demonstrated. This chapter also goes into detail explaining the use of Annotation scales and how they can be used for managing viewport scales. A tutorial exercise follows to let the user practice this advanced layout technique.

CHAPTER 20 – SOLID MODELING FUNDAMENTALS

This chapter begins with a comparison between isometric, extruded, wireframe, surfaced, and solid model drawings. The chapter continues with a discussion on the use of the 3D Modeling workspace. Creating User Coordinate Systems and how they are positioned to construct objects in 3D is a key concept to master in this chapter. Creating User Coordinate Systems dynamically is also emphasized. The display of 3D images through the View Cube, Steering Wheel, and the 3DORBIT command are discussed along with the creation of visual styles. Creating various solid primitives such as boxes, cones, and cylinders is discussed in addition to the ability to construct complex solid objects through the use of the Boolean operations of union, subtraction, and intersection. The chapter continues by discussing extruding, rotating, sweeping, and lofting operations for creating solid models in addition to filleting and chamfering solid models. Tutorial exercises follow at the end of this chapter.

CHAPTER 21 – CONCEPT MODELING AND EDITING SOLIDS

This chapter begins with a detailed study on how concept models can easily be created by dragging on grips located at key locations of a solid primitive. The ability to press and pull faces of a solid model and easily change its shape is also discussed. The 3DALIGN and 3DROTATE commands are discussed as a means of introducing the editing capabilities of AutoCAD on solid models. Modifications can also be made to a solid model through the use of the SOLIDEDIT command, which is discussed in this chapter. The ability to extrude existing faces, imprint objects, and create thin wall shells is demonstrated throughout this chapter. A tutorial exercise can be found at the end of this chapter.

CHAPTER 22 – CREATING 2D MULTIVIEW DRAWINGS FROM A SOLID MODEL

Once the solid model is created, the SOLVIEW command is used to lay out 2D views of the model, and the SOLDRAW command is used to draw the 2D views. Layers are automatically created to assist in the annotation of the drawing through the use of dimensions. The use of the FLATSHOT command is also explained as another means of projecting 2D geometry from a 3D model. A tutorial exercise is available at the end of this chapter, along with instructions on how to apply the techniques learned in this chapter to other solid models.

CHAPTER 23 – PRODUCING RENDERINGS AND MOTION STUDIES

This chapter introduces you to the uses and techniques of producing renderings from 3D models in AutoCAD. A brief overview of the rendering process is covered, along with detailed information about placing lights in your model, producing scenes from selected lighting arrangements, loading materials through the materials library supplied in AutoCAD, attaching materials to your 3D models, creating your own custom materials, applying a background to your rendered image, and experimenting with the use of motion path animations for creating walkthroughs of 3D models.

HOW THE BOOK WAS PRODUCED

The following hardware and software tools were used to create this version of the AutoCAD Tutor Book:

Hardware: Precision Workstation by Dell Computer Corporation

CAD Software: AutoCAD 2009 by Autodesk, Inc

Word Processing: Microsoft Word by Microsoft Corporation

Screen Capture Software: SnagIt! By TechSmith

Image Manipulation Software: Paint Shop Pro by Jasc Software, Inc.

Page Proof Review Software: Acrobat 7.0 by Adobe Corporation

ONLINE COMPANION

This new edition contains a special Internet companion piece. The Online Companion is your link to AutoCAD on the Internet. Monthly updates include a command of the month, FAQs, and tutorials. You can find the Online Companion at:

http://www.autodeskpress.com/resources/olcs/index.aspx

ACKNOWLEDGMENTS

I wish to thank the staff at Autodesk Press for their assistance with this document, especially John Fisher, Sandy Clark, and Stacy Masucci.

I would also like to thank Kevin Lang, Engineering Design Graphics, Trident Technical College, Charleston, SC for performing the technical edit on various chapters.

The publisher and author would like to thank and acknowledge the many professionals who reviewed the manuscript to help us publish this AutoCAD text. A special acknowledgment is due to the following instructors, who reviewed the chapters in detail:

Wen. M. Andrews, J. Sargeant Reynolds Community College, Richmond, VA

Kirk Barnes, Ivy Tech State College, Bloomington, IN

Paul Ellefson, Hennepin Technical College, Brooklyn Park, MN

Rajit Gadh, University of Wisconsin, Madison, WI

Gary J. Masciadrelli, Springfield Technical Community College, Springfield, MA

ABOUT THE AUTHOR

Alan J. Kalameja is the Department Head of Design and Construction at Trident Technical College, located in Charleston, South Carolina. He has been at the college for more than 27 years and has been using AutoCAD since 1984. He has authored the AutoCAD Tutor for Engineering Graphics in Release 10, 12, 14, and AutoCAD 2000, 2002, 2004, 2006, 2007, and 2008. He is also one of the principal authors involved in creating the Autodesk Inventor 2009 Essentials PLUS book. All of his books are published by Thomson Delmar Learning.

CONVENTIONS

All tutorials in this publication use the following conventions in the instructions:

Whenever you are told to enter text, the text appears in boldface type. This may take the form of entering an AutoCAD command or entering such information as absolute, relative, or polar coordinates. You must follow these and all text inputs by pressing the ENTER key to execute the input. An icon for most commands is also shown to assist in activating a command. For example, to draw a line using the LINE command from point 3,1 to 8,2, the sequence would look like the following:

```
Command: L (For line)
Specify first point: 3,1
Specify next point or [Undo]: 8,2
Specify next point or [Undo]: (Press ENTER to exit this command)
```

Instructions for selecting objects are in italic type. When instructed to select an object, move the pickbox on the object to be selected and press the pick button on the mouse.

If you enter the wrong command for a particular step, you may cancel the command by pressing the ESC key. This key is located in the upper left-hand corner of any standard keyboard.

Instructions in some tutorials are designed to enter all commands, options, coordinates, and so forth, from the keyboard. You may use the same commands by selecting them from the pull-down menu area, or from one of the floating toolbars.

Other tutorial exercises are provided with minimal instructions to test your ability to complete the exercise.

NOTES TO THE STUDENT AND INSTRUCTOR CONCERNING THE USE OF TUTORIAL EXERCISES

Various tutorial exercises have been designed throughout this book and can be found at the end of each chapter. The main purpose of each tutorial is to follow a series of steps toward the completion of a particular problem or object. Performing the tutorial will also prepare you to undertake the numerous drawing problems also found at the end of each chapter.

As you work on the tutorials, you should follow the steps very closely, taking care not to make a mistake. However, most individuals rush through the tutorials to get the correct solution in the quickest amount of time, only to forget the steps used to complete the tutorial. A typical comment made by many is "I completed the tutorial . . . but I don't understand what I did to get the correct solution."

It is highly recommended to both student and instructor that all tutorial exercises be performed two or even three times. Completing the tutorial the first time will give you the confidence that it can be done; however, you may not understand all the steps involved. Completing the tutorial a second time will allow you to focus on where certain operations are performed and why things behave the way they do. This still may not be enough. More complicated tutorial exercises may need to be performed a third time. This will allow you to anticipate each step and have a better idea what operation to perform in each step. Only then will you be comfortable and confident to attempt the many drawing problems that follow the tutorial exercises.

The CD-ROM in the back of the book contains AutoCAD drawing files for the Try It! exercises. To use drawing files, copy files to your hard drive, then remove their read-only attribute. Files cannot be used without AutoCAD. Files are located in the /Drawing Files/ directory.

SUPPLEMENTS

e.resource™--This is an educational resource that creates a truly electronic classroom. It is a CD-ROM containing tools and instructional resources that enrich your classroom and make your preparation time shorter. The elements of *e.resource* link directly to the text and tie together to provide a unified instructional system. Spend your time teaching, not preparing to teach.

ISBN 1-4354-4225-3

Features contained in *e.resource* include:

Instructor Guide

- **Syllabus:** Lesson plans created by chapter. You have the option of using these lesson plans with your own course information.
- **Chapter Hints:** Objectives and teaching hints that provide the basis for a lecture outline that helps you to present concepts and material.
- **PowerPoint® Presentation:** These slides provide the basis for a lecture outline that helps you to present concepts and material. Key points and concepts can be graphically highlighted for student retention. There are more than 300 slides, covering every chapter in the text.
- **Exam View Computerized Test Bank:** More than 600 questions of varying levels of difficulty are provided in true/false and multiple-choice formats. Exams can be generated to assess student comprehension or questions can be made available to the student for self-evaluation.
- **Video and Animation Resources:** These AVI files graphically depict the execution of key concepts and commands in drafting, design, and AutoCAD and let you bring multimedia presentations into the classroom.

Spend your time teaching, not preparing to teach!

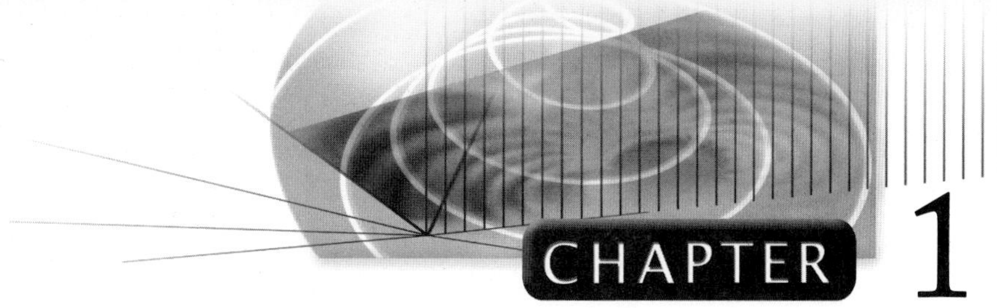

Getting Started with AutoCAD

This chapter begins with an explanation of the components that make up a typical AutoCAD display screen. You will learn various methods of selecting commands: some from the pull-down menu, others from toolbars, some from the Ribbon, and still others by entering the command from the keyboard. You will learn how to begin a new drawing. Once in a drawing, you will construct line segments with the LINE command in addition to drawing lines accurately with the Direct Distance mode, absolute coordinates, relative coordinates, and polar coordinates. Object Snap will be discussed as a way to construct all types of objects with greater precision. Additional drawing aids such as Object Snap Tracking and Polar Tracking will be discussed. Other basic drawing commands such as constructing circles and polylines will be shown. You will also be introduced to the ERASE command for removing drawing objects.

THE 2D DRAFTING & ANNOTATION WORKSPACE

The initial load of AutoCAD displays in a workspace. Workspaces are considered task-oriented environments that use a default drawing template and even launch such items as toolbars and palettes, depending on the workspace. By default, AutoCAD loads the 2D Drafting & Annotation workspace as shown in the following image. This workspace controls the display of what is called the Ribbon, which is used for accessing the more popular commands in AutoCAD. This workspace contains other items such as the Menu Browser, the Quick Access Toolbar, the graphic cursor, and the InfoCenter for getting information about commands as shown in the following image. Two other workspaces are supplied with AutoCAD, namely AutoCAD Classic and 3D Modeling. The 3D Modeling workspace will be discussed in greater detail in chapter 20.

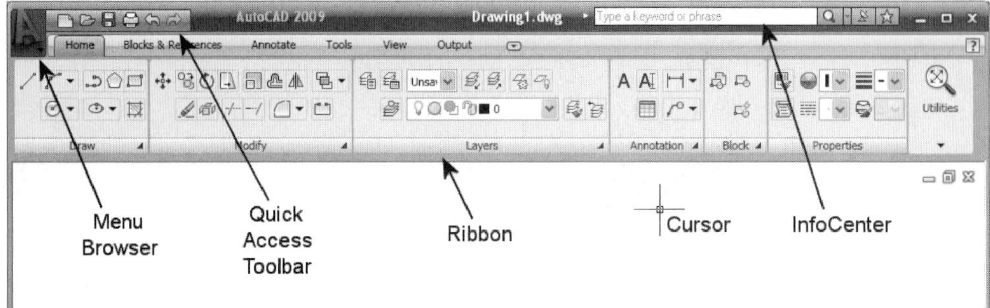

Figure 1.1

THE AUTOCAD CLASSIC WORKSPACE

A second workspace is provided, namely AutoCAD Classic, which is shown in the following image. This workspace differs from the previous workspace in that a number of extra toolbars display in the upper, left, and right vertical portions of the display screen. Also, the Tool Palette displays, which allows you to drag and drop commands for use. This workspace does not display the Ribbon although this can be turned on if you want to use it.

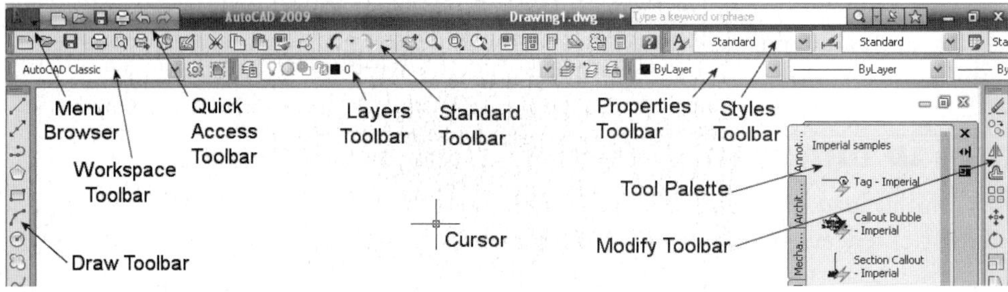

Figure 1.2

While major differences occur at the top of the display screen when you are activating different workspaces, most of the tools available at the bottom of the screen are common to both workspaces. Study the difference between the AutoCAD Classic and the 2D Drafting and Annotation workspaces as shown in the following image.

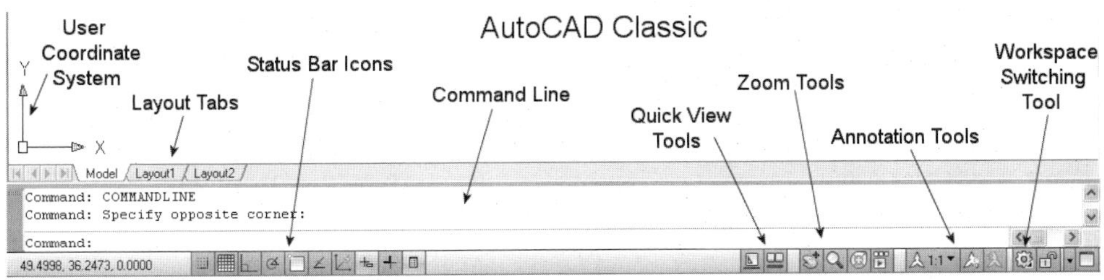

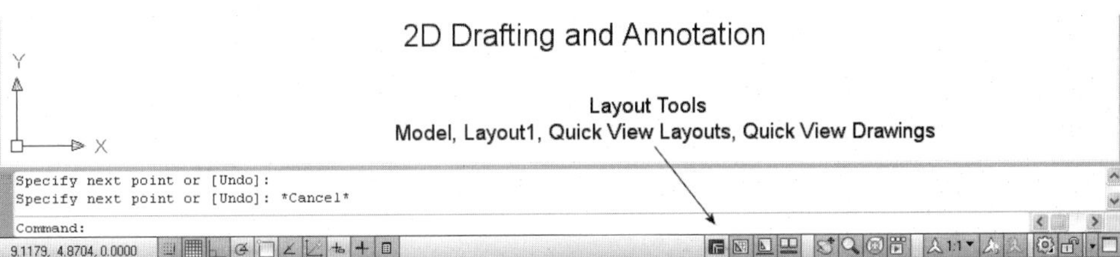

Figure 1.3

ACCESSING WORKSPACES

How you switch to different workspaces depends on the current workspace that you are in. For instance, if the current workspace is AutoCAD Classic, you can click in the Workspaces toolbar, as shown on the left in the following image, and choose a different workspace. If however, the current workspace is 2D Drafting and Annotation, you would click on the Workspace Switching Icon located in the bottom right of the display screen as shown on the right in the following image. This would activate a menu that displays all available workspaces to choose from. The presence of the checkmark alongside the 2D Drafting & Annotation workspace signifies that it is current. In addition to using these pre-existing workspaces, you can also arrange your screen to your liking and save these screen changes as your own custom workspace.

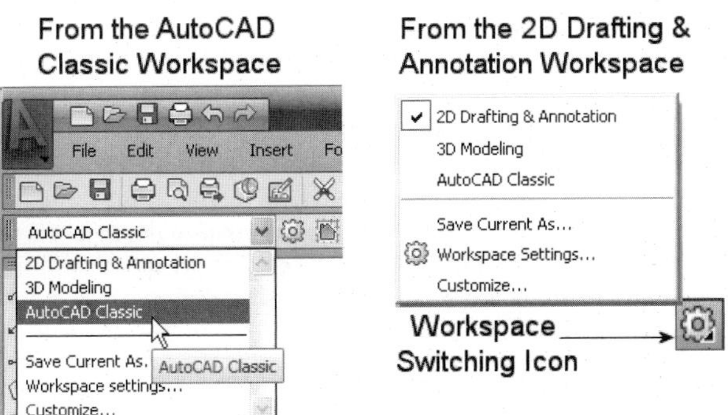

Figure 1.4

Note: A Default AutoCAD workspace may be present in the list shown in the previous image. This workspace is automatically created if you are upgrading from a previous version of AutoCAD.

THE STATUS BAR

The status bar, illustrated in the following image, is used to toggle ON or OFF the following modes: Coordinate Display, Snap, Grid, Ortho, Polar Tracking, Object Snap (OSNAP), Object Snap Tracking (OTRACK), Dynamic User Coordinate System (DUCS), Dynamic Input (DYN), Line Weight (LWT), and Quick Properties (QP). Click the button once to turn the mode on or off. A button with a blue color indicates that the mode is on. For example, the following image illustrates Ortho turned off (gray color) and Polar turned on (blue color). Right-clicking on any button in the status bar activates the menu shown in the following image. Clicking on Use Icons will change the graphic icons to text mode icons.

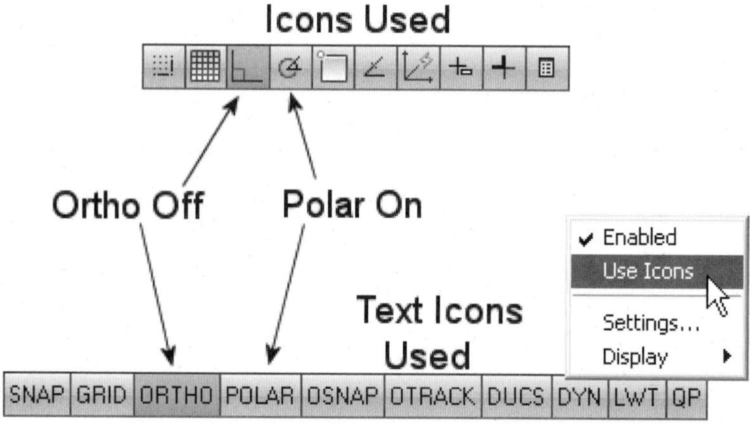

Figure 1.5

The following table gives a brief description of each component located in the status bar:

Button	Tool	Description
5.5000, 3.0000	Coordinate Display	Toggles the coordinate display, located in the lower-left corner of the status bar, ON or OFF. When the coordinate display is off, the coordinates are updated when you pick an area of the screen with the cursor. When the coordinate display is on, the coordinates dynamically change with the current position of the cursor.
	SNAP	Toggle Snap mode ON or OFF. The SNAP command forces the cursor to align with grid points. The current snap value can be modified and can be related to the spacing of the grid.
	GRID	Toggles the display of the grid ON or OFF. The actual grid spacing is set by the DSETTINGS OR GRID command and not by this function key.

Button	Tool	Description
	ORTHO	Toggles Ortho mode ON or OFF. Use this key to force objects such as lines to be drawn horizontally or vertically.
	POLAR	Toggles the Polar Tracking ON or OFF. Polar Tracking can force lines to be drawn at any angle, making it more versatile than Ortho mode. The Polar Tracking angles are set through a dialog box. Also, if you turn Polar Tracking on, Ortho mode is disabled, and vice versa.
	OSNAP	Toggles the current Object Snap settings ON or OFF. This will be discussed later in this chapter.
	OTRACK	Toggles Object Snap Tracking ON or OFF. This feature will also be discussed later in this chapter.
	DUCS	Toggles the Dynamic User Coordinate System ON or OFF. This feature is used mainly for modeling in 3D.
	DYN	Toggles Dynamic Input ON or OFF. When turned on, your attention is directed to your cursor position as commands and options are executed. When turned off, all commands and options are accessed through the Command prompt at the bottom of the display screen.
	LWT	Toggles Lineweight ON or OFF. When turned off, no lineweights are displayed. When turned on, lineweights that have been assigned to layers are displayed in the drawing.
	Quick Properties	When turned on, this tool will list the most popular properties of a selected object.

Right-clicking one of the buttons displays the shortcut menu in the following image. Choose Settings to access various dialog boxes that control certain features associated with the button. These controls will be discussed later in this chapter and also in chapter 2. Depending on which button is clicked on, additional tools may be available as shown by right-clicking on SNAP, POLAR, or OSNAP.

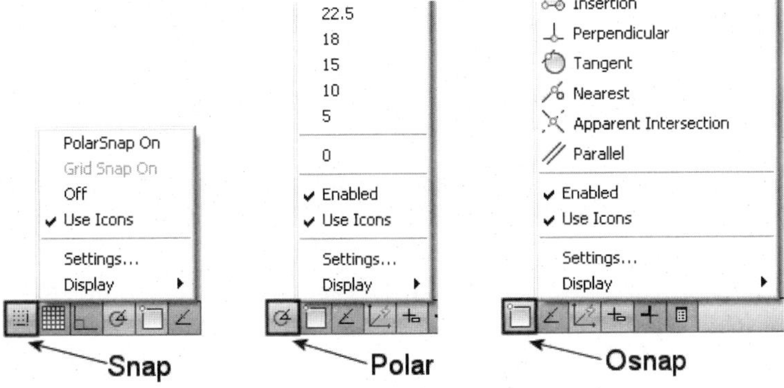

Figure 1.6

You can also access most tools located in the status bar through the function keys located at the top of any standard computer keyboard. The following table describes each function key.

Function Key	Definitions
F1	Displays AutoCAD Help Topics
F2	Toggle Text/Graphics Screen
F3	Object Snap settings ON/OFF
F4	Toggle Tablet Mode ON/OFF
F5	Toggle Isoplane Modes
F6	Toggle Dynamic UCS ON/OFF
F7	Toggle Grid Mode ON/OFF
F8	Toggle Ortho Mode ON/OFF
F9	Toggle Snap Mode ON/OFF
F10	Toggle Polar Mode ON/OFF
F11	Toggle Object Snap Tracking ON/OFF
F12	Toggle Dynamic Input (DYN) ON/OFF

Note: Most of the function keys are similar in operation to the modes found in the status bar except for the following:
When you press F1, the AutoCAD Help Topics dialog box is displayed.
Pressing F2 takes you to the text screen consisting of a series of previous prompt sequences. This may be helpful for viewing the previous command sequence in text form.
Use F4 to toggle Tablet mode ON or OFF. This mode is only activated when the digitizing tablet has been calibrated for the purpose of tracing a drawing into the computer.
Pressing F5 scrolls you through the three supported Isoplane modes used to construct isometric drawings (Right, Left, and Top).
Pressing CTRL+SHIFT+P toggles Quick Properties mode ON or OFF.

ADDITIONAL STATUS BAR CONTROLS

Located at the far right end of the status bar are additional buttons separated into four distinct groups used to manage the appearance of the AutoCAD display screen and the annotation scale of a drawing. These items include Quick View Layouts and Drawings, Display tools, Annotation Scale tools, the Workspace Switching tool, the Toolbar Unlocking tool, the Status Bar menu tool, and the Clean Screen tool. When annotative objects such as text and dimensions are created, they are scaled based on the current annotation scale and automatically displayed at the correct size. This feature will be discussed in greater detail in chapter 19. The following table gives a brief description of the remaining buttons found in this area.

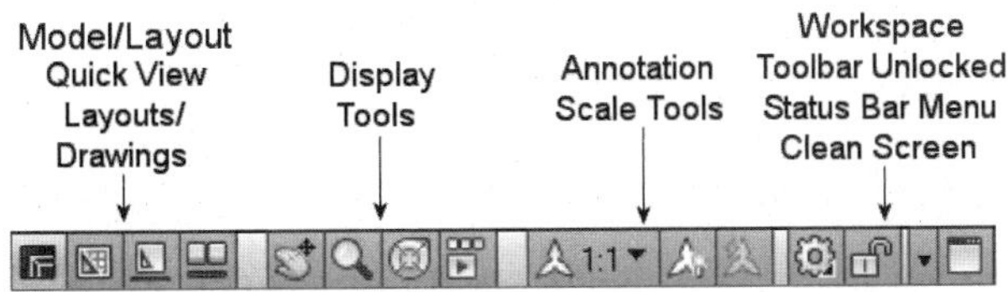

Figure 1.7

Button	Tool	Description
	Workspace Switching	Allows you to switch between the workspaces already defined in the drawing.
	Toolbar/Window Positions Toggle	Locks the position of all toolbars on the display screen.
	Status Bar Menu Controls	Activates a menu used for turning on or off certain status bar buttons.
	Clean Screen	Removes all toolbars from the screen, giving your display an enlarged appearance. Click this button again to return the toolbars to the screen.

COMMUNICATING WITH AUTOCAD

THE COMMAND LINE

How productive the user becomes in using AutoCAD may depend on the degree of understanding of the command execution process with AutoCAD. One of the means of command execution is through the command prompt that is located at the bottom of the display screen. As a command is selected from a toolbar, AutoCAD prompts the user with a series of steps needed to complete this command. In the following image, the CIRCLE command is chosen as the command. The next series of lines in the command line prompts the user to first specify or locate a center point for the circle. After this is accomplished, you are then prompted to specify the radius of the circle. These actions are also called the default prompt.

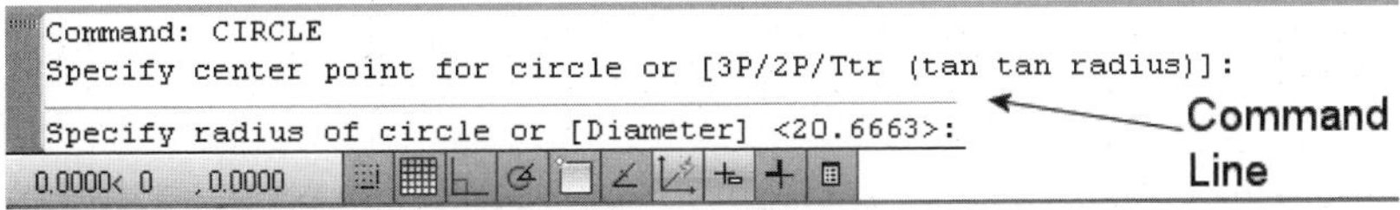

Figure 1.8

UNDERSTANDING THE COMMAND PROMPT

In the previous image of the command line, notice the string of CIRCLE command options displayed as the following:

```
[3P/2P/Ttr (tan tan radius)]
```

Items identified inside the square brackets are referred to as options. Typing in this option from the keyboard activates it.

DYNAMIC INPUT

Yet another more efficient means of command execution within AutoCAD is through the Dynamic Input feature, which is activated by clicking the DYN button located in the status bar at the bottom of your display screen. Whether a command is picked from the Ribbon, toolbar, or pull-down menu or entered from the keyboard, you see immediate feedback at your cursor location. The following image illustrates how the CIRCLE command prompts display at the cursor location. As the cursor is moved around, the Specify center point for circle prompt also moves. Also notice that the current screen position is displayed. If the down directional arrow is typed on the keyboard, options of the CIRCLE command display. Typing the down arrow cycles through the available options in executing the CIRCLE command.

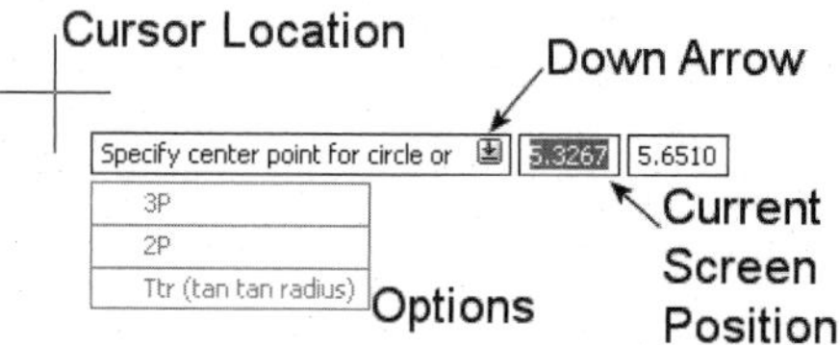

Figure 1.9

THE MENU BROWSER

The Menu Browser provides you with the ability to access most AutoCAD commands. Clicking on the Icon in the upper left corner will display the various categories. Clicking on a category will display the individual commands. In the following image, the Insert category is picked, which displays all commands native to this category. At the bottom left corner of the Menu Browser are a series of document controls used for opening up existing drawing files. At the very bottom of the Menu Browser are two buttons, one called Options for launching the Options dialog box and the other called Exit AutoCAD used for exiting the AutoCAD environment. The Options dialog box controls various settings internal to AutoCAD and is considered an advanced feature.

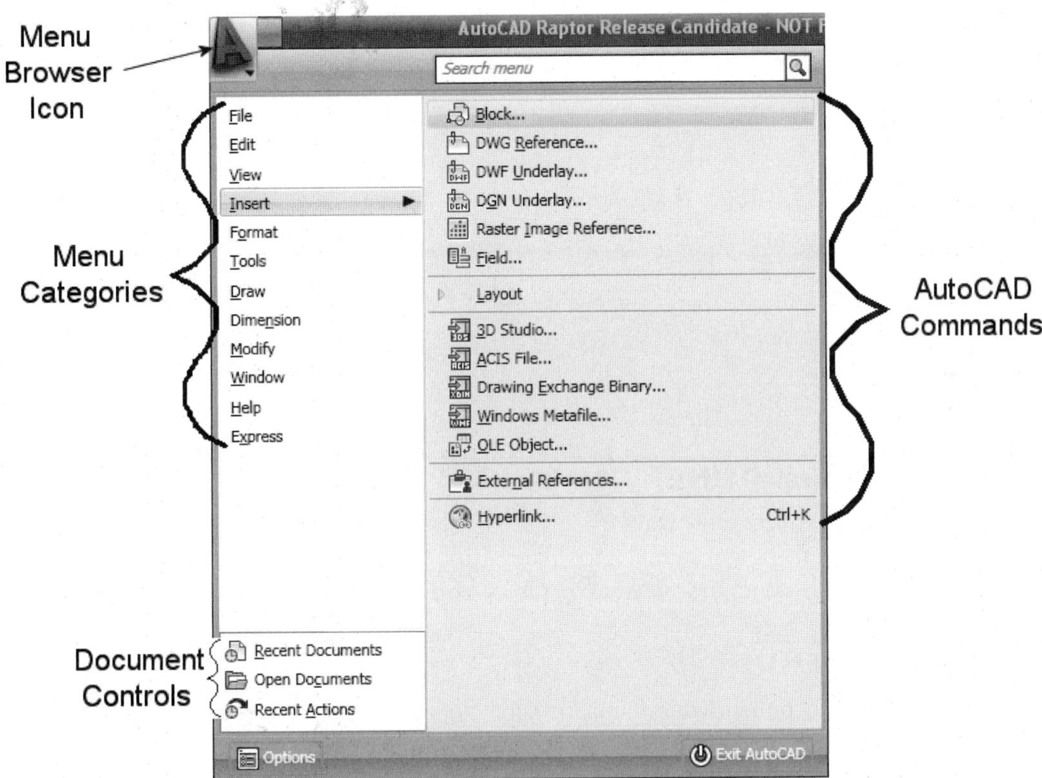

Figure 1.10

DOCUMENT CONTROLS

Clicking on the three document control items in the lower-left corner of the Menu Browser displays a series of panes used for viewing recent documents or actions. These three panes are illustrated in the following image. The Recent Documents mode allows you to display existing files in an ordered list or group them by date or file type. When you move your cursor over one of these files, a preview image automatically appears in addition to information about the document. You can also view a list of the most recent actions that you performed through the Recent Actions pane.

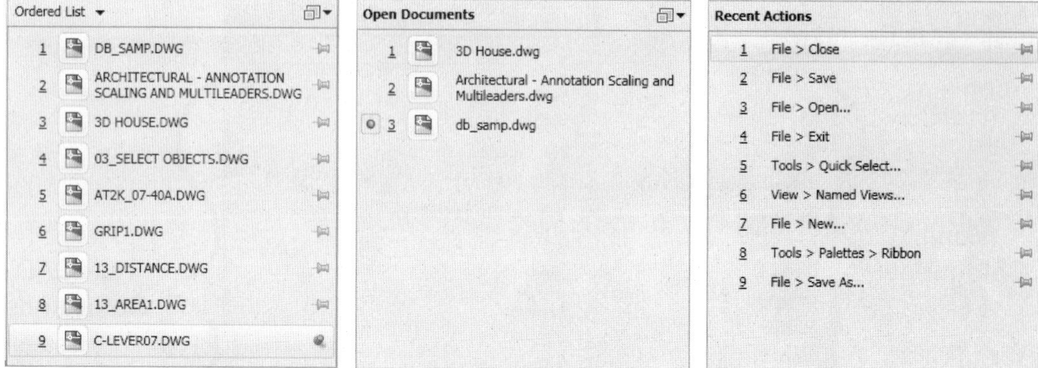

Figure 1.11

METHODS OF CHOOSING COMMANDS

AutoCAD provides numerous ways to enter or choose commands for constructing or editing drawings: pull-down menus, toolbars, Ribbon display, dialog boxes, right-click shortcut menus, and command aliases for direct entry from the keyboard.

PULL-DOWN MENUS

In either workspace, pull-down menus provide an easy way to access most AutoCAD commands. The pull-down menu area is displayed when you click on the Menu Browser icon located in the upper-left corner of the AutoCAD display screen. Here, various categories exist such as File, Edit, View, Insert and so on. Clicking one of these category headings pulls down a menu consisting of commands related to this heading. For example, clicking the Format heading followed by Layer tools in the following image pulls down the menu containing extra commands used for managing layers in a drawing.

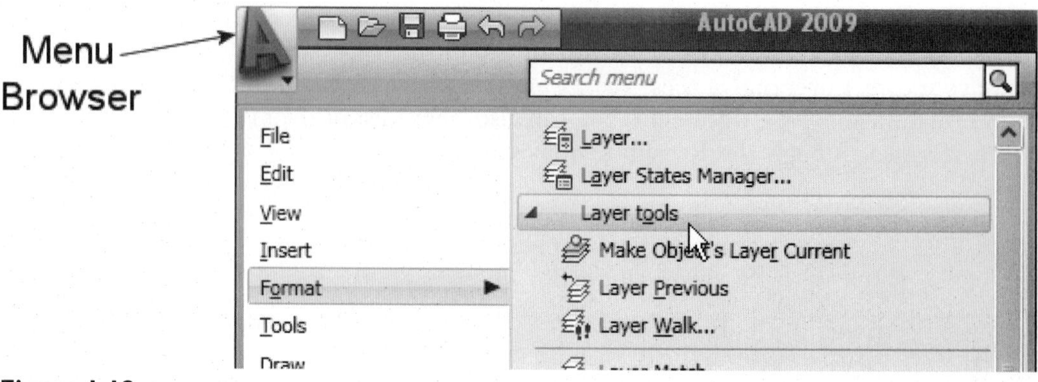

Figure 1.12

In the 2D Drafting & Annotation workspace, you can activate the pull-down menus in the upper part of the display screen by right-clicking on any button located in the Quick Access menu and choosing Show Menu Bar from the menu, as shown in the following image.

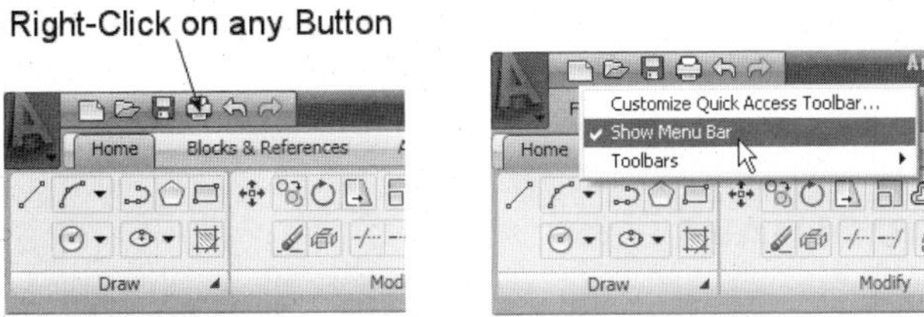

Figure 1.13

This will display the pull-down menu bar at the top of the screen, which may be a more convenient location to choose commands from as shown in the following image.

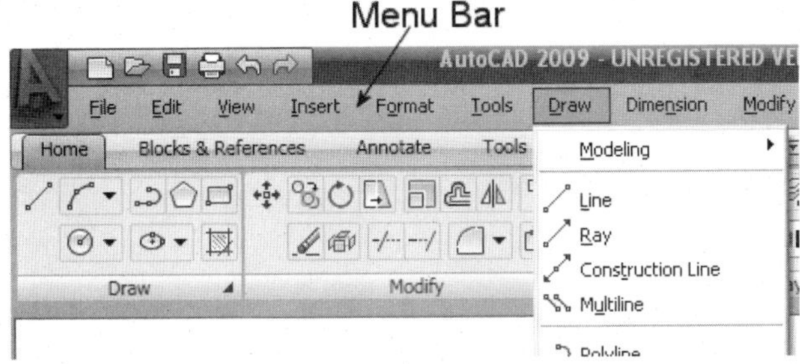

Figure 1.14

TOOLBARS FROM THE AUTOCAD CLASSIC WORKSPACE

Activating the AutoCAD Classic workspace will automatically display toolbars. The following image shows the Standard, Workspaces, Layers, and Draw toolbars displayed.

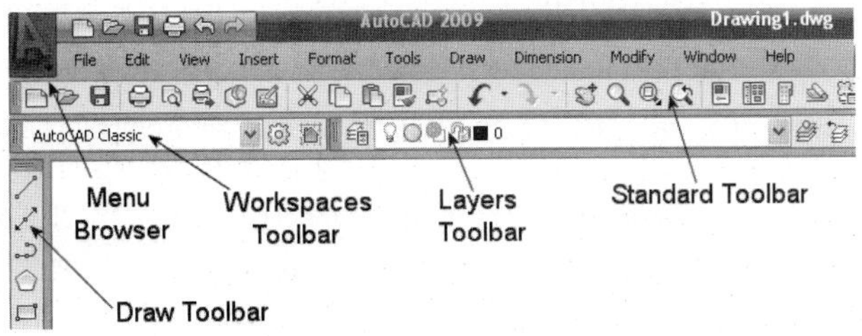

Figure 1.15

The toolbar in this example allows you access to most ZOOM command options. When the cursor rolls over a tool, a 3D border is displayed, along with a tooltip that explains the purpose of the command, as shown on the right in the following image.

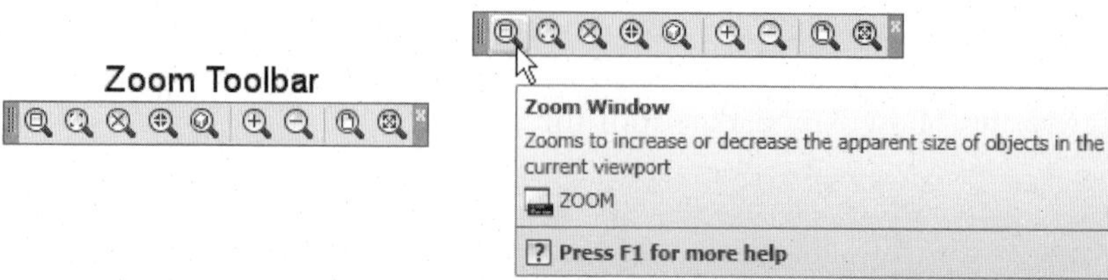

Figure 1.16

ACTIVATING TOOLBARS

Many toolbars are available to assist the user in executing other types of commands. When working in the AutoCAD Classic workspace, six toolbars are already active or displayed in all drawings: Draw, Layers, Modify, Properties, Standard, and Styles. To activate a different toolbar, move the cursor over the top of any command button and press the right mouse button. A shortcut menu appears that displays all toolbars, as shown in the following image. In this example, placing a check beside Text displays this toolbar.

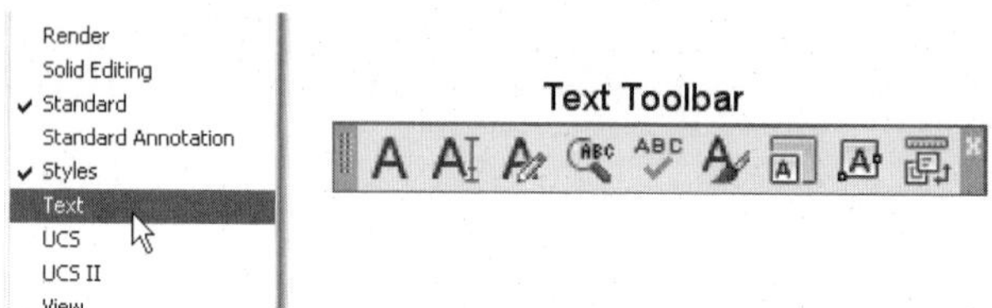

Figure 1.17

DOCKING TOOLBARS

It is considered good practice to line the top or side edges of the display screen with toolbars. The method of moving toolbars to the sides of your screen is called docking. Press down on the toolbar title strip and slowly drag the toolbar to the top of the screen until the toolbar appears to jump. Letting go of the mouse button docks the toolbar to the top of the screen as shown in the following image. Practice this by docking various toolbars to your screen.

Tip: To prevent docking, press the CTRL key as you drag the toolbar. This allows you to move the toolbar into the upper or lower portions of the display screen without the toolbar docking. Also, if a toolbar appears to disappear, it might actually be alongside or below toolbars that already exist. Closing toolbars will assist in finding the missing one.

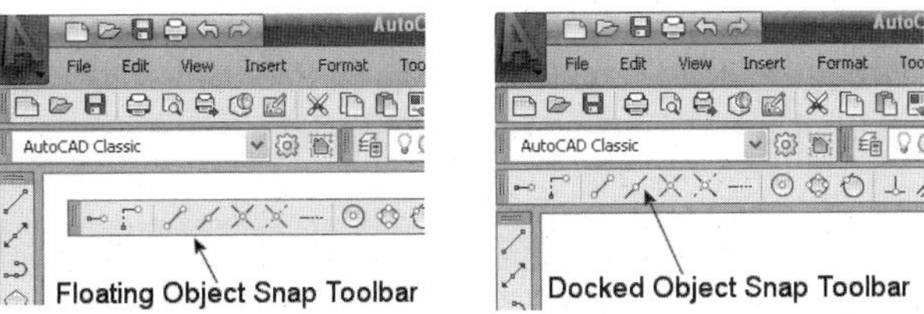

Figure 1.18

TOOLBARS FROM THE 2D DRAFTING & ANNOTATION WORKSPACE

Inside of the 2D Drafting & Annotation workspace, it is possible to display toolbars as in the AutoCAD Classic workspace. To accomplish this, right-click on the Quick Access Toolbar and pick AutoCAD, as shown in the following image. This will display all of the toolbars similar to those that are present in the AutoCAD Classic Workspace. Pick one of the names from the list to show the toolbar on the display screen.

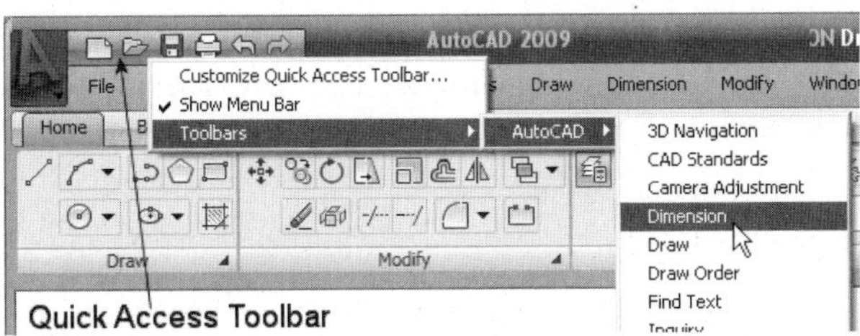

Figure 1.19

CUSTOMIZING THE QUICK ACCESS TOOLBAR

One of the advantages of the Quick Access Toolbar is the ability to display the AutoCAD commands that you frequently use. By Default, the following commands are already provided: NEW, OPEN, SAVE, PLOT, UNDO, and REDO. In this example, the CLOSE command will be added to the Quick Access Toolbar. In order to add this or other commands, first right-click on any command in the Quick Access Toolbar and choose Customize Quick Access Toolbar as shown in the following image.

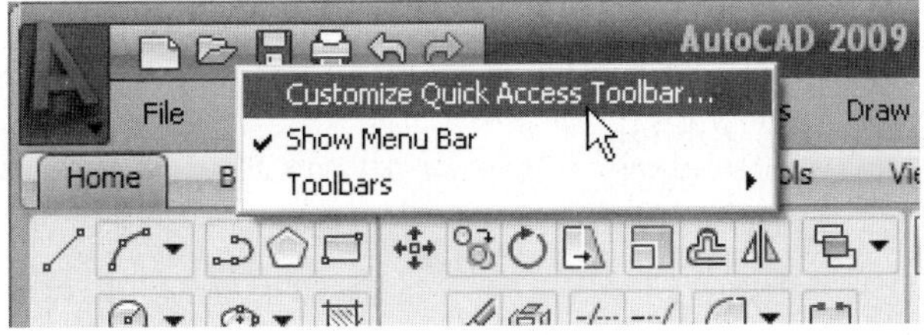

Figure 1.20

The Customize User Interface dialog box will appear. Scroll through the list of all commands and locate the CLOSE command. Then drag and drop this command into the

Quick Access Toolbar after the OPEN command but before the SAVE command as shown in the following image. It will appear as part of the Quick Access Toolbar. When finished, click the Apply button followed by the OK button.

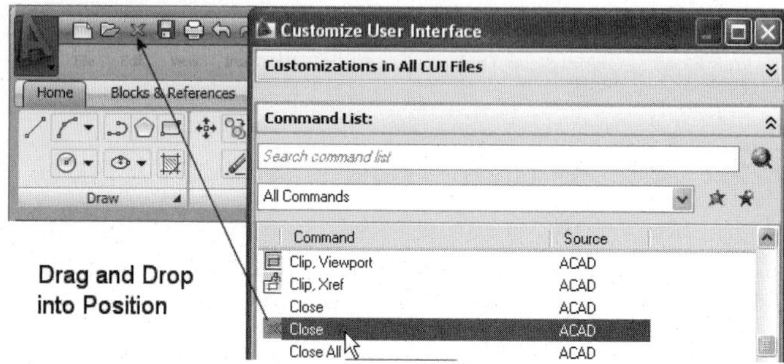

Figure 1.21

The Quick Access Toolbar is now populated with the CLOSE command from the Customize User Interface as shown in the following image. This provides a quick and efficient means of closing your drawings instead of looking for this command from the other AutoCAD menus.

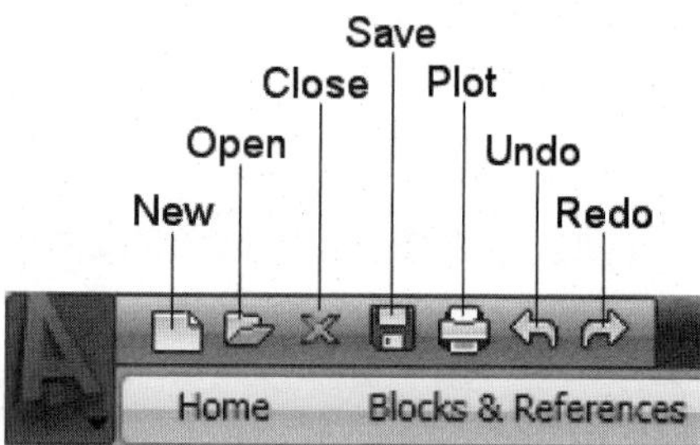

Figure 1.22

RIBBON DISPLAY MODES

A small button with an arrow is displayed at the end of the Ribbon tabs. This button allows you to minimize the Ribbon and display more of your screen. Three modes are available, as shown in the following image, that allow you to minimize to the title panels, minimize to the tabs, or show the full Ribbon.

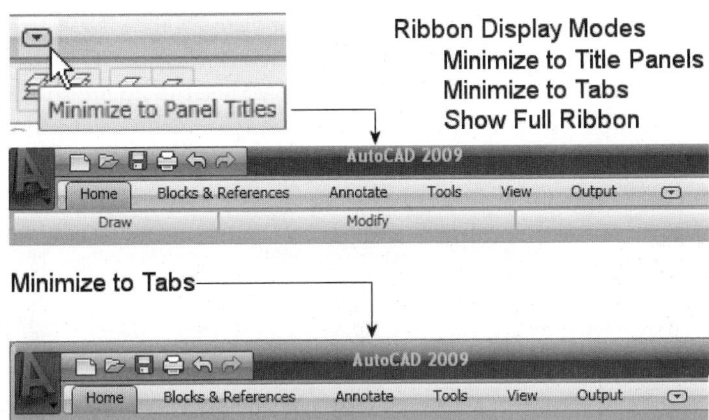

Figure 1.23

DIALOG BOXES AND ICON MENUS

Settings and other controls can be changed through dialog boxes. Illustrated on the left in the following image is the Drawing Units dialog box, which will be discussed in chapter 2. Illustrated on the right in the following image is the Hatch Pattern Palette. This palette provides an icon menu that makes it easy to choose the desired hatch pattern. Simply select the pattern by reviewing the small images (icons) and click it. Palettes are similar to dialog boxes with the exception that palettes allow for the display of small images. Certain dialog boxes can be increased in size by moving your cursor over their borders; this is true of the Hatch Pattern Palette dialog box. When two arrows appear, hold down the pick button of the mouse (usually the left button) and stretch the dialog box in that direction. If the cursor is moved to the corner of the dialog box, the box is stretched in two directions. These methods can also be used to make the dialog box smaller, although there is a default size for the dialog boxes, which limits smaller sizes. The dialog box cannot be stretched if no arrows appear when you move your cursor over the border of the dialog box; this is true of the Drawing Units dialog box as well.

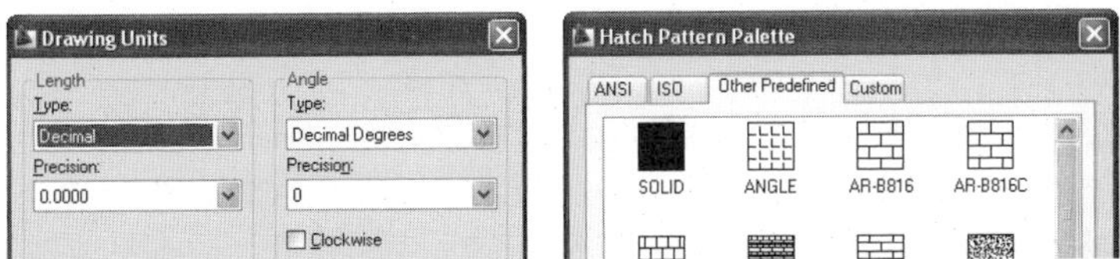

Figure 1.24

TOOL PALETTES

Commands can also be accessed from a number of tool palettes. To launch the Tool Palette, click the Tool Palettes Window button, which is located in the Standard toolbar of the AutoCAD Classic workspace, as shown in the following image. If you are in the 2D Drafting & Annotation workspace, first click on the View tab and select Tool Palettes. The

Tool Palette is a long, narrow bar that consists of numerous tabs. Three tabs, namely, Modify (A), Draw (B), and Architectural (C) are illustrated below. Use these tabs to access the more popular drawing and modify commands. While this image shows three palettes, in reality only one will be present on your screen at any one time. Simply click a different tab to display the commands associated with the tab.

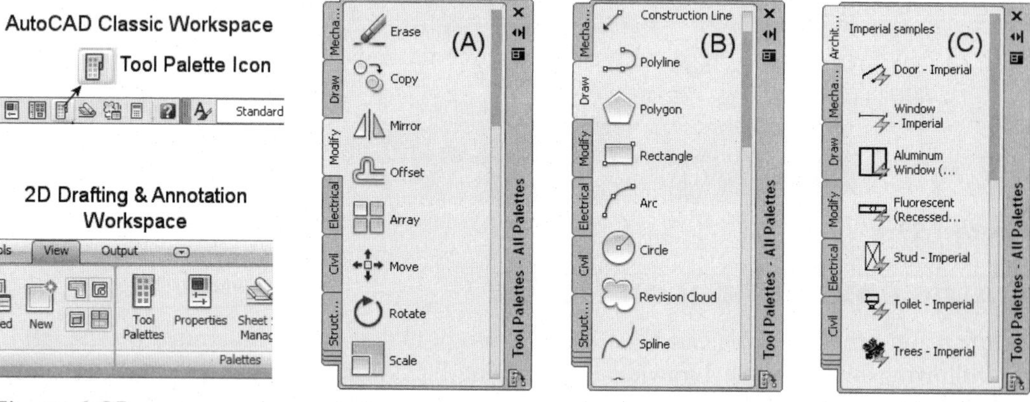

Figure 1.25

RIGHT-CLICK SHORTCUT MENUS

Many shortcut or cursor menus have been developed to assist with the rapid access to commands. Clicking the right mouse button activates a shortcut menu that provides access to these commands. The Default shortcut menu is illustrated on the left in the following image. It is displayed whenever you right-click in the drawing area and no command or selection set is in progress.

On the right in the following image is an example of the Edit shortcut menu. This shortcut consists of numerous editing and selection commands. This menu activates whenever you right-click in the display screen with an object or group of objects selected but no command is in progress.

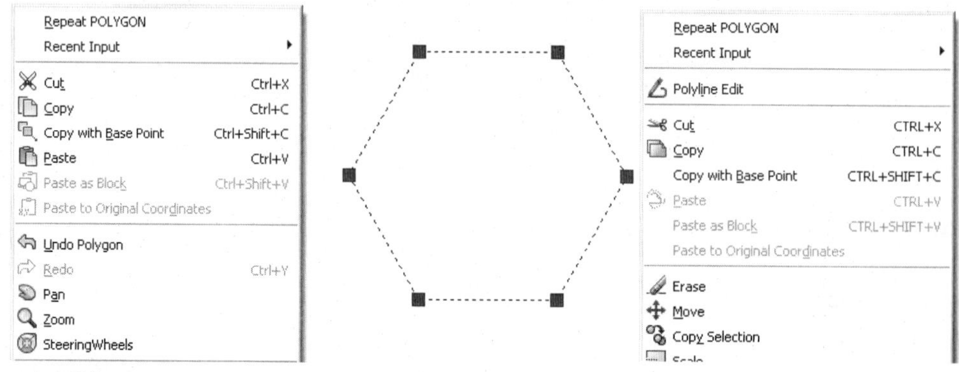

Figure 1.26

Right-clicking in the COMMAND prompt area of the display screen activates the shortcut menu, as shown on the left in the following image. This menu provides quick access to the Options dialog box, which is used to control various settings in AutoCAD. Also, a record of the six most recent commands is kept, which allows the user to select from this group of previously used commands.

Illustrated on the right in the following image is an example of a Command-Mode shortcut menu. When you enter a command and right-click, this menu displays options of the command. This menu supports a number of commands. In the figure, the 3P, 2P, and Ttr (tan tan radius) listings are all options of the CIRCLE command.

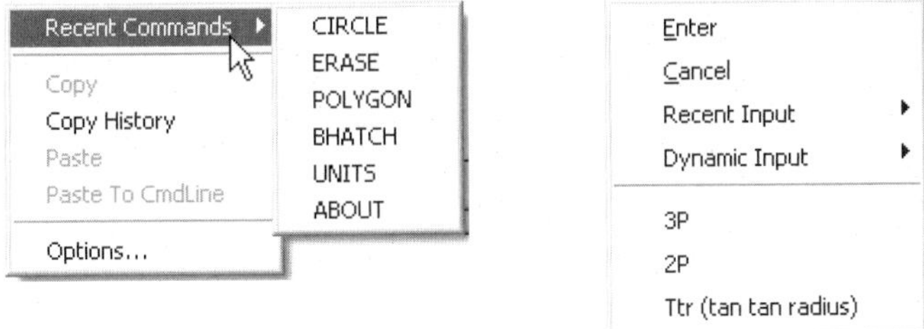

Figure 1.27

COMMAND ALIASES

Commands can be executed directly through keyboard entry. This practice is popular for users who are already familiar with the commands. However, users must know the command name, including its exact spelling. To assist with the entry of AutoCAD commands from the keyboard, numerous commands are available in shortened form, referred to as aliases. For example, instead of typing in LINE, all that is required is L. The letter E can be used for the ERASE command, and so on. These command aliases are listed throughout this book. The complete list of all command aliases can be found under the Express pull-down menu by clicking Tools and Command Alias Editor as shown in the following image. The AutoCAD Command Alias Editor will appear, displaying all of the commands that have their names shortened. Once you are comfortable with the keyboard, command aliases provide a fast and efficient method of activating AutoCAD commands.

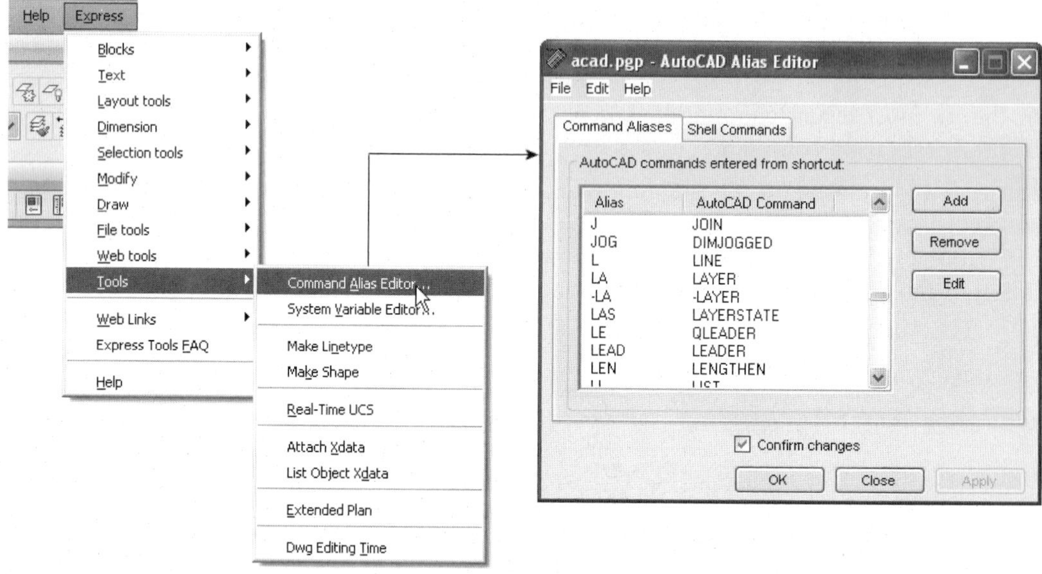

Figure 1.28

STARTING A NEW DRAWING

To begin a new drawing file, select the NEW command using one of the following methods:

- From the Quick Access toolbar
- From the Standard toolbar of the AutoCAD Classic workspace
- From the Menu Browser (File > New)
- From the keyboard (New)

```
Command: NEW
```

Entering the NEW command displays the dialog box illustrated in the following image. This dialog box provides a list of templates to use for starting a new drawing.

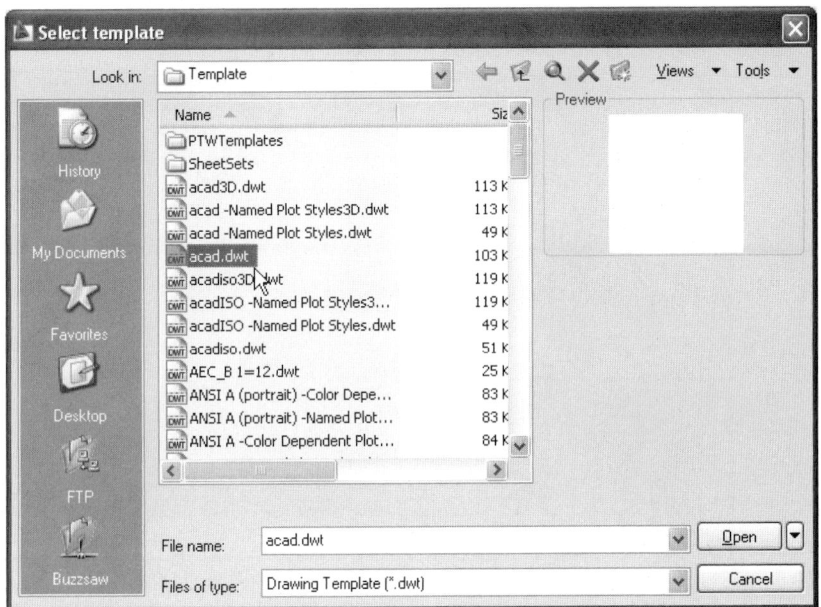

Figure 1.29

Note: A QNEW command is also available for creating new drawings. This command is similar to the NEW command but provides the option of starting with a pre-selected template.

OPENING AN EXISTING DRAWING

The OPEN command is used to edit a drawing that has already been created. Select this command from one of the following:

- From the Quick Access toolbar
- From the Standard toolbar of the AutoCAD Classic workspace
- From the Menu Browser (File > Open)
- From the keyboard (Open)

When you select this command, a dialog box appears similar to the following image. Listed in the field area are all files that match the type shown at the bottom of the dialog box. Because the file type is .DWG, all drawing files supported by AutoCAD are listed. To choose a different folder, use standard Windows file management techniques by clicking in the Look in field. This displays all folders associated with the drive. Clicking the folder displays any drawing files contained in it.

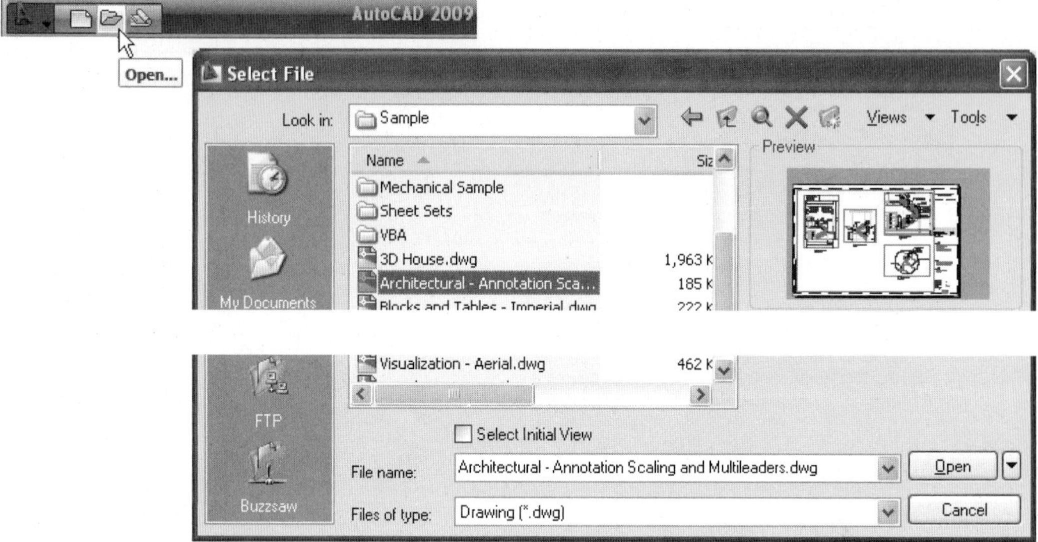

Figure 1.30

Additional tools are available in the Menu Browser to assist in locating drawings to open. These tools include Recent Documents, Open Documents, and Recent Actions. Illustrated in the following image is an example of clicking on Recent Documents, which is located in the lower left corner of the Menu Browser. Notice the ordered list of all drawings that were recently opened enabling you to select these more efficiently.

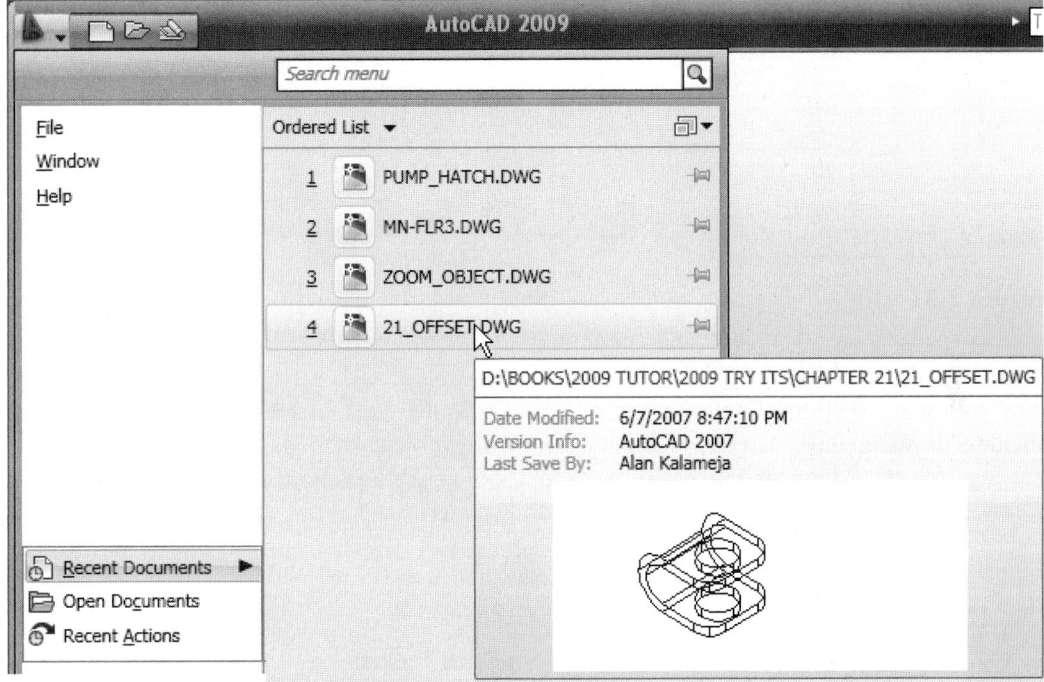

Figure 1.31

When viewing files from the Ordered List Pane, clicking on the Ordered List icon will expand the menu to include a number of options to display files. The default setting to display files is through Icons as shown on the left in the following image. Changing to Small Images will change the Ordered List to small images of each file as shown on the right. You can also change the Ordered List types to Medium or even Large Images.

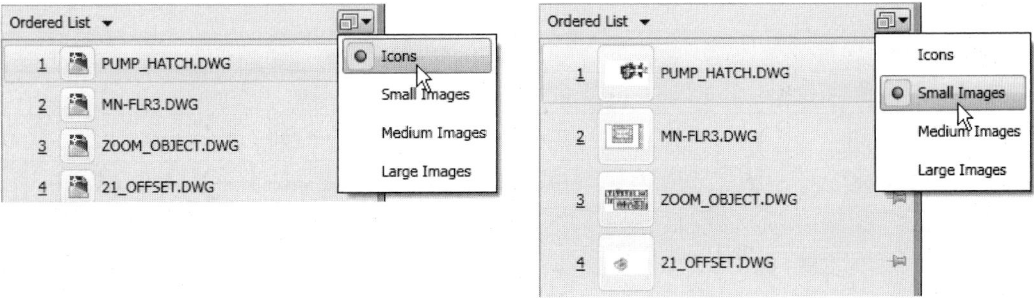

Figure 1.32

Because the Ordered List can get large, certain drawing names drop off, which means you need to look for the drawing again. For drawings that are used most frequently, you can click on the pin icon to change the orientation of the pin as shown on the right in the following image. The presence of this pin means that this drawing will always be displayed in the Ordered List.

Figure 1.33

BASIC DRAWING COMMANDS

The following sections discuss some basic techniques used in creating drawings. These include drawing lines, circles, and polylines, using object snap modes and tracking, and erasing objects. Many of the basic drawing tools can be easily accessed using either the Draw toolbar or Ribbon, as shown in the following image.

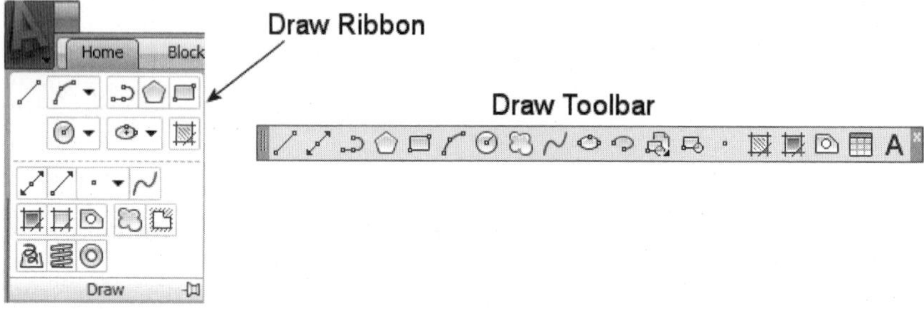

Figure 1.34

The following table gives a brief description of the LINE, CIRCLE, and PLINE commands:

Button	Tool	Key-In	Function
	Line	L	Draws individual or multiple line segments
	Circle	C	Constructs circles of specified radius or diameter
	Pline	PL	Used to construct a polyline, which is similar to a line except that all segments made with the PLINE command are considered a single object

THE LINE COMMAND

Use the LINE command to construct a line from one endpoint to the other. Choose this command from one of the following:

- From the Draw toolbar of the AutoCAD Classic workspace

- From the Ribbon > Home Tab > Draw Panel
- From the Menu Browser (Draw > Line)
- From the keyboard (L or Line)

As the first point of the line is marked, the rubber-band cursor is displayed along with the normal crosshairs to assist in locating where the next line segment will be drawn. The LINE command stays active until the user either executes the Close option or issues a null response by pressing ENTER at the prompt "To point."

Try It!: Create a new drawing from scratch. Study the following image on the left and follow the command sequence for using the LINE command.

```
Command: L (For LINE)
Specify first point: (Pick a point at "A")
Specify next point or [Undo]: (Pick a point at "B")
Specify next point or [Undo]: (Pick a point at "C")
Specify next point or [Close/Undo]: (Pick a point at "D")
Specify next point or [Close/Undo]: (Pick a point at "E")
Specify next point or [Close/Undo]: (Pick a point at "F")
Specify next point or [Close/Undo]: C (To close the shape and
    exit the command)
```

If a mistake is made in drawing a segment, as illustrated in the following image, the user can correct the error without exiting the LINE command. The built-in UNDO option within the LINE command removes the previously drawn line while still remaining in the LINE command. Refer to the following image on the right and prompts to use the Undo option of the LINE command.

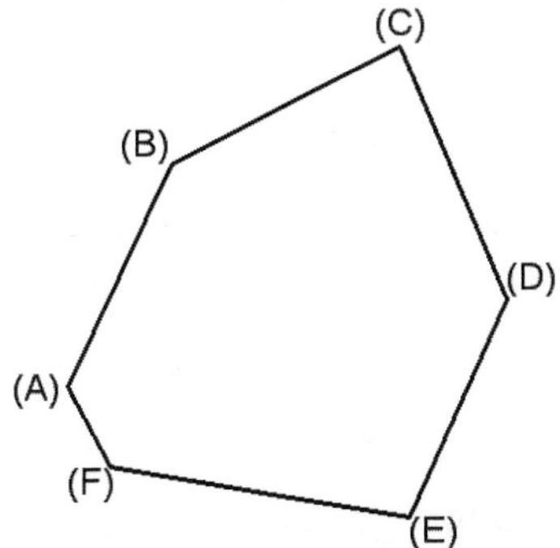

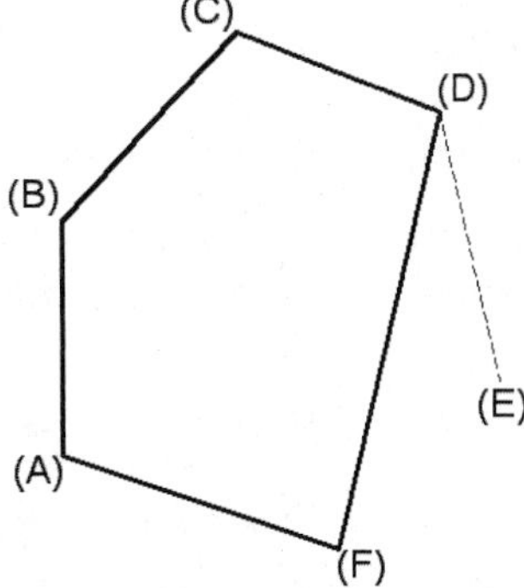

Figure 1.35

```
Command: L (For LINE)
Specify first point: (Pick a point at "A")
Specify next point or [Undo]: (Pick a point at "B")
Specify next point or [Undo]: (Pick a point at "C")
Specify next point or [Close/Undo]: (Pick a point at "D")
Specify next point or [Close/Undo]: (Pick a point at "E")
```

```
Specify next point or [Close/Undo]: U (To undo or remove the
    segment from "D" to "E" and still remain in the LINE command)
Specify next point or [Close/Undo]: (Pick a point at "F")
Specify next point or [Close/Undo]: End (For Endpoint mode)
of (Select the endpoint of the line segment at "A")
Specify next point or [Close/Undo]: (Press ENTER to exit this
    command)
```

CONTINUING LINES

Another option of the LINE command is the Continue option. The dashed line segment in the following image was the last segment drawn before the LINE command was exited. To pick up at the last point of a previously drawn line segment, type the LINE command and press ENTER. This activates the Continue option of the LINE command.

```
Command: L (For LINE)
Specify first point: (Press ENTER to activate Continue Mode)
Specify next point or [Undo]: (Pick a point at "B")
Specify next point or [Undo]: (Pick a point at "C")
Specify next point or [Close/Undo]: End (For Endpoint mode)
of (Select the endpoint of the vertical line segment at "D")
Specify next point or [Close/Undo]: (Press ENTER to exit this
    command)
```

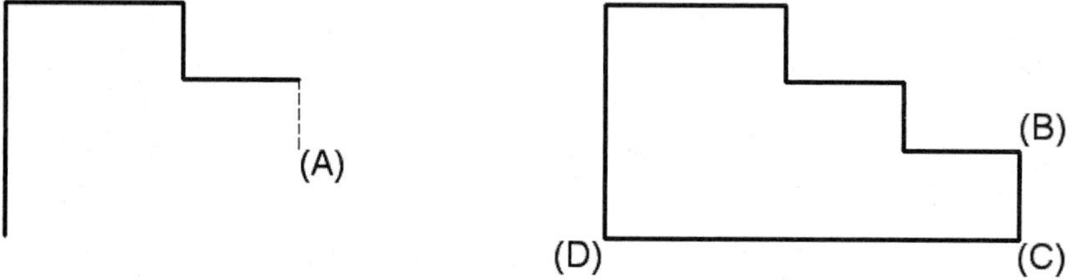

Figure 1.36

DYNAMIC INPUT AND LINES

With Dynamic Input turned on in the status bar, additional feedback can be obtained when drawing line segments. In addition to the command prompt and down arrow being displayed at your cursor location, a dynamic distance and angle are displayed to assist you in the construction of the line segment, as shown in the following image.

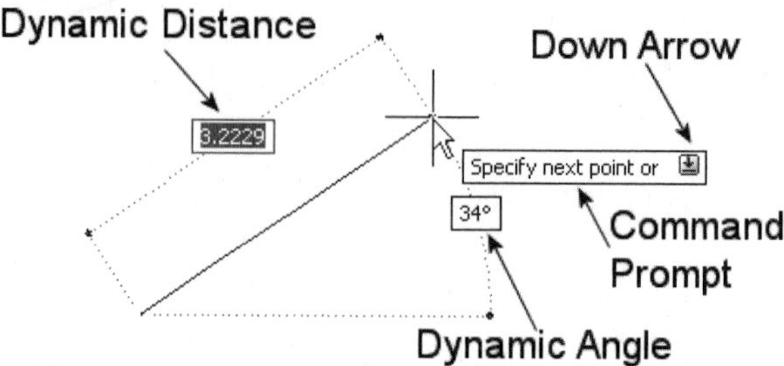

Figure 1.37

Command prompts for using Dynamic Input now appear in the drawing window next to the familiar AutoCAD cursor.

- When constructing line segments, dynamic dimensions in the form of a distance and an angle appear on the line. If the distance dimension is highlighted, entering a new value from your keyboard will change its value.
  - Pressing the TAB key switches you from the distance dimension to the angle dimension, where you can change its value.
- By default in Dynamic Input, coordinates for the second point of a line are considered relative. In other words, you do not need to type the @ symbol in front of the coordinate. The @ symbol means "last point" and will be discussed in greater detail later in this chapter.
- You can still enter relative and polar coordinates as normal using the @ symbol if you desire. These older methods of coordinate entry override the default dynamic input setting.
- The appearance of an arrow symbol in the Dynamic Input prompt area indicates that this command has options associated with it. To view these command options, press the DOWN ARROW key on your keyboard. These options will display on your screen. Continue pressing the DOWN ARROW until you reach the desired command option and then press the ENTER key to select it.
- Dynamic Input can be toggled ON or OFF in the status bar by clicking the DYN button or by pressing the F12 function key.

THE DIRECT DISTANCE MODE FOR DRAWING LINES

Another method is available for constructing lines, and it is called drawing by Direct Distance mode. In this method, the direction a line will be drawn in is guided by the location of the cursor. You enter a value, and the line is drawn at the specified distance at the angle specified by the cursor. This mode works especially well for drawing horizontal and vertical lines. The following image illustrates an example of how the Direct Distance mode is used.

Try It!: Create a new drawing from scratch. Turn Ortho mode on in the status bar. Then use the following command sequence to construct the line segments using the Direct Distance mode of entry.

```
Command: L (For LINE)
Specify first point: 2.00,2.00
Specify next point or [Undo]: (Move the cursor to the right and
    enter a value of 7.00 units)
Specify next point or [Undo]: (Move the cursor up and enter a
    value of 3.00 units)
Specify next point or [Close/Undo]: (Move the cursor to the left
    and enter a value of 4.00 units)
```

```
Specify next point or [Close/Undo]: (Move the cursor down and
    enter a value of 1.00 units)
Specify next point or [Close/Undo]: (Move the cursor to the left
    and enter a value of 2.00 units)
Specify next point or [Close/Undo]: C (To close the shape and
    exit the command)
```

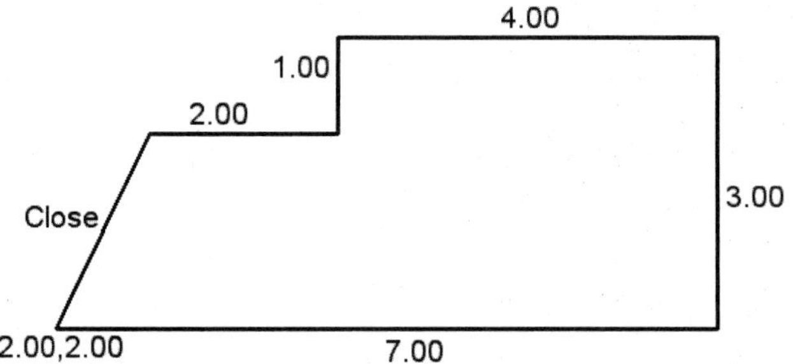

Figure 1.38

Tip: If Ortho mode is currently turned on, you can temporarily turn Ortho off while in the LINE command by pressing the SHIFT key as you drag your cursor to draw the next line.

The following image shows another example of an object drawn with Direct Distance mode. Each angle was constructed from the location of the cursor. In this example, Ortho mode is turned off.

Try It!: Create a new drawing from scratch. Be sure Ortho mode is turned off.

Then use the following command sequence to construct the line segments using the direct distance mode of entry.

```
Command: L (For LINE)
Specify first point: (Pick a point at "A")
Specify next point or [Undo]: (Move the cursor and enter 3.00)
Specify next point or [Undo]: (Move the cursor and enter 2.00)
Specify next point or [Close/Undo]: (Move the cursor and
    enter 1.00)
Specify next point or [Close/Undo]: (Move the cursor and
    enter 4.00)
Specify next point or [Close/Undo]: (Move the cursor and
    enter 2.00)
Specify next point or [Close/Undo]: (Move the cursor and
    enter 1.00)
Specify next point or [Close/Undo]: (Move the cursor and
    enter 1.00)
Specify next point or [Close/Undo]: C (To close the shape and
    exit the command)
```

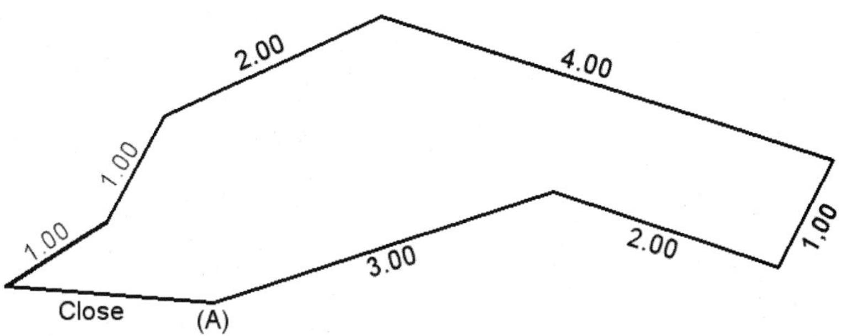

Figure 1.39

USING OBJECT SNAP FOR GREATER PRECISION

A major productivity tool that allows locking onto key locations of objects is Object Snap. The following image is an example of the construction of a vertical line connecting the endpoint of the fillet with the endpoint of the line at "A." The LINE command is entered and the Endpoint mode activated. When the cursor moves over a valid endpoint, an Object Snap symbol appears along with a tooltip indicating which OSNAP mode is currently being used. Another example of the use of Object Snap is in dimensioning applications where exact endpoints and intersections are needed.

Try It!: Open the drawing file 01_Endpoint. Use the illustration in the following image and the command sequence below to draw a line segment from the endpoint of the arc to the endpoint of the line.

```
Command: L (For LINE)
Specify first point: End (For Endpoint mode)
of (Pick the endpoint of the fillet at "A" illustrated in the
    following image)
Specify next point or [Undo]: End (For Endpoint mode)
of (Pick the endpoint of the line at "B")
Specify next point or [Undo]: (Press ENTER to exit this command)
```

Perform the same operation to the other side of this object using the Endpoint mode of OSNAP. The results are shown on the right in the following image.

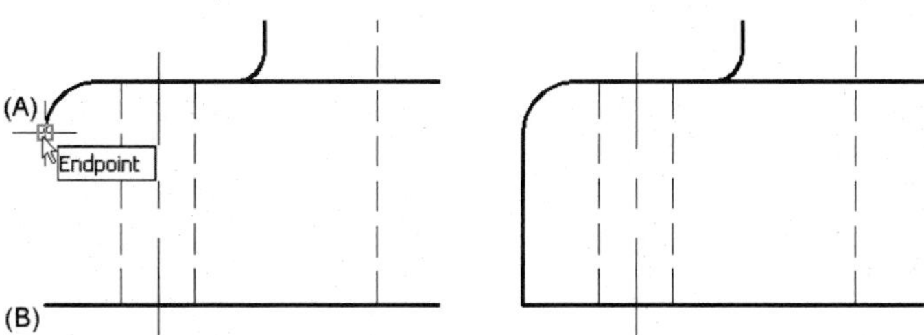

Figure 1.40

Object Snap modes can be selected in a number of different ways. Illustrated in the following image is the standard Object Snap toolbar. The following table gives a brief description of each Object Snap mode. In this table, notice the Key-In column. When the Object Snap modes are executed from keyboard input, only the first three letters are required.

Figure 1.41

The following table gives a brief description of each Object Snap mode:

Button	Tool	Key-In	Function
	Center	Cen	Snaps to the centers of circles and arcs
	Endpoint	End	Snaps to the endpoints of lines and arcs
	Extension	Ext	Creates a temporary extension line or arc when your cursor passes over the endpoint of objects; you can specify new points along the temporary line
	From	Fro	Snaps to a point at a specified distance and direction from a selected reference point
	Insert	Ins	Snaps to the insertion point of blocks and text
	Intersection	Int	Snaps to the intersections of objects
	Apparent Intersection	Int	Mainly used in creating 3D wireframe models; finds the intersection of points not located in the same plane
	Midpoint	Mid	Snaps to the midpoint of lines and arcs
	Midpoint Between 2 Points	M2P	Snaps to the middle of two selected points
	Nearest	Nea	Snaps to the nearest point found along any object
	Node	Nod	Snaps to point objects (including dimension definition points) and text objects (including multiline text and dimension text)
	None	Non	Disables object snap
	Osnap Settings	Osnap	Launches the Drafting Settings dialog box and activates the Object Snap tab
	Parallel	Par	Draws an object parallel to another object
	Perpendicular	Per	Snaps to a perpendicular location on an object
	Quadrant	Qua	Snaps to four key points located on a circle
	Tangent	Tan	Snaps to the tangent location of arcs and circles

The following image shows the Object Snap modes that can be activated when you hold down SHIFT or CTRL and press the right mouse button while within a command such as LINE or MOVE. This shortcut menu will appear wherever the cursor is currently positioned.

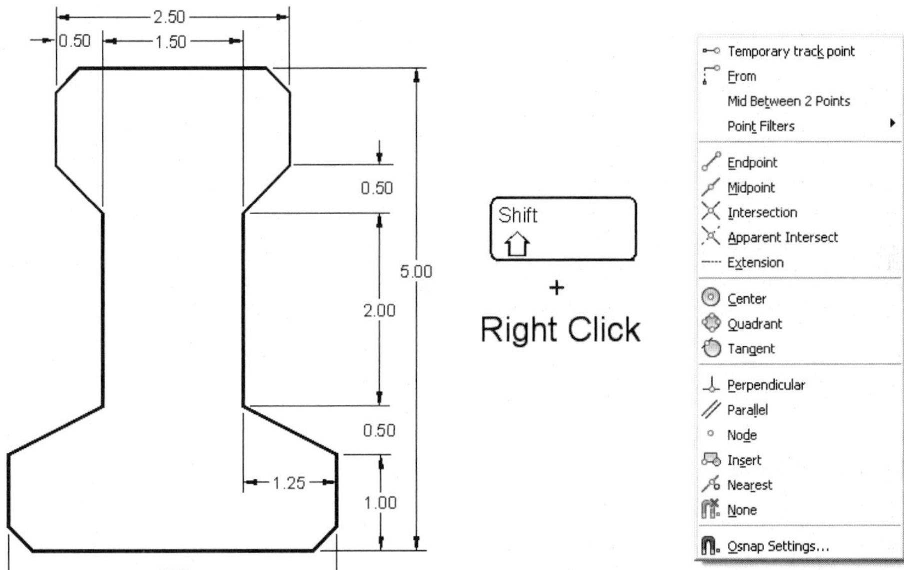

Figure 1.42

OBJECT SNAP MODES

CENTER (CEN)

Use the Center mode to snap to the center of a circle or arc. To accomplish this, activate the mode by clicking the Center button and moving the cursor along the edge ofthe circle or arc, as shown in the following image. Notice the AutoSnap symbol appearing at the center of the circle or arc.

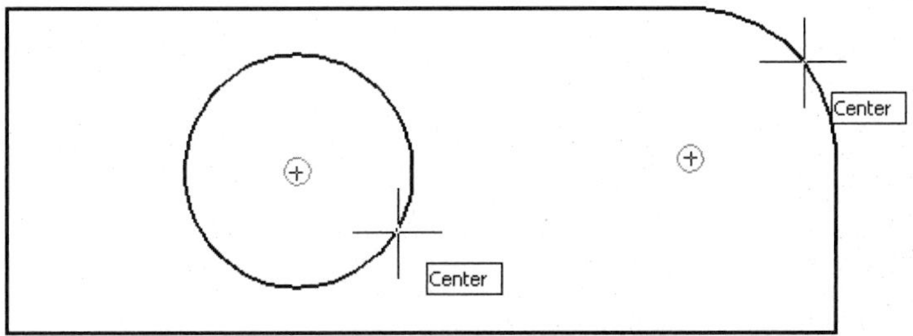

Figure 1.43

ENDPOINT (END)

The Endpoint mode is one of the more popular Object Snap modes; it is helpful in snapping to the endpoints of lines or arcs as shown in the following image. One application of Endpoint is during the dimensioning process, where exact distances are needed to produce the desired dimension. Activate this mode by clicking the Endpoint button, and then move the cursor along the edge of the object to snap to the endpoint. In the case of the line or arc shown in the following image, the cursor does not actually have to be positioned at the endpoint; favoring one end automatically snaps to the closest endpoint.

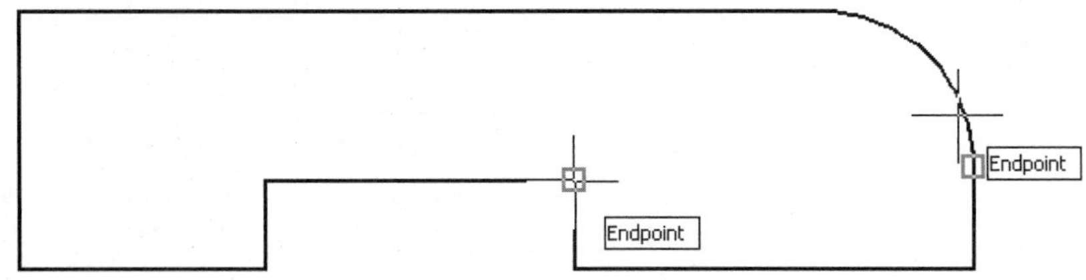

Figure 1.44

EXTENSION (EXT)

When you acquire a line or an arc, the Extension mode creates a temporary path that extends from the object. Once the Extension object snap is selected, move your cursor over the end of the line at "A," as shown in the following image, to acquire it. Moving your cursor away provides an extension at the same angle as the line. To un-acquire an extension, simply move your cursor over the end of the line again. A tooltip displays the current extension distance and angle. Acquiring the end of the arc at "B" provides the radius of the arc and displays the current length in the tooltip.

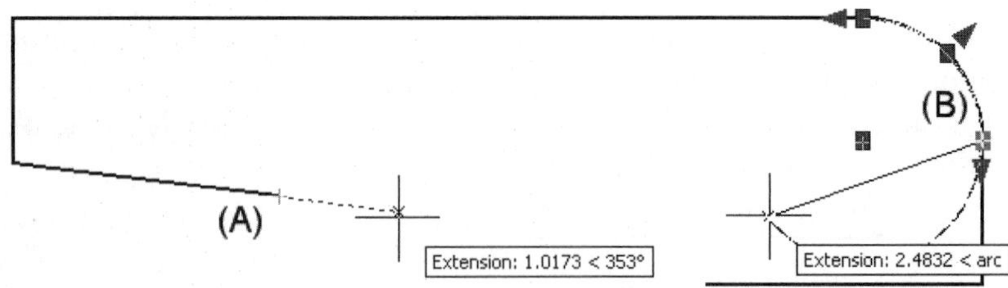

Figure 1.45

FROM (FRO)

Use the From mode along with a secondary Object Snap mode to establish a reference point and construct an offset from that point. Open the drawing file 01_Osnap From. In the following image, the circle needs to be drawn 1.50 units in the X and Y directions from point "A." The CIRCLE command is activated and the Object Snap From mode is used in combination with the Object Snap Intersection mode. The From option requires a base point. Identify the base point at the intersection of corner "A." The next prompt asks for an offset value; enter the relative coordinate value of @1.50,1.50 (this identifies a point

1.50 units in the positive X direction and 1.50 units in the positive Y direction). This completes the use of the From option and identifies the center of the circle at "B." Study the following command sequence to accomplish this operation:

Try It!: Open the drawing file 01_Osnap From. Use the illustration and prompt sequence below for constructing a circle inside the shape with the aid of the Object Snap From mode.

```
Command: C (For CIRCLE)
Specify center point for circle or [3P/2P/Ttr (tan tan
    radius)]: From
Base point: Int (For Intersection Mode)
of (Select the intersection at "A" in the following image)
<Offset>: @1.50,1.50
Specify radius of circle or [Diameter]: D (For Diameter)
Specify diameter of circle: 1.25
```

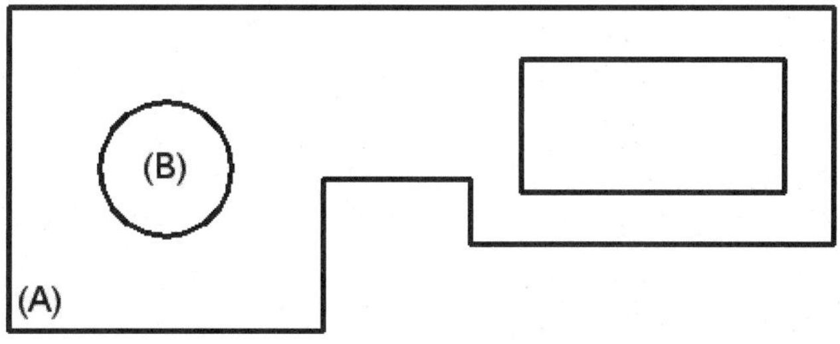

Figure 1.46

INSERT (INS)

The Insert mode snaps to the insertion point of an object. In the case of the text object in the following image, activating the Insert mode and positioning the cursor anywhere on the text snaps to its insertion point, in this case at the lower-left corner of the text at "A." The other object illustrated in the following image is called a block. It appears to be constructed of numerous line objects; however, all objects that make up the block are considered to be a single object. Blocks can be inserted in a drawing. Typical types of blocks are symbols such as doors, windows, bolts, and so on—anything that is used several times in a drawing. In order for a block to be brought into a drawing, it needs an insertion point, or a point of reference. The Insert mode, when you position the cursor on a block, will snap to the insertion point at "B" of that block.

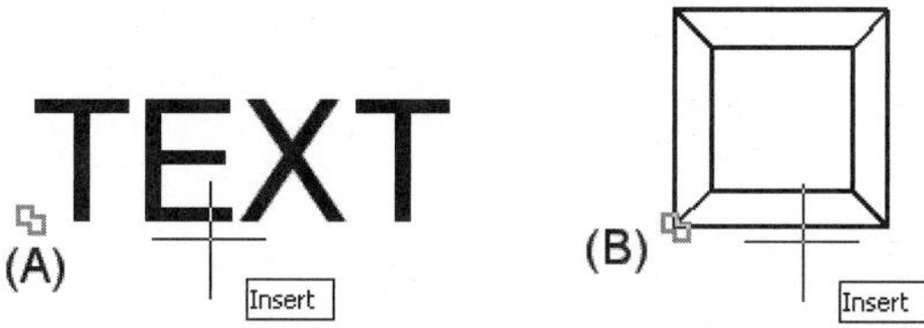

Figure 1.47

INTERSECTION (INT)

Another popular Object Snap mode is Intersection. Use this mode to snap to the intersection of two objects. Position the cursor anywhere near the intersection of two objects and the intersection symbol appears. See the following image.

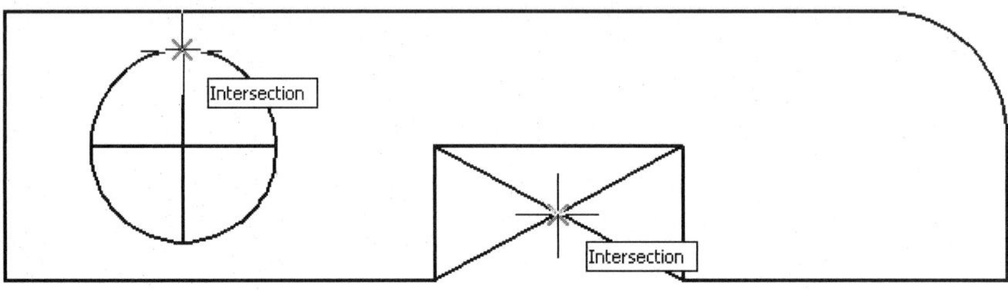

Figure 1.48

EXTENDED INTERSECTION (INT)

Another type of intersection snap is the Extended Intersection mode, which is used to snap to an intersection not considered obvious from the previous example. The same Object Snap Intersection button is utilized for performing an extended intersection operation. The following image shows two lines that do not intersect. Activate the Extended Intersection mode and pick both lines. Notice the intersection symbol present where the two lines, if extended, would intersect.

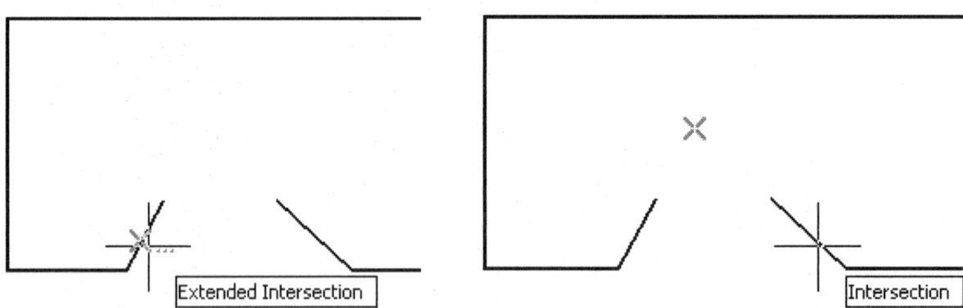

Figure 1.49

MIDPOINT (MID)

The Midpoint mode snaps to the midpoint of objects. Line and arc examples are shown in the following image. When activating the Midpoint mode, touch the object anywhere with some portion of the cursor; the midpoint symbol appears at the exact midpoint of the object.

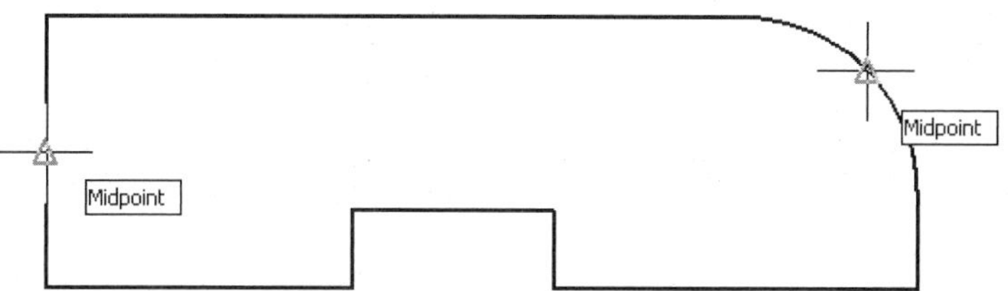

Figure 1.50

MIDPOINT OF TWO SELECTED POINTS (M2P)

This Object Snap mode snaps to the midpoint of two selected points. To access this mode, type either M2P or MTP at the command prompt. While this Object Snap mode is not accessible through a toolbar button, it can be found by pressing SHIFT + Right Mouse Button to display the Object Snap menu, as shown on the left in the following image.

The following command sequence and illustration in the following image show the construction of a circle at the midpoint of two selected points.

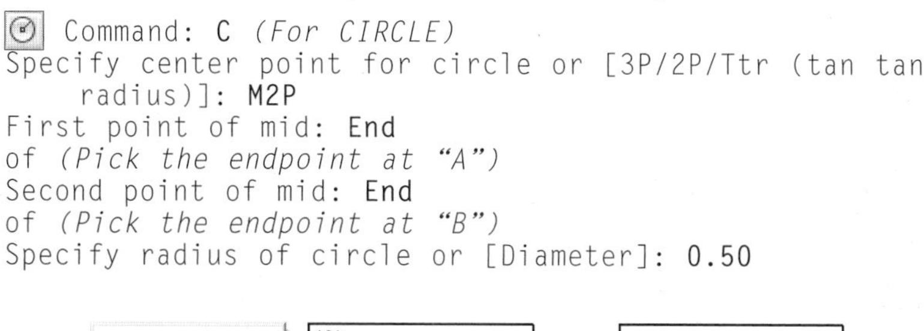

```
Command: C (For CIRCLE)
Specify center point for circle or [3P/2P/Ttr (tan tan
    radius)]: M2P
First point of mid: End
of (Pick the endpoint at "A")
Second point of mid: End
of (Pick the endpoint at "B")
Specify radius of circle or [Diameter]: 0.50
```

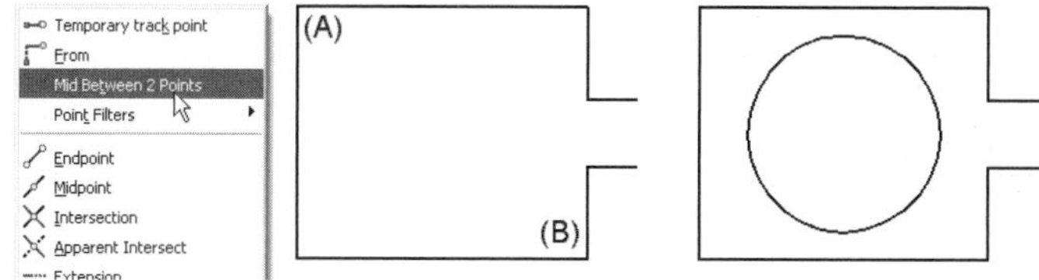

Figure 1.51

NEAREST (NEA)

The Nearest mode snaps to the nearest point it finds on an object. Use this mode when a point on an object needs to be selected and an approximate location on the object is sufficient. The nearest point is calculated based on the closest distance from the intersection of the crosshairs perpendicular to the object or the shortest distance from the crosshairs to the object. In the following image, the appearance of the Nearest symbol helps to show where the point identified by this mode is actually located.

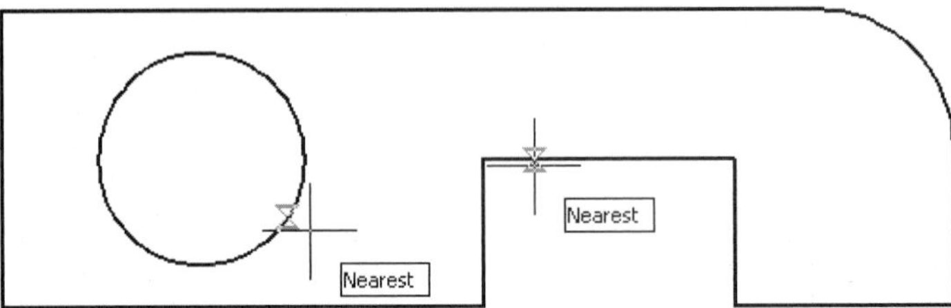

Figure 1.52

NODE (NOD)

The Node mode snaps to a node or point. Picking the point in the following image snaps to its center.

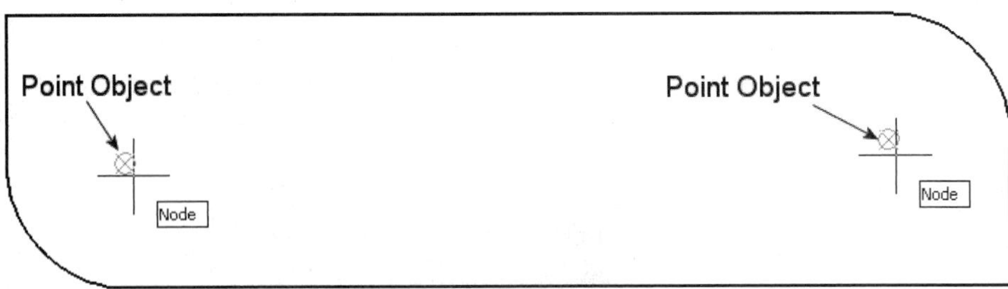

Figure 1.53

PARALLEL (PAR)

Use the Parallel mode to construct a line parallel to another line. In the following image, the LINE command is started and a beginning point of the line is picked. The Parallel mode is activated by selecting the Parallel icon and then hovering the cursor over the existing line. The existing line is highlighted at "A" and the Parallel symbol appears. Finally, moving the cursor to the approximate position that makes the new line parallel to the one just selected allows the Parallel mode to construct a parallel line, the tracking path and the tooltip giving the current distance and angle. The result of this mode is illustrated on the right in the following image.

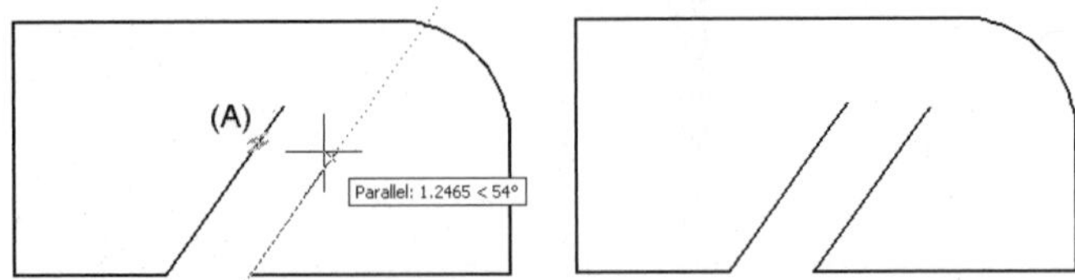

Figure 1.54

PERPENDICULAR (PER)

The Perpendicular mode is helpful for snapping to an object normal (or perpendicular) from a previously identified point. The following image shows a line segment drawn perpendicular from the point at "A" to the inclined line "B." A 90° angle is formed with the perpendicular line segment and the inclined line "B." With this mode, the user can also construct lines perpendicular to circles.

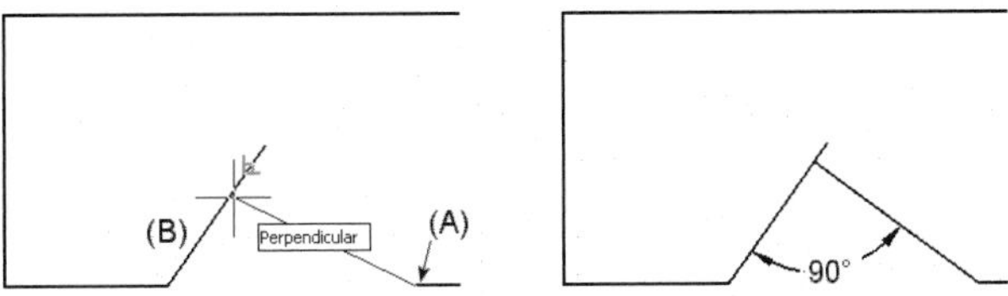

Figure 1.55

QUADRANT (QUA)

Circle quadrants are defined as points located at the 0°, 90°, 180°, and 270° positions of a circle, as in the following image. Using the Quadrant mode will snap to one of these four positions as the edge of a circle or arc is selected. In the example of the circle in the following image, the edge of the circle is selected by the cursor location. The closest quadrant to the cursor is selected.

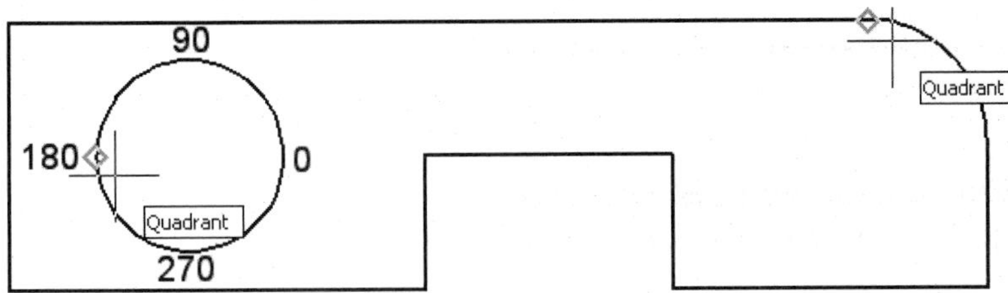

Figure 1.56

TANGENT (TAN)

The Tangent mode is helpful in constructing lines tangent to other objects such as the circles in the following image. In this case, the Deferred Tangent mode is being used in conjunction with the LINE command. The point at "A" is first picked at the bottom of the circle using the Tangent mode. When dragged to the next location, the line will be tangent at point "A." Then, with Tangent mode activated and the location at "B" picked, the line will be tangent to the large circle near "B." The results are illustrated on the right in the following image.

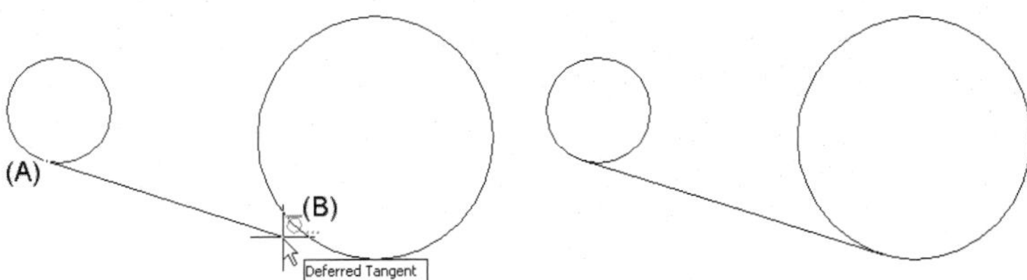

Figure 1.57

Try It!: Open the drawing file 01_Tangent. Follow this command sequence for constructing a line segment tangent to two circles:

```
Command: L (For LINE)
Specify first point: Tan (For Tangent mode)
from (Select the circle near "A")
Specify next point or [Undo]: Tan (For Tangent mode)
to (Select the circle near "B")
```

Try It!: Open the drawing file 01_Osnap. Various objects consisting of lines, circles, arcs, points, and blocks need to be connected with line segments at their key locations. Use the prompt sequence and the following image for performing this operation.

```
Command: L (For LINE)
Specify first point: End
of (Pick the endpoint at "A")
Specify next point or [Undo]: Nod
of (Pick the node at "B")
Specify next point or [Undo]: Tan
to (Pick the circle at "C")
Specify next point or [Close/Undo]: Int
of (Pick the intersection at "D")
Specify next point or [Close/Undo]: Int
of (Pick the line at "E")
and (Pick the horizontal line at "F")
Specify next point or [Close/Undo]: Qua
of (Pick the circle at "G")
Specify next point or [Close/Undo]: Cen
of (Pick the arc at "H")
Specify next point or [Close/Undo]: Mid
of (Pick the line at "J")
Specify next point or [Close/Undo]: Per
to (Pick the line at "K")
Specify next point or [Close/Undo]: Ins
of (Pick on the I-Beam symbol near "L")
Specify next point or [Close/Undo]: Nea
to (Pick the circle at "M")
Specify next point or [Close/Undo]: (Press ENTER to exit this
    command)
```

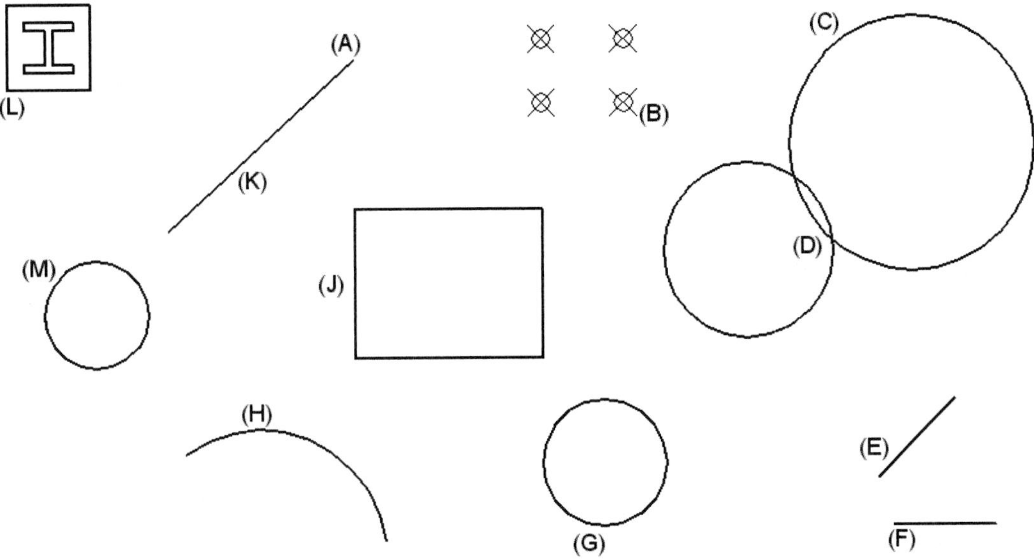

Figure 1.58

CHOOSING RUNNING OBJECT SNAP

So far, all Object Snap modes have been selected from the shortcut menu or entered at the keyboard. The problem with these methods is that if you're using a certain mode over a series of commands, you have to select the Object Snap mode every time. It is possible to make the Object Snap mode or modes continuously present through Running Osnap. Right-click the OSNAP button located in the status bar at the bottom of the drawing area, as shown in the following image. Selecting Settings from the shortcut menu activates the Drafting Settings dialog box illustrated in the following image. By default, the Endpoint, Center, Intersection, and Extension modes are automatically selected. Whenever the cursor lands over an object supported by one of these four modes, a yellow symbol appears to alert the user to the mode. It is important to know that when changes are made inside this dialog box, the changes are automatically saved to the system. Even if a drawing session is terminated without saving, the changes made to the Object Snap modes are automatic. Other Object Snap modes can be selected by checking their appropriate boxes in the dialog box; removing the check disables the mode.

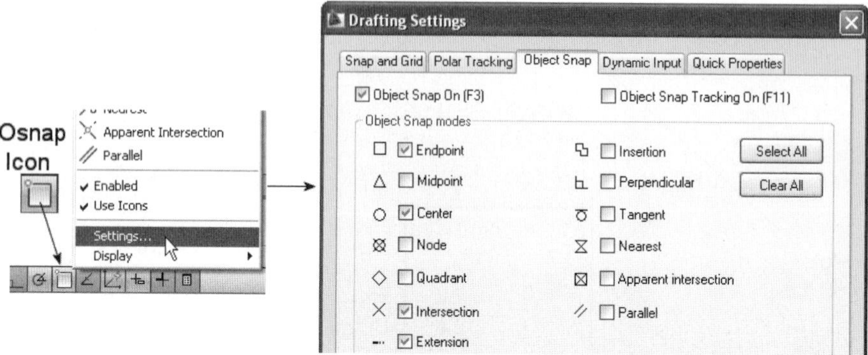

Figure 1.59

These Object Snap modes remain in effect during drawing until you click the OSNAP button illustrated in the status bar in the following image; this turns off the current running Object Snap modes. To reactivate the running Object Snap modes, click the OSNAP button again and the previous set of Object Snap modes will be back in effect. Object Snap can also be activated and reactivated by pressing the F3 function key. It is also important to note that anytime you select an Object Snap from the toolbar, cursor menu, or by typing it in, it overrides the Running Osnap for that single operation. This ensures that you only snap to that specific mode and not accidentally to one of the set Running Osnaps.

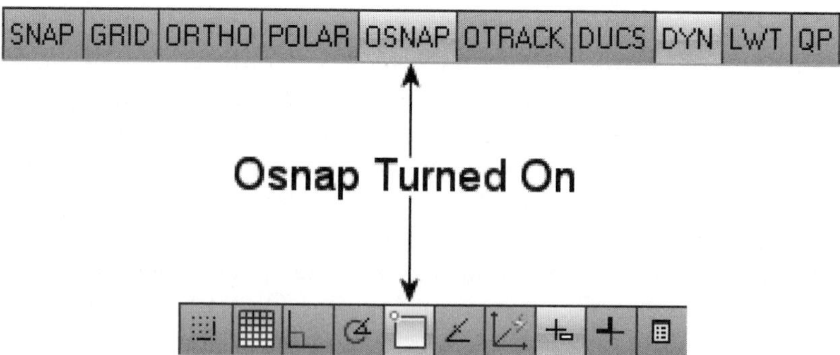

Figure 1.60

POLAR TRACKING

Previously in this chapter, the Direct Distance mode was highlighted as an efficient means of constructing orthogonal lines without the need for using one of the coordinate modes of entry. However, it did not lend itself to lines at angles other than 0°, 90°, 180°, and 270°. Polar Tracking has been designed to remedy this. This mode allows the cursor to follow a tracking path that is controlled by a preset angular increment. The POLAR button located at the bottom of the display in the Status area turns this mode on or off. Right-clicking

POLAR in the status bar at the bottom of the screen and picking Settings will display the Drafting Settings dialog box shown in the following image and the Polar Tracking tab.

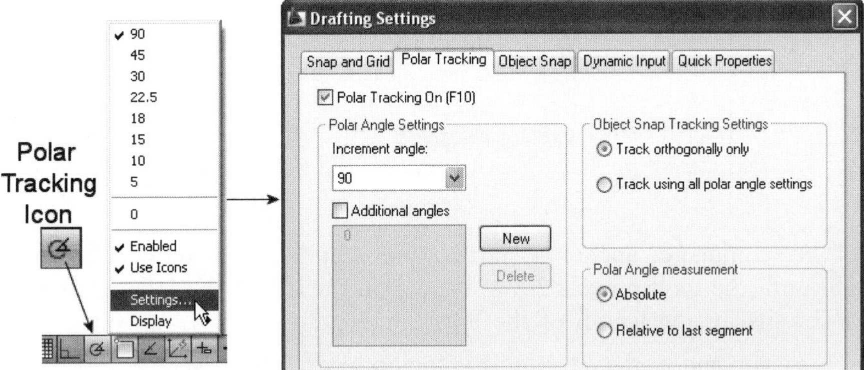

Figure 1.61

A few general terms are defined before continuing:

Tracking Path—This is a temporary dotted line that can be considered a type of construction line. Your cursor will glide or track along this path (see the following image).

Tooltip—This displays the current cursor distance and angle away from the tracking point (see the following image).

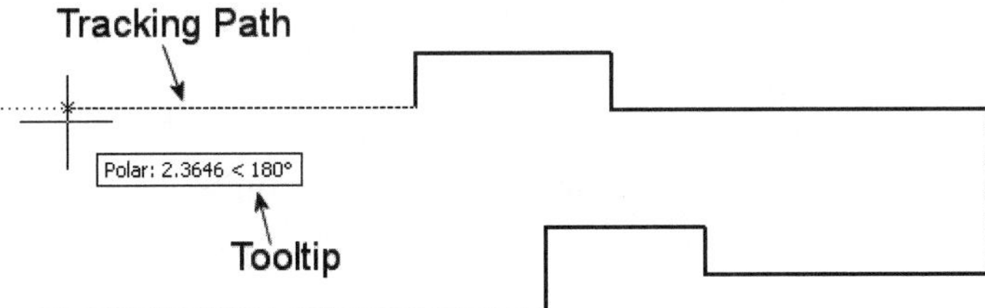

Figure 1.62

Tip: Both POLAR and ORTHO modes cannot be turned on at the same time. Once you turn POLAR on, ORTHO automatically turns off, and vice versa.

Try It!: To see how the Polar Tracking mode functions, construct an object that consists of line segments at 10° angular increments. Create a new drawing starting from scratch. Then, set the angle increment through the Polar Tracking tab of the Drafting Settings dialog box to 10°, as shown in the following image.

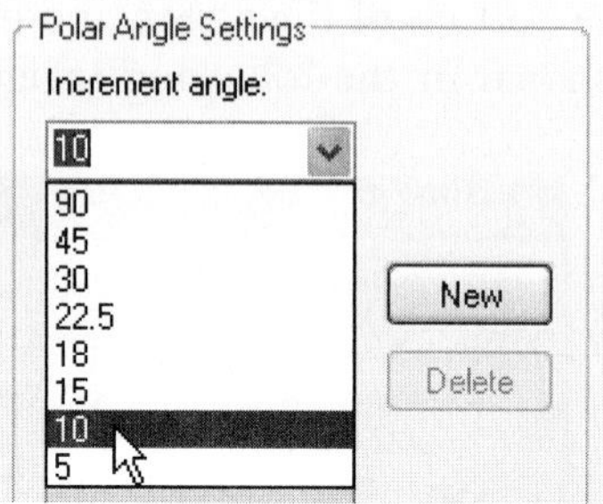

Figure 1.63

Start the LINE command, anchor a starting point at "A," and move the cursor to the upper-right until the tooltip reads 20° as shown on the left in the following image. Enter a value of 2 units for the length of the line segment.

Move the cursor up and to the left until the tooltip reads 110°, as shown on the right in the following image, and enter a value of 2 units. (This will form a 90° angle with the first line.)

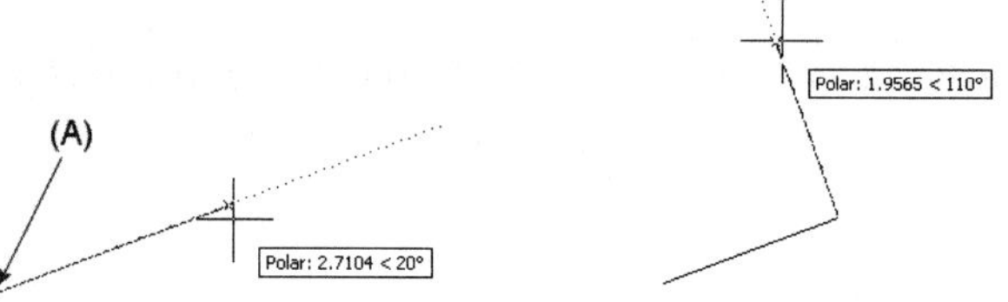

Figure 1.64

Move the cursor until the tooltip reads 20°, as shown on the left in the following image, and enter a value of 1 unit.

Move the cursor until the tooltip reads 110°, as shown on the right in the following image, and enter a value of 1 unit.

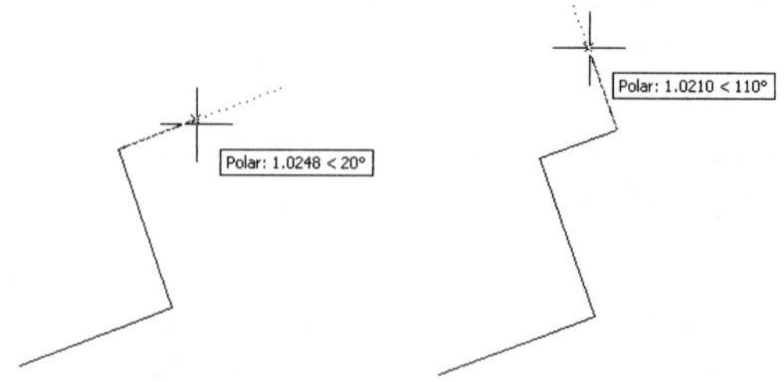

Figure 1.65

Move the cursor until the tooltip reads 200°, as shown on the left in the following image, and enter a value of 3 units.

Move the cursor to the endpoint, as shown on the right in the following image, or use the Close option of the LINE command to close the shape and exit the command.

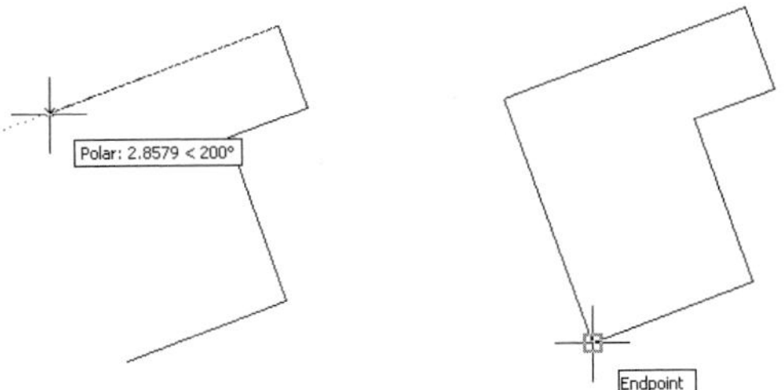

Figure 1.66

SETTING A POLAR SNAP VALUE

An additional feature of using polar snap is illustrated in the following image. Clicking the Snap and Grid tab of the Drafting Settings dialog box displays the dialog box in the following image. Clicking the Polar Snap option found along the lower-left corner of the dialog box allows the user to enter a polar distance. When SNAP and POLAR are both turned on and the cursor is moved to draw a line, not only will the angle be set but the cursor will also jump to the next increment set by the polar snap value.

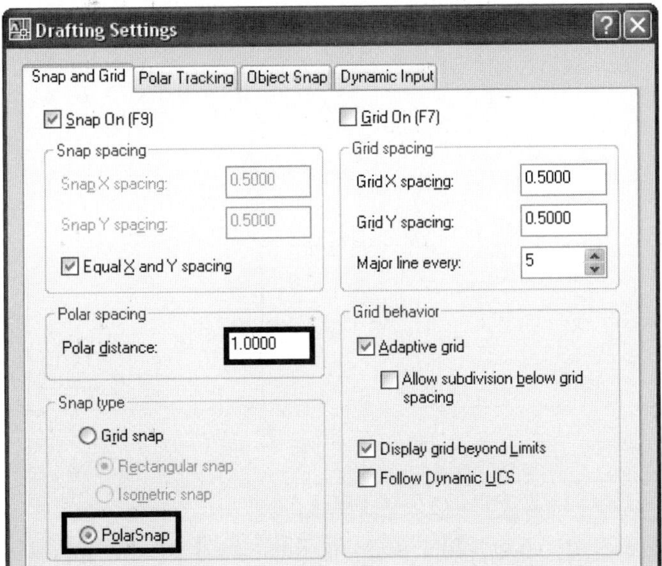

Figure 1.67

Try It!: Open the drawing file 01_Polar. Set the polar angle to 30 degrees and a polar snap distance to 0.50 unit increments. Be sure POLAR and SNAP are both turned on in your status bar and that all other modes are turned off. Begin constructing the object in the following image using the command prompt sequence below as a guide.

```
Command: L (For LINE)
Specify first point: 7.00,4.00
```

```
Specify next point or [Undo]: (Move your cursor down until the
    tooltip reads Polar: 2.5000<270 and pick a point)
Specify next point or [Undo]: (Move your cursor right until the
    tooltip reads Polar: 1.5000<0 and pick a point)
Specify next point or [Close/Undo]: (Polar: 2.0000<30)
Specify next point or [Close/Undo]: (Polar: 2.0000<60)
Specify next point or [Close/Undo]: (Polar: 2.5000<90)
Specify next point or [Close/Undo]: (Polar: 3.0000<150)
Specify next point or [Close/Undo]: (Polar: 1.5000<180)
Specify next point or [Close/Undo]: (Polar: 2.5000<240)
Specify next point or [Close/Undo]: (Polar: 2.5000<120)
Specify next point or [Close/Undo]: (Polar: 1.5000<180)
Specify next point or [Close/Undo]: (Polar: 3.0000<210)
Specify next point or [Close/Undo]: (Polar: 2.5000<270)
Specify next point or [Close/Undo]: (Polar: 2.0000<300)
Specify next point or [Close/Undo]: (Polar: 2.0000<330)
Specify next point or [Close/Undo]: (Polar: 1.5000<0)
Specify next point or [Close/Undo]: (Polar: 2.5000<90)
Specify next point or [Close/Undo]: C (To close the shape)
```

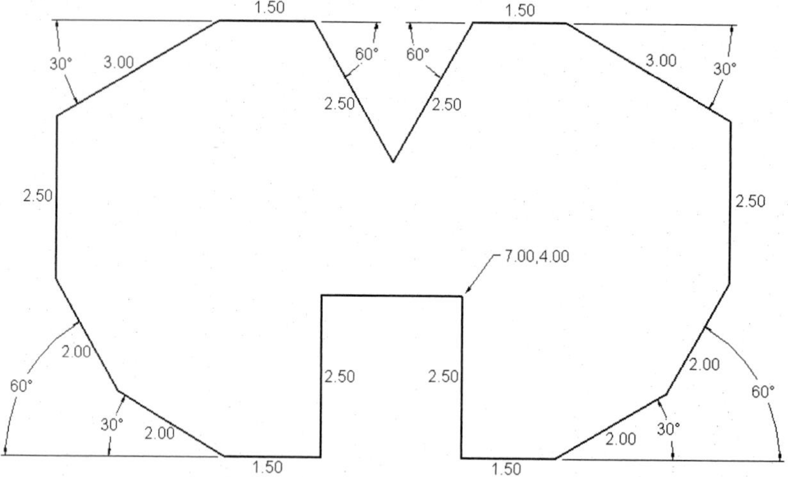

Figure 1.68

SETTING A RELATIVE POLAR ANGLE

An additional feature of using polar snap is illustrated in the following image. When activating the Polar tab of the Drafting Settings dialog box, located in the lower-right corner are two settings that deal with the Polar Angle measurement; they are Absolute and Relative to last segment.

> **Absolute**—This is the default setting when dealing with Polar Angle measurement. This setting controls all angle measurements based on the position of the current user coordinate system, the icon located in the lower-left corner of all AutoCAD drawing screens.

Relative to last segment—When changing the Polar Angle measurement to Relative to last segment, the Polar Tracking angle is based on the last line segment drawn.

Figure 1.69

To get a better understanding of the two different Polar Angle measurement settings, study the following image. The illustration on the left is an example of the Absolute setting. The 150° angle was drawn from point (A) to point (B). This angle is derived from the absolute position of angle 0 (zero) set by default to the 3 o'clock position as defined in the Drawing Units dialog box. In the illustration on the right, the same line segment is drawn. However, this time the Relative to last segment setting is used. Notice how the 120° angle is calculated. The angle is based on the last line segment, not on an angle calculated in relation to 0 (zero). This is the reason for the parallel line segment at (C).

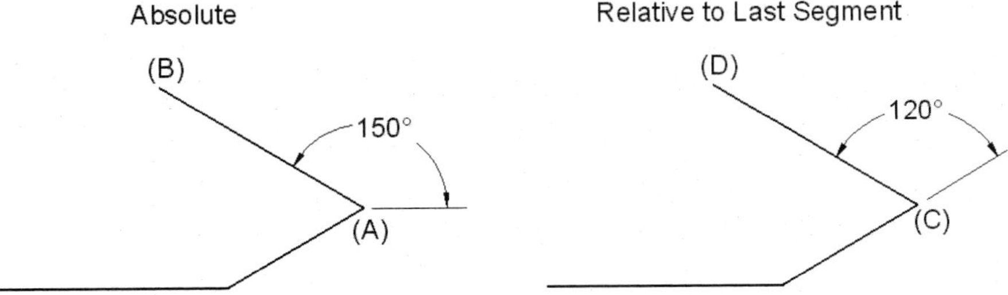

Figure 1.70

OBJECT SNAP TRACKING MODE

Object Snap Tracking works in conjunction with Object Snap. Before you can track from an Object Snap point, you must first set an Object Snap mode or modes from the Object Snap tab of the Drafting Settings dialog box. Object Snap Tracking can be toggled ON or OFF with the OTRACK button, which is located in the status bar shown in the following image.

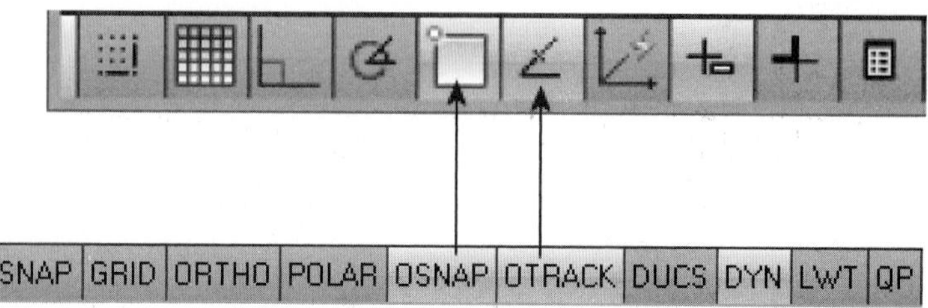

Figure 1.71

The advantage of using Object Snap Tracking is in the ability to choose or acquire points to be used for construction purposes. Acquired points are temporarily selected by hovering the cursor over the point versus selecting with the mouse. Care must be taken when acquiring points that the points are in fact not picked. They are used only for construction purposes. For example, two line segments need to be added to the object, as shown on the left in the following image, to form a rectangle. Here is how to perform this operation using Polar Tracking.

Try It!: Open the drawing file 01_Otrack Lines. Notice in the status bar that POLAR, OSNAP, and OTRACK are all turned on. Be sure that Running Osnap is set to Endpoint mode. Enter the LINE command and pick a starting point for the line at "A." Then, move the cursor directly to the left until the tooltip reads 180° as shown on the left in the following image. The starting point for the next line segment is considered acquired.

Rather than enter the length of this line segment, move your cursor over the top of the corner at "B" to acquire this point (be careful not to pick the point here). Then move your cursor up until the tooltip reads 90° as shown on the right in the following image.

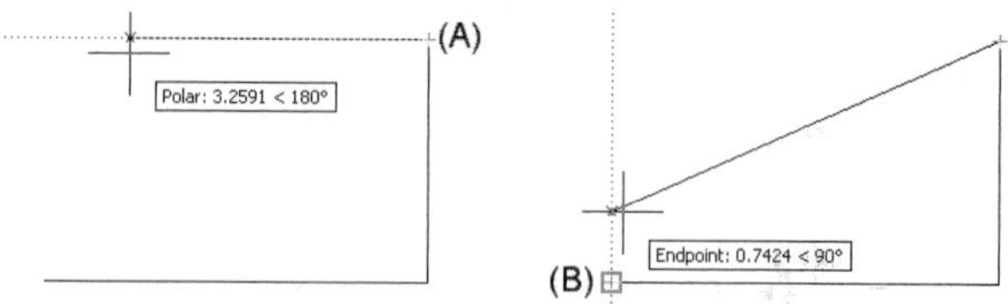

Figure 1.72

Move your cursor up until the tooltip now reads angles of 90° and 180°. Also notice the two tracking paths intersecting at the point of the two acquired points. Picking this point at "C" will construct the horizontal line segment as shown on the left in the following image. Finally, slide your cursor to the endpoint at "D" to complete the rectangle as shown on the right in the following image.

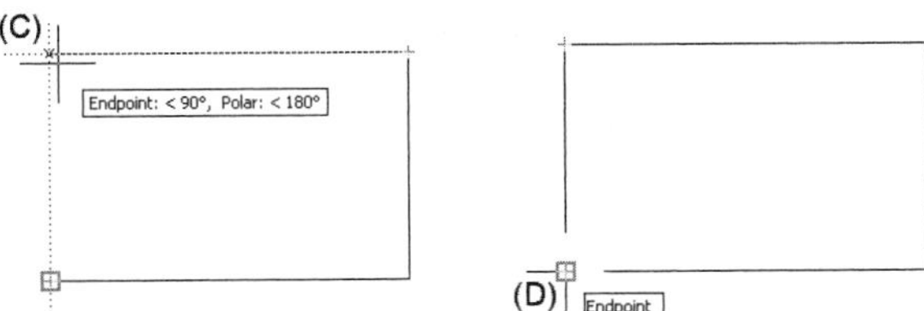

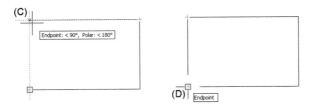

Figure 1.73

Tip: Pausing the cursor over an existing acquired point a second time removes the tracking point from the object.

Try It!: Open the drawing file 01_Otrack Pipes. Set Running Osnap mode to Midpoint. Set the polar angle to 90 degrees. Be sure POLAR, OSNAP, and OTRACK are all turned on. Enter the line command and connect all fittings with lines illustrated in the following image.

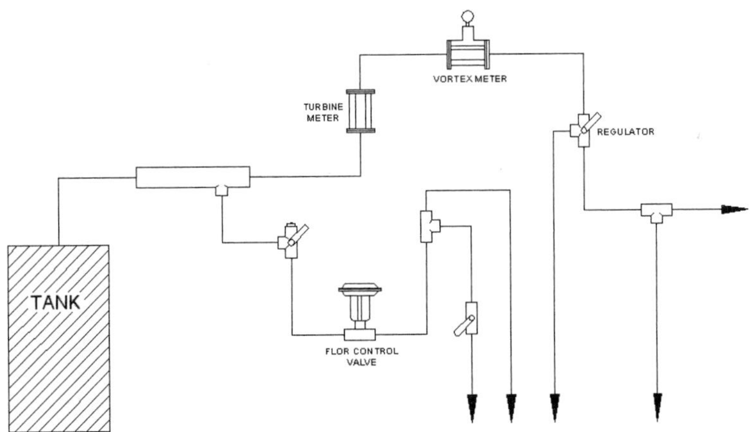

Figure 1.74

USING TEMPORARY TRACKING POINTS

A very powerful construction tool includes the ability to use an extension path along with the Temporary Tracking Point tool to construct objects under difficult situations, as illustrated in the following image.

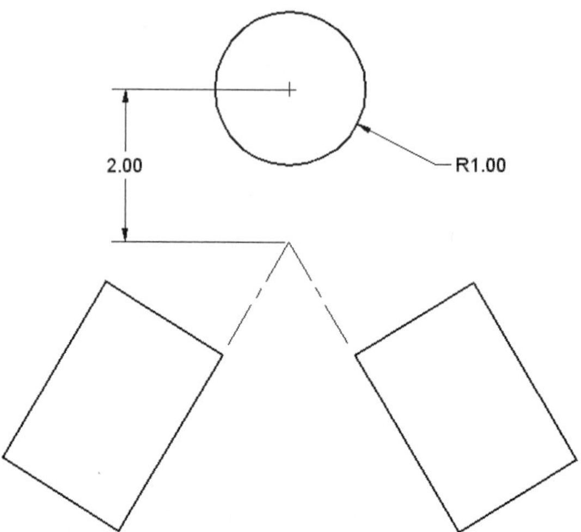

Figure 1.75

An extension path is similar to a tracking path except that it is present when the Object Snap Extension mode is activated. To construct the circle in relation to the two inclined rectangles, follow the next series of steps.

Try It!: Open the drawing file 01_Temporary Point. Set Running Osnap to Endpoint, Intersection, Center, and Extension. Check to see that OSNAP and OTRACK are turned on and all other modes are turned off. Activate the circle command; this prompts you to specify the center point for the circle. Move the cursor over the corner of the right rectangle at "A" to acquire this point as shown on the left in the following image. Move the cursor up and to the left, making sure the tooltip lists the Extension mode.

With the point acquired at "A," move the cursor over the corner of the left rectangle at "B" and acquire this point. Move the cursor up and to the right, making sure the tooltip lists the Extension mode, as shown on the right in the following image.

Move the cursor until both acquired points intersect as shown on the right in the following image. The center of the circle is located 2 units above this intersection. Click the Object Snap Temporary Tracking button and pick this intersection as shown on the right in the following image.

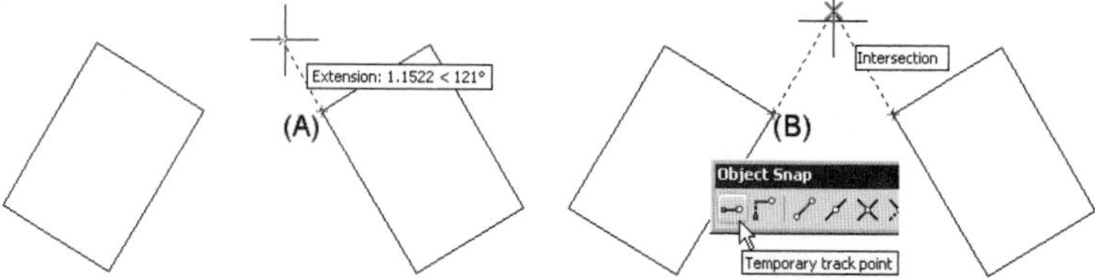

Figure 1.76

Next, move the cursor directly above the temporary tracking point, as shown on the left in the following image. The tooltip should read 90°. Entering a value of 2 units identifies the center of the circle.

The completed construction operation is illustrated as the right in the following image.

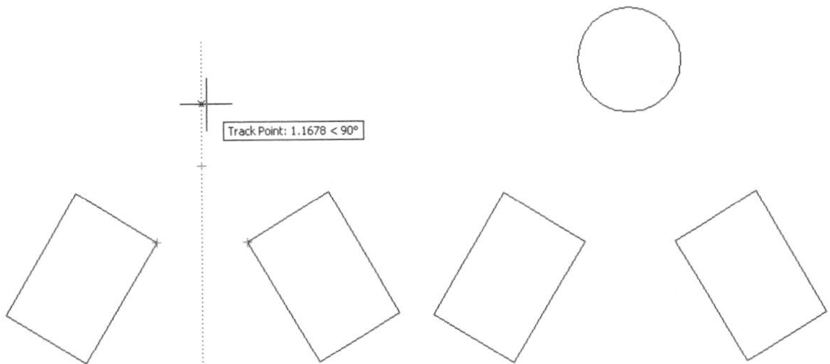

Figure 1.77

ALTERNATE METHODS USED FOR PRECISION DRAWING: CARTESIAN COORDINATES

Before drawing precision geometry such as lines and circles, it is essential to have an understanding of coordinate systems. The Cartesian or rectangular coordinate system is a system constructed of an orthogonal axis intersecting at an origin that creates four quadrants, allowing location of any point by specifying the coordinates. A coordinate is made up of a horizontal and vertical pair of numbers identified as X and Y. The coordinates are then plotted on a type of graph or chart. An example of a rectangular coordinate system is shown in the following image. The coordinates of the origin are 0,0. From the origin, all positive directions move up and to the right. All negative directions move down and to the left.

The coordinate axes are divided into four quadrants that are labeled I, II, III, and IV, as shown in the following image. In Quadrant I, all X and Y values are positive. Quadrant II has a negative X value and positive Y value. Quadrant III has negative values for X and Y. Quadrant IV has positive X values and negative Y values.

Note: When you begin a drawing in AutoCAD, the screen display reflects Quadrant I of the Cartesian coordinate system, as shown in the following image. The origin 0,0 is located in the lower-left corner of the drawing screen.

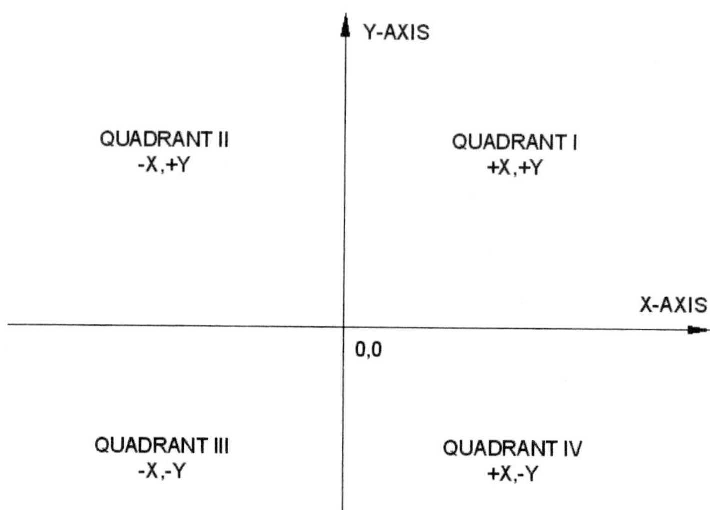

Figure 1.78

For each set of (X,Y) coordinates, X values represent distances from the origin horizontally to the right if positive and horizontally to the left if negative. Y values represent distances from the origin vertically up if positive and vertically down if negative. The following image shows a series of coordinates plotted on the number lines. One coordinate is identified in each quadrant to show the positive and negative values. As an example, coordinate 3,2 in Quadrant I represents a point 3 units to the right and 2 units vertically up from the origin. The coordinate –5,3 in Quadrant II represents a point 5 units to the left and 3 units vertically up from the origin. Coordinate –2,–2 in Quadrant III represents a point 2 units to the left and 2 units vertically down from the origin. Lastly, coordinate 2,–4 in Quadrant IV represents a point 2 units to the right and 4 units vertically down from the origin.

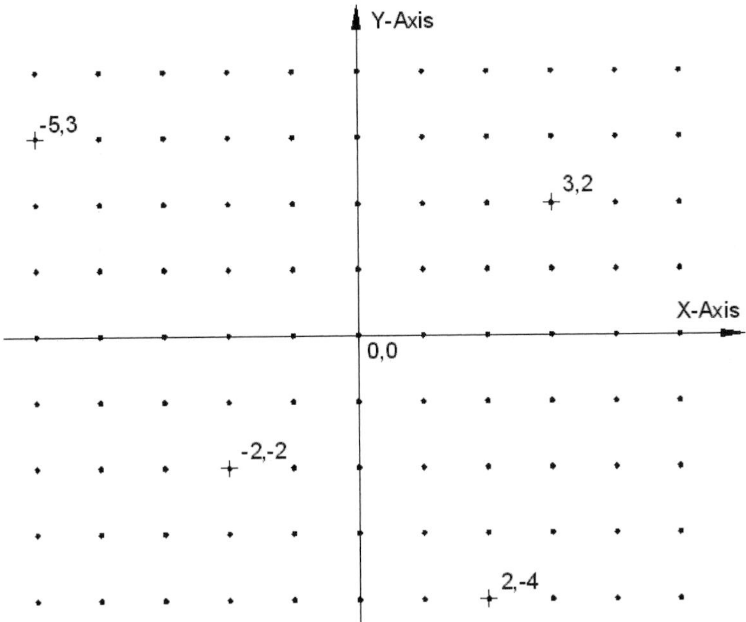

Figure 1.79

ABSOLUTE COORDINATE MODE FOR DRAWING LINES

When drawing geometry such as lines, the user must use a method of entering precise distances, especially when accuracy is important. This is the main purpose of using coordinates. The simplest and most elementary form of coordinate values is absolute coordinates. Absolute coordinates conform to the following format:

X,Y

One problem with using absolute coordinates is that all coordinate values refer back to the origin 0,0. This origin on the AutoCAD screen is usually located in the lower-left corner when a new drawing is created. The origin will remain in this corner unless it is altered with the LIMITS command.

Try It!: Create a new drawing file starting from scratch. For this exercise, turn off Dynamic Input by clicking the DYN button located in the status bar. Then use the line command prompts below and the following image to construct the shape.

```
Command: L (For LINE)
Specify first point: 2,2 (at "A")
Specify next point or [Undo]: 2,7 (at "B")
Specify next point or [Undo]: 5,7 (at "C")
Specify next point or [Close/Undo]: 7,4 (at "D")
Specify next point or [Close/Undo]: 10,4 (at "E")
Specify next point or [Close/Undo]: 10,2 (at "F")
Specify next point or [Close/Undo]: C (To close the shape and
    exit the command)
```

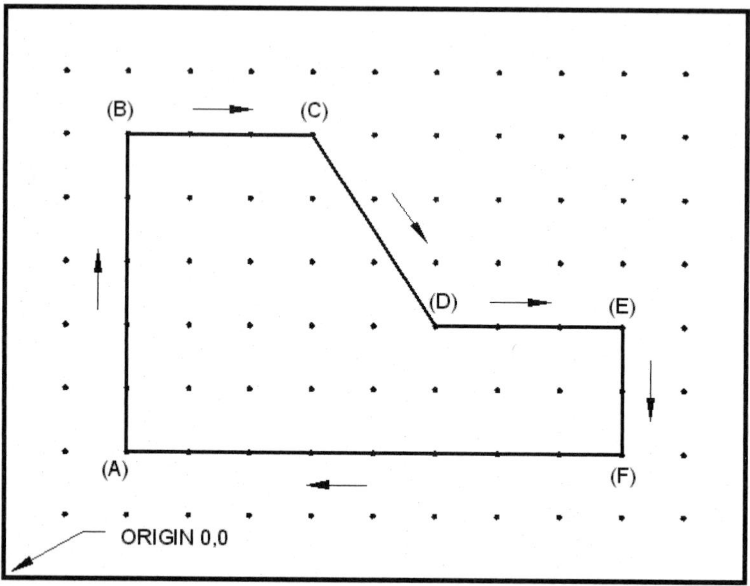

Figure 1.80

As shown in the previous example, all points on the object make reference to the origin at 0,0. Even though absolute coordinates are useful in starting lines, there are more efficient ways to construct lines and draw objects.

RELATIVE COORDINATE MODE FOR DRAWING LINES

With absolute coordinates, the horizontal and vertical distance from the origin at 0,0 must be kept track of at all times in order for the correct coordinate to be entered. With complicated objects, this is difficult to accomplish, and as a result, the wrong coordinate may be entered. It is possible to reset the last coordinate to become a new origin or 0,0 point. The new point would be relative to the previous point, and for this reason, this point is called a relative coordinate. The format is as follows:

@X,Y

In this format, we use the same X and Y values with one exception: the At symbol or @ resets the previous point to 0,0 and makes entering coordinates less confusing.

Try It!: Create a new drawing file starting from scratch. Use the LINE command prompts below and the following image to construct the shape.

```
Command: L (For LINE)
Specify first point: 2,2 (at "A")
Specify next point or [Undo]: @0,4 (to "B")
Specify next point or [Undo]: @4,2 (to "C")
Specify next point or [Close/Undo]: @3,0 (to "D")
Specify next point or [Close/Undo]: @3,-4 (to "E")
Specify next point or [Close/Undo]: @-3,-2 (to "F")
```

```
Specify next point or [Close/Undo]: @-7,0 (back to "A")
Specify next point or [Close/Undo]: (Press ENTER to exit this
    command)
```

Tip: In each command, the @ symbol resets the previous point to 0,0.

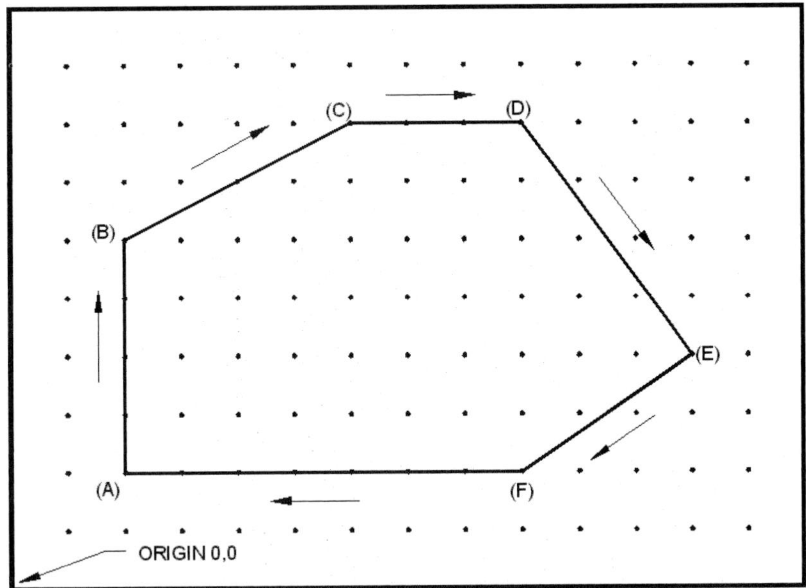

Figure 1.81

POLAR COORDINATE MODE FOR DRAWING LINES

Another popular method of entering coordinates is the polar coordinate mode. The format is as follows:

@Distance<Direction

As the preceding format implies, the polar coordinate mode requires a known distance and a direction. The @ symbol resets the previous point to 0,0. The direction is preceded by the < symbol, which reads the next number as a polar or angular direction. The following image illustrates the directions supported by the polar coordinate mode.

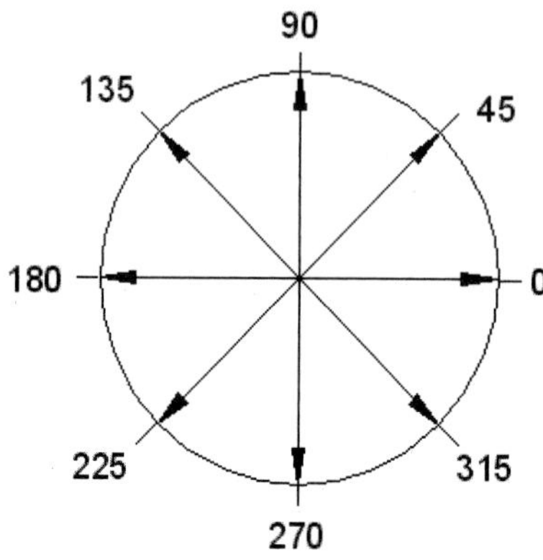

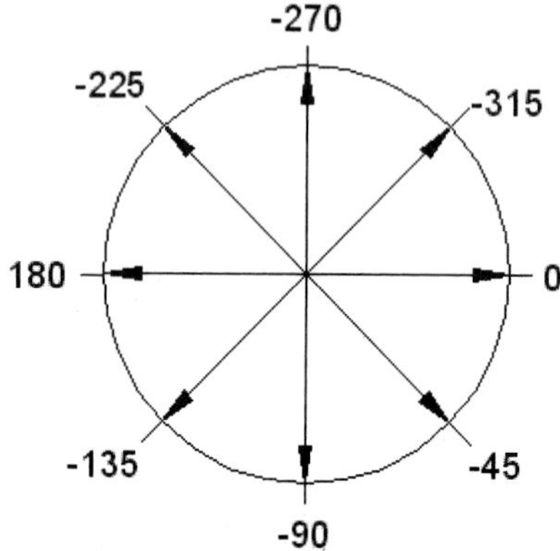

Figure 1.82

Try It!: Create a new drawing file starting from scratch. Use the LINE command prompts below and the following image to construct the shape.

```
Command: L (For LINE)
Specify first point: 3,2 (at "A")
Specify next point or [Undo]: @8<0 (to "B")
Specify next point or [Undo]: @5<90 (to "C")
Specify next point or [Close/Undo]: @5<180 (to "D")
Specify next point or [Close/Undo]: @4<270 (to "E")
Specify next point or [Close/Undo]: @2<180 (to "F")
Specify next point or [Close/Undo]: @2<90 (to "G")
Specify next point or [Close/Undo]: @1<180 (to "H")
Specify next point or [Close/Undo]: @3<270 (to close back
    to "A")
Specify next point or [Close/Undo]: (Press ENTER to exit this
    command)
```

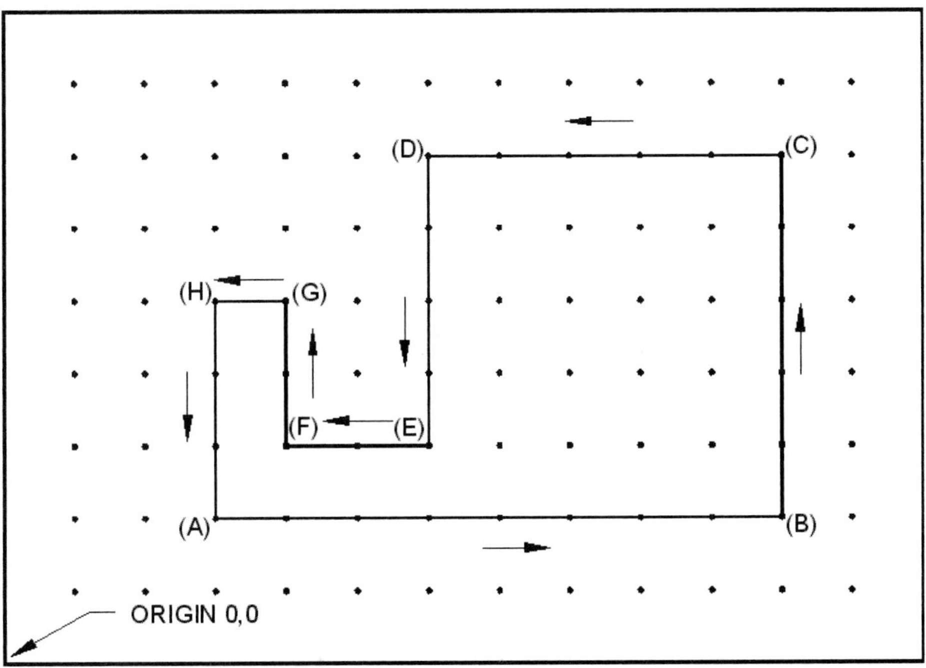

Figure 1.83

COMBINING COORDINATE MODES FOR DRAWING LINES

So far, the preceding pages concentrated on using each example of coordinate modes (absolute, relative, and polar) separately to create geometry. While the examples focused on each individual mode, it is important to note that maximum productivity is usually obtained through use of a combination of modes during a drawing session. It is fairly common to use one, two, or three coordinate modes in combination with one another. In the following image, the drawing starts with an absolute coordinate, changes to a polar coordinate, and changes again to a relative coordinate. The user should develop proficiency in each mode in order to be most productive.

Try It!: Create a new drawing file starting from scratch. Use the LINE command prompts below and the following image to construct the shape.

```
Command: L (For LINE)
Specify first point: 2,2 (at "A")
Specify next point or [Undo]: @3<90 (to "B")
Specify next point or [Undo]: @2,2 (to "C")
Specify next point or [Close/Undo]: @6<0 (to "D")
Specify next point or [Close/Undo]: @5<270 (to "E")
Specify next point or [Close/Undo]: @3<180 (to "F")
Specify next point or [Close/Undo]: @3<90 (to "G")
Specify next point or [Close/Undo]: @2<180 (to "H")
Specify next point or [Close/Undo]: @-3,-3 (back to "A")
```

```
Specify next point or [Close/Undo]: (Press ENTER to exit this
    command)
```

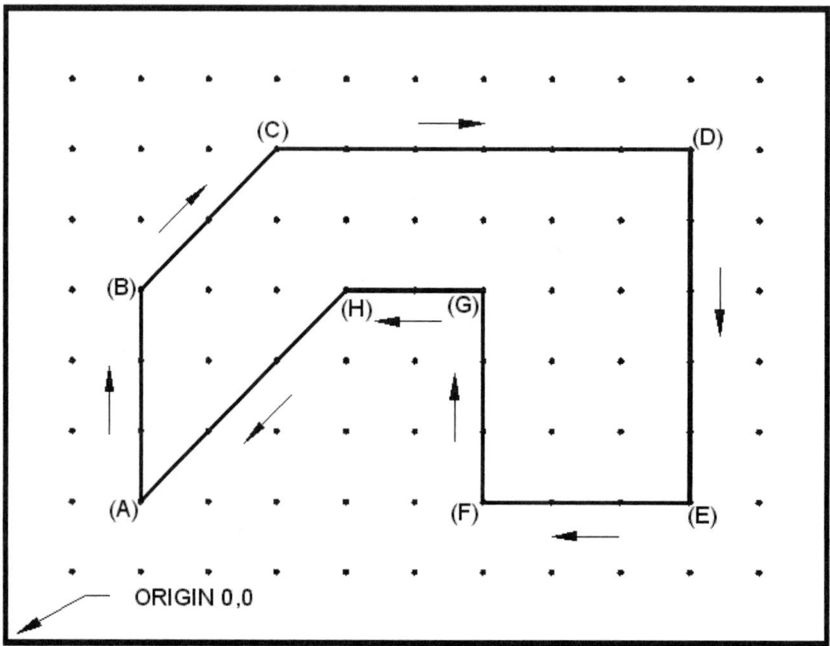

Figure 1.84

THE CIRCLE COMMAND

The CIRCLE command constructs circles of various radii or diameter. This command can be selected from any of the following:

- From the Draw toolbar of the AutoCAD Classic workspace
- From the Ribbon > Home Tab > Draw Panel
- From the Menu Browser (Draw > Circle)
- From the keyboard (C or Circle)

Choosing Circle from the Draw pull-down menu displays the cascading menu shown in the following image. All supported methods of constructing circles are displayed in the list. Circles may be constructed by radius or diameter. This command also supports circles defined by two or three points and construction of a circle tangent to other objects in the drawing. These last two modes will be discussed in chapter 5, "Performing Geometric Constructions."

Figure 1.85

CIRCLE BY RADIUS MODE

Use the CIRCLE command and the Radius mode to construct a circle by a radius value that you specify. After selecting a center point for the circle, the user is prompted to enter a radius for the desired circle.

Try It!: Create a new drawing file starting from scratch. Use the CIRCLE command prompts below and the illustration on the left in the following image to construct a circle by radius.

```
Command: C (For CIRCLE )
Specify center point for circle or [3P/2P/Ttr (tan tan radius)]:
    (Mark the center at “A”)
Specify radius of circle or [Diameter]: 1.50
```

CIRCLE BY DIAMETER MODE

Use the CIRCLE command and the Diameter mode to construct a circle by a diameter value that you specify. After selecting a center point for the circle, you are prompted to enter a diameter for the desired circle.

Try It!: Create a new drawing file starting from scratch. Use the CIRCLE command prompts below and the illustration on the right in the following image to construct a circle by diameter.

```
Command: C (For CIRCLE)
Specify center point for circle or [3P/2P/Ttr (tan tan radius)]:
    (Mark the center at “A”)
Specify radius of circle or [Diameter]: D (For Diameter)
Specify diameter of circle: 3.00
```

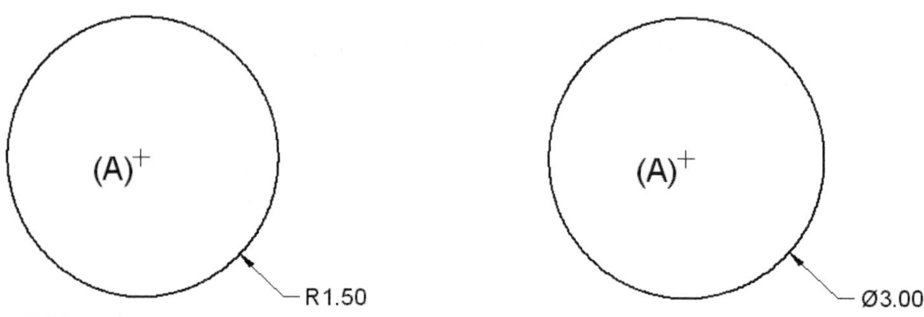

Figure 1.86

DYNAMIC INPUT AND CIRCLES

When using Dynamic Input mode for constructing circles, notice the appearance of a dynamic radius readout when you drag your cursor as shown in the following image. Simply enter the desired radius or press the Down Arrow key twice on your keyboard to select the Diameter option, press ENTER, and provide a diameter value.

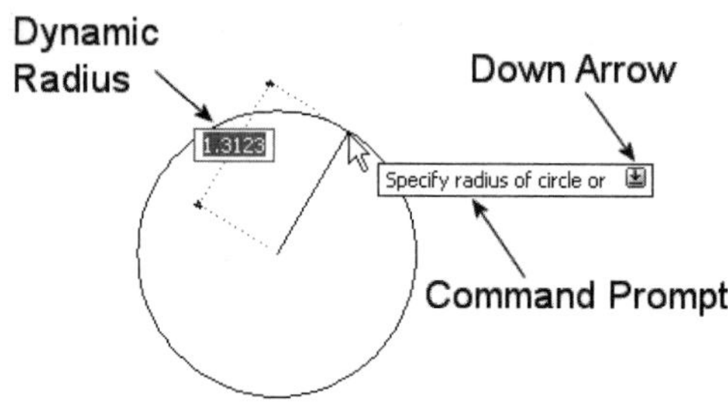

Figure 1.87

USING THE PLINE COMMAND

Polylines are similar to individual line segments except that a polyline can consist of numerous segments and still be considered a single object. Width can also be assigned to a polyline, unlike regular line segments; this makes polylines perfect for drawing borders and title blocks. Polylines can be constructed by selecting any of the following:

- From the Draw toolbar of the AutoCAD Classic workspace
- From the Ribbon > Home Tab > Draw Panel
- From the pull-down menu (Draw > Polyline)
- From the keyboard (PL or Pline)

Study the following images and both command sequences that follow to use the PLINE command.

Try It!: Create a new drawing file starting from scratch. Follow the command prompt sequence and illustration below to construct the polyline.

```
Command: PL (For PLINE)
Specify start point: (Pick a point at "A" in the
    following image)
Current line-width is 0.0000
Specify next point or [Arc/Close/Halfwidth/Length/Undo/Width]: W
    (For Width)
Specify starting width <0.0000>: 0.10
Specify ending width <0.1000>: (Press ENTER to accept the default)
Specify next point or [Arc/Close/Halfwidth/Length/Undo/Width]:
    (Pick a point at "B")
Specify next point or [Arc/Close/Halfwidth/Length/Undo/Width]:
    (Pick a point at "C")
Specify next point or [Arc/Close/Halfwidth/Length/Undo/Width]:
    (Pick a point at "D")
Specify next point or [Arc/Close/Halfwidth/Length/Undo/Width]:
    (Pick a point at "E")
Specify next point or [Arc/Close/Halfwidth/Length/Undo/Width]:
    (Press ENTER to exit this command)
```

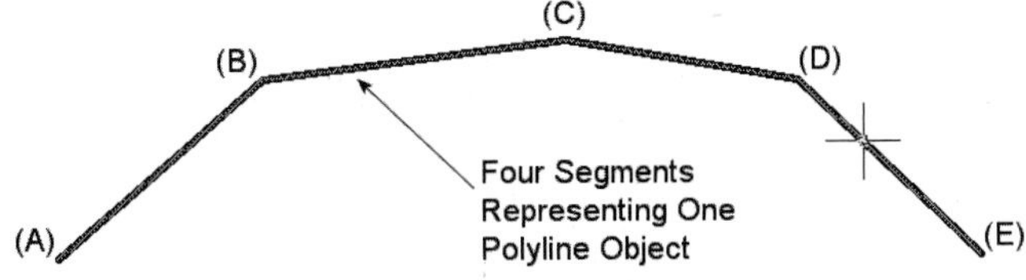

Figure 1.88

Try It!: Create a new drawing file starting from scratch. Follow the command prompt sequence and illustration below to construct the polyline object. The Direct Distance mode of entry could also be used to construct this object from the dimensions given on the object.

```
Command: PL (For PLINE)
Specify start point: 2,2.5 Current line-width is 0.0000
Specify next point or [Arc/Close/Halfwidth/Length/Undo/Width]:
    Move cursor to the right and type 8
(To "B")
Specify next point or [Arc/Close/Halfwidth/Length/Undo/Width]:
    Move cursor up and type 1
(To "C")
Specify next point or [Arc/Close/Halfwidth/Length/Undo/Width]:
    Move cursor to the right and type 1
(To "D")
Specify next point or [Arc/Close/Halfwidth/Length/Undo/Width]:
    Move cursor up and type 2
(To "E")
Specify next point or [Arc/Close/Halfwidth/Length/Undo/Width]:
    Move cursor to the left and type 2
```

```
(To "F")
Specify next point or [Arc/Close/Halfwidth/Length/Undo/Width]:
    Move cursor down and type 2
(To "G")
Specify next point or [Arc/Close/Halfwidth/Length/Undo/Width]:
    Move cursor to the left and type 6
(To "H")
Specify next point or [Arc/Close/Halfwidth/Length/Undo/Width]:
    Move cursor up and type 2
(To "I")
Specify next point or [Arc/Close/Halfwidth/Length/Undo/Width]:
    Move cursor to the left and type 2
(To "J")
Specify next point or [Arc/Close/Halfwidth/Length/Undo/Width]:
    Move cursor down and type 2
(To "K")
Specify next point or [Arc/Close/Halfwidth/Length/Undo/Width]:
    Move cursor to the right and type 1
(To "L")
Specify next point or [Arc/Close/Halfwidth/Length/Undo/Width]: C
    (To close the shape and exit the PLINE command)
```

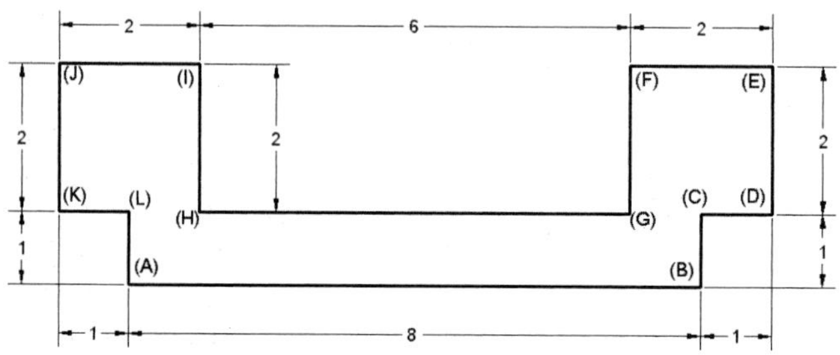

Figure 1.89

DYNAMIC INPUT AND PLINES

Using Dynamic Input for polylines is similar to lines. As you move your cursor while in the PLINE command, you can observe the appearance of your pline through the dynamic distance and angle features, as shown in the following image. As with all dynamic input modes, pressing the down arrow on your keyboard displays options for the PLINE command that you can cycle through.

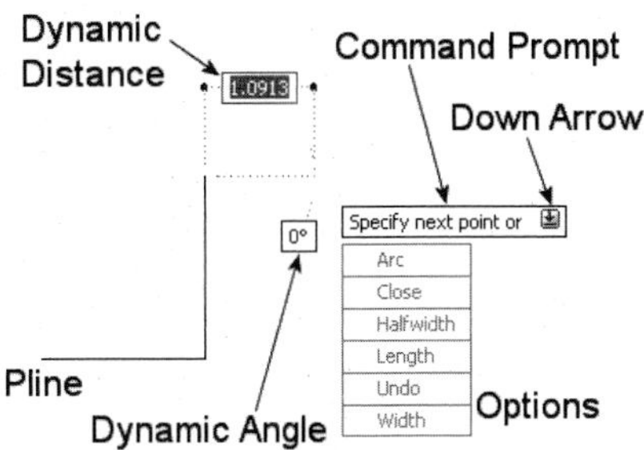

Figure 1.90

ERASING OBJECTS

Throughout the design process, as objects such as lines and circles are placed in a drawing, changes in the design will require the removal of objects. The ERASE command deletes objects from the database. The ERASE command is selected from any of the following:

- From the Modify toolbar of the AutoCAD Classic workspace
- From the Ribbon > Home Tab > Modify Panel
- From the pull-down menu (Modify > Erase)
- From the keyboard (E or Erase)

On the left in the following image, line segments "A" and "B" need to be removed in order for a new line to be constructed, closing the shape. Two ways of erasing these lines will be introduced here.

When first entering the ERASE command, you are prompted to "Select objects:" to erase. Notice that your cursor changes in appearance from crosshairs to a pickbox, as shown on the right in the following image. Move the pickbox over the object to be selected and pick this item. Notice that it will be highlighted as a dashed object to signify it is now selected. At this point, the "Select objects:" prompt appears again. Additional objects may be selected at this point. Once all the objects are selected, pressing ENTER performs the erase operation.

```
Command: E (For ERASE)
Select objects: (Pick line "A," as shown on the right in the
    following image)
Select objects: (Press ENTER to perform the erase operation)
```

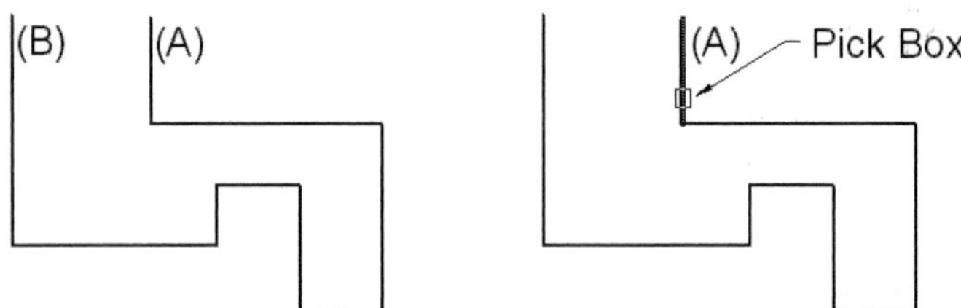

Figure 1.91

Another method of erasing is illustrated on the left in the following image. Instead of using the ERASE command, highlight the line by selecting it with the cursor crosshairs from the "command:" prompt. Square boxes also appear at the endpoints and midpoints of the line. With this line segment highlighted, press DELETE on the keyboard, resulting in removal of the line from the drawing.

With both line segments erased, a new line is constructed from the endpoint at "A" to the endpoint at "B," as shown on the right in the following image.

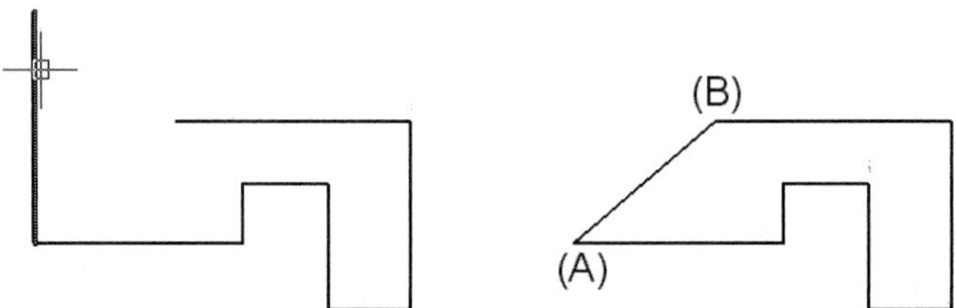

Figure 1.92

SAVING A DRAWING FILE

You can save drawings using the QSAVE and SAVEAS commands. The QSAVE command can be selected from the following:

- From the Quick Access Toolbar
- From the Standard toolbar of the AutoCAD Classic workspace
- From the pull-down menu (File > Save)
- From the keyboard (Qsave)

The SAVEAS command can be selected from any of the following:

- From the pull-down menu (File > Save As)
- From the keyboard (Saveas)

These commands are found on the File menu, as shown on the left in the following image.

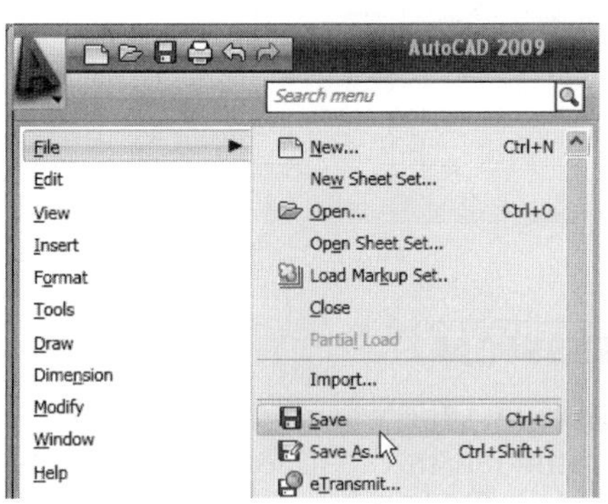

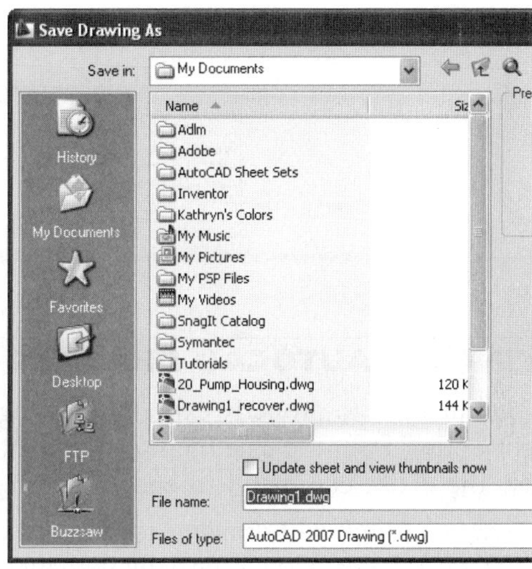

Figure 1.93

SAVE

Selecting Save from the File pull-down menu, as shown in the previous image, activates the QSAVE command, which stands for Quick Save. If a drawing file has never been saved and this command is selected, the dialog box shown on the right in the previous image is displayed. Once a drawing file has been initially saved, selecting this command causes an automatic save and the Save Drawing As dialog box is no longer displayed.

SAVE AS

Using the SAVEAS command always displays the dialog box shown on the right in the previous image. Simply click the Save button or press ENTER to save the drawing under the current name, which is displayed in the field. This command is more popular for saving the current drawing under an entirely different name. Simply enter the new name in place of the highlighted name in the field. Once a drawing is given a new name through this command, it also becomes the new current drawing file.

The ability to exchange drawings with past releases of AutoCAD is still important to many industry users. When the Files of type field is selected in the following image, a drop-down list appears. Use this list to save a drawing file in AutoCAD 2004, 2000, and even R14 formats. The user can also save a drawing file as a Drawing Standard (.dws), a Drawing Template (.dwt), and a Drawing Interchange Format (.dxf). The Drawing Interchange Format is especially useful with opening up an AutoCAD drawing in a competitive CAD system.

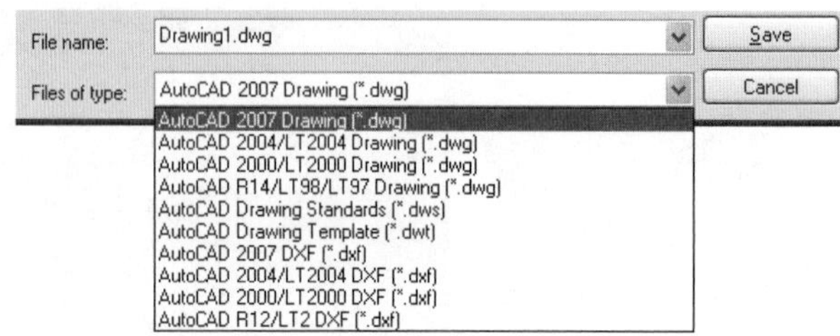

Figure 1.94

EXITING AN AUTOCAD DRAWING SESSION

It is good practice to properly exit any drawing session. One way of exiting is by choosing the Exit option from the File menu, as shown on the left in the following image. You can also use the CLOSE command to end the current AutoCAD drawing session.

Whenever an AutoCAD drawing session is exited, a built-in safeguard provides a second chance to save the drawing, especially if changes were made and a Save was not performed. You may be confronted with three options, illustrated in the AutoCAD alert dialog box shown on the right in the following image. By default, the Yes button is highlighted.

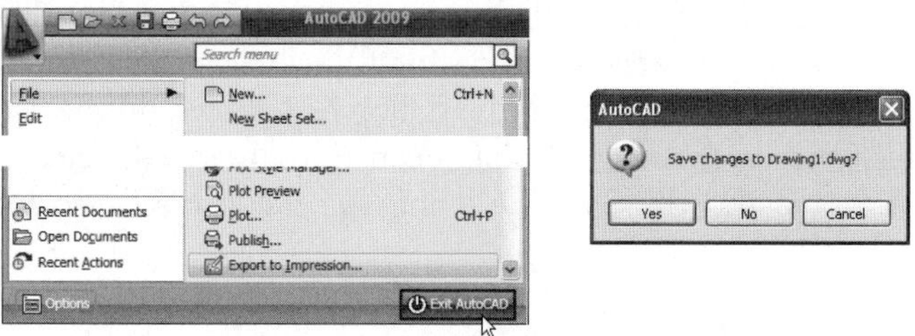

Figure 1.95

If changes were made to a drawing but no Save was performed, the user can now save them by clicking the Yes button before exiting the drawing. Changes to the drawing will be saved and the software exits back to the operating system.

If changes were made but the user does not want to save them, clicking the No button is appropriate. Changes to the drawing will not be saved and the software exits back to the operating system.

If changes are made to the drawing and the Exit option is chosen mistakenly, clicking the Cancel button cancels the Exit option and returns the user to the current drawing.

TUTORIAL EXERCISE: 01_GAGE BLOCK.DWG

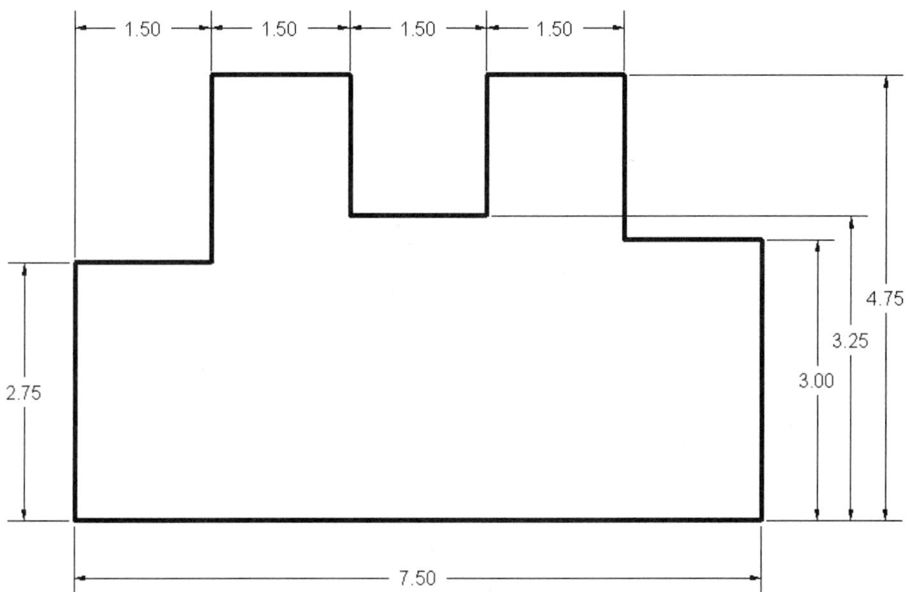

Figure 1.96

PURPOSE

This tutorial is designed to allow you to construct a one-view drawing of the Gage Block using Polar Tracking and Direct Distance mode.

SYSTEM SETTINGS

Use the current default settings for the limits of this drawing, (0,0) for the lower-left corner and (12,9) for the upper-right corner.

SUGGESTED COMMANDS

Open the drawing file called 01_Gage Block. The LINE command will be used entirely for this tutorial in addition to the Polar Tracking and Direct Distance modes. Running Object Snap should already be set to the following modes: Endpoint, Center, Intersection, and Extension.

Step 1

Begin this tutorial by first turning Polar Tracking on. This can be accomplished by clicking Polar in the status bar located at the bottom of your display screen. Then activate the LINE command and follow the next series of prompt sequences to complete this object.

```
Command: L (For LINE)
Specify first point: 1,1
Specify next point or [Undo]: Move cursor to the right and
    type 7.5
```

```
Specify next point or [Undo]: Move cursor up and type 3
Specify next point or [Close/Undo]: Move cursor to the left and
    type 1.5
Specify next point or [Close/Undo]: Move cursor up and type 1.75
Specify next point or [Close/Undo]: Move cursor to the left and
    type 1.5
Specify next point or [Close/Undo]: Move cursor down and
    type 1.5
Specify next point or [Close/Undo]: Move cursor to the left and
    type 1.5
Specify next point or [Close/Undo]: Move cursor up and type 1.5
Specify next point or [Close/Undo]: Move cursor to the left and
    type 1.5
Specify next point or [Close/Undo]: Move cursor down and type 2
Specify next point or [Close/Undo]: Move cursor to the left and
    type 1.5
Specify next point or [Close/Undo]: C (To close the shape)
```

TUTORIAL EXERCISE: 01_ANGLE BLOCK.DWG

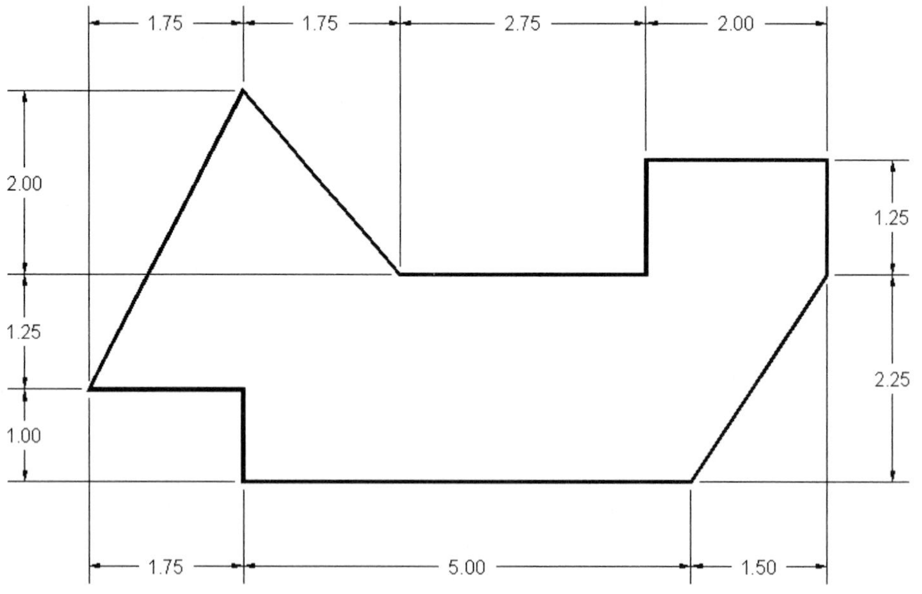

Figure 1.97

This tutorial is designed to allow you to construct a one-view drawing of the Angle Block using a combination of Polar Tracking, Direct Distance mode, and relative coordinates.

SYSTEM SETTINGS

Use the current default settings for the limits of this drawing, (0,0) for the lower-left corner and (12,9) for the upper-right corner.

SUGGESTED COMMANDS

Open the drawing file called 01_Angle Block. The LINE command will be used entirely for this tutorial in addition to the Polar Tracking and Direct Distance modes. Running Object Snap should already be set to the following modes: Endpoint, Center, Intersection, and Extension.

Step 1

Begin this tutorial by first checking that Polar Tracking is turned on. Then activate the LINE command and follow the next series of prompt sequences to complete this object.

```
Command: L (For LINE)
Specify first point: 3,1
Specify next point or [Undo]: Move cursor to the right and
    type 5
Specify next point or [Undo]: @1.5,2.25
Specify next point or [Close/Undo]: Move cursor up and
    type 1.25
Specify next point or [Close/Undo]: Move cursor to the left and
    type 2
Specify next point or [Close/Undo]: Move cursor down and
    type 1.25
Specify next point or [Close/Undo]: Move cursor to the left and
    type 2.75
Specify next point or [Close/Undo]: @-1.75,2
Specify next point or [Close/Undo]: @-1.75,-3.25
Specify next point or [Close/Undo]: Move cursor to the right and
    type 1.75
Specify next point or [Close/Undo]: C (To close the shape)
```

TUTORIAL EXERCISE: 01_ANGLE PLATE.DWG

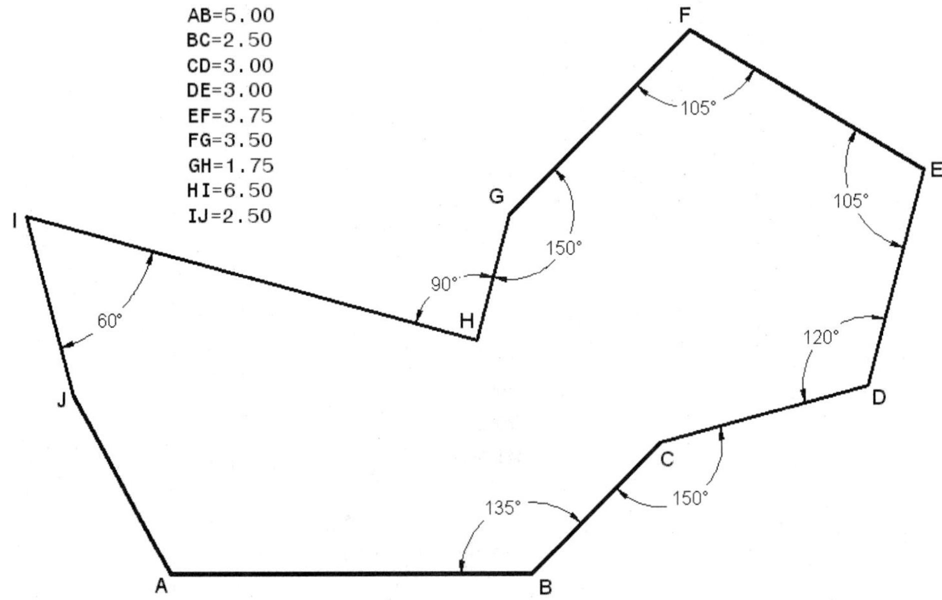

Figure 1.98

This tutorial is designed to allow you to construct a one-view drawing of the Angle Plate using Polar Tracking set to relative mode and Direct Distance mode.

SYSTEM SETTINGS

Use the current default settings for the limits of this drawing, (0,0) for the lower-left corner and (12,9) for the upper-right corner.

SUGGESTED COMMANDS

Open the drawing file called 01_Angle Plate. The LINE command will be used entirely for this tutorial in addition to the Polar Tracking and Direct Distance modes. Polar Tracking will need to be set to a new incremental angle of 15°. Also, Polar Tracking will need to be set to relative mode. Running Object Snap should already be set to the following modes: **End**point, Center, Intersection, and Extension.

Step 1

Right-click POLAR in the status bar at the bottom of the display screen and pick Settings from the menu. When the Drafting Settings dialog box appears, click the Polar Tracking tab. While in this tab, set the Incremental Angle to 15° under Polar Angle Settings area. Then set the Polar Angle measurement to Relative to last segment, as shown in the following image. Click the OK button to save the settings and exit the Drafting Settings dialog box.

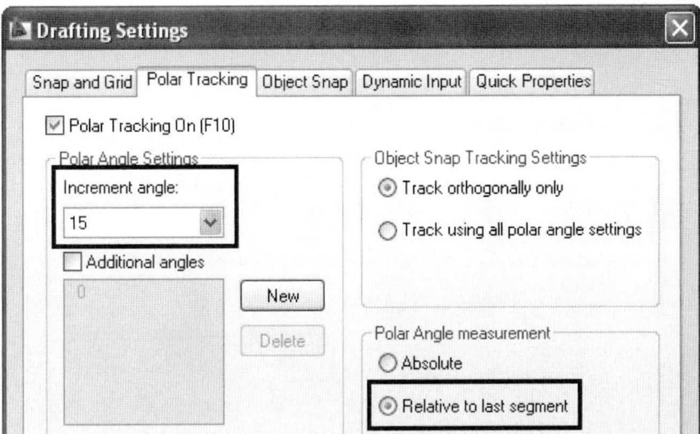

Figure 1.99

Step 2

After setting the angle increment and relative mode under the Polar Tracking tab of the Drafting Settings dialog box, activate the LINE command and follow the next series of prompt sequences to complete this object.

```
Command: L (For LINE)
Specify first point: 3,1
Specify next point or [Undo]: (Move your cursor to the right
    until the polar tooltip reads 0° and enter 5)
Specify next point or [Undo]: (Move your cursor until the polar
    tooltip reads 45° and enter 2.5)
Specify next point or [Close/Undo]: (Move your cursor until the
    polar tooltip reads 330° and enter 3)
Specify next point or [Close/Undo]: (Move your cursor until the
    polar tooltip reads 60° and enter 3)
Specify next point or [Close/Undo]: (Move your cursor until the
    polar tooltip reads 75° and enter 3.75)
Specify next point or [Close/Undo]: (Move your cursor until the
    polar tooltip reads 75° and enter 3.5)
Specify next point or [Close/Undo]: (Move your cursor until the
    polar tooltip reads 30° and enter 1.75)
Specify next point or [Close/Undo]: (Move your cursor until the
    polar tooltip reads 270° and enter 6.5)
Specify next point or [Close/Undo]: (Move your cursor until the
    polar tooltip reads 120° and enter 2.5)
Specify next point or [Close/Undo]: C (To close the shape)
```

Step 3

When finished with this problem, change the Increment angle under Polar Angle Settings back to 90° and the Polar Angle measurement back to Absolute, as shown in the following image.

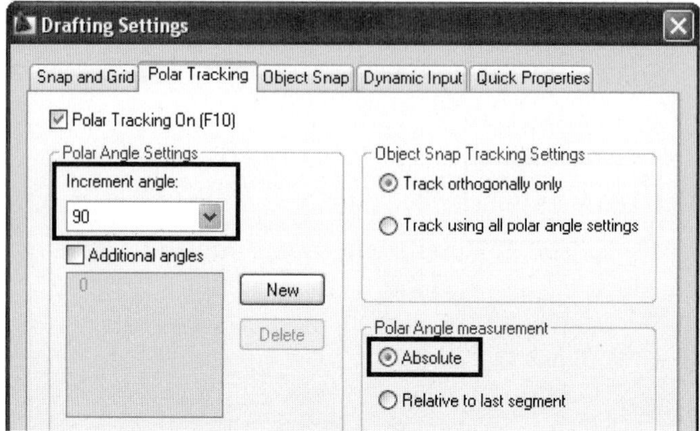

Figure 1.100

TUTORIAL EXERCISE: 01_PATTERN.DWG

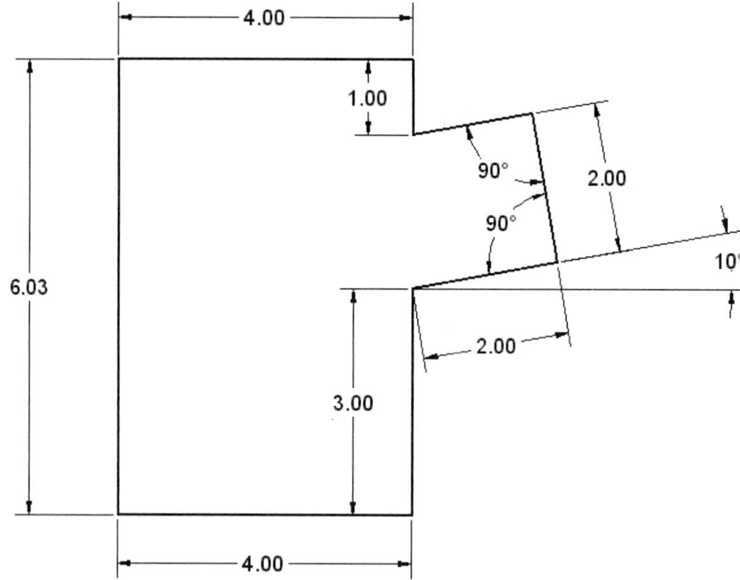

Figure 1.101

PURPOSE

This tutorial is designed to allow you to construct a one-view drawing of the Pattern using Polar Tracking techniques.

SYSTEM SETTINGS

Use the current default settings for the limits of this drawing, (0,0) for the lower-left corner and (12,9) for the upper-right corner.

LAYERS

Create the following layer with the format:

Name	Color	Linetype
Object	Green	Continuous

SUGGESTED COMMANDS

Open the drawing file called 01_Pattern. The LINE command will be used entirely for this tutorial in addition to the Polar Tracking mode. Running Object Snap should already be set to the following modes: **End**point, Center, Intersection, and Extension. Dynamic Input has been turned off for this exercise.

Step 1

Open the drawing file 01_Pattern. Activate the Drafting Settings dialog box, click the Polar Tracking tab, and change the Increment angle setting to 10° as shown on the left in the following image. Verify that POLAR, OSNAP, and OTRACK are all turned on.

Step 2

Activate the LINE command, select a starting point, move your cursor to the right, and enter a value of 4 units as shown on the right in the following image. Notice that your line is green because the current layer, Object, has been assigned the green color.

```
Command: L (For LINE)
Specify first point: 3,1
Specify next point or [Undo]: (Move your cursor to the right and
    enter 4)
```

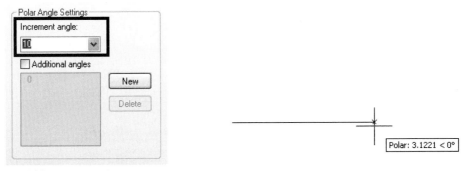

Figure 1.102

Step 3

While still in the LINE command, move your cursor directly up, and enter a value of 3 units as shown on the left in the following image.

```
Specify next point or [Undo]: (Move your cursor up and enter 3)
```

Step 4

While still in the LINE command, move your cursor up and to the right until the tooltip reads 10°, and enter a value of 2 units as shown on the right in the following image.

```
Specify next point or [Close/Undo]: (Move your cursor up and to
    the right at a 10° angle and enter 2)
```

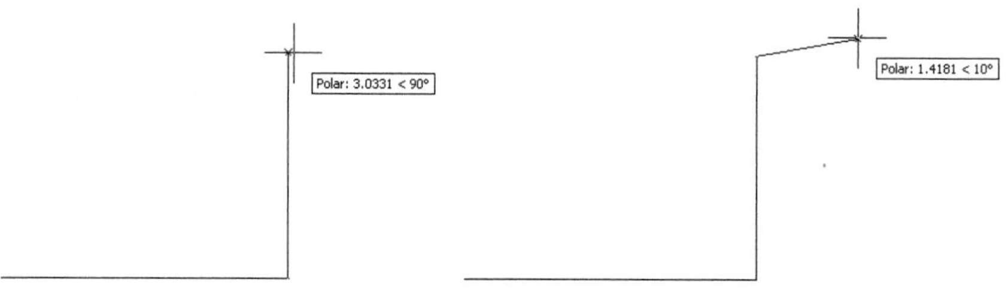

Figure 1.103

Step 5

While still in the LINE command, move your cursor up and to the left until the tooltip reads 100°, and enter a value of 2 units as shown on the left in the following image.

```
Specify next point or [Close/Undo]: (Move your cursor up and to
    the left at a 100° angle and enter 2)
```

Step 6

While still in the LINE command, first acquire the point at "A." Then move your cursor below and to the left until the Polar value in the tooltip reads 190°, and pick the point at "B" as shown on the right in the following image.

```
Specify next point or [Close/Undo]: (Acquire the point at "A"
    and pick the new point at "B")
```

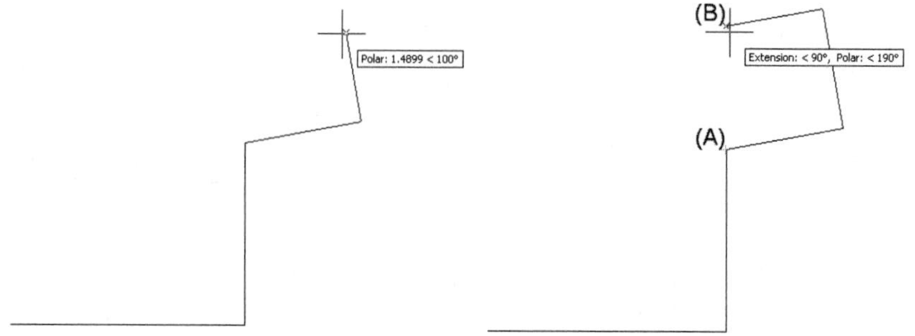

Figure 1.104

Step 7

While still in the LINE command, move your cursor directly up, and enter a value of 1 unit as shown on the left in the following image.

```
Specify next point or [Close/Undo]: (Move your cursor up and
    enter 1)
```

Step 8

While still in the LINE command, first acquire the point at "C." Then move your cursor to the left until the tooltip reads Polar: < 180°, and pick the point at "D" as shown on the right in the following image.

```
Specify next point or [Close/Undo]: (Acquire the point at "C"
    and pick the new point at "D")
```

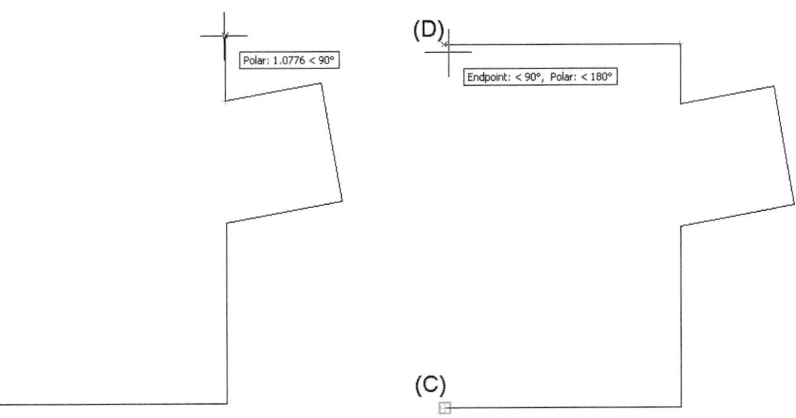

Figure 1.105

Step 9

While still in the LINE command, complete the object by closing the shape as shown in the following image.

```
Specify next point or [Close/Undo]: C (To close the shape and
    exit the LINE command)
```

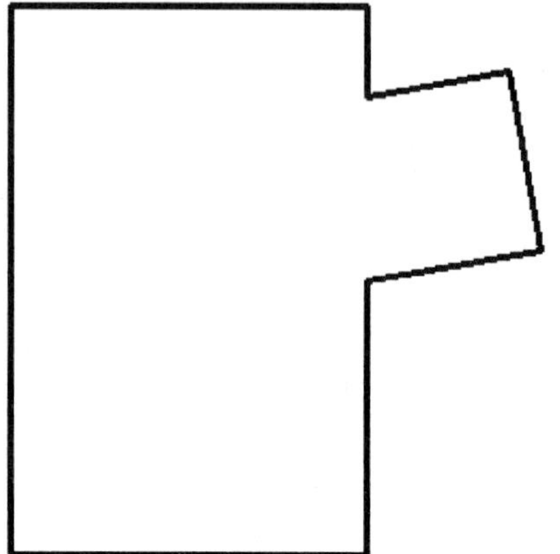

Figure 1.106

TUTORIAL EXERCISE: 01_TEMPLATE.DWG

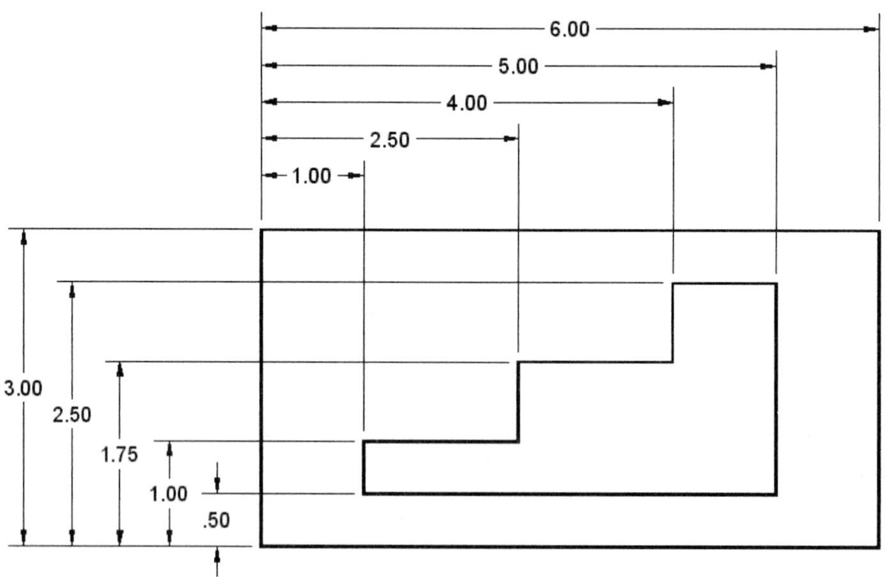

Figure 1.107

PURPOSE

This tutorial is designed to allow you to construct a one-view drawing of the template using Relative Coordinate mode in combination with the Direct Distance mode. The Direct Distance mode can also be used to perform this exercise.

SYSTEM SETTINGS

Use the current default settings for the limits of this drawing, (0,0) for the lower-left corner and (12,9) for the upper-right corner.

LAYERS

The following layer has already been created:

Name	Color	Linetype
Object	Green	Continuous

SUGGESTED COMMANDS

The LINE command will be used entirely for this tutorial, in addition to a combination of coordinate systems. The ERASE command could be used (however, using this command will force the user to exit out of the LINE command), although a more elaborate method of correcting mistakes while using the LINE command is to execute the Undo option. This option allows the user to delete (or undo) previously drawn lines without having to exit the LINE command. The Object Snap From mode will also be used to construct lines from a point of reference. The coordinate mode of entry and the Direct Distance mode will be used throughout this tutorial exercise.

Step 1

Open the drawing file 01_Template.dwg. Then use the LINE command to draw the outer perimeter of the box using the Direct Distance mode. Because the box consists of horizontal and vertical lines, Ortho mode is first turned on; this forces all movements to be in the horizontal or vertical direction. To construct a line segment, move the cursor in the direction in which the line is to be drawn and enter the exact value of the line. The line is drawn at the designated distance in the current direction of the cursor. Repeat this procedure for the other lines that make up the box, as shown in the following image.

```
Command: L (For LINE)
Specify first point: 2,2
Specify next point or [Undo]: (Move the cursor to the right and
    enter a value of 6.00 units)
Specify next point or [Undo]: (Move the cursor up and enter a
    value of 3.00 units)
Specify next point or [Close/Undo]: (Move the cursor to the left
    and enter a value of 6.00 units)
Specify next point or [Close/Undo]: C (To close the shape)
```

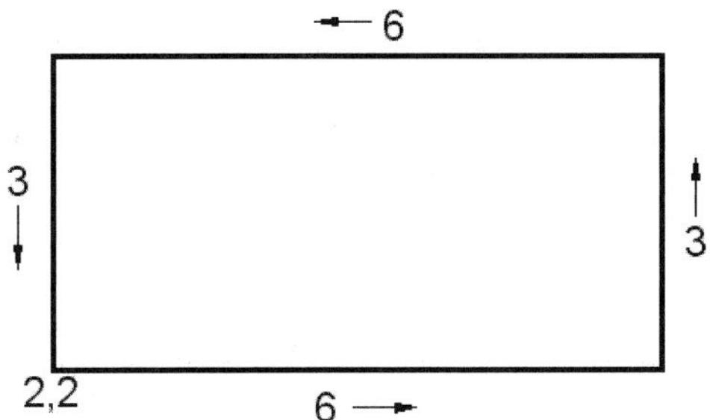

Figure 1.108

Step 2

The next step is to draw the stair step outline of the template using the LINE command again. However, we first need to identify the starting point of the template.

Absolute coordinates could be calculated, but in more complex objects this would be difficult. A more efficient method is to use the Object Snap From mode along with the Object Snap Intersection mode to start the line relative to another point. Both Object Snap selections are found on the Object Snap toolbar. Use the following command sequence and image as guides for performing this operation.

```
Command: L (For LINE)
Specify first point: From
Base point: Int
of (Pick the intersection at "A" as shown in the following
    image)
<Offset>: @1.00,0.50
```

The relative coordinate offset value begins a new line a distance of 1.00 units in the X direction and 0.50 units in the Y direction.

Continue with the LINE command to construct the stair step outline shown in the following image. Use the Direct Distance mode to accomplish this task. In this example, Direct Distance mode is a good choice to use, especially since all lines are either horizontal or vertical. Use the following command sequence to construct the object with this alternate method.

```
Specify next point or [Undo]: (Move the cursor to the right and
    enter a value of 4.00 units)
Specify next point or [Undo]: (Move the cursor up and enter a
    value of 2.00 units)
Specify next point or [Close/Undo]: (Move the cursor to the left
    and enter a value of 1.00 units)
Specify next point or [Close/Undo]: (Move the cursor down and
    enter a value of 0.75 units)
Specify next point or [Close/Undo]: (Move the cursor to the left
    and enter a value of 1.50 units)
Specify next point or [Close/Undo]: (Move the cursor down and
    enter a value of 0.75 units)
Specify next point or [Close/Undo]: (Move the cursor to the left
    and enter a value of 1.50 units)
Specify next point or [Close/Undo]: C (To close the shape)
```

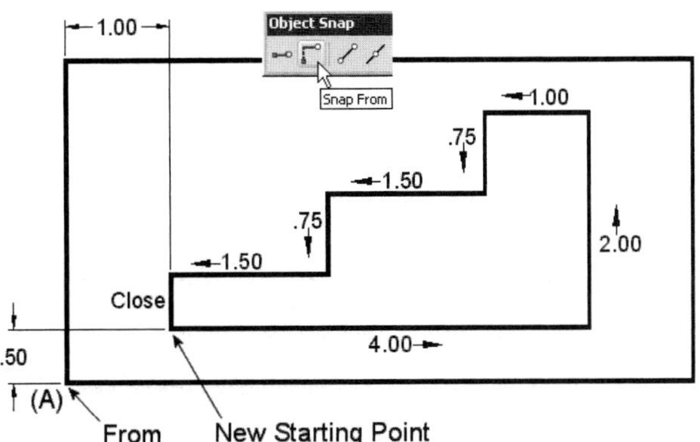

Figure 1.109

PROBLEMS FOR CHAPTER 1

DIRECTIONS FOR PROBLEMS 1-1 THROUGH 1-8

Construct one-view drawings from the following figures using the LINE command along with Direct Distance mode, relative coordinates, and/or polar coordinates.

PROBLEM 1-1

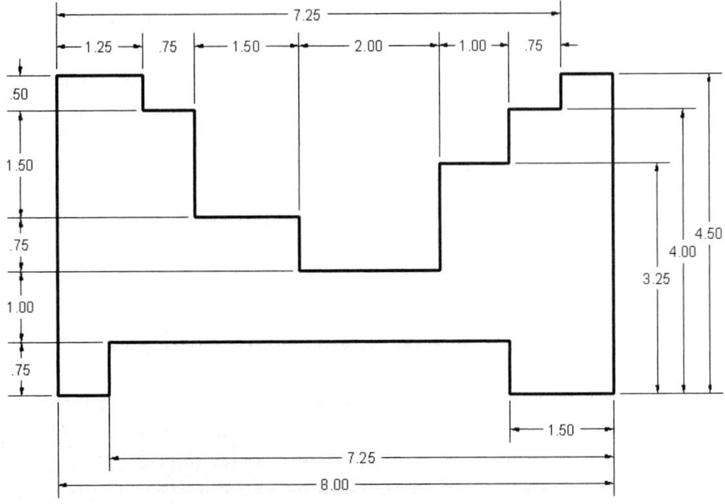

PROBLEM 1-2

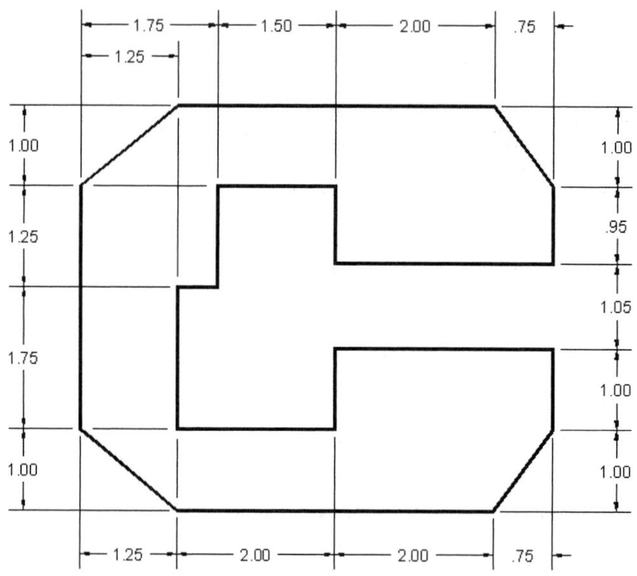

PROBLEM 1-3

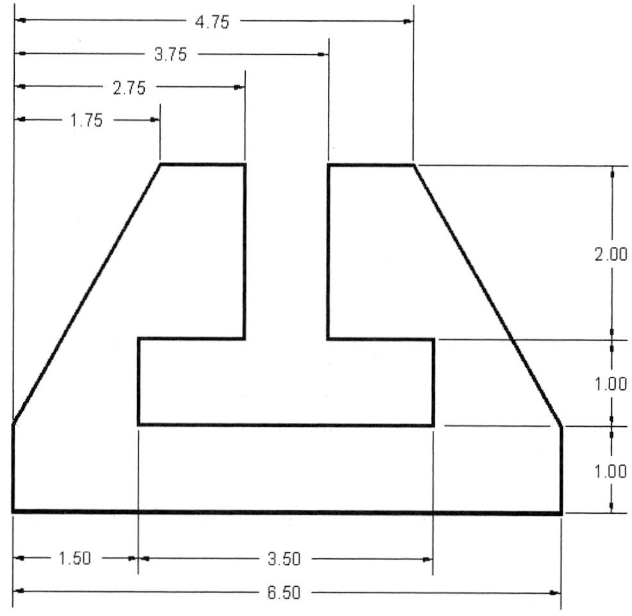

PROBLEM 1-4

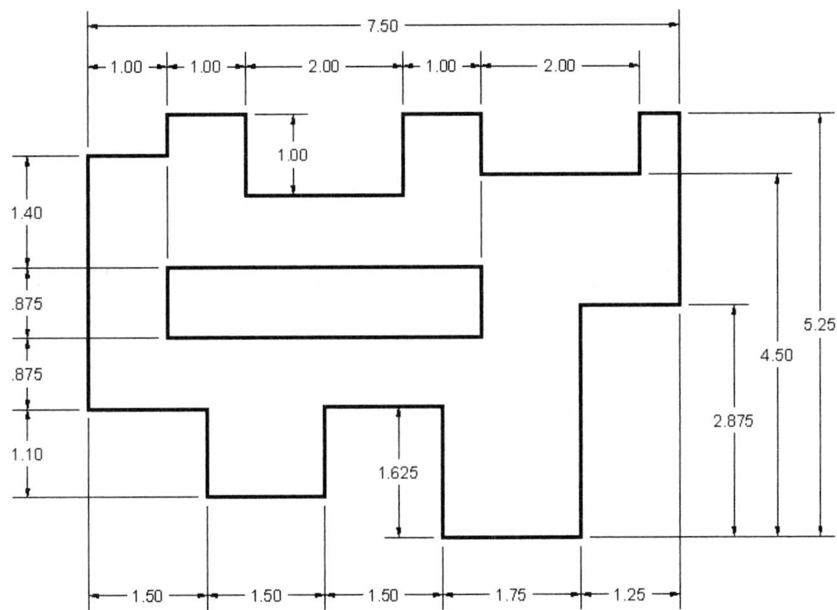

PROBLEM 1-5

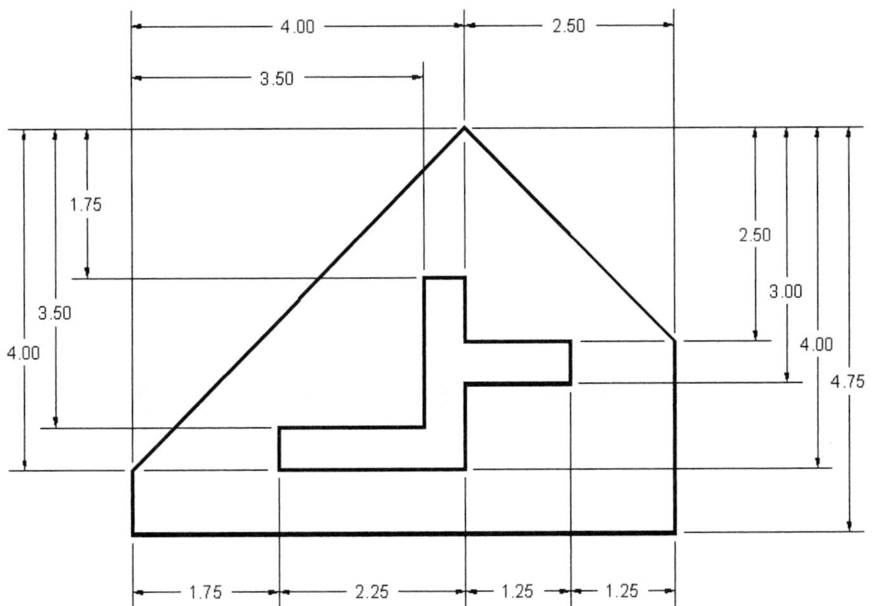

PROBLEM 1-6

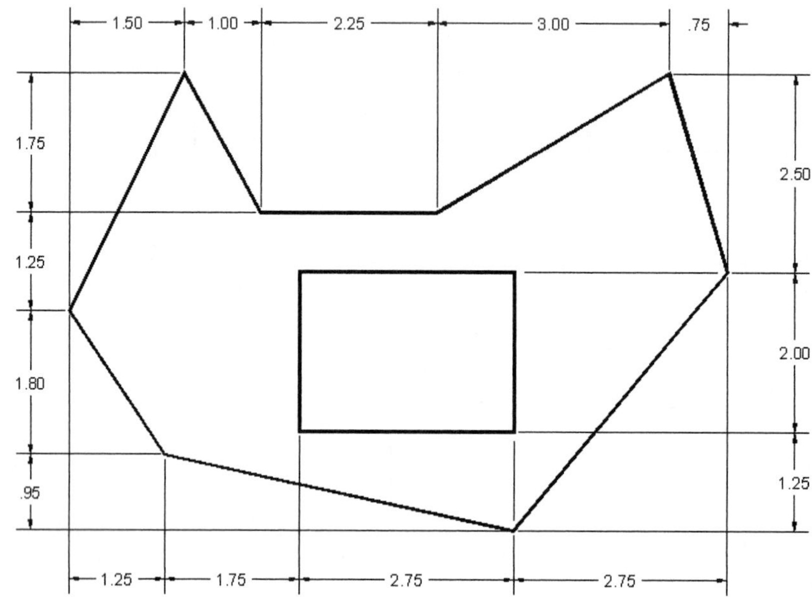

PROBLEM 1-7

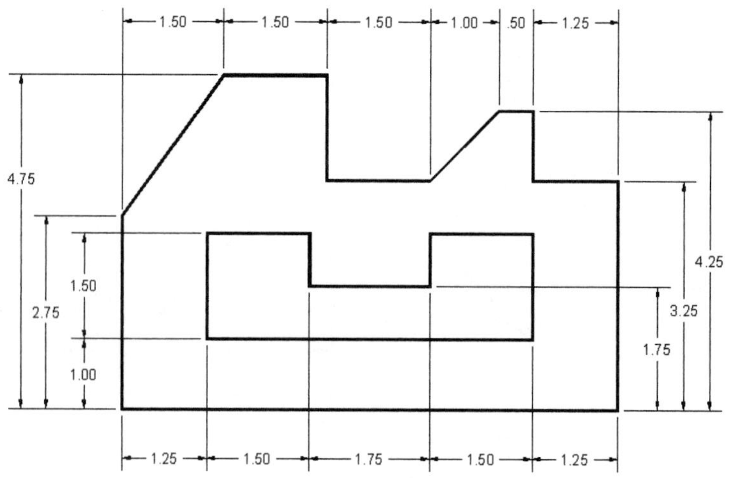

PROBLEM 1-8

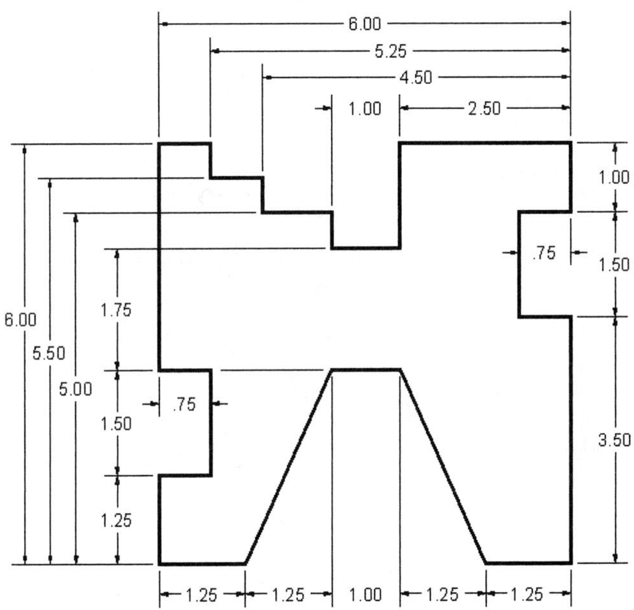

DIRECTIONS FOR PROBLEM 1-9

Supply the appropriate absolute, relative, and/or polar coordinates for this figure in the table that follows the object.

PROBLEM 1-9

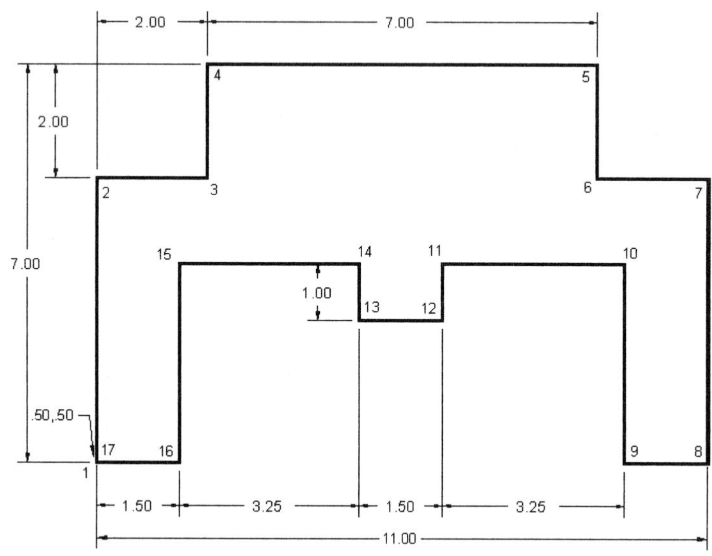

	Absolute	Relative	Polar
From Pt (1)	.50,.50	.50,.50	.50,.50
To Pt (2)	___,___	___,___	___,___
To Pt (3)	___,___	___,___	___,___
To Pt (4)	___,___	___,___	___,___
To Pt (5)	___,___	___,___	___,___
To Pt (6)	___,___	___,___	___,___
To Pt (7)	___,___	___,___	___,___
To Pt (8)	___,___	___,___	___,___
To Pt (9)	___,___	___,___	___,___
To Pt (10)	___,___	___,___	___,___
To Pt (11)	___,___	___,___	___,___
To Pt (12)	___,___	___,___	___,___
To Pt (13)	___,___	___,___	___,___
To Pt (14)	___,___	___,___	___,___
To Pt (15)	___,___	___,___	___,___

	Absolute	Relative	Polar
To Pt (16)	___ , ___	___ , ___	___ , ___
To Pt (17)	___ , ___	___ , ___	___ , ___
To Pt	Enter	Enter	Enter

CHAPTER 2

Drawing Setup and Organization

Chapter 2 covers a number of drawing setup commands. The user will learn how to assign different units of measure with the UNITS command. The default sheet size can also be increased on the display screen with the LIMITS command. Controlling the grid and snap will be briefly discussed through the Snap and Grid tab located in the Drafting Settings dialog box. The major topic of this chapter is the discussion of layers. All options of the Layer Properties Manager palette will be demonstrated, along with the ability to assign color, linetype, and lineweight to layers. The Layer Control box and Properties toolbar will provide easy access to all layers, colors, linetypes, and lineweights used in a drawing. Controlling the scale of linetypes through the LTSCALE command will also be discussed. Advanced layer tools such as Filtering Layers and creating Layer States will be introduced. This chapter concludes with a section on creating template files.

SETTING DRAWING UNITS

The Drawing Units dialog box is available for interactively setting the units of a drawing. Choosing Units from the Format heading of the Menu Browser activates the dialog box illustrated in the following image.

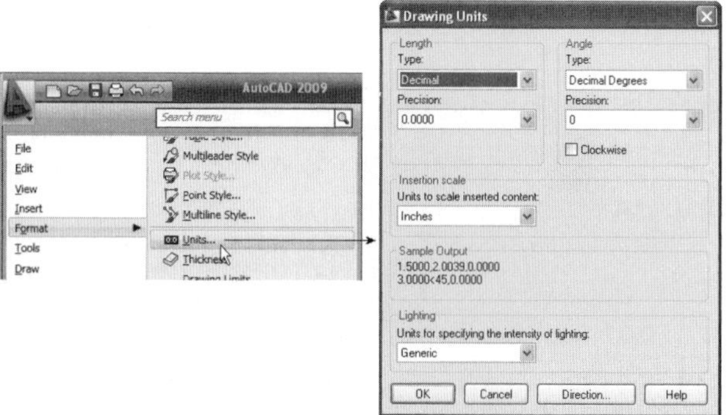

Figure 2.1

By default, decimal units are set along with four-decimal-place precision. The following systems of units are available: Architectural, Decimal, Engineering,

Fractional, and Scientific (see the left side of the following image). Architectural units are displayed in feet and fractional inches. Engineering units are displayed in feet and decimal inches. Fractional units are displayed in fractional inches. Scientific units are displayed in exponential format.

Methods of measuring angles supported in the Drawing Units dialog box include Decimal Degrees, Degrees/Minutes/Seconds, Grads, Radians, and Surveyor's Units (see the middle of the following image). Accuracy of decimal degree for angles may be set between zero and eight places.

Selecting Direction... in the main Drawing Units dialog box displays the Direction Control dialog box shown on the right in the following image. This dialog box is used to control the direction of angle zero in addition to changing whether angles are measured in the counterclockwise or clockwise direction. By default, angles are measured from zero degrees in the east and in the counterclockwise direction.

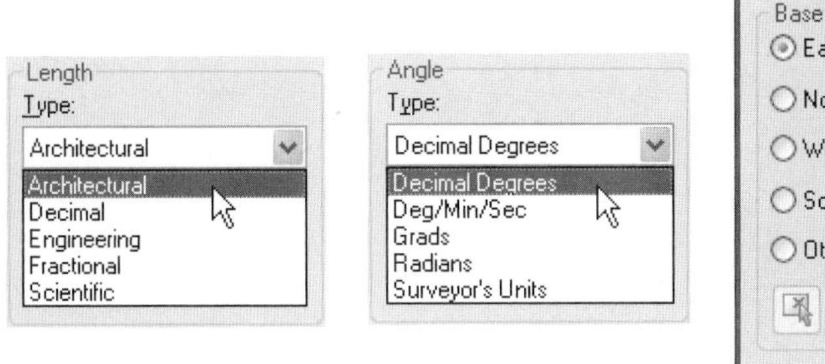

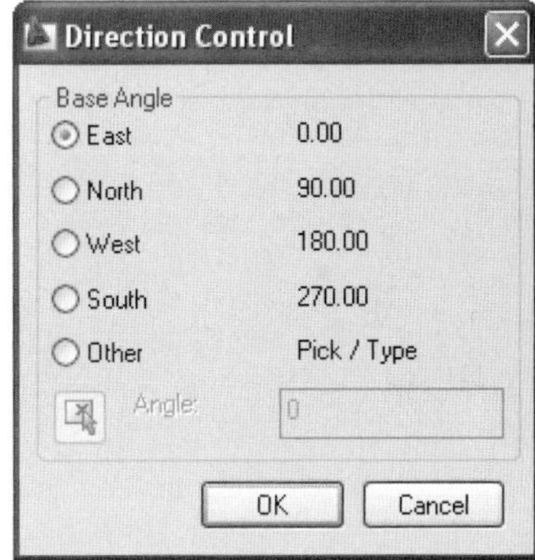

Figure 2.2

ENTERING ARCHITECTURAL VALUES FOR DRAWING LINES

The method of entering architectural values in feet and inches is a little different from the method for entering them in decimal places. To designate feet, you must enter the apostrophe symbol (') from the keyboard after the number. For example, "ten feet" would be entered as (10'), as shown in the following image. When feet and inches are necessary, you cannot use the Spacebar to separate the inch value from the foot value. For example, thirteen feet seven inches would be entered as (13'7), as shown in the following image. If you do use the Spacebar after the (13') value, this is interpreted as the ENTER key and your value is accepted as (13'). If you have to enter feet, inches, and fractions of an inch, use the hyphen (-) to separate the inch value from the fractional value. For example, to draw a line seventeen feet eleven and one-quarter inches, you would enter the following value in at the keyboard: (17'11-1/4). See the following image. Placing the inches symbol (") is not required since all numbers entered without the foot symbol are interpreted as inches.

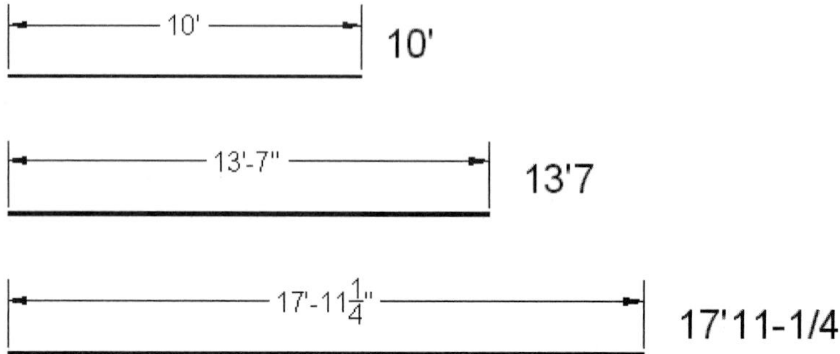

Figure 2.3

Try It!: Open the drawing file 02_Architectural. Verify that the units setting is in architectural units by activating the Drawing Units dialog box. With the following image as a guide, use the Direct Distance mode of entry to construct the shape using architectural values..

```
Command: L (For LINE)
Specify first point: 4',2'
Specify next point or [Undo]: (Move your cursor to the right and
    enter a value of 10')
Specify next point or [Undo]: (Move your cursor up and enter a
    value of 4'6
Specify next point or [Close/Undo]: (Move your cursor to the
    right and enter a value of 13'7-1/2)
Specify next point or [Close/Undo]: (Move your cursor up and
    enter a value of 4'9-1/2)
Specify next point or [Close/Undo]: (Move your cursor to the
    left and enter a value of 7')
Specify next point or [Close/Undo]: (Move your cursor up and
    enter a value of 3')
Specify next point or [Close/Undo]: (Move your cursor to the
    left and enter a value of 16'7-1/2)
Specify next point or [Close/Undo]: C (To close the shape and
    exit the LINE command)
```

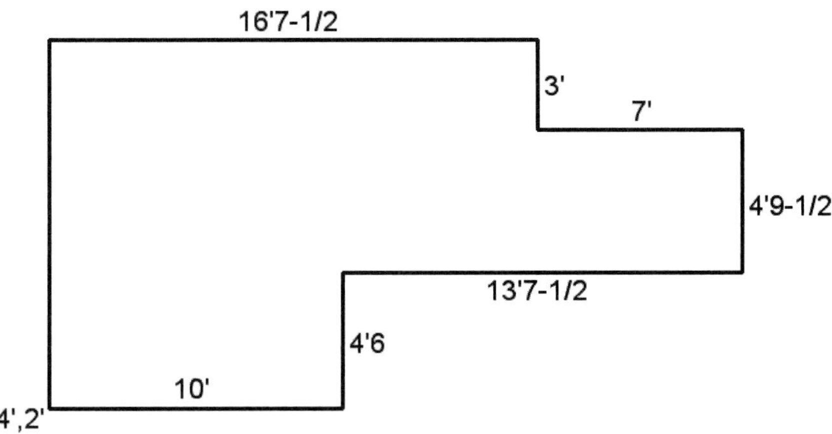

Figure 2.4

SETTING THE LIMITS OF THE DRAWING

By default, the size of the drawing screen in a new drawing file measures 12 units in the X direction and 9 units in the Y direction. This size may be ideal for small objects, but larger drawings require more drawing screen area. Use the LIMITS command for increasing the size of the drawing area. Select this command by picking Drawing Limits from the Format heading of the Menu Browser as shown in the following image; you can also enter this command directly at the command prompt by typing "Limits." Illustrated in the following image is a single-view drawing that fits on a screen size of 24 units in the X direction and 18 units in the Y direction. Follow the next command sequence to change the limits of a drawing.

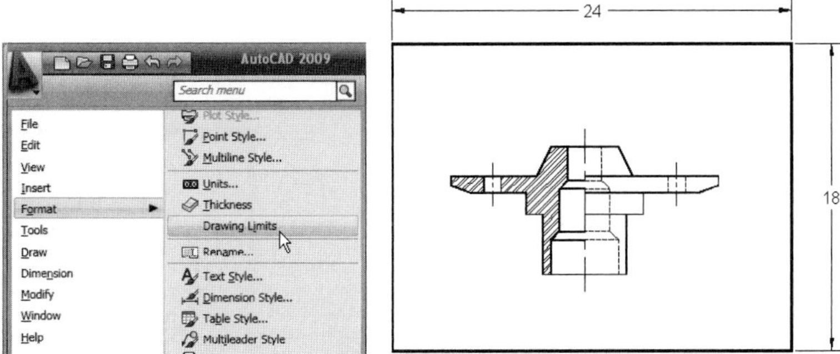

Figure 2.5

```
Command: LIMITS
Reset Model space limits:
Specify lower left corner or [ON/OFF] <0.0000,0.0000>: (Press
    ENTER to accept this value)
Specify upper right corner <12.0000,9.0000>: 24,18
```

Changing the limits does not change the current viewing area in the display screen. Before continuing, perform a ZOOM-All to change the size of the display screen to reflect the changes in the limits of the drawing. You can find ZOOM-All under View in the Menu Browser. It can also be accessed from the Zoom icon on the status bar or through the Ribbon from the Home tab and Utilities panel.

```
Command: Z (For Zoom)
All/Center/Dynamic/Extents/Previous/Scale/Window/Object<real
    time>: A (For All)
```

USING GRID IN A DRAWING

Use grid to get a relative idea as to the size of objects. Grid is also used to define the size of the display screen originally set by the LIMITS command. The dots that make up the grid will never plot out on paper even if they are visible on the display screen. You can turn grid dots on or off by using the GRID command or by pressing F7, or by single-clicking the GRID icon, located in the status bar at the bottom of the display screen. By default, the grid is displayed in 0.50-unit intervals similar to the following image on the left. Illustrated in the following image on the right is a grid that has been set to a value of 0.25, or half its original size.

Try It!: Open the drawing file 02_Grid. Use the following command sequence and illustrations below for using the GRID command.

```
Command: GRID
Specify grid spacing(X) or [ON/OFF/Snap/Major/aDaptive/Limits/
    Follow/Aspect] <0.5000>: On
Command: GRID
Specify grid spacing(X) or [ON/OFF/Major/aDaptive/Limits/Follow/
    Snap/Aspect] <0.5000>: 0.25
```

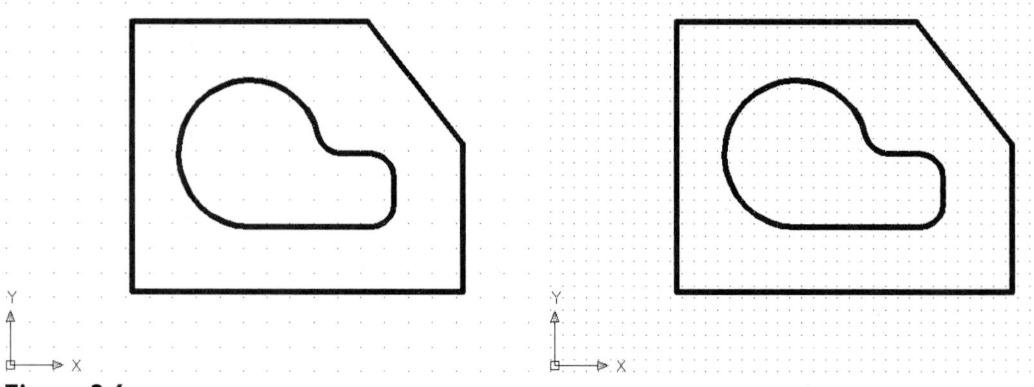

Figure 2.6

Note: One advantage of using a grid is the ability to create objects where scale is not important such as electrical symbols. You only want to make sure that all of the symbols are proportional to each other. This technique of using a grid will be covered in chapter 16.

SETTING A SNAP VALUE

It is possible to have the cursor lock on to or snap to a grid dot, as illustrated in the following image on the left; this is the purpose of the SNAP command. By default, the current snap spacing is 0.50 units. Even though a value is set, the snap must be turned on for the cursor to be positioned on a grid dot. You can accomplish this by using the SNAP command (as shown in the following sequence), by pressing F9, or by single-clicking SNAP icon in the status bar at the bottom of the display screen.

Some drawing applications require that the snap be rotated at a specific angular value (see the following image on the right). Changing the snap in this fashion also affects the cursor. Use the following command sequence for rotating the snap.

Try It!: Open the drawing file 02_Snap. Use the command sequence below and the illustration in the following image for using the SNAP command.

```
Command: SN (For SNAP)
Specify snap spacing or [ON/OFF/Aspect/Style/Type] <0.5000>: On
Command: SN (For SNAP)
Specify snap spacing or [ON/OFF/Aspect/Style/Type] <0.5000>: R
    (For Rotate)
Specify base point <0.0000,0.0000>: (Press ENTER to accept this
    value)
Specify rotation angle <0>: 30
```

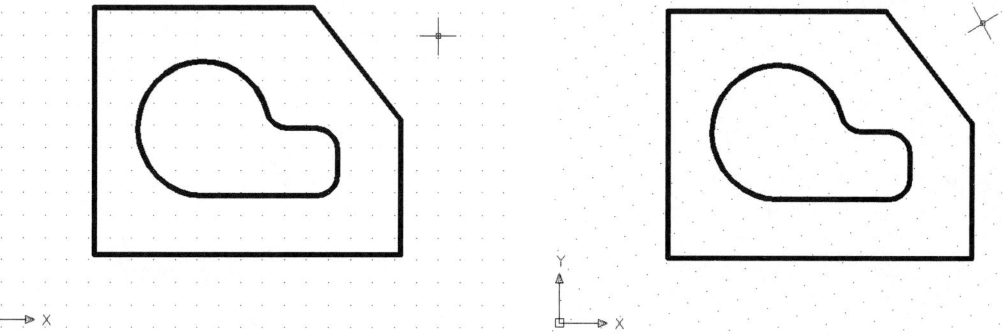

Figure 2.7

Tip: To affect both the grid and the snap, set the grid to a value of zero (0). When setting a new snap value, this value is also used for the spacing of the grid.

CONTROLLING SNAP AND GRID THROUGH THE DRAFTING SETTINGS DIALOG BOX

Right-clicking the SNAP or GRID icons in the status bar displays the menu as shown in the following image on the left. Clicking Settings displays the Drafting Settings dialog box shown in the image on the right. Use this dialog box for making changes to the grid and snap settings. The Snap type area controls whether the isometric grid is present or not.

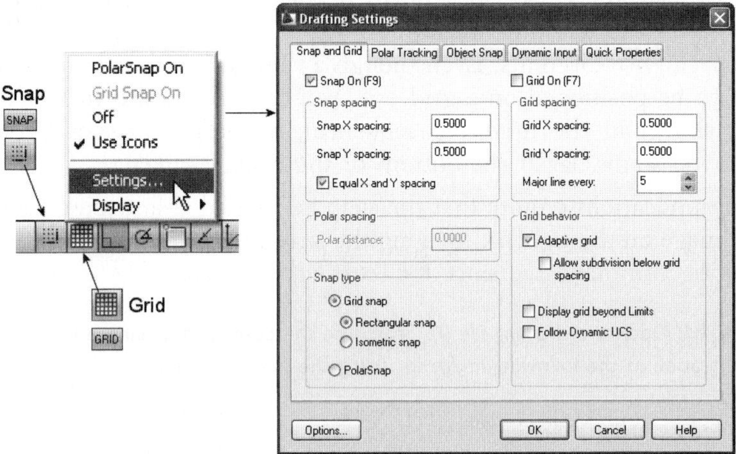

Figure 2.8

CONTROLLING DYNAMIC INPUT

Right-clicking the Dynamic Input icon, located in the status bar, and picking Settings launches the Dynamic Input tab of the Drafting Settings dialog box, as shown in the following image. Various checkboxes are available to turn on or off the Pointer Input (the absolute coordinate display of your cursor position when you are inside a command) and the Dimension Input (the display of distance and angle information for commands that support this type of input). You can also control the appearance of the Dynamic Prompts. This can take the form of changing the background color or even assigning a level of transparency to the display of the dynamic input.

Clicking the Settings button under Pointer Input and Dimension Input launches the dialog boxes that allow you to change settings to further control Pointer and Dimension Inputs.

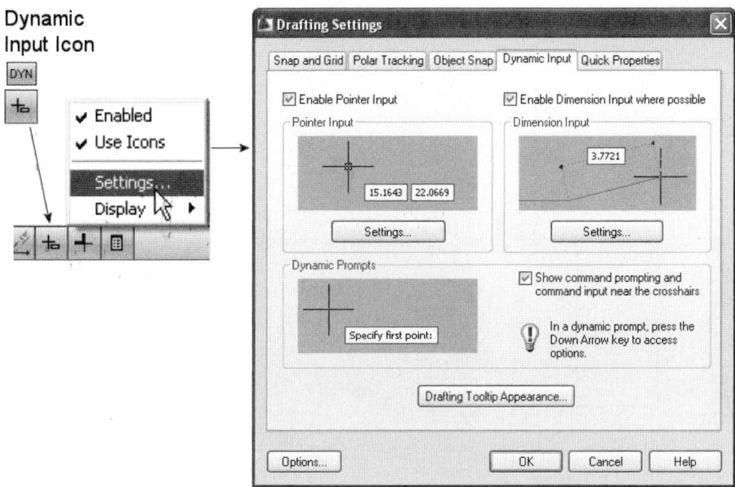

Figure 2.9

THE ALPHABET OF LINES

Engineering drawings communicate information through the use of lines and text, which, if used appropriately, accurately convey a project from design to construction. Before you construct engineering drawings, the quality of the lines that make up the drawing must first be discussed. Some lines of a drawing should be made thick; others need to be made thin. This is to emphasize certain parts of the drawing and it is controlled through a line quality system. Illustrated in the following image is a two-view drawing of an object complete with various lines that will be explained further.

The most important line of a drawing is the object line, which outlines the basic shape of the object. Because of their importance, object lines are made thick and continuous so they stand out among the other lines in the drawing. It does not mean that the other lines are considered unimportant; rather, the object line takes precedence over all other lines.

The cutting plane line is another thick line; it is used to determine the placement in the drawing where an imaginary saw will cut into the drawing to expose interior details. It stands out by being drawn as a series of long dashes separated by spaces. Arrowheads determine the viewing direction for the adjacent view. This line will be discussed in greater detail in chapter 9, "Creating Section Views."

The hidden line is a thin weight line used to identify edges that become invisible in a view. It consists of a series of dashes separated by spaces. Whether an edge is visible or invisible, it still must be shown with a line.

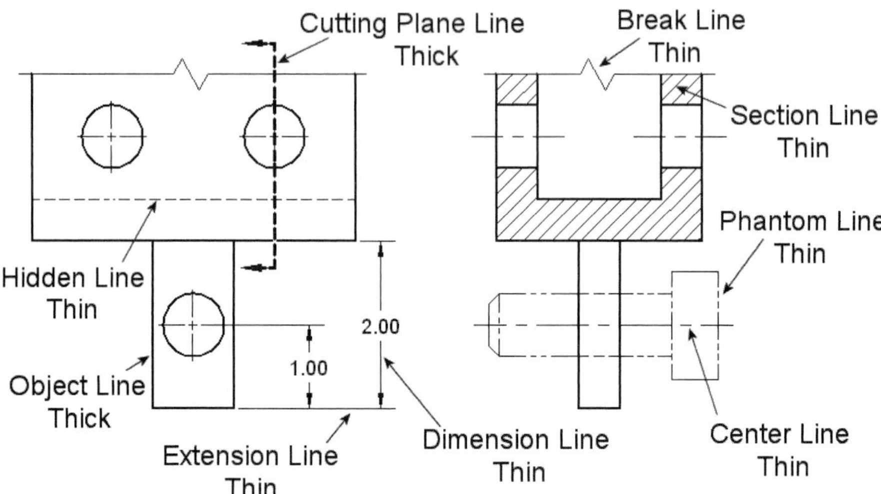

Figure 2.10

The dimension line is a thin line used to show the numerical distance between two points. The dimension text is placed within the dimension line, and arrowheads are placed at opposite ends of the dimension line.

The extension line is another thin continuous line, used as a part of the overall dimension. Extension lines provide a means to move dimension lines away from the object into a clear area where they can easily be seen and interpreted.

When you use the cutting Plane line to create an area to cut or slice, the surfaces in the adjacent view are section lined using the section line, a thin continuous line.

Another important line used to identify the centers of symmetrical objects such as cylinders and holes is the centerline. It is a thin line consisting of a series of long and short dashes. It is a good practice to dimension to centerlines; for this reason centerlines, extension lines, and dimensions are made the same line thickness.

The phantom line consists of a thin line made with a series of two short dashes and one long dash. It is used to simulate the placement or movement of a part or component without actually detailing the component.

The long break line is a thin line with a "zigzag" symbol used to establish where an object is broken to simulate a continuation of the object.

ORGANIZING A DRAWING THROUGH LAYERS

As a means of organizing objects, a series of layers should be devised for every drawing. You can think of layers as a group of transparent sheets that combine to form the completed drawing. The illustration on the left in the following image displays a drawing consisting of object lines, dimension lines, and border. An example of organizing these three drawing components by layers is illustrated on the right in the following image. Only the drawing border occupies a layer called "Border." The object lines occupy a layer called

"Object," and the dimension lines are drawn on a layer called "Dimension." At times, it may be necessary to turn off the dimension lines for a clearer view of the object. Creating all dimensions on a specific layer allows you to turn off the dimensions while viewing all other objects on layers that are still turned on.

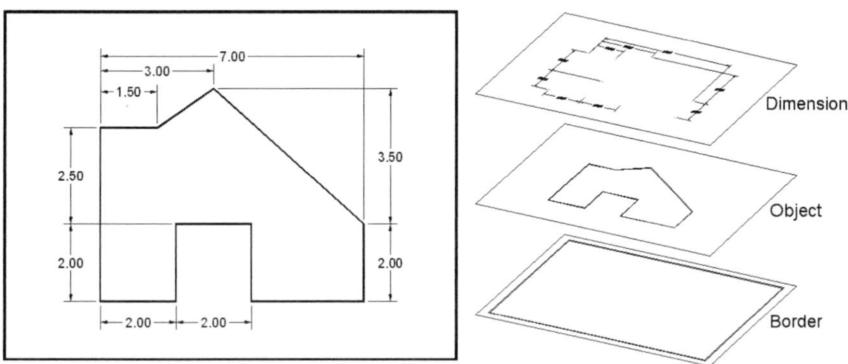

Figure 2.11

THE LAYER PROPERTIES MANAGER PALETTE

The Layer Properties Manager palette is the tool used to create and manage layers. This palette is activated through the following methods: clicking the Layer Properties Manager button from the toolbar located in the AutoCAD Classic workspace, as shown on the left in the following image; choosing Format followed by Layer... from the Menu Browser, as shown in the middle in the following image; or clicking the Layer Properties Manager button from the Ribbon, as shown on the right in the following image.

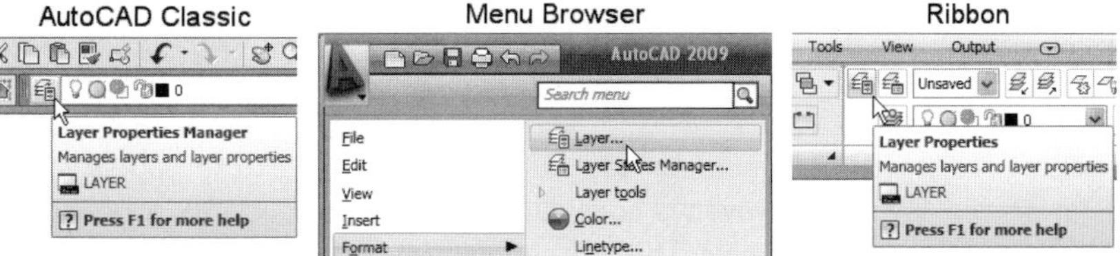

Figure 2.12

The Layer Properties Manager palette, illustrated in the following image, is divided into two separate panes. The first pane on the left is the tree view pane used for displaying layer filter, group, or state information. The main body of the Layer Properties Manager is the list view pane. This area lists the individual layers that currently exist in the drawing.

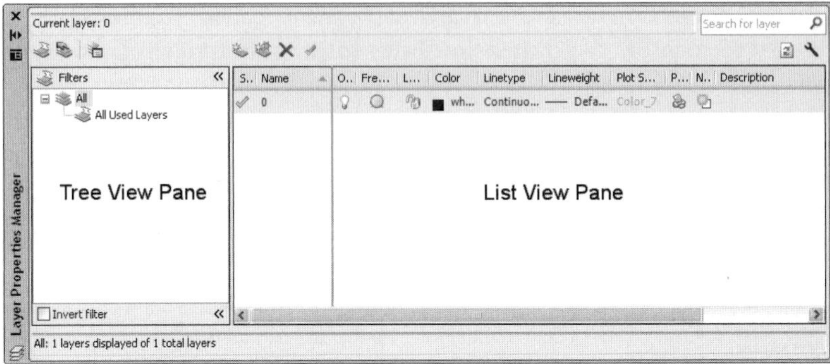

Figure 2.13

The layer information located in the list view pane is sometimes referred to as layer states that allow you to perform the following operations: turning layers on or off, freezing or thawing layers, locking or unlocking layers, assigning a color, linetype, lineweight, and plot style to a layer or group of layers, as shown in the following image. A brief explanation of each layer state is provided below:

Status	Name	On	Freeze	Lock	Color	Linetype	Lineweight	Plot Style	Plot	New VP Freeze	Description
	0				white	Continuous	—— Default	Color_7			

Figure 2.14

Status – When a green checkmark is displayed, this layer is considered current.

Name – Displays the name of the layer.

On/Off – Makes all objects created on a certain layer visible or invisible on the display screen. The On state is symbolized by a yellow light bulb. The Off state has a light bulb icon shaded black.

Freeze – This state is similar to the Off mode; objects frozen appear invisible on the display screen. Freeze, however, is considered a major productivity tool used to speed up the performance of a drawing. This is accomplished by not calculating any frozen layers during drawing regenerations. A snowflake icon symbolizes this layer state.

Thaw – This state is similar to the On mode; objects on frozen layers reappear on the display screen when they are thawed. The sun icon symbolizes this layer state.

Lock – This state allows objects on a certain layer to be visible on the display screen while protecting them from accidentally being modified through an editing command. A closed padlock icon symbolizes this layer state.

Unlock – This state unlocks a previously locked layer and is symbolized by an open padlock icon.

Color – This state displays a color that is assigned to a layer and is symbolized by a square color swatch along with the name of the color. By default, the color white is assigned to a layer.

Linetype – This state displays the name of a linetype that is assigned to a layer. By default, the Continuous linetype is assigned to a layer.

Lineweight – This state sets a lineweight to a layer. An image of this lineweight value is visible in this layer state column.

Plot Style – A plot style allows you to override the color, linetype, and lineweight settings made in the Layer Properties Manager dialog box. Notice how this area is grayed out. When working with a plot style that is color dependent, you cannot change the plot style. Plot styles will be discussed in greater detail later in this book.

Plot – This layer state controls which layers will be plotted. The presence of the printer icon symbolizes a layer that will be plotted. A printer icon with a red circle and diagonal slash signifies a layer that will not be plotted.

New VP Freeze – Creates a new layer and automatically freezes this layer in any new viewport.

Description – This state allows you to enter a detailed description for a layer.

THE STATUS OF LAYERS

An additional layer property called Status is illustrated in the following image. It is used to identify visually which layers contain or hold drawing objects and which are considered empty or unused. If the Status icon is solid blue, objects such as lines, circles, arcs, and so forth reside on this layer. If the Status icon is faded gray in color, the layer is empty of any objects.

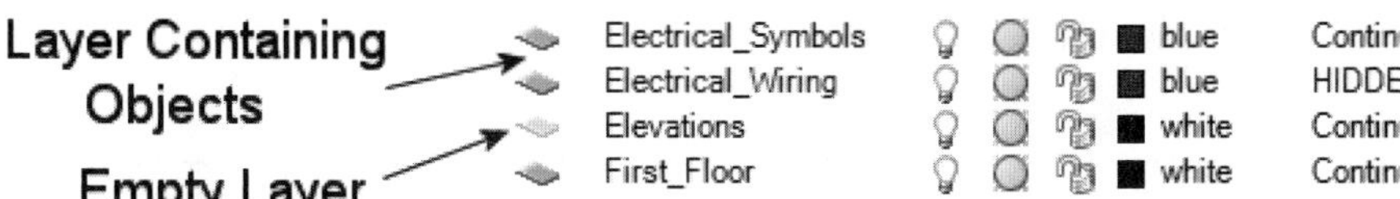

Figure 2.15

CREATING NEW LAYERS

Four buttons used for creating and deleting layers and for making a layer current are available at the top of the Layer Properties Manager palette. Most of these layer options can also be activated by pressing the ALT key along with a specific letter, as shown in the following image.

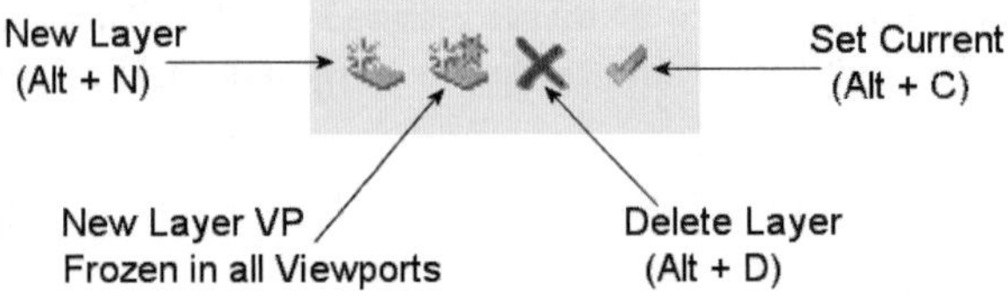

Figure 2.16

Clicking the New button of the Layer Properties Manager palette creates a new layer called Layer1, which displays in the layer list box, as shown in the following image. Since this layer name is completely highlighted, you may elect to change its name to something more meaningful, such as a layer called Object or Hidden to hold all object or hidden lines in a drawing.

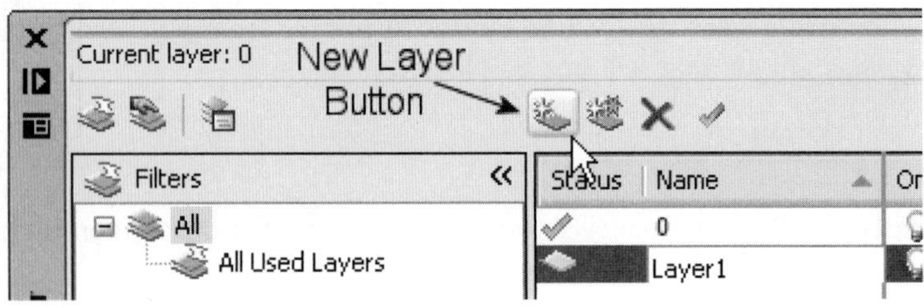

Figure 2.17

Illustrated in the following image is the result of changing the name of the layer from Layer1 to Object. You could also have entered "OBJECT" or "object" and these names would appear in the palette.

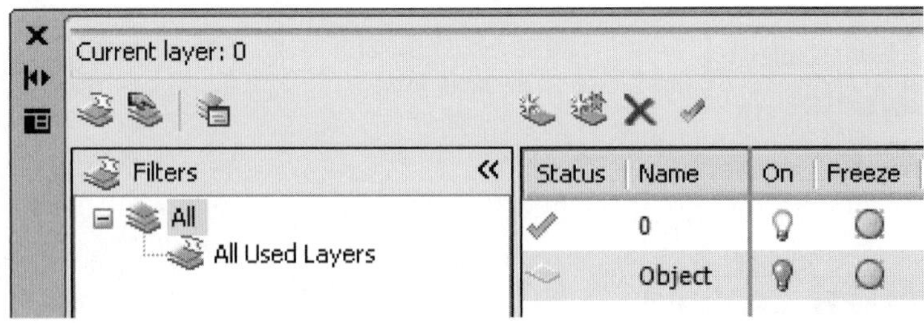

Figure 2.18

You can also be descriptive with layer names. In the following image, a layer has been created called "Section (this layer is designed to control section lines)." You are allowed to add spaces and other characters in the naming of a layer. Because of space limitations, the entire layer name may not display until you move your cursor over the top of the layer name. As a result, the layer name will appear truncated as in the layer Section (... in the following image).

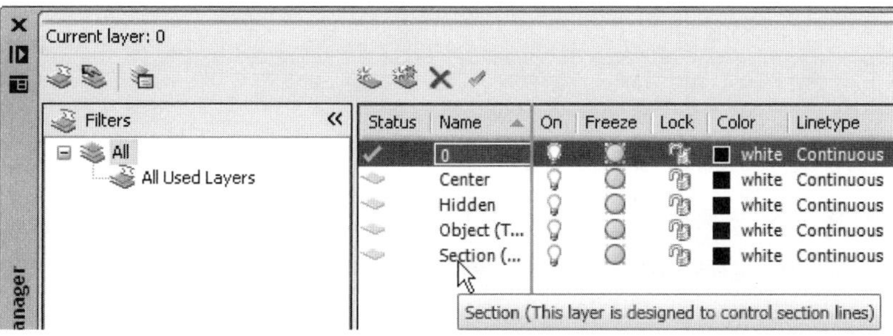

Figure 2.19

If more than one layer needs to be created, it is not necessary to continually click the New button. A more efficient method would be to perform either of the following operations: After creating a new Layer1, change its name to Dimension followed by a comma (,). This automatically creates a new Layer1 as shown in the following image. Continue creating new layer names followed by the comma. You could also press ENTER twice, which would create another new layer.

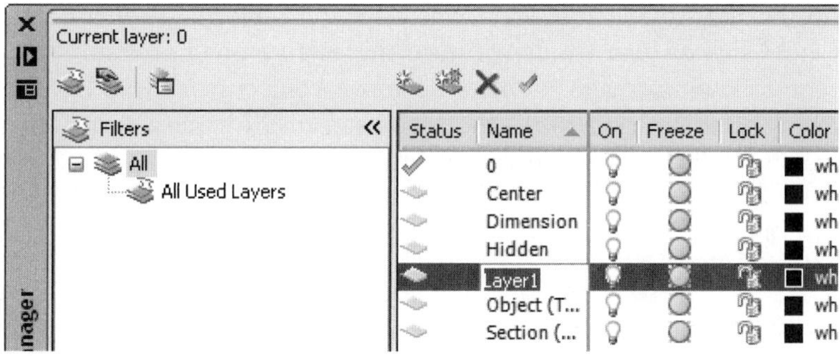

Figure 2.20

DELETING LAYERS

To delete a layer or group of layers, highlight the layers for deletion and click on the delete button as shown on the left in the following image. The results are displayed on the right in the following image with the selected layers being deleted from the palette.

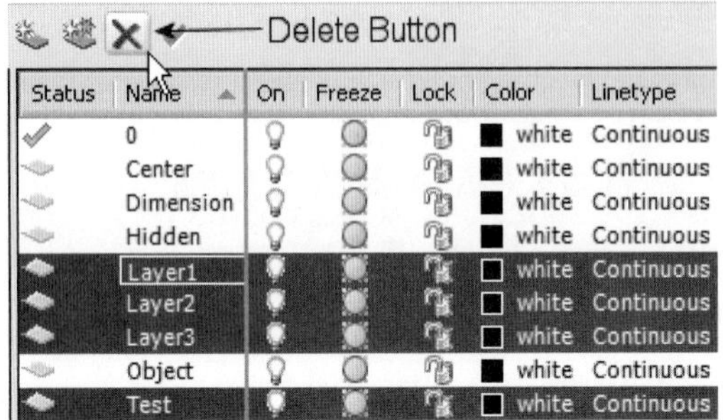

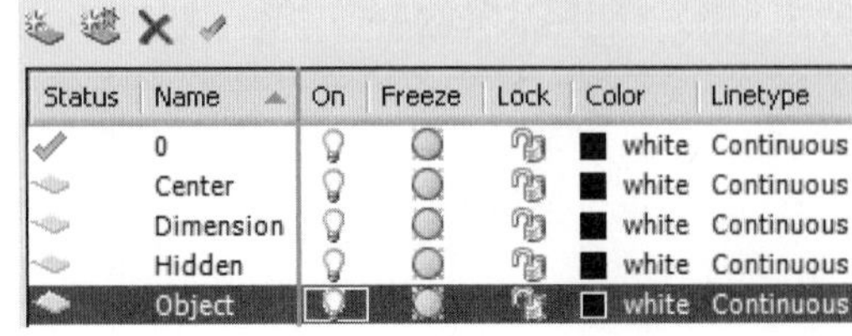

Figure 2.21

Note: Only layers that do not contain any drawing geometry or objects can be deleted.

AUTO-HIDING THE LAYER PROPERTIES MANAGER PALETTE

While the Layer Properties Manager palette can display on the screen preventing you from working on detail segments of your drawing, it is possible to collapse or Auto-hide the palette. Right clicking on the title strip will display a menu as shown on the left in the following image. Click on Auto-hide will turn this feature on. The results are displayed on the right in the following image. When you move your cursor away from the palette, it collapses allowing your drawing to fill the entire screen. Moving your cursor over the title strip of the palette will display it in its entirety.

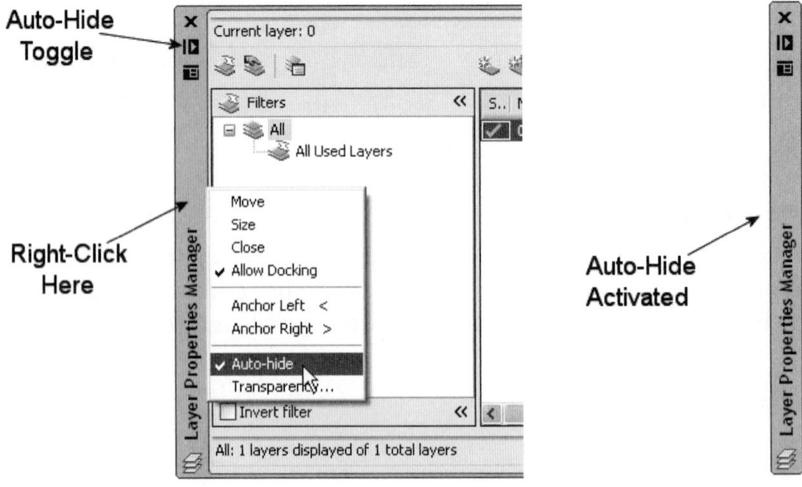

Figure 2.22

ASSIGNING COLOR TO LAYERS

Once you select a layer from the list box of the Layer Properties Manager palette and the color swatch is selected in the same row as the layer name, the Select Color dialog box shown in the following image is displayed. Three tabs allow you to select three different color groupings. The three groupings (Index Color, True Color, and Color Books) are described as follows.

INDEX COLOR TAB

This tab, shown in the following image, allows you to make color settings based on 255 AutoCAD Color Index (ACI) colors. Standard, Gray Shades, and Full Color Palette areas are available for you to choose colors from.

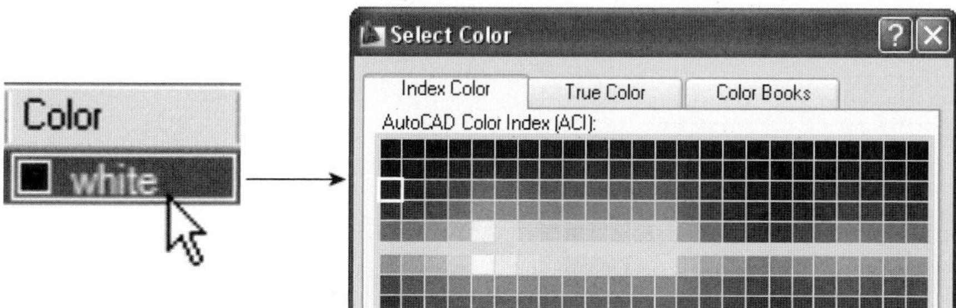

Figure 2.23

TRUE COLOR TAB

Use this tab to make color settings using true colors, also known as 24-bit color. Two color models are available for you to choose from, namely Hue, Saturation, and Luminance (HSL), or Red, Green, and Blue (RGB). Through this tab, you can choose from over sixteen million colors, as shown on the left in the following image.

COLOR BOOKS TAB

Use the Color Books tab to select colors that use third-party color books (such as Pantone) or user-defined color books. You can think of a color book as similar to those available in hardware stores when selecting household interior paints. When you select a color book, the name of the selected color book will be identified in this tab, as shown on the right in the following image.

Note: The Index tab will be used throughout this text. However, you are encouraged to experiment with the True Color and Color Books tabs.

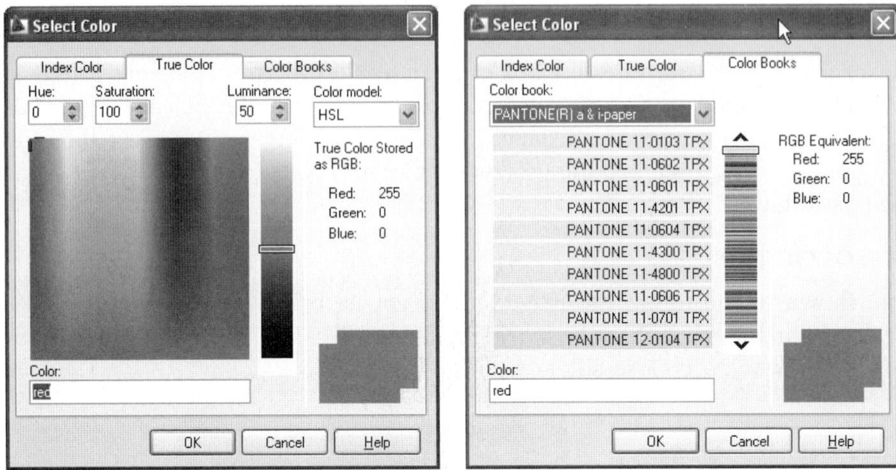

Figure 2.24

ASSIGNING LINETYPES TO LAYERS

Selecting the name Continuous next to the highlighted layer activates the Select Linetype dialog box, as shown in the following image. Use this dialog box to dynamically select preloaded linetypes to be assigned to various layers. By default, the Continuous linetype is loaded for all new drawings.

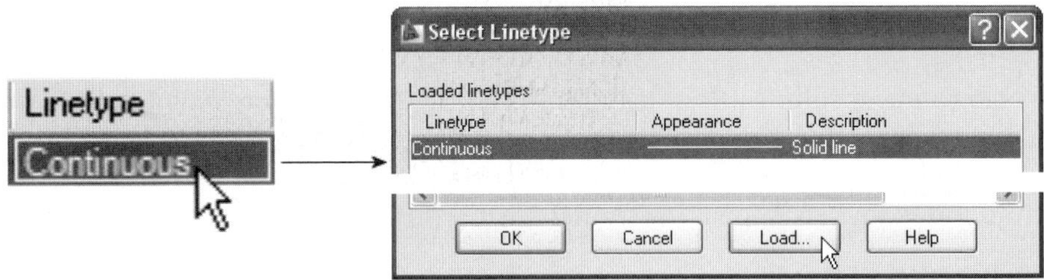

Figure 2.25

To load other linetypes, click the Load button of the Select Linetype dialog box; this displays the Load or Reload Linetypes dialog box, as shown in the following image.

Use the scroll bars to view all linetypes contained in the file ACAD.LIN. Notice that, in addition to standard linetypes such as HIDDEN and PHANTOM, a few linetypes are provided that have text automatically embedded in the linetype. As the linetype is drawn, the text is placed depending on how it was originally designed. Notice also three variations of Hidden linetypes; namely HIDDEN, HIDDEN2, and HIDDENX2. The HIDDEN2 represents a linetype where the distances of the dashes and spaces in between each dash are half of the original HIDDEN linetype. HIDDENX2 represents a linetype where the distances of the dashes and spaces in between each dash of the original HIDDEN linetype are doubled. Click the desired linetypes to load. When finished, click the OK button.

The loaded linetypes now appear in the Select Linetype dialog box, as shown in the following image. It must be pointed out that the linetypes in this list are only loaded into the drawing and are not assigned to a particular layer. Clicking the linetype in this dialog box assigns this linetype to the layer currently highlighted in the Layer Properties Manager palette.

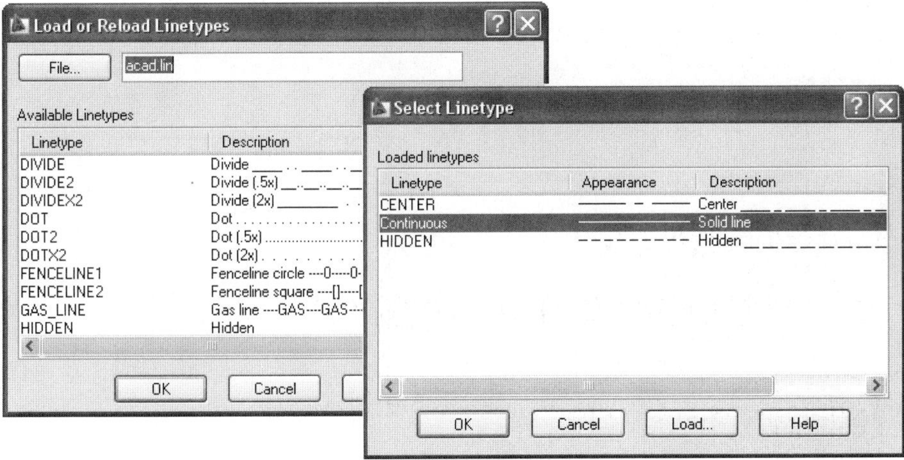

Figure 2.26

ASSIGNING LINEWEIGHT TO LAYERS

Selecting the name Default under the Lineweight heading of the Layer Properties Manager dialog box activates the Lineweight dialog box, as shown in the following image. Use this to attach a lineweight to a layer. Lineweights are very important to a drawing file—they give contrast to the drawing. As stated earlier, the object lines should stand out over all other lines in the drawing. A thick lineweight would then be assigned to the object line layer.

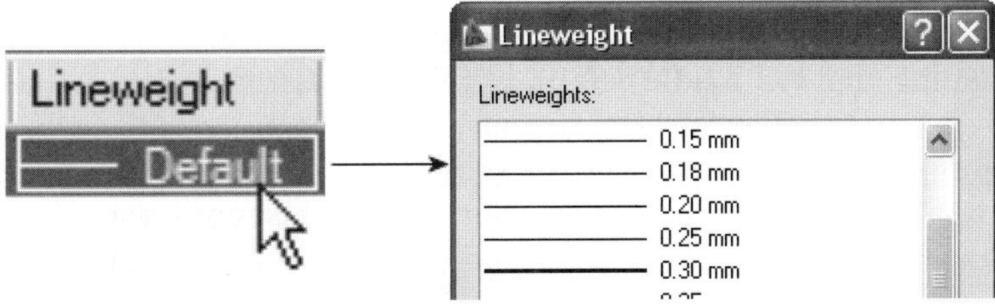

Figure 2.27

In the following image on the left, a lineweight of 0.50 mm has been assigned to all object lines and a lineweight of 0.30 has been assigned to hidden lines. However, all lines in this figure appear to consist of the same lineweight. This is due to the lineweight feature being turned off. Notice in this figure the Lineweight icon in the Status bar. Use this button to toggle ON or OFF the display of assigned lineweights. Clicking the Lineweight icon, as shown on the right in the following image, turns the lineweight function on (the icon will turn a blue color.) It should be noted that the assigned lineweight will be plotted whether or not the LWT button is activated.

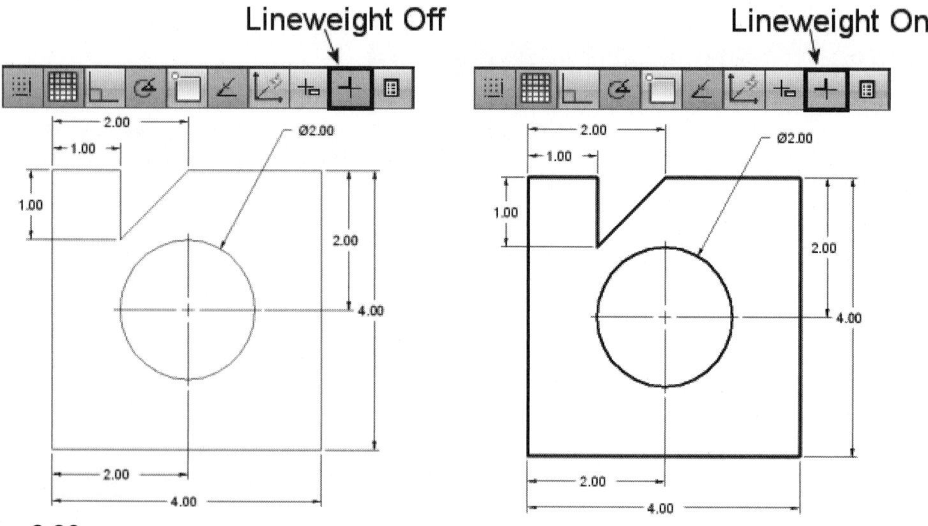

Figure 2.28

THE LINEWEIGHT SETTINGS DIALOG BOX

You have probably already noticed that when you turn on the lineweights through the LWT button, all lineweights may appear thick. In a complicated drawing, the lineweights could be so thick that it would be difficult to interpret the purpose of the drawing. To give your lineweights a more pleasing appearance, a dialog box is available to control the display of your lineweights. This control does not affect the plotted lineweights assigned; only what is displayed on the screen. Clicking Format followed by Lineweight, found in the Menu Browser, as shown on the left in the following image, displays the Lineweight Settings dialog box. Notice the position of the slider bar in the Adjust Display Scale area of the dialog box. Sliding the bar to the left near the Min setting reduces the width of all lineweights. Try experimenting with this on any lineweight assigned to a layer. If you think your lineweights appear too thin, slide the bar to the right near the Max setting and observe the results. Continue to adjust your lineweights until they have a pleasing appearance in your drawing.

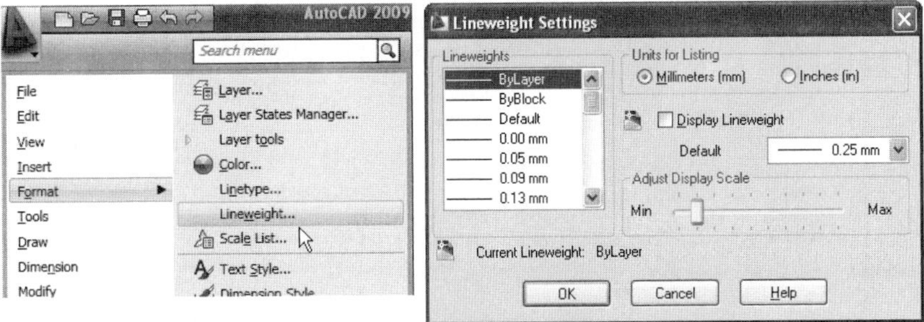

Figure 2.29

THE LINETYPE MANAGER DIALOG BOX

Choosing Format followed by Linetype, found in the pull-down menu, as shown on the left in the following image (or found in the Menu Browser), displays the Linetype Manager dialog box. This dialog box is designed mainly to preload linetypes.

Clicking the Load... button activates the Load or Reload Linetypes dialog box illustrated in a previous segment of this chapter. You can select individual linetypes in this dialog box or load numerous linetypes at once by pressing CTRL and clicking each linetype. Clicking the OK button loads the selected linetypes into the dialog box.

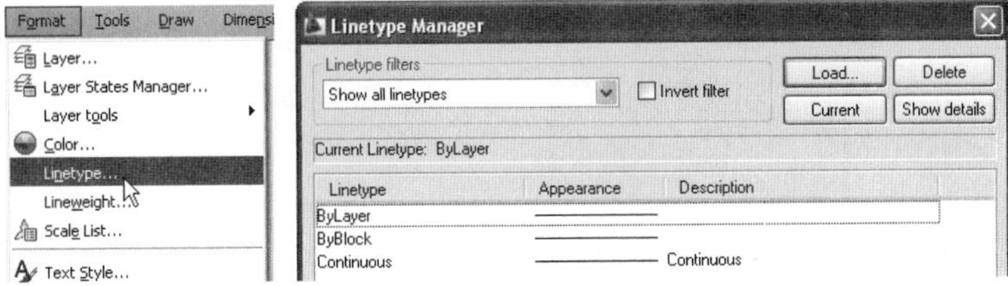

Figure 2.30

The Linetype Manager dialog box, as shown in the following image, also displays a number of extra linetype details (click Show Details). The Global scale factor value has the same effect as the LTSCALE command. Also, setting a different value in the Current object scale box can scale the linetype of an individual object.

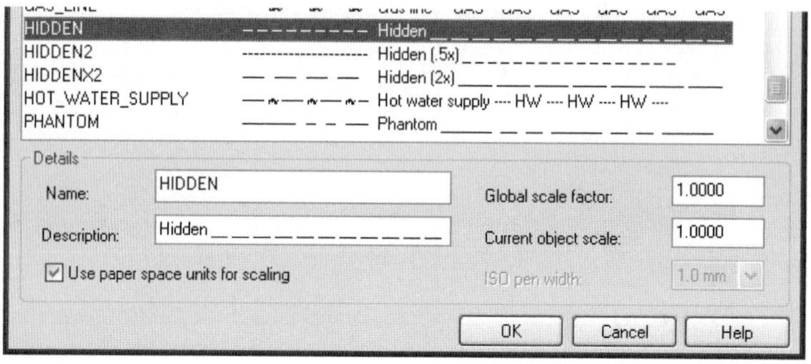

Figure 2.31

LOCKED LAYERS

Objects on locked layers remain visible on a drawing; they cannot, however, be selected when performing editing operations. Quite often, it can be difficult to distinguish objects on layers that are locked from normal layers. To assist with the identification of locked objects, a lock icon appears when you move your cursor over an object considered locked, as shown in the following image.

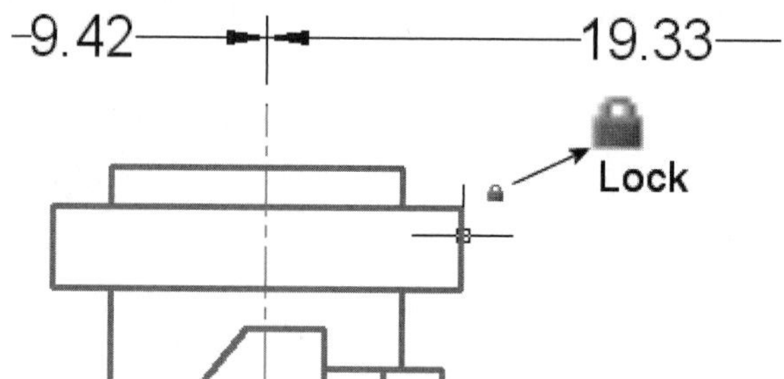

Figure 2.32

Another tool used to assist with identifying locked layers can be found by activating the Home tab of the Ribbon. Expanding the Layers panel will display the Layer Lock Fading tool, as shown in the following image. A slider bar that will dim or fade an object on a locked layer is present. Notice that the image on the left has objects on a locked layer faded to 50% while the image on the right is faded 80%. This allows you to better distinguish objects on regular layers or on locked layers.

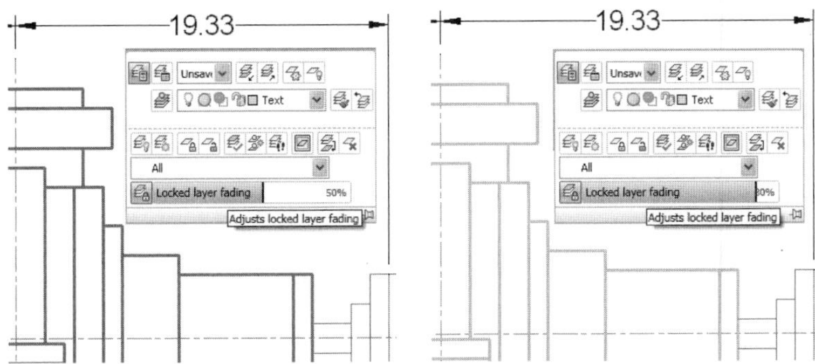

Figure 2.33

THE LAYERS TOOLBAR

The Layers toolbar provides an area to better control the layer properties or states. A toolbar is illustrated on the left in the following image. You can also access this feature through the Ribbon as shown on the right in the following image. The Layer Control area will now be discussed in greater detail.

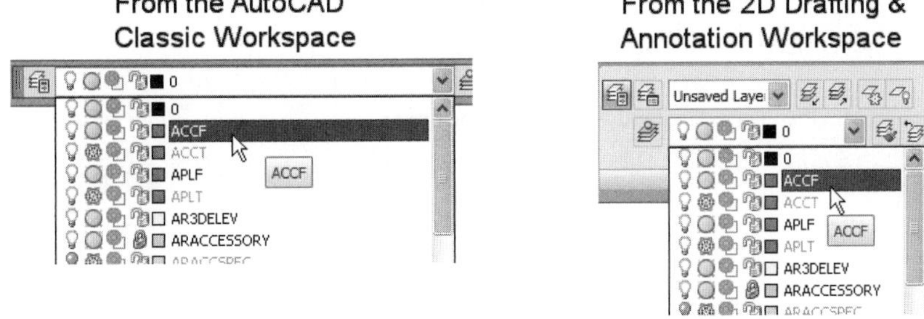

Figure 2.34

CONTROL OF LAYER PROPERTIES

Expanding the drop down list cascades all layers defined in the drawing in addition to their properties, identified by symbols (see the following image). The presence of the light bulb signifies that the layer is turned on. Clicking the lightbulb symbol turns the layer off. The sun symbol signifies that the layer is thawed. Clicking the sun turns it into a snowflake symbol, signifying that the layer is now frozen. The padlock symbol controls whether a layer is locked or unlocked. By default, all layers are unlocked. Clicking the padlock changes the symbol to display the image of a locked symbol, signifying that the layer is locked.

Study the following image for a better idea of how the symbols affect the state of certain layers.

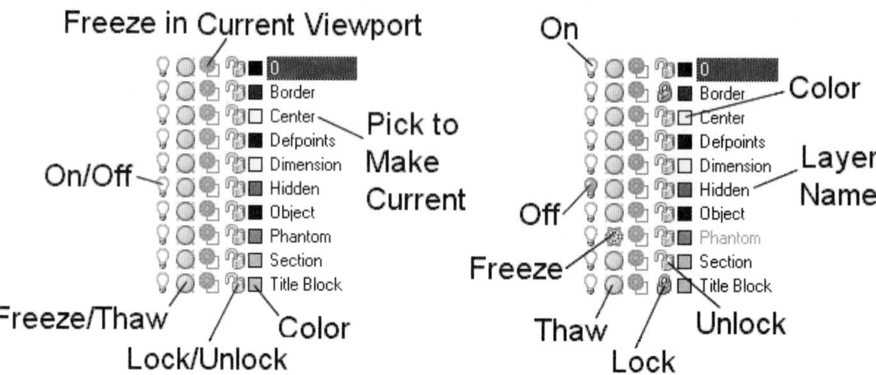

Figure 2.35

THE PROPERTIES TOOLBAR

In addition to the Layers Property Manager Palette, a number of other toolbars supporting layer properties are also available. The Properties toolbar, as shown on the left in the following image provides three areas to better access options for the control of Colors, Linetypes, and Lineweights. A fourth area controls plot styles and is not discussed in this book. You can also access these same properties through the Properties pane of the Ribbon as shown on the right in the following image.

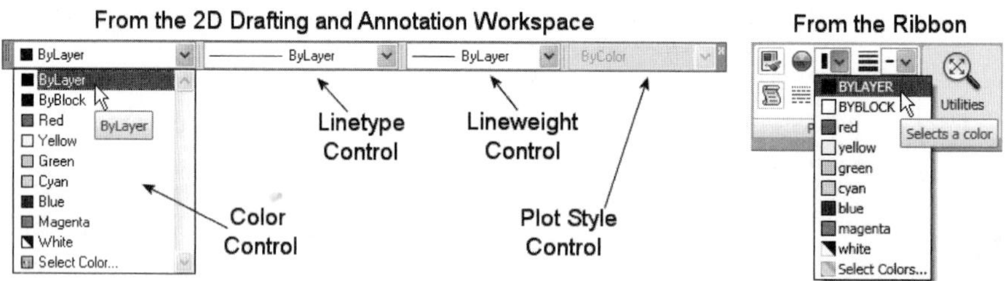

Figure 2.36

Note: While the Properties toolbar and pane allow you to change color, linetype, and lineweight on the fly, it is poor practice to do so. Check to see that each category in the Properties toolbar reads "ByLayer," which means that layers control this category.

MAKING A LAYER CURRENT

Various methods can be employed to make a layer current to draw on. Select a layer in the Layer Properties Manager palette and then click the Current button to make the layer current.

Picking a layer name from the Layer Control box, as shown on the left in the following image, will also make the layer current.

The Make Object's Layer Current button, located in the Ribbon as shown on the right in the following image, allows you to make a layer the new current layer by clicking an object in the drawing. The layer is now made current based on the layer of the selected object.

```
Command: LAYMCUR
Select object whose layer will become current: (Pick an object)
```

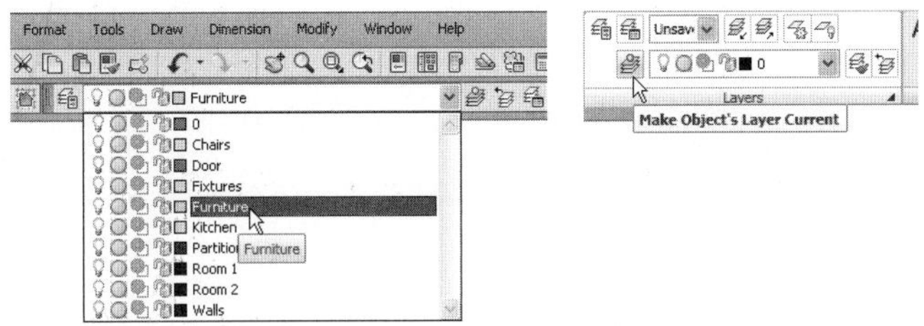

Figure 2.37

USING THE LAYER PREVIOUS COMMAND

The Layer Previous button is used to undo changes that you have made to layer settings such as color or lineweight. This button can be found on the left in the Layers toolbar (2D Drafting & Annotation Workspace) or on the right in the Layers pane of the Ribbon as shown in the following image. You could even turn a number of layers off and use the Layer Previous command to have the layers turned back on in a single step. This command can also be entered from the keyboard in the following prompt sequence:

```
Command: LAYERP
```

There are a few exceptions to the use of the Layer Previous command:

- If you rename a layer and change its properties, such as color or lineweight, issuing the Layer Previous command restores the original properties but does not change the layer name back to the original.
- Purged or deleted layers will not be restored by using Layer Previous.
- If you add a new layer, issuing the Layer Previous command will not remove the new layer.

Figure 2.38

RIGHT-CLICK SUPPORT FOR LAYERS

While inside the Layer Properties Manager, right-clicking inside the layer information area displays the shortcut menu in the following image. Use this menu to make the selected layer current, to make a new layer based on the selected layer, to select all layers in the dialog box, or to clear all layers. You can even select all layers except for the current layer. This shortcut menu provides you with easier access to commonly used layer manipulation tools.

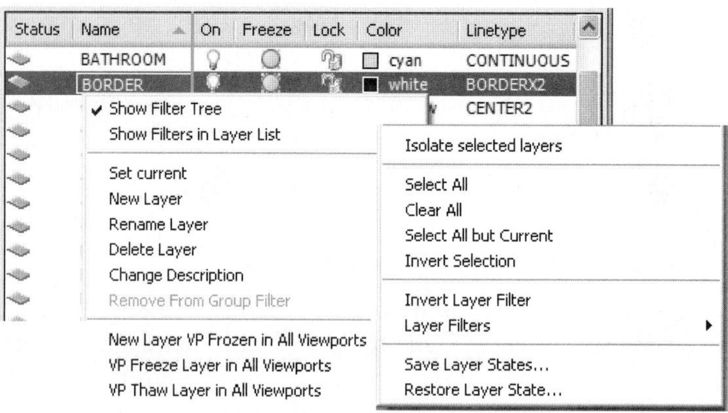

Figure 2.39

OTHER RIGHT-CLICK LAYER CONTROLS

If you right-click in one of the layer header names (Name, On, Freeze, etc.), you get the menu illustrated in the following image. This menu allows you to turn off header names as a means of condensing the list of headers and making it easier to interpret the layers that you are using. Other areas of this menu allow you to maximize all columns in order to view all information in full regarding layers. By default, the Name column is frozen. This means when you scroll to the left or right to view the other layer properties, the Name column does not scroll and allows you to view information at the end of the Layer Properties Manager palette.

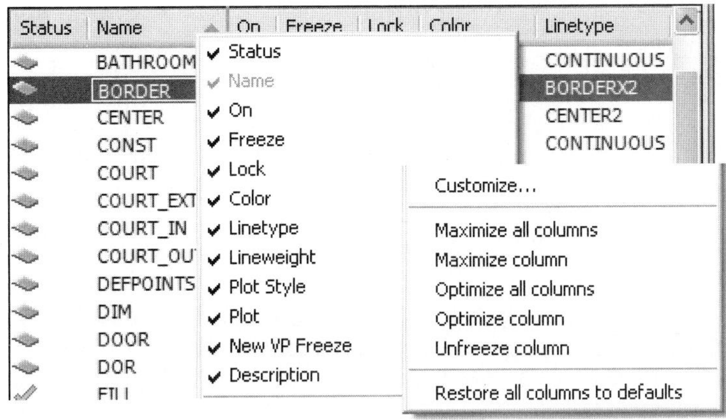

Figure 2.40

CONTROLLING THE LINETYPE SCALE

Once linetypes are associated with layers and placed in a drawing, a LTSCALE command is available to control their scale. On the left side of the following image, the default linetype scale value of 1.00 is in effect. This scale value acts as a multiplier for all linetype distances. In other words, if the hidden linetype is designed to have dashes 0.25 units long, a linetype scale value of 1.00 displays the dash of the hidden line at a value of 0.25 units. The LTSCALE command displays the following command sequence:

```
Command: LTS (For LTSCALE)
Enter new linetype scale factor <1.0000>: (Press ENTER to accept
    the default or enter another value)
```

Try It!: Open the drawing file 02_LTScale. Follow the directions, command prompts, and illustrations below for using the LTSCALE command.

In the middle of the following image, a linetype scale value of 0.50 units has been applied to all linetypes. As a result of the 0.50 multiplier, instead of all hidden line dashes measuring 0.25 units, they now measure 0.125 units.

```
Command: LTS (For LTSCALE)
Enter new linetype scale factor <1.0000>: 0.50
```

On the right side of the following image, a linetype scale value of 2.00 units has been applied to all linetypes. As a result of the 2.00 multiplier, instead of all hidden line dashes measuring 0.25 units, they now measure 0.50 units. Notice how a large multiplier displays the centerlines as what appears to be a continuous linetype. The lines are not long enough to display the size of dashes specified.

```
Command: LTS (For LTSCALE)
Enter new linetype scale factor <0.5000>: 2.00
```

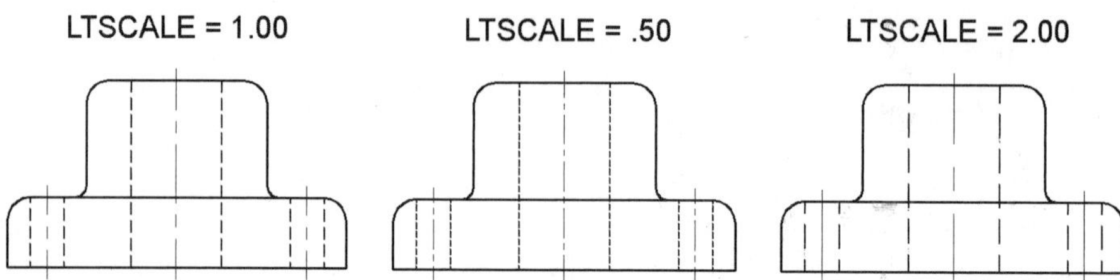

Figure 2.41

Try It!: Open the drawing file 02_Floor Plan illustrated in the following image on the left. This floor plan is designed to plot out at a scale of 1/8" = 1'0". This creates a scale factor of 96 (found by dividing 1' by 1/8"). A layer called "Dividers" was created and assigned the Hidden linetype to show all red hidden lines as potential rooms in the plan. However, the hidden lines do not display. For all linetypes to show as hidden, this multiplier should be applied to the drawing through the LTSCALE command.

Since the drawing was constructed in real-world units or full size, the linetypes are converted to these units beginning with the multiplier of 96, as shown in the following image on the right, through the LTSCALE command.

```
Command: LTS (For LTSCALE)
Enter new linetype scale factor <1.0000>: 96
```

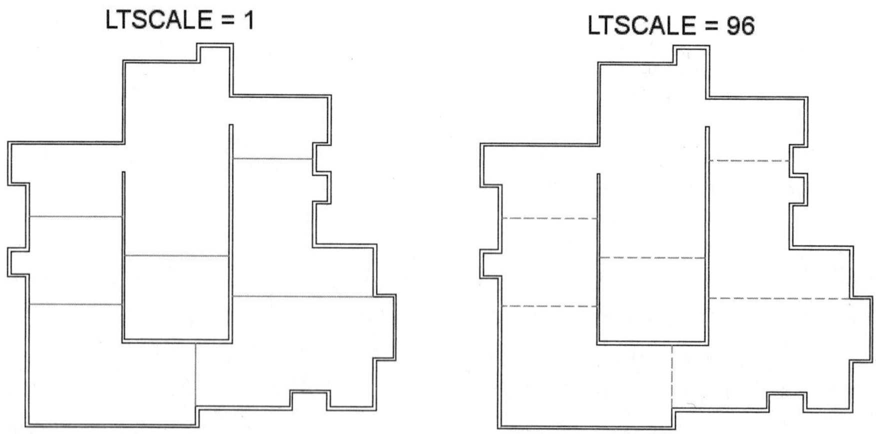

Figure 2.42

Note: Changing the default linetype scale value from 1 to another value affects the Annotation Scale when switching to a drawing layout. Annotation Scales will be discussed in greater detail in chapter 19. In the meantime, use linetype scales to change their values and observe how the linetype will appear in your drawing. When finished, change the linetype scale back to 1.

ADVANCED LAYER TOOLS

LAYER FILTERS

The first button located above the Tree View pane allows you to create a New Property Filter, as shown at the left in the following image. Clicking this button displays the Layer Filter Properties dialog box. You enter information in the Filter definition area (upper half) of the dialog box and observe the results of the filter in the Filter preview area (lower half). In this example, a layer filter has been created in the definition area to identify all layers that begin with AR*. The results show a number of layers in the preview area. Notice at the top of the dialog box that this filter has been given the name AR. You can build various named filters as a means of further organizing your layers by function, color, name, and linetype, or even through a combination of these states.

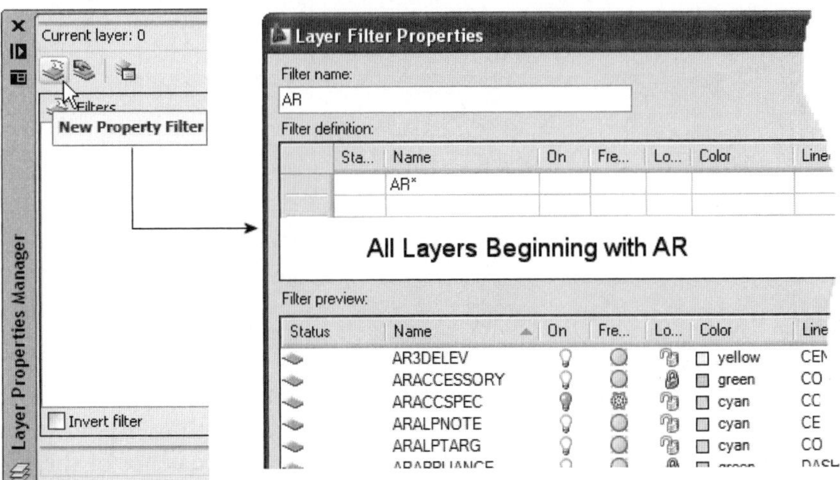

Figure 2.43

LAYER GROUPS

Layer groups allow you to collect a number of layers under a unique name. For instance, you may want to group a number of layers under the name Foundation; or in a mechanical application, you may want to group a number of layers under the name Fasteners. Choose this command by clicking the second button illustrated on the left in the following image. You could also move your cursor in the Tree View pane and right-click to display the menu shown on the right in the following image. Then click New Group Filter.

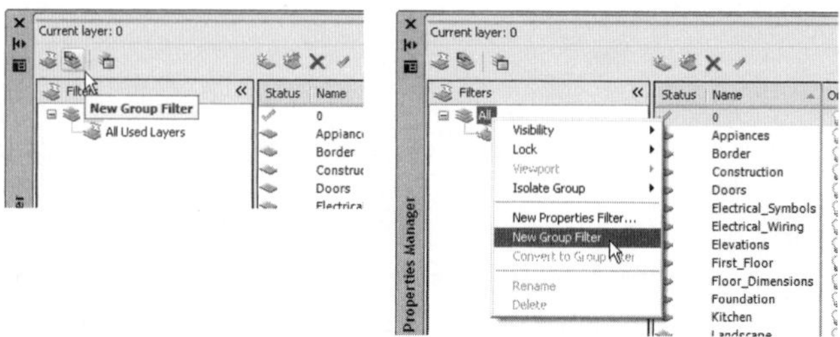

Figure 2.44

As shown on the left in the following image, a new layer group called Electrical Plan is created and located in the Tree View pane. To associate layers with this group, select the layers in the List View pane and drag the layers to the layer group. You are not moving these layers; rather, you are grouping them under a unique name.

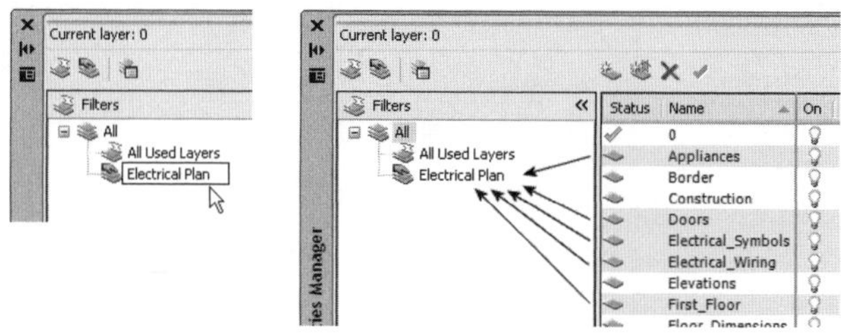

Figure 2.45

Clicking the Electrical Plan layer group displays only those layers common to this group in the List View pane of the Layer Properties Manager dialog box, as shown on the left in the following image. Notice also in the Tree View pane the existence of the Floor_Plan, Foundation Plan, and Unused Layers groups.

Note: Clicking All in the Tree View pane lists all layers defined in the drawing.

Additional controls allow you to further manipulate layers. With the Electrical Plan group selected, it is possible to invert this filter. This means to select all layers not part of the Electrical Plan group. To perform this operation, place a check in the box next to Invert filter, located in the lower-left corner of the Layer Properties Manager palette. The result of this inverted list is shown on the right in the following image.

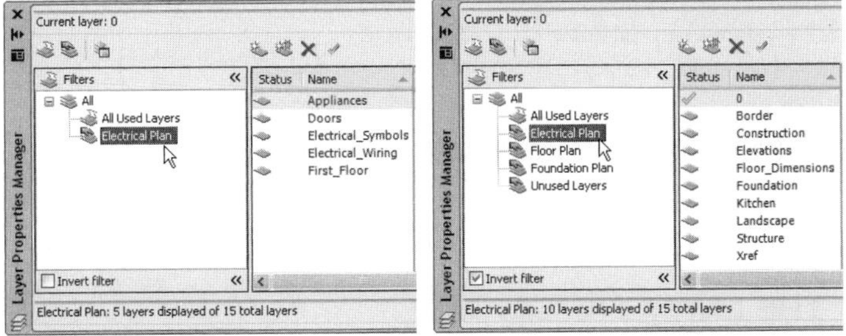

Figure 2.46

LAYER STATES MANAGER

The Layer States Manager allows you to group a number of layer settings under a unique name and then retrieve this name later to affect the display of a drawing. This layer tool can be selected from the Layer Properties Manager palette, as shown on the left in the following image. This launches the Layer States Manager dialog box. First, you arrange your drawing in numerous states by turning off certain layers. As you do this, you assign a unique name in the Layer States Manager dialog box.

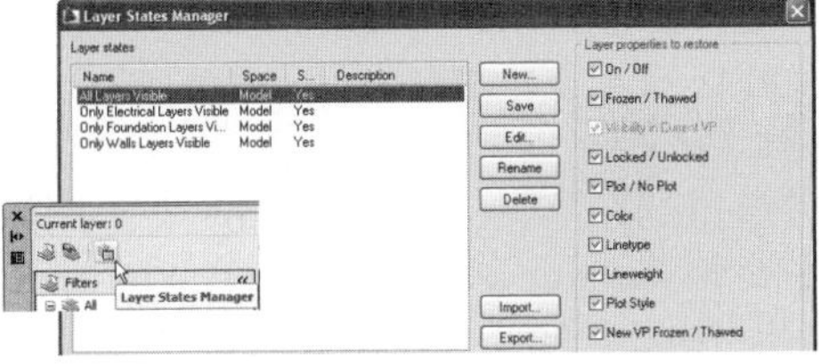

Figure 2.47

Once you have created numerous states, you test them out by double-clicking on a name such as "Display Walls Only" as shown on the left in the following image. Your drawing should update to only display the walls as shown in the right in the following image.

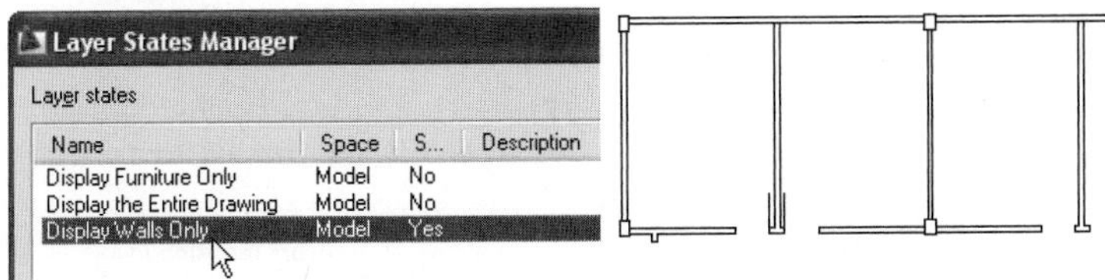

Figure 2.48

It is considered good practice to create a layer state that displays all layers. That way, when you call up various states, you can always get all layers back by calling up this all-layers state. On the left in the following image is the "Display the Entire Drawing" state. Double-clicking on this state will display all layers in the drawing as shown on the right in the following image.

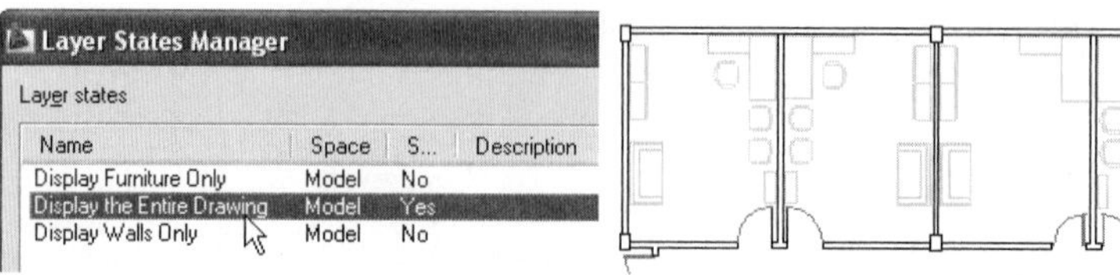

Figure 2.49

COLLAPSING THE LAYER FILTERS TREE

In an effort to provide more room to display your layer information, you can click on the double arrows as shown on the left in the following image. This results in collapsing the layer filter tree. To re-display the filter tree, click on the double arrows again as shown on the right in the following image.

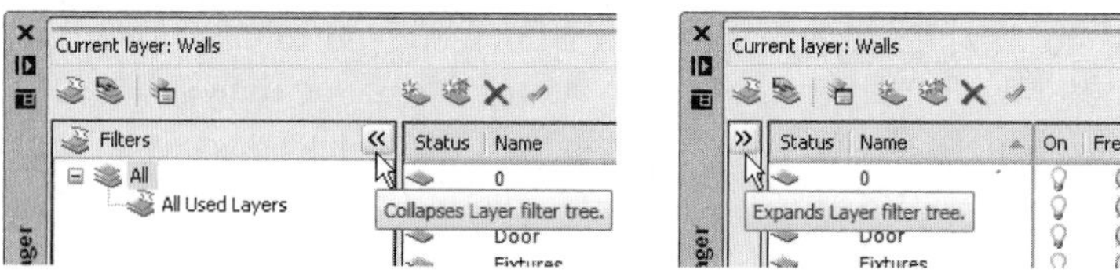

Figure 2.50

ADDITIONAL LAYER TOOLS

Clicking Format in the Menu Browser, followed by Layer Tools, displays more layer commands that can be used to further manage layers. The display of all of these tools is shown on the left in the following image. A special toolbar called Layers II is also available, as shown on the right in the following image.

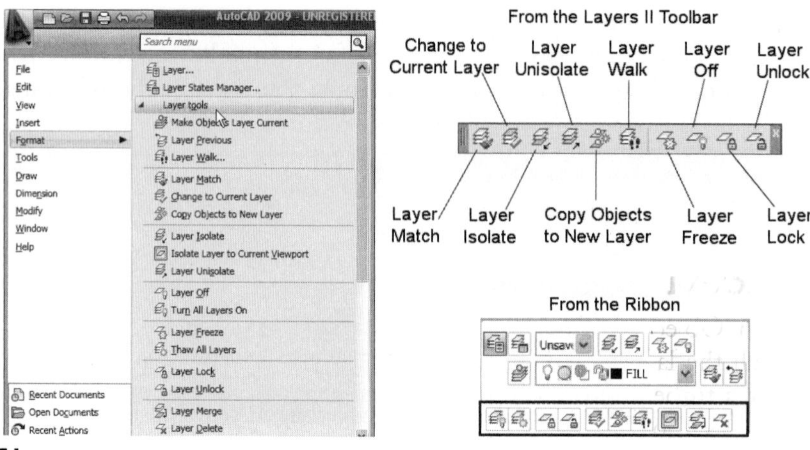

Figure 2.51

Pull-Down Menu Title	Command	Description
Make Objects Layer Current	LAYMCUR	Makes the layer of a selected object current
Layer Previous	LAYERP	Undoes any changes made to the settings of layers such as color, linetype, or lineweight
Layer Walk	LAYWALK	Activates a dialog box that isolates layers in sequential order
Layer Match	LAYMCH	Changes the layer of selected objects to that of a selected object
Change to Current Layer	LAYCUR	Changes the layer of selected objects to the current layer
Copy Objects to New Layer	COPYTOLAYER	Copies objects to other layers
Layer Isolate	LAYISO	Isolates layers of selected objects by turning all other layers off
Isolate Layer to Current Viewport	LAYVPI	Isolates an object's layer in a viewport
Layer Unisolate	LAYUNISO	Turns on layers that were turned off with the last LAYISO command
Layer Off	LAYOFF	Turns off the layers of selected objects
Turn All Layers On	LAYON	Turns all layers on
Layer Freeze	LAYFRZ	Freezes the layers of selected objects
Thaw All Layers	LAYTHW	Thaws all layers
Layer Lock	LAYLCK	Locks the layers of selected objects
Layer Unlock	LAYULK	Unlocks the layers of selected objects
Layer Merge	LAYMRG	Merges two layers, and then removes the first layer from the drawing
Layer Delete	LAYDEL	Permanently deletes layers from drawings

Note: The file 02_Facilities Plan.dwg is available in the Try It! folder for you to use on the following additional layer tools.

MATCH OBJECT'S LAYER (LAYMCH)

The Match Object's Layer command allows you to change the layers of selected objects to match the layer of a selected destination object. The command is LAYMCH; first select the objects to be changed, such as the three chairs shown in the following image. Next select the object on the destination layer, as in the shelf at "D." The three chairs now belong to the same layer as the shelf.

```
Command: LAYMCH
Select objects to be changed:
```

```
Select objects: (Select the three chairs labeled "A," "B,"
    and "C")
Select objects: (Press ENTER to continue)
Select object on destination layer or [Name]: (Pick the shelf
    at "D")
3 objects changed to layer "Furniture."
```

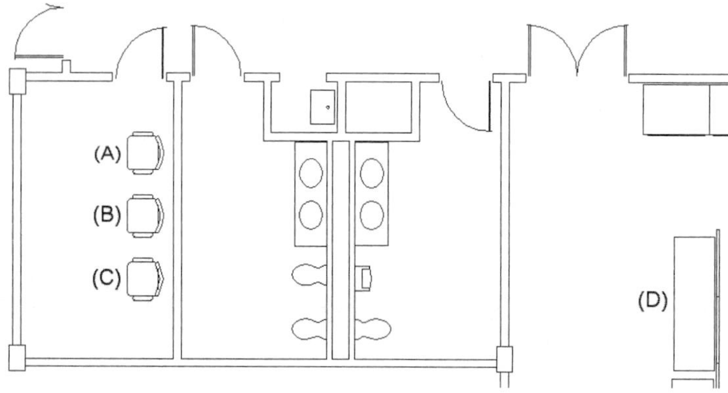

Figure 2.52

CHANGE TO CURRENT LAYER (LAYCUR)

The Change to Current Layer command is used to change the layer of one or more objects to the current layer. The command is LAYCUR. This command is particularly helpful if objects were constructed on the wrong layer. In the following image, the three chairs labeled "A," "B," and "C" were drawn on the wrong layer. First make the desired layer current; in the following image, the current layer is Furniture. Selecting the three chairs after issuing the LAYCUR command changes the three chairs to the Furniture layer.

```
Command: LAYCUR
Select objects to be changed to the current layer: (Select the
    three chairs "A," "B," and "C")
Select objects: (Press ENTER to change the three objects to the
    current layer)
3 objects changed to layer "Furniture" (the current layer)
```

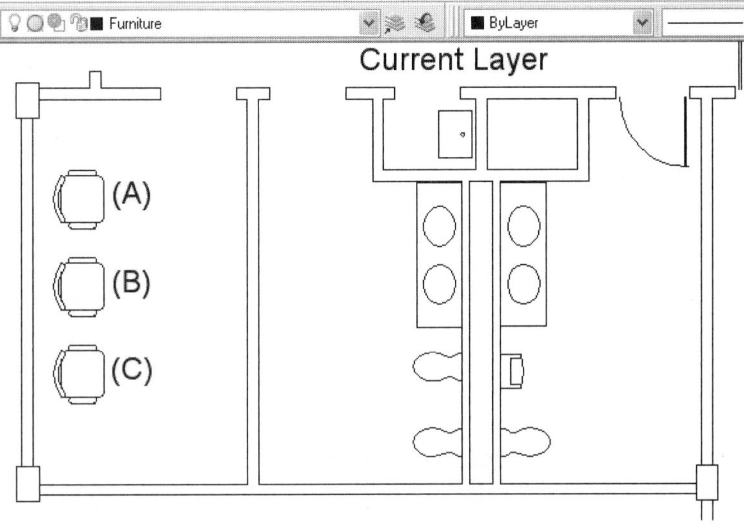

Figure 2.53

ISOLATE OBJECT'S LAYER (LAYISO)

The Isolate Object's Layer command isolates the layer or layers of one or more selected objects by turning all other layers off or by locking/fading them. A Settings option in the command allows you to toggle between the two choices. The effects of this command are the same as using the Layer Properties Manager palette and picking the layer names to turn off or lock. The LAYISO command is yet another tool that manipulates layers in a more efficient manner than through conventional methods. After issuing the command, you are prompted to select the object or objects on the layer or layers to be isolated. On the top in the following image, pick any outside wall and press the ENTER key to execute the command. A message in the command prompt area alerts you that layer Walls is isolated. The option for turning off the layers was selected for the example shown below. All objects on the Walls layer are visible, as shown on the bottom in the following image; all other layers are turned off.

```
Command: LAYISO
Current setting: Hide Layers, Viewports=Off
Select objects on the layer(s) to be isolated or [Settings]:
     (Pick one of the walls)
Select objects: (Press ENTER to isolate the layer based on the
     object selected)
Layer "Walls" has been isolated.
```

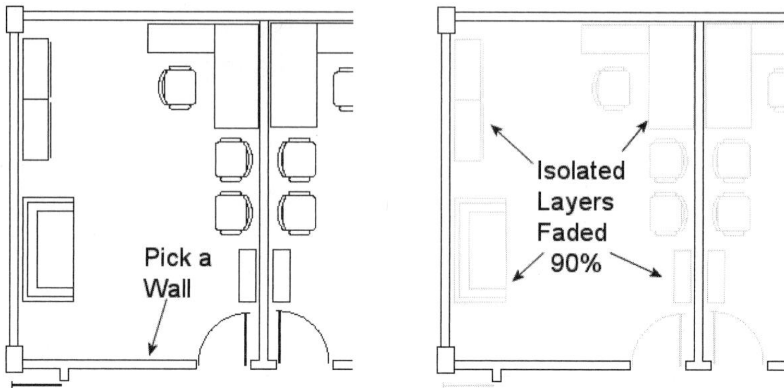

Figure 2.54

UNISOLATE OBJECT'S LAYER (LAYUNISO)

Use this command to restore layers that were turned off or locked with the last usage of the Layer Isolate command (LAYISO).

COPY OBJECTS TO NEW LAYER (COPYTOLAYER)

The Copy Objects to New Layer command copies selected objects to a different layer while leaving the original objects on the original layer. In the following image, three chairs on layer Room 1 need to be copied and have their layer changed to Room 2.

```
Command: COPYTOLAYER
Select objects to copy: (Select the three chairs illustrated in
    the following image)
Select objects to copy: (Press ENTER to continue)
Select object on destination layer or [Name] <Name>: (Press ENTER.
    When the Copy to Layer dialog box appears, select Room 2)
3 objects copied and placed on layer "Room 2"
Specify base point or [Displacement/eXit] <exit>: (Pick the
    endpoint at "A")
Specify second point of displacement or <use first point as
    displacement>: (Pick a point at "B")
```

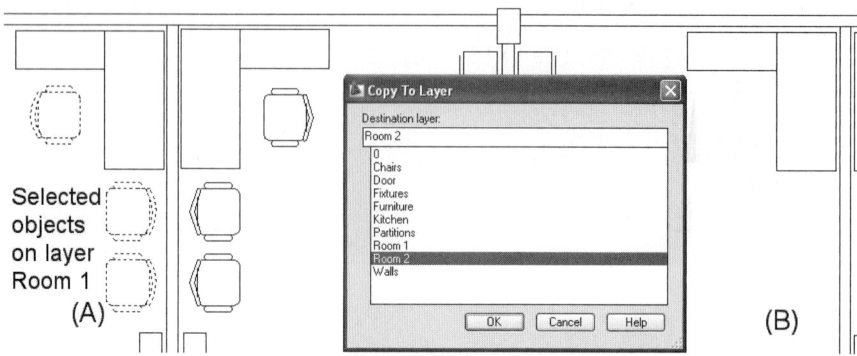

Figure 2.55

The results are illustrated in the following image. The three original objects remain on layer Room 1, while the new set of objects has been copied to a new location and to a new layer (Room 2).

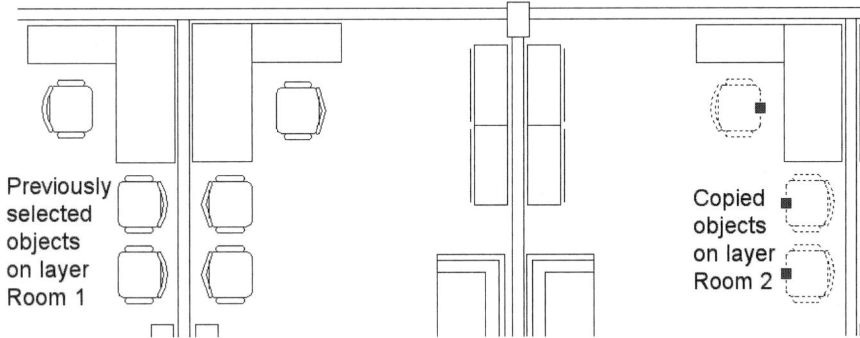

Figure 2.56

LAYER WALK (LAYWALK)

The Layer Walk command displays a dialog box containing all layers in the present drawing. Clicking a layer in this dialog box turns off all other layers. If layers are detected as unreferenced (unused), a purge button activates, allowing you to delete the layer.

In the illustration of the office plan shown on the left in the following image, the LAYWALK command has activated the Layer Walk dialog box. Notice in this dialog box that all layers are highlighted, signifying that all layers in the drawing are visible.

Clicking the layer Walls, as shown on the right in the following image, turns off all layers except for the Walls layer. You can press the CTRL or SHIFT key while selecting layers in the Layer Walk dialog box to select multiple layers.

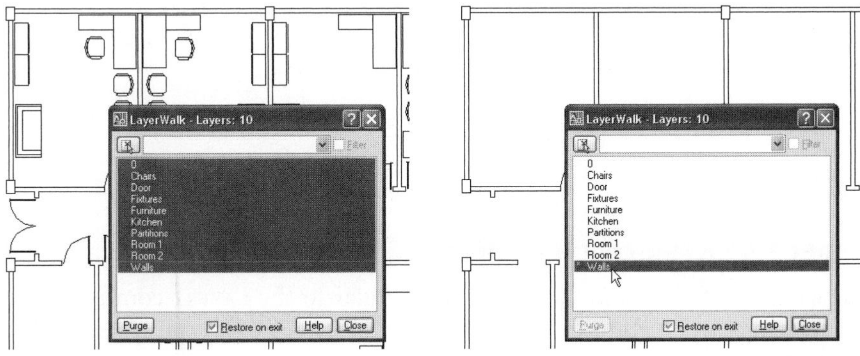

Figure 2.57

FREEZE OBJECT'S LAYER (LAYFRZ)

The Freeze Object's Layer command freezes layers by picking objects to control the layers to be frozen. Again, the Layer Properties Manager palette and the Layer Control box are normally used to perform all layer freeze operations.

```
Command: LAYFRZ
Current settings: Viewports=Vpfreeze, Block nesting level=Block
Select an object on the layer to be frozen or [Settings/Undo]:
    (Pick a chair)
Layer "Chairs" has been frozen.
Select an object on the layer to be frozen or [Settings/Undo]:
    (Pick a shelf)
Layer "Furniture" has been frozen.
Select an object on the layer to be frozen or [Settings/Undo]:
    (Pick a sink)
Layer "Fixtures" has been frozen.
Select an object on the layer to be frozen or [Settings/Undo]:
    (Pick a refrigerator)
Layer "Kitchen" has been frozen.
Select an object on the layer to be frozen or [Settings/Undo]:
    (Pick a door)
Layer "Doors" has been frozen.
Select an object on the layer to be frozen or [Settings/Undo]:
    (Press ENTER to complete this command. The results should be
    similar to the following image.)
```

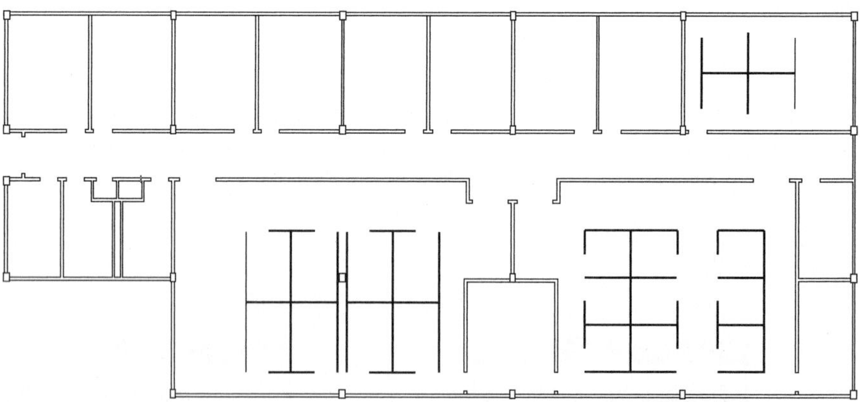

Figure 2.58

TURN OBJECT'S LAYER OFF (LAYOFF)

The Turn Object's Layer Off command is similar to the LAYFRZ command except that, instead of freezing layers, you can turn off a layer or group of layers by selecting an object or group of objects.

```
Command: LAYOFF
Current settings: Viewports=Vpfreeze, Block nesting level=Block
Select an object on the layer to be turned off or [Settings/
    Undo]: (Pick a shelf)
```

```
Layer "Furniture" has been turned off.
Select an object on the layer to be turned off or [Settings/
    Undo]: (Pick a partition)
Layer "Partitions" has been turned off.
Select an object on the layer to be turned off or [Settings/
    Undo]: (Pick a chair)
Layer "Chairs" has been turned off.
Select an object on the layer to be turned off or [Settings/
    Undo]: (Press ENTER to exit this command. The results should
    be similar to the following image.)
```

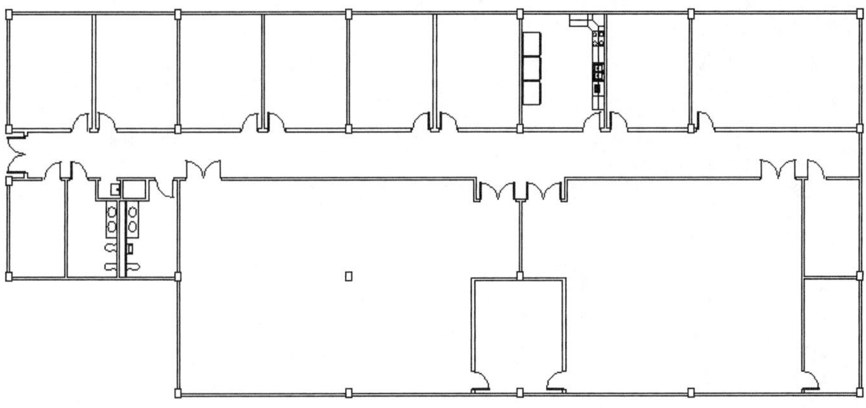

Figure 2.59

LOCK OBJECT'S LAYER (LAYLCK)

The Lock Object's Layer command locks the layer of a selected object. A locked layer is visible on the display screen; however, any object associated with a locked layer is non-selectable. The command to lock a layer is LAYLCK.

```
Command: LAYLCK
Select an object on the layer to be locked: (Pick an exterior
    wall in the previous image)
Layer "Walls" has been locked.
```

UNLOCK OBJECT'S LAYER (LAYULK)

The Unlock Object's Layer command unlocks the layer of a selected object. Objects on an unlocked layer can now be selected whenever the prompt "Select objects:" appears. The command to unlock a layer is LAYULK.

```
Command: LAYULK
Select an object on the layer to be unlocked: (Pick an exterior
    wall in the previous image)
Layer "Walls" has been unlocked.
```

MERGING LAYERS INTO ONE (LAYMRG)

The Merging Layers into One command (LAYMRG) merges the contents of one layer with a target layer. The layer containing all chairs (Chairs layer) illustrated in the following image will be merged with the Furniture layer. One of the chairs is first selected as the layer to merge. Then one of the cabinets at "A," as shown in the following image, is selected as the target layer. Before the layers are merged, a warning message appears in the command prompt alerting you that the Chairs layer will be permanently merged into the Furniture layer. After you perform this operation, the Chairs layer automatically is purged from the drawing.

```
Command: LAYMRG
Select object on layer to merge or [Name]: (Pick a chair, as
    shown in the following image)
Selected layers: "Chairs"
Select object on layer to merge or [Name/Undo]: (Press ENTER to
    continue)
Select object on target layer or [Name]: (Pick the cabinet at
    "A" in the following image)
******** WARNING ********
There are 1 block definition(s) which reference the layer(s) you
    are merging.
The block(s) will be redefined and the entities referencing the
    layer(s)
will be changed to reference layer "Furniture."
You are about to merge layer "Chairs" into layer "Furniture."
Do you wish to continue? [Yes/No] <No>: Y
Redefining block "CHAIR7."
Deleting layer "Chairs."
1 layer deleted.
```

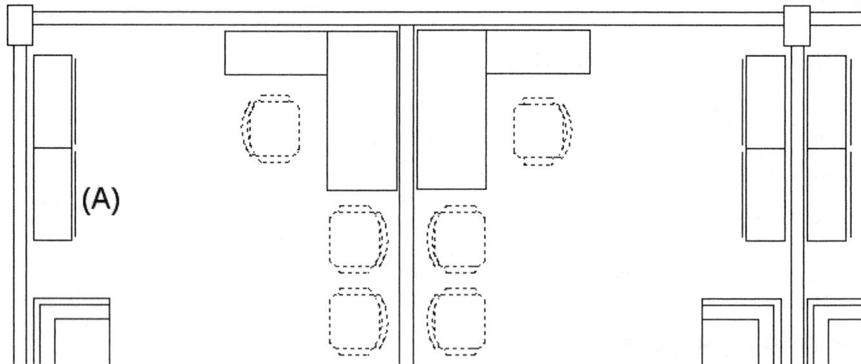

Figure 2.60

If you are not sure about the layer to merge, you can use the Name option, which launches the Merge Layers dialog box, as shown in the following image. You can then choose the layer by name. The same warning appears in the command prompt area that alerts you that the layer name and all of the objects on the layer (Chairs) will be merged with the target.

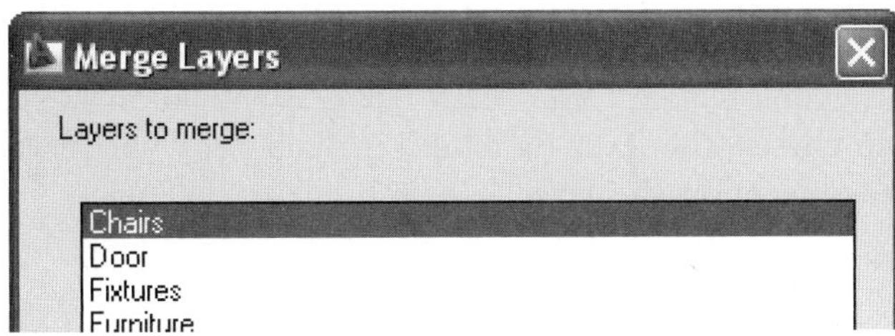

Figure 2.61

DELETING ALL OBJECTS ON A LAYER (LAYDEL)

Use the Layer Delete command (LAYDEL) to delete all objects on a specified layer. First select one of the chairs, as shown on the left in the following image. The chair is located on the Chairs layer. Before all objects on this layer are deleted, a message appears in the command prompt, asking whether you really want to perform this operation. Answering "Yes" deletes all objects assigned to the layer in addition to purging the layer from the drawing. Although this tool has many beneficial advantages, you must exercise care in deciding when to use it and when not to use it.

```
Command: LAYDEL
Select object on layer to delete or [Name]: (Pick the chair at
    "A" in the following image)
Selected layers: Chairs
Select object on layer to delete or [Name/Undo]: (Press ENTER to
    continue)

******** WARNING ********

There are 1 block definition(s) which reference the layer(s) you
    are deleting.
The block(s) will be redefined and the entities referencing the
    layer(s) will be removed from the block definition(s).
You are about to delete layer "Chairs" from this drawing.
Do you wish to continue? [Yes/No] <No>: Y
Redefining block "CHAIR7".
Deleting layer "Chairs".
1 layer deleted.
```

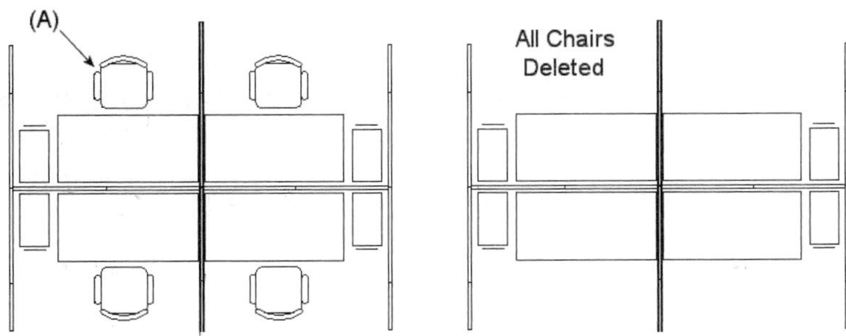

Figure 2.62

As with merging layers, if you are not sure about the layer to delete, you can use the Name option, which launches the Delete Layers dialog box, as shown on the left in the following image. You can then choose the layer by name. A dialog box will alert you that the layer name and all of the objects on the layer (Chairs) will be deleted.

```
Command: LAYDEL
Select object on layer to delete or [Name]: N (For Name)
```

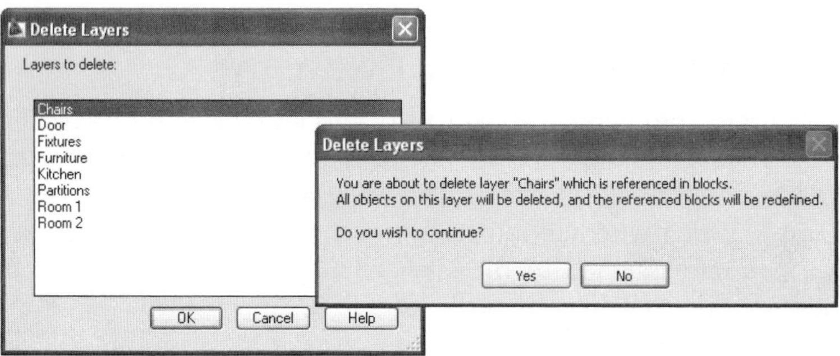

Figure 2.63

CREATING TEMPLATE FILES

While layers are one productive way to work on drawings, another way is through template files. Template files have settings such as units, limits, and layers already created, which enhances the productivity of the user. By default, when you start a new drawing from scratch, AutoCAD uses a template. Two templates are available, depending on whether you are drawing in inches or millimeters; they are Acad.Dwt (inches) and Acadiso.Dwt (millimeters). Various settings are made in a template file. For instance, in the Acad.Dwt template file, the drawing units are set to decimal units with four-decimal-place precision. The drawing limits are set to 12,9 units for the upper-right corner. Only one layer exists, namely layer 0. The grid in this template is set to a spacing of 0.50 units.

If you have to create several drawings that have the same limits and units settings and use the same layers, rather than start with the Acad.Dwt file and make changes to the settings at the start of each new drawing file, a better technique would be to create your own template. Templates can be made up of entire drawings or just have a company border already placed in the drawing. Other templates may appear empty of objects such as lines or circles. However, the limits of the drawing may already be set along with grid and snap settings. The following settings are usually assigned to a drawing template:

- Drawing units and precision
- Drawing limits
- Grid and Snap Settings
- Layers, Linetypes, and Lineweights

- Linetype scale factor depending on the final plot scale of the drawing

Try It!: Create a new drawing file starting from scratch. You will be creating a template designed to draw an architectural floor plan. Make changes to the following settings:

- In the Drawing Units dialog box, change the units type from Decimal to Architectural.
- Set the limits of the drawing to 17',11' for the upper-right corner and perform a ZOOM All. Change the grid and snap to 3" for the X and Y values.
- Create the following layer names:
  - Floor Plan, Hidden, Center, Dimension, Doors, Windows
- Make your own color assignments to these layers.
- Load the Hidden and Center linetypes and assign these to the appropriate layers.
- Assign a lineweight of 0.70mm to the Floor Plan layer.
- Finally, use the LTSCALE command and set the linetype scale to 12.

Now it is time to save the settings in a template format. Follow the usual steps for saving your drawing. However, when the Save Drawing As dialog box appears, first click the arrow on the other side of "Files of type": in the following image to expose the file types.

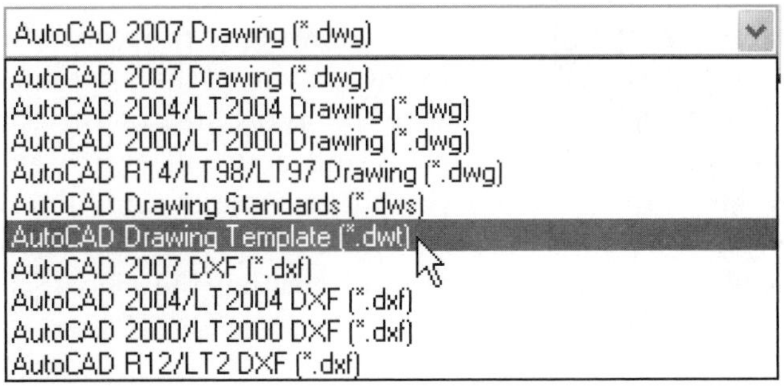

AutoCAD 2007 Drawing (*.dwg)

AutoCAD 2007 Drawing (*.dwg)
AutoCAD 2004/LT2004 Drawing (*.dwg)
AutoCAD 2000/LT2000 Drawing (*.dwg)
AutoCAD R14/LT98/LT97 Drawing (*.dwg)
AutoCAD Drawing Standards (*.dws)
AutoCAD Drawing Template (*.dwt)
AutoCAD 2007 DXF (*.dxf)
AutoCAD 2004/LT2004 DXF (*.dxf)
AutoCAD 2000/LT2000 DXF (*.dxf)
AutoCAD R12/LT2 DXF (*.dxf)

Figure 2.64

Click the field that states, "AutoCAD Drawing Template (*.dwt)." AutoCAD will automatically take you to the folder, shown in the following image, that holds all template information. It is at this time that you enter the name of the template as AEC_B_1=12. The name signifies an architectural drawing for the B-size drawing sheet at a scale of 1" = 1'-0".

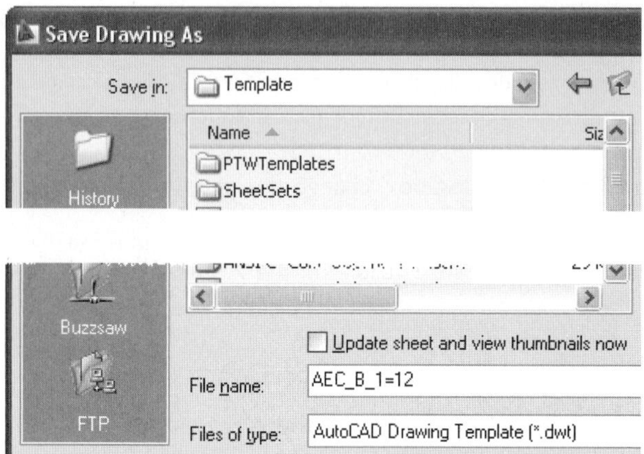

Figure 2.65

Before the template file is saved, you have the opportunity to document the purpose of the template. It is always good practice to create this documentation, especially if others will be using your template.

RECORDING A MACRO

Another way of capturing repetitive tasks in a drawing is to record a series of commonly used tasks and then play these tasks back in any drawing. This is accomplished through the Action Recorder. This tool is located under the Tools area of the Ribbon. The Action Recorder will allow you to replay any command and will even wait for user input if necessary. Since we are on the topic of layers, an example of using the Action Recorder to record the creation of a number of layers and then playing the macro back in a new drawing will be shown. To begin the process of creating the macro, click on the Record button as shown in the following image.

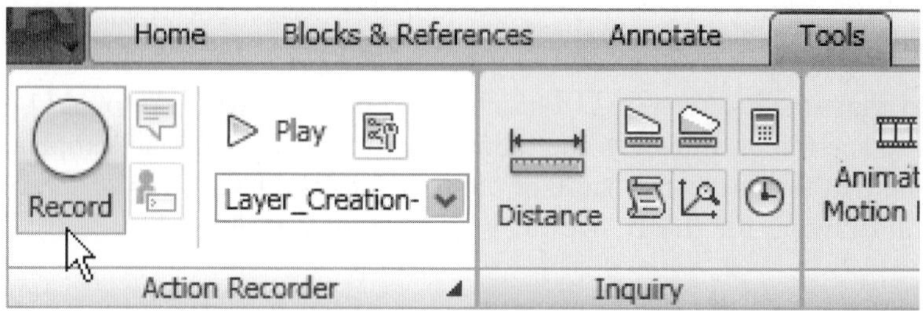

Figure 2.66

As you move your cursor around, you will notice the appearance of a red dot that signifies that anything you do from now on will be recorded. In this example, the Layer Properties Manager palette is active and layers are created as shown on the right in the following image. When you are satisfied with the information you created, click on the Stop button to end the macro.

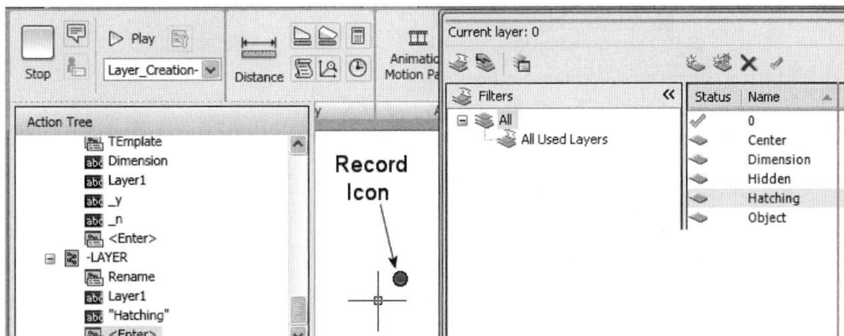

Figure 2.67

When stopping the macro, the Action Macro dialog box will appear and you will be prompted to create a name for the macro as shown in the following image. This represents a file that you can later use in other drawings.

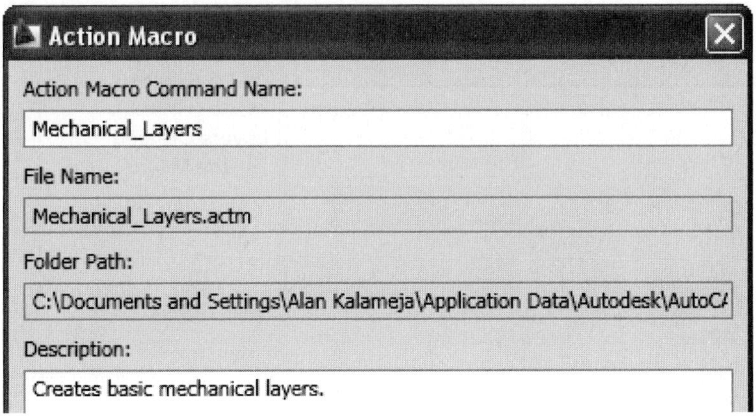

Action Macro

Action Macro Command Name:

Mechanical_Layers

File Name:

Mechanical_Layers.actm

Folder Path:

C:\Documents and Settings\Alan Kalameja\Application Data\Autodesk\AutoC/

Description:

Creates basic mechanical layers.

Figure 2.68

Create a new drawing and in the drop-down list, select the macro that was just saved. Then click on the play button as shown on the right in the following image.

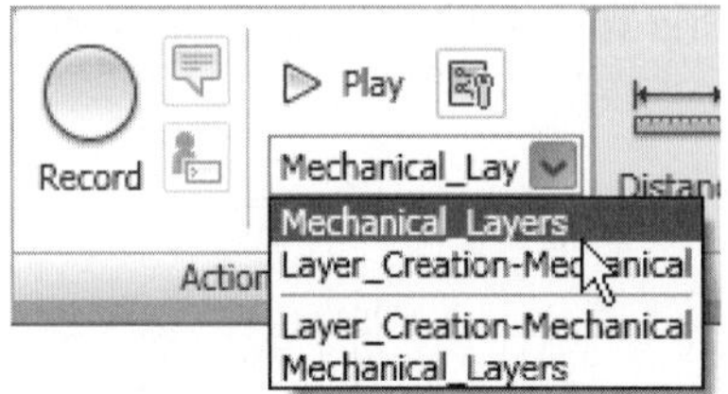

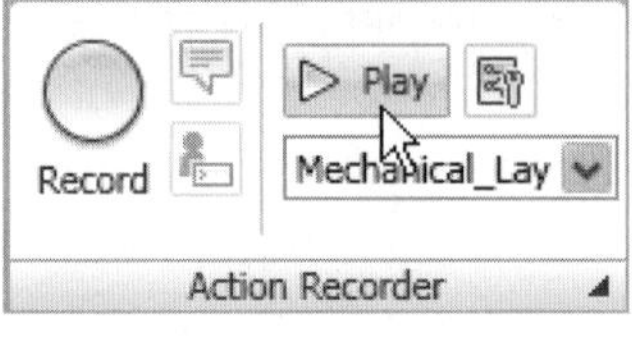

Figure 2.69

A message will appear alerting you that the action macro is complete. Activating the Layer Properties Manager palette will show the layers automatically created in the new drawing as a result of using the action macro.

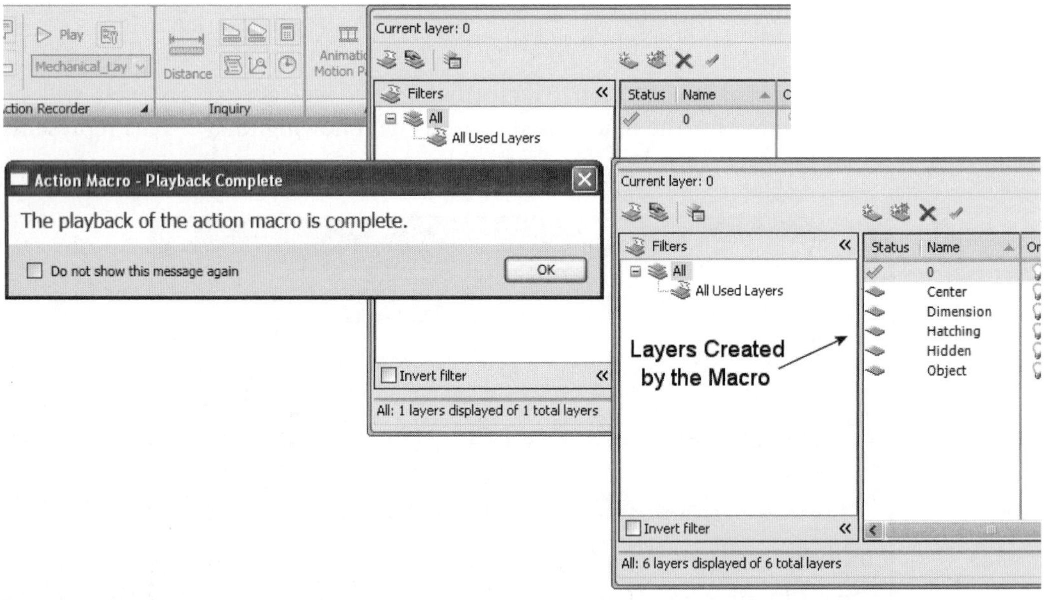

Figure 2.70

TUTORIAL EXERCISE: CREATING LAYERS USING THE LAYER PROPERTIES MANAGER PALETTE

Layer Name	Color	Linetype	Lineweight
Object	White	Continuous	0.70
Hidden	Red	Hidden	0.30
Center	Yellow	Center	Default
Dimension	Yellow	Continuous	Default

Layer Name	Color	Linetype	Lineweight
Section	Blue	Continuous	Default

PURPOSE

Use the following steps to create layers according to the above specifications.

Step 1

Activate the Layer Properties Manager palette by clicking the Layer button located on the Layers toolbar, as shown on the left in the following image or in the Ribbon as shown on the right in the following image.

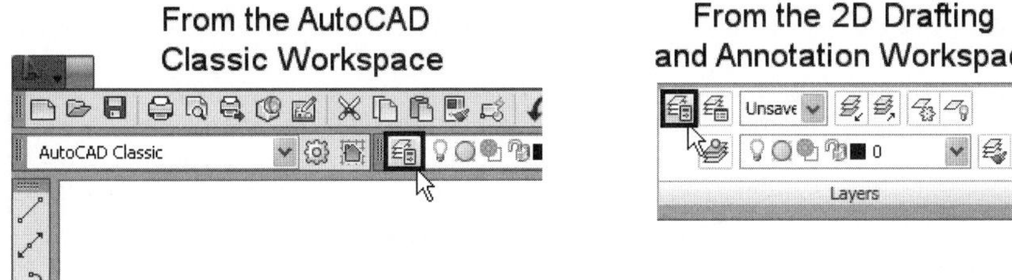

Figure 2.71

Step 2

Once the Layer Properties Manager palette displays, as shown in the following image, notice on your screen that only one layer is currently listed, namely Layer 0. This happens to be the default layer, which is the layer that is automatically available to all new drawings.

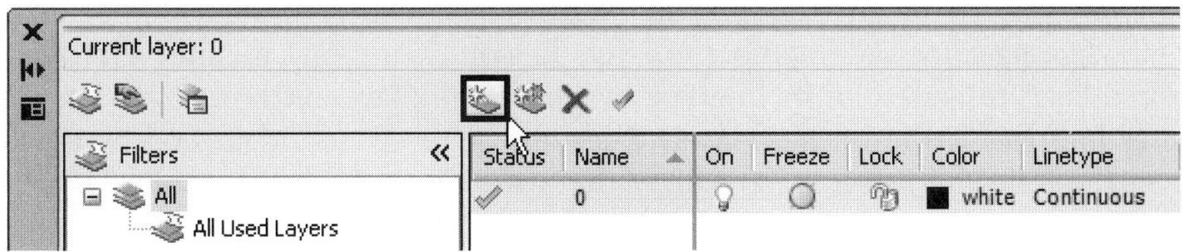

Figure 2.72

Since it is poor practice to construct any objects on Layer 0, new layers will be created not only for object lines but for hidden lines, centerlines, dimension objects, and section lines as well. To create these layers, click the New button and notice that a layer is automatically added to the list of layers. This layer is called Layer1 (see the following image). While this layer is highlighted, enter the first layer name, "Object."

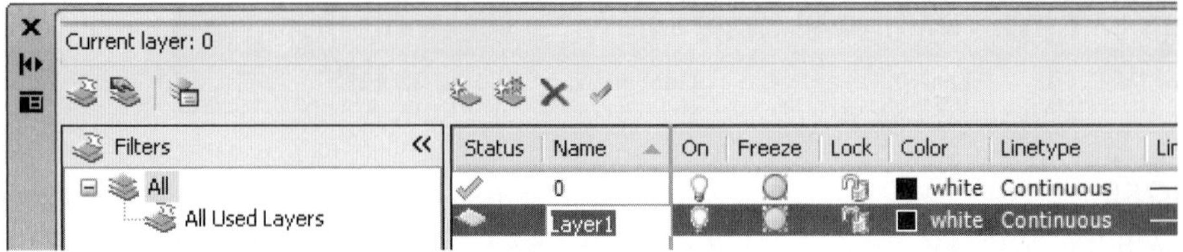

Figure 2.73

Entering a comma after the name of the layer allows more layers to be added to the listing of layers without having to click the New button. Once you have entered the comma after the layer "Object" and the new layer appears, enter the new name of the layer as "Hidden." Repeat this procedure of using the comma to create multiple layers for "Center," "Dimension," and "Section." Press ENTER after typing in "Section." The complete list of all layers created should be similar to the illustration shown in the following image.

Status	Name	On	Freeze	Lock	Color	Linetype	Lineweight	Plot Style	Plot
	0				white	Continuous	—— Default	Color_7	
	Center				white	Continuous	—— Default	Color_7	
	Dimension				white	Continuous	—— Default	Color_7	
	Hidden				white	Continuous	—— Default	Color_7	
	Object				white	Continuous	—— Default	Color_7	
	Section				white	Continuous	—— Default	Color_7	

Figure 2.74

Step 3

As all new layers are displayed, the names may be different, but they all have the same color and linetype assignments (see the following image). At this point, the dialog box comes in handy in assigning color and linetypes to layers in a quick and easy manner. First, highlight the desired layer to add color or linetype by picking the layer. A horizontal bar displays, signifying that this is the selected layer. Click the color swatch identified by the box, as shown in the following image, and follow the next step to assign the color "Red" to the "Hidden" layer name.

Status	Name	On	Freeze	Lock	Color	Linetype	Lineweight	Plot Style	Plot
	0				white	Continuous	—— Default	Color_7	
	Center				white	Continuous	—— Default	Color_7	
	Dimension				white	Continuous	—— Default	Color_7	
	Hidden				white	Continuous	—— Default	Color_7	
	Object				white	Continuous	—— Default	Color_7	
	Section				white	Continuous	—— Default	Color_7	

Figure 2.75

Step 4

Clicking the color swatch in the previous step displays the Select Color dialog box, as shown in the following image. Select the desired color from one of the following areas: Standard Colors, Gray Shades, or Full Color Palette. The standard colors represent colors 1 (Red) through 9 (Gray). On display terminals with a high-resolution graphics card, the full color palette displays different shades of the standard colors. This gives you a greater variety of colors to choose from. For the purposes of this tutorial, the color "Red" will be assigned to the "Hidden" layer. Select the box displaying the color red; a box outlines the color and echoes the color in the bottom portion of the dialog box. Click the OK button to complete the color assignment; if you select Cancel, the color assignment will be removed, requiring this operation to be used again. Continue with this step by assigning the color Yellow to the Center and Dimension layers; assign the color Blue to the Section layer.

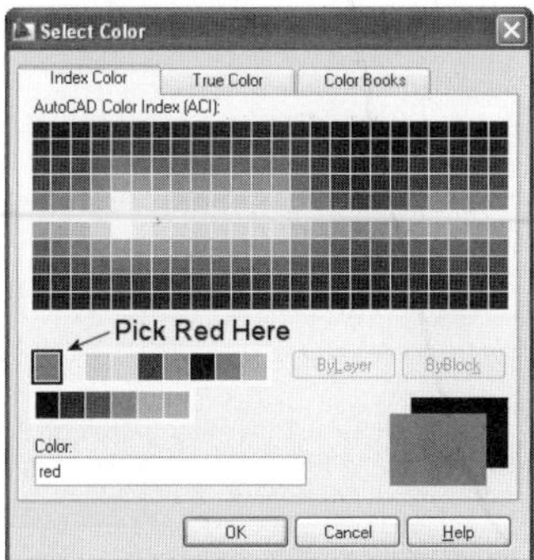

Figure 2.76

Step 5

Once the color has been assigned to a layer, the next step is to assign a linetype, if any, to the layer. The "Hidden" layer requires a linetype called "HIDDEN." Click the "Continuous" linetype to display the Select Linetype dialog box illustrated in the following image. By default, and to save space in each drawing, AutoCAD initially only loads the Continuous linetype. Click the Load button to load more linetypes.

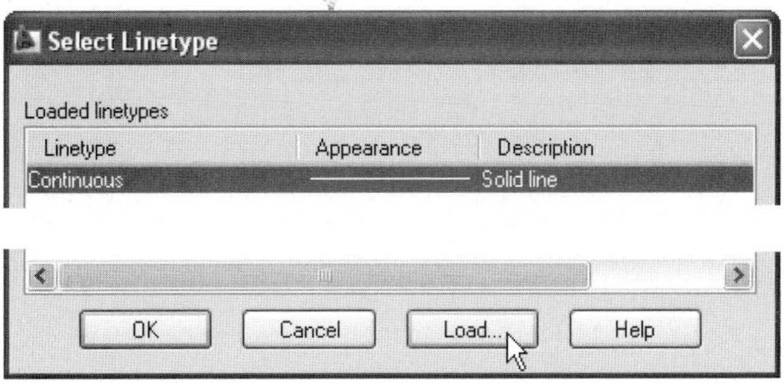

Figure 2.77

Choose the desired linetype to load from the Load or Reload the Linetype dialog box, as shown in the following image. Scroll through the linetypes until the "HIDDEN" linetype is found. Click the OK button to return to the Select Linetype dialog box.

Once you are back in the Select Linetype dialog box, as shown in the following image, notice the "HIDDEN" linetype listed along with "Continuous." Because this linetype has just been loaded, it still has not been assigned to the "Hidden" layer. Click the Hidden linetype listed in the Select Linetype dialog box, and click the OK button. Once the Layer Properties Manager palette reappears, notice that the "HIDDEN" linetype has been assigned to the "Hidden" layer. Repeat this procedure to assign the "CENTER" linetype to the "Center" layer.

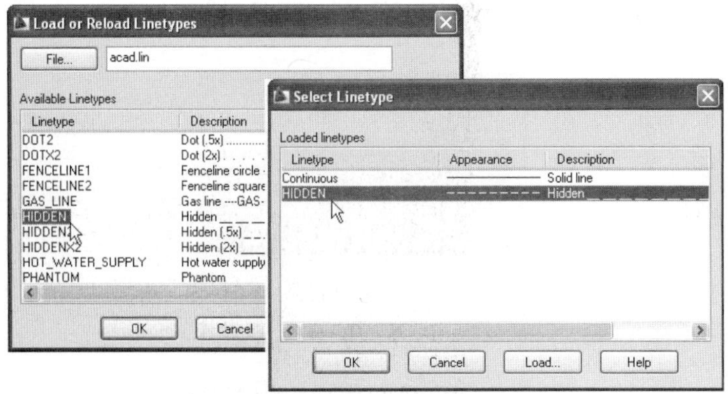

Figure 2.78

Step 6

Another requirement of the "Hidden" layer is that it uses a lineweight of 0.30 mm to have the hidden lines stand out when compared with the object in other layers. Clicking the highlighted default lineweight displays the Lineweight dialog box shown in the image on the right. Click the 0.30 mm lineweight followed by the OK button to assign this lineweight to the "Hidden" line. Use the same procedure to assign a lineweight of 0.70 mm to the Object layer.

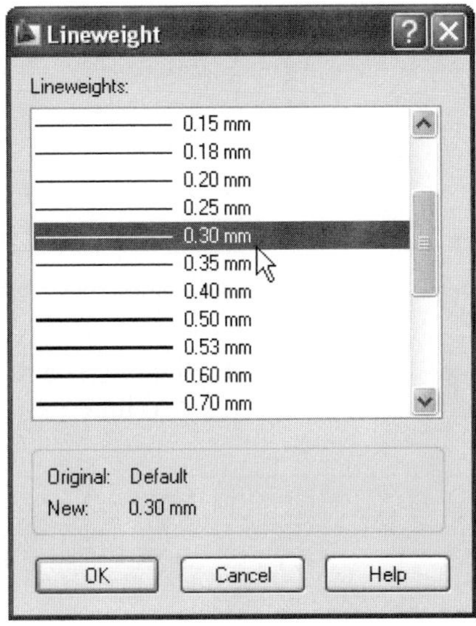

Figure 2.79

Step 7

Once you have completed all color and linetype assignments, the Layer Properties Manager palette should appear similar to the following image. Click the OK button to save all layer assignments and return to the drawing editor.

Status	Name	On	Freeze	Lock	Color	Linetype	Lineweight	Plot Style	Plot
	0				white	Continuous	Default	Color_7	
	Center				yell...	CENTER	Default	Color_2	
	Dimension				yell...	Continuous	Default	Color_2	
	Hidden				red	HIDDEN	0.30 ...	Color_1	
	Object				white	Continuous	0.70 ...	Color_7	
	Section				blue	Continuous	Default	Color_5	

Figure 2.80

Step 8

Clicking in the Name box at the top of the list of layers reorders the layers. Now, the layers are listed in reverse alphabetical order (see the following image). This same effect occurs when you click in the On, Freeze, Color, Linetype, Lineweight, and Plot header boxes.

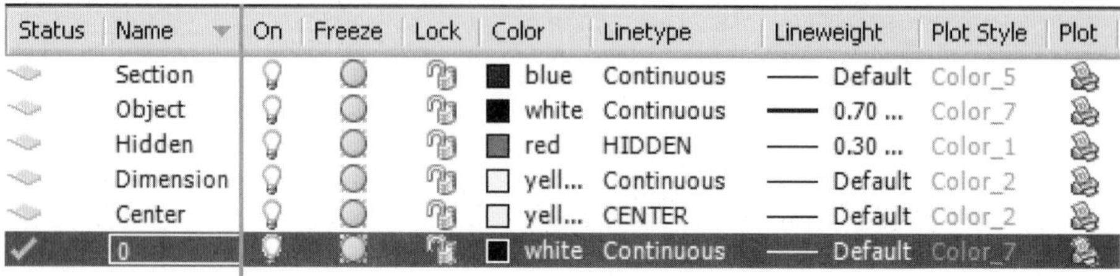

Status	Name	On	Freeze	Lock	Color	Linetype	Lineweight	Plot Style	Plot
	Section				blue	Continuous	Default	Color_5	
	Object				white	Continuous	0.70 ...	Color_7	
	Hidden				red	HIDDEN	0.30 ...	Color_1	
	Dimension				yell...	Continuous	Default	Color_2	
	Center				yell...	CENTER	Default	Color_2	
✓	0				white	Continuous	Default	Color_7	

Figure 2.81

The following image displays the results of clicking the Color header: all colors are reordered, starting with Red (1) and followed by Yellow (2), Blue (5), and White (7).

Status	Name	On	Freeze	Lock	Color	Linetype	Lineweight	Plot Style	Plot
	Hidden				red	HIDDEN	0.30 ...	Color_1	
	Center				yell...	CENTER	Default	Color_2	
	Dimension				yell...	Continuous	Default	Color_2	
	Section				blue	Continuous	Default	Color_5	
✓	0				white	Continuous	Default	Color_7	
	Object				white	Continuous	0.70 ...	Color_7	

Figure 2.82

TEMPLATE CREATION EXERCISES FOR CHAPTER 2

ARCHITECTURAL APPLICATION

Create an Architectural template (dwt) that contains the following setup information.

Use the Drawing Units dialog box to set the units of the drawing to Architectural. Keep the remaining default values for all other unit settings.

Use the Layer Properties Manager palette to create the layers contained in the table below:

Layer Name	Color	Linetype	Lineweight
Border	Blue (5)	Continuous	1.00
Center	Green (3)	Center	0.25
Dimension	Green (3)	Continuous	0.25
Doors	Red (2)	Continuous	0.50
Electrical	Green (3)	Continuous	0.50
Elevations	White (7)	Continuous	0.70
Foundation	Cyan (4)	Continuous	0.50
Furniture	Red (1)	Continuous	0.50
Hatching	Magenta (6)	Continuous	0.50

Layer Name	Color	Linetype	Lineweight
Hidden	Red (1)	Hidden	0.50
HVAC	Cyan (4)	Continuous	0.50
Misc	Green (3)	Continuous	0.50
Plumbing	Cyan (4)	Continuous	0.50
Text	Green (3)	Continuous	0.50
Viewports	Gray (9)	Continuous	0.25
Walls	White (7)	Continuous	0.70
Windows	Yellow (2)	Continuous	0.50

When finished making the changes to the drawing units and layer assignments, save this file as a new AutoCAD template called ARCH-Imperial.DWT (imperial units are in feet and inches).

MECHANICAL APPLICATION

Create a Mechanical template (dwt) that contains the following setup information.

Use the Drawing Units dialog box to set the units of the drawing to Mechanical. Also change the precision to two decimal places. Keep the remaining default values for all other unit settings.

Use the Layer Properties Manager palette to create the layers contained in the table below:

Layer Name	Color	Linetype	Lineweight
Border	Blue (5)	Continuous	1.00
Center	Green (3)	Center	0.25
Construction	Gray (8)	Continuous	0.25
Dimension	Green (3)	Continuous	0.25
Hatching	Magenta (6)	Continuous	0.50
Hidden	Red (1)	Hidden	0.50
Misc	Green (3)	Continuous	0.50
Object	White (7)	Continuous	0.70
Phantom	Cyan (4)	Phantom	0.50
Text	Green (3)	Continuous	0.50
Viewports	Gray (9)	Continuous	0.25

When you are finished making the changes to the drawing units and layer assignments, save this file as a new AutoCAD template called MECH-Imperial.DWT (imperial units are in inches).

CIVIL APPLICATION

Create a Civil template (dwt) that contains the following setup information.

Use the Drawing Units dialog box to set the units of the drawing to Decimal. Also, change the precision to two decimal places. Keep the remaining default values for all other unit settings.

Use the Layer Properties Manager palette to create the layers contained in the table below:

Layer Name	Color	Linetype	Lineweight
Border	Blue (5)	Continuous	1.00
Building Outline	White (7)	Continuous	0.70
Center	Green (3)	Center	0.25
Contour Lines – Existing	White (7)	Dashed	0.50
Contour Lines – New	Magenta (6)	Continuous	0.50
Dimension	Green (3)	Continuous	0.25
Drainage	Cyan (4)	Divide	0.50
Fire Line	Red (1)	Continuous	0.50
Gas Line	Red (1)	Gas_Line	0.50
Hatching	Magenta (6)	Continuous	0.50
Hidden	Red (1)	Hidden	0.50
Misc	Green (3)	Continuous	0.50
Parking	Yellow (2)	Continuous	0.50
Property Line	Green (3)	Phantom	0.80
Sewer	White (7)	Continuous	0.50
Text	Green (3)	Continuous	0.50
Viewports	Gray (9)	Continuous	0.25
Water Line	Cyan (4)	Continuous	0.50
Wetlands	Green (3)	Dashdot	0.50

When finished making the changes to the drawing units and layer assignments, save this file as a new AutoCAD template called CIVIL-Imperial.DWT (imperial units are in feet and inches).

PROBLEMS FOR CHAPTER 2

PROBLEM 2–1

Open the existing drawing called 02-Storage.Use the Layer Properties Manager along with the Layer Filter Properties dialog box to answer the following layer-related questions:

1. The total number of layers frozen in the database of this drawing is ________________ .
2. The total number of layers locked in the database of this drawing is ________________ .
3. The total number of layers turned off in the database of this drawing is ________________ .
4. The total number of layers assigned the color red and frozen in the database of this drawing is ________________ .
5. The total number of layers that begin with the letter "L," are assigned the color yellow, and are turned off in the database of this drawing is ________________ .

PROBLEM 2–2

Open the existing drawing called 02_I_Pattern.Dwg. Use the Layer Properties Manager along with the Layer Filter Properties dialog box to answer the following layer-related questions:

1. The total number of layers assigned the color red and turned off in the database of this drawing is ________________ .
2. The total number of layers assigned the color red and locked in the database of this drawing is ________________ .
3. The total number of layers assigned the color white and frozen in the database of this drawing is ________________ .
4. The total number of layers assigned the color yellow and assigned the hidden linetype in the database of this drawing is ________________ .
5. The total number of layers assigned the color red and assigned the phantom linetype in the database of this drawing is ________________ .
6. The total number of layers assigned the center linetype in the database of this drawing is ________________ .

CHAPTER 3

AutoCAD Display and Basic Selection Operations

This chapter introduces you to the various ways of viewing your drawing. A number of options of the ZOOM command will be explained first, followed by a review of panning and understanding how a wheel mouse is used with AutoCAD. User-defined portions of a drawing screen called views will also be introduced in this chapter. You will be introduced to various Object Selection methods such as window, crossing, crossing polygon, window polygon, fence, and previous, to name just a few.

VIEWING YOUR DRAWING WITH ZOOM

The ability to magnify details in a drawing or reduce the drawing to see it in its entirety is a function of the ZOOM command. It does not take much for a drawing to become very busy, complicated, or dense when displayed in the drawing editor. Therefore, use the ZOOM command to work on details or view different parts of the drawing. Zoom can be selected through the following menus as shown in the following image: Zoom toolbar, Menu Browser, pull-down menu, and Ribbon.

The Zoom toolbar contains the more popular tools of the ZOOM command.

Another way to choose options of the ZOOM command is through the Menu Browser found along the left side of the display screen. These options include zooming in real time, zooming to the previous display, using a window to define a boxed area to zoom to, dynamic zooming, zooming to a user-defined scale factor, zooming based on a center point and a scale factor, or performing routine operations such as zooming in or out, zooming all, or zooming to the extents of the drawing. All these modes will be discussed in the pages that follow.

Choosing Zoom from the pull-down menu displays the various options of the ZOOM command that can be found in the Menu Browser.

ZOOM commands can also be selected from the Ribbon when operating in the 2D Drafting & Annotation Workspace.

You can also activate the ZOOM command from the keyboard by entering either ZOOM or the letter Z, which is its command alias.

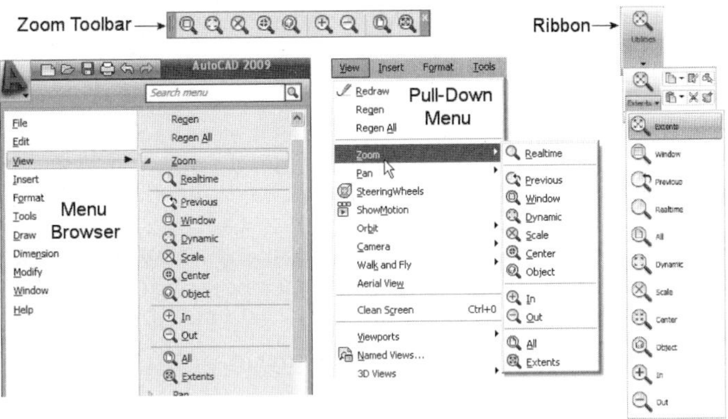

Figure 3.1

The following table illustrates each ZOOM command function and the icon or button that it relates to.

Button	Tool	Function
	ZOOM-Window	Zooms to a rectangular area by specifying two diagonal points
	ZOOM-Dynamic	Uses a viewing box to zoom in to a portion of a drawing
	ZOOM-Scale	Zooms the display based on a specific scale factor
	ZOOM-Center	Zooms to an area of the drawing based on a center point and magnification value
	ZOOM-Object	Zooms to the largest possible magnification based on the objects selected
	ZOOM-In	Magnifies the display screen by a scale factor of 2
	ZOOM-Out	De-magnifies the display screen by a scale factor of .5
	ZOOM-All	Zooms to the largest possible magnification based on the current grid setting or drawing extents, whichever is larger
	ZOOM-Extents	Zooms to the largest possible magnification based on all objects in the drawing
	ZOOM-Realtime	Zooms dynamically when the user moves the cursor up or down
	ZOOM-Previous	Zooms to the previous view

The following image illustrates a floor plan. To work on details of this and other drawings, use the ZOOM command to magnify or reduce the display screen. The following are the options of the ZOOM command (note that nX and nXP refer to the relative scale factors, which will be explained later in the chapter):

```
Command: Z (For ZOOM)
Specify corner of window, enter a scale factor (nX or nXP), or
[All/Center/Dynamic/Extents/Previous/Scale/Window/Object] <real
    time>: (enter one of the listed options)
```

Executing the ZOOM command and picking a blank part of the screen places you in automatic ZOOM-Window mode. Selecting another point zooms in to the specified area. Refer to the following command sequence to use this mode of the ZOOM command on the floor plan as shown on the left in the following image.

```
Command: Z (For ZOOM)
Specify corner of window, enter a scale factor (nX or nXP), or
[All/Center/Dynamic/Extents/Previous/Scale/Window/Object] <real
    time>: (Mark a point at "A")
Other corner: (Mark a point at "B")
```

The ZOOM-Window option is automatically invoked once you select a blank part of the screen and then pick a second point. The resulting magnified portion of the screen appears as shown on the right in the following image.

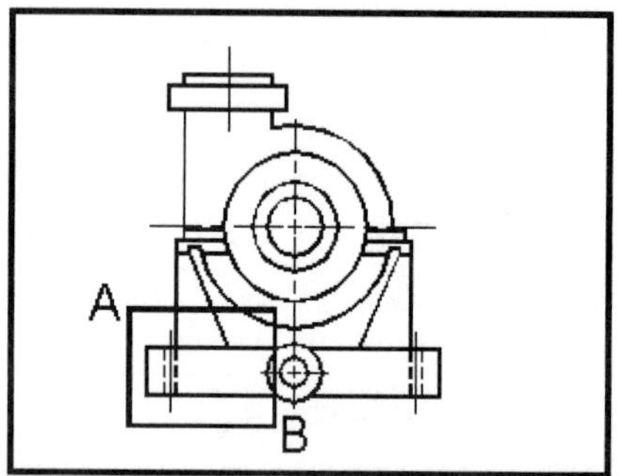

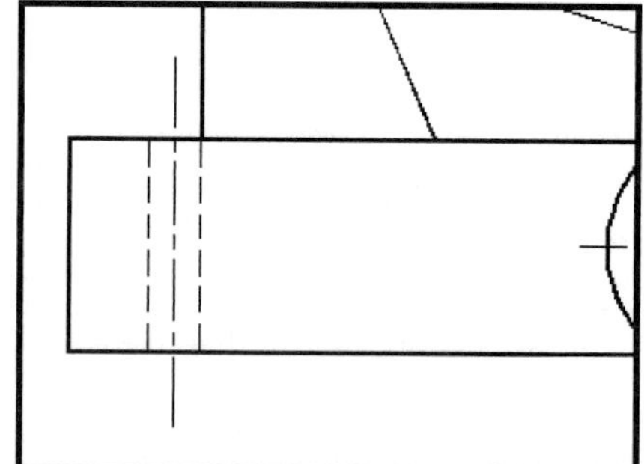

Figure 3.2

ZOOMING WITH A WHEEL MOUSE

One of the easiest ways for performing zooming and panning operations is through the mouse. Most computers are equipped with a standard Microsoft two-button mouse with the addition of a wheel, as illustrated in the following image. Rolling the wheel forward zooms in to, or magnifies, the drawing. Rolling the wheel backward zooms out of, or reduces, the drawing.

Pressing and holding the wheel down, as shown on the right in the following image, places you in Realtime PAN mode. The familiar hand icon on the display screen identifies this mode.

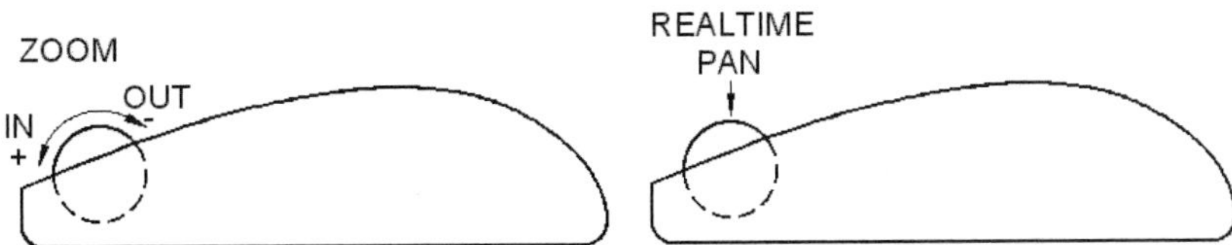

Figure 3.3

Pressing CTRL and then depressing the wheel places you in Joystick Pan mode (see the following image). This mode is identified by a pan icon similar to that shown in the middle of the following image. This icon denotes all directions in which panning may occur. Moving the mouse in any direction with the wheel depressed displays the icon on the right in the following image, which shows the direction of the pan. The farther away from the center dot you move your cursor, the faster the panning takes place.

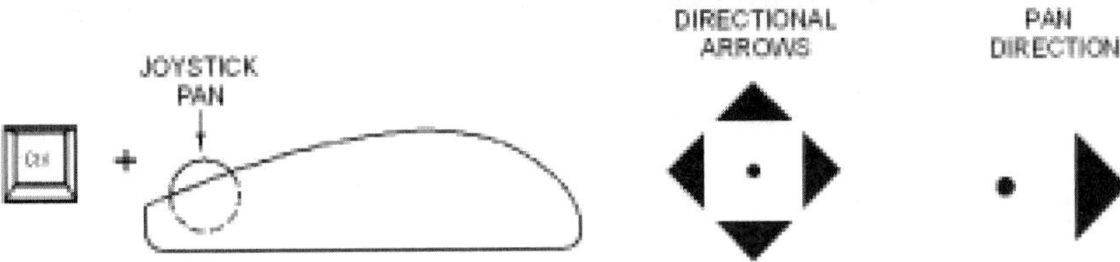

Figure 3.4

The wheel can also function like a mouse button. Double-clicking on the wheel, as in the following image, performs a ZOOM-Extents and is extremely popular for viewing the entire drawing on the display screen.

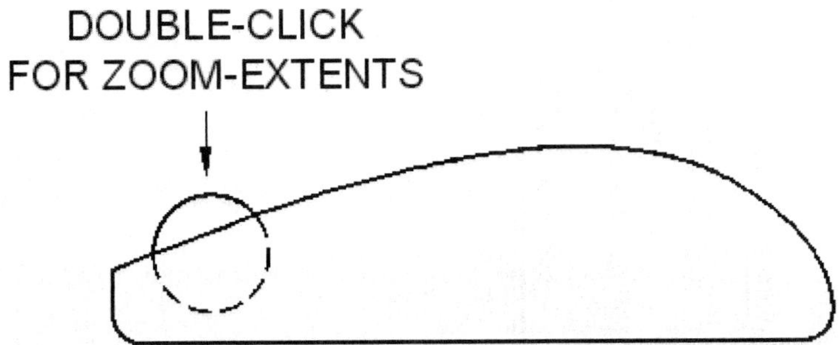

Figure 3.5

ZOOMING IN REAL TIME

A powerful option of the ZOOM command is performing screen magnifications or reductions in real time. This is the default option of the command. Issuing the Realtime option of the ZOOM command displays a magnifying glass icon with a positive sign and a negative sign above the magnifier icon. Identify a blank part of the drawing editor, press down the Pick button of the mouse (the left mouse button), and move in an upward direction to zoom in to the drawing in real time. Identify a blank part of the drawing editor, press down the Pick button of the mouse, and move in a downward direction to zoom out of the drawing in real time. Use the following command sequence and illustration on the left in the following image for performing this task.

```
Command: Z (For ZOOM)
Specify corner of window, enter a scale factor (nX or nXP), or
[All/Center/Dynamic/Extents/Previous/Scale/Window/Object] <real
    time>: (Press ENTER to accept Realtime as the default)
```

Identify the lower portion of the drawing editor, press and hold down the Pick button of the mouse, and move the Realtime cursor up; notice the image zooming in.

Once you are in the Realtime mode of the ZOOM command, press the right mouse button to activate the shortcut menu, as shown on the right in the following image. Use this menu to switch between Realtime ZOOM and Realtime PAN, which gives you the ability to pan across the screen in real time. The ZOOM-Window, Original (Previous), and Extents options are also available in the cursor menu.

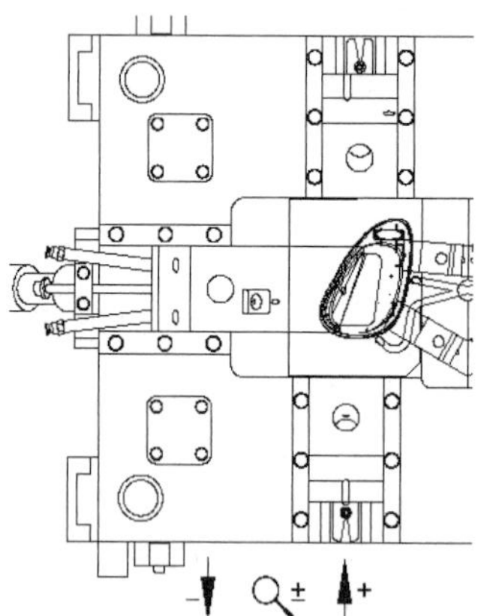
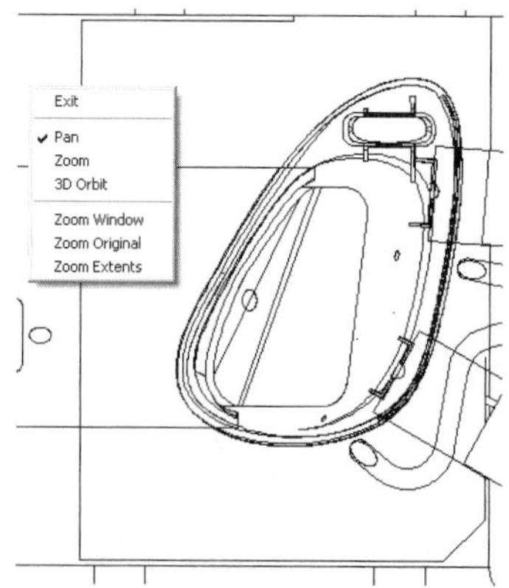

Figure 3.6

USING ZOOM-ALL

Another option of the ZOOM command is All. Use this option to zoom to the current limits of the drawing as set by the LIMITS command. In fact, right after the limits of a drawing have been changed, issuing a ZOOM-All updates the drawing file to reflect the latest screen size. To use the ZOOM-All option, refer to the following command sequence.

```
Command: Z (For ZOOM)
Specify corner of window, enter a scale factor (nX or nXP), or
[All/Center/Dynamic/Extents/Previous/Scale/Window/Object] <real
    time>: A (For All)
```

The illustration on the left in the following image shows a zoomed-in portion of a part. Use the ZOOM-All option to zoom to the drawing's current limits, as shown on the right in the following image.

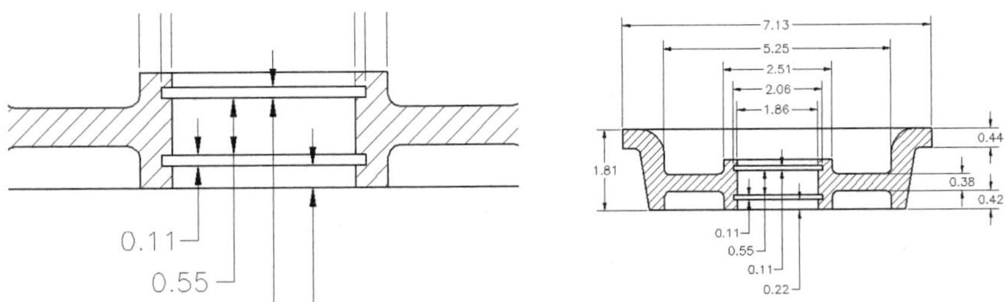

Figure 3.7

USING ZOOM-CENTER

The ZOOM-Center option allows you to specify a new display based on a selected center point as shown on the left in the following image. A window height controls whether the image on the display screen is magnified or reduced. If a smaller value is specified for the magnification or height, the magnification of the image is increased (you zoom in to the object). If a larger value is specified for the magnification or height, the image gets smaller, or a ZOOM-Out is performed.

Try It!: Open the drawing file 03_Zoom Center. Follow the illustrations and command sequence below to perform a zoom based on a center point.

```
Command: Z (For ZOOM)
Specify corner of window, enter a scale factor (nX or nXP), or
[All/Center/Dynamic/Extents/Previous/Scale/Window/Object] <real
    time>: C (For Center)
Center point: (Mark a point at the center of circle "A" as shown
    on the left)
Magnification or Height <7.776>: 2
```

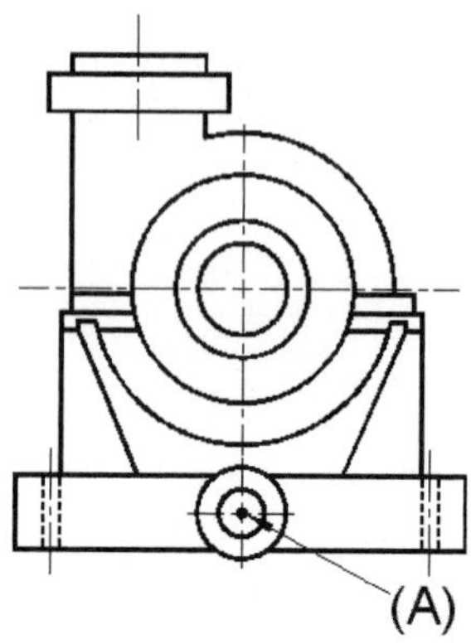

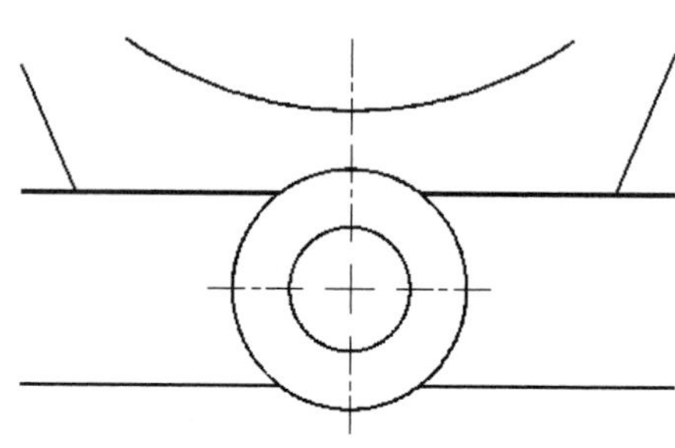

Figure 3.8

USING ZOOM-EXTENTS

The left image of the pump in the following image reflects a ZOOM-All operation. Use this option to display the entire drawing area based on the drawing limits even if the objects that make up the image appear small. Instead of performing a zoom based on the drawing limits, ZOOM-Extents uses the extents of the image on the display screen to perform the zoom. The right image in the following image shows the largest possible image displayed as a result of using the ZOOM command and the Extents option.

Try It!: Open the drawing file 03_Zoom Extents. Follow the illustrations and command sequence below to perform a zoom based on the drawing limits (All) and the objects in the drawing (Extents).

```
Command: Z (For ZOOM)
Specify corner of window, enter a scale factor (nX or nXP), or
[All/Center/Dynamic/Extents/Previous/Scale/Window/Object] <real
    time>: A (For All)
Command: Z (For ZOOM)
Specify corner of window, enter a scale factor (nX or nXP), or
[All/Center/Dynamic/Extents/Previous/Scale/Window/Object] <real
    time>: E (For Extents)
```

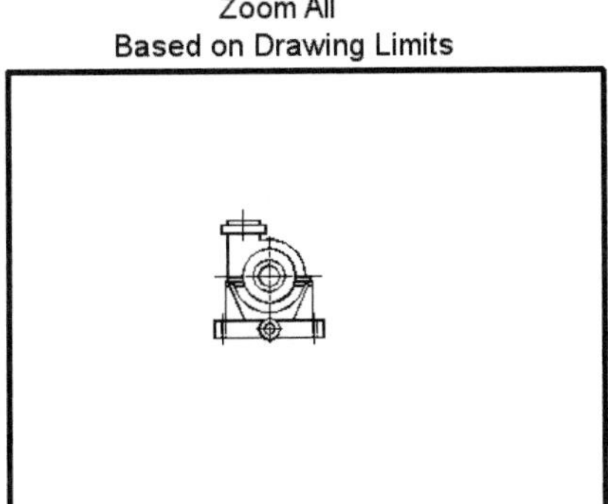

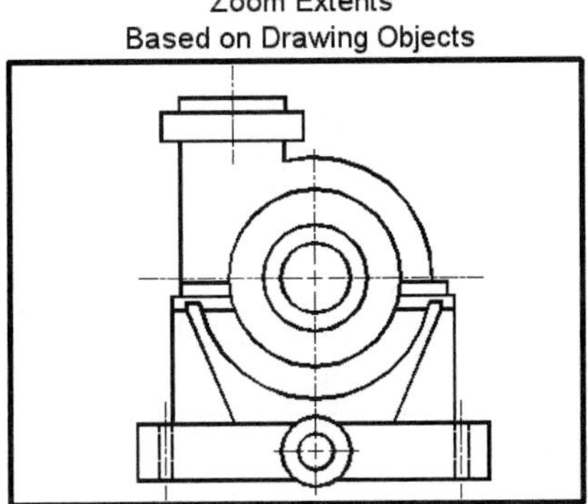

Figure 3.9

USING ZOOM-WINDOW

The ZOOM-Window option allows you to specify the area to be magnified by marking two points representing a rectangle, as shown on the left in the following image. The center of the rectangle becomes the center of the new image display; the image inside the rectangle is either enlarged, as shown on the right in the following image, or reduced.

Try It!: Open the drawing file 03_Zoom Window. Follow the illustrations and command sequence below to perform a zoom based on a window.

```
Command: Z (For ZOOM)
Specify corner of window, enter a scale factor (nX or nXP), or
[All/Center/Dynamic/Extents/Previous/Scale/Window/Object] <real
    time>: W (For Window)
First corner: (Mark a point at "A")
Other corner: (Mark a point at "B")
```

By default, the Window option of zoom is automatic; in other words, without entering the Window option, the first point you pick identifies the first corner of the window box. The prompt "Other corner:" completes ZOOM-Window, as indicated in the following prompts:

```
Command: Z (For ZOOM)
Specify corner of window, enter a scale factor (nX or nXP), or
[All/Center/Dynamic/Extents/Previous/Scale/Window/Object] <real
    time>: (Mark a point at "A")
Other corner: (Mark a point at "B")
```

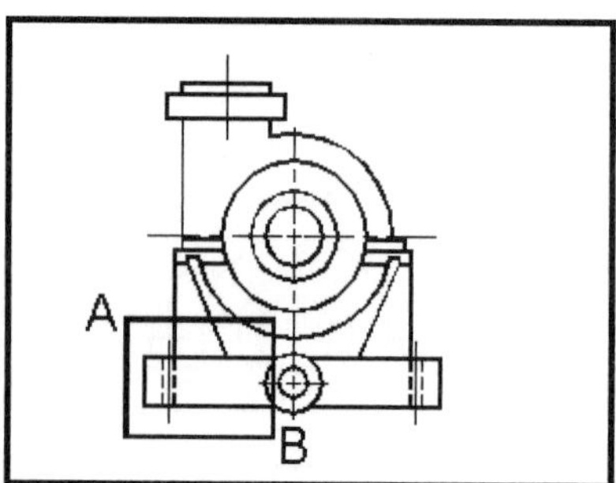

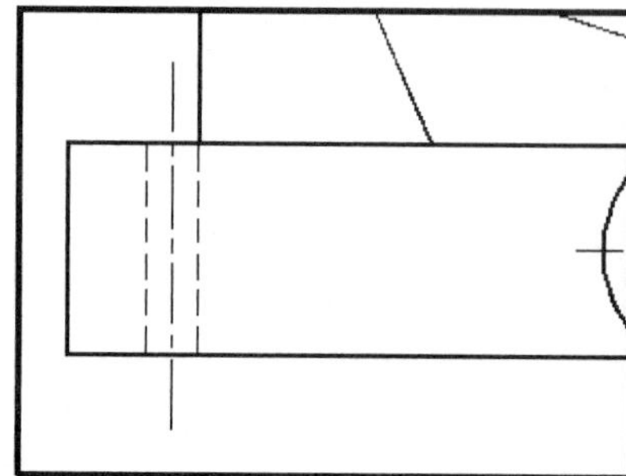

Figure 3.10

USING ZOOM-PREVIOUS

After magnifying a small area of the display screen, use the Previous option of the ZOOM command to return to the previous display. The system automatically saves up to ten views when zooming. This means you can begin with an overall display, perform two zooms, and use the ZOOM-Previous command twice to return to the original display, as shown in the following image.

```
Command: Z (For ZOOM)
Specify corner of window, enter a scale factor (nX or nXP), or
[All/Center/Dynamic/Extents/Previous/Scale/Window/Object] <real
    time>: P (For Previous)
```

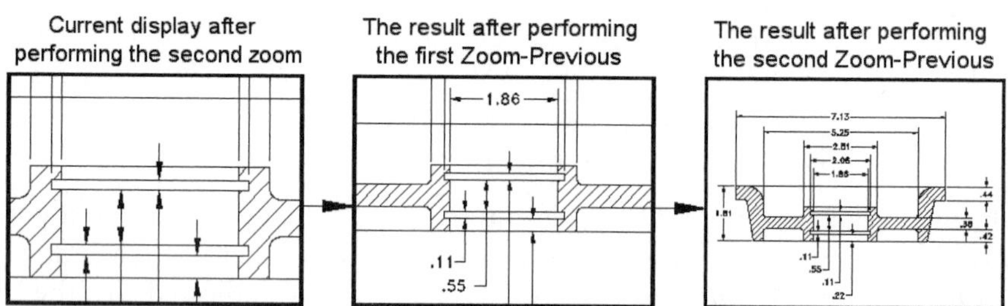

Figure 3.11

USING ZOOM-OBJECT

A zooming operation can also be performed based on an object or group of objects that you select in the drawing. The illustration on the left in the following image displays a facilities plan. The ZOOM-Object option is used to magnify one of the chairs using the following command sequence:

```
Command: Z (For ZOOM)
Specify corner of window, enter a scale factor (nX or nXP), or
[All/Center/Dynamic/Extents/Previous/Scale/Window/Object] <real
    time>: O (For Object)
Select objects: (Pick the chair)
Select objects: (Press ENTER to perform a ZOOM-Extents based on
    the chair)
```

The results are shown on the right in the following image where the chair is magnified to fit in the display screen.

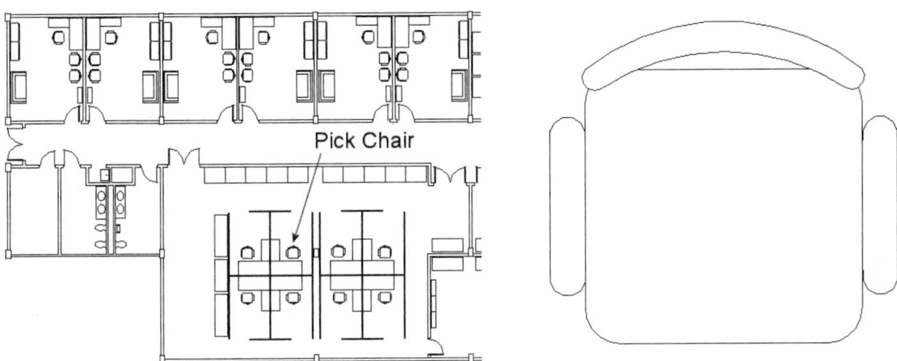

Figure 3.12

USING ZOOM-SCALE

In addition to performing zoom operations by picking points on your display screen for such options as zooming to a window or center point, you can also fine-tune your zooming by entering zoom scale factors directly from your keyboard. For example, entering a scale factor of 2 will zoom in to your drawing at a factor of 2 times. Entering a scale factor of 0.50 will zoom out of your drawing.

Try It!: Open the drawing file 03_Zoom Scale. Use the command sequences and illustration in the following image for performing a zoom based on a scale factor.

If a scale factor of 0.50 is used, the zoom is performed in the drawing at a factor of 0.50, based on the original limits of the drawing. Notice that the image gets smaller.

```
Command: Z (For ZOOM)
Specify corner of window, enter a scale factor (nX or nXP), or
[All/Center/Dynamic/Extents/Previous/Scale/Window/Object] <real
    time>: 0.50
```

If a scale factor of 0.50X is used, the zoom is performed in the drawing again at a factor of 0.50; however, the zoom is based on the current display screen. The image gets even smaller.

```
Command: Z (For ZOOM)
Specify corner of window, enter a scale factor (nX or nXP), or
[All/Center/Dynamic/Extents/Previous/Scale/Window/Object] <real
    time>: 0.50X
```

Enter a scale factor of 0.90. The zoom is again based on the original limits of the drawing. As a result, the image displays larger.

```
Command: Z (For ZOOM)
Specify corner of window, enter a scale factor (nX or nXP), or
[All/Center/Dynamic/Extents/Previous/Scale/Window/Object] <real
    time>: 0.90
```

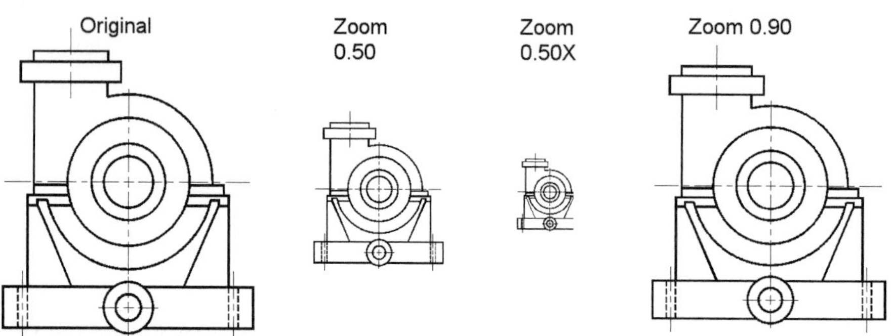

Figure 3.13

USING ZOOM-IN

Clicking on this button automatically performs a Zoom-In operation at a scale factor of 2X; the "X" uses the current screen to perform the Zoom-In operation.

USING ZOOM-OUT

Clicking on this button automatically performs a Zoom-Out operation at a scale factor of 0.5X; the "X" uses the current screen to perform the zoom-out operation.

THE PAN COMMAND

As you perform numerous ZOOM-Window and ZOOM-Previous operations, it becomes apparent that it would be nice to zoom in to a detail of a drawing and simply slide the drawing to a new area without changing the magnification; this is the purpose of the PAN command. In the following image, the ZOOM-Window option is used to construct a rectangle around the Top view to magnify it.

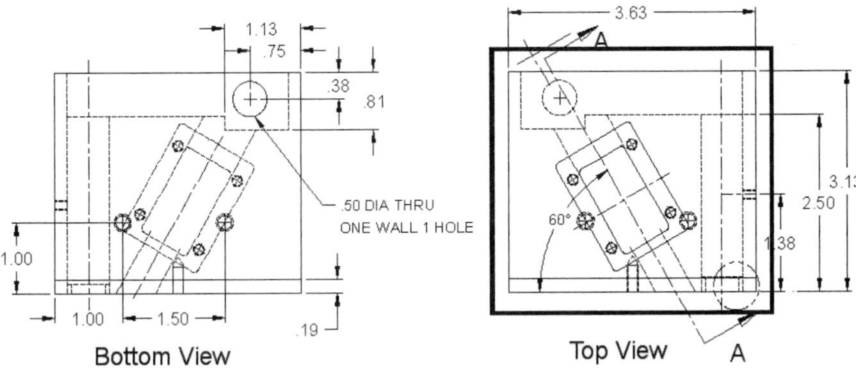

Figure 3.14

The result is shown on the right in the following image. Now the Bottom view needs to be magnified to view certain dimensions. Rather than use ZOOM-Previous and then ZOOM-Window again to magnify the Bottom view, use the PAN command.

```
Command: P (For PAN)
Press ESC or ENTER to exit, or right-click to display
    shortcut menu.
```

Issuing the PAN command displays the Hand symbol. Pressing the Pick button down at "A" and moving the Hand symbol to the right at "B" as shown on the right in the following image pans the screen and displays a new area of the drawing in the current zoom magnification.

The Bottom view is now visible after the drawing is panned from the Top view to the Bottom view, with the same display screen magnification as shown on the left in the following image. PAN can also be used transparently; that is, while in a current command, you can select the PAN command, which temporarily interrupts the current command, performs the pan, and then restores the current command.

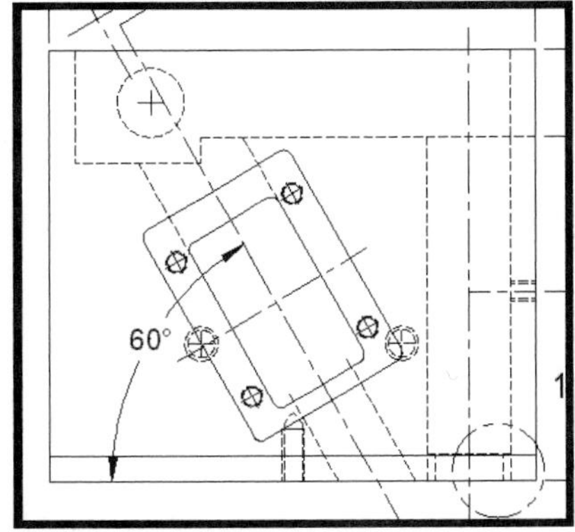

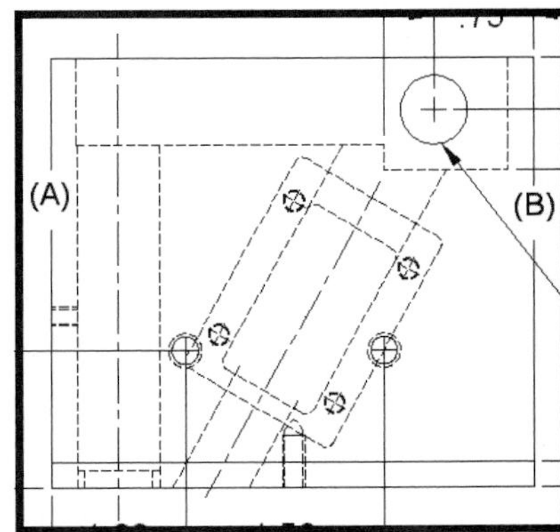

Figure 3.15

CREATING NAMED VIEWS

An alternate method of performing numerous zooms is to create a series of views of key parts of a drawing. Then, instead of using the ZOOM command, restore the named view to perform detail work. This named view is saved in the database of the drawing for use in future editing sessions. Named Views can be found in the View pull-down menu, as shown on the left in the following image, or from the Menu Browser on the right in the following image. You can activate this same dialog box through the keyboard by entering the following at the command prompt:

```
Command: V (For VIEW)
```

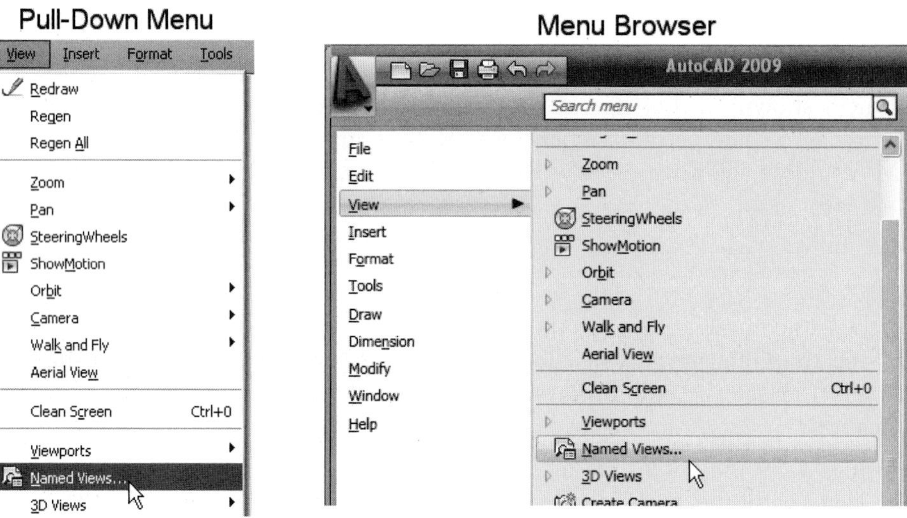

Figure 3.16

Clicking on either one of these items will activate the View Manager dialog box as shown in the following image.

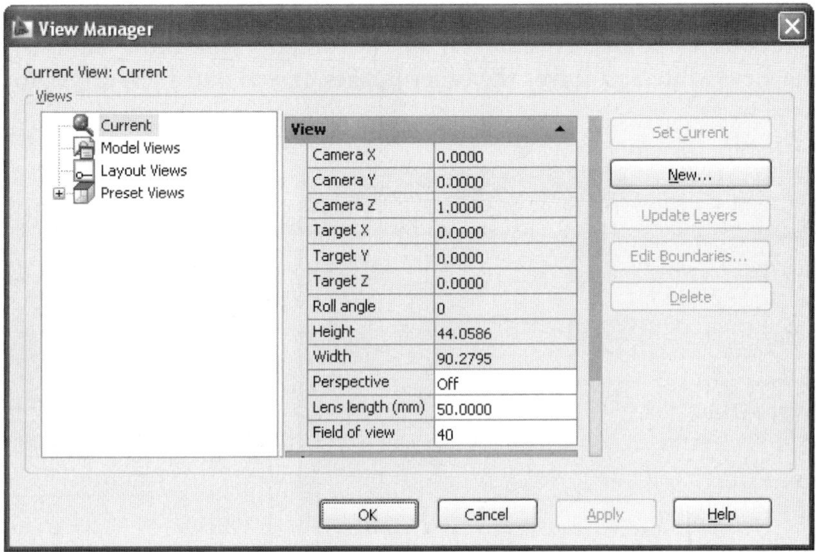

Figure 3.17

The following table describes each view type:

View Type	Description
Current	Displays the current view, view information, and clipping properties.
Model Views	Displays a list of all views created along with general, camera, and clipping properties.
Layout Views	Displays a list of views created in a layout in addition to general and view information (height and width).
Preset Views	Displays a list of orthogonal and isometric views, and lists the general properties for each preset view.

Try It!: Open the drawing file 03_Views. Follow the next series of steps and illustrations used to create a view called "FRONT."

Clicking the New button in the View Manager dialog box activates the New View/Shot Properties dialog box, as shown on the left in the following image. Use this dialog box to guide you in creating a new view. By definition, a view is created from the current display screen. This is the purpose of the Current Display radio button. Many views are created with the Define Window radio button, which creates a view based on the contents of a window that you define. Choosing the Define View Window button returns you to the display screen where the drawing appears grayed out. You are prompted for the two corners required to create a new view by window.

As illustrated on the right in the following image, a rectangular window is defined around the Front view using points "A" and "B" as the corners. This turns the image captured inside of the view white and leaves the other images grayed out. Press Enter to accept this as the new view.

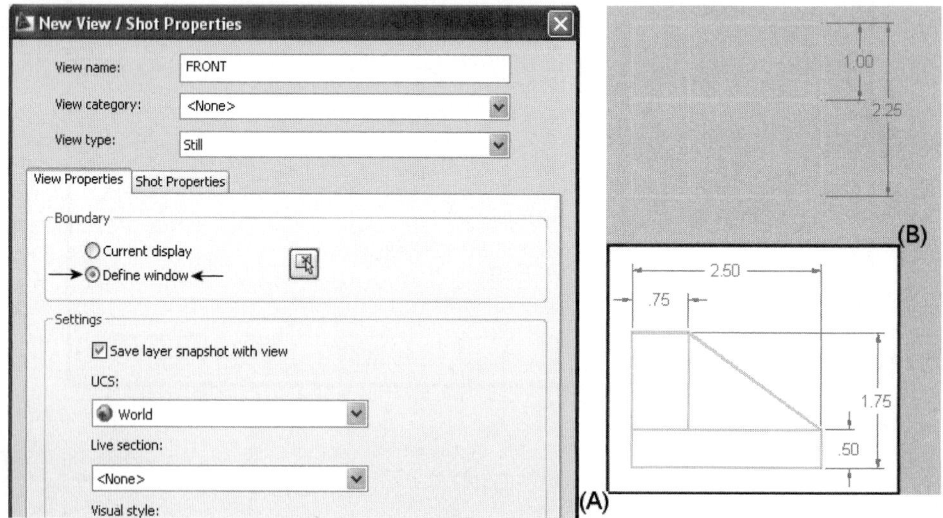

Figure 3.18

Accepting the window created in the previous image redisplays the New View/Shot Properties dialog. Clicking OK saves the view name FRONT under the Model Views heading in the View Manager dialog box, as shown in the following image.

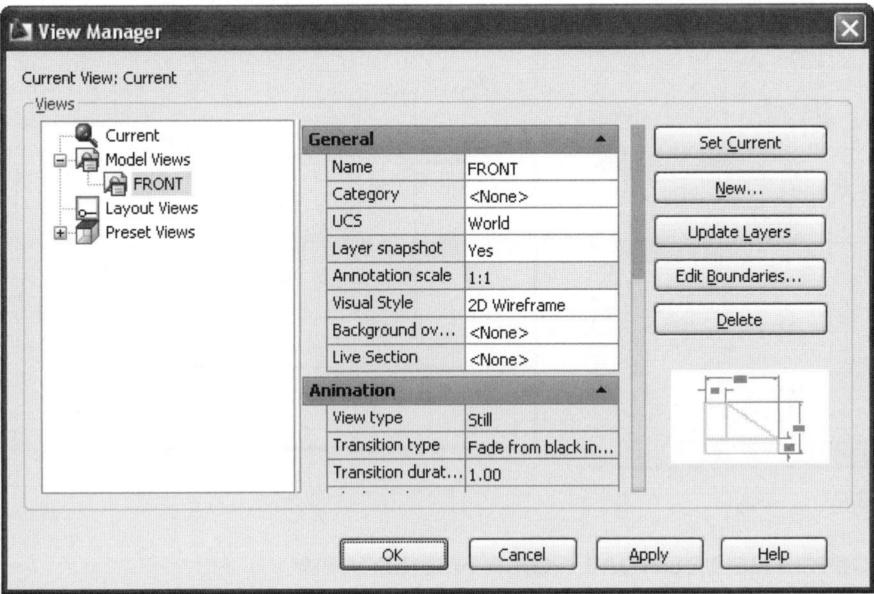

Figure 3.19

Try It!: Open the drawing file 03_Views Complete. A series of views have already been created inside this drawing. Activate the View Manager dialog box and experiment with restoring a number of these views.

The following image illustrates numerous views of a drawing already created, namely FRONT, ISO, OVERALL, SIDE, and TOP.

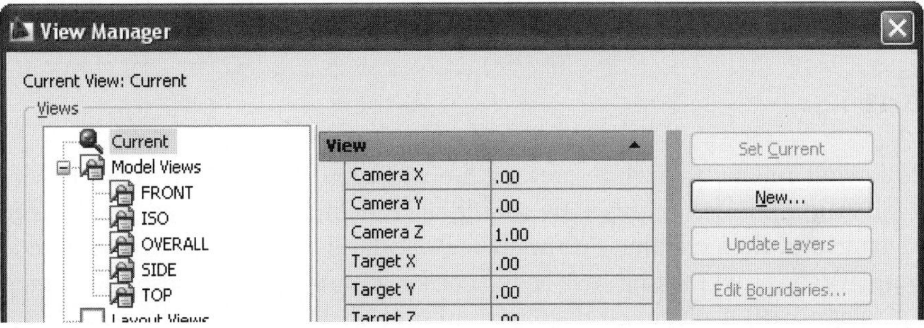

Figure 3.20

Clicking on a defined view name (FRONT), as shown on the left in the following image, and right-clicking the mouse, displays the shortcut menu used to set the view current. You can also create a new view, update layers, edit the boundaries of the view, and delete the view through this shortcut menu. Clicking on Set Current and clicking OK exits the View Manager dialog box displays the Front view, as shown on the right in the following image.

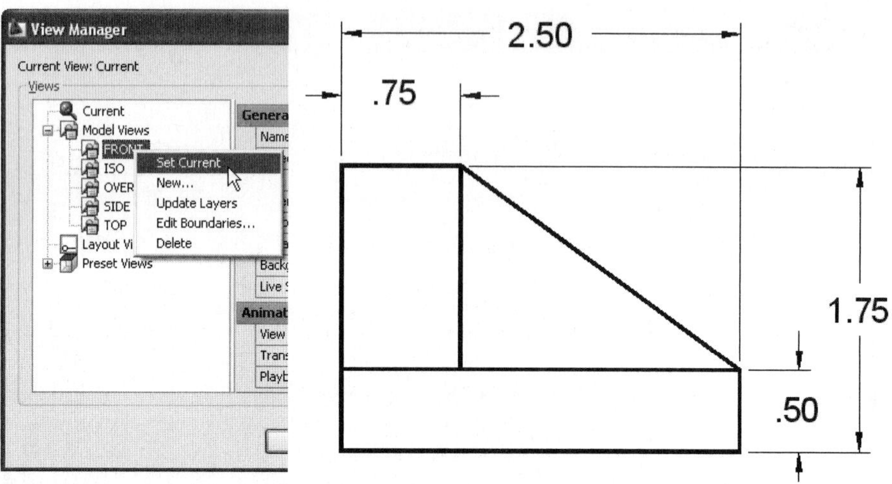

Figure 3.21

CHANGING THE BOUNDARY OF A NAMED VIEW

The View Manager dialog box has an Edit Boundaries button.

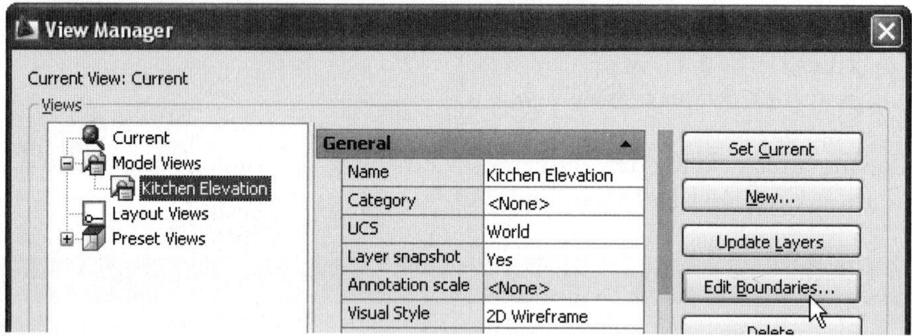

Figure 3.22

In the following image, the white area of the drawing illustrates the current view called Kitchen Elevation. However, this view should also include the cabinet details. Clicking on the view name and then clicking on the Edit Boundaries button allows you to redefine the view boundary box by picking two new corner points. This is also illustrated in the following image with a rectangle displayed around all kitchen elevations.

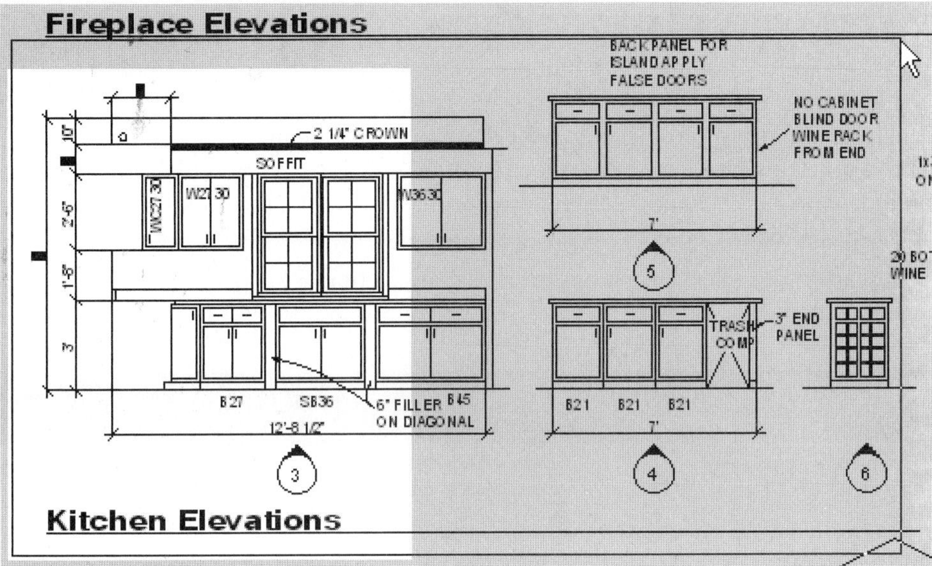

Figure 3.23

After you construct this new boundary, the results should be similar to the following image. The new view is displayed in white. Non-view components have the color gray in the background.

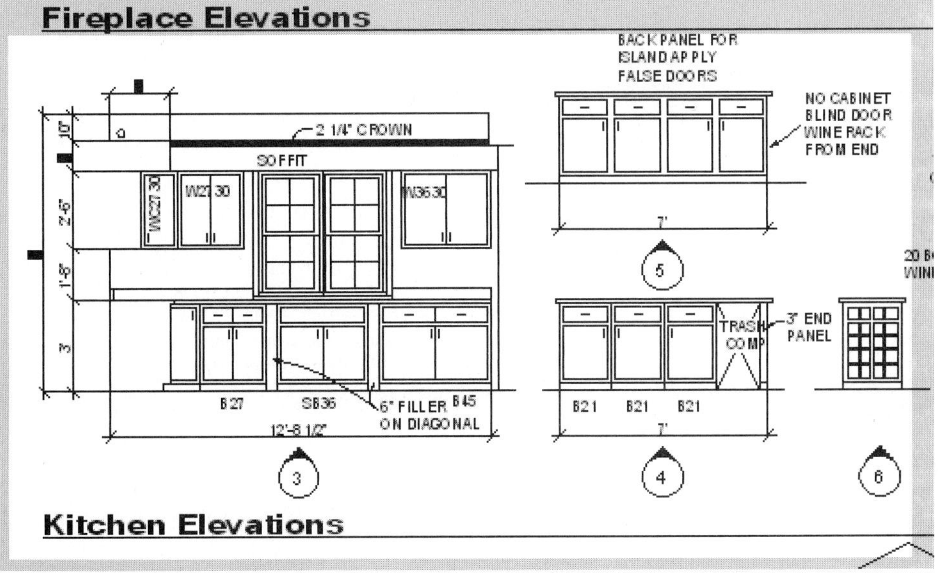

Figure 3.24

CREATING OBJECT SELECTION SETS

Selection sets are used to group a number of objects together for the purpose of modifying. Applications of selection sets are covered in the following pages. Once a selection set has been created, the group of objects may be modified by moving, copying, mirroring, and so on. The operations supported by selection sets will be covered in chapter 4, "Modify Commands." An object manipulation command supports the creation of selection sets if it prompts you to "Select objects."

The commonly used selection set options (how a selection set is made) are briefly described in the following table.

Selection Tool	Selection Key-In	Function
All	All	Selects all objects in the database of the drawing. This mode even selects objects on layers that are turned off. Objects that reside on frozen layers are not selected.
Crossing	C	You create a rectangular box by dragging your cursor from right to left (transparent green window). This rectangular box appears as dotted line segments. All objects completely enclosed and touching this rectangular box are selected.
Crossing Polygon	CP	You create an irregular closed shape. The polygon formed appears as dotted line segments. All objects completely enclosed and touching this irregular shape are selected.
Fence	F	You create a series of line segments. These line segments appear dotted. All objects touching these fence lines are selected. You cannot close the fence shape; it must remain open.
Last	L	Selects the last object you created.
Previous	P	Selects the object or objects from the most recently created selection set.
Window	W	You create a rectangular box by dragging your cursor from left to right (transparent blue window). This rectangular box appears as solid lines. All objects completely enclosed in this rectangular box are selected. Objects touching the edges of the box are ignored.
Window Polygon	WP	You create an irregular closed shape. The polygon formed appears as solid line segments. All objects completely enclosed and touching this irregular shape are selected.

OBJECT PRE-HIGHLIGHT

Whenever you move your cursor on an object, the object pre-highlights. This provides an easy way to select the correct object, especially in busy drawings. In the following image, the cursor is moved over the rectangle that surrounds the 128 text object. This action pre-highlights this rectangle until you move your cursor off this object.

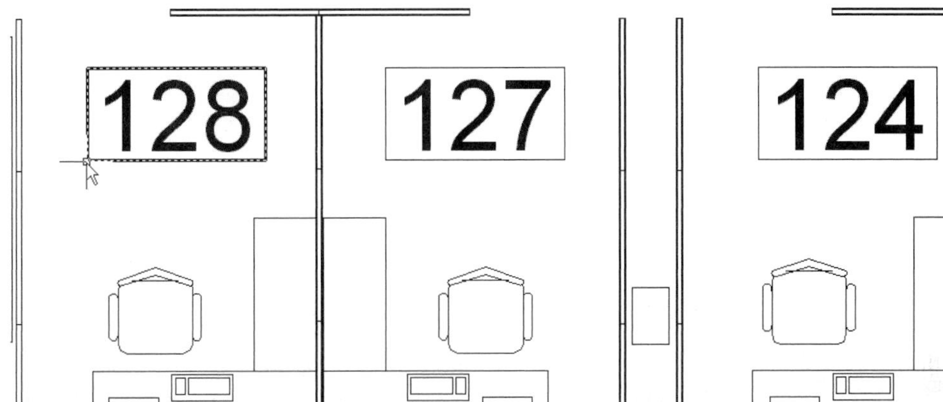

Figure 3.25

SELECTING OBJECTS BY INDIVIDUAL PICKS

When AutoCAD prompts you with "Select objects," a pickbox appears as the cursor is moved across the display screen, as shown on the left in the following image. As you move your cursor over an object, the object highlights and turns bold to signify that it is selectable, as shown on the right in the following image. An object remains highlighted once it is selected.

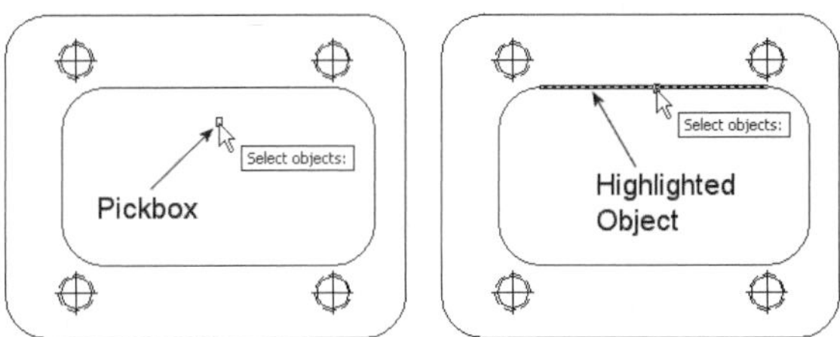

Figure 3.26

Try It!: Open the drawing file 03_Select. Enter the ERASE command and at the Select objects prompt, pick the arc segment labeled "A" as shown on the left in the following image. To signify that the object is selected the rectangle highlights.

```
Command: E (For ERASE)
Select objects: (Pick the object at "A")
Select objects: (Press ENTER to execute the ERASE command. The
    rectangle disappears as shown on the right in the
    following image)
```

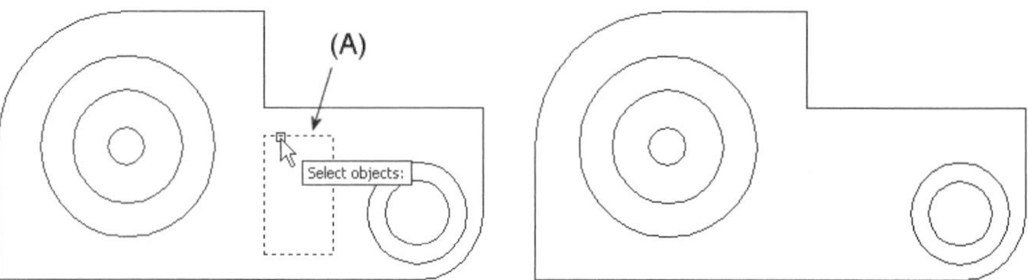

Figure 3.27

SELECTING OBJECTS BY WINDOW

The individual pick method previously mentioned works fine for small numbers of objects. However, when numerous objects need to be edited, selecting each individual object could prove time-consuming. Instead, you can select all objects that you want to become part of a selection set by using the Window selection mode. This mode requires you to create a rectangular box by picking two diagonal points. In the following image, a selection window has been created with point "A" as the first corner and "B" as the other corner. As an additional aid in selecting, the rectangular box you construct is displayed as a transparent blue box by default. When you use this selection mode, only those objects completely enclosed by the blue window box are selected. Even though the window touches other objects, they are not completely enclosed by the window and therefore are not selected.

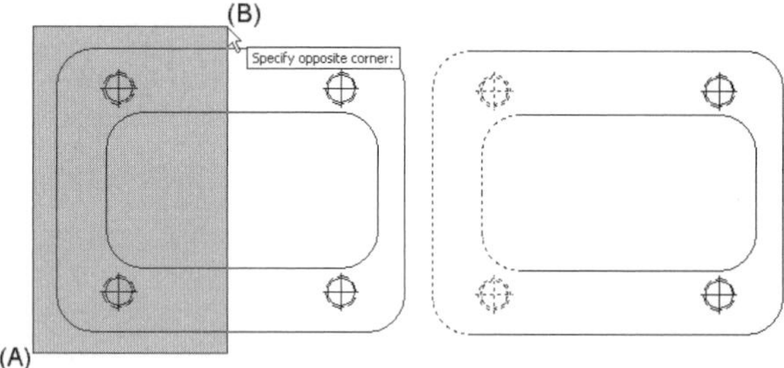

Figure 3.28

SELECTING OBJECTS BY CROSSING WINDOW

In the previous example of producing a selection set by a window, the window selected only those objects completely enclosed by it. The following image is an example of selecting objects by a crossing window. The Crossing Window option requires two points to define a rectangle, as does the Window selection option. In the following image, a transparent green rectangle with dashed edges is used to select objects, using "A" and "B" as corners for the rectangle; however, this time the crossing window was used. The highlighted objects illustrate the results. All objects that are touched by or enclosed by the crossing rectangle are selected. Because the transparent green crossing rectangle passes through the three circles without enclosing them, they are still selected by this object selection mode.

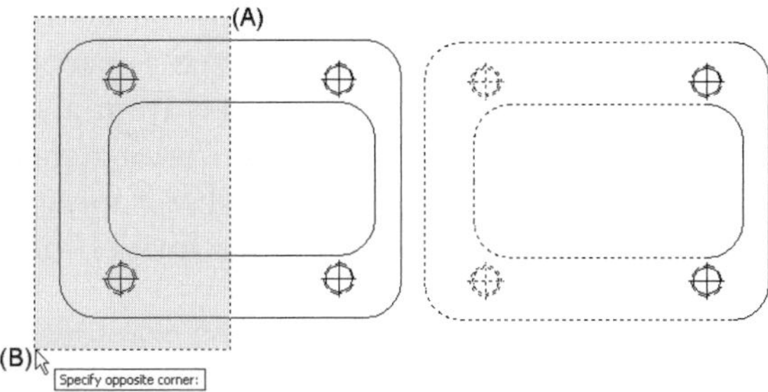

Figure 3.29

SELECTING OBJECTS BY A FENCE

Use this mode to create a selection set by drawing a line or group of line segments called a fence. Any object touched by the fence is selected. The fence does not have to end exactly where it was started. In the following image, all objects touched by the fence are selected, as represented by the dashed lines.

Try It!: Open the drawing file 03_Fence. Follow the command sequence below and the illustration to select a group of objects using a fence.

```
Command: E (For ERASE)
Select objects: F (For Fence)
First fence point: (Pick a first fence point)
Specify endpoint of line or [Undo]: (Pick a second fence point)
Specify endpoint of line or [Undo]: (Pick a third fence point)
Specify endpoint of line or [Undo]: (Press ENTER to exit
    fence mode)
Select objects: (Press ENTER to execute the ERASE command)
```

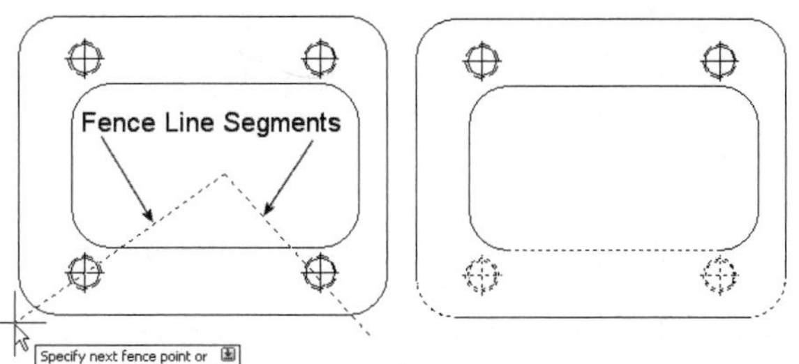

Figure 3.30

REMOVING OBJECTS FROM A SELECTION SET

All the previous examples of creating selection sets have shown you how to create new selection sets. What if you select the wrong object or objects? Instead of canceling out of the command and trying to select the correct objects, you can use the Remove option to remove objects from an existing selection set. As illustrated on the left in the following image, a selection set has been created and made up of all the highlighted objects. However, the outer lines and arcs were mistakenly selected as part of the selection set. The Remove option allows you to remove highlighted objects from a selection set. To activate Remove, press SHIFT and pick the object or objects you want removed; this works only if the Select objects prompt is present. When the highlighted objects are removed from the selection set, as shown on the right in the following image, they regain their original display intensity.

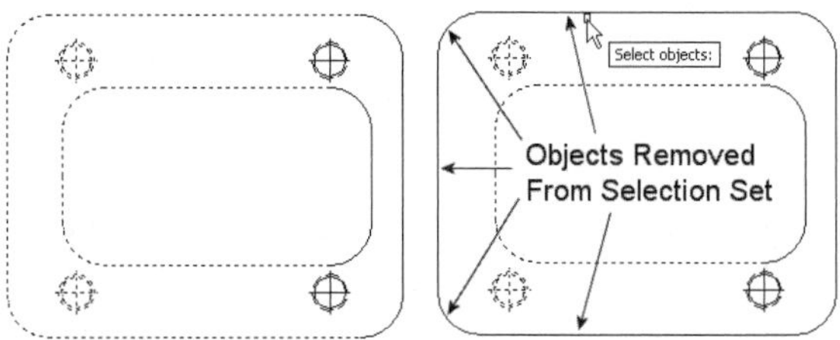

Figure 3.31

SELECTING THE PREVIOUS SELECTION SET

When you create a selection set of objects, this grouping is remembered until another selection set is made. The new selection set replaces the original set of objects. Let's say you moved a group of objects to a new location on the display screen. Now you want to rotate these same objects at a certain angle. Rather than select the same set of objects to rotate, you would pick the Previous option or type P at the Select objects prompt. This selects the previous selection set. The buffer holding the selection set is cleared whenever you use the U command to undo the previous command.

SELECTING OBJECTS BY A CROSSING POLYGON

When you use the Window or Crossing Window mode to create selection sets, two points specify a rectangular box for selecting objects. At times, it is difficult to select objects by the rectangular window or crossing box because in more cases than not, extra objects are selected and have to be removed from the selection set.

Try It!: Open the drawing file 03_Select CP. The following image shows a mechanical part with a "C"-shaped slot. Rather than use Window or Crossing Window modes, you can pick the Crossing Polygon mode (CPolygon) or type CP at the Select objects prompt. You simply pick points representing a polygon. Any object that touches or is inside the polygon is added to a selection set of objects. As illustrated on the left in the following image, the crossing polygon is constructed using points "1" through "5." A similar but different selection set mode is the Window Polygon (WPolygon). Objects are selected using this mode when they lie completely inside the Window Polygon, which is similar to the regular Window mode.

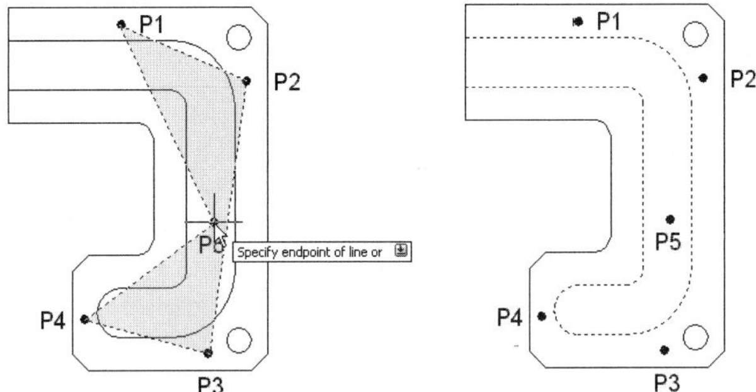

Figure 3.32

CYCLING THROUGH OBJECTS

At times, the process of selecting objects can become quite tedious. Often, objects lie directly on top of each other. As you select the object to delete, the other object selects instead. To remedy this, press and hold the SHIFT key and press the SPACEBAR when prompted to "Select objects." This activates Object Selection Cycling and enables you to scroll through all objects in the vicinity of the pickbox by pressing the SPACEBAR. A message appears in the prompt area when you begin scrolling, alerting you that cycling is on; you can now pick objects until the desired object is highlighted. Pressing ENTER not only accepts the highlighted object but toggles cycling off. In the following image and with selection cycling on, the first pick selects the circle, the second pick selects the line segment, and the third pick selects the rectangle. Keep picking until the desired object highlights.

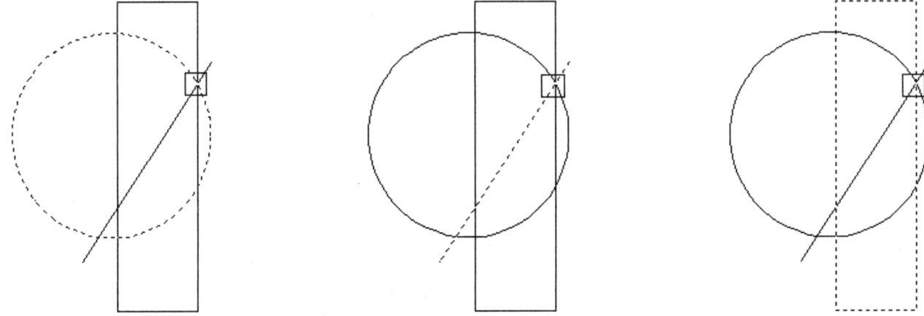

Figure 3.33

Try It!: Open the drawing file 03_Cycle and experiment using this feature on the object illustrated in the previous image.

PRE-SELECTING OBJECTS

You can bypass the Select objects prompt by pre-selecting objects at the Command prompt. Notice that when you pick an object, not only does it highlight, but a series of blue square boxes also appears. These objects are called grips and will be discussed in greater detail in chapter 7. To cancel or de-select the object, press the ESC key and notice that even the grips disappear. You could also pick a blank part of your screen at the Command prompt. Moving your cursor to the right and picking a point has the same effect as using the Window option of Select objects. Only objects completely inside the window are selected. If you pick a blank part of your screen at the Command prompt and move your cursor to the left, this has the same effect as using the Crossing option of Select objects in which any items touched by or completely enclosed by the box are selected.

Tip: Pressing CTRL + A at the Command prompt selects all objects in the entire drawing and displays the blue grip boxes. This even select objects that are on a layer that has been turned off, but does not select those on a layer that is frozen or locked.

THE QSELECT COMMAND

Yet another way of creating a selection set is by matching the object type and property with objects currently in use in a drawing. This is the purpose of the QSELECT (Quick Select) command. This command can be chosen from the Tools pull-down menu, as shown on the left in the following image, which launches the dialog box on the right.

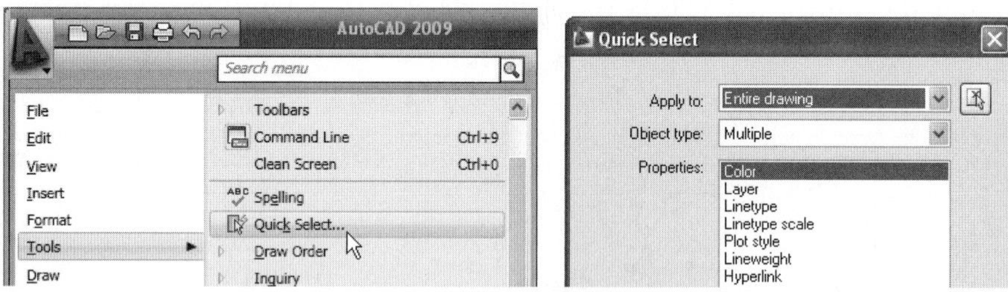

Figure 3.34

Try It!: Open the drawing file 03_Qselect and activate the Quick Select dialog box. This command works only if objects are defined in a drawing; the Quick Select dialog box does not display in a drawing file without any objects drawn. Clicking in the Object type field displays all object types currently used in the drawing. This enables you to create a selection set by the object type. For instance, to select all line segments in the drawing file, click on Line in the Object type field, as illustrated on the left in the following image.

Clicking the OK button at the bottom of the dialog box returns you to the drawing and highlights all the line segments in the drawing, as shown on the right in the following image.

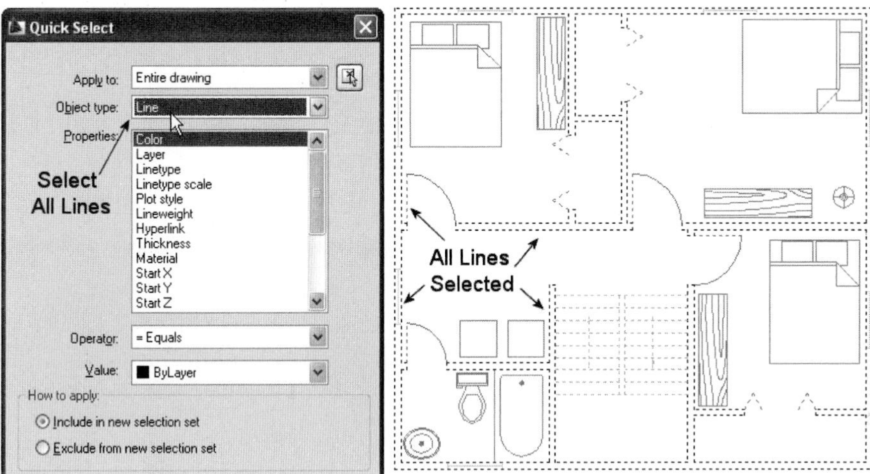

Figure 3.35

Other controls of Quick Select include the ability to select the object type from the entire drawing or from just a segment of the drawing. You can narrow the selection criteria by adding various properties to the selection mode such as Color, Layer, and Linetype, to name a few. You can also create a reverse selection set. The Quick Select dialog box lives up to its name—it enables you to create a quick selection set.

TUTORIAL EXERCISE: 03_SELECT OBJECTS.DWG

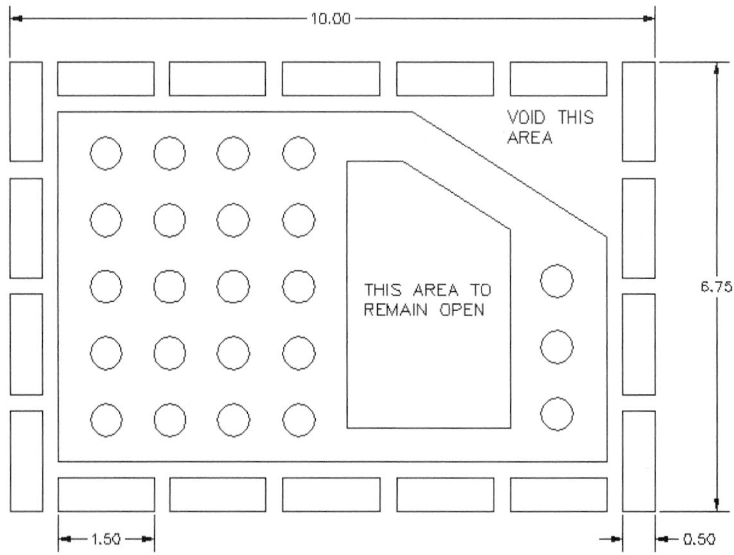

Figure 3.36

PURPOSE

The purpose of this tutorial is to experiment with the object's selection modes on the drawing shown in the previous image.

SYSTEM SETTINGS

Keep the current limits settings of (0,0) for the lower-left corner and (16,10) for the upper-right corner. Keep all remaining system settings.

SUGGESTED COMMANDS

This tutorial utilizes the ERASE command as a means of learning the basic object selection modes. The following selection modes will be highlighted for this tutorial: Window, Crossing, Window Polygon, Crossing Polygon, Fence, and All. The effects of locking layers and selecting objects will also be demonstrated.

Step 1

Open the drawing 03_Select Objects and activate ERASE at the Command prompt. At the Select objects prompt, pick a point on the blank part of your screen at "A." At the next Select objects prompt, pick a point on your screen at "B." Notice that a solid blue transparent box is formed as shown in the following image. The presence of this box signifies the Window option of selecting objects.

```
Command: E (For ERASE)
Select objects: (Pick a point at "A")
Specify opposite corner: (Pick a point at "B")
```

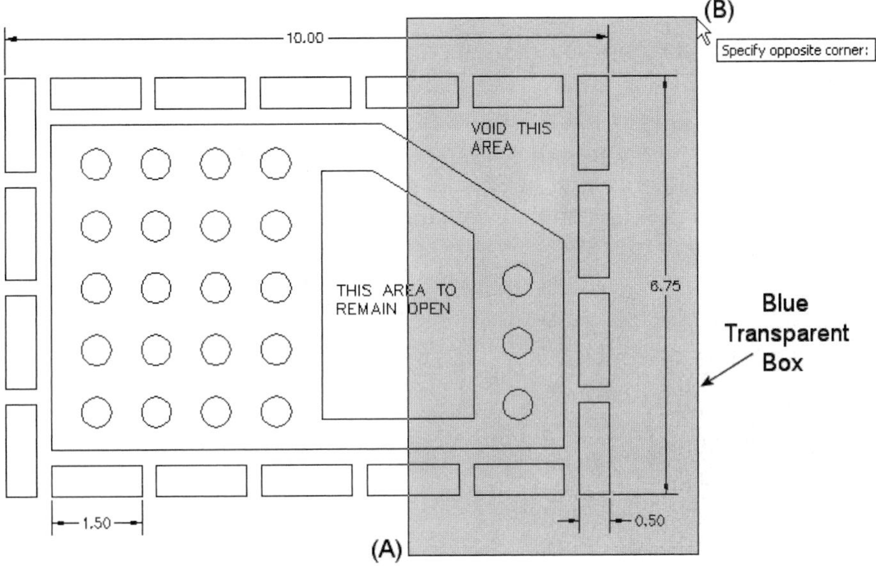

Figure 3.37

The result of selecting objects by a window is illustrated in the following image. Only those objects completely surrounded by the window are highlighted. Pressing ENTER performs the erase operation. Before continuing on to the next step, issue the U (UNDO) command, which negates the previous ERASE operation.

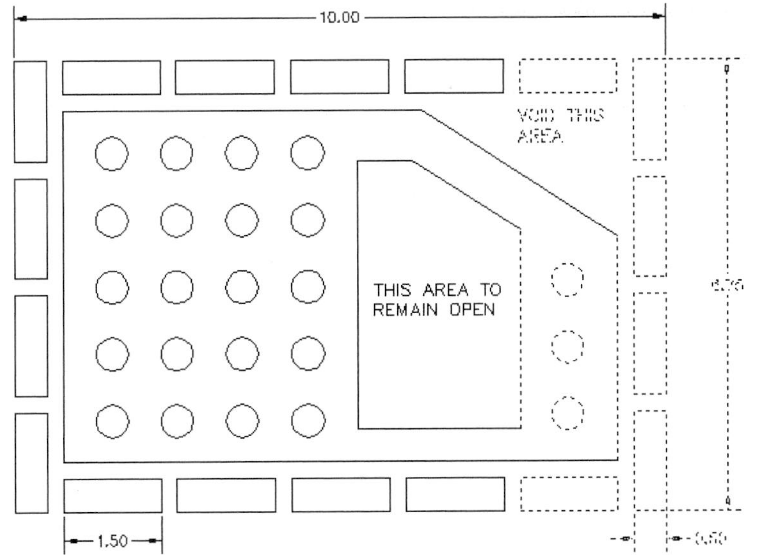

Figure 3.38

Step 2

With the entire object displayed on the screen, activate the ERASE command again. At the Select objects prompt, pick a point on the blank part of your screen at "A." At the Specify opposite corner prompt, pick a point on your screen at "B." Notice that the transparent box is now dashed and green in appearance as shown in the following image. The presence of this box signifies the Crossing option of selecting objects.

```
Command: E (For ERASE)
Select objects: (Pick a point at "A")
Specify opposite corner: (Pick a point at "B")
```

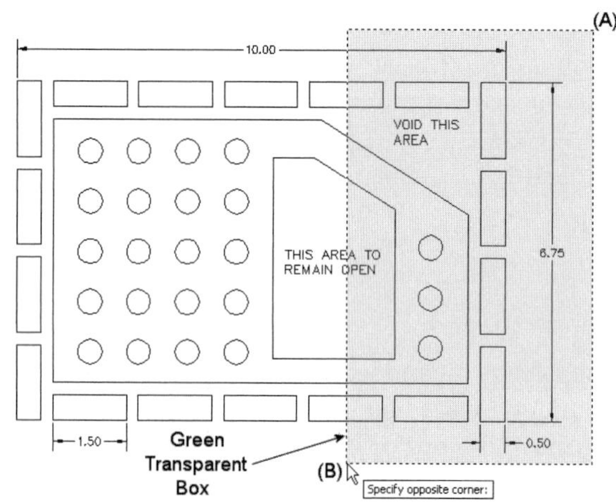

Figure 3.39

The result of selecting objects by a crossing window is illustrated in the following image. Notice that objects touched by the crossing box, as well as those objects completely surrounded by the crossing box, are highlighted. Pressing ENTER performs the ERASE operation. Before continuing on to the next step, issue the U (UNDO) command, which negates the previous ERASE operation.

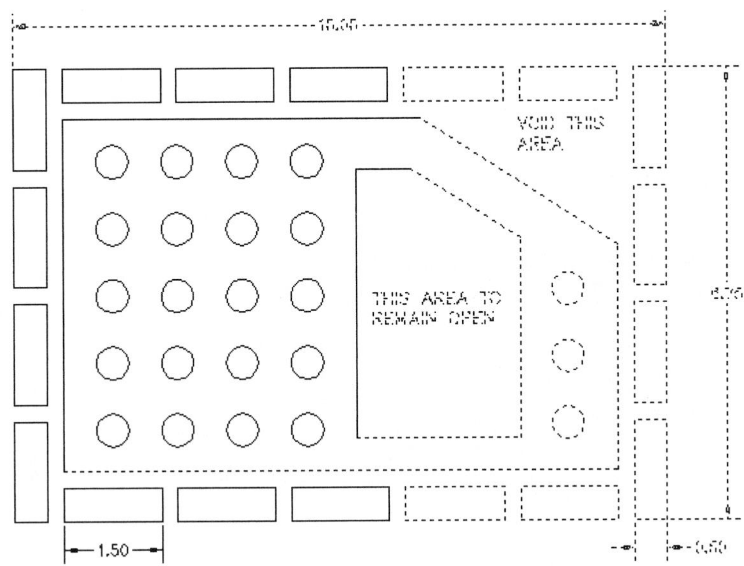

Figure 3.40

Step 3

Activate the ERASE command again. At the Select objects prompt, pick a point on the blank part of your screen at "A." At the Specify opposite corner prompt, pick a point on your screen at "B" as shown in the following image. The presence of this transparent blue box signifies the Window option of selecting objects.

```
Command: E (For ERASE)
Select objects: (Pick a point at "A")
Specify opposite corner: (Pick a point at "B")
```

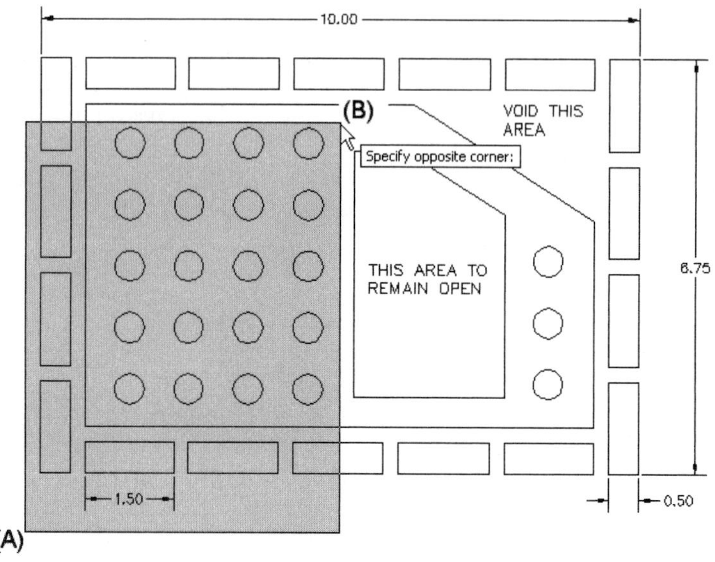

Figure 3.41

The result of selecting objects by a window is illustrated in the following image on the left. Notice that a number of circles and rectangles have been selected along with a single dimension. Unfortunately, the rectangles and the dimension were selected by mistake. Rather than cancel the ERASE command and select the objects again, press and hold down the SHIFT key and click on the highlighted rectangles and dimension. Notice how this operation de-selects the objects (see the following image on the right). Pressing ENTER performs the ERASE operation. Before continuing on to the next step, issue the U (UNDO) command, which negates the previous ERASE operation.

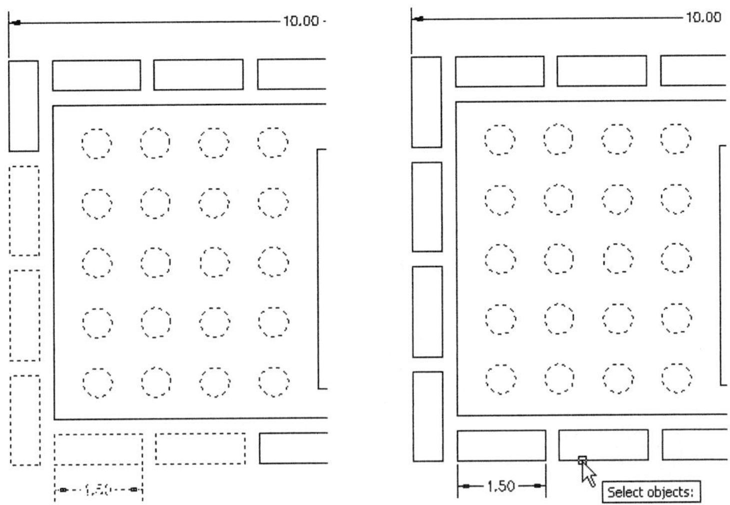

Figure 3.42

Step 4

Yet another way to select objects is by using a Window Polygon. This method allows you to construct a closed irregular shape for selecting objects. Again, activate the ERASE command. At the Select objects prompt, enter WP for Window Polygon. At the First polygon point prompt, pick a point on your screen at "A." Continue picking other points until the desired polygon, in transparent blue, is formed, as shown in the following image.

Note: The entire polygon must form a single closed shape; edges of the polygon cannot cross each other.

```
Command: E (For ERASE)
Select objects: WP (For Window Polygon)
First polygon point: (Pick a point at "A")
Specify endpoint of line or [Undo]: (Pick at "B")
Specify endpoint of line or [Undo]: (Pick at "C")
Specify endpoint of line or [Undo]: (Pick at "D")
Specify endpoint of line or [Undo]: (Pick at "E")
Specify endpoint of line or [Undo]: (Pick at "F")
Specify endpoint of line or [Undo]: (Pick at "G")
Specify endpoint of line or [Undo]: (Press ENTER)
11 found
```

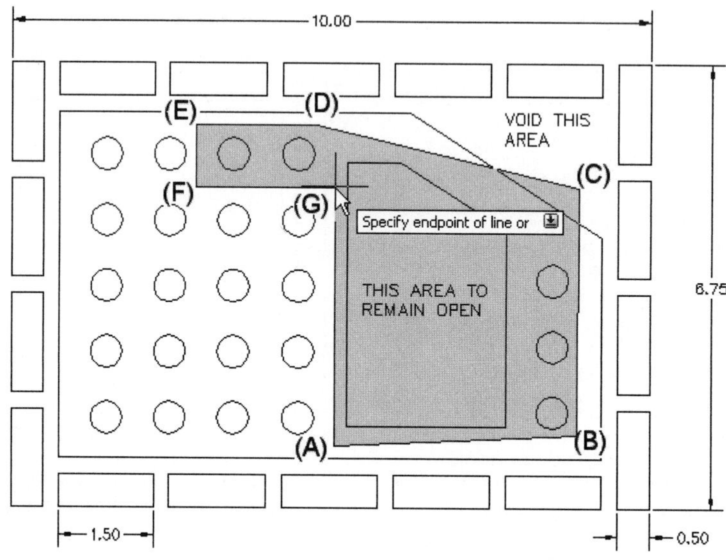

Figure 3.43

The result of selecting objects by a window polygon is illustrated in the following image. As with the window selection mode, objects must lie entirely inside the window polygon in order for them to be selected. Pressing ENTER performs the ERASE operation. Before continuing on to the next step, issue the U (UNDO) command, which negates the previous ERASE operation.

Note: A Crossing Polygon (CP) mode is also available to select objects that touch the crossing polygon or are completely surrounded by the polygon. A green polygonal shape displays as it is being created.

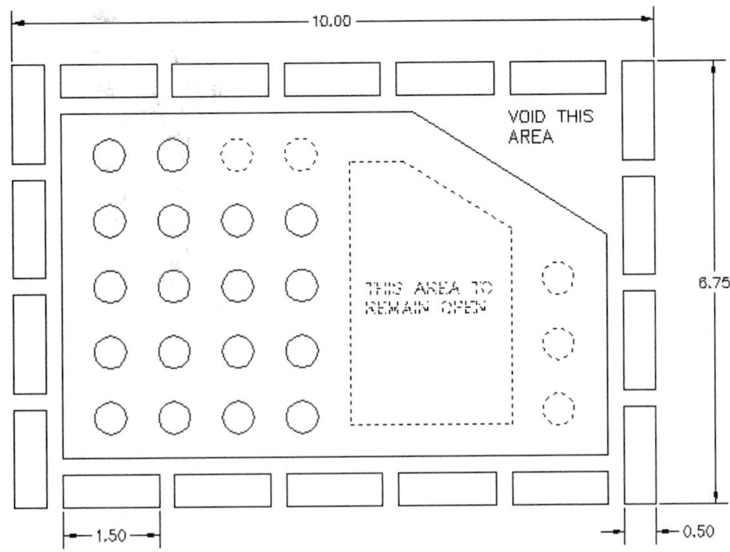

Figure 3.44

Step 5

Objects can also be selected by a fence. This is represented by a crossing line segment. Any object that touches the crossing line is selected. You can construct numerous crossing line segments. You cannot use the Fence mode to surround objects as with the Window or Crossing modes. Activate the ERASE command again. At the Select objects prompt, enter F, for Fence. At the First fence point prompt, pick a point on your screen at "A." Continue constructing crossing line segments until you are satisfied with the objects being selected (see the following image).

```
Command: E (For ERASE)
Select objects: F (For Fence)
First fence point: (Pick at "A")
Specify endpoint of line or [Undo]: (Pick at "B")
Specify endpoint of line or [Undo]: (Pick at "C")
Specify endpoint of line or [Undo]: (Pick at "D")
Specify endpoint of line or [Undo]: (Pick at "E")
Specify endpoint of line or [Undo]: (Press ENTER to create the
    selection set)
```

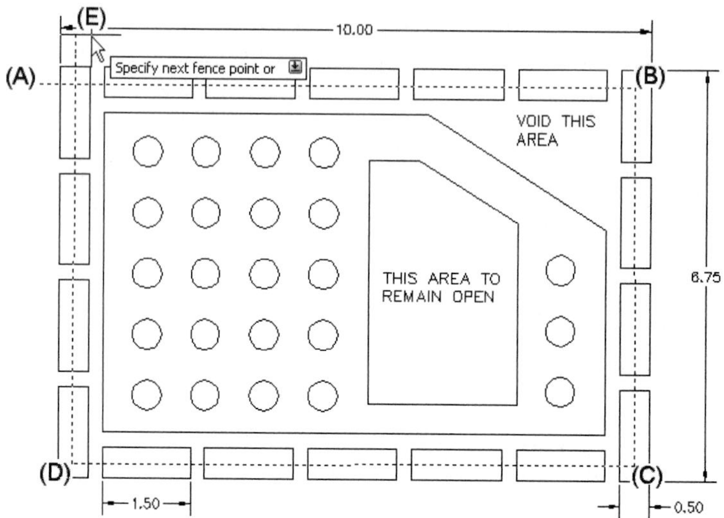

Figure 3.45

The result of selecting objects by a fence is illustrated in the following image. Notice that objects touched by the fence are highlighted. Pressing ENTER performs the ERASE operation. Before continuing on to the next step, issue the U (UNDO) command, which negates the previous ERASE operation.

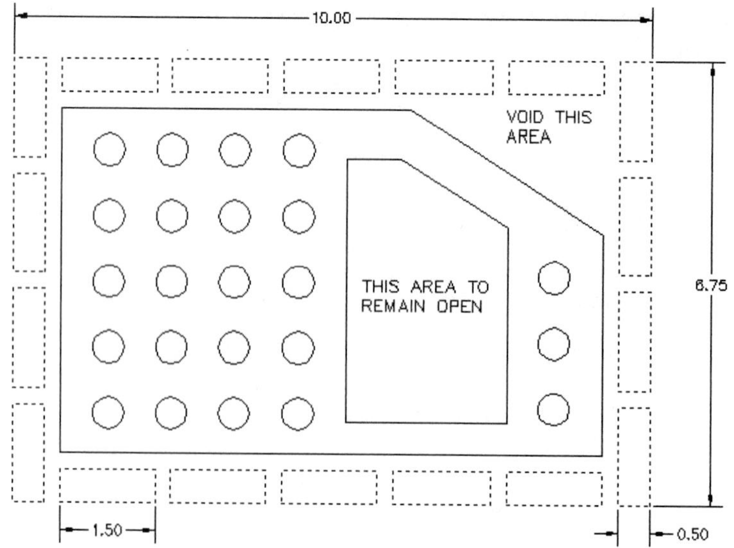

Figure 3.46

Step 6

When situations require you to select all objects in the entire database of a drawing, the All option is very efficient. Activate the ERASE command again. At the Select objects prompt, enter All. Notice that in the following image all objects are selected by this option.

```
Command: E (For ERASE)
Select objects: All
```

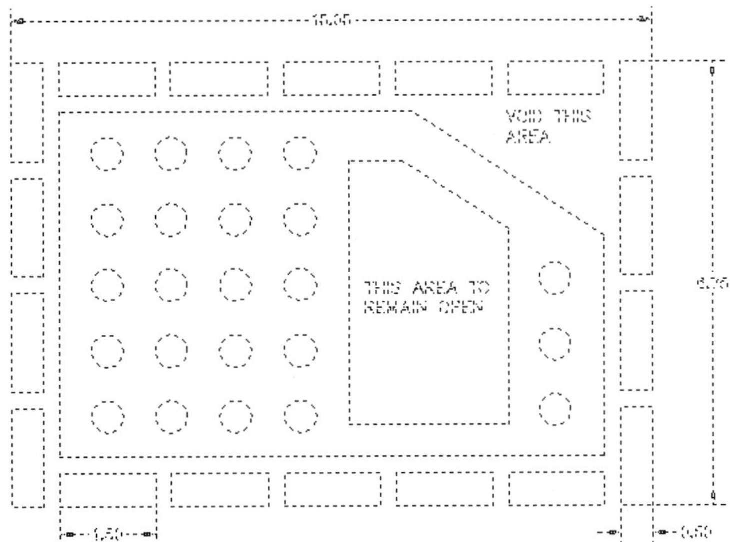

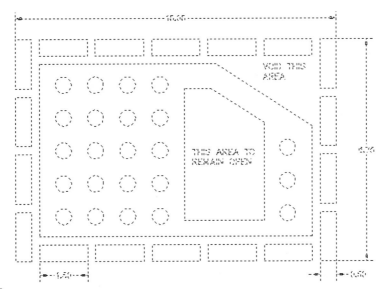

Figure 3.47

Pressing ENTER performs the ERASE operation. Before continuing on to the next step, issue the U (UNDO) command, which negates the previous ERASE operation.

Note: The All option also selects objects even on layers that are turned off. The All option does not select objects on layers that are frozen.

Step 7

Before performing another ERASE operation on a number of objects, activate the Layer Properties Manager dialog box, select the "Circles" layer, and click the lock icon, as shown in the following image. This operation locks the "Circles" layer.

Status	Name	On	Freeze	Lock	Color	Linetype	Lineweight	Plot Style	Plot	New VF
	0				white	CONTINUOUS	—— Default	Color_7		
	Circles				white	CONTINUOUS	—— Default	Color_7		
	Defpoints				white	CONTINUOUS	—— Default	Color_7		
	Dimension				white	CONTINUOUS	—— Default	Color_7		
	Lines				white	CONTINUOUS	—— Default	Color_7		
	OBJECT				white	CONTINUOUS	—— Default	Color_7		
	Rectangles				white	CONTINUOUS	—— Default	Color_7		
	Text				white	CONTINUOUS	—— Default	Color_7		

Figure 3.48

To see what effect this has on selecting objects, activate the ERASE command again. At the Select objects prompt, pick a point on the blank part of your screen at "A." At the Specify opposite corner prompt, pick a point on your screen at "B." You have once again selected a number of objects by a green crossing box (see the following image).

```
Command: E (For ERASE)
Select objects: (Pick a point at "A")
Specify opposite corner: (Pick a point at "B")
20 were on a locked layer.
```

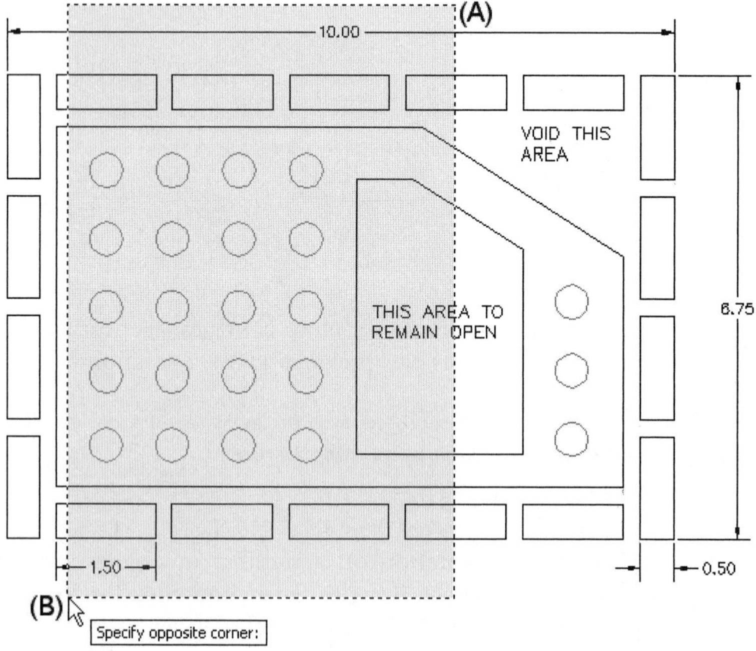

Figure 3.49

The result of selecting objects by a crossing window is illustrated in the following image. Notice that even though a group of circles was completely surrounded by the crossing window, the circles do not highlight because they belong to a locked layer. Pressing ENTER performs the ERASE operation. Before continuing on to the next step, issue the U (UNDO) command, which negates the previous ERASE operation.

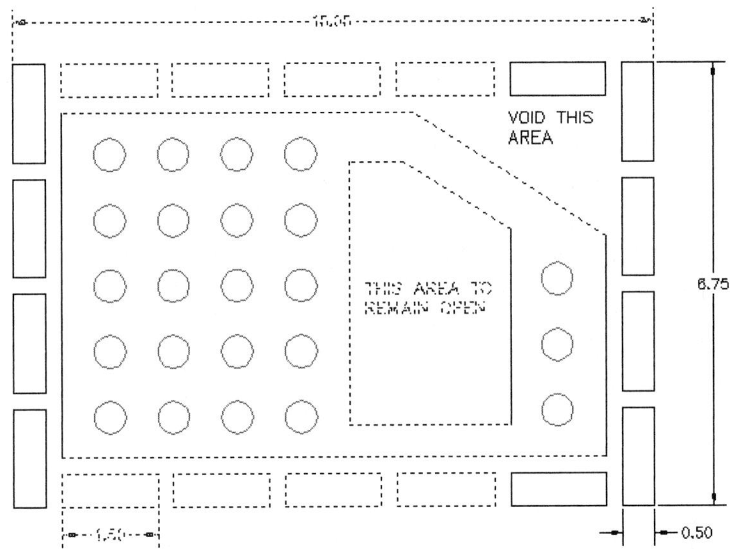

Figure 3.50

Step 8

This final step illustrates the use of the ZOOM-All and ZOOM-Extents commands and how they differ. First activate the ZOOM command and use the All option (a button is available for performing this operation).

```
Command: Z (For ZOOM)
Specify corner of window, enter a scale factor (nX or nXP), or
[All/Center/Dynamic/Extents/Previous/Scale/Window/Object] <real
    time>: A (For All)
```

The result is illustrated in the following image. Performing a ZOOM-All zooms the display based on the current drawing limits (the limits of this drawing have been established as 0,0 for the lower-left corner and 16,10 for the upper-right corner).

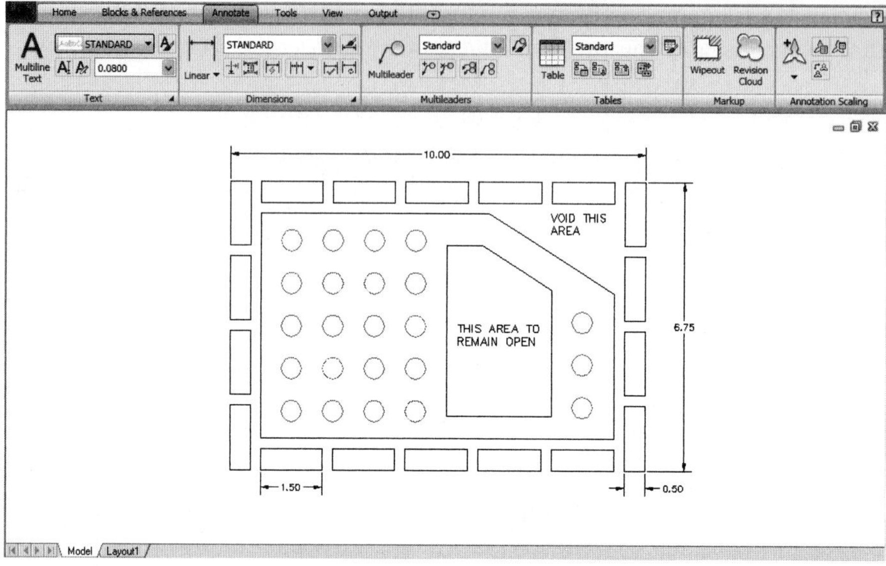

Figure 3.51

Now activate the ZOOM command and use the Extents option (a button is also available for performing this operation).

```
Command: Z (For ZOOM)
Specify corner of window, enter a scale factor (nX or nXP), or
[All/Center/Dynamic/Extents/Previous/Scale/Window/Object] <real
    time>: E (For Extents)
```

The result is illustrated in the following image. This zoom is performed to the limits that contain all the objects currently in the drawing.

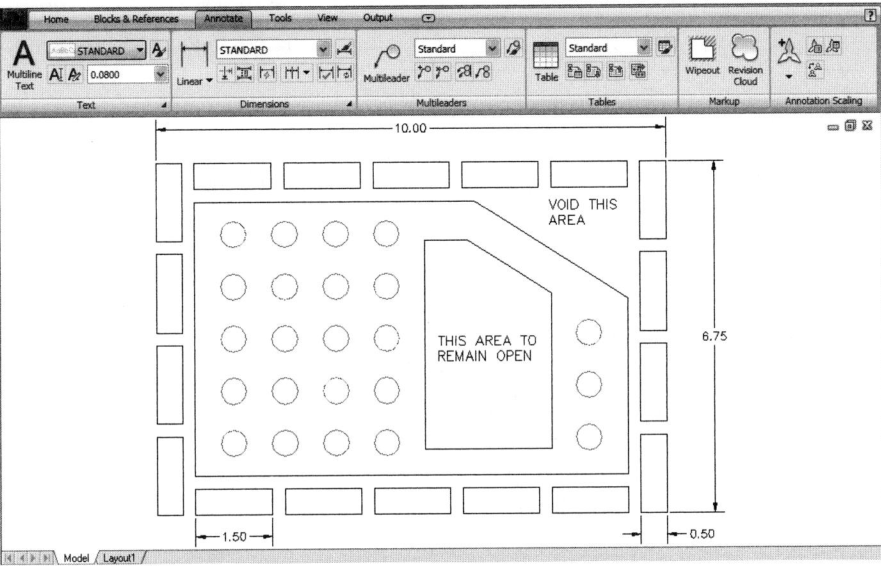
10.00
VOID THIS AREA
THIS AREA TO REMAIN OPEN
6.75
1.50
0.50

Figure 3.52

Modifying Your Drawings

The heart of any CAD system is its ability to modify and manipulate existing geometry, and AutoCAD is no exception. Many modify commands relieve the designer of drudgery and mundane tasks, and this allows more productive time for conceptualizing the design. This chapter will break all AutoCAD modify commands down into two separate groupings. The first grouping is called Level I and will cover the MOVE, COPY, SCALE, ROTATE, OFFSET, FILLET, CHAMFER, TRIM EXTEND, and BREAK commands. The second grouping is called Level II and will cover the ARRAY, MIRROR, STRETCH, PEDIT, EXPLODE, LENGTHEN, JOIN, UNDO, and REDO commands. A number of small exercises accompany each command in order to reinforce the importance of its use.

METHODS OF SELECTING MODIFY COMMANDS

As with all commands, you can find the main body of modify commands on the Menu Browser, as shown on the left in the following image, or through the Ribbon and pull-down menu (of the AutoCAD Classic workspace), both shown on the right in the following image.

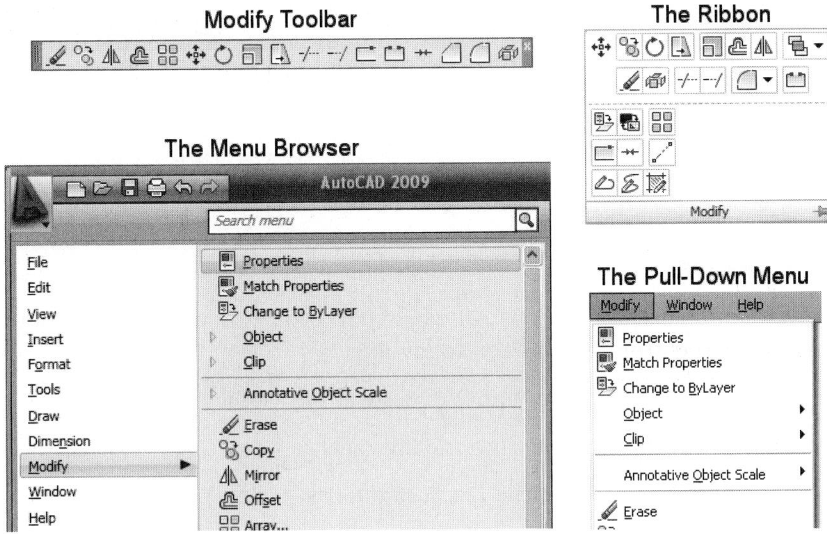

Figure 4.1

You can also enter all modify commands directly from the keyboard either using their entire name or through command aliasing, as in the following examples:

Enter F for the FILLET command

Enter M for the MOVE command

LEVEL I MODIFY COMMANDS

With all the modify commands available in AutoCAD, the following represent beginning, or Level I commands, which you will find yourself using numerous times as you make changes to your drawing. These commands are briefly described in the following table:

Button	Tool	Shortcut	Function
	Move	M	Used for moving objects from one location to another
	Copy	Cp or Co	Used for copying objects from one location to another
	Scale	Sc	Used for increasing or reducing the size of objects
	Rotate	Ro	Used for rotating objects to a different angle
	Fillet	F	Used for rounding off the corners of objects at a specified radius
	Chamfer	Cha	Used to connect two objects with an angled line forming a bevel
	Offset	O	Used for copying objects parallel to one another at a specified distance
	Trim	Tr	Used for partially deleting objects based on a cutting edge

Button	Tool	Shortcut	Function
	Extend	Ex	Used for extending objects based on a boundary edge
	Break	Br	Creates a gap in an object between two specified points
	Break at Point	Br	Breaks an object into two objects at a specified point without a gap present

MOVING OBJECTS

The MOVE command repositions an object or group of objects at a new location. Choose this command from one of the following:

- The Modify toolbar of the AutoCAD Classic workspace
- From the Ribbon > Home Tab > Modify Panel
- The pull-down menu or Menu Browser (Modify > Move)
- The keyboard (M or MOVE)
- By right-clicking the mouse after selecting object

Once the objects to move are selected, AutoCAD prompts the user to select a base point of displacement (where the object is to move from). Next, AutoCAD prompts the user to select a second point of displacement (where the object is to move to), as shown in the following image.

```
Command: M (For MOVE)
Select objects: (Select the bed, as shown in the following image)
Select objects: (Press ENTER to continue)
Specify base point or [Displacement] <Displacement>: (Select the
    endpoint of the bed at "A")
Specify second point or <use first point as displacement>: (Mark
    a point at "B")
```

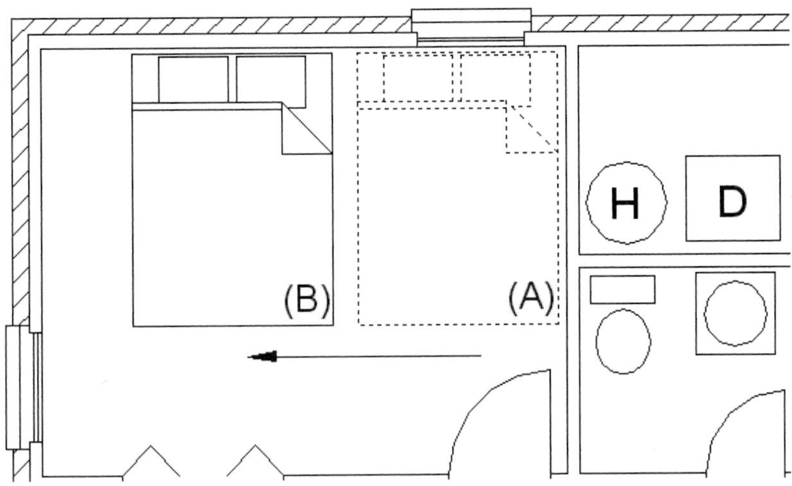

Figure 4.2

Try It!: Open the drawing file 04_Move. The slot as shown on the left in the following image is incorrectly positioned; it needs to be placed 1.00 unit away from the left edge of the object. You can use the MOVE command in combination with a polar coordinate or Direct Distance mode to perform this operation. Use this illustration and the following command sequence for performing this operation.

```
Command: M (For MOVE)
Select objects: (Select the slot and all centerlines)
Select objects: (Press ENTER to continue)
Specify base point or [Displacement] <Displacement>: Cen (For
    Osnap Center)
of (Select the edge of arc "A")
Specify second point or <use first point as displacement>: (Turn
    Polar on, move your cursor to the right, and type .50 at the
    command prompt)
```

As the slot is moved to a new position with the MOVE command, the horizontal dimension will reflect the correct distance from the edge of the object to the centerline of the arc, as shown in the following image.

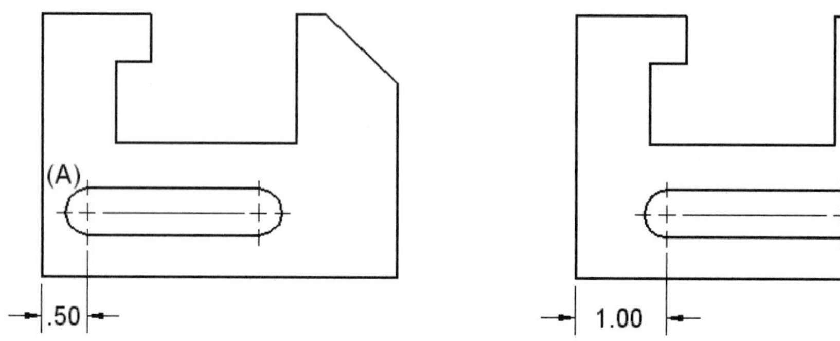

Figure 4.3

PRESS AND DRAG MOVE

If accuracy is not important and you simply need to move an object or group of objects to a new approximate location, you can use a press and drag technique. First, select the objects at the command prompt. Then, press and hold down the left mouse button on one of the highlighted objects (not one of the blue grips), and drag the objects to the new location.

COPYING OBJECTS

The COPY command is used to duplicate an object or group of objects. Choose this command from one of the following:

- The Modify toolbar of the AutoCAD Classic workspace
- From the Ribbon > Home Tab > Modify Panel
- The pull-down menu or Menu Browser (Modify > Copy)
- The keyboard (CP or COPY)
- By right-clicking the mouse after selecting object

Once the COPY command is executed, the multiple copy mode is on. To duplicate numerous objects while staying inside the COPY command, simply keep picking new second points of displacement and the objects will copy to these new locations, as shown in the following image. Once the copy is completed, press ENTER or ESC to exit.

```
Command: CP (For COPY)
Select objects: (Select the chair to copy)
Select objects: (Press ENTER to continue)
Specify base point or [Displacement/mode} <Displacement>: (Pick a
    reference point for the copy operation)
Specify second point or <use first point as displacement>: (Pick
    a location for the first copy)
Specify second point or [Exit/Undo] <Exit>: (Pick a location for
    the second copy)
Specify second point or [Exit/Undo] <Exit>: (Pick a location for
    the third copy)
Specify second point or [Exit/Undo] <Exit>: (Press ENTER or ESC to
    exit this command)
```

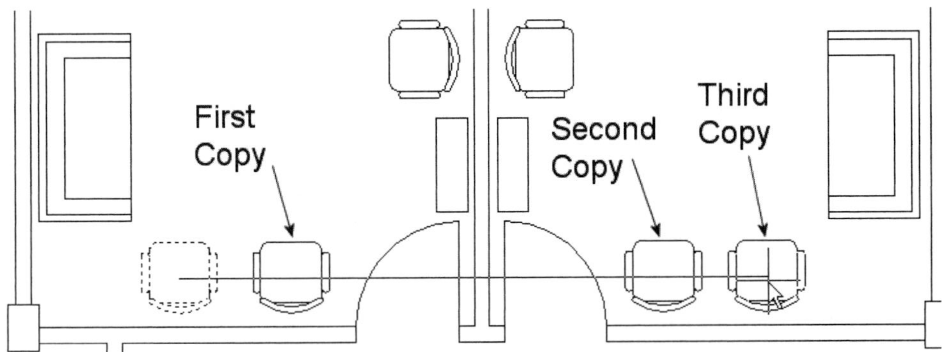

Figure 4.4

Try It!: Open the drawing file 04_Copy Multiple. Follow the command sequence in the previous example to copy the three holes multiple times. Use the intersection of "A" as the base point for the copy. Then copy the three holes to the intersections located at "B," "C," "D," "E," "F," "G," "H," and "J," as shown on the left in the following image. The results are illustrated on the right in the following image.

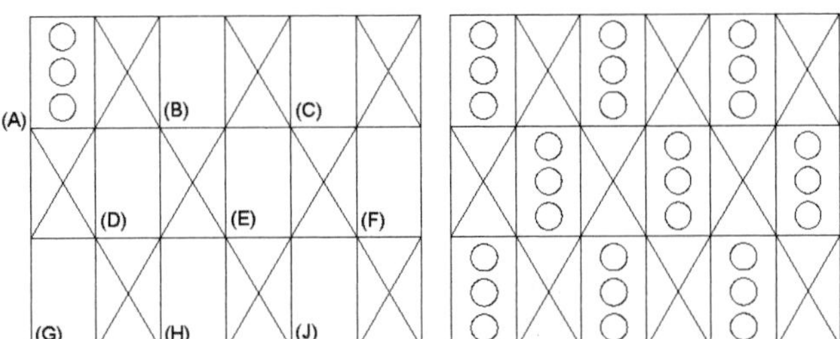

Figure 4.5

PRESS AND DRAG COPY

As with the MOVE command, you can also use the press and drag technique to copy objects to a new approximate location. First, select the objects at the command prompt. Then, press and hold down the right mouse button on one of the highlighted objects (not one of the blue grips), and drag the objects to the new location. When you release the mouse button a menu displays, as shown in the following image. Select the Copy Here item to copy the item or group of items.

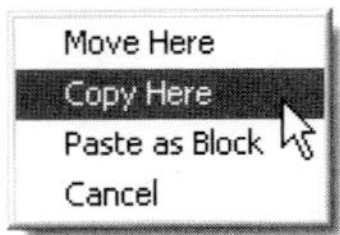

Figure 4.6

SCALING OBJECTS

Use the SCALE command to change the overall size of an object. The size may be larger or smaller in relation to the original object or group of objects. The SCALE command requires a base point and scale factor to complete the command. Choose this command from one of the following:

- The Modify toolbar of the AutoCAD Classic workspace
- From the Ribbon > Home Tab > Modify Panel
- The pull-down menu or Menu Browser (Modify > Scale)
- The keyboard (SC or SCALE)
- By right-clicking the mouse after selecting object

Try It!: Open the drawing file 04_Scale1. With a base point at "A" and a scale factor of 0.50, the results of using the SCALE command on a group of objects are shown in the following image.

```
Command: SC (For SCALE)
Select objects: All
Select objects: (Press ENTER to continue)
Specify base point: (Select the endpoint of the line at "A")
Specify scale factor or [Copy/Reference] <1.0000>: 0.50
```

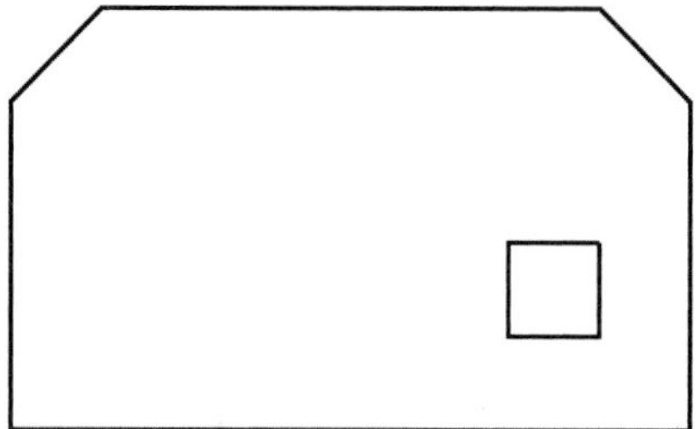

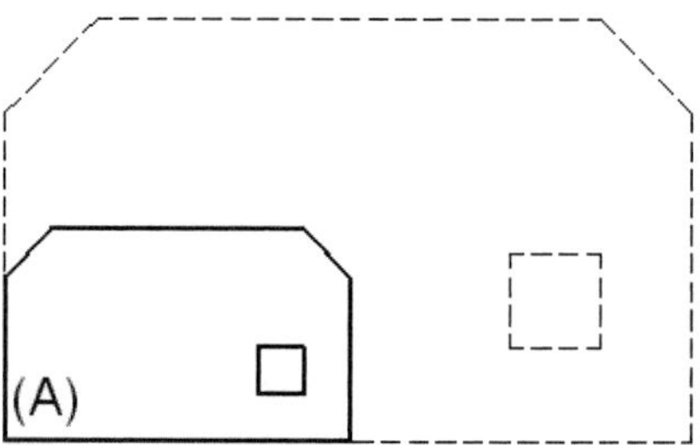

Figure 4.7

Try It!: Open the drawing file 04_Scale2. The example in the following image shows the effects of identifying a new base point in the center of the object.

```
Command: SC (For SCALE)
Select objects: All
Select objects: (Press ENTER to continue)
Specify base point: (Pick a point near "A")
Specify scale factor or [Copy/Reference] <1.0000>: 0.40
```

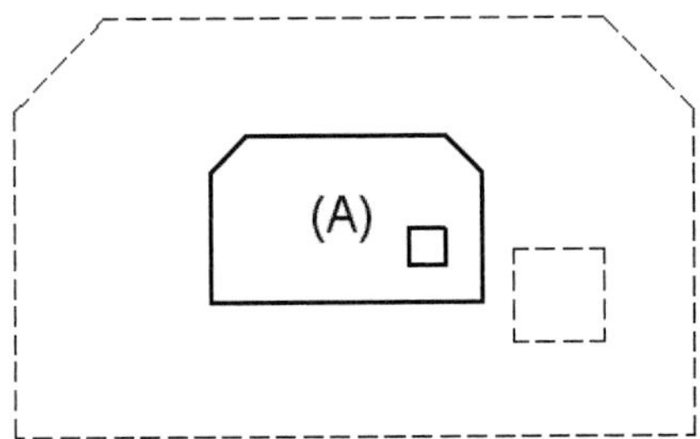

Figure 4.8

Note: After identifying the base point for the scaling operation, you can use the Copy option to create a scaled copy of the objects you are scaling.

SCALE—REFERENCE

Suppose you are given a drawing that has been scaled down in size. However, no one knows what scale factor was used. You do know what one of the distances should be. In this special case, you can use the Reference option of the SCALE command to identify endpoints of a line segment that act as a reference length. Entering a new length value could increase or decrease the entire object proportionally.

Try It!: Open the drawing file 04_Scale Reference. Study the following image and the following prompts for performing this operation.

```
Command: SC (For SCALE)
Select objects: (Pick a point at "A")
Specify opposite corner: (Pick a point at "B")
Select objects: (Press ENTER to continue)
Specify base point: (Select the edge of the circle to identify
    its center)
Specify scale factor or [Copy/Reference] <1.0000>: R (For
    Reference)
Specify reference length <1.0000>: (Select the endpoint of the
    line at "C")
Specify second point: (Select the endpoint of the line at "D")
Specify new length or [Points] <1.0000>: 2.00
```

Because the length of line "CD" was not known, the endpoints were picked after the Reference option was entered. This provided the length of the line to AutoCAD. The final step to perform was to make the line 2.00 units, which increased the size of the object while also keeping its proportions.

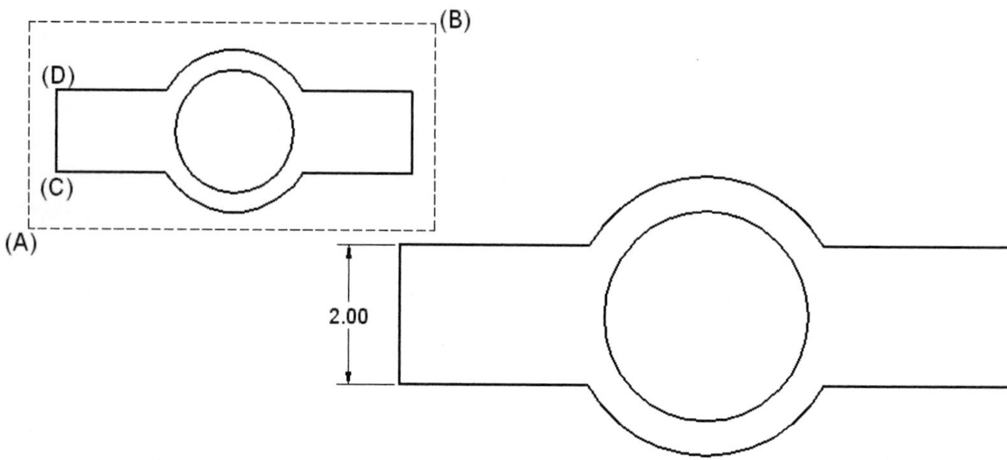

Figure 4.9

ROTATING OBJECTS

The ROTATE command changes the orientation of an object or group of objects by identifying a base point and a rotation angle that completes the new orientation. Choose this command from one of the following:

- The Modify toolbar of the AutoCAD Classic workspace
- From the Ribbon > Home Tab > Modify Panel
- The pull-down menu or Menu Browser (Modify > Rotate)
- The keyboard (RO or ROTATE)
- By right-clicking the mouse after selecting object

The following image shows an object, complete with crosshatch pattern, that needs to be rotated to a 30° angle using point "A" as the base point.

Try It!: Open the drawing file 04_Rotate. Use the following prompts and image to perform the rotation.

```
Command: RO (For ROTATE)
Current positive angle in UCS: ANGDIR=counterclockwise ANGBASE=0
Select objects: All
Select objects: (Press ENTER to continue)
Specify base point: (Select the endpoint of the line at "A")
Specify rotation angle or [Copy/Reference] <0>: 30
```

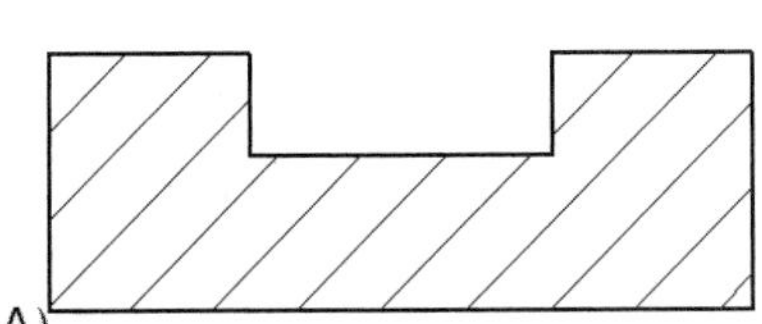

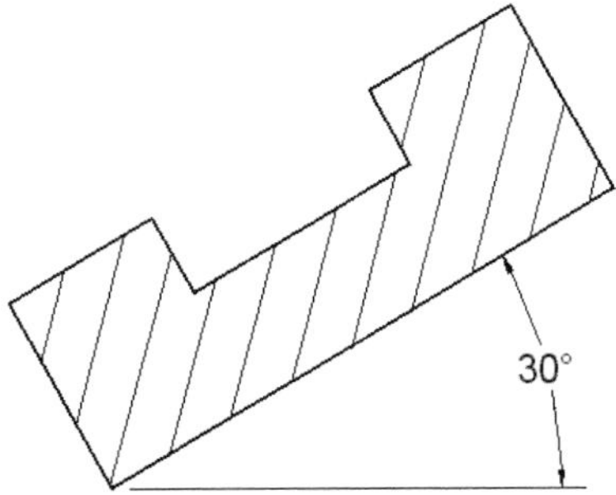

Figure 4.10

Note: After identifying the base point for the rotating operation, you can use the Copy option to create a rotated copy of the objects you are rotating.

ROTATE—REFERENCE

At times it is necessary to rotate an object to a desired angular position. However, this must be accomplished even if the current angle of the object is unknown. To maintain the accuracy of the rotation operation, use the Reference option of the ROTATE command. The following image shows an object that needs to be rotated to the 30°-angle position. Unfortunately, we do not know the angle in which the object currently lies. Entering the Reference angle option and identifying two points create a known angle of reference. Entering a new angle of 30° rotates the object to the 30° position from the reference angle.

Try It!: Open the drawing file 04_Rotate Reference. Use the following prompts and image to accomplish this.

```
Command: RO (For ROTATE)
Current positive angle in UCS: ANGDIR=counterclockwise ANGBASE=0
Select objects: (Select the object in the following image)
Select objects: (Press ENTER to continue)
Specify base point: (Pick either the edge of the circle or two
    arc segments to locate the center)
Specify rotation angle or [Copy/Reference] <0>: R (For Reference)
Specify the reference angle <0>: (Pick either the edge of the
    circle or two arc segments to locate the center)
Specify second point: Mid
of (Select the line at "A" to establish the reference angle)
Specify the new angle or [Points] <0>: 30
```

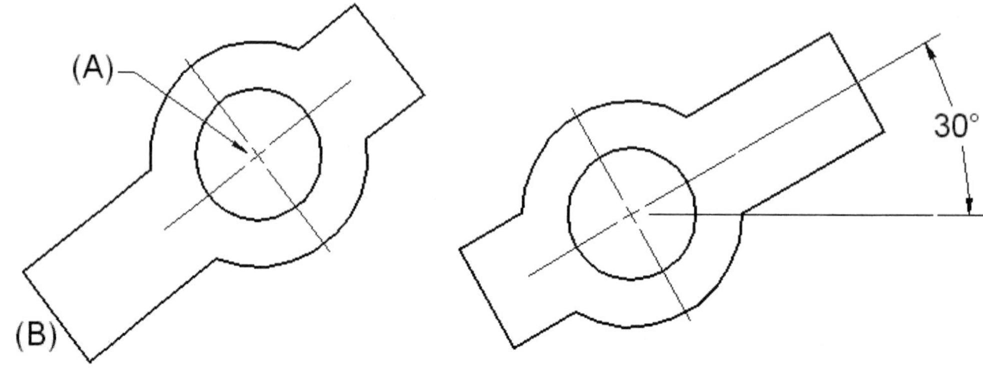

Figure 4.11

CREATING FILLETS AND ROUNDS

Many objects require highly finished and polished surfaces consisting of extremely sharp corners. Fillets and rounds represent the opposite case, where corners are rounded off, either for ornamental purposes or as required by design. Generally a fillet consists of a rounded edge formed in the corner of an object, as illustrated in the following image. A round is formed at an outside corner. Fillets and rounds are primarily used where objects are cast or made from poured metal. The metal forms more easily around a shape that has rounded corners instead of sharp corners, which usually break away. Some drawings have so many fillets and rounds that a note is used to convey the size of them all, similar to "All Fillets and Rounds .125 Radius."

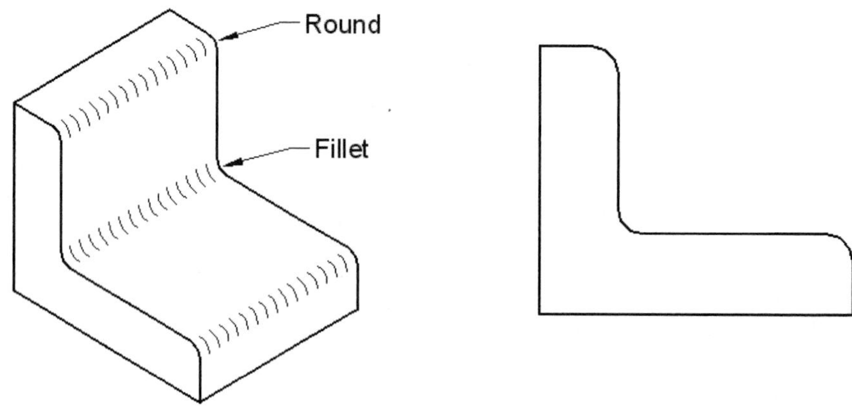

Figure 4.12

AutoCAD provides the FILLET command, which allows you to enter a radius followed by the selection of two lines. The result is a fillet of the specified radius to the two lines selected. The two lines are also automatically trimmed, leaving the radius drawn from the endpoint of one line to the endpoint of the other line. Choose this command from one of the following:

- The Modify toolbar of the AutoCAD Classic workspace
- From the Ribbon > Home Tab > Modify Panel
- The pull-down menu or Menu Browser (Modify > Fillet)
- The keyboard (F or FILLET)

FILLETING BY RADIUS

Illustrated in the following image is an example of setting a radius in the FILLET command for creating rounded-off corners.

Try It!: Open the drawing file 04_Fillet. Follow the illustration on the left in the following image and command sequence below to place fillets at the three corner locations.

```
Command: F (For FILLET)
Current settings: Mode = TRIM, Radius = 0.0000
Select first object or [Undo/Polyline/Radius/Trim/Multiple]: R
    (For Radius)
Specify fillet radius <0.0000>: 0.25
Select first object or [Undo/Polyline/Radius/Trim/Multiple]:
    (Select at "A")
Select second object or shift-select to apply corner: (Select
    at "B")
Command:
(Press ENTER to re-execute this command)
```

Repeat this procedure for creating additional fillets using lines "BC" and "CD." When finished, your display should appear similar to the illustration on the right in the following image.

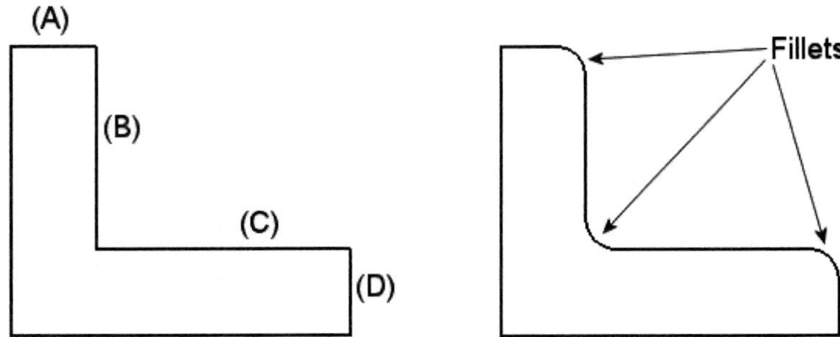

Figure 4.13

FILLET AS A CORNERING TOOL

A very productive feature of the FILLET command is its use as a cornering tool. To accomplish this, set the fillet radius to a value of 0. This produces a corner out of two non-intersecting objects.

Try It!: Open the drawing file 04_Fillet Corner1. Follow the illustration on the left in the following image and the command sequence below for performing this task.

```
Command: F (For FILLET)
Current settings: Mode = TRIM, Radius = 0.5000
Select first object or [Undo/Polyline/Radius/Trim/Multiple]: R
    (For Radius)
Enter fillet radius <0.5000>: 0
Select first object or [Undo/Polyline/Radius/Trim/Multiple]:
    (Select line "A")
Select second object or shift-select to apply corner: (Select
    line "B")
```

Repeat the procedure for the remaining two corners using lines "BC" and "CD." When finished, your display should appear similar to the illustration on the right in the following image.

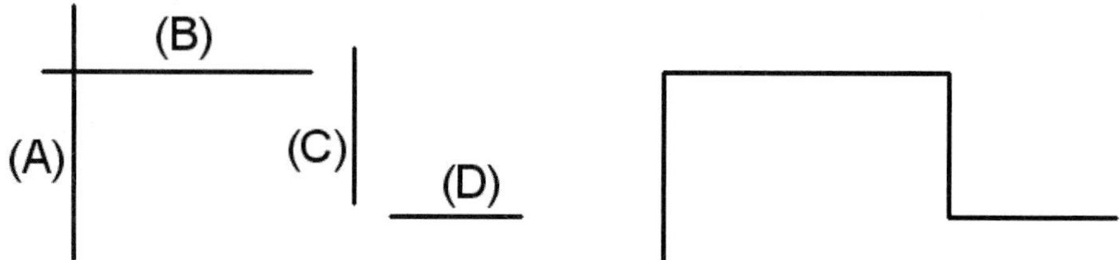

Figure 4.14

Tip: Even with a fillet radius set to a positive value, such as .50, you can easily apply a corner to two lines by holding down the SHIFT key when picking the second object. The use of the SHIFT key in this example temporarily sets the fillet radius to zero. Releasing the SHIFT key sets the fillet radius back to the current value, in this case .50.

PERFORMING MULTIPLE FILLETS

Since the filleting of lines is performed numerous times, a multiple option is available that automatically repeats the FILLET command. This option can be found by picking Multiple from the Cursor menu or by typing M for Multiple at the command prompt.

Try It!: Open the drawing file 04_Fillet Multiple. Activate the FILLET command and verify that the radius is set to 0. Use the Multiple option to make the command repeat. Then click on the corners until your object appears similar to the illustration on the right in the following image. If you make a mistake by picking two lines incorrectly, type U to undo this operation and still remain in the FILLET command.

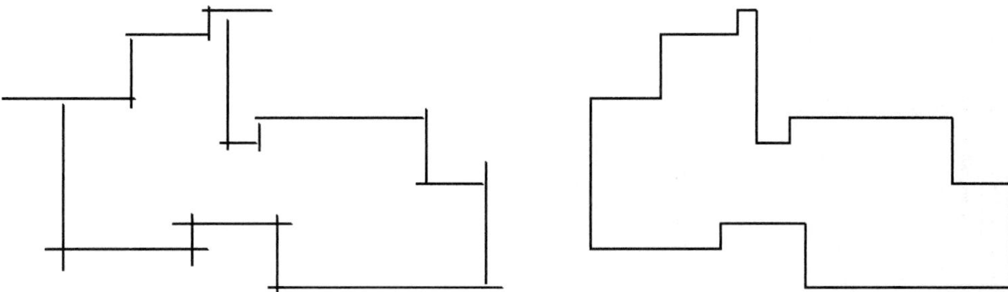

Figure 4.15

FILLETING POLYLINES

In the previous examples of using the FILLET command, you had to pick individual line segments in order to produce one rounded corner. You also had to repeat these picks for additional rounded corners. If the object you are filleting is a polyline, you can have the FILLET command round off all corners of this polyline in a single pick.

Try It!: Open the drawing file 04_Fillet Pline. Using the FILLET command on a polyline object produces rounded edges at all corners of the polyline in a single operation. Follow the illustration in the following image and the command sequence below for performing this task.

```
Command: F (For FILLET)
Current settings: Mode = TRIM, Radius = 0.0000
Select first object or [Undo/Polyline/Radius/Trim/Multiple]: R
    (For Radius)
Enter fillet radius <0.0000>: 0.25
Select first object or [Undo/Polyline/Radius/Trim/Multiple]: P
    (For Polyline)
Select 2D polyline: (Select the polyline at "A")
```

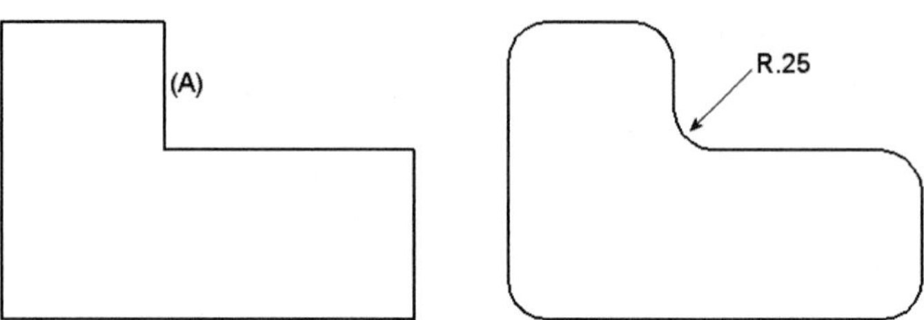

Figure 4.16

FILLETING PARALLEL LINES

Filleting two parallel lines, as shown in the following image, automatically constructs a semicircular arc object connecting both lines at their endpoints. When performing this operation, it does not matter what the radius value is set to.

Try It!: Open the drawing file 04_Fillet Parallel. Use the illustration on the left in the following image and the command sequence below for performing this task.

```
Command: F (For FILLET)
Current settings: Mode = TRIM, Radius = 1.0000
Select first object or [Undo/Polyline/Radius/Trim/Multiple]:
     (Select line "A")
Select second object or shift-select to apply corner: (Select
     line "B")
```

Continue filleting the remaining parallel lines to complete all slots. Quicken the process by using the Multiple option when prompted to "Select first object."

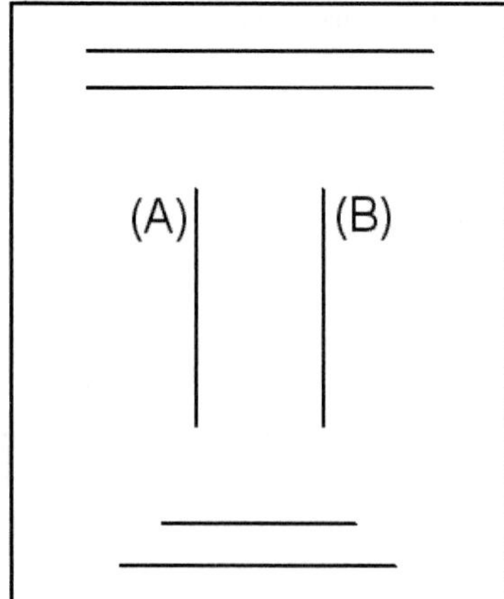

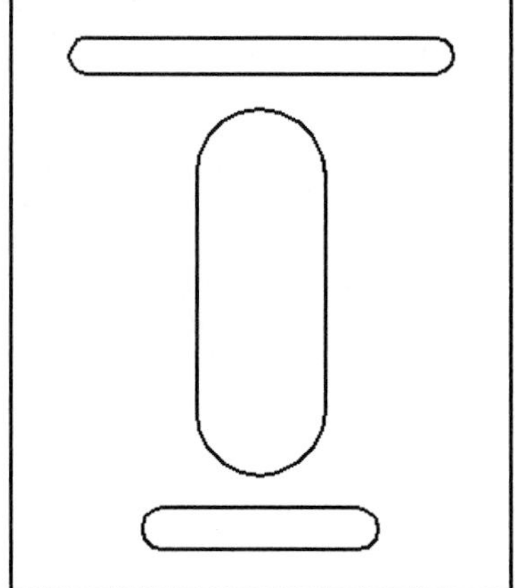

Figure 4.17

CREATING CHAMFERS

Chamfers represent a way to finish a sharp corner of an object. The CHAMFER command produces an inclined surface at an edge of two intersecting line segments. Distances determine how far from the corner the chamfer is made. The following image illustrates two examples of chamfered edges; one edge is created from unequal distances, while the other uses equal distances.

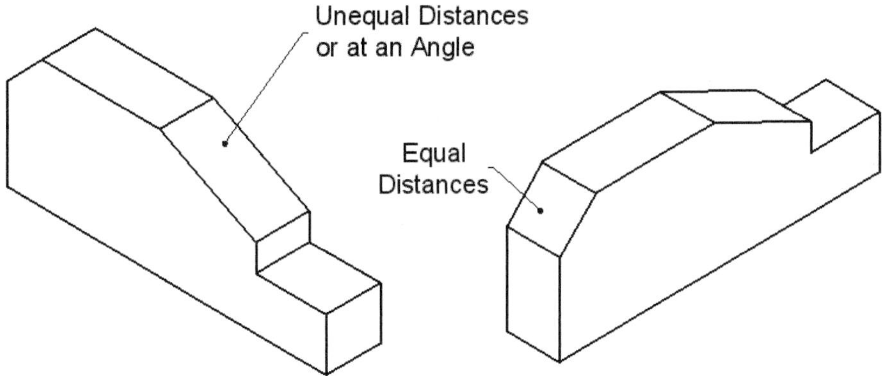

Figure 4.18

The CHAMFER command is designed to draw an angle across a sharp corner given two chamfer distances. Choose this command from one of the following:

- The Modify toolbar of the AutoCAD Classic workspace
- From the Ribbon > Home Tab > Modify Panel
- The pull-down menu or Menu Browser (Modify > Chamfer)
- The keyboard (CHA or CHAMFER)

CHAMFER BY EQUAL DISTANCES

The most popular chamfer involves a 45° angle, which is illustrated in the following image. You can control this angle by entering two equal distances.

Try It!: Open the drawing file 04_Chamfer Distances. In the example in the following image, if you specify the same numeric value for both chamfer distances, a 45°-angled chamfer is automatically formed. As long as both distances are the same, a 45° chamfer will always be drawn. Study the illustration in the following image and the following prompts:

```
Command: CHA (For CHAMFER)
(TRIM mode) Current chamfer Dist1 = 0.5000, Dist2 = 0.5000
Select first line or [Undo/Polyline/Distance/Angle/Trim/mEthod/
    Multiple]: D (For Distance)
Specify first chamfer distance <0.5000>: 0.15
```

```
Specify second chamfer distance <0.1500>: (Press ENTER to accept
    the default)
Select first line or [Undo/Polyline/Distance/Angle/Trim/mEthod/
    Multiple]: (Select the line at "A")
Select second line or shift-select to apply corner: (Select the
    line at "B")
```

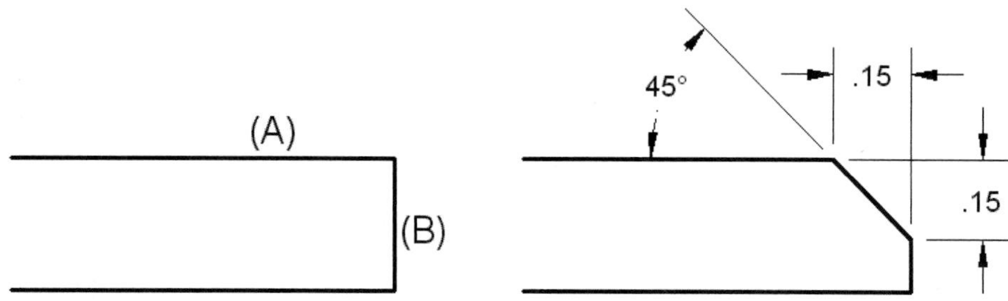

Figure 4.19

Note: When both chamfer distances are set to 0 (zero) and two edges are selected, the effects are identical to setting the fillet radius to 0 (zero); the CHAMFER command is used here as a cornering tool.

CHAMFER BY ANGLE

Another technique of constructing a chamfer is when one distance and the angle are given. When the chamfer is made up of an angle other than 45°, it is commonly referred to as a beveled edge.

Try It!: Open the drawing file 04_Chamfer Angle. The following image illustrates the use of the CHAMFER command by setting one distance and identifying an angle.

```
Command: CHA (For CHAMFER)
(TRIM mode) Current chamfer Dist1 = 0.5000, Dist2 = 0.5000
Select first line or [Undo/Polyline/Distance/Angle/Trim/mEthod/
    Multiple]: A (For Angle)
Specify chamfer length on the first line <0.1500>: 0.15
Specify chamfer angle from the first line <60>: 60
Select first line or [Undo/Polyline/Distance/Angle/Trim/mEthod/
    Multiple]: (Select the line at "A")
Select second line or shift-select to apply corner: (Select the
    line at "B")
```

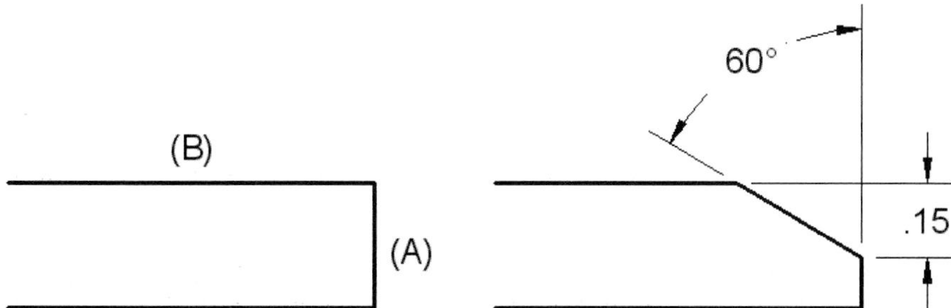

Figure 4.20

CHAMFERING A POLYLINE

When working with polyline objects, you have the opportunity to select only one of the edges of the polyline. All edges of the polyline will be chamfered to the specified distances or angle.

Try It!: Open the drawing file 04_Chamfer Pline. Because a polyline consists of numerous segments representing a single object, using the CHAMFER command with the Polyline option produces corners throughout the entire polyline, as shown in the following image.

```
Command: CHA (For CHAMFER)
(TRIM mode) Current chamfer Dist1 = 0.00, Dist2 = 0.00
Select first line or [Undo/Polyline/Distance/Angle/Trim/mEthod/
    Multiple]: D (For Distance)
Specify first chamfer distance <0.00>: 0.50
Specify second chamfer distance <0.50>: (Press ENTER to accept the
    default)
Select first line or [Undo/Polyline/Distance/Angle/Trim/mEthod/
    Multiple]: P (For Polyline)
Select 2D Polyline: (Select the Polyline at "A")
```

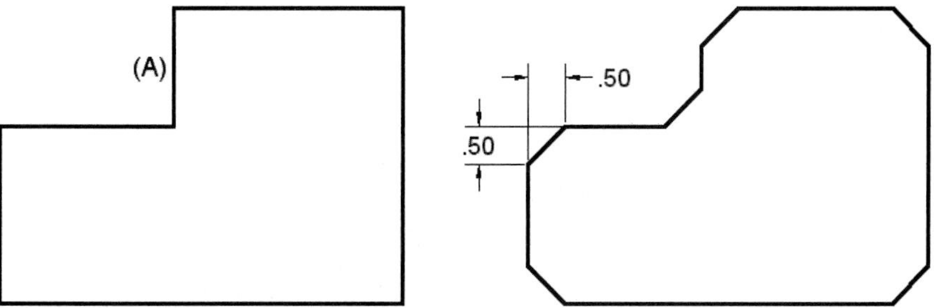

Figure 4.21

Note: A Multiple option of the CHAMFER command allows you to chamfer edges that share the same chamfer distances without exiting and reentering the command.

CHAMFER PROJECT—BEAM

The following Try It! exercise involves the chamfering of various corners at different distance settings.

Try It!: Open the drawing file 04_Chamfer Beam. Using the illustration provided in the following image, follow these directions: Apply equal chamfer distances of 0.25 units to corners "AB," "BC," "DE," and "EF." Set new equal chamfer distances to 0.50 units and apply these distances to corners "GH" and "JK." Set a new first chamfer distance to 1.00; set a second chamfer distance to 0.50 units. Apply the first chamfer distance to line "L" and the second chamfer distance to line "H." Complete this object by applying the first chamfer distance to line "M" and the second chamfer distance to line "K."

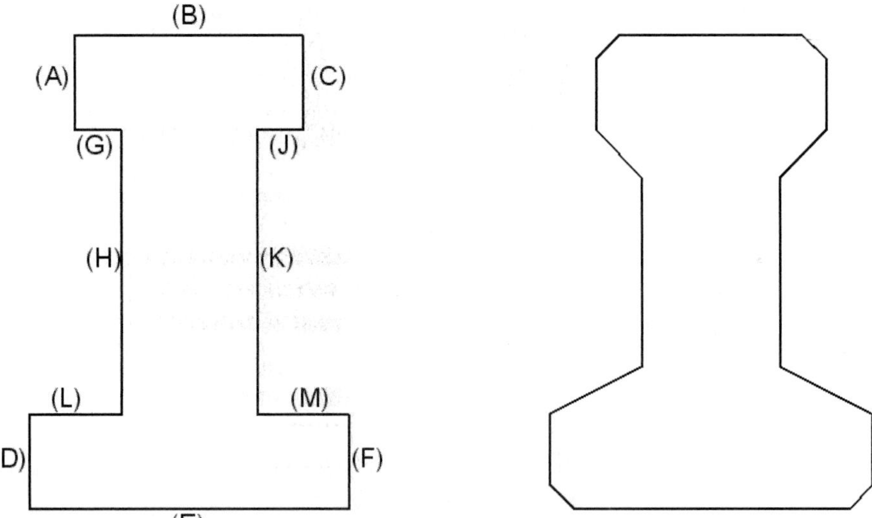

Figure 4.22

OFFSETTING OBJECTS

The OFFSET command is commonly used for creating a copy of one object that is parallel to another. Choose this command from one of the following:

- The Modify toolbar of the AutoCAD Classic workspace
- From the Ribbon > Home Tab > Modify Panel
- The pull-down menu or Menu Browser (Modify > Offset)
- The keyboard (O or OFFSET)

OFFSETTING USING A THROUGH POINT

One method of offsetting is to identify a point to offset through, called a through point. Once an object is selected to offset, a through point is identified. The selected object offsets to the point shown in the following image.

Try It!: Open the drawing file 04_Offset Through. Refer to the following image and command sequence to use this method of the OFFSET command.

```
Command: O (For OFFSET)
Current settings: Erase source=No Layer=Source OFFSETGAPTYPE=0
Specify offset distance or [Through/Erase/Layer] <Through>: T
    (For Through)
Select object to offset or [Exit/Undo] <Exit>: (Select the line
    at "A")
Specify through point or [Exit/Multiple/Undo] <Exit>: Nod (For
    Osnap Node)
of (Select the point at "B")
Select object to offset or [Exit/Undo] <Exit>: (Press ENTER to
    exit this command)
```

Figure 4.23

OFFSETTING BY A DISTANCE

Another method of offsetting is by a specified offset distance, as shown in the following image, where the objects need to be duplicated at a set distance from existing geometry. The COPY command could be used for this operation; a better command would be OFFSET. This allows you to specify a distance and a side for the offset to occur. The result is an object parallel to the original object at a specified distance. All objects in the following image need to be offset 0.50 toward the inside of the original object.

Try It!: Open the drawing file 04_Offset Shape. See the command sequence and following image to perform this operation.

```
Command: O (For OFFSET)
Current settings: Erase source=No Layer=Source OFFSETGAPTYPE=0
Specify offset distance or [Through/Erase/Layer] <Through>: 0.50
Select object to offset or [Exit/Undo] <Exit>: (Select the
    horizontal line at "A")
Specify point on side to offset or [Exit/Multiple/Undo] <Exit>:
    (Pick a point anywhere on the inside near "B")
```

Repeat the preceding procedure for the remaining lines by offsetting them inside the shape, as shown on the left in the following image.

Notice that when all lines were offset, the original lengths of all line segments were maintained. Because all offsetting occurs inside, the segments overlap at their intersection points, as shown in the middle of the following image. In one case, at "A" and "B," the lines did not meet at all. The FILLET command is used to edit all lines to form a sharp corner. You can accomplish this by setting the fillet radius to 0.

```
Command: F (For FILLET)
Current settings: Mode = TRIM, Radius = 0.5000
Select first object or [Undo/Polyline/Radius/Trim/Multiple]: R
    (For Radius)
Enter fillet radius <0.5000>: 0
Select first object or [Undo/Polyline/Radius/Trim/Multiple]:
    (Select line "A")
Select second object: (Select line "B")
```

Repeat the above procedure for the remaining lines, as shown in the middle in the following image.

Using the OFFSET command along with the FILLET command produces the result shown on the right the following image. The fillet radius must be set to a value of 0 for this special effect.

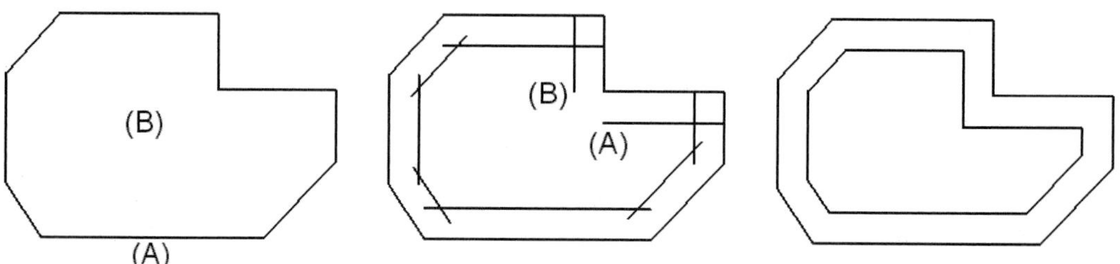

Figure 4.24

PERFORMING MULTIPLE OFFSETS

If you know ahead of time that you will be offsetting the same object the same preset distance, you can use the Multiple option of the OFFSET command to work more efficiently.

Try It!: Open the drawing file 04_Offset Multiple. See the command sequence and following image to perform this operation.

```
Command: O (For OFFSET)
Current settings: Erase source=No Layer=Source OFFSETGAPTYPE=0
Specify offset distance or [Through/Erase/Layer] <Through>: 0.40
Select object to offset or [Exit/Undo] <Exit>: (Pick vertical
    line "A")
Specify point on side to offset or [Exit/Multiple/Undo] <Exit>: M
    (For Multiple)
```

```
Specify point on side to offset or [Exit/Undo] <next object>:
    (Pick at "B")
Specify point on side to offset or [Exit/Undo] <next object>:
    (Pick at "B")
Specify point on side to offset or [Exit/Undo] <next object>:
    (Pick at "B")
Specify point on side to offset or [Exit/Undo] <next object>:
    (Press enter)
Select object to offset or [Exit/Undo] <Exit>: (Press ENTER to
    exit)
```

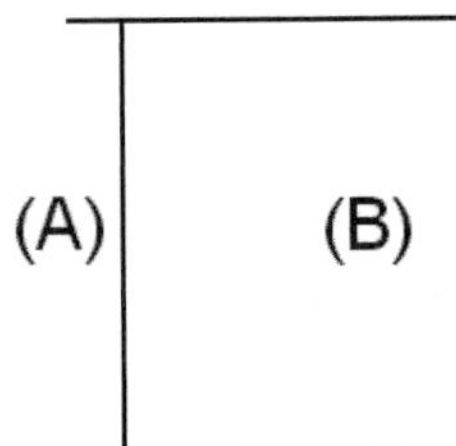

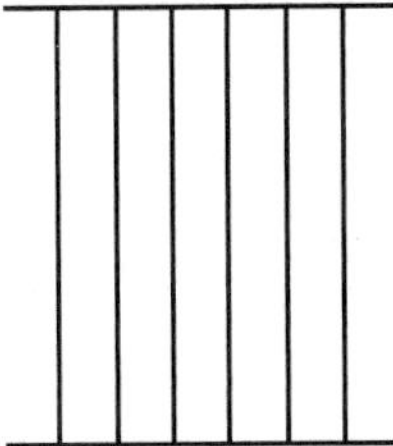

Figure 4.25

OTHER OFFSET OPTIONS

Other options of the OFFSET command include Erase and Layer. When the Erase option is used, the original object you select to offset is erased after the offset copy is made. When using the Layer option, the object being offset can take on the layer properties of the source object or can be based on the current layer. When using the source, the offset copy takes on the same layer as the source object you pick with offsetting. You could also make a new layer current. Using the Current Layer mode when offsetting changes all offset copies to the current layer. These extra offset modes allow you more flexibility with using this command.

TRIMMING OBJECTS

Use the TRIM command to partially delete an object or a group of objects based on a cutting edge. Choose this command from one of the following:

- The Modify toolbar of the AutoCAD Classic workspace
- From the Ribbon > Home Tab > Modify Panel
- The pull-down menu or Menu Browser (Modify > Trim)
- The keyboard (TR or TRIM)

SELECTING INDIVIDUAL CUTTING EDGES

As illustrated on the left in the following image, the four dashed lines are selected as cutting edges. Next, segments of the circles are selected to be trimmed between the cutting edges.

Try It!: Open the drawing file 04_Trim Basics. Use the following image and the command sequence below to perform this task.

```
Command: TR (For TRIM)
Current settings: Projection=UCS Edge=None
Select cutting edges ...
Select objects or <select all>: (Select the four dashed lines on
    the left in the following image)
Select objects: (Press ENTER to continue)
Select object to trim or shift-select to extend or [Fence/
    Crossing/Project/Edge/eRase/Undo]: (Select the circle areas
    at "A" through "D")
Select object to trim or shift-select to extend or [Fence/
    Crossing/Project/Edge/eRase/Undo]: (Press ENTER to exit this
    command)
```

The results of performing trim on this object are illustrated on the right in the following image.

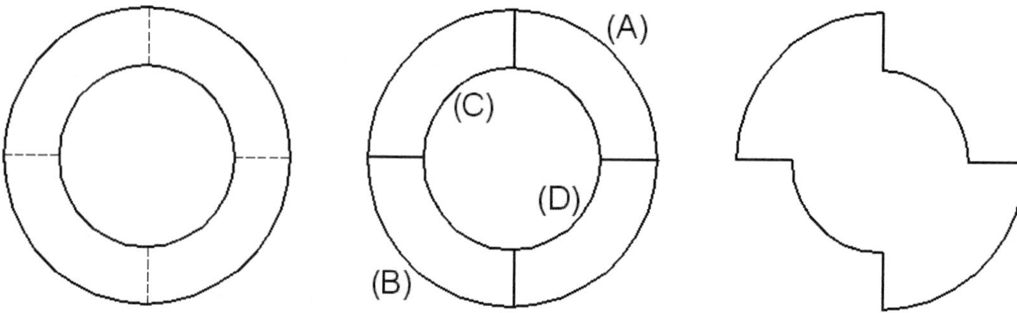

Figure 4.26

SELECTING ALL OBJECTS AS CUTTING EDGES

An alternate method of selecting cutting edges is to press ENTER in response to the prompt "Select objects." This automatically creates cutting edges out of all objects in the drawing. When you use this method, the cutting edges do not highlight. This is a very efficient means of trimming out unnecessary objects. You must, however, examine what you are trimming before using this method.

Try It!: Open the drawing file 04_Trim All. Enter the TRIM command; press ENTER at the Select Objects prompt to select all objects as cutting edges. In the following image, pick the lines at "A," "B," and "E," and the arc segments at "C" and "D" as the objects to trim.

```
Command: TR (For TRIM)
Current settings: Projection=UCS Edge=None
Select cutting edges ...
Select objects or <select all>: (Press ENTER to select all objects
    as cutting edges)
```

```
Select object to trim or shift-select to extend or [Fence/
    Crossing/Project/Edge/eRase/Undo]: (Select segments "A"
    through "E")
Select object to trim or shift-select to extend or [Fence/
    Crossing/Project/Edge/eRase/Undo]: (Press ENTER to exit this
    command)
```

The results of performing trim on this object are illustrated on the right in the following image.

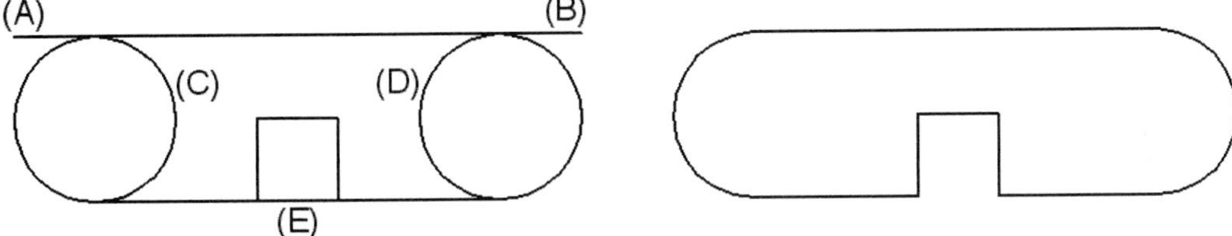

Figure 4.27

TRIMMING BY A CROSSING BOX

When trimming objects, you do not have to pick objects individually. You can erect a crossing box and trim objects out more efficiently.

Try It!: Open the drawing file 04_Trim Crossing. Yet another application of the TRIM command uses the Crossing option of "Select objects." First, invoke the TRIM command and select the small circle as the cutting edge. Begin the response to the prompt of "Select object to trim" by clicking a blank part of your screen; this automatically activates Crossing mode. See the illustration in the middle of the following image. Turn off Running OSNAP before conducting this exercise.

```
Command: TR (For TRIM)
Current settings: Projection=UCS Edge=None
Select cutting edges ...
Select objects or <select all>: (Select the small circle, as
    shown on the left in the following image)
Select objects: (Press ENTER to continue)
Select object to trim or shift-select to extend or [Fence/
    Crossing/Project/Edge/eRase/Undo]: (Pick a corner in the
    middle image)
Specify opposite corner: (Pick an opposite corner in the middle
    image)
Select object to trim or shift-select to extend or [Fence/
    Crossing/Project/Edge/eRase/Undo]: (Press ENTER to exit this
    command)
```

The power of the Crossing option of "Select objects" is shown on the right in the following image. Eliminating the need to select each individual line segment inside the small circle to trim, the Crossing mode trims all objects it touches in relation to the cutting edge.

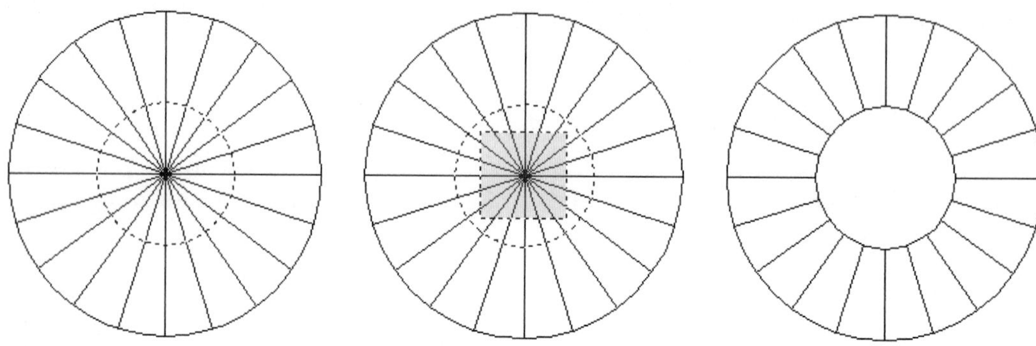

Figure 4.28

MORE ABOUT INDIVIDUAL CUTTING EDGES

Care must be taken to decide when it is appropriate to press ENTER and select all objects in your drawing as cutting edges using the TRIM command. To see this in effect, try the next exercise.

Try It!: Open the drawing file 04_Trim Cut. You need to remove the six vertical lines from the inside of the object. However, if you press ENTER to select all cutting edges, each individual segment would need to be trimmed, which is considered unproductive. Select lines "A" and "B" as cutting edges in the following image and select the inner vertical lines as the objects to trim. This is considered a more efficient way of using this command.

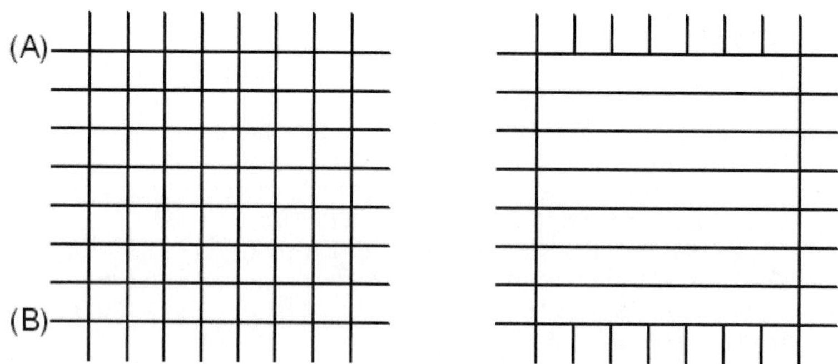

Figure 4.29

TRIMMING EXERCISE—FLOOR PLAN

Use any technique that you have learned to trim the walls of the floor plan in order for the object to appear similar to the illustration in the following image.

Try It!: Open the drawing file 04_Trim Walls. Using the following image as a guide, use the TRIM command to trim away the extra overshoots and complete the floor plan illustrated in the figure.

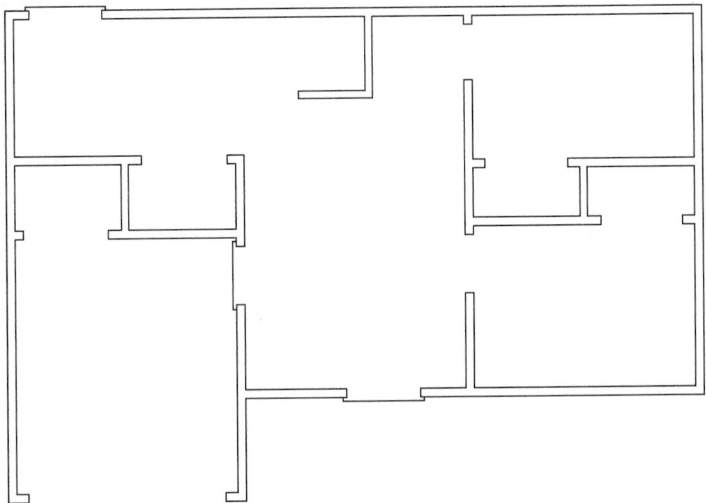

Figure 4.30

Tip: While inside the TRIM command, you can easily toggle to the EXTEND command by holding down the SHIFT key at the following command prompt:

```
Select object to trim or shift-select to extend or [Fence/
    Crossing/Project/Edge/eRase/Undo]:
```

EXTENDING OBJECTS

The EXTEND command is used to extend objects to a specified boundary edge. Choose this command from one of the following:

- The Modify toolbar of the AutoCAD Classic workspace
- From the Ribbon > Home Tab > Modify Panel
- The pull-down menu or Menu Browser (Modify > Extend)
- The keyboard (EX or EXTEND)

SELECTING INDIVIDUAL BOUNDARY EDGES

In the following image, select all dashed objects as the boundary edges. After pressing ENTER to continue with the command, select the lines at "A," "B," "C," and "D" to extend these objects to the boundary edges. If you select the wrong end of an object, use the Undo feature, which is an option of the command, to undo the change and repeat the procedure at the correct end of the object.

Try It!: Open the drawing file 04_Extend Basics. Use the following illustration and command sequence for accomplishing this task.

```
Command: EX (For EXTEND)
Current settings: Projection=UCS Edge=None
Select boundary edges ...
Select objects or <select all>: (Select the objects represented
    by dashes)
Select objects: (Press ENTER to continue)
Select object to extend or shift-select to trim or [Fence/
    Crossing/Project/Edge/Undo]: (Select the ends of the lines
    at "A" through "D")
Select object to extend or shift-select to trim or [Fence/
    Crossing/Project/Edge//Undo]: (Press ENTER to exit this
    command)
```

Tip: An alternate method of selecting boundary edges is to press ENTER in response to the "Select objects" prompt. This automatically creates boundary edges out of all objects in the drawing. When you use this method, however, the boundary edges do not highlight.

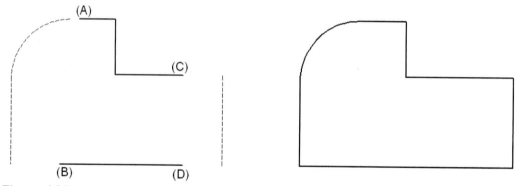

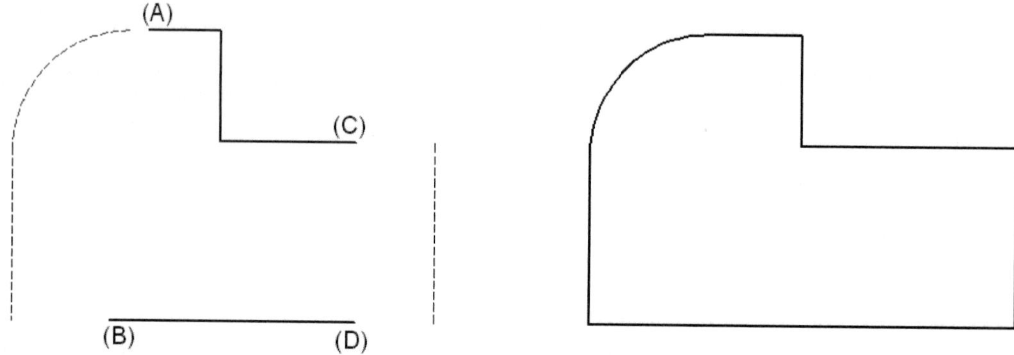

Figure 4.31

EXTENDING MULTIPLE OBJECTS

The EXTEND command can be used for extending multiple objects. After identifying the boundary edges, you can use a number of crossing boxes to identify numerous items to extend. This is a very productive method of using this command.

Try It!: Open the drawing file 04_Extend Multiple. To extend multiple objects such as the line segments shown in the following image, select the lines at "A" and "B" as the boundary edge and use the Crossing mode to create two crossing boxes, represented by the dashed rectangles. This extends all line segments to intersect with the boundaries.

```
Command: EX (For EXTEND)
Current settings: Projection=UCS Edge=None
Select boundary edges ...
Select objects or <select all>: (Select the lines at "A"
    and "B")
```

```
Select objects: (Press ENTER to continue)
Select object to extend or shift-select to trim or [Fence/
    Crossing/Project/Edge/Undo]: C (For Crossing or you can
    click a blank area as you did earier with the TRIM command)
Specify first corner: (Pick a point to start the first
    rectangle)
Specify opposite corner: (Pick a second point for the first
    rectangle)
Select object to extend or shift-select to trim or [Fence/
    Crossing/Project/Edge/Undo]: (Press ENTER to exit this
    command)
```

Repeat this procedure to extend the lines on the other side of the concrete block.

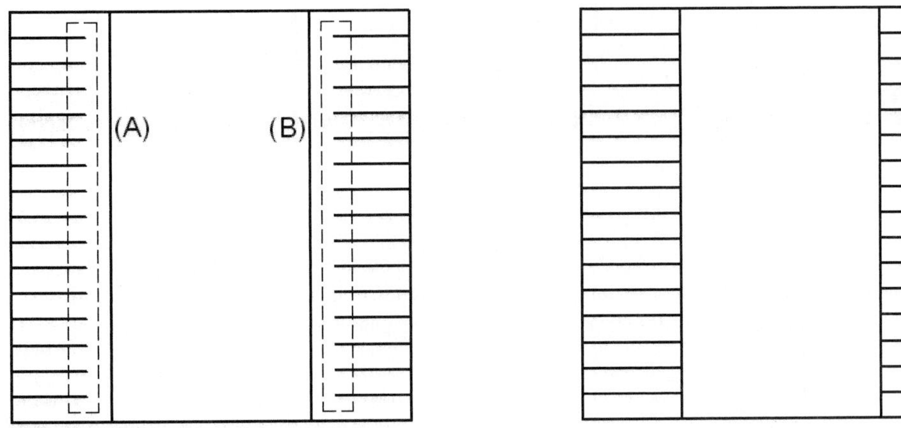

Figure 4.32

TOGGLING FROM EXTEND TO TRIM

While inside the EXTEND command, you can easily toggle to the TRIM command by holding down the SHIFT key at the following command prompt:

```
Select object to extend or shift-select to trim or [Fence/
    Crossing/Project/Edge/Undo]: (Pressing SHIFT while picking
    objects activates the TRIM command.)
```

The next Try It! exercise illustrates this technique.

Try It!: Open the drawing file 04_Extend and Trim. First, activate the EXTEND command and press ENTER, which selects all edges of the object in the following image as boundary edges. When picking the edges to extend, click on the ends of the lines from "A" through "J," as shown in the following image. At this point, do not exit the command. Press and hold down the SHIFT key; this activates the TRIM command. Now pick all of the ends of the lines until your shape appears like the illustration on the right in the following image. The use of the SHIFT key when trimming or extending provides a quick means of switching between commands.

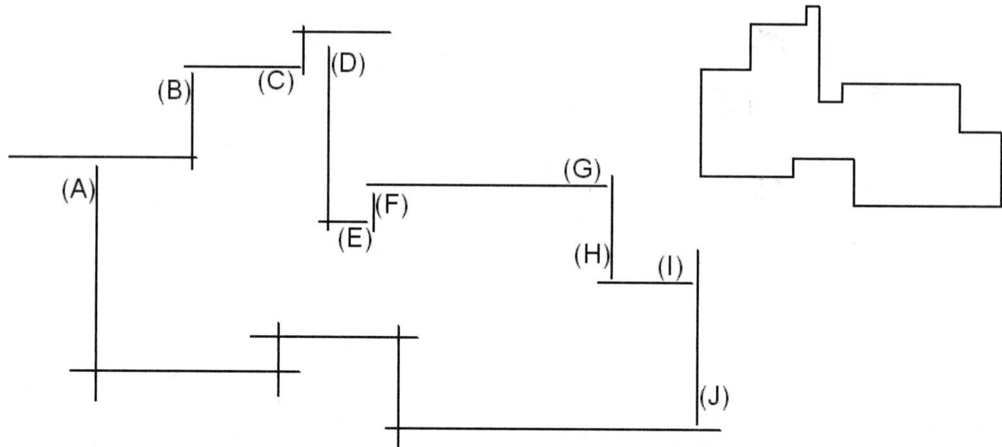

Figure 4.33

EXTEND EXERCISE—PIPING DIAGRAM

Try It!: Open the drawing file 04_Extend Pipe. Enter the EXTEND command and press ENTER when the "Select objects:" prompt appears. This selects all objects as boundary edges. Select the ends of all magenta lines representing pipes as the objects to extend. They will extend to intersect with the adjacent pipe fitting. Your finished drawing should appear similar to the following image.

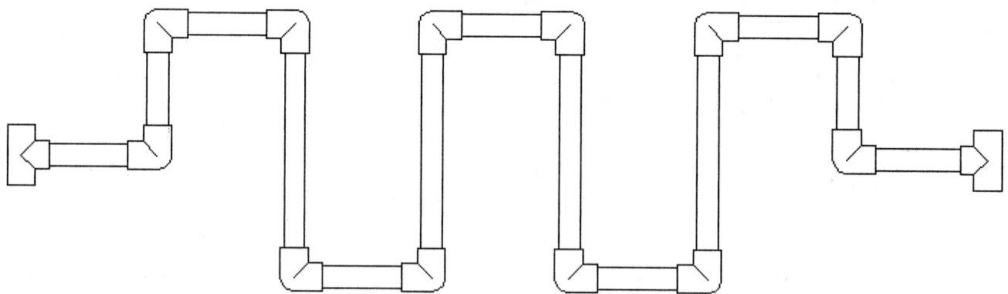

Figure 4.34

BREAKING OBJECTS

The BREAK command is used to partially delete a segment of an object. Choose this command from one of the following:

- The Modify toolbar of the AutoCAD Classic workspace
- From the Ribbon > Home Tab > Modify Panel
- The pull-down menu or Menu Browser (Modify > Break)
- The keyboard (BR or BREAK)

BREAKING AN OBJECT

The following command sequence and image show how the BREAK command is used.

Try It!: Open the drawing file 04_Break Gap. Turn off Running OSNAP prior to conducting this exercise. Use the following prompts and illustrations to break the line segment.

```
Command: BR (For BREAK)
Select objects: (Select the line at "A")
Specify second break point or [First point]: (Pick the line
    at "B")
```

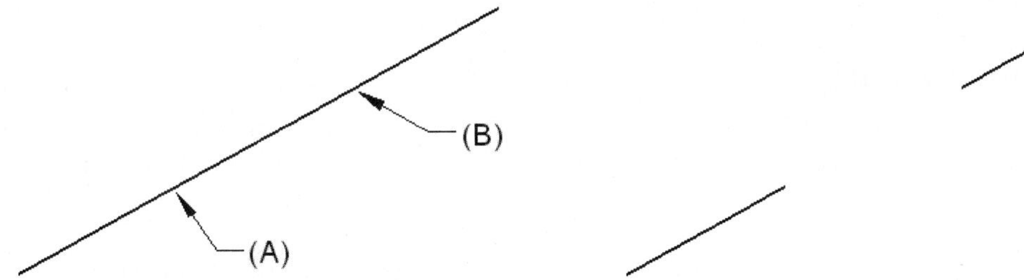

Figure 4.35

IDENTIFYING A NEW FIRST BREAK POINT

In the previous example of using the BREAK command, the location where the object was selected became the first break point. You can select the object and then be prompted to pick a new first point. This option resets the command and allows you to select an object to break followed by two different points that identify the break. The following exercise illustrates this technique.

Try It!: Open the drawing file 04_Break First. Utilize the First option of the BREAK command along with OSNAP options to select key objects to break. The following command sequence and image demonstrate using the First option of the BREAK command:

```
Command: BR (For BREAK)
Select object: (Select the line)
Specify second break point or [First point]: F (For First)
Specify first break point: Int
of (Pick the intersection of the two lines at "A")
Specify second break point: End
of (Pick the endpoint of the line at "B")
```

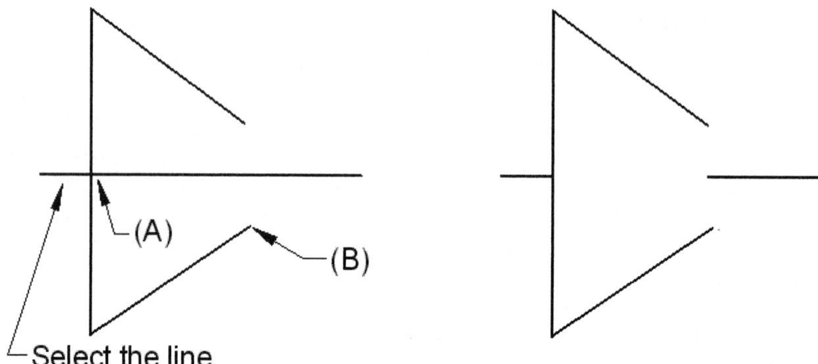

Figure 4.36

BREAKING CIRCLES

Circles can also be broken into arc segments with the BREAK command. There is only one rule to follow when breaking circles: you must pick the two break points in a counter-clockwise direction when identifying the endpoints of the segment to be removed.

Try It!: Open the drawing file 04_Break Circle. Study the following command sequence and image for breaking circles.

```
Command: BR (For BREAK)
Select objects: (Select the circle)
Specify second break point or [First point]: F (For First)
Specify first break point: (Pick at "First Point")
Specify second break point: (Pick at "Second Point")
```

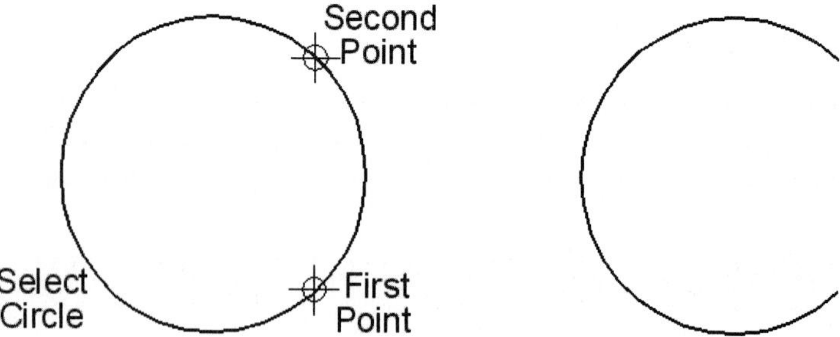

Figure 4.37

BREAK AT POINT

You can also break an object at a selected point. The break is so small that you cannot find it no matter how much you zoom in to the break point. As illustrated on the left in the following image, the line is highlighted to prove that it consists of a continuous object. Clicking on the Break at Point tool in the Modify toolbar activates the following command sequence:

```
Command: BR (For BREAK)
Select object: (Select the line anywhere)
Specify second break point or [First point]: F (For First point)
Specify first break point: Mid
of (Pick the midpoint of the line)
Specify second break point: @ (For previous point)
```

The results are illustrated on the right in the following image. Here the line is again selected. Notice that only half of the line selects because the line was broken at its midpoint.

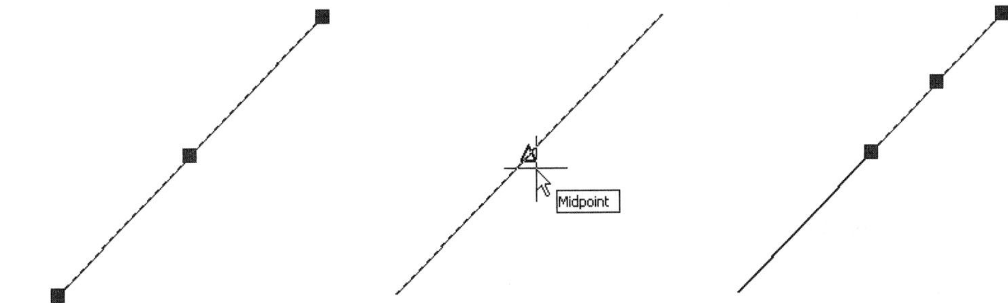

Figure 4.38

Note: The Break At tool automates the command sequence listed above. For this tool to operate correctly you may have to turn off running OSNAP.

LEVEL II MODIFY COMMANDS

This second grouping of modify commands is designed to perform more powerful editing operations compared with the Level I modify commands already discussed in this chapter. These commands are briefly described in the following table:

Button	Tool	Key-In	Function
	Array	AR	Creates multiple copies of objects in a rectangular or circular pattern
	Mirror	MI	Creates a mirror image of objects based on an axis of symmetry
	Stretch	S	Used for moving or stretching the shape of an object
	Pedit	PE	Used for editing polylines
	Explode	X	Breaks a compound object such as a polyline, block, or dimension into individual objects
	Lengthen	LEN	Changes the length of lines and arcs

Button	Tool	Key-In	Function
	Join	JO	Joins collinear objects to form a single unbroken object
	Undo	U	Used for backtracking or reversing the action of the previously used command
	Redo	RD	Reverses the effects of the previously used UNDO command operation

CREATING ARRAYS

If you need to create copies of objects that form rectangular or circular patterns, the ARRAY command is available to help with this task. This is a very powerful command that is dialog-box driven. If performing a rectangular array, you will need to supply the number of rows and columns for the pattern in addition to the spacing between these rows and columns. When performing a circular or polar array, you need to supply the center point of the array, the number of items to copy, and the angle to fill. The next series of pages documents both methods of performing arrays. Choose this command from one of the following:

- The Modify toolbar of the AutoCAD Classic workspace
- From the Ribbon > Home Tab > Modify Panel
- The pull-down menu or Menu Browser (Modify > Array)
- The keyboard (AR or ARRAY)

CREATING RECTANGULAR ARRAYS

CREATING RECTANGULAR PATTERNS WITH POSITIVE OFFSET

The Array dialog box allows you to arrange multiple copies of an object or group of objects in a rectangular or polar (circular) pattern. When creating a rectangular array, you are prompted to enter the number of rows and columns for the array. A row is a group of objects that are copied vertically in the positive or negative direction. A column is a group of objects that are copied horizontally, also in the positive or negative direction.

Try It!: Open the drawing file 04_Array Rectangular Positive. Suppose the object illustrated on the left in the following image needs to be copied in a rectangular pattern consisting of three rows and three columns. You also want a row spacing of .50 units and a column spacing of 1.25 units. The row offset would be 1.00 and the column offset would be 2.00 (measured center to center or from the same location on each object) in order to space the rectangular shapes away from one another. The result is illustrated on the right in the following image.

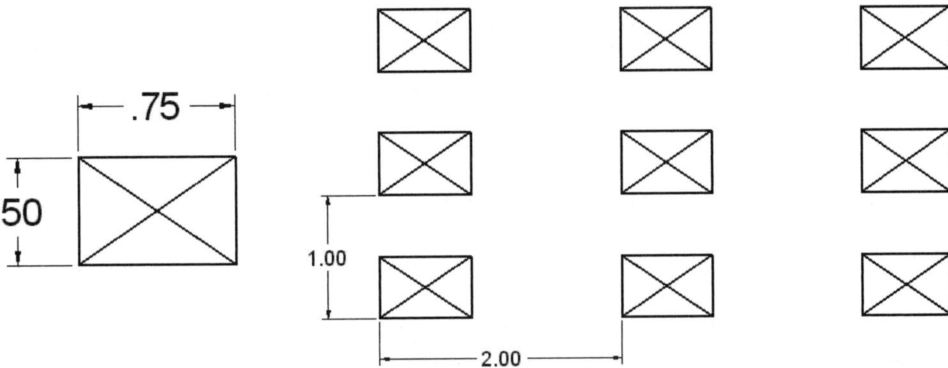

Figure 4.39

Clicking Array in the Modify pull-down menu displays the Array dialog box. When it appears, make the following changes. Be sure the Rectangular Array option is selected at the top left corner of the dialog box. Enter 3 for the number of rows and 3 for the number of columns. For Row Offset, enter a value of 1.00 units. For Column Offset, enter a value of 2.00 units. Click the Select objects button in the upper-right corner of the dialog box. This returns you to the drawing. Pick the rectangle and the two diagonal line segments. Pressing ENTER when finished returns you to the Array dialog box. The Array dialog box should now appear similar to the following image.

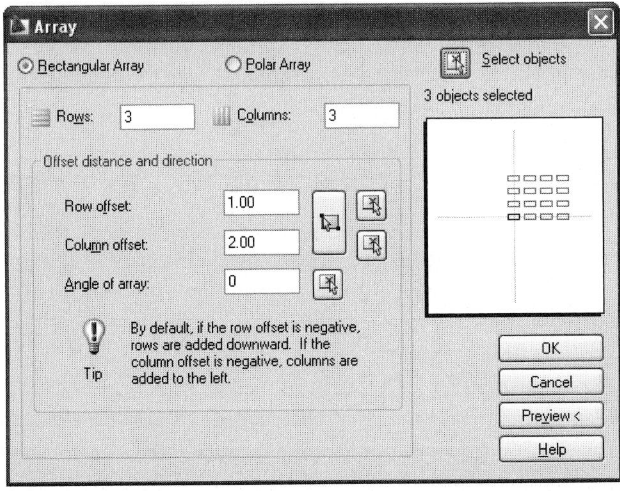

Figure 4.40

Observe the pattern in the preview image and notice that it has been updated to reflect three rows and three columns. Notice also that the Preview < button in the lower-right corner of the dialog box is active. Click this button to preview what the rectangular pattern will look like in your drawing. Also notice the following prompt will appear in the Command prompt or attached to your cursor location: "Pick or press ESC to return to dialog or <Right-click to accept array>:" Right clicking will accept the array pattern as is. If you notice an error in the array pattern, press the ESC key. This returns you to the Array dialog box and allows you to make changes to any value. The results of performing the rectangular array are illustrated in the following image.

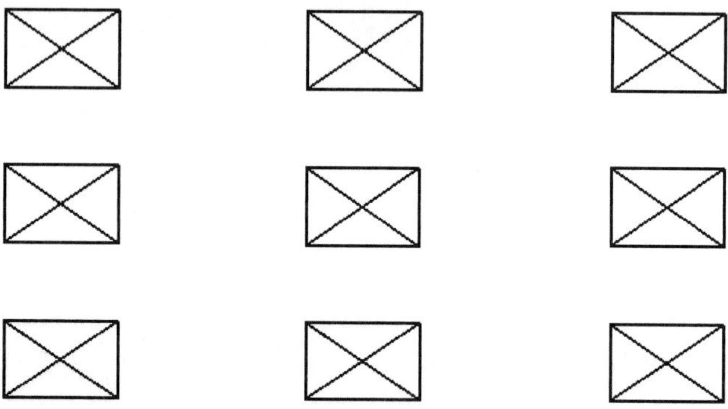

Figure 4.41

Tip: When performing rectangular arrays, a reference point must be established from the object to be arrayed and the distance between the objects determined. Not only must the spacing distance be used, but also the overall size of the object plays a role in determining the offset distances. With the total height of the original object at 0.50 and a required spacing between rows of 0.50, both object height and spacing result in a distance of 1.00 from one reference point on an object to the next. A center-to-center distance is often specified on engineering drawings, which becomes the row or column offset required.

CREATING RECTANGULAR PATTERNS WITH NEGATIVE OFFSETS

In the previous array example, the rectangular array illustrates a pattern that runs to the right of and above the original figure. At times these directions change to the left of and below the original object. The only change occurs in the distances between rows and columns, where negative values dictate the direction of the rectangular array, as shown on the right in the following image.

Try It!: Open the drawing file 04_Array Rectangular Negative. Follow the illustration in the image and the following command prompt sequence for performing this operation.

Activate the Array dialog box, as shown on the right in the following image. While in Rectangular Array mode, set the number of rows to 3 and the number of columns to 2. Because the directions of the array will be to the left of and below the original object, both row and column distances will be negative values. When finished, preview the array; if the results are similar to the illustration on the left in the following image, right-click to accept the results and create the rectangular array pattern.

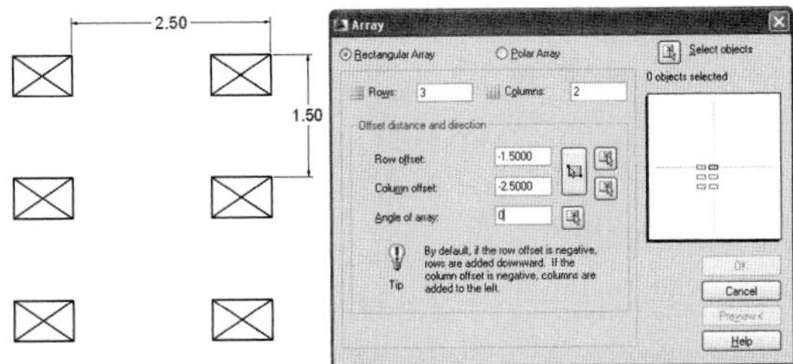

Figure 4.42

Try It!: Open the drawing 04_Array I-Beam and create a rectangular pattern consisting of 4 rows and 5 columns. Create a space of 20' between both rows and columns. Dimensions may be added at a later time. The results are displayed in the following image.

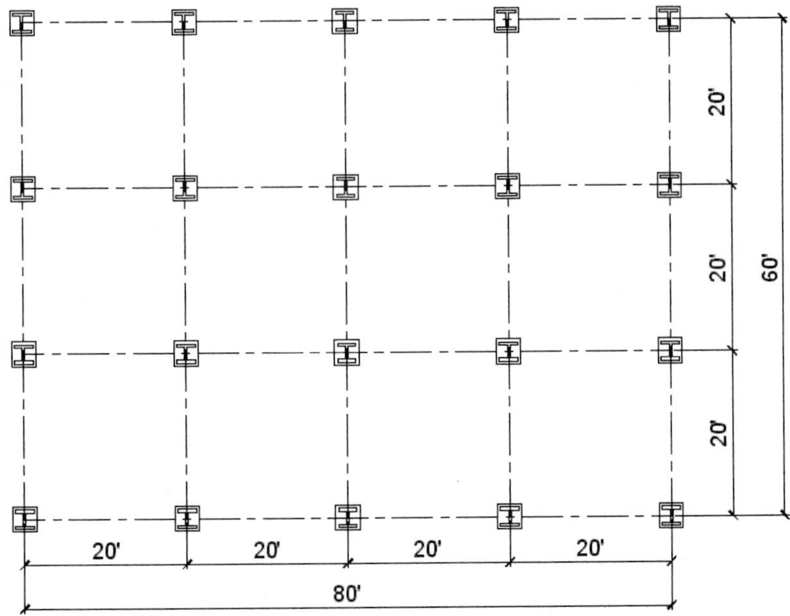

Figure 4.43

Note: The I-beam and footing were converted to a block object with the insertion point located at the center of the footing. For this reason, the I-beams can be arrayed from their centers.

CREATING POLAR ARRAYS

Polar arrays allow you to create multiple copies of objects in a circular or polar pattern. After selecting the objects to array, you pick a center point for the array in addition to the number of items to copy and the angle to fill.

Try It!: Open the drawing file 04_Array Polar and activate the Array dialog box. Be sure to select the Polar Array option at the top of the dialog box. Click the Select objects button at the top of the dialog box and pick the circle at "A" and the center line at "B" as the objects to array. For the Center point of the array, click the Select objects button and pick the intersection at "C" (this works only if the Intersection OSNAP mode is checked). Enter 4 for the Total number of items and be sure the Angle to fill is set to 360 (degrees). Also, verify that the box is checked next to Rotate items as copied. The Array dialog box should appear similar to the illustration shown on the right in the following image. Observe the results in the image icon and click the Preview < button to see the results. Right-click to accept the results and complete the array operation.

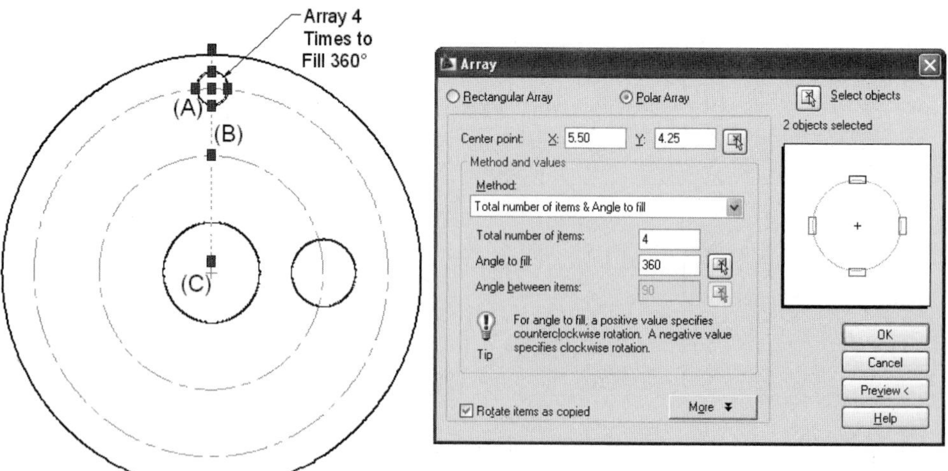

Figure 4.44

The results are illustrated on the left in the following image. This image also illustrates the creation of 4 holes that fill an angle of 180° in the clockwise direction. Notice in the Array dialog box shown on the right that an angle of −180° is given. This negative value drives the array in the clockwise direction. Remember that positive angles drive polar arrays in the counterclockwise direction.

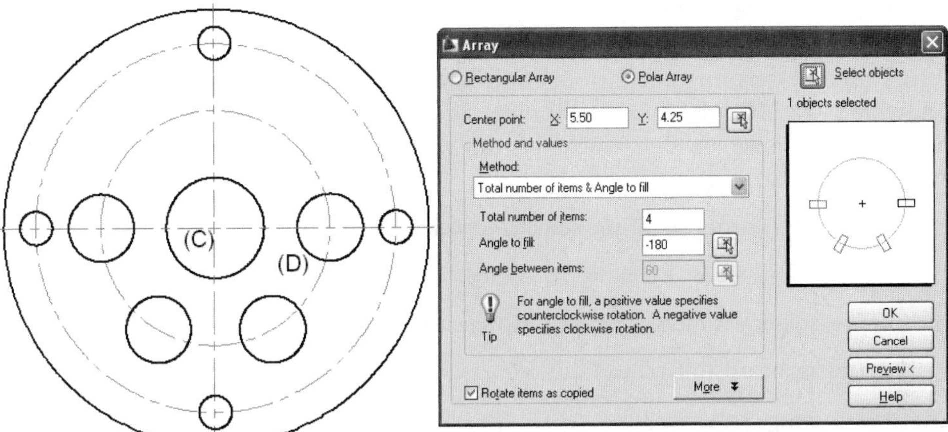

Figure 4.45

Try It!: Open the drawing file 04_Array Polar Rotate. Illustrated in the following image are three different results for arraying noncircular objects. The image on the right illustrates a polar array formed by rotating the square object as it is being copied. In the middle image, the square object is not being rotated as it is being copied. This results in the reference point at "B" pulling the arrayed pattern slightly to the right and off the main circular center line. To array rectangular or square objects in a polar pattern without rotating the objects, first convert the square or rectangle to a block with an insertion point located in the center of the square, as shown on the right in the following image (this process will be covered in detail in chapter 16). Now all squares lie an equal distance from their common center.

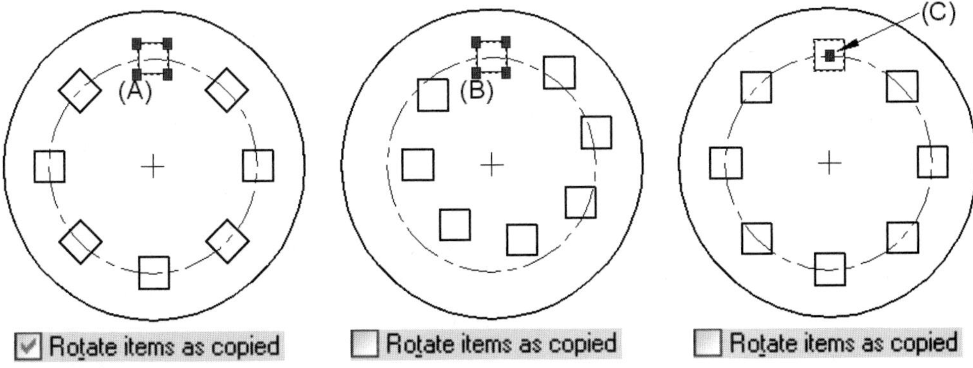

Figure 4.46

MIRRORING OBJECTS

The MIRROR command is used to create a mirrored copy of an object or group of objects. When performing a mirror operation, you have the option of deleting the original object, which would be the same as flipping the object, or keeping the original object along with the mirror image, which would be the same as flipping and copying. Choose this command from one of the following:

- The Modify toolbar of the AutoCAD Classic workspace
- From the Ribbon > Home Tab > Modify Panel
- The pull-down menu or Menu Browser (Modify > Mirror)
- The keyboard (MI or MIRROR)

MIRRORING AND COPYING

The default action of the MIRROR command is to copy and flip the set of objects you are mirroring. After selecting the objects to mirror, you identify the first and second points of a mirror line. You then decide to keep or delete the source objects. The mirror operation is performed in relation to the mirror line. Usually, most mirroring is performed in relation to horizontal or vertical lines. As a result, it is usually recommended to turn on ORTHO mode to force orthogonal mirror lines (horizontal or vertical).

Try It!: Open the drawing file 04_Mirror Copy. Refer to the following prompts and image for using the MIRROR command:

```
Command: MI (For MIRROR)
Select objects: (Select a point near "X")
Specify opposite corner: (Select a point near "Y")
Select objects: (Press ENTER to continue)
Specify first point of mirror line: (Select the endpoint of the
    centerline at "A")
Specify second point of mirror line: (Select the endpoint of the
    centerline at "B")
Erase source objects? [Yes/No] <N>: (Press ENTER for default)
```

Because the original object needed to be retained by the mirror operation, the image result is shown on the right in the following image. The MIRROR command works well when symmetry is required.

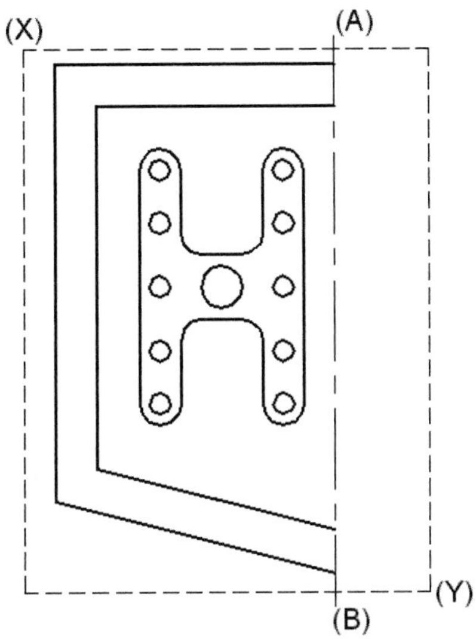

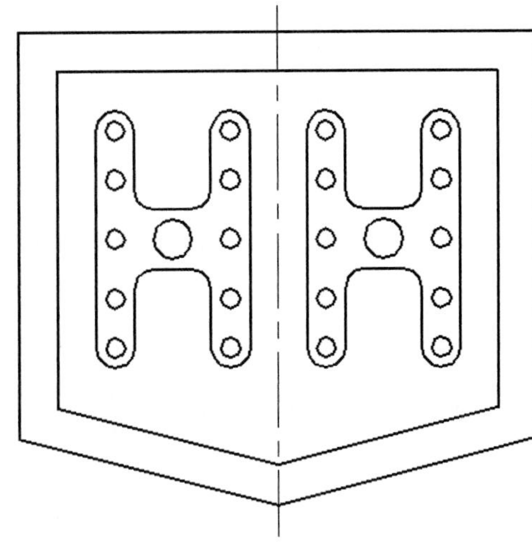

Figure 4.47

MIRRORING BY FLIPPING

The illustration in the following image is a different application of the MIRROR command. It is required to have all items that make up the bathroom plan flip but not copy to the other side. This is a typical process involving "what if" scenarios.

Try It!: Open the drawing file 04_Mirror Flip. Use the following command prompts to perform this type of mirror operation. The results are displayed on the right in the following image.

```
Command: MI (For MIRROR)
Select objects: All (This selects all objects)
Select objects: (Press ENTER to continue)
Specify first point of mirror line: Mid
of (Select the midpoint of the line at "A")
Specify second point of mirror line: Per
to (Select line "B," which is perpendicular to point "A")
Erase source objects? [Yes/No] <N>: Y (For Yes)
```

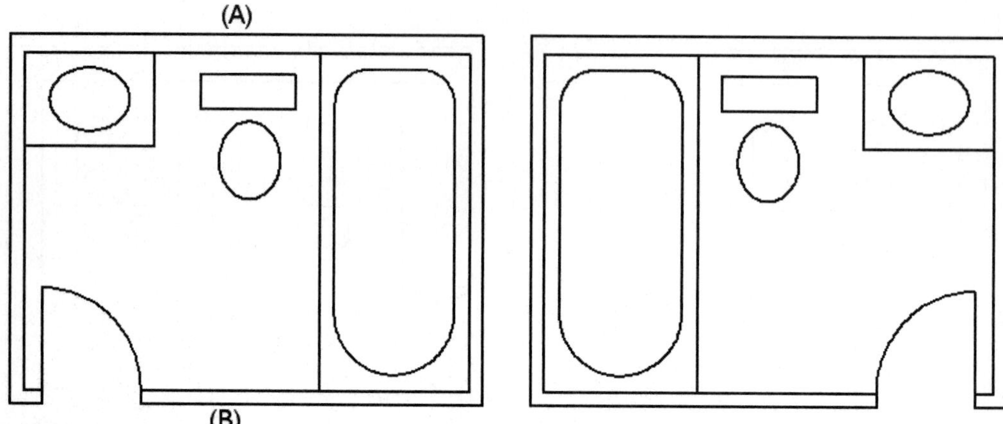

Figure 4.48

MIRRORING TEXT

In addition to mirroring other object types, text can also be mirrored, as in the following image of the duplex complex. Rather than copying and moving the text into position for the matching duplex half, you can include the text in the mirroring operation.

Try It!: Open the drawing file 04_Mirror Duplex. Use the MIRROR command and create a mirror image of the Duplex floor plan using line "AB" as the points for the mirror line. Do not delete the source objects. Your finished results should be similar to the following image.

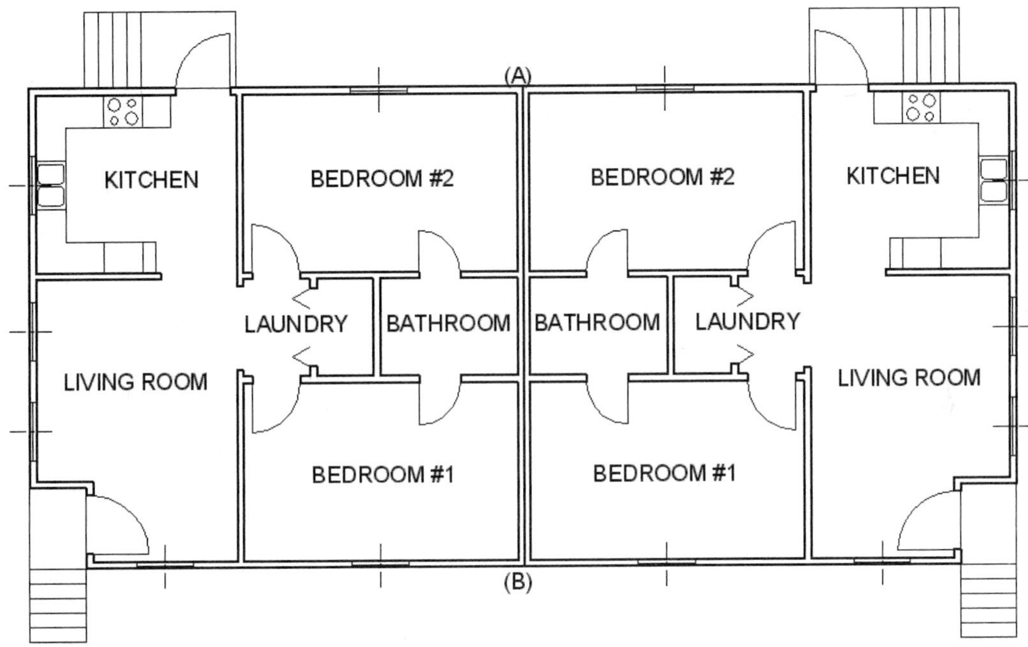

Figure 4.49

MORE INFORMATION ON MIRRORING TEXT

There may be times when you mirror text, but the text is not right-reading. The mirror operation makes the text backwards and difficult to read. This usually occurs with older AutoCAD drawings.

If this happens, a special system variable called MIRRTEXT is available. This variable must be entered at the Command prompt. If this variable is set to a value of 1 (or on), change the value to 0 (Zero, or off).

```
Command: MIRRTEXT
Enter new value for MIRRTEXT <1>: 0
```

STRETCHING OBJECTS

Use the STRETCH command to move a portion of a drawing while still preserving the connections to parts of the drawing remaining in place. Choose this command from one of the following:

- The Modify toolbar of the AutoCAD Classic workspace
- From the Ribbon > Home Tab > Modify Panel
- The pull-down menu or Menu Browser (Modify > Stretch)
- The keyboard (S or STRETCH)

THE BASICS OF STRETCHING

To perform this type of operation, you must use the Crossing option of "Select objects." Items inside the crossing box will move, while those touching the box will change length. In the following image, a group of objects is selected with the crossing box. Next, a base point is identified at the approximate location of "C." Finally, a second point of displacement is identified with a polar coordinate or by using the Direct Distance mode. Once the objects selected in the crossing box are stretched, the objects not only move to the new location but also mend themselves.

Try It!: Open the drawing file 04_Stretch. Use the following image and command sequence to perform this task.

```
Command: S (For STRETCH)
Select objects to stretch by crossing-window or crossing-
    polygon...
Select objects: (Pick a point at "A")
Specify opposite corner: (Pick a point at "B")
Select objects: (Press ENTER to continue)
Specify base point or [Displacement] <Displacement>: (Select a
    point at "C")
Specify second point or <use point place as displacement>: (With
    Polar turned on, move your cursor to the left and enter a
    value of .75)
```

The results of performing this stretching operation are illustrated on the right in the following image.

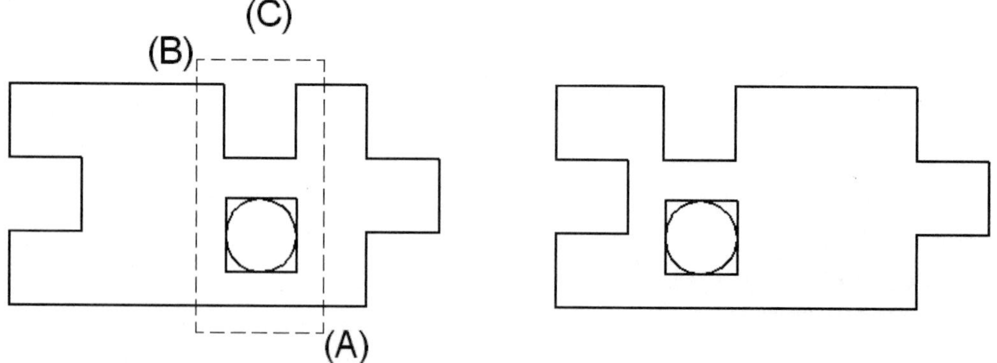

Figure 4.50

HOW STRETCHING AFFECTS DIMENSIONS

Applications of the STRETCH command are illustrated in the following image, in which a number of architectural features need to be positioned at a new location. The whole success of using the STRETCH command on these features is in the selection of the objects to stretch through the crossing box. Associative dimensions will automatically move with the objects.

Try It!: Open the drawing file 04_Stretch Arch. Use the following command sequence and image to stretch the window, wall, and door the designated distances using the Direct Distance mode or by entering a polar coordinate at the keyboard. Since each of the stretch distances is a different value, the STRETCH command must be used three separate times.

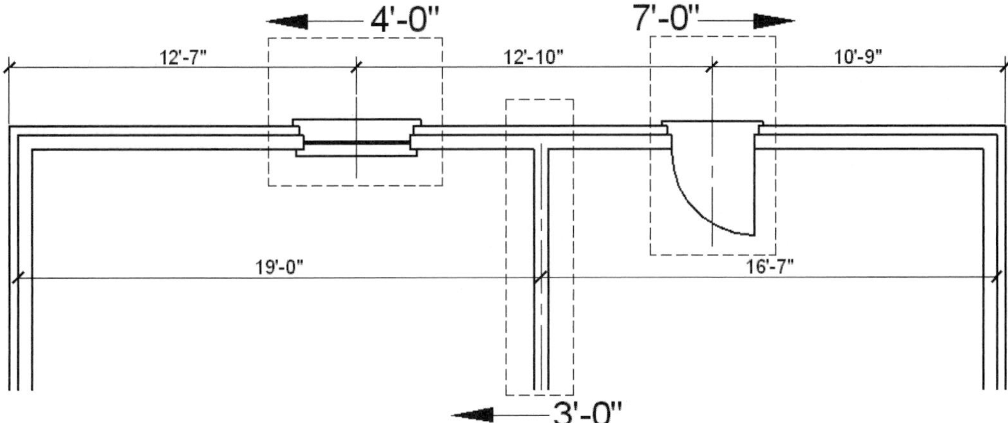

Figure 4.51

STRETCHING USING MULTIPLE CROSSING WINDOWS

When identifying items to stretch by crossing box, you are not limited to a single crossing box. You can surround groups of objects with multiple crossing boxes. All items selected in this manner will be affected by the stretching operation.

Try It!: Open the drawing file 04_Stretch Fence. Enter the STRETCH command and construct three separate crossing boxes to select the top of the fence boards at "A," "B," and "C," as shown in the following image. When prompted for a base point or displacement, pick a point on a blank part of your screen. Stretch these boards up at a distance of 12' or 1'-0".

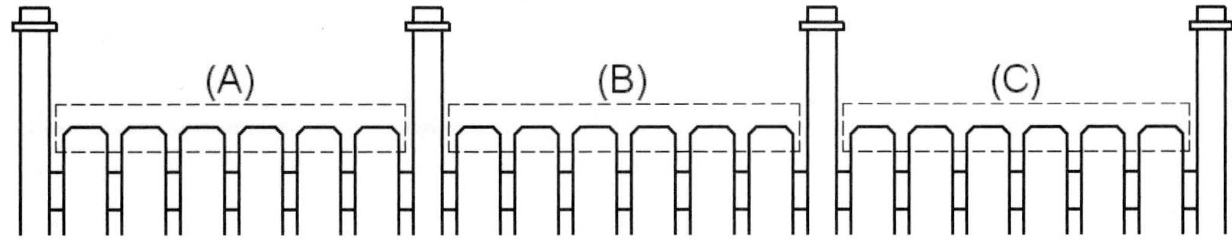

Figure 4.52

The results of performing the stretch operation on the fence are illustrated in the following image. When you have to stretch various groups of objects in a single operation, you can create numerous crossing boxes to better perform this task.

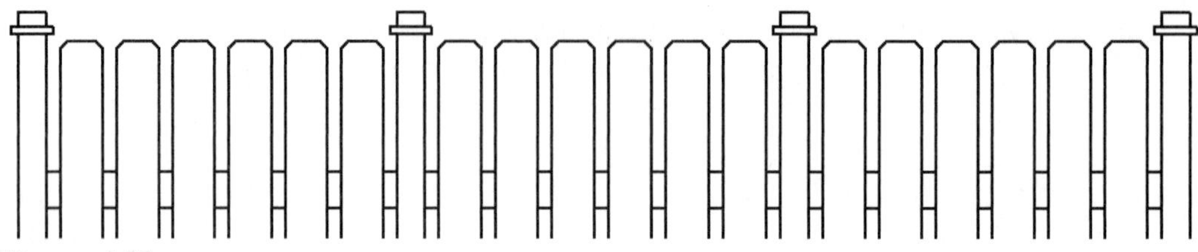

Figure 4.53

EDITING POLYLINES

Editing polylines can lead to interesting results. A few of these options will be explained in the following pages. Choose this command from one of the following:

- The Modify II toolbar of the AutoCAD Classic workspace
- From the Ribbon > Home Tab > Modify Panel (Expanded)
- The pull-down menu or Menu Browser (Modify > Object > Polyline)
- The keyboard (PE or PEDIT)

CHANGING THE WIDTH OF A POLYLINE

Illustrated on the left in the following image is a polyline of width 0.00. The PEDIT command is used to change the width of the polyline to 0.10 units, as shown on the right in the following image.

Try It!: Open the drawing file 04_Pedit Width. Refer to the following command sequence to use the PEDIT command with the Width option.

```
Command: PE (For PEDIT)
Select polyline or [Multiple]: (Select the polyline)
Enter an option [Open/Join/Width/Edit vertex/Fit/Spline/Decurve/
    Ltype gen/Undo]: W (For Width)
Specify new width for all segments: 0.10
Enter an option [Open/Join/Width/Edit vertex/Fit/Spline/Decurve/
    Ltype gen/Undo]: (Press ENTER to exit this command)
```

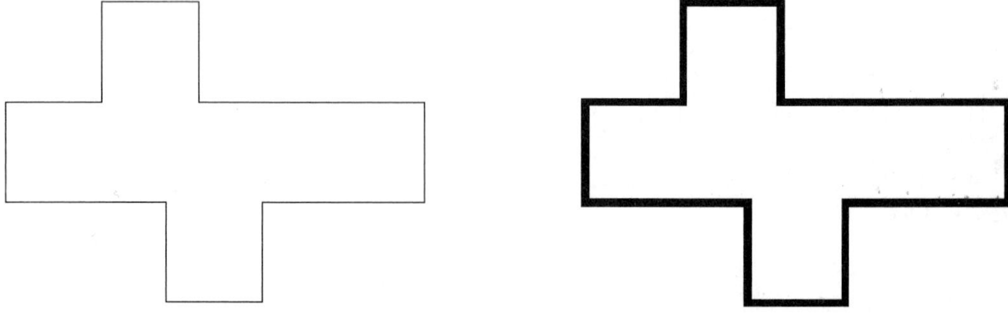

Figure 4.54

JOINING OBJECTS INTO A SINGLE POLYLINE

It is very easy to convert regular objects such as lines and arcs into polylines (circles cannot be converted). As you can convert individual objects into polylines, the results are a collection of individual polylines. As long as the polyline endpoints match with one another, these polylines can be easily joined into one single polyline object using the Join option of the PEDIT command.

Try It!: Open the drawing file 04_Pedit Join. Refer to the following command sequence and image to use this command.

```
Command: PE (For PEDIT)
Select polyline or [Multiple]: (Select the line at "A")
Object selected is not a polyline
Do you want to turn it into one? <Y> (Press ENTER)
Enter an option [Close/Join/Width/Edit vertex/Fit/Spline/Decurve/
    Ltype gen/Undo]: J (For Join)
Select objects: (Pick a point at "B")
Specify opposite corner: (Pick a point at "C")
Select objects: (Press ENTER to join the lines)
56 segments added to polyline
Enter an option [Open/Join/Width/Edit vertex/Fit/Spline/Decurve/
    Ltype gen/Undo]: (Press ENTER to exit this command)
```

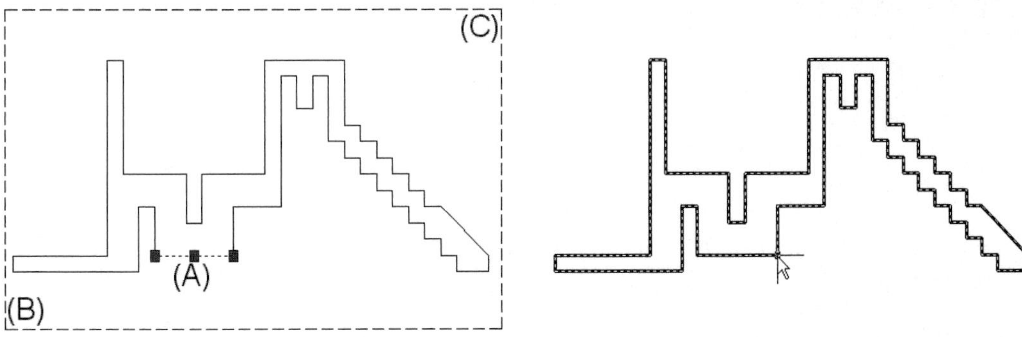

Figure 4.55

CURVE GENERATION

Polylines can be edited to form various curve-fitting shapes. Two curve-fitting modes are available, namely, Splines, and Fit Curves.

GENERATING SPLINES

The Spline option produces a smooth-fitting curve based on control points in the form of the vertices of the polyline.

Try It!: Open the drawing file 04_Pedit Spline Curve. Refer to the following command sequence and image to use this command.

```
Command: PE (For PEDIT)
Select polyline or [Multiple]: (Select the polyline frame)
Enter an option [Close/Join/Width/Edit vertex/Fit/Spline/Decurve/
    Ltype gen/Undo]: S (For Spline)
Enter an option [Close/Join/Width/Edit vertex/Fit/Spline/Decurve/
    Ltype gen/Undo]: (Press ENTER to exit this command)
```

The results of creating a spline curve from a polyline are shown on the right in the following image.

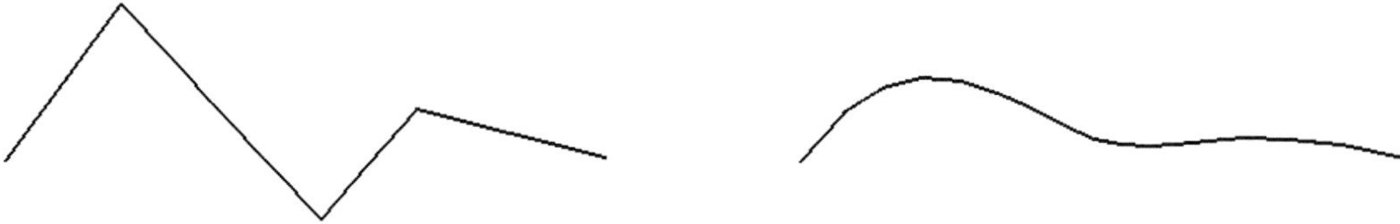

Figure 4.56

GENERATING FIT CURVES

The Fit Curve option passes entirely through the control points, producing a more exaggerated curve.

Try It!: Open the drawing file 04_Pedit Fit Curve. Refer to the following command sequence and image to use this command.

```
Command: PE (For PEDIT)
Select polyline or [Multiple]: (Select the polyline)
Enter an option [Close/Join/Width/Edit vertex/Fit/Spline/Decurve/
    Ltype gen/Undo]: F (For Fit)
Enter an option [Close/Join/Width/Edit vertex/Fit/Spline/Decurve/
    Ltype gen/Undo]: (Press ENTER to exit this command)
```

The results of creating a fit curve from a polyline are shown on the right in the following image.

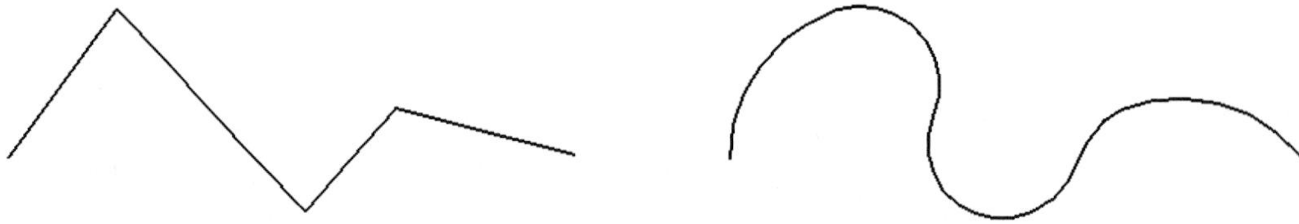

Figure 4.57

LINETYPE GENERATION OF POLYLINES

The Linetype Generation option of the PEDIT command controls the pattern of the linetype from polyline vertex to vertex. In the polyline illustrated on the left in the following image, the hidden linetype is generated from the first vertex to the second vertex. An entirely different pattern is formed from the second vertex to the third vertex, and so on. The polyline illustrated on the right in the following image has the linetype generated throughout the entire polyline. In this way, the hidden linetype is smoothed throughout the polyline.

Try It!: Open the drawing file 04_Pedit Ltype Gen. Refer to the following command sequence and image to use this command.

```
Command: PE (For PEDIT)
Select polyline or [Multiple]: (Select the polyline)
Enter an option [Close/Join/Width/Edit vertex/Fit/Spline/Decurve/
    Ltype gen/Undo]: L (For Ltype gen)
Enter polyline linetype generation option [ON/OFF] <Off>: On
Enter an option [Close/Join/Width/Edit vertex/Fit/Spline/Decurve/
    Ltype gen/Undo]: (Press ENTER to exit this command)
```

Figure 4.58

OFFSETTING POLYLINE OBJECTS

Once a group of objects has been converted to and joined into a single polyline object, the entire polyline can be copied at a parallel distance using the OFFSET command.

Try It!: Open the drawing file 04_Pedit Offset. Use the OFFSET command to copy the shape in the following image a distance of 0.50 units to the inside.

```
Command: O (For OFFSET)
Specify offset distance or [Through/Erase/Layer] <Through>: 0.50
Select object to offset or [Exit/Undo] <Exit>: (Select the
    polyline at "A")
Specify point on side to offset or [Exit/Multiple/Undo] <Exit>:
    (Select a point anywhere near "B")
Select object to offset or [Exit/Undo] <Exit>: (Press ENTER to
    exit this command)
```

Because the object was converted to a polyline, all objects are offset at the same time, as shown on the right in the following image.

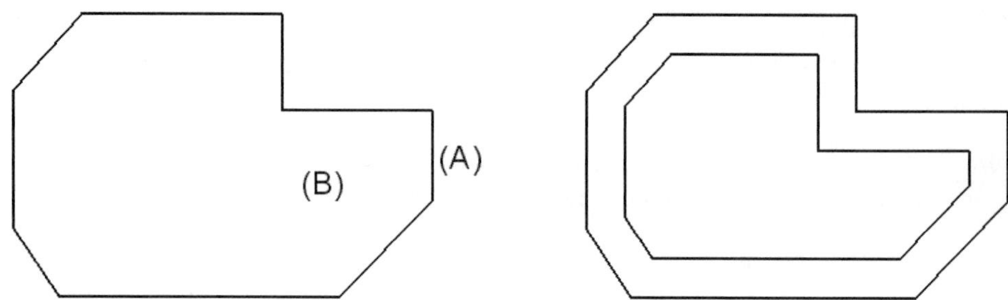

Figure 4.59

MULTIPLE POLYLINE EDITING

Multiple editing of polylines allows for multiple objects to be converted to polylines. This is accomplished with the PEDIT command and the Multiple option. Illustrated on the left in the following image is a rectangle and four slots, all considered individual objects. When you run the Multiple option of the PEDIT command, not only can you convert all objects at once into individual polylines, but you can join the endpoints of common shapes as well. The result of editing multiple polylines and then joining them is illustrated on the right in the following image.

Try It!: Open the drawing file 04_Pedit Multiple1. Use the prompt sequence below and the following image to illustrate how the MPEDIT command is used.

```
Command: PE (For PEDIT)
Select polyline or [Multiple]: M (For Multiple)
Select objects: All
Select objects: (Press ENTER to continue)
Convert Lines and Arcs to polylines [Yes/No]? <Y> (Press ENTER)
Enter an option [Close/Open/Join/Width/Fit/Spline/Decurve/Ltype
    gen/Undo]: J (For Join)
Join Type = Extend
Enter fuzz distance or [Jointype] <0.0000>: (Press ENTER to accept
    this default value)
15 segments added to 5 polylines
Enter an option [Close/Open/Join/Width/Fit/Spline/Decurve/Ltype
    gen/Undo]: (Press ENTER to exit)
```

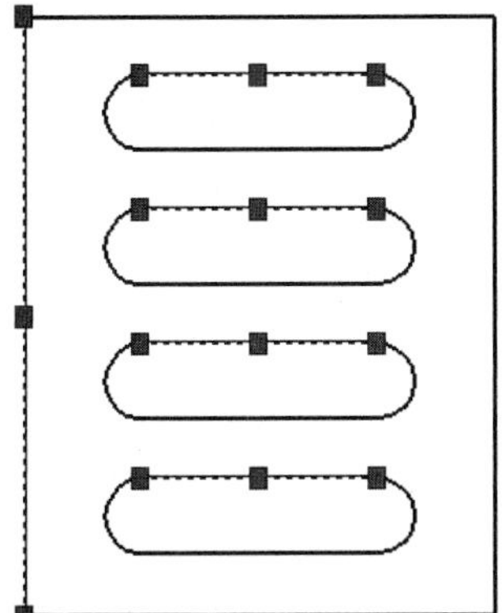

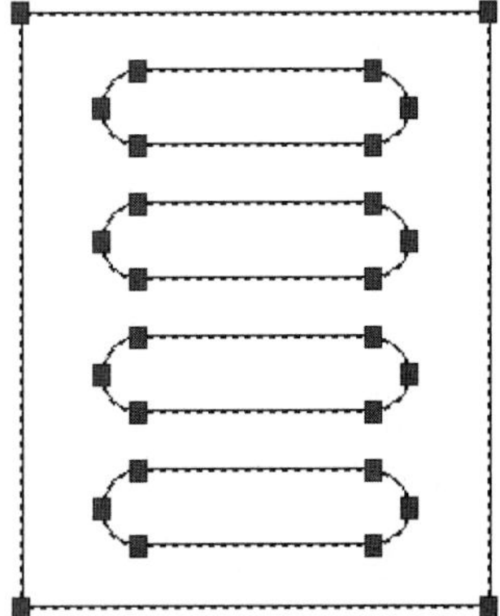

Figure 4.60

As a general rule when joining polylines, you cannot have gaps present or overlapping occurring when performing this operation. This is another feature of using the Multiple option of the PEDIT command. This option works best when joining two objects that have a gap or overlap. After selecting the two objects to join, you will be asked to enter a fuzz factor. This is the distance used by this command to bridge a gap or trim overlapping lines. You could measure the distance between two objects to determine this value. Study the following example for automatically creating corners in objects using a fuzz factor.

Try It!: Open the drawing file 04_Pedit Multiple2. In this example, one of the larger gaps was measured to be .12 units in length. As a result, a fuzz factor slightly larger than this calculated value is used (.13 units). The completed object is illustrated on the right in the following image. You may have to experiment with various fuzz factors before you arrive at the desired results.

```
Command: PE (For PEDIT)
Select polyline or [Multiple]: M (For Multiple)
Select objects: All
Select objects: (Press ENTER to continue)
Convert Lines and Arcs to polylines [Yes/No]? <Y> (Press ENTER)
Enter an option [Close/Open/Join/Width/Fit/Spline/Decurve/Ltype
    gen/Undo]: J (For Join)
Join Type = Extend
Enter fuzz distance or [Jointype] <0.25>: .13
11 segments added to polyline
Enter an option [Close/Open/Join/Width/Fit/Spline/Decurve/Ltype
    gen/Undo]: (Press ENTER to exit)
```

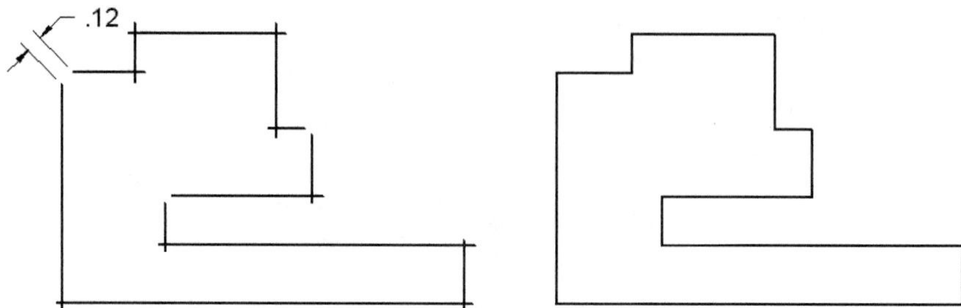

Figure 4.61

EXPLODING OBJECTS

Using the EXPLODE command on a polyline, dimension, or block separates the single object into its individual parts. Choose this command from one of the following:

- The Modify toolbar of the AutoCAD Classic workspace
- From the Ribbon > Home Tab > Modify Panel
- The pull-down menu or Menu Browser (Modify > Explode)
- The keyboard (X or EXPLODE)

Illustrated on the left in the following image is a polyline that is considered one object. Using the EXPLODE command and selecting the polyline breaks the polyline into numerous individual objects, as shown on the right in the following image.

```
Command: X (For EXPLODE)
Select objects: (Select the polyline)
Select objects: (Press ENTER to perform the explode operation)
```

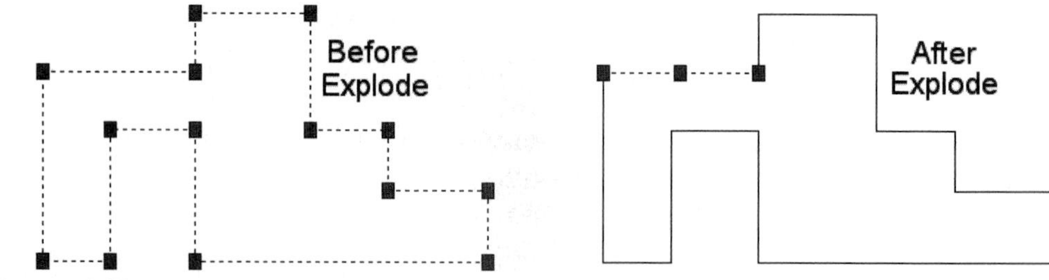

Figure 4.62

Note: Dimensions consist of extension lines, dimension lines, dimension text, and arrowheads, all grouped into a single object. (Dimensions are covered in chapter 11, "Adding Dimensions to Your Drawing.") Using the EXPLODE command on a dimension breaks the dimension down into individual extension lines, dimension lines, arrowheads, and dimension text. This is not advisable, because the dimension value will not be updated if the dimension is stretched along with the object being dimensioned. Also, the ability to manipulate dimensions with a feature called grips is lost. Grips are discussed in chapter 7.

LENGTHENING OBJECTS

The LENGTHEN command is used to change the length of a selected object without disturbing other object qualities such as angles of lines or radii of arcs. Choose this command from one of the following:

- The pull-down menu or Menu Browser (Modify > Lengthen)
- From the Ribbon > Home Tab > Modify Panel (Expanded)The keyboard (LEN or LENGTHEN)

Try It!: Open the drawing file 04_Lengthen1. Use the illustration in the following image and the command sequence below for performing this task.

```
Command: LEN (For LENGTHEN)
Select an object or [DElta/Percent/Total/DYnamic]: (Select line
    "A")
Current length: 12.3649
Select an object or [DElta/Percent/Total/DYnamic]: T (For Total)
Specify total length or [Angle] (1.0000)>: 20
Select an object to change or [Undo]: (Select the line at "A")
Select an object to change or [Undo]: (Press ENTER to exit)
```

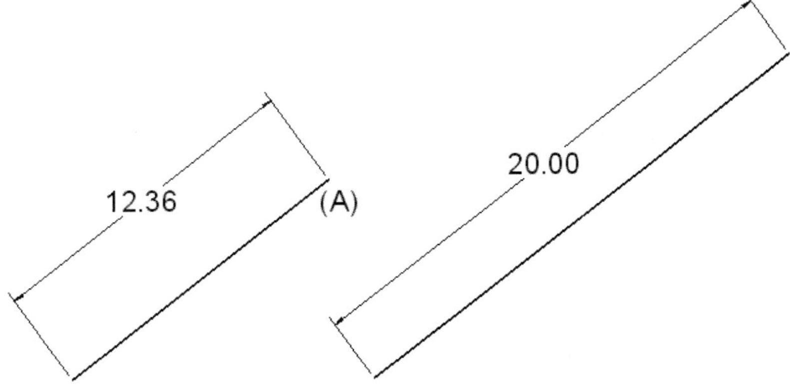

Figure 4.63

Tip: After supplying the new total length of any object, be sure to select the desired end to lengthen when you are prompted to select an object to change.

JOINING OBJECTS

For special cases in which individual line segments need to be merged together as a single segment, the JOIN command can be used to accomplish this task. Usually this occurs when gaps occur in line segments and all segments lie in the same line of sight. This condition is sometimes referred to as collinear. Rather than connect the gaps with additional individual line segments, considered very unproductive and poor in practice, use JOIN to connect all segments as one. Choose this command from one of the following:

- The Modify toolbar of the AutoCAD Classic workspace
- From the Ribbon > Home Tab > Modify Panel (Expanded)
- The pull-down menu or Menu Browser (Modify > Join)
- The keyboard (J or JOIN)

Try It!: Open the drawing file 04_Join. Pick one line segment at "A" as the source. Then select the other line segment at "B" to join to the source. Use the following command prompt and image to join various segments using the JOIN command.

```
Command: J (For Join)
Select source object: (Select line "A")
Select lines to join to source: (Select line "B")
Select lines to join to source: (Press ENTER to join the segments)
1 line joined to source
```

Continue using the JOIN command on the other line segments. A different JOIN command must be used for each group of line segments that appears broken in the following image.

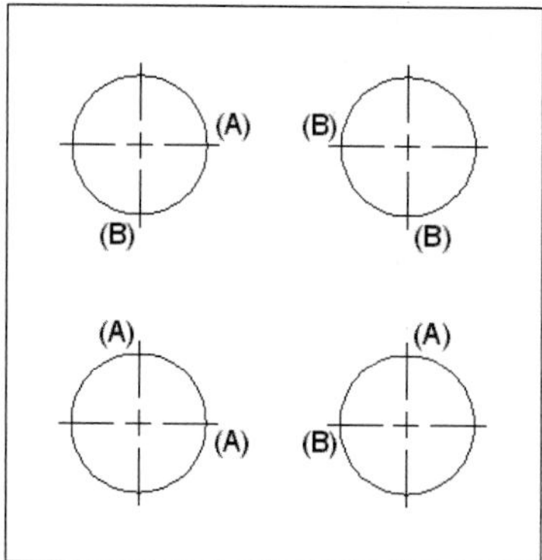

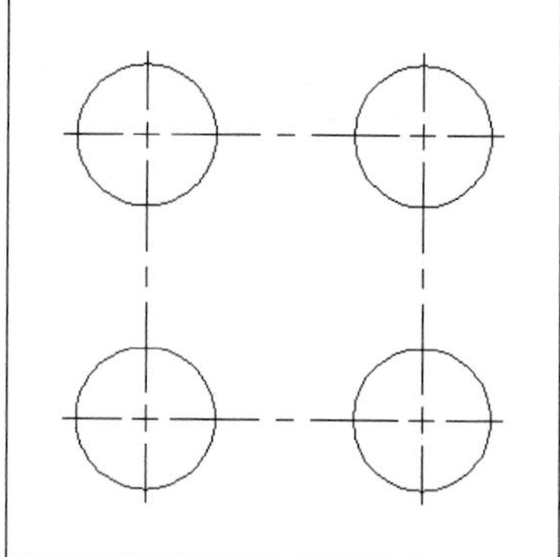

Figure 4.64

UNDOING AND REDOING OPERATIONS

The UNDO command can be used to undo the previous task or command action while the REDO command reverses the effects of any previous undo. Choose these commands from one of the following:

- The Standard toolbar of the AutoCAD Classic workspace
- The Quick Access Toolbar
- The pull-down menu or Menu Browser (Edit > Undo) or (Edit > Redo)
- The keyboard (U or UNDO) or (REDO)
- Select anywhere in the drawing and right-click

For example, if you draw an arc followed by a line followed by a circle, issuing the UNDO command will undo the action caused by the most recent command; in this case, the circle would be removed from the drawing database. This represents one of the easiest ways to remove data or backtrack the drawing process.

Expanding the Undo list found in the Standard toolbar, shown on the left in the following image, allows you to undo several actions at once. From this example, notice that the Rectangle, Line, and Circle actions are highlighted for removal.

You can also reverse the effect of the UNDO command by using REDO immediately after the UNDO operation.

Clicking the REDO command button from the Standard toolbar negates one UNDO operation. You can click on this button to cancel the effects of numerous UNDO operations.

As with UNDO, you can also REDO several actions at once through the Redo list shown on the right in the following image. This list can be accessed from the Standard toolbar.

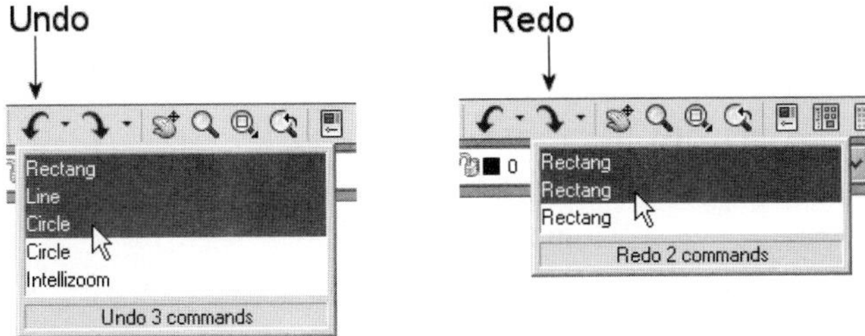

Figure 4.65

Note: When grouping actions to be undone, you cannot, in the previous image, for example, highlight Line, skip Trim, and highlight Fillet to be removed. The groupings to undo must be strung together in this dialog box.

REDO only works if you have undone a previous operation. Otherwise, REDO remains inactive.

TUTORIAL EXERCISE: ANGLE.DWG

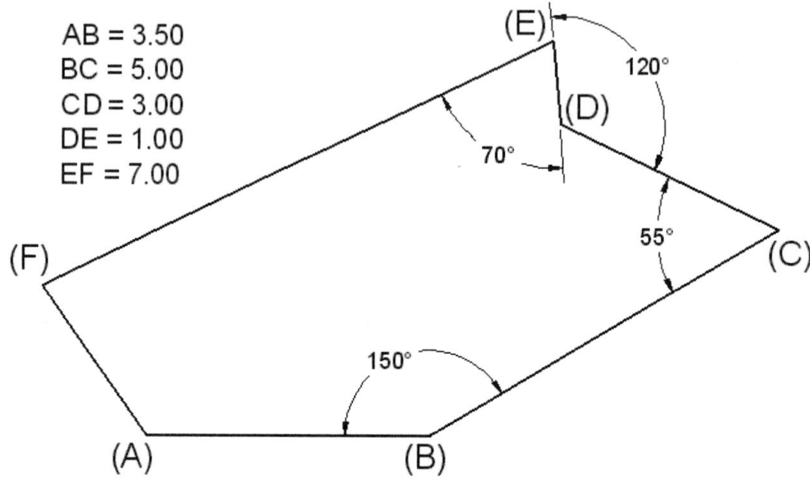

Figure 4.66

PURPOSE

This tutorial is designed to allow you to construct a one-view drawing of the angle as shown in the previous image using the ARRAY and LENGTHEN commands.

SYSTEM SETTINGS

Use the Drawing Units dialog box and change the precision of decimal units from 4 to 2 places. Use the current default settings for the limits of this drawing, (0,0) for the lower-left corner, and (12,9) for the upper-right corner. Check to see that the following Object Snap modes are already set: Endpoint, Extension, Intersection, Center.

LAYERS

Create the following layer with the format:

Name	Color	Linetype
Object	White	Continuous

SUGGESTED COMMANDS

Make the Object layer current. Begin this drawing by constructing line "AB," which is horizontal. Use the ARRAY command to copy and rotate line "AB" at an angle of 150° in the clockwise direction. Once the line is copied and rotated, use the LENGTHEN command and modify the new line to the designated length. Repeat this procedure for lines "CD," "DE," and "EF." Complete the drawing by constructing a line segment from the endpoint at vertex "F" to the endpoint at vertex "A." This object could also have been constructed using a polar angle setting of 5° and a polar setting relative to the previous line.

Step 1

Draw line "AB" using the Polar coordinate or Direct Distance mode, as shown in the following image. (Line "AB" is considered a horizontal line.)

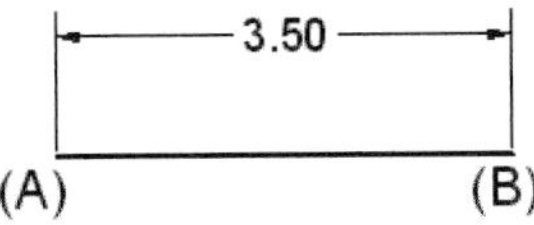

Figure 4.67

Step 2

One technique of constructing the adjacent line at 150° from line "AB" is to use the Array dialog box, as shown on the left in the following image, and perform a polar operation. Select line "AB" as the object to array, pick the endpoint at "B" as the center of the array, and enter a value of -150° for the angle to fill. Entering a negative angle copies the line in the clockwise direction. The value of your array center point will be different from the value displayed in the dialog box.

The result is shown on the right in the following image.

Rectangular Array
Polar Array
Center point: X: 20.8434 Y: 55.5073
Method and values
Method:
Total number of items & Angle to fill
Total number of items: 2
Angle to fill: -150
Angle between items: 150
For angle to fill, a positive value specifies counterclockwise rotation. A negative value specifies clockwise rotation.
Tip
Rotate items as copied
More

(A)
(B)

Figure 4.68

Step 3

The array operation allowed line "AB" to be rotated and copied at the correct angle, namely -150°. However, the new line is the same length as line "AB." Use the LENGTHEN command to increase the length of the new line to a distance of 5.00 units. Use the Total option, specify the new total length of 5.00, and select the end of the line at "1" as the object to change, as shown on the left in the following image.

```
Command: LEN (For LENGTHEN)
Select an object or [DElta/Percent/Total/DYnamic]: T (For Total)
Specify total length or [Angle] <1.00)>: 5.00
```

```
Select an object to change or [Undo]: (Pick the end of the line
    at "1")
Select an object to change or [Undo]: (Press ENTER to exit this
    command)
```

The result is shown on the right in the following image.

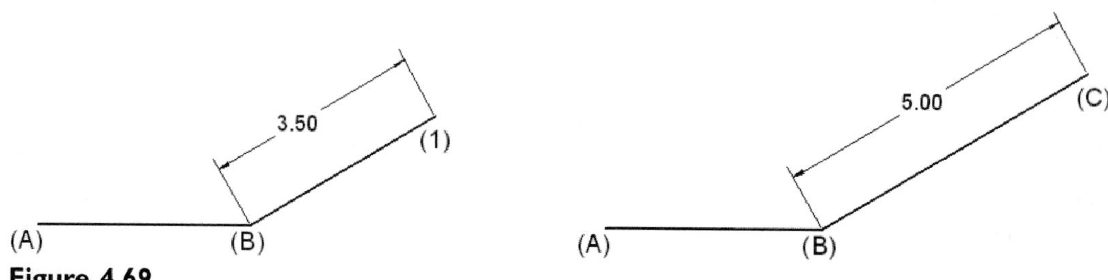

Figure 4.69

Step 4

Use the Array dialog box, shown on the left in the following image, and perform a polar operation. Select line "BC" as the object to array, pick the endpoint at "C" as the center of the array, and enter a value of –55° for the angle to fill. Entering a negative angle copies the line in the clockwise direction.

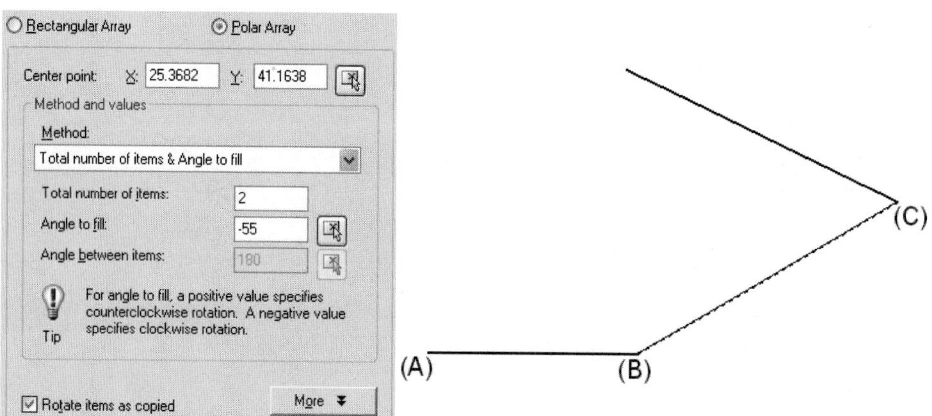

Figure 4.70

Step 5

Then use the LENGTHEN command to reduce the length of the new line from 5.00 units to 3.00 units. Use the Total option, specify the new total length of 3.00, and select the end of the line at "1" as the object to change, as shown on the left in the following image.

```
Command: LEN (For LENGTHEN)
Select an object or [DElta/Percent/Total/DYnamic]: T (For Total)
Specify total length or [Angle] <5.00)>: 3.00
```

```
Select an object to change or [Undo]: (Pick the end of the line
    at "1")
Select an object to change or [Undo]: (Press ENTER to exit this
    command)
```

The result is shown on the right in the following image.

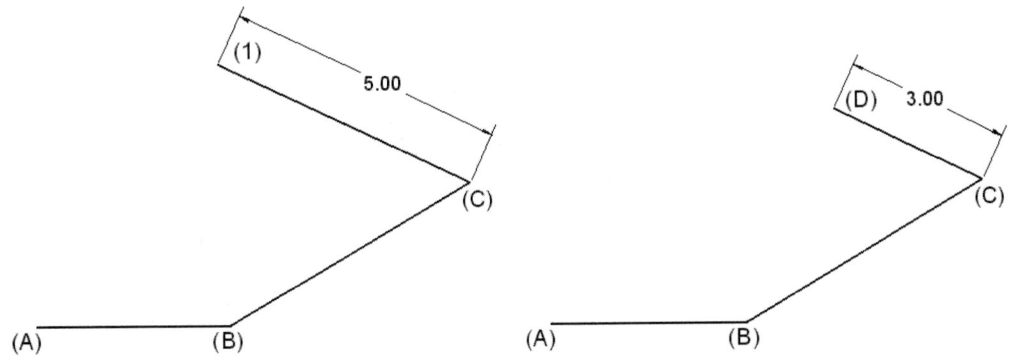

Figure 4.71

Step 6

Use the Array dialog box and perform another polar operation. Select line "CD" as the object to array, as shown on the left in the following image, pick the endpoint at "D" as the center of the array, and enter a value of 120° for the angle to fill. Entering a positive angle copies the line in the counterclockwise direction.

Then use the LENGTHEN command to reduce the length of the new line from 3.00 units to 1.00 unit, as shown on the right in the following image.

```
Command: LEN (For LENGTHEN)
Select an object or [DElta/Percent/Total/DYnamic]: T (For Total)
Specify total length or [Angle] <3.00)>: 1.00
Select an object to change or [Undo]: (Pick the end of the line
    at "1")
Select an object to change or [Undo]: (Press ENTER to exit this
    command)
```

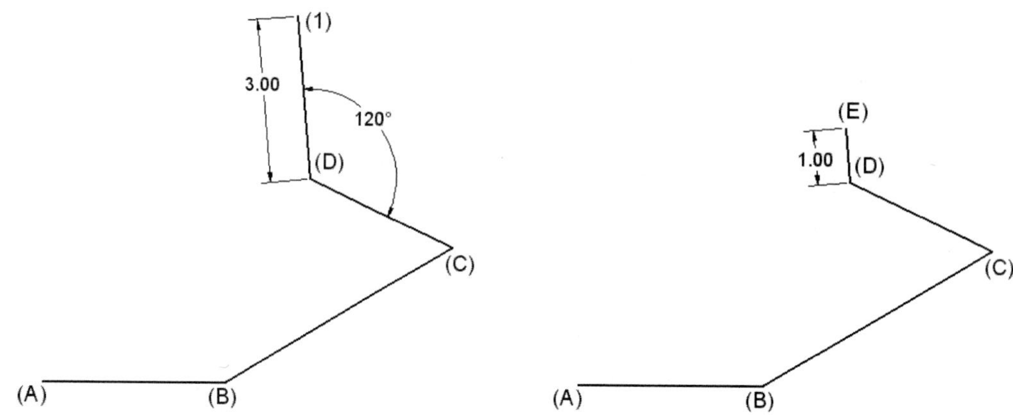

Figure 4.72

Step 7

Use the Array dialog box and perform the final polar operation. Select line "DE" as the object to array, pick the endpoint at "E" as the center of the array, and enter a value of –70° for the angle to fill. Entering a negative angle copies the line in the clockwise direction. Then use the LENGTHEN command to increase the length of the new line from 1.00 unit to 7.00 units, as shown on the right in the following image.

```
Command: LEN (For LENGTHEN)
Select an object or [DElta/Percent/Total/DYnamic]: T (For Total)
Specify total length or [Angle] <1.00)>: 7.00
Select an object to change or [Undo]: (Pick the end of the line
    at "1")
Select an object to change or [Undo]: (Press ENTER to exit this
    command)
```

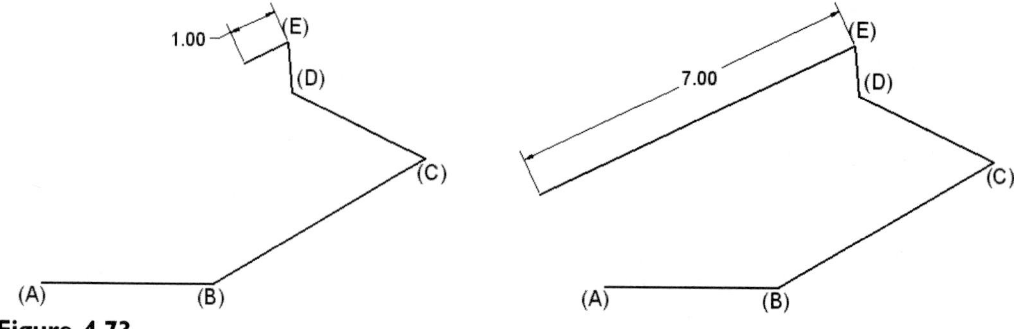

Figure 4.73

Step 8

Connect endpoints "F" and "A" with a line as shown in the following image.

```
Command: L (For LINE)
Specify first point: (Pick the endpoint of the line at "F")
Specify next point or [Undo]: (Pick the endpoint of the line
    at "A")
Specify next point or [Undo]: (Press ENTER to exit this command)
```

The completed drawing is illustrated on the right in the following image. You may add dimensions at a later date.

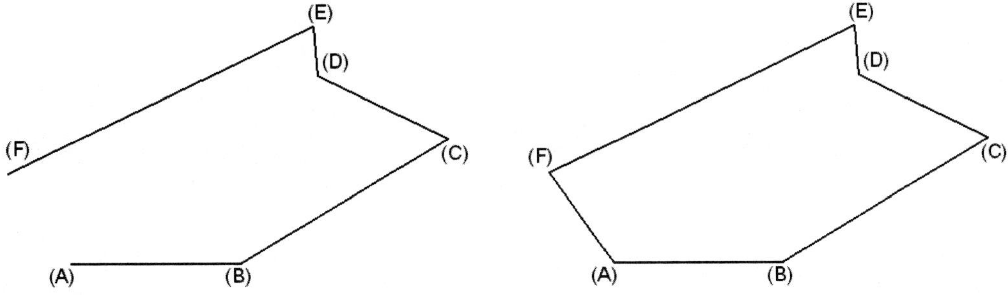

Figure 4.74

TUTORIAL EXERCISE: GASKET.DWG

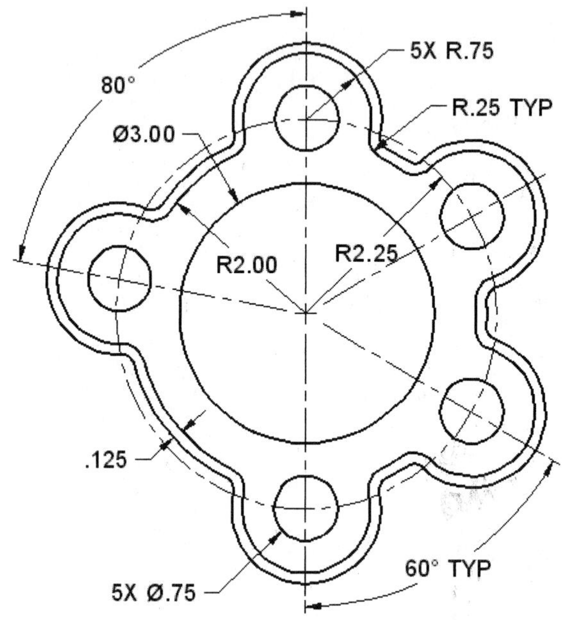

Figure 4.75

PURPOSE

This tutorial is designed to allow you to construct a one-view drawing of the gasket using the Array dialog box.

SYSTEM SETTINGS

Use the current default settings for the units and limits of this drawing, (0,0) for the lower-left corner and (12,9) for the upper-right corner. Check to see that the following Object Snap modes are already set: Endpoint, Extension, Intersection, Center, and Quadrant.

LAYERS

Create the following layers with the format:

Name	Color	Linetype
Object	White	Continuous
Center	Yellow	Center

SUGGESTED COMMANDS

Draw the basic shape of the object using the LINE and CIRCLE commands. Lay out a centerline circle; draw one of the gasket tabs at the top of the center circle. Use the Array dialog box to create four copies in the –180° direction and two copies in the 80° direction.

Trim out the excess arc segments to form the gasket. Convert the outer profile of the gasket into one continuous polyline object and offset this object .125 units to the outside of the gasket.

Step 1

Use existing unit and limit settings. Create an Object and a Center layer. Assign the Center linetype to the Center layer.

Make the Object layer current and construct a circle of 2.00 radius with its center at absolute coordinate 4.00,4.00, as shown on the left in the following image.

Make the Center layer current and construct a circle of 2.25 radius using the previous center point, as shown on the right in the following image. Change the linetype scale from 1.00 to 0.50 units using the LTSCALE (LTS) command.

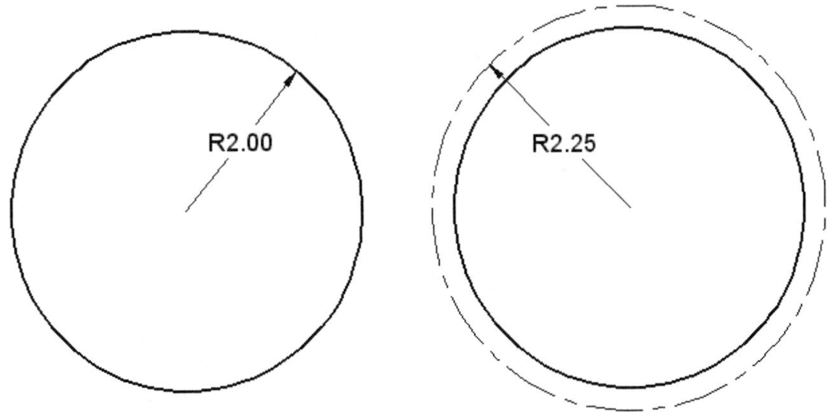

Figure 4.76

Step 2

Make the Object layer current again and construct a circle of radius .75' from the quadrant at the top of the centerline circle, as shown on the left in the following image. Also construct a circle of .75' in diameter from the same quadrant at the top of the centerline circle, as shown on the right in the following image.

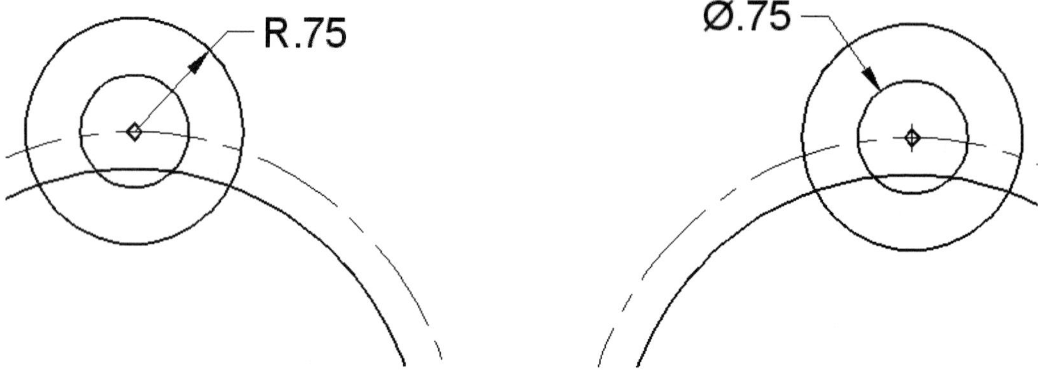

Figure 4.77

Step 3

Copy the two top circles just created in a polar (circular) pattern using the Array dialog box. Use the center of the 2.00-radius circle as the center point of the array. Change the total number of items to 4 and the angle to fill to –180°, as shown on the left in the following image. The negative angle drives the array in the clockwise direction, as shown on the right in the following image.

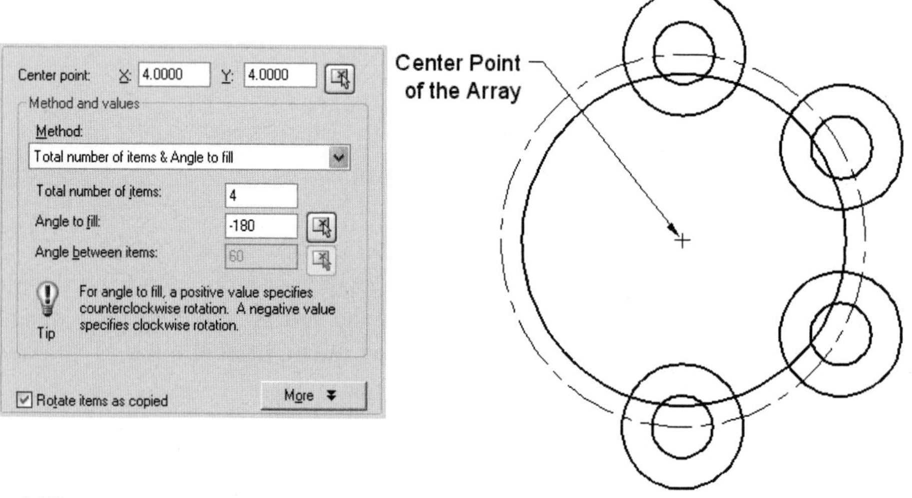

Figure 4.78

Step 4

Perform another array operation on the top two circles. Again, use the center of the 2.00-radius circle as the center point of the array. Change the total number of items to 2 and the angle to fill to 80°, as shown on the left in the following image. The positive angle drives the array in the counterclockwise direction, as shown on the right in the following image.

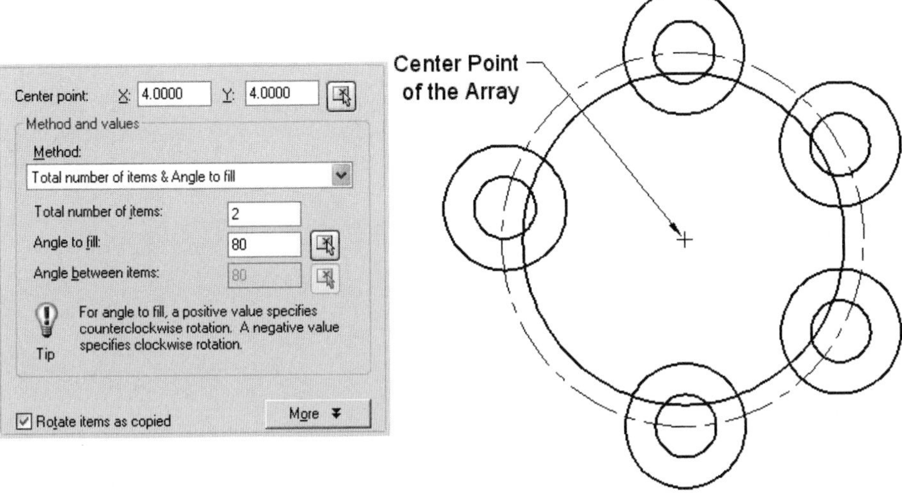

Figure 4.79

Step 5

Trim out the inside edges of the five circles labeled "A" through "E" using the dashed circle as the cutting edge, as shown on the left in the following image. The results are displayed on the right.

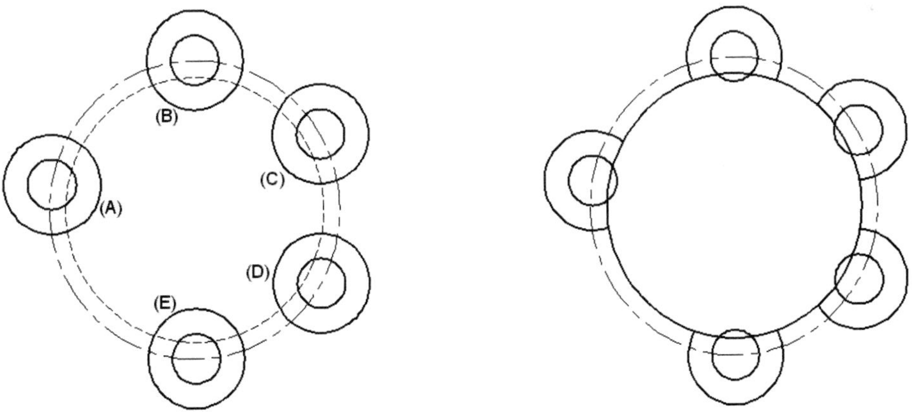

Figure 4.80

Step 6

Use the TRIM command again. Select the five dashed arc segments as cutting edges and trim away the portions labeled "A" through "E," as shown on the left in the following image. The results are displayed on the right.

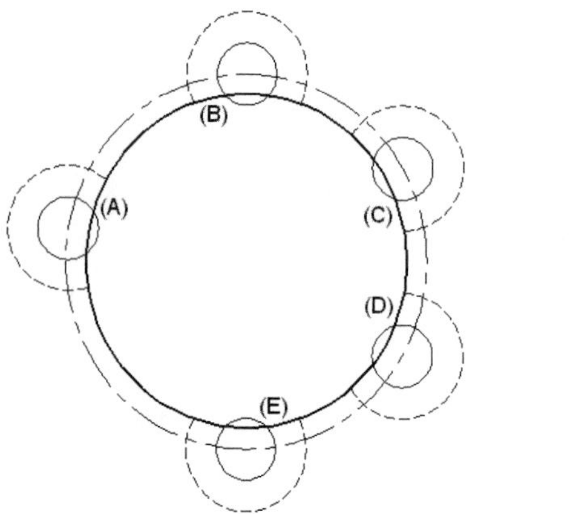

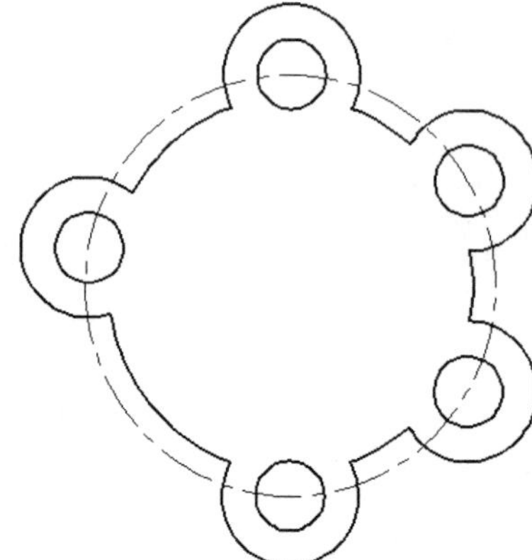

Figure 4.81

Step 7

Fillet the inside corners of the .75-radius arcs with the 2.00-radius arc using the FILLET command and a radius set to .25'. These corners are marked by a series of points, shown on the left in the following image. The results are displayed on the right.

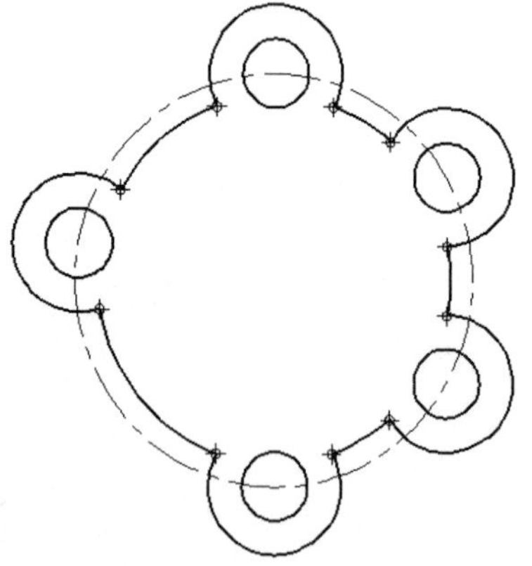
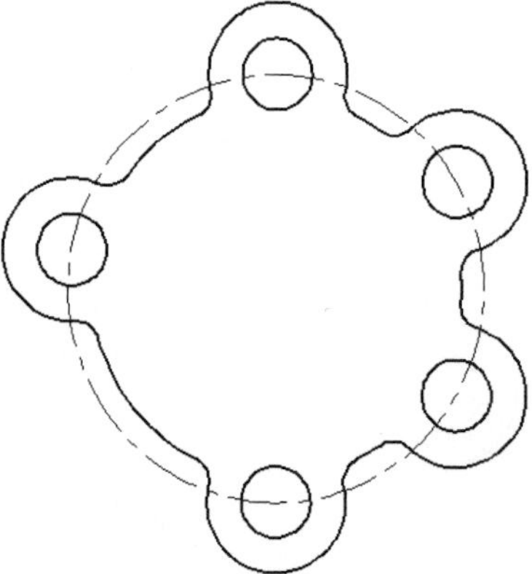

Figure 4.82

Step 8

Create the large 1.50-radius hole in the center of the gasket using the CIRCLE command. Change the outer perimeter of the gasket into one continuous polyline object using the PEDIT command. Use the Join option of this command to accomplish this task. Finally, create a copy of the outer profile of the gasket a distance of .125' using the OFFSET command. Offset the profile to the outside at "A" as shown on the left in the following image. The completed gasket is displayed on the right.

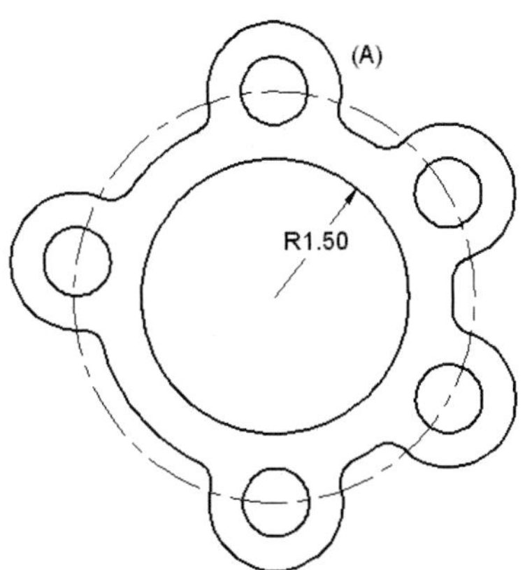

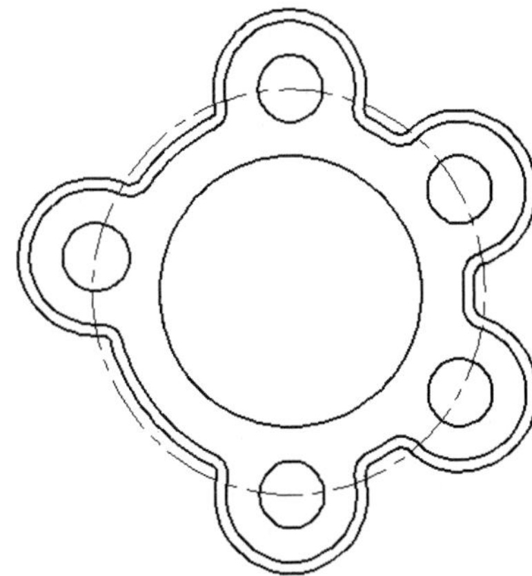

Figure 4.83

TUTORIAL EXERCISE: TILE.DWG

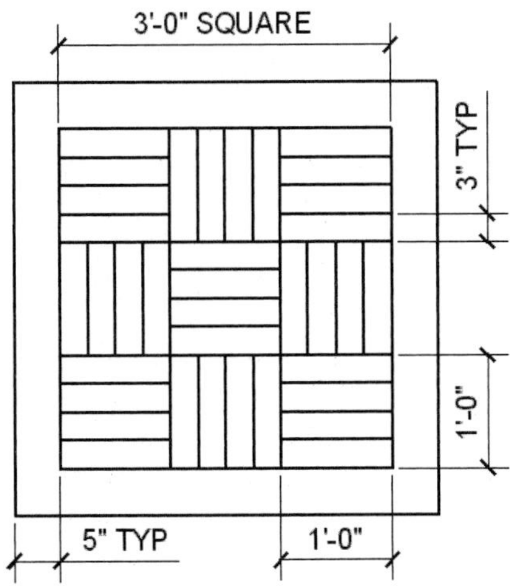

Figure 4.84

PURPOSE

This tutorial is designed to use the OFFSET and TRIM commands to complete the drawing of the floor tile shown in the previous image.

SYSTEM SETTINGS

Use the Drawing Units dialog box and change the units of measure from decimal to architectural units. Keep the remaining default settings. Use the LIMITS command and change the limits of the drawing to (0,0) for the lower-left corner and (10',8') for the upper-right corner. Use the ZOOM command and the All option to fit the new drawing limits to the display screen.

Check to see that the following Object Snap modes are currently set: Endpoint, Extension, Intersection, and Center.

LAYERS

Create the following layer with the format:

Name	**Color**	**Linetype**
Object	White	Continuous

SUGGESTED COMMANDS

Make the Object layer current. Use the RECTANGLE command to begin the inside square of the tile. The OFFSET command is used to copy the inner square a distance of 5' to form the outer square. The ARRAY command is used to copy selected line segments in a rectangular pattern at a specified distance. The TRIM command is then used to form the inside tile patterns.

Step 1

Verify that the current units are set to architectural and that the drawing limits set to 10', 8' for the upper-right corner. Be sure to perform a ZOOM-All on your screen. Draw the inner 3'-0" square using the RECTANG command, as shown on the left in the following image. Then offset the square 5' to the outside using the OFFSET command, as shown on the right in the following image. Because the square was drawn as a polyline, the entire shape offsets to the outside.

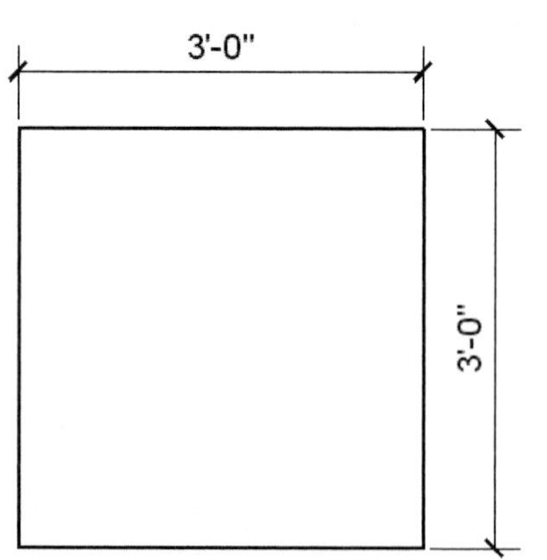

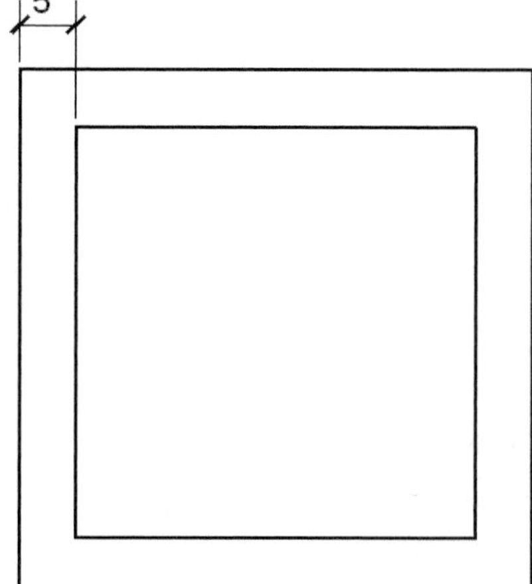

Figure 4.85

Step 2

Notice that when you click on the inner square on the left in the following image, the entire object highlights because it consists of a single polyline object. Use the EXPLODE command to break up the inner square into individual line segments. Now when you click on a line that is part of the inner square, only that line highlights, as shown on the right in the following image. This procedure is required in order to perform the next step.

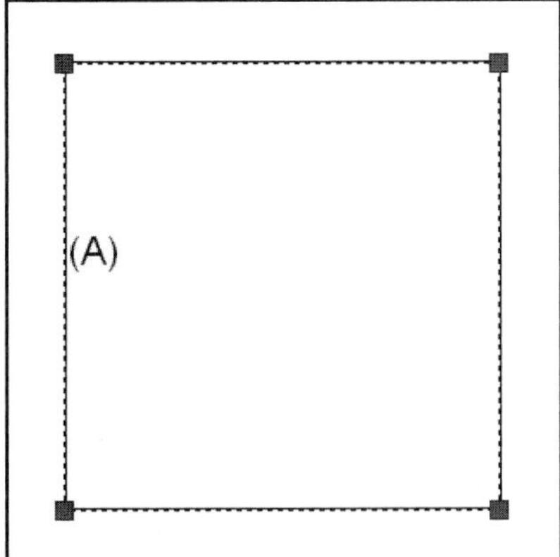

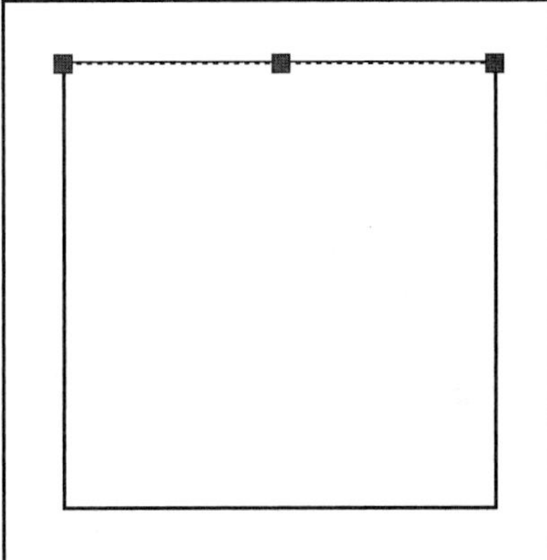

Figure 4.86

Step 3

You will now begin laying out the individual tiles with a spacing of 3" between each. The Array dialog box will be used to accomplish this. Activate the Array dialog box, change the number of rows to 1 and the number of columns to 12. Also change the distance between columns to 3". Finally, pick the line at "A" as the object to array, as shown on the left in the following image. Your display should appear similar to the illustration on the right in the following image.

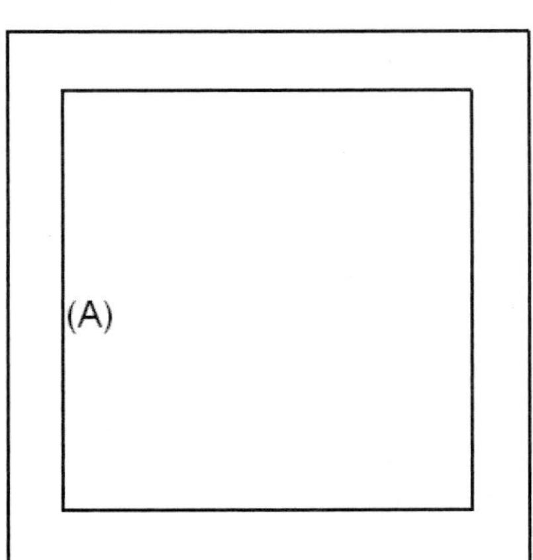

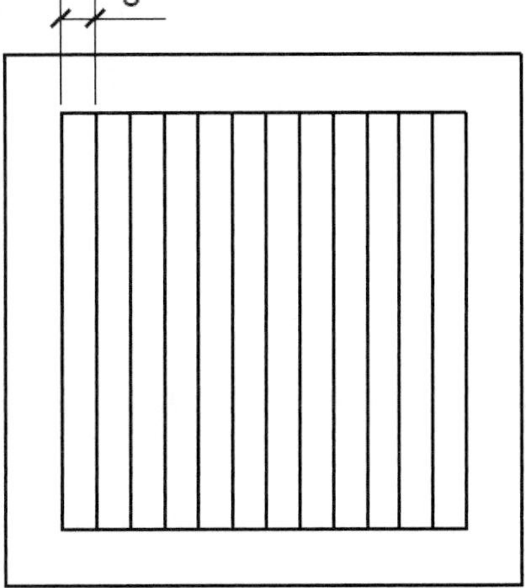

Figure 4.87

Step 4

The bottom horizontal line needs to be copied multiple times vertically. Activate the Array dialog box; change the number of rows to 12 and the number of columns to 1. Also change the distance between rows to 3". Finally, pick the line at "A" as the object to array, as shown on the left in the following image. Your display should appear similar to the illustration on the right in the following image.

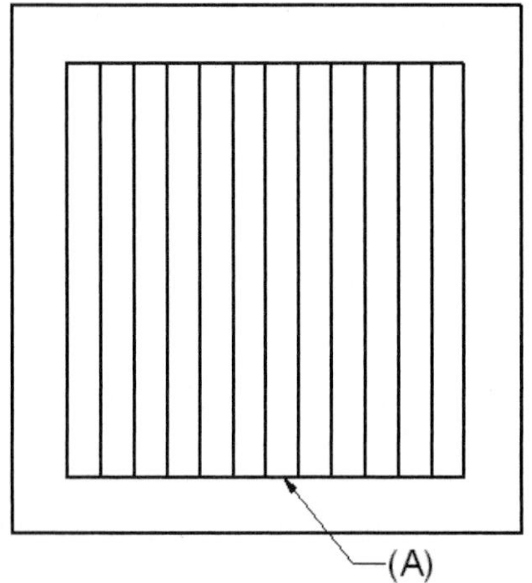

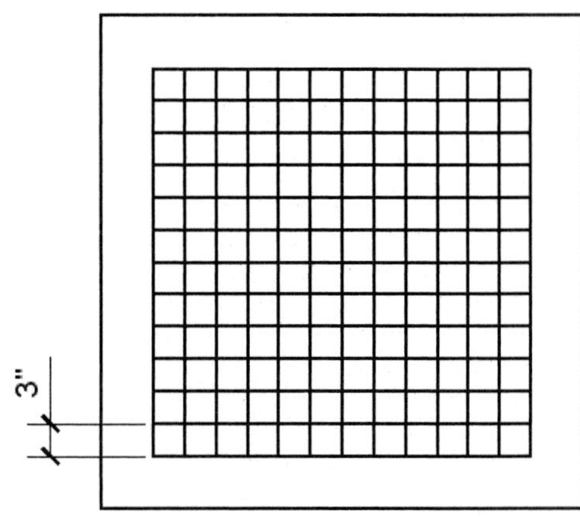

Figure 4.88

Step 5

The TRIM command will now be used to clean up the inner lines and form the 3' tiles. When using TRIM, do not press ENTER and select all cutting edges. This would be counterproductive. Instead, select the two vertical dashed lines, as shown on the left in the following image. Then trim away the horizontal segments in zones "A" through "D," as shown on the right in the following image. Remember that if you make a mistake and trim the wrong line, you can enter U in the command line to restore the previous trimmed line and pick the correct line.

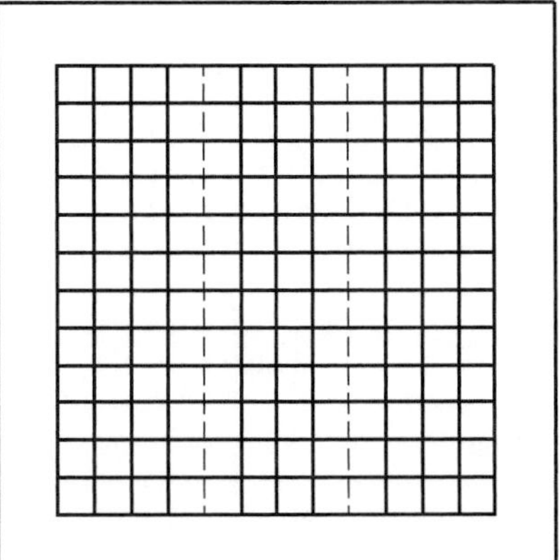

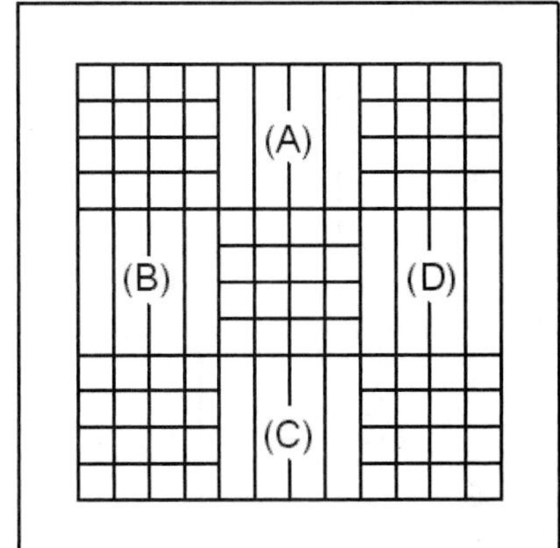

Figure 4.89

Step 6

Use the TRIM command again to finish cleaning up the object. Select the two horizontal dashed lines as cutting edges, as shown on the left in the following image. Then trim away the vertical segments in zones "A" through "E," as shown on the right in the following image. This completes this exercise on creating the tile.

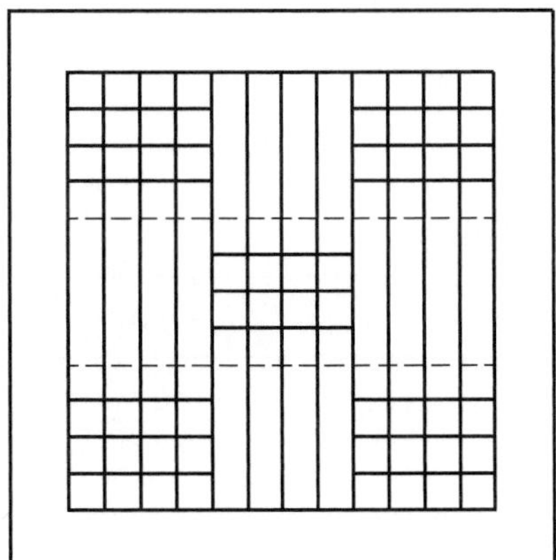

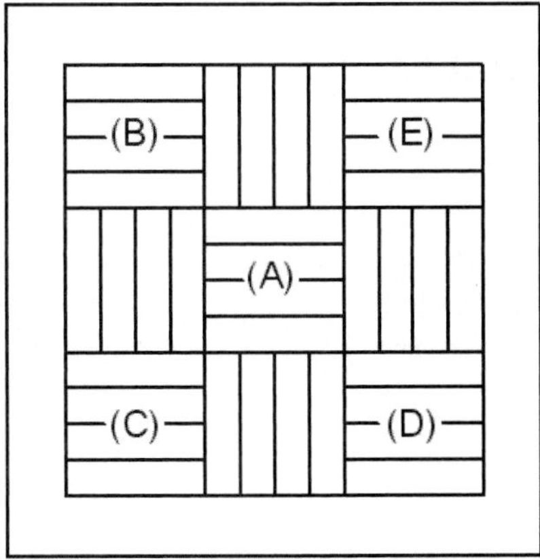

Figure 4.90

PROBLEMS FOR CHAPTER 4

DIRECTIONS FOR PROBLEMS 4–1 THROUGH 4–14

Construct each one-view drawing using the appropriate coordinate mode or Direct Distance mode. Utilize advanced commands such as ARRAY and MIRROR whenever possible.

PROBLEM 4–1

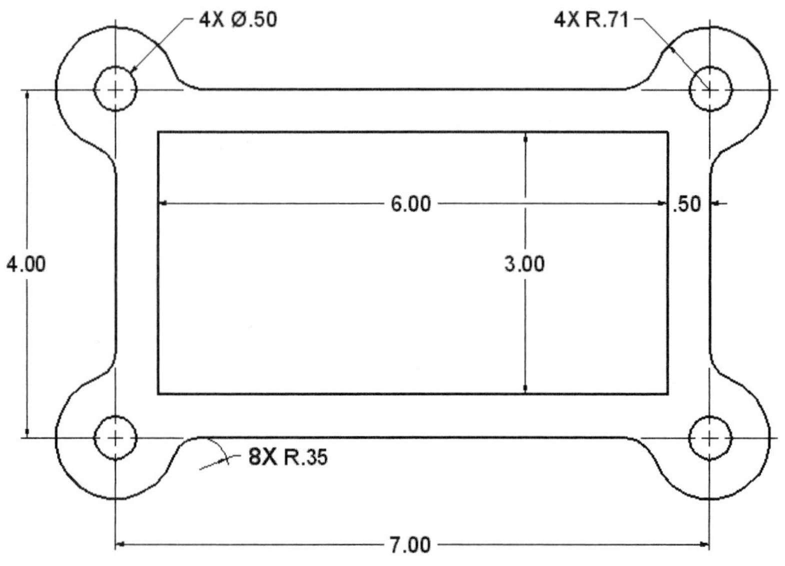

PROBLEM 4–2

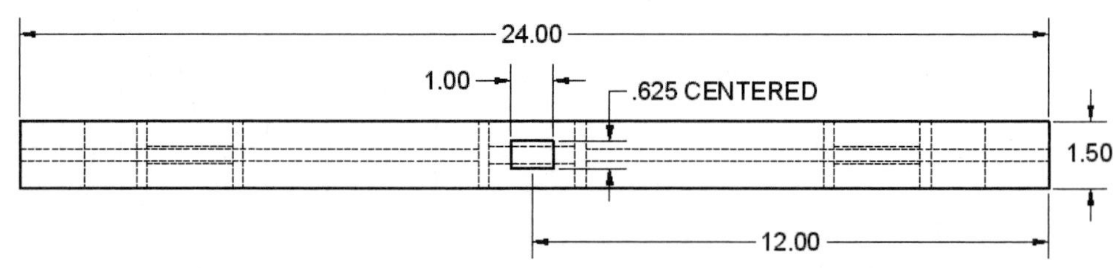

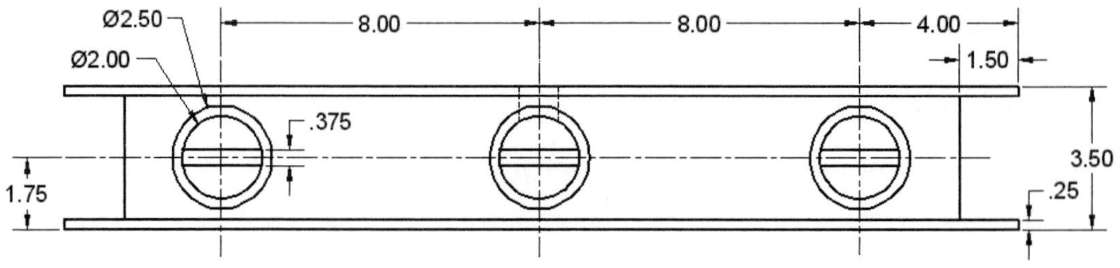

PROBLEM 4–3

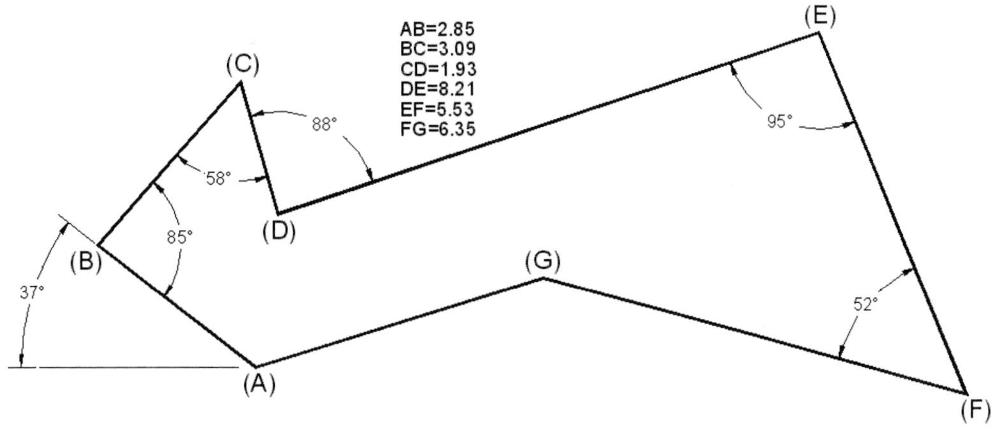

PROBLEM 4–4

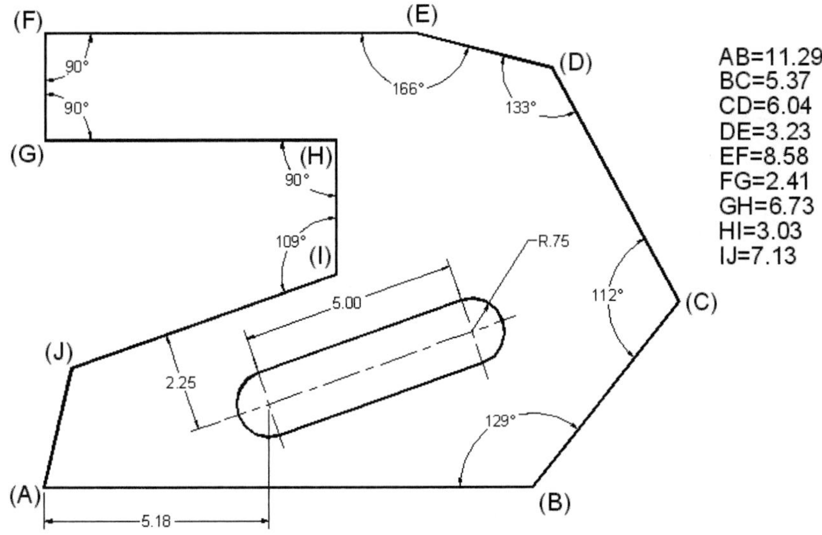

PROBLEM 4–5

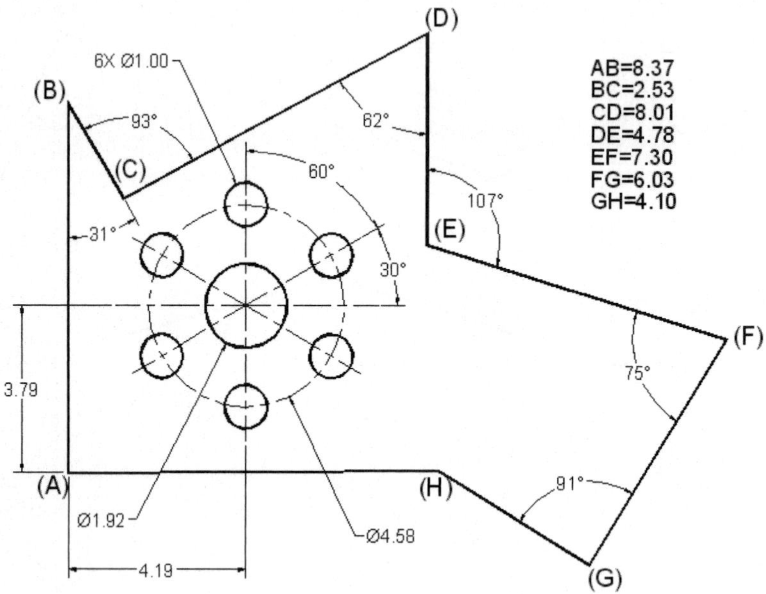

PROBLEM 4–6

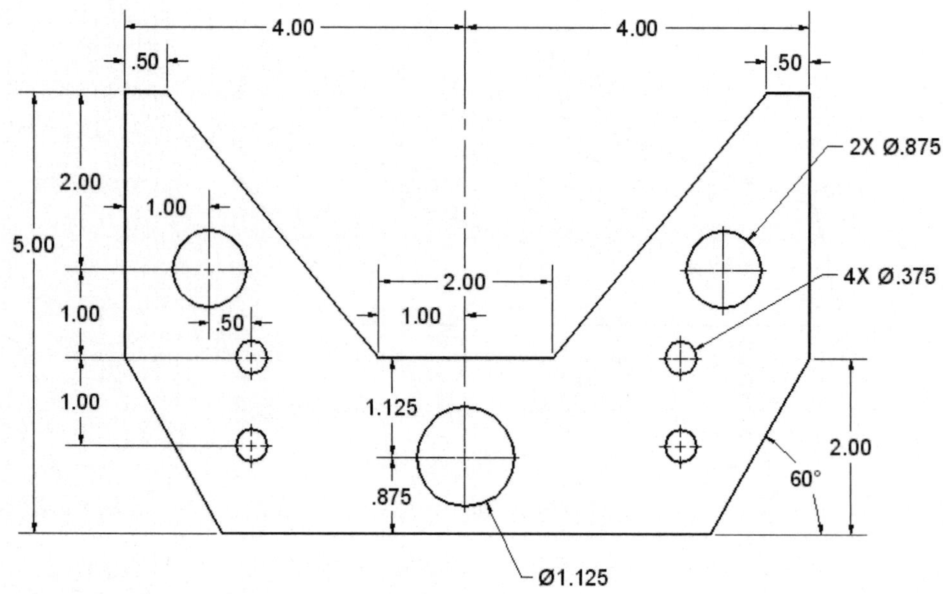

PROBLEM 4–7

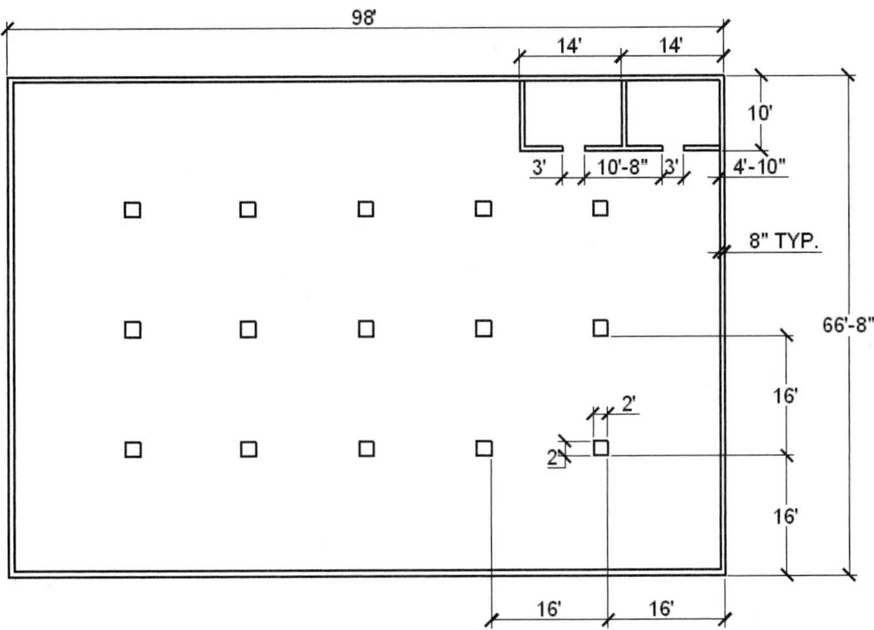

PROBLEM 4–8

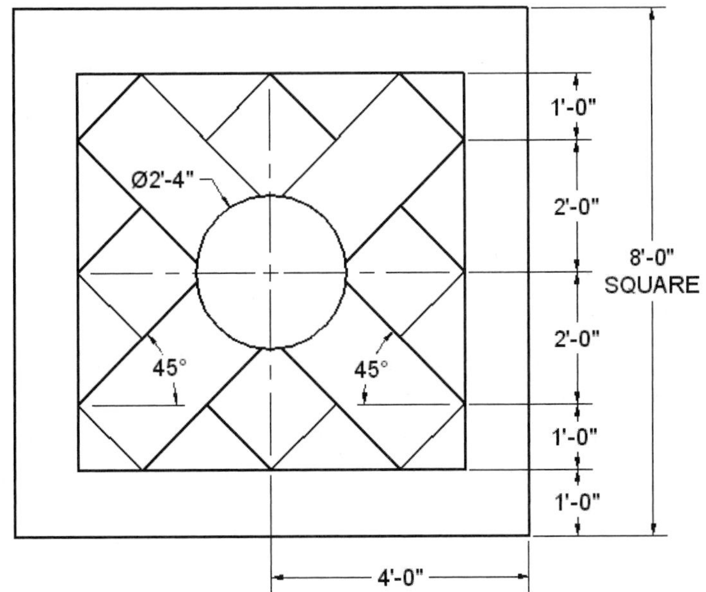

PROBLEM 4–9

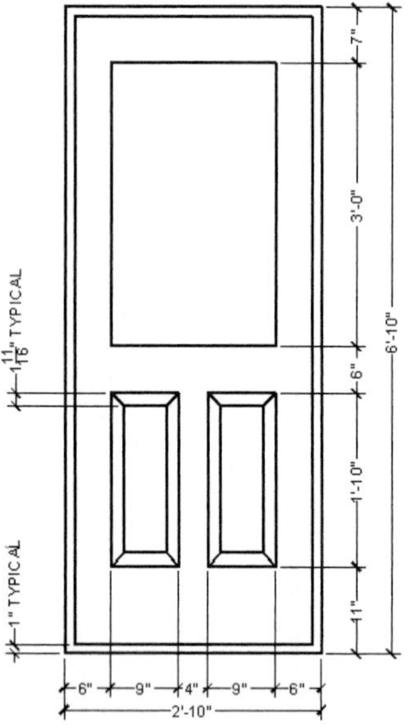

PROBLEM 4–10

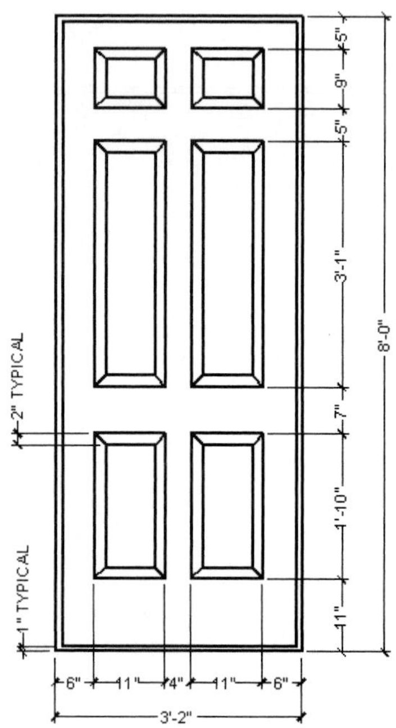

PROBLEM 4–11

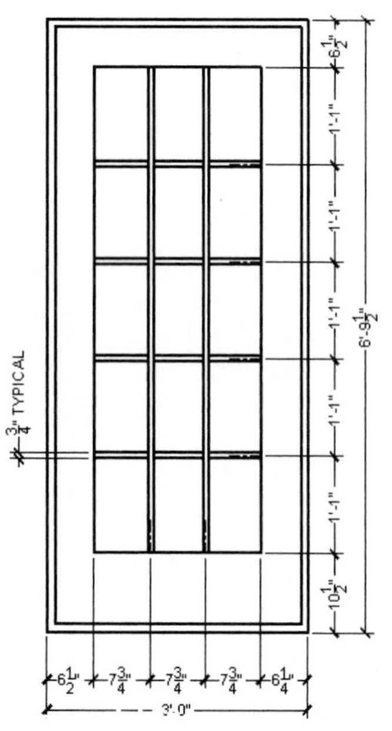

PROBLEM 4–12

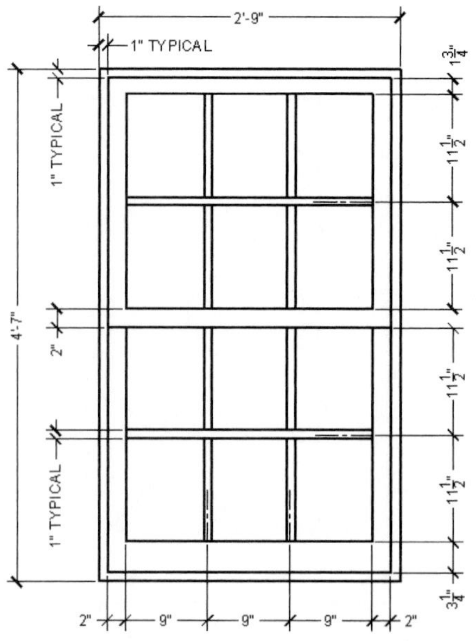

PROBLEM 4–13

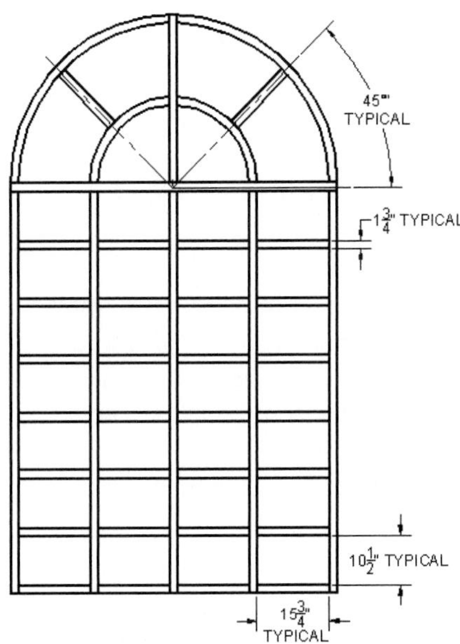

PROBLEM 4–14

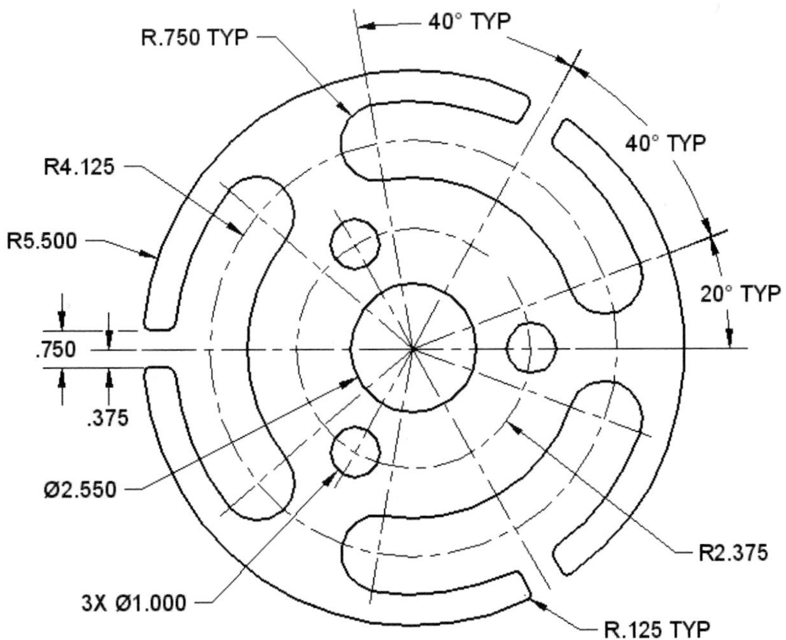

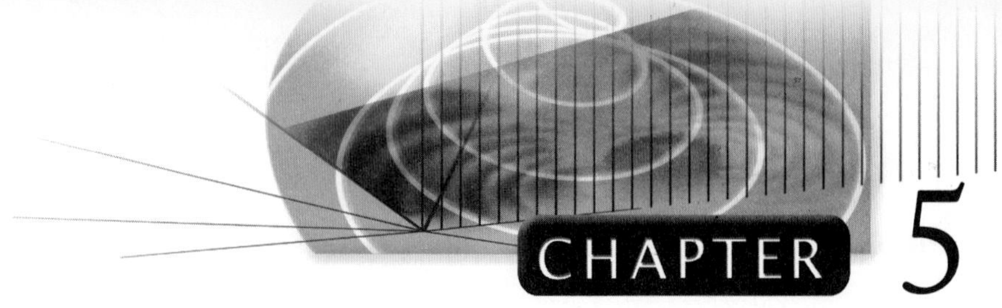

Performing Geometric Constructions

In chapter 1, the LINE, CIRCLE, and PLINE commands were introduced. The remainder of the drawing commands used for object creation are introduced in this chapter and include following: ARC, BOUNDARY, CIRCLE-2P, CIRCLE-3P, CIRCLE-TTR, CIRCLE-TTT, DONUT, ELLIPSE, POINT, DIVIDE, MEASURE, POLYGON, RAY, RECTANG, REVCLOUD, SPLINE, WIPEOUT, and XLINE.

Various scenarios of the CIRCLE-TTR command will be used to show the power of how AutoCAD can be used to create circles tangent to other object types. When covering the POINT command, the Point Style dialog box will also be discussed as a means of changing the appearance of points in your drawing. Multilines will also be covered, as will the ability to create a multiline style and then edit any intersections of a multiline.

METHODS OF SELECTING OTHER DRAW COMMANDS

You can find the main body of draw commands on the Menu Browser, Draw Toolbar, Tool Palette, pull-down menu, and Ribbon as shown in the following image.

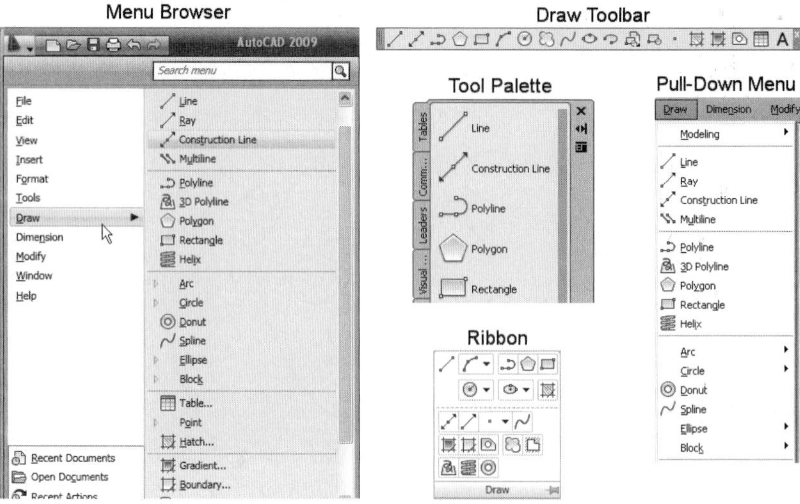

Figure 5.1

Refer to the following table for a complete listing of all buttons, tools, shortcuts, and functions of most drawing commands.

Button	Tool	Shortcut	Function
	Arc	A	Constructs an arc object using a number of command options
	Boundary	BO	Traces a polyline boundary over the top of an existing closed group of objects
	Circle	C	Constructs a circle by radius, diameter, 2 points, 3 points, 2 tangent points and a radius, and 3 tangent points
	Donut	DO	Creates a filled-in circular polyline object
	Ellipse	EL	Creates an elliptical object
None	Mline	ML	Creates multiline objects that consist of multiple parallel lines
	Point	PO	Creates a point object
None	Divide	DIV	Inserts evenly spaced points or blocks along an object's length or perimeter
None	Measure	ME	Inserts points or blocks along an object's length or perimeter at designated increments
	Polygon	POL	Creates an equilateral polygon shape consisting of various side combinations
	Ray	RAY	Creates a semi-infinite line used for construction purposes
	Rectang	REC	Creates a rectangular polyline object
	Revcloud	REVCLOUD	Creates a revision cloud consisting of a polyline made up of sequential arc segments
	Spline	SPL	Creates a smooth curve from a sequence of points
None	Wipeout	WIPEOUT	Use to cover or hide objects with a blank area
	Xline	XL	Creates an infinite line used for construction purposes

CONSTRUCTING ARCS

Use the ARC command to construct portions of circular shapes by radius, diameter, arc length, included angle, and direction. Choose this command from one of the following:

- The Draw toolbar of the AutoCAD Classic workspace
- From the Ribbon > Home Tab > Draw Panel

- The pull-down menu or Menu Browser (Draw > Arc)
- The keyboard (A or ARC)

Choosing Arc from the Draw pull-down menu displays the cascading menu shown in the following image. Choosing the Arc icon from the Ribbon displays an enlarged menu dealing with arc objects. Arcs can also be selected from the Menu Browser also shown in the following image. By default, the 3 Points Arc mode supports arc constructions in the clockwise as well as the counterclockwise direction. Providing a negative or positive angle in the Angle option also allows a clockwise or counterclockwise direction. All other arc modes support the ability to construct arcs only in the counterclockwise direction. Three arc examples will be demonstrated in the next series of pages, namely, how to construct an arc by 3 Points, by a Starting point, Center point, and Ending point, and how to Continue arcs.

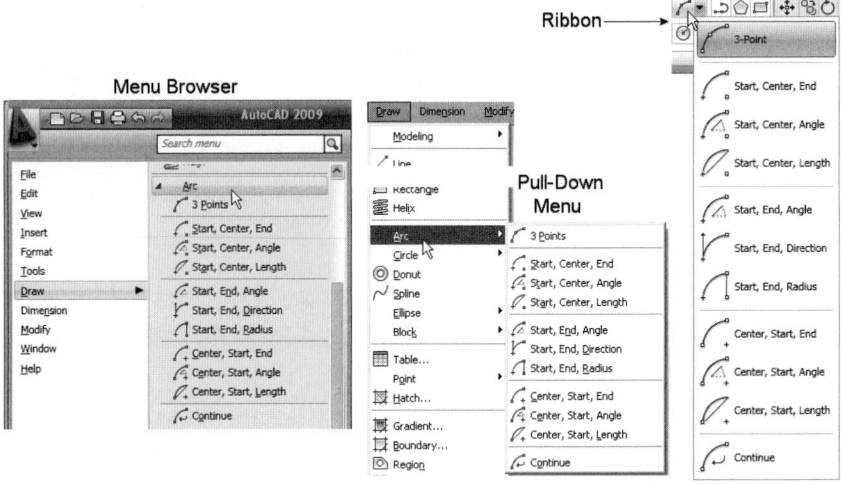

Figure 5.2

3 POINTS ARC MODE

By default, arcs are drawn with the 3 Points method. The first and third points identify the endpoints of the arc. This arc may be drawn in either the clockwise or counterclockwise direction.

Try It!: Create a new drawing file starting from scratch. Use the following command sequence and image to construct a 3-point arc.

```
Command: A (For ARC)
Specify start point of arc or [Center]: (Pick a point at "A")
Specify second point of arc or [Center/End]: (Pick a point at "B")
Specify end point of arc: (Pick a point at "C")
```

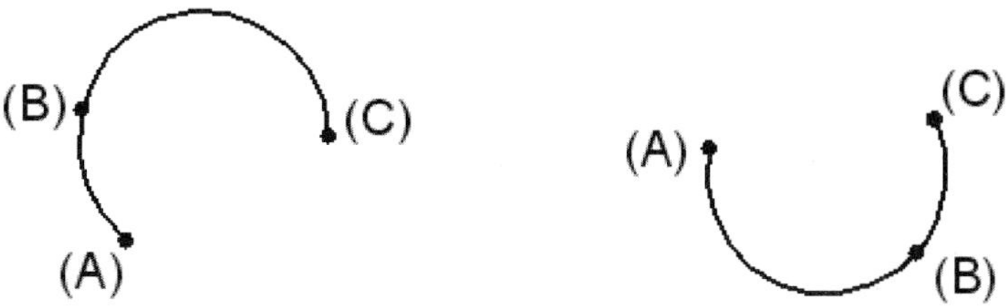

Figure 5.3

START, CENTER, END MODE

Use this ARC mode to construct an arc by defining its start point, center point, and endpoint. This arc will always be constructed in a counterclockwise direction.

Try It!: Open the drawing file 05_Door Swing. Use the following command sequence and image for constructing an arc representing the door swing by start, center, and end points.

```
Command: A (For ARC)
Specify start point of arc or [Center]: (Pick the endpoint at "A")
Specify second point of arc or [Center/End]: C (For Center)
Specify center point of arc: (Pick the endpoint at "B")
Specify end point of arc or [Angle/chord Length]: (Pick a point
    at "C")
```

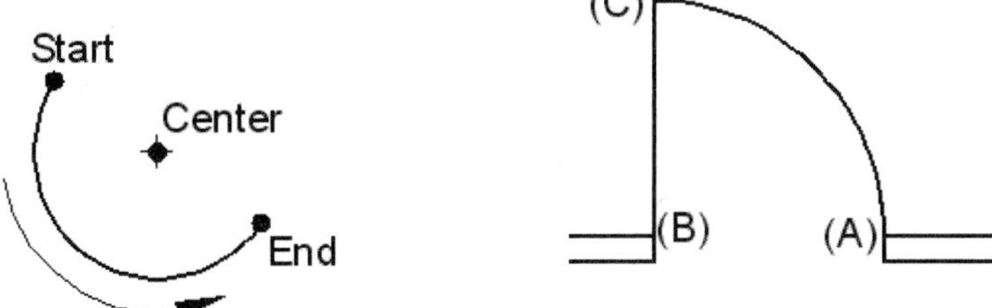

Figure 5.4

CONTINUE MODE

Use this mode to continue a previously drawn arc. All arcs drawn through Continue mode are automatically constructed tangent to the previous arc. Activate this mode from the Arc menu shown in the following image. The new arc begins at the last endpoint of the previous arc.

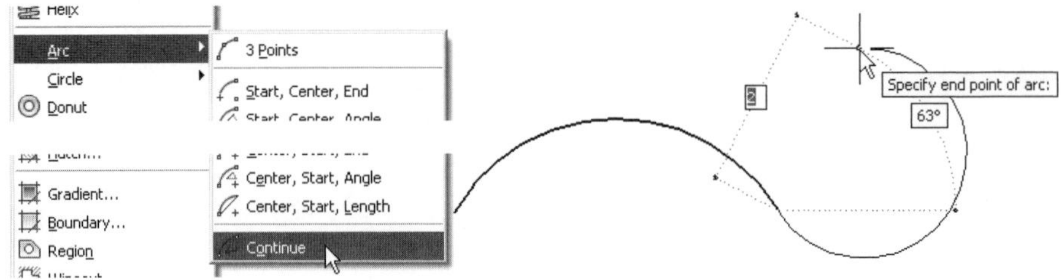

Figure 5.5

CREATING A BOUNDARY

The BOUNDARY command is used to create a polyline boundary around any closed shape. Choose this command from one of the following:

- The pull-down menu or Menu Browser (Draw > Boundary)
- From the Ribbon > Home Tab > Draw Panel
- The keyboard (BO or BOUNDARY)

It has already been demonstrated that the Join option of the PEDIT command is used to join object segments into one continuous polyline. The BOUNDARY command automates this process even more. Start this command by choosing Boundary from the Menu Browser, as shown on the left in the following image. This activates the Boundary Creation dialog box, shown on the right in the following image.

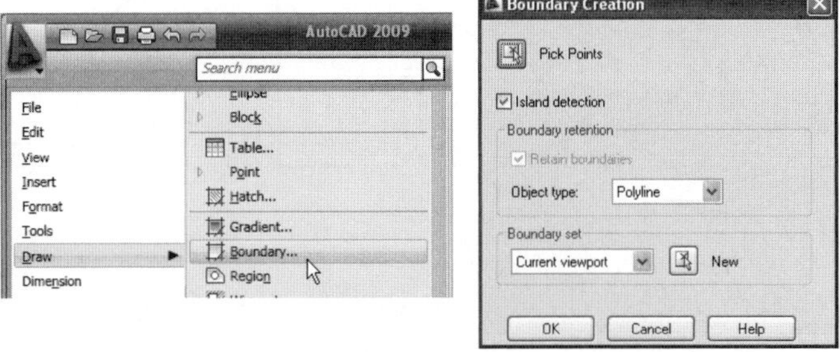

Figure 5.6

Before you use this command, it is good practice to create a separate layer to hold the polyline object; this layer could be called "Boundary" or "BP," for Boundary Polyline. Unlike the Join option of the PEDIT command, which converts individual objects to polyline objects, the BOUNDARY command traces a polyline in the current layer on top of individual objects.

Try It!: Open the drawing file 05_Boundary Extrusion. Activate the Boundary dialog box and click the Pick Points button. Then pick a point inside the object illustrated in the following image. Notice how the entire object is highlighted. To complete the command, press ENTER when prompted to select another internal point, and the polyline will be traced over the top of the existing objects.

```
Command: BO (For BOUNDARY)
(The Boundary Creation dialog box appears. Click the Pick Points
    button.)
Pick internal point: (Pick a point at "A" in the following
    image)
Selecting everything...
Selecting everything visible...
Analyzing the selected data...
Analyzing internal islands...
Pick internal point: (Press ENTER to construct the boundary)
BOUNDARY created 1 polyline
```

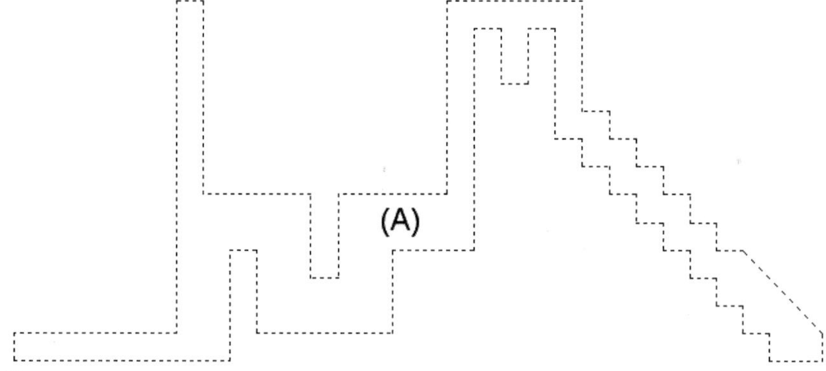

Figure 5.7

Once the boundary polyline is created, the boundary may be relocated to a different position on the screen with the MOVE command. The results are illustrated in the following image. The object on the left consists of the original individual objects. When the object on the right is selected, all objects highlight, signifying that the object is made up of a polyline object made through the use of the BOUNDARY command.

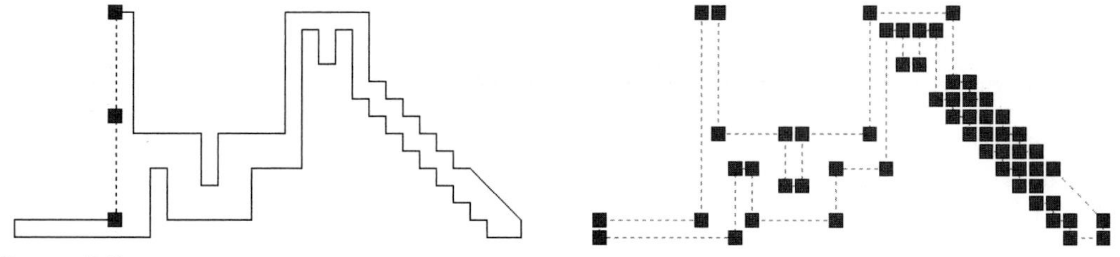

Figure 5.8

When the BOUNDARY command is used on an object consisting of an outline and internal islands similar to the drawing in the following image, a polyline object is also traced over these internal islands.

Try It!: Open the drawing file 05_Boundary Cover in the following image. Notice that the current layer is Boundary and the color is Magenta. Issue the BOUNDARY command and pick a point inside the middle of the object at "A" without OSNAP turned on. Notice that the polyline is constructed on the top of all existing objects. Turn the Object layer off to display just the Boundary layer.

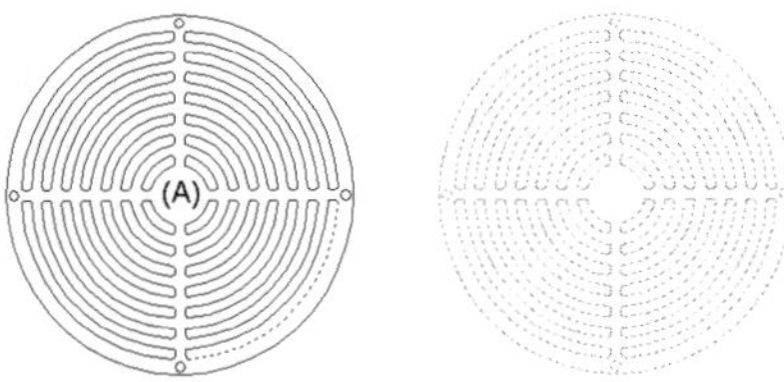

Figure 5.9

In addition to creating a special layer to hold the boundary polyline, another important rule to follow when using the BOUNDARY command is to be sure there are no gaps in the object. In the following image, when the BOUNDARY command encounters the gap at "A," a dialog box informs you that no internal boundary could be found. This is likely because the object is not completely closed. In this case, you must exit the command, close the object, and activate the BOUNDARY command again. It is acceptable, however, to have lines cross at the intersection at "B." While these "overshoots," as they are called, work well when using the Boundary Creation dialog box, this may not be considered a good drawing practice.

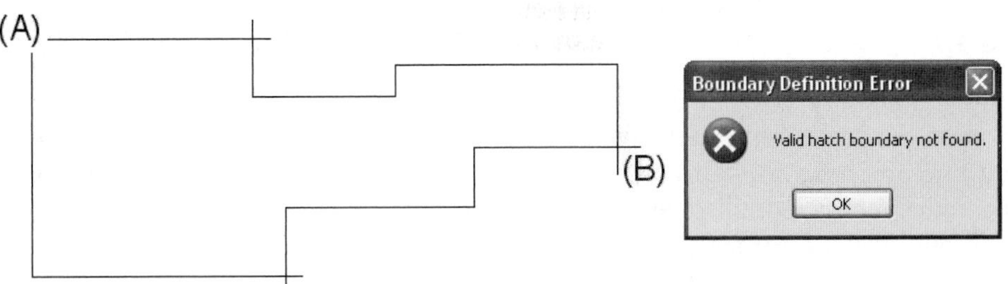

Figure 5.10

ADDITIONAL OPTIONS FOR CREATING CIRCLES

Additional options of the CIRCLE command may be selected from the Menu Browser under the Circle heading, as shown on the left in the following image, or from the Circle icon that is found under the Home tab of the Ribbon, as shown on the right in the following image. The 2 Points, 3 Points, Tan Tan Radius, and Tan Tan Tan modes will be explained in the next series of examples.

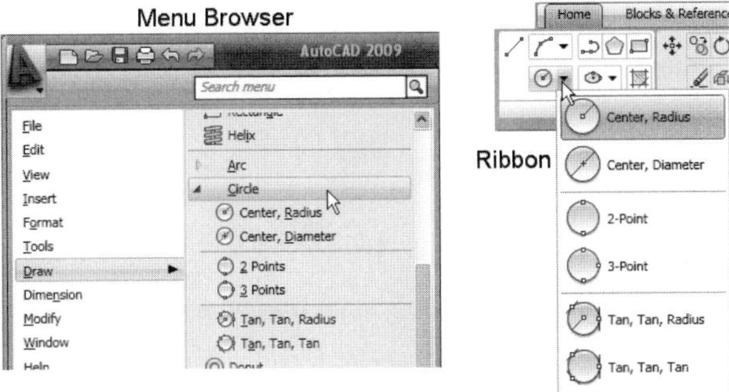

Figure 5.11

3 POINTS CIRCLE MODE

Use the CIRCLE command and the 3 Points mode to construct a circle by three points that you identify. No center point is utilized when you enter the 3 Points mode. Simply select three points and the circle is drawn.

Try It!: Open the drawing file 05_Circle_3P. Study the following prompts and image for constructing a circle using the 3 Points mode.

```
Command: C (For CIRCLE)
Specify center point for circle or [3P/2P/Ttr (tan tan
    radius)]: 3P
Specify first point on circle: (Pick a midpoint at "A")
Specify second point on circle: (Pick a midpoint at "B")
Specify third point on circle: (Pick a midpoint at "C")
```

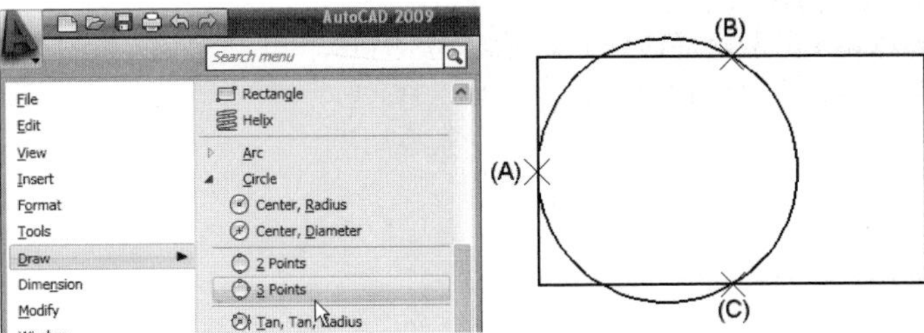

Figure 5.12

2 POINTS CIRCLE MODE

Use the CIRCLE command and the 2 Points mode to construct a circle by selecting two points. Choose this command from the Menu Browser as shown in the following image. These points form the diameter of the circle. No center point is utilized when you use the 2 Points mode.

Try It!: Open the drawing file 05_Circle_2P. Study the following prompts and image for constructing a circle using the 2 Points mode.

```
Command: C (For CIRCLE)
Specify center point for circle or [3P/2P/Ttr (tan tan
    radius)]: 2P
Specify first end point of circle's diameter: (Pick a midpoint
    at "A")
Specify second end point of circle's diameter: (Pick a midpoint
    at "B")
```

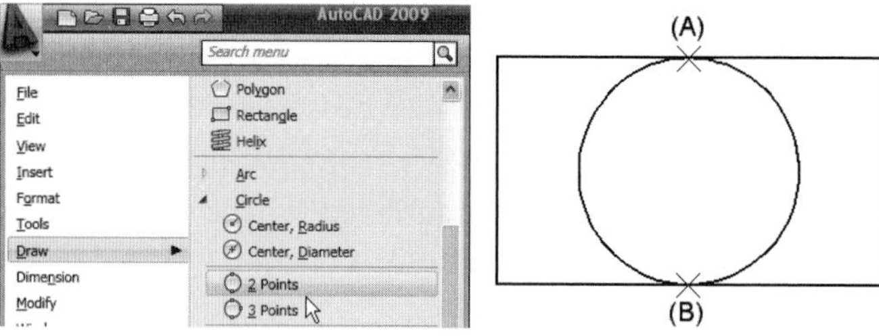

Figure 5.13

CONSTRUCTING AN ARC TANGENT TO TWO LINES USING CIRCLE-TTR

Illustrated on the left in the following image are two inclined lines. The purpose of this example is to connect an arc tangent to the two lines at a specified radius. The CIRCLE-TTR (Tangent-Tangent-Radius) command will be used here along with the TRIM command to clean up the excess geometry. To assist with this operation, the OSNAP-Tangent mode is automatically activated when you use the TTR option of the CIRCLE command.

First, use the CIRCLE-TTR command to construct an arc tangent to both lines, as shown in the middle in the following image.

```
Command: C (For CIRCLE)
Specify center point for circle or [3P/2P/Ttr (tan tan radius)]:
    T (For TTR)
Specify point on object for first tangent of circle: (Select the
    line at "A")
Specify point on object for second tangent of circle: (Select
    the line at "B")
Specify radius of circle: (Enter a radius value)
```

Use the TRIM command to clean up the lines and arc. The completed result is illustrated on the right in the following image. The FILLET command could also have been used for this procedure. Not only will the curve be drawn, but this command also automatically trims the lines.

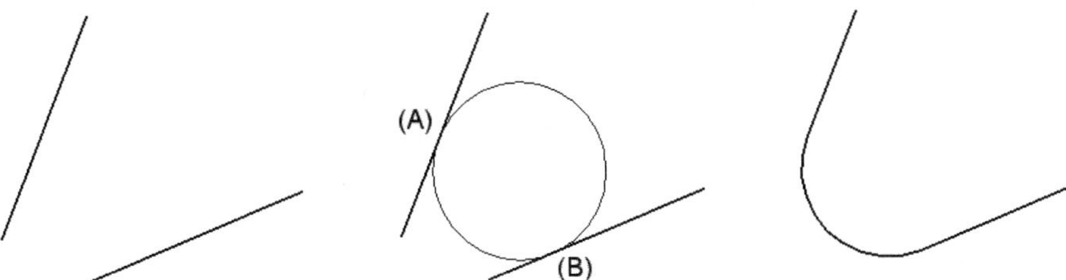

Figure 5.14

Try It!: Open the drawing file 05_TTR1. Use the CIRCLE command and the TTR option to construct a circle tangent to lines "A" and "B" in the following image. Use a circle radius of 0.50 units. With the circle constructed, use the TRIM command and select the circle as well as lines "C" and "D" as cutting edges. Trim the circle at "E" and the lines at "C" and "D." Observe the final results of this operation shown in the following image.

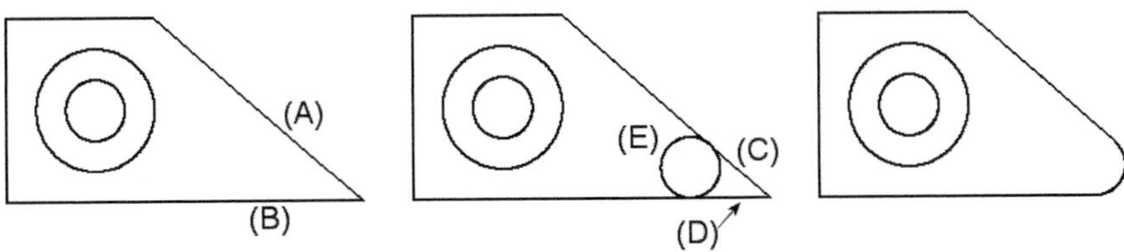

Figure 5.15

CONSTRUCTING AN ARC TANGENT TO A LINE AND ARC USING CIRCLE-TTR

Illustrated on the left in the following image is an arc and an inclined line. The purpose of this example is to connect an additional arc tangent to the original arc and line at a specified radius. The CIRCLE-TTR command will be used here, along with the TRIM command to clean up the excess geometry.

First, use the CIRCLE-TTR command to construct an arc tangent to the arc and inclined line, as shown on the right in the following image.

```
[Tan, Tan, Radius] Command: C (For CIRCLE)
Specify center point for circle or [3P/2P/Ttr (tan tan radius)]:
    T (For TTR)
Specify point on object for first tangent of circle: (Select the
    arc at "A")
Specify point on object for second tangent of circle: (Select
    the line at "B")
Specify radius of circle: (Enter a radius value)
```

Take note that the radius must be greater than or equal to half the distance between the circle and the line; otherwise, the second circle cannot be constructed. Use the TRIM command to clean up the arc and line. The completed result is illustrated at the bottom in the following image.

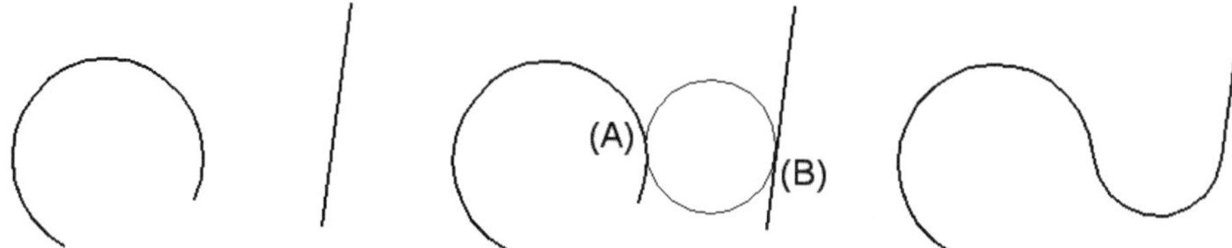

Figure 5.16

Try It!: Open the drawing file 05_TTR2. Using the following image as a guide, use the CIRCLE command and the TTR option to construct a circle tangent to the line at "A" and arc at "B." Use a circle radius value of 0.50 units. Construct a second circle tangent to the arc at "C" and line at "D" using the default circle radius value of 0.50 units. With the circles constructed, use the TRIM command, press ENTER to select all cutting edges, and trim the lines at "E" and "J" and the arc at "F" and "H," in addition to the circles at "G" and "K." Observe the final results of this operation shown in the following image.

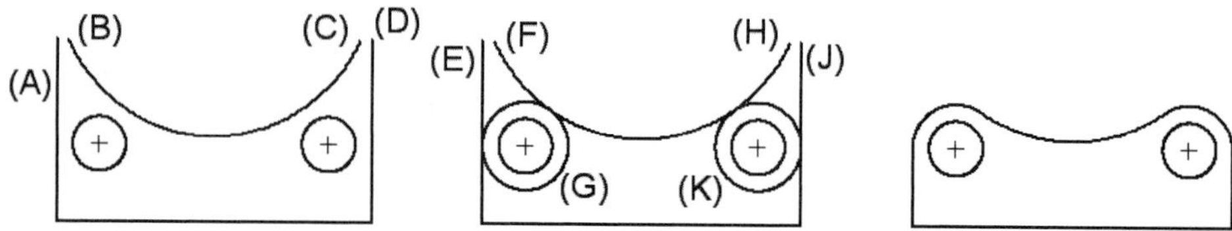

Figure 5.17

CONSTRUCTING AN ARC TANGENT TO TWO ARCS USING CIRCLE-TTR

Method #1

Illustrated on the left in the following image are two arcs. The purpose of this example is to connect a third arc tangent to the original two at a specified radius. The CIRCLE-TTR command will be used here along with the TRIM command to clean up the excess geometry.

Use the CIRCLE-TTR command to construct an arc tangent to the two original arcs, as shown in the middle in the following image.

```
[Tan, Tan, Radius] Command: C (For CIRCLE)
Specify center point for circle or [3P/2P/Ttr (tan tan radius)]:
    T (For TTR)
Specify point on object for first tangent of circle: (Select the
    arc at "A")
Specify point on object for second tangent of circle: (Select
    the arc at "B")
Specify radius of circle: (Enter a radius value)
```

Use the TRIM command to clean up the two arcs, using the circle as a cutting edge. The completed result is illustrated on the right in the following image.

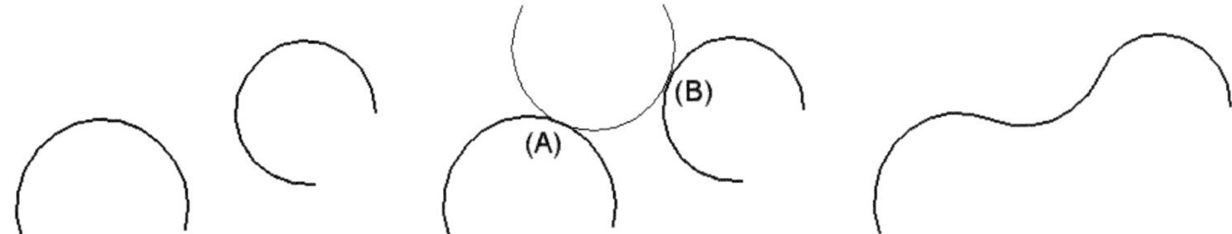

Figure 5.18

Try It!: Open the drawing file 05_TTR3. Use the CIRCLE command and the TTR option to construct a circle tangent to the circles at "A" and "B." Construct a second circle tangent to the circles at "C" and "D" in the following image. Use a circle radius of 4.50 units for both circles. With the circles constructed, use the TRIM command and select both small circles as cutting edges. Trim the circles at "E" and "F." Observe the final results of this operation shown in the following image.

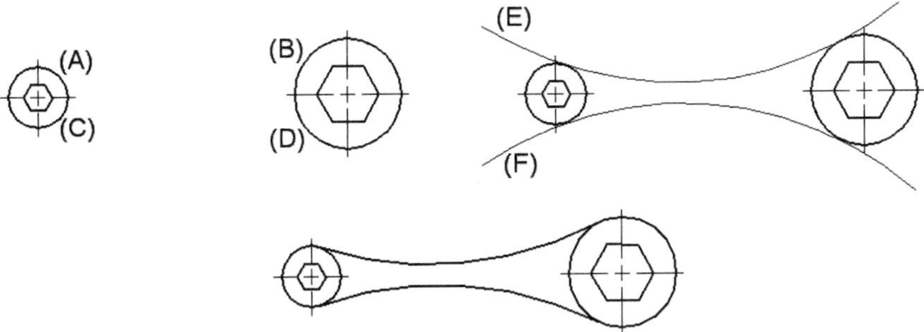

Figure 5.19

CONSTRUCTING AN ARC TANGENT TO TWO ARCS USING CIRCLE-TTR

Method #2

Illustrated on the left in the following image are two arcs. The purpose of this example is to connect an additional arc tangent to and enclosing both arcs at a specified radius. The CIRCLE-TTR command will be used here along with the TRIM command.

First, use the CIRCLE-TTR command to construct an arc tangent to and enclosing both arcs, as shown in the middle in the following image.

```
Command: C (For CIRCLE)
Specify center point for circle or [3P/2P/Ttr (tan tan radius)]:
    T (For TTR)
Specify point on object for first tangent of circle: (Select the
    arc at "A")
Specify point on object for second tangent of circle: (Select
    the arc at "B")
Specify radius of circle: (Enter a radius value)
```

Use the TRIM command to clean up all arcs. The completed result is illustrated on the right in the following image.

Figure 5.20

Try It!: Open the drawing file 05_TTR4. Use the CIRCLE command and the TTR option to construct a circle tangent to the circles at "A" and "B." Construct a second circle tangent to the circles at "C" and "D" in the following image. Use a circle radius of 1.50 units for both circles. With the circles constructed, use the TRIM command and trim all circles until you achieve the final results of this operation shown in the following image.

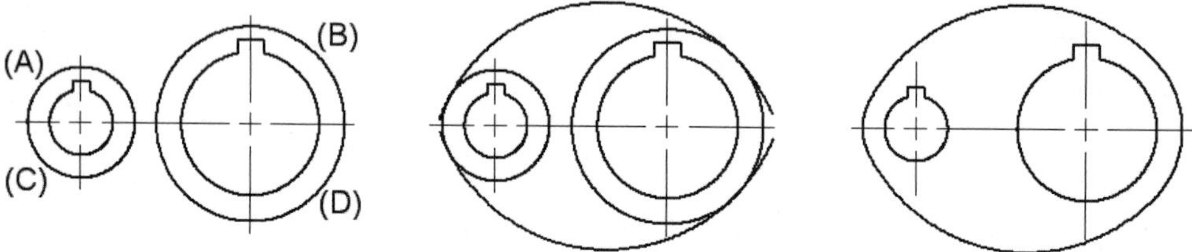

Figure 5.21

CONSTRUCTING AN ARC TANGENT TO TWO ARCS USING CIRCLE-TTR

Method #3

Illustrated on the left in the following image are two arcs. The purpose of this example is to connect an additional arc tangent to one arc and enclosing the other. The CIRCLE-TTR command will be used here along with the TRIM command to clean up unnecessary geometry.

First, use the CIRCLE-TTR command to construct an arc tangent to the two arcs. Study the illustration in the middle in the following image and the following prompts to understand the proper pick points for this operation. It is important to select the tangent points in the approximate location of the tangent to each circle in order to obtain the desired arc tangent orientation.

```
Tan, Tan, Radius  Command: C (For CIRCLE)
Specify center point for circle or [3P/2P/Ttr (tan tan radius)]:
    T (For TTR)
Specify point on object for first tangent of circle: (Select the
    arc at "A")
Specify point on object for second tangent of circle: (Select
    the arc at "B")
Specify radius of circle: (Enter a radius value)
```

Use the TRIM command to clean up the arcs. The completed result is illustrated on the right in the following image.

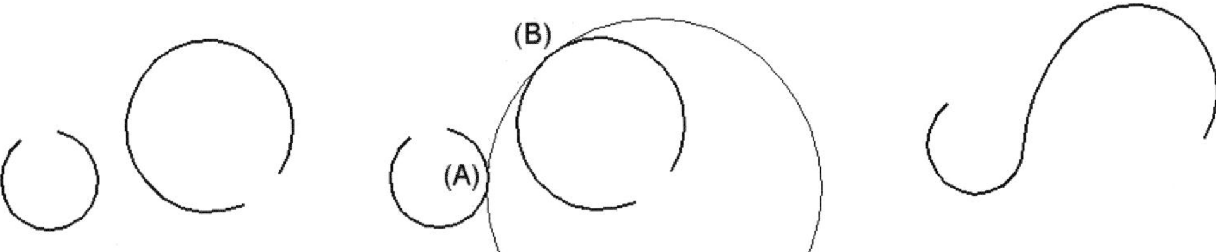

Figure 5.22

Try It!: Open the drawing file 05_TTR5. Use the CIRCLE command and the TTR option to construct a circle tangent to the circles at "A" and "B." Construct a second circle tangent to the circles at "C" and "D" in the following image. Use a circle radius of 3.00 units for both circles. With the circles constructed, use the TRIM command and trim all circles until you achieve the final results of this operation shown in the following image.

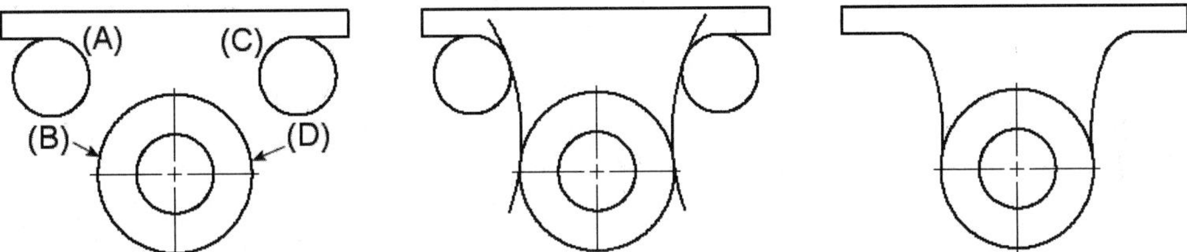

Figure 5.23

CONSTRUCTING A LINE TANGENT TO TWO ARCS OR CIRCLES

Illustrated on the left in the following image are two circles. The purpose of this example is to connect the two circles with two tangent lines. This can be accomplished with the LINE command and the OSNAP-Tangent option.

Use the LINE command to connect two lines tangent to the circles, as shown in the middle in the following image. The following procedure is used for the first line. Use the same procedure for the second.

```
Command: L (For LINE)
Specify first point: Tan
to (Select the circle near "A")
Specify next point or [Undo]: Tan
to (Select the circle near "B")
Specify next point or [Undo]: (Press ENTER to exit this command)
```

Use the TRIM command to clean up the circles so that the appearance of the object is similar to the illustration on the right in the following image.

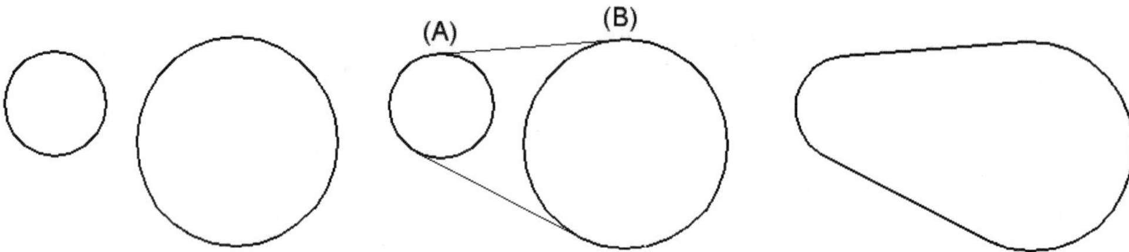

Figure 5.24

Try It!: Open the drawing file 05_Tangent Lines. Construct a line tangent to the two circles at "A" and "B." Repeat this procedure for the circles at "C" and "D" in the following image. Use the TRIM command and trim all circles until you achieve the final results of this operation shown in the following image.

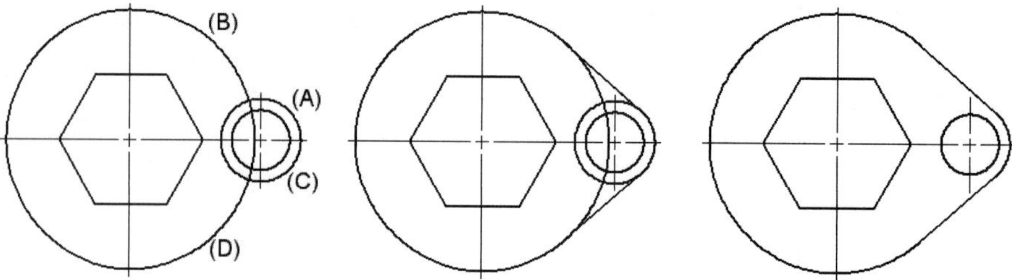

Figure 5.25

CIRCLE—TAN TAN TAN

Yet another mode of the CIRCLE command allows you to construct a circle based on three tangent points. This mode is actually a variation of the 3 Points mode together with using the OSNAP-Tangent mode three times. This mode requires you to select three objects, as in the example of the three line segments shown on the right in the following image.

```
Tan, Tan, Tan  Command: C (For CIRCLE)
Specify center point for circle or [3P/2P/Ttr (tan tan
    radius)]: 3P
Specify first point on circle: Tan
to (Select the line at "A")
Specify second point on circle: Tan
to (Select the line at "B")
Specify third point on circle: Tan
to (Select the line at "C")
```

The result is illustrated on the right in the following image, with the circle being constructed tangent to the edges of all three line segments.

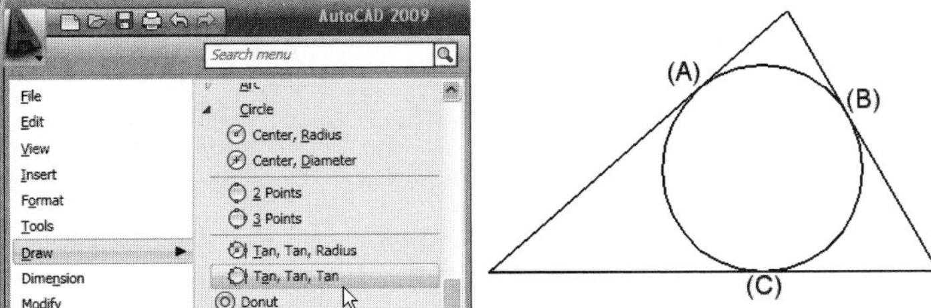

Figure 5.26

Try It!: Open the drawing file 05_TTT. Use the illustration in the previous image and the previous command sequence for constructing a circle tangent to three lines.

QUADRANT VERSUS TANGENT OSNAP OPTION

Various examples have been given on previous pages concerning drawing lines tangent to two circles, two arcs, or any combination of the two. The object on the left in the following image illustrates the use of the OSNAP-Tangent option when used with the LINE command.

```
Command: L (For LINE)
Specify first point: Tan
to (Select the arc near "A")
Specify next point or [Undo]: Tan
to (Select the arc near "B")
Specify next point or [Undo]: (Press ENTER to exit this command)
```

Note that the angle of the line formed by points "A" and "B" is neither horizontal nor vertical. This is a typical example of the capabilities of the OSNAP-Tangent option.

The object illustrated on the right in the following image is a modification of the example on the left with the inclined tangent lines changed to horizontal and vertical tangent lines. This example is to show that two OSNAP options are available for performing tangencies, namely OSNAP-Tangent and OSNAP-Quadrant. However, it is up to you to evaluate under what conditions to use these OSNAP options. In the example on the right, you can use the OSNAP-Tangent or OSNAP-Quadrant option to draw the lines tangent to the arcs. The Quadrant option could be used only because the lines to be drawn are perfectly horizontal or vertical. Usually it is impossible to know this ahead of time, and in this case, the OSNAP-Tangent option should be used whenever possible.

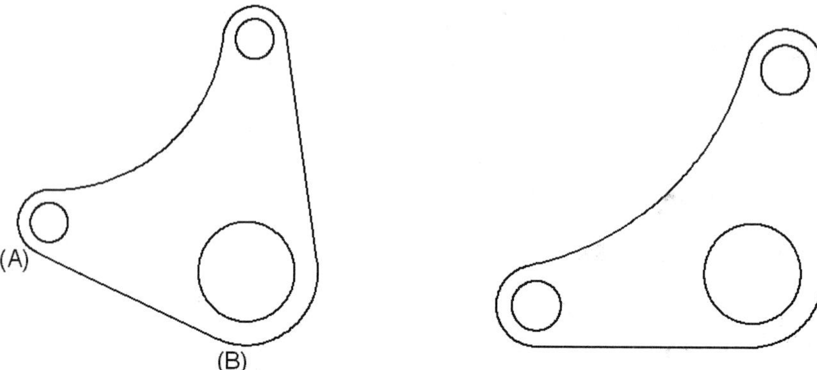

Figure 5.27

CREATING FILLED-IN DOTS (DONUTS)

Use the DONUT command to construct a filled-in circle. Choose this command from one of the following:

- The pull-down menu or Menu Browser (Draw > Donut)
- From the Ribbon > Home Tab > Draw Panel (Expanded)
- The keyboard (DO or DONUT)

This object belongs to the polyline family. The donut illustrated on the left in the following image has an inside diameter of 0.50 units and an outside diameter of 1.00 units. When you place donuts in a drawing, the Multiple option is automatically invoked. This means you can place as many donuts as you like until you exit the command.

Try It!: Create a new drawing file starting from scratch. Use the following command sequence for constructing this type of donut.

```
Command: DO (For DONUT)
Specify inside diameter of donut <0.50>: (Press ENTER to accept
    the default)
Specify outside diameter of donut <1.00>: (Press ENTER to accept
    the default)
Specify center of donut or <exit>: (Pick a point to place the
    donut)
Specify center of donut or <exit>: (Pick a point to place
    another donut or press ENTER to exit this command)
```

Setting the inside diameter of a donut to a value of zero (0) and an outside diameter to any other value constructs a donut representing a dot, as shown on the right in the following image.

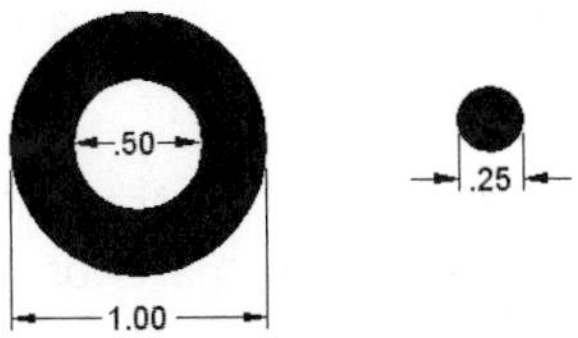

Figure 5.28

Try It!: Create a new drawing file starting from scratch. Use the following command sequence for constructing this type of donut.

```
Command: DO (For DONUT)
Specify inside diameter of donut <0.50>: 0
Specify outside diameter of donut <1.00>: 0.25
Specify center of donut or <exit>: (Pick a point to place the
    donut)
Specify center of donut or <exit>: (Pick a point to place
    another donut or press ENTER to exit this command)
```

Try It!: Open the drawing file 05_Donut. Activate the DONUT command and set the inside diameter to 0 and the outside diameter to .05. Place four donuts at the intersections of the electrical circuit, as shown in the following image.

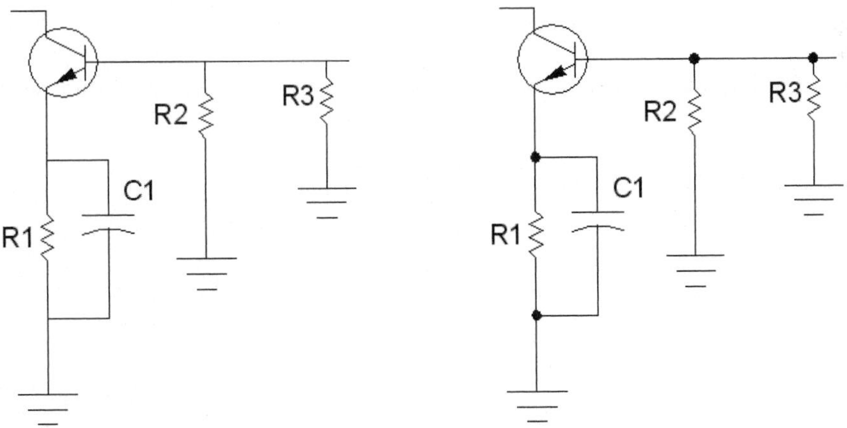

Figure 5.29

CONSTRUCTING ELLIPTICAL SHAPES

Use the ELLIPSE command to construct a true elliptical shape. Choose this command from one of the following:

- The Draw toolbar of the AutoCAD Classic workspace
- From the Ribbon > Home Tab > Draw Panel

- The pull-down menu or Menu Browser (Draw > Ellipse)
- The keyboard (EL or ELLIPSE)

Before studying the three examples for ellipse construction, see the illustration on the right in the following image to view two important parts of any ellipse, namely its major and minor diameters.

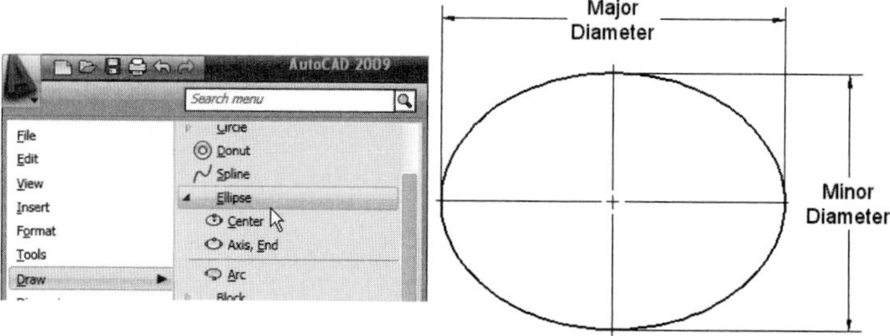

Figure 5.30

You can construct an ellipse by marking two points, which specify one of its axes, as shown in the following image. These first two points also identify the angle with which the ellipse will be drawn. Responding to the prompt "Specify distance to other axis or [Rotation]" with another point identifies half of the other axis. The rubber-banded line is added to assist you in this ellipse construction method.

Try It!: Create a new drawing file starting from scratch. Use the illustration in the following image and the command sequence to construct an ellipse by locating three points.

```
Command: EL (For ELLIPSE)
Specify axis endpoint of ellipse or [Arc/Center]: (Pick a point
    at "A")
Specify other endpoint of axis: (Pick a point at "B")
Specify distance to other axis or [Rotation]: (Pick a point
    at "C")
```

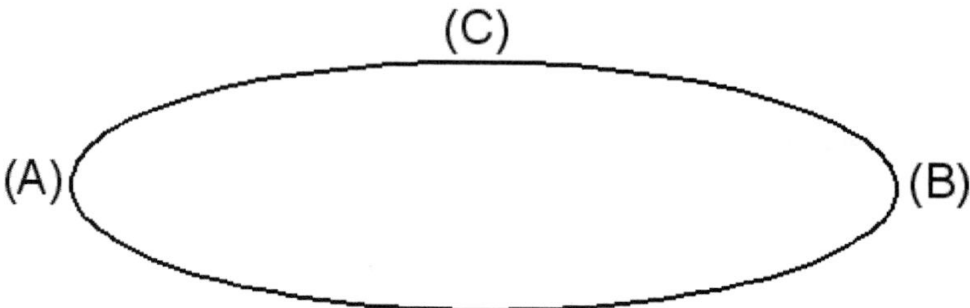

Figure 5.31

You can also construct an ellipse by first identifying its center. You can pick points to identify its axes or use polar coordinates to accurately define the major and minor diameters of the ellipse, as shown on the left in the following image.

Try It!: Create a new drawing file starting from scratch. Use the illustration on the left in the following image and the following command sequence for constructing an ellipse based on a center. Use a polar coordinate or the Direct Distance mode for locating the two axis endpoints of the ellipse. For this example turn Polar or Ortho modes on.

```
Command: EL (For ELLIPSE)
Specify axis endpoint of ellipse or [Arc/Center]: C (For Center)
Specify center of ellipse: (Pick a point at "A")
Specify endpoint of axis: (Move your cursor to the right and
    enter a value of 2.50)
Specify distance to other axis or [Rotation]: (Move your cursor
    straight up and enter a value of 1.50)
```

The last method of constructing an ellipse, as shown on the right in the following image, illustrates constructing an ellipse by way of rotation. Identify the first two points for the first axis. Reply to the prompt "Specify distance to other axis or [Rotation]" with Rotation. The first axis defined is now used as an axis of rotation that rotates the ellipse into a third dimension.

Try It!: Create a new drawing file starting from scratch. Use the illustration on the right in the following image and the following command sequence for constructing an ellipse based on a rotation around the major axis.

```
Command: EL (For ELLIPSE)
Specify axis endpoint of ellipse or [Arc/Center]: C (For Center)
Specify center of ellipse: (Pick a point at "A")
Specify endpoint of axis: (Pick a point at "B")
Specify distance to other axis or [Rotation]: R (For Rotation)
Specify rotation around major axis: 80
```

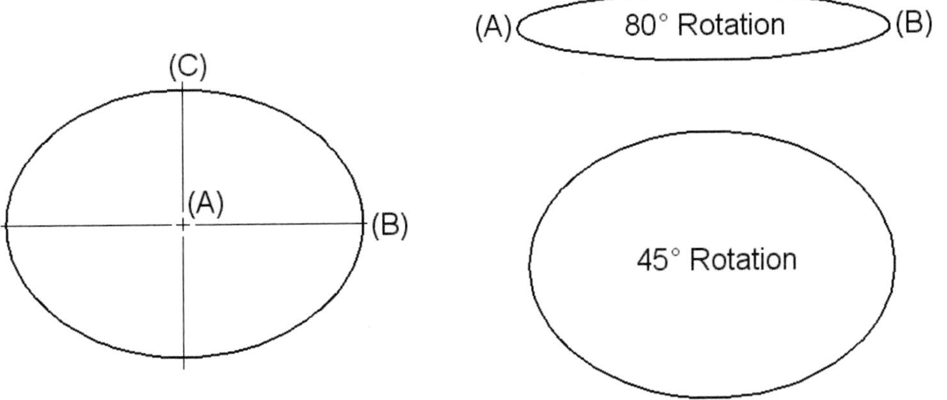

Figure 5.32

CREATING POINT OBJECTS

Use the POINT command to identify the location of a point on a drawing, which may be used for reference purposes. Choose this command from one of the following:

- The Draw toolbar of the AutoCAD Classic workspace
- From the Ribbon > Home Tab > Draw Panel (Expanded)
- The pull-down menu or Menu Browser (Draw > Point)
- The keyboard (PO or POINT)

Choose Point from the Menu Browser, as shown on the left in the following image. Clicking on the Single Point option allows for the creation of one point. If numerous points need to be constructed in the same operation, choose the Multiple Point option. The OSNAP Node or Nearest option is used to snap to points. By default, a point is displayed as a dot on the screen. Illustrated on the right in the following image is an object constructed using points that have been changed in appearance to resemble Xs.

```
Command: PO (For POINT)
Current point modes: PDMODE=3 PDSIZE=0.0000
Specify a point: (Pick the new position of a point)
Specify a point: (Either pick another point location or press ESC
    to exit this command)
```

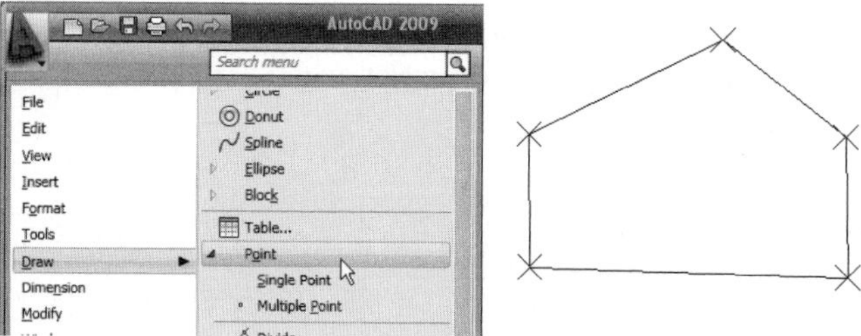

Figure 5.33

SETTING A NEW POINT STYLE

Because the appearance of the default point as a dot may be confused with the existing grid dots already on the screen, a mechanism is available to change the appearance of the point. Choosing Point Style from the Format heading in the Menu Browser, as shown on the left in the following image, displays the Point Style dialog box shown on the right. Use this icon menu to set a different point mode and point size.

Tip: Only one point style may be current in a drawing. Once a point is changed to a current style, the next drawing regeneration updates all points to this style.

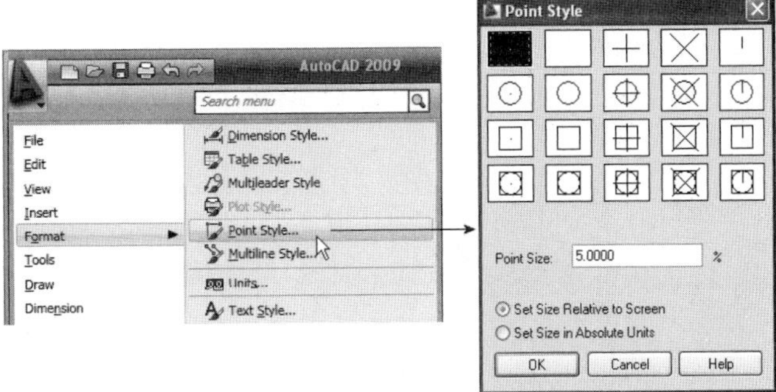

Figure 5.34

DIVIDING OBJECTS INTO EQUAL SPACES

Illustrated in the following image is an arc constructed inside a rectangular object. The purpose of this example is to divide the arc into an equal number of parts. This was a tedious task with manual drafting methods, but thanks to the DIVIDE command, this operation is much easier to perform. The DIVIDE command instructs you to supply the number of divisions and then performs the division by placing a point along the object to be divided. Choose this command from one of the following:

- The pull-down menu or Menu Browser (Draw > Point > Divide)
- From the Ribbon > Home Tab > Draw Panel (Expanded – under Point)
- The keyboard (DIV or DIVIDE)

The Point Style dialog box controls the point size and shape. Be sure the point style appearance is set to produce a visible point. Otherwise, the results of the DIVIDE command will not be obvious.

Try It!: Open the drawing file 05_Divide. Use the DIVIDE command and select the arc as the object to divide, as illustrated on the left in the following image, and enter a value of 6 for the number of segments. The command divides the object by a series of points, as shown in the middle in the following image. Then construct circles with a diameter of .25 using each point as its center, as shown on the right in the following image. The OSNAP-Node mode works with point objects.

```
Command: DIV (For DIVIDE)
Select object to divide: (Select the arc)
Enter the number of segments or [Block]: 6
```

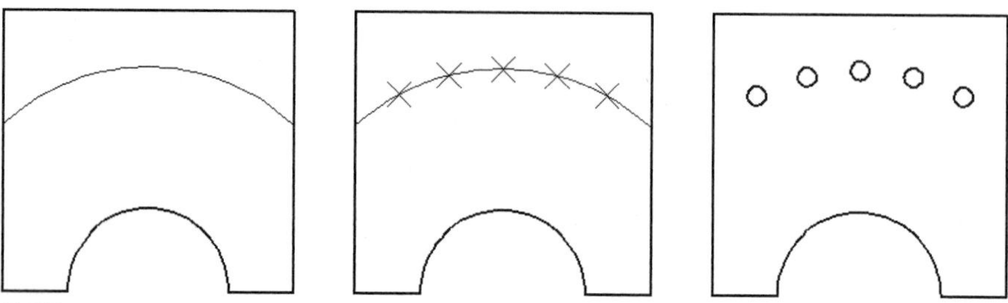

Figure 5.35

For polyline objects, the DIVIDE command divides the entire polyline object into an equal number of segments. This occurs even if the polyline consists of a series of line and arc segments, as in the following image.

Try It!: Open the drawing file 05_Divide Pline. Use the DIVIDE command and select the outer polyline as the object to divide, as shown on the left in the following image, and enter a value of 9 for the number of segments. The command divides the entire polyline by a series of points, as shown on the right.

```
Command: DIV (For DIVIDE)
Select object to divide: (Select the outer polyline shape)
Enter the number of segments or [Block]: 9
```

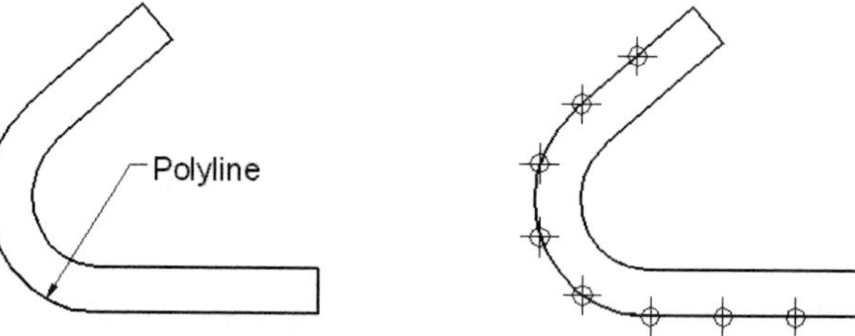

Figure 5.36

MEASURING OBJECTS

The MEASURE command takes an object such as a line or an arc and measures along it depending on the length of the segment. The MEASURE command, similar to the DIVIDE command, places a point on the object at a specified distance given in the MEASURE command. Choose this command from one of the following:

- The pull-down menu or Menu Browser (Draw > Point > Measure)
- From the Ribbon > Home Tab > Draw Panel (Expanded – under Point)
- The keyboard (ME or MEASURE)

It is important to note that as a point is placed along an object during the measuring process, the object is not automatically broken at the location of the point. Rather, the point is commonly used to construct from, along with the OSNAP-Node option. Also, the measuring starts at the endpoint closest to the point you used to select the object.

Choose Point from the Draw pull-down menu or Menu Browser, and then choose the MEASURE command.

Try It!: Open the drawing file 05_Measure. Use the illustration in the following image and the command sequence below to perform this operation.

```
Command: ME (For MEASURE)
Select object to measure: (Select the left end of the diagonal
    line, as shown on the left in the following image)
Specify length of segment or [Block]: 1.25
```

The results, shown in the following image, illustrate various points placed at 1.25 increments. As with the DIVIDE command, the appearance of the points placed along the line is controlled through the Point Style dialog box.

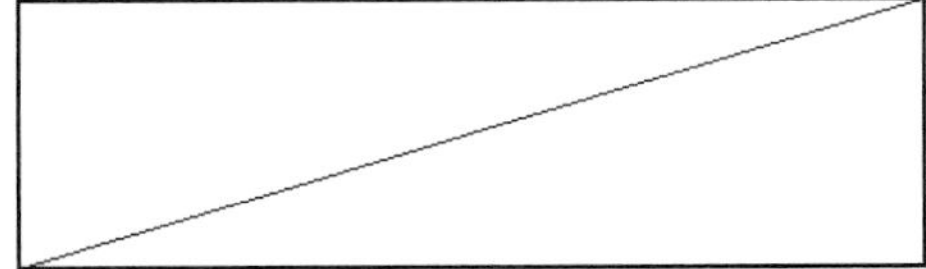

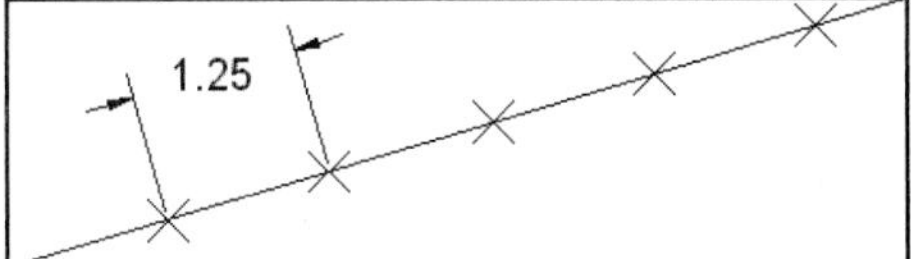

Figure 5.37

CREATING POLYGONS

The POLYGON command is used to construct a regular polygon. Choose this command from one of the following:

- The Draw toolbar of the AutoCAD Classic workspace
- From the Ribbon > Home Tab > Draw Panel
- The pull-down menu or Menu Browser (Draw > Polygon)
- The keyboard (POL or POLYGON)

You create polygons by identifying the number of sides for the polygon, locating a point on the screen as the center of the polygon, specifying whether the polygon is inscribed or circumscribed, and specifying a circle radius for the size of the polygon. Polygons consist of a closed polyline object with width set to zero.

Try It!: Create a new drawing file starting from scratch. Use the following command sequence to construct the inscribed polygon, as shown on the left in the following image.

```
Command: POL (For POLYGON)
Enter number of sides <4>: 6
Specify center of polygon or [Edge]: (Mark a point at "A," as
    shown on the left in the following image)
```

```
Enter an option [Inscribed in circle/Circumscribed about circle]
    <I>: I (For Inscribed)
Specify radius of circle: 1.00
```

Try It!: Create a new drawing file starting from scratch. Use the following command sequence to construct the circumscribed polygon, as shown on the right in the following image.

```
Command: POL (For POLYGON)
Enter number of sides <4>: 6
Specify center of polygon or [Edge]: (Pick a point at "A," as
    shown on the right in the following image)
Enter an option [Inscribed in circle/Circumscribed about circle]
    <I>: C (For Circumscribed)
Specify radius of circle: 1.00
```

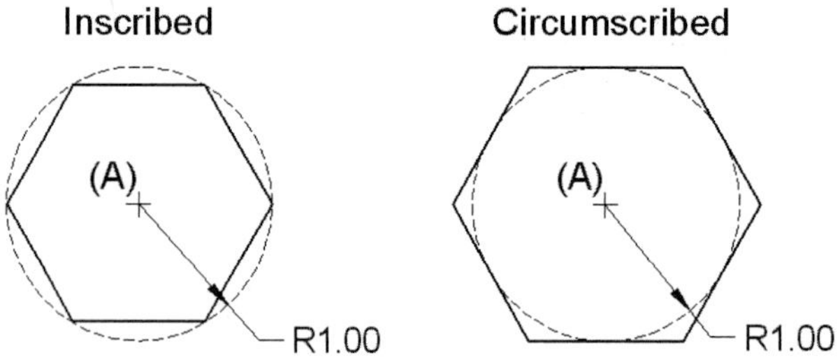

Figure 5.38

A third method for creating polygons can be accomplished by locating the endpoints of one of its edges. The polygon is then drawn in a counterclockwise direction.

Try It!: Create a new drawing file starting from scratch. Study the following image and the command sequence to construct a polygon by edge.

```
Command: POL (For POLYGON)
Enter number of sides <4>: 5
Specify center of polygon or [Edge]: E (For Edge)
Specify first endpoint of edge: (Mark a point at "A")
Specify second endpoint of edge: (Mark a point at "B")
```

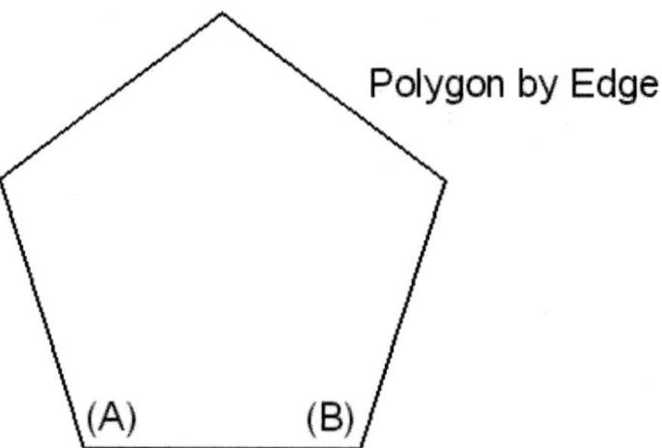

Figure 5.39

CREATING A RAY CONSTRUCTION LINE

A ray is a type of construction line object that begins at a user-defined point and extends to infinity in only one direction. Choose this command from one of the following:

- The pull-down menu or Menu Browser (Draw > Ray)
- From the Ribbon > Home Tab > Draw Panel (Expanded)
- The keyboard (RAY)

In the following image, the quadrants of the circles identify all points where the ray objects begin and are drawn to infinity to the right. You should organize ray objects on specific layers. You should also exercise care in the editing of rays, and take special care not to leave segments of objects in the drawing database as a result of breaking ray objects. Breaking the ray object at "A" in the following image converts one object to an individual line segment; the other object remains a ray. Study the following command sequence for constructing a ray.

```
Command: RAY
Specify start point: (Pick a point on an object)
Specify through point: (Pick an additional point to construct the
    ray object)
Specify through point: (Pick another point to construct the ray
    object or press ENTER to exit this command)
```

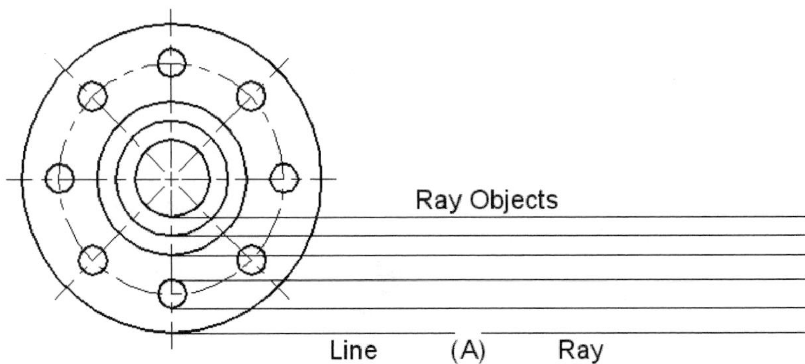

Figure 5.40

CREATING RECTANGLE OBJECTS

Use the RECTANG command to construct a rectangle by defining two points. Choose this command from one of the following:

- The Draw toolbar of the AutoCAD Classic workspace
- From the Ribbon > Home Tab > Draw Panel
- The pull-down menu or Menu Browser (Draw > Rectangle)
- The keyboard (REC or RECTANG or RECTANGLE)

RECTANGLE BY PICKING TWO DIAGONAL POINTS

As illustrated in the following image, two diagonal points are picked to define the rectangle. The rectangle is drawn as a single polyline object.

```
Command: REC (For RECTANG)
Specify first corner point or [Chamfer/Elevation/Fillet/
    Thickness/Width]: (Pick a point at "A")
Specify other corner point or [Area/Dimensions/Rotation]: (Pick a
    point at "B")
```

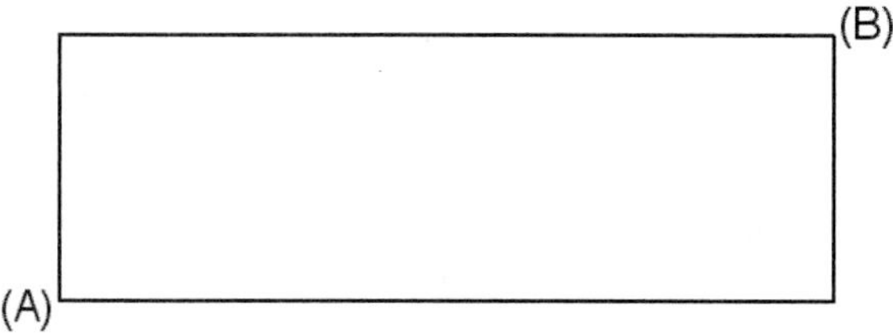

Figure 5.41

CHANGING RECTANGLE PROPERTIES

Other options of the RECTANG command enable you to construct a chamfer or fillet at all corners of the rectangle, to assign a width to the rectangle, and to have the rectangle drawn at a specific elevation and at a thickness for 3D purposes. As illustrated in the following image, a rectangle is constructed with a chamfer distance of .20 units; the width of the rectangle is also set at .05 units. A relative coordinate value of 4.00,1.00 is used to construct the rectangle 4 units in the X direction and 1 unit in the Y direction. The @ symbol resets the previous point at "A" to zero.

Try It!: Create a new AutoCAD drawing starting from scratch. Use the following command sequence image for constructing a wide rectangle with its corners chamfered.

```
Command: REC (For RECTANGLE)
Specify first corner point or [Chamfer/Elevation/Fillet/
    Thickness/Width]: C (For Chamfer)
Specify first chamfer distance for rectangles <0.0000>: .20
Specify second chamfer distance for rectangles <0.2000>: (Press
    ENTER to accept this default value)
Specify first corner point or [Chamfer/Elevation/Fillet/
    Thickness/Width]: W (For Width)
Specify line width for rectangles <0.0000>: .05
Specify first corner point or [Chamfer/Elevation/Fillet/
    Thickness/Width]: (Pick a point at "A")
Specify other corner point or [Area/Dimensions/Rotation]:
    @4.00,1.00 (To identify the other corner at "B")
```

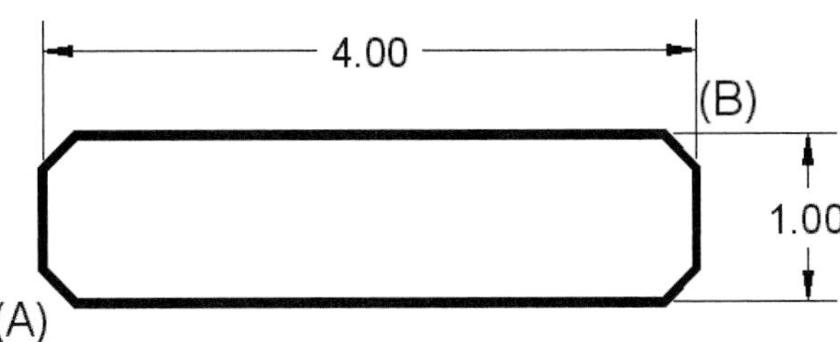

Figure 5.42

RECTANGLE BY DIMENSIONS

The Dimensions option provides another method of constructing a rectangle in which you supply the length and width dimensions for the rectangle. Before the rectangle is constructed, you indicate a point where the opposite corner of the rectangle will be positioned. There are four possible positions or quadrants the rectangle can be drawn in using this method: upper-right, upper-left, lower-right, and lower-left.

RECTANGLE BY AREA

The final method of constructing a rectangle is by area. In this method, you pick a first corner point, as in previous rectangle modes. After entering the Area option, you enter the area of the rectangle based on the current drawing units. You then enter the length or width of the rectangle. If, for instance, you enter the length, the width of the rectangle will automatically be calculated based on the area. In the following command sequence, an area of 250 is entered, along with a length of 20 units. A width of 12.50 units is automatically calculated based on the other two numbers.

```
Command: REC (For RECTANGLE)
Specify first corner point or [Chamfer/Elevation/Fillet/
    Thickness/Width]: (Pick a point to start the rectangle)
Specify other corner point or [Area/Dimensions/Rotation]: A
    (For Area)
Enter area of rectangle in current units <100.0000>: 250
Calculate rectangle dimensions based on [Length/Width] <Length>:
    L (For Length)
Enter rectangle length <10.0000>: 20
```

Try It!: Create a new AutoCAD drawing starting from scratch. You can also specify the dimensions (length and width) of the rectangle by using the following command sequence and image.

```
Command: REC (For RECTANGLE)
Specify first corner point or [Chamfer/Elevation/Fillet/
    Thickness/Width]: (Pick a point at "A" on the screen)
Specify other corner point or [Area/Dimensions/Rotation]: D (For
    Dimensions)
Specify length for rectangles <0.0000>: 3.00
Specify width for rectangles <0.0000>: 1.00
Specify other corner point or [Area/Dimensions/Rotation]: (Moving
    your cursor around positions the rectangle in four possible
```

```
    positions. Click the upper-right corner of your screen at
    "B" to anchor the upper- right corner of the rectangle)
```

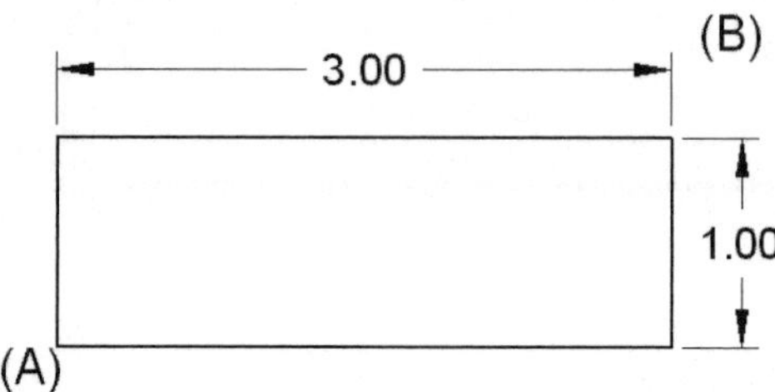

Figure 5.43

Note: Rectangles can also be constructed based on a user-specified angle.

CREATING A REVISION CLOUD

The REVCLOUD command creates a polyline object consisting of arc segments in a sequence. This feature is commonly used in drawings to identify areas where the drawing is to be changed or revised. Choose this command from one of the following:

- The Draw toolbar of the AutoCAD Classic workspace
- From the Ribbon > Home Tab > Draw Panel (Expanded)
- The pull-down menu or Menu Browser (Draw > Revision Cloud)
- The keyboard (REVCLOUD)

In the following image, an area of the house is highlighted with the revision cloud. By default, each arc segment has a minimum and maximum arc length of 0.50 units. In a drawing such as a floor plan, the arc segments would be too small to view. In cases in which you wish to construct a revision cloud and the drawing is large, the arc length can be increased through a little experimentation. Once the arc segment is set, pick a point on your drawing to begin the revision cloud. Then move your cursor slowly in a counter-clockwise direction. You will notice the arc segments of the revision cloud being constructed. Continue surrounding the object, while at the same time heading back to the origin of the revision cloud. Once you hover near the original start, the end of the revision cloud snaps to the start and the command ends, leaving the revision cloud as the polyline object. Study the command prompt sequence below and the illustration in the following image for the construction of a revision cloud.

```
Command: REVCLOUD
Minimum arc length: 1/2" Maximum arc length: 1/2" Style: Normal
Specify start point or [Arc length/Object/Style] <Object>: A (For
    Arc length)
Specify minimum length of arc <1/2">: 24
Specify maximum length of arc <2'>: (Press ENTER)
```

```
Specify start point or [Arc length/Object/Style] <Object>: (Pick
    a point on your screen to start the revision cloud. Surround
    the item with the revision cloud by moving your cursor in a
    counterclockwise direction. When you approach the start
    point, the revision cloud closes and the command exits.)
Guide crosshairs along cloud path...
Revision cloud finished.
```

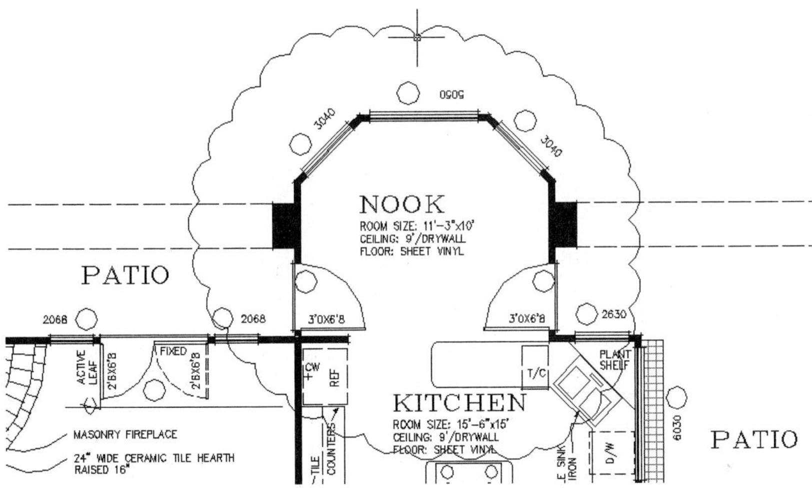

Figure 5.44

CALLIGRAPHY REVISION CLOUDS

The Calligraphy option of creating revision clouds allows you to enhance the use of this command and make your drawings more dramatic. Use the following command sequence and image, which illustrate this feature.

```
Command: REVCLOUD
Minimum arc length: 24.0000 Maximum arc length: 24.0000 Style:
    Normal
Specify start point or [Arc length/Object/Style] <Object>: S (For
    Style)
Select arc style [Normal/Calligraphy] <Normal>: C (For
    Calligraphy)
Arc style = Calligraphy
Specify start point or [Arc length/Object/Style] <Object>: (Pick
    a starting point)
Guide crosshairs along cloud path...
Revision cloud finished.
```

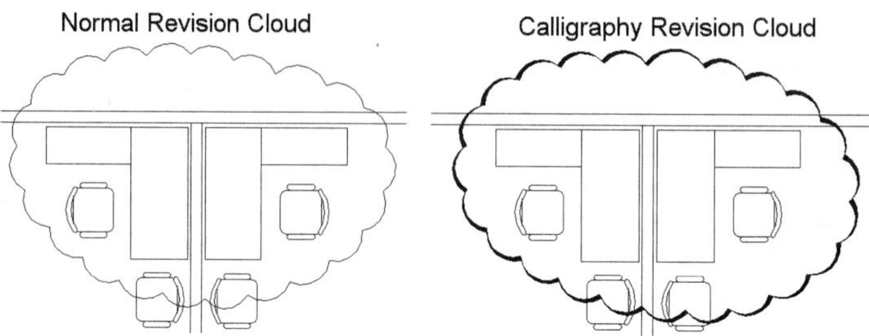

Figure 5.45

CREATING SPLINES

Use the SPLINE command to construct a smooth curve given a sequence of points. Choose this command from one of the following:

- The Draw toolbar of the AutoCAD Classic workspace
- From the Ribbon > Home Tab > Draw Panel (Expanded)
- The pull-down menu or Menu Browser (Draw > Spline)
- The keyboard (SPL or SPLINE)

You have the option of changing the accuracy of the curve given a tolerance range. The basic command sequence follows, which constructs the spline segment shown on the left in the following image.

```
Command: SPL (For SPLINE)
Specify first point or [Object]: (Pick a first point)
Specify next point: (Pick another point)
Specify next point or [Close/Fit tolerance] <start tangent>:
    (Pick another point)
Specify next point or [Close/Fit tolerance] <start tangent>:
    (Pick another point)
Specify next point or [Close/Fit tolerance] <start tangent>:
    (Press ENTER to continue)
Specify start tangent: (Press ENTER to accept)
Specify end tangent: (Press ENTER to accept the end tangent
    position, which exits the command and places the spline)
```

The spline may be closed to display a continuous segment, as shown on the right in the following image. Entering a different tangent point at the end of the command changes the shape of the curve connecting the beginning and end of the spline.

```
Command: SPL (For SPLINE)
Specify first point or [Object]: (Pick a first point)
Specify next point: (Pick another point)
Specify next point or [Close/Fit tolerance] <start tangent>:
    (Pick another point)
Specify next point or [Close/Fit tolerance] <start tangent>:
    (Pick another point)
Specify next point or [Close/Fit tolerance] <start tangent>: C
    (To Close)
```

```
Specify tangent: (Press ENTER to exit the command and place the
    spline)
```

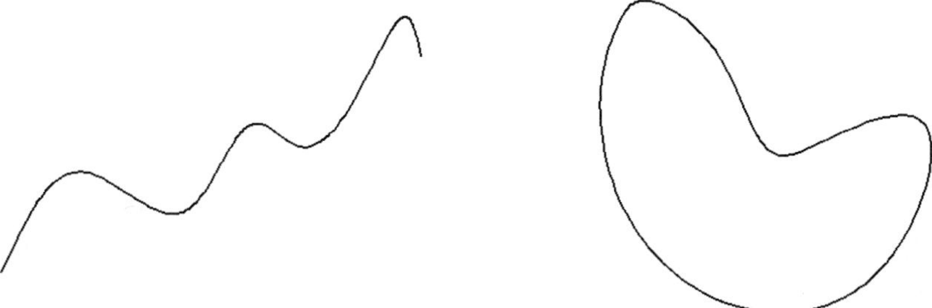

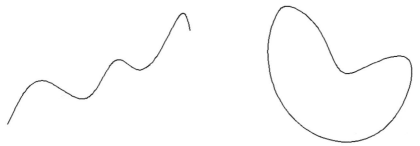

Figure 5.46

Try It!: Open the drawing file 05_ Spline Rasp Handle, as shown on the left in the following image. Turn OSNAP on and set Running OSNAP to Node. This allows you to snap to points when drawing splines or other objects. Construct a spline by connecting all points between "A" and "B." Construct an arc with its center at "C," the start point at "D," and the ending point at "B." Connect points "D" and "E" with a line segment. Construct another spline by connecting all points between "E" and "F." Construct another arc with its center at "G," the start point at "H," and the ending point at "F." Connect points "H" and "A" with a line segment.

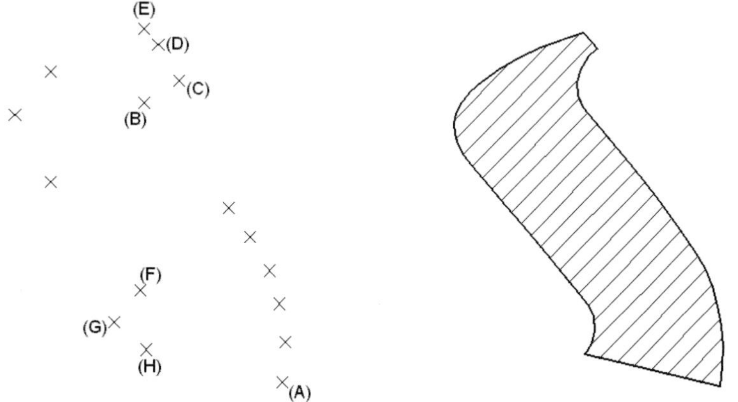

Figure 5.47

MASKING TECHNIQUES WITH THE WIPEOUT COMMAND

To mask or hide objects in a drawing without deleting them or turning off layers, the WIPEOUT command could be used. Choose this command from one of the following:

- The pull-down menu or Menu Browser (Draw > Wipeout)
- From the Ribbon > Home Tab > Draw Panel (Expanded)
- The keyboard (WIPEOUT)

This command reads the current drawing background color and creates a mask over anything defined by a frame. In the illustration on the left in the following image, a series of text objects needs to be masked over. The middle image shows a four-sided frame that was created over the text using the following command prompt sequence:

```
Command: WIPEOUT
Specify first point or [Frames/Polyline] <Polyline>: (Pick a
    first point)
Specify next point: (Pick a second point)
Specify next point or [Undo]: (Pick a third point)
Specify next point or [Close/Undo]: (Pick a fourth point)
Specify next point or [Close/Undo]: (Press ENTER to exit the
    command and create the wipeout)
```

As the text seems to disappear, the wipeout frame is still visible. A visible frame is important if you would like to unmask or delete the wipeout. If you want to hide all wipeout frames, use the following command sequence:

```
Command: WIPEOUT
Specify first point or [Frames/Polyline] <Polyline>: F (For
    Frames)
Enter mode [ON/OFF] <ON>: Off
```

Now all wipeout frames in the current drawing are turned off, as shown in the illustration on the right in the following image.

Note: You could also create a predefined polyline object and then convert it into a wipeout using the Polyline option of the WIPEOUT command.

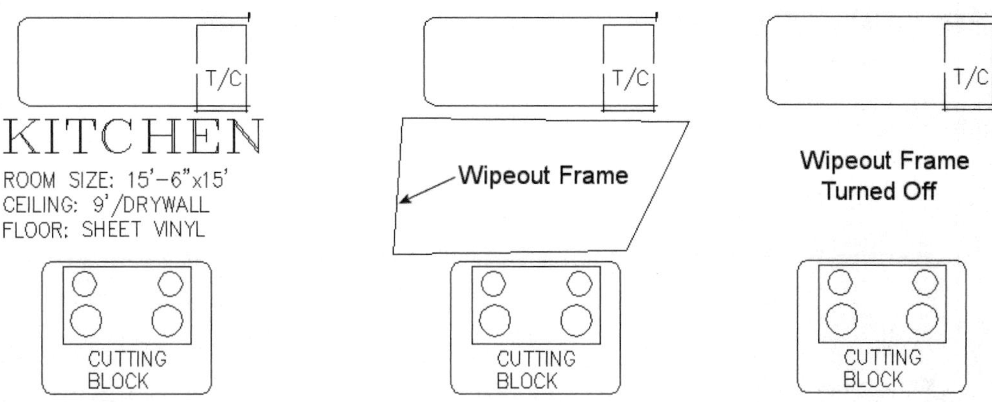

Figure 5.48

CREATING CONSTRUCTION LINES WITH THE XLINE COMMAND

Xlines are construction lines drawn from a user-defined point. Choose this command from one of the following:

- The Draw toolbar of the AutoCAD Classic workspace
- From the Ribbon > Home Tab > Draw Panel (Expanded)
- The pull-down menu or Menu Browser (Draw > Construction Line)
- The keyboard (XL or XLINE)

You are not prompted for any length information because the Xline extends an unlimited length, beginning at the user-defined point and going off to infinity in opposite directions from the point. Xlines can be drawn horizontal, vertical, and angular. You can bisect an angle using an Xline or offset the Xline at a specific distance. Xlines are particularly useful in constructing orthogonal drawings by establishing reference lines between views. As illustrated on the left in the following image, the circular view represents the Front view of a flange. To begin the creation of the Side views, lines are usually projected from key features on the adjacent view. In the case of the Front view, the key features are the top of the plate in addition to the other circular features. In this case, the Xlines were drawn with the Horizontal mode from the Quadrant of all circles. The following prompts outline the XLINE command sequence:

```
Command: XL (For XLINE)
Specify a point or [Hor/Ver/Ang/Bisect/Offset]: H (For
    Horizontal)
Specify through point: (Pick a point on the display screen to
    place the first Xline)
Specify through point: (Pick a point on the display screen to
    place the second Xline)
```

Since the Xlines continue to be drawn in both directions, care must be taken to manage these objects. Construction management techniques of Xlines could take the form of placing all Xlines on a specific layer to be turned off or frozen when not needed. When editing Xlines (especially with the BREAK command), you need to take special care to remove all excess objects that remain on the drawing screen. As illustrated on the right in the following image, breaking the Xline converts the object to a ray object. Use the ERASE command to remove any excess Xlines.

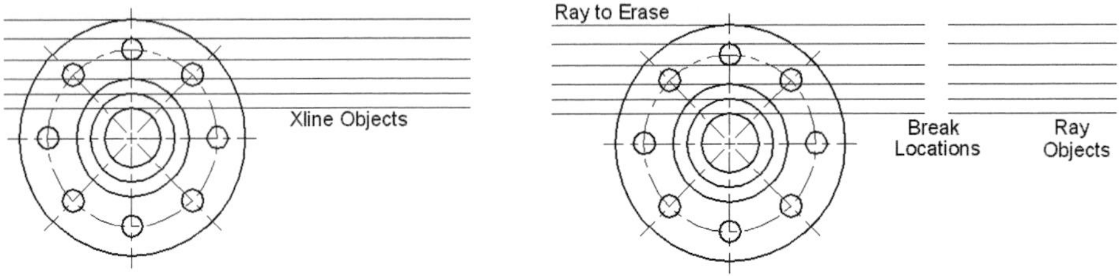

Figure 5.49

Try It!: Open the drawing file 05_Xline. Another application of Xlines is illustrated on the left in the following image. Three horizontal and vertical Xlines are constructed. You must corner the Xlines to create the object illustrated on the right in the following image. Use the FILLET command and set the radius to zero. Pick the lines at "A" and "B" to create the first corner. Repeat this sequence on the remaining lines to form the other corners.

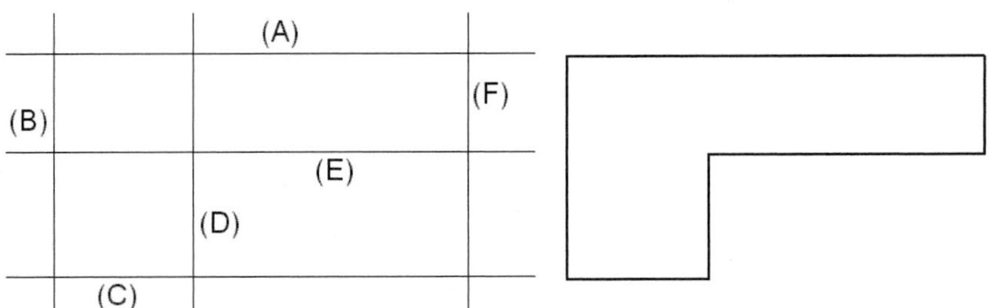

Figure 5.50

OGEE OR REVERSE CURVE CONSTRUCTION

An ogee curve connects two parallel lines with a smooth, flowing curve that reverses itself in a symmetrical form.

Try It!: Open the drawing file 05_Ogee. To begin constructing an ogee curve to line segments "AB" and "CD," a line was drawn from "B" to "C," which connects both parallel line segments, as shown on the left in the following image.

Use the DIVIDE command to divide line segment "BC" into four equal parts. Be sure to set a new point mode by picking a new point from the Point Style dialog box. Construct vertical lines from "B" and "C." Complete this step by constructing line segment "XY," which is perpendicular to line "BC," as shown on the right in the following image. Do not worry about where line "XY" is located at this time.

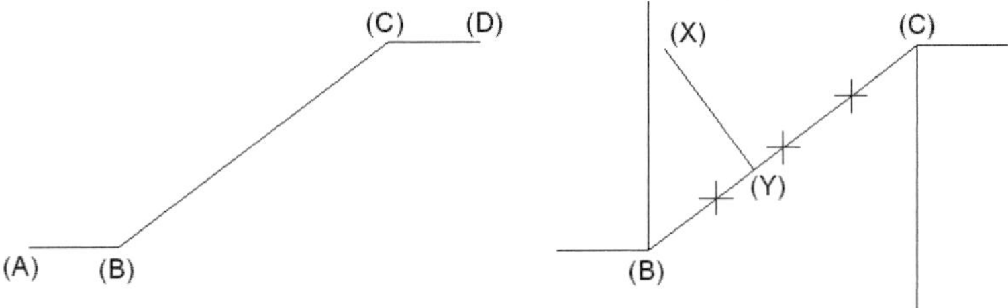

Figure 5.51

Move line "XY" to the location identified by the point, as shown on the left in the following image. Complete this step by copying line "XY" to the location identified by point "Z," as shown on the left in the following image.

Construct two circles with centers located at points "X" and "Y," as shown on the right in the following image. Use the OSNAP-Intersection mode to accurately locate the centers. If an intersection is not found from the previous step, use the EXTEND command to find the intersection and continue with this step. The radii of both circles are equivalent to distances "XB" and "YC."

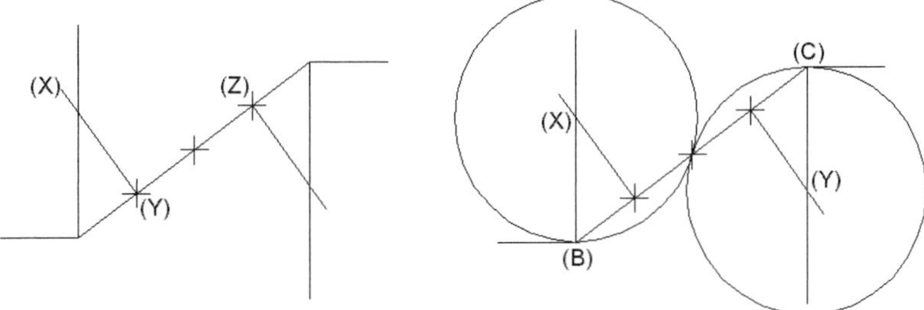

Figure 5.52

Use the TRIM command to trim away any excess arc segments to form the ogee curve, as shown on the left in the following image.

This forms the frame of the ogee for the construction of objects such as the wrench illustrated on the right in the following image.

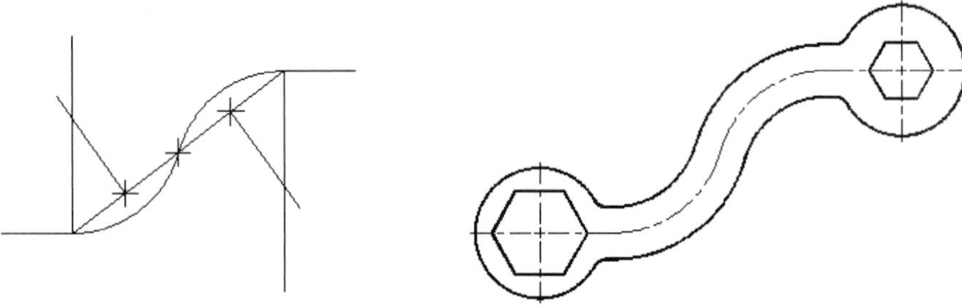

Figure 5.53

TUTORIAL EXERCISE: GEAR-ARM.DWG

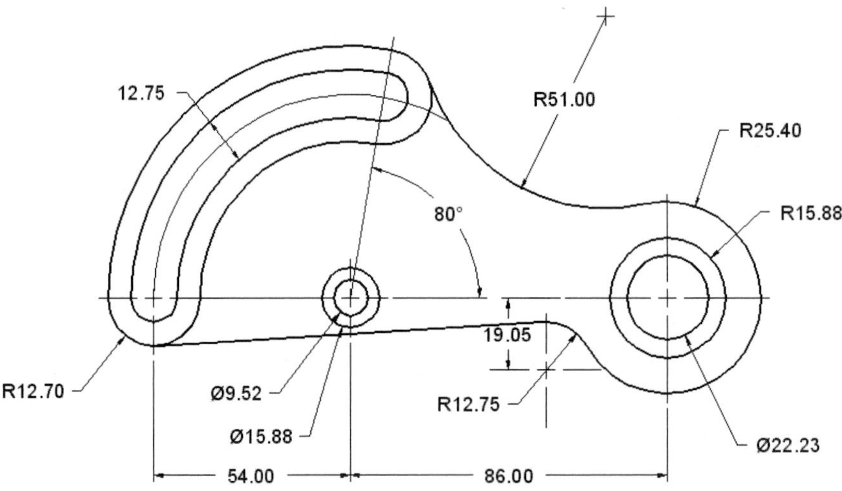

Figure 5.54

PURPOSE

This tutorial is designed to use geometric commands to construct a one-view drawing of the gear-arm in metric format, as illustrated in the previous image.

SYSTEM SETTINGS

Use the Drawing Units dialog box to change the number of decimal places past the zero from four to two. Keep the remaining default unit values. Using the LIMITS command, keep (0,0) for the lower-left corner and change the upper-right corner from (12,9) to (265.00,200.00). Perform a ZOOM-All after changing the drawing limits. Since a layer called "Center" must be created to display centerlines, use the LTSCALE command and change the default value of 1.00 to 15.00. This makes the long and short dashes of the centerlines appear on the display screen. Check to see that the following Object Snap modes are already set: Endpoint, Extension, Intersection, Center.

LAYERS

Create the following layers with the format:

Name	Color	Linetype
Center	Yellow	Center
Construction	Gray	Center
Dimension	Yellow	Continuous
Object	White	Continuous

SUGGESTED COMMANDS

Begin a new drawing called Gear-arm. The object consists of a combination of circles and arcs along with tangent lines and arcs. Before beginning, be sure to set Object as the new current layer. Construction lines will be used as a layout tool to mark the centers of key circles and arcs. The CIRCLE-TTR command will be used for constructing tangent arcs to existing geometry. Use the ARC command to construct a series of arcs for the left side of the gear-arm. The TRIM command will be used to trim circles, lines, and arcs to form the basic shape.

Step 1

Begin constructing the gear arm by creating construction geometry that will be used to create all circles. First make the Construction layer current and use the LINE command to lay out the long horizontal line and middle vertical line. Then use the OFFSET command to copy the vertical line at the specified distances. Also, create a line 70 units long at an 80° angle, as shown in the following image.

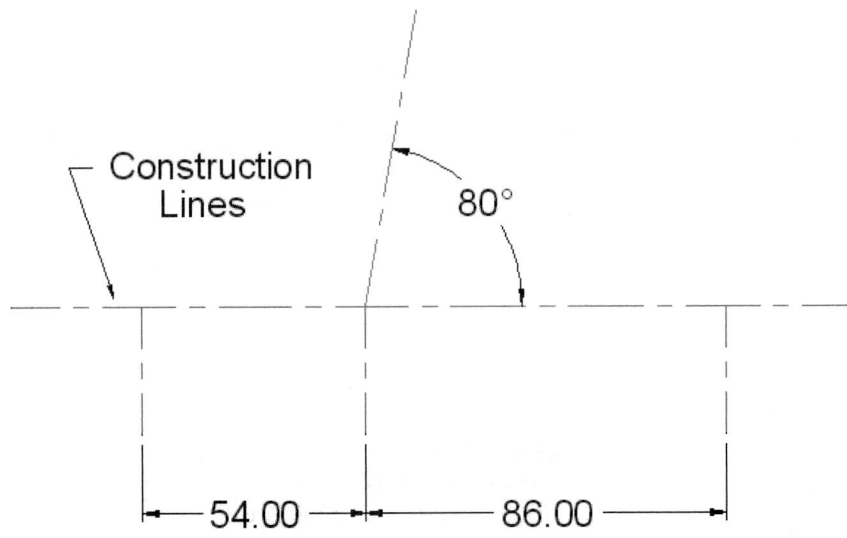

Figure 5.55

Step 2

Continue using the Construction layer to create an arc that will be used to locate the slot detail. This arc is to be constructed using the Center, Start, Angle mode of the arc command, as shown on the left in the following image. Pick "A" as the center of the arc, "B" as the start of the arc, as shown on the right in the following image, and for angle, enter a value of −110°. The negative degree value is needed to construct the arc in the clockwise direction.

Create another construction object; this time use the OFFSET command to offset the long horizontal construction line 19.05 units down. The line shown on the right in the following image was shortened on each end using the BREAK command.

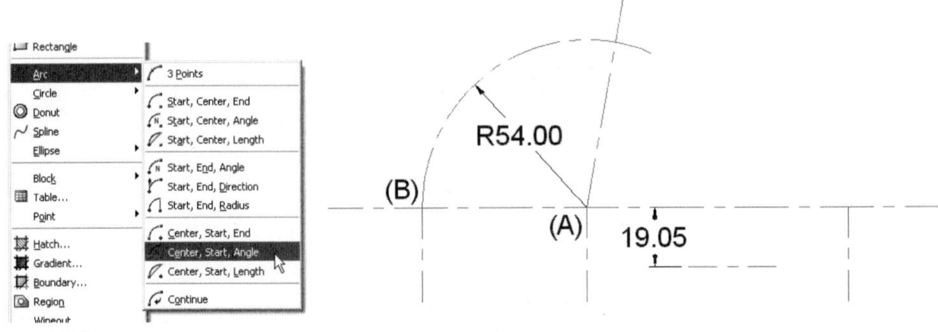

Figure 5.56

Step 3

One final item of construction geometry needs to be created, namely, the circle at a diameter of 76.30. Construct this circle from the intersection of the construction lines at "A," as shown on the left in the following image. Convert the circle to an arc segment by using the BREAK command and picking the first break point at "B" and the second break point at "C." This converts the circle of diameter 76.30 to an arc of radius 38.15, as shown on the right in the following image.

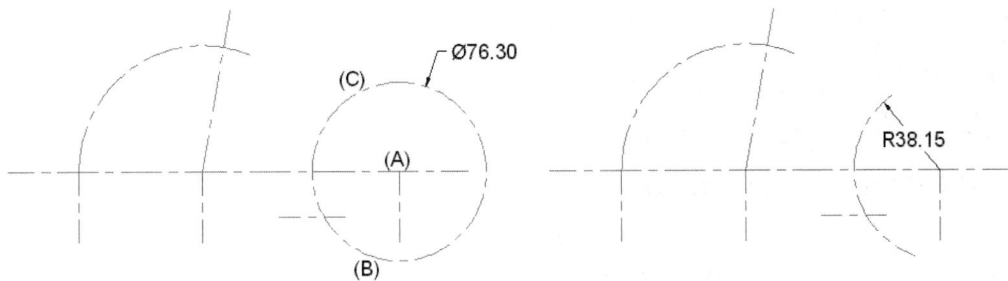

Figure 5.57

Step 4

Using the following image as a guide, create all circles using the intersection of the construction line geometry as the centers for the circles. All circles are given with diameter dimensions.

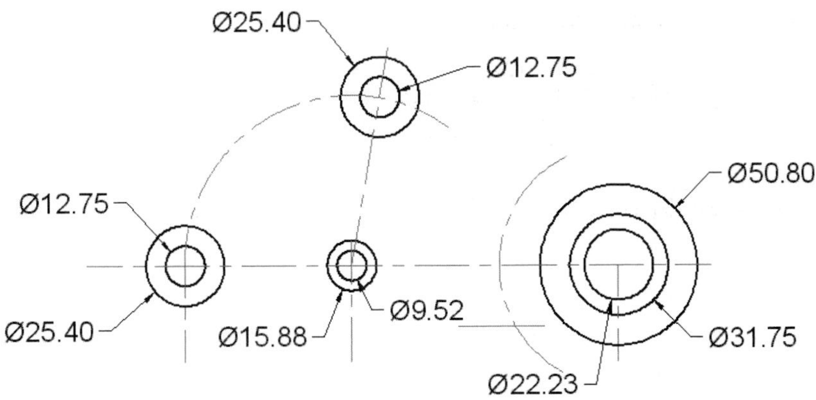

Figure 5.58

Step 5

Use the Arc Center, Start, End command to create the arc shown on the left in the following image. Pick "A" as the center of the arc, "B" as the start, and "C" as the end of the arc. You must use a combination of Object Snaps in order for the arcs to be drawn correctly. Following these steps constructs the outer arc of the gear arm. Use the same Arc Center, Start, End command to create the inside arc using a different set of points.

When finished creating the outer arc segments, create the two inner arc segments, as shown on the right in the following image, using the same Arc Center, Start, End command. Continue to use Object Snap to pick valid intersections in order to create the most accurate set of arcs.

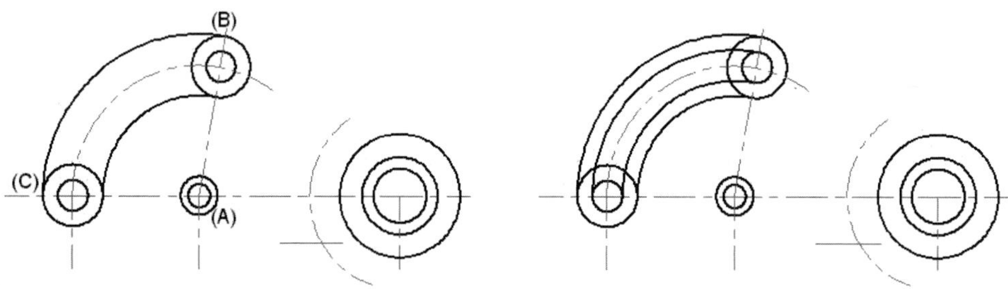

Figure 5.59

Step 6

Use the CIRCLE-TTR command to construct a circle tangent to two existing circles. Pick the two dashed circles as the objects to be made tangent, as shown on the left in the following image, and enter a radius value of 51.

After the circle is created, enter the TRIM command, pick the two dashed circles as cutting edges, and pick the upper edge of the large 51-radius circle. The results of this operation are illustrated on the right in the following image.

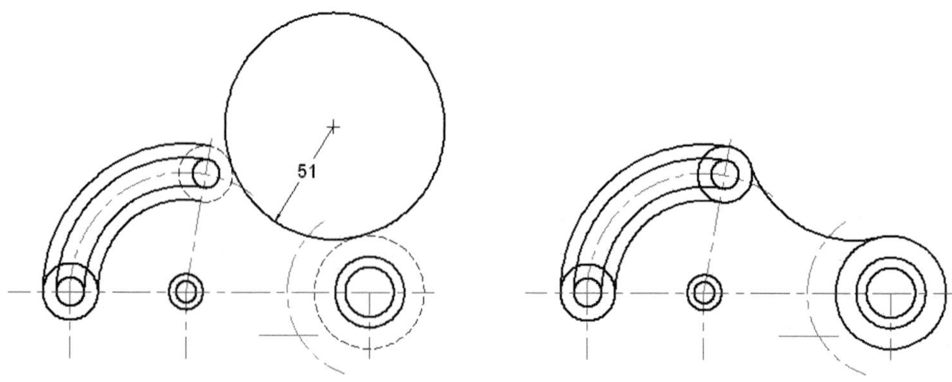

Figure 5.60

Step 7

Enter the TRIM command again, pick the two dashed arcs as cutting edges, and pick the two inside edges of the 25.4-diameter circles, as shown on the left in the following image. The results of this operation are illustrated on the right in the following image.

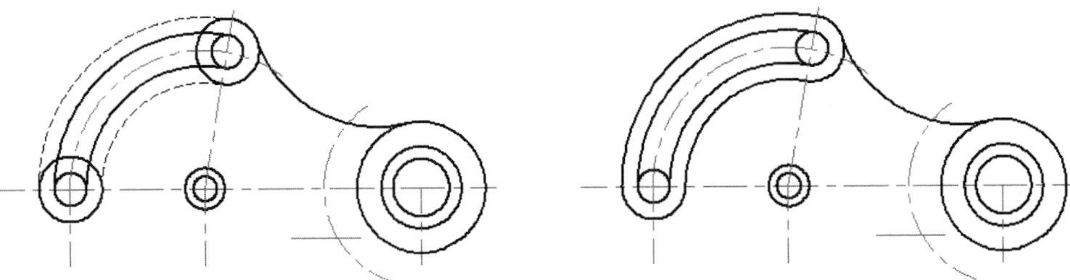

Figure 5.61

Step 8

Do the same procedure to clean up the geometry that makes up the inner slot. Enter the TRIM command, pick the two dashed arcs as cutting edges, and pick the two inside edges of the 12.75-diameter circles, as shown on the left in the following image. The results of this operation are illustrated on the right in the following image.

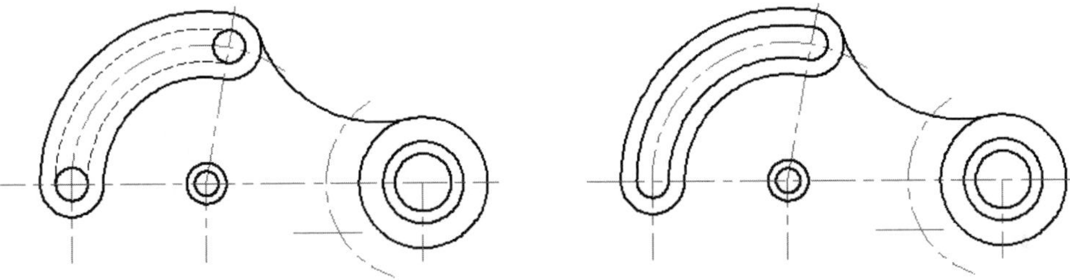

Figure 5.62

Step 9

Construct a new circle of diameter 25.50 from the intersection of the construction line and arc, as shown on the left in the following image. Then draw a line segment that is tangent with the bottom of the arc and top of the circle just constructed, as shown on the right in the following image.

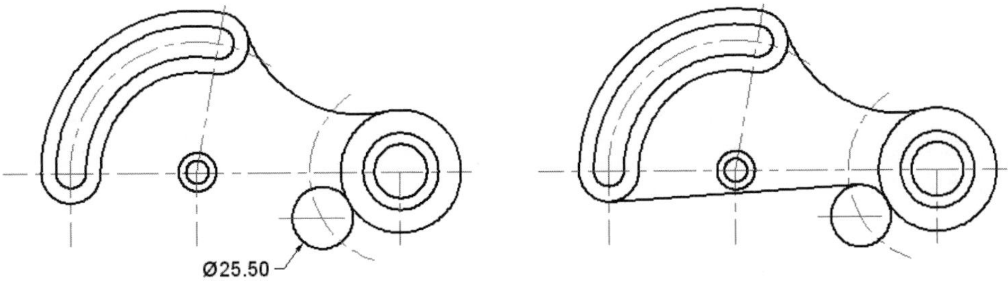

Figure 5.63

Step 10

Use the TRIM command to clean up all unnecessary segments of geometry. Select all dashed objects as cutting edges, as shown on the left in the following image. Trim objects until your display appears similar to the illustration on the right in the following image.

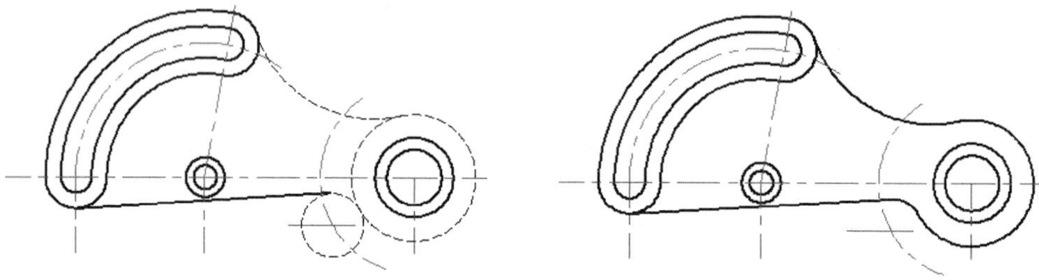

Figure 5.64

Step 11

Erase all unnecessary construction geometry, as shown on the left in the following image. Notice that the arc and 70° angle line both remain and will be used later to define the centers of the gear arm.

Next create centerlines to mark the centers of all circles. Switch to the Center layer. A system variable that controls centerlines and marks called DIMCEN must first be entered from the keyboard and changed to a value of –.05. The negative value constructs the small dash and long segment, all representing the centerline. Entering a value of .05 would only mark the center of circles by constructing two short intersecting line segments. Then use the DIMCENTER (or DCE for short) command and pick the edge of a circle or arc segment to place the centerline. The results should appear similar to the illustration on the right in the following image.

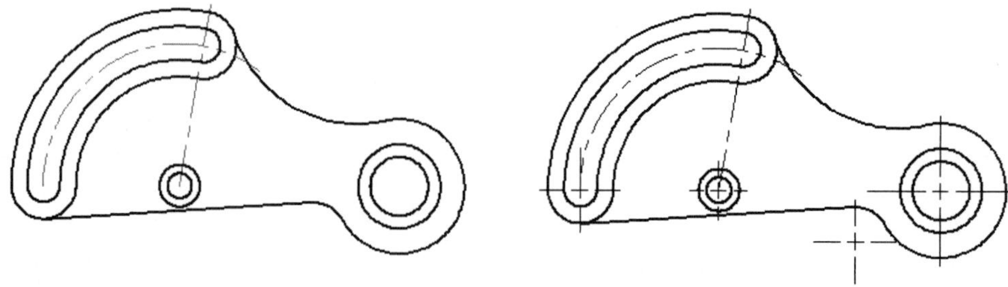

Figure 5.65

Step 12

The completed gear arm drawing is illustrated as shown on the left in the following image. The finished object may be dimensioned as an optional step, as shown on the right in the following image.

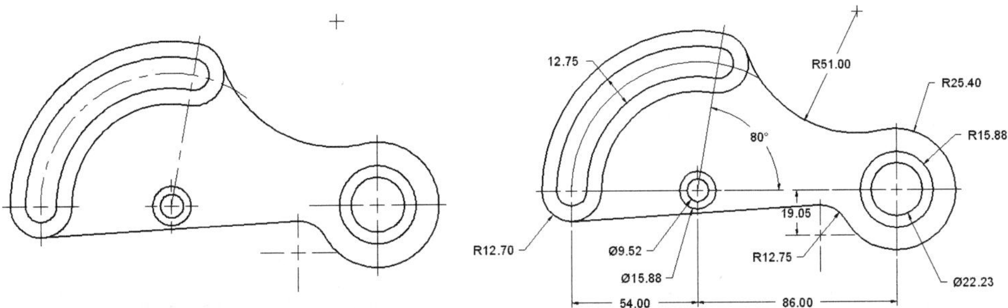

Figure 5.66

TUTORIAL EXERCISE: PATTERN1.DWG

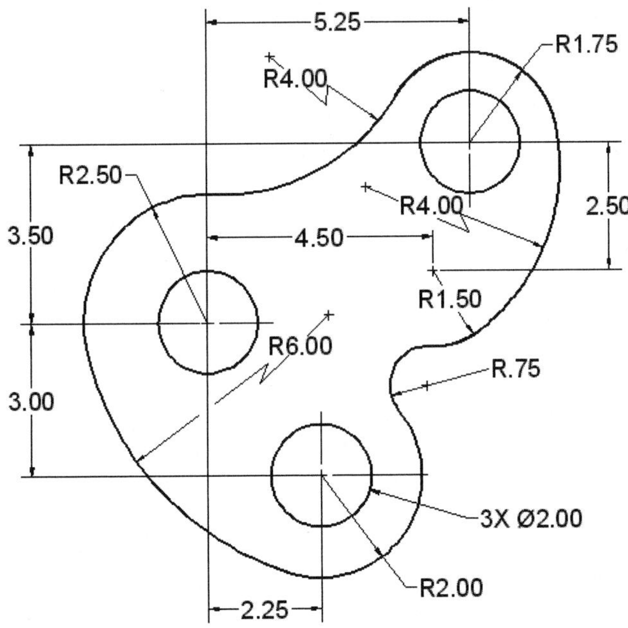

Figure 5.67

PURPOSE

This tutorial is designed to use various Draw commands to construct a one-view drawing of Pattern1, as shown in the previous image. Refer to the following special system settings and suggested command sequences.

SYSTEM SETTINGS

Use the Drawing Units dialog box to change the number of decimal places past the zero from four to two. Keep the remaining default unit values. Using the LIMITS command, keep (0,0) for the lower-left corner and change the upper-right corner from (12,9) to

(21.00,16.00). Perform a ZOOM-All after changing the drawing limits. Check to see that the following Object Snap modes are already set: Endpoint, Extension, Intersection, and Center.

Setting OSNAP Mode would also be helpful with completing this exercise.

LAYERS

Create the following layers with the format:

Name	Color	Linetype
Object	White	Continuous
Center	Yellow	Center
Dimension	Yellow	Continuous

SUGGESTED COMMANDS:

Begin constructing Pattern1 by first creating four circles by relative coordinate mode. Then use the CIRCLE-TTR command/option to draw tangent arcs to the circles already drawn. Use the TRIM command to clean up and partially delete circles to obtain the outline of the pattern. Then add the 2.00-diameter holes followed by the center markers, using the DIMCENTER (DCE) command.

Step 1

Check that the current layer is set to Object. See the following image for the dimensions and table for locating the centers of all circles. Use the CIRCLE command to create the circle at "A" whose center point is located at coordinate 9.50, 4.50 and which has a diameter of 4.00 units. Locate circle "B" at relative coordinate @-2.25,3.00 and assign a diameter of 5.00 units. The relative coordinate is based on the previous circle. Continue by drawing circle "C" using the center point at relative coordinate @5.25,3.50 and assign a diameter of 3.50 units. Construct the last circle, "D," with center point @–.75,–2.50 and a diameter of 3.00 units. Notice that when using relative coordinate mode, you do not need to draw construction geometry and then erase it later.

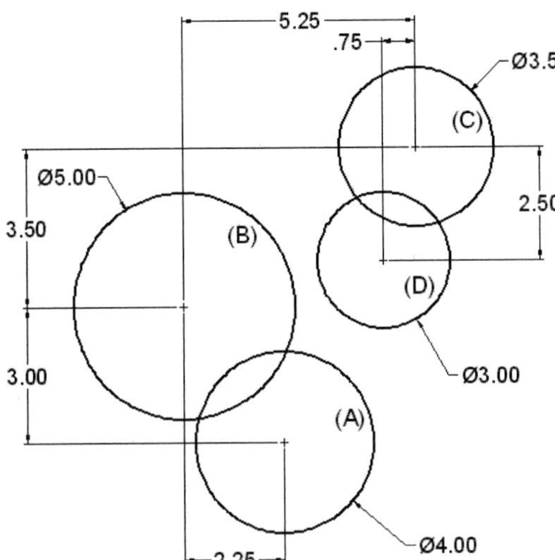

CIRCLE	Center Point Coordinate
A	9.50,4.50
B	@-2.25,3.00
C	@5.25,3.50
D	@-.75,-2.50

Figure 5.68

Step 2

Use the CIRCLE-TTR command/option to construct a 4.00-radius circle tangent to the two dashed circles, as shown on the left in the following image. Then use the TRIM command to trim away part of the circle so your display appears similar to the illustration on the right in the following image.

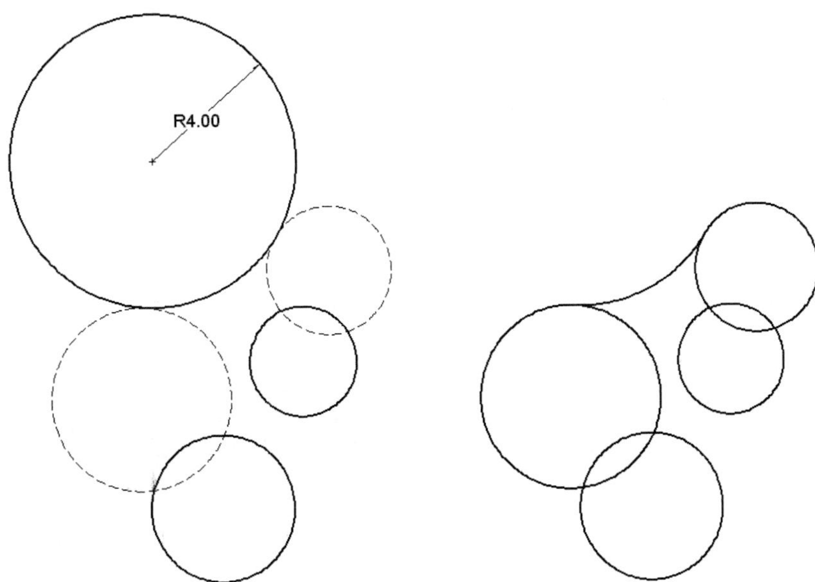

Figure 5.69

Step 3

Use the CIRCLE-TTR command/option to construct another 4.00-radius circle tangent to the two dashed circles, as shown on the left in the following image. Then use the TRIM command to trim away part of the circle so your display appears similar to the illustration on the right in the following image.

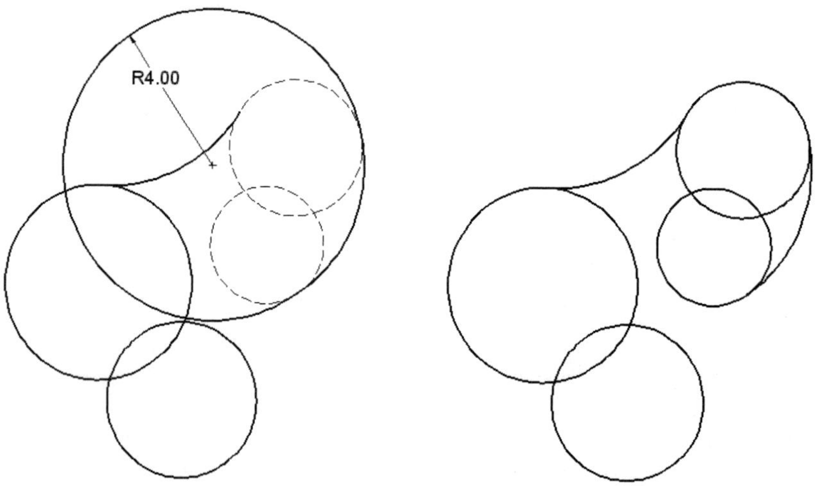

Figure 5.70

Step 4

Use the CIRCLE-TTR command/option to construct a 6.00-radius circle tangent to the two dashed circles, as shown on the left in the following image. Then use the TRIM command to trim away part of the circle so your display appears similar to the illustration on the right in the following image.

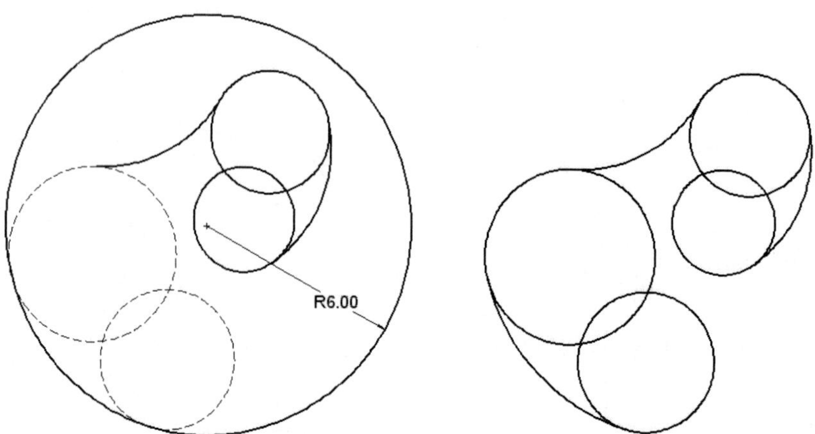

Figure 5.71

Step 5

Use the CIRCLE-TTR command/option to construct a .75-radius circle tangent to the two dashed circles, as shown on the left in the following image. Then use the TRIM command to trim away part of the circle so your display appears similar to the illustration on the right in the following image.

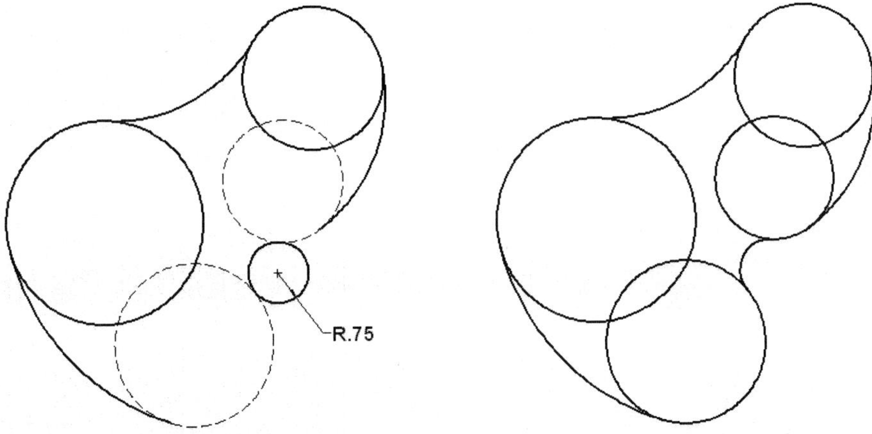

Figure 5.72

Step 6

Use the TRIM command, select all dashed arcs, shown on the left in the following image, as cutting edges, and trim away the circular segments to form the outline of the Pattern1 drawing. When finished, your display should appear similar to the illustration on the right in the following image.

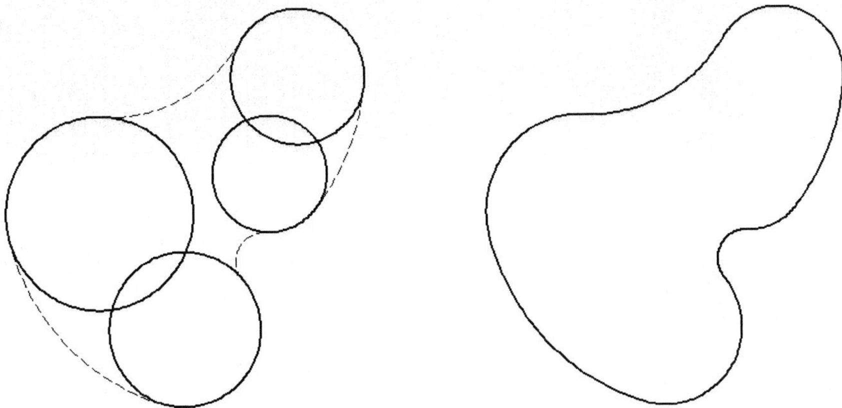

Figure 5.73

Step 7

Use the CIRCLE command to construct a total of three circles of 2.00 diameter. Use the centers of arcs "A," "B," and "C," as shown on the left in the following image, as the centers for the circles. Switch to the Center layer, change the DIMCEN system variable to a value of .07, and use the DIMCENTER command to place center marks identifying the centers of all circles and arcs, as shown on the right in the following image.

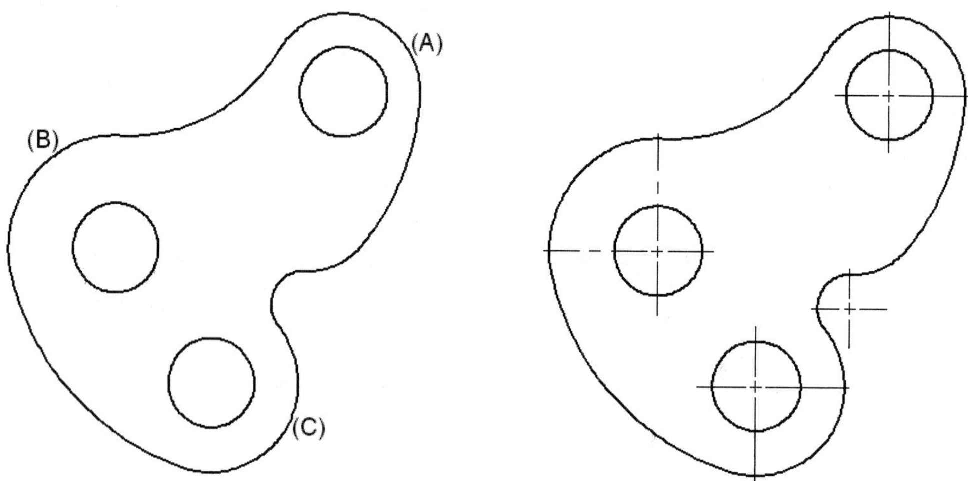

Figure 5.74

PROBLEMS FOR CHAPTER 5

DIRECTIONS FOR PROBLEMS 5–1 THROUGH 5–14:

Construct these geometric construction figures using existing AutoCAD commands.

PROBLEM 5–1

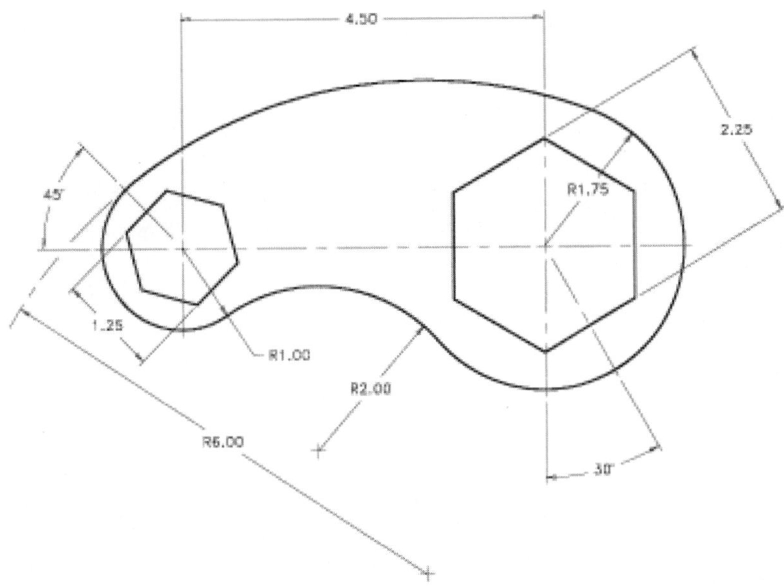

PROBLEM 5–2

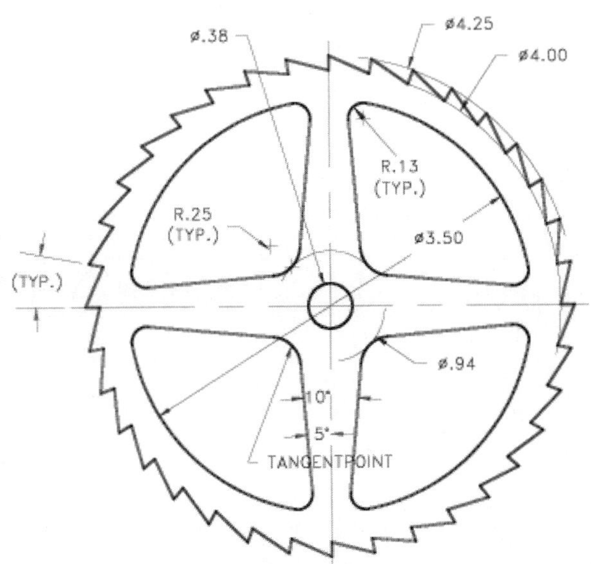

PROBLEM 5–3

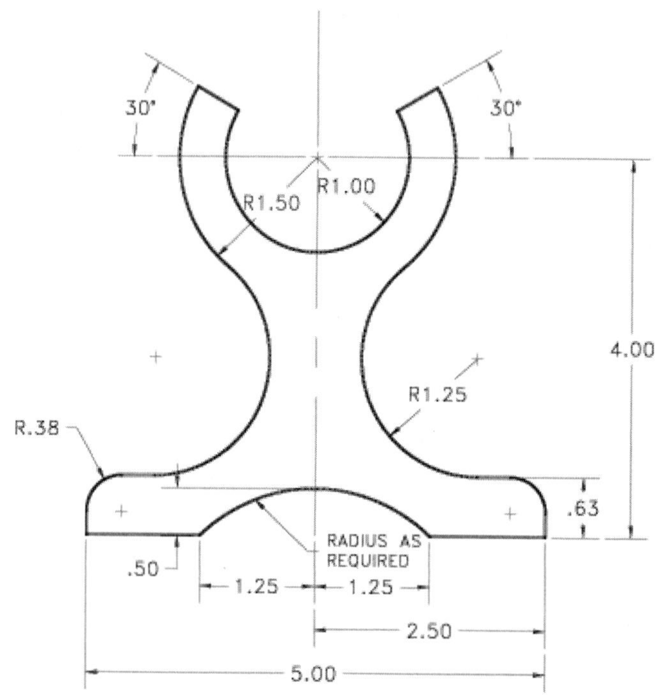

PROBLEM 5–4

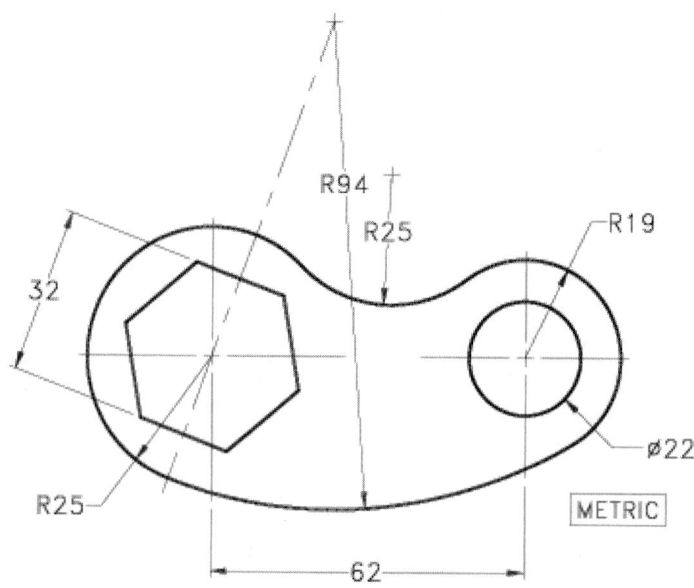

PROBLEM 5–5

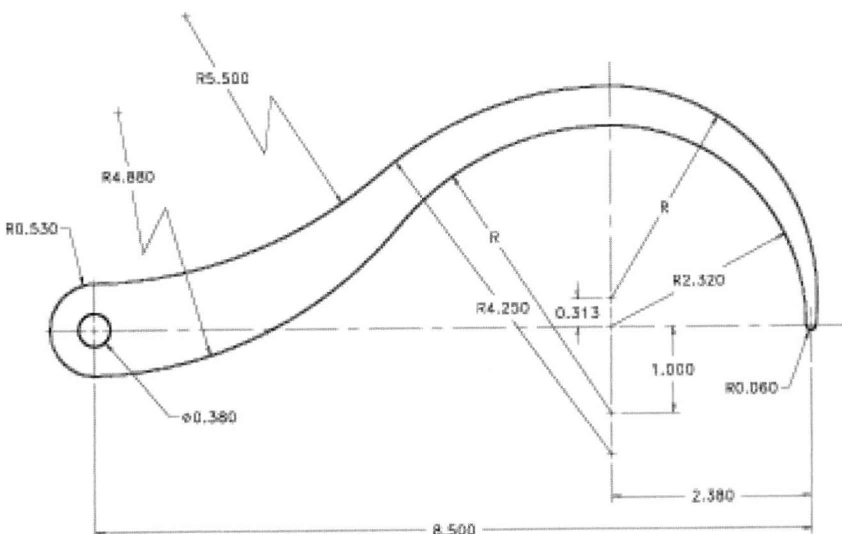

PROBLEM 5–6

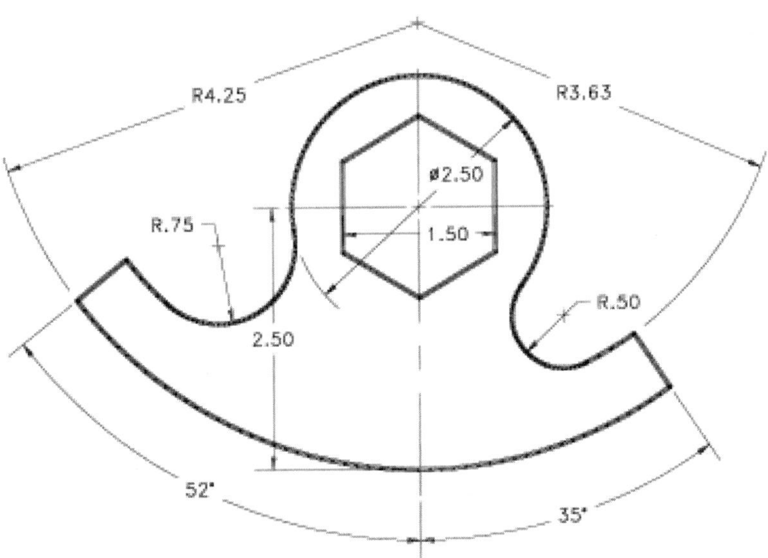

PROBLEM 5–7

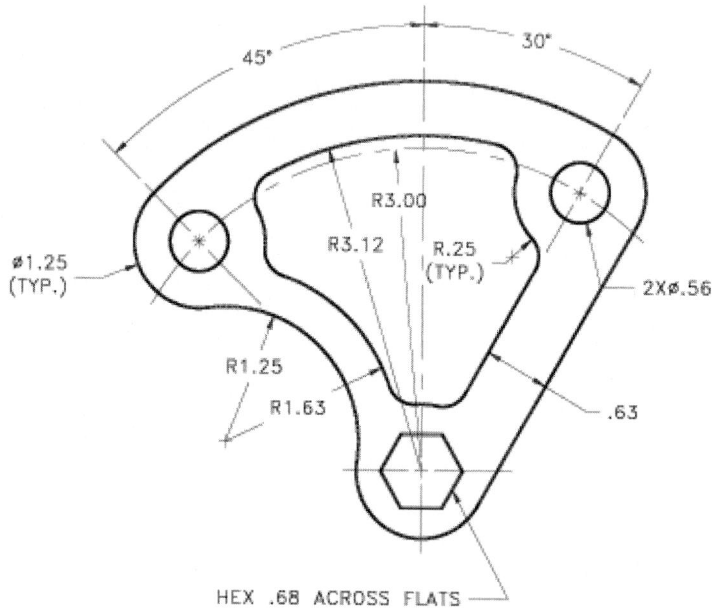

PROBLEM 5–8

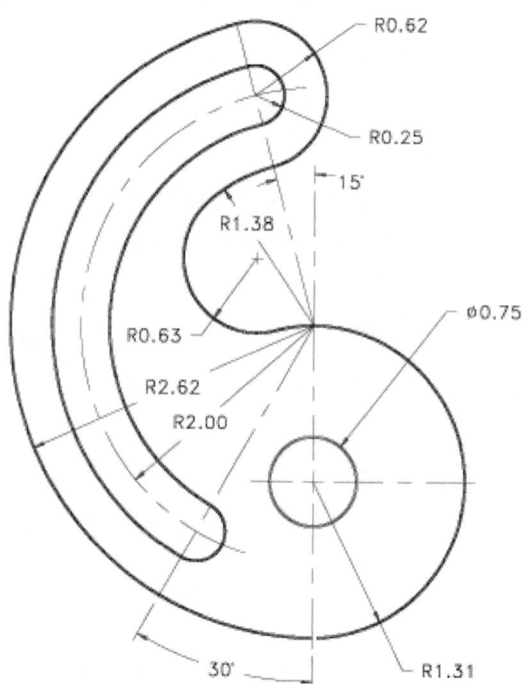

PROBLEM 5–9

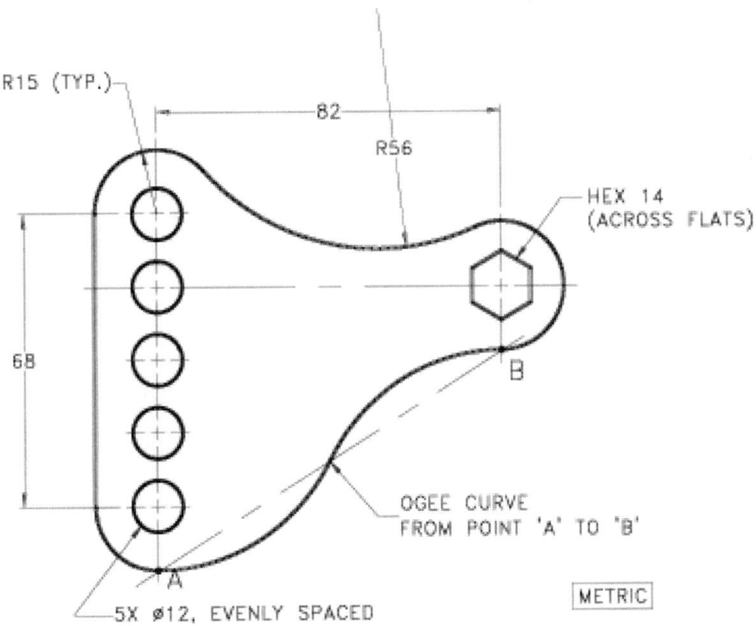

R15 (TYP.)
82
R56
HEX 14
(ACROSS FLATS)
68
B
OGEE CURVE
FROM POINT 'A' TO 'B'
A
METRIC
5X ø12, EVENLY SPACED

PROBLEM 5–10

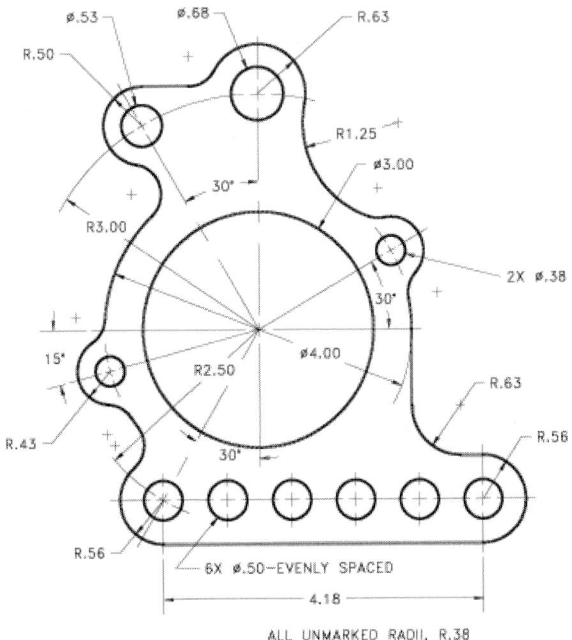

ø.53
ø.68
R.63
R.50
R1.25
ø3.00
30°
R3.00
2X ø.38
30°
15°
ø4.00
R2.50
R.63
R.43
R.56
30°
R.56
6X ø.50–EVENLY SPACED
4.18
ALL UNMARKED RADII, R.38

PROBLEM 5–11

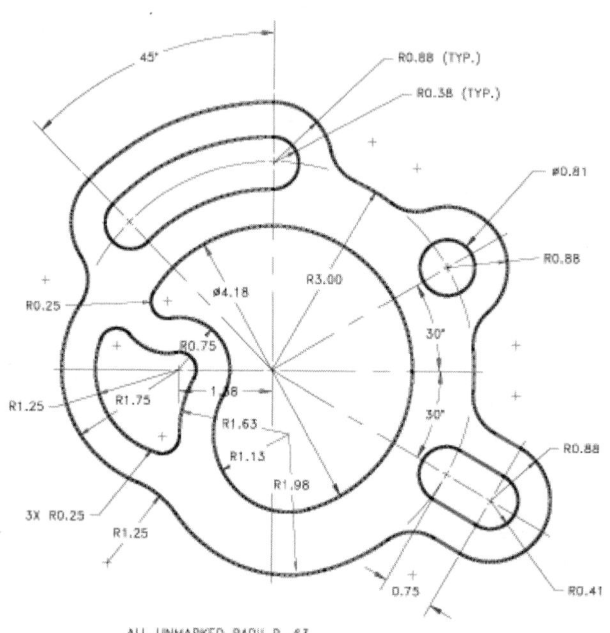

PROBLEM 5–12

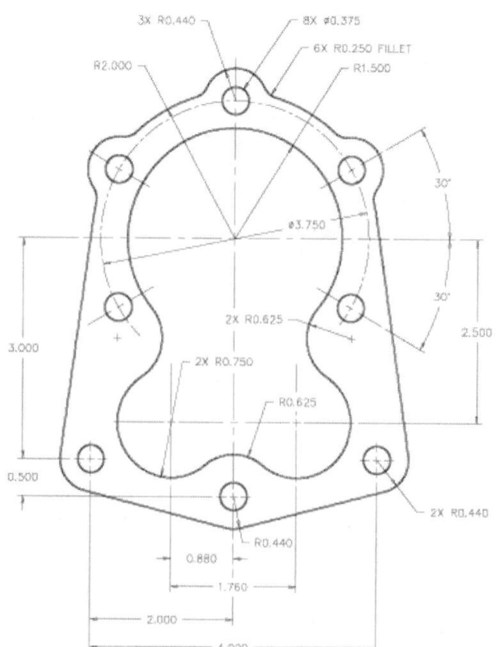

PROBLEM 5–13

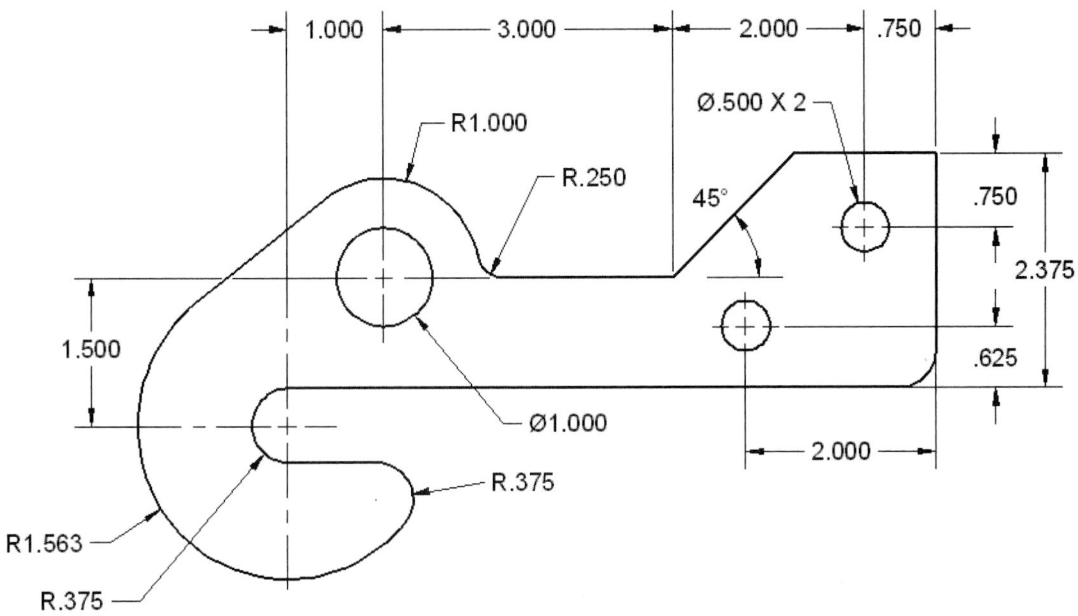

PROBLEM 5–14

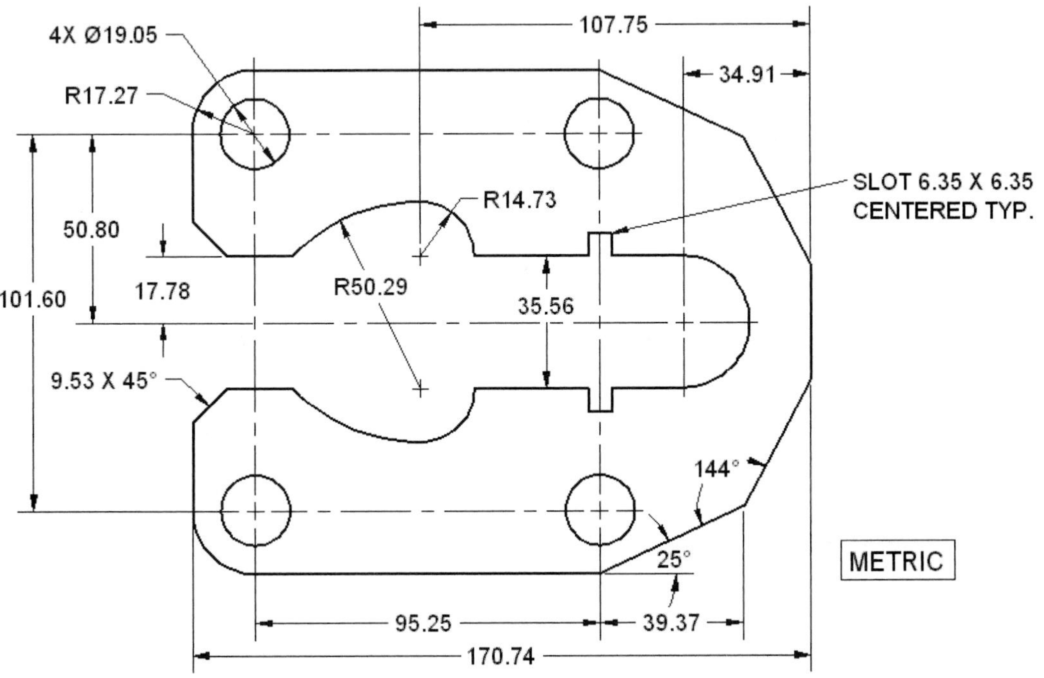

CHAPTER 6

Working with Text, Fields, and Tables

AutoCAD provides a robust set of text commands that allow you to place different text objects, edit those text objects, and even globally change the text height without affecting the justification. The heart of placing text in a drawing is through the MTEXT command. This chapter discusses the uses of the DTEXT command, which stands for Dynamic Text or Single Line Text. In this chapter, you will also be given information on how to edit text created with the MTEXT and DTEXT commands. The SCALETEXT and JUSTIFYTEXT commands will be explained in detail, as will the ability to create custom text styles and assign different text fonts to these styles. Intelligent text in the form of fields will also be discussed in this chapter, as will the creation of tables.

AUTOCAD TEXT COMMANDS

All text commands can be located easily by launching the Text toolbar or Ribbon, illustrated in the following image. The two main modes for entering text are Multiline Text (the MTEXT command) and Single Line Text (the DTEXT command). You can also find text commands by choosing the Text cascading menu from the Draw pull-down menu or the Menu Browser, as shown in the following image.

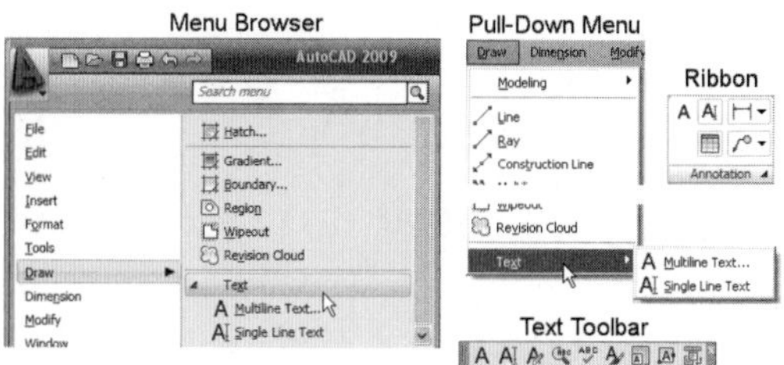

Figure 6.1

Study the information in the following table for a brief description of each text command.

Button	Tool	Shortcut	Function
	Multiline Text	MT	Creates paragraphs of text as a single object
	Single Line Text	DT	Creates one or more lines of text; each line of text is considered a separate object
	Edit Text	ED	Used for editing multiline and single line text in addition to dimension and attribute text
	Find and Replace	FIND	Used for finding, replacing, selecting, or zooming in to a particular text object
	Spell Check	SP	Performs a spell check on a drawing
	Text Style	ST	Launches the Text Style dialog box used for creating different text styles
	SCALETEXT	None	Used for scaling selected text objects without affecting their locations
	JUSTIFYTEXT	None	Used for justifying selected text objects without affecting their locations
	SPACETRANS	None	Converts text heights between Model Space and Paper Space (Layout mode)

ADDING MULTILINE TEXT IN THE AUTOCAD CLASSIC WORKSPACE

The MTEXT command allows for the placement of text in multiple lines. Entering the MTEXT command from the command prompt displays the following prompts:

```
Command: MT (For MTEXT)
Current text style: "GENERAL NOTES" Text height: 0.2000
    Annotative: No
Specify first corner: (Pick a point at "A" to identify one
    corner of the Mtext box)
Specify opposite corner or [Height/Justify/Line spacing/Rotation/
    Style/Width/Columns]: (Pick another corner forming a box)
```

Picking a first corner displays a user-defined box with an arrow at the bottom, as shown in the following image. This box defines the area where the multiline text will be placed. If the text cannot fit on one line, it will wrap to the next line automatically.

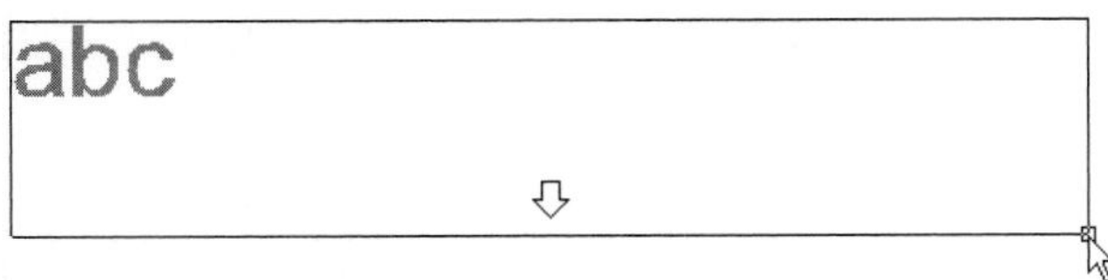

Figure 6.2

After you click a second point marking the other corner of the insertion box, the Text Formatting toolbar appears, along with the Multiline Text Editor. The parts of these items are shown in the following image. One of the advantages of this tool is the transparency associated with the Multiline Text Editor. As you begin entering text, you can still see your drawing in the background.

Notice in the following image other important areas of the Text Formatting toolbar. The current text style and font are displayed, in addition to the text height. You can make the text bold, italicized, or underlined by using the B, I, and U buttons present in the toolbar. When you have finished entering the text, click the OK button to dismiss the Text Formatting toolbar and text editor and place the text in the drawing.

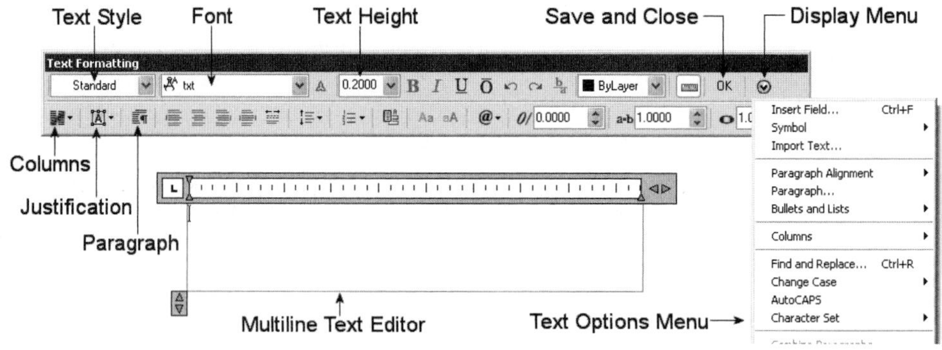

Figure 6.3

As you begin to type, your text appears in the Text Editor box shown in the following image. As the multiline text is entered, it automatically wraps to the next line depending on the size of the initial bounding box. Tabs and indents can be utilized as necessary, and even the size of the box can be modified, if required.

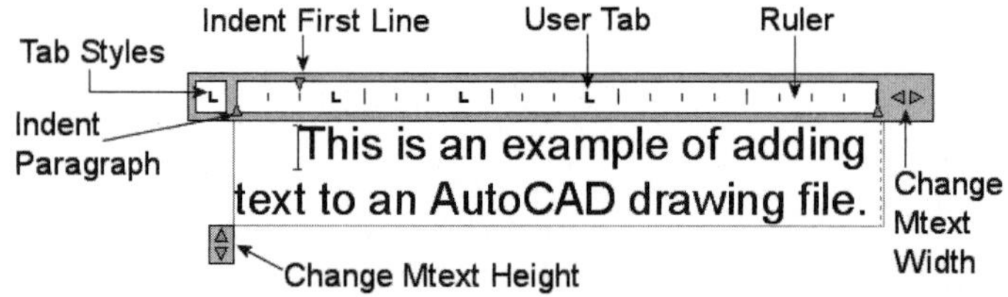

Figure 6.4

Note: Clicking a blank part of your screen after you are finished entering text in the previous example will also exit the Text Formatting toolbar.

Another method of displaying additional options of the Multiline Text Editor is to right-click inside the text box. This displays the menu shown in the following image.

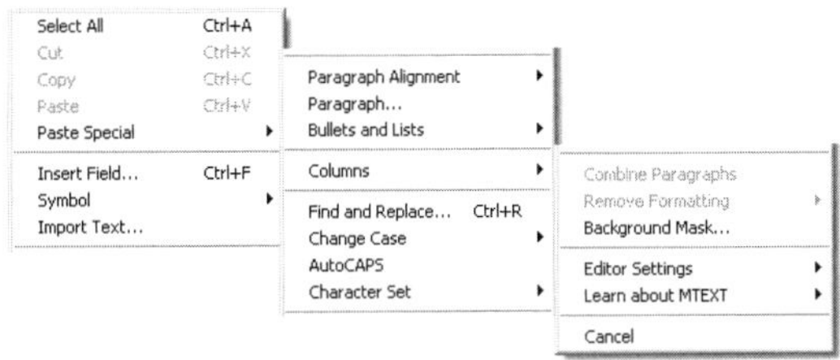

Figure 6.5

TEXT FORMATTING TOOLBAR BUTTONS – UPPER ROW

The Text Formatting toolbar has a wealth of controls available to manipulate text items. These buttons will be explained in two separate groups: buttons located on the top and bottom rows of the toolbar. The following table will explain the top row controls and buttons.

Button	Tool	Function
Standard	Text Style Name	Displays the current text style. Click the down arrow to change to a different text style.
Txt	Text Font Name	Displays the current text font
	Annotative	Turns annotative scale on or off
0.2000	Text Height	Displays the current text height
B	Bold	Used to change highlighted text to bold
I	Italics	Used to change highlighted text to italics
U	Underline	Used to underline highlighted text
O	Overline	Used to place a line over the top of highlighted text
	Undo	Used to undo the previous text option and remain in the Text Formatting toolbar
	Redo	Used to redo the previous undo
	Fraction Stack	Used for turning stacked fraction mode on or off
	Color	Used to change the color of highlighted text
	Ruler	Displays a ruler used for setting tabs and indents
OK	OK	Used to accept the current multiline text and exit the Text Formatting toolbar
	Options	Used to display additional text options

TEXT FORMATTING TOOLBAR BUTTONS – LOWER ROW

The following table explains the top row controls and buttons located in the Text Formatting toolbar.

Button	Tool	Function
	Columns	Displays a menu used to control the type of text column generated
	Mtext Justification	Displays a menu consisting of nine text justification modes
	Paragraph	Displays the Paragraph dialog box used for changing such items as indentations that affect paragraph text
	Left	Justifies the left edge of text with the left edge of the margin
	Center	Justifies mtext objects to the center
	Right	Justifies the right edge of text with the right edge of the margin
	Justify	Adjusts the horizontal spacing of text so the text is aligned evenly along the right and left margins
	Distribute	Distributes mtext across the width of the mtext box and adjusts the spacing in between individual letters
	Line Spacing	Displays a menu used for changing the spacing in between lines of text
	Numbering	Displays a menu used for applying bullets and numbers to text
	Field	Used for inserting a field
	Uppercase	Used for changing all highlighted text to uppercase
	Lowercase	Used for changing all highlighted text to lowercase
	Symbol	Displays a menu for selecting special text characters and symbols
	Oblique Angle	Determines the forward or backward slant of text
	Tracking	Used to increase or decrease the space between selected text characters
	Width Factor	Used to create wide or narrow text characters

TEXT JUSTIFICATION MODES

The following image illustrates a sample text item and the various locations by which the text can be justified. By default, when you place text, it is left justified. These justification modes are designed to work in combination with one another. For example, one common justification is to have the text middle centered. Options in text commands allow you to pick Middle Center from a menu or type MC in at the keyboard. The same type of combination is available for the other justification modes.

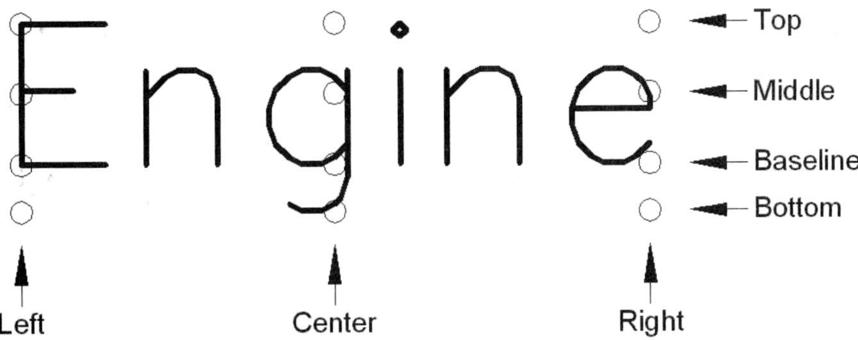

Figure 6.6

JUSTIFYING MULTILINE TEXT

Various justification modes are available to change the appearance of the text in your drawing. Justification modes can be retrieved by clicking the Justification button found in the second row of the Text Formatting toolbar as shown in the following image. Clicking this button displays the nine modes.

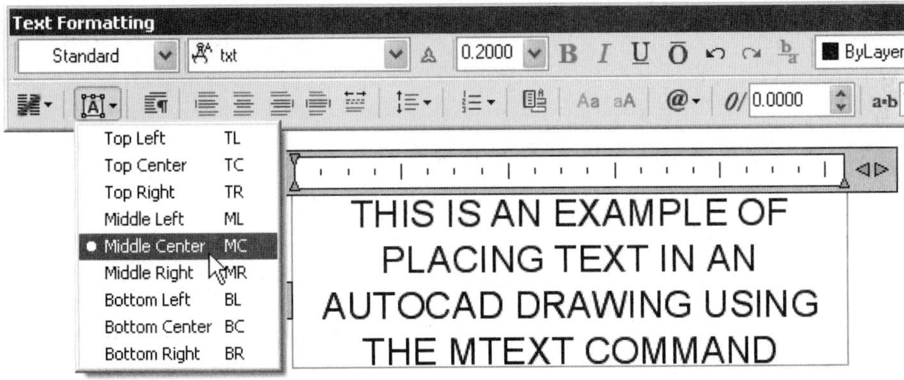

Figure 6.7

Additionally, Paragraph Alignment modes are also available in the second row of the Text Formatting dialog box, as shown in the following image. These justification modes are similar to those found in word processors and allow you to align individual paragraphs in the text editor. Paragraphs are separated by hard returns (pressing the ENTER key).

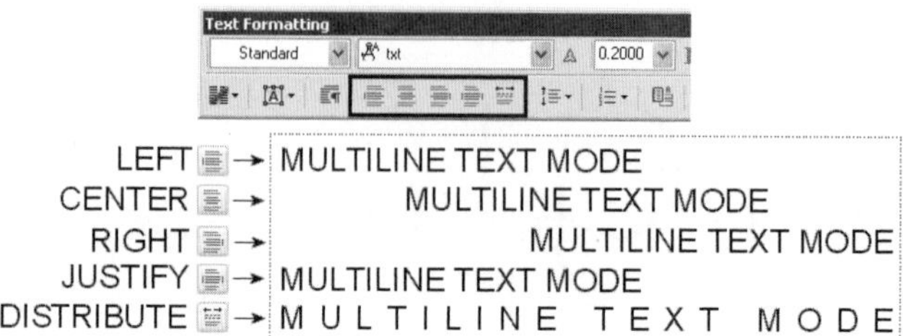

Figure 6.8

INDENTING TEXT

Multiline text can be indented in order to align text objects to form a list or table. A ruler, illustrated in the following image, consisting of short and longer tick marks displays the current paragraph settings.

The tabs and indents that you set before you start to enter text apply to the whole multiline text object. If you want to set tabs and indents to individual paragraphs, click in a paragraph or select multiple paragraphs to apply the indentations.

Two sliders are available on the left side of the ruler to show the amount of indentation applied to the various multiline text parts. In the following image, the top slider is used to indent the first line of a paragraph. A tooltip is available to remind you of this function.

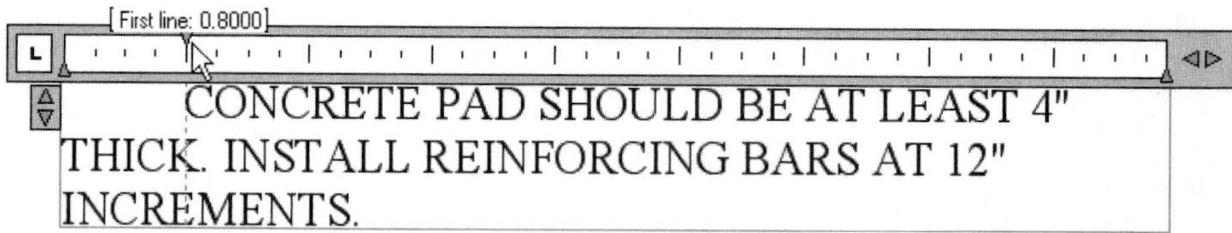

Figure 6.9

In the following image, a bottom slider is also available. This slider controls the amount of indentation applied to the other lines of the paragraph. You can see the results of using this slider in the following image.

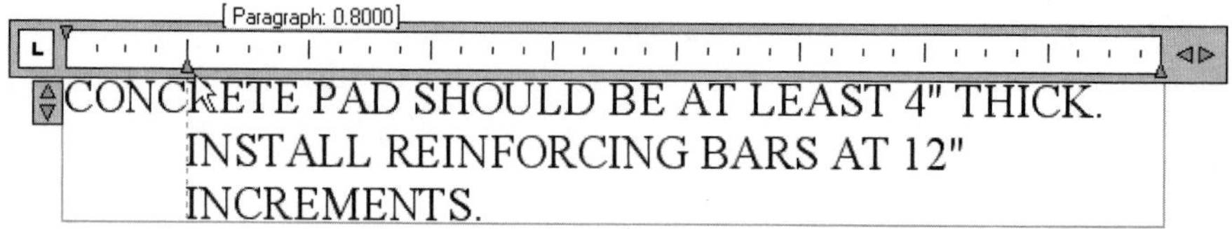

Figure 6.10

Note: If you set tabs and indents prior to starting to enter multiline text, the tabs are applied to the whole multiline object.

FORMATTING WITH TABS

You can further control your text by adding tabs. The long tick marks on the ruler show the default tab stops. Clicking the Tab Style button toggles to four different tab modes, as shown on the left in the following image. When you click the ruler to set your own tabs, the ruler displays a small, default L-shaped marker at each custom tab stop, as shown in the following image. When you anchor your cursor in front of a sentence and press the TAB key on your keyboard, the text aligns to the nearest tab marker. To remove a custom tab stop, drag the marker off the ruler.

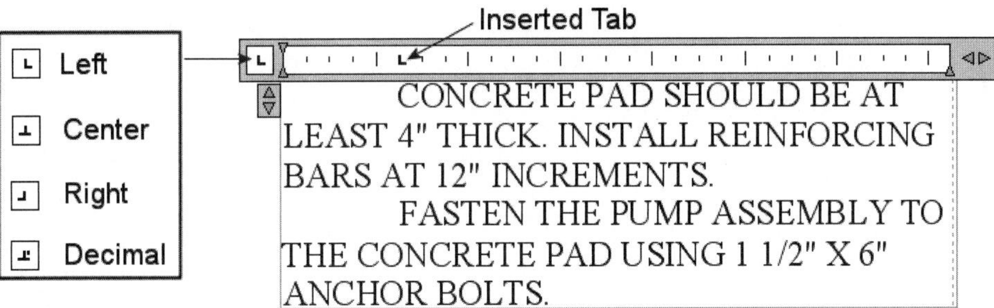

Figure 6.11

BULLETING AND NUMBERING TEXT

Multiline text can also be formatted in a numbered, bulleted, or alphabetic character list, either uppercase or lowercase. The four types of lists are illustrated in the following image.

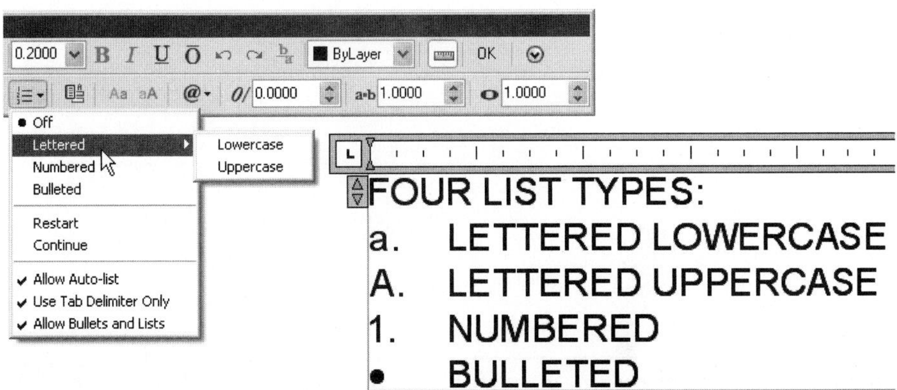

Figure 6.12

As lists are created, times occur when you need to change the format for new lists generated in the same multiline text command. In the following image on the left, an alphanumeric list was interrupted to generate a numbered list. The menu shown on the right can assist with this operation. When going from the lettered to the numbered list, the Restart mode was utilized. Notice also in the illustration on the left that a paragraph was added with more numbers continuing beneath it. This demonstrates the ability to continue a list.

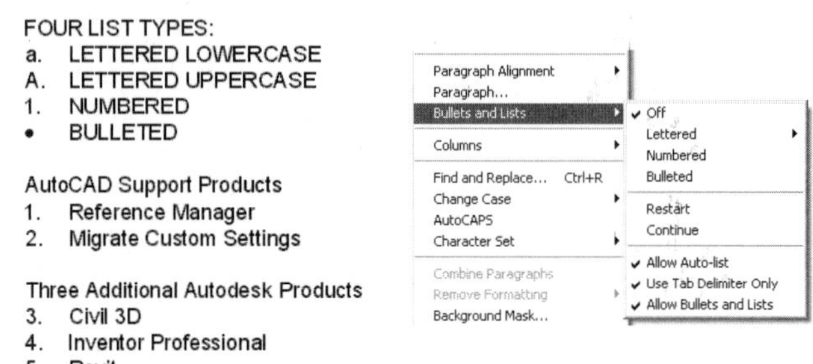

Figure 6.13

FORMATTING FRACTIONAL TEXT

Fractional text can be reformatted through the Multiline Text Editor. Entering a space after the fraction value in the following image displays the AutoStack Properties dialog box, also displayed in the following image. Use this dialog box to enable the autostacking of fractions. You can also remove the leading space in between the whole number and the fraction. In the following image, the fraction on the left was converted to the fraction on the right using the AutoStack Properties dialog box. A Fraction Stack button, which can be used to stack or un-stack a fraction, is also provided in the top line of the Text Formatting Toolbar.

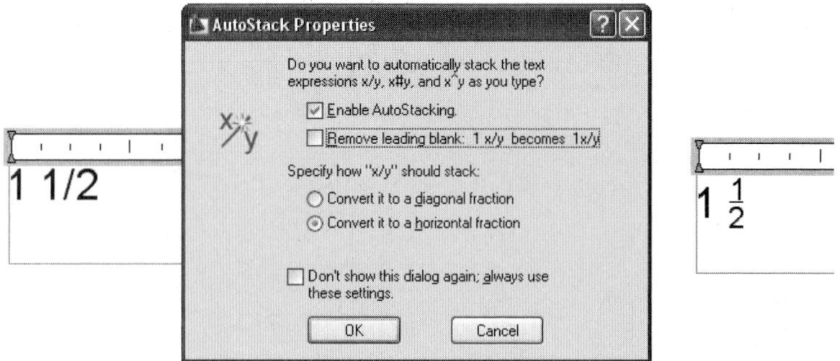

Figure 6.14

CHANGING THE MTEXT WIDTH AND HEIGHT

When you first construct a rectangle that defines the boundary of the mtext object, you are not locked into this boundary. Right-clicking inside the tab marker area, as illustrated in the following image, activates the menu used for displaying either Set Mtext Width or Height dialog boxes. The current value in the Width field is the current width of the rectangle used to construct the mtext object. Enter a larger or smaller value to expand or contract the mtext object. You can also use the two sets of arrows at either ends of the ruler to change the mtext width or height.

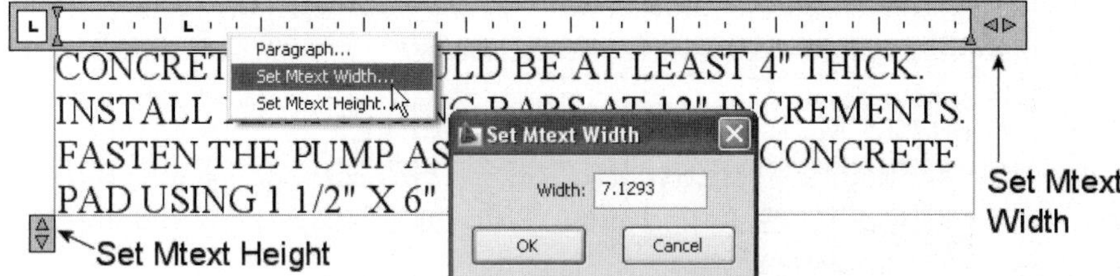

Figure 6.15

Note: Grips provide another method for changing the size of the text box once the text has already been created. Clicking the multiline text object as it appears in your drawing causes blue boxes called grips to appear in the four corners of the text box. Picking one of these grips and moving to the left or right increases or decreases the size of the text box. Grips will be covered in greater detail in chapter 7.

MULTILINE TEXT AND THE RIBBON

All of the text examples presented up to this point illustrated placing text through the AutoCAD Classic workspace. Text can also be placed through the 2D Drafting and Annotation workspace that utilizes the Annotation tab of the Ribbon as shown in the following image. Basic text commands are contained in this area.

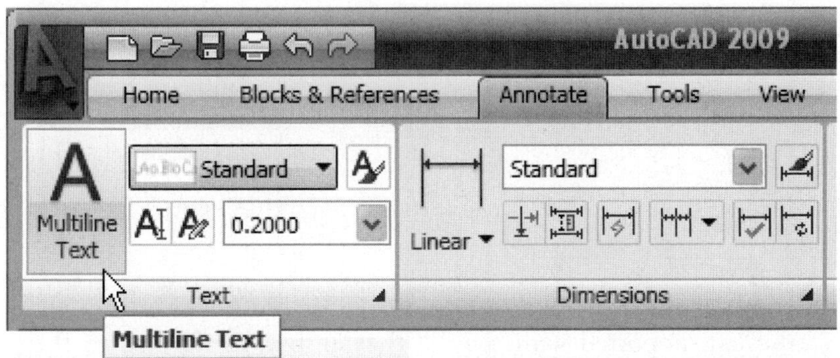

Figure 6.16

When you activate the Multiline Text command and generate the rectangular text box, a new menu appears in the Ribbon as shown in the following image. These headings allow you to control the text style, formatting, paragraph, text insertion, and other text options. This gives you a more robust set of text related commands all at your fingertips.

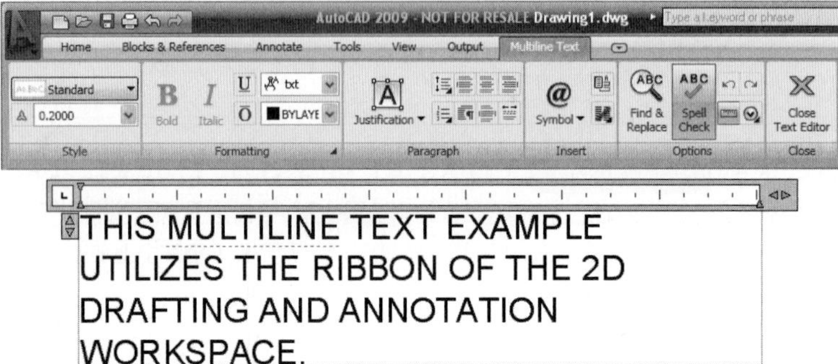

Figure 6.17

IMPORTING TEXT INTO YOUR DRAWING

If you have existing files created in a word processor, you can import these files directly into AutoCAD and have them converted into mtext objects. As illustrated in the following image, right-click in the Multiline Text Editor box to display the menu; then pick Import Text. An open file dialog box displays, allowing you to pick the desired file to import. The Import Text feature of the Multiline Text Editor supports two file types: those ending in a TXT extension and those ending in the RTF (rich text format) extension. Text with the TXT extension conforms to the current AutoCAD text style regarding font type. Text with the RTF extension remembers formatting and font settings from the word-processing document that it was originally created in.

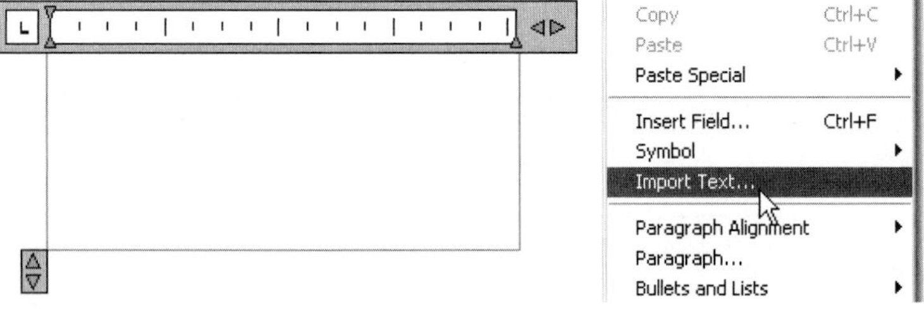

Figure 6.18

The finished result of this text-importing operation is illustrated in the following image. The text on the left, imported with the TXT extension, conformed to the present text style, which has a font of Arial assigned to it. The text on the right was imported with the RTF extension. Even with the text style being set to Arial, the RTF-imported text displays in the Times New Roman font. This is the font in which it was originally created in the word-processing document.

TXT Extension	RTF Extension
The Import Text feature of the Multiline Text supports two file types; those ending in a TXT extension and the other in the RTF (Rich Text Format) extension.	The Import Text feature of the Multiline Text supports two file types; those ending in a TXT extension and the other in the RTF (Rich Text Format) extension.

Figure 6.19

MULTILINE TEXT SYMBOLS

While in the Multiline Text Editor, right-clicking in the text window and selecting Symbol, as shown in the following image, displays various symbols that can be incorporated into your drawing.

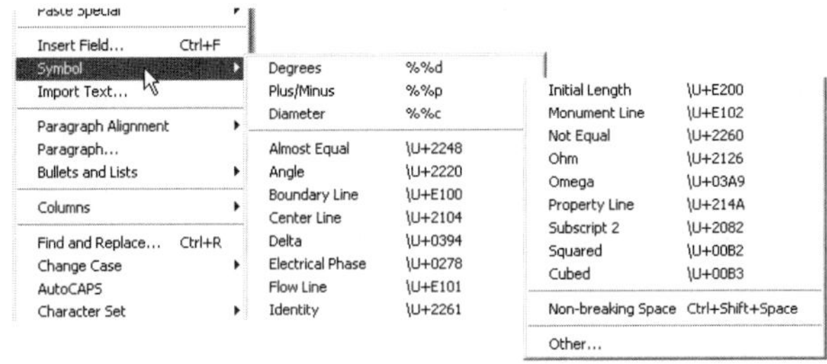

Figure 6.20

The table illustrated in the following image shows all available multiline text symbols along with their meanings.

MTEXT SYMBOLS			
SYMBOL	DESCRIPTION	SYMBOL	DESCRIPTION
°	Degrees	≡	Identity
±	Plus/Minus	O͢	Initial Length
Ø	Diameter	ℳ	Monument Line
≈	Almost Equal	≠	Not Equal
∠	Angle	Ω	Ohm
ℬ	Boundary Line	Ω	Omega
℄	Center Line	℘	Property Line
Δ	Delta	H_2O	Subscript 2
ϕ	Electrical Phase	4^2	Squared
ℱ	Flow Line	4^3	Cubed

Figure 6.21

CREATING PARAGRAPHS OF TEXT

Clicking the Paragraph button of the Text formatting toolbar opens the Paragraph dialog box. This feature allows you to set indentations for the first lines of paragraphs and for the entire paragraph. In addition to indents, you can also specify the type of tab stop, paragraph spacing, and the spacing of lines in a paragraph.

The Tab area of the dialog box allows you to set left, center, right, and decimal tabs. You can also add or remove tabs using the appropriate control buttons.

The Left Indent area allows you to set the indentation value for the first line of text. The Hanging indent allows you to set indentations to the selected or current paragraphs.

The Right Indent area applies an indentation to the entire selected or current paragraph.

Placing a check in the Paragraph Alignment box activates five alignment properties for the selected or current paragraph.

Placing a check in the Paragraph Spacing box activates controls for setting the spacing before or after the selected or current paragraph. The spacing between two paragraphs is calculated from the total of the After value and the Before value.

Placing a check in the Paragraph Line Spacing box allows you to set the spacing between individual lines in the selected or current paragraph.

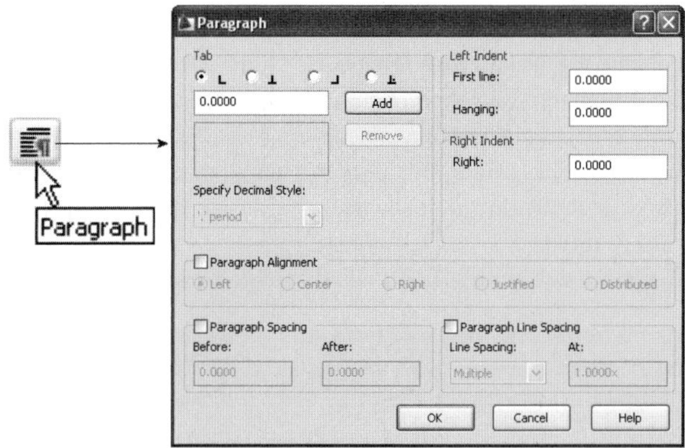

Figure 6.22

ORGANIZING TEXT BY COLUMNS

This feature in the Text Formatting toolbar allows you to create columns of an mtext object. You begin by first creating an mtext object consisting of a single column. You then choose one of the two different column methods, (Dynamic Columns and Static Columns) from the Columns menu, as shown on the left in the following image.

Dynamic Columns allow columns of text to flow, causing columns to be added or removed. You can control this flow through the Auto height or Manual height modes.

Static Columns allow you to specify the width, height, and number of columns. In this way, all columns share the same height.

Other column options include the following:

No Columns specifies no columns for the current mtext object.

Insert Column Break ALT + ENTER inserts a manual column break.

Column Settings displays the Column Settings dialog box shown on the right in the following image. Through this dialog box, you specify the column and gutter width, height, and number of columns. Grips are used to edit the width and height of the column.

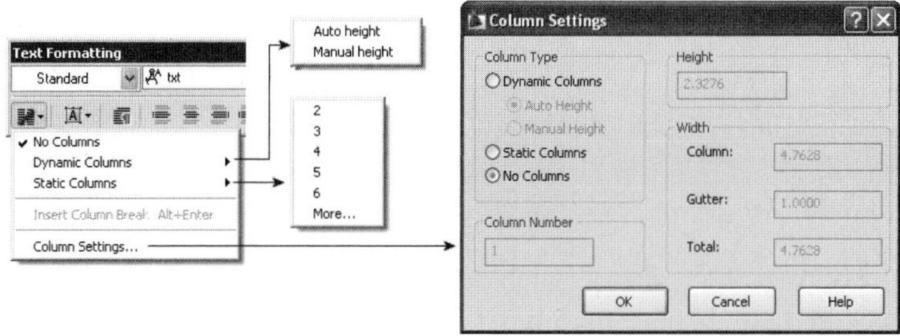

Figure 6.23

Try It!: Open the drawing file 06_Specifications. The text in this drawing consists of one long column. You will arrange this text in three columns, which will allow all text to be visible on the drawing sheet. Activate the Mutliline Text Editor by entering the ED command and picking any text object. Click the Columns button located in the Text Formatting toolbar. Click the Dynamic Columns from the menu followed by Manual Height as shown in the following image. Click the OK button to dismiss the Text Formatting toolbar.

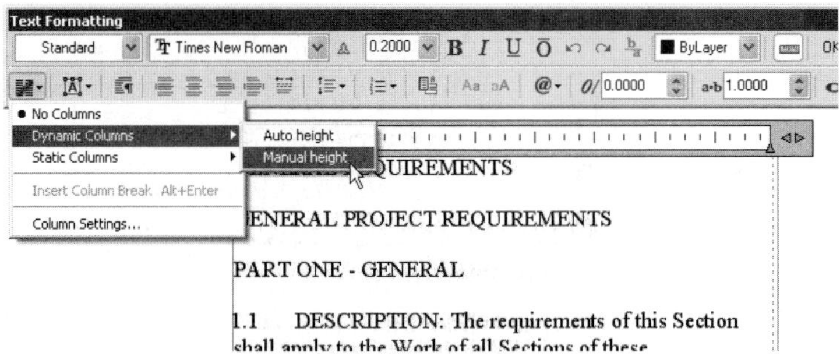

Figure 6.24

When you return to the drawing, click the multiline text object and notice an arrow grip present at the very bottom of the text paragraph as shown on the left in the following image. Pick this grip and move the bottom of the paragraph up. As you do this, two or three columns appear, as shown on the right in the following image.

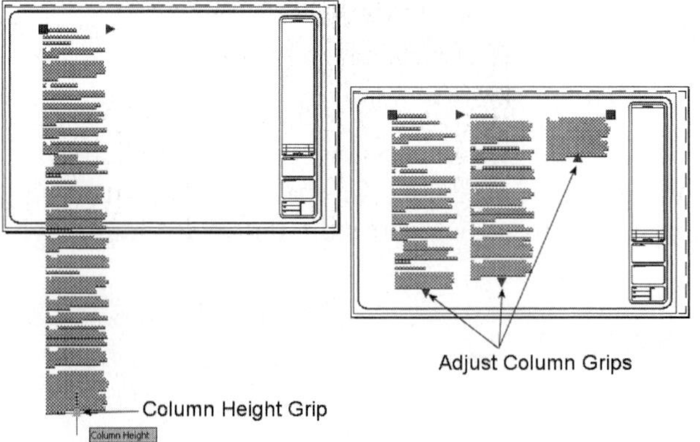

Figure 6.25

Zoom in to the columns of text and keep adjusting the arrow grips at the bottom of each paragraph until your text display matches the following image. In this image, notice that Part Two and Part Three are at the top of their respective columns.

GENERAL REQUIREMENTS

GENERAL PROJECT REQUIREMENTS

PART ONE - GENERAL

1.1 DESCRIPTION: The requirements of this Section shall apply to the Work of all Sections of these Specifications.

1.2 QUALITY ASSURANCE: In procuring all items used in the Work, its is the Contractor's responsibility to verify the detailed requirements of the specifically named codes and standards and to verify that the items procured for use in this Work meet or exceed the specified requirements.

1.3 PRODUCT HANDLING

1.3.1 Delivery of materials: Deliver all materials to the job site in original, new and unopened containers bearing the manufacturer's name and label.

1.3.2 Storage of materials: Provide proper storage to prevent damage to and deterioration of materials.

1.3.3 Protection: Use all means necessary to protect the materials of these Sections before, during and after installation and to protect the work and materials of all other trades.

1.3.4 Replacements: In the event of damage, immediately make all repairs and replacements necessary at not additional cost.

1.4 ALLOWANCES: Where the work of certain Sections is to be performed under specified allowances, the Contractor shall submit the following for the work of each given Section:

(a) the initial bid price
(b) quantity "take-offs
(c) material invoices as selected and installed
(d) invoices for labor (where applicable)

This information will be used to determine the appropriateness of extras or credits due as applied for by the Contractor.

PART TWO - PRODUCTS

2.1 COLORS AND PATTERNS OF MATERIALS: Unless the precise color and pattern are specifically described in the Contract Documents, and whenever a choice of color or pattern is available in a specified product, submit accurate color and pattern charts for review and selection.

2.2 SUBSTITUTIONS: Do not substitute materials, equipment or methods unless such substitution has been specifically approved for this Work. Where the phrase "or equal" occurs in the Contract Documents, do not assume that materials, equipment or methods will be approved as equal unless the item has been specifically approved for this Work.

2.3 CLEANING MATERIALS AND EQUIPMENT: Provide all required personnel, equipment and materials needed to maintain the specified standard of cleanliness, as recommended by the manufacturer of the material.

2.4 OTHER MATERIALS: All other materials, not specifically described but required for a complete and proper installation as indicated on the Drawings, shall be new, suitable for the intended use and subject to approval.

PART THREE - EXECUTION

3.1 INSPECTION: Each trade shall examine the areas and conditions under which their work will be performed. Correct conditions detrimental to the proper and timely completion of the Work. Do not proceed until unsatisfactory conditions have been corrected.

3.2 COORDINATION: Each trade shall be responsible to carefully coordinate with all other trades to ensure proper and adequate interface of the work of other trades with their work.

3.3 DELIVERIES: Stockpile all materials sufficiently in advance of need to insure their availability in a timely manner for this Work.

3.4 COMPLIANCE OF MATERIALS: Do not permit materials not complying with the various provisions of these Specifications to be brought onto or to be stored at the job site. Immediately remove from the job site all non-complying materials and replace them with materials meeting the requirements of this Section.

3.5 TIMIMG OF SUBMITTALS: Make all submittals far enough in advance of scheduled dates for installation to provide all time required for reviews, for securing necessary approvals, for possible revisions and re-submittals, and for placing orders and securing delivery.

3.6 CLEANING: Retain all stored items in an orderly arrangement allowing maximum access, not impeding drainage or traffic and providing the required protection of materials. Do not allow the accumulation of scrap, debris, waste material and other items not required for construction of the Work. Prior to completion of the Work, remove from the job site all tools, surplus materials, equipment, scrap, debris and waste. Visually inspect all surfaces and remove all traces of soil, waste material, smudges and other foreign matter. Remove all traces of splashed materials from adjacent surfaces. Remove all paint drippings, spots, stains and dirt from finished surfaces.

Figure 6.26

CREATING SINGLE LINE TEXT

Single line text is another method of adding text to your drawing. The actual command used to perform this operation is DTEXT, which stands for Dynamic Text mode and which allows you to place a single line of text in a drawing and view the text as you type it. This command can be found on the Ribbon in either the Home or Annotate tabs. You might also choose the command from the Menu Browser or the Draw pull-down menu, as shown in the following image.

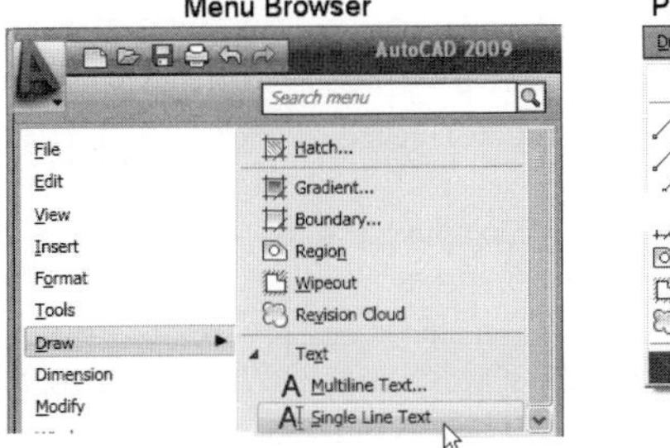

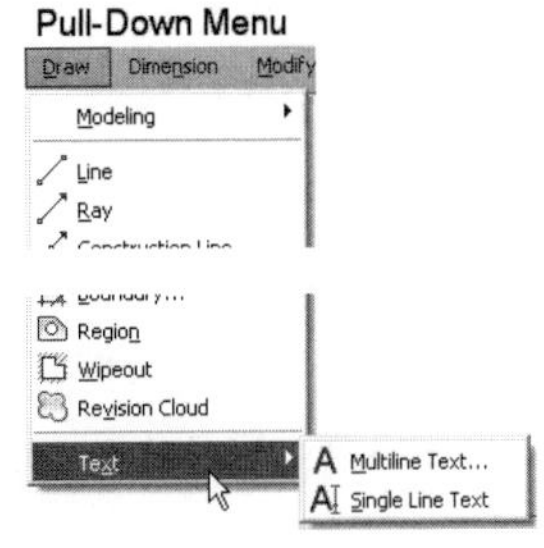

Figure 6.27

When using the DTEXT command, you are prompted to specify a start point, height, and rotation angle. You are then prompted to enter the actual text. As you do this, each letter displays on the screen. When you are finished with one line of text, pressing ENTER drops the Insert bar to the next line, where you can enter more text. Pressing ENTER again drops the Insert bar down to yet another line of text, as shown on the left in the following image. Pressing ENTER at the "Enter text" prompt exits the DTEXT command and permanently adds the text to the database of the drawing, as shown on the right in the following image.

```
Command: DT (For DTEXT)
Current text style: "Standard" Text height: 0.2000 Annotative: No
Specify start point of text or [Justify/Style]: (Pick a point
    at "A")
Specify height <0.2000>: 0.50
Specify rotation angle of text <0>: (Press ENTER to accept this
    default value)
(Enter text:) ENGINEERING (After this text is entered, press ENTER
    to drop to the next line of text)
(Enter text:) GRAPHICS (After this text is entered, press ENTER to
    drop to the next line of text)
(Enter text:) (Either add more text or press ENTER to exit this
    command and place the text)
```

ENGINEERING GRAPHICS | ENGINEERING GRAPHICS

Figure 6.28

ADDITIONAL SINGLE LINE TEXT APPLICATIONS

When you place a line of text using the DTEXT command and press ENTER, the Insert bar drops down to the next line of text. This sequence continues until you press ENTER at the "Enter text" prompt, which exits the command and places the text permanently in the drawing. You can also control the placement of the Insert bar by clicking a new location in response to the "Enter text" prompt. In the following image, various labels need to be placed in the pulley assembly. The first label, "BASE PLATE," is placed with the DTEXT command. Without pressing ENTER, immediately pick a new location at "B" and place the text "SUPPORT." Continue this process with the other labels. The Insert bar at "E" denotes the last label that needs to be placed. When you perform this operation, pressing ENTER one last time at the "Enter text:" prompt places the text and exits the command. Follow the command sequence below for a better idea of how to perform this operation.

Try It!: Open the drawing file 06_Pulley Text. Use the following image and command sequence to perform this operation.

```
Command: DT (For DTEXT)
Current text style: "Standard" Text height: 0.2500 Annotative: No
Specify start point of text or [Justify/Style]: (Pick a point
    at "A")
Specify height <0.2500>: (Press ENTER to accept this default
    value)
Specify rotation angle of text <0>: (Press ENTER to accept this
    default value)
Enter text: BASE PLATE (After this text is entered, pick
    approximately at "B")
```

```
Enter text: SUPPORT (After this text is entered, pick
    approximately at "C")
Enter text: SHAFT (After this text is entered, pick approximately
    at "D")
Enter text: PULLEY (After this text is entered, pick
    approximately at "E")
Enter text: (Enter BUSHING for this part and then press ENTER)
Enter text: (Press ENTER to exit this command)
```

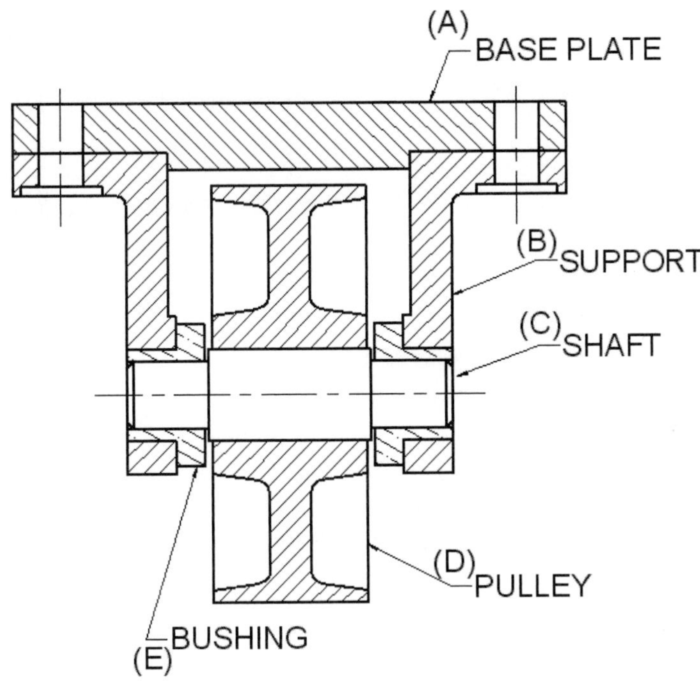

Figure 6.29

EDITING TEXT

Text constructed with the MTEXT and DTEXT commands is easily modified with the DDEDIT command. Start this command by choosing Edit..., which is found in the Object and Text cascades of the Modify pull-down menu, as shown on the left in the following image. Selecting the multiline text object displays the Multiline Text Editor, as shown on the right in the following image. This is the same dialog box as that used to initially create text through the MTEXT command. Use this dialog box to change the text height, font, color, and justification. The DDEDIT command can also be accessed from the Menu Browser if you are not utilizing the AutoCAD Classic workspace.

Note: A quicker way to edit multiline text is to double-click the text object. This launches the Multiline Text Editor box and Text Formatting toolbar.

Try It!: Open the drawing file 06_Edit Text1. Find the Edit text command in the Modify pull-down menu or use the following command sequence to display the Multiline Text Editor.

```
Command: ED (For DDEDIT)
Select an annotation object or [Undo]: (Select the Mtext object
    and the Multiline Text Editor appears. Perform any text
    editing task and click the OK button)
Select an annotation object or [Undo]: (Pick another Mtext object
    to edit or press ENTER to exit this command)
```

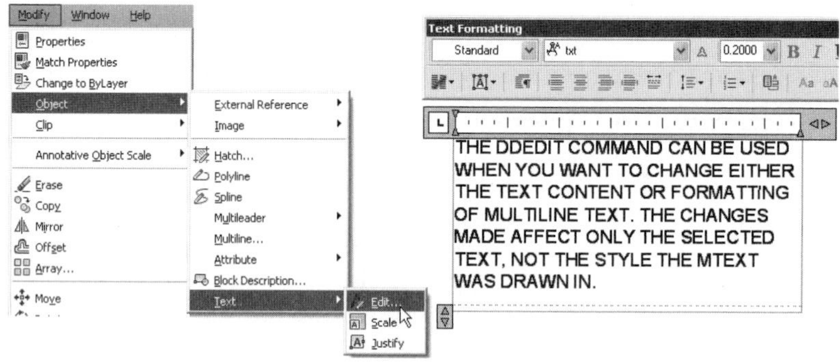

Figure 6.30

Try It!: Open the drawing file 06_Edit Text2. Activate the Multiline Text Editor by entering the ED command and picking any text object. In this dialog box, the text is currently drawn in the RomanD font and needs to be changed to Arial. To accomplish this, first highlight all the text. Next, change to the desired font, as shown in the following image.

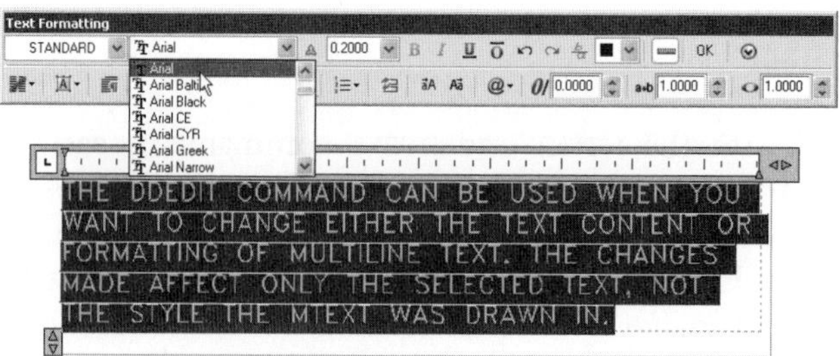

Figure 6.31

Since all text was highlighted, changing the font to Arial updates all text, as shown in the following image. Clicking the OK button dismisses the dialog box and changes the text font in the drawing.

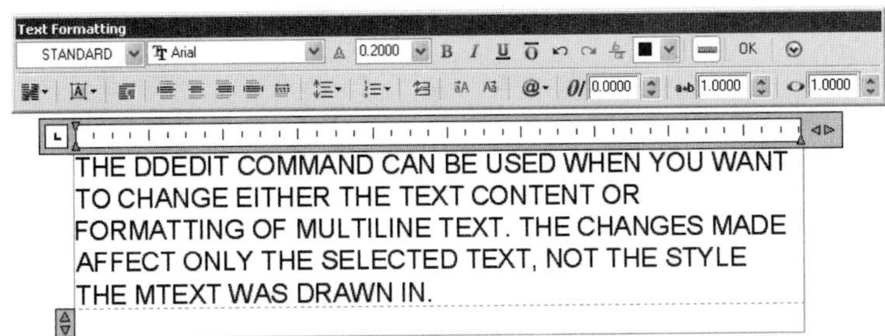
Text Formatting
STANDARD
Arial
0.2000
OK
0.0000
1.0000
1.0000
THE DDEDIT COMMAND CAN BE USED WHEN YOU WANT
TO CHANGE EITHER THE TEXT CONTENT OR
FORMATTING OF MULTILINE TEXT. THE CHANGES MADE
AFFECT ONLY THE SELECTED TEXT, NOT THE STYLE
THE MTEXT WAS DRAWN IN.

Figure 6.32

When editing Mtext objects, you can selectively edit only certain words that are part of a multiline text string. To perform this task, highlight the text to change and apply the new formatting style, such as font, underscore, or height, to name a few.

Try It!: Open the drawing file 06_Edit Text3. Activate the Multiline Text Editor by entering the ED command and picking any text object. In the following image, the text "PLACING" needs to be underscored, the text "AutoCAD" needs to be changed to a Swiss font, and the text "MTEXT" needs to be increased to a text height of 0.30 units. When performing this type of operation, you need to highlight only the text object you want to change. Clicking the OK button dismisses the dialog box and changes the individual text objects in the drawing.

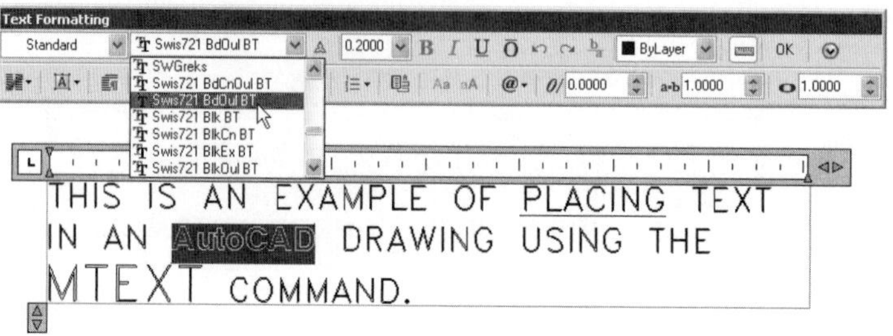

Figure 6.33

Using the DDEDIT (ED) command or double-clicking a text object created with the DTEXT command, as shown on the left in the following image, displays a field, as shown on the right in the following image. Use this to change text in the field provided. Font, justification, and text height are not supported in this field.

MECHANICAL MECHANICAL

Figure 6.34

GLOBALLY MODIFYING THE HEIGHT OF TEXT

The height of a text object can be easily changed through a command found in the Modify pull-down menu or Menu Browser. Click Object, followed by Text, and then Scale. This command allows you to pick a text object and change its text height. The important feature of this command is that scaling has no effect on the justification of the text. This means that after scaling the text, you should not have to move each individual text item to a new location.

Try It!: Open the drawing file 06_Scale Text, illustrated in the following image. All offices in this facilities plan have been assigned room numbers. One of the room numbers (ROOM 114) has a text height of 12", while all other room numbers are 8" in height. You could edit each individual room number until all match the height of ROOM 114, or you could use the SCALETEXT command by using the following command sequence:

```
Command: SCALETEXT
Select objects: (Pick all 16 mtext objects except for ROOM 114)
Select objects: (Press ENTER to continue)
Enter a base point option for scaling
[Existing/Left/Center/Middle/Right/TL/TC/TR/ML/MC/MR/BL/BC/BR]
    <Existing>: (Press ENTER to accept this value)
Specify new model height or [Paper height/Match object/Scale
    factor] <8.0000>: M (For Match object)
Select a text object with the desired height: (Pick the text
    identified by ROOM 114)
Height=12.0000
```

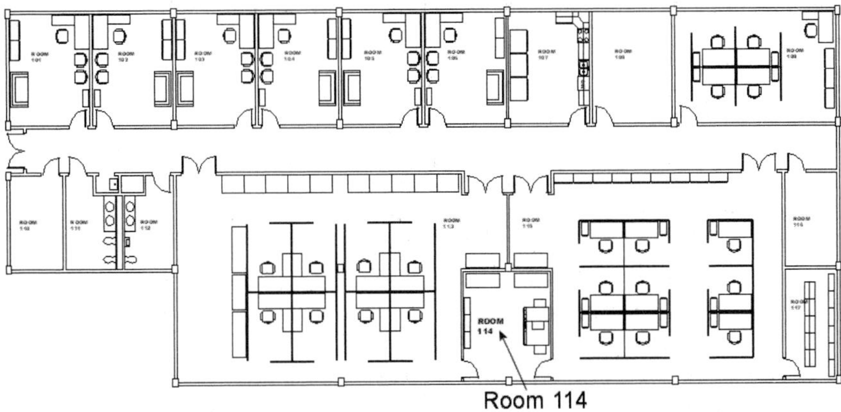

Figure 6.35

The results are displayed in the following image. All the text was properly scaled without affecting the justification locations.

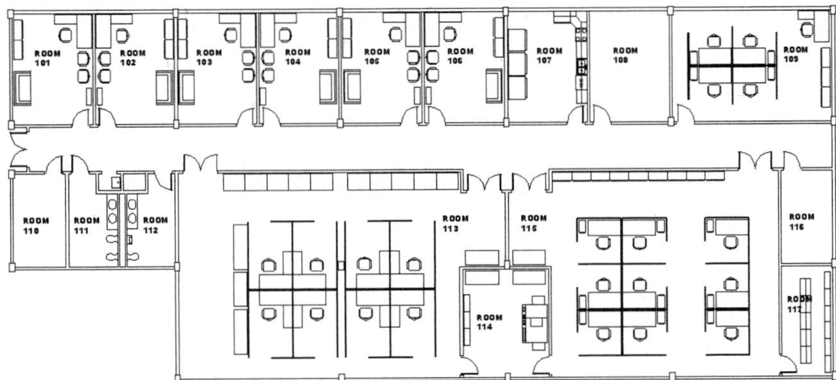

Figure 6.36

Note: The Quick Select dialog box could be used to select all mtext objects. You could also use a Window to select the mtext since the SCALETEXT command will only select text objects.

GLOBALLY MODIFYING THE JUSTIFICATION OF TEXT

Yet another text editing mode can be found in the Modify pull-down menu or Menu Browser. Click Object, followed by Text, and then Justify. This command allows you to pick a text item and change its current justification. If you pick a multiline text object, all text in this mtext object will have its justification changed. In the case of regular text placed with the DTEXT command, you would have to select each individual line of text for it to be justified.

Try It!: Open the drawing file 06_Justify Text, illustrated in the following image. All the text justification points in this facilities drawing need to be changed from left justified to top center justified. Use the following command prompt and illustration in the following image to accomplish this task.

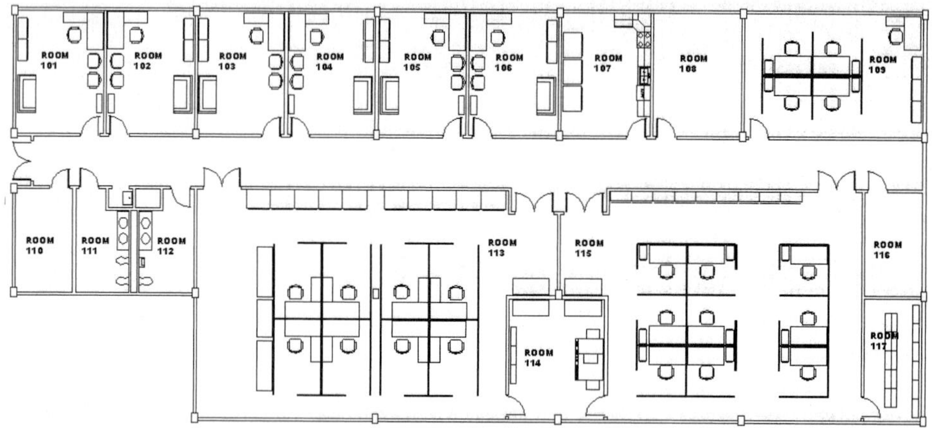

Figure 6.37

```
Command: JUSTIFYTEXT
Select objects: (Select all 17 text objects)
Select objects: (Press ENTER to continue)
Enter a justification option
[Left/Align/Fit/Center/Middle/Right/TL/TC/TR/ML/MC/MR/BL/BC/BR]
     <Left>: TC (For Top Center)
```

The results are illustrated in the following image. All text objects have been globally changed from left justified to top center justified.

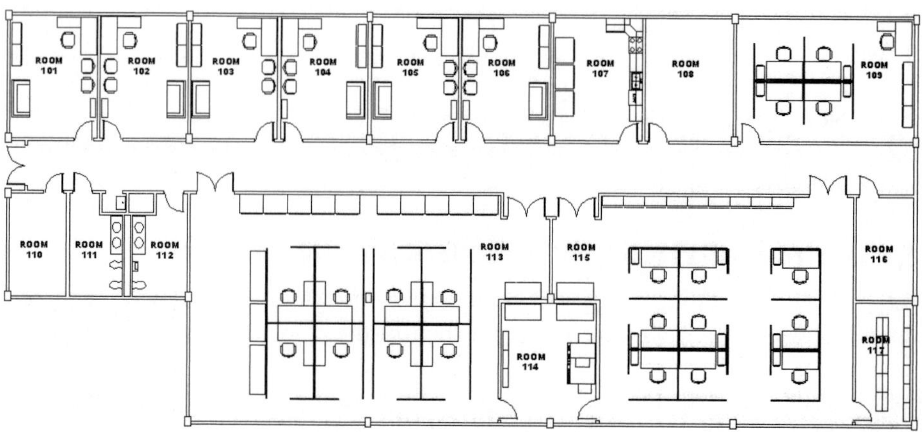

Figure 6.38

SPELL-CHECKING TEXT

Spell-checking is an important part of the drawing documentation process. Words, notes, or engineering terms that are spelled correctly elevate the drawing to a higher level of professionalism. Spell-checking a drawing can be activated through the Menu Browser, Ribbon, or Text toolbar, as shown in the following image.

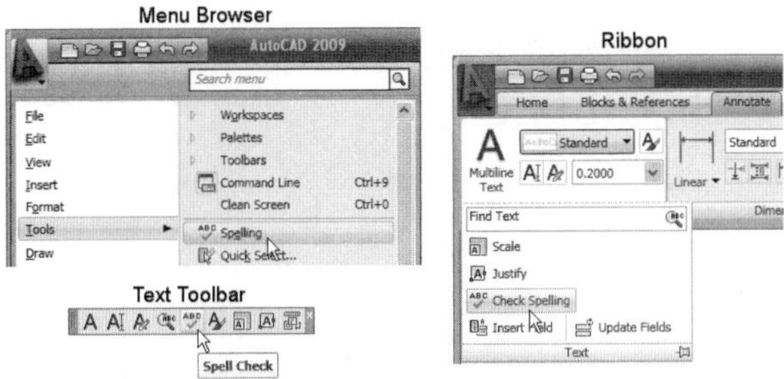

Figure 6.39

Issue the SPELL command and click the Start button to begin the spell-check. The "Where to Check" drop-down list provides the option of checking the entire drawing or limiting the check to the current space/layout or to only selected text items. The "Not in Dictionary" box displays the first word that is suspected of being spelled wrong. It wasn't found in the current dictionary. A drop-down list and Dictionaries button is provided to select other dictionaries. The Suggestions area displays all possible alternatives to the word identified as being misspelled. The Ignore button allows you to skip the current word; this would be applicable especially in the case of acronyms such as CAD and GDT. Clicking Ignore All skips all remaining words that match the current word. The Change button replaces the word "Not in the dictionary" with the word in the Suggestions box. The "Add to Dictionary" button adds the current word to the current dictionary. In the following image, the word "DIRECION" was identified as being misspelled. Notice that the correct word, "DIRECTION," is listed as a suggested correction. Clicking the Change button replaces the misspelled word and continues with the spell-checking operation until completed.

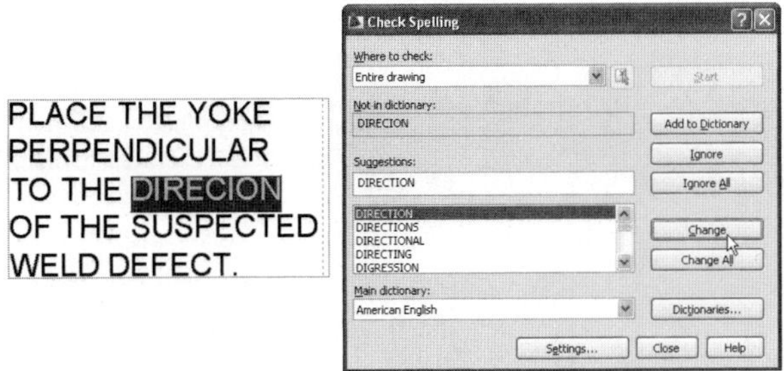

Figure 6.40

After completing the spell-checking operation, the mtext object is displayed in the following image.

PLACE THE YOKE
PERPENDICULAR
TO THE DIRECTION
OF THE SUSPECTED
WELD DEFECT.

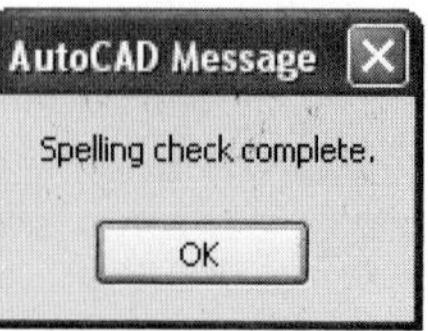

Figure 6.41

Try It!: Open the drawing 06_Spell Check1 and perform a spell-check operation on this mtext object.

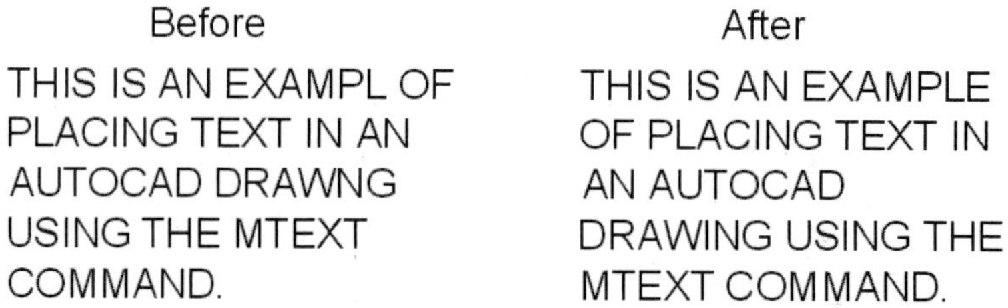

Before	After
THIS IS AN EXAMPL OF PLACING TEXT IN AN AUTOCAD DRAWNG USING THE MTEXT COMMAND.	THIS IS AN EXAMPLE OF PLACING TEXT IN AN AUTOCAD DRAWING USING THE MTEXT COMMAND.

Figure 6.42

CREATING DIFFERENT TEXT STYLES

A text style is a collection of settings that are applied to text placed with the DTEXT or MTEXT command. These settings could include presetting the text height and font, in addition to providing special effects, such as an oblique angle for inclined text. Choose Text Style... from the Menu Browser, Annotation tab of the Ribbon, Format pull-down menu, or from the Text toolbar, shown in the following image.

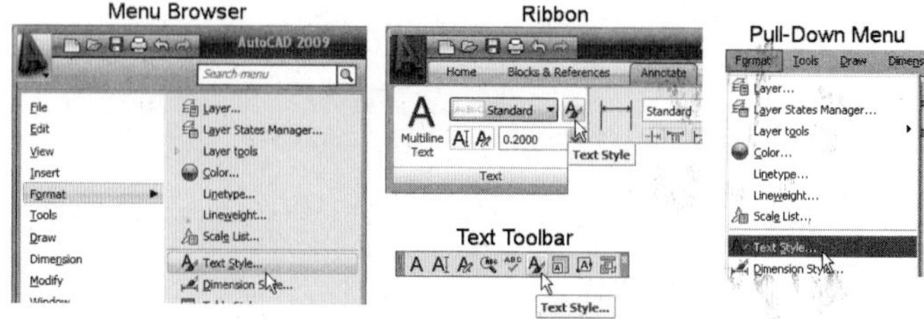

Figure 6.43

Initiating the STYLE command will launch the Text style dialog box as shown in the following image, which is used to create new text styles. As numerous styles are created, this dialog box can also be used to make an existing style current in the drawing. By default, when you first begin a drawing, the current Style Name is STANDARD. It is considered good practice to create your own text style and not rely on STANDARD. Once a new style is created, a font name is matched with the style. Clicking in the field for Font Name displays a list of all text fonts supported by the operating system. These fonts have different extensions, such as SHX and TTF. TTF or TrueType Fonts are especially helpful in AutoCAD because these fonts display in the drawing in their true form. If the font is bold and filled-in, the font in the drawing file displays as bold and filled-in.

When a Font Name is selected, it displays in the Preview area located in the lower-right corner of the dialog box. The Effects area allows you to display the text upside down, backwards, or vertically. Other effects include a width factor, explained later, and the oblique angle for text displayed at a slant.

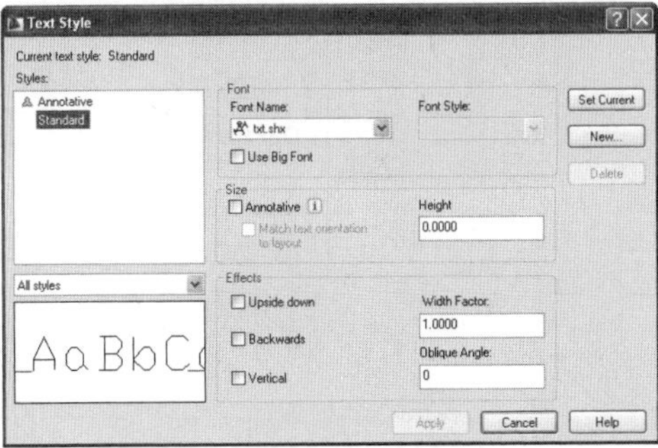

Figure 6.44

Clicking the New... button of the Text Style dialog box displays the New Text Style dialog box, as shown in the following image. A new style is created called General Notes. Clicking the OK button returns you to the Text Style dialog box. Clicking in the Font Name field displays all supported fonts. Clicking Arial assigns the font to the style name General Notes. Clicking the Apply button saves the font to the database of the current drawing file.

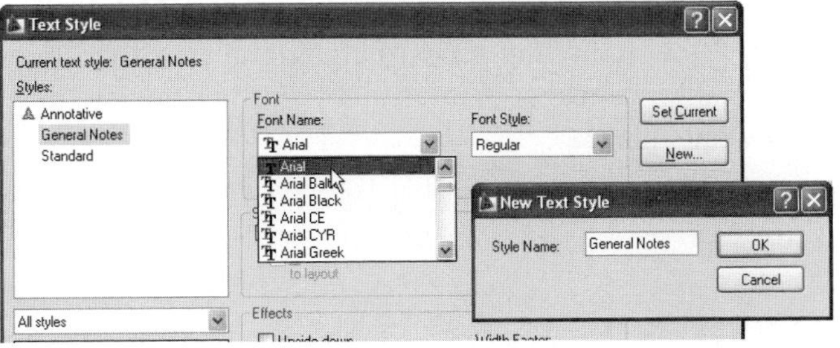

Figure 6.45

THE TEXT STYLE CONTROL BOX

The Styles toolbar provides a convenient method for changing styles. Use the Text Style Control box illustrated on the left in the following image to make an existing text style current. This control box allows you to change from one text style to another easily. The Text panel of the Ribbon also provides a drop-down list for selecting the current text style in a more graphical form, as shown on the right in the following image.

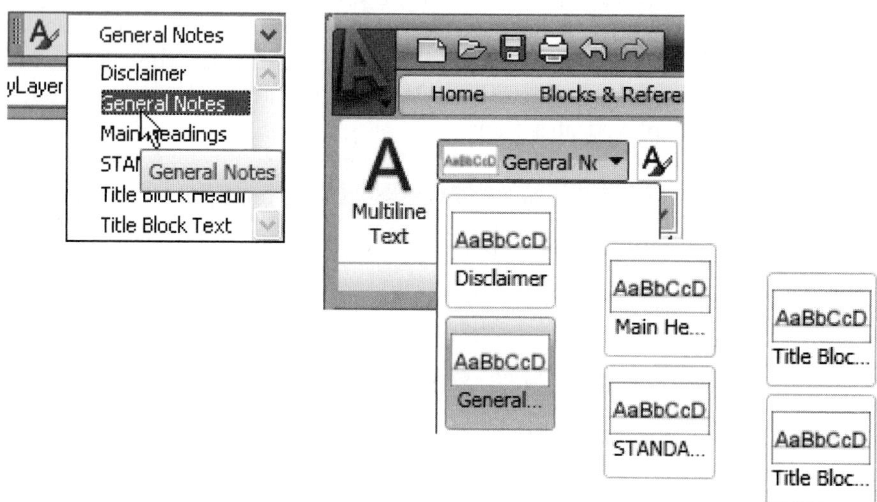

Figure 6.46

Note: You can also highlight a text object and use the Text Style Control box to change the selected text to a different text style. This action is similar to changing an object from one layer to another through the Layer Control box.

FIELDS

Fields are a type of intelligent text that you can add to your drawings. In the following image, fields are identified by a distinctive gray background, which is not present on regular text. Fields are intended for items that tend to change during the life cycle of a drawing, such as the designer's name, checker, creation date of the drawing, and drawing number, as shown in the following image.

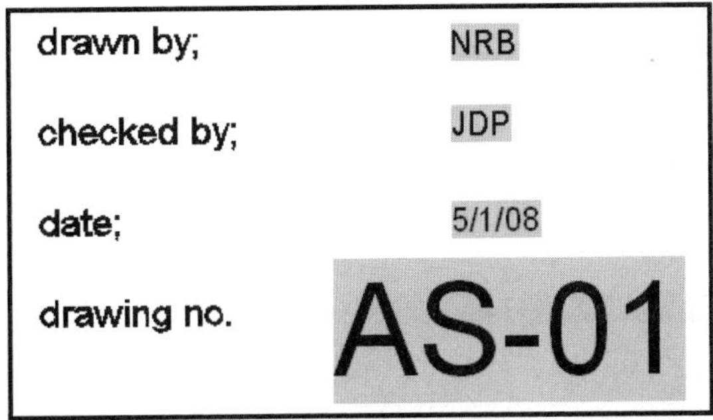

Figure 6.47

The FIELD command can be selected from the Insert pull-down menu, as shown on the left in the following image. This activates the Field dialog box, which consists of Field categories, Field names, Author, and Format information. Rather than cycle through all field names, you can pick a field category, which limits the amount of items in the Field names area.

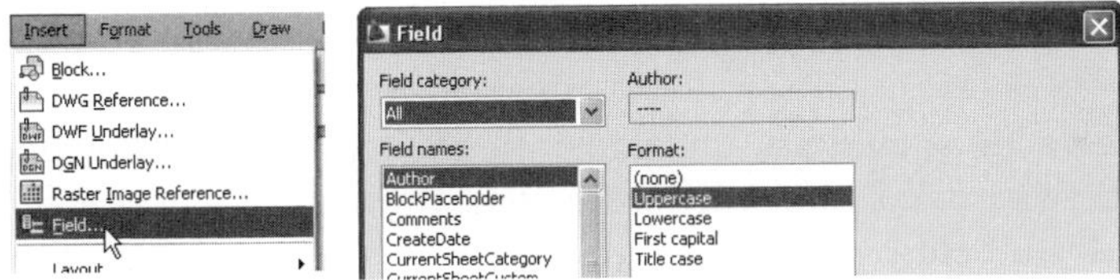

Figure 6.48

By default, all fields are available under the Field names area. You can specify a certain Field category to narrow down the list of field names. Illustrated on the left in the following image is a list of all field categories; the Date & Time category is selected. This activates specific field names only related to this category, as shown on the right in the following image. With the Field name SaveDate highlighted, notice all the examples that pertain to when the drawing was last saved. There is also a Hints area that gives examples of various month, day, and year notations.

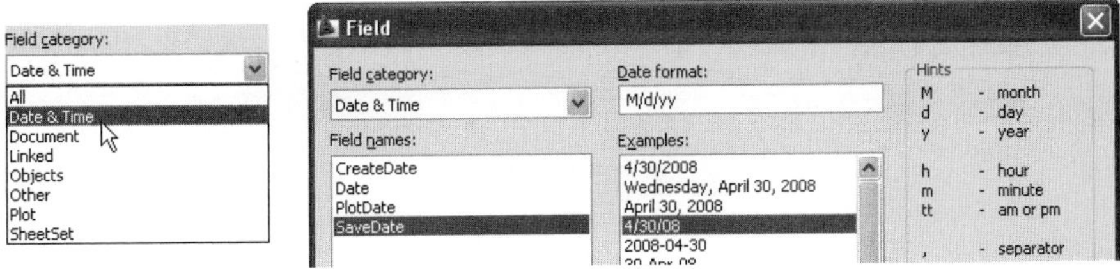

Figure 6.49

FIELDS AND MULTILINE TEXT

While placing multiline text, you can create a field through the Text Formatting toolbar, as shown in the following image. This allows you to place numerous fields at one time. Right-clicking in the text box and then picking Insert Field from the menu also activates the Field dialog box.

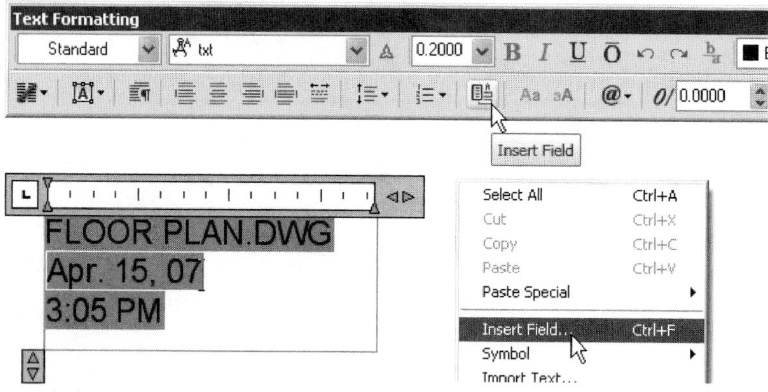

Figure 6.50

The results of creating a field are illustrated on the left in the following image. In this example, the following field categories were used: Filename and Date. The time entry is a subset of the Date field.

Whenever field information changes, such as the date and time, you may need to update the field in order to view the latest information. Updating fields is accomplished by clicking Update Fields, which is located under the Tools pull-down menu area, as shown in the middle in the following image. When picking this command, you are prompted to select the field. In the following image on the right, the date automatically changed. Whenever you open a drawing that contains fields, the fields will automatically update to reflect the most up-to-date information.

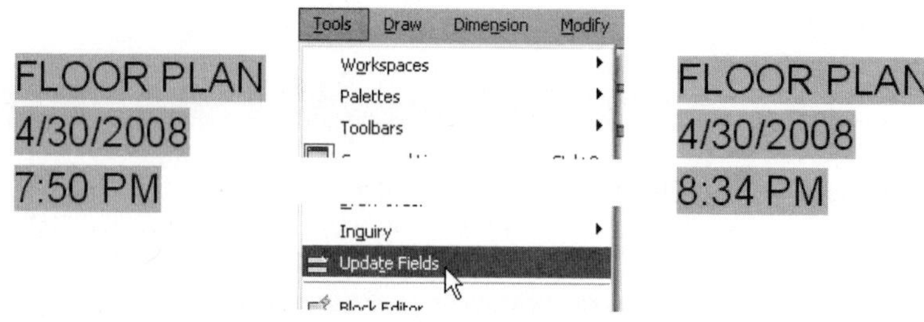

Figure 6.51

FIELDS AND DRAWING PROPERTIES

When creating some fields, information is displayed as a series of four dashed lines. This is to signify that the information contained in the field name either has not been entered or has not been updated. In the following image, the Author, Subject, and Title fields need to be completed before the correct information displays.

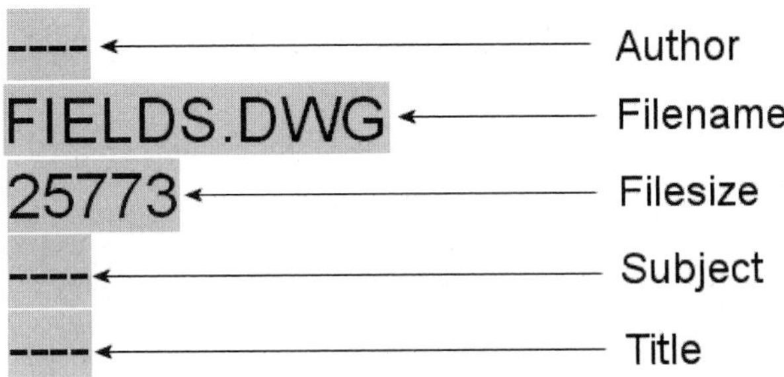

Figure 6.52

In the previous illustration of the Author, Subject, and Title field names, activating the Drawing Properties dialog box (pick Drawing Properties from the File pull-down menu), as shown on the left in the following image, allows you to fill in this information. When returning to the drawing, click the Tools pull-down menu, select Update Fields, and pick the field that needs updating. The information listed in the Drawing Properties dialog box will be transferred to the specific fields, as shown on the right in the following image.

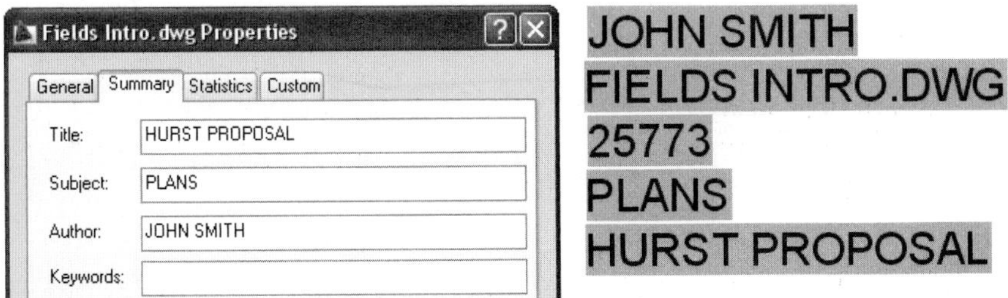

Figure 6.53

CREATING TABLES

As a means of further organizing your work, especially text, you can create cells organized in rectangular patterns that consist of rows and columns. The TABLE command is used to accomplish this and can be selected using one of the following methods, as shown in the following image: Menu Browser, pull-down menu, Ribbon, or Draw toolbar.

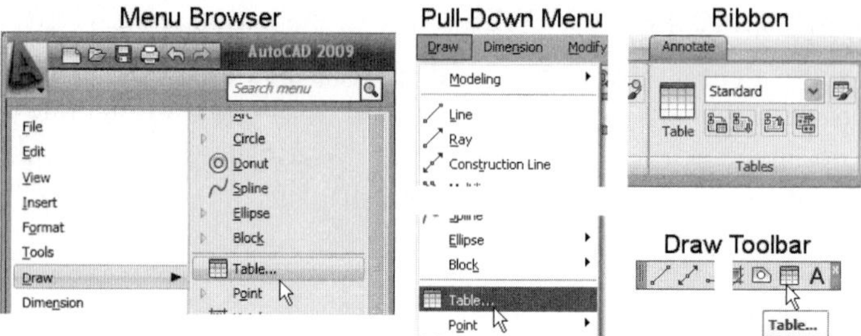

Figure 6.54

Using one of the methods shown in the previous illustration will activate the Insert Table dialog box, as shown in the following image. It is here that you specify the number of rows and columns that make up the table, in addition to the column width and row height. You can also set the type of styles that will be applied to a cell and then get a preview of what the table will look like.

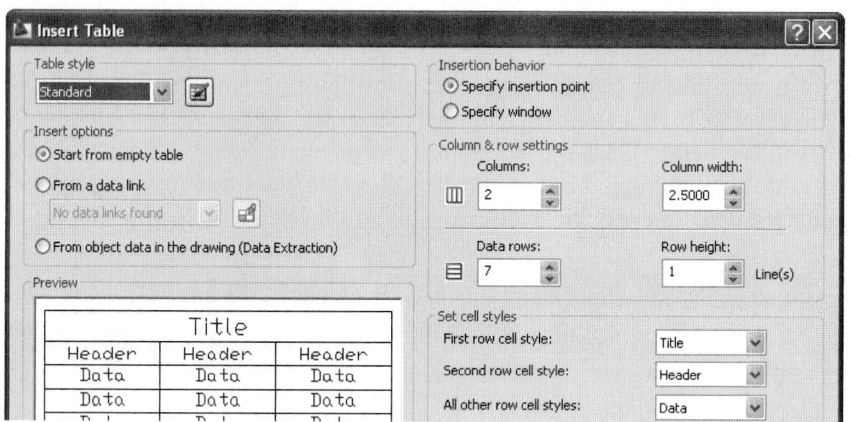

Figure 6.55

Once you decide on the number of rows and columns for the table, clicking the OK button returns you to the drawing editor. Here is where you locate the position of the table in your drawing. This is a similar operation to inserting a block. After the table is positioned, the table appears in the drawing editor and the Text Formatting toolbar displays above the table, as shown in the following image. In this example, the text "WINDOW SCHEDULE" was added as the title of the table. Notice also that the text height automatically adjusts to the current table style. This will be discussed later in this chapter.

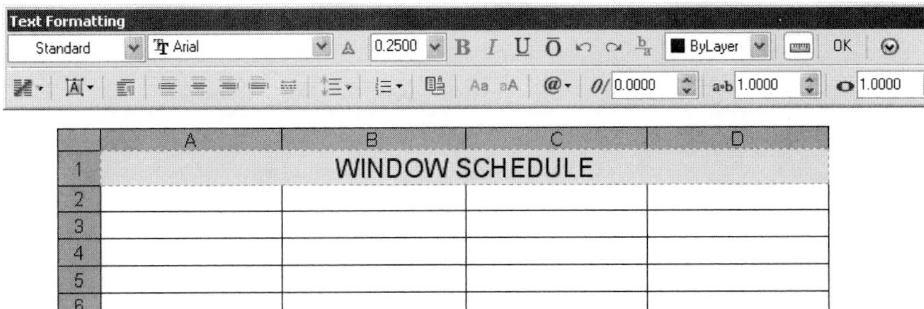

Figure 6.56

After you have entered text into the title cell, pressing the TAB key moves you from one cell to another, as shown in the following image. In this example, cell headers are added to identify the various categories of the table.

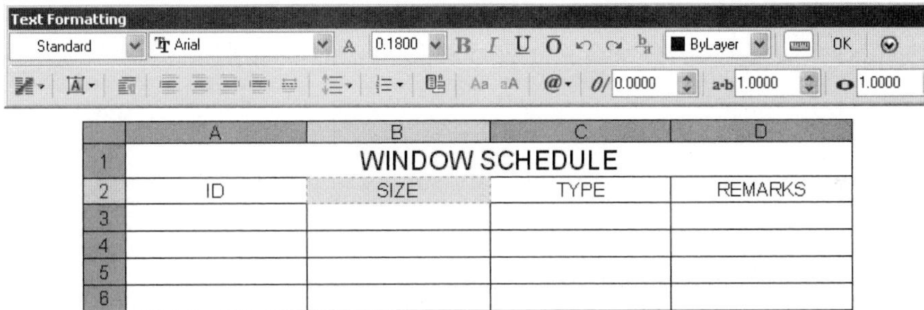

Figure 6.57

Note: Pressing the TAB key moves you from one cell to another; however, you cannot reverse this direction through the TAB key. Instead, press SHIFT + TAB to reverse the direction in moving from one cell to another inside a table.

THE TABLE TOOLBAR

Clicking a table entry highlights the cell and displays the Table toolbar as shown in the following image. This toolbar contains numerous commands used for manipulating tables.

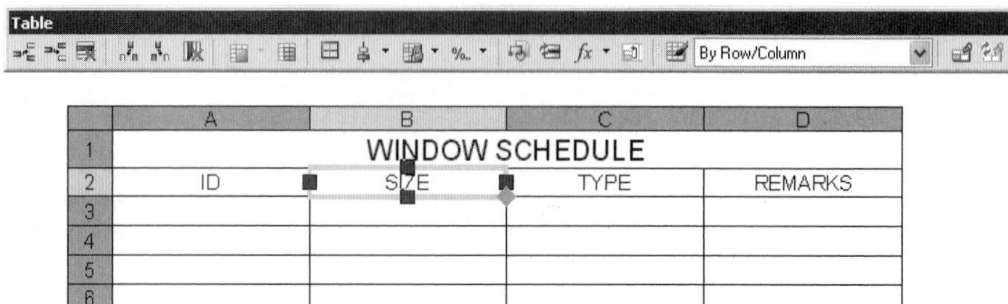

Figure 6.58

Note: If the Ribbon menu is present, clicking on a table cell will activate the following menu that deals specifically with table rows, columns, merge, cell styles, cell format, insert, and data headings.

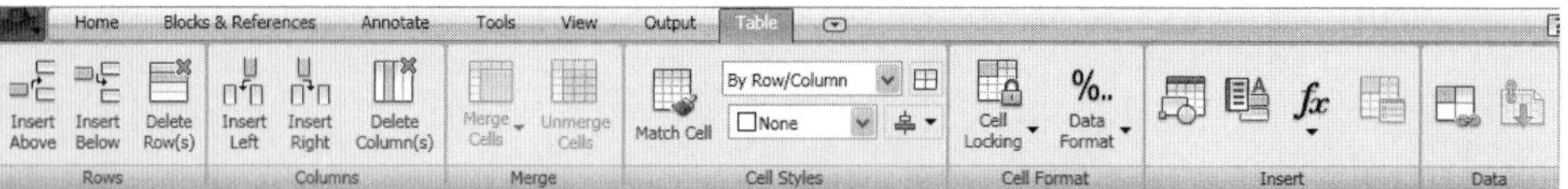

Figure 6.59

The Table toolbar has numerous controls for manipulating tables. The following chart will explain these buttons.

Button	Tool	Function
	Insert Row Above	Inserts a row above a selected cell
	Insert Row Below	Inserts a row below a selected cell
	Delete Row	Deletes an entire row
	Insert Column Left	Inserts a column to the left of a selected cell
	Insert Column Right	Inserts a column to the right of a selected cell
	Delete Column	Deletes an entire column
	Merge Cells	Merges all selected cells or cells by row or column
	Unmerge Cells	Unmerges selected cells
	Cell Borders	Launches the Cell Border Properties dialog box designed to control the display of cell borders
	Middle Center Align	Displays nine text alignment modes that deal with tables
	Locking	Displays a menu used for locking the cell content, cell format, or both. Also used for unlocking cells
%..	Data Format	Launches the Table Cell Format dialog box used for defining the data type of a cell or group of cells

Button	Tool	Function
	Insert Block	Launches the Insert a Block in a Table dialog box used for placing and fitting blocks in table cells
	Insert Field	Launches the Field dialog box designed for creating a field in a table cell
fx	Insert Formula	Displays a menu used for performing summations and other calculations on a group of cells
	Manage Cell Content	Displays the Manage Cell Content dialog box used for changing the order of cell content as well as changing the direction in which cell content will display
	Match Cell	Matches the contents of a source cell with destination cell
By Row/Column	Cell Style	Displays a drop-down list that allows you to change to a different cell style
	Link Cell	Displays the Select Data Link dialog box that lists the current links with Excel
	Download Changes	Used for downloading changes from a source file to the table

MERGING CELLS IN A TABLE

There are times when you need to merge cells in the rows or columns of a table. To accomplish this, first pick inside a cell, then hold down the SHIFT key as you pick the cells you want to merge with it. Then pick the Merge Cells button followed by All from the menu, as shown in the following image. The final example of merging the cells that contain the "FLOOR" text is also shown in the following image.

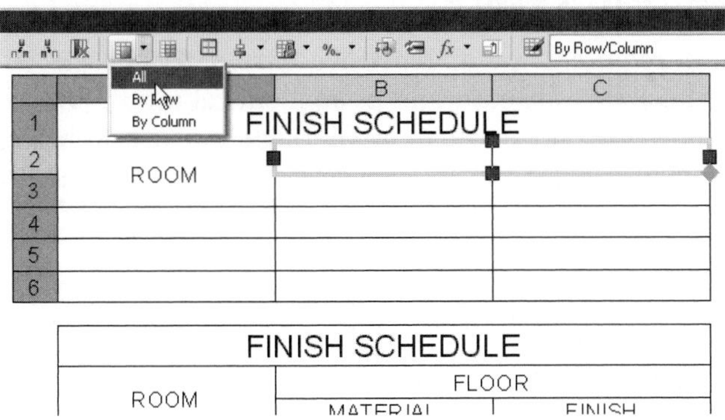

Figure 6.60

MODIFYING A TABLE

The modification of table cells, table rows, and even the overall size of the table is greatly enhanced through the use of grips. You can achieve different results depending on the table grip that is selected. For instance, first select the table to display all grips. Then, select the grip in the upper-left corner of the table and move your cursor. This action moves the table to a new location. Selecting the grip in the upper-right corner of the table and then dragging your cursor left or right increases or decreases the spacing of all table columns.

Selecting the grip in the lower-left corner of the table increases the spacing between rows. Study the following image to see how other grips affect the sizing of a table. Grips will be covered in greater detail in chapter 7.

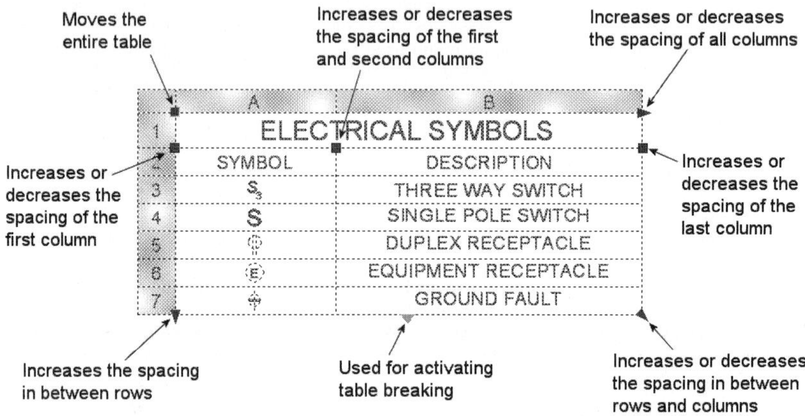

Figure 6.61

Note: When using grips to stretch rows or columns in a table, you may want to turn running OSNAP off. Otherwise, if OSNAP Endpoint is set, your table may stretch back to its original location, giving the appearance that the operation did not work.

AUTOFILLING CELLS IN A TABLE

Tables can, at times, require a lot of repetitive data entry. In the table example in the following image, you could easily copy and paste identical information from one cell to the next. Another very powerful tool exists in tables that allow text to be created incrementally such as the room numbers in the image. This is accomplished through the Autofill feature of tables.

ROOM	NAME OF SPACE	FLR	BASE	MATL	FIN	COLOR	MATL	FIN
100	CLASSROOM	CPT	V	GBW	PT		GBW	PT
101	CLASSROOM	CPT	V	GBW	PT		GBW	PT
102	CLASSROOM	CPT	V	GBW	PT		GBW	PT
103	CLASSROOM	CPT	V	GBW	PT		GBW	PT
104	HALLWAY	CPT	V	GBW	PT		GBW	PT
105	PASTOR	CPT	V	GBW	PT		GBW	PT
106	RESTROOM	CPT	V	GBW	PT		GBW	PT
107	OFFICE	CPT	V	GBW	PT		GBW	PT
108	OFFICE	CPT	V	GBW	PT		GBW	PT

Figure 6.62

Cell 3A illustrated on the left in the following image is populated with the number "1." If you know the next number below the "1" will be "2" and so on, first click in cell 3A. Notice the light blue diamond shaped grip in the lower-right corner of the cell. Click on this grip and move your cursor in a downward direction as shown in the middle in the following image. You will notice your cursor changing to different numbers the more you drag the cursor. This image illustrates how the numbers "1" through "8" can be autofilled. The results are shown on the right in the following image with the Autofill feature creating the proper incremental data.

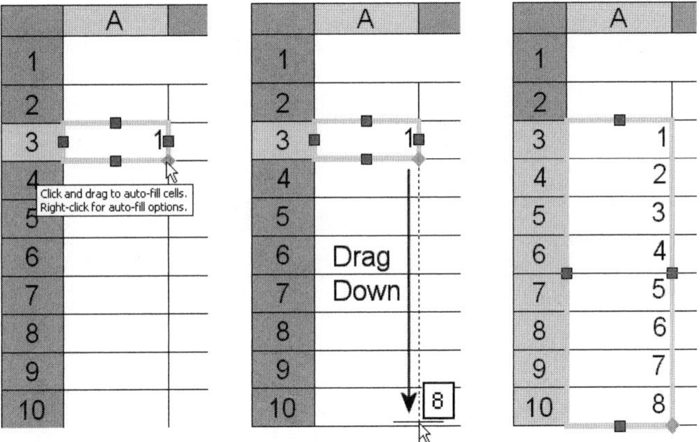

Figure 6.63

Autofill also works with regular text entries. If you want to reproduce the word "VINYL" in the remaining cells below, click in the cell with the word. When the light blue diamond grip appears, drag your cursor to the cells that will contain the same text. The results are shown on the right in the following image with the Autofill feature being used to automatically place identical text entries into table cells.

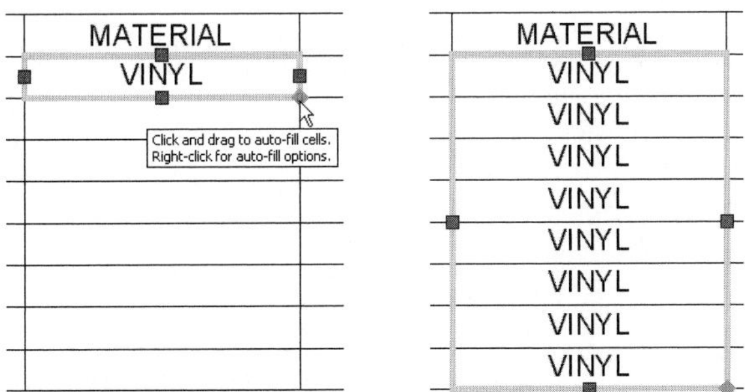

Figure 6.64

TABLES AND MICROSOFT EXCEL

Microsoft Excel spreadsheets can easily be imported into AutoCAD drawings as tables. The illustration on the left in the following image represents information organized in an Excel spreadsheet. With the spreadsheet open, copy the appropriate data to the Windows Clipboard. Switch to AutoCAD and while in the drawing editor, click the Edit pull-down menu and pick Paste Special, as shown in the middle of the following image. This displays the Paste Special dialog box, as shown on the right in the following image. Click AutoCAD Entities to merge the spreadsheet into an AutoCAD drawing as a table.

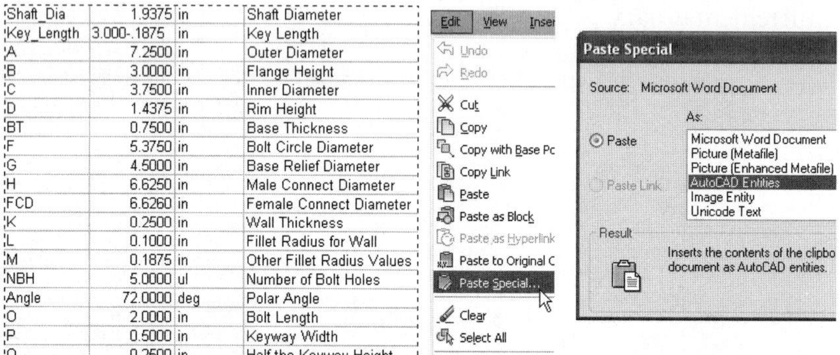

Shaft_Dia	1.9375	in	Shaft Diameter
Key_Length	3.000-.1875	in	Key Length
A	7.2500	in	Outer Diameter
B	3.0000	in	Flange Height
C	3.7500	in	Inner Diameter
D	1.4375	in	Rim Height
BT	0.7500	in	Base Thickness
F	5.3750	in	Bolt Circle Diameter
G	4.5000	in	Base Relief Diameter
H	6.6250	in	Male Connect Diameter
FCD	6.6260	in	Female Connect Diameter
K	0.2500	in	Wall Thickness
L	0.1000	in	Fillet Radius for Wall
M	0.1875	in	Other Fillet Radius Values
NBH	5.0000	ul	Number of Bolt Holes
Angle	72.0000	deg	Polar Angle
O	2.0000	in	Bolt Length
P	0.5000	in	Keyway Width
Q	0.2500	in	Half the Keyway Height

Figure 6.65

The initial results of importing an Excel spreadsheet into an AutoCAD drawing are illustrated on the left in the following image. The Excel spreadsheet was converted into an AutoCAD drawing object. However, a few columns need to be lengthened and the rows shortened in order to organize the data in a single line. Also, headings need to be added to the top of the table by inserting a row above the top line of data. The completed table is illustrated on the right in the following image.

Shaft_Dia	1.9375	in	Shaft Diameter
Key_Leng th	3.000-.18 75	in	Key Length
A	7.2500	in	Outer Diameter
B	3.0000	in	Flange Height
C	3.7500	in	Inner Diameter
D	1.4375	in	Rim Height
BT	0.7500	in	Base Thickness
F	5.3750	in	Bolt Circle Diameter
G	4.5000	in	Base Relief Diameter
H	6.6250	in	Male Connect Diameter
FCD	6.6260	in	Female Connect Diameter
K	0.2500	in	Wall Thickness
L		in	Fillet Radius for

PARAMETER	VALUE	UNITS	DESCRIPTION
Shaft_Dia	1.9375	in	Shaft Diameter
Key_Length	3.000-.1875	in	Key Length
A	7.2500	in	Outer Diameter
B	3.0000	in	Flange Height
C	3.7500	in	Inner Diameter
D	1.4375	in	Rim Height
BT	0.7500	in	Base Thickness
F	5.3750	in	Bolt Circle Diameter
G	4.5000	in	Base Relief Diameter
H	6.6250	in	Male Connect Diameter
FCD	6.6260	in	Female Connect Diameter
K	0.2500	in	Wall Thickness
L	0.1000	in	Fillet Radius for Wall
M	0.1875	in	Other Fillet Radius Values
NBH	5.0000	ul	Number of Bolt Holes
Angle	72.0000	deg	Polar Angle
O	2.0000	in	Bolt Length
P	0.5000	in	Keyway Width
Q	0.2500	in	Half the Keyway Height

Figure 6.66

CREATING TABLE STYLES

Table styles, as with text styles, can be used to organize different properties that make up a table. Table styles can be accessed through the Menu Browser, pull-down menu, Ribbon, and Styles toolbar as shown in the following image.

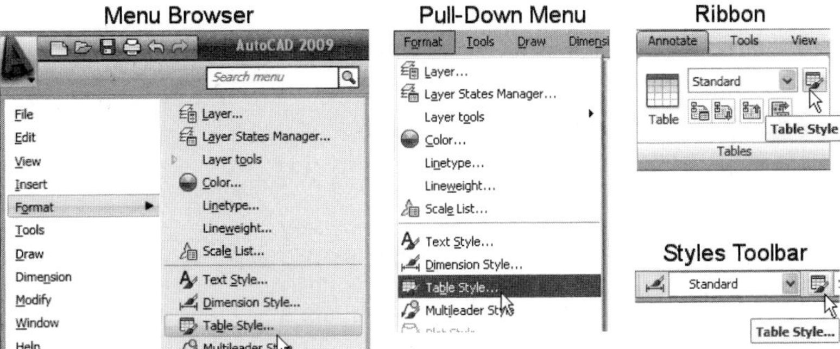

Figure 6.67

Clicking on either one of the Table Style listings shown in the previous image will display the Table Style dialog box, as shown in the following image. Clicking the New button displays the Create New Table Style dialog box, in which you enter the name of the table style.

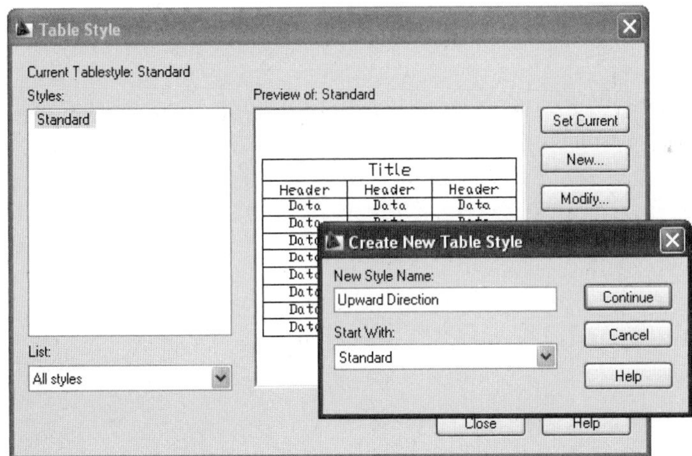

Figure 6.68

Clicking Continue allows you to make various changes to a table based on the type of cell. For example, in the following image three tabs can be used to make changes to the text that occupies a cell in a table based on whether the text is cell data, the column heads of a cell, or even the title cell of the table. These text settings can be used to emphasize the information in the cells. For example, you would want the title of the table to stand out with a larger text height than the column headings. You might also want the column headings to stand out over the data in the cells below. Also, as in the following image, you can change the direction of the table. By default, a table is created in the downward direction. You may want to change this direction to upward.

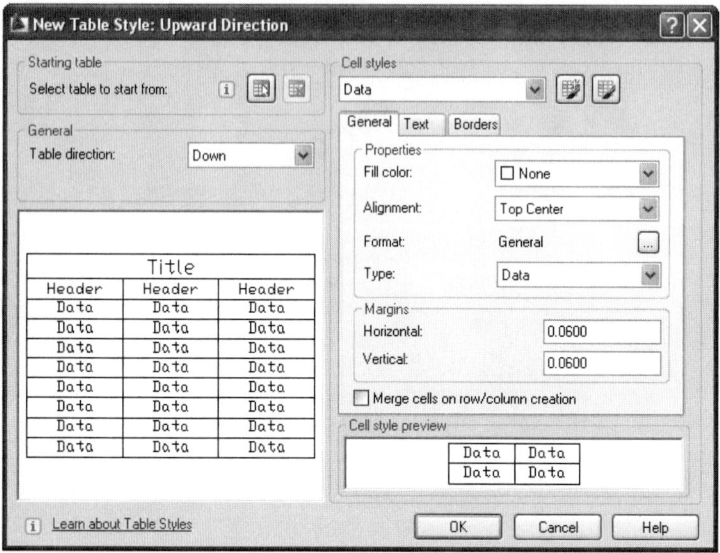

Figure 6.69

The following image illustrates three tabs containing information that can be modified for the cell type. In this image, the Data cell type is selected.

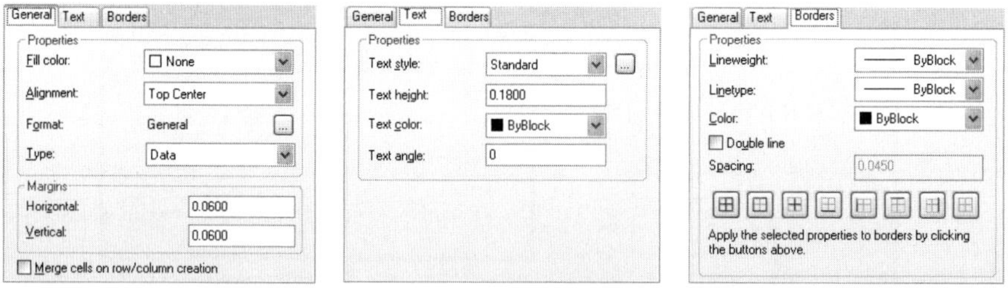

Figure 6.70

To make changes to the Header or Title cells, select the cell type in the Cell Styles field, as shown in the following image, and modify the information in the tabs as desired.

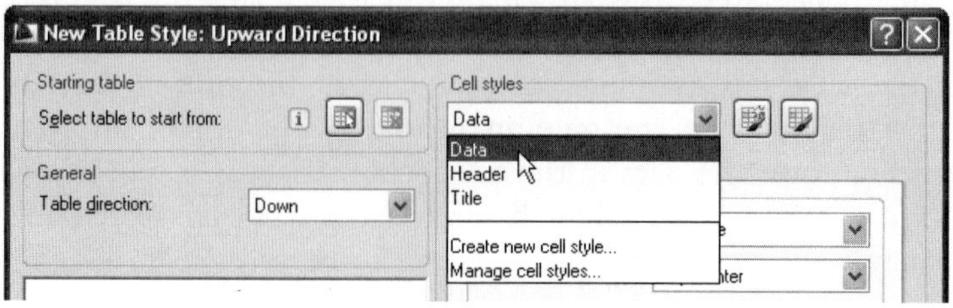

Figure 6.71

TUTORIAL EXERCISE: 06_PUMP_ASSEMBLY.DWG

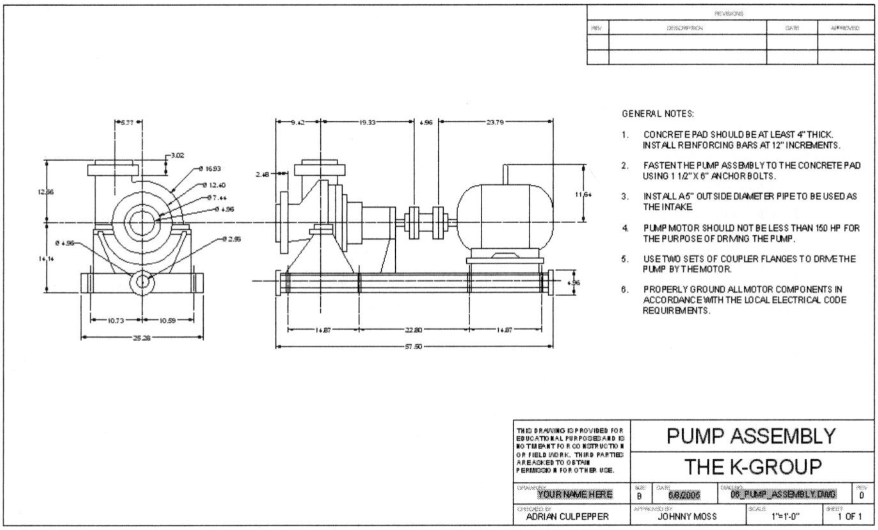

Figure 6.72

PURPOSE

This tutorial is designed to create numerous text styles and add different types of text objects to the title block illustrated in the previous image.

SYSTEM SETTINGS

Since this drawing is already provided on the CD, open an existing drawing file called 06_Pump_Assembly. Follow the steps in this tutorial for creating a number of text styles and then placing text in the title block area. Check to see that the following Object Snap modes are already set: Endpoint, Extension, Intersection, Center.

LAYERS

Various layers have already been created for this drawing. Since this tutorial covers the topic of text, the current layer is Text.

SUGGESTED COMMANDS

Open the drawing called 06_Pump_Assembly. In this tutorial you will create four text styles called Company Info, Title Block Text, Disclaimer, and General Notes. The text style Company Info will be applied to one part of the title block. Title Block Text will be applied to a majority of the title block, where questions such as Drawn By, Date, and Scale are asked. Also, the Drawn By, Date, and Drawing Number will be supplied by creating fields. A disclaimer will be imported into the title block in the Disclaimer text style. Finally, a series of notes will be imported in the General Notes text style. Once these general notes are imported, a spell-check operation will be performed on the notes.

Step 1

Opening the drawing file 06_Pump _Assembly displays an image similar to the following image. The purpose of this tutorial is to fill in the title block area with a series of text, multiline text, and field objects. Also, a series of general notes will be placed to the right of the pump assembly. Before you place the text, four text styles will be created to assist with the text creation. The four text styles are also identified in the following image. The text style Company Info will be used to place the name of the company and drawing title. Information in the form of scale, date, and who performed the drawing will be handled by the text style Title Block Text. A disclaimer will be imported from an existing .TXT file available on the CD; this will be accomplished in the Disclaimer text style. A listing of six general notes is also available on the CD. It is in .RTF format, originally created in Microsoft Word. It will be imported into the drawing and placed in the General Notes text style.

Since you will be locating various justification points for the placement of text, turn OSNAP off by single-clicking OSNAP in the Status bar at the bottom of the display screen.

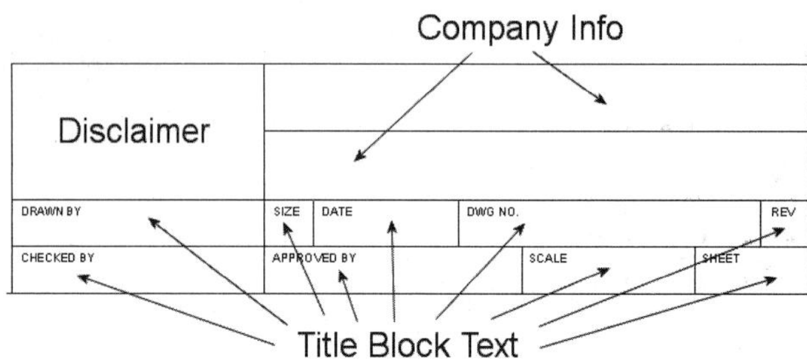

Figure 6.73

Step 2

Before creating the first text style, first see what styles are already defined in the drawing by choosing Text Style... from the Format pull-down menu, as shown on the left in the following image. This displays the Text Style dialog box, as shown on the right in the following image. (You could also choose Text Style from the Menu Browser.) The Styles text box displays the current text styles defined in the drawing. Every AutoCAD drawing contains the STANDARD text style. This is created by default and cannot be deleted. Also, the Title Block Headings text style was already created. This text style was used to create the headings in each title block box, for example, "Scale" and "Date."

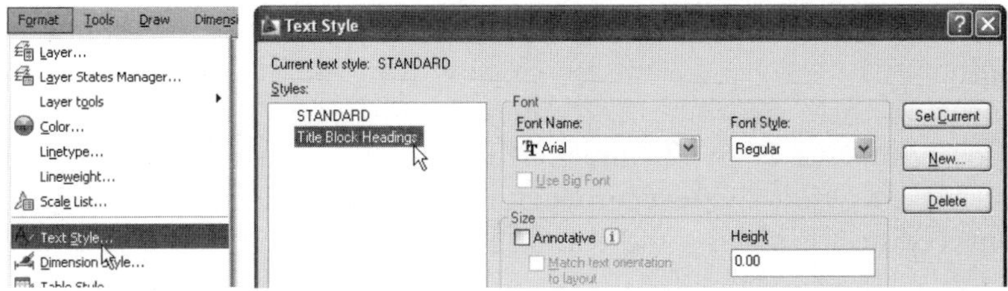

Figure 6.74

Step 3

Create the first text style by clicking the New... button, as shown in the following image. This activates the New Text Style dialog box, also shown in the following image. For Style Name, enter Company Info. When finished, click the OK button. This takes you back to the Text Style dialog box, as shown on the right in the following image. In the Font area, verify that the name of the font is Arial. Notice the font appearing in the Preview area. If necessary, click the Apply button to complete the text style creation process.

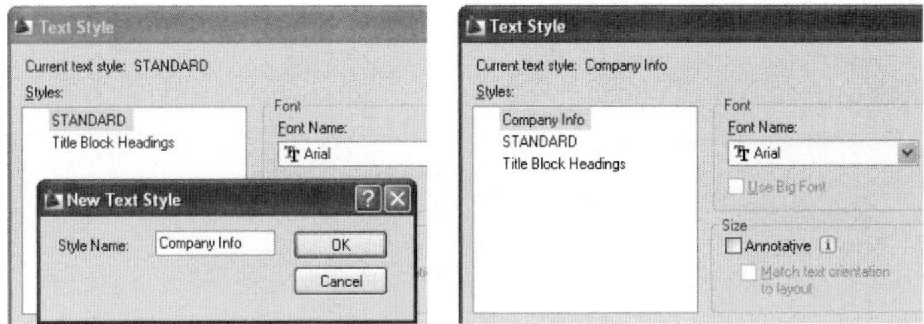

Figure 6.75

Step 4

Using the same procedure outlined in Step 3, create the following new text styles: Disclaimer, General Notes, and Title Block Text. Verify that the Arial font is assigned to all of these style names. Keep all other remaining default text settings. When finished, your display should appear similar to the following image.

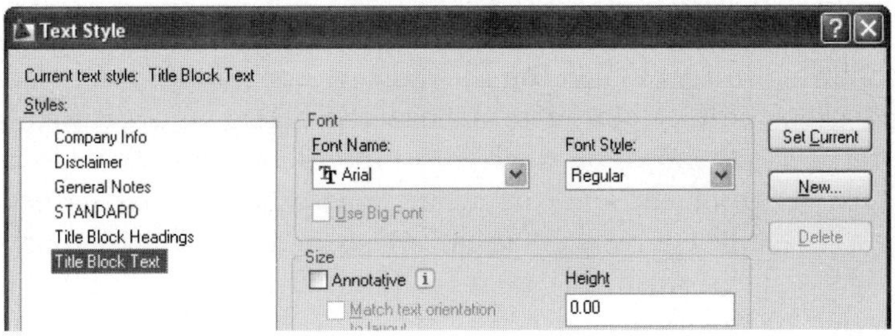

Figure 6.76

Step 5

Make the Company Info style current by selecting it in the Style text box and clicking the Set Current button. Add multiline text under the Company Info text style. Change the text height for both entries to .25 units. Middle-Center justify the text. Add the text PUMP ASSEMBLY inside the text formatting box. Click the OK button to place the text. Repeat this procedure for placing the text THE K-GROUP in the next title block space.

```
Command: MT (For MTEXT)
MTEXT Current text style: "Company Info" Text height: 0.20
    Annotative: No
Specify first corner: (Pick a point at "A")
Specify opposite corner or [Height/Justify/Line spacing/Rotation/
    Style/Width/Columns]: H (For Height)
Specify height <0.20>: .25
Specify opposite corner or [Height/Justify/Line spacing/Rotation/
    Style/Width/Columns]: J (For Justify)
Enter justification [TL/TC/TR/ML/MC/MR/BL/BC/BR] <TL>: MC (For
    Middle Center)
Specify opposite corner or [Height/Justify/Line spacing/Rotation/
    Style/Width/Columns]: (Pick a point at "B")
```

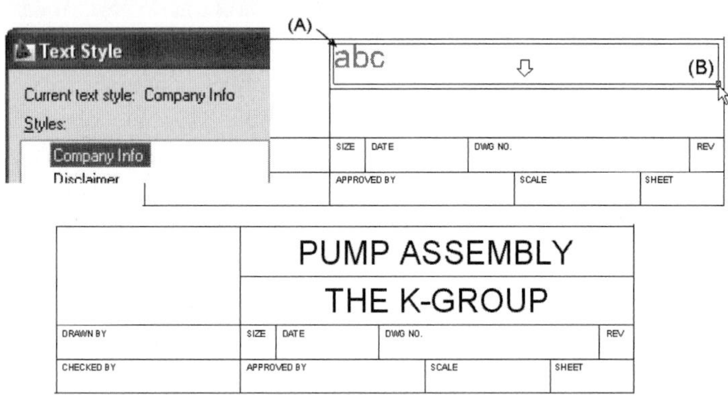

Figure 6.77

Step 6

Change the current text style to Title Block Text, as shown in the following image. Begin placing fields into the title block. Click Insert Field under the Insert pull-down menu, as shown in the following image. When the Field dialog box appears, pick Author under the Field name and click the OK button to dismiss the dialog box. Place this in the space identified as Drawn By, as shown on the right in the following image. Use a height of .10 for all of the next series of entries. Notice the appearance of four dashes inside a gray rectangle. This will be filled in when you enter your name in the appropriate area of the Properties dialog box later on in this exercise. Use the following command sequence to assist in the placing of this first field.

```
Command: FIELD
MTEXT Current text style: "Title Block Text" Text height: 0.20
Specify start point or [Height/Justify]: H (For Height)
Specify height <0.20>: .10
Specify start point or [Height/Justify]: (Pick a point inside of
    the area identified as Drawn By)
```

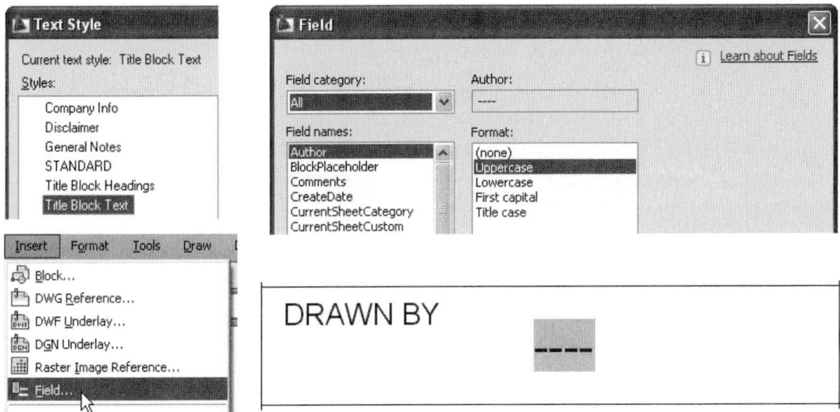

Figure 6.78

Step 7

Place a new field for the Date category. Activate the Field dialog box, as shown in the following image, and pick the Date Field name. Also select the format of the date from the list of examples located in this dialog box. Place this field under the DATE heading, as shown in the following image. Notice that the date is automatically calculated based on the current date setting on your computer.

```
Command: FIELD
MTEXT Current text style: "Title Block Text" Text height: 0.10
Specify start point or [Height/Justify]: (Pick a point inside of
    the area identified as Date)
```

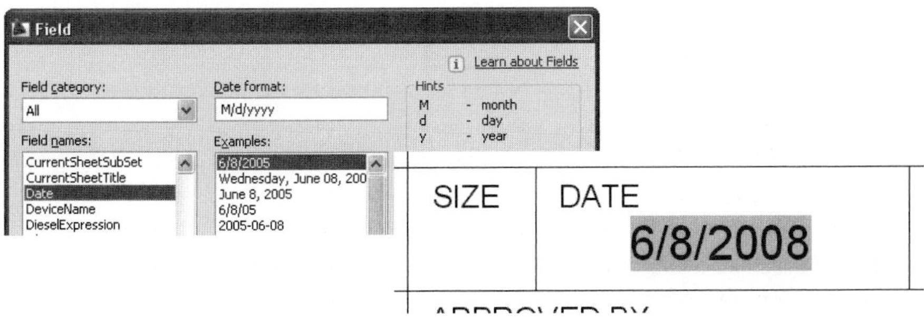

Figure 6.79

Step 8

Place a new field for the Filename category. Activate the Field dialog box, as shown in the following image, and pick the Filename Field name. Pick Uppercase for the Format. Also click the radio button next to Filename only. This limits the number of letters in the drawing name. Locate this field under the DWG NO. heading name, as shown in the following image.

```
Command: FIELD
MTEXT Current text style: "Title Block Text" Text height: 0.10
Specify start point or [Height/Justify]: (Pick a point inside of
    the area identified as DWG NO.)
```

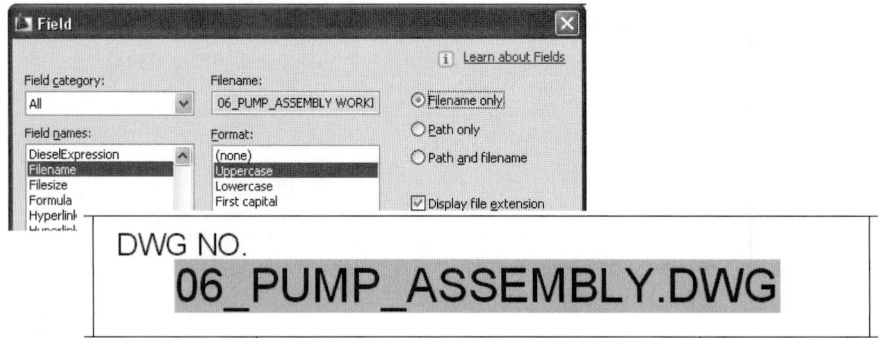

Figure 6.80

Step 9

While still in the Title Block Text style, use the Single Line text command and add the following entries to the title block using the following prompts and image as guides.

```
Command: DT (For DTEXT)
Current text style: "Title Block Text" Text height: 0.10
    Annotative: No
Specify start point of text or [Justify/Style]: (Pick a point
    approximately at "A")
Specify height <0.10>: (Press ENTER to accept this default value)
Specify rotation angle of text <0d0'>: (Press ENTER to accept this
    default value)
Enter text: ADRIAN CULPEPPER (After this text is entered, pick
    approximately at "B")
Enter text: (Use the table and figure below to enter the
    remaining text; when complete, press ENTER twice to exit this
    command)
```

Text	Title Block Area
B	SIZE (at "B")
JOHNNY MOSS	APPROVED BY (at "C")

Text	Title Block Area
1"=1'-0"	SCALE (at "D")
0	REV (at "E")
1 OF 1	SHEET (at "F")

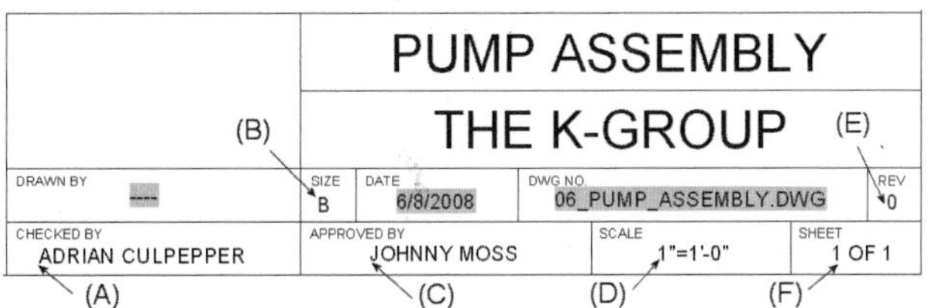

Figure 6.81

Step 10

Open the Text Style dialog box and double-click the Disclaimer text style, as shown in the following image. Activate the MTEXT command, pick a first corner at "A," and pick a second corner at "B." While in the Text Formatting dialog box, use the toolbar to change the text height to .08, right-click in the text editor, and pick Import Text from the menu, as shown in the following image.

```
Command: MT (For MTEXT)
MTEXT Current text style: "Disclaimer" Text height: 0.20
    Annotative: No
Specify first corner: (Pick a point at "A")
Specify opposite corner or [Height/Justify/Line spacing/Rotation/
    Style/Width/Columns]: (Pick a point at "B")
```

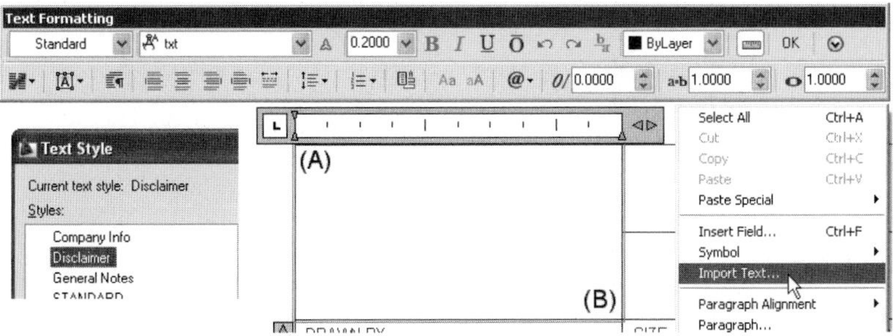

Figure 6.82

When the Select File dialog box appears, click the location of your CD and pick the file 06_Disclaimer.txt, as shown on the left in the following image. Since this information was already created in an application outside AutoCAD, it will be imported into the Multiline

Text Editor. Click the OK button to place the text in the title block, as shown on the right in the following image.

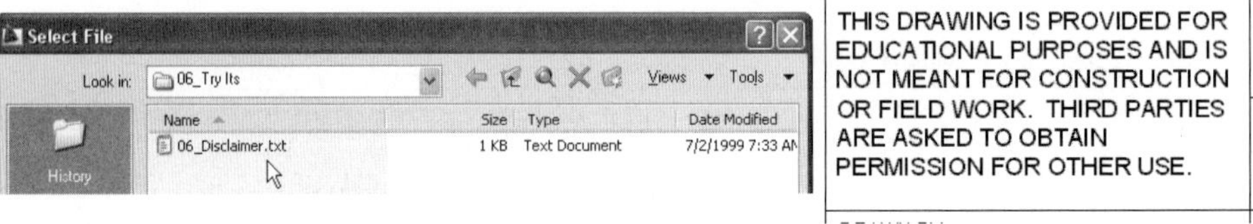

Figure 6.83

Step 11

Before placing the last text object, zoom to the extents of the drawing. Then click the General Notes text style located in the Text Style Control box (located on the Styles toolbar or the Annotate tab of the Ribbon, as shown on the left in the following image. Use the MTEXT command to create a rectangle to the right of the Pump Assembly, as shown on the right in the following image. This will be used to hold a series of general notes in multiline text format.

```
Command: MT (For MTEXT)
Current text style: "General Notes" Text height: 0.12
    Annotative: No
Specify first corner: (Pick a point at "A")
Specify opposite corner or [Height/Justify/Line spacing/Rotation/
    Style/Width/Columns]: (Pick a point at "B")
```

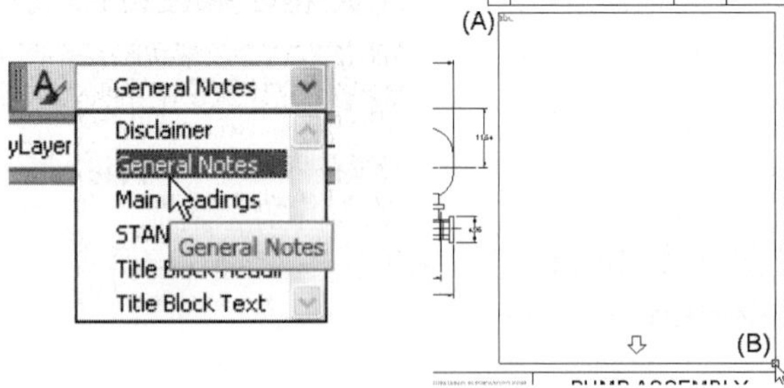

Figure 6.84

When the Multiline Text Editor appears, click the Import Text... button, as shown on the left in the following image. Then click the location of your CD and select the file 06_GENERAL_NOTES.rtf, as shown on the right in the following image. Make sure the file type you are searching for in the Select File dialog box is set to RTF. This RTF file (rich text file) was created outside AutoCAD in Microsoft Word.

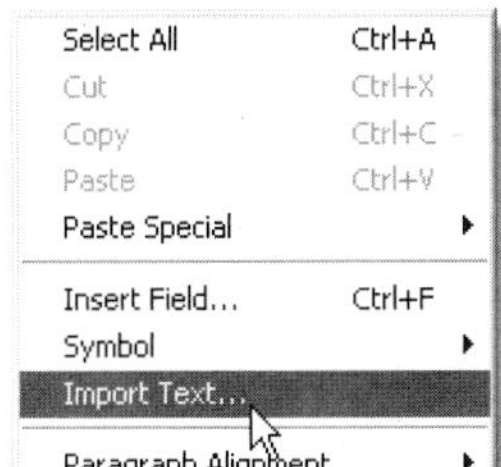

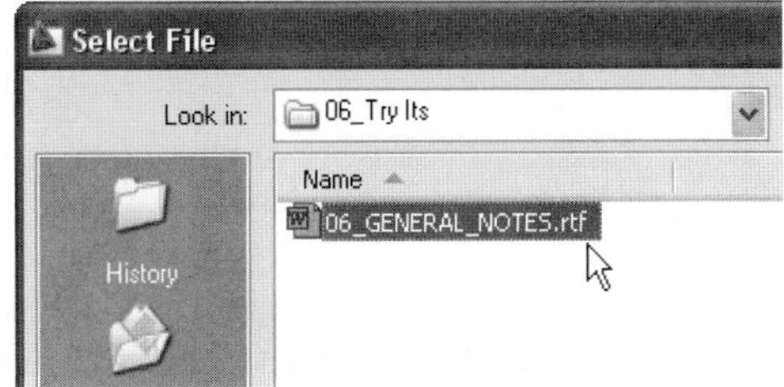

Figure 6.85

Step 12

The results of this operation are displayed on the left in the following image. However, the text appears to be displayed in a different font. This is due to the fact that when files are imported in RTF format, the original format of the text is kept. This means that the Times New Roman font was used even though the current text style uses the Arial font.

While still inside the Text Formatting toolbar, highlight all the general notes and change the font to Arial, as shown on the right in the following image.

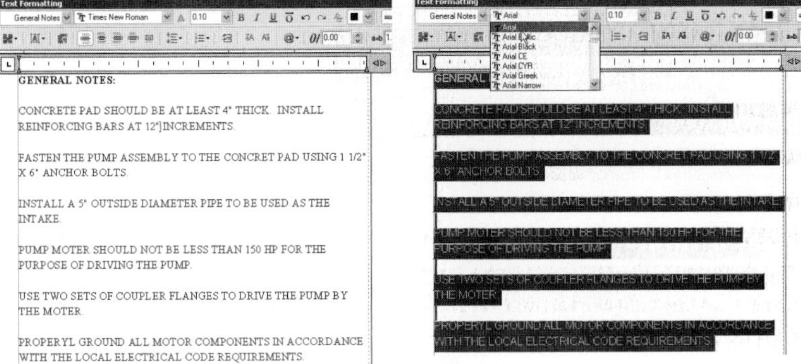

Figure 6.86

One more item needs to be taken care of before you leave the Text Formatting toolbar. Highlight all the text directly under the heading of GENERAL NOTES, as shown in the following image. Click the Numbering button. This creates a numbered list based on all highlighted lines of text, as shown in the following image. If necessary, change the tab and paragraph indent to match the following image. When finished, click the OK button to return to the drawing and place the general notes.

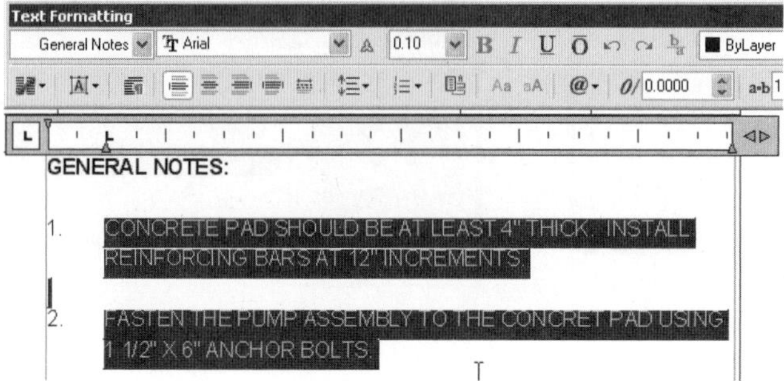

Figure 6.87

Step 13

Use the SPELL command and pick "Selected objects" from the Where to check field. Pick the Select objects button and click the multiline text object that holds the six general notes. Click the Start button and the Check Spelling dialog box appears, as shown in the following image; make the corrections to the words CONCRETE, MOTOR, and PROPERLY.

Figure 6.88

Step 14

In a previous step, you created a field for the author of the drawing. However, this field displayed as empty. You will now complete the title block by filling in the information and completing the Author field. Click Drawing Properties, found under the File pull-down menu, as shown on the left in the following image. When the 06_Pump Assembly.dwg Properties dialog box appears, enter your name in the Author text box, as shown in the middle in the following image.

When you return to your drawing, the Author field located in the title block is still not reflecting your name. Click Update Fields, found under the Tools pull-down menu area, and select the field to place your name in the title block.

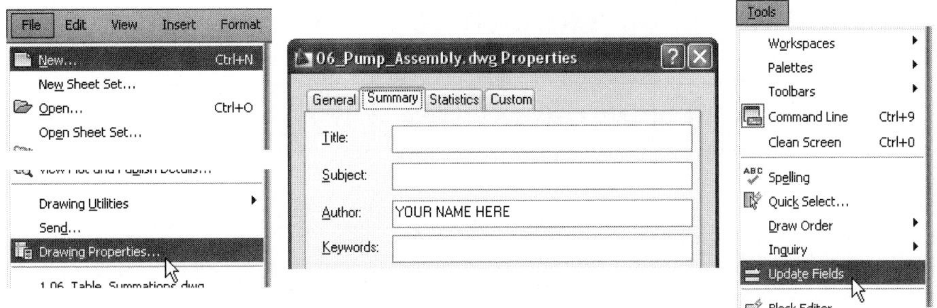

Figure 6.89

Step 15

The completed title block and general notes area are displayed in the following image.

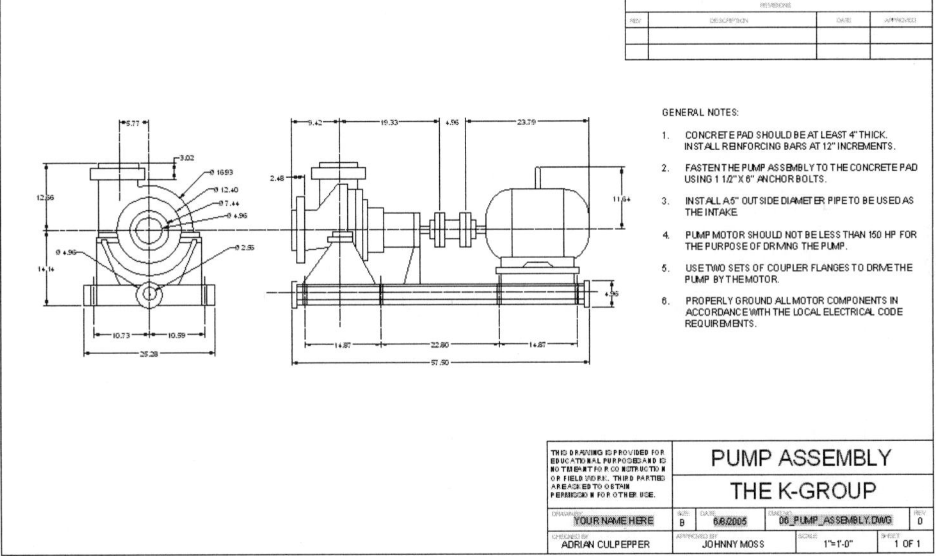

Figure 6.90

TUTORIAL EXERCISE: 06_TABLE.DWG

DOOR SCHEDULE					
MARK	WIDTH	HEIGHT	THICKNESS	MATERIAL	NOTES
1	9'-0"	8'-0"	1 3/4"	VINYL	~~
2	9'-0"	8'-0"	1 3/4"	VINYL	~~
3	9'-0"	8'-0"	1 3/4"	VINYL	~~
4	3'-0"	6-8"	1 3/8"	HCW	~~
5	2'-4"	6-8"	1 3/8"	HCW	~~
6	1'-6"	6-8"	1 3/8"	HCW	~~
7	5-0"	6-8"	1 3/4"	METAL CLAD	FRENCH DOOR
8	5'-0"	6-8"	1 3/4"	METAL CLAD	FRENCH DOOR
9	2-8"	6-8"	1 3/4"	METAL	SCREENED DOOR

Figure 6.91

PURPOSE

This tutorial is designed to create a new table style and then a new table object, as shown in the previous image.

SYSTEM SETTINGS

Since this drawing is already provided on the CD, open an existing drawing file called 06_Table. Two text styles are already created for this exercise; namely Architectural and Title.

LAYERS

A Table layer is already created for this drawing.

SUGGESTED COMMANDS

Open the drawing called 06_Table. You will create a new table style called Door Schedule. Changes will be made to the Data, Heading, and Data fields through the Table Style dialog box. Next, you will create a new table consisting of six columns and eleven rows. Information related to the door schedule will be entered in at the Title, Heading, and Data fields.

Step 1

Open the existing drawing file 06_Table. Activate the Text Style dialog box (STYLE command) and notice the two text styles already created, as shown on the left in the following image, namely, Title and Architectural. The Title text style uses the Arial font while Architectural uses the CityBlueprint font.

Next, activate the Table Style dialog box and create a new table style called Door Schedule, as shown on the right in the following image.

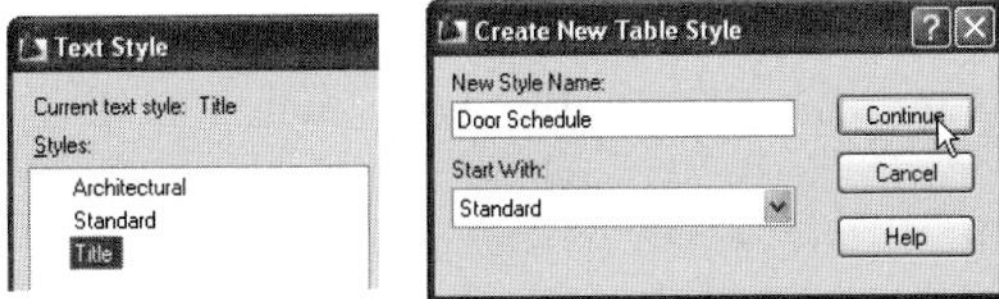

Figure 6.92

Step 2

Clicking the Continue button located in the previous image takes you to the main New Table Style dialog box. Use the information in the following image to change the properties with the Cell styles set to Data.

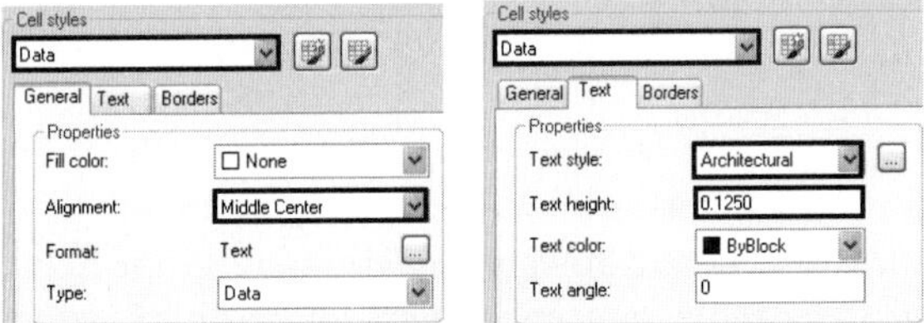

Figure 6.93

Use the information in the following image to change the properties under the Header cell style.

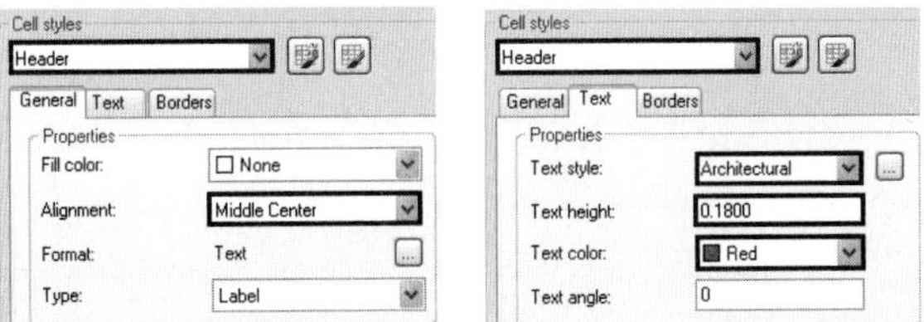

Figure 6.94

Use the information in the following image to change the properties under the Title cell style.

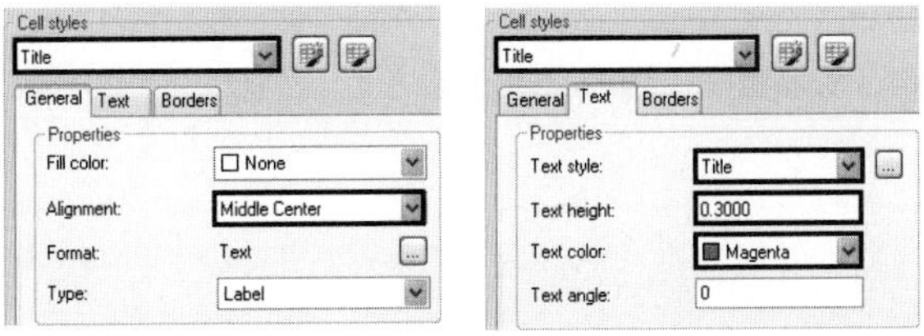

Figure 6.95

Step 3

Click the OK button to return to the Table Style dialog box. Verify the Door Schedule style is the current table style, as shown in the following image (you can double-click any table style name from the list to make it current). When finished, close the Table Style dialog box.

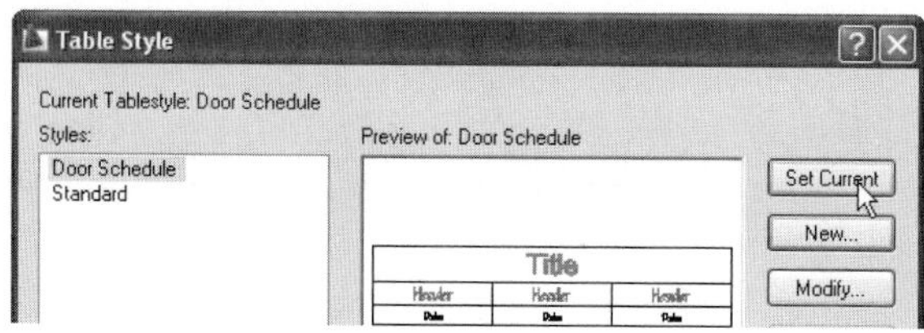

Figure 6.96

Step 4

Activate the TABLE command and while in the Insert Table dialog box, create a table consisting of six columns and eleven rows using the new Door Schedule table style. Click the OK button to place the table in a convenient location on your screen.

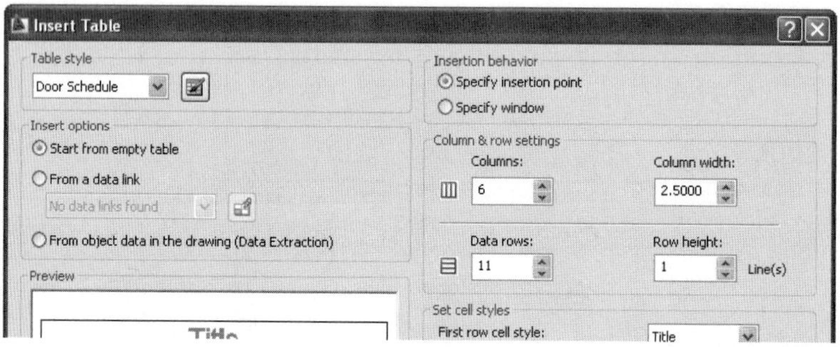

Figure 6.97

Step 5

When the Text Formatting toolbar appears, fill out the title of the table as DOOR SCHEDULE, as shown in the following image. Click OK.

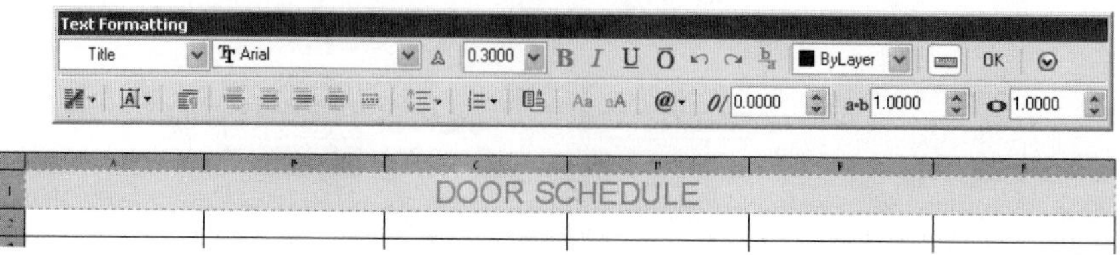

Figure 6.98

Step 6

Next, double-click inside the first column on the next row, and when the Text Formatting toolbar appears, label this cell MARK. Then press the TAB key to advance to the next column header and label this WIDTH. Repeat this procedure for the next series of column headers, namely HEIGHT, THICKNESS, MATERIAL, and NOTES, as shown in the following image. Click OK.

DOOR SCHEDULE					
MARK	WIDTH	HEIGHT	THICKNESS	MATERIAL	NOTES

Figure 6.99

Step 7

As the table appears to be stretched too long, click the table and pick the grip in the upper-right corner. Then stretch the table to the left, which shortens the length of all rows, as shown in the following image.

DOOR SCHEDULE					
MARK	WIDTH	HEIGHT	THICKNESS	MATERIAL	NOTES

Figure 6.100

Step 8

Complete the table by filling in all rows and columns that deal with the data. Utilize the Autofill technique for incrementing the numbers under the Mark column. To save time, you can also utilize Autofill to copy the "like data" values. The final results are shown in the following image.

DOOR SCHEDULE					
MARK	WIDTH	HEIGHT	THICKNESS	MATERIAL	NOTES
1	9'-0"	8'-0"	1 3/4"	VINYL	~ ~
2	9'-0"	8'-0"	1 3/4"	VINYL	~ ~
3	9'-0"	8'-0"	1 3/4"	VINYL	~ ~
4	3'-0"	6-8"	1 3/8"	HCW	~ ~
5	2'-4"	6-8"	1 3/8"	HCW	~ ~
6	1'-6"	6-8"	1 3/8"	HCW	~ ~
7	5-0"	6-8"	1 3/4"	METAL CLAD	FRENCH DOOR
8	5'-0"	6-8"	1 3/4"	METAL CLAD	FRENCH DOOR
9	2-8"	6-8"	1 3/4"	METAL	SCREENED DOOR

Figure 6.101

TUTORIAL EXERCISE: 06_TABLE_SUMMATIONS.DWG

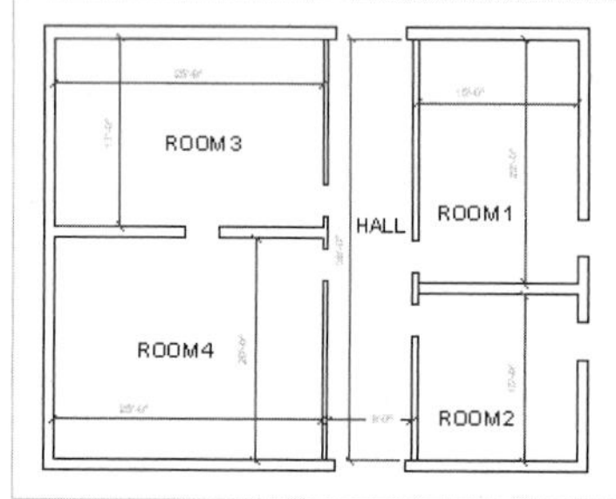

ROOM AREA CHART			
ROOM NUMBER	LENGTH	WIDTH	ROOM AREA (SQ FT)
HALLWAY	38	8	
ROOM 1	22	15	
ROOM 2	15	15	
ROOM 3	17	25	
ROOM 4	20	25	
		TOTAL AREA	

Figure 6.102

PURPOSE

This tutorial is designed to perform mathematical calculations and summations on a table.

SYSTEM SETTINGS

No special system settings need to be made for this drawing.

LAYERS

Layers are already created for this drawing.

SUGGESTED COMMANDS

Open the drawing file 06_Table_Summations, as shown in the previous image. Notice how the drawing opens up in a layout called Floor Plan, as shown in the following image. In addition to the floor plan being displayed, a table is present. Basic mathematical calculations such as addition, subtraction, multiplication, and division can be made on information in a table. Order of operations can also be used on the basic math calculations. You can also perform sum, average, and counting operations. You will insert formulas into this table that will calculate the area of each room. You will then perform a summation on all room areas.

Step 1

Zoom in to the table. You will first assign a formula that calculates the area of the Hallway. Click the cell that will hold the area of the Hallway; the Table toolbar displays as shown in the following image. Click the Insert Formula button, and when the menu displays, as shown in the following image, click Equation.

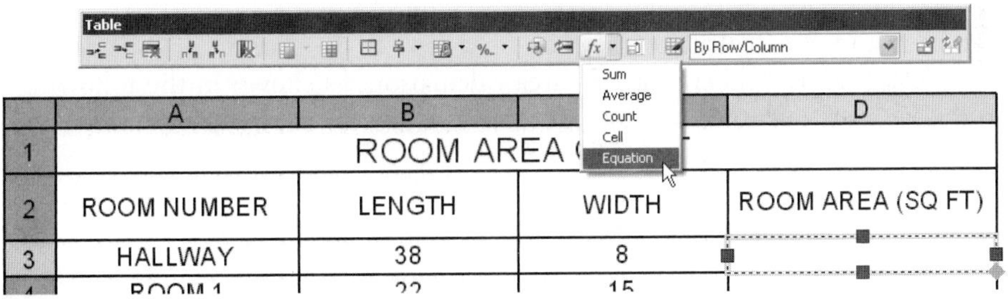

	A	B	C	D
1	ROOM AREA CHART			
2	ROOM NUMBER	LENGTH	WIDTH	ROOM AREA (SQ FT)
3	HALLWAY	38	8	

Figure 6.103

Step 2

Clicking the Equation in the previous step displays the Text Formatting toolbar, as shown in the following image. In the highlighted cell, you will need to multiply the Hallway length (38) by the width (8). You should not enter these numbers directly into the equation cell. Rather, enter the column and row number of each value. For example, the value 38 is identified by B3 and the value 8 is identified by C3. Enter these two cell identifiers separated by the multiplication symbol (*). Your display should appear similar to the following image.

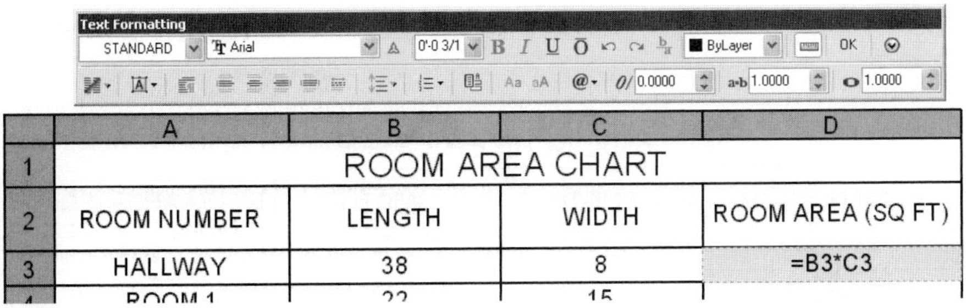

	A	B	C	D
1	ROOM AREA CHART			
2	ROOM NUMBER	LENGTH	WIDTH	ROOM AREA (SQ FT)
3	HALLWAY	38	8	=B3*C3

Figure 6.104

Step 3

Exiting the Text Formatting dialog box creates a field in the cell the calculation was being performed in, as shown in the following image. This field is easily recognized by the typical gray text background. Placing information as a field means if the information in the length and width cells changes, the field information will automatically recalculate itself.

ROOM AREA CHART			
ROOM NUMBER	LENGTH	WIDTH	ROOM AREA (SQ FT)
HALLWAY	38	8	304
ROOM 1	22	15	

Figure 6.105

Step 4

Follow steps 1 and 2 to calculate the area of Room 1. Be careful to use the correct cell identifying letters and numbers for this area calculation, as shown in the following image. You could also use the Autofill feature of tables as shown in earlier portions of this chapter for creating incremental copies of cells.

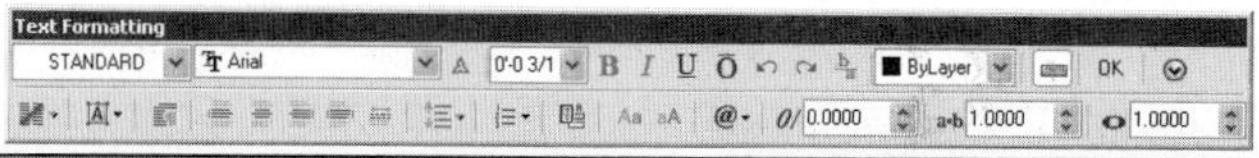

	A	B	C	D
1	ROOM AREA CHART			
2	ROOM NUMBER	LENGTH	WIDTH	ROOM AREA (SQ FT)
3	HALLWAY	38	8	304
4	ROOM 1	22	15	=B4*C4

Figure 6.106

Note: An equation can be started by simply clicking in a cell and typing the equal "=" sign.

Step 5

Continue inserting equations to calculate the area of Rooms 2, 3, and 4, as shown in the following image.

ROOM AREA CHART			
ROOM NUMBER	LENGTH	WIDTH	ROOM AREA (SQ FT)
HALLWAY	38	8	304
ROOM 1	22	15	330
ROOM 2	15	15	225
ROOM 3	17	25	425
ROOM 4	20	25	500
		TOTAL AREA	

Figure 6.107

Step 6

With the individual room areas calculated, you will now perform a summation that will add all current area fields. First, click in the cell in which the summation will be performed. When the Table toolbar appears, click the Insert Formula button followed by Sum, as shown in the following image.

Table
By Row/Column
Sum
Average
Count
Cell
Equation

	A	B		D
1	ROOM AREA (			
2	ROOM NUMBER	LENGTH	WIDTH	ROOM AREA (SQ FT)
3	HALLWAY	38	8	304
4	ROOM 1	22	15	330
5	ROOM 2	15	15	225
6	ROOM 3	17	25	425
7	ROOM 4	20	25	500
8			TOTAL AREA	

Figure 6.108

Step 7

After selecting Sum from the menu in the previous step, you will be prompted for the following:

- Select first corner of table cell range:
  - Answer this prompt by picking a point inside the upper-left corner of the cell that displays an area of 304. You will be prompted a second time.
- Select second corner of table cell range:
  - Answer this prompt by picking a point inside the lower-right corner of the cell that displays an area of 500.
  - This action identifies the cells to perform the summations.

	A	B	C	D
1	ROOM AREA CHART			
2	ROOM NUMBER	LENGTH	WIDTH	ROOM AREA (SQ FT)
3	HALLWAY	38	8	304
4	ROOM 1	22	15	330
5	ROOM 2	15	15	225
6	ROOM 3	17	25	425
7	ROOM 4	20	25	500
8			TOTAL AREA	

Figure 6.109

Step 8

After you select the second corner of the table cell range in the previous step, the equation is automatically created to add the values of cells D3 through D7, as shown in the following image.

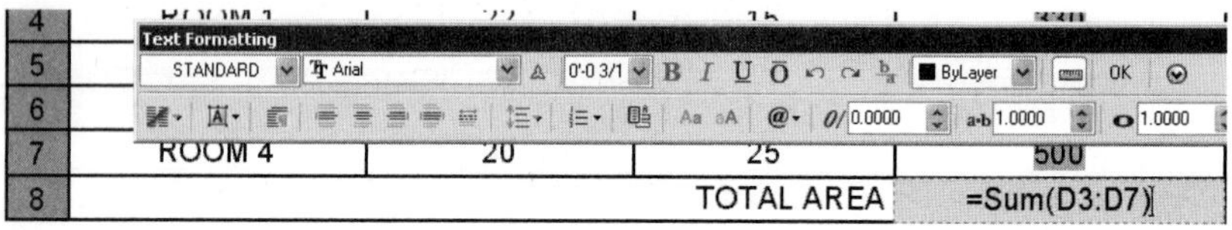

Figure 6.110

Step 9

The final results are displayed in the following image. This completes this tutorial exercise.

ROOM AREA CHART			
ROOM NUMBER	LENGTH	WIDTH	ROOM AREA (SQ FT)
HALLWAY	38	8	304
ROOM 1	22	15	330
ROOM 2	15	15	225
ROOM 3	17	25	425
ROOM 4	20	25	500
		TOTAL AREA	1784

Figure 6.111

CHAPTER 7

Object Grips and Changing the Properties of Objects

This chapter begins with a discussion of what grips are and how they are used to edit portions of your drawing. Various Try It! exercises are available to practice using grips. This chapter continues by examining a number of methods used to modify objects. These methods are different from the editing commands learned in chapter 4. Sometimes you will want to change the properties of objects such as layer, color, and even linetype. This is easily accomplished through the Properties Palette. The Match Properties command will also be introduced in this chapter. This powerful command allows you to select a source object and have the properties of the source transferred to other objects that you pick.

USING OBJECT GRIPS

An alternate method of editing is to use object grips. A grip is a small box appearing at key object locations such as the endpoints and midpoints of lines and arcs or the center and quadrants of circles. Once grips are selected, the object may be stretched, moved, rotated, scaled, or mirrored. Grips are at times referred to as visual Object Snaps because your cursor automatically snaps to all grips displayed along an object.

Try It!: Open the drawing file called 07_Grip Objects. While in the Command: prompt, click on each object type displayed in the following image to activate the grips. Examine the grip locations on each object.

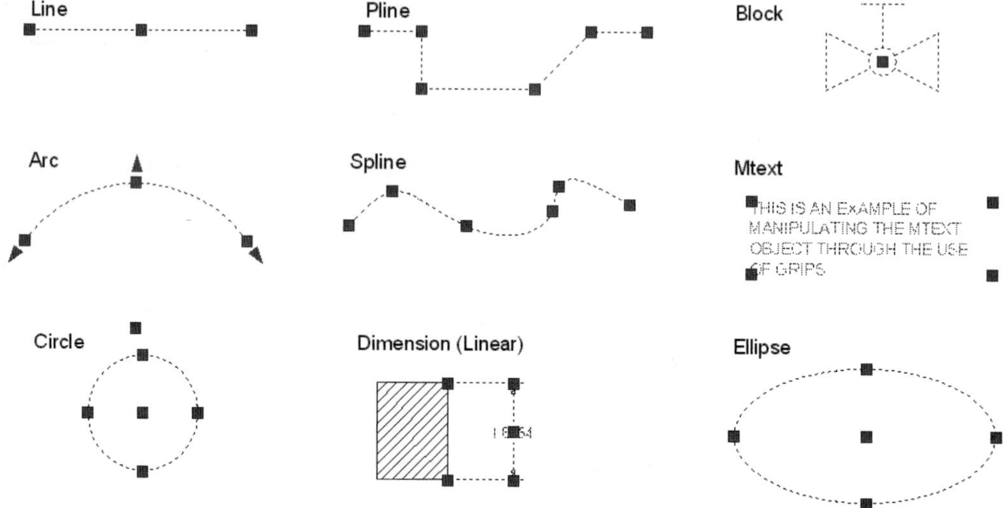

Figure 7.1

Changing the settings, color, and size of grips is accomplished under the Selection tab of the Options dialog box. Right-clicking on a blank part of your screen and choosing Options... from the menu, as shown on the left in the following image, displays the main Options dialog box. Click on the Selection tab to display the grip settings, as shown on the right in the following image.

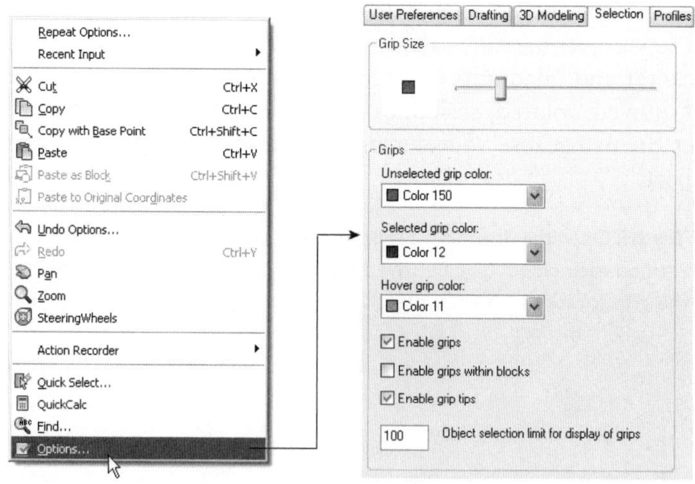

Figure 7.2

The following grip settings are explained as follows:

Grip size—Use the Grip Size area to move a slider bar to the left to make the grip smaller or to the right to make the grip larger.

Unselected grip color—Controls the color for an unselected grip. It is good practice to change the Unselected grip color to a light color if you are using a black screen background.

Selected grip color—Controls the color for a grip that is selected.

Hover grip color—Controls the color of an unselected grip when your cursor pauses over it.

Enable grips—By default, grips are enabled; a check in the Enable grips box means that grips will display when you select an object.

Enable grips within blocks—Normally a single grip is placed at the insertion point when a block is selected. Check the Enable grips within blocks box if you want grips to be displayed along with all individual objects that make up the block.

Enable grip tips—Some custom objects support grip tips. This controls the display of these tips when your cursor hovers over the grip on a custom object that supports grip tips.

Object selection limit for display of grips—By default, this value is set to 100. This means that when you select less than 100 objects, grips will automatically appear on the highlighted objects. When you select more than 100 objects, the display of grips is suppressed.

OBJECT GRIP MODES

The following image shows the main types of grips. When an object is first selected, the object highlights and the square grips are displayed in the default color of blue. In this example of a line, the grips appear at the endpoints and midpoint. The entire object is subject to the many grip edit commands. As you pause your cursor over a grip, the grip turns green. You will also see the angle and distance of this line, as shown in the middle of the following image, provided that dynamic input is turned on. When one of the grips is selected, it turns red by default, as shown on the right side of the following image. As you move your cursor, dynamic input appears in the form of direction and angle fields used to assist in making changes to this line. If dynamic input is turned off, distance and angle feedback do not appear. Once a grip is selected, the following prompts appear in the command prompt area:

```
** STRETCH **
Specify stretch point or [Base point/Copy/Undo/eXit]: (Press the
    SPACEBAR)
** MOVE **
Specify move point or [Base point/Copy/Undo/eXit]: (Press the
    SPACEBAR)
** ROTATE **
Specify rotation angle or [Base point/Copy/Undo/Reference/eXit]:
    (Press the SPACEBAR)
** SCALE **
Specify scale factor or [Base point/Copy/Undo/Reference/eXit]:
    (Press the SPACEBAR)
```

```
** MIRROR **
Specify second point or [Base point/Copy/Undo/eXit]: (Press the
    SPACEBAR to begin STRETCH mode again or enter X to exit
    grip mode)
```

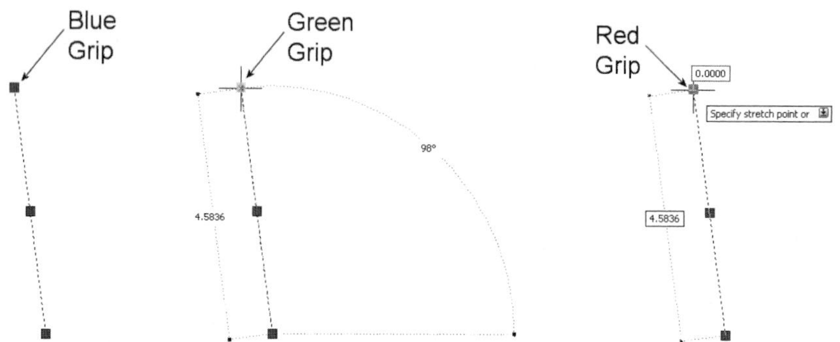

Figure 7.3

Note: When you hover your cursor over an unselected grip, the grip turns green underneath your cursor to signify that it is selectable.

To move from one edit command mode to another, press the Spacebar. Once an editing operation is completed, pressing ESC removes the highlight and removes the grips from the object. The following image shows various examples of each editing mode.

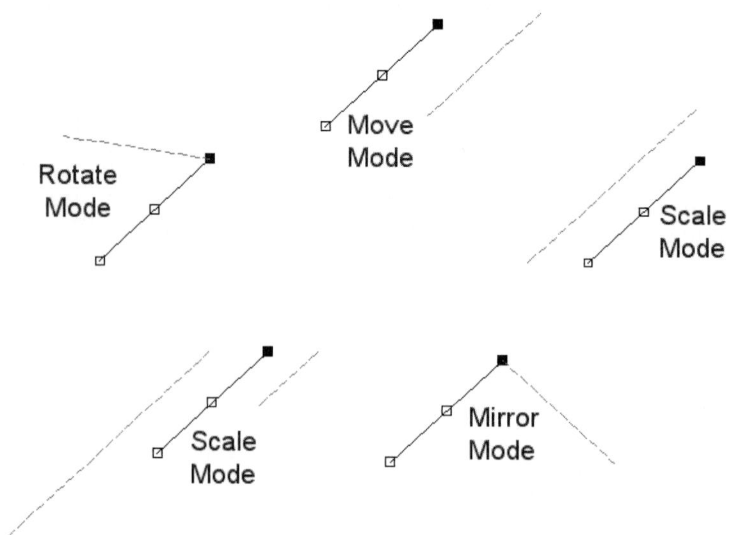

Figure 7.4

In the case of using grips with arcs, directional arrows display when grips appear. Use these arrows to guide you through the various directions in which the grips can be stretched, as shown in the following image.

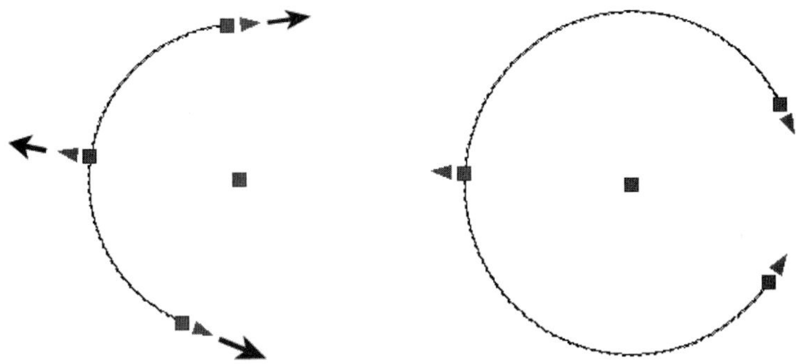

Figure 7.5

ACTIVATING THE GRIP SHORTCUT MENU

In the following image, a horizontal line has been selected and grips appear. The rightmost endpoint grip has been selected. Rather than use the Spacebar to scroll through the various grip modes, click the right mouse button. Notice that a shortcut menu on grips appears. This provides an easier way of navigating from one grip mode to another.

Try It!: Open the drawing file 07_Grip Shortcut. Click on any object, pick on a grip, and then press the right mouse button to activate the grip shortcut menu.

USING THE GRIP—STRETCH MODE

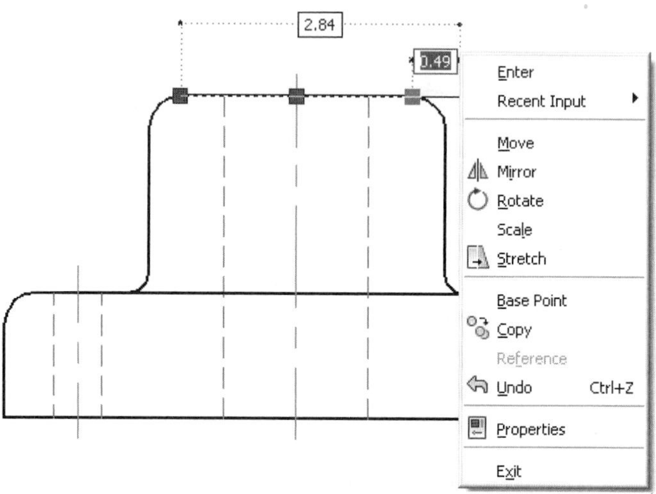

Figure 7.6

The STRETCH mode of grips operates similarly to the normal STRETCH command. Use STRETCH mode to move an object or group of objects and have the results mend themselves similar to the following image. The line segments "A" and "B" are both too long by two units. To decrease these line segments by two units, use the STRETCH mode by selecting lines "A," "B," and "C" at the command prompt. Next, while holding down SHIFT, select the grips "D," "E," and "F." This selects multiple grips and turns the grips red and ready for the stretch operation. Release SHIFT and pick the grip at "E" again. The STRETCH mode appears in the command prompt area. The last selected grip is considered the base point. Moving your cursor to the left and entering a value of 2.00 stretches the three highlighted grip objects to the left a distance of two units. To remove the object highlight and grips, press ESC at the command prompt.

Try It!: Open the drawing file 07_Grip Stretch. Use the illustration in the following image and the prompt sequence below to perform this task.

```
Command: (Select the three dashed lines labeled "A", "B", and
    "C" as shown in the following image. Then, while holding
    down SHIFT, select the grips at "D," "E," and "F." Release the
    SHIFT key and pick the grip at "E" again)
**STRETCH**
Specify stretch point or [Base point/Copy/Undo/eXit]: (Turn on
    Ortho or Polar mode, move your cursor to the left, and
    type 2)
Command: (Press ESC to remove the object highlight and grips)
```

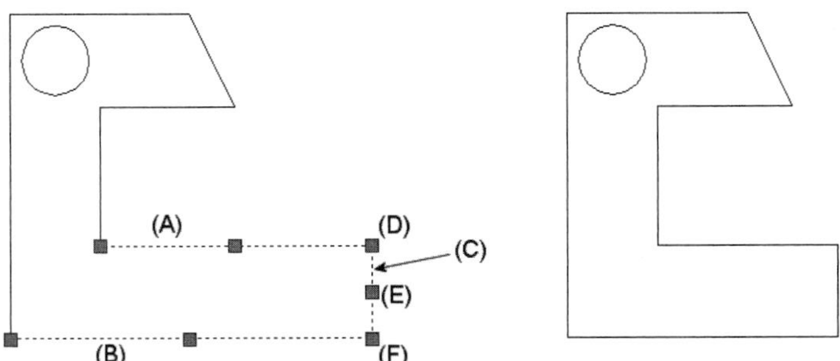

Figure 7.7

USING THE GRIP—SCALE MODE

Using the SCALE mode of object grips allows an object to be uniformly scaled in both the X and Y directions. This means that a circle, such as the one shown in the following image, cannot be converted to an ellipse through different X and Y values. As the grip is selected, any cursor movement drags the scale factor until a point is marked where the object will be scaled to that factor. The following image and the following prompt illustrate the use of an absolute value to perform the scaling operation of half the circle's normal size.

Try It!: Open the drawing file 07_Grip Scale. Use the illustration in the following image and the prompt sequence below to perform this task.

```
Command: (Select the circle to display grips, and then select
    the grip at the center of the circle at "A". Press the
    SPACEBAR until the SCALE mode appears at the bottom of the
    prompt line or press the right mouse button to activate the
    grip cursor menu to choose Scale)
**SCALE**
Specify scale factor or [Base point/Copy/Undo/Reference/eXit]: 0.50
Command: (Press ESC to remove the object highlight and grips)
```

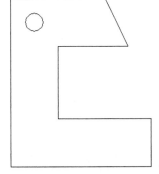

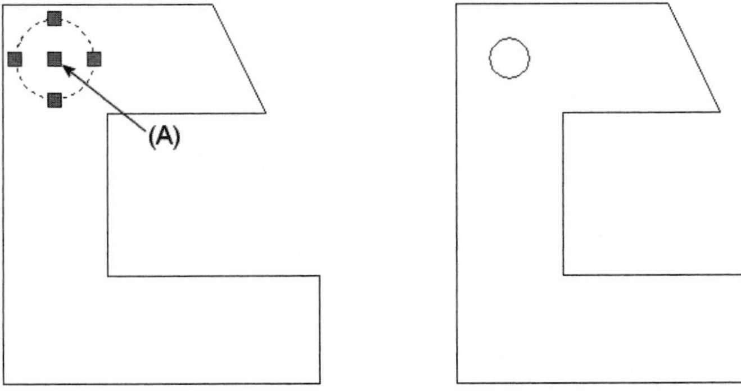

Figure 7.8

USING THE GRIP—MOVE/COPY MODE

The Multiple Copy option of the MOVE mode is demonstrated with the circle shown on the left in the following image. The circle is copied using Direct Distance mode at distances 2.50 and 5.00. This Multiple Copy option is actually disguised under the command options of object grips.

First, select the circle to display the grips at the circle center and quadrants, and then select the center grip. Use the Spacebar to scroll past the STRETCH mode to the MOVE mode. Issue a Copy within the MOVE mode to be placed in Multiple MOVE mode as shown in the following image.

Try It!: Open the drawing file 07_Grip Copy. Use the illustration in the middle of the following image and the prompt sequence below to perform this task.

```
Command: (Select the circle to activate the grips at the center
    and quadrants; select the grip at the center of the circle
    at "A". Then press the SPACEBAR until the MOVE mode appears at
    the bottom of the prompt line or press the right mouse
    button to activate the grip cursor menu to choose Move)
**MOVE**
Specify move point or [Base point/Copy/Undo/eXit]: C (For Copy)
**MOVE (multiple)**
Specify move point or [Base point/Copy/Undo/eXit]: (Move your
    cursor straight down and enter a value of 2.50)
**MOVE (multiple)**
Specify move point or [Base point/Copy/Undo/eXit]: (Move your
    cursor straight down and enter a value of 5.00)
**MOVE (multiple)**
Specify move point or [Base point/Copy/Undo/eXit]: X (For exit)
Command: (Press ESC to remove the object highlight and grips)
```

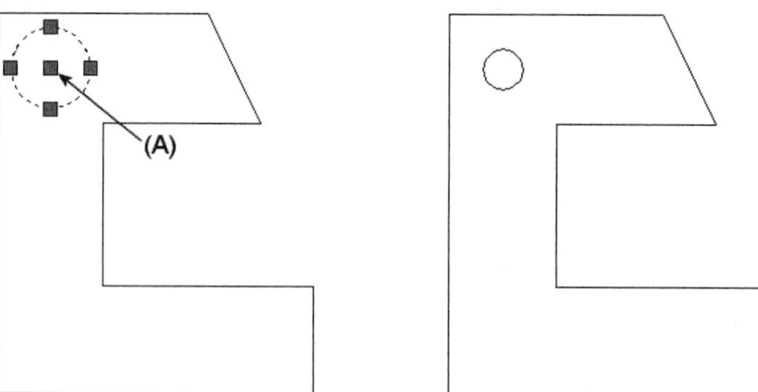

Figure 7.9

USING THE GRIP—MIRROR MODE

Use the grip-MIRROR mode to flip an object along a mirror line similar to the one used in the regular MIRROR command. Follow the prompts for the MIRROR option if an object needs to be mirrored but the original does not need to be saved. This performs the mirror operation but does not produce a copy of the original. If the original object needs to be saved during the mirror operation, use the Copy option of MIRROR mode. This places you in multiple MIRROR mode. Locate a base point and a second point to perform the mirror operation. See the following image.

Try It!: Open the drawing file 07_Grip Mirror. Use the illustration in the following image and the prompt sequence below to perform this task.

```
Command: (Select the circle in the following image to enable
    grips, and then select the grip at the center of the circle.
    Press the SPACEBAR until the MIRROR mode appears at the bottom
    of the prompt line or press the right mouse button to
    activate the grip cursor menu to choose Mirror)
**MIRROR**
Specify second point or [Base point/Copy/Undo/eXit]: C (For Copy)
**MIRROR (multiple)**
Specify second point or [Base point/Copy/Undo/eXit]: B (For
    Base Point)
Base point: Mid
of (Pick the midpoint at "A")
**MIRROR** (multiple)
Specify second point or [Base point/Copy/Undo/eXit]: (Move your
    cursor straight up and enter a value of 1.00)
**MIRROR** (multiple)
Specify second point or [Base point/Copy/Undo/eXit]: X (For Exit)
Command: (Press ESC to remove the object highlight and grips)
```

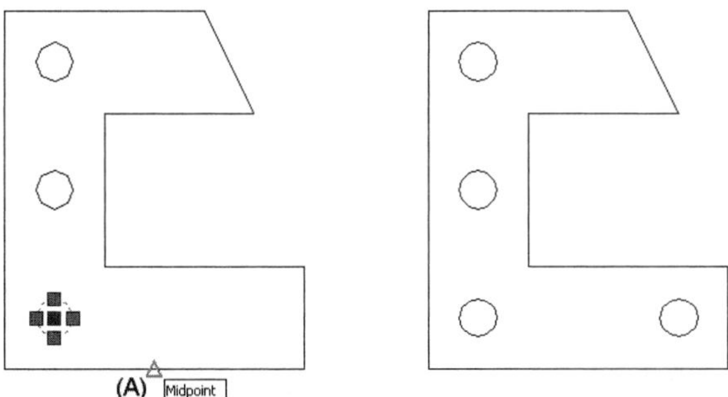

Figure 7.10

USING THE GRIP—ROTATE MODE

Numerous grips may be selected with window or crossing boxes. At the command prompt, pick a blank part of the screen; this should place you in Window/Crossing selection mode. Picking up or below and to the right of the previous point places you in Window selection mode; picking up or below and to the left of the previous point places you in Crossing selection mode. This method is used on all objects shown in the following image. Selecting the lower-left grip and using the Spacebar to advance to the ROTATE option allows all objects to be rotated at a defined angle in relation to the previously selected grip.

Try It!: Open the drawing file 07_Grip Rotate. Use the illustration in the following image and the prompt sequence below to perform this task.

```
Command: (Pick near "X," then near "Y" to create a window
    selection set and enable all grips in all objects. Select
    the grip at the lower-left corner of the object. Then press
    the SPACEBAR until the ROTATE mode appears at the bottom of the
    prompt line or click the right mouse button to activate the
    grip cursor menu to choose Rotate)
**ROTATE**
Specify rotation angle or [Base point/Copy/Undo/Reference/eXit]: 30
Command: (Press ESC to remove the object highlight and grips)
```

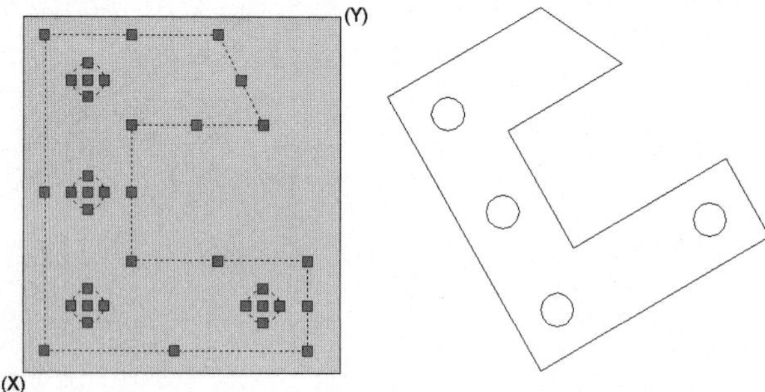

Figure 7.11

USING THE GRIP—MULTIPLE ROTATE MODE

With object grips, you may use ROTATE mode to rotate and copy an object. The following image illustrates a line that needs to be rotated and copied at a 40° angle. With a positive angle, the direction of the rotation is counterclockwise.

Selecting the line on the left in the following image enables grips located at the endpoints and midpoint of the line. Select the grip at "B" in the middle of the following image. This grip also locates the vertex of the required angle. Distance and angle feedback will appear if dynamic input is turned on. Press the Spacebar until the ROTATE mode is reached. Enter Multiple ROTATE mode by entering C for Copy when you are prompted in the following command sequence. Finally, enter a rotation angle of 40 to produce a copy of the original line segment at a 40° angle in the counterclockwise direction, as shown on the right in the following image.

Try It!: Open the drawing file 07_Grip Multiple Rotate. Use the illustration in the following image and the prompt sequence below to perform this task.

```
Command: (Select line segment "A"; then select the grip at "B."
    Press the SPACEBAR until the ROTATE mode appears at the bottom
    of the prompt line or click the right mouse button to
    activate the grip cursor menu to choose Rotate)
```

```
**ROTATE**
Specify rotation angle or [Base point/Copy/Undo/Reference/eXit]:
    C (For Copy)
**ROTATE (multiple)**
Specify rotation angle or [Base point/Copy/Undo/Reference/eXit]: 40
Specify rotation angle or [Base point/Copy/Undo/Reference/eXit]:
    X (For Exit)
Command: (Press ESC to remove the object highlight and grips)
```

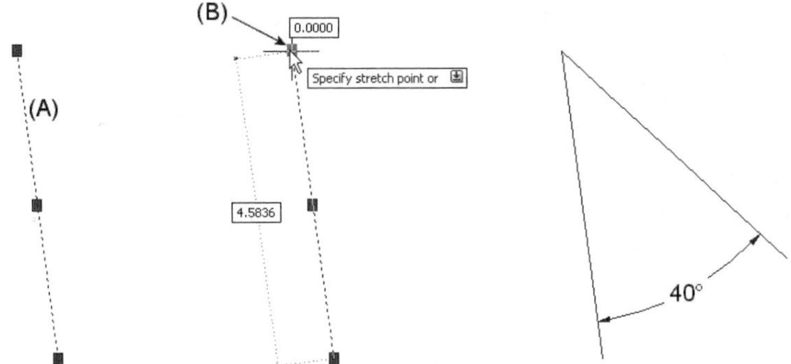

Figure 7.12

USING GRIP OFFSET SNAP FOR ROTATIONS

All Multiple Copy modes within grips may be operated in a snap location mode while you hold down the CTRL key. Here's how it works. On the left in the following image, the vertical centerline and circle are selected. The objects highlight and the grips appear. A multiple copy of the selected objects needs to be made at an angle of 45°. The grip ROTATE option is used in Multiple Copy mode.

Try It!: Open the drawing file 07_Grip Offset Snap Rotate. Use the illustration in the following image and the prompt sequence below to perform this task.

```
Command: (Select centerline segment "A" and circle "B"; and then
    select the grip near the center of circle "C." Press the
    SPACEBAR until the ROTATE mode appears at the bottom of the
    prompt line)
**ROTATE**
Specify rotation angle or [Base point/Copy/Undo/Reference/eXit]:
    C (For Copy)
**ROTATE (multiple)**
Specify rotation angle or [Base point/Copy/Undo/Reference/eXit]:
    B (For Base Point)
Base point: Cen
of (Select the circle at "C" to snap to the center of the circle)
**ROTATE (multiple)**
Specify rotation angle or [Base point/Copy/Undo/Reference/eXit]: 45
```

Rather than enter another angle to rotate and copy the same objects, hold down the CTRL key, which places you in offset snap location mode. Moving the cursor snaps the selected objects to the angle just centered, namely 45°, as shown in the middle of the following image.

```
**ROTATE (multiple)**
Specify rotation angle or [Base point/Copy/Undo/Reference/eXit]:
    (Hold down the CTRL key and move the circle and centerline
    until it snaps to the next 45° position shown in the middle
    of the following image)
**ROTATE (multiple)**
Specify rotation angle or [Base point/Copy/Undo/Reference/eXit]:
    X (For Exit)
Command: (Press ESC to remove the object highlight and grips)
```

The Rotate-Copy-Snap Location mode could allow you to create the circles illustrated on the right side of the following image without the aid of the ARRAY command. Since all angle values are 45°, continue holding down CTRL to snap to the next 45° location, and mark a point to place the next group of selected objects.

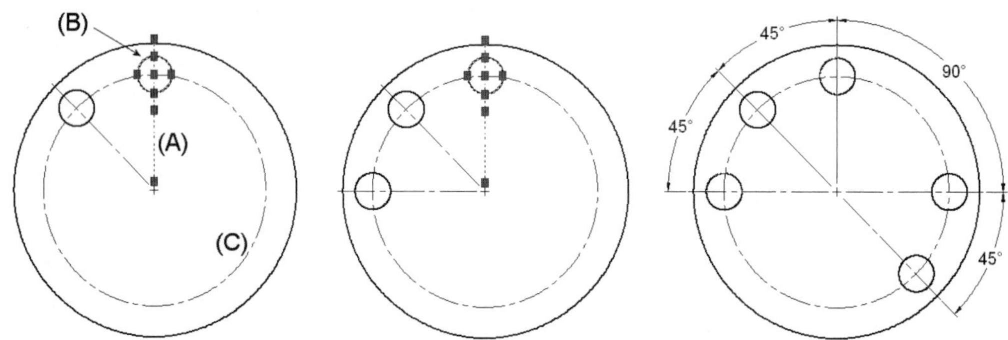

Figure 7.13

USING GRIP OFFSET SNAP FOR MOVING

As with the previous example of using Offset Snap Locations for ROTATE mode, these same snap locations apply to MOVE mode. The illustration on the left in the following image shows two circles along with a common centerline. The circles and centerline are first selected, which highlights these three objects and activates the grips. The intent is to move and copy the selected objects at 2-unit increments.

Try It!: Open the drawing file 07_Grip Offset Snap Move. Use the illustration in the following image and the prompt sequence below to perform this task.

```
Command: (Select the two circles and centerline to activate the
    grips; select the grip at the midpoint of the centerline.
    Then press the SPACEBAR until the MOVE mode appears at the
    bottom of the prompt line or click the right mouse button to
    activate the grip cursor menu to choose Move)
**MOVE**
Specify move point or [Base point/Copy/Undo/eXit]: C (For Copy)
```

```
**MOVE (multiple)**
Specify move point or [Base point/Copy/Undo/eXit]: (Move your
    cursor straight down and enter a value of 2.00)
```

In the middle of the following image, instead of remembering the previous distance and entering it to create another copy of the circles and centerline, hold down CTRL and move the cursor down again to see the selected objects snap to the distance already specified.

```
**MOVE (multiple)**
Specify move point or [Base point/Copy/Undo/eXit]: (Hold down the
    CTRL key and move the cursor down to have the selected
    objects snap to another 2.00-unit distance)
**MOVE (multiple)**
Specify move point or [Base point/Copy/Undo/eXit]: X (To exit)
Command: (Press ESC to remove the object highlight and grips)
```

The illustration on the right side of the following image shows the completed hole layout, the result of using the Offset Snap Location method of object grips.

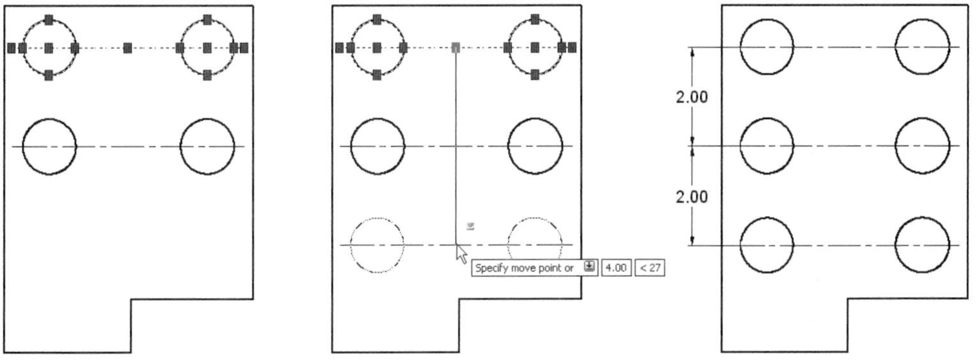

Figure 7.14

GRIP PROJECT—LUG.DWG

Try It!: Open the drawing file 07_Lug. Use the illustration in the following image and the prompt sequence below to perform this task.

Using the illustration in the following image as a guide, make the following changes to the lug:

1. Use the Grip-Stretch mode to join the endpoint of the inclined line at (A) with the endpoint of the horizontal line also at (A).
2. Use the Grip-Stretch mode to join the endpoint of the inclined line at (B) with the endpoint of the horizontal line also at (B).
3. Use the Grip-Stretch mode, pick the 1.50 diameter dimension, make the grip located on the dimension text hot, and position the dimension text in the lower-right corner of the object.

4. Use the Grip-Scale mode, pick the 1.50 diameter dimension and the circle, make the grip located at the center of the circle hot, and use a scale factor of .50 to reduce the size of the circle by half. Notice that the dimension recalculates to reflect this change.
5. Use the Grip-Mirror mode, and pick the circle, 1.50 diameter dimension, two vertical lines, both fillets, and the horizontal line at (E). Pick one of the grips to make it hot. Use the midpoint of the long horizontal line of the object as the new base point. Turn Ortho on and move your cursor down to establish the mirror line.
6. Use the Grip-Stretch mode to increase the size of the .50 throat at (F) to a new distance of 1.25. Pick both fillets and the three line segments attached to the fillets. Hold down the SHIFT key and select the grips at the endpoints of both fillets and the middle of the horizontal line. Release the SHIFT key and pick the grip at the midpoint of the horizontal line again. Move your cursor up (either Ortho or Polar mode needs to be turned on) and type a value of .75.
7. The completed exercise is illustrated on the right in the following image.

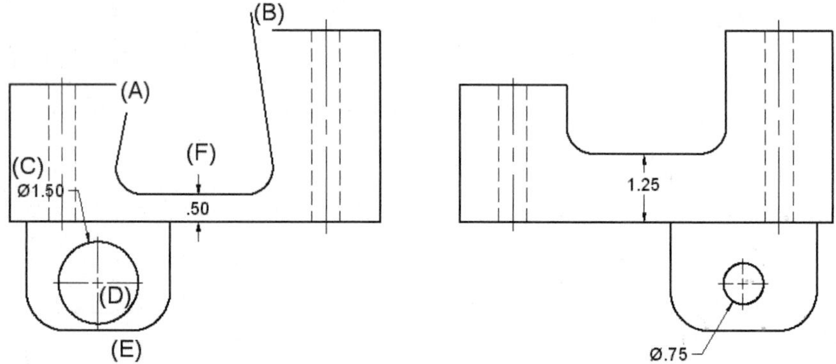

Figure 7.15

MODIFYING THE PROPERTIES OF OBJECTS

At times, objects are drawn on the wrong layer, color, or even in the wrong linetype. The lengths of line segments are incorrect, or the radius values of circles and arcs are incorrect. Eliminating the need to erase these objects and reconstruct them to their correct specifications, a series of tools are available to modify the properties of these objects. One such tool is the Properties Palette which can be selected from either the Menu Browser or the Standard Toolbar as shown on the left in the following image. This will display the Properties Palette as shown on the right in the following image.

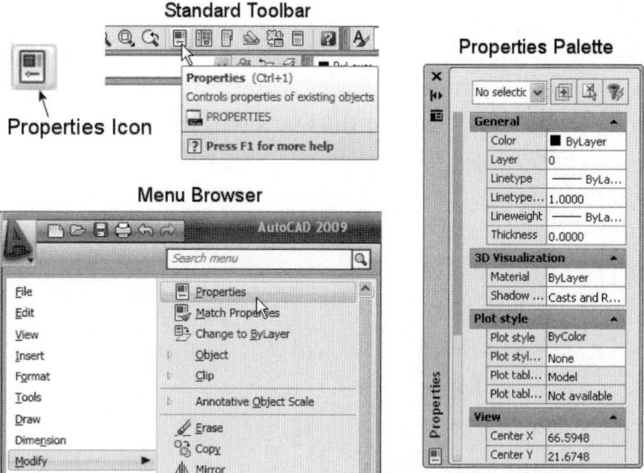

Figure 7.16

Illustrated in the following image are two line segments. One of the line segments has been pre-selected, as shown by the highlighted appearance and the presence of grips.

Clicking on the Properties button on the Standard toolbar in the following image displays the Properties Palette. This palette displays information about the object already selected; in this case the information is about the line segment, which is identified at the top of the palette.

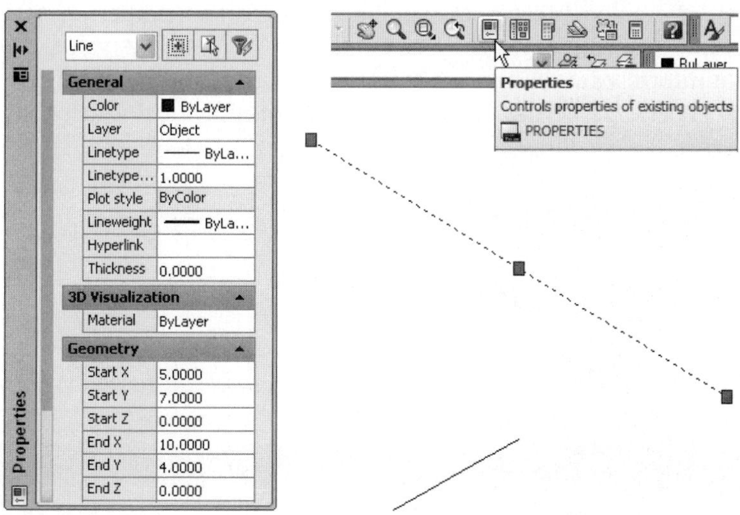

Figure 7.17

When changing a selected object to a different layer, click in the layer field on the left in the following image and the current layer displays (in this example, Layer Object). Clicking the down arrow displays all layers defined in the drawing. Clicking on one of these layers changes the selected object to a different layer.

A number of options that control the Properties Palette are illustrated on the right in the following image. You can elect to move, size, or close the Properties Palette from this menu. You can allow or prevent the Properties Palette from docking. Auto-hiding displays the Properties Palette when your cursor lies anywhere inside the window. When your cursor moves off of the Properties Palette, the window collapses to display only the blue side strip. Checking Description controls the display of a description area located at the bottom of the palette.

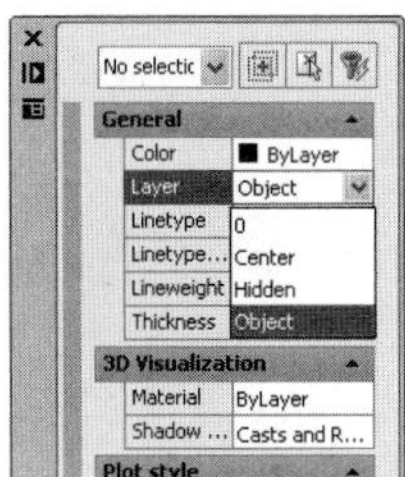

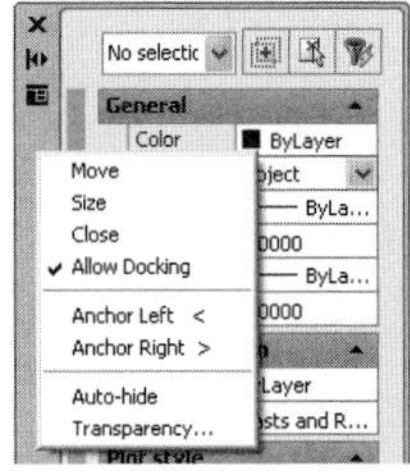

Figure 7.18

Three line segments are illustrated in the following image and all three segments have been pre-selected, as shown by their highlighted appearance and the presence of grips.

Clicking on the Properties button on the Standard toolbar displays the Properties Palette, as shown in the following image. At the top of the palette, the three lines are identified. You can change the color, layer, linetype, and other general properties of all three lines. However, you are unable to enter the Start and End X, Y, and Z values, and there is no length or angle information. Whenever you select more than one of the same object, you can change the general properties of the object but not individual values that deal with the object's geometry.

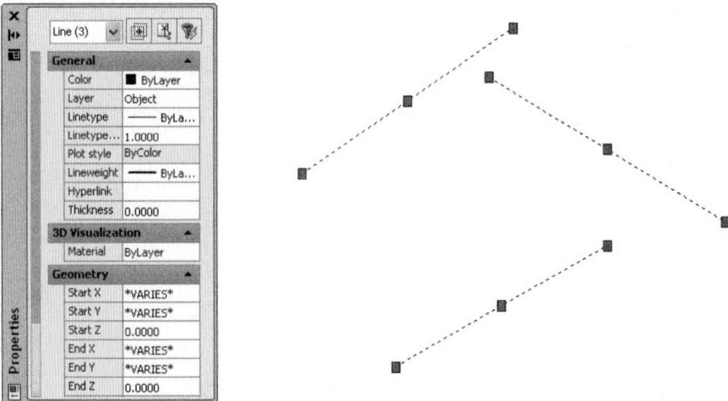

Figure 7.19

On the left side of the following image, an arc, circle, and line are pre-selected as identified by the appearance of grips. Activating the Properties Palette shown on the right side of the following image displays a number of object types at the top of the dialog box. You can click which object or group of objects to modify. With "All (3)" highlighted, you can change the general properties, such as layer and linetype, but not any geometry settings.

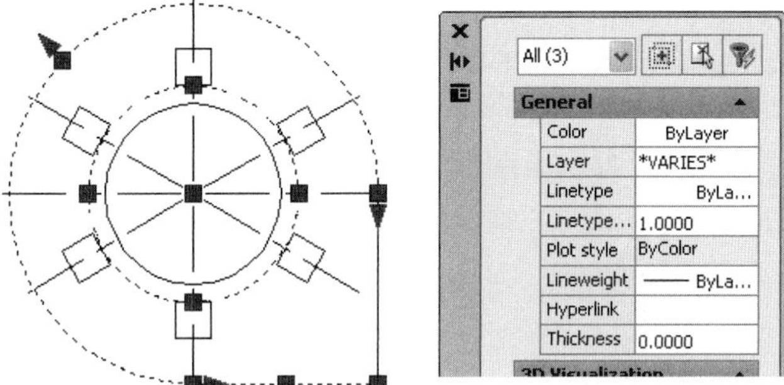

Figure 7.20

Try It!: Open the drawing file 07_Clutch Properties. What if you need to increase the radius of the circle to 1.25 units? Click on the inner circle and then select the Properties button. Since only one object was selected, the full complement of general and geometry settings is present for you to modify. Click in the Radius field and change the current value to 1.25 units. Pressing ENTER automatically updates the other geometry settings in addition to the actual object in the drawing, as shown in the following image. When finished, dismiss this dialog box to return to the drawing.

USING THE QUICK PROPERTIES TOOL

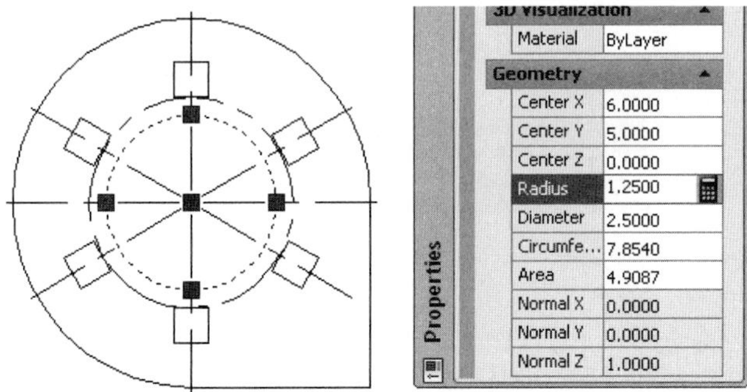

Figure 7.21

Located in the status bar at the bottom of every display screen is a Quick Properties button as shown on the left in the following image. When this tool is turned on, you can either list or make changes to properties depending on what you select. For example, if you hover your cursor over the line or block in the following examples, information is listed regarding the color, layer, and linetype of the selected objects. If you select a text object while Quick Properties is turned on as shown on the right in the following image, a panel appears that allows you to make changes to items such as layer, contents, and style. This panel can also be expanded to display more information that can be changed. This makes Quick Properties a very useful tool especially since it is virtually at your fingertips through the use of a basic wheel mouse.

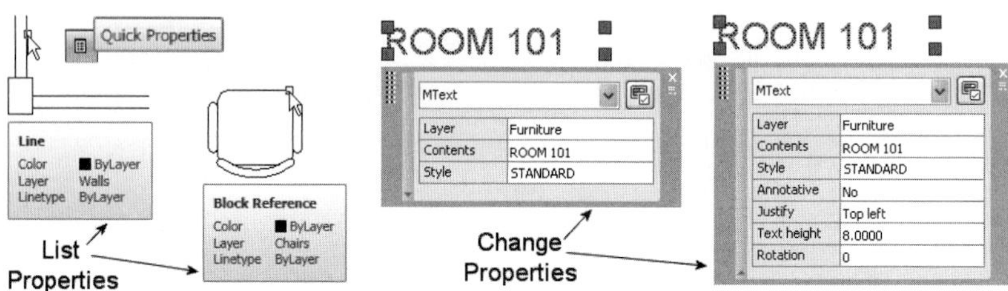

Figure 7.22

By default, when you click on an object with Quick Properties turned on, the Quick Properties panel locates itself at a set distance away from the cursor. Whenever you pick a new location on an object, the panel follows along with the cursor location at this set distance. Right clicking on the Quick Properties palette displays the menu as shown in the following image. If you change the Location Mode of the panel from Cursor to Float, the panel will remain in its position no matter where you select the object.

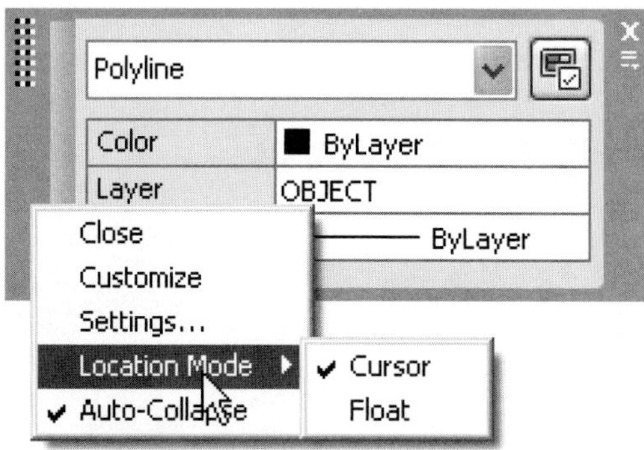

Figure 7.23

Another feature of Quick Properties is to display more settings found under the Quick Properties tab of the Drafting Settings dialog box as shown in the following image. Experiment with the various settings under this dialog box to see how it affects the display of the Quick Properties panel.

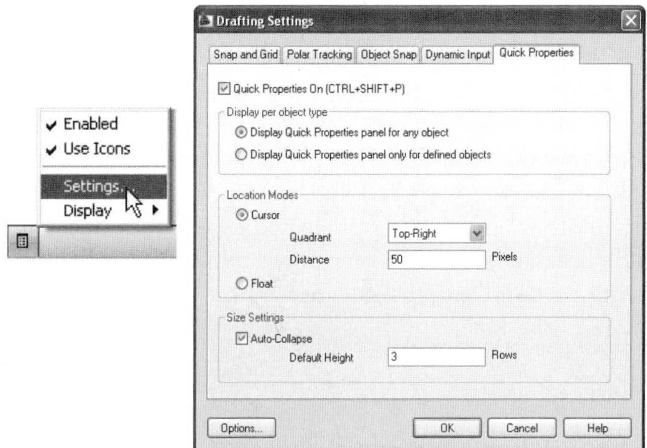

Figure 7.24

USING THE QUICK SELECT DIALOG BOX

At various times in a drawing, you need to build a selection set and modify the properties of these selected objects. However, there can be numerous objects in a drawing and if you attempt to select objects through the conventional methods such as window or crossing, you may accidentally ignore a few objects. A tool is available in which you enter certain parameters and AutoCAD selects the objects based on these parameters; this tool is called Quick Select. This tool can be activated by clicking on Quick Select from the Menu Browser or Tools pull-down menu in the following image. It displays the Quick Select dialog box, also shown in the following image.

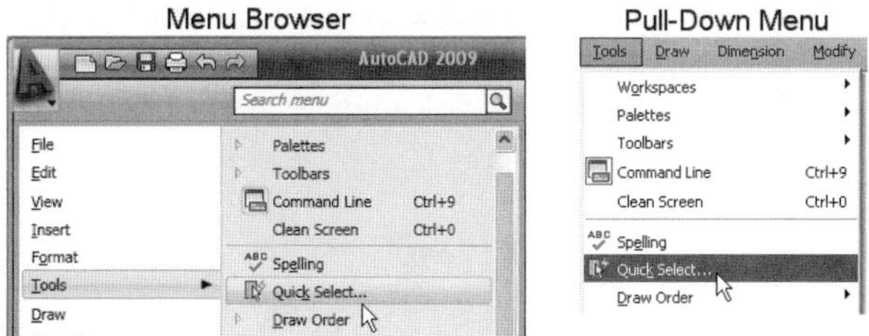

Figure 7.25

Clicking on Quick Select from either one of the menus launches the Quick Select dialog box as shown in the following image. To have AutoCAD create a selection set by object type and property, first choose the appropriate object type in the dialog box. Notice in the following example that you can choose from Block Reference or Line. Once you determine the object type, the property area of this dialog box changes depending on the object. In the following example, study the two rows of properties on the right in the following image. Notice that the properties for the Block Reference object type greatly differ from those for the Line object type. This, however, is what makes Quick Select so powerful. Not only can you create a selection set based on object type, but you can also narrow your search down based on layer and color for most objects or even by block name.

By default, whenever you apply the object type and property information, this information becomes the basis for the new selection set. You could also create the inverse of this information; this occurs when you click on the radio button next to Exclude from new selection set entry. If this button is picked and you build a selection set based on a certain object and property, all other objects will be selected except this object and property. As you can see, the Quick Select dialog box is a very powerful tool that actually works for you and saves you from having to build tedious selection sets.

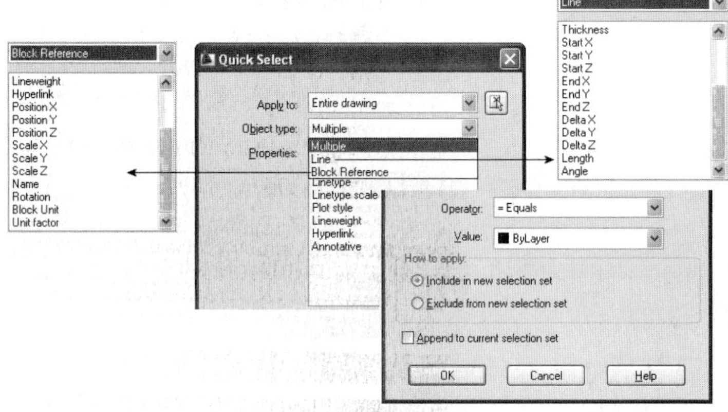

Figure 7.26

Try It!: Open the drawing file 07_Change Text Height. In the following image, the room numbers in the rectangular boxes are currently set to a height of 18". All text items need to be changed to a new height of 12". Rather than individually changing each text item, use the Quick Select dialog box to assist with this operation.

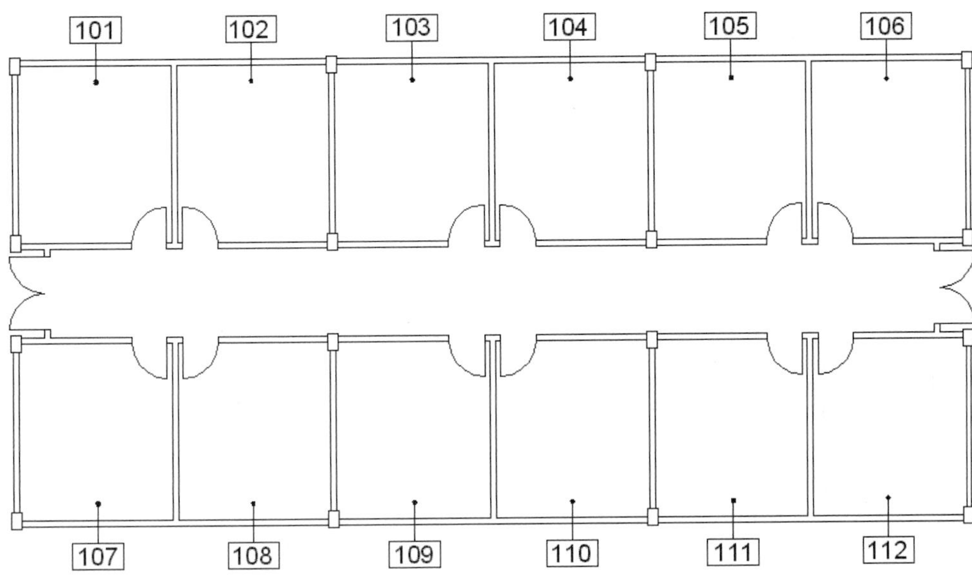

Figure 7.27

First, activate the Properties Palette and click on the Quick Select button, as shown on the left in the following image. This represents another method of activating Quick Select. Once inside the Quick Select dialog box, click in the Object type window and select "Text," as shown on the right side of the following image.

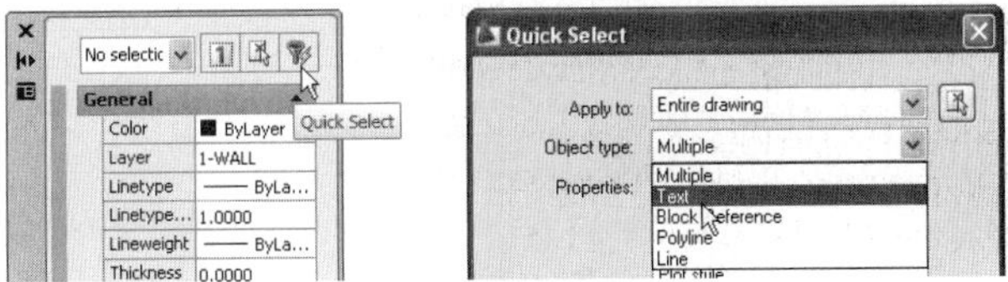

Figure 7.28

Clicking the OK button at the bottom of the Quick Select dialog box returns you to your drawing. Notice that all text is highlighted and a text height of 18.0000 is listed in the Properties Palette. Change this value to 12, as shown in the following image.

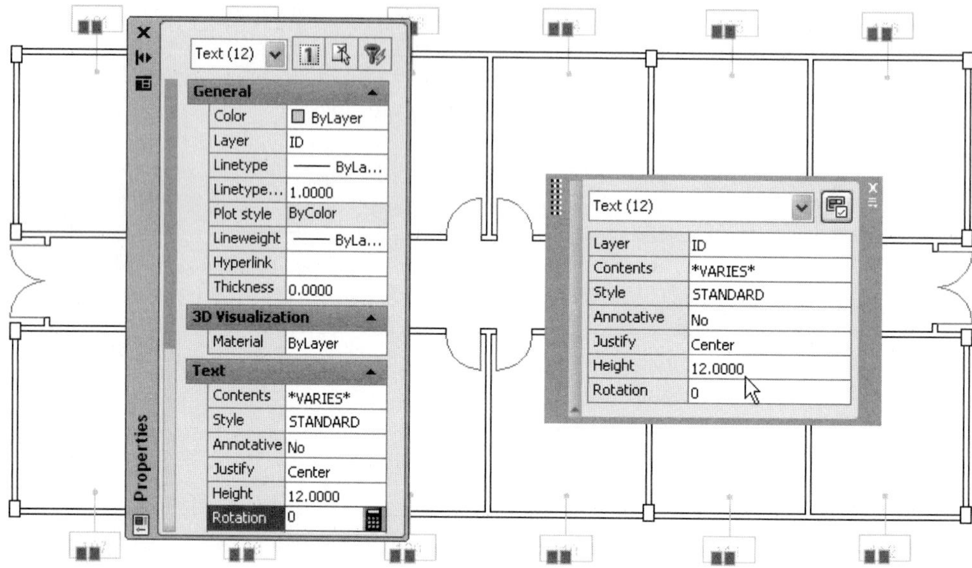

Figure 7.29

When finished changing the text height in the Properties Palette, press the ENTER key followed by the ESC key and notice that all text in the following image has been changed to the new height of 12".

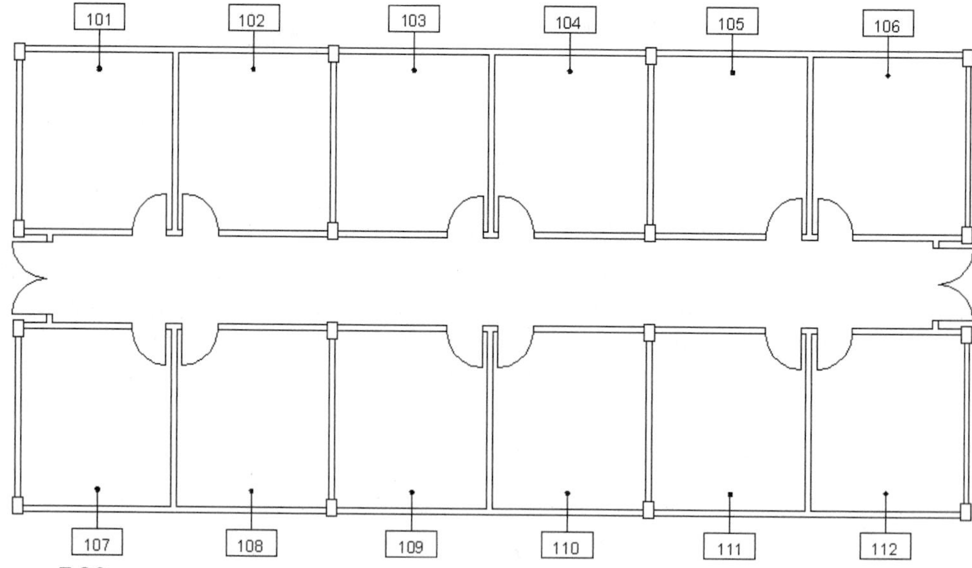

Figure 7.30

USING THE PICKADD FEATURE OF THE PROPERTIES PALETTE

Another very interesting feature is available to you inside the Properties Palette; it is the PICKADD button located in the following image. To see how it operates, follow the next exercise.

Try It!: Open the drawing file 07_PickAdd. A number of object types ranging from lines to text to multiline text and polylines with dimensions are displayed. Activate the Properties Palette and pick on one object. Notice that it highlights, the grips appear, and information about the object is displayed in the Properties Palette. Suppose, however, that you want to list information about another object. You must first press ESC to deselect the original object. Now select a different object and this information is displayed in the Properties Palette. This time, select the PICKADD button on the left in the following image. Notice that the button changes in appearance. Instead of the "plus" sign, a "1" is displayed on the right in the following image. Click on any object and notice the information displayed in the Properties Palette. Without pressing ESC, click on another object. The original object deselects and the new object highlights with its information displayed in the Properties Palette. Very simply, the PICKADD button eliminates the need for the ESC key when displaying information about individual objects.

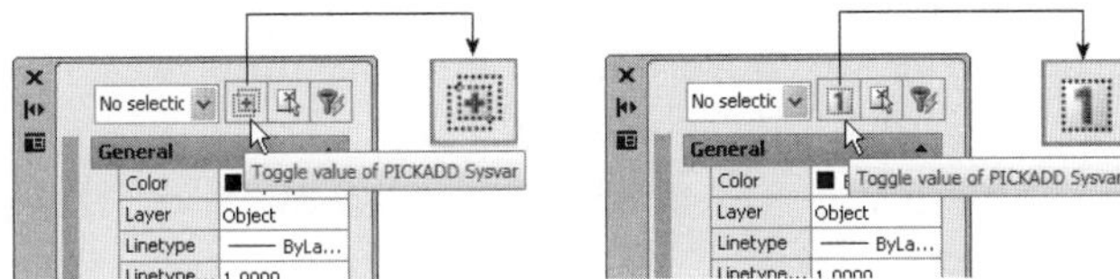

Figure 7.31

Note: Changing the PICKADD button affects all drawings. Since this change is global, click the PICKADD button to set it back to the "plus" sign before continuing on.

PERFORMING MATHEMATICAL CALCULATIONS

A mathematical calculator is available to assist in making a variety of calculations while still remaining in an AutoCAD drawing. The calculator is activated by clicking on QuickCalc, found under Tools in the Menu Browser, as shown on the left in the following image. This launches the calculator, as shown on the right in the following image. The calculator resembles the Properties Palette and can be manipulated in the same fashion. When you perform calculations throughout the design cycle, these calculations are stored in a History area. Calculations in this area can be retrieved for later use. As you enter numbers and mathematical functions such as addition and subtraction from the supplied keypads, they are displayed in the Input area of the calculator. Four tabbed areas control the methods by which calculations are made. They are Number Pad, Scientific, Units Conversion, and Variables.

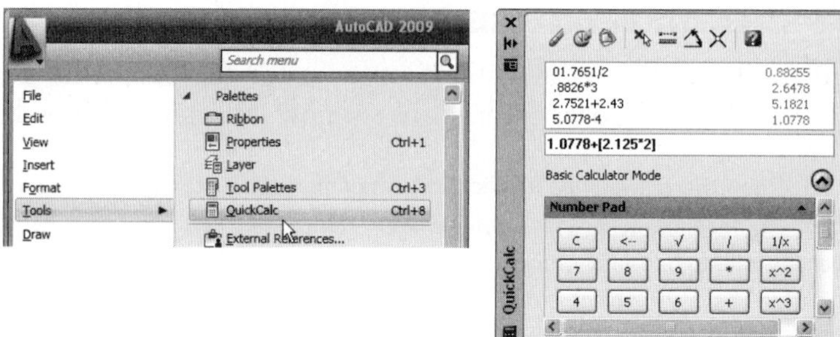

Figure 7.32

Each of the four tabbed areas of the calculator are illustrated in the following image. A brief description of each follows:

Numeric Pad: This area of the calculator is used to perform the most basic of mathematical functions. Clicking on a number or function places this information in the Input area of the calculator.

Scientific: Use this tabbed area to enter trigonometric and other advanced functions.

Units Conversion: This area is used to convert units from one system to another. You enter the units type, the unit type to convert from, and the units to convert to. Entering a numeric value to convert displays the converted value.

Variables: This area contains a series of shortcut commands that can be used for performing specialized operations.

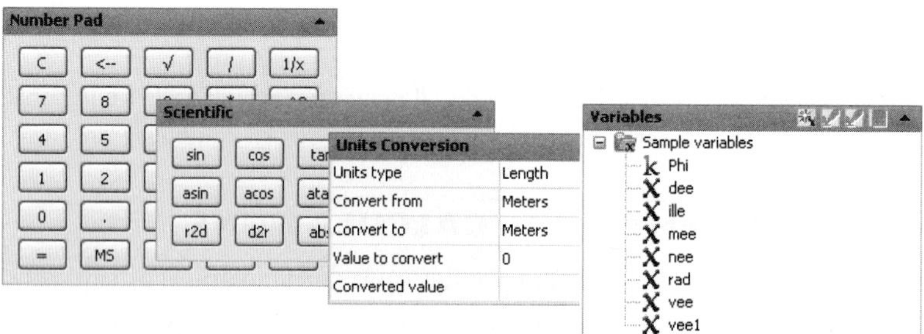

Figure 7.33

Table 7.1 *The following table illustrates the variable names and their functions*

Variable	Function
Phi	Golden Ratio (1.61803399)
dee	Finds the distance between two endpoints
illi	Finds the intersection of two lines defined by four endpoints
mee	Finds the midpoint between two endpoints

Table 7.1 *The following table illustrates the variable names and their functions (continued)*

Variable	Function
nee	Finds the unit vector in the XY plane normal (perpendicular) to two endpoints
rad	Obtains the radius from a circle, arc, or polyline arc segment
vee	Finds a vector from two endpoints
Vee1	Finds a unit vector from two endpoints

When you are using the Properties Palette and you need to change the value of one of the Geometry listings, a calculator icon is present. Clicking this icon launches the QuickCalc dialog box, as shown on the right in the following image. Entering a mathematical formula in the Input area and clicking the Apply button sends the calculated value back to the Properties Palette and change the highlighted object.

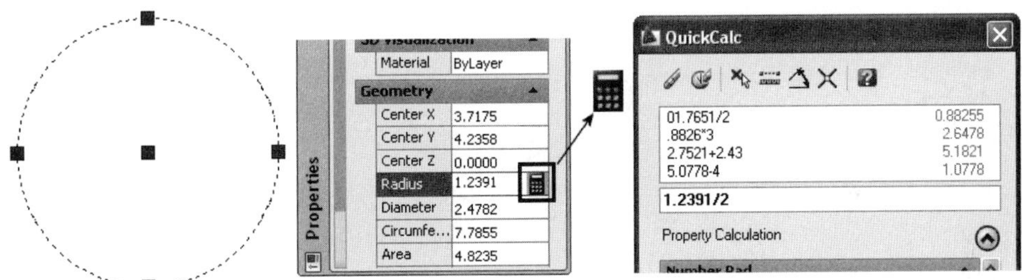

Figure 7.34

USING THE LAYER CONTROL BOX TO MODIFY OBJECT PROPERTIES

If all you need to do is to change an object or a group of objects from one layer to another, the Layer Control box can easily perform this operation.

Try It!: Open the drawing file 07_Change Layer. In the following image, select the arc and two line segments. Notice that the current layer is 0 in the Layer Control box. These objects need to be on the OBJECT layer.

Click in the Layer Control box to display all layers defined in the drawing. Then click on the desired layer for all highlighted objects (in this case, the OBJECT layer in the following image).

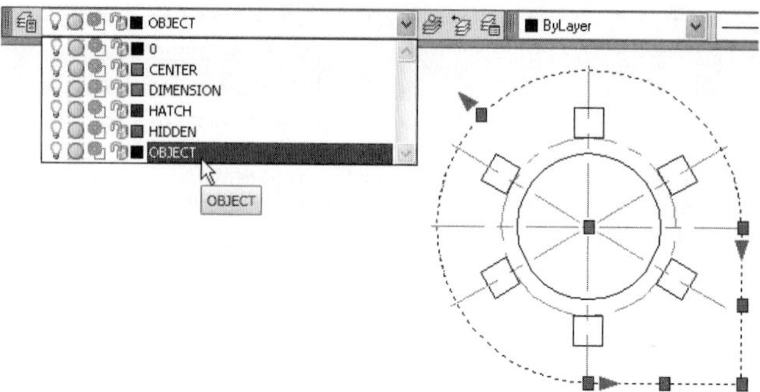

Figure 7.35

Notice that in the following image, with the objects still highlighted, the layer listed is OBJECT. This is one of the quickest and most productive ways of changing an object from one layer to another.

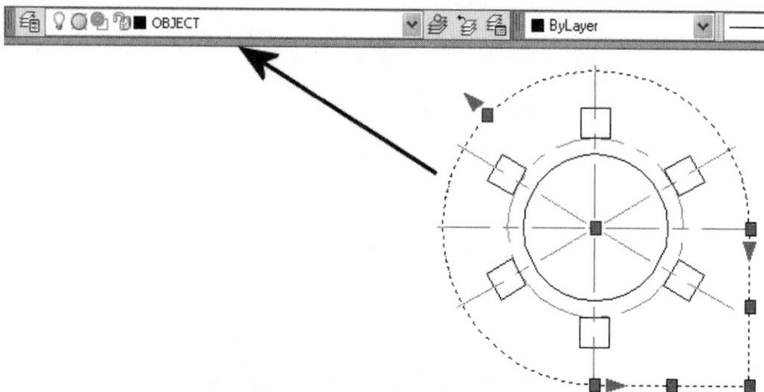

Figure 7.36

Try It!: Open the drawing file 07_Mosaic. In the following image, this drawing consists of 5 different types of objects all drawn on Layer 0: circles, squares drawn as closed polylines, lines, text (the letter X), and text (the letter Y). The Quick Select dialog box will be used to select these object types individually.

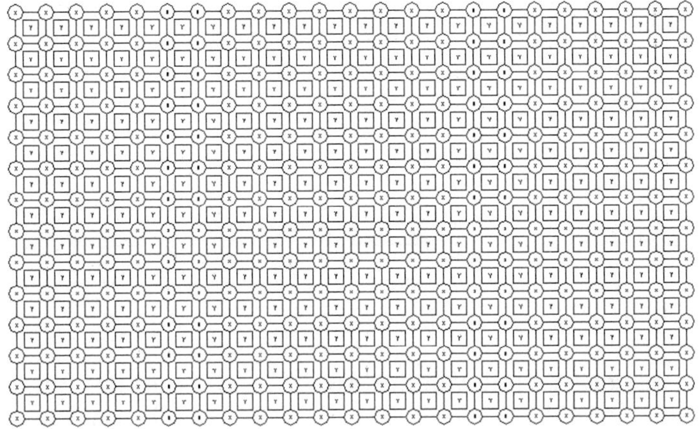

Figure 7.37

Once you have selected them, change the objects to the correct layers, which are also supplied with this drawing and identified on the left in the following image.

First, activate the Quick Select dialog box from the Tools pull-down menu, shown in the middle in the following image. In the Object type box, select Circle, as you see on the right in the following image, and click the OK button.

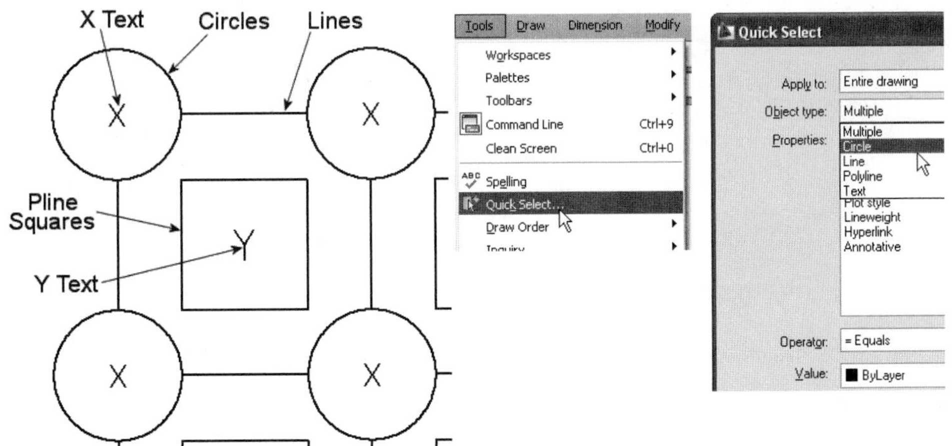

Figure 7.38

With all circles selected, click the Layer Control box and pick the Circles layer as shown in the following image. All circles in the mosaic pattern should turn red. Press ESC to remove the object highlight.

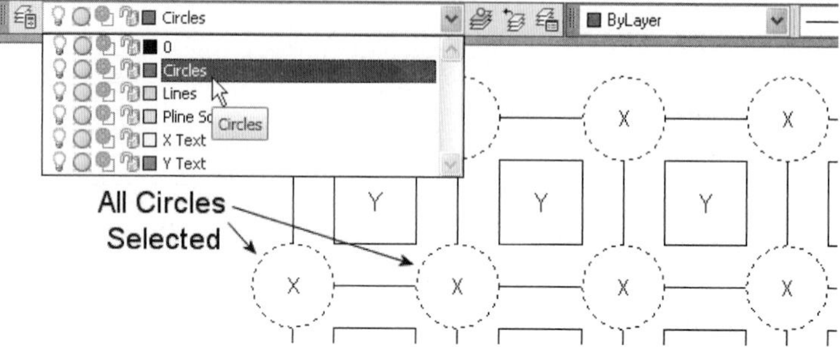

Figure 7.39

Next, you need to select all the text with the letter "X" and change this text to the layer called X Text. Activate the Quick Select dialog box and make the following changes: Change the Object type to Text, set the Properties to Contents, and enter "X" as the value. Your display should be similar to the dialog box on the left in the following image. Click the OK button to dismiss the Quick Select dialog box and select all text with the letter "X."

Change all letters to the X Text layer in the Layer Control box, as shown on the right in the following image. Press ESC to remove the object highlight and grips. Follow the same procedures for changing all line segments to the Lines layer, all polylines to the Pline Squares layer, and all "Y" text to the Y Text layer.

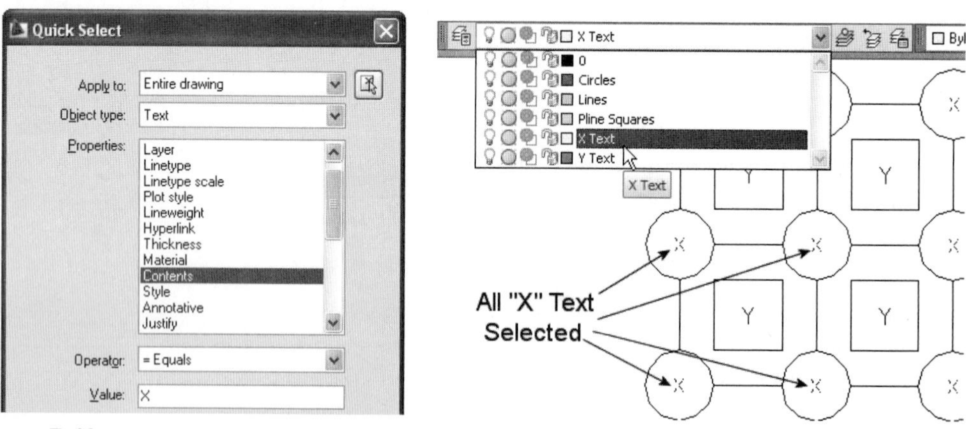

Figure 7.40

Try It!: Open the drawing file 07_Qselect Duplex. Use the illustration in the following image and the information below to change the items to their proper layers.

```
Change all lines to the Walls layer
Change all text to the Room Labels layer
Change the blocks "Door" and "Louver" to the Door layer
Change the block "Window" to the Window layer
Change the blocks "Countertop," "Range," "Sink," and
    "Refrigerator" to the Countertop layer
```

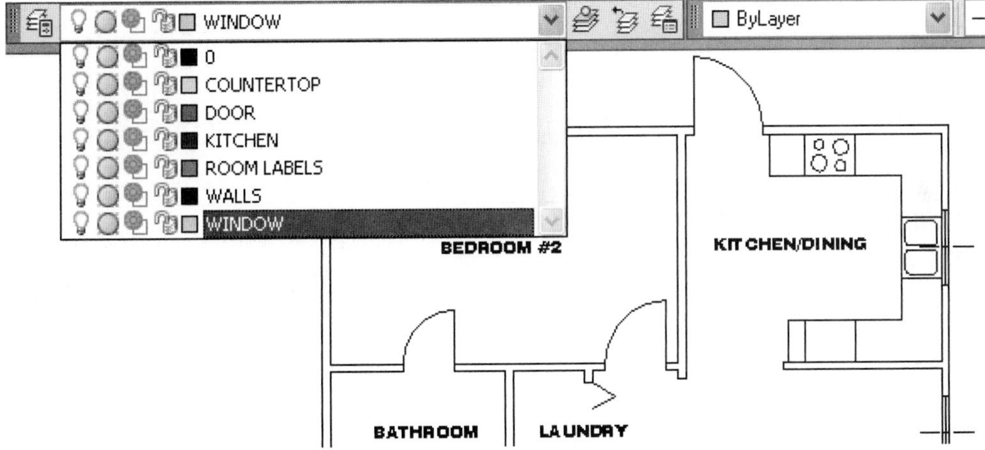

Figure 7.41

Tip: For the blocks, see the illustration in the following image for supplying the correct information in the Quick Select dialog box; change the Object type to Block Reference. Change the Properties to Name. Click the down arrow next to the Value heading and select the desired block from the list provided. Then, after clicking the OK button and all desired blocks are selected, change the highlighted blocks to the correct layer. Follow this procedure for selecting the Window, Countertop, Range, Sink, Door, Louver, and Refrigerator blocks.

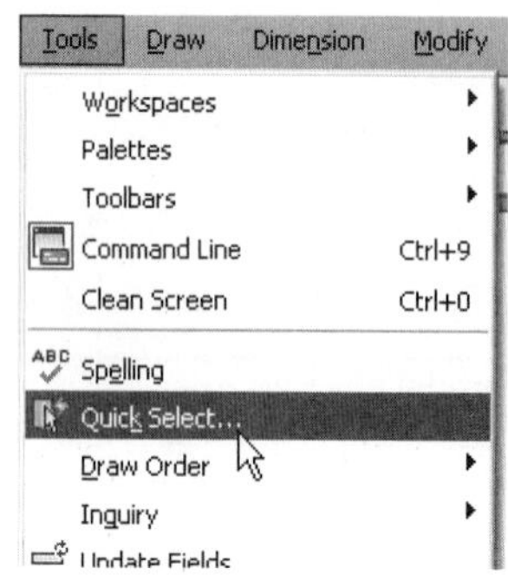

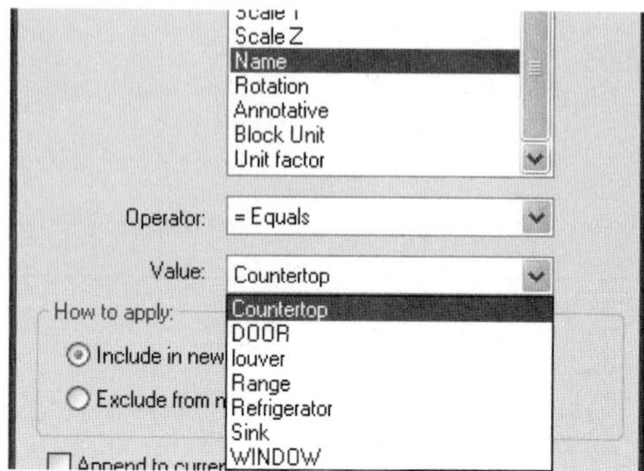

Figure 7.42

DOUBLE-CLICK EDIT ON ANY OBJECT

Double-clicking on any object provides you with a quick way of launching the Properties Palette or other related dialog boxes depending on the object type. For instance, double-clicking on a line segment launches the Properties Palette, which displays information about the line. Double-clicking on a text object displays the Edit Text dialog box, allowing you to enter or delete words. Whenever the Properties Palette is launched and you want to modify a different object, first press ESC to deselect the current object. You may also want to dismiss the Properties Palette.

Try It!: Open the drawing file 07_Double Click Edit. In the following image, double-click on the magenta centerline and the Properties Palette launches with information about the line. Press ESC and dismiss the Properties Palette. Double-click on the word "BLOCK" to launch the Edit Text function. Press ESC to exit the edit function when finished. Double-click on the sentence "THE OBJECT SHOWN ABOVE IS A WINDOW SYMBOL" to launch the Multiline Text Editor dialog box. Press ESC to dismiss the dialog box when finished. Double-click on the hatch pattern and to launch the Hatch Edit dialog box. Pressing ESC dismisses this dialog box when finished. Continue by double-clicking on the circle, dimension, and rectangle and observe the type of dialog box launched through this method. Press ESC to dismiss the dialog boxes when finished with each operation.

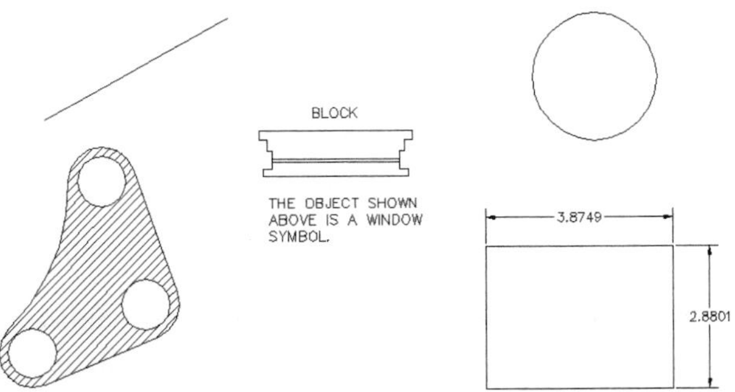

Figure 7.43

MATCHING THE PROPERTIES OF OBJECTS

At times objects are drawn on the wrong layers or the wrong color scheme is applied to a group of objects. Text objects are sometimes drawn with an incorrect text style. You have just seen how the Properties Palette and the Layer Control box provide quick ways to fix such problems. Yet another tool is available for changing the properties of objects—the MATCHPROP command. Choose this command from the Standard toolbar, as shown on the left in the following image, or from the Modify pull-down menu, as shown on the right in the following image.

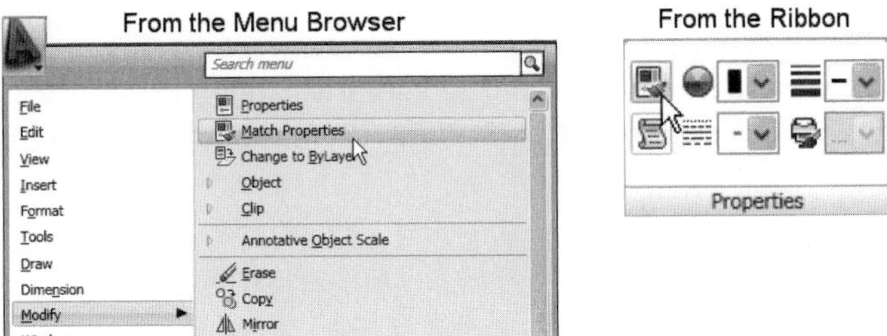

Figure 7.44

Try It!: Open the drawing file 07_Matchprop Flange. Use the illustration in the following image and the command sequence below for performing this task.

When you start the command, a source object is required. This source object transfers all its current properties to other objects designated as "Destination Objects." As shown on the left in the following image, the flange requires the object lines located at "B," "C," "D," and "E" to be converted to hidden lines. Using the MATCHPROP command, select the existing hidden line "A" as the source object. Notice the appearance of the Match Properties icon. Select lines "B" through "E" as the destination objects using this icon.

```
Command: MA (For MATCHPROP)
Select source object: (Select the hidden line at "A")
Current active settings: Color Layer Ltype Ltscale Lineweight
    Thickness
```

```
PlotStyle Text Dim Hatch Polyline Viewport Table Material Shadow
    display Multileader
Select destination object(s) or [Settings]: (Select line "B")
Select destination object(s) or [Settings]: (Select line "C")
Select destination object(s) or [Settings]: (Select line "D")
Select destination object(s) or [Settings]: (Select line "E")
Select destination object(s) or [Settings]: (Press ENTER to exit
    this command)
```

The results appear in the flange illustrated on the right in the following image, where the continuous object lines were converted to hidden lines. Not only did the linetype property get transferred, but the color, layer, lineweight, and linetype scale information did as well.

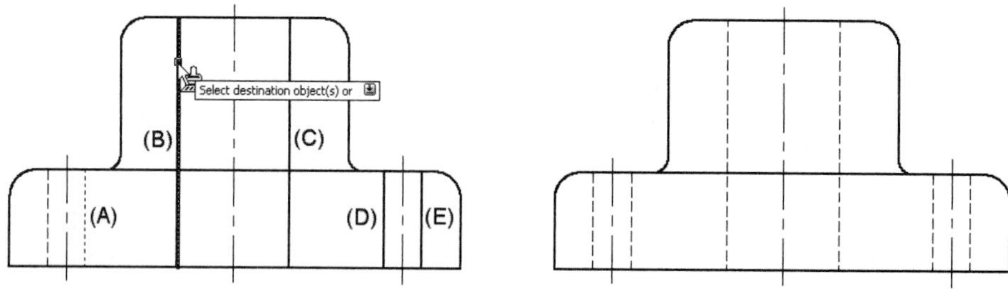

Figure 7.45

To get a better idea of what object properties are affected by the MATCHPROP command, reenter the command, pick a source object, and instead of picking a destination object immediately, enter S for Settings. This displays the Property Settings dialog box in the following image.

```
Command: MA (For MATCHPROP)
Select source object: (Select the hidden line at "A" in the
    previous image)
Current active settings: Color Layer Ltype Ltscale Lineweight
    Thickness
PlotStyle Dim Text Hatch Polyline Viewport Table Material Shadow
    display Multileader
Select destination object(s) or [Settings]: S (For Settings; this
    displays the Property Settings dialog box in the following
    image)
```

Any box with a check displayed in it transfers that property from the source object to all destination objects. If you need to transfer only the layer information and not the color and linetype properties of the source object, remove the checks from the Color and Linetype properties before you select the destination objects. This prevents these properties from being transferred to any destination objects.

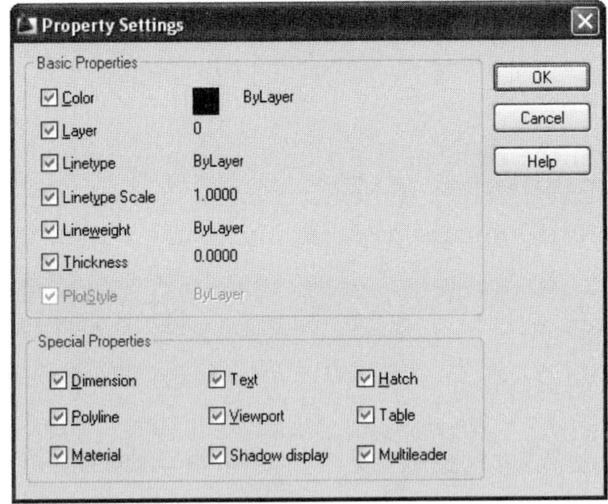

Figure 7.46

MATCHING DIMENSION PROPERTIES

The MATCHPROP command can control special properties of dimensions, text, hatch patterns, polylines, viewports, and tables. The Dimension property will be featured next as shown in the following image. (Even though the topic of dimensions will not be covered until chapter 11, this concept is introduced here.)

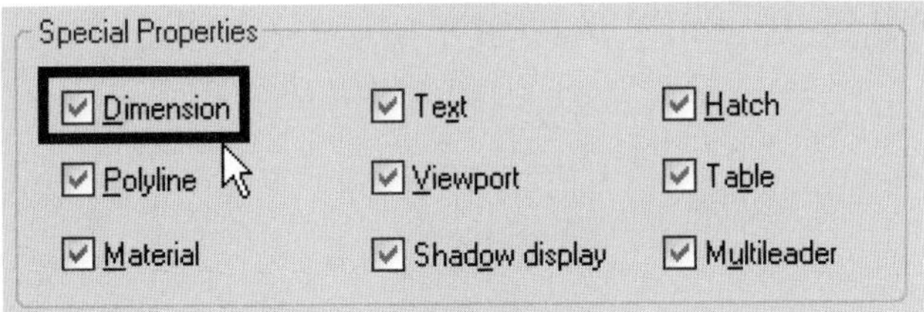

Figure 7.47

Try It!: Open the drawing file 07_Matchprop Dim. The following image shows two blocks: the block assigned a dimension value of 46.6084 was dimensioned with the METRIC dimension style with the Arial font applied. The block assigned a dimension value of 2.3872 was dimensioned with the STANDARD dimension style with the TXT font applied. Both blocks need to be dimensioned with the METRIC dimension style. Issue the MATCHPROP command and select the 46.6084 dimension as the source object and then select the 2.3872 dimension as the destination object.

```
Command: MA (For MATCHPROP)
Select source object: (Select the dimension at "A" in the
    following image)
Current active settings: Color Layer Ltype Ltscale Lineweight
    Thickness
PlotStyle Dim Text Hatch Polyline Viewport Table Material Shadow
    display Multileader
```

```
Select destination object(s) or [Settings]: (Select the dimension
    at "B")
Select destination object(s) or [Settings]: (Press ENTER to exit
    this command)
```

The results are shown on the right in the following image, with the METRIC dimension style applied to the STANDARD dimension style through the use of the MATCHPROP command. Because the text font was associated with the dimension style, it also changed in the destination object.

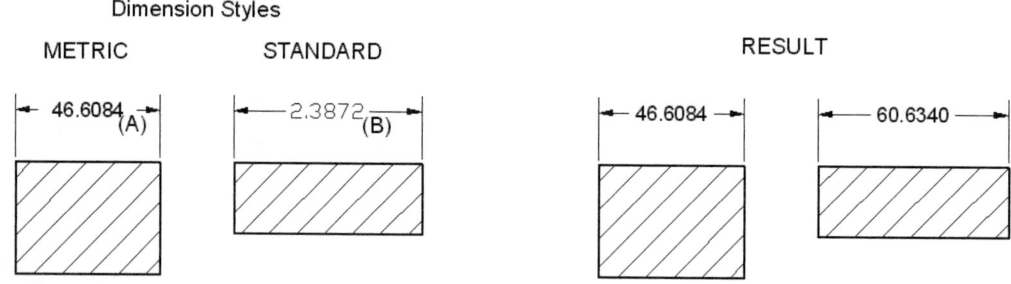

Figure 7.48

MATCHING TEXT PROPERTIES

The following is an example of how the MATCHPROP command affects a text object with the Text property shown in the following image.

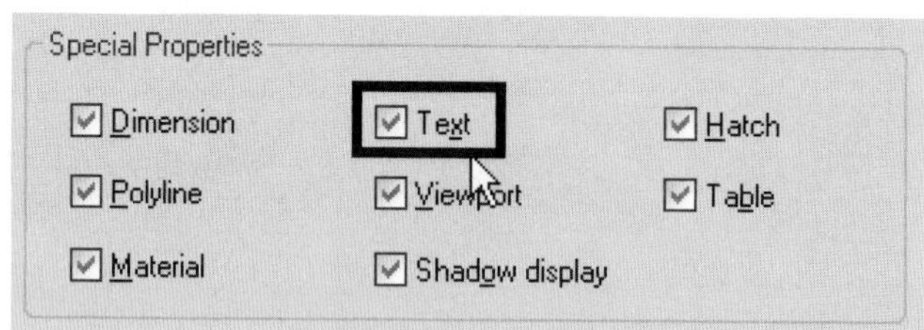

Figure 7.49

Try It!: Open the drawing file 07_Matchprop Text. The following image shows two text items displayed in different fonts. The text "Coarse Knurl" at "A" was constructed with a text style called Arial. The text "Medium Knurl" at "B" was constructed with the default text style called STANDARD. Use the following command sequence to match the STANDARD text style with the Arial text style using the MATCHPROP command.

```
Command: MA (For MATCHPROP)
Select source object: (Select the text at "A")
Current active settings: Color Layer Ltype Ltscale Lineweight
    Thickness
PlotStyle Dim Text Hatch Polyline Viewport Table Material Shadow
    display Multileader
```

```
Select destination object(s) or [Settings]: (Select the text
     at "B")
Select destination object(s) or [Settings]: (Press ENTER to exit
     this command)
```

The result is shown on the right in the following image. Both text items now share the same text style. Notice that the text string stays intact when text properties are matched. Only the text style of the source object is applied to the destination object.

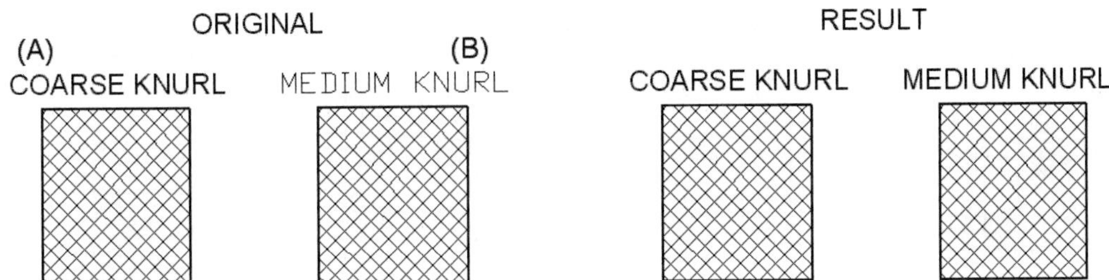

Figure 7.50

MATCHING HATCH PROPERTIES

A source hatch object can also be matched to a destination pattern with the MATCHPROP command and the Hatch property as shown in the following image (Even though the topic of hatching will not be covered until chapter 9, this concept is introduced here.)

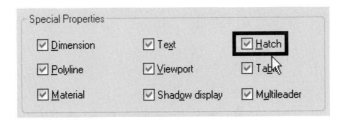

Special Properties

Dimension | Text | Hatch
Polyline | Viewport | Table
Material | Shadow display | Multileader

Figure 7.51

Try It!: Open the drawing file 07_Matchprop Hatch. In the following image, the crosshatch patterns at "B" and "C" are at the wrong angle and scale. They should reflect the pattern at "A" because it is the same part. Use the MATCHPROP command, select the hatch pattern at "A" as the source object, and select the patterns at "B" and "C" as the destination objects.

```
Command: MA (For MATCHPROP)
Select source object: (Select the hatch pattern at "A")
Current active settings: Color Layer Ltype Ltscale Lineweight
     Thickness
PlotStyle Dim Text Hatch Polyline Viewport Table Material Shadow
     display Multileader
Select destination object(s) or [Settings]: (Select the hatch
     pattern at "B")
```

```
Select destination object(s) or [Settings]: (Press ENTER to exit
    this command)
```

The results appear on the right in the following image, where the source hatch pattern property was applied to all destination hatch patterns.

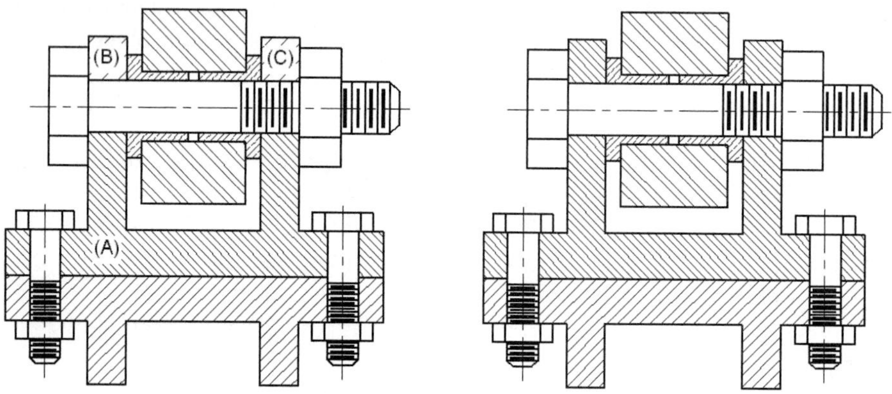

Figure 7.52

MATCHING POLYLINE PROPERTIES

A source polyline object can also be matched to a destination pattern with the MATCHPROP command and the Polyline property as shown in the following image.

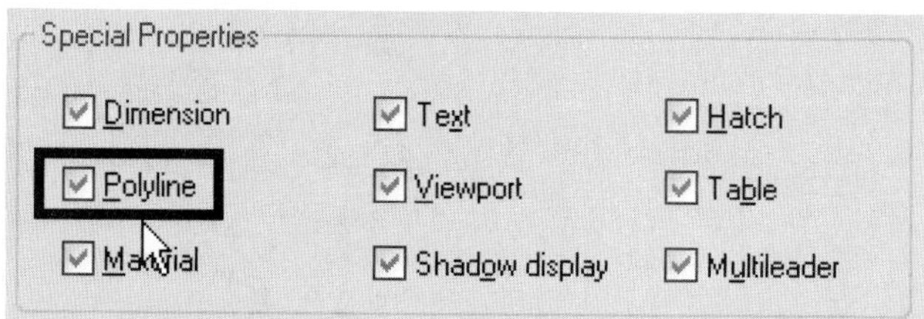

Figure 7.53

Try It!: Open the drawing file 07_Matchprop Pline. In the following image, the polyline at "B" is at the wrong width. It should match the width of the polyline at "A." Using the MATCHPROP command, select the polyline at "A" as the source object, and select the polyline at "B" as the destination object.

```
Command: MA (For MATCHPROP)
Select source object: (Select the polyline at "A")
Current active settings: Color Layer Ltype Ltscale Lineweight
    Thickness
PlotStyle Dim Text Hatch Polyline Viewport Table Material Shadow
    display Multileader
Select destination object(s) or [Settings]: (Select the polyline
    at "B")
```

```
Select destination object(s) or [Settings]: (Press ENTER to exit
    this command)
```

The results appear on the right in the following image, where the source polyline property (width, in this example) was applied to the destination polyline.

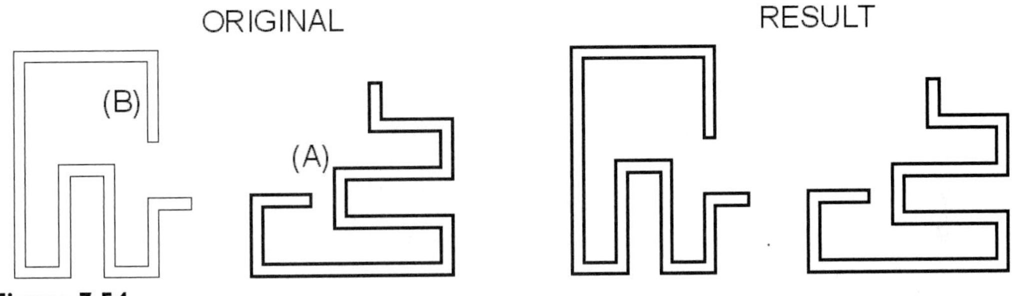

Figure 7.54

MATCHING VIEWPORT PROPERTIES

A source viewport object can also be matched to a destination viewport with the MATCHPROP command and the Viewport property as shown in the following image. Even though the topic of viewports will not be covered until chapter 19, this concept is introduced here. Viewports are used to arrange images of a drawing in layout mode before they are plotted out. Matching the properties of a viewport can transfer the scale of one viewport to another. The following Try It! exercise illustrates this.

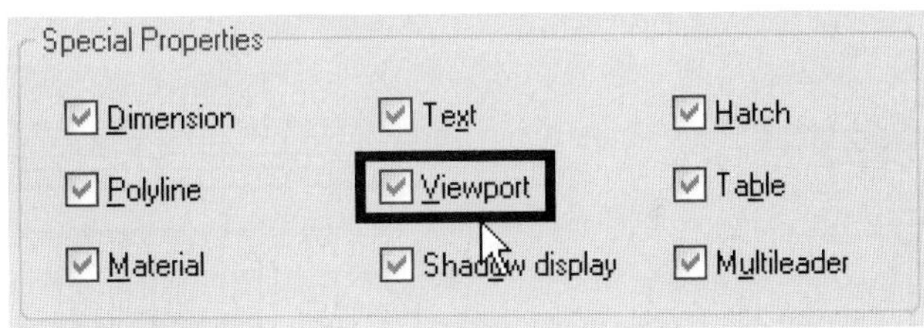

Figure 7.55

Try It!: Open the drawing file 07_Matchprop Viewport. In the following image, the viewport image at "A" is at the correct scale. The viewport at "B" should have the same scale as that of viewport "A." Using the MATCHPROP command, select the viewport at "A" as the source object, and select the viewport at "B" as the destination object.

```
Command: MA (For MATCHPROP)
Select source object: (Select the edge of the viewport at "A")
Current active settings: Color Layer Ltype Ltscale Lineweight
    Thickness
PlotStyle Dim Text Hatch Polyline Viewport Table Material Shadow
    display Multileader
Select destination object(s) or [Settings]: (Select the viewport
    at "B")
```

`Select destination object(s) or [Settings]:` *(Press* ENTER *to exit this command)*

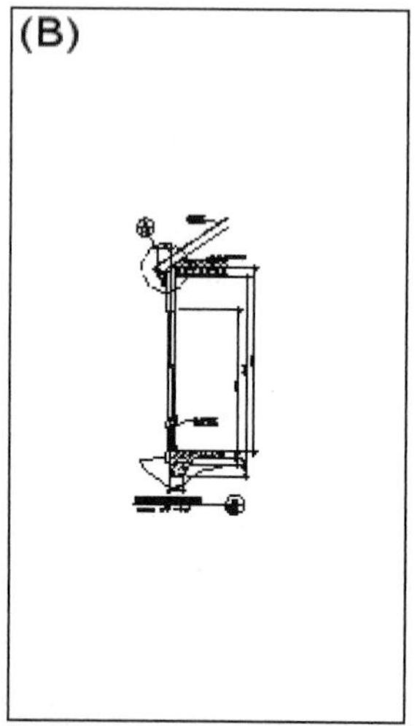

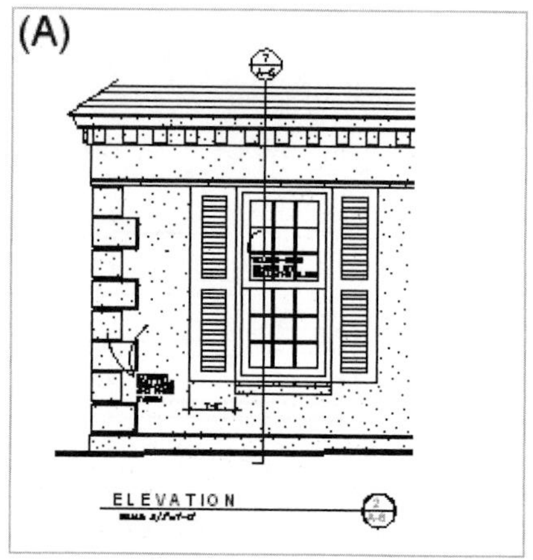

Figure 7.56

The results appear in the following image, where the source viewport property (scale of the image) was applied to the destination viewport. Notice also that the layer of the viewport "B" was changed from 0 to Vports.

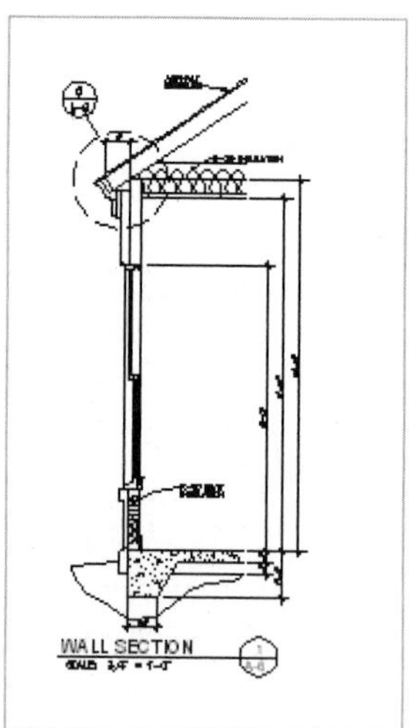

Figure 7.57

MATCH PROPERTIES AND TABLES

A source table object can be matched to a destination table using the Table property, as shown in the following image. When one table is matched with another, the table style of the source table updates the table style of the destination table.

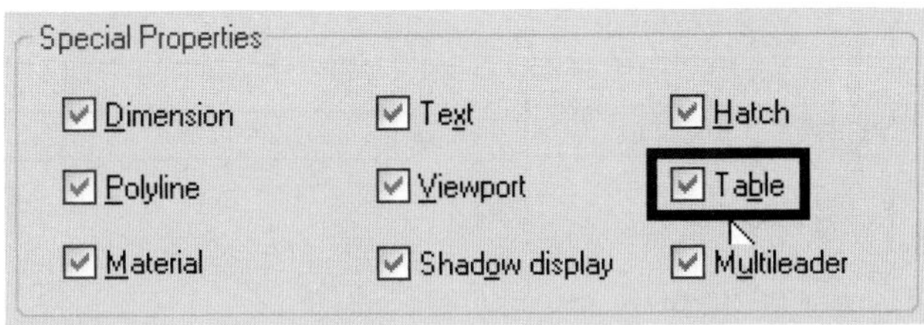

Figure 7.58

Try It!: Open the drawing file 07_Matchprop Table. In the following image, one table has the title strip originating at the top and the other table title originates at the bottom. Using the MATCHPROP command, select the table on the left as the source object, and select the table on the right as the destination object.

```
Command: MA (For MATCHPROP)
Select source object: (Select the table on the left in the
    following image)
Current active settings: Color Layer Ltype Ltscale Lineweight
    Thickness
PlotStyle Dim Text Hatch Polyline Viewport Table Material Shadow
    display Multileader
Select destination object(s) or [Settings]: (Select the table on
    the right in the following image)
Select destination object(s) or [Settings]: (Press ENTER to exit
    this command)
```

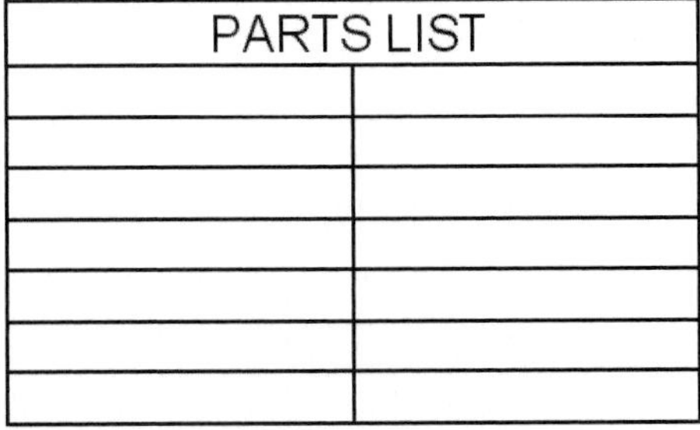

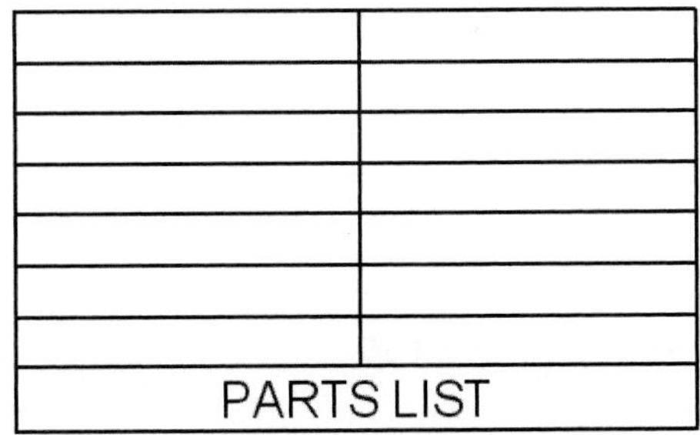

Figure 7.59

In the following image, notice how the table on the right changes to the downward direction arrangement as controlled by the table style.

PARTS LIST	

PARTS LIST	

Figure 7.60

Note: You may have to move the table into position.

TUTORIAL EXERCISE: 07_MODIFY-EX.DWG

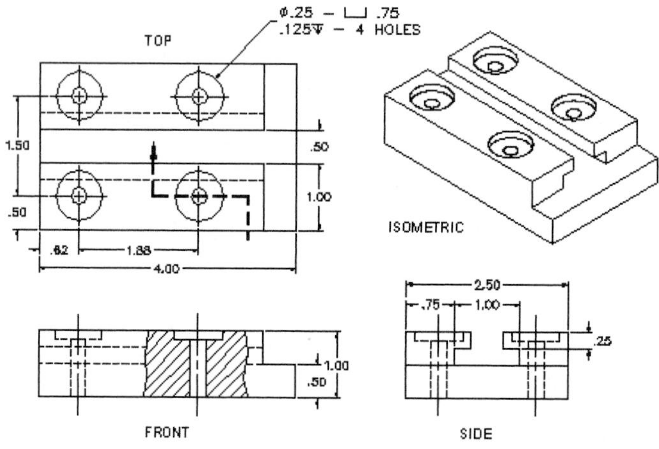

Figure 7.61

PURPOSE

This tutorial exercise is designed to change the properties of existing objects displayed in the above image.

SYSTEM SETTINGS

Since this drawing is provided on the CD, open an existing drawing file called "07_Modify-Ex." Follow the steps in this tutorial for changing various objects to the correct layer, text style, and dimension style.

LAYERS

Layers have already been created in this drawing.

SUGGESTED COMMANDS

Begin this tutorial by using the Properties Palette to change the isometric object to a different layer. Continue by changing the text height and layer of the view identifiers (FRONT, TOP, SIDE, ISOMETRIC). The MATCHPROP command will be used to transfer the properties from one dimension to another, one text style to another, and one hatch pattern to another. The Layer Control box will be used to change the layer of various objects located in the Front and Top views.

Step 1

Loading this drawing displays the objects in a page layout called "Orthographic Views." Page layout is where the drawing will be plotted out in the future. This name is present next to the Model tab in the bottom portion of the drawing screen. Since a majority of changes will be made in Model mode, click on the Model tab near the bottom of the display screen. Your drawing will appear similar to the following image.

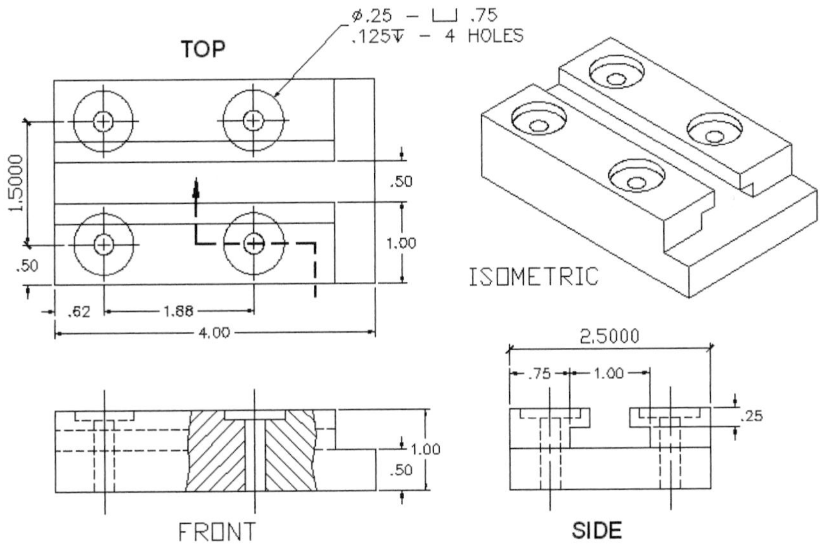

Figure 7.62

Step 2

While in the Model environment, select all lines that make up the isometric view in the following image. You can accomplish this by using the Window mode from the Command prompt. If you accidentally select the word ISOMETRIC, de-select this word. Activate the Properties Palette and click in the Layer field. This displays the current layer the objects are drawn on (DIM) in the following image. Click the down arrow to display the other layers and pick the OBJECT layer. This changes all selected objects that make up the isometric view to the OBJECT layer. Press ESC to remove the object highlight and the grips from the drawing.

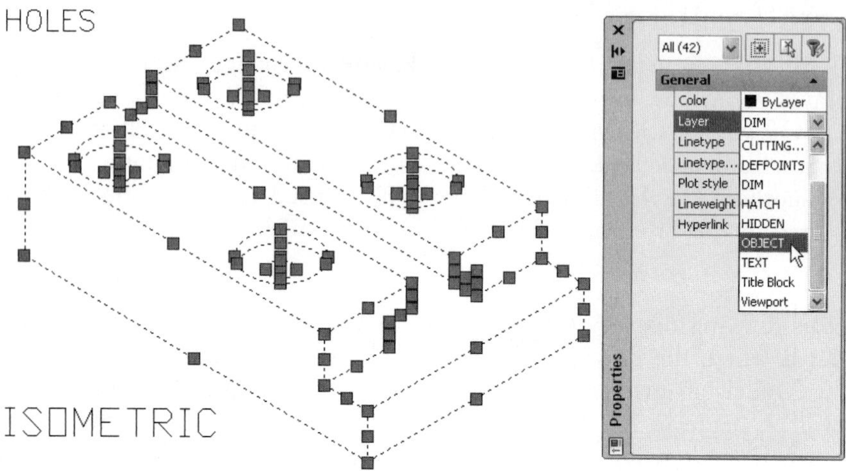

Figure 7.63

Step 3

Select the view titles (FRONT, TOP, SIDE, ISOMETRIC). These text items need to be changed to a height of 0.15 and the TEXT layer. With all four text objects highlighted, and the Properties Palette already active, click in the Layer field, and click the down arrow to change the selected objects to the TEXT layer in the following image.

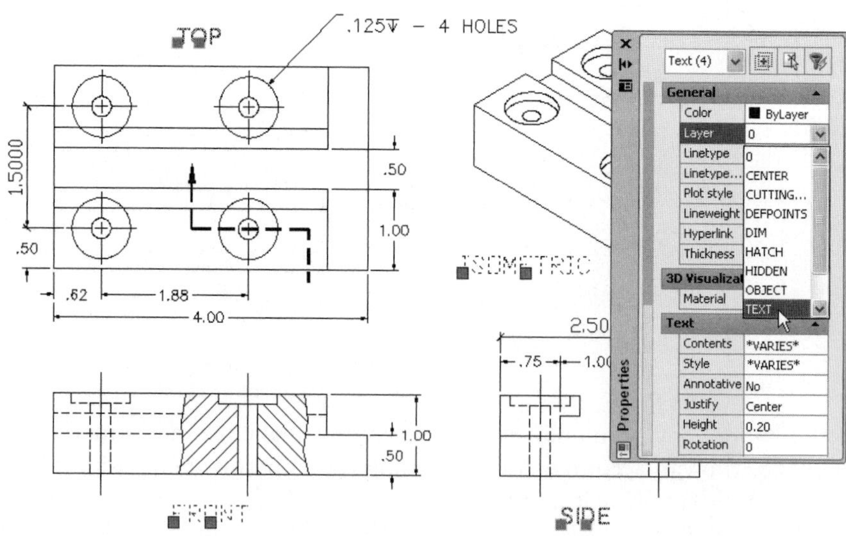

Figure 7.64

With the text still highlighted, change the text height in the Properties Palette from 0.20 to a new value of 0.15. Pressing ENTER after making this change automatically updates all selected text objects to this new height in the following image. Dismiss this dialog box when finished. Press ESC to remove the object highlight and the grips from the drawing.

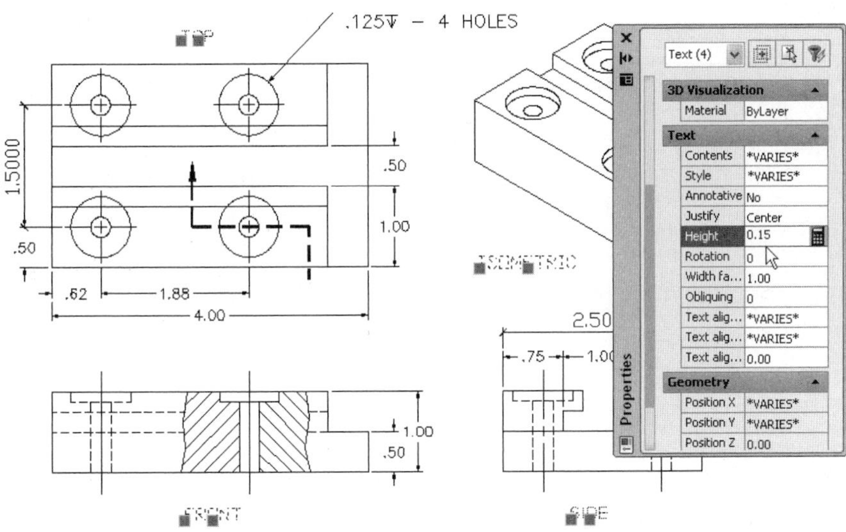

Figure 7.65

Step 4

When you examine the view titles, TOP and SIDE are in one text style while FRONT and ISOMETRIC are in another. To remain consistent in the design process, you should make sure all text identifying the view titles has the same text style as TOP. Activate the MATCHPROP command. Click the text object TOP as the source object. When the Match Property icon appears in the following image, select ISOMETRIC and FRONT as the destination objects. All properties associated with the TOP text object are transferred to ISOMETRIC and FRONT, including the text style.

```
Command: MA (for MATCHPROP)
Select source object: (Select the text object "TOP," which should
    highlight in the following image)
Current active settings: Color Layer Ltype Ltscale Lineweight
    Thickness
PlotStyle Dim Text Hatch Polyline Viewport Table Material Shadow
    display Multileader
Select destination object(s) or [Settings]: (Select the text
    object "ISOMETRIC")
Select destination object(s) or [Settings]: (Select the text
    object "FRONT")
Select destination object(s) or [Settings]: (Press ENTER to exit
    this command)
```

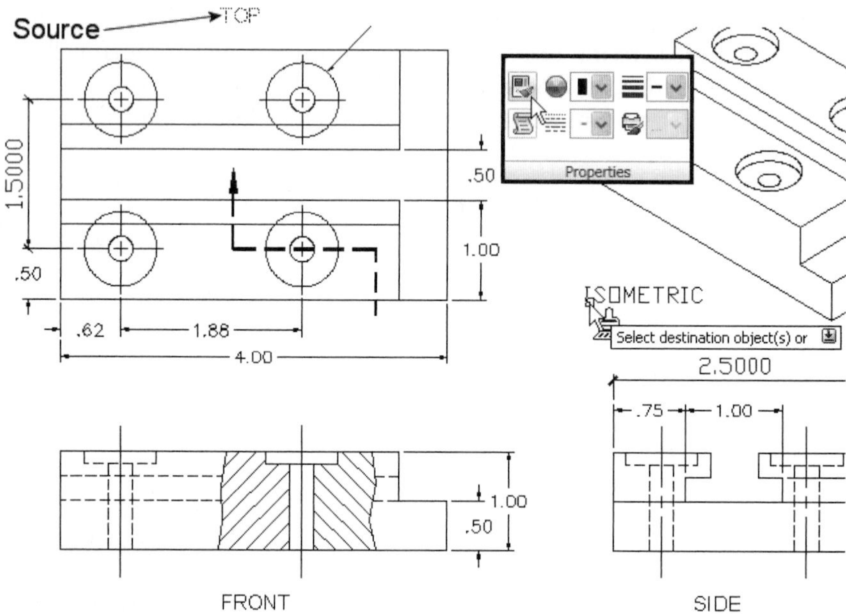

Figure 7.66

Step 5

Notice the dimensions in this drawing. Two dimensions stand out above the rest (the 1.5000 vertical dimension in the Top view and the 2.5000 horizontal dimension in the Side view). Again, to remain consistent in the design process, you should make sure all dimensions have the same appearance (dimension text height, number of decimal places, whether they are broken inside instead of placed above the dimension line). Activate the MATCHPROP command again by clicking on the button in the Ribbon. Click the 4.00 horizontal dimension in the Top view as the source object. When the Match Property icon appears, as in the following image, select the 1.5000 and 2.5000 vertical dimensions as the destination objects. All dimension properties associated with the 4.00 dimension are transferred to the 1.5000 and 2.5000 dimensions.

```
Command: MA (for MATCHPROP)
Select source object: (Select the 4.00 dimension, which should
    highlight)
Current active settings: Color Layer Ltype Ltscale Lineweight
    Thickness
PlotStyle Dim Text Hatch Polyline Viewport Table Material Shadow
    display Multileader
Select destination object(s) or [Settings]: (Select the 2.5000
    dimension at "A")
Select destination object(s) or [Settings]: (Select the 1.5000
    dimension at "B")
Select destination object(s) or [Settings]: (Press ENTER to exit
    this command)
```

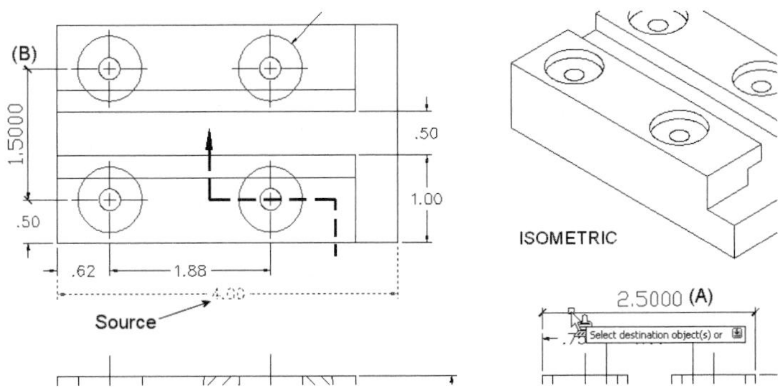

Figure 7.67

Step 6

In the Front view, an area is crosshatched. However, both sets of crosshatching lines need to be drawn in the same direction, rather than opposing each other. Activate the MATCHPROP command again by clicking on the button in the Ribbon. Click the left hatch pattern as the source object. When the Match Property icon appears, as in the following image, select the right hatch pattern as the destination object. All hatch properties associated with the left hatch pattern are transferred to the right hatch pattern.

```
Command: MA (for MATCHPROP)
Select source object: (Select the left hatch pattern, which
    should highlight)
Current active settings: Color Layer Ltype Ltscale Lineweight
    Thickness
PlotStyle Dim Text Hatch Polyline Viewport Table Material Shadow
    display Multileader
Select destination object(s) or [Settings]: (Pick the right hatch
    pattern)
Select destination object(s) or [Settings]: (Press ENTER to exit
    this command.)
```

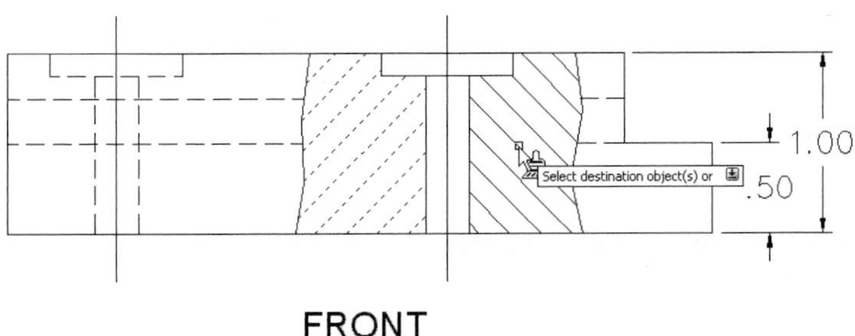

Figure 7.68

Step 7

Your display should appear similar to the following image. All view titles (FRONT, TOP, SIDE, ISOMETRIC) share the same text style and are at the same height. All dimensions share the same parameters and text orientation. Both crosshatch patterns are drawn in the same direction.

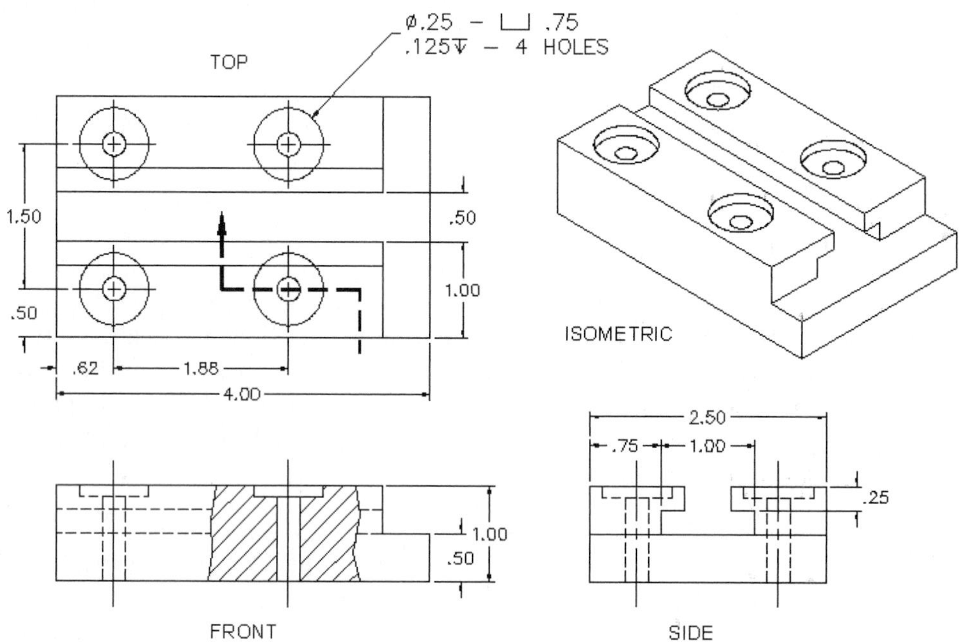

Figure 7.69

Step 8

A different method will now be used to change the specific layer properties of objects. The two highlighted lines in the Top view in the following image were accidentally drawn on Layer OBJECT and need to be transferred to the HIDDEN layer. With the lines highlighted, click in the Layer Control box to display all layers. Click on the HIDDEN layer to change the highlighted lines to the HIDDEN layer. Press ESC to remove the object highlight and the grips from the drawing.

Note: You may need to Regen the drawing in order to get the dashes to appear in the hidden lines.

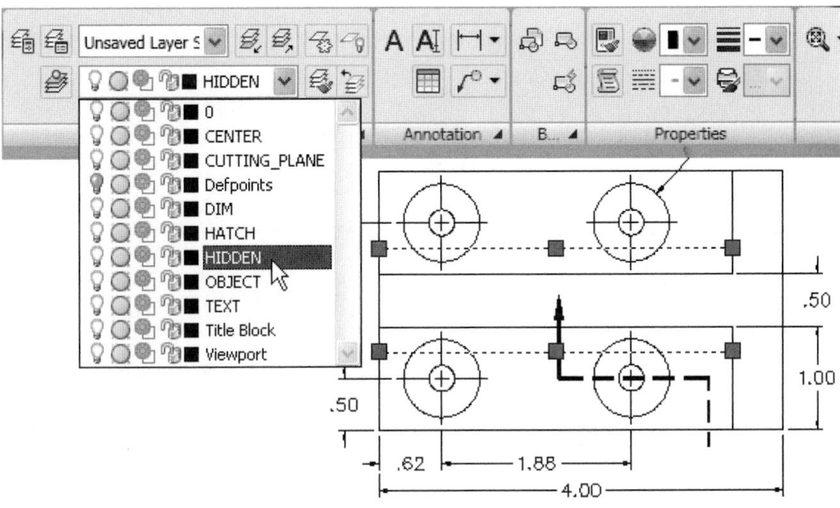

Figure 7.70

Step 9

The two highlighted lines in the Front view in the following image were accidentally drawn on the TEXT Layer and need to be transferred to the CENTER layer. With the lines highlighted, click in the Layer Control box to display all layers. Click on the CENTER layer to change the highlighted lines to the CENTER layer. Press ESC to remove the object highlight and the grips from the drawing.

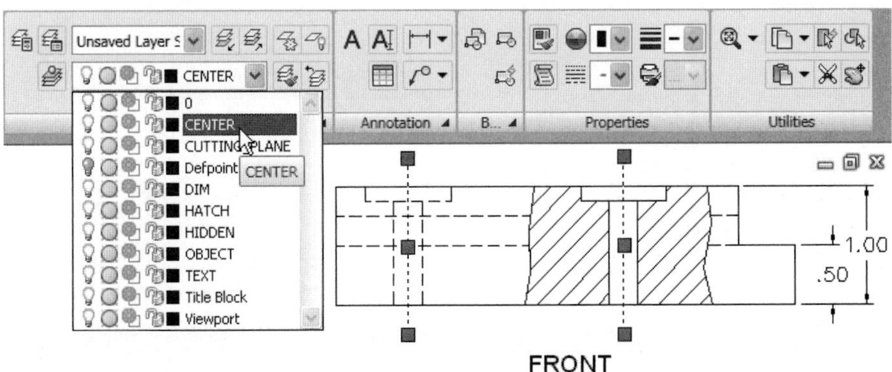

Figure 7.71

Step 10

Your display should appear similar to the following image. Notice the hidden lines in the Top view and the centerlines in the Front view.

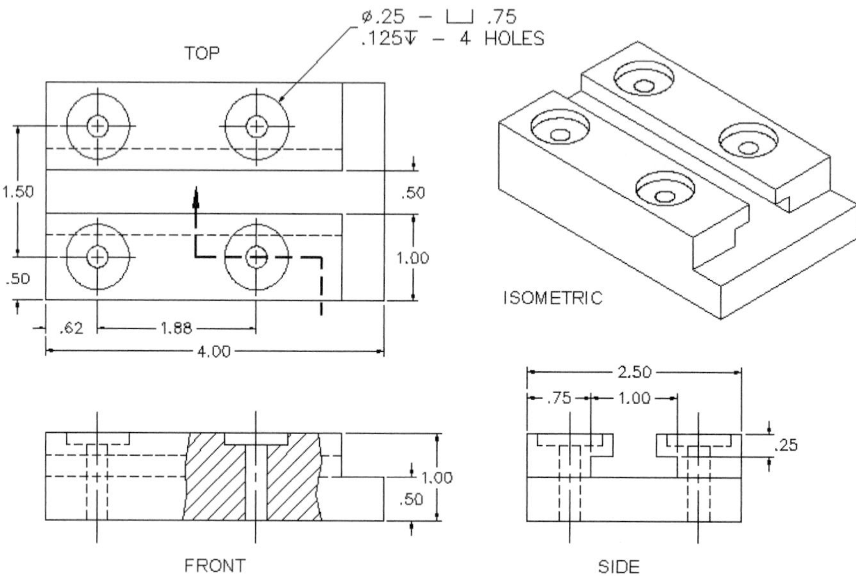

Figure 7.72

Step 11

Click on the Orthographic Views tab. Select the rectangular viewport and change this object's layer to "Viewport" in the following image.

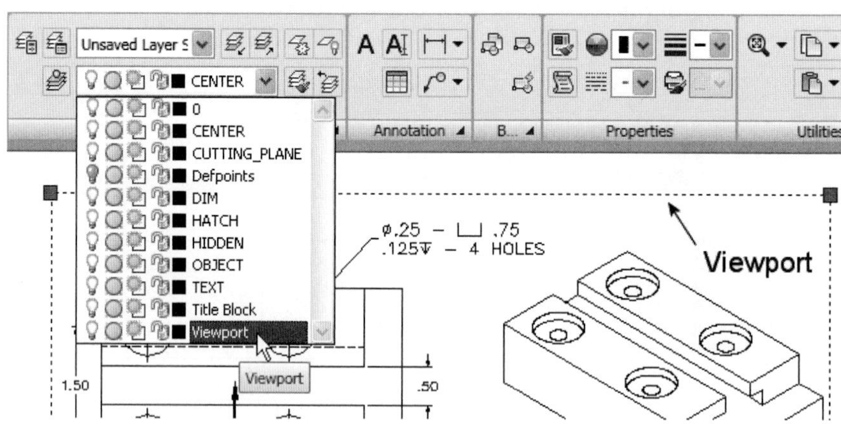

Figure 7.73

Step 12

Turn off the Viewport layer. Your display should appear similar to the following image. This completes this tutorial exercise.

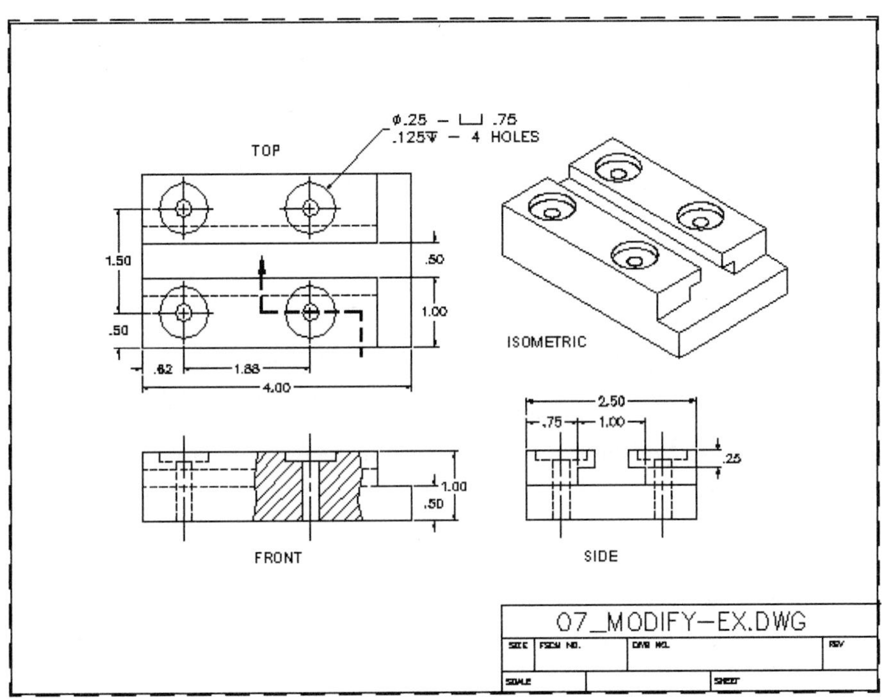
Ø.25 – ⌴ .75
.125↧ – 4 HOLES
TOP
1.50
.50
.50
1.00
.62
1.88
4.00
ISOMETRIC
2.50
.75
1.00
.25
1.00
.50
FRONT
SIDE
07_MODIFY-EX.DWG

Figure 7.74

CHAPTER 8

Shape Description/Multiview Projection

Before any object is made in production, some type of drawing needs to be created. This is not just any drawing, but rather an engineering drawing consisting of overall object sizes with various views of the object organized on the computer screen. This chapter introduces the topic of shape description or how many views are really needed to describe an object. The art of multiview projection includes methods of constructing one-view, two-view, and three-view drawings using AutoCAD commands.

SHAPE DESCRIPTION

Begin constructing an engineering drawing by first analyzing the object being drawn. To accomplish this, describe the object by views or by how an observer looks at the object. The illustration on the left in the following image shows a simple wedge; this object can be viewed at almost any angle to get a better idea of its basic shape. However, to standardize how all objects are to be viewed, and to limit the confusion usually associated with complex multiview drawings, some standard method of determining how and where to view the object must be exercised.

Even though the simple wedge is easy to understand because it is currently being displayed in picture or isometric form, it would be difficult to produce this object because it is unclear what the sizes of the front and top faces are. Illustrated on the right in the following image are six primary ways or directions in which to view an object. The Front view begins the shape description and is followed by the Top view and Right Side view. Continuing on, the Left Side view, Back view, and Bottom view complete the primary ways to view an object.

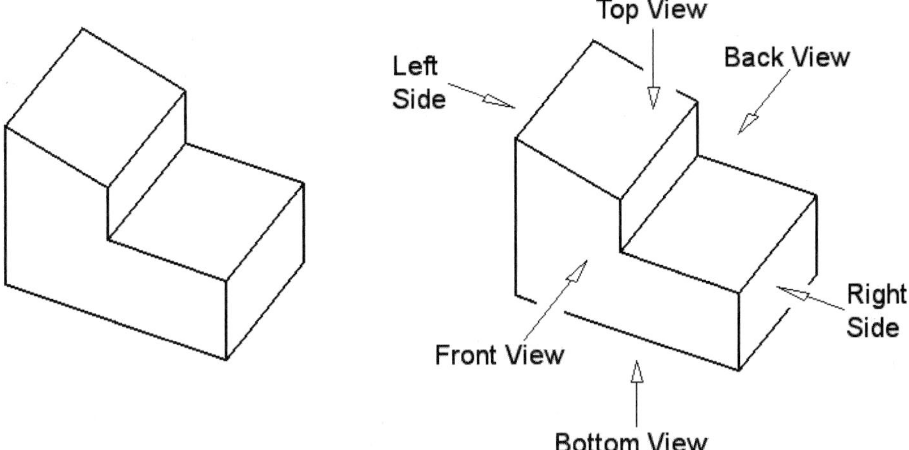

Figure 8.1

Now that the primary ways of viewing an object have been established, the views need to be organized to promote clarity and have the views reference one another. Imagine the simple wedge positioned in a transparent glass box similar to the illustration on the left in the following image. With the entire object at the center of the box, the sides of the box represent the ways to view the object. Images of the simple wedge are projected onto the glass surfaces of the box.

With the views projected onto the sides of the glass box, we must now prepare the views to be placed on a two-dimensional (2D) drawing screen. For this to be accomplished, the glass box, which is hinged, is unfolded as shown on the right in the following image. All folds occur from the Front view, which remains stationary.

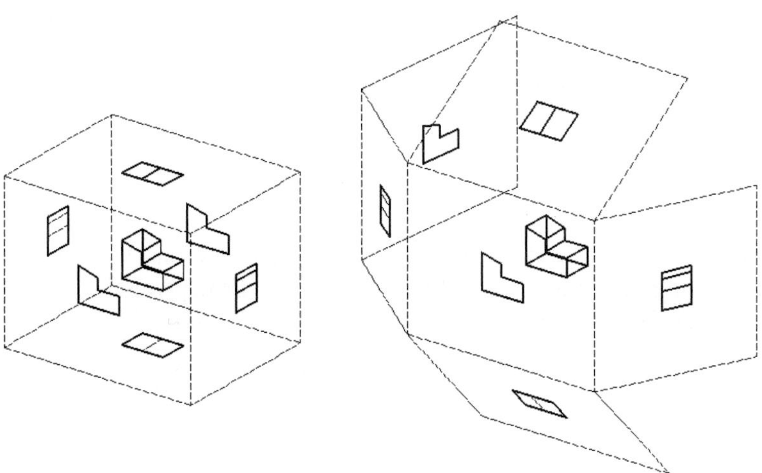

Figure 8.2

The following image shows the views in their proper alignment to one another. However, this illustration is still in a pictorial view. These views need to be placed flat before continuing.

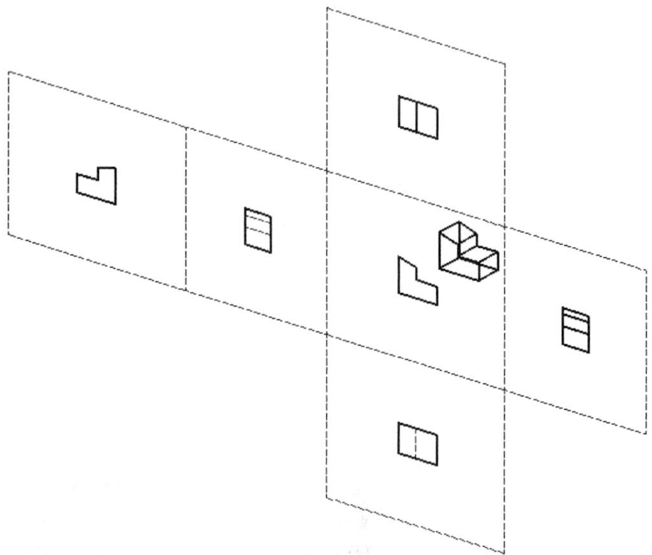

Figure 8.3

The result of the glass box being completely unfolded and laid flat is illustrated on the left in the following image. The Front view becomes the main view, with other views placed in relation to it. Above the Front view is the Top view. To the right of the Front view is the Right Side view. To the left of the Front view is the Left Side view, followed by the Back view. Underneath the Front view is the Bottom view. This becomes the standard method of laying out the views needed to describe an object. But are all the views necessary? Upon closer inspection we find that, except for being a mirror image, the Front and Back views are identical. The Top and Bottom views appear similar, as do the Right and Left Side views. One very important rule to follow in multiview objects is to select only those views that accurately describe the object and discard the remaining views.

The complete multiview drawing of the simple wedge is illustrated on the right in the following image. Only the Front, Top, and Right Side views are needed to describe this object. When laying out views, remember that the Front view is usually the most important—it holds the basic shape of the object being described. Directly above the Front view is the Top view, and to the right of the Front view is the Right Side view. All three views are separated by a space of varying size. This space is commonly called a dimension space because it is a good area in which to place dimensions describing the size of the object. The space also acts as a separator between views; without it, the views would touch one another, making them difficult to read and interpret. The minimum distance of this space is usually 1.00 unit.

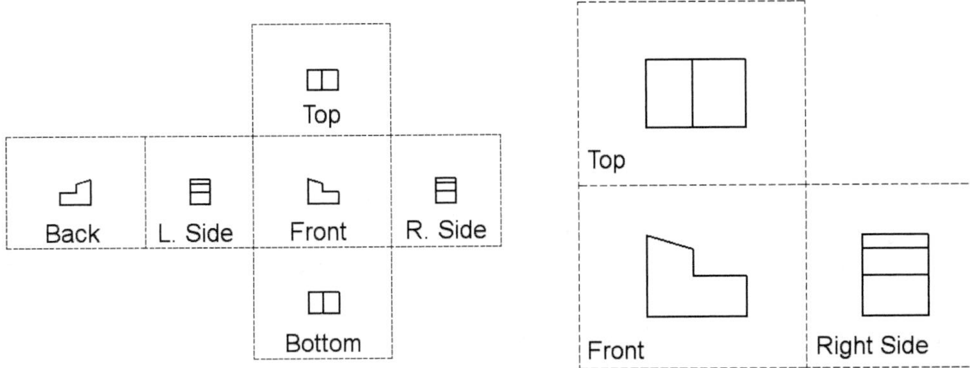

Figure 8.4

RELATIONSHIPS BETWEEN VIEWS

Some very interesting and important relationships are set up when the three views are placed in the configuration illustrated on the left in the following image. Notice that because the Top view is directly above the Front view, both share the same width dimension. The Front and Right Side views share the same height. The relationship of depth on the Top and Right Side views can be explained with the construction of a 45°-projection line at "A" and the projection of the lines over and down or vice versa to get the depth. Another principle illustrated by this example is that of projecting lines up, over, across, and down to create views. Editing commands such as ERASE and TRIM are then used to clean up unnecessary lines.

With the three views identified as shown on the right in the following image, the only other step—and it is an important one—is to annotate the drawing, or add dimensions to the views. With dimensions, the object drawn in multiview projection can now be produced. Even though this has been a simple example, the methods of multiview projection work even for the most difficult and complex objects.

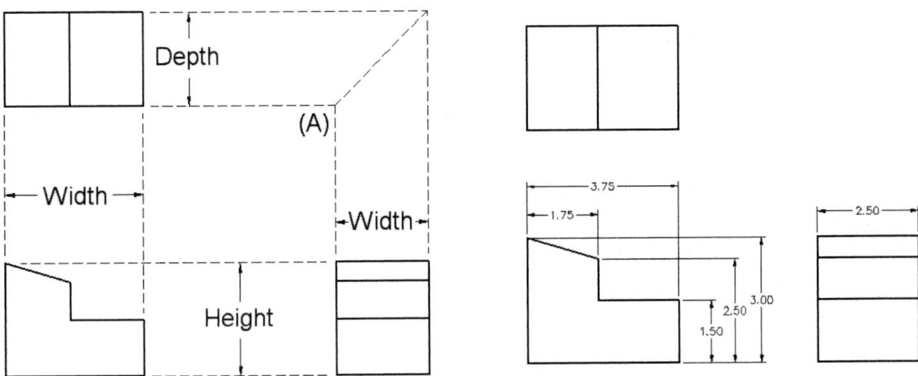

Figure 8.5

REVIEW OF LINETYPES AND CONVENTIONS

At the heart of any engineering drawing is the ability to assign different types of lines to convey meaning to the drawing. When plotted, all lines of a drawing are dark; border and title block lines are the thickest. Object lines outline visible features of a drawing and are made thick and dark, but not as thick as a border line. Features that are behind a surface but are important to the description of the object are identified by a hidden line. This line is a series of dashes 0.12 units in length with a spacing of 0.06 units. Centerlines identify the centers of circular features, such as holes, or show that a hidden feature in one view is circular in another. The centerline consists of a series of long and short dashes. The short dash measures approximately 0.12 units, whereas the long dash may vary from 0.75 to 1.50 units. A gap of 0.06 units separates dashes. Study the examples of these lines, shown on the left in the following image.

The object shown on the right in the following image is a good illustration of the use of phantom lines in a drawing. Phantom lines are especially useful where motion is applied. The arm on the right is shown by standard object lines consisting of the continuous linetype. To show that the arm rotates about a center pivot, the identical arm is duplicated through the ARRAY command, and all lines of this new element are converted to phantom lines through the Properties dialog box, MATCHPROP command, or Layer Control box. Notice that the smaller segments are not shown as phantom lines due to their short lengths.

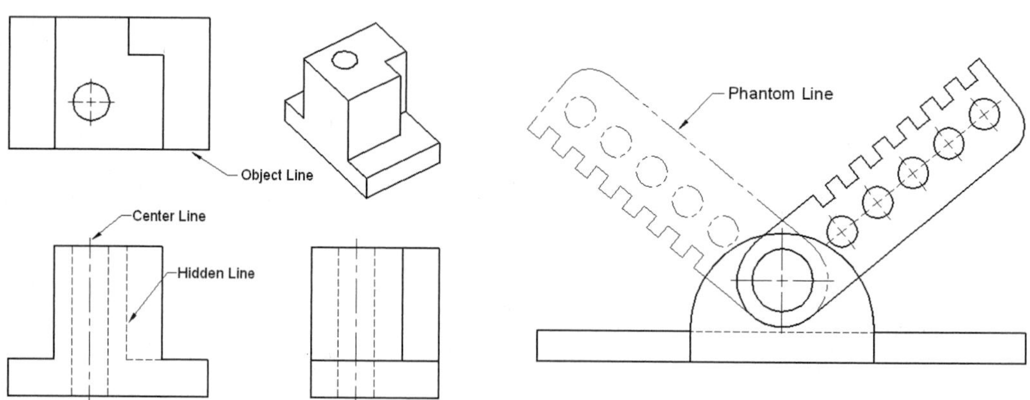

Figure 8.6

Use of linetypes in a drawing is crucial to the interpretation of the views and the final design before the object is actually made. Sometimes the linetype appears too long; in other cases, the linetype does not appear at all, even though using the LIST command on the object will show the proper layer and linetype. The LTSCALE or Linetype Scale command is used to manipulate the size of all linetypes loaded into a drawing. By default, all linetypes are assigned a scale factor of 1.00. This means that the actual dashes and/or spaces of the linetype are multiplied by this factor. The view illustrated on the left in the following image shows linetypes that use the default value of 1.00 from the LTSCALE command.

If a linetype dash appears too long, use the LTSCALE command and set a new value to less than 1.00. If a linetype appears too short, use the LTSCALE command and set a new value to greater than 1.00. The view illustrated in the middle of the following image shows the effects of the LTSCALE command set to a new value of 0.75. Notice that this view has one more series of dashes than the same object illustrated previously. The 0.75 multiplier affects all dashes and spaces defined in the linetype.

The view on the right in the following image shows how using the LTSCALE command with a new value of 0.50 shortens the linetype dashes and spaces even more. Now even the center marks identifying the circles have been changed to centerlines. When you use the LTSCALE command, the new value, whether larger or smaller than 1.00, affects all linetypes visible on the display screen.

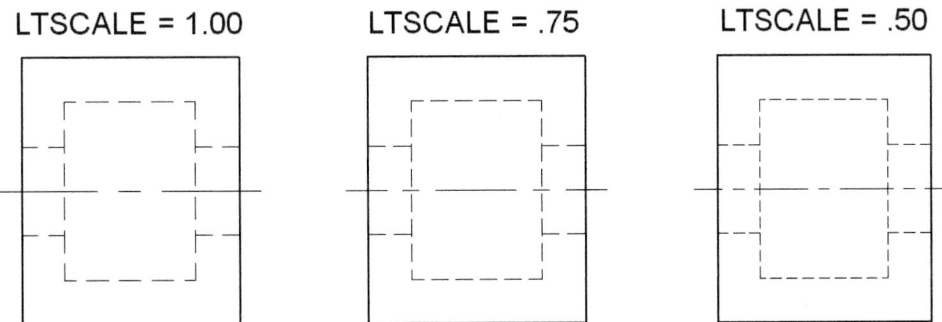

Figure 8.7

ONE-VIEW DRAWINGS

An important rule to remember concerning multiview drawings is to draw only enough views to accurately describe the object. In the drawing of the gasket in the following image, Front and Side views are shown. However, the Side view is so narrow that it is difficult to interpret the hidden lines drawn inside. A better approach would be to leave out the Side view and construct a one-view drawing consisting of just the Front view.

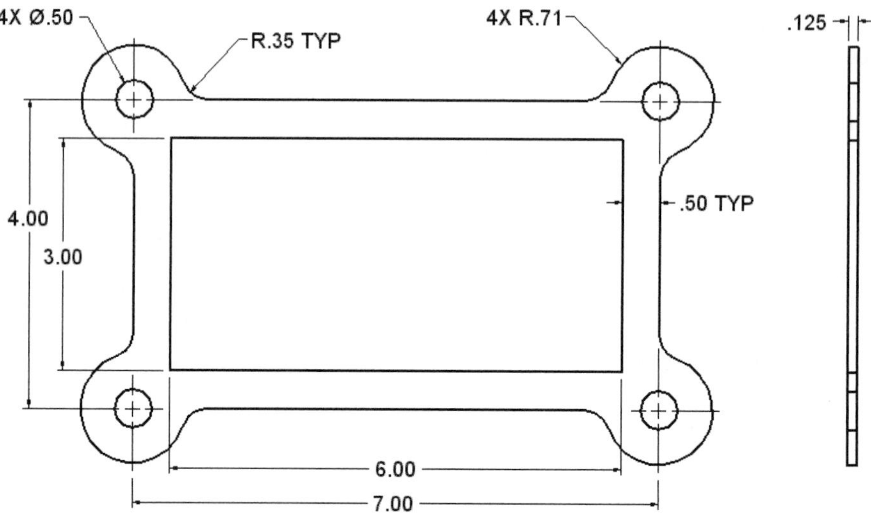

Figure 8.8

Begin the one-view drawing of the gasket by first laying out centerlines marking the centers of all circles and arcs, as shown on the left in the following image. A layer containing centerlines could be used to show all lines as centerlines.

Use the CIRCLE command to lay out all circles representing the bolt holes of the gasket, as shown on the right in the following image. A layer containing continuous object lines could be used for these circles. You could use the OFFSET command to form the large rectangle on the inside of the gasket. If lines of the rectangle extend past each other, use the FILLET command set to a value of 0. Selecting two lines of the rectangle will form a corner. Repeat this procedure for any other lines that do not form exact corners.

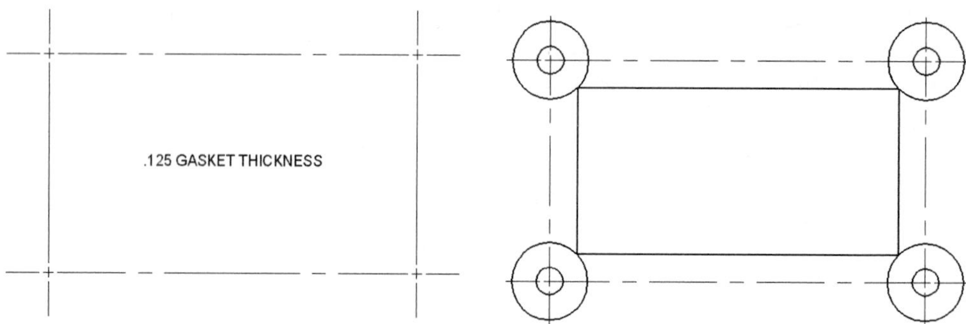

Figure 8.9

Use the TRIM command to begin forming the outside arcs of the gasket, as shown on the left in the following image.

Use the FILLET command set to the desired radius to form a smooth transition from the arcs to the outer rectangle, as shown on the right in the following image.

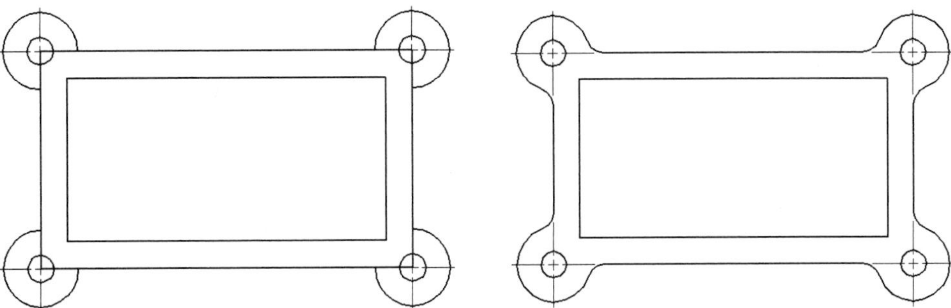

Figure 8.10

TWO-VIEW DRAWINGS

Before attempting any drawing, determine how many views need to be drawn. A minimum number of views are needed to describe an object. Drawing extra views is not only time-consuming, but it may result in two identical views with mistakes in each view. You must interpret which is the correct set of views. The illustration on the left in the following image is a three-view multiview drawing of a coupler. The circles and hidden circle identify the Front view. Except for their rotation angles, the Top and Right Side views are identical. In this example, or for other symmetrical objects, only two views are needed to accurately describe the object being drawn. The Top view has been deleted to leave the Front and Right Side views. The Side view could have easily been deleted in favor of leaving the Front and Top views. This decision is up to the designer, depending on sheet size and which views are best suited for the particular application. The desired two view drawing is displayed as shown on the right in the following image.

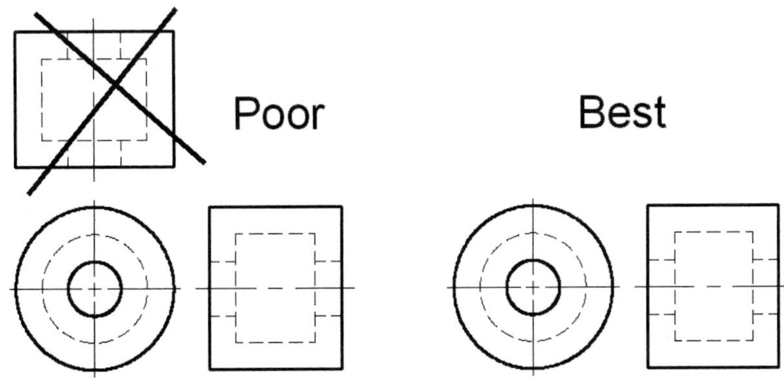

Figure 8.11

Try It!: Open the drawing file 08_R-Guide. Use the following images and descriptions for constructing a two-view drawing of this object.

To illustrate how AutoCAD is used as the vehicle for creating a two-view engineering drawing, study the two-view drawing along with the pictorial drawing illustrated in the following image to get an idea of how the drawing will appear.

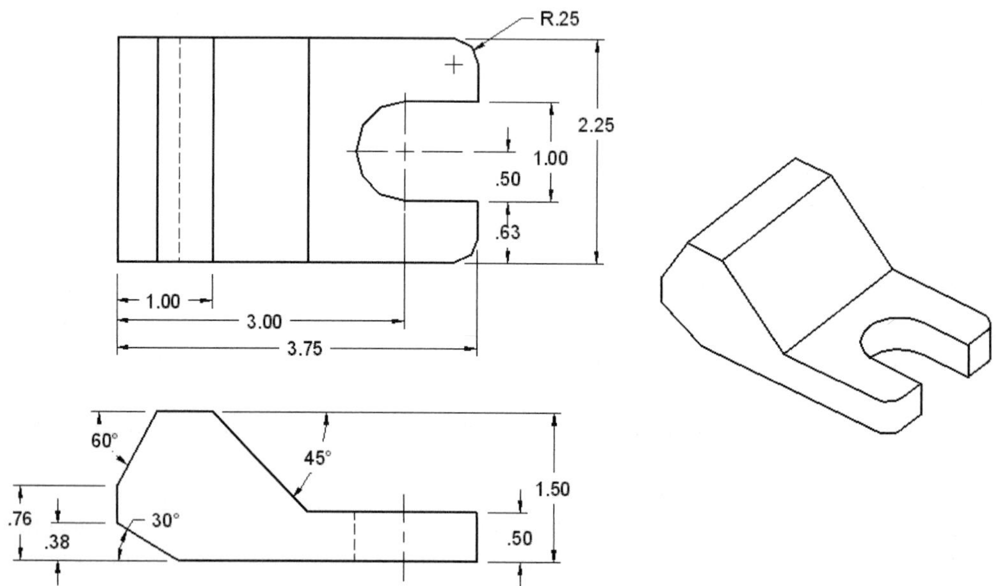

Figure 8.12

Begin the two-view drawing by using the LINE command to lay out the Front and Top views. You can find the width of the Top view by projecting lines up from the front because both views share the same width, as shown on the left in the following image. Provide a space of 1.50 units between views to act as a separator and allow for dimensions at a later time. With the two views laid out, use the TRIM command to trim unnecessary lines in order for your drawing to appear similar to the illustration on the right in the following image.

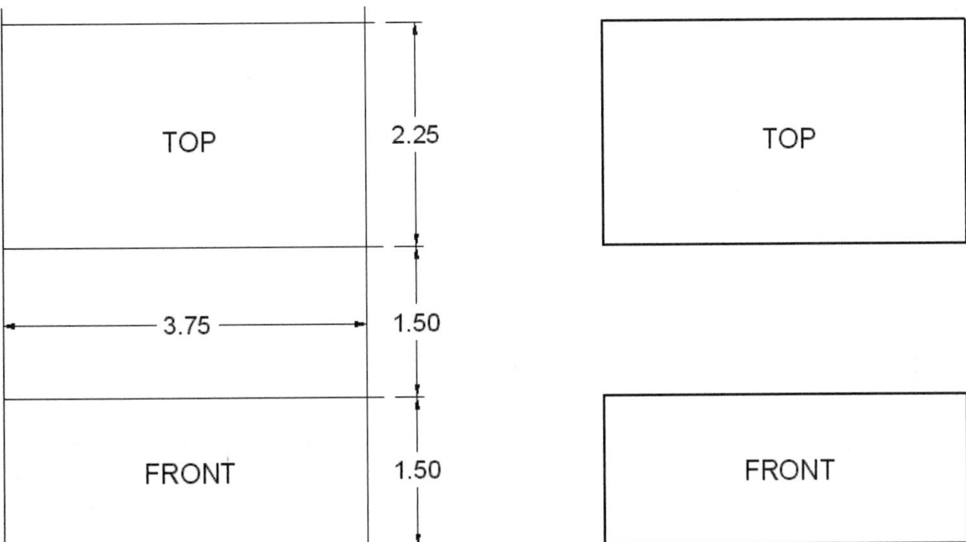

Figure 8.13

Next, add visible details to the views, such as arcs, filleted corners, and angles, as shown on the left in the following image. Use various editing commands such as TRIM, EXTEND, and OFFSET to clean up unnecessary geometry.

From the Front view, project corners up to the Top view, as shown on the right in the following image. These corners form visible edges in the Top view.

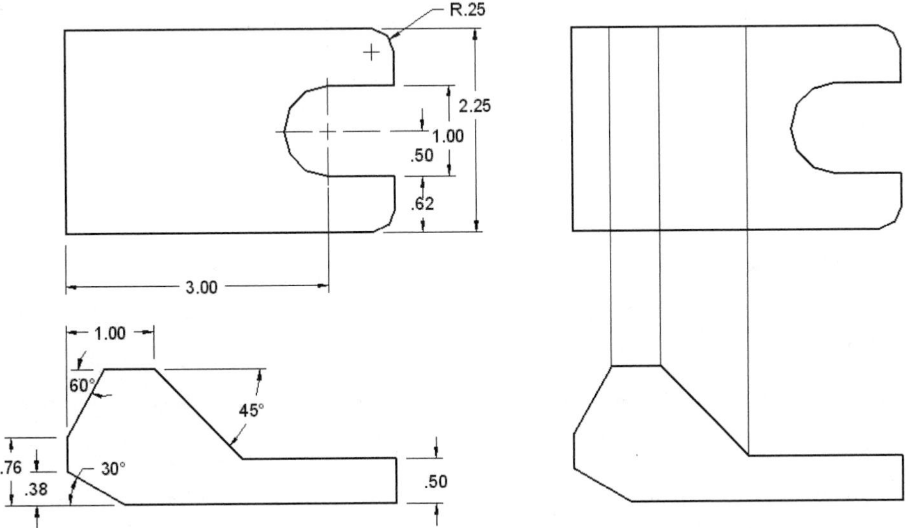

Figure 8.14

Use the same projection technique to project features from the Top view to the Front view, as shown on the left in the following image. Then use the TRIM command to delete any geometry that appears in the 1.50 dimension space. The views now must conform to engineering standards by showing which lines are visible and which are invisible, as shown on the right in the following image. Use the Layer Control box to change the line in the Top view from the Object layer to the Hidden layer. In the same manner, the slot visible in the Top view is hidden in the Front view. Again change the line in the Front view to the Hidden layer. Since the slot in the Top view represents a circular feature, use the DIMCENTER command to place a center marker at the center of the semicircle. To show in the Front view that the hidden line represents a circular feature, add one centerline consisting of one short dash and two longer dashes. Be sure this line is drawn on the Center layer. If the slot in the Top view were square instead of circular, centerlines would not be necessary.

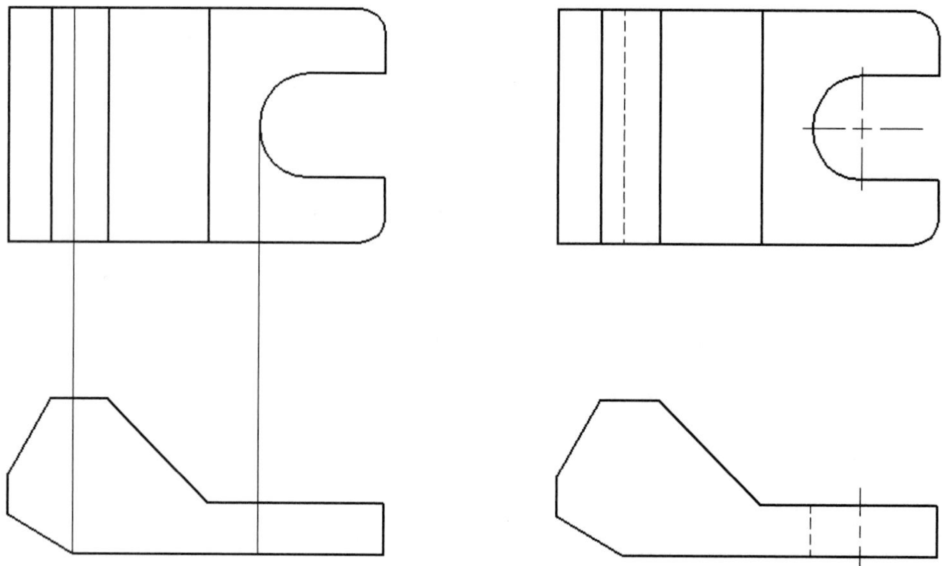

Figure 8.15

THREE-VIEW DRAWINGS

If two views are not enough to describe an object, draw three views: Front, Top, and Right Side views.

Try It!: Open the drawing file 08_Guide Block. Use the following images and descriptions for constructing a three-view drawing of this object.

A three-view drawing of the guide block, as illustrated in orthographic and pictorial formats in the following image, will be the focus of this segment. Notice the broken section exposing the Spotface operation above a drill hole.

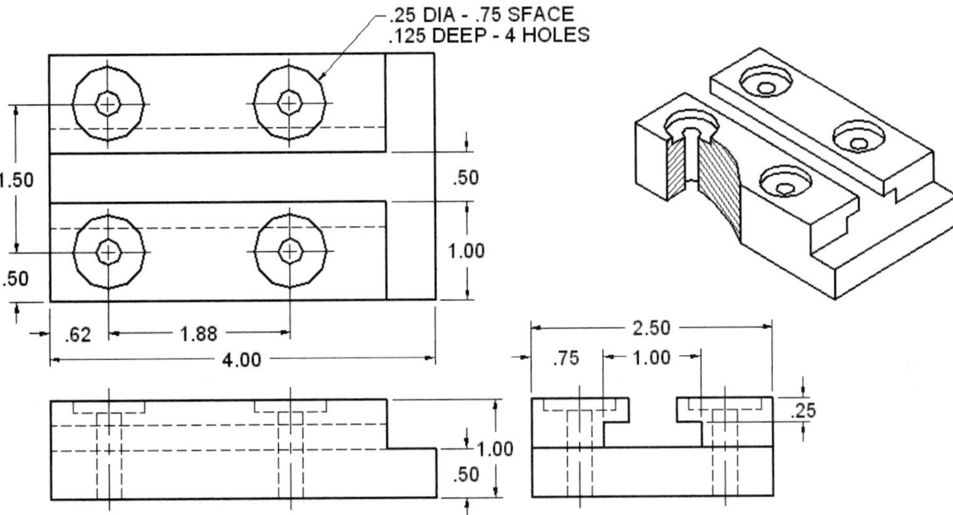

Figure 8.16

Begin this drawing by laying out all views using overall dimensions of width, depth, and height, as shown in the following image. The LINE and OFFSET commands are popular commands used to accomplish this. Provide a space between views to accommodate dimensions at a later time.

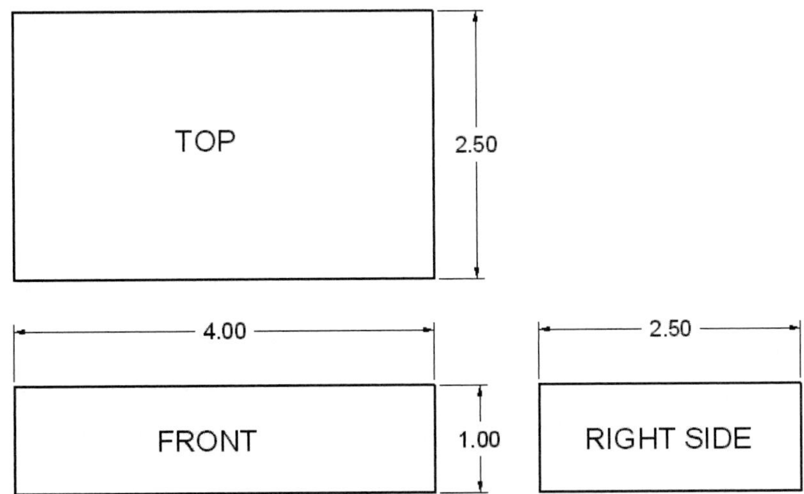

Figure 8.17

Next, draw features in the views where they are visible, as shown in the following image. Since the Spotface holes appear above, draw these in the Top view. The notch appears in the Front view; draw it there. A slot is visible in the Right Side view and is drawn there.

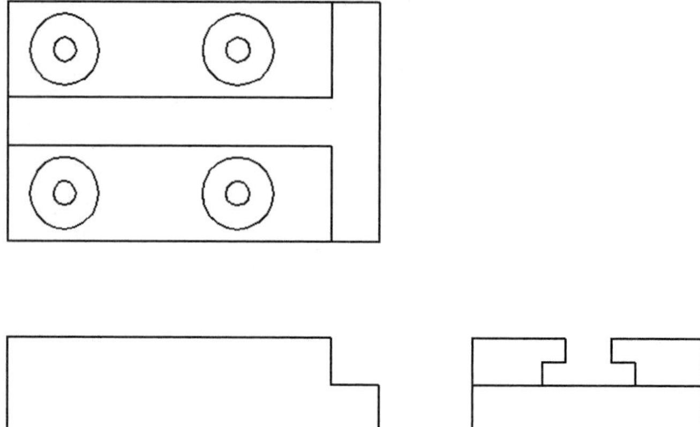

Figure 8.18

As in two-view drawings, all features are projected down from the Top to the Front view. To project depth measurements from the Top to the Right Side view, construct the 45° line as shown in the following image.

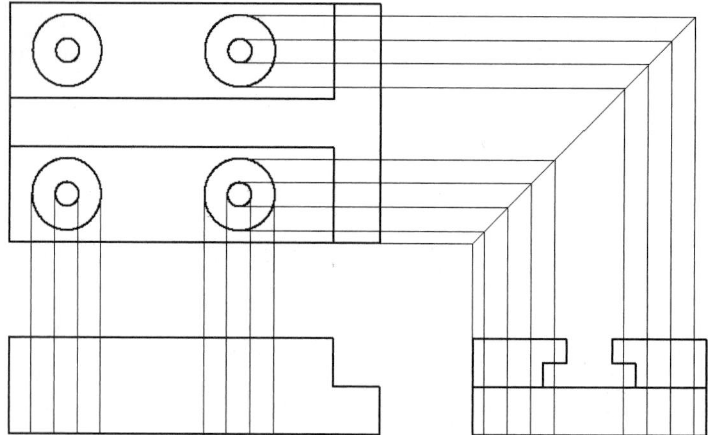

Figure 8.19

Use the 45° line to project the slot from the Right Side view to the Top view, as shown in the following image. Project the height of the slot from the Right Side view to the Front view.

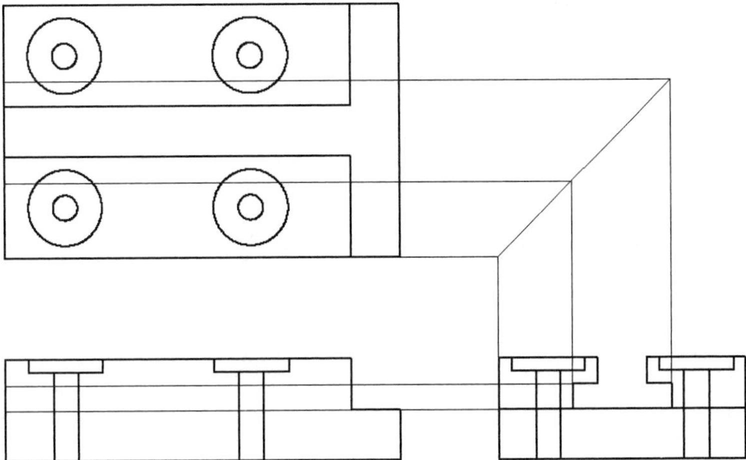

Figure 8.20

Change the continuous lines to the Hidden layer where features appear invisible, such as the holes in the Front and Right Side views as shown in the following image. Change the remaining lines to the Hidden layer. Erase any construction lines, including the 45°-projection line.

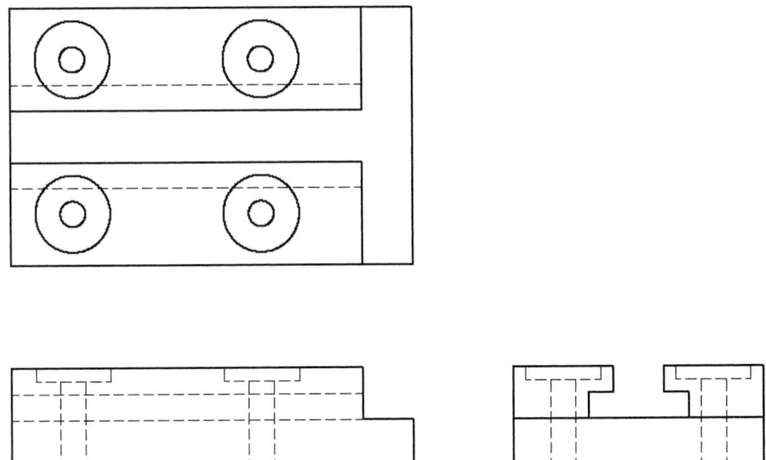

Figure 8.21

Begin adding centerlines to label circular features, as shown in the following image. Either construct these centerlines on the Center layer or change them later on to the Center layer. The DIMCENTER (type DCE at the Command prompt) command is used where the circles are visible. When features are hidden but represent circular features, the single centerline consisting of one short dash and two long dashes is used.

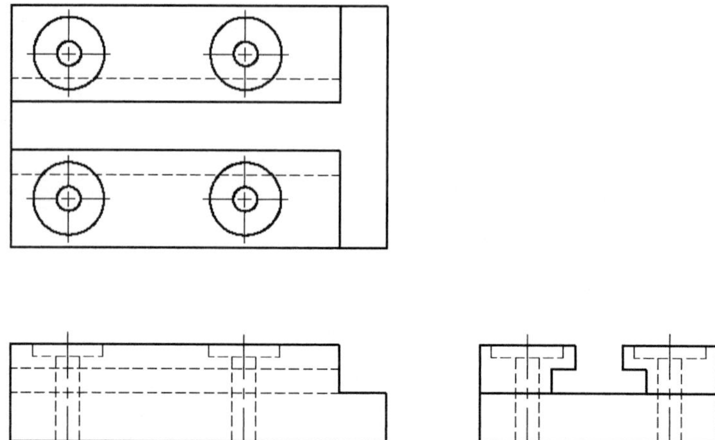

Figure 8.22

RUNOUTS

Where flat surfaces become tangent to cylinders, there must be some method of accurately representing this using fillets. In the object illustrated on the left in the following image, the Front view shows two cylinders connected to each other by a tangent slab. The Top view is complete; the Front view has all the geometry necessary to describe the object, with the exception of the exact intersection of the slab with the cylinder.

The illustration on the right in the following image displays the correct method for finding intersections or runouts—areas where surfaces intersect others and blend in, disappear, or simply run out. A point of intersection is found at "A" in the Top view with the intersecting slab and the cylinder. This actually forms a 90° angle with the line projected from the center of the cylinder and the angle made by the slab. A line is projected from "A" in the Top view to intersect with the slab found in the Front view.

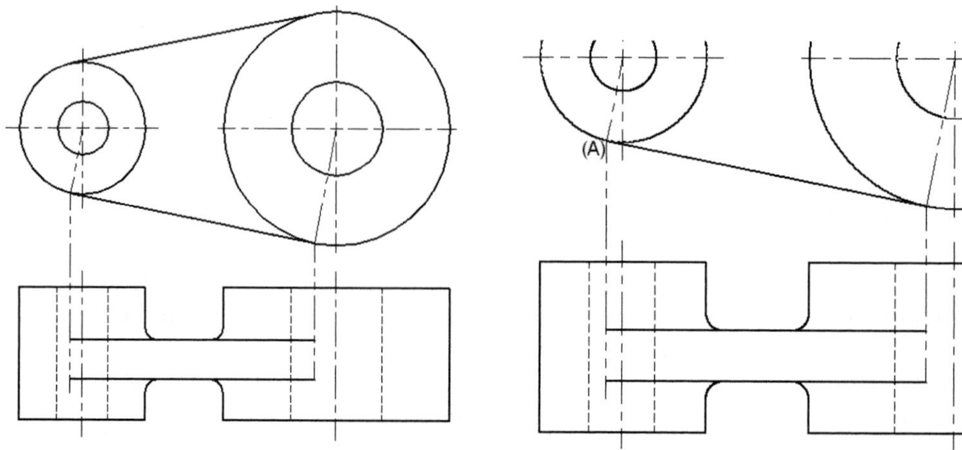

Figure 8.23

As illustrated on the left in the following image, fillets are copied to represent the slab and cylinder intersections. This forms the runout.

The resulting two-view drawing, complete with runouts, is illustrated on the right in the following image.

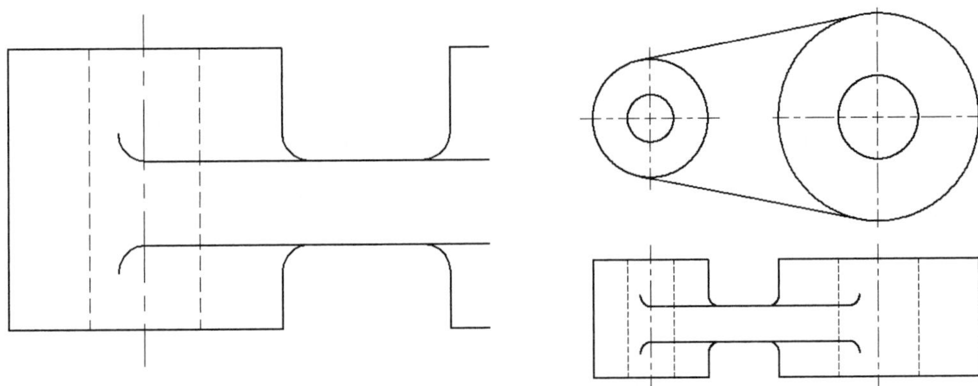

Figure 8.24

TUTORIAL EXERCISE: ORTHOGRAPHIC BLOCK.DWG

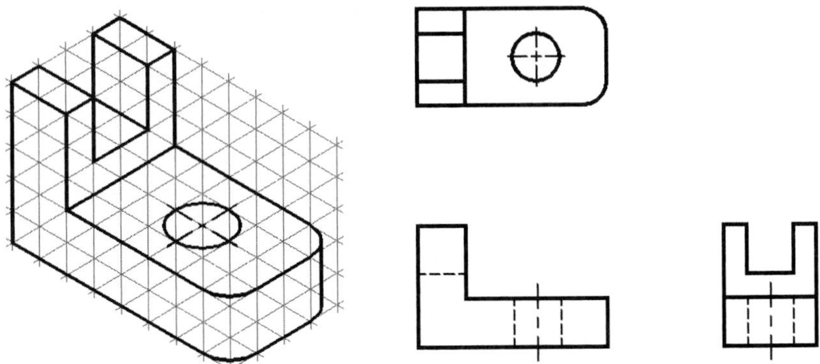

Figure 8.25

PURPOSE

Using an isometric grid, construct the Front, Top, and Right Side views of the Orthographic Block, as shown in the previous image.

SYSTEM SETTINGS

Begin a new drawing called Orthographic Block.dwg. Use the Drawing Units dialog box to change the precision from four to two decimal places. Keep the remaining default unit values.

Using the LIMITS command, keep (0,0) for the lower-left corner and change the upper-right corner from (12,9) to (15.50,9.50).

It is very important to check and see that the following Object Snap modes are already set: Endpoint, Extension, Intersection, and Center.

LAYERS

Create the following layers with the format:

Name	Color	Linetype
Object	Green	Continuous
Hidden	Red	Hidden
Center	Yellow	Center

SUGGESTED COMMANDS

Begin this tutorial by laying out the three primary views using the LINE and OFFSET commands. Use a grid spacing of 0.25 units for all distance calculations.

Use the TRIM command to clean up any excess line segments. As an alternative projection method, use temporary tracking points in combination with Object Snap options to add features in other views.

Step 1

Begin constructing the Orthographic Block by laying out the Front, Top, and Right Side views using only the overall dimensions. Do not be concerned about details such as holes or slots; these will be added to the views in a later step. The length of the object shown on the left in the following image is 8 grid units (2 inches). The height of the object is 5 grid units (1.25 inches); the depth of the object is 4 units (1 inch). The distance between views is 5 grid units (1.25 inches).

Step 2

Once the overall dimensions have been used to lay out the Front, Top, and Right Side views, begin adding visible details to the views. The "L" shape is added to the Front view; the hole and corner fillets are added to the Top view; the rectangular slot is added to the Right Side view, as shown on the right in the following image. Refer to the isometric view of this object at the beginning of this tutorial for the dimensions of the "L" shape, the hole, corner fillets, and the rectangular slot.

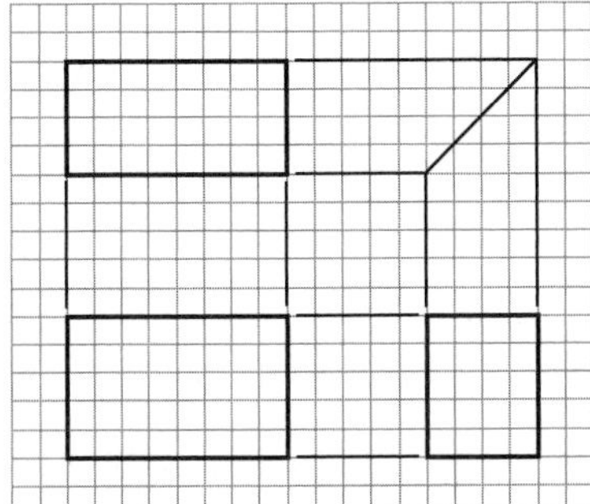

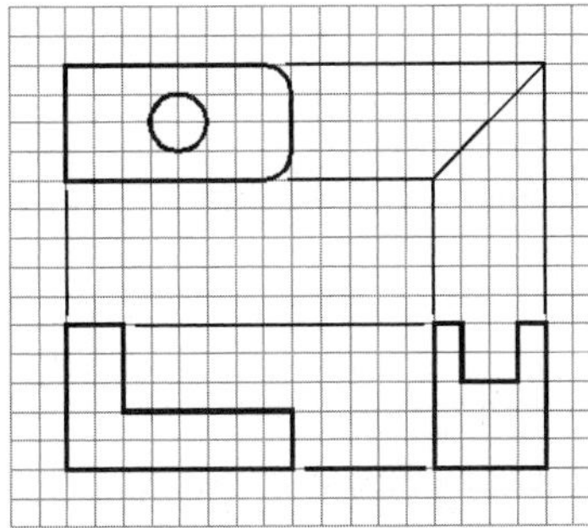

Figure 8.26

Step 3

Begin projecting the visible edges from hole and slot features onto other views. Slot information is added to the Top view and height information is projected onto the Right Side view from the Front view. At this point, only add visible information to other views where required, as shown on the left in the following image.

Step 4

Now project all hidden features to the other views. The hole projection is hidden in the Front view along with the slot visible in the Right Side view. The hole is also hidden in the Right Side view, as shown on the right in the following image. Notice how the 45° angle is used to project the hole from the Top view to the Right Side view.

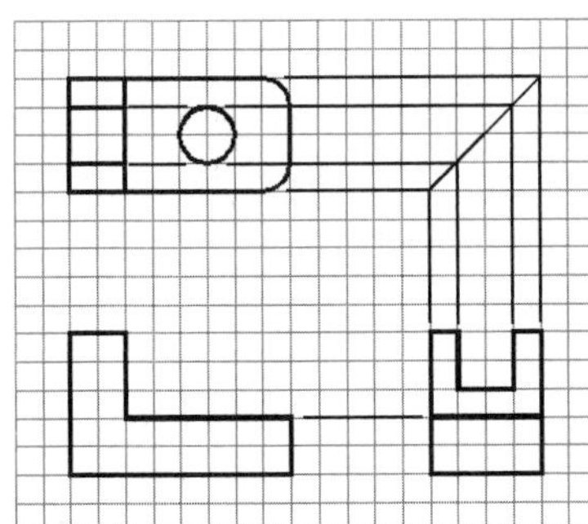

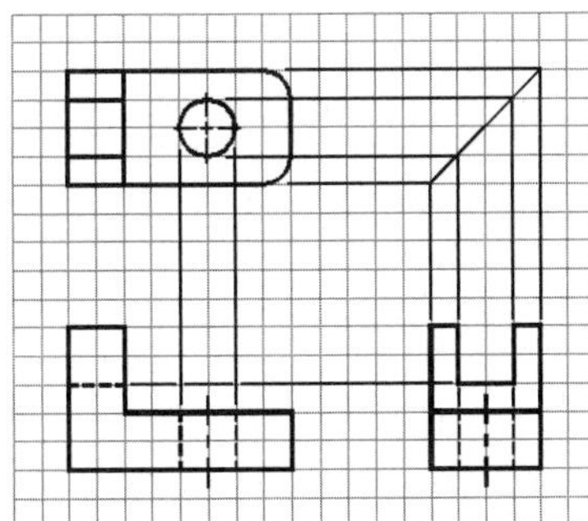

Figure 8.27

Step 5

The completed multiview drawing solution is illustrated on the left in the following image. Dimensions are added to document the exact size of the object, as shown on the right in the following image. Proper placement of dimensions will be discussed in chapter 11.

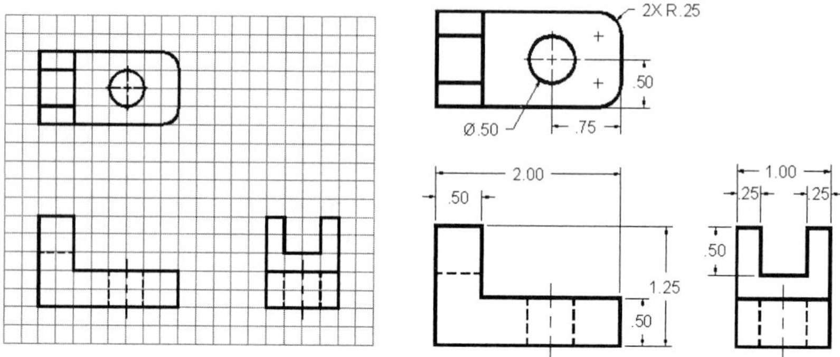

Figure 8.28

TUTORIAL EXERCISE: 08_SHIFTER.DWG

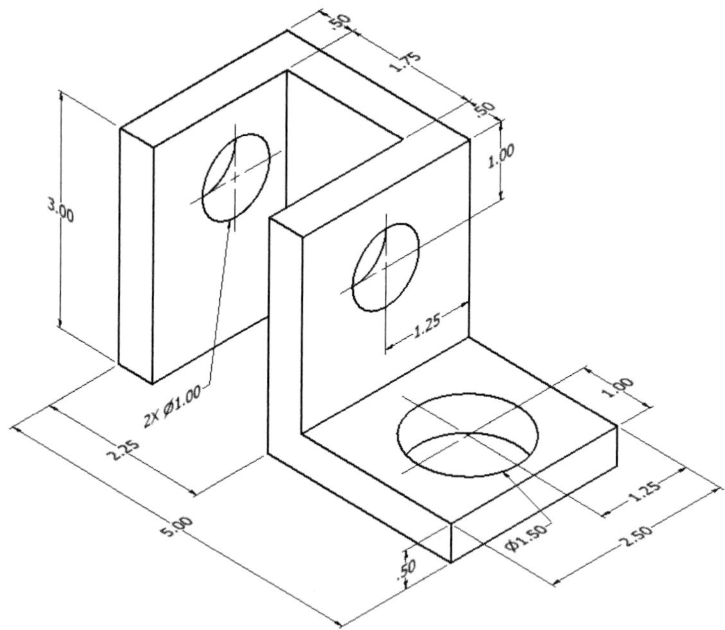

Figure 8.29

PURPOSE

This tutorial is designed to allow the user to construct a three-view drawing of the 08_Shifter as shown in the previous image.

SYSTEM SETTINGS

Use the settings from the previous tutorial and check to see that the following Object Snap modes are already set: Endpoint, Extension, Intersection, and Center.

LAYERS

Create two additional layers:

Name	Color	Linetype
Dimension	Yellow	Continuous
Projection	Cyan	Continuous

SUGGESTED COMMANDS

The primary commands used during this tutorial are OFFSET and TRIM. The OFFSET command is used for laying out all views before the TRIM command is used to clean up excess lines. Since different linetypes represent certain features of a drawing, the Layer Control box is used to convert to the desired linetype needed as set in the Layer Properties Manager dialog box. Once all visible details are identified in the primary views, project the visible features to the other views using the LINE command. A 45° inclined line is constructed to project lines from the Top view to the Right Side view and vice versa.

Step 1

Open the drawing file 08_Shifter. Make the Object layer current. Then use the LINE and OFFSET commands and create the three views displayed in the following image. The dimensions in this view represent the overall width, height, and depth of the Shifter. Space the views a distance of 1.50 units away from each other.

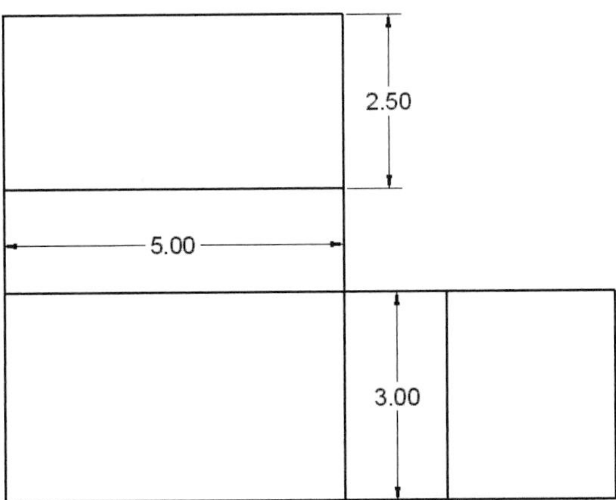

Figure 8.30

Step 2

Continue using the OFFSET command to add the various features to the Front, Top, and Right Side views using the dimensions provided in the following image. Use the TRIM command to clean up all intersections in order for your display to appear similar to the following image.

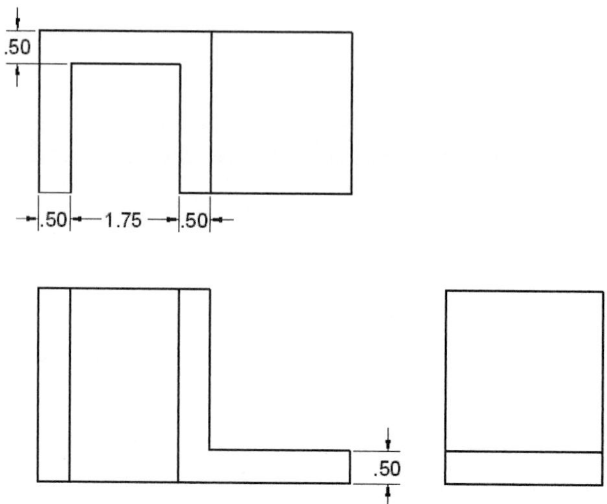

Figure 8.31

Step 3

Add both circles to their respective views, as shown in the following image. One method of finding the circle centers is to use the OFFSET command. Another method would be to use the OSNAP-From mode and the dimensions displayed in the following image.

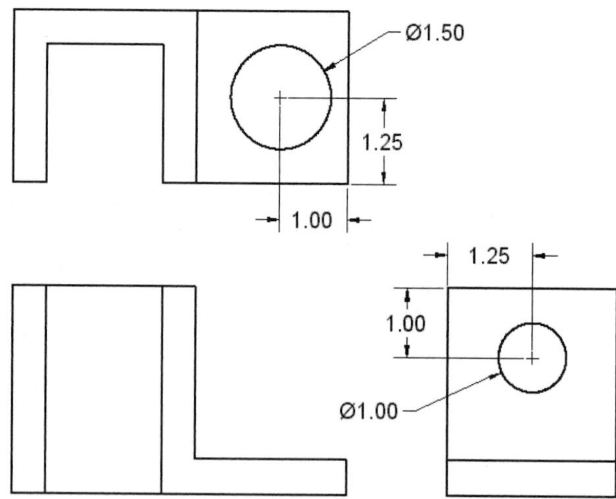

Figure 8.32

Step 4

Project lines from the quadrants of the circles from the Top and Side views into the Front view, as shown in the following image. These projected lines in the Front view represent the hidden edges of the circles.

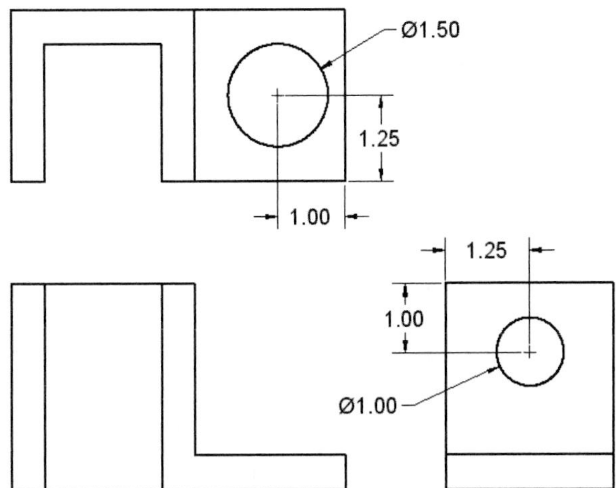

Figure 8.33

Step 5

Use the following image to guide you along the trimming of unnecessary projected lines representing the circle edges in the Front view. Then change the highlighted lines in the following image to the Hidden layer.

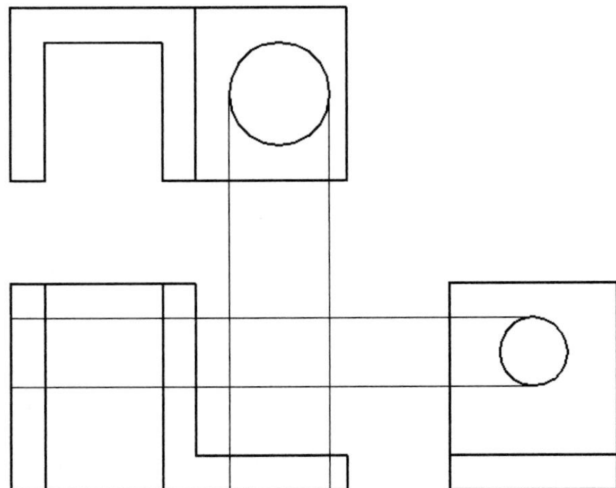

Figure 8.34

Step 6

With the Center layer current, add center marks to both circles using the Dimension Center (DCE) command by touching the edge of each circle to place the center mark, as shown in the following image.

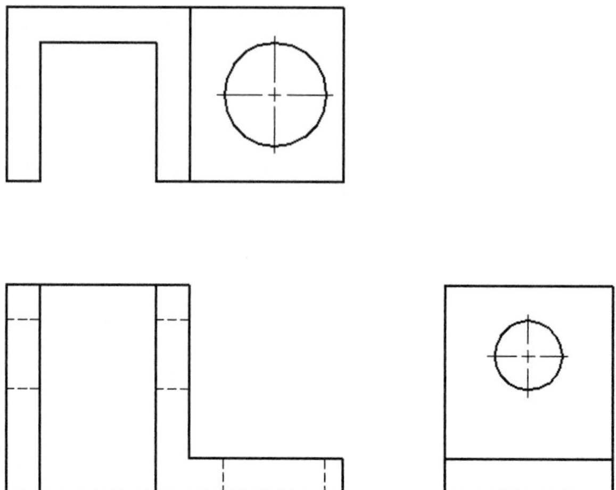

Figure 8.35

Step 7

Project the endpoints of the center marks into the Front view, as shown in the following image.

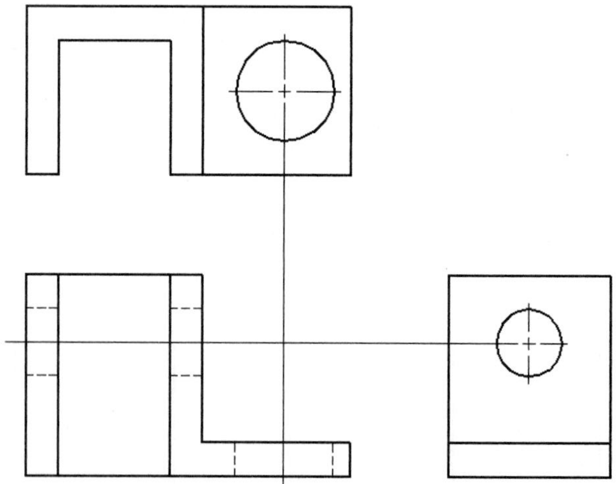

Figure 8.36

Step 8

Use the BREAK command to form the centerlines in the Front view. Then change these lines to the Center layer, as shown in the following image.

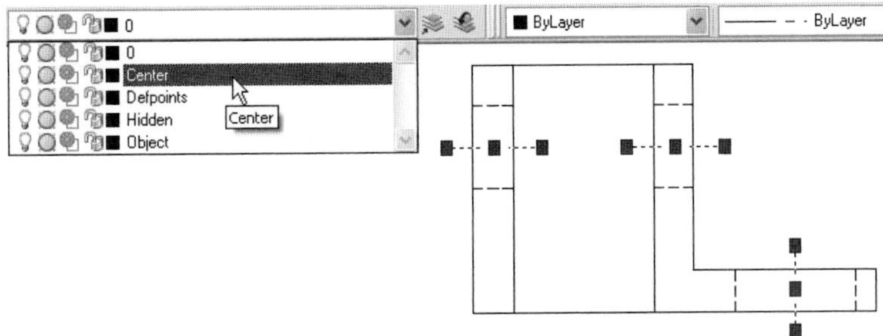

Figure 8.37

Step 9

Use the FILLET command to create a corner between the Top and Side views, as shown in the following image. Then create a 45° line from the corner at "A" at a distance of 4 units. This distance is an approximation and will be used to project edges to the Top and Side views.

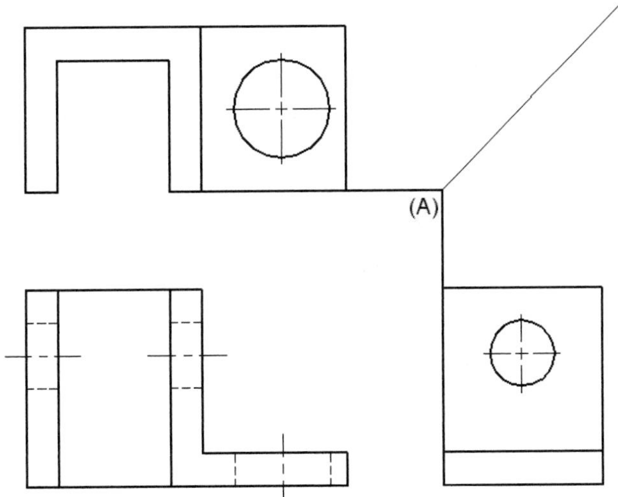

Figure 8.38

Step 10

Project lines from the edges of the circles in the Top view to intersect with the 45° angle. Then continue constructing lines into the Side view, as shown in the following image.

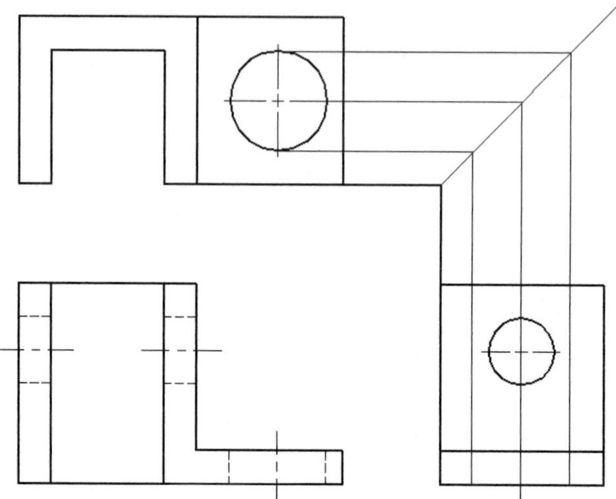

Figure 8.39

Step 11

Use the TRIM and BREAK commands to clean up lines in the Side view, as shown in the following image. Then change these lines to the Hidden and Center layers.

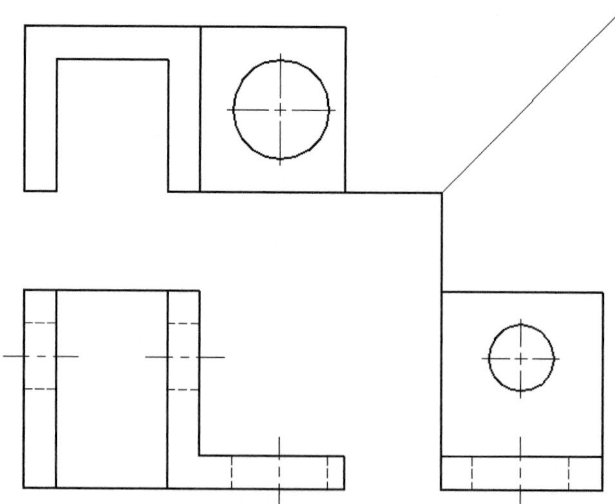

Figure 8.40

Step 12

Project lines from the edges of the circles in the Side view to intersect with the 45° angle. Then continue constructing lines into the Top view, as shown in the following image.

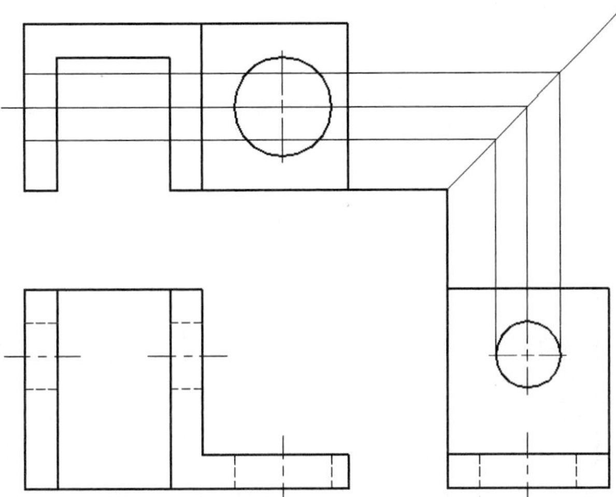

Figure 8.41

Step 13

Use the TRIM and BREAK commands to clean up lines in the Top view, as shown in the following image. Then change these lines to the Hidden and Center layers.

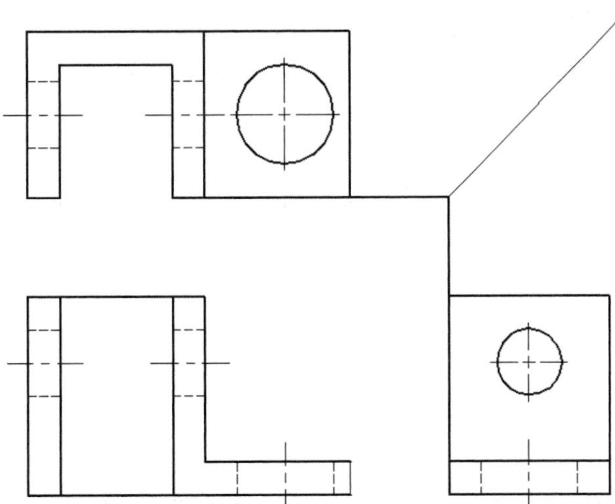

Figure 8.42

Step 14

Erase the 45° angle line, trim the corner connecting the Top and Side views, and extend the hidden line in the Side view to the top of the line at "A." This line represents a hidden edge. The completed orthographic drawing of the Shifter is shown in the following image.

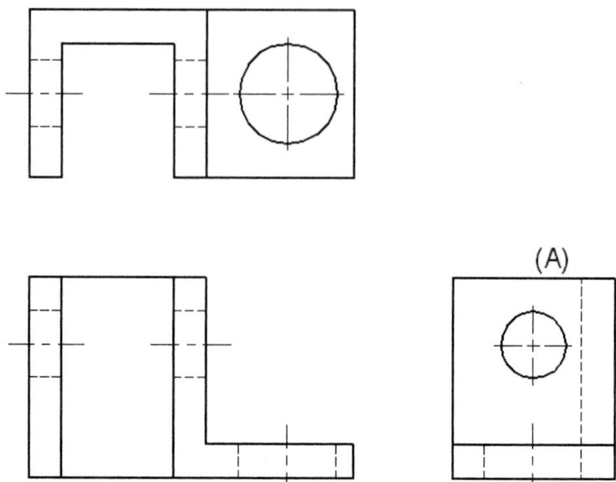

Figure 8.43

PROBLEMS FOR CHAPTER 8

DIRECTIONS FOR PROBLEMS 8–1 AND 8–2:

Open each drawing file individually (Pr08-01.dwg and Pr08-02.dwg). Create two new layers called Missing Visible Lines and Missing Hidden Lines. Assign the color red to both of these layers. Assign the Hidden linetype to the Missing Hidden Lines layer. Using projection techniques, find the missing lines for each object and draw in the correct solution on the proper layer.

PROBLEM 8–1

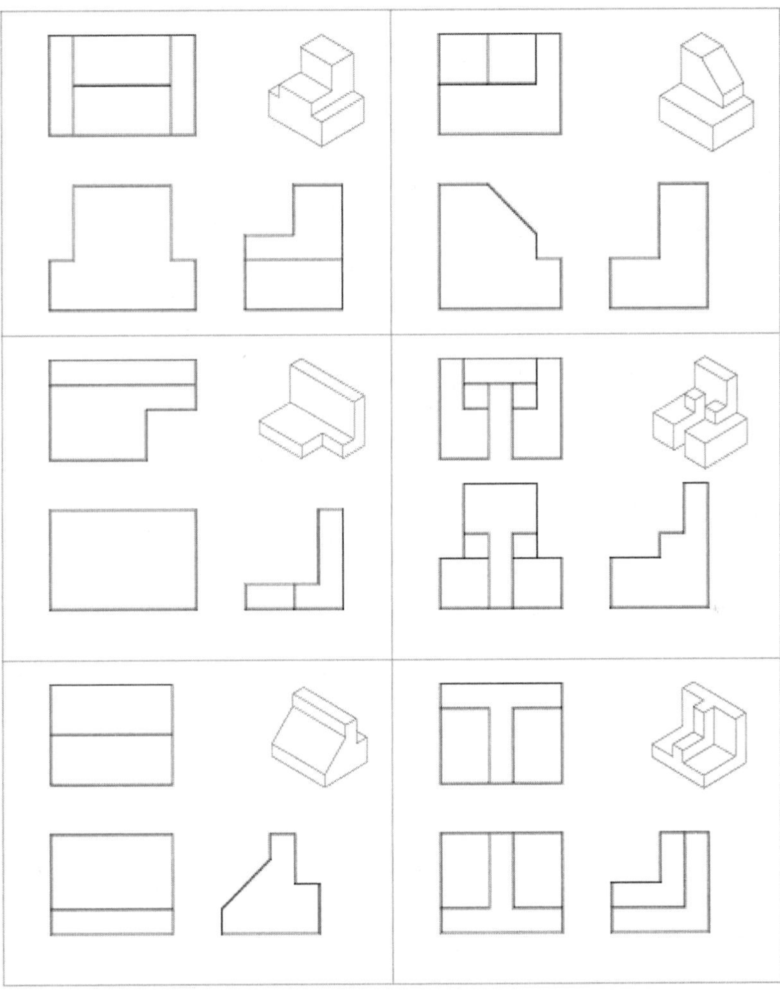

PROBLEM 8–2

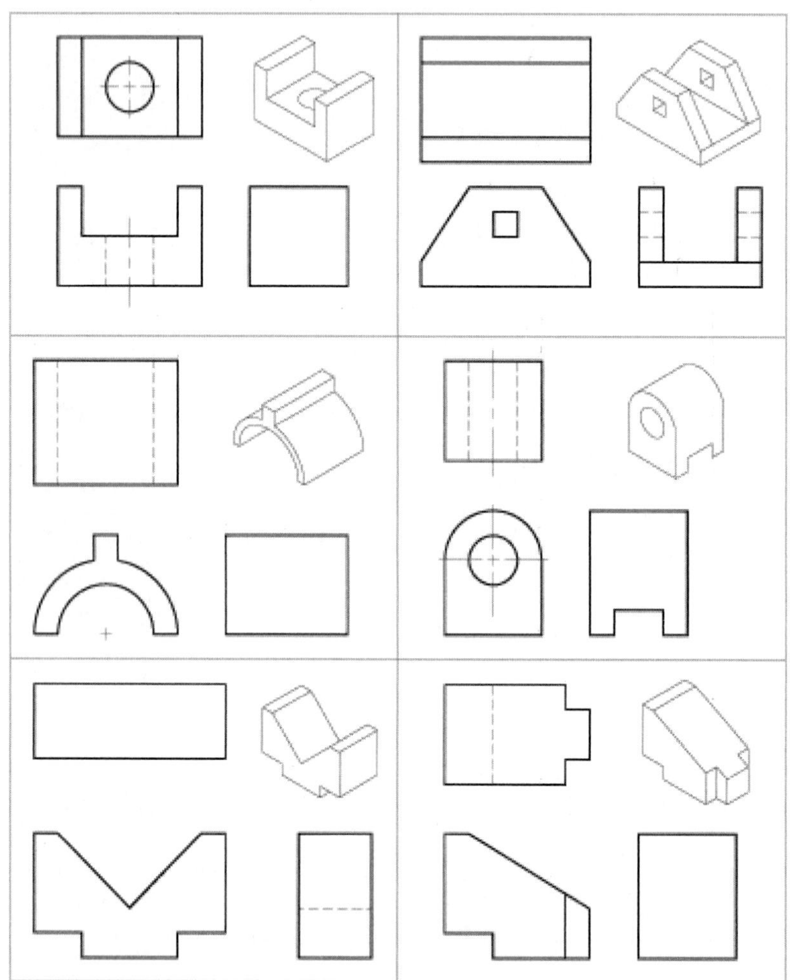

DIRECTIONS FOR PROBLEMS 8–3 THROUGH 8–20:

Construct a multiview drawing by sketching the Front, Top, and Right Side views. Use a grid of 0.25 units to assist in the construction of the sketches. These sketching problems can also be constructed directly in AutoCAD using typical commands and projection techniques.

PROBLEM 8–3

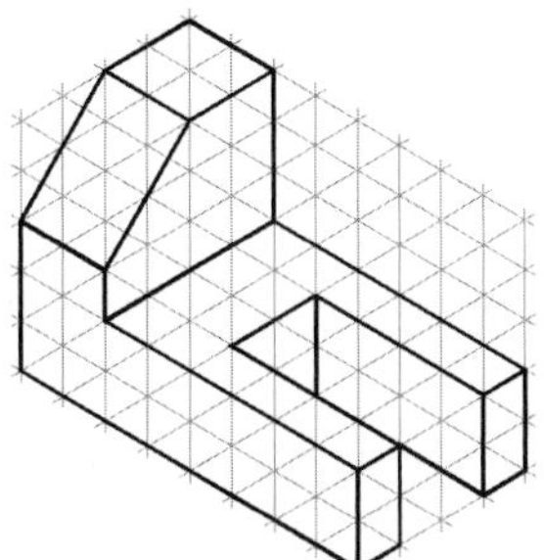

PROBLEM 8-4

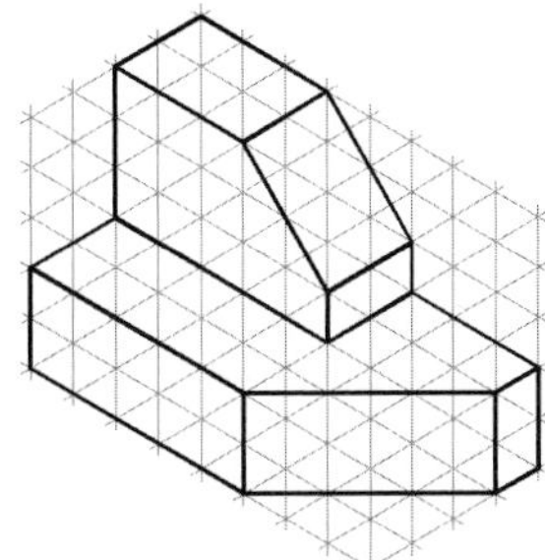

PROBLEM 8–5

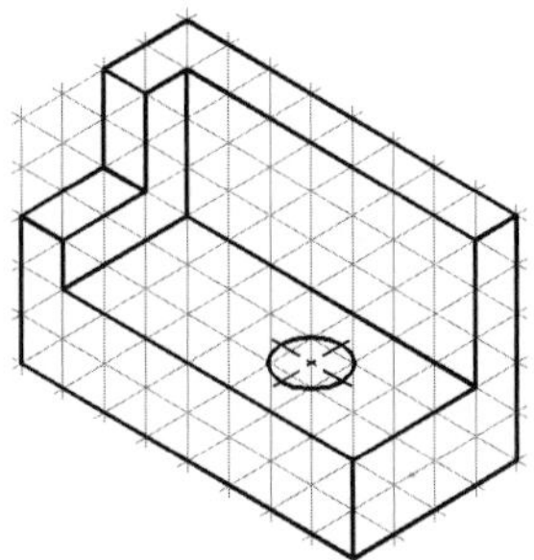

PROBLEM 8–6

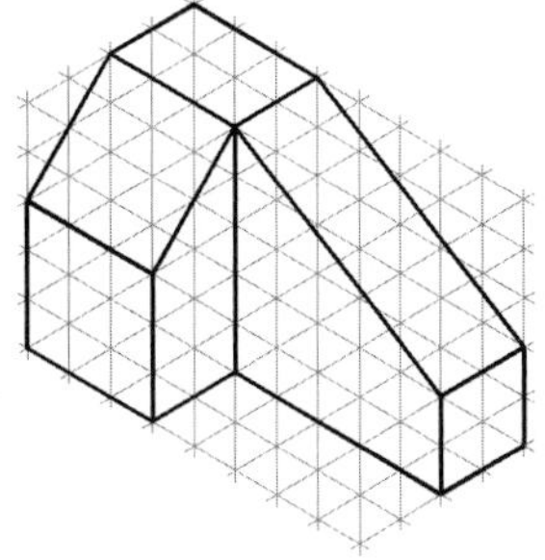

PROBLEM 8–7

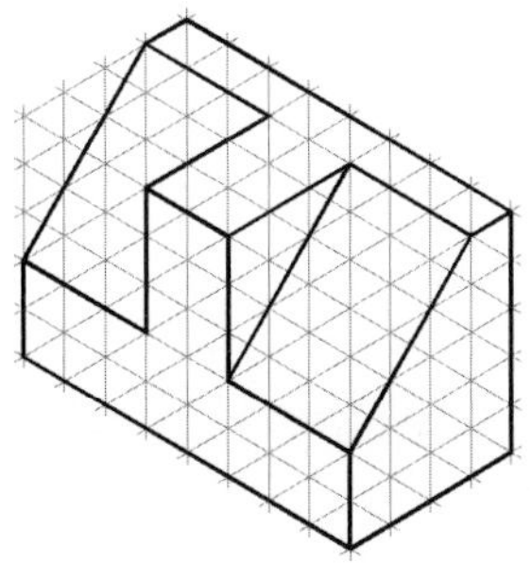

PROBLEM 8–8

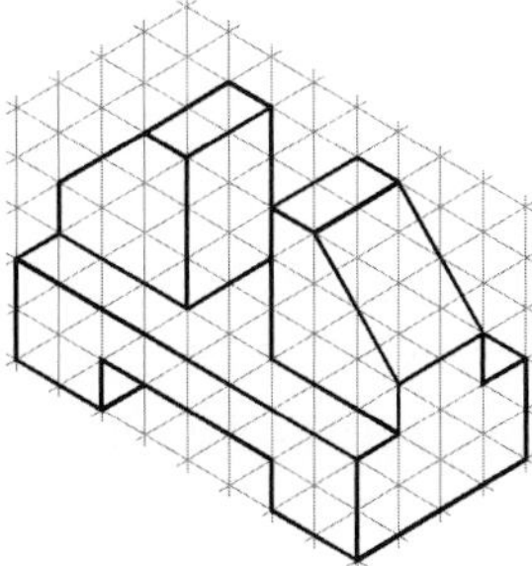

PROBLEM 8–9

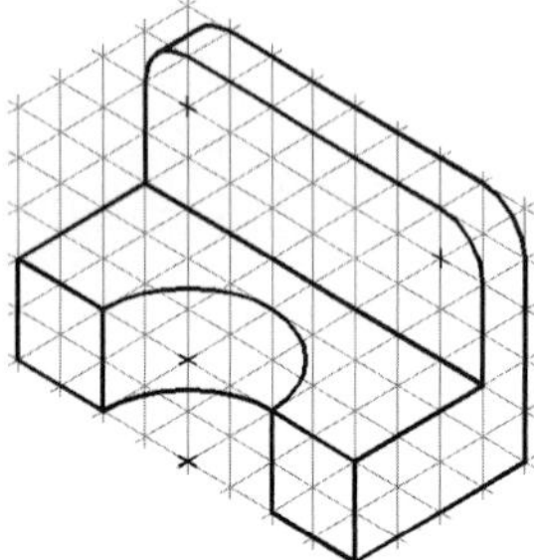

PROBLEM 8–10

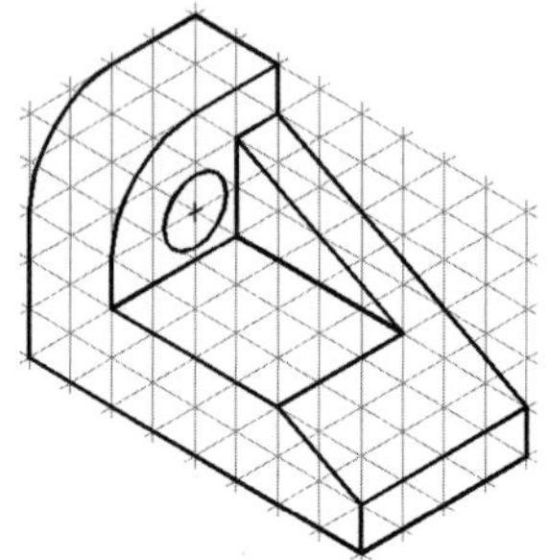

PROBLEM 8–11

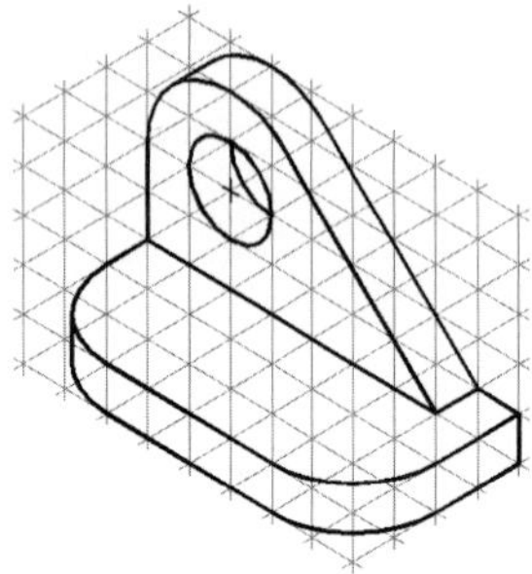

PROBLEM 8–12

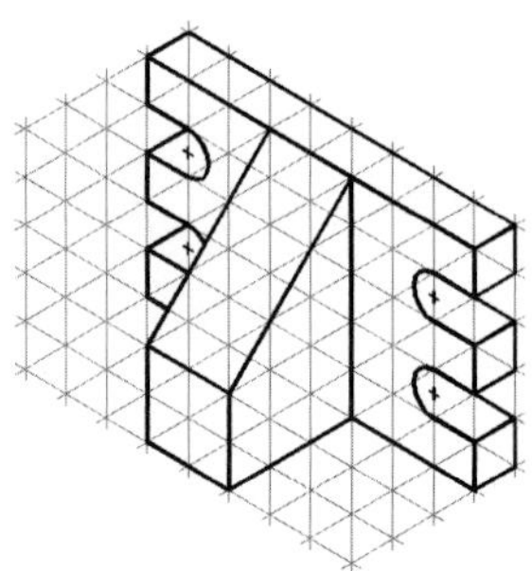

PROBLEM 8–13

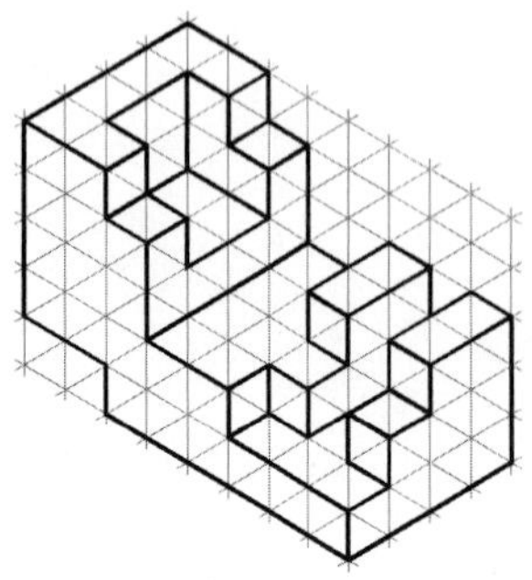

PROBLEM 8–14

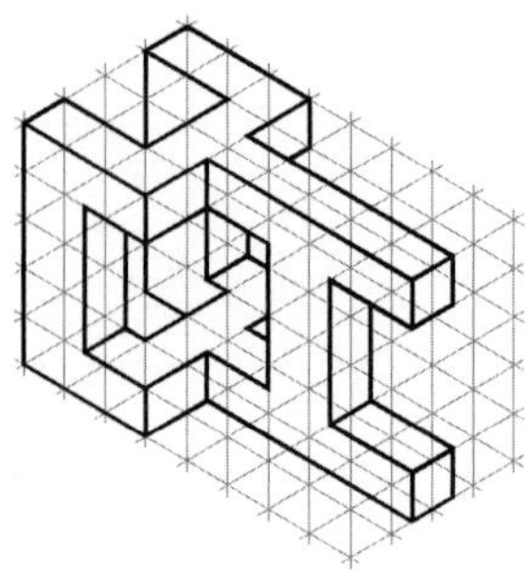

PROBLEM 8–15

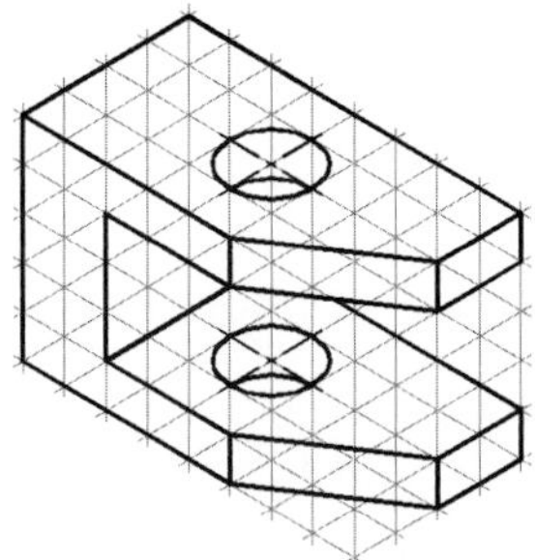

PROBLEM 8–16

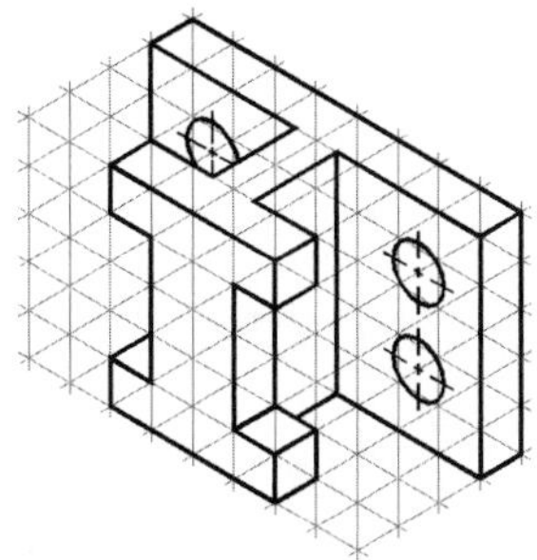

PROBLEM 8–17

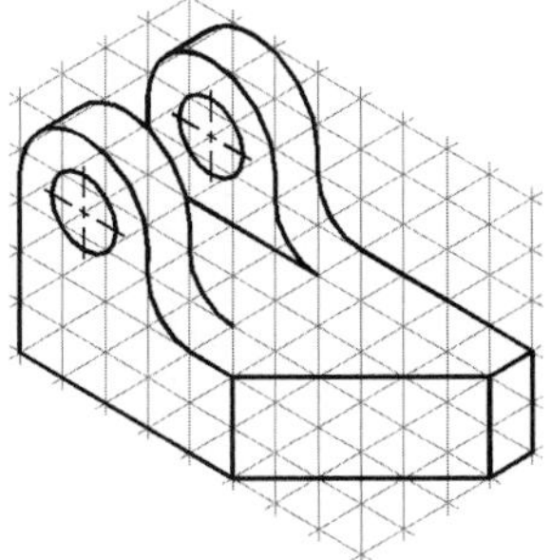

PROBLEM 8–18

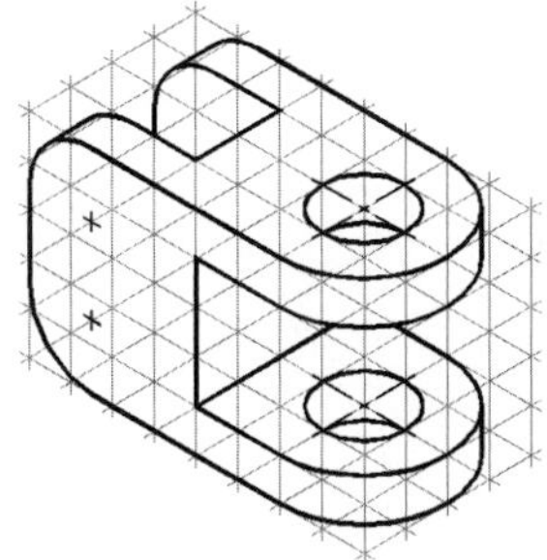

PROBLEM 8–19

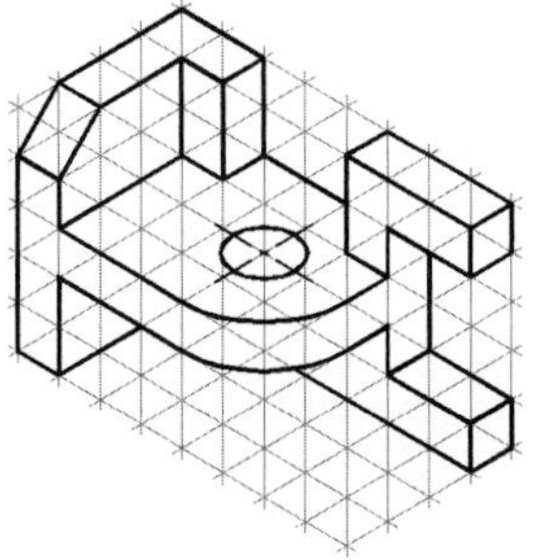

PROBLEM 8–20

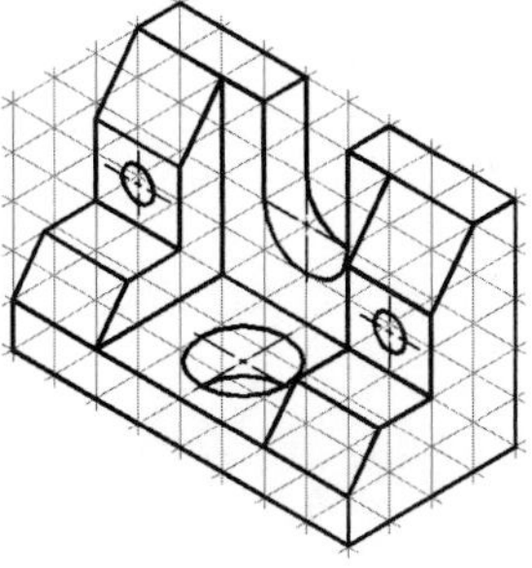

DIRECTIONS FOR PROBLEMS 8–21 THROUGH 8–35:

Construct a multiview drawing of each object. Be sure to construct only those views that accurately describe the object.

PROBLEM 8–21

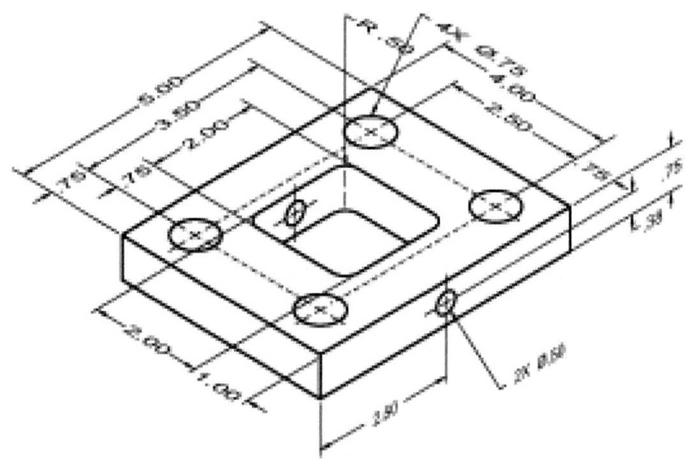

PROBLEM 8–22

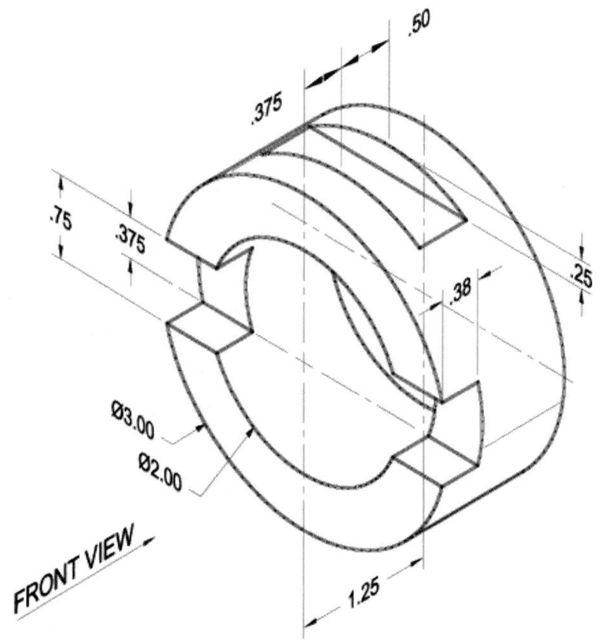

PROBLEM 8–23

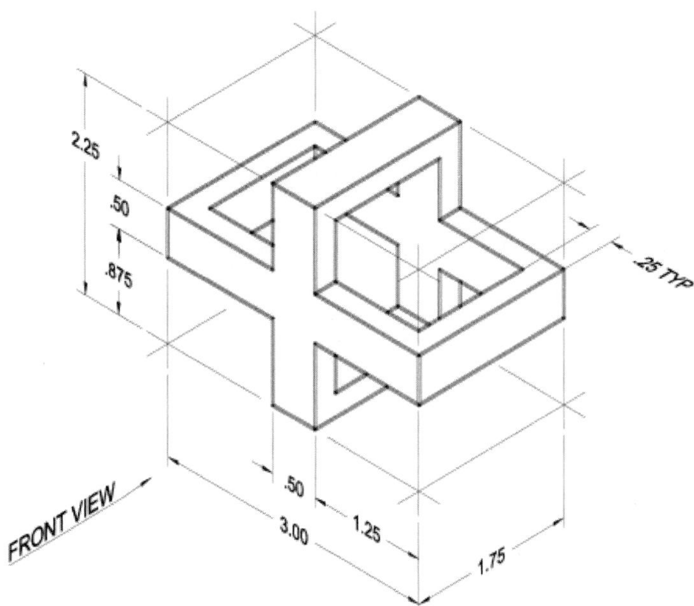

PROBLEM 8–24

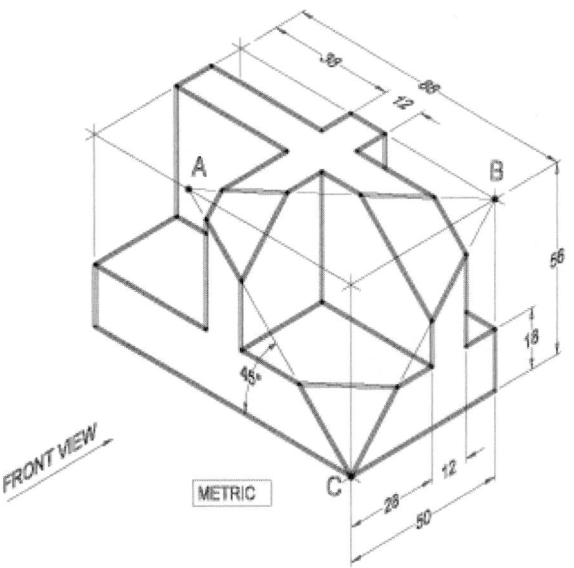

PROBLEM 8–25

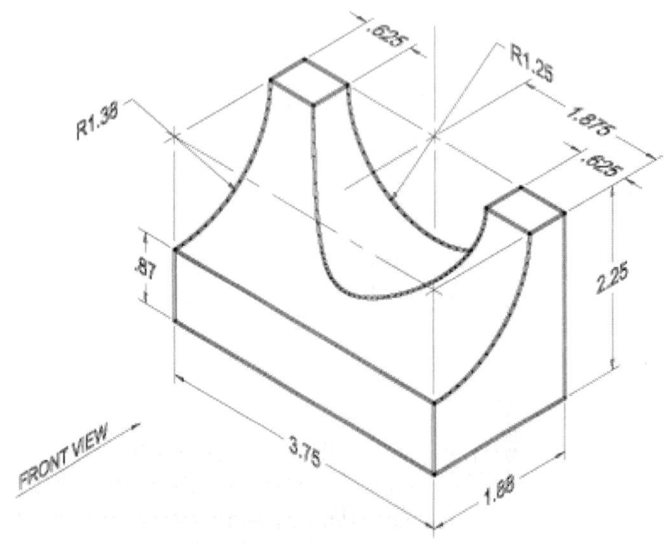

PROBLEM 8–26

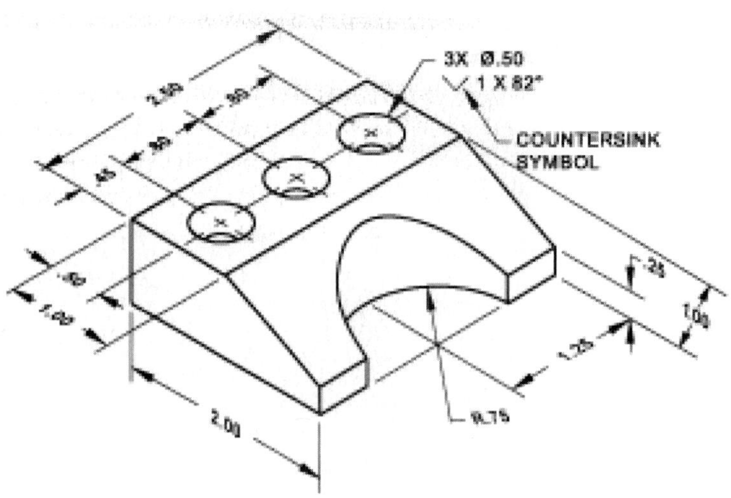

PROBLEM 8–27

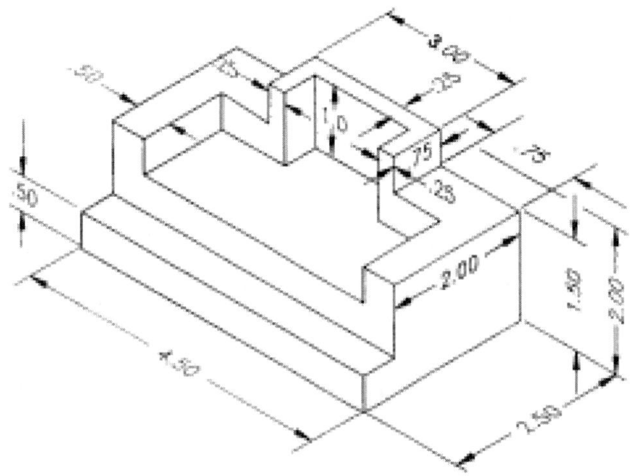

PROBLEM 8–28

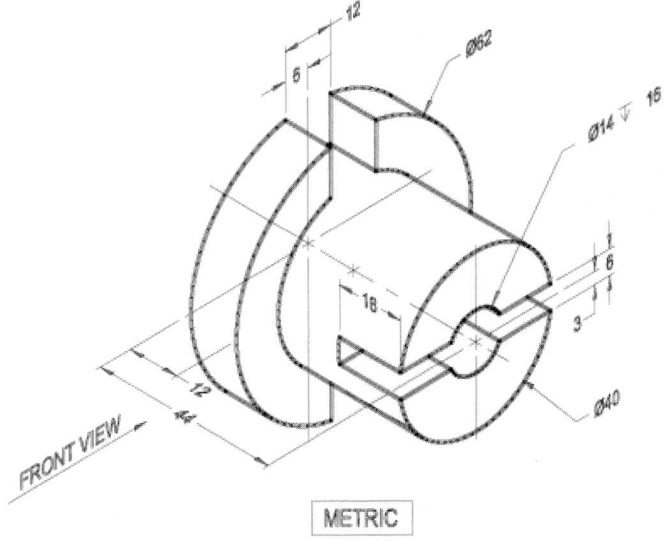

PROBLEM 8–29

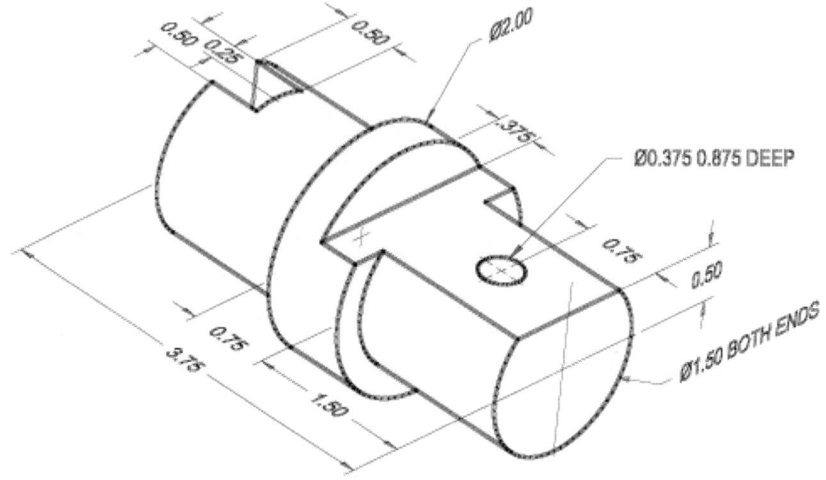

PROBLEM 8–30

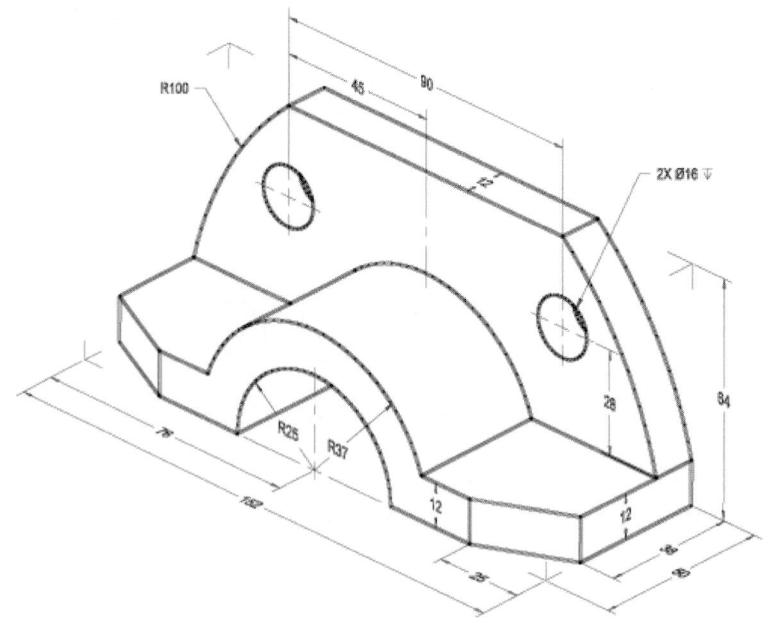

PROBLEM 8–31

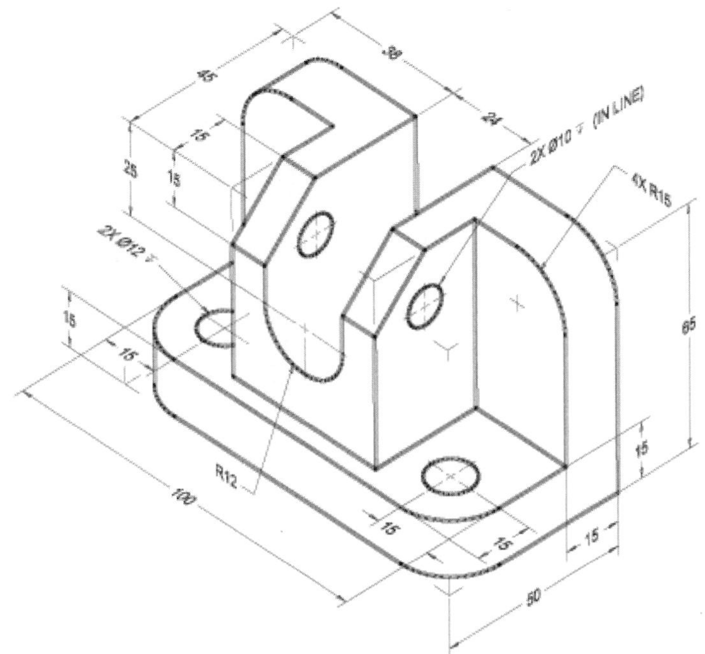

PROBLEM 8–32

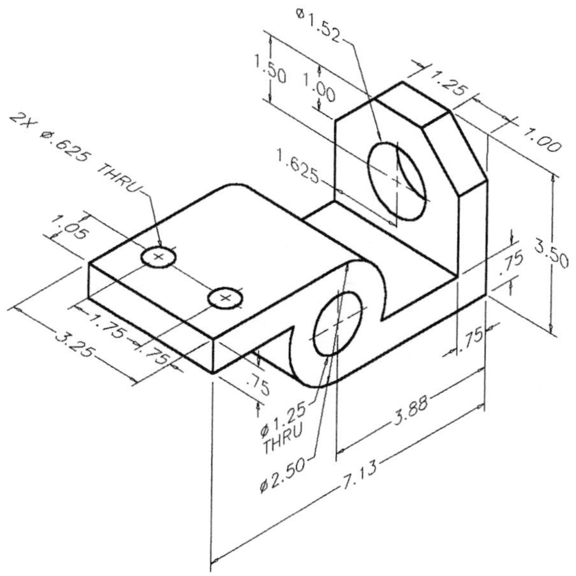

PROBLEM 8–33

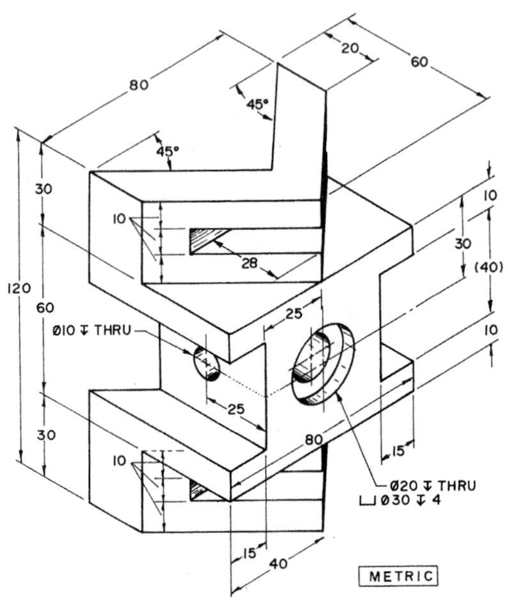

PROBLEM 8–34

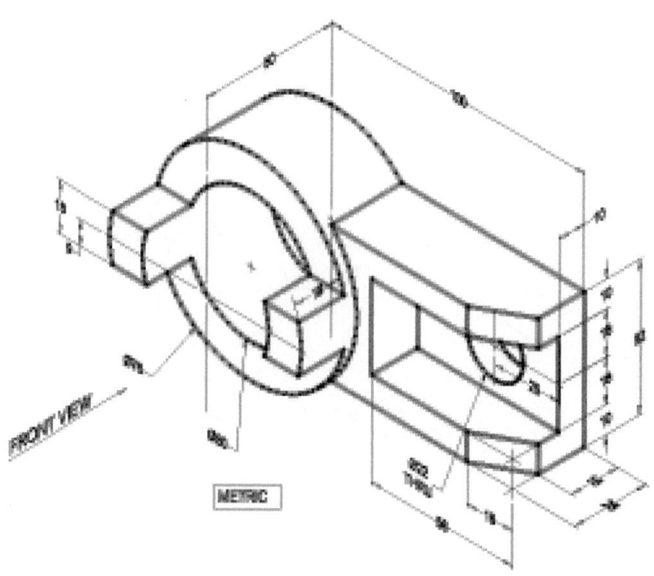

PROBLEM 8–35

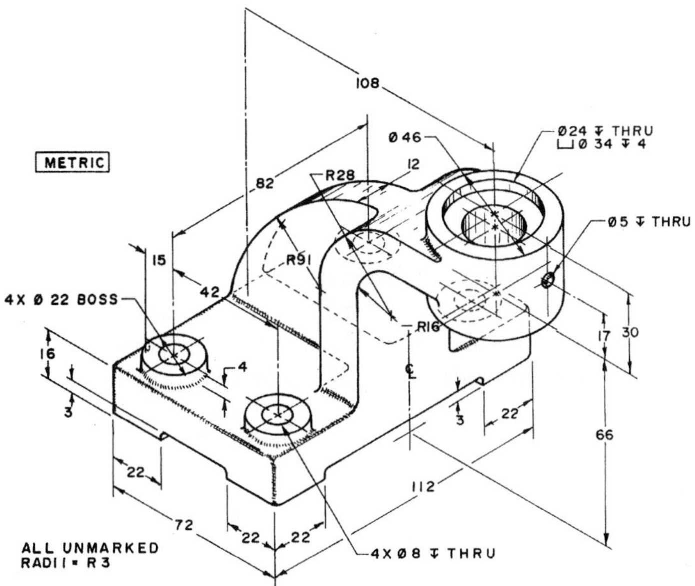

METRIC
108
82
Ø46
Ø24 ↧ THRU
⌴Ø34 ↧ 4
R28
12
Ø5 ↧ THRU
15
R91
42
4X Ø22 BOSS
R16
16
4
17
30
3
3
22
66
22
112
72
22
22
4X Ø8 ↧ THRU
ALL UNMARKED
RADII = R3

CHAPTER 9

Creating Section Views

This chapter will cover section views, which are created by slicing an object along a cutting plane in order to view its interior details. The basics of section views will be covered first. Full, half, assembly, aligned, and offset sections will be shown. Other special section topics include sectioning ribs, creating broken sections, revolved sections, removed sections, and isometric sections. After covering the basics of sections, the chapter will shift to a discussion on how AutoCAD crosshatches objects through the Hatch and Gradient dialog box (BHATCH command). You will be able to select from a collection of many hatch patterns including gradient patterns for special effects when adding hatch patterns to your drawings. Hatch scaling and angle considerations will also be covered along with associative crosshatching. Once hatching exists in a drawing, it can be easily modified through the Hatch Edit dialog box. Automatic hatch scaling in a drawing layout (Paper Space) will also be covered.

SECTION VIEW BASICS

Principles of orthographic projections remain the key method for the production of engineering drawings. As these drawings get more complicated in nature, the job of the designer becomes more challenging in the interpretation of views, especially where hidden features are involved. The concept of slicing a view to expose these interior features is the purpose of performing a section. On the left in the following image is a pictorial representation of a typical flange consisting of eight bolt holes and a counterbore hole in the center.

The drawings on the right in the following image show a typical solution to a multiview problem complete with Front and Side views. The Front view displaying the eight bolt holes is obvious to interpret; however, the numerous hidden lines in the Side view make the drawing difficult to understand, and this is considered a relatively simple drawing. To relieve the confusion associated with a drawing too difficult to understand because of numerous hidden lines, a section drawing is made for the part.

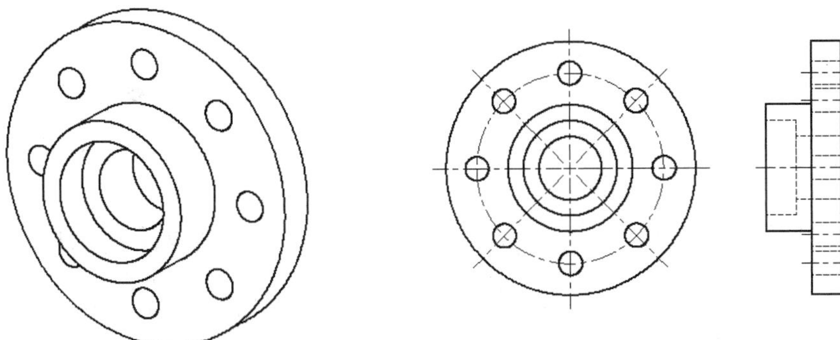

Figure 9.1

To understand section views better, see the illustration on the left in the following image. Creating a section view slices an object in such a way as to expose what used to be hidden features and convert them to visible features. This slicing or cutting operation can be compared to that of using a glass plate or cutting plane to perform the section. In the object shown on the left in the following image, the glass plate cuts the object in half. It is the responsibility of the designer or CAD operator to convert one half of the object to a section and to discard the other half. Surfaces that come in contact with the glass plane are crosshatched to show where material was cut through.

A completed section view drawing is shown on the right in the following image. Two new types of lines are also shown, namely, a cutting plane line and section lines. The cutting plane line performs the cutting operation on the Front view. In the Side view, section lines show the surfaces that were cut. The counterbore edge is shown because this corner would be visible in the section view. Notice that holes are not section lined because the cutting plane passes across the center of the hole. Notice also that hidden lines are not displayed in the Side view. It is poor practice to merge hidden lines into a section view, although there are always exceptions. The arrows of the cutting plane line tell the designer to view the section in the direction of the arrows and discard the other half.

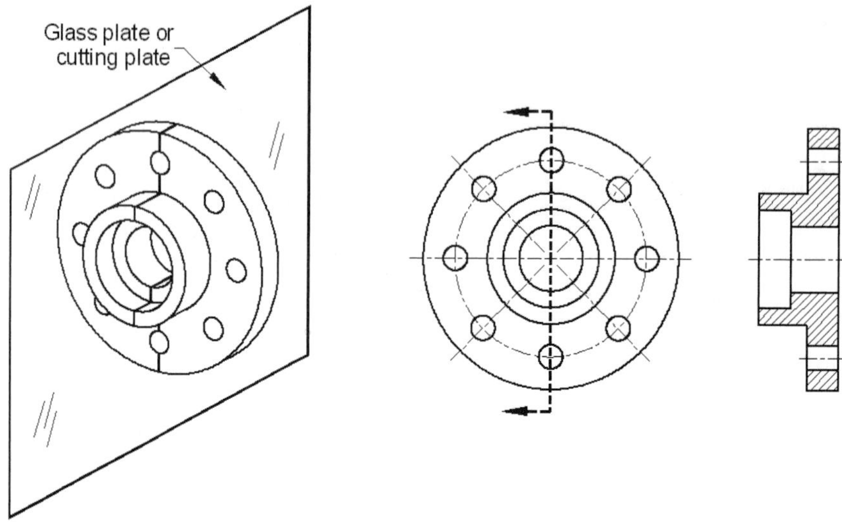

Figure 9.2

The cutting plane line consists of a very thick line with a series of dashes approximately 0.25" in length, as shown on the left in the following image. A polyline may be used to create this line of 0.05" thickness. The arrows point in the direction of sight used to create the section, with the other half generally discarded. Assign this line one of the dashed linetypes; the hidden linetype is reserved for detailing invisible features in views. The section line, in contrast with the cutting plane line, is a very thin line, as shown on the right in the following image. This line identifies the surfaces being cut by the cutting plane line. The section line is usually drawn at an angle and at a specified spacing.

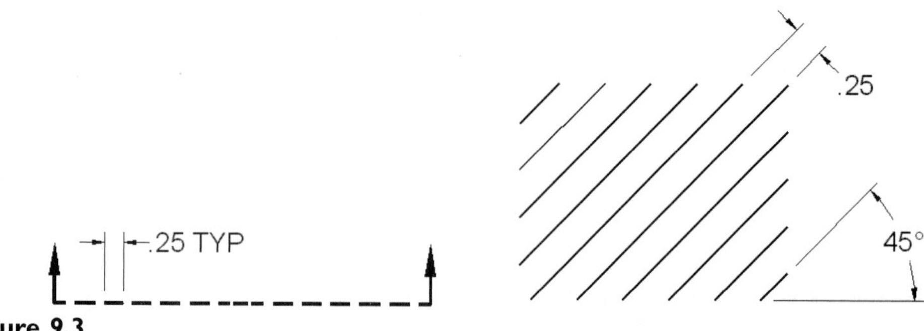

Figure 9.3

A wide variety of hatch patterns is supplied with the software. One of these patterns, ANSI31, is displayed on the left in the following image. This is one of the more popular patterns, with lines spaced 0.125" apart and at a 45° angle.

The object on the right in the following image illustrates proper section lining techniques. Much of the pain of spacing the section lines apart from each other and at angles has been eased considerably with the use of the computer as a tool. However, the designer must still practice proper section lining techniques at all times for clarity of the section.

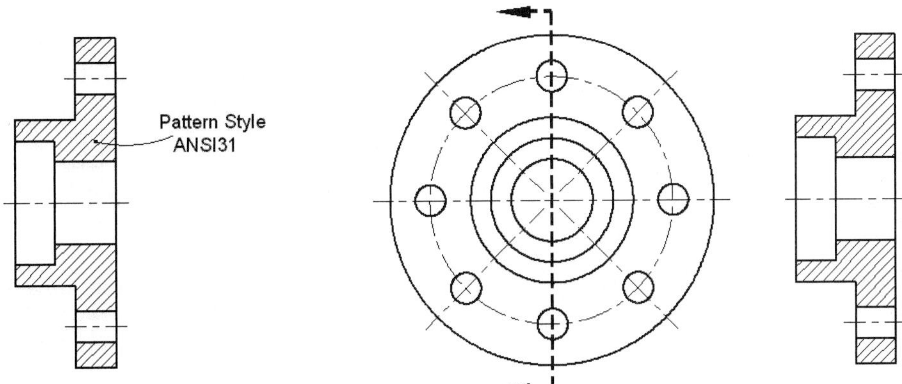

Figure 9.4

The following image illustrates common errors involved when applying section lines to a cutaway view. At "A," the section lines run in the correct direction and at the same angle; however, the hidden lines have not been converted to object lines. This will confuse the more experienced designer because the presence of hidden lines in the section means more complicated invisible features.

At "B" is another error, encountered when sections are created. Again, the section lines are properly placed; however, all surfaces representing holes have been removed, which displays the object as a series of sectioned blocks unconnected, implying four separate parts.

At "C," the object appears to be properly section lined; however, upon closer inspection, notice that the angle of the crosshatch lines in the upper half differs from the angle of the same lines in the lower-left half. This suggests two different parts when, in actuality, it is the same part.

At "D," all section lines run in the correct direction. The problem is that the lines run through areas that were not sliced by the cutting plane line. These areas in "D" represent drill and counterbore holes and are typically left unsectioned.

We use standard orthographic projection to show features of the object in areas where we are not cutting through material.

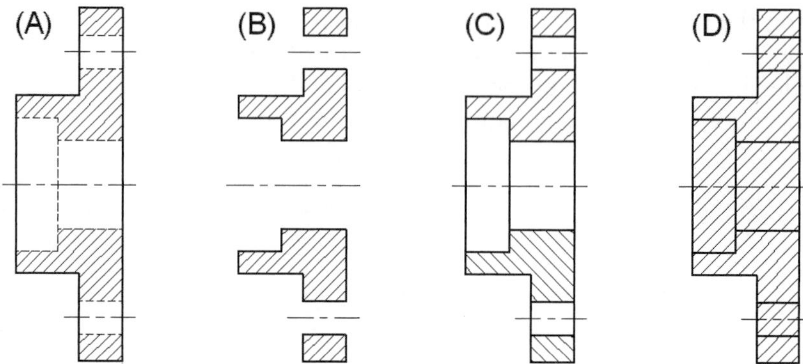

Figure 9.5

These have been identified as the most commonly made errors when an object is crosshatched. Remember just a few rules to follow:

- Section lines are present only on surfaces that are cut by the cutting plane line.
- Section lines are drawn in one direction when the same part is crosshatched.
- Hidden lines are usually omitted when a section view is created.
- Areas such as holes are not sectioned because the cutting line only passes across this feature.

FULL SECTIONS

When the cutting plane line passes through the entire object, a full section is formed. As illustrated on the left in the following image, a full section would be the same as taking an object and cutting it completely down the middle. Depending on the needs of the designer, one half is kept while the other half is discarded. The half that is kept is shown with section lines.

The multiview solution to this problem is shown on the right in the following image. The Front view is drawn with lines projected across to form the Side view. To show that the Side view is in section, a cutting plane line is added to the Front view. This line performs the physical cut. You have the option of keeping either half of the object. This is the purpose of adding arrowheads to the cutting plane line. The arrowheads define the direction of sight in which you view the object to form the section. You must then interpret what surfaces are being cut by the cutting plane line in order to properly add crosshatching lines to the section that is located in the Right Side view. Hidden lines are not necessary once a section has been made.

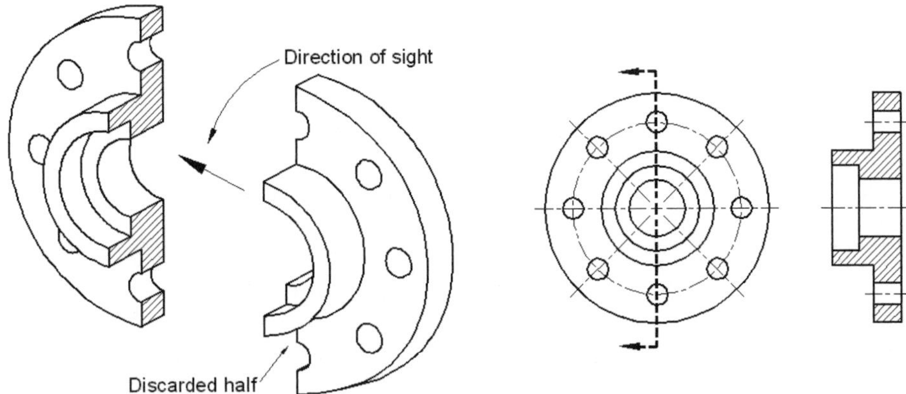

Figure 9.6

Numerous examples illustrate a cutting plane line with the direction of sight off to the left. This does not mean that a cutting plane line cannot have a direction of sight going to the right, as shown in the following image. In this example, the section is formed from the Left Side view if the circular features are located in the Front view.

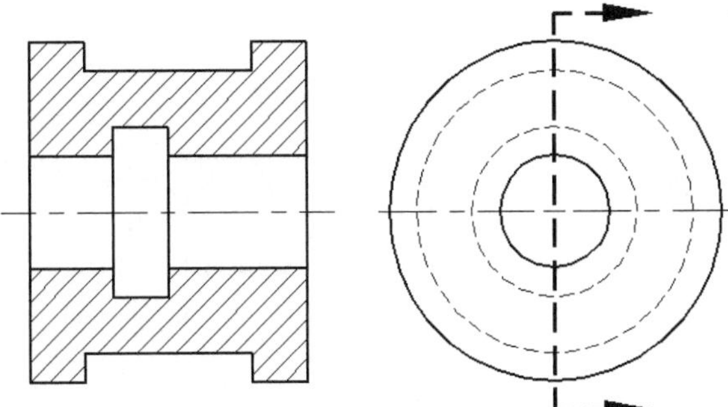

Figure 9.7

HALF SECTIONS

When symmetrical-shaped objects are involved, sometimes it is unnecessary to form a full section by cutting straight through the object. Instead, the cutting plane line passes only halfway through the object, which makes the pictorial drawing as shown on the left in the following image a half section. The rules for half sections are the same as for full sections; namely, once a direction of sight is established, part of the object is kept, and part is discarded.

The views are laid out, as shown on the right in the following image, in the usual multiview format. To prepare the object as a half section, the cutting plane line passes halfway through the Front view before being drawn off to the right. Notice that there is only one direction-of-sight arrow and a centerline is used to depict where the sectioned portion of the object ends in the Right Side view. The Right Side view is converted to a half section by the crosshatching of the upper half of the Side view while hidden lines remain in the lower half.

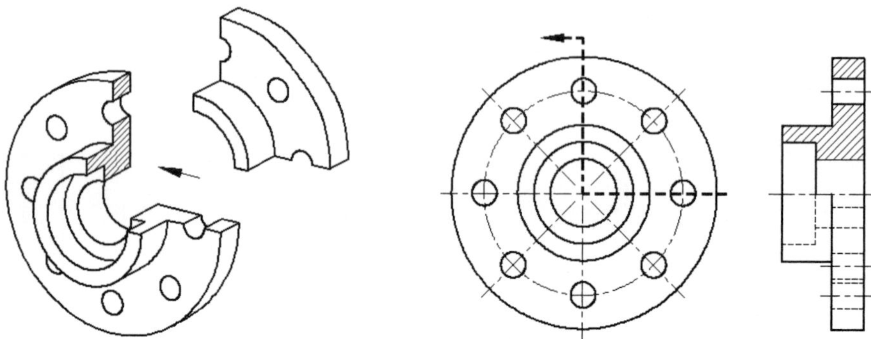

Figure 9.8

Depending on office practices, some designers prefer to omit hidden lines entirely from the Side view, similar to the drawing on the left in the following image. In this way, the lower half is drawn showing only what is visible.

The illustration on the right in the following image shows another way of drawing the cutting plane line to conform to the Right Side view drawn in section. Hidden lines have been removed from the lower half; only those lines visible are displayed. This is consistent with the practice of omitting hidden lines from section views.

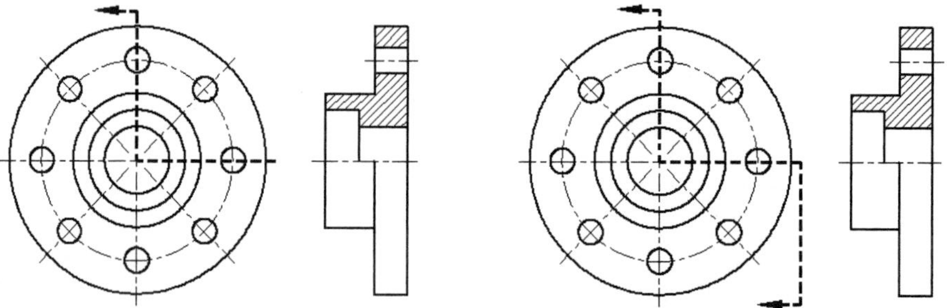

Figure 9.9

ASSEMBLY SECTIONS

It would be unfair to give designers the impression that section views are used only for displaying internal features of individual parts. Another advantage of using section views is that they permit the designer to create numerous objects, assemble them, and then slice the assembly to expose internal details and relationships of all parts. This type of section is an assembly section similar to the illustration on the left in the following image. For all individual parts, notice the section lines running in the same direction. This follows one of the basic rules of section views: keep section lines at the same angle for each individual part.

The illustration on the right in the following image shows the difference between assembly sections and individual parts that have been sectioned. For parts in an assembly that contact each other, it is good practice to alternate the directions of the section lines to make the assembly much clearer and to distinguish the parts from each other. You can accomplish this by changing the angle of the hatch pattern or even the scale of the pattern.

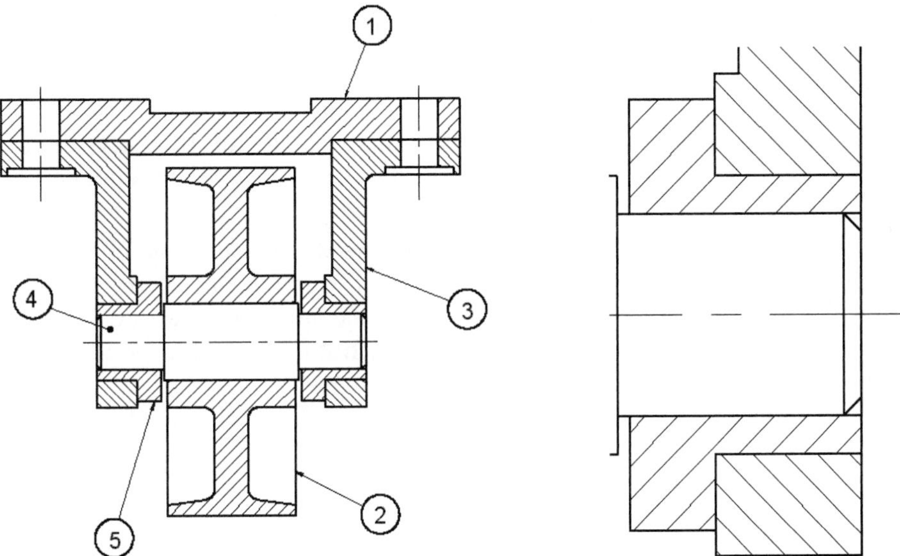

Figure 9.10

To identify parts in an assembly, an identifying part number along with a circle and arrowhead line are used, as shown on the left in the following image. The line is very similar to a leader line used to call out notes for specific parts on a drawing. The addition of the circle highlights the part number. Sometimes this type of callout is referred to as a "bubble."

In the enlarged assembly shown on the right in the following image, the large area in the middle is actually a shaft used to support a pulley system. With the cutting plane passing through the assembly including the shaft, refrain from crosshatching features such as shafts, screws, pins, or other types of fasteners. The overall appearance of the assembly is actually enhanced when these items are not crosshatched.

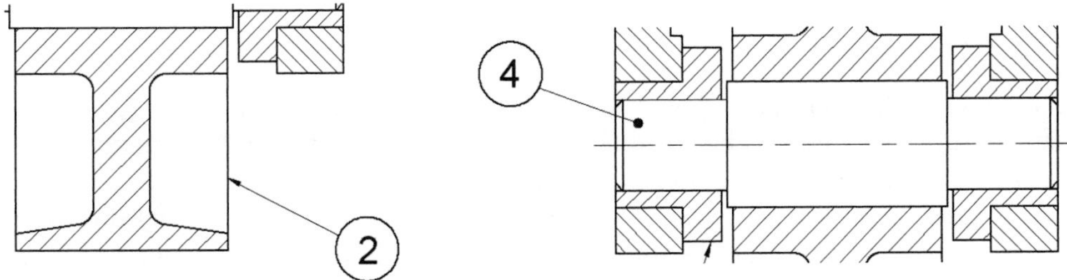

Figure 9.11

ALIGNED SECTIONS

Aligned sections take into consideration the angular position of details or features of a drawing. Instead of the cutting plane line being drawn vertically through the object, as shown on the left in the following image, the cutting plane is angled or aligned with the same angle as the elements it is slicing. Aligned sections are also made to produce better clarity of a drawing. As illustrated on the left in the following image, with the cutting plane forming a full section of the object, it is difficult to obtain the true size of the angled elements. In the Side view, they appear foreshortened or not to scale. Hidden lines were added as an attempt to better clarify the view.

Instead of the cutting plane line being drawn all the way through the object, the line is bent at the center of the object before being drawn through one of the angled legs, as shown on the right in the following image. The direction of sight arrows on the cutting plane line not only determines the direction in which the view will be sectioned, but also shows another direction for rotating the angled elements so they line up with the upper elements. This rotation is usually not more than 90°. As lines are projected across to form the Side view, the section appears as if it were a full section. This is only because the features were rotated and projected in section for greater clarity of the drawing. It is interesting to note that the aligned section is drawn higher than the Front view from which it is drawn. Standard orthographic projection is usually not applicable to aligned sections.

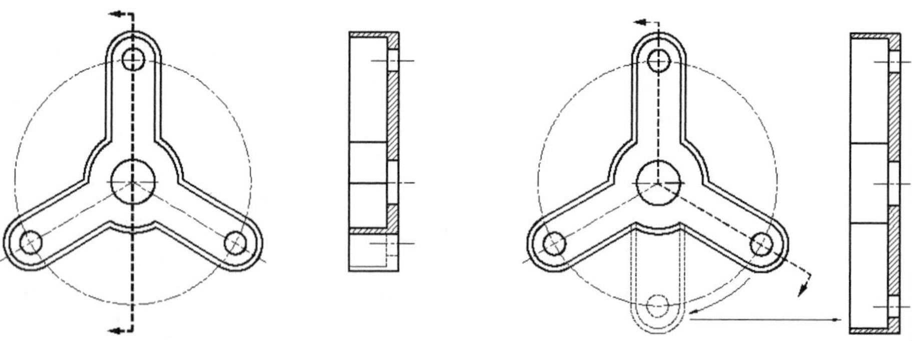

Figure 9.12

OFFSET SECTIONS

Offset sections take their name from the offsetting of the cutting plane line to pass through certain features in a view, as shown in the following image. If the cutting plane line passes straight through any part of the object, some features would be exposed while others would remain hidden. By offsetting the cutting plane line, the designer controls its direction and which features of a part it passes through. The view to section follows the basic section rules. Notice that the changes in direction of the cutting plane as it passes through the object are shown in the Top view but not in the Front view.

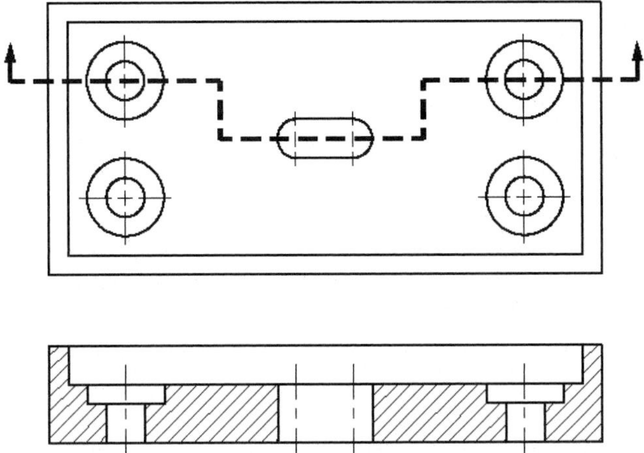

Figure 9.13

SECTIONING RIBS

Parts made out of cast iron with webs or ribs used for reinforcement do not follow the basic rules of sections. As illustrated on the left in the following image, the Front view has the cutting plane line passing through the entire view; the Side view at "A" is crosshatched according to the basics of section views. However, it is difficult to read the thickness of the base because the crosshatching includes the base along with the web. A more efficient method is to leave off crosshatching webs, as in "B." In this example, eliminating the crosshatching of the web exposes other important details such as thickness of bases and walls.

The object shown on the right in the following image is another example of performing a full section on an area consisting of webbed or ribbed features. If you do not crosshatch the webbed areas, more information is available, such as the thickness of the base and wall areas around the cylindrical hole. While this may not be considered true projection, it is good practice.

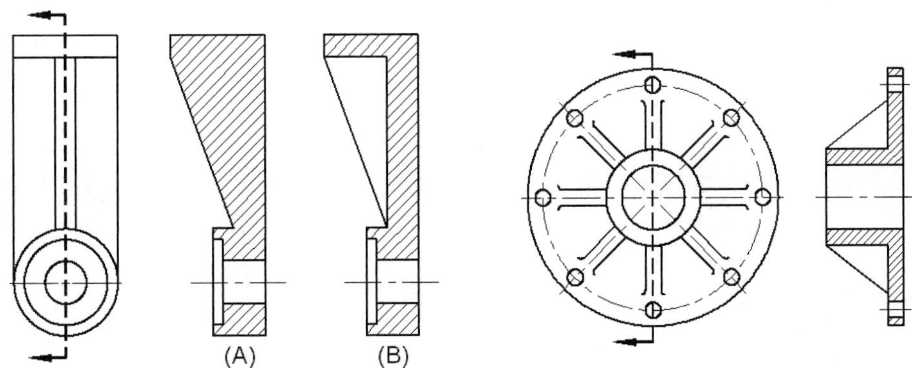

Figure 9.14

BROKEN SECTIONS

At times, only a partial section of an area needs to be created. For this reason, a broken section might be used. The object illustrated on the left in the following image at "A" shows a combination of sectioned areas and conventional areas outlined by the hidden lines. When converting an area to a broken section, you create a break line, crosshatch one area, and leave the other area in a conventional drawing. Break lines may take the form of short freehanded line segments, as shown at "B" in the following image, or a series of long lines separated by break symbols, as shown at "C" in the following image. You can use the AutoCAD LINE command with Ortho off to produce the desired effect, as in "A" and "B."

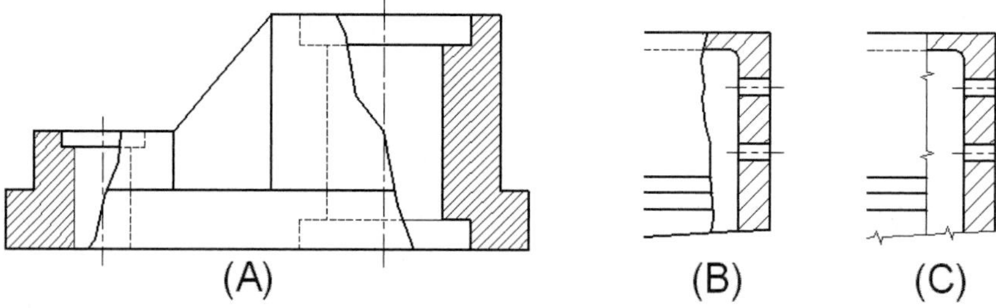

Figure 9.15

REVOLVED SECTIONS

Section views may be constructed as part of a view with revolved sections. As illustrated on the left in the following image, the elliptical shape is constructed while it is revolved into position and then crosshatched. The crosshatched shape represents the cross section shape of the arm.

Illustrated on the right in the following image is another example of a revolved section where a cross section of the C-clamp was cut away and revolved to display its shape.

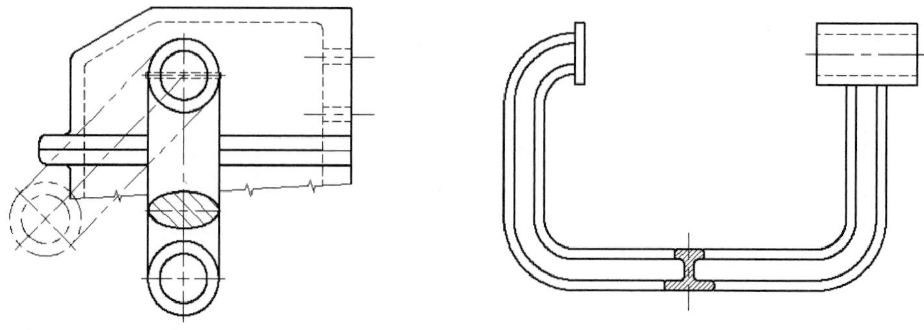

Figure 9.16

REMOVED SECTIONS

Removed sections are very similar to revolved sections except that instead of the section being drawn somewhere inside the object, as is the case with a revolved section, the section is placed elsewhere or removed to a new location in the drawing. The cutting plane line is present, with the arrows showing the direction of sight. Identifying letters are placed on the cutting plane and underneath the section to keep track of the removed sections, especially when there are a number of them on the same drawing sheet. See the illustration on the left in the following image.

Another way of displaying removed sections is to use centerlines as a substitute for the cutting plane line. As illustrated on the right in the following image, the centerlines determine the three shapes of the chisel and display the basic shapes. Identification numbers are not required in this particular example.

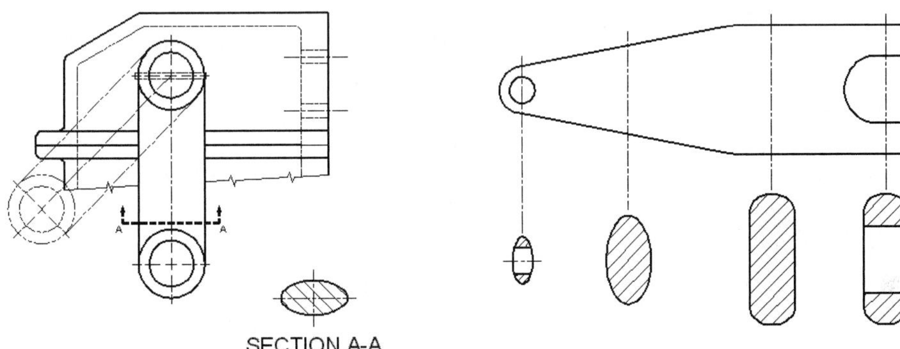

Figure 9.17

ISOMETRIC SECTIONS

Section views may be incorporated into pictorial drawings that appear in the following image. The illustration on the left is an example of a full isometric section with the cutting plane passing through the entire object. In keeping with basic section rules, only those surfaces sliced by the cutting plane line are crosshatched. Isometric sections make it easy to view cut edges compared to holes or slots. The illustration on the right is an example of an isometric drawing converted to a half section.

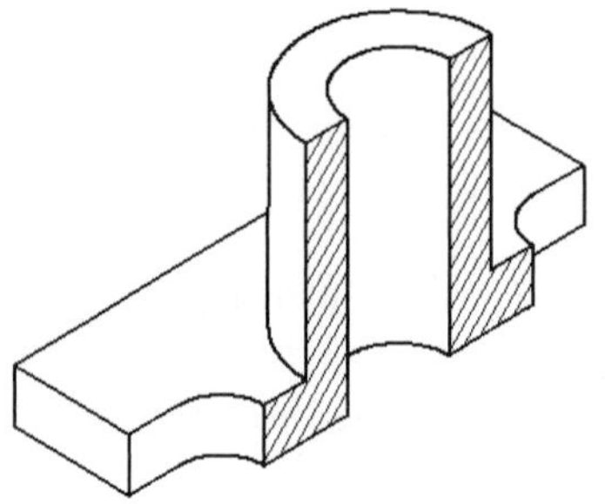
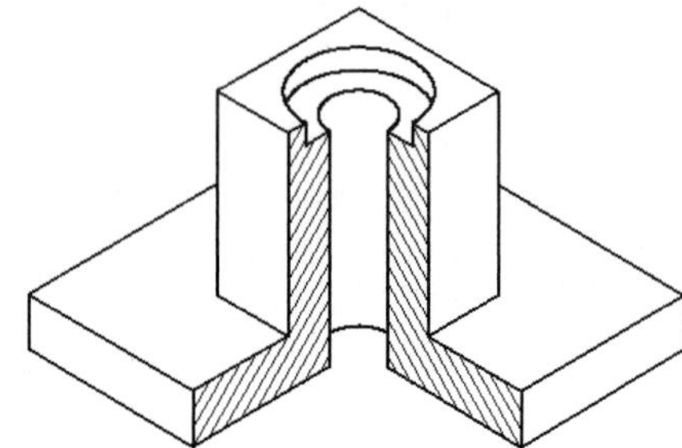

Figure 9.18

ARCHITECTURAL SECTIONS

Mechanical representations of machine parts are not the only type of drawings where section views are used. Architectural drawings rely on sections to show the building materials and construction methods for the construction of foundation plans, roof details, or wall sections, as shown in the following image. Here, numerous types of crosshatching symbols are used to call out the different types of building materials, such as brick veneer at "A," insulation at "B," finished flooring at "C," floor joists at "D," concrete block at "E," earth at "F," and poured concrete at "G." Section hatch patterns that are provided with AutoCAD may be used to crosshatch most building components.

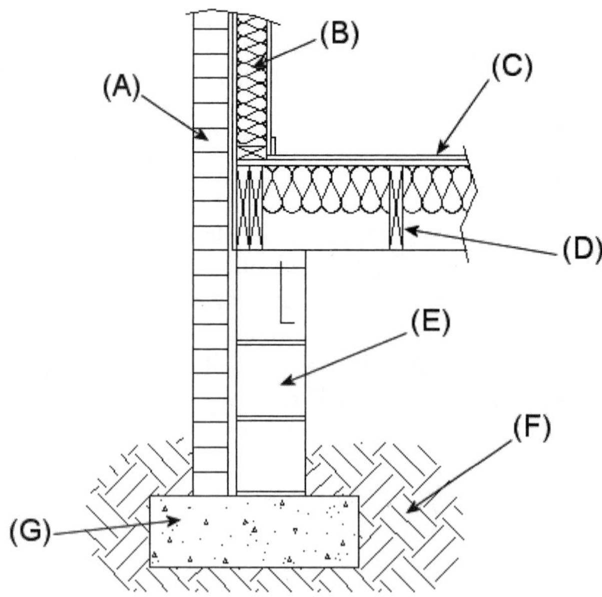

Figure 9.19

THE BHATCH COMMAND

With the basics of sections having been covered, it is time now to see how AutoCAD is used to generate section lines. This is accomplished by using the Boundary Hatch, or BHATCH, command. This command can be chosen from any one of the following menus: the Menu Browser, pull-down menu, Ribbon, or Draw Toolbar as shown in the following image.

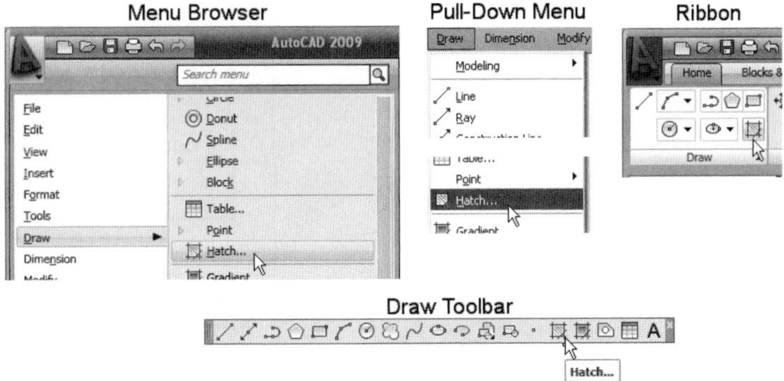

Figure 9.20

Clicking either one of the Hatch buttons in the previous image activates the Hatch and Gradient dialog box, as shown in the following image. Use the Hatch tab to pick points identifying the area or areas to be crosshatched, select objects using one or more of the popular selection set modes, select hatch patterns supported in the software, and change the scale and angle of the hatch pattern. Any areas grayed out in this dialog box are considered inactive. Only when additional parameters have been satisfied will these buttons activate for use.

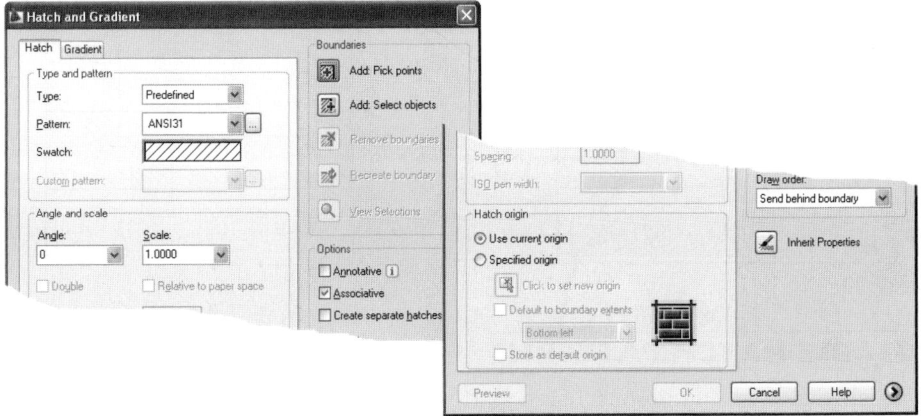

Figure 9.21

Try It!: Open the drawing file 09_Hatch Basics. The object shown on the left in the following image will be used to demonstrate the boundary crosshatching method. The object needs to have areas "A," "B," and "C" crosshatched.

Areas "A," "B," and "C" first need to be identified as the boundaries to be hatched. The BHATCH command is a very efficient method used for automatically highlighting areas to be crosshatched, as shown on the right in the following image.

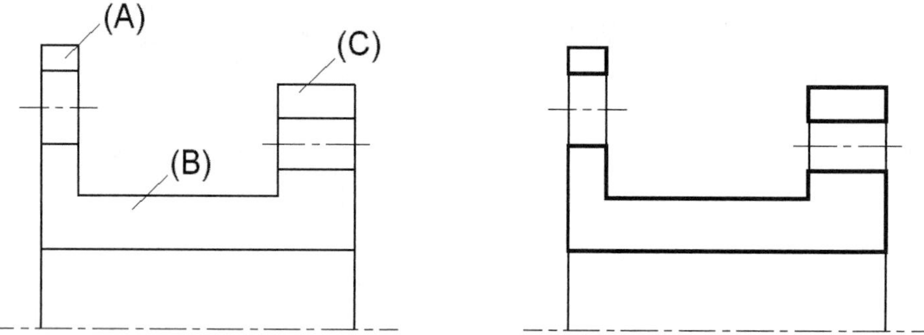

Figure 9.22

The BHATCH command opens the Hatch tab of the Hatch and Gradient dialog box. Click the Add: Pick points button found in the upper-right corner of this dialog box. The illustration on the left in the following image shows three areas with internal points identified by the Xs. The dashed areas identify the highlighted areas to be crosshatched. In the final step of the BHATCH command, the crosshatch pattern is applied to the areas, and the highlighted outlines deselect, leaving the crosshatch patterns. See the illustration on the right in the following image.

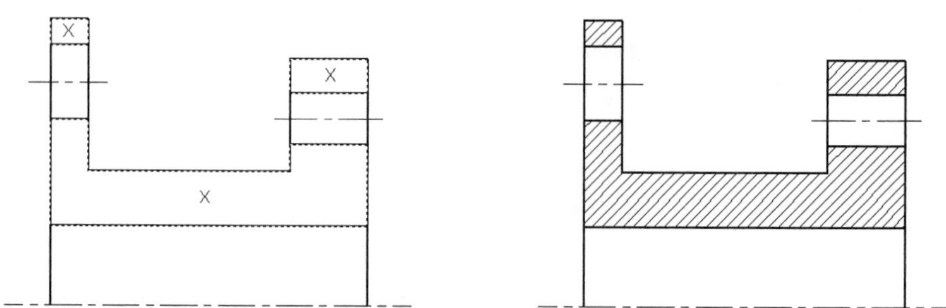

Figure 9.23

AVAILABLE HATCH PATTERNS

The Hatch tab holds numerous hatch patterns in a drop-down list box, shown in the following image. Clicking on a pattern in this listing makes the pattern current.

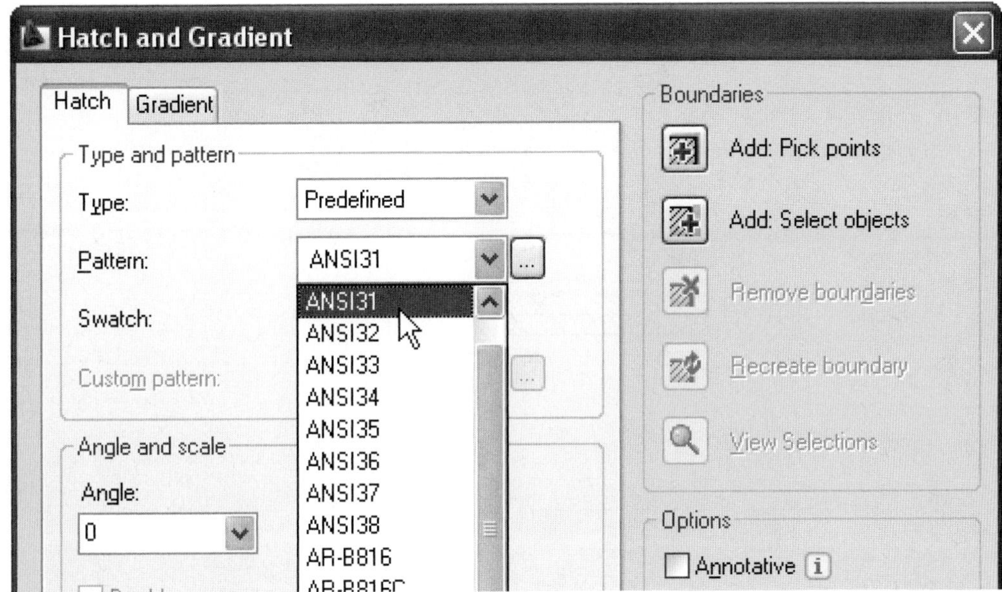

Figure 9.24

Selecting the pattern (...) button or ellipses next to the pattern name, as shown on the left in the following image, activates the Hatch Pattern Palette dialog box. Here, a series of tabs are present that display a number of crosshatching patterns already created and ready for use, as shown on the right in the following image. Select a particular pattern by clicking on the pattern itself. If the wrong pattern was selected, simply choose the correct one or move to another tab to view other patterns for use.

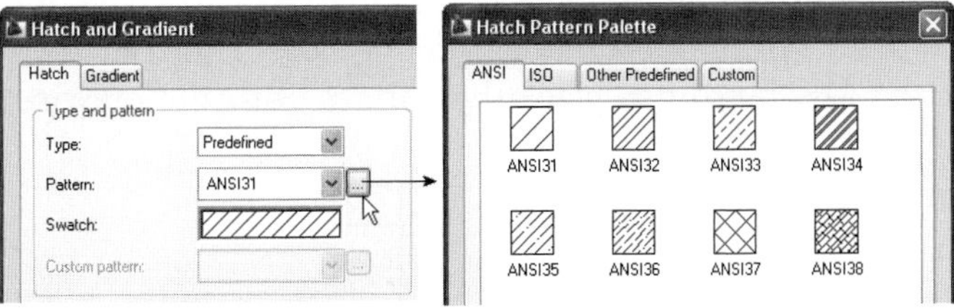

Figure 9.25

Once a hatch pattern is selected, the name is displayed in the Hatch and Gradient dialog box, as shown in the following image. If the scaling and angle settings look favorable, click the Add: Pick points button. The command prompt requires you to select an internal point or points.

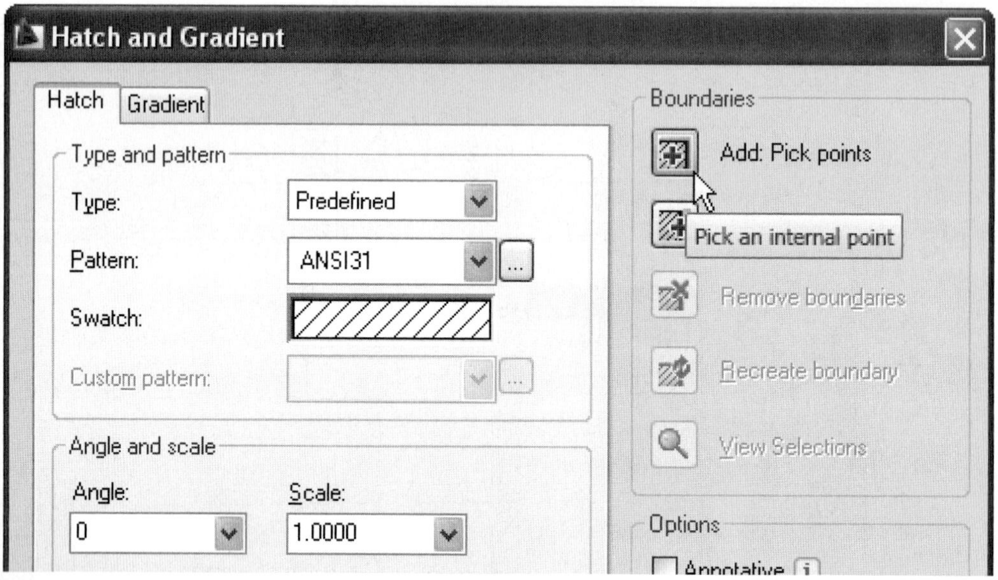

Hatch and Gradient

Hatch | Gradient

Type and pattern
Type: Predefined
Pattern: ANSI31
Swatch:
Custom pattern:

Angle and scale
Angle: 0
Scale: 1.0000

Boundaries
Add: Pick points
Pick an internal point
Remove boundaries
Recreate boundary
View Selections

Options
Annotative

Figure 9.26

Continue working with the drawing file 09_Hatch Basics. Click in the three areas marked by the X, as shown on the left in the following image. Notice that these areas highlight. If these areas are correct, press ENTER to return to the Hatch and Gradient dialog box.

When you are back in the Hatch and Gradient dialog box, you now have the option of first previewing the hatch to see if all settings and the appearance of the pattern are desirable. Click the Preview button in the lower-left corner of the dialog box to accomplish this. The results appear on the right in the following image. At this point in the process, the pattern is still not permanently placed on the object. Pressing the ENTER key to exit Preview mode returns you to the main Hatch and Gradient dialog box. If the hatch pattern is correct in appearance, click the OK button to place the pattern with the drawing.

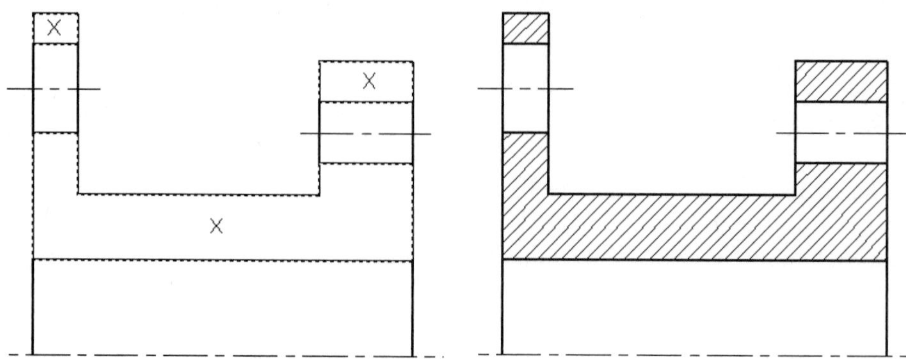

Figure 9.27

SOLID FILL HATCH PATTERNS

Activating the Hatch and Gradient dialog box and clicking on the pattern (...) button or ellipses followed by the Other Predefined tab displays the dialog box as shown on the left in the following image. Additional patterns are available and include brick, roof, and other construction-related materials. One powerful pattern is called SOLID. This pattern is used to fill in an enclosed area with a solid color based on the current layer. The illustration on the right in the following image shows a typical application of filling in the walls of a floor plan with a solid color.

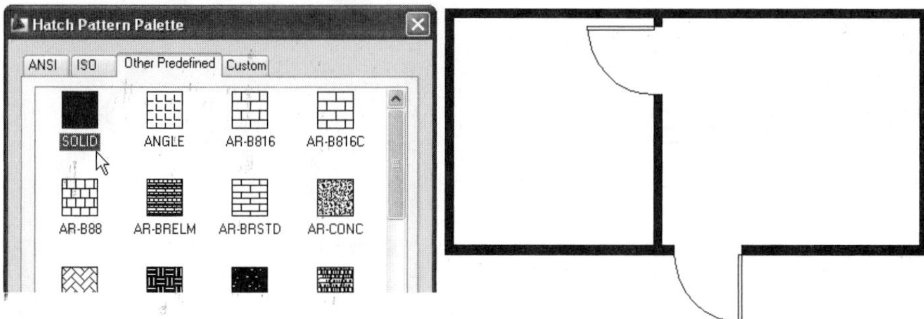

Figure 9.28

Try It!: Open the drawing file 09_Hatch Solid1. The object shown on the left in the following image illustrates a thin wall that needs to be filled in. The BHATCH command used with the SOLID hatch pattern can perform this task very efficiently. When you pick an internal point, the entire closed area is filled with a solid pattern and placed on the current layer with the same color and other properties of the current layer.

The solid hatch pattern shown on the right in the following image was placed with one internal point pick.

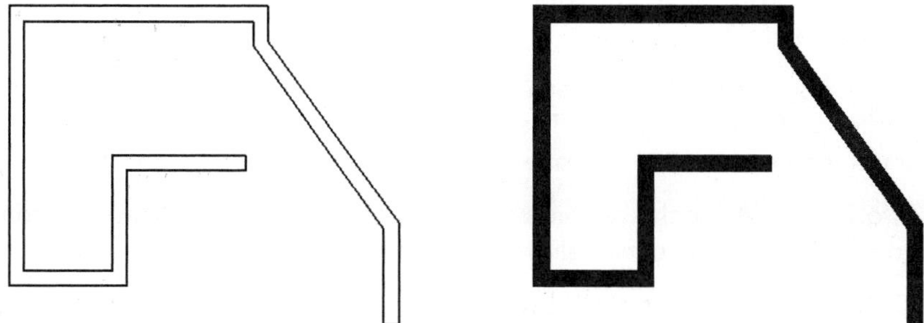

Figure 9.29

Try It!: Open the drawing file 09_Hatch Solid2. Applying the SOLID hatch pattern to the object, as shown on the left in the following image, can easily be accomplished with the BHATCH command. This object also is a thin wall example, with the addition of various curved edges and fillets.

The solid hatch pattern also works for curved outlines. The illustration on the right in the following image displays the solid pattern, completely filling in the outline where all corners consist of some type of curve generated by the FILLET command.

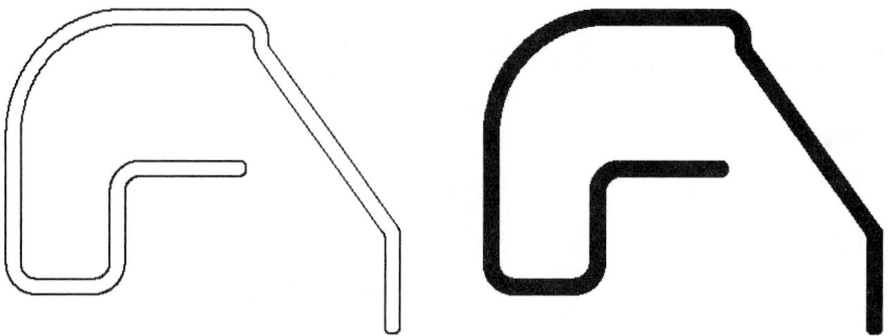

Figure 9.30

GRADIENT PATTERNS

A gradient hatch pattern is a solid hatch fill that makes a smooth transition from a lighter shade to a darker shade. Predefined patterns such as linear, spherical, and radial sweep are available to provide different effects. As with the vector hatch patterns that have always been supplied with AutoCAD, the angle of the gradient patterns can also be controlled. Gradient patterns can also be associative; the pattern updates if the boundary is changed. To edit a gradient pattern, simply double-click it to launch the Hatch Edit dialog box. As with hatch patterns, clicking on Gradient in any of the following menu items will launch the Gradient tab of the Hatch and Gradient dialog box: Menu Browser, Draw Toolbar, pull-down menu, or Ribbon. A few of the controls for this tab of the dialog box are explained as follows:

- One Color—When this option is selected, a color swatch with a Browse button and a Shade and Tint slider appears. The One Color option designates a fill that uses a smooth transition between darker shades and lighter tints of a single color.
- Two Color—When this option is selected, a color swatch and Browse button display for colors 1 and 2. This option allows the fill to transition smoothly between two colors.
- Color Swatch—This is the default color displayed as set by the current color in the drawing. This option specifies the color to be used for the gradient fill. The presence of the three dots, called ellipses (...), next to the color swatch signifies the Browse button. Use this to display the Select Color dialog box, similar to the dialog used for assigning colors to layers. Use this dialog box to select color based on the AutoCAD Index Color, True Color, or Color Book.

- Shade and Tint Slider—This option designates the amount of tint and shade applied to the gradient fill of one color. Tint is defined as the selected color mixed with white. Shade is defined as the selected color mixed with black.
- Centered Option—Use this option for creating special effects with gradient patterns. When this option is checked, the gradient fill appears symmetrical. If the option is not selected, AutoCAD shifts the gradient fill up and to the left. This position creates the illusion that a light source is located to the left of the object.
- Angle—Set an angle that affects the gradient pattern fill. This angle setting is independent of the angle used for regular hatch patterns.
- Gradient Patterns—Nine gradient patterns are available for you to apply. These include linear sweep, spherical, and parabolic.

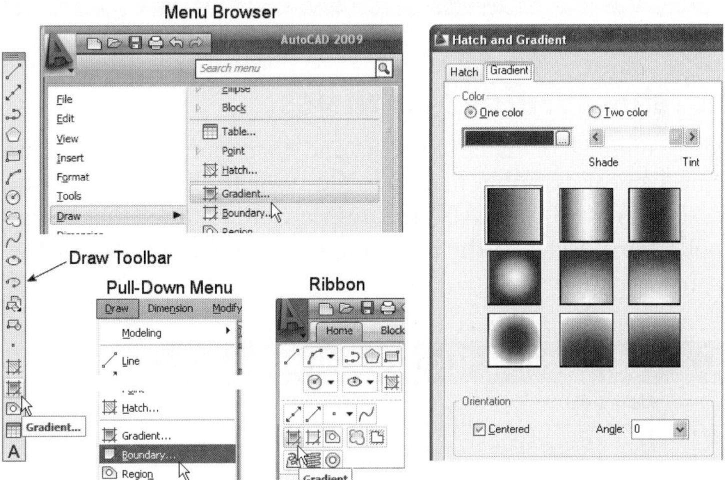

Figure 9.31

The example of the house elevation in the following image illustrates parabolic linear sweep being applied above and below the house.

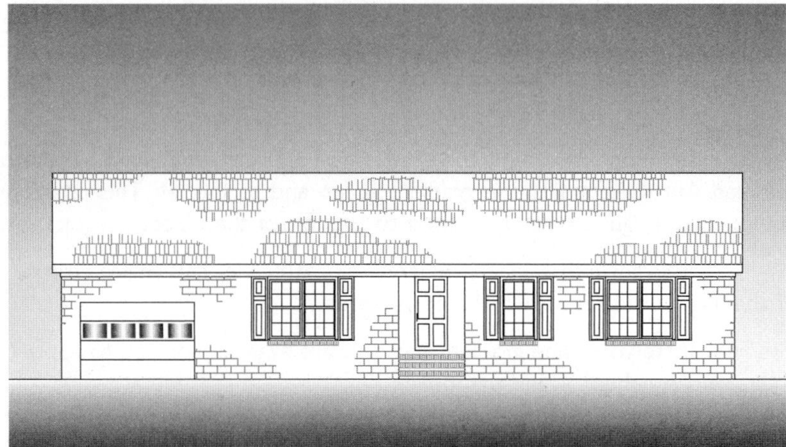

Figure 9.32

HATCH PATTERN SYMBOL MEANINGS

Listed below are a number of hatch patterns and their purposes. For example, if you are constructing a mechanical assembly in which you want to distinguish plastic material from steel, use the patterns ANSI34 and ANSI32, respectively. Patterns that begin with AR- have architectural applications. Refer to the following list for the purposes of other materials and their associated hatch patterns.

ANSI31	ANSI Iron, Brick, Stone masonry
ANSI32	ANSI Steel
ANSI33	ANSI Bronze, Brass, Copper
ANSI34	ANSI Plastic, Rubber
ANSI35	ANSI Fire brick, Refractory material
ANSI36	ANSI Marble, Slate, Glass
ANSI37	ANSI Lead, Zinc, Magnesium, Sound/Heat/Elec Insulation
ANSI38	ANSI Aluminum
AR-B816	8x16 Block elevation stretcher bond
AR-B816C	8x16 Block elevation stretcher bond with mortar joints
AR-B88	8x8 Block elevation stretcher bond
AR-BRELM	Standard brick elevation English bond with mortar joints
AR-BRSTD	Standard brick elevation stretcher bond

ANSI31	ANSI Iron, Brick, Stone masonry
AR-CONC	Random dot and stone pattern
AR-HBONE	Standard brick herringbone pattern @ 45 degrees
AR-PARQ1	2x12 Parquet flooring: pattern of 12x12

HATCHING ISLANDS

Try It!: Open the drawing file 09_Hatch Islands. When charged with the task of crosshatching islands, first activate the Hatch and Gradient dialog box and select a pattern. Select the Add: Pick points button in the upper-right corner of the dialog box. Next pick a point at "A" on the left in the following image, which will define not only the outer perimeter of the object but the inner shapes as well. This result is due to Flood mode in the Island Detection area being activated in the Advanced tab of the Hatch and Gradient dialog box.

Selecting the Preview button displays the hatch pattern shown on the right in the following image. If changes need to be made, such as a change in the hatch scale or angle, the preview allows these changes to be made. After making changes, be sure to preview the pattern once again to check whether the results are desirable. Choosing OK places the pattern and exits to the command prompt.

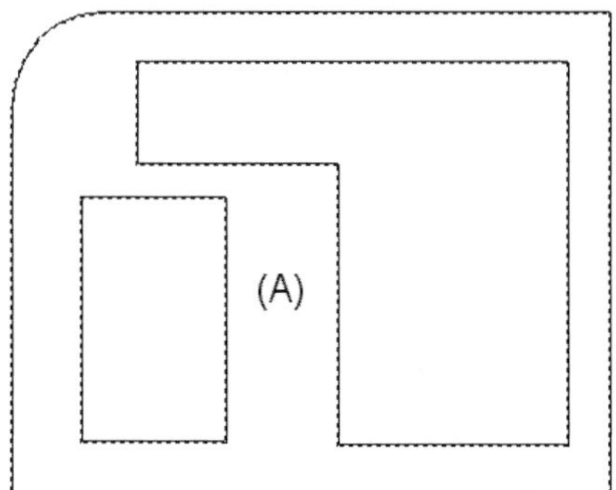

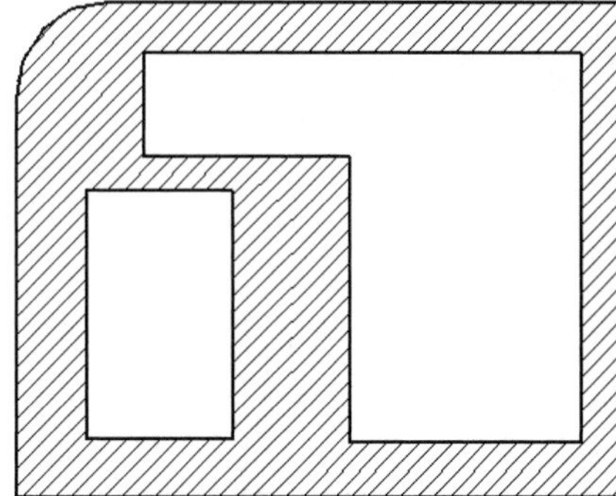

Figure 9.33

HATCH PATTERN SCALING

Hatching patterns are predefined in size and angle. When you use the BHATCH command, the pattern used is assigned a scale value of 1.00, which will draw the pattern exactly the way it was originally created.

Try It!: Open the drawing file 09_Hatch Scale. Activate the Hatch and Gradient dialog box, accept all default values, and pick internal points to hatch the object as shown on the left in the following image.

Entering a different scale value for the pattern in the Hatch and Gradient dialog box either increases or decreases the spacing between crosshatch lines. The illustration on the right in the following image is an example of the ANSI31 pattern with a new scale value of 0.50.

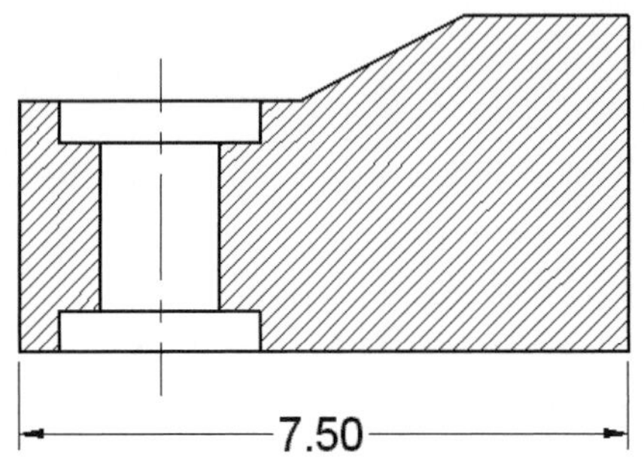

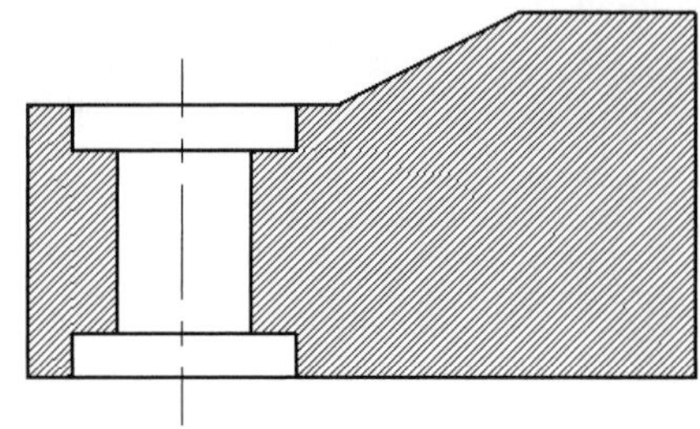

Figure 9.34

As you can decrease the scale of a pattern to hatch small areas, you can also scale the pattern up for large areas. The illustration in the following image has a hatch scale of 2.00, which doubles all distances between hatch lines.

Use care when hatching large areas. In this image, the total length of the object measures 190.50 millimeters. If the hatch scale of 1.00 were used, the pattern would take on a filled appearance. This results in numerous lines being generated, which increases the size of the drawing file. A value of 25.4, the number of millimeters in an inch, is used to scale hatch lines for metric drawings.

Try It!: Open the drawing file 09_Hatch Scale mm. Activate the Hatch and Gradient dialog box and change the pattern scale to 25.4 units. Click the Add: Pick points button to pick internal points to crosshatch the object illustrated in the following image.

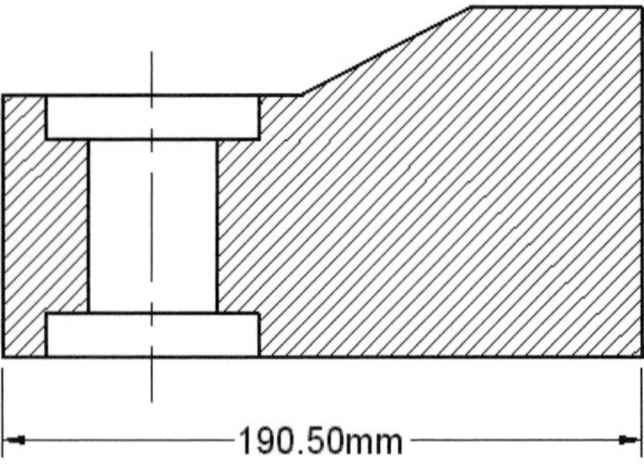

Figure 9.35

HATCH PATTERN ANGLE MANIPULATION

As with the scale of the hatch pattern, depending on the effect, you can control the angle for the hatch pattern within the area being hatched. By default, the BHATCH command displays a 0° angle for all patterns.

Try It!: Open the drawing file 09_Hatch Angle. Activate the Hatch and Gradient dialog box and hatch the object as shown on the left in the following image, keeping all default values. The angle for ANSI31 is drawn at 45°—the angle in which the pattern was originally created.

Experiment with the angle setting of the Hatch and Gradient dialog box by entering any angle different from the default value of "0" to rotate the hatch pattern by that value. This means that if a pattern were originally designed at a 45° angle, like ANSI31, entering a new angle for the pattern would begin rotating the pattern starting at the 45° position.

Entering an angle other than the default rotates the pattern from the original angle to a new angle. In the illustration on the right in the following image, an angle of 90° has been applied to the ANSI31 pattern in the Hatch and Gradient dialog box. Providing different angles for patterns is useful when you create section assemblies in which different parts are in contact with each other because it makes the different parts easy to see.

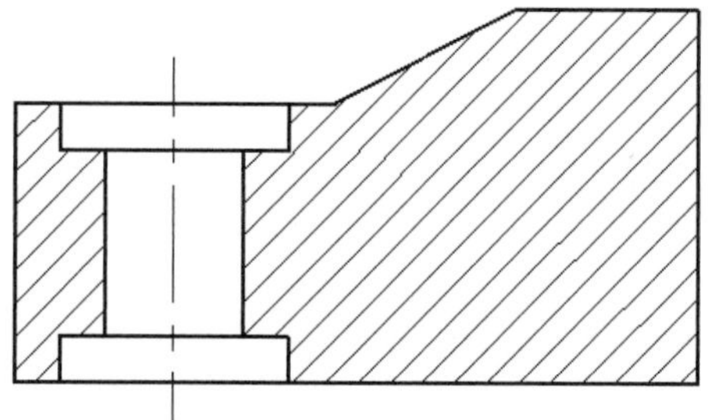

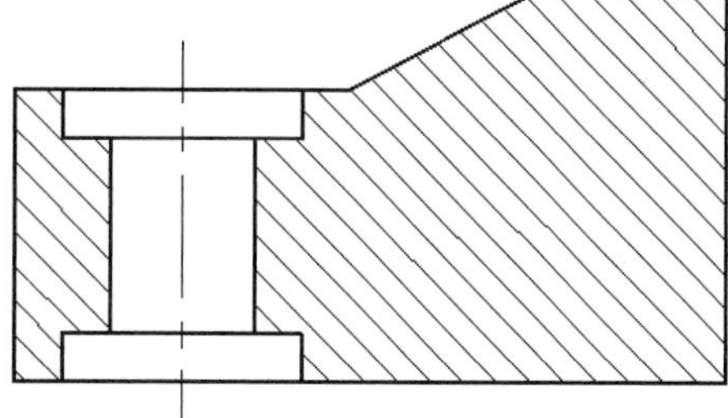

Figure 9.36

TRIMMING HATCH AND FILL PATTERNS

Hatch patterns can also be trimmed when you add objects to a drawing. In the illustration on the left in the following image, the rectangle is constructed and hatched with the ANSI31 pattern. In the image on the right, two circles have been added to the ends of the rectangle. You now want to remove the hatch pattern from both halves of the circles.

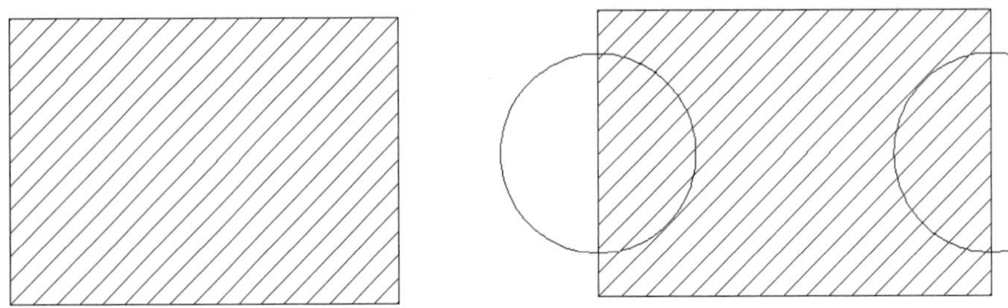

Figure 9.37

To accomplish this, enter the TRIM command and select the edges of both circles as cutting edges, as shown on the left in the following image. Then pick the hatch patterns inside both circles. The results are illustrated on the right, with the hatch pattern trimmed out of both circles.

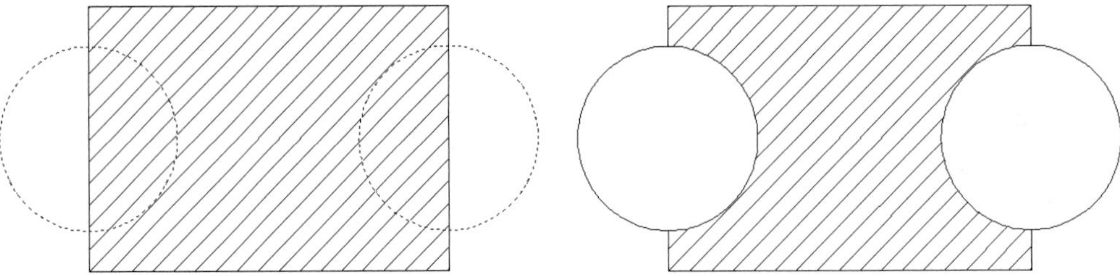

Figure 9.38

MODIFYING ASSOCIATIVE HATCHES

Try It!: Open the drawing file 09_Hatch Assoc. The following is a review of associative hatching. In the following image, the plate needs to be hatched in the ANSI31 pattern at a scale of one unit and an angle of zero; the two slots and three holes are to be considered islands. Enter the BHATCH command, make the necessary changes in the Hatch and Gradient dialog box (including making sure Associative is picked in the Options part of the dialog box), click the Add: Pick points button, and mark an internal point somewhere inside the object, such as at "A," as shown in the following image.

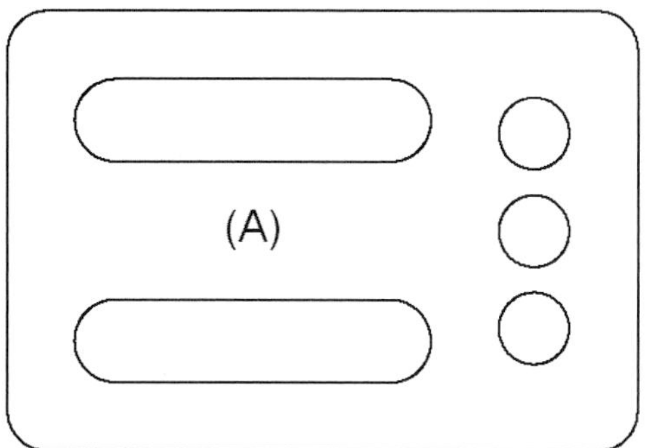

Figure 9.39

When all objects inside and including the outline highlight, press ENTER to return to the Hatch and Gradient dialog box. It is very important to realize that all objects highlighted are tied to the associative crosshatch object. Each shape works directly with the hatch pattern to ensure that the outline of the object is being read by the hatch pattern and that the hatching is performed outside the outline. You have the option of first previewing the hatch pattern or applying the hatch pattern. In either case, the results should appear similar to the object on the left in the following image.

Associative hatch objects may be edited, and the results of this editing have an immediate impact on the appearance of the hatch pattern. For example, the two outer holes need to be increased in size by a factor of 1.5; also, the middle hole needs to be repositioned to the other side of the object, as shown on the right in the following image. Not only does using the MOVE command allow you to reposition the hole, but when the move is completed, the hatch pattern mends itself automatically to the moved circle. In the same manner, using the SCALE command to increase the size of the two outer circles by a value of 1.5 units makes the circles larger and automatically mends the hatch pattern.

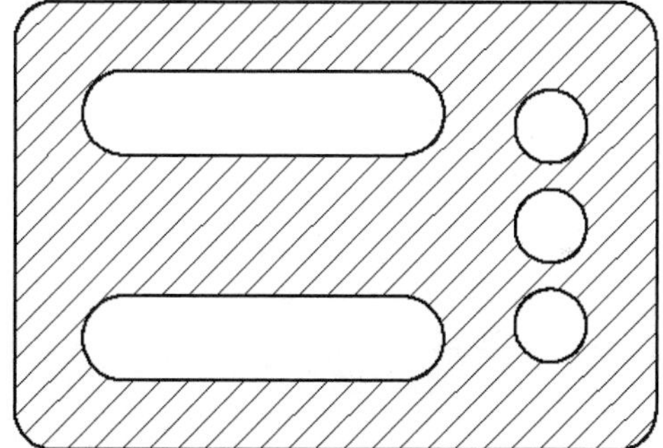

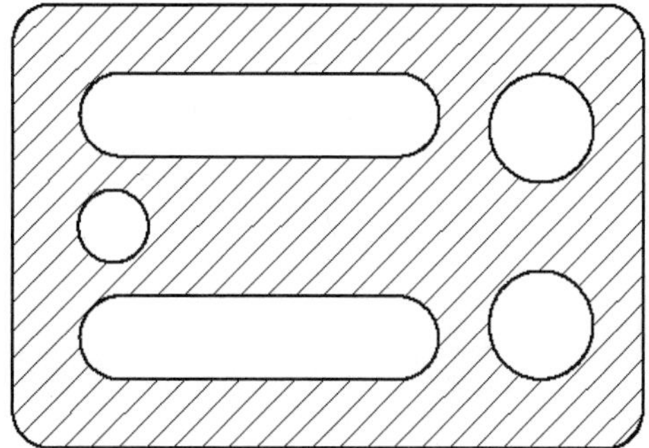

Figure 9.40

On the left in the following image, the length of the slots needs to be shortened. Also, hole "A" needs to be deleted. Use the STRETCH command to shorten the slots. Use the crossing box at "B" to select the slots and stretch them one unit to the right. Use the ERASE command to delete hole "A."

The result of the editing operations is shown on the right in the following image. In this figure, associative hatching has been enhanced for members to be erased while the hatch pattern still maintains its associativity. The inner hole was deleted with the ERASE command. After the ERASE command was executed, the hatch pattern updated itself and hatched through the area once occupied by the hole.

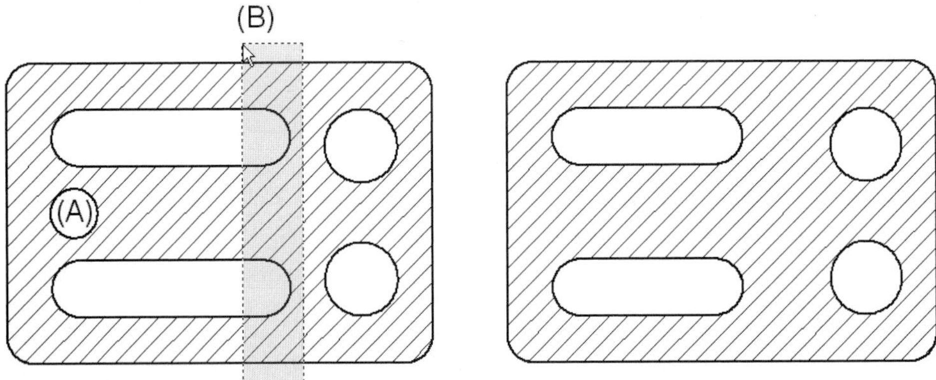

Figure 9.41

Once a hatch pattern is placed in the drawing, it does not update itself to any new additions in the form of closed shapes in the drawing. On the left in the following image, a rectangle was added at the center of the object. Notice, however, that the hatch pattern cuts directly through the rectangle. The hatch pattern does not have the intelligence to recognize the new boundary.

On the other hand, the TRIM command could be used to select the rectangle as the cutting edge and select the hatch pattern inside the rectangle to remove it, as shown on the right in the following image. After this trimming operation is performed, the rectangle is now recognized as a valid member of the associative hatch.

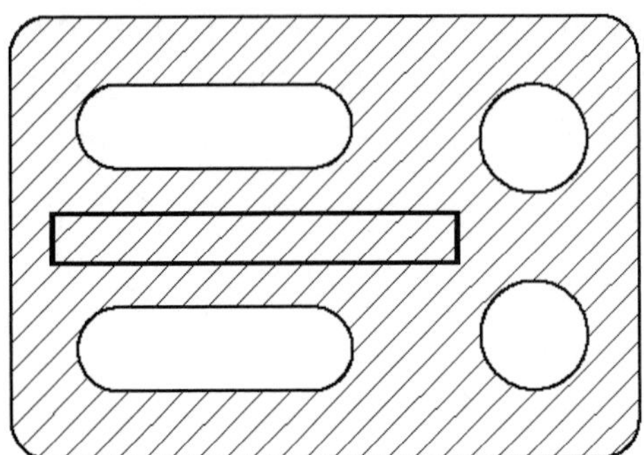

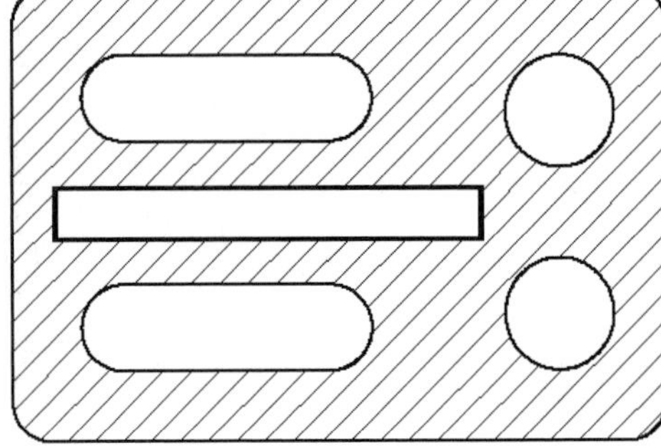

Figure 9.42

EDITING HATCH PATTERNS

Try It!: Open the drawing file 09_Hatchedit. In the illustration on the left in the following image, the pattern needs to be increased to a new scale factor of 3 units and the angle of the pattern needs to be rotated 90°. The current scale value of the pattern is 1 unit and the angle is 0°. Issuing the HATCHEDIT command produces a prompt for you to select the hatch pattern to edit. Clicking on the hatch pattern anywhere inside the object as shown on the left in the following image displays the Hatch Edit dialog box, as shown on the right in the following image.

With the dialog box displayed, click in the Scale field to change the scale from the current value of 1 unit to the new value of 3 units. Next, click in the Angle field and change the angle of the hatch pattern from the current value of 0° to the new value of 90°. This increases the spacing between hatch lines and rotates the pattern 90° in the counterclockwise direction.

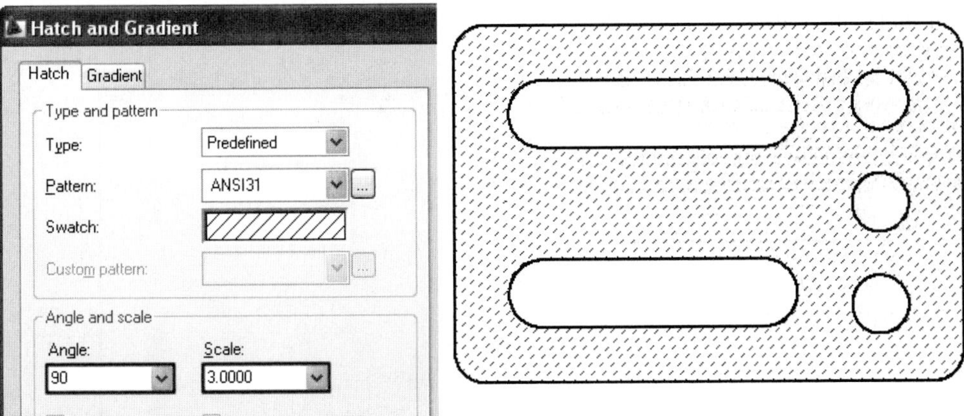

Figure 9.43

Clicking the OK button in the Hatch Edit dialog box returns you to the drawing editor and updates the hatch pattern to these changes, as shown in the following image. Be aware of the following rules for the HATCHEDIT command to function: it only works on an associative hatch pattern. If the pattern loses associativity through the use of the MIRROR command or if the hatch pattern is exploded, the HATCHEDIT command will cease to function.

Note: Double-clicking on any associative hatch pattern automatically launches the Hatch Edit dialog box. You could also pre-select the hatch pattern, right-click, and click Hatch edit... to launch the Hatch Edit dialog box.

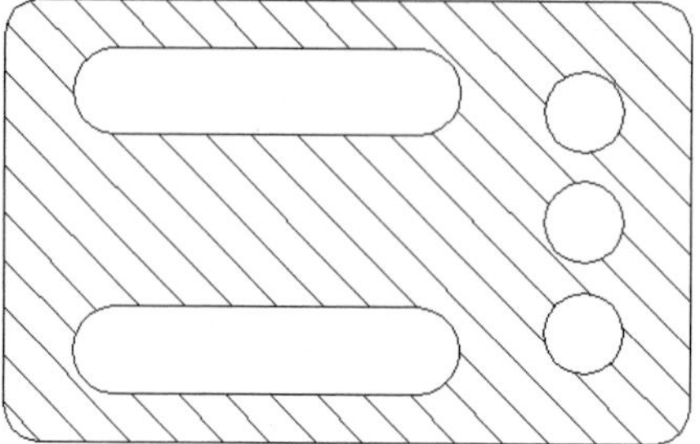

Figure 9.44

ADVANCED HATCHING TECHNIQUES

Advanced features are available inside the Hatch and Gradient dialog box. To view these features, click the arrow located in the lower-right corner of the dialog box, as shown in the following image. This expands the Hatch and Gradient dialog box to include the advanced features, which include the following areas: Islands, Boundary Retention, Boundary Set, Gap Tolerance, and Inherit Options.

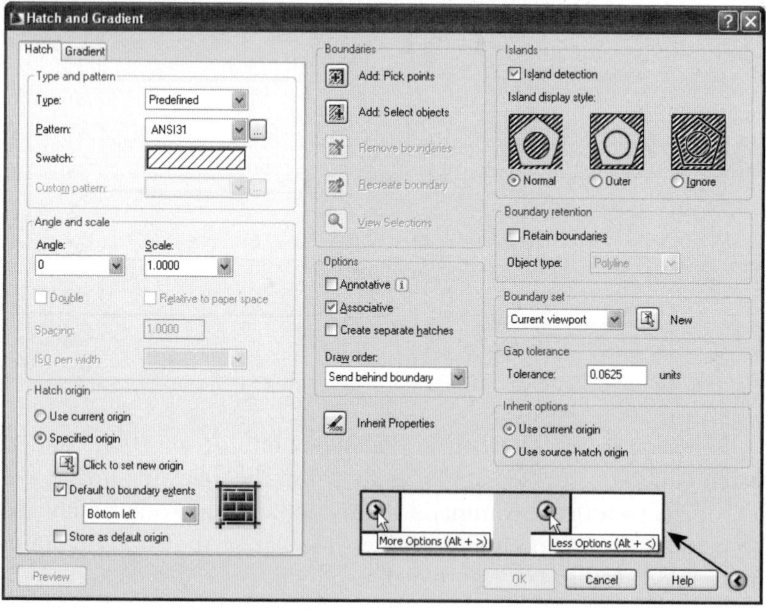

Figure 9.45

The illustration on the left in the following image shows three boundary styles and how they affect the crosshatching of Islands. The Normal boundary style is the default hatching method in which, when you select all objects within a window, the hatching begins with the outermost boundary, skips the next inside boundary, hatches the next innermost boundary, and so on. The Outer boundary style hatches only the outermost boundary of the object. The Ignore style ignores the default hatching methods of alternating hatching and hatches the entire object.

You have the option of retaining the hatch boundary by clicking in the checkbox for Retain boundaries, as shown on the right in the following image.

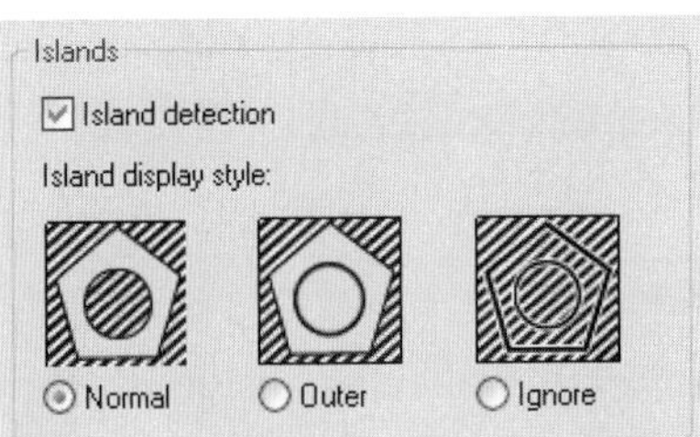

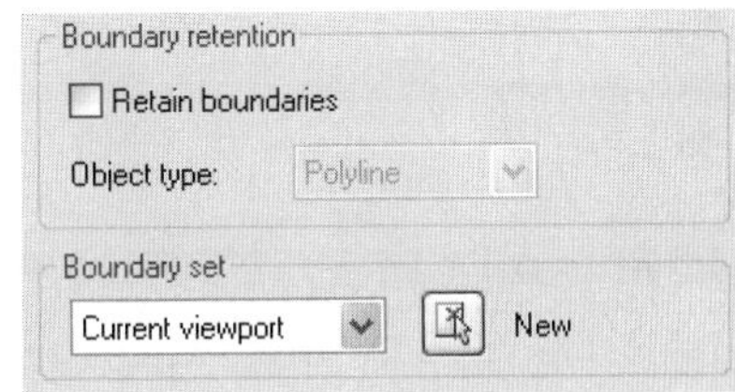

Figure 9.46

GAP TOLERANCE

Your success in hatching an object relies heavily on making sure the object being hatched is completely closed. This requirement can be relaxed with the addition of a Gap tolerance field, as shown on the left in the following image. In this example, a gap tolerance of .20 has been entered. However, the object in the middle has a gap of .40. Since the object's gap is larger than the tolerance value, an alert dialog box appears, as shown on the right in the following image. You must set the gap tolerance larger than the gap in the object or completely close the shape.

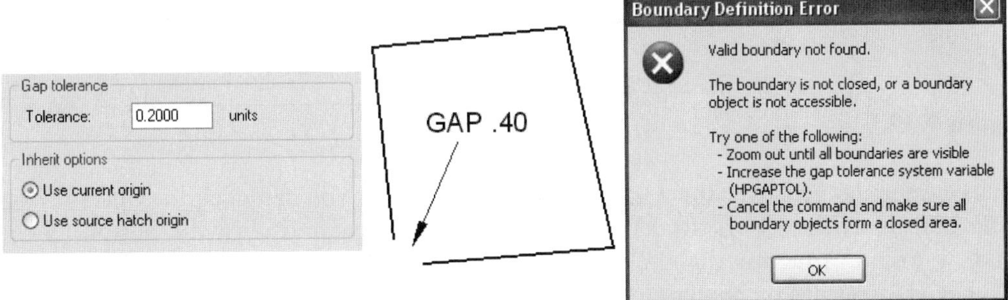

Figure 9.47

The illustration on the left in the following image shows a gap of .19. With a gap tolerance setting of .20, another alert dialog box warns you that the shape is not closed but if the gap setting is larger than the gap in the object, the object will be hatched. The results are shown on the right in the following image.

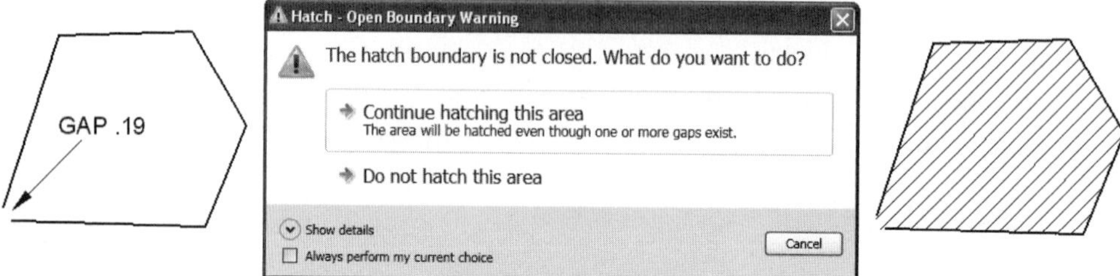

Figure 9.48

PRECISION HATCH PATTERN PLACEMENT

At times, you want to control where the hatch pattern begins inside a shape. In the following image, a brick pattern was applied without any special insertion base points. In this image, the pattern seems to be laid out based on the center of the rectangle. As the brick pattern reaches the edges of the rectangle, the pattern just ends. This occurs because the hatch pattern uses a current origin point to be constructed from. This control can be found in the main Hatch and Gradient dialog box under the Hatch origin area, as shown on the left in the following image.

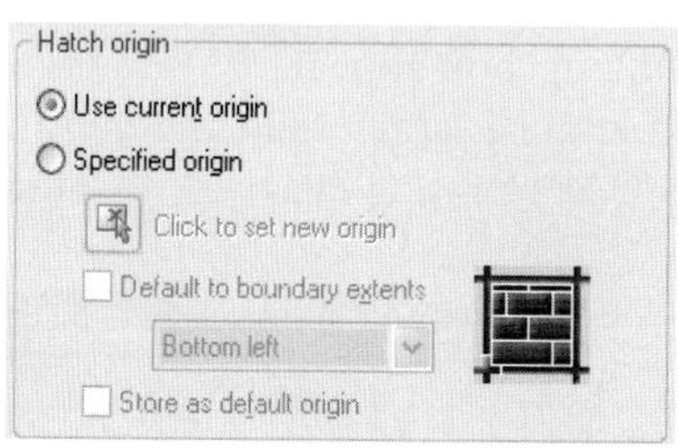

Figure 9.49

To have more control over the origin of the hatch pattern, you can click the Specific origin radio button, as shown on the left in the following image, and click the button Click to set new origin. You can then select the new origin of the hatch pattern back in your drawing. The results are shown on the right in the following image, with the hatch pattern origin located at the lower-left corner of the rectangle.

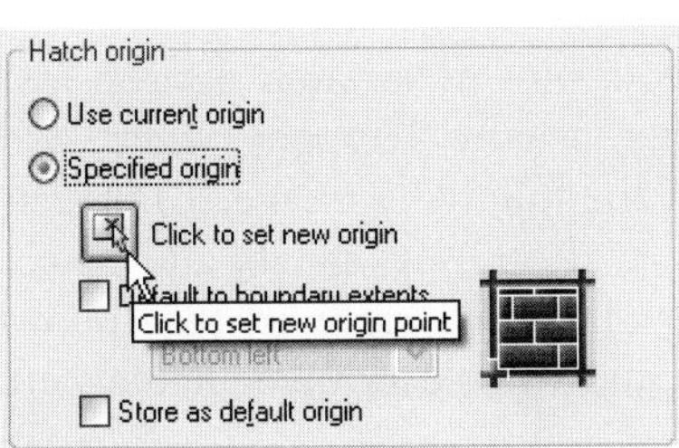

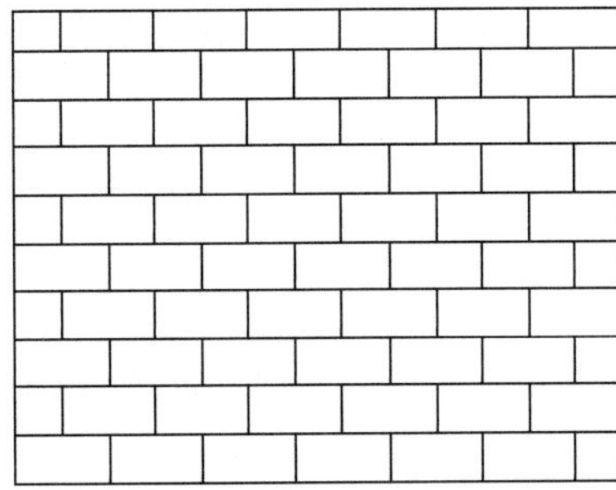

Figure 9.50

INHERIT HATCHING PROPERTIES

Try It!: Open the drawing file 09_Hatch Inherit. The illustration on the left in the following image consists of a simple assembly drawing. At least three different hatch patterns are displayed to define the different parts of the assembly. Unfortunately, a segment of one of the parts was not hatched, and it is unclear what pattern, scale, and angle were used to place the pattern.

Whenever you are faced with this problem, click the Inherit Properties button in the main Hatch and Gradient dialog box, as shown in the following image. Clicking on the pattern at "A" sets the pattern, scale, and angle to match the selected pattern.

To complete the hatch operation, click an internal point in the empty area. Right-click and select ENTER from the shortcut menu. The hatch pattern is placed in this area, and it matches that of the other patterns to identify the common parts, as shown on the right in the following image.

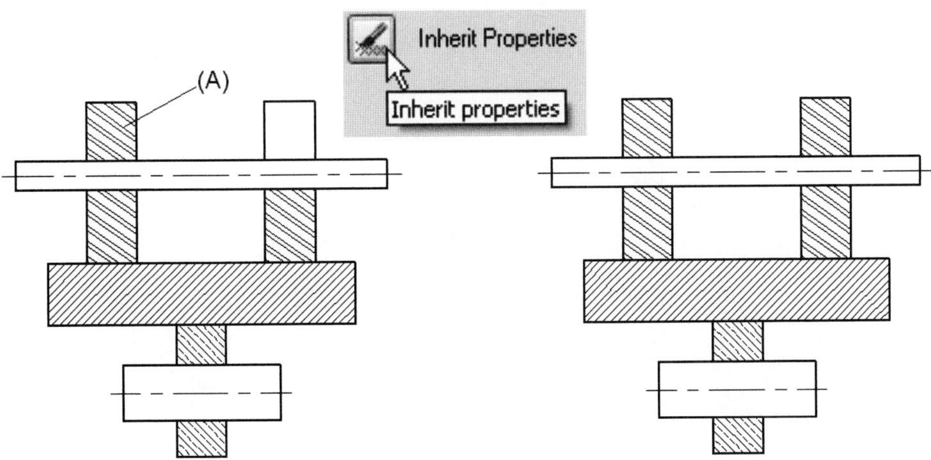

Figure 9.51

EXCLUDE HATCH OBJECTS FROM OBJECT SNAPS

At times, it is easy to mistakenly snap to the endpoint of a hatch pattern instead of the endpoint of the perimeter of an object. This usually happens when adding linear dimensions to a drawing, as shown on the left in the following image. To prevent this from happening, activate the Options dialog box, select the Drafting tab, and place a check in the box next to Ignore Hatch Objects, as shown on the right in the following image. Object Snap modes will now ignore endpoints and intersections of hatch patterns.

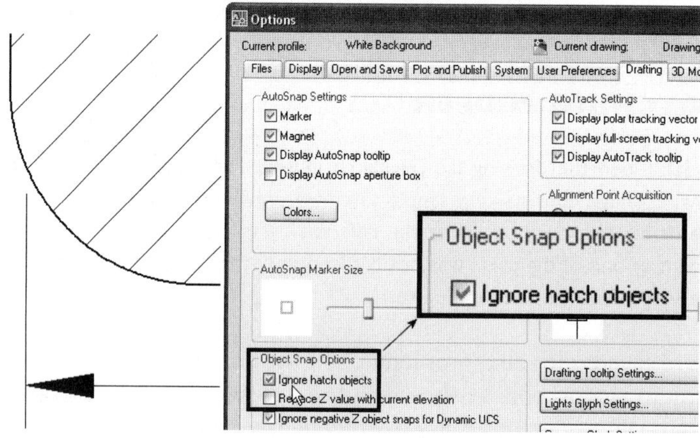

Figure 9.52

TUTORIAL EXERCISE: 09_COUPLER.DWG

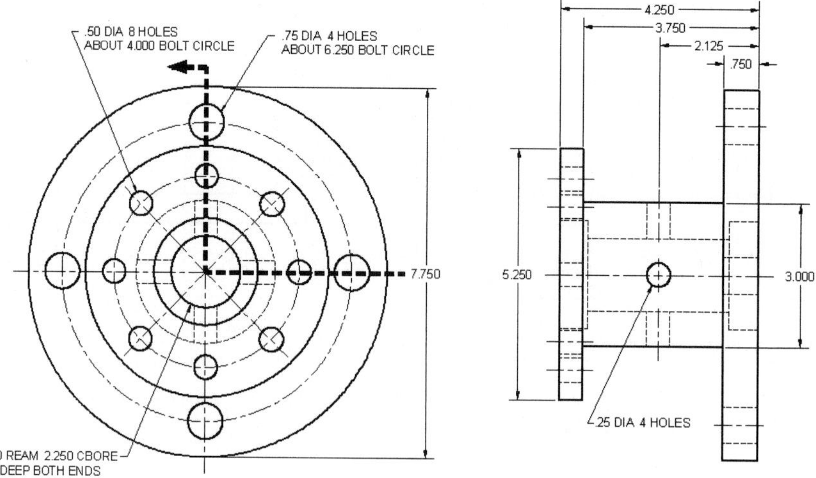

Figure 9.53

PURPOSE

This tutorial is designed to use the MIRROR and BHATCH commands to convert 09_Coupler to a half section, as shown in the previous image.

SYSTEM SETTINGS

Since this drawing is provided on CD, edit an existing drawing called 09_Coupler. Follow the steps in this tutorial for converting the upper half of the object to a half section. All Units, Limits, Grid, and Snap values have been previously set.

LAYERS

The following layers are already created:

Name	Color	Linetype
Object	White	Continuous
Center	Yellow	Center
Hidden	Red	Hidden
Hatch	Magenta	Continuous
Cutting Plane Line	Yellow	Dashed
Dimension	Yellow	Continuous

SUGGESTED COMMANDS

Convert one-half of the object to a section by erasing unnecessary hidden lines. Use the Layer Control box and change the remaining hidden lines to the Object layer. Issue the BHATCH command and use the ANSI31 hatch pattern to hatch the upper half of the 09_Coupler on the Hatch layer.

Step 1

Use the MIRROR command to copy and flip the upper half of the Side view and form the lower half. When in Object Selection mode, use the Remove option to deselect the main centerline, hole centerlines, and hole as shown on the left in the following image. If these objects are included in the mirror operation, a duplicate copy of these objects will be created.

```
Command: MI (For MIRROR)
Select objects: (Pick a point at "A")
Specify opposite corner: (Pick a point at "B")
Select objects: (Hold down the SHIFT key and pick centerlines "C"
    and "D" to remove them from the selection set)
Select objects: (With the SHIFT key held down, pick a point at "E")
Specify opposite corner: (With the SHIFT key held down, pick a
    point at "F")
Select objects: (Press ENTER to continue)
```

```
Specify first point of mirror line: (Select the endpoint of the
    centerline near "C")
Specify second point of mirror line: (Select the endpoint of the
    centerline near "D")
Erase source objects? [Yes/No] <N>: (Press ENTER to perform the
    mirror operation)
```

Begin converting the upper half of the Side view to a half section by using the ERASE command to remove any unnecessary hidden lines and centerlines from the view as shown on the right in the following image.

```
Command: E (For ERASE)
Select objects: (Carefully select the hidden lines labeled "G,"
    "H," "J," and "K")
Select objects: (Select the centerline labeled "L")
Select objects: (Press ENTER to execute the ERASE command)
```

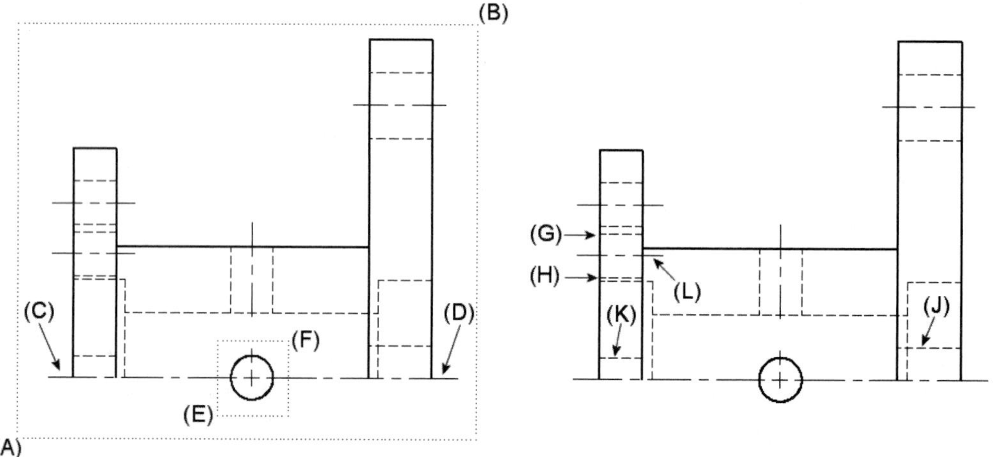

Figure 9.54

Step 2

Since the remaining hidden lines actually represent object lines when shown in a full section, use the Layer Control box in the following image to convert all highlighted hidden lines from the Hidden layer to the Object layer.

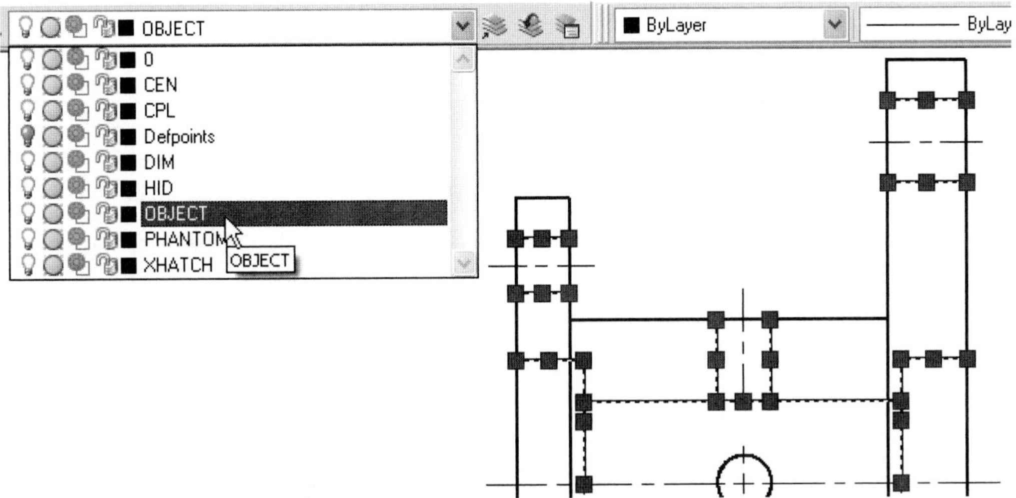

Figure 9.55

Step 3

Remove unnecessary line segments from the upper half of the converted section using the TRIM command. Use the horizontal line at "A" as the cutting edge, and select the two vertical segments at "B" and "C" as the objects to trim, as shown on the left in the following image.

Make the Hatch layer the current layer. Then, use the BHATCH command to display the Hatch and Gradient dialog box. Use the pattern "ANSI31" and keep all default settings. Click the Add: Pick points button. When the drawing reappears, click inside areas "D," "E," "F," and "G," as shown on the right in the following image. When finished selecting these internal points, press ENTER to return to the Hatch and Gradient dialog box.

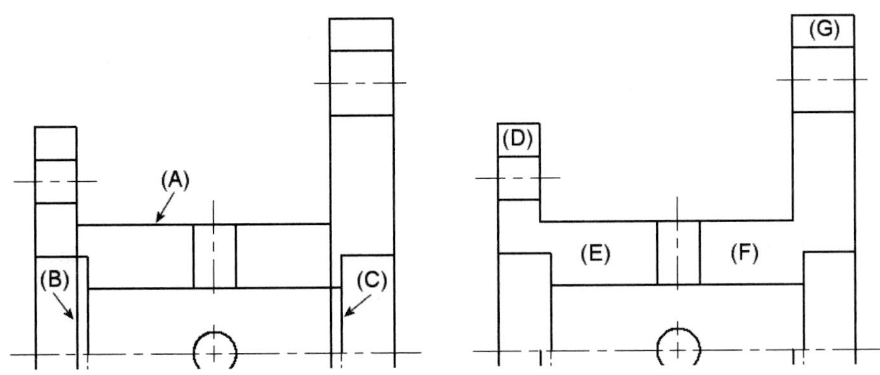

Figure 9.56

Step 4

The complete hatched view is shown in the following image.

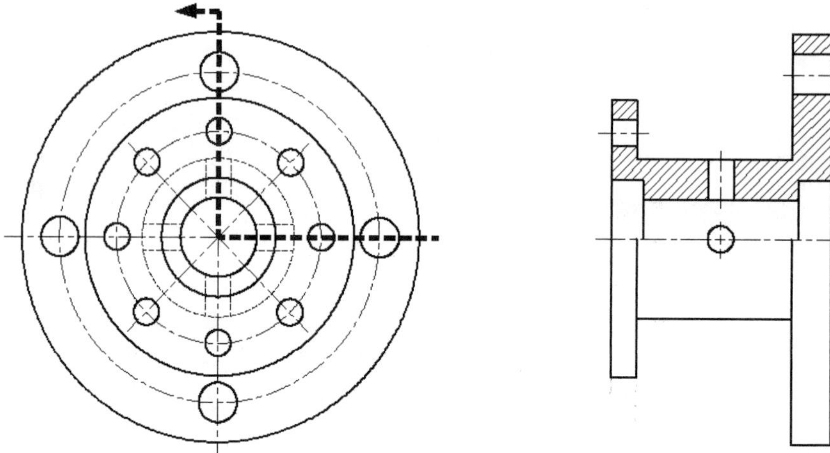

Figure 9.57

TUTORIAL EXERCISE: 09_ELEVATION.DWG

Figure 9.58

PURPOSE

This tutorial is designed to use the Inherit Properties button of the Hatch and Gradient dialog box on 09_Elevation. Gradient hatch patterns will also be applied to this drawing for presentation purposes.

SYSTEM SETTINGS

Since this drawing is provided on CD, edit an existing drawing called 09_Elevation. Follow the steps in this tutorial for creating an elevation of the residence.

LAYERS

The following layers are already created:

Name	Color	Linetype
Elevations	White	Continuous
Exterior Brick	Red	Continuous
Gradient Background	Blue	Continuous
Roof Boundaries	153	Continuous
Roof Hatch	73	Dashed
Sill	30	Continuous
Wall Boundaries	Magenta	Continuous

SUGGESTED COMMANDS

Open up the existing drawing file 09_Elevation and notice the appearance of existing roof and brick patterns. Use the Inherit Properties feature of the Hatch and Gradient dialog box to transfer the roof and brick patterns to the irregular-shaped areas of this drawing. Next, a gradient pattern will be added to the upper part of the drawing for the purpose of creating a background. Another gradient pattern will be added to the lower portion of the drawing. Finally, the last gradient pattern will be added to the garage door windows. In this way, the gradient patterns will be used to enhance the elevation of the house.

Step 1

Open the drawing 09_Elevation. Notice the appearance of a roof and brick hatch patterns in the following image. The roof and brick patterns need to be applied to the other irregular areas of this house elevation. First, make the Roof Hatch layer current through the Layer Control box.

Figure 9.59

Step 2

Activate the Hatch and Gradient dialog box and click the Inherit Properties button, as shown on the left in the following image. This button allows you to click an existing hatch pattern and have the pattern name, scale, and angle transfer to the proper fields in the dialog box. In other words, you do not have to figure out the name, scale, or angle of the pattern. The only requirement is that a hatch pattern already exists in the drawing.

When you return to the drawing, shown on the right in the following image, notice the appearance of a glyph that is similar to the Match Properties icon. Click the existing roof pattern in the figure.

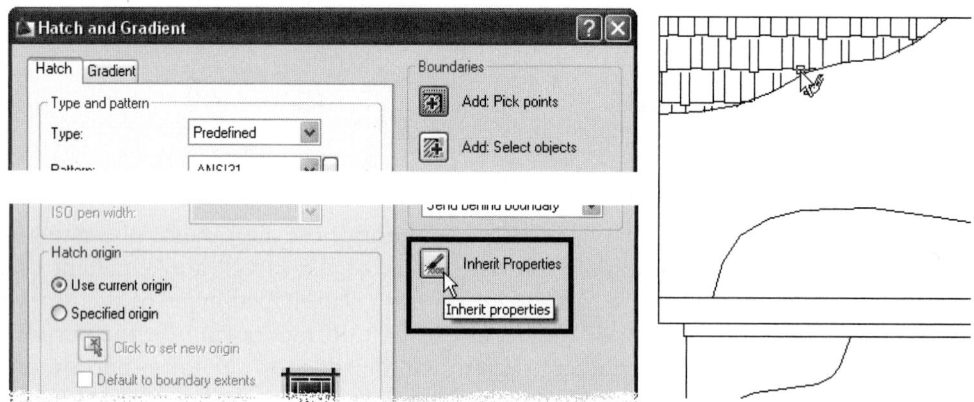

Figure 9.60

Step 3

After you pick the existing hatch pattern in the previous image, your cursor changes appearance again. Now pick internal points inside every irregular shape in the following image. Notice that each one highlights. When you're finished, press ENTER to return to the Hatch and Gradient dialog box. Click the Preview button to examine the results. If the hatch pattern appearance is correct, right-click to place the roof hatch patterns.

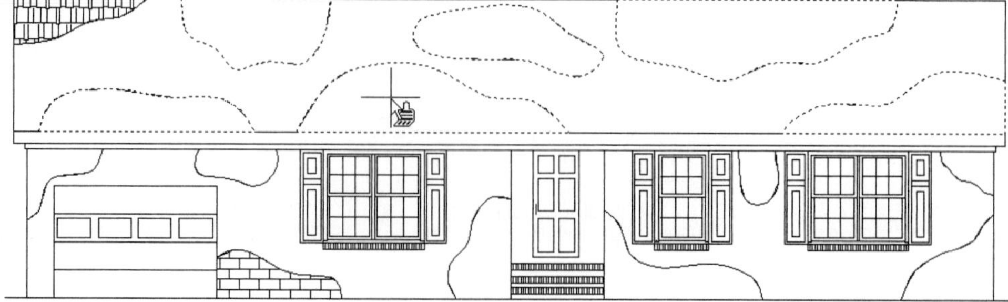

Figure 9.61

Step 4

Next, make the Exterior Brick layer current using the Layer Control box. Then activate the Hatch and Gradient dialog box, click the Inherit Properties button, pick the existing brick pattern, and click inside every irregular shape in the following image until each one highlights. Preview the hatch pattern and right-click if the results are correct.

Figure 9.62

Step 5

To give the appearance that the hatch patterns are floating on the roof and wall, use the Layer Control box to turn off the two layers that control the boundaries of these patterns, namely, Roof Boundaries and Wall Boundaries. Your display should appear similar to the following image.

Figure 9.63

Step 6

A number of gradient hatch patterns will now be applied to the outer portions of the elevation. First perform a ZOOM-EXTENTS. Then change the current layer to Gradient Background using the Layer Control box, as shown in the following image.

```
Command: Z (For ZOOM)
Specify corner of window, enter a scale factor (nX or nXP), or
[All/Center/Dynamic/Extents/Previous/Scale/Window/Object]
    <real time>:
E (For Extents)
```

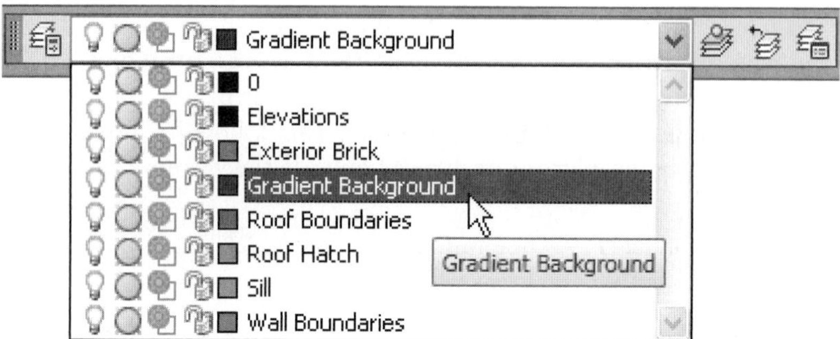

Figure 9.64

Step 7

You will now apply a gradient hatch pattern to the upper portion of the elevation plan. Activate the Hatch and Gradient dialog box, click the Gradient tab, and pick the pattern as shown on the left in the following image. Then click the Add: Pick points button and click an internal point in the figure of the elevation. Once the boundary highlights, press ENTER, preview the hatch, and right-click to place the hatch pattern.

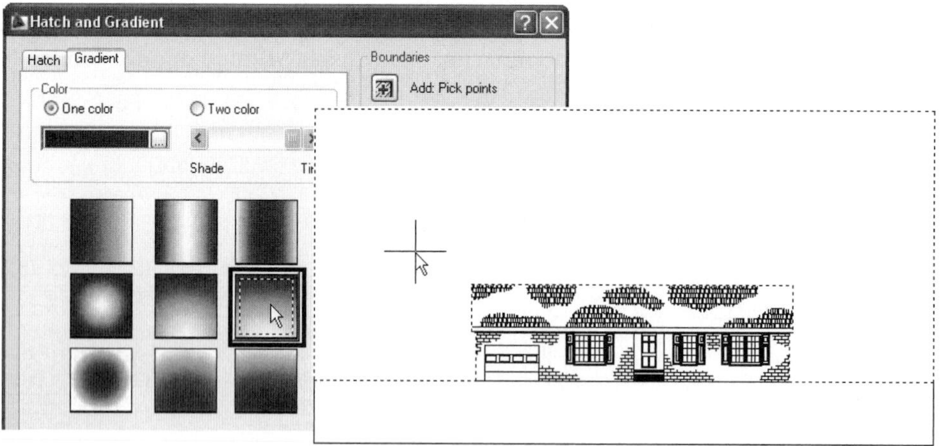

Figure 9.65

Step 8

Apply another gradient hatch pattern to the lower portion of the elevation plan. Activate the Hatch and Gradient dialog box, click the Gradient tab, and pick the pattern as shown on the left in the following image. Then click the Add: Pick points button and click an internal point in the figure of the elevation. Once the boundary highlights, press ENTER, preview the hatch, and right-click to place the hatch pattern.

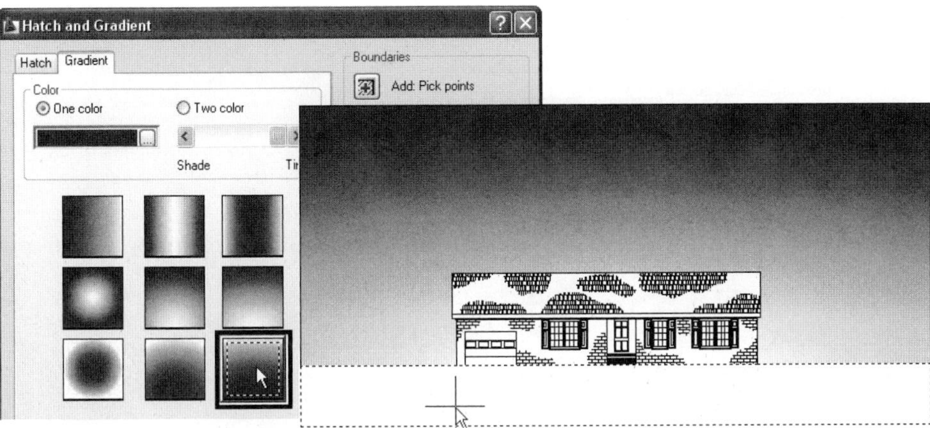

Figure 9.66

Step 9

Apply gradient patterns to each of the garage windows, as shown in the following image, by using the BHATCH command for each of the four windows.

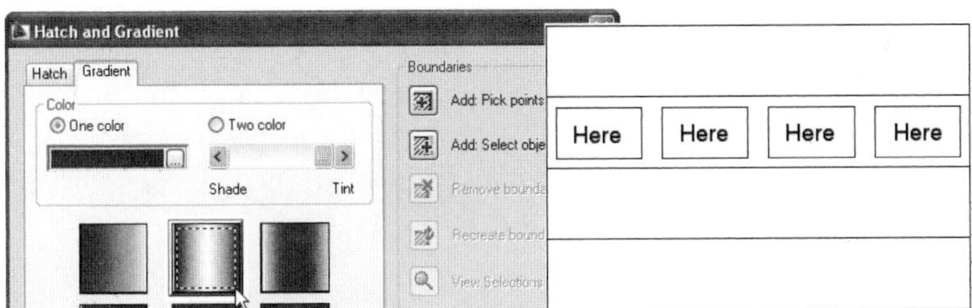

Figure 9.67

Step 10

If the boundary lines that define the elevation disappear, send the gradient patterns back using the Draw Order feature, as shown in the following image.

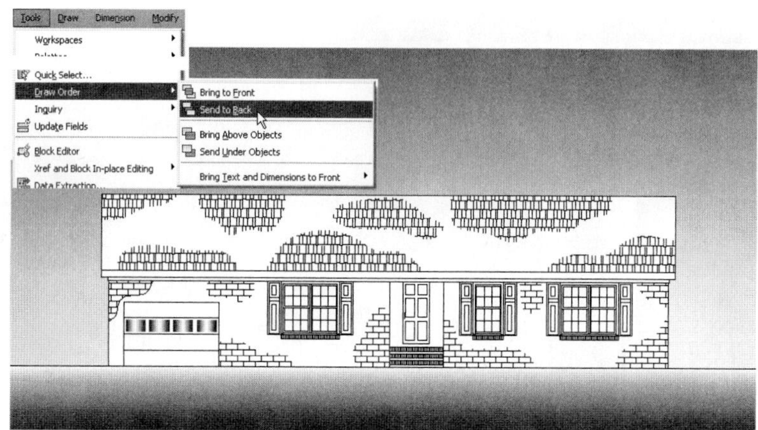

Figure 9.68

PROBLEMS FOR CHAPTER 9

PROBLEM 9–1

Center a three-view drawing and make the Front view a full section.

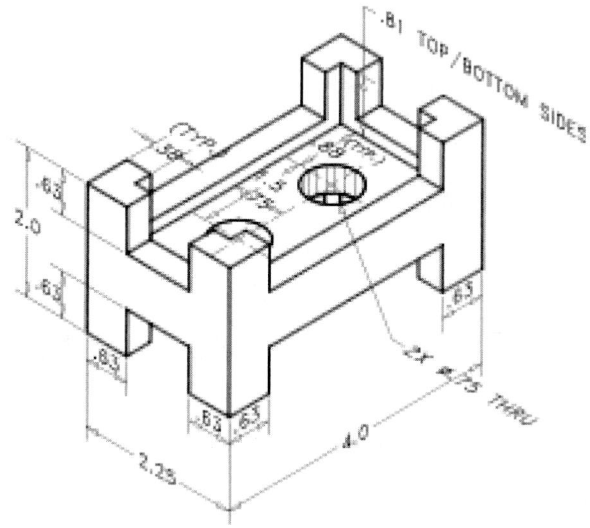

PROBLEM 9–2

Center two views within the work area and make one view a full section. Use correct drafting practices for the ribs.

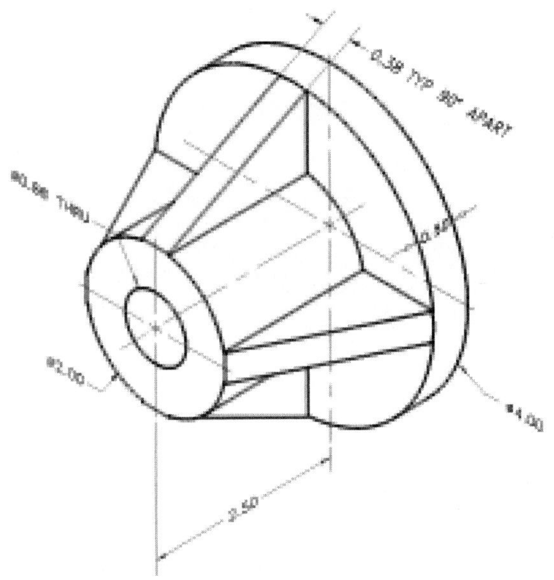

PROBLEM 9–3

Center two views within the work area and make one view a half section.

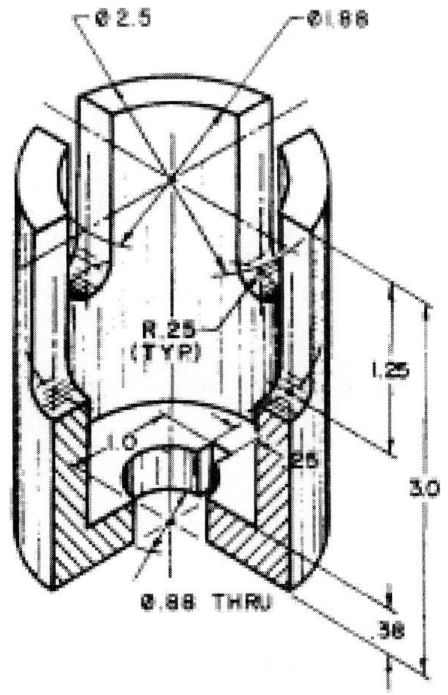

PROBLEM 9–4

Center two views within the work area and make one view a half section.

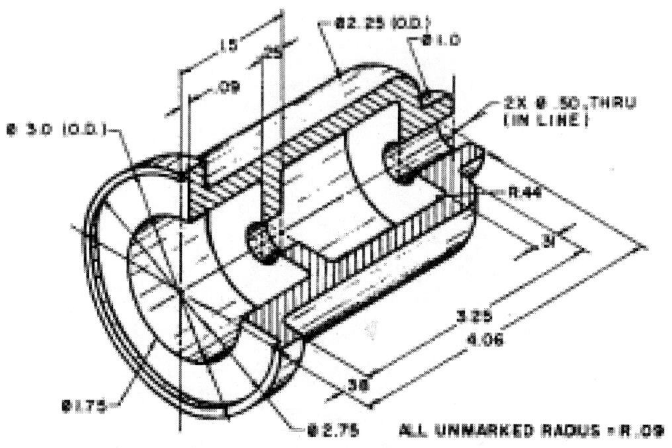

PROBLEM 9–5

Center three views within the work area and make one view an offset section.

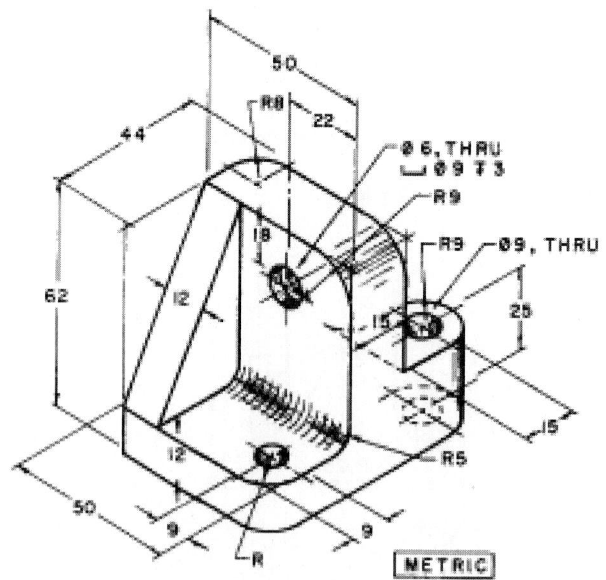

PROBLEM 9–6

Center three views within the work area and make one view an offset section.

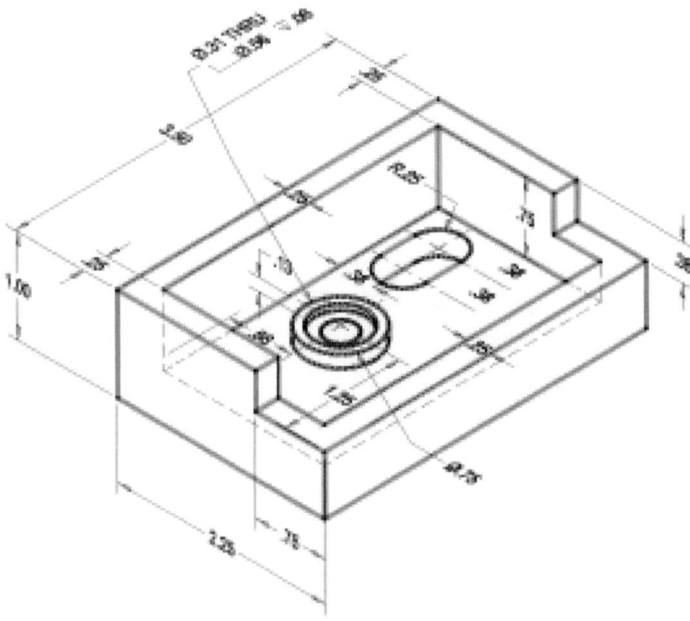

PROBLEM 9–7

Center three views within the work area and make one view an offset section.

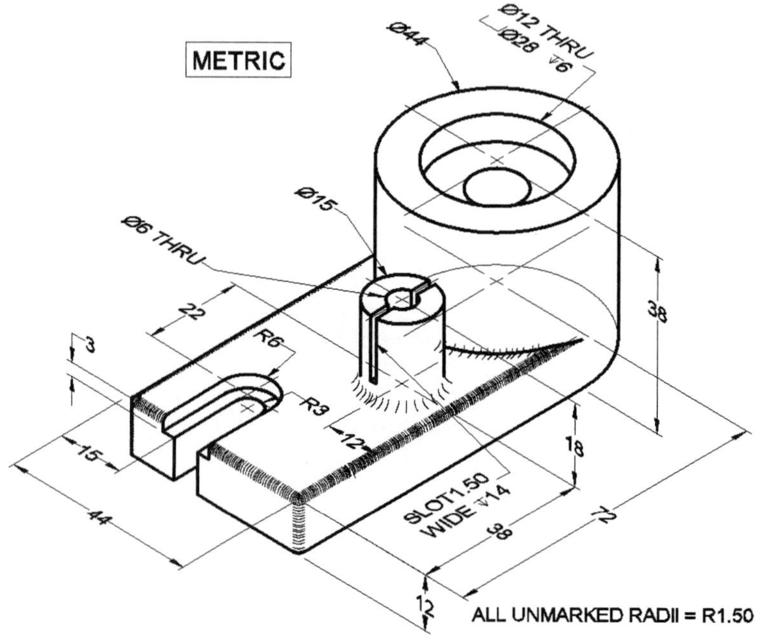

PROBLEM 9–8

Center two views within the work area and make one view a half section.

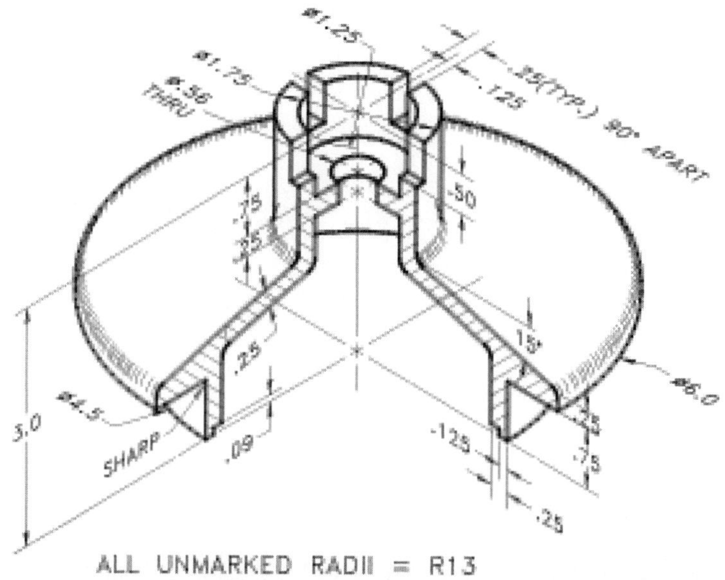

PROBLEM 9–9

Center the required views within the work area and make one view a broken-out section to illustrate the complicated interior area.

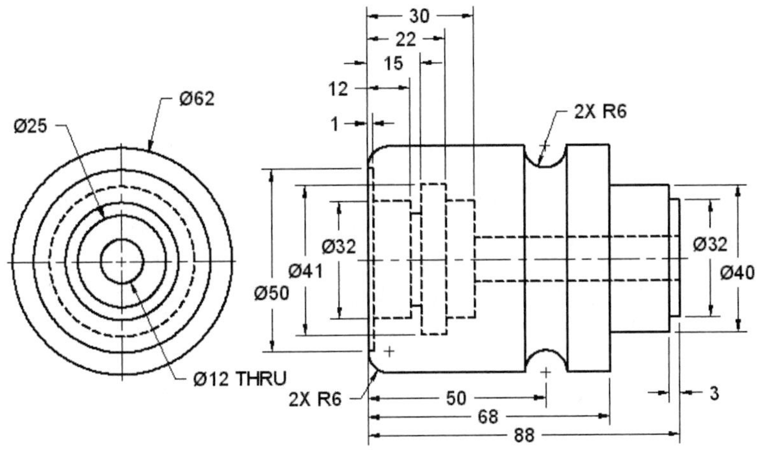

PROBLEM 9–10

Create the required views within the work area and add a full removed section called A-A.

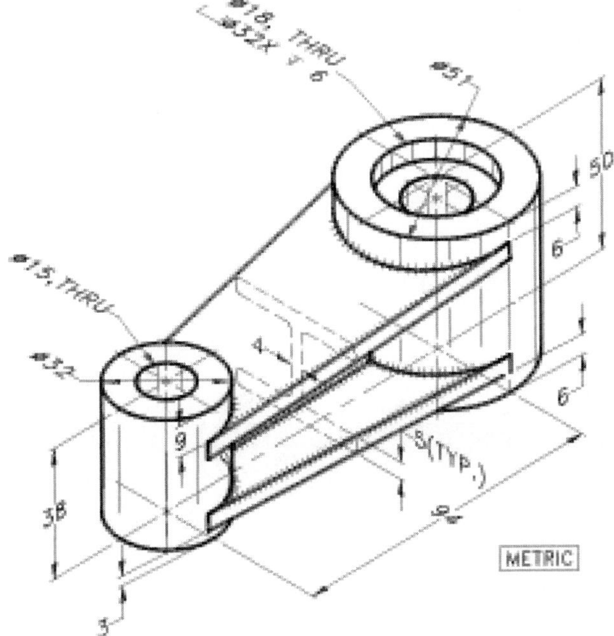

PROBLEM 9–11

Center the required views within the work area and add a full removed section called A-A.

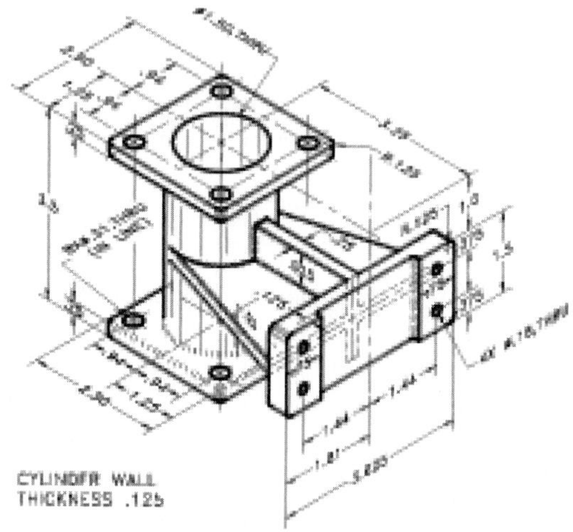

ALL UNMARKED RADII = R.06

DIRECTIONS FROM PROBLEMS 9–12 THROUGH 9–14

Center the required views within the work area. Leave a 1" or 25-mm space between the views. Make one view a full section to fully illustrate the object. Consult your instructor if you need to add dimensions.

PROBLEM 9–12

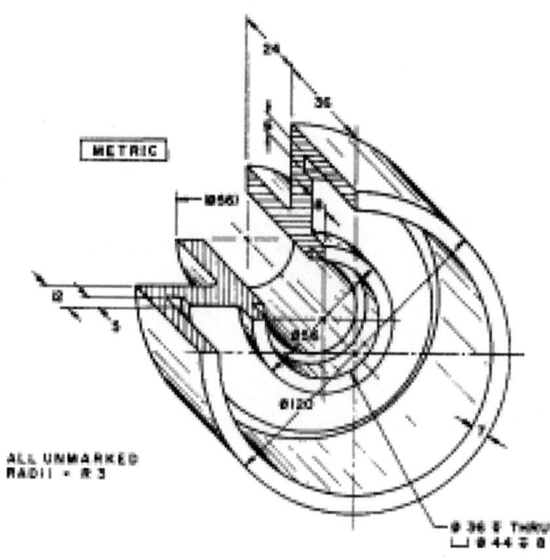

PROBLEM 9–13

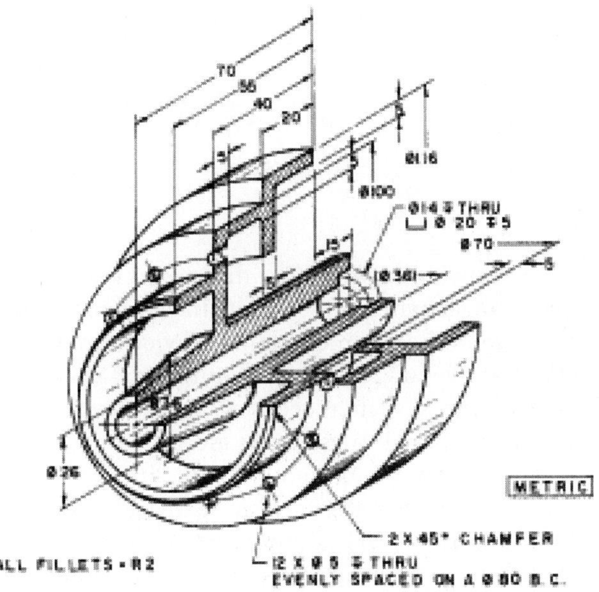

PROBLEM 9–14

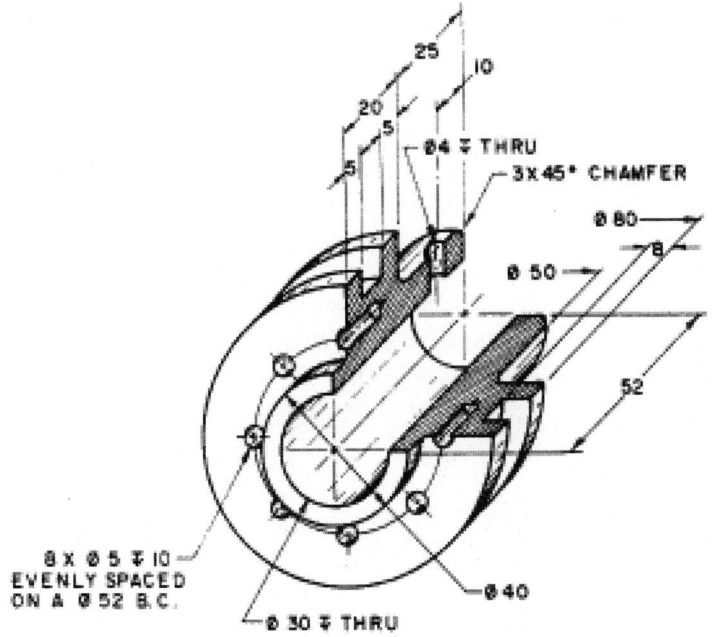

Creating Auxiliary Views

During the discussion of multiview drawings, we discovered that you need to draw enough views of an object to accurately describe it. In most cases, this requires a Front, Top, and Right Side view. Sometimes additional views are required, such as Left Side, Bottom, and Back views, to show features not visible in the three primary views. Other special views, such as sections, are created to expose interior details for better clarity. Sometimes all these views are still not enough to describe the object, especially when features are located on an inclined surface. To produce a view perpendicular to this inclined surface, an auxiliary view is drawn. This chapter will describe where auxiliary views are used and how they are projected from one view to another. A tutorial exercise is presented to show the steps in the construction of an auxiliary view.

AUXILIARY VIEW BASICS

The illustration on the left in the following image presents interesting results if constructed as a multiview drawing or orthographic projection.

The illustration on the right in the following image should be quite familiar; it represents the standard glass box with objects located in the center. Again, the purpose of this box is to prove how orthographic views are organized and laid out. This illustration is no different. First the primary views (Front, Top, and Right Side) are projected from the object to intersect perpendicularly with the glass plane. Under normal circumstances, this procedure would satisfy most multiview drawing cases. Remember, only those views necessary to describe the object are drawn. However, upon closer inspection, we notice that the object in the Front view consists of an angle forming an inclined surface.

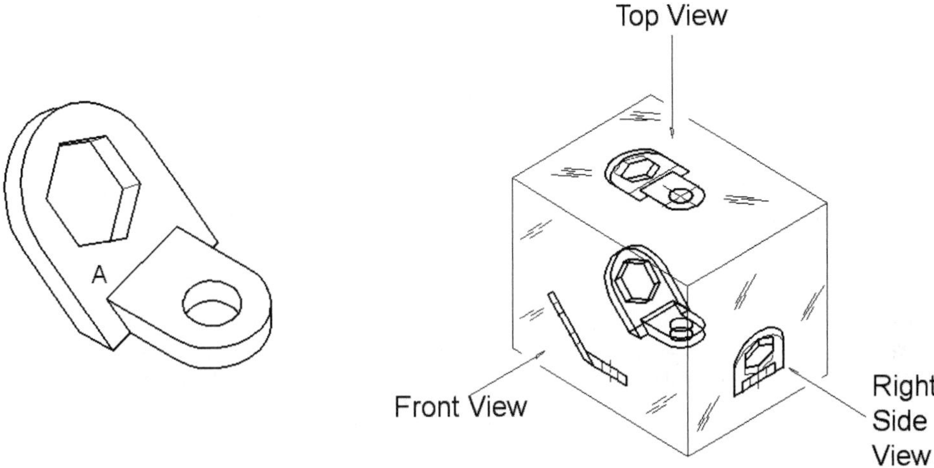

Figure 10.1

When you lay out the Front, Top, and Right Side views, a problem occurs. The Front view shows the basic shape of the object and the angle of the inclined surface, as shown in the following image. The Top view shows the true size and shape of the surface formed by the circle and arc. The Right Side view shows the true thickness of the hole from information found in the Top view. However, the true size and shape of the features found in the inclined surface at "A" do not exist. We see the hexagonal hole going through the object in the Top and Right Side views. These views, however, do not show the detail in actual size; instead, the views are foreshortened. For that matter, the entire inclined surface is foreshortened in these views. Dimensioning the size and location of the hexagonal hole would be challenging with these three views. This is one case in which the Front, Top, and Right Side views are not enough to describe the object. An additional view, or auxiliary view, is used to display the true shape of surfaces along an incline.

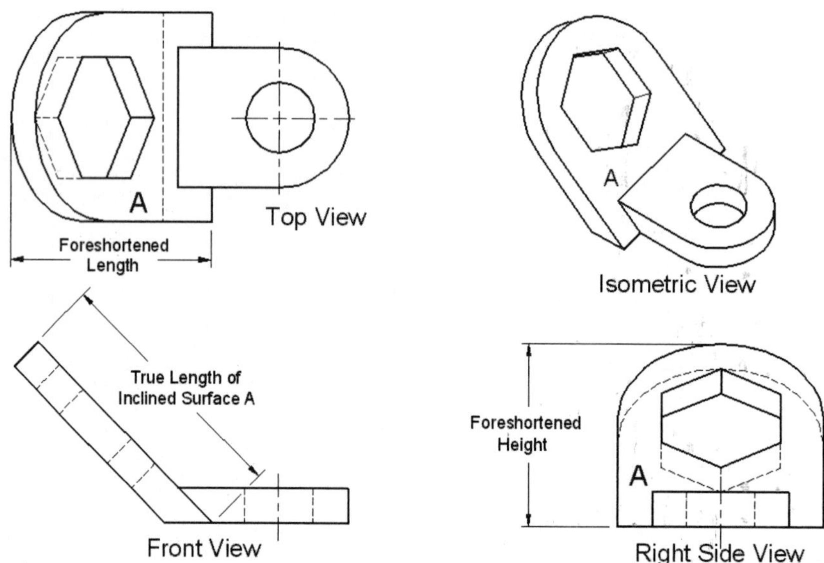

Figure 10.2

To demonstrate the creation of an auxiliary view, let's create another glass box, this time with an inclined plane on one side. This plane is always parallel to the inclined surface of the object. Instead of just the Front, Top, and Right Side views being projected, the geometry describing the features along the inclined surface is projected to the auxiliary plane, as shown on the left in the following image.

As in multiview projection, the edges of the glass box are unfolded, with the edges of the Front view plane as the pivot, as shown on the right in the following image.

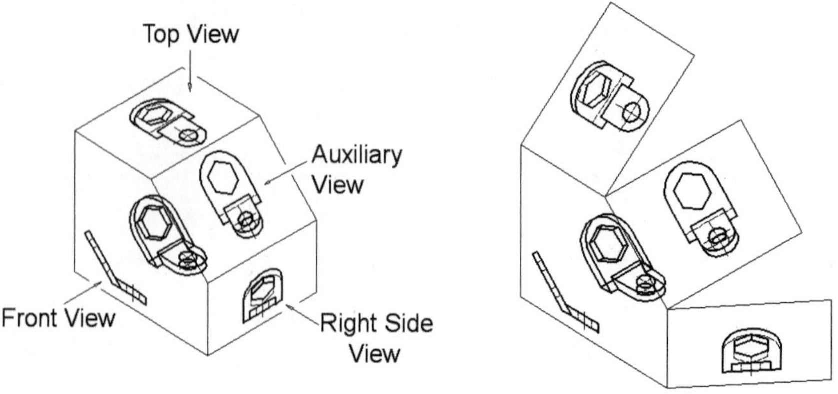

Figure 10.3

All planes are extended perpendicular to the Front view, where the rotation stops. The result is the organization of the multiview drawing, complete with an auxiliary viewing plane, as shown on the left in the following image.

The illustration on the right in the following image represents the final layout, complete with auxiliary view. This figure shows the auxiliary being formed as a result of the inclined surface present in the Front view. An auxiliary view may be created in relation to any inclined surface located in any view. Also, the figure displays circles and arcs in the Top view that appear as ellipses in the auxiliary view.

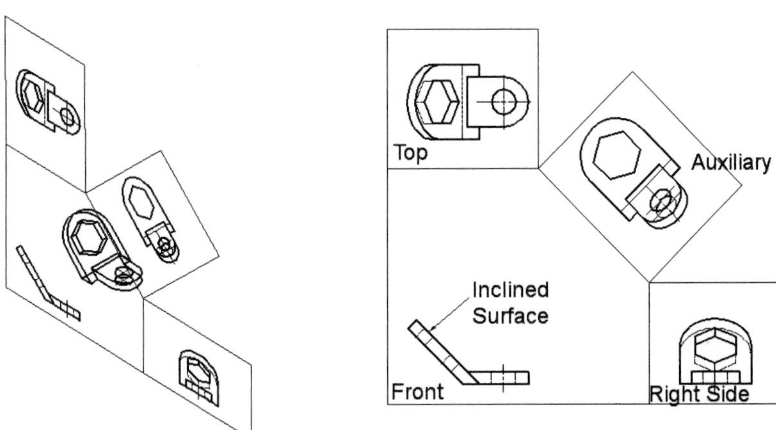

Figure 10.4

It is usually not required that elliptical shapes be drawn in one view where the feature is shown at true size and shape in another. The resulting view minus these elliptical features is called a partial view, used extensively in auxiliary views. An example of the Top view converted to a partial view is shown on the left in the following image.

A few rules to follow when constructing auxiliary views are displayed pictorially on the right in the following image. First, the auxiliary plane is always constructed parallel to the inclined surface. Once this is established, visible as well as hidden features are projected from the incline to the auxiliary view. These projection lines are always drawn perpendicular to the inclined surface and perpendicular to the auxiliary view.

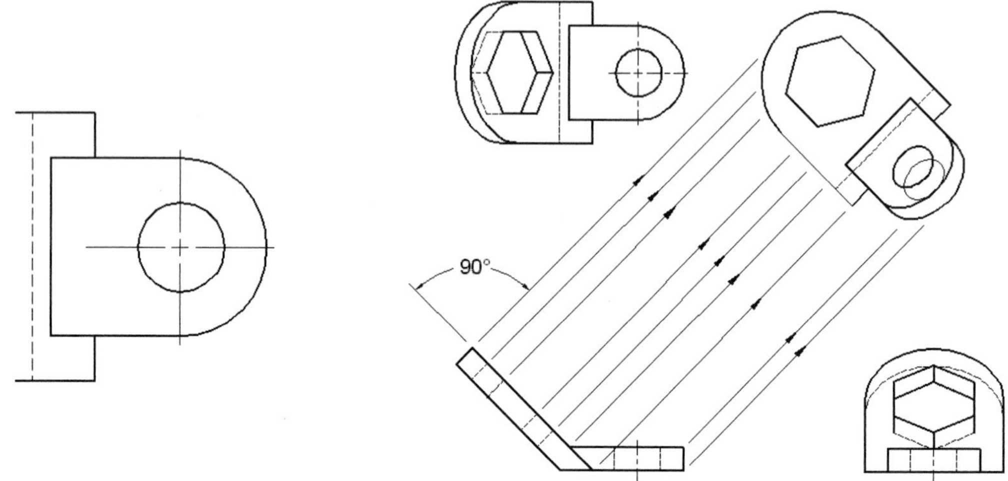

Figure 10.5

CREATING AUXILIARY VIEWS USING XLINES

The following image shows an object consisting of front and right side views in addition to an isometric or pictorial view. To construct the perpendicular projector lines that will be used to create the auxiliary view, the XLINE command is used. Xlines are considered construction lines and were discussed in chapter 5. The Angle and Reference options of the XLINE command provide an easy way to construct the perpendicular projection lines. You first enter the Angle option followed by the Reference option. You will be prompted to select an object; pick line segment "AE," as shown in the following image. Then enter an angle of 90° as the angle of the xline, and pick the endpoints at "A" through "E" to construct the xline objects. By default, all xlines are drawn infinitely in two directions. You can trim, fillet, and even break xlines.

```
Command: XL (For XLINE)
XLINE Specify a point or [Hor/Ver/Ang/Bisect/Offset]: A (For Ang)
Enter angle of xline (0) or [Reference]: R (For Reference)
Select a line object: (Select line segment "AE")
Enter angle of xline <0>: 90
Specify through point: (Pick the endpoints of points "A"
    through "E")
```

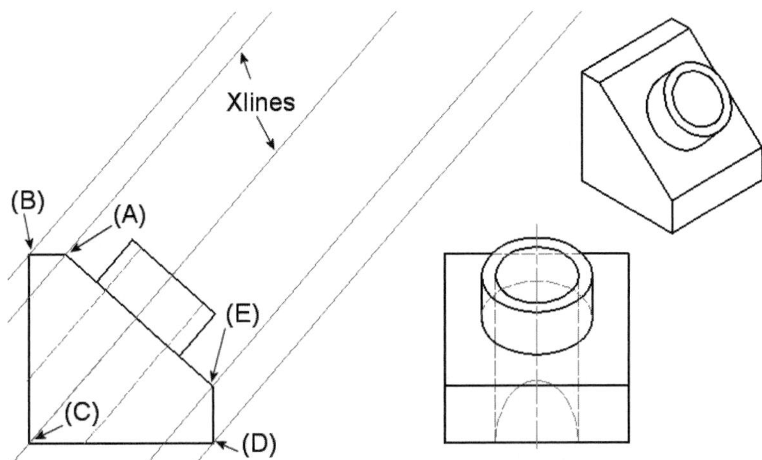

Figure 10.6

With the perpendicular projection lines constructed, next construct a base edge that is perpendicular to the projection lines. This base edge normally forms one of the edges of the auxiliary view. In the following example, line segment "1-2" located in the right side view coincides with xline segment "1-2" found in the auxiliary view.

```
Command: XL (For XLINE)
XLINE Specify a point or [Hor/Ver/Ang/Bisect/Offset]: A (For Ang)
Enter angle of xline (0) or [Reference]: R (For Reference)
Select a line object: (Pick the xline at "1")
Enter angle of xline <0>: 90
Specify through point: (Pick an approximate location in the
    drawing to locate the base edge of the auxiliary view.)
```

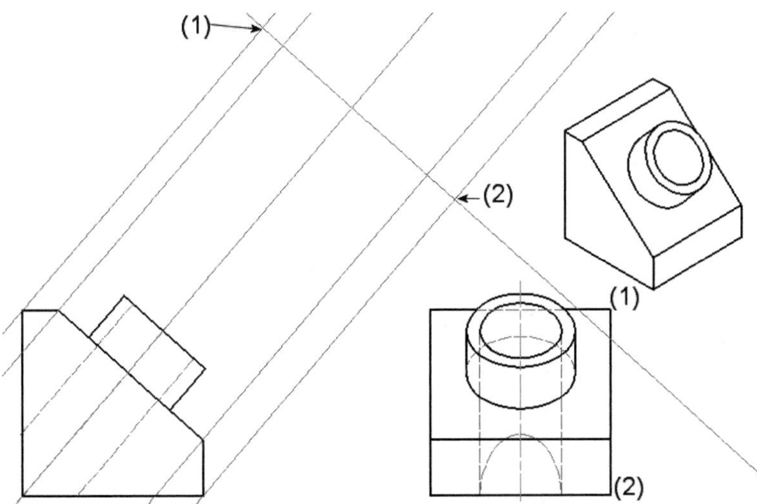

Figure 10.7

TRANSFERRING DISTANCES WITH THE OFFSET COMMAND

With the perpendicular projectors constructed along with the base edge using xlines, the next step is to transfer distances from one view and create the auxiliary view. The auxiliary view in our example is a depth type. All views projected from the front view show depth. This means that the distances along the projectors can be transferred from the depth dimensions in a top or right side view. In this example the depth will be transferred from the right side view using the OFFSET command. To accomplish this, enter OFFSET and for the offset distance, pick the endpoint at "A" in the right side view, as shown in the following image. For the second point, pick the endpoint at "B." These two endpoints form the depth of the right side view. To transfer this distance to the auxiliary view, pick line "C" as the object to offset and then pick a location at "D" as the side to perform the offset shown in the following image.

```
Command: O (For OFFSET)
Current settings: Erase source=No Layer=Source OFFSETGAPTYPE=0
Specify offset distance or [Through/Erase/Layer] <Through>: (Pick
    the endpoint at "A")
Specify second point: (Pick the endpoint at "B")
Select object to offset or [Exit/Undo] <Exit>: (Pick line "C")
Specify point on side to offset or [Exit/Multiple/Undo] <Exit>:
    (Pick the location at "D")
Select object to offset or [Exit/Undo] <Exit>: (Press ENTER to
    exit this command)
```

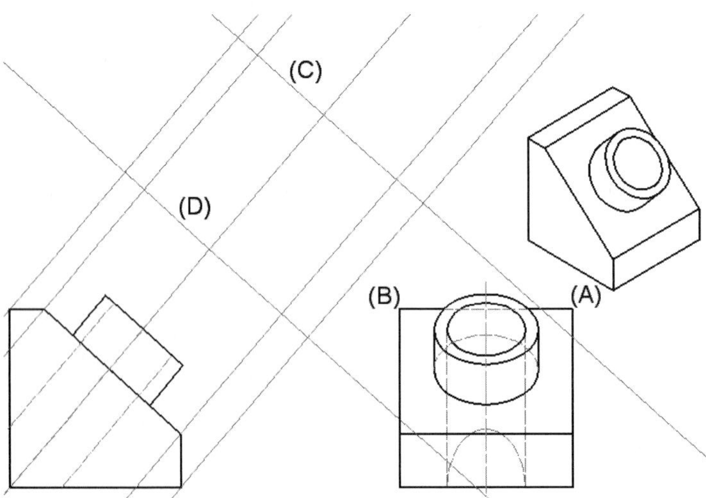

Figure 10.8

Continue using the OFFSET command to transfer more distances, such as the hole location, from the right side view to the auxiliary view, as shown in the following image. To complete the auxiliary view use the TRIM and/or FILLET commands to remove excess lines.

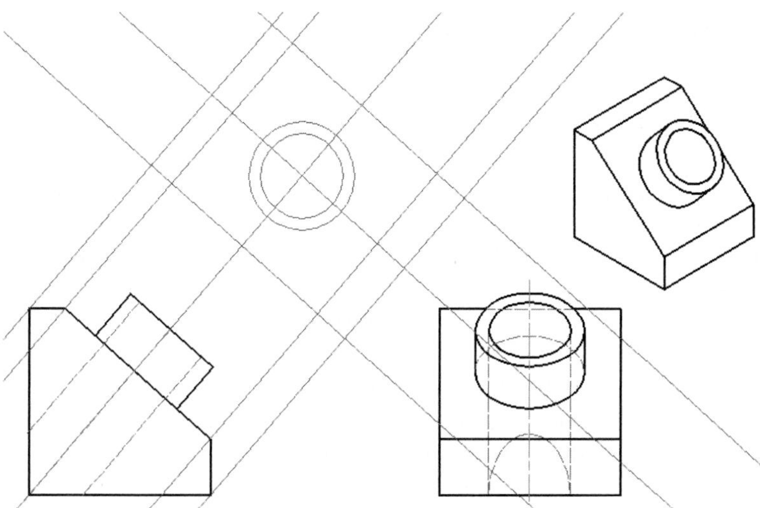

Figure 10.9

The completed auxiliary view is illustrated in the following image.

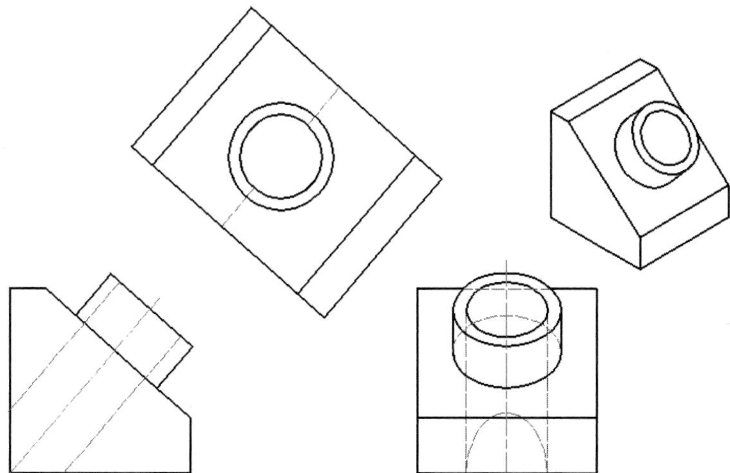

Figure 10.10

CONSTRUCTING AN AUXILIARY VIEW

Try It!: Open the drawing file 10_Aux Basics. Illustrated on the left in the following image is a basic multiview drawing consisting of Front, Top, and Right Side views. The inclined surface in the Front view is displayed in the Top and Right Side views; however, the surface appears foreshortened in both adjacent views. An auxiliary view of the incline needs to be made to show its true size and shape. Follow the next series of images that illustrate one suggested method for projecting to find auxiliary views. Notice that the current layer is Construction. You will create all lines in this layer. Later you will change lines to their correct layer to indicate their purpose, such as Object and Hidden.

Construct the perpendicular projection lines that make up the auxiliary view. Use the XLINE command to create all infinite lines perpendicular to the incline in the front view at the five endpoint or intersection locations shown on the right in the following image. These form the projection lines used for beginning the creation of the auxiliary view.

```
Command: XL (For XLINE)
Specify a point or [Hor/Ver/Ang/Bisect/Offset]: A (For Ang)
Enter angle of xline (0) or [Reference]: R (For Reference)
Select a line object: (Select the inclined line in the front
    view at "A")
Enter angle of xline <0>: 90
Specify through point: (Pick the endpoints at the five locations)
```

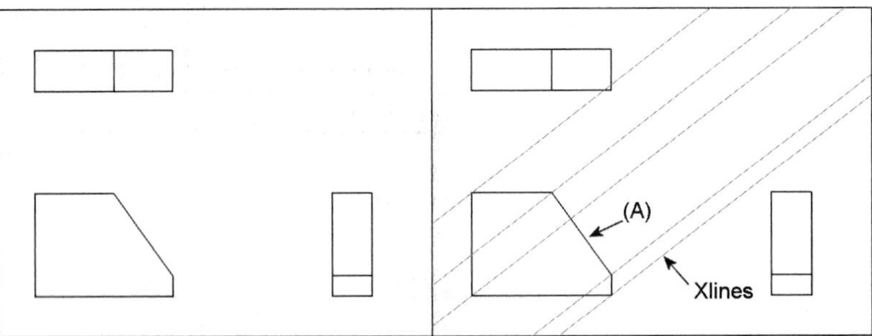

Figure 10.11

Create another xline; however, this time the infinite line will be constructed to one of the perpendicular projection lines at "A" created in the previous step. This line will form the front edge of the finished auxiliary view identified by points 1 and 2, as shown on the left in the following image.

```
Command: XL (For XLINE)
Specify a point or [Hor/Ver/Ang/Bisect/Offset]: A (For Ang)
Enter angle of xline (0) or [Reference]: R (For Reference)
Select a line object: (Select the perpendicular projection line
    at "A")
```

```
Enter angle of xline <0>: 90
Specify through point: (Pick a convenient location on your
    screen)
```

As shown on the right in the following image, use the OFFSET command to first set an offset distance from "A" to "B" in the side view. This distance represents the depth of the side view and will be transferred to the auxiliary view. Offset xline "C" located in the auxiliary view this distance in the direction indicated at "D".

```
Command: O (For OFFSET)
Current settings: Erase source=No Layer=Source OFFSETGAPTYPE=0
Specify offset distance or [Through/Erase/Layer] <Through>: (Pick
    the endpoint at "A")
Specify second point: (Pick the endpoint at "B")
Select object to offset or [Exit/Undo] <Exit>: (Pick xline "C")
Specify point on side to offset or [Exit/Multiple/Undo] <Exit>:
    (Pick a point on your screen at "D")
Select object to offset or [Exit/Undo] <Exit>: (Press ENTER to
    exit this command and perform the operation.)
```

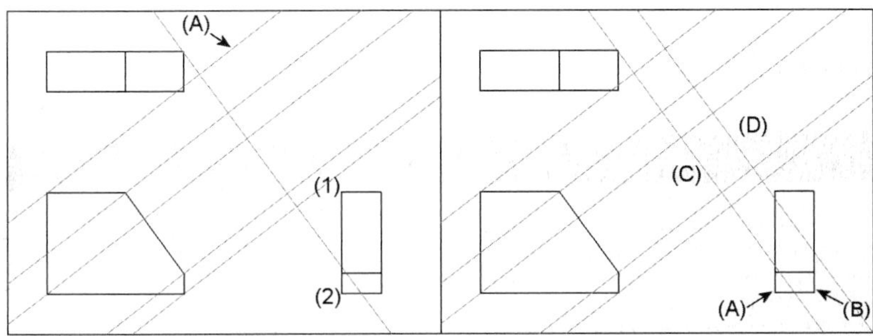

Figure 10.12

Trim all excess lines using the four projection lines labeled "A" through "D" as cutting edges, as shown on the left in the following image. Your display should appear similar to the example shown on the right in the following image.

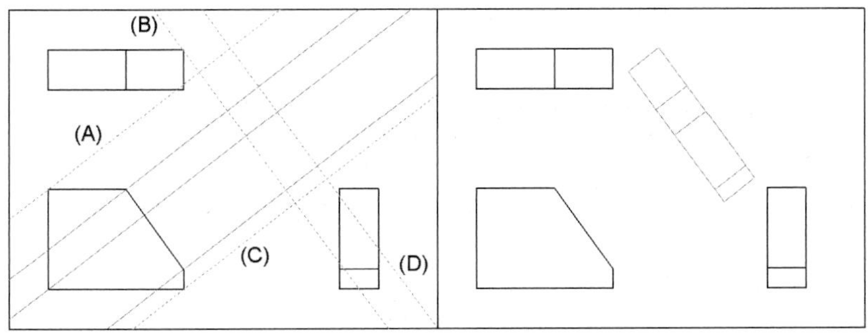

Figure 10.13

Change all lines to the Object layer. Change the single line segment to the Hidden layer, as shown on the left in the following image. The result is a multiview drawing complete with auxiliary view displaying the true size and shape of the inclined surface, as shown on the right in the following image. For dimensioning purposes, aligned dimensions could be used to annotate the distances located in the auxiliary view.

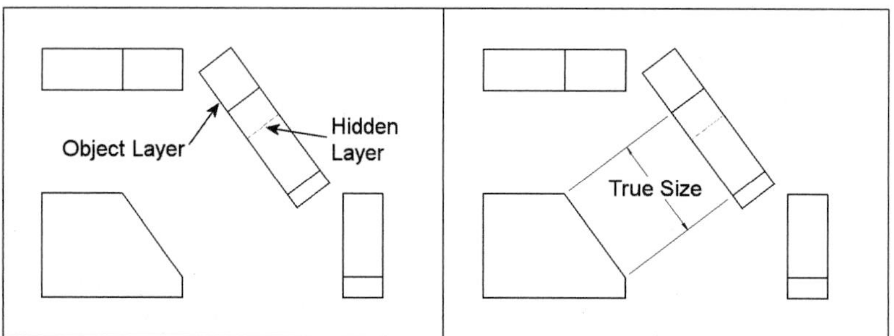

Figure 10.14

CONSTRUCTING THE TRUE SIZE OF A CURVED SURFACE

Try It!: Open the file 10_Aux Curve.dwg as shown on the left in the following image. From front and right side views, an auxiliary view will be created to display the true size and shape of the inclined surface. Notice that a layer called Construction has already been created and is current. This layer will be used throughout this tutorial exercise.

Construct the perpendicular projection lines that make up the auxiliary view. Use the XLINE command to create three infinite lines perpendicular to the incline in the front view at the three locations shown on the right in the following image. These form the projection lines used for beginning the auxiliary view.

```
Command: XL (For XLINE)
Specify a point or [Hor/Ver/Ang/Bisect/Offset]: A (For Ang)
Enter angle of xline (0) or [Reference]: R (For Reference)
Select a line object: (Select the inclined line in the front view)
Enter angle of xline <0>: 90
Specify through point: (Pick the three locations)
```

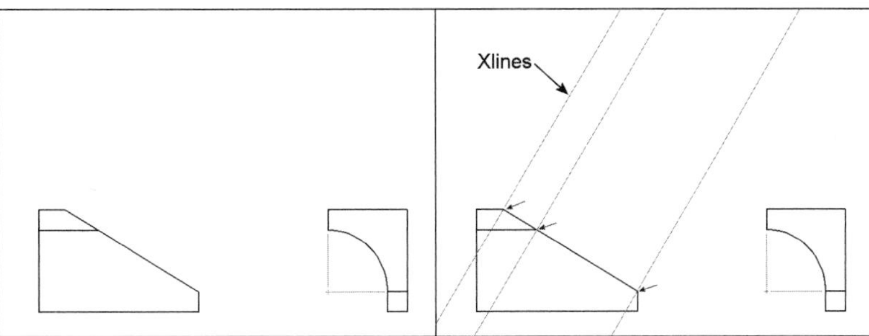

Figure 10.15

Create another xline; however, this time the infinite line will be constructed to one of the perpendicular projection lines created in the previous step. This line will form the back edge of the finished auxiliary view identified by points 1 and 2 shown on the left in the following image.

```
Command: XL (For XLINE)
Specify a point or [Hor/Ver/Ang/Bisect/Offset]: A (For Ang)
Enter angle of xline (0) or [Reference]: R (For Reference)
Select a line object: (Select one of the perpendicular projection
    lines)
Enter angle of xline <0>: 90
Specify through point: (Pick a location)
```

Trim the back corners of the projection lines. Your display should appear as shown on the right in the following image.

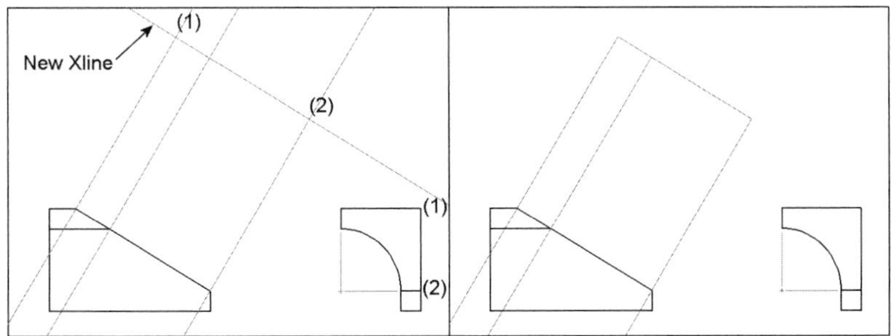

Figure 10.16

Set an offset distance from "A" to "B" in the side view. This distance represents the depth of the object and will be transferred to the auxiliary view. Offset the xline located in the auxiliary view this distance in the direction indicated on the left in the following image.

```
Command: O (For OFFSET)
Current settings: Erase source=No Layer=Source OFFSETGAPTYPE=0
Specify offset distance or [Through/Erase/Layer] <Through>: (Pick
    the endpoint at "A")
```

```
Specify second point: (Pick the endpoint at "B")
Select object to offset or [Exit/Undo] <Exit>: (Pick xline "C")
Specify point on side to offset or [Exit/Multiple/Undo] <Exit>:
    (Pick a point on your screen at "D")
Select object to offset or [Exit/Undo] <Exit>: (Press ENTER to
    exit this command and perform the operation.)
```

Set another offset distance from "A" to "B" in the side view; then offset back line "C" located in the auxiliary view this distance to "D," as shown on the right in the following image.

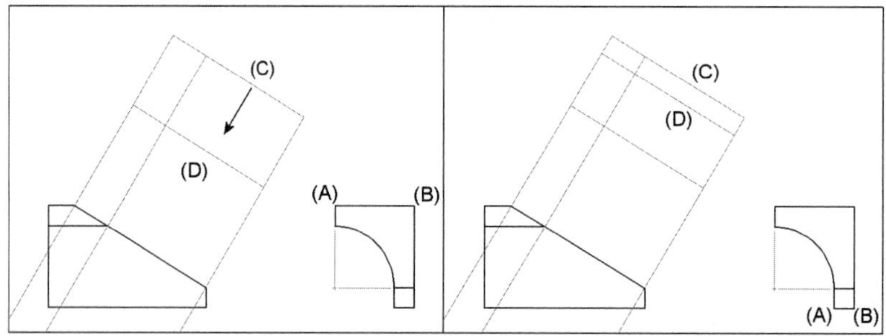

Figure 10.17

Trim and erase lines until your display appears as shown on the left in the following image. Construct a horizontal xline from the lower point of the incline. Then offset this xline at .25 increments. Your display should appear as shown on the right in the following image.

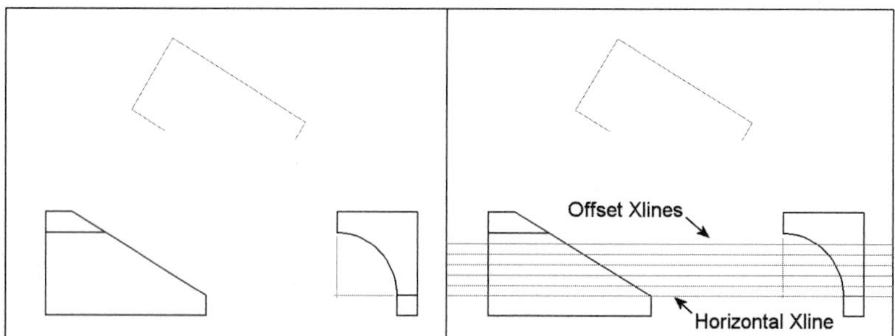

Figure 10.18

Construct an angled xline perpendicular to every intersection along the incline. You should have 5 xlines constructed from these points. Your display should appear as shown on the left in the following image. Create an offset distance from "A" to "B" along line segment 1 in the side view. Then offset the back line "C" this distance at "D" in the auxiliary view, as shown on the right in the following image.

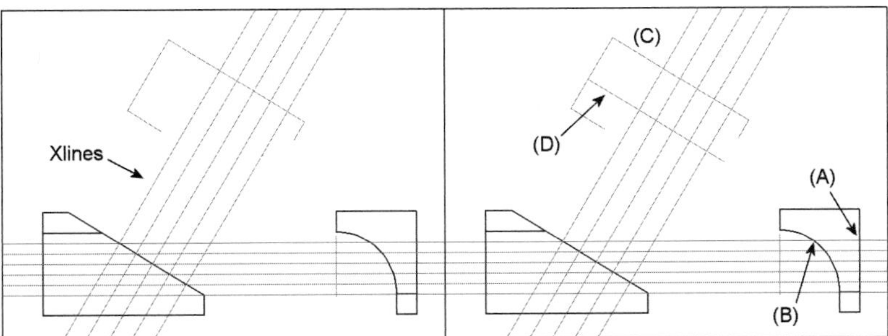

Figure 10.19

Place points at their respective intersections, as shown on the left in the following image. The purpose of the points is to identify a segment of the arc to be constructed in the auxiliary view. Perform the same series of steps to locate the remaining points that make up the curve in the auxiliary view, as shown on the right in the following image.

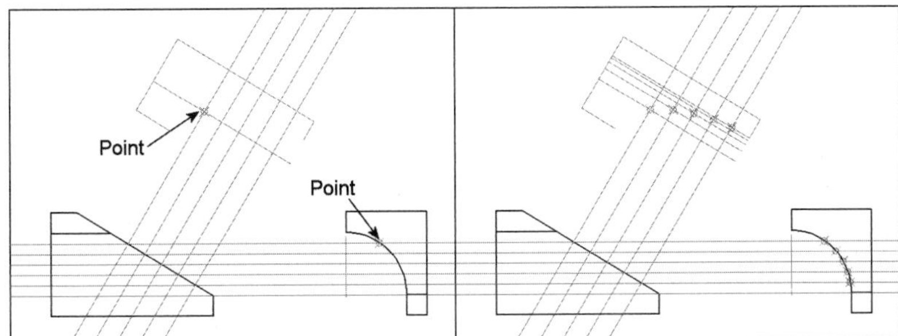

Figure 10.20

Use the SPLINE command to create a spline connecting all points along the auxiliary view, as shown on the left in the following image. Use the Osnap-Node mode for locking onto each point. Make the Object layer current. Then change the visible lines of the auxiliary view to the Object layer. Turn off the Construction layer. Your display should appear similar to the example on the right in the following image.

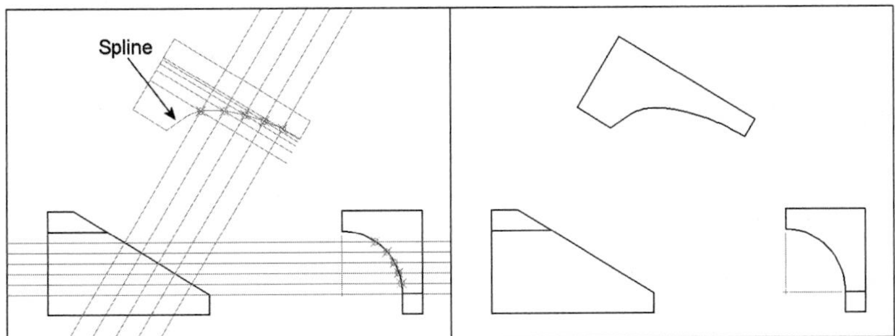

Figure 10.21

TUTORIAL EXERCISE: 10_BRACKET.DWG

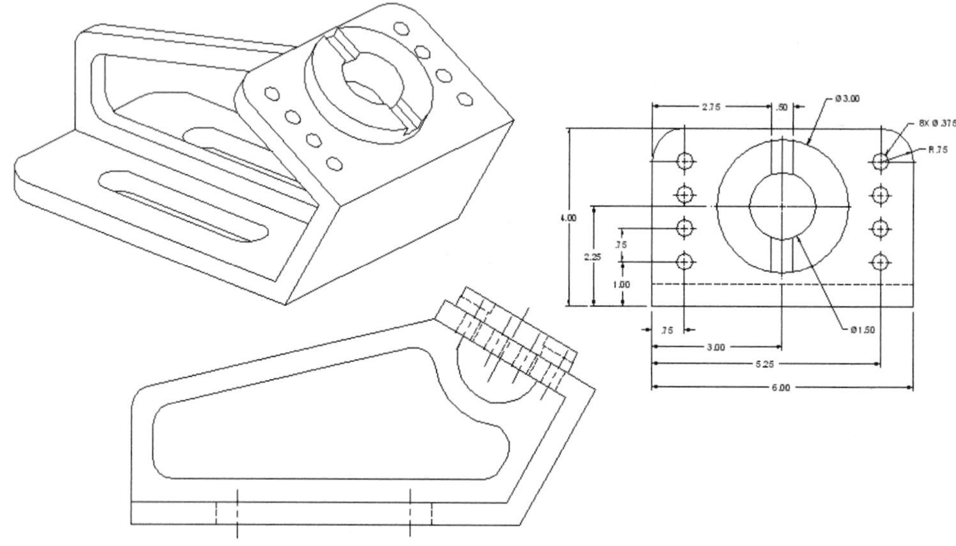

Figure 10.22

PURPOSE

This tutorial is designed to allow you to construct an auxiliary view of the inclined surface for the bracket shown in the previous image.

SYSTEM SETTINGS

Since this drawing is provided on CD, edit an existing drawing called 10_Bracket. Follow the steps in this tutorial for the creation of an auxiliary view.

LAYERS

The following layers have already been created with the following format:

Name	Color	Linetype
CEN	Yellow	Center
DIM	Yellow	Continuous
HID	Red	Hidden
OBJ	Cyan	Continuous

SUGGESTED COMMANDS

Begin this tutorial by using the OFFSET command to construct a series of lines parallel to the inclined surface containing the auxiliary view. Next construct lines perpendicular to the inclined surface. Use the CIRCLE command to begin laying out features that lie in the auxiliary view. Use ARRAY to copy the circle in a rectangular pattern. Add centerlines using the DIMCENTER command. Insert a predefined view called Top. A three-view drawing consisting of Front, Top, and auxiliary views is completed.

Step 1

Begin the construction of the auxiliary view by using the OFFSET command to copy a line parallel to the inclined line located in the Front view, as shown on the left in the following image. Use an offset distance of 8.50, pick line "A" as the object to offset, and pick near "B" as the side to perform the offset. Then use the OFFSET command again to offset line "B" to the side at "C" at a distance of 6.00 units.

The previous two lines formed using the OFFSET command define the depth of the auxiliary view. To determine the width of the auxiliary view, use the XLINE command to construct the two perpendicular projection lines, as shown in the following image. Project the two xlines from the endpoints of the Front view at "D" and "E" at an angle perpendicular to line "A," as shown on the right in the following image.

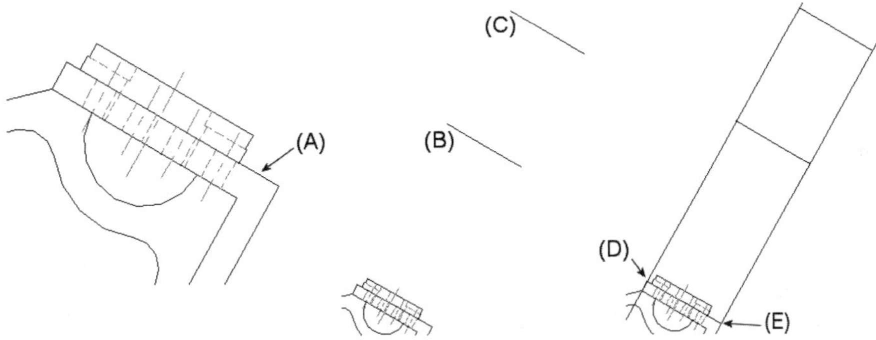

Figure 10.23

Step 2

Use the ZOOM-Window option to magnify the display of the auxiliary view similar to the illustration on the left in the following image. Then use the FILLET command to create four corners using the lines labeled "A" through "D." The fillet radius should be set to 0 in order to form the corners. The results are displayed on the right in the following image.

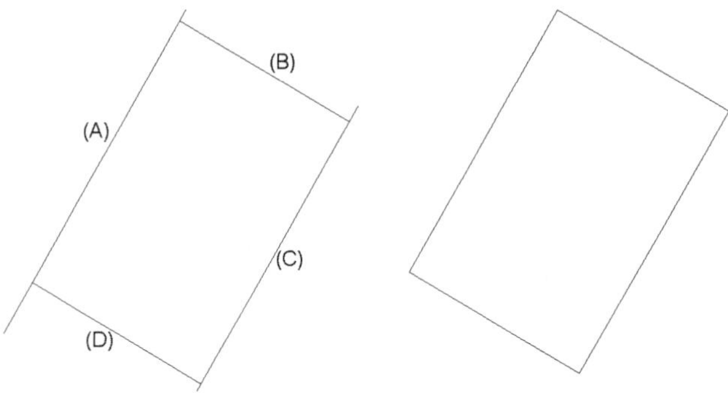

Figure 10.24

Step 3

When you are finished with the filleting operation, activate the View Manager dialog box, as shown on the left in the following image, and click the New button. When the New View/Shot Properties dialog box appears, as shown on the right in the following image, verify that the view will be defined based on the Current display; a radio button should be active for this mode. Then save the display to a new name, AUX.

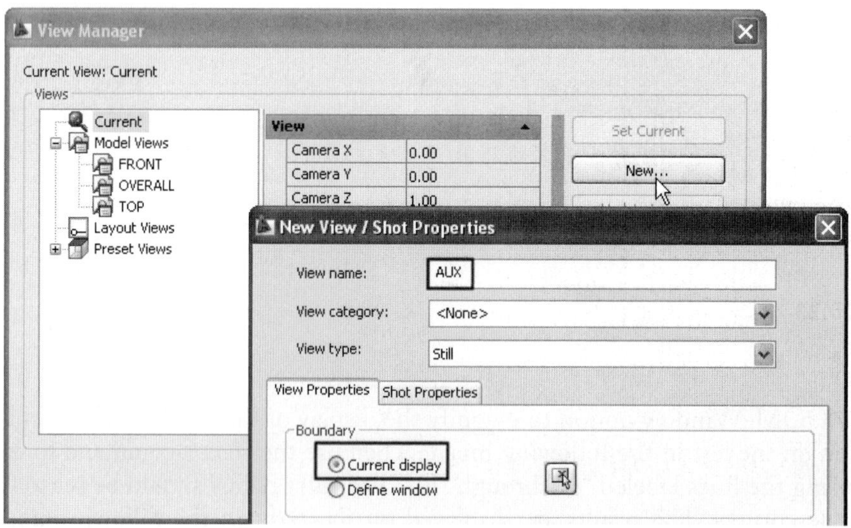

Figure 10.25

Step 4

Use the ZOOM-Previous option or other ZOOM mode to demagnify the screen back to the original display. Use XLINE to create a perpendicular projection from the endpoint of the centerline at "A," as shown on the left in the following image. Use the Reference option and pick the centerline. Keep the default angle of 0.

This is one of the construction lines that will be used for finding the centers of the circular features located in the auxiliary view.

Then use the OFFSET command to create the other construction line, as shown on the right in the following image. Use an offset distance of 3.00, pick line "B" as the object to offset, and pick a point near "C," as shown on the right in the following image.

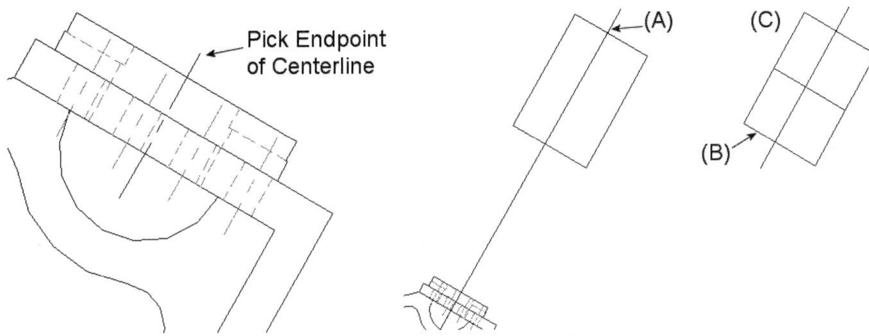

Figure 10.26

Step 5

Activate the View Manager dialog box and set the view AUX current, as shown on the left in the following image. Then draw two circles of diameters 3.00 and 1.50 from the center at "A," as shown on the right in the following image, using the CIRCLE command. For the center of the second circle, you can use the @ option to pick up the previous point that was the center of the first circle.

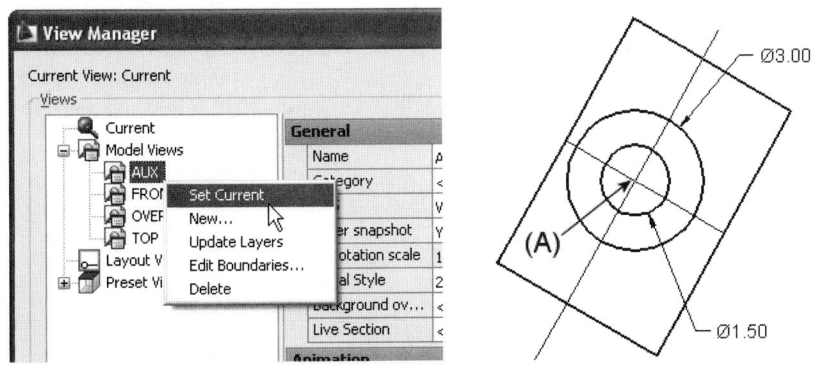

Figure 10.27

Step 6

Use the OFFSET command to offset the centerline the distance of .25 units, as shown on the left in the following image. Perform this operation on both sides of the centerline. Both offset lines form the width of the .50 slot.

Then use the TRIM command to trim away portions of the offset lines, using the two circles as cutting edges. When finished, your display should resemble the illustration on the right in the following image.

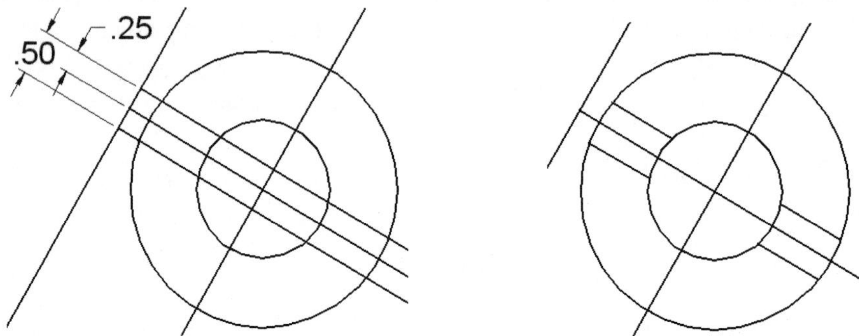

Figure 10.28

Step 7

Use the ERASE command to delete the two lines at "A" and "B," as shown on the left in the following image. Standard centerlines will be placed here later, marking the center of both circles.

Two more construction lines need to be made. These lines will identify the center of the small .37-diameter circle. Use the OFFSET command to create offsets for the lines illustrated on the right in the following image. Offset distances of .75 and 1.00 are used to perform this task. Also be careful to offset the correct line at the specified distance.

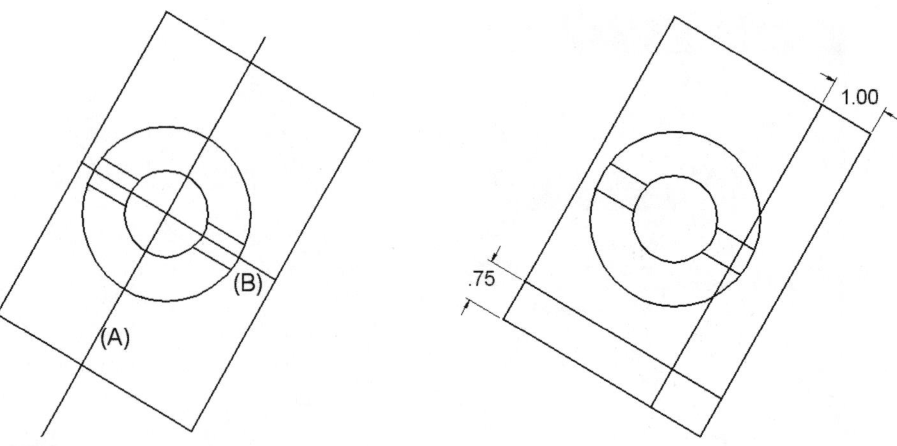

Figure 10.29

Step 8

Draw a circle of .37 units in diameter from the intersection of the two lines created in the last OFFSET command, as shown on the left in the following image. Then use the ERASE command to delete the two lines labeled "A" and "B." A standard center marker will be placed later at the center of this circle.

For the next phase of this step, make the CEN layer current through the Layer Control Box. The DIMCENTER command (or DCE for short) will be used to create the center mark for the .37-diameter circle. Type DCE at the command prompt and pick the edge of the circle at "C," as shown in the middle of the following image, to place this center mark.

Next, use the ROTATE command to rotate the center marker parallel to the edges of the auxiliary view. Activate the ROTATE command and pick the centerlines. For the base point of the rotation, pick a point at the center of the circle at "D" using OSNAP Center or Intersection mode. For the rotation angle, use OSNAP Perpendicular and pick the line near "E." The centerlines should rotate parallel to the edge of the auxiliary, as shown on the right in the following image.

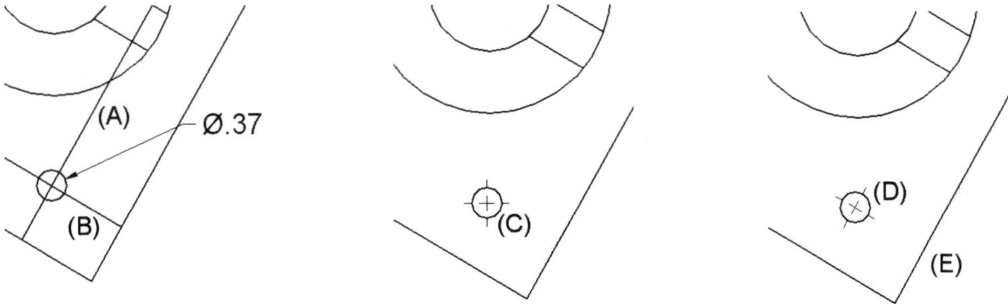

Figure 10.30

Step 9

Since the remaining seven holes form a rectangular pattern, use the Array dialog box, as shown on the left in the following image, and perform a rectangular array based on the .37-diameter circle and center marks. The number of rows is two and the number of columns is four. The distance between rows is 4.50 units and between columns is -0.75 units. Set the angle of the array by picking two points. To accomplish this, click the Pick angle of array button as shown in the following image and pick the endpoint at "A" first and the endpoint at "B" next. This creates the angle of the array at 330°. Also be sure to select the circle and centerlines as the objects to array. Performing a preview results in the array displayed on the right in the following image.

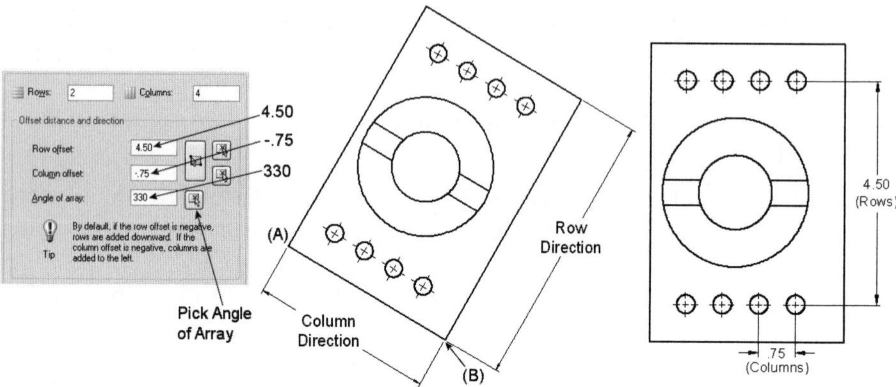

Figure 10.31

Step 10

Use the FILLET command set to a radius of .75 to place a radius along the two corners of the auxiliary, as shown on the left in the following image.

Place a center mark in the center of the two large circles using the DIMCEN (or DCE for short) command. When prompted to select the arc or circle, pick the edge of the large circle. The results are displayed in the middle of the following image.

Then rotate the center mark using the ROTATE command. Use OSNAP Perpendicular to set the rotation angle. The results are displayed on the right in the following image.

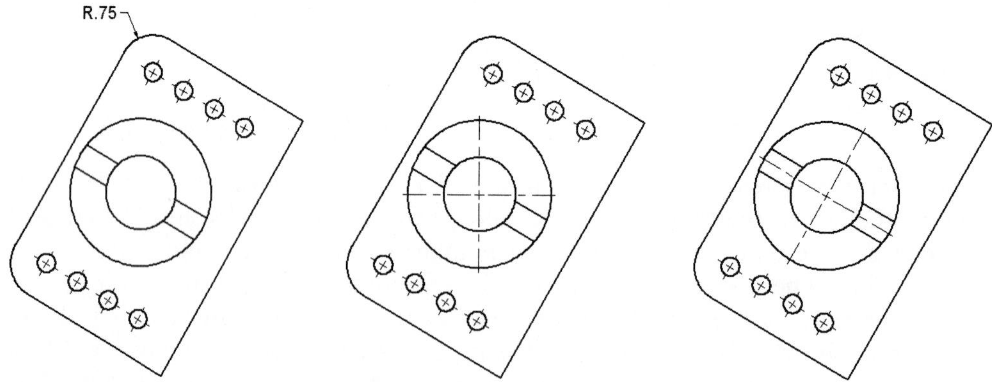

Figure 10.32

Step 11

Activate the View Manager dialog box, as shown in the following image, and set the view named OVERALL current. This should return your display to view all views of the Bracket.

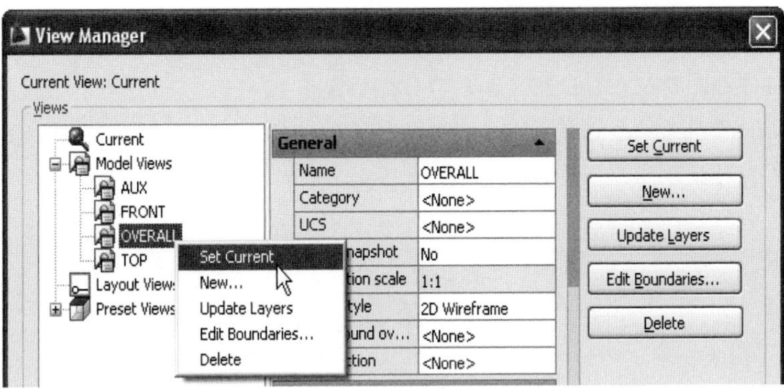

Figure 10.33

Step 12

Complete the drawing of the bracket by activating the Insert dialog box as shown in the following image and inserting an existing block called TOP into the drawing. This block represents the complete Top view of the drawing. Use an insertion point of 0,0 for placing this view in the drawing. Your display should appear similar to the following image.

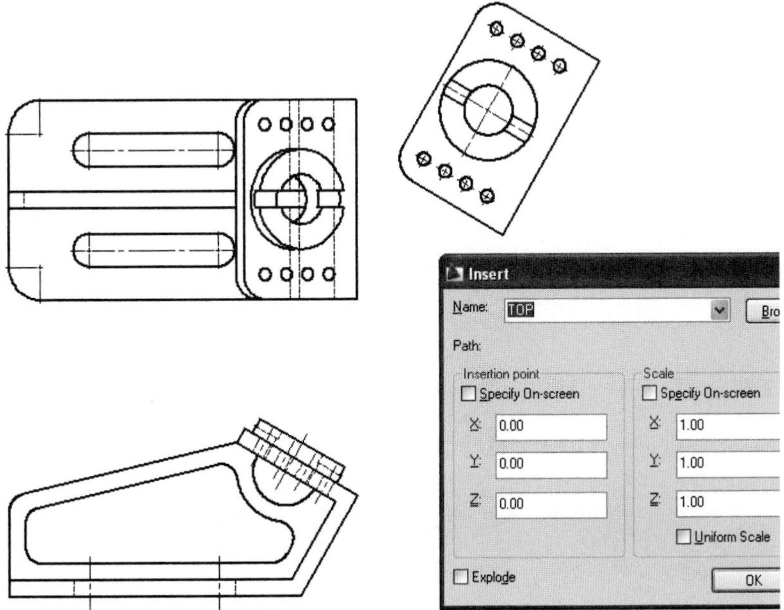

Figure 10.34

PROBLEMS FOR CHAPTER 10

DIRECTIONS FOR PROBLEMS 10–1 THROUGH 10–10

Draw the required views to fully illustrate each object. Be sure to include an auxiliary view.

PROBLEM 10–1

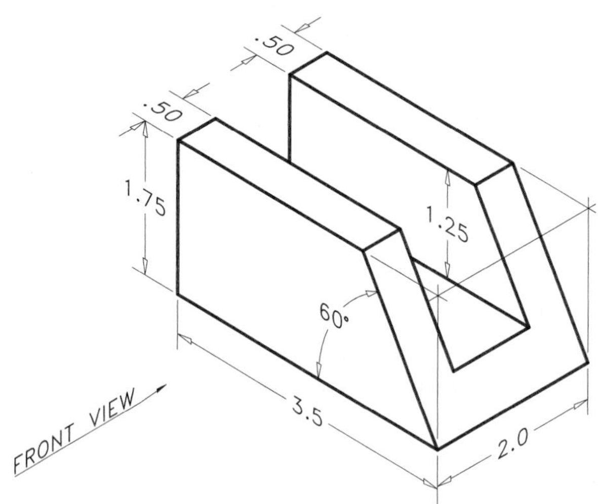

PROBLEM 10–2

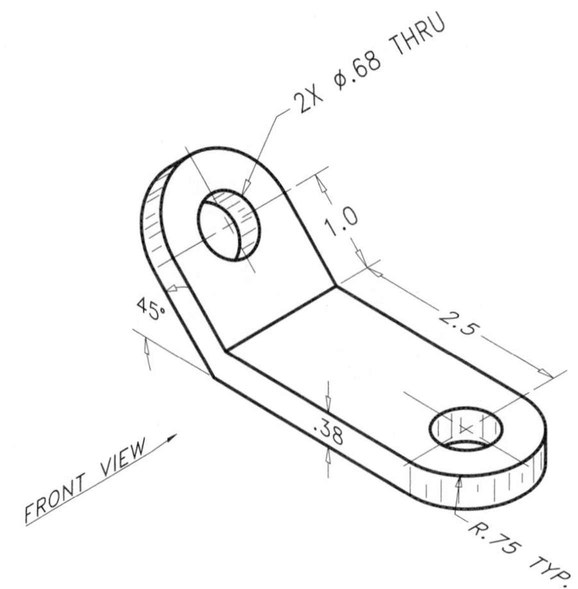

PROBLEM 10–3

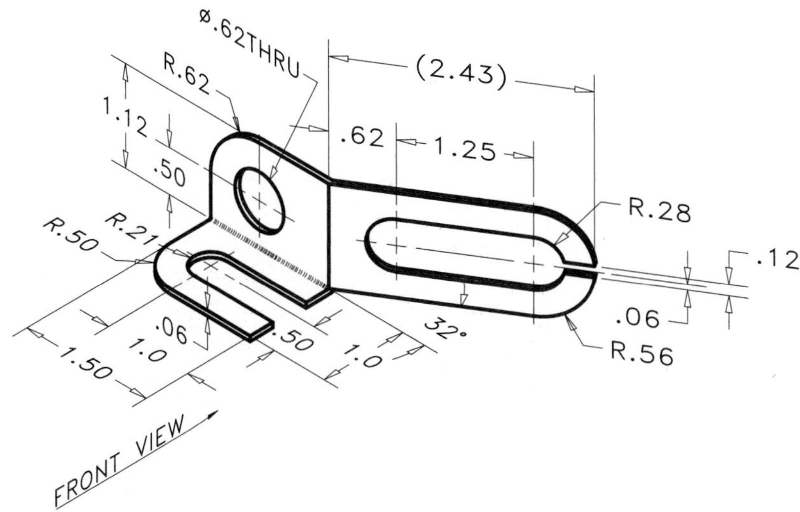

PROBLEM 10–4

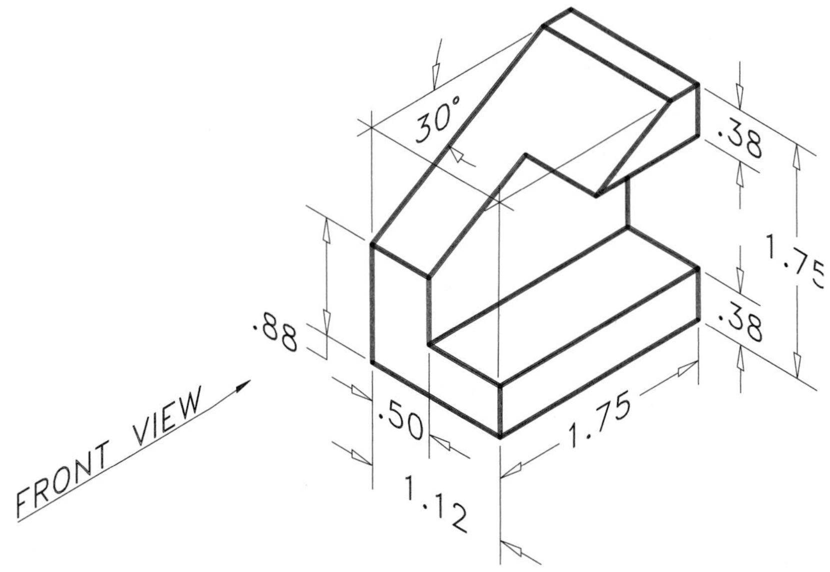

PROBLEM 10–5

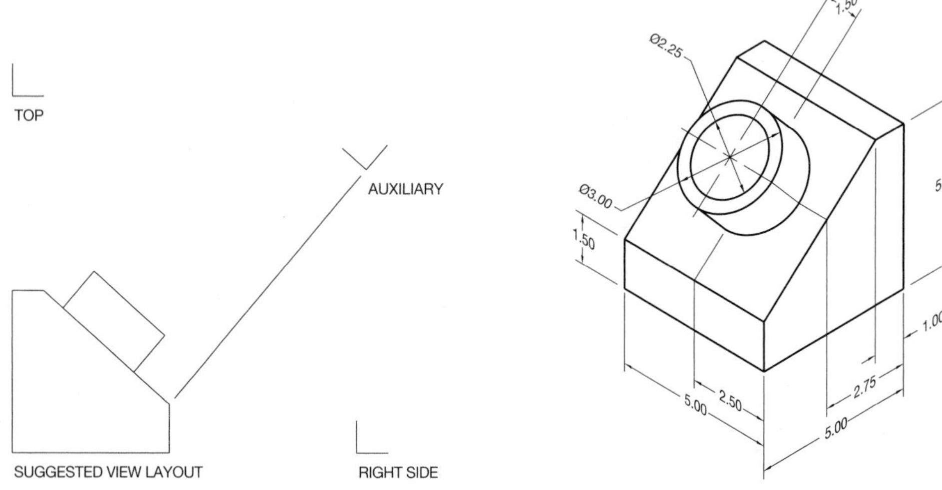

PROBLEM 10–6

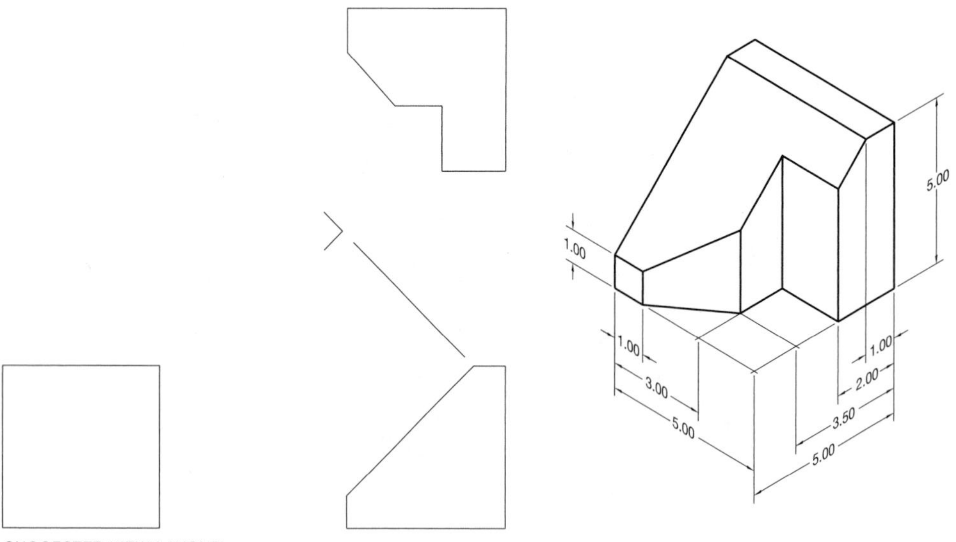

PROBLEM 10–7

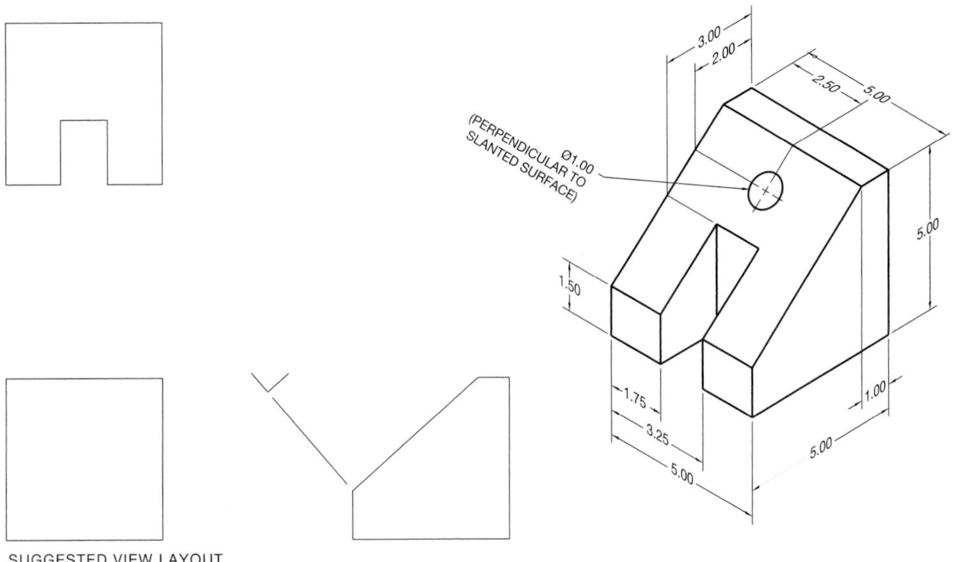

PROBLEM 10–8

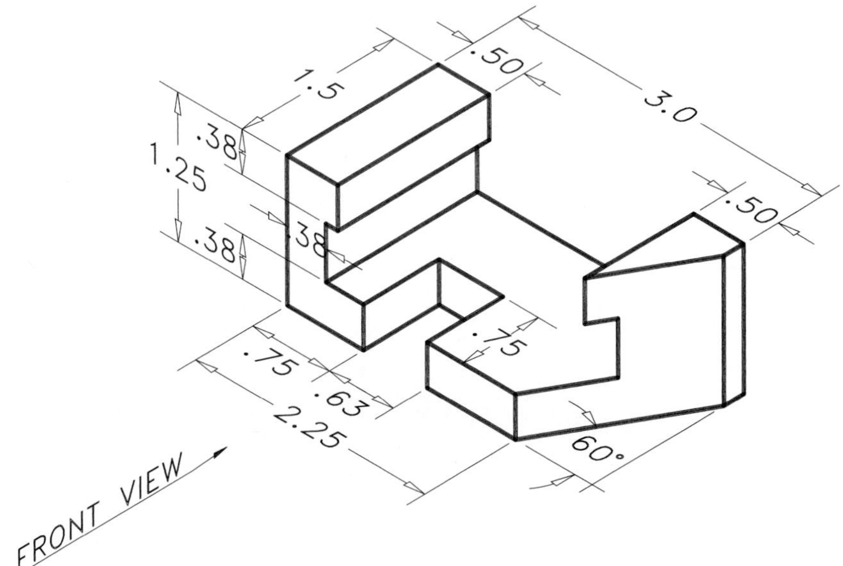

PROBLEM 10–9

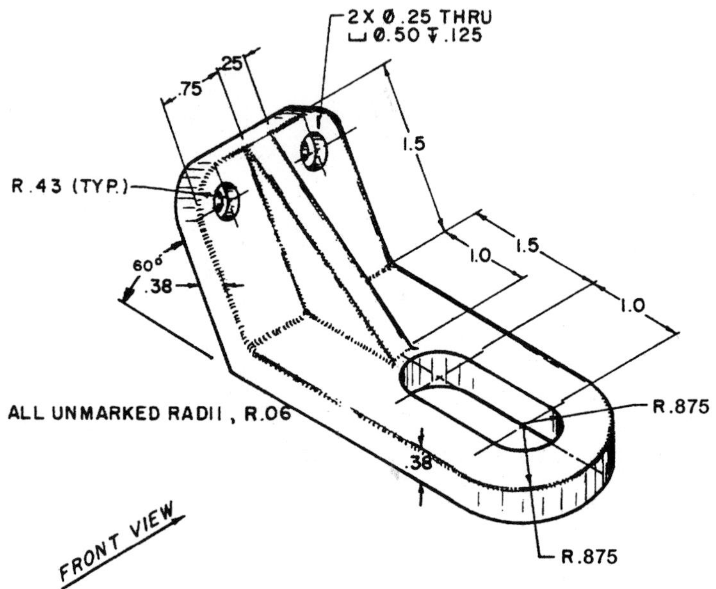

PROBLEM 10–10

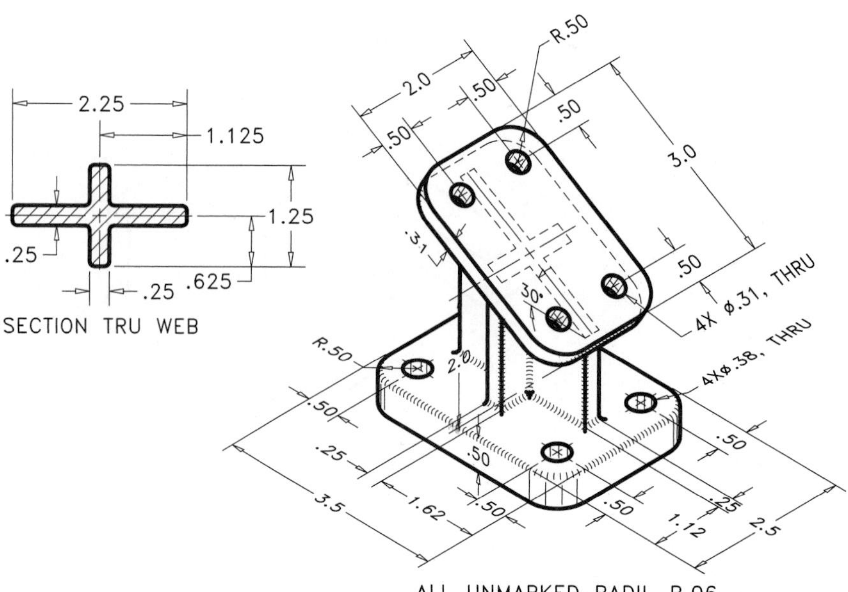

DIRECTIONS FOR PROBLEMS 10–11 THROUGH 10–13

Create the missing Auxiliary view through projection methods.

PROBLEM 10–11

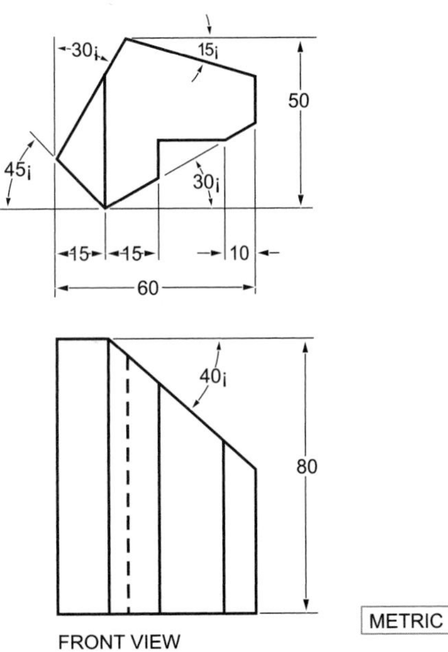

PROBLEM 10–12

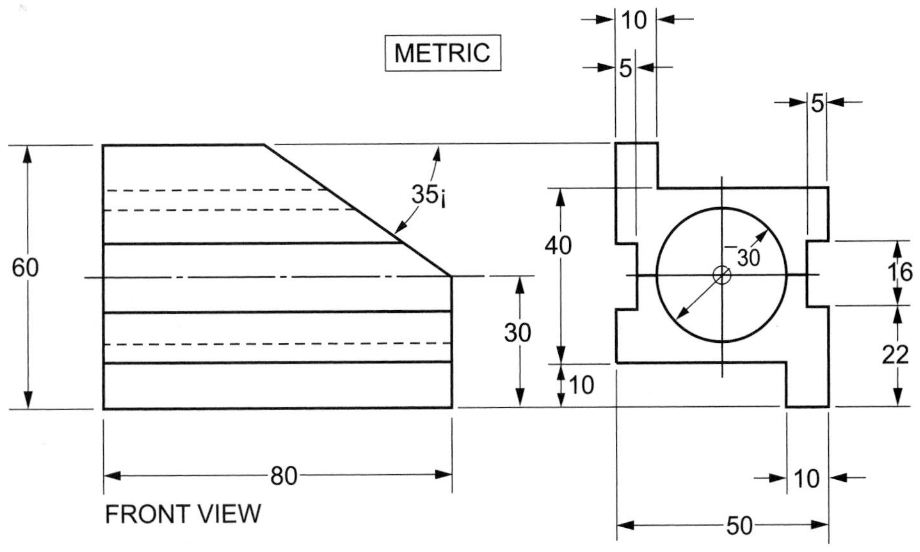

PROBLEM 10–13

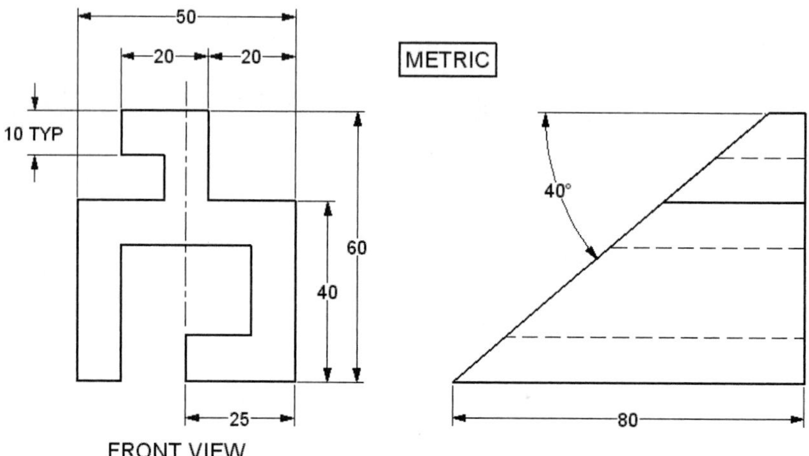

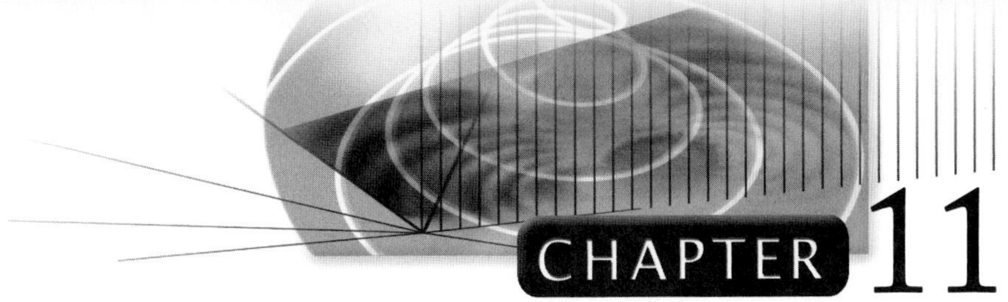

Adding Dimensions to Your Drawing

Once orthographic views have been laid out, a design is not ready for the production line until dimensions describing the width, height, and depth of the object are added to the drawing. However, these numbers must be added in a certain organized fashion; otherwise, the drawing becomes difficult to read. That may lead to confusion and the possible production of a part that is incorrect according to the original design. This chapter begins with an introduction to the proper placement of dimensions in a drawing. The chapter continues with a discussion of the commands used in AutoCAD for creating dimensions. These commands include Linear, Aligned, Ordinate, Radius, Diameter, Angular, Quick, Baseline, and Continue dimensions. A section in this chapter deals with the placement of geometric dimensioning and tolerancing symbols. A short segment on the effects grips have on dimensions is included in this chapter.

DIMENSIONING BASICS

Before discussing the components of a dimension, remember that object lines (at "A") continue to be the thickest lines of a drawing, with the exception of border or title blocks (see the following image). To promote contrasting lines, dimensions become visible, yet thin, lines. The heart of a dimension is the dimension line (at "B"), which is easily identified by the location of arrow terminators at both ends (at "D"). In mechanical cases, the dimension line is broken in the middle, which provides an excellent location for the dimension text (at "E"). For architectural applications, dimension text is usually placed above an unbroken dimension line. The extremities of the dimension lines are limited by the placement of lines that act as stops for the arrow terminators. These lines, called extension lines (at "C"), begin close to the object without touching the object. Extension lines will be discussed in greater detail in the pages that follow. For placing diameter and radius dimensions, a leader line consisting of an inclined line with a short horizontal shoulder is used (at "F"). Other applications of leader lines are for adding notes to drawings.

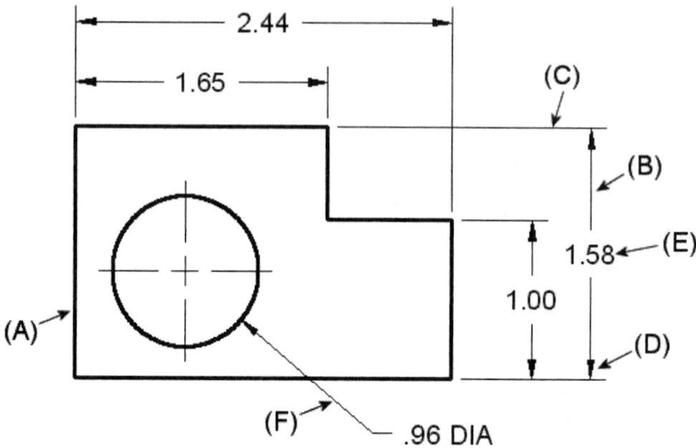

Figure 11.1

When you place dimensions in a drawing, provide a spacing of at least 0.38 units between the first dimension line and the object being dimensioned (at "A" on the left in the following image). If you're placing stacked or baseline dimensions, provide a minimum spacing of at least 0.25 units between the first and second dimension line (at "B") or any other dimension lines placed thereafter. This prevents dimensions from being placed too close to each other.

It is recommended that extensions never touch the object being dimensioned and that they begin between approximately 0.03 and 0.06 units away from the object (at "C" on the right in the following image). As dimension lines are added, extension lines should extend no further than 0.12 units beyond the arrow or any other terminator (at "D"). The height of dimension text is usually 0.125 units (at "E"). This value also applies to notes placed on objects with leader lines. Certain standards may require a taller lettering height. Become familiar with your office's standard practices, which may deviate from these recommended values.

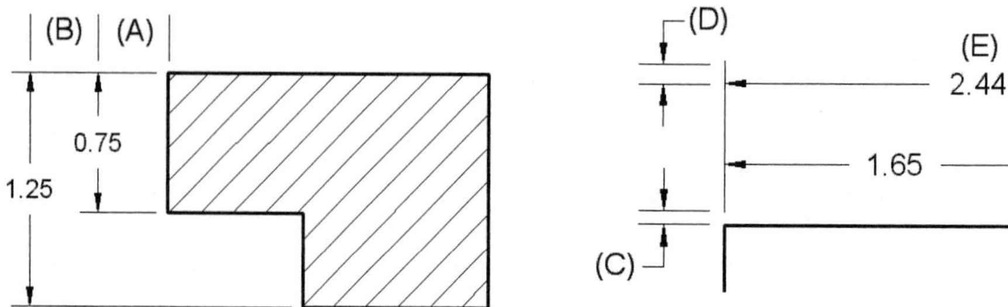

Figure 11.2

PLACEMENT OF DIMENSIONS

When placing multiple dimensions on one side of an object, place the shorter dimension closest to the object, followed by the next larger dimension, as shown on the left in the following image. When you are placing multiple horizontal and vertical dimensions involving extension lines that cross other extension lines, do not place gaps in the extension lines at their intersection points.

It is acceptable for extension lines to cross each other, but it is considered unacceptable practice for extension lines to cross dimension lines, as in the example on the right in the following image. The shorter dimension should be placed closest to the object, followed by the next largest dimension.

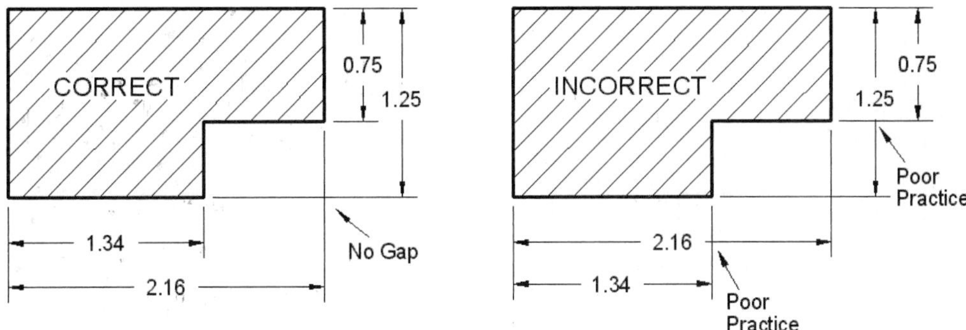

Figure 11.3

Dimensions should be added to the outside areas of the view being dimensioned, as shown on the left in the following image. It is considered poor practice to dimension inside an object when there is sufficient room to place dimensions on the outside; however, there may be exceptions to this rule. It is also considered poor practice to cross dimension lines, because this may render the drawing confusing and possibly result in the inaccurate interpretation of the drawing. It is also considered poor practice to superimpose extension lines over object lines. See the illustration on the right in the following image.

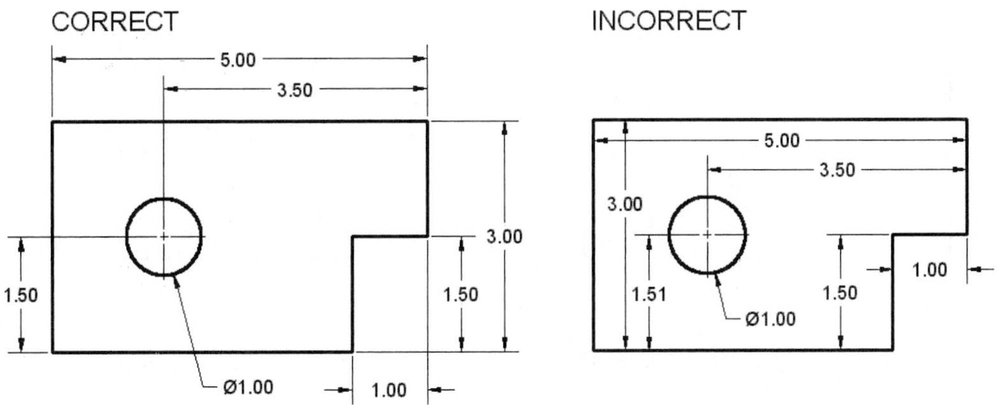

Figure 11.4

PLACING EXTENSION LINES

Because two extension lines may intersect without providing a gap, so also may extension lines and object lines intersect with each other without the need for a gap between them. The practice is the same when centerlines extend beyond the object without gapping, as shown on the left in the following image.

On the right in the following image, the gaps in the extension lines may appear acceptable; however, in a very complex drawing, gaps in extension lines would render a drawing confusing. Draw extension lines as continuous lines without providing breaks in the lines.

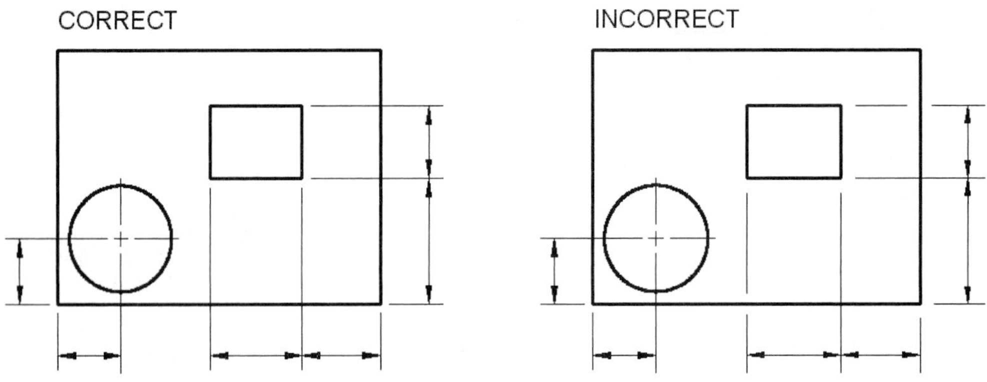

Figure 11.5

In the same manner, when centerlines are used as extension lines for dimension purposes, no gap is provided at the intersection of the centerline and the object, as shown on the left in the following image. As with extension lines, the centerline should extend no further than 0.125 units past the arrow terminator.

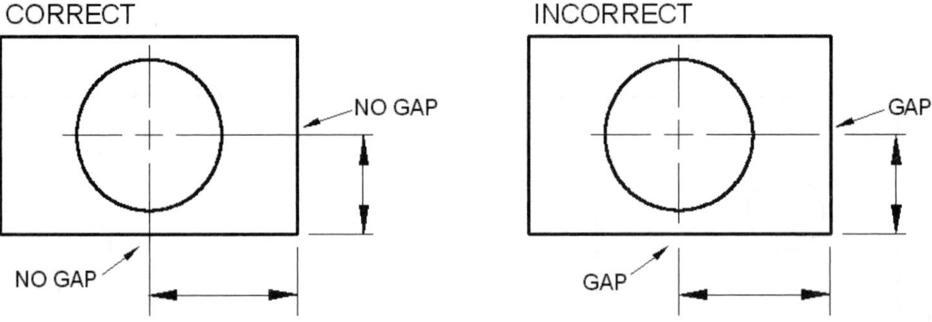

Figure 11.6

GROUPING DIMENSIONS

To promote ease of reading and interpretation, group dimensions whenever possible, as shown on the left in the following image. In addition to making the drawing and dimensions easier to read, this promotes good organizational skills and techniques. As in previous examples, always place the shorter dimensions closest to the drawing, followed by any larger or overall dimensions.

Avoid placing dimensions to an object line substituting for an extension line, as shown on the right in the following image. The drawing is more difficult to follow, with the dimensions placed at different levels instead of being grouped. In some cases, however, this practice of dimensioning is unavoidable.

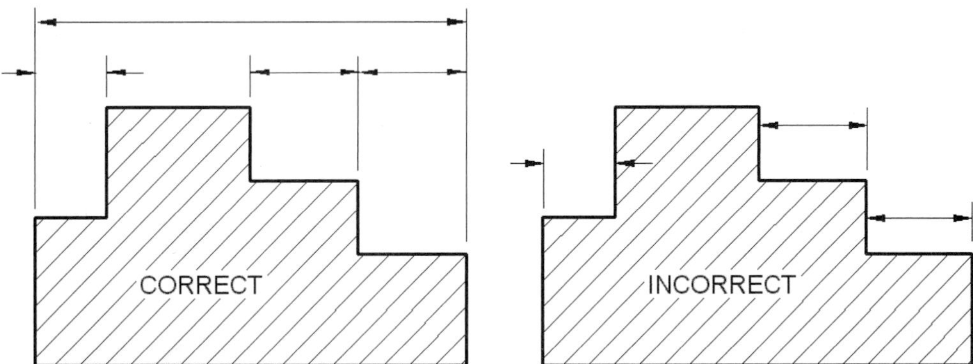

Figure 11.7

DIMENSIONING TO VISIBLE FEATURES

On the left in the following image, the object is dimensioned correctly. However, the problem is that dimensions have been added to hidden features. Although there are always exceptions, try to avoid dimensioning to any hidden surfaces or features.

The object shown on the right in the following image has been converted to a full section. Surfaces that were previously hidden are now exposed. This example illustrates a better way to dimension details that were previously hidden.

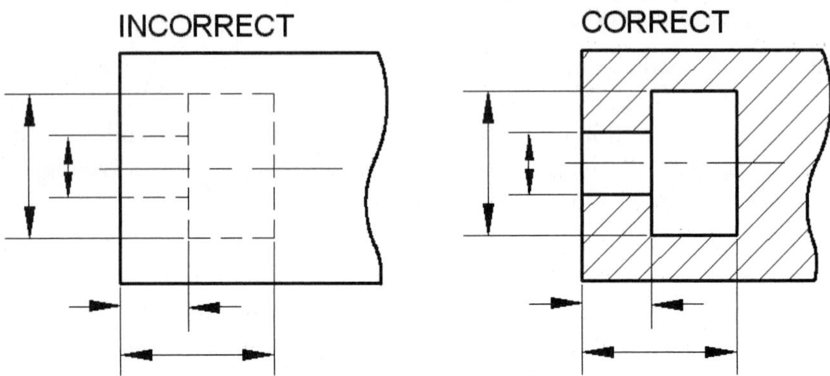

Figure 11.8

DIMENSIONING TO CENTERLINES

Centerlines are used to identify the center of circular features, as in "A" in the following image. You can also use centerlines to indicate an axis of symmetry, as in "B." Here, the centerline, consisting of a short dash flanked by two long dashes, signifies that the feature is circular in shape and form. Centerlines may take the place of extension lines when dimensions are placed in drawings.

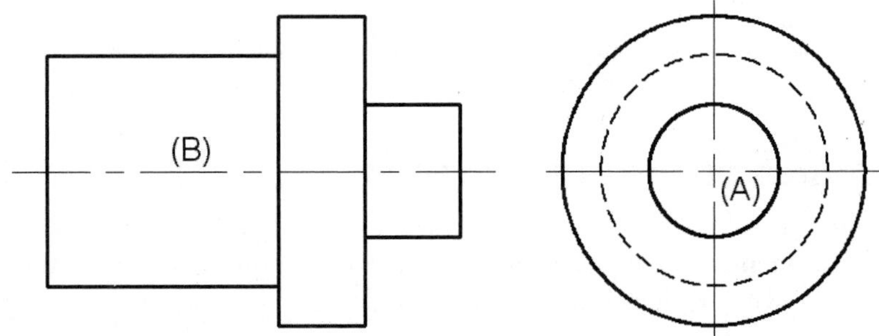

Figure 11.9

The following image on the left represents the top view of a U-shaped object with two holes placed at the base of the U. It also represents the correct way of utilizing centerlines as extension lines when dimensioning to holes. What makes this example correct is the rule of always dimensioning to visible features. The example in this image uses centerlines to dimension to holes that appear as circles.

The following image on the right represents the front view of the U-shaped object. The hidden lines display the circular holes passing through the object along with centerlines. Centerlines are being used as extension lines for dimensioning purposes; however, it is considered poor practice to dimension to hidden features or surfaces. Always attempt to dimension to a view where the features are visible before dimensioning to hidden areas.

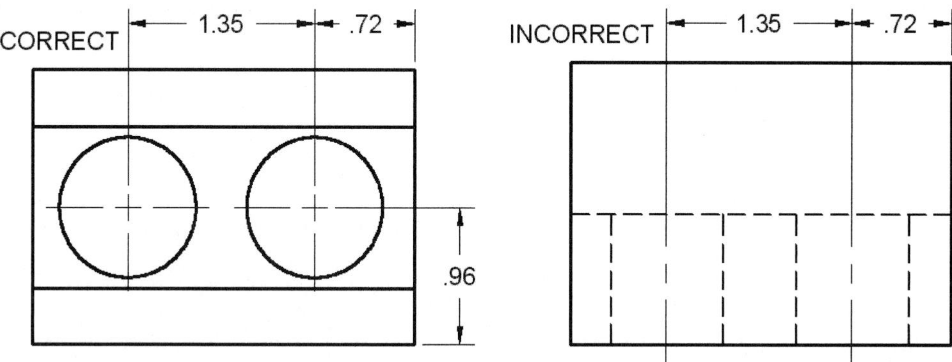

Figure 11.10

CONTINUE AND BASELINE DIMENSIONS

Additional dimensioning techniques are illustrated in the following image. In the illustration on the left, the dimensions are arranged alongside each other. This technique is commonly referred to as continue or chain dimensions. In the illustration on the right, a series of dimensions can be placed by stacking method. Also, all dimensions begin from the left edge of the part. This technique is called baseline dimensioning.

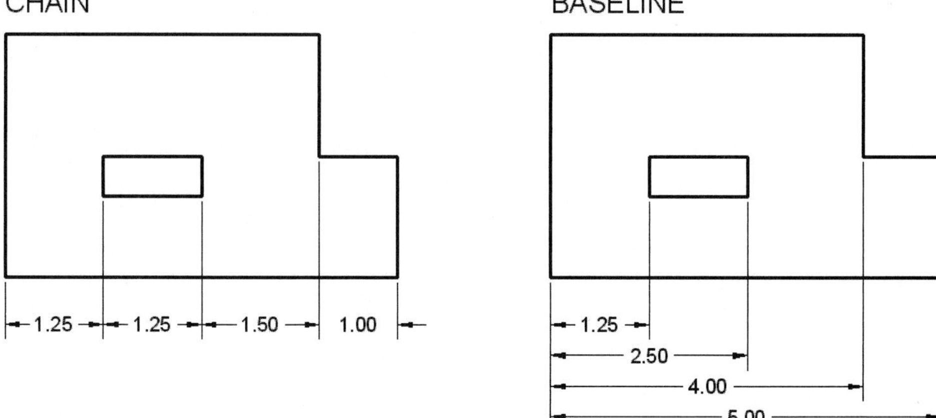

Figure 11.11

DIMENSIONING CYLINDERS

When dimensioning the diameters of cylinders, it is preferable to add these dimensions to the noncircular view, as shown on the right in the following image. Notice also that these dimensions are placed between views for clarity. While the illustration on the left is dimensionally correct, it is difficult to interpret the diameter dimensions placed in the Top view. Also, radius dimensions are rarely, used to dimension full diameters, since measurement tools such as micrometers and calipers are designed to check diameters.

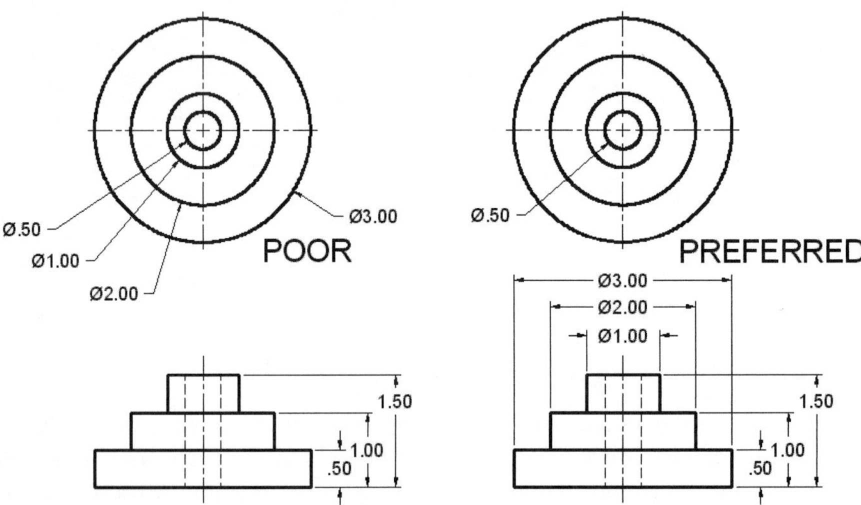

Figure 11.12

DIMENSIONING SYSTEMS

A typical example of placing dimensions using AutoCAD is illustrated on the left in the following image. Here, all text items are aligned horizontally. This applies to all vertical, aligned, angular, and diameter dimensions. When all dimension text can be read from one direction, this is the Unidirectional Dimensioning System. By default, all AutoCAD settings are set to dimension in the Unidirectional System.

The same object is also illustrated on the right in the following image. Notice that all horizontal dimensions have the text positioned in the horizontal direction, as in the unidirectional system. However, vertical and aligned dimension text is rotated or aligned with the direction being dimensioned. This is the most notable feature of the Aligned Dimensioning System. Text along vertical dimensions is rotated in such a way that the drawing must be read from the right.

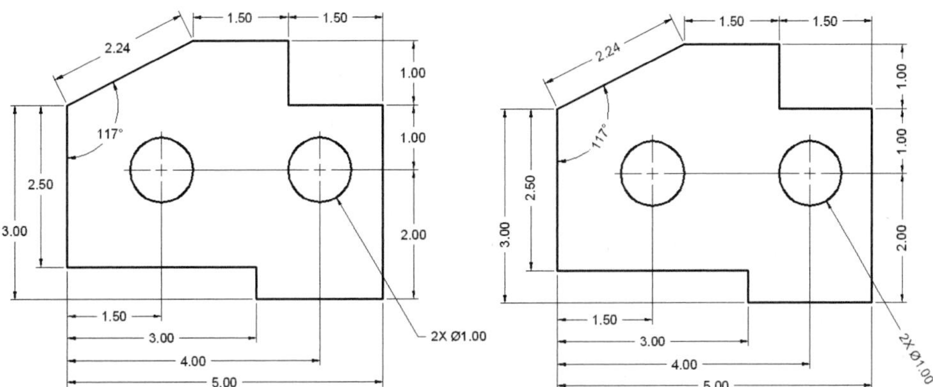

Figure 11.13

ORDINATE DIMENSIONS

A different type of dimensioning system is illustrated in the following image. Notice how all hole locations are designated by two numbers. Actually, these sets of numbers are coordinates. One of the numbers measures a distance in the X direction, while the other number measures a distance in the Y direction. No arrows or dimension lines are used. This type of dimensioning technique is referred to as Ordinate dimensioning. All hole location coordinates reference a known 0,0 location in the drawing. As shown in the following image, this type of dimensioning is well suited for locating hole positions.

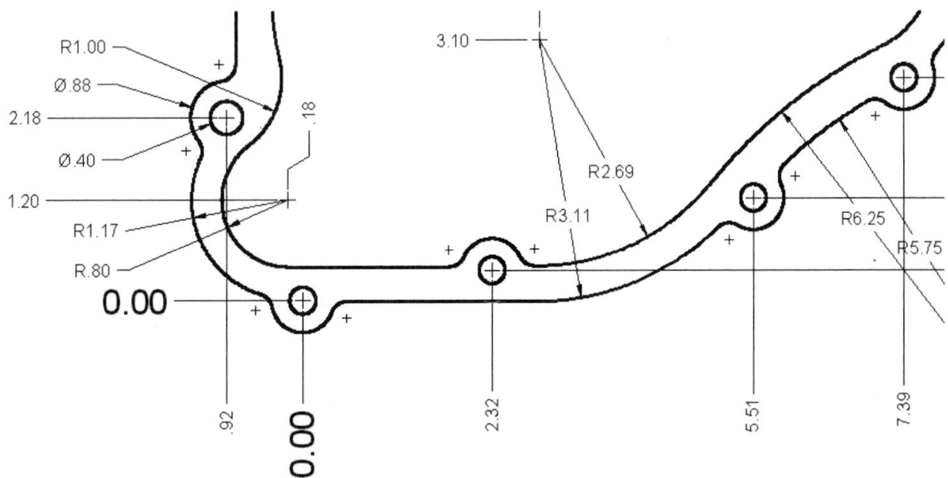

Figure 11.14

ARCHITECTURAL DIMENSIONS

The following image illustrates another type of dimensioning technique. Architectural dimensions usually have their text drawn directly above the dimension line. Instead of an arrow, architectural dimensions are terminated with a tick mark. Architectural dimension text for vertical dimensions is read from the left side; this makes all vertical architectural dimensions aligned or parallel to the dimension line.

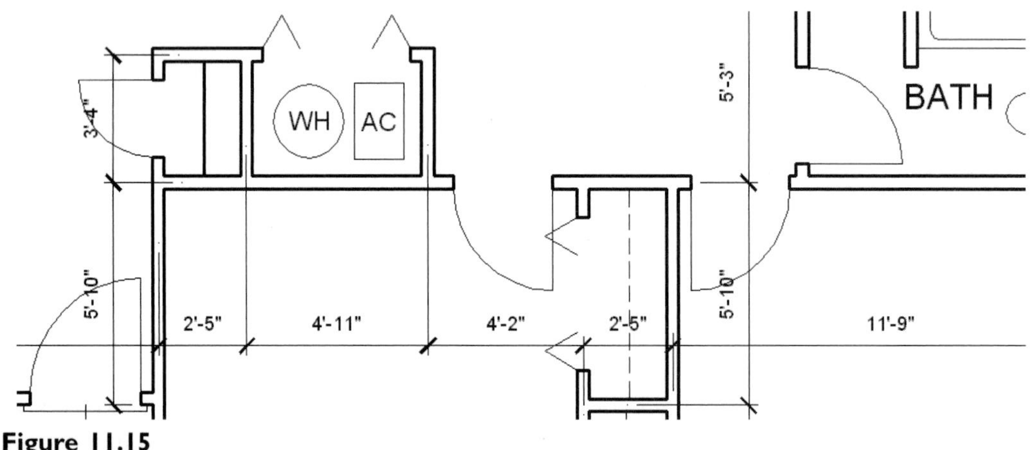

Figure 11.15

REPETITIVE DIMENSIONS

Throughout this chapter, numerous methods of dimensioning are discussed, supported by many examples. Just as it was important in multiview projection to draw only those views that accurately describe the object, it is also important to dimension these views. In this manner, the actual production of the part may start. However, take care when placing dimensions. It takes planning to better place a dimension. The problem with the illustration on the left in the following image is that, even though the views are correct and the dimensions call out the overall sizes of the object, there are too many cases in which dimensions are duplicated. Once a feature has been dimensioned, such as the overall width of 3.50 units shown between the Front and Top view, this dimension does not need to be repeated in the Top view. This is the purpose of understanding the relationship between views and what dimensions the views have in common with each other. Adding unnecessary dimensions also makes the drawing very busy and cluttered in addition to being very confusing to read. Also, if a size changes in production and a dimension needs to change on the drawings, more instances of the dimension must be found and corrected. Compare the illustration on the left with that on the right in the following image, which shows only those dimensions needed to describe the size of the object.

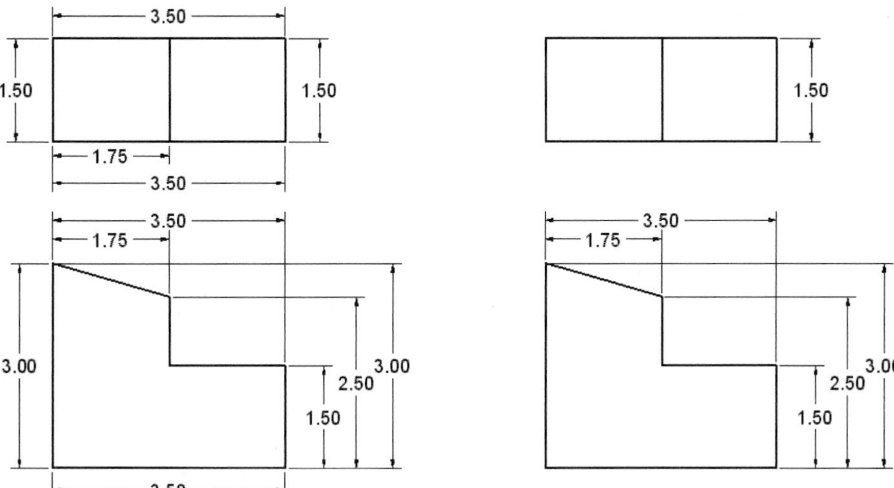

Figure 11.16

METHODS OF CHOOSING DIMENSION COMMANDS

A number of tools are available for choosing dimension-related commands. The top item in the following image illustrates the Dimension toolbar. Also illustrated are the Menu Browser, pull-down menu, and Ribbon that are used for selecting dimension commands. Another way of activating dimension commands is through the keyboard. These commands tend to get long; the DIMLINEAR command is one example. To spare you the effort of entering the entire command, all dimension commands have been abbreviated to three letters. For example, DLI is the alias for the DIMLINEAR command.

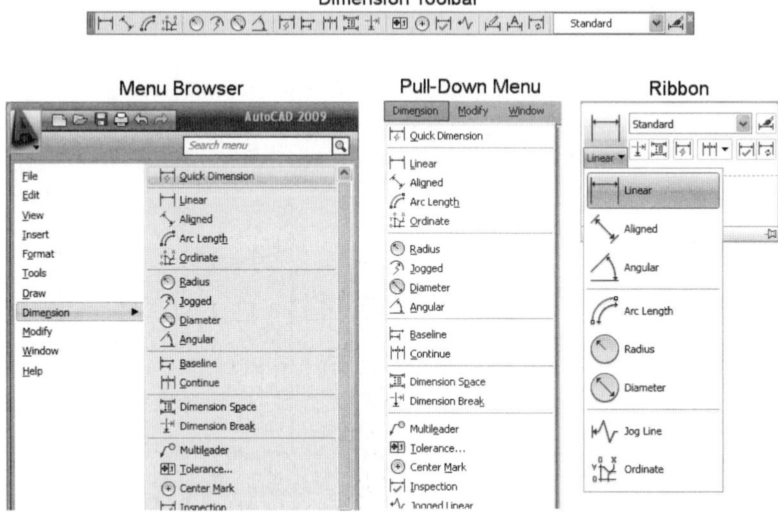

Figure 11.17

Study the following table for a brief description of each tool.

Button	Tool	Shortcut	Function
	Linear	DLI	Creates a horizontal, vertical, or rotated dimension
	Aligned	DAL	Creates a linear dimension parallel to an object
	Arc Length	DAR	Dimensions the total length of an arc
	Ordinate	DOR	Creates an ordinate dimension based on the current position of the User Coordinate System (UCS)
	Radius	DRA	Creates a radius dimension
	Jogged	DJO	Creates a jogged dimension
	Diameter	DIA	Creates a diameter dimension
	Angular	DAN	Creates an angular dimension
	Quick Dimension	QDIM	Creates a quick dimension
	Continue	DCO	Creates a continued dimension
	Baseline	DBA	Creates a baseline dimension
	Dimension Space	—	Adjusts the spacing of parallel linear and angular dimensions to be equal
	Dimension Break	—	Creates breaks in dimension, extension, and leader lines
	Tolerance	TOL	Activates the Tolerance dialog box

Button	Tool	Shortcut	Function
	Center Mark	DCE	Places a center mark inside a circle or an arc
	Inspection	—	Creates an inspection dimension
	Jogged Linear	DJL	Creates a jogged linear dimension
	Edit	DED	Used to edit a dimension
	Text Edit	—	Used to edit the text of a dimension
	Update	—	Used for updating existing dimensions to changes in the current dimension style
	Dimstyle Dialog	D	Activates the Dimension Style Manager dialog box

BASIC DIMENSION COMMANDS

LINEAR DIMENSIONS

The Linear Dimensioning mode generates either a horizontal or vertical dimension, depending on the location of the dimension. The following prompts illustrate the generation of a horizontal dimension with the DIMLINEAR command. Notice that identifying the dimension line location at "C" in the following image automatically generates a horizontal dimension.

Try It!: Open the drawing file 11_Dim Linear1. Verify that OSNAP is on and set to Endpoint. Use the following command sequence and image for performing this dimensioning task.

```
Command: DLI (For DIMLINEAR)
Specify first extension line origin or <select object>: (Select
    the endpoint of the line at "A")
Specify second extension line origin: (Select the other endpoint
    of the line at "B")
Specify dimension line location or [Mtext/Text/Angle/Horizontal/
    Vertical/Rotated]: (Pick a point near "C" to locate the
    dimension)
Dimension text = 8.00
```

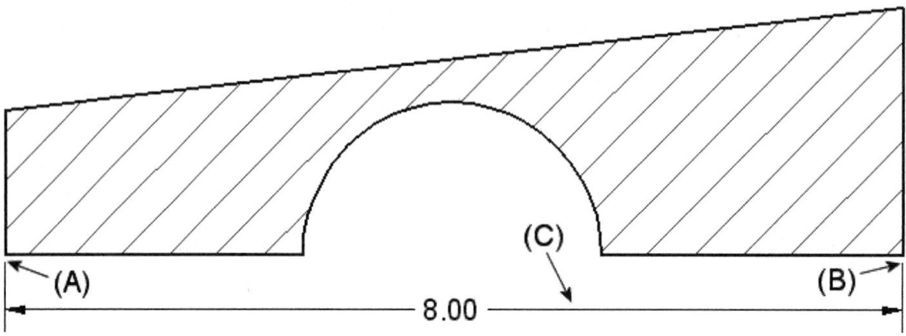

Figure 11.18

The linear dimensioning command is also used to generate vertical dimensions. The following prompts illustrate the generation of a vertical dimension with the DIMLINEAR command. Notice that identifying the dimension line location at "C" in the following image automatically generates a vertical dimension.

Try It!: Open the drawing file 11_Dim Linear2. Verify that OSNAP is on and set to Endpoint. Use the following command sequence and image for performing this dimensioning task.

```
Command: DLI (For DIMLINEAR)
Specify first extension line origin or <select object>: (Select
    the endpoint of the line at "A")
Specify second extension line origin: (Select the endpoint of the
    line at "B")
Specify dimension line location or [Mtext/Text/Angle/Horizontal/
    Vertical/Rotated]: (Pick a point near "C" to locate the
    dimension)
Dimension text = 1.14
```

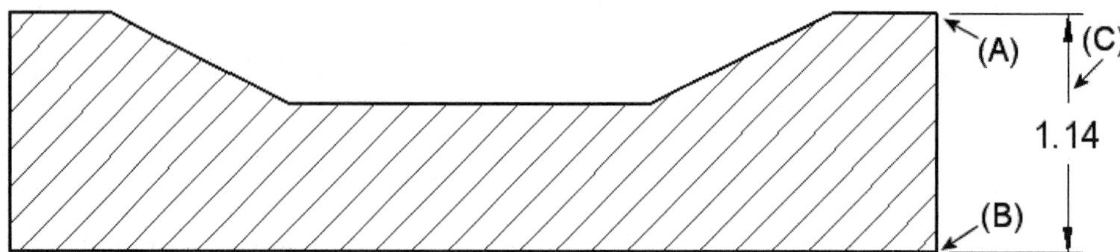

Figure 11.19

Rather than select two separate endpoints to dimension to, certain situations allow you to press ENTER and select the object (in this case, the line). This selects the two endpoints and prompts you for the dimension location. The completed dimension is illustrated in the following image.

Try It!: Open the drawing file 11_Dim Linear3. Use the following command sequence and image for performing this dimensioning task.

```
Command: DLI (For DIMLINEAR)
Specify first extension line origin or <select object>: (Press
    ENTER to select an object)
Select object to dimension: (Select the line at "A")
Specify dimension line location or [Mtext/Text/Angle/Horizontal/
    Vertical/Rotated]: (Pick a point near "B" to locate the
    dimension)
Dimension text = 9.17
```

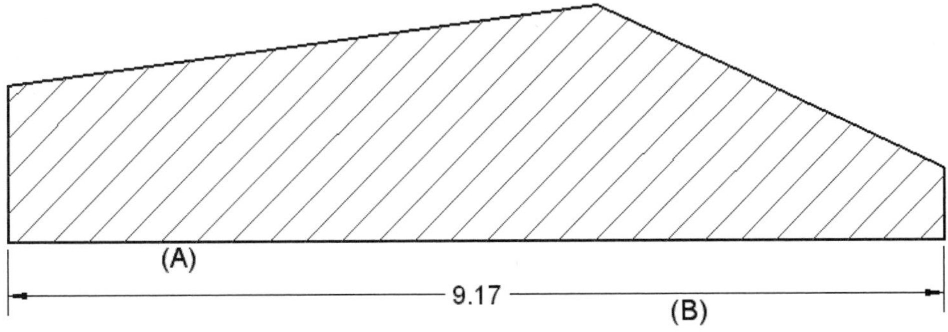

Figure 11.20

ALIGNED DIMENSIONS

The Aligned Dimensioning mode generates a dimension line parallel to the distance specified by the location of two extension line origins, as shown in the following image.

Try It!: Open the drawing file 11_Dim Aligned. Verify that OSNAP is on and set to Endpoint. The following prompts and image illustrate the creation of an aligned dimension with the DIMALIGNED command.

```
Command: DAL (For DIMALIGNED)
Specify first extension line origin or <select object>: (Select
    the endpoint of the line at "A")
Specify second extension line origin: (Select the endpoint of the
    line at "B")
Specify dimension line location or [Mtext/Text/Angle]: (Pick a
    point at "C" to locate the dimension)
Dimension text = 12.06
```

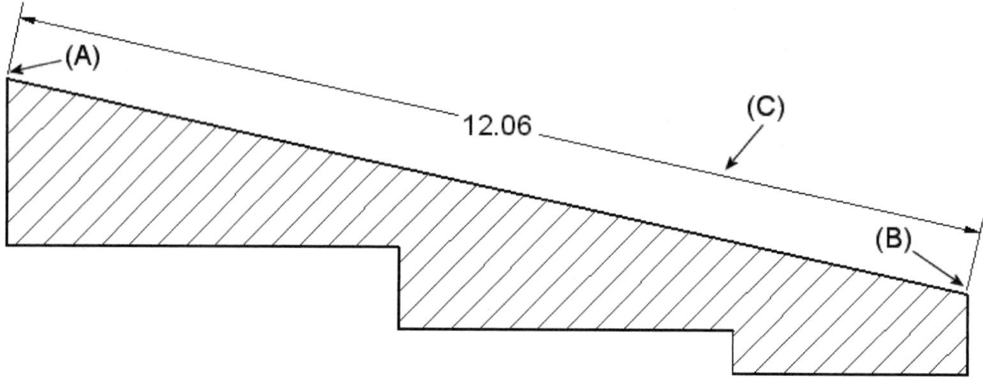

Figure 11.21

ROTATED LINEAR DIMENSIONS

This Linear Dimensioning mode can also create a dimension line rotated at a specific angle, as shown in the following image. The following prompts illustrate the generation of a rotated dimension with the DIMLINEAR command and a known angle of 45°. If you do not know the angle, you could easily establish the angle by clicking the endpoints at "A" and "D" in the following image when prompted to "Specify angle of dimension line <0>."

Try It!: Open the drawing file 11_Dim Rotated. Verify that OSNAP is on and set to Endpoint. Use the following command sequence and image for creating a rotated dimension.

```
Command: DLI (For DIMLINEAR)
Specify first extension line origin or <select object>: (Select
    the endpoint of the line at "A")
Specify second extension line origin: (Select the endpoint of the
    line at "B")
Specify dimension line location or [Mtext/Text/Angle/Horizontal/
    Vertical/Rotated]: R (For Rotated)
Specify angle of dimension line <0>: 15
Specify dimension line location or [Mtext/Text/Angle/Horizontal/
    Vertical/Rotated]: (Pick a point at "C" to locate the
    dimension)
Dimension text = 6.00
```

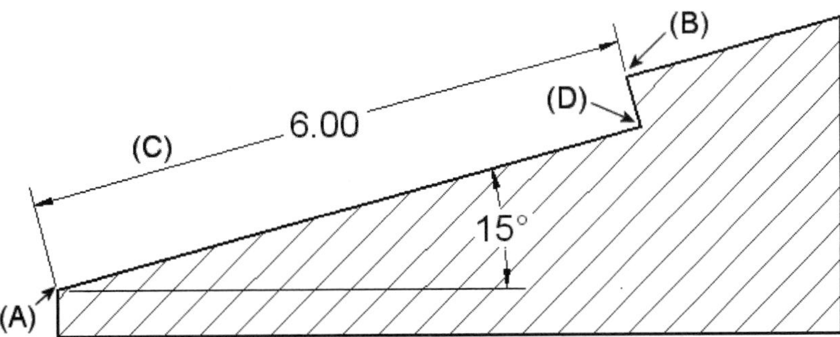

Figure 11.22

CONTINUE DIMENSIONS

The power of grouping dimensions for ease of reading has already been explained. The following image shows yet another feature of dimensioning in AutoCAD: the practice of using continue dimensions. With one dimension already placed with the DIMLINEAR command, you issue the DIMCONTINUE command, which prompts you for the second extension line location. Picking the second extension line location strings the dimensions next to each other or continues the dimension.

Try It!: Open the drawing file 11_Dim Continue. Verify that Running OSNAP is on and set to Endpoint. Use the following command sequence and image for creating continue dimensions.

```
Command: DLI (For DIMLINEAR)
Specify first extension line origin or <select object>: (Select
    the endpoint of the line at "A")
Specify second extension line origin: (Select the endpoint of the
    line at "B")
Specify dimension line location or [Mtext/Text/Angle/Horizontal/
    Vertical/Rotated]: (Locate the 1.75 horizontal dimension)
Dimension text = 1.75

Command: DCO (For DIMCONTINUE)
Specify a second extension line origin or [Undo/Select] <Select>:
    (Select the endpoint of the line at "C")
Dimension text = 1.25
Specify a second extension line origin or [Undo/Select] <Select>:
    (Select the endpoint of the line at "D")
Dimension text = 1.50
Specify a second extension line origin or [Undo/Select] <Select>:
    (Select the endpoint of the line at "E")
Dimension text = 1.00
Specify a second extension line origin or [Undo/Select] <Select>:
    (Press ENTER when finished)
Select continued dimension: (Press ENTER to exit this command)
```

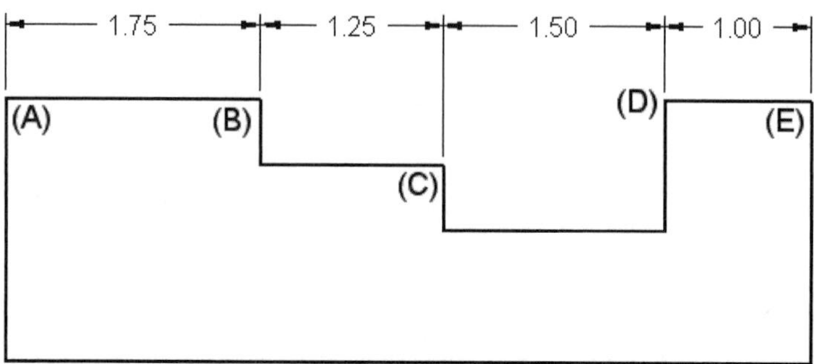

Figure 11.23

BASELINE DIMENSIONS

Yet another aid in grouping dimensions is the DIMBASELINE command. Continue dimensions place dimensions next to each other; baseline dimensions establish a base or starting point for the first dimension, as shown in the following image. Any dimensions that follow in the DIMBASELINE command are calculated from the common base point already established. This is a very popular mode to use when one end of an object acts as a reference edge. When you place dimensions using the DIMBASELINE command, a default baseline spacing setting of 0.38 units controls the spacing of the dimensions from each other. The DIMBASELINE command is initiated after the DIMLINEAR command.

Try It!: Open the drawing file 11_Dim Baseline. Verify that Running OSNAP is on and set to Endpoint. Use the following command sequence and image for creating baseline dimensions.

```
Command: DLI (For DIMLINEAR)
Specify first extension line origin or <select object>: (Select
    the endpoint of the line at "A")
Specify second extension line origin: (Select the endpoint of the
    line at "B")
Specify dimension line location or [Mtext/Text/Angle/Horizontal/
    Vertical/Rotated]: (Locate the 1.75 horizontal dimension)
Dimension text = 1.75

Command: DBA (For DIMBASELINE)
Specify a second extension line origin or [Undo/Select] <Select>:
    (Select the endpoint of the line at "C")
Dimension text = 3.00
Specify a second extension line origin or [Undo/Select] <Select>:
    (Select the endpoint of the line at "D")
Dimension text = 4.50
Specify a second extension line origin or [Undo/Select] <Select>:
    (Select the endpoint of the line at "E")
Dimension text = 5.50
Specify a second extension line origin or [Undo/Select] <Select>:
    (Press ENTER when finished)
Select base dimension: (Press ENTER to exit this command)
```

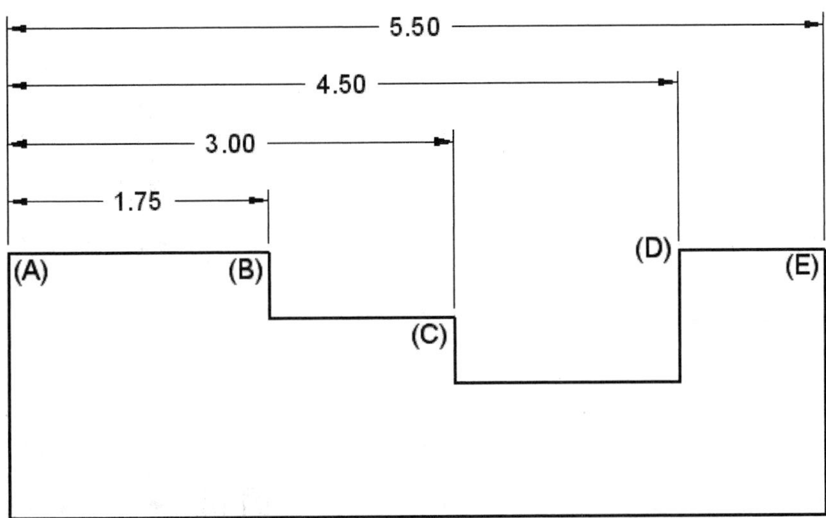

Figure 11.24

THE QDIM COMMAND

A more efficient means of placing a series of continued or baseline dimensions is the QDIM, or Quick Dimension, command. You identify a number of valid corners representing intersections or endpoints of an object and all dimensions are placed.

CONTINUOUS MODE

In the following image, a crossing box is used to identify all corners to dimension to.

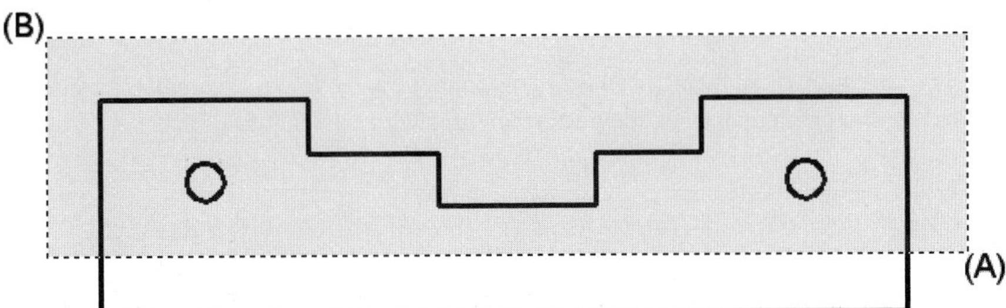

Figure 11.25

When you have finished identifying the crossing box and press the ENTER key, a preview of the dimensioning mode appears, as in the following image. During this preview mode, you can right-click and have a shortcut menu appear. This allows you to select other dimension modes.

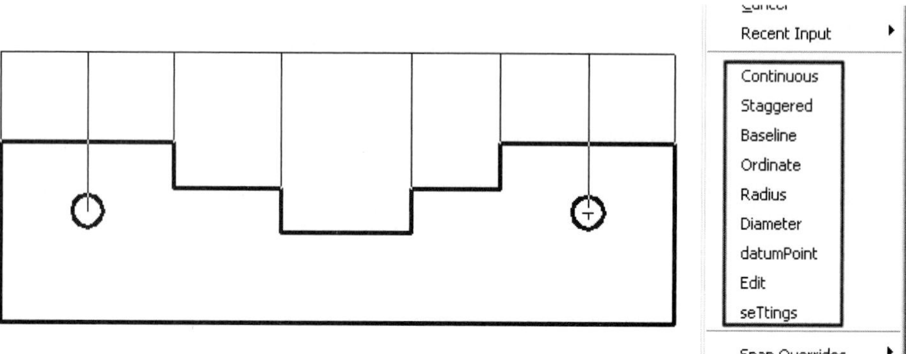

Figure 11.26

When the dimension line is identified, as in the following image, all continued dimensions are placed.

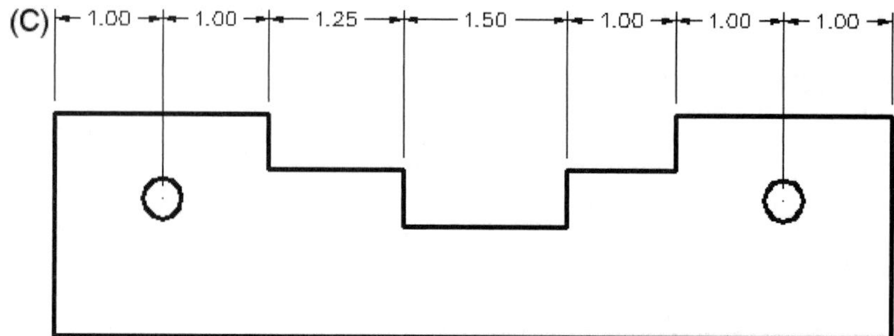

Figure 11.27

Try It!: Open the drawing file 11_Qdim Continuous. Follow the command sequence below and previous images for performing this dimensioning task.

```
Command: QDIM
Select geometry to dimension: (Pick a point at "A")
Specify opposite corner: (Pick a point at "B")
Select geometry to dimension: (Press ENTER to continue)
Specify dimension line position, or
[Continuous/Staggered/Baseline/Ordinate/Radius/Diameter/
    datumPoint/Edit/seTtings]
<Continuous>: (Change to a different mode or locate the dimension
    line at "C")
```

STAGGERED MODE

By default, the QDIM command places continued dimensions. Before locating the dimension line, you have the option of placing the staggered dimensions in the following image. The process with this style begins with adding dimensions to inside details and continuing outward until all features are dimensioned.

Try It!: Open the drawing file 11_Qdim Staggered. Activate the QDIM command and pick the lines labeled "A" through "H," as shown on the left in the following image. Change to the Staggered mode. Your display should appear similar to the illustration on the right in the following image.

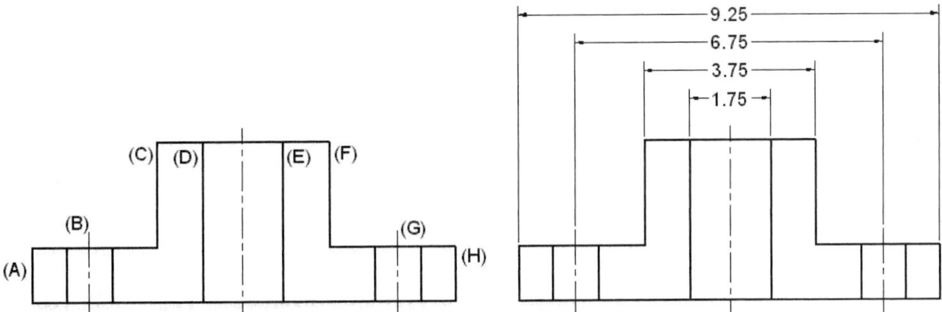

Figure 11.28

BASELINE MODE

Another option of the QDIM command is the ability to place baseline dimensions. As with all baseline dimensions, an edge is used as the baseline, or datum. All dimensions are calculated from the left edge, as shown in the following image. The datumPoint option of the QDIM command allows you to select the right edge of the object in the following image as the new datum or baseline.

Try It!: Open the drawing file 11_Qdim Baseline. Activate the QDIM command and identify the same set of objects as in the first Qdim exercise. Change to the Baseline mode. Your display should appear similar to the illustration in the following image.

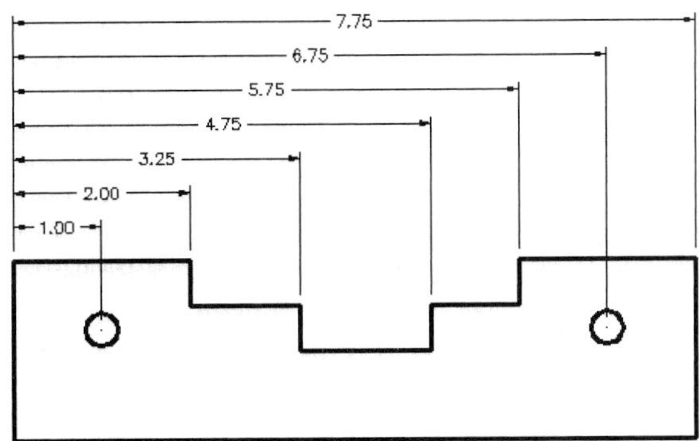

Figure 11.29

ORDINATE MODE

The Ordinate option of the QDIM command calculates dimensions from a known 0,0 corner. The main Ordinate dimensioning topic will be discussed in greater detail later in this chapter. A new User Coordinate System must first be created in the corner of the object. Then the QDIM command is activated along with the Ordinate options to display the results, as shown in the following image. Dimension lines are not used in ordinate dimensions, in order to simplify the reading of the drawing.

Try It!: Open the drawing file 11_Qdim Ordinate. Activate the QDIM command and identify the same set of objects as in the first Qdim exercise. Change to the Ordinate mode. Your display should appear similar to the illustration in the following image.

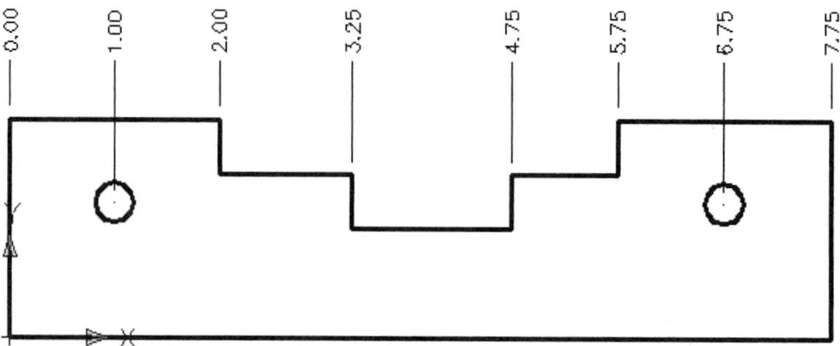

Figure 11.30

Tip: The QDIM command can also be used to edit an existing group of dimensions. Activate the command, select all existing dimensions with a crossing box, and enter an option to change the style of all dimensions.

DIAMETER AND RADIUS DIMENSIONING

Arcs and circles should be dimensioned in the view where their true shape is visible. The mark in the center of the circle or arc indicates its center point, as shown in the following image. You may place the dimension text either inside or outside the circle; you may also use grips to aid in the dimension text location of a diameter or radius dimension. When dimensioning a small radius, an arc extension line is formed, depending on where you locate the radius dimension text.

Try It!: Open the drawing file 11_Dim Radial. Use the following command sequence and image for placing diameter and radius dimensions.

```
Command: DDI (For DIMDIAMETER)
Select arc or circle: (Select the edge of the large circle)
Dimension text = 2.50
Specify dimension line location or [Mtext/Text/Angle]: (Pick a
    point to locate the diameter dimension)
```

```
Command: DDI (For DIMDIAMETER)
Select arc or circle: (Select the edge of the small circle)
Dimension text = 1.00
Specify dimension line location or [Mtext/Text/Angle]: (Pick a
    point to locate the diameter dimension)
```

```
Command: DRA (For DIMRADIUS)
Select arc or circle: (Select the edge of the large arc)
Dimension text = 1.00
Specify dimension line location or [Mtext/Text/Angle]: (Pick a
    point to locate the radius dimension)
```

```
Command: DRA (For DIMRADIUS)
Select arc or circle: (Select the edge of the small arc)
Dimension text = .50
Specify dimension line location or [Mtext/Text/Angle]: (Pick a
    point to locate the radius dimension)
```

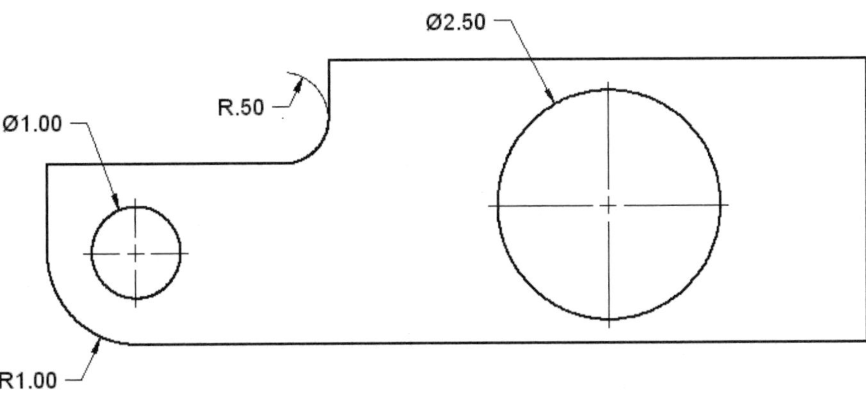

Figure 11.31

USING QDIM FOR RADIUS AND DIAMETER DIMENSIONS

As with linear dimensions, the QDIM command has options that can be applied to radius and diameter dimensions. In the following image, all arcs and circles are selected. Activating either the Radius or Diameter option displays the results in the illustrations. When locating the radius or diameter dimensions, you can specify the angle of the leader. A predefined leader length is applied to all dimensions. Grips could be used to relocate the dimensions to better places.

Try It!: Open the drawing file 11_Qdim Radius. Use the illustration on the left in the following image and the command sequence below for placing a series of radius dimensions using the QDIM command.

```
Command: QDIM
Select geometry to dimension: (Select the four arcs on the left
    in the following image)
Select geometry to dimension: (Press ENTER to continue)
```

```
Specify dimension line position, or
[Continuous/Staggered/Baseline/Ordinate/Radius/Diameter/
    datumPoint/Edit/seTtings]
<Continuous>: R (For Radius)
```

Try It!: Open the drawing file 11_Qdim Diameter. Use the illustration on the right in the following image and the command sequence below for placing a series of diameter dimensions using the QDIM command.

```
Command: QDIM
Select geometry to dimension: (Select the four circles on the
    right in the following image)
Select geometry to dimension: (Press ENTER to continue)
Specify dimension line position, or
[Continuous/Staggered/Baseline/Ordinate/Radius/Diameter/
    datumPoint/Edit/seTtings]
<Continuous>: D (For Diameter)
Specify dimension line position, or
[Continuous/Staggered/Baseline/Ordinate/Radius/Diameter/
    datumPoint/Edit/seTtings]
<Diameter>: (Pick a point to locate the diameter dimension)
```

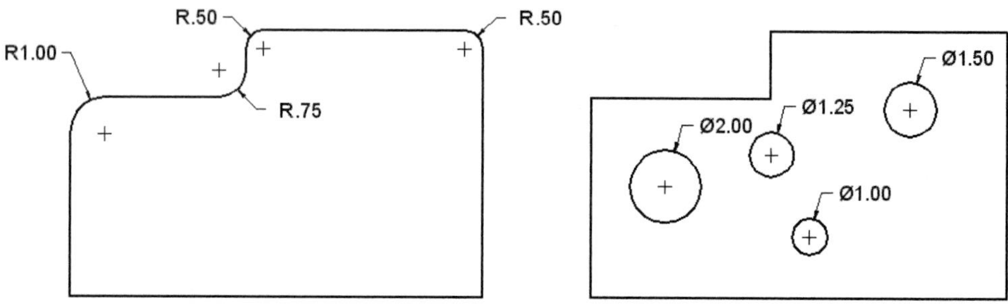

Figure 11.32

SPACING DIMENSIONS

This tool is used to adjust the spacing between parallel linear and angular dimensions to be equal. You can automatically have this tool calculate the dimension spacing distance based on the dimension text height. You can also specify a value for the dimensions to be separated by. These parameters work well for baseline dimensions. In the case of continue dimensions, entering a value of 0 lines up all continue dimensions.

Try It!: Open the drawing file 11_Dim Spacing. Use the following command sequence and image for performing this operation on both baseline and continue dimensions.

```
Command: DIMSPACE
Select base dimension: (Select dimension "A")
Select dimensions to space: (Select dimensions "B", "C", and "D")
Select dimensions to space: (Press ENTER to continue)
Enter value or [Auto] <Auto>: A (For Auto)
```

```
Command: DIMSPACE
Select base dimension: (Select dimension "E")
Select dimensions to space: (Select dimensions "F", "G", and "H")
Select dimensions to space: (Press ENTER to continue)
Enter value or [Auto] <Auto>: 0
```

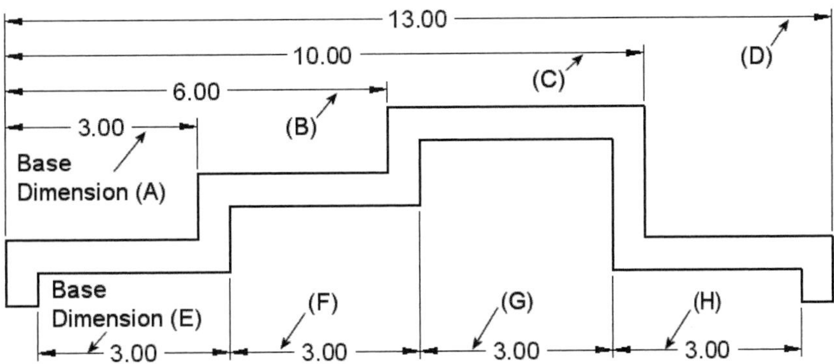

Figure 11.33

APPLYING BREAKS IN DIMENSIONS

Use this tool to break dimension, extension, or leader lines. Dimension breaks can be added to the following dimension types; Linear, Angular, Diameter, Radius, Jogged, Ordinate, and Multileaders. The following objects act as cutting edges when producing dimension breaks: Arcs, Circles, Dimensions, Ellipses, Leaders, Lines, Mtext, Polylines, Splines, and Text. Dimension breaks cannot be placed on an arrowhead or dimension text.

An Auto option is available when using the Dimension Break command. The size of the break is controlled through the Symbols and Arrows tab of the Dimension Style Manager dialog box, which will be explained in chapter 12.

Try It!: Open the drawing file 11_Dim Break Mech. Use the following command sequence and image for creating numerous breaks in a dimensions extension line.

```
Command: DIMBREAK
Select a dimension or [Multiple]: (Pick the 2.00 vertical
    dimension on the right)
Select object to break dimension or [Auto/Restore/Manual] <Auto>:
    (Press ENTER to automatically break this dimension)

Command: DIMBREAK
Select a dimension or [Multiple]: (Pick the dimension at "A")
Select object to break dimension or [Auto/Restore/Manual] <Auto>:
    (Pick the dimension at "B")
Select object to break dimension: (Press ENTER to exit this
    command)
```

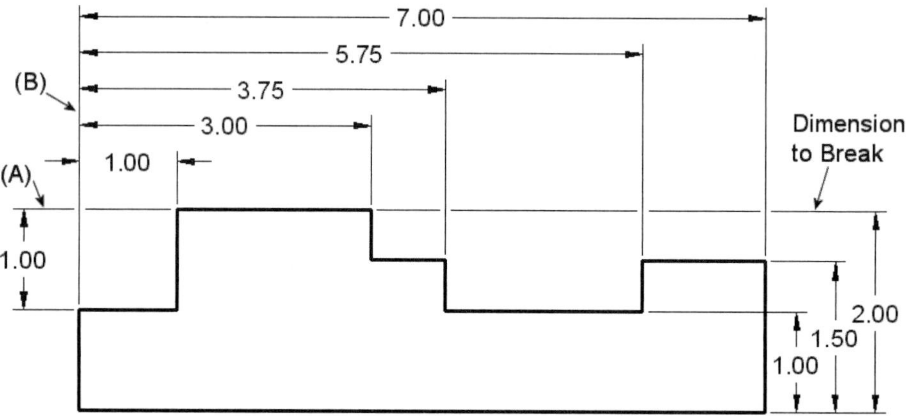

Figure 11.34

INSPECTION DIMENSIONS

An inspection dimension identifies a distance that needs to be checked to ensure the value is within a specific range such as a tolerance. Part of the inspection dimension includes a parameter stating how often the dimension should be tested. Inspection dimensions are created from an already existing dimension.

Before converting an existing dimension to an inspection dimension, a dialog box displays, as shown in the following image. The dialog box displays the shape, label, and inspection rate. You can choose from a round or angular frame; you can even have no frame applied to the inspection dimension. The inspection label is present at the leftmost portion of the inspection dimension; the actual dimension value is located in the center section of the inspection dimension. The inspection rate located at the rightmost section of the inspection dimension is used to indicate the frequency at which the dimension value is inspected.

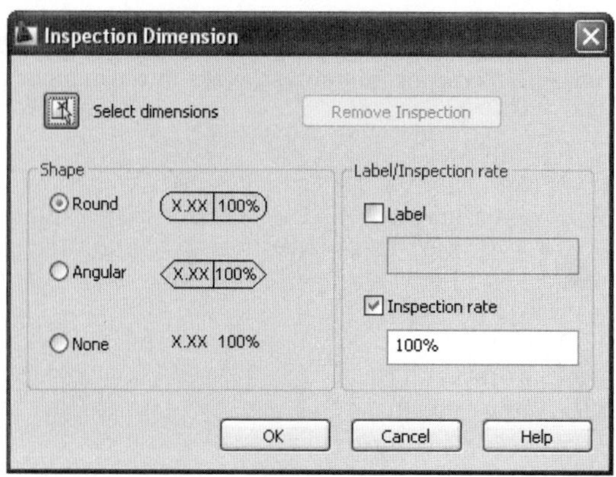

Figure 11.35

Try It!: Open the drawing file 11_Dim Inspect. Use the following command sequence and image for creating an inspection dimension.

```
Command: DIMINSPECT
(When the Inspection Dimension dialog box displays, change the
    shape to Angular and place checks in the label and
    Inspection rate boxes, as shown on the left in the following
    image. Enter "A" as the label designation and click the
    Select dimensions button)
Select dimensions: (Pick the 8.00 vertical dimension on the
    right)
Select dimensions: (Press ENTER to return to the dialog box; click
    OK)
```

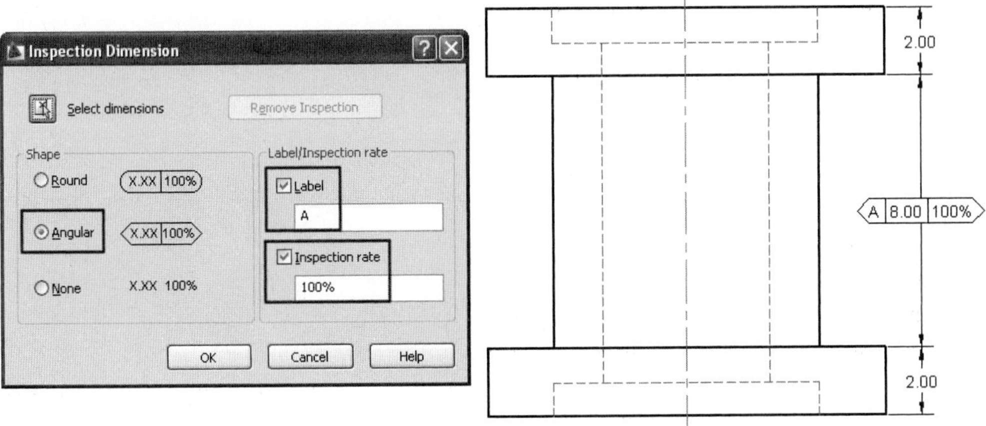

Figure 11.36

ADDING JOGGED DIMENSIONS

At times, you need to add a radius dimension to a large arc. When creating the radius dimension, the center for the dimension is placed outside the boundaries of the drawing. To give better control over these situations, you can create a jog in the radius. This is represented by a zigzag appearance, or jog, in the leader holding the radius dimension. Use the following command sequence and image for creating a jogged radius dimension.

```
Command: DJO (For DIMJOGGED)
Select arc or circle: (Select the arc in the following image)
Specify center location override: (Locate the center at "A")
Dimension text = 23.1588
Specify dimension line location or [Mtext/Text/Angle]: (Locate
    the dimension line)
Specify jog location: (Pick a point at "B")
```

ADDING ARC DIMENSIONS

You can also dimension the length of an arc using the Dimension Arc command (DIMARC, or the shortcut DAR). After you select the arc, extension lines are created at the endpoints of the arc and a dimension arc is constructed parallel to the arc being dimensioned. The arc symbol is placed with the dimension text. This symbol can also be located above the dimension text. This technique will be covered in the next chapter.

```
Command: DAR (For DIMARC)
Select arc or polyline arc segment: (Select the arc)
Specify arc length dimension location, or [Mtext/Text/Angle/
    Partial]: (Locate the dimension)
Dimension text = 10.8547
```

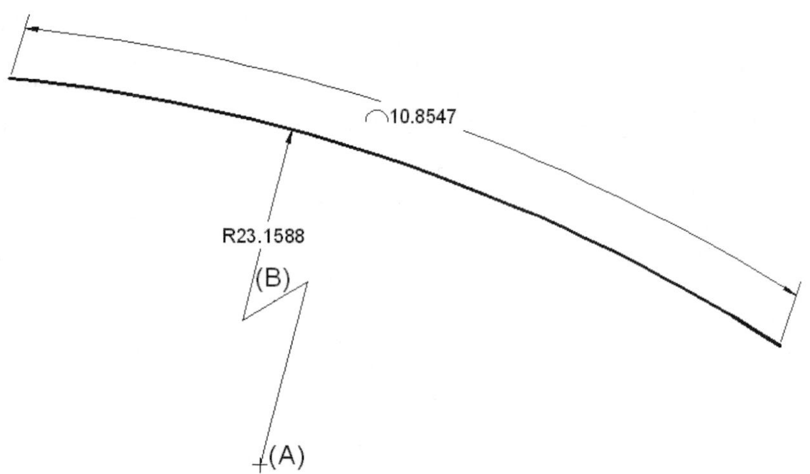

Figure 11.37

LINEAR JOG DIMENSIONS

Jog lines are used to represent a dimension value that does not display the actual measurement. Typically, the actual measurement value of the dimension is smaller than the displayed value. The Linear Jog dimension tool is used for creating this type of jogged dimension.

Try It!: Open the drawing file 11_Dim Linear Jog. Use the following command sequence and image for creating a jog along a linear dimension.

```
Command: DJL (For DIMJOGLINE)
Select dimension to add jog or [Remove]: (Select dimension "A")
Specify jog location (or press ENTER): (Pick the jog location
    at "B")
```

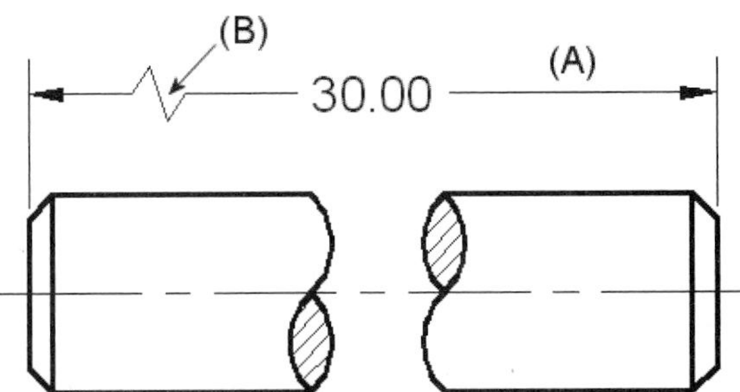

Figure 11.38

LEADER LINES

A leader line is a thin, solid line leading from a note or dimension and ending with an arrowhead, as illustrated at "A" in the following image. The arrowhead should always terminate at an object line such as the edge of a hole or an arc. A leader to a circle or arc should be radial; this means it is drawn so that if extended, it would pass through the center of the circle, as illustrated at "B." Leaders should cross as few object lines as possible and should never cross each other. The short horizontal shoulder of a leader should meet the dimension illustrated at "A." It is poor practice to underline the dimension with the horizontal shoulder, as illustrated at "C." Example "C" also illustrates a leader not lined up with the center or radial. This may affect the appearance of the leader. Again, check your company's standard practices to ensure that this example is acceptable.

Yet another function of a leader is to attach notes to a drawing, as illustrated at "D." Notice that the two notes attached to the view have different terminators: arrows and dots. Dots are sometimes used as terminators when the note is referring to an entire surface. It is good practice to be consistent with the type of terminators for the duration of the drawing.

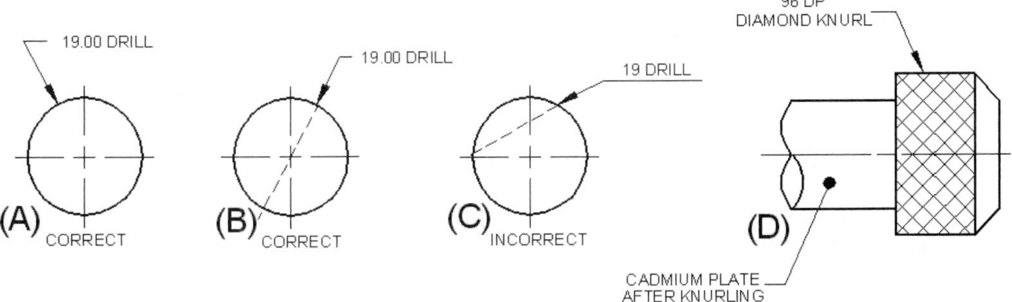

Figure 11.39

THE QLEADER COMMAND

The QLEADER command, or Quick Leader, provides numerous controls for placing leaders in your drawing. The QLEADER command is actually Leader and must be entered in at the command prompt with LE.

Try It!: Open the drawing file 11_Qleader. Study the prompt sequence below and the following image for this command.

```
Command: LE (For QLEADER)
Specify first leader point, or [Settings]<Settings>: Nea
to (Pick the point nearest at "A")
Specify next point: (Pick a point at "B")
Specify next point: (Press ENTER to continue)
Specify text width <0.00>: (Press ENTER to accept this default)
Enter first line of annotation text <Mtext>: (Press ENTER to
    display the Multiline Text Editor dialog box. Enter "2X
    R.20". Click the OK button to place the leader)
```

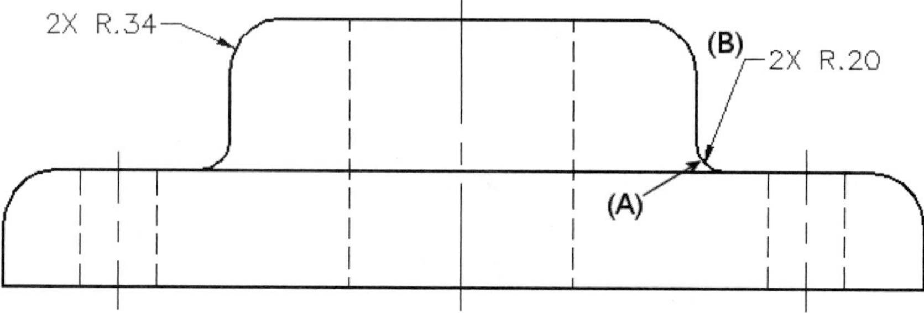

Figure 11.40

Pressing ENTER at the first Quick Leader prompt displays the Leader Settings dialog box that consists of three tabs used for controlling leaders. The Annotation tab deals with the object placed at the end of the leader. By default, the MText radio button is selected, allowing you to add a note through the Multiline Text Editor dialog box. You could also copy an object at the end of the leader, have a geometric tolerancing symbol placed in the leader, have a pre-defined block placed in the leader, or leave the leader blank.

```
Command: LE (For QLEADER)
Specify first leader point, or [Settings]<Settings>: (Press ENTER
    to accept the default value and display the Leader Settings
    dialog box in the following image)
```

The Leader Line & Arrow tab in the following image allows you to draw a leader line consisting of straight segments or in the form of a spline object. You can control the number of points used to define the leader; a maximum of three points is more than enough to create your leader. You can even change the arrowhead type.

The Attachment tab in the following image allows you to control how text is attached to the end of the leader through various justification modes. The settings in the figure are the default values.

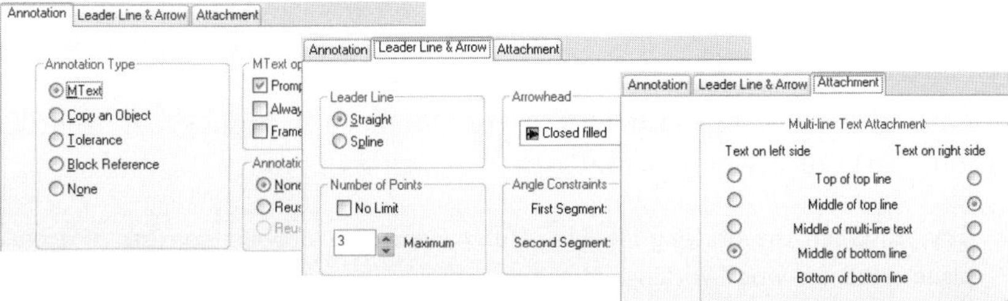

Figure 11.41

ANNOTATING WITH MULTILEADERS

Multileaders allow for leader lines to be aligned, collected, added, or removed. These types of leaders allow for more control than traditional leaders. Choose multileader tools from the Menu Browser, Multileader Toolbar, Ribbon, or pull-down menu as shown in the following image.

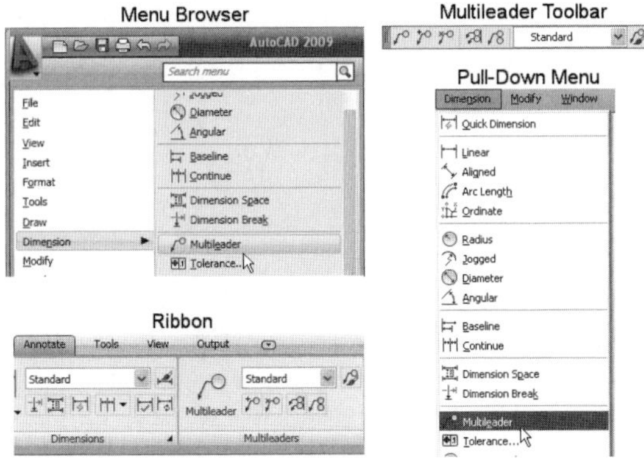

Figure 11.42

The following table illustrates each multileader tool along with a brief description.

Button	Tool	Function
	MLEADER	Creates a multileader
	--	Adds a multileader
	--	Removes a multileader
	MLEADERALIGN	Aligns a number of multileaders

Button	Tool	Function
	MLEADERCOLLECT	Collects a number of multileaders
	MLEADERSTYLE	Launches the Multileader Style Properties dialog box

CREATING MULTILEADERS

The following Try It! exercise allows you to place a number of multileaders that identify various parts of a wood plane.

Try It!: Open the drawing file 11 _Wood_Plane. This drawing consists of a typical plane used in woodworking. Use the MLEADER command to place a number of multileaders that identify various parts of the wood plane.

```
Command: MLEADER
Specify leader arrowhead location or [leader Landing first/
    Content first/Options] <Options>: (Pick a point at "A")
Specify leader landing location: (Pick a point at "B." When the
    Text Formatting toolbar appears, enter the name of the wood
    plane part. When finished, click the OK button to dismiss
    the Text Formatting toolbar and place the multileader as
    shown on the left in the following image)
```

Continue placing additional multileaders to identify additional wood plane parts, as shown on the right in the following image.

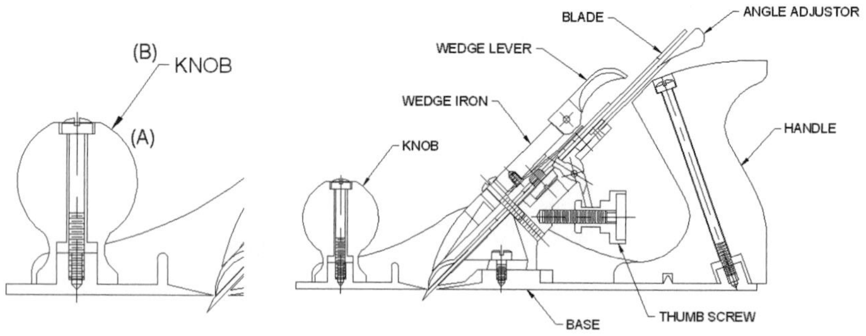

Figure 11.43

ALIGNING AND ADDING MULTILEADERS

As numerous leaders are placed in a drawing, the time comes when they need to be aligned with each other to promote a neat and organized drawing. The next Try It! exercise will show how to align leaders. Then you will add additional leaders to an existing leader line.

Try It!: Open the drawing file 11_Multileaders Align. A number of multileaders are already created; however, notice that each leader is out of alignment with the others. Begin the alignment process by activating the Ribbon and clicking the Align Multileaders button found under the Annotate tab, as shown in the following image. First select all of the multileaders present in the following image. You could select Baseboard Molding as the multileader to align to. Instead, type O for Options. Then type D for Distribute and pick the two points, as shown in the following image.

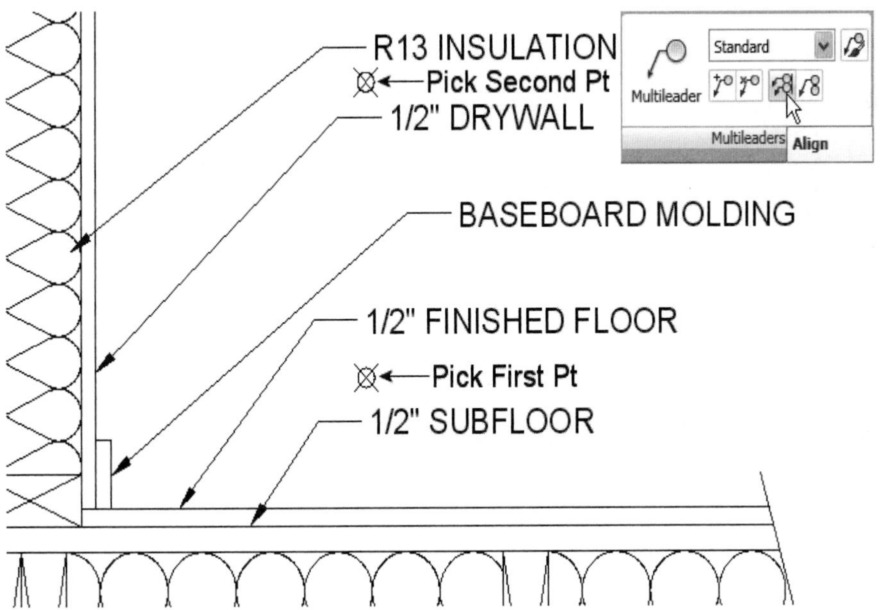

Figure 11.44

The results are illustrated in the following image with all multileaders aligned with each other and equally spaced due to the distribute option. This makes the leaders more presentable in the drawing.

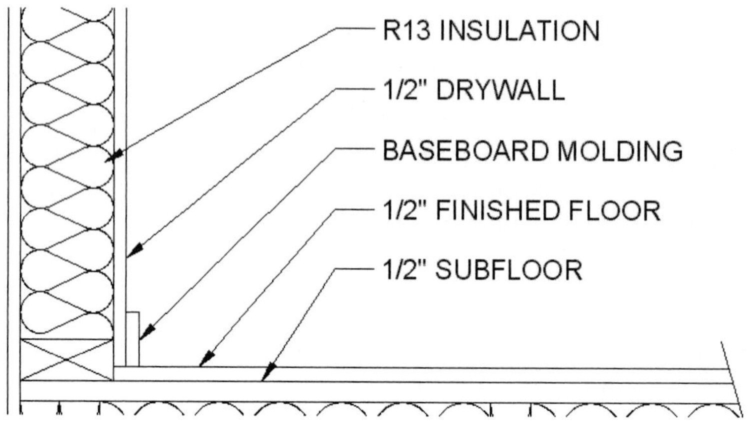

Figure 11.45

Continue with this Try It! exercise by panning to the lower portion of the architectural detail until your image appears similar to the following image. One concrete block wall is already called out with a multileader. With the Ribbon still present, click the add leader button, as shown in the following image. When prompted to select a multileader, click the existing leader identified by concrete block. When prompted to specify the leader arrowhead location, use Osnap Nearest to pick the edges of the other two concrete block symbols, as shown in the following image. The additional leaders will be added to the existing concrete block leader.

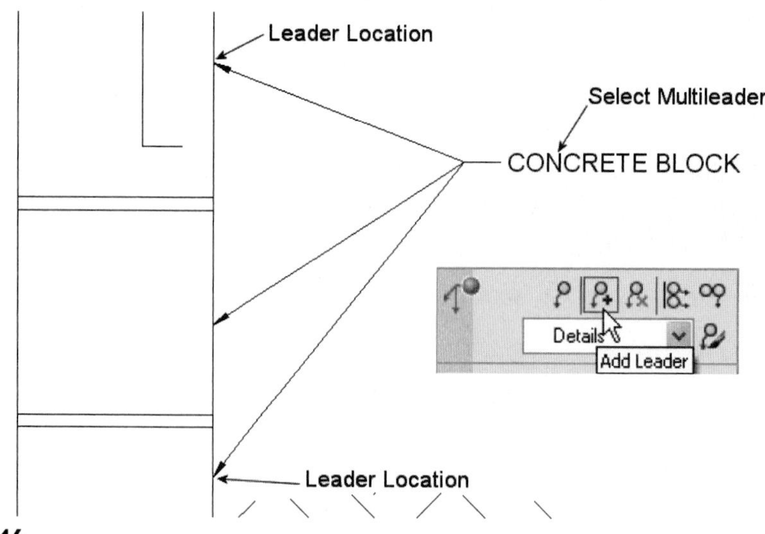

Figure 11.46

COLLECTING MULTILEADERS

In the previous Try It! exercise, typical wall section elements were identified by multileaders with text. Multileaders can also take the form of blocks or symbols. Blocks will be covered in great detail in chapter 16. Typical examples of multileader blocks include circles, boxes, and triangles, to name a few. Text is typically placed inside the circles that are usually called balloons and are used to identify items located in a parts list. These multileader blocks can either be displayed individually or can be grouped or collected using a single multileader. The next Try It! exercise illustrates this.

Try It!: Open the drawing file 11_Multileaders_Collect. Four multileaders have been placed using circles as blocks. Begin by clicking the Collect Multileaders button found under the Annotate tab of the Ribbon, as shown in the following image. When prompted to select the multileaders to group, pick items 1 through 4 in order.

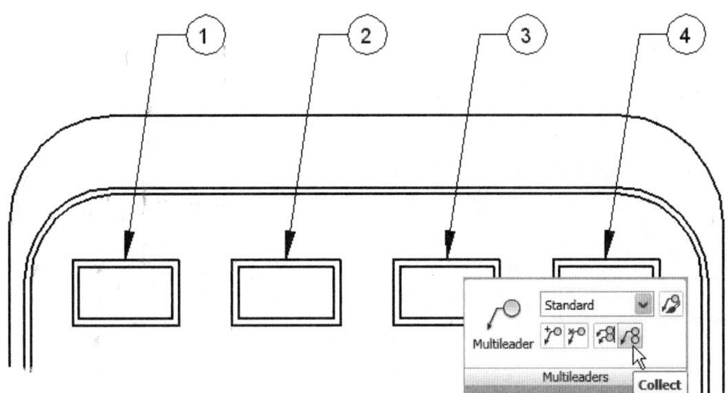

Figure 11.47

When prompted to select a new location, pick the location as shown in the following image. Notice how all four blocks are collected under the common leader.

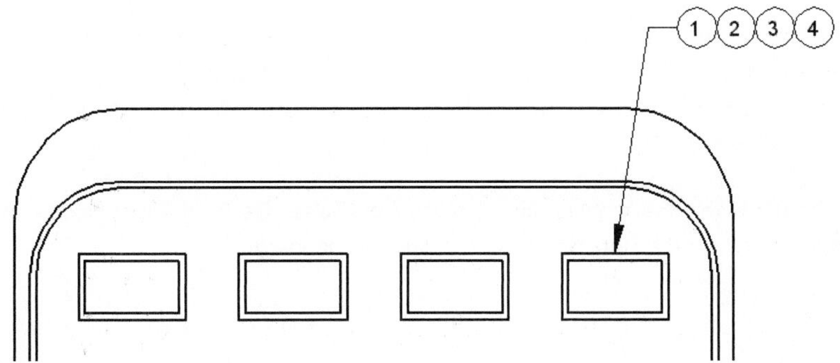

Figure 11.48

DIMENSIONING ANGLES

Dimensioning angles requires two lines forming the angle in addition to the location of the vertex of the angle along with the dimension arc location. Before going any further, understand how the curved arc for the angular dimension is derived. At "A" in the following image, the dimension arc is struck with its center at the vertex of the object.

Try It!: Open the drawing file 11_Dim Angle. Use the following command sequence and image for performing this operation on both angles.

```
Command: DAN (For DIMANGULAR)
Select arc, circle, line, or <specify vertex>: (Select line "A")
Select second line: (Select line "B")
Specify dimension arc line location or [Mtext/Text/Angle/
    Quadrant]: (Pick a point at "C" to locate the dimension)
Dimension text = 53
```

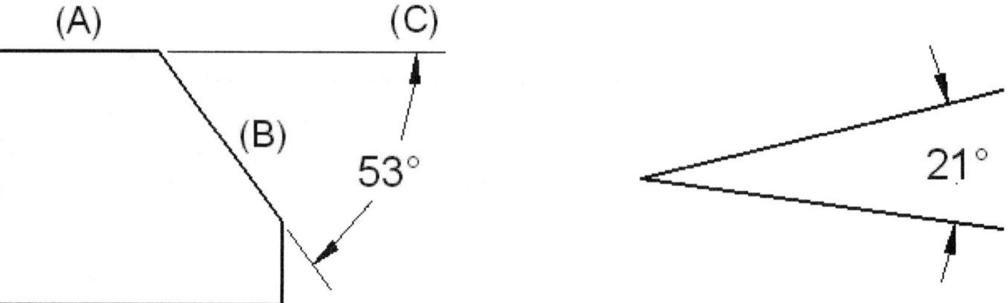

Figure 11.49

DIMENSIONING SLOTS

For slots, first select the view where the slot is visible. Two methods of dimensioning the slot are illustrated in the following image. With the first method, you call out a slot by locating the center-to-center distance of the two semicircles, followed by a radius dimension to one of the semicircles; which radius dimension is selected depends on the available room to dimension. A second method involves the same center-to-center distance followed by an overall distance designating the width of the slot. This dimension is the same as the diameter of the semicircles. It is good practice to place this dimension inside the slot if room is available.

Try It!: Open the drawing file 11_Dim Slot1. Use this file for practice in placing dimensions on the slot, as shown in the following image.

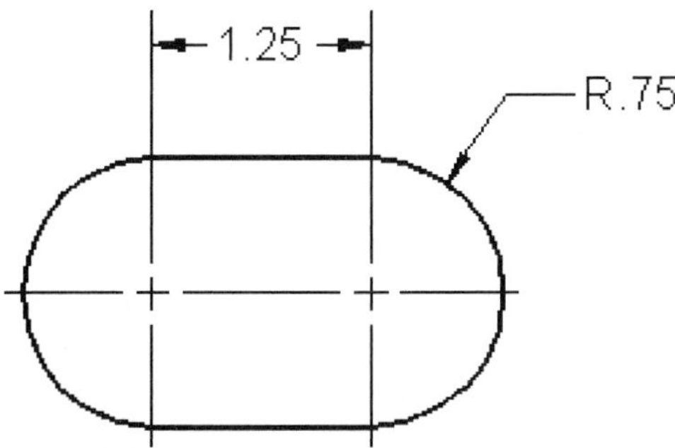

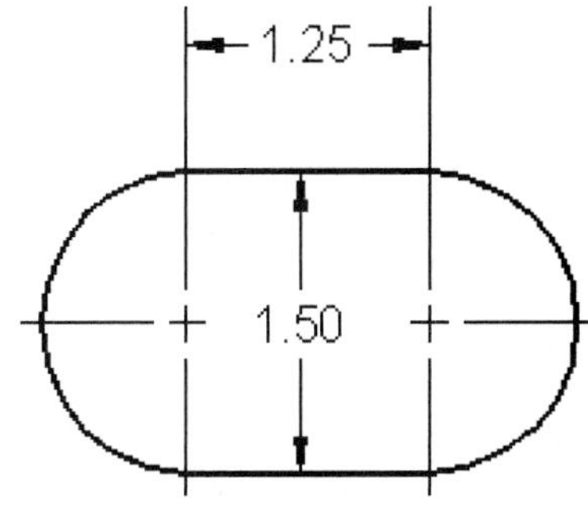

Figure 11.50

A more complex example in the following image involves slots formed by curves and angles. Here, the radius of the circular center arc is called out. Angles reference each other for accuracy. The overall width of the slot is dimensioned, which happens to be the diameter of the semicircles at opposite ends of the slot.

Try It!: Open the drawing file 11_Dim Slot2. Use this file for practice in placing dimensions on the angular slot in the following image. Use the DIMALIGN command to place the .48 dimension. Use OSNAP-Nearest to identify the location at "A" and OSNAP-Perpendicular to identify the location at "B."

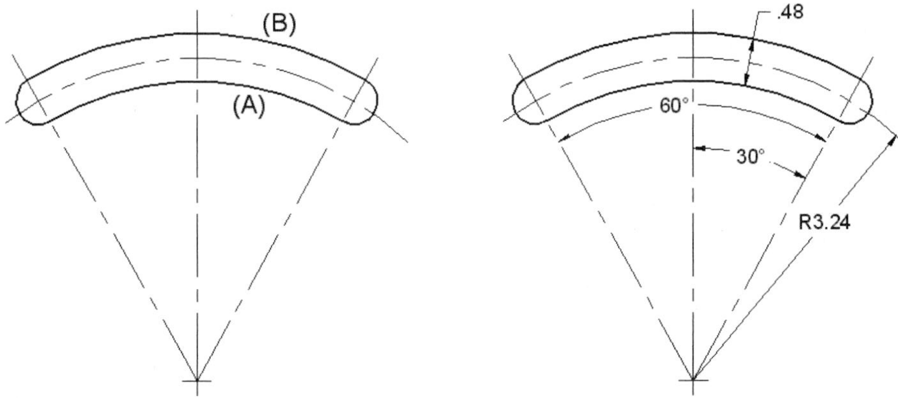

Figure 11.51

ADDING ORDINATE DIMENSIONS

The plate on the left in the following image consists of numerous drill holes with a few slots, in addition to numerous 90° angle cuts along the perimeter. This object is not considered difficult to draw or make because it consists mainly of drill holes. However, conventional dimensioning techniques make the plate appear complex because a dimension is required for the location of every hole and slot in both the X and Y directions. Add standard dimension components such as extension lines, dimension lines, and arrowheads, and it is easy to get lost in the complexity of the dimensions even on this simple object. A better dimensioning method, called ordinate, or datum, dimensioning, is illustrated on the right in the following image. Here, dimension lines or arrowheads are not drawn; instead, one extension line is constructed from the selected feature to a location specified by you. A dimension is added to identify this feature in either the X or Y direction. It is important to understand that all dimension calculations occur in relation to the current User Coordinate System (UCS), or the current 0,0 origin. In the following image, with the 0,0 origin located in the lower-left corner of the plate, all dimensions in the horizontal and vertical directions are calculated in relation to this 0,0 location. Holes and slots are called out with the DIMDIAMETER command. The following illustrates a typical ordinate dimensioning command sequence:

```
Command: DOR (For DIMORDINATE)
Specify feature location: (Select a feature using an Osnap
    option)
Specify leader endpoint or [Xdatum/Ydatum/Mtext/Text/Angle]:
    (Locate a point outside of the object)
Dimension text = Calculated value
```

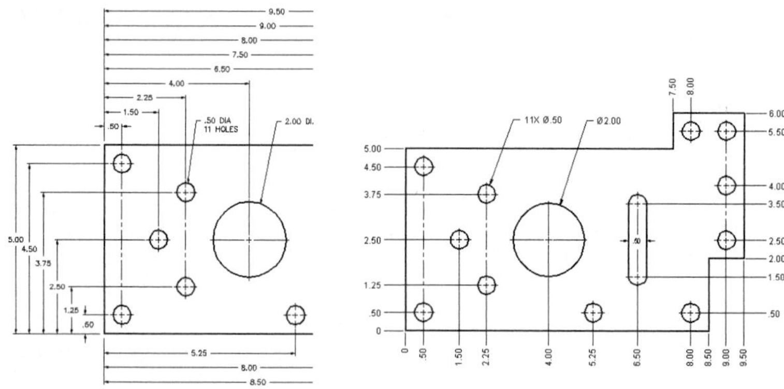

Figure 11.52

To understand how to place ordinate dimensions, see the example in the following image and the prompt sequence below. Before you place any dimensions, a new User Coordinate System must be moved to a convenient location on the object with the UCS command and the Origin option. All ordinate dimensions will reference this new origin because it is located at coordinate 0,0. At the command prompt, enter DOR (for DIMORDINATE) to begin ordinate dimensioning. Select the quadrant of the arc at "A" as the feature. For the leader endpoint, pick a point at "B." Be sure Ortho mode is on. It is also helpful to snap to a convenient snap point for this and other dimensions along the direction. This helps in keeping all ordinate dimensions in line with one another.

Try It!: Open the drawing file 11_Dim Ordinate. Follow the next series of figures and the following command sequences to place ordinate dimensions.

```
Command: DOR (For DIMORDINATE)
Specify feature location: Qua
of (Select the quadrant of the slot at "A" as shown on the left
    in the following image)
Specify leader endpoint or [Xdatum/Ydatum/Mtext/Text/Angle]:
    (Locate a point at "B" as shown on the left in the following
    image)
Dimension text = 1.50
```

With the previous example highlighting horizontal ordinate dimensions, placing vertical ordinate dimensions is identical, as shown on the right in the following image. With the UCS still located in the lower-left corner of the object, select the feature at "A," using either the Endpoint or Quadrant mode. Pick a point at "B" in a convenient location on the drawing. Again, it is helpful if Ortho is on and you snap to a grid dot.

```
Command: DOR (For DIMORDINATE)
Specify feature location: Qua
of (Select the quadrant of the slot at "A," as shown on the
    right in the following image)
Specify leader endpoint or [Xdatum/Ydatum/Mtext/Text/Angle]:
    (Locate a point at "B," as shown on the right in the
    following image)
Dimension text = 3.00
```

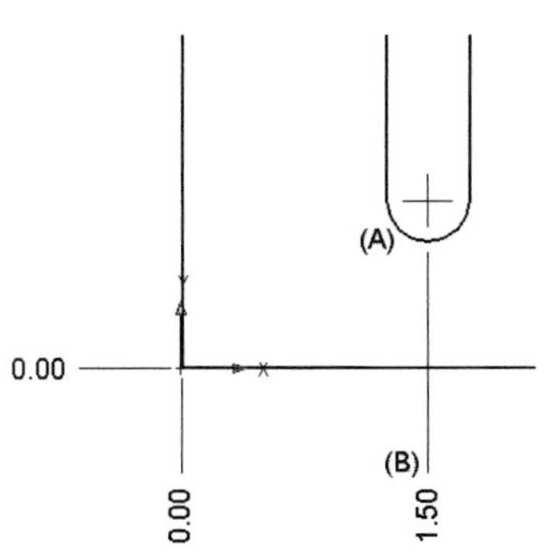

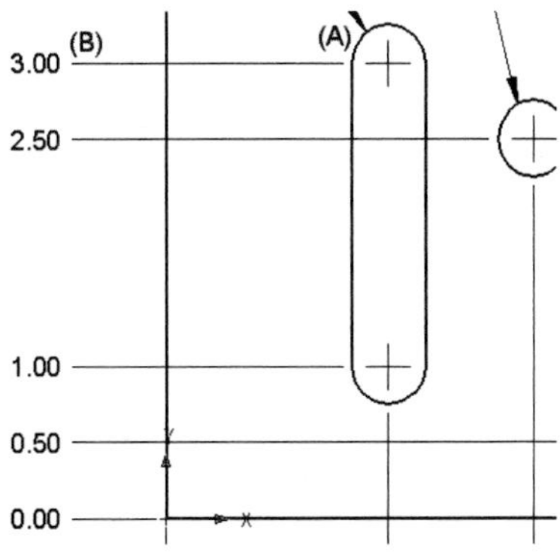

Figure 11.53

When spaces are tight to dimension to, two points not parallel to the X or Y axis will result in a "jog" being drawn, as shown on the left in the following image. It is still helpful to snap to a grid dot when performing this operation; however, be sure Ortho is turned off.

```
Command: DOR (For DIMORDINATE)
Specify feature location: End
of (Select the endpoint of the line at "A," as shown on the left
    in the following image)
```

```
Specify leader endpoint or [Xdatum/Ydatum/Mtext/Text/Angle]:
    (Locate a point at "B," as shown on the left in the
    following image)
Dimension text = 2.00
```

Ordinate dimensioning provides a neat and easy way of organizing dimensions for machine tool applications. Only two points are required to place the dimension that references the current location of the UCS, as shown on the right in the following image.

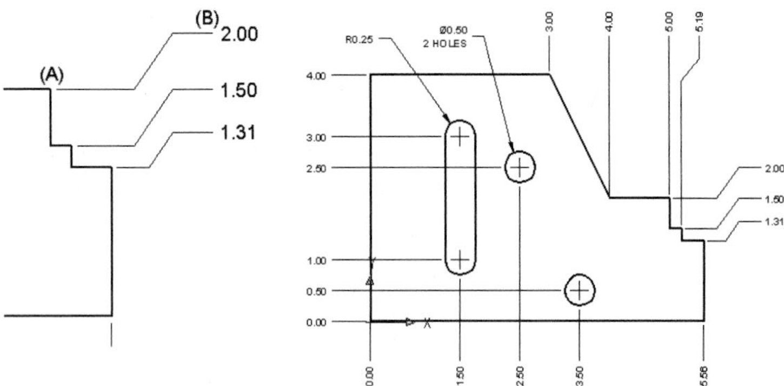

Figure 11.54

EDITING DIMENSIONS

Use the DIMEDIT command to add text to the dimension value, rotate the dimension text, rotate the extension lines for an oblique effect, or return the dimension text to its home position. The following image shows the effects of adding text to a dimension and rotating the dimension to a user-specified angle.

Try It!: Open the drawing file 11_Dim Dimedit. Follow the next series of figures and command prompt sequences for accomplishing this task.

```
Command: DED (For DIMEDIT)
Enter type of dimension editing [Home/New/Rotate/Oblique] <Home>:
    N (For New. This displays the Multiline Text Editor dialog
    box. Add the text "TYPICAL" on the other side of the <>.
    When finished, click the OK button)
Select objects: (Select the 5.00 dimension)
Select objects: (Press ENTER to perform the dimension edit
    operation)
```

```
Command: DED (For DIMEDIT)
Enter type of dimension editing [Home/New/Rotate/Oblique] <Home>:
    R (For Rotate)
Specify angle for dimension text: 10
Select objects: (Select the 5.00 dimension)
Select objects: (Press ENTER to perform the dimension edit
    operation)
```

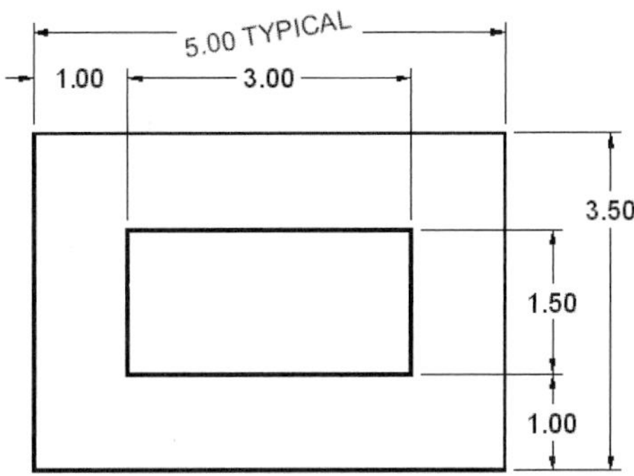

Figure 11.55

Regular AutoCAD modify commands can also affect dimensions. When you use the STRETCH command on the object on the left in the following image, points "A" and "B" identify a crossing window. Point "C" is the base point of displacement. The results are displayed on the right in the following image. Not only did the object lines stretch to the new position, but the dimensions also all updated themselves to new values.

```
Command: S (For STRETCH)
Select objects to stretch by crossing-window or crossing-
    polygon...
Select objects: (Pick a point at "A")
Specify opposite corner: (Pick a point at "B" to activate the
    crossing window)
Select objects: (Press ENTER to continue)
Specify base point or [Displacement] <Displacement>: (Pick a
    point at "C"; it could also be anywhere on the screen)
Specify second point or <use first point as displacement>: (Use
    the Direct Distance mode; with ORTHO mode on, move your
    cursor to the left and type .75 to perform the stretch)
```

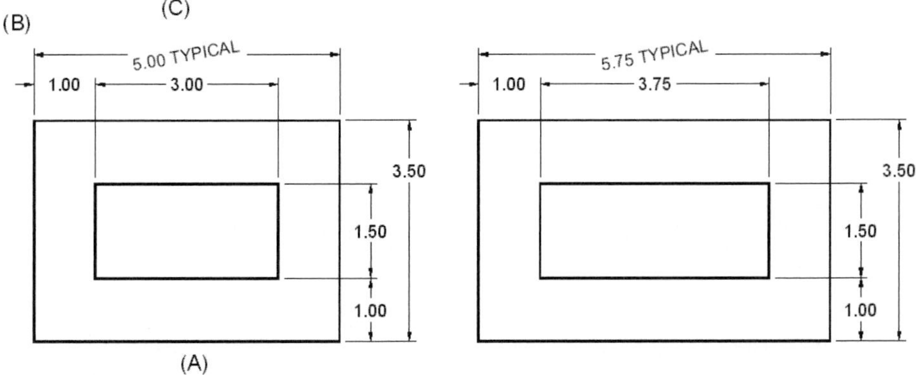

Figure 11.56

One of the other options of the DIMEDIT command is the ability for you to move and rotate the dimension text and still have the text return to its original or home location. This is the purpose of the Home option. Selecting the 5.75 dimension returns it to its original position, as shown in the following image. However, you would have to use the New option of the DIMEDIT command to remove the text "TYPICAL."

```
Command: DED (For DIMEDIT)
Enter type of dimension editing [Home/New/Rotate/Oblique] <Home>:
    H (Press ENTER to continue)
Select objects: (Select the 5.00 dimension)
Select objects: (Press ENTER to perform the dimension edit
    operation)
```

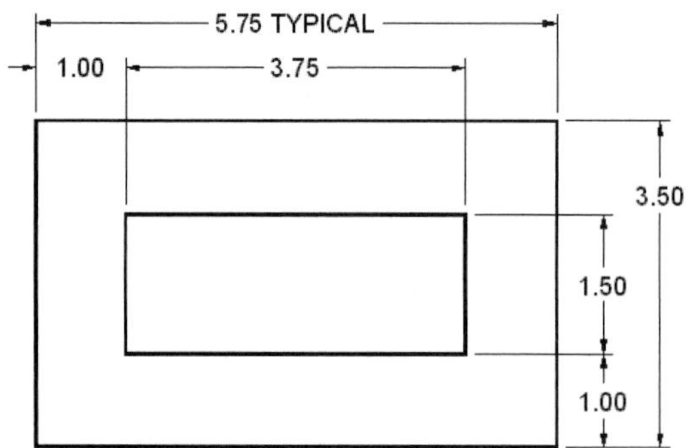

Figure 11.57

The DIMEDIT command also has an Oblique option that allows you to enter an obliquing angle, which rotates the extension lines and repositions the dimension line. This option is useful if you are interested in placing dimensions on isometric drawings.

Try It!: Open the drawing file 11_Dim Oblique. Follow the next series of images and command prompt sequence for accomplishing this task.

```
Command: DED (For DIMEDIT)
Enter type of dimension editing [Home/New/Rotate/Oblique] <Home>:
    O (For Oblique)
Select objects: (Select the 2.00 and 1.00 dimensions on the left
    in the following image)
Select objects: (Press ENTER to continue)
Enter obliquing angle (Press ENTER for none): 150
```

The results are illustrated on the right in the following image, with both dimensions being repositioned with the Oblique option of the DIMEDIT command. Notice that the extension and dimension lines were affected; however, the dimension text remained the same.

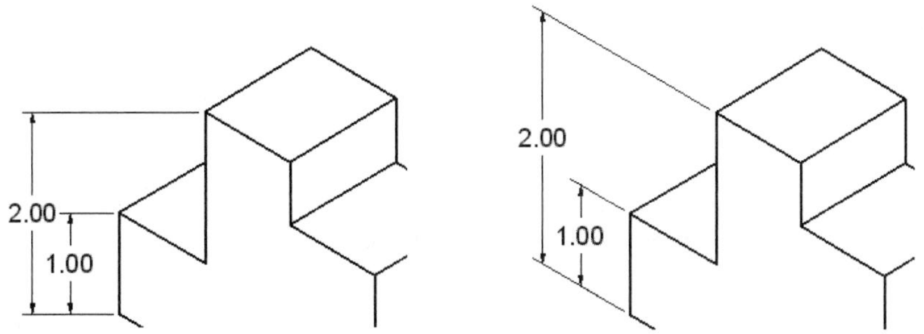

Figure 11.58

Use the Oblique option to complete the editing of this drawing by rotating the dimension at "A" at an obliquing angle of 210°. An obliquing angle of –30° was used to rotate the dimensions at "B" and "C." The dimensions at "D" require an obliquing angle of 90°, as shown in the following image. This represents proper isometric dimensions, except for the orientation of the text.

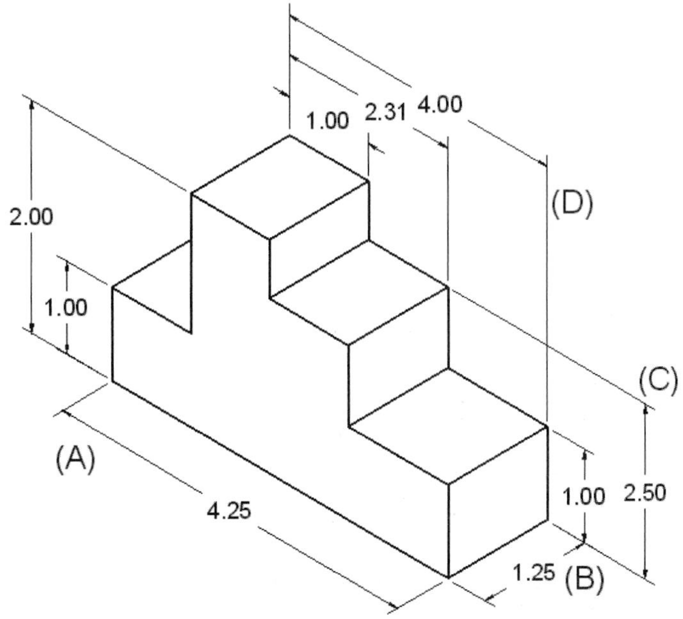

Figure 11.59

FLIPPING DIMENSION ARROWS

You can control the placement of dimension arrows by first selecting the dimensions and then right-clicking one of the arrows to display the menu in the following image. Clicking Flip Arrow from this menu flips this arrowhead. The other arrowhead remains normal. In the illustration on the right, the Flip Arrow mode was used twice to flip both arrowheads to the outside of the extension lines.

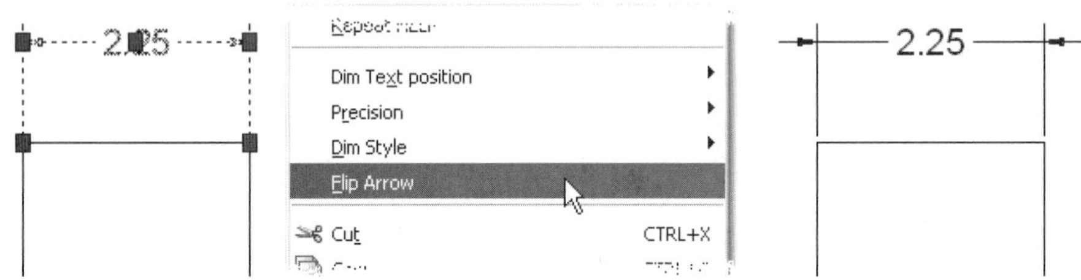

Figure 11.60

GEOMETRIC DIMENSIONING AND TOLERANCING (GDT)

In the object in the following image, approximately 1000 items need to be manufactured based on the dimensions of the drawing. Also, once constructed, all objects need to be tested to be sure the 90° surface did not deviate over 0.005 units from its base. Unfortunately, the wrong base was selected as the reference for all dimensions. Which should have been the correct base feature? As all items were delivered, they were quickly returned because the long 8.80 surface drastically deviated from the required 0.005 unit deviation or zone. This is one simple example that demonstrates the need for using geometric dimensioning and tolerancing techniques. First, this method deals with setting tolerances to critical characteristics of a part. Of course, the function of the part must be totally understood by the designer in order for tolerances to be assigned. The problem with the object was the note, which did not really specify which base feature to choose as a reference for dimensioning.

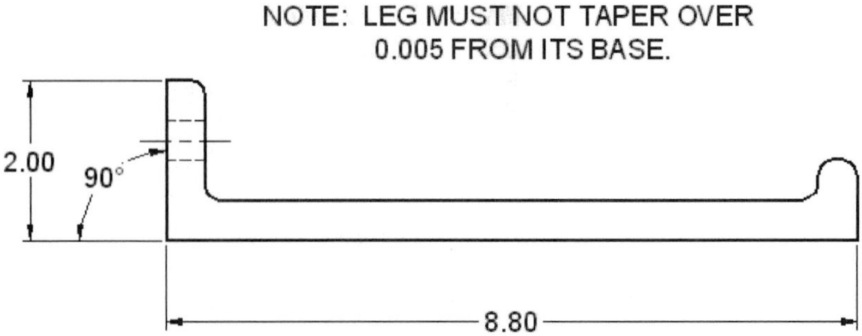

Figure 11.61

The following image shows the same object complete with dimensioning and tolerancing symbols. The letter "A" inside the rectangle identifies the datum or reference surface. The tolerance symbol at the end of the long edge tells the individual making this part that the long edge cannot deviate more than 0.005 units using surface "A" as a reference. The datum triangles that touch the part are a form of dimension arrowhead.

Using geometric tolerancing symbols ensures more accurate parts with less error in interpreting the dimensions.

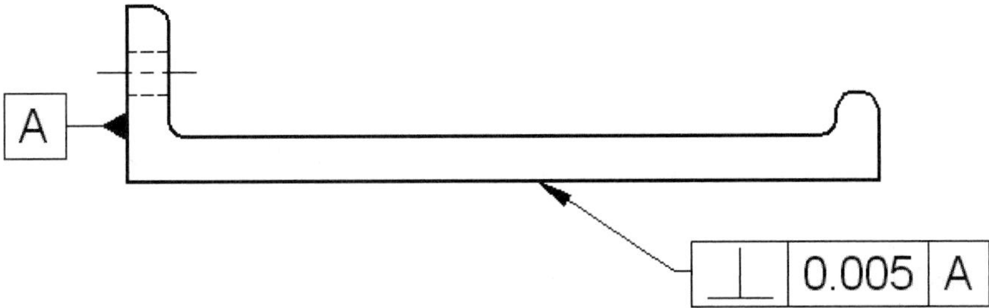

Figure 11.62

GDT SYMBOLS

Entering TOL at the command prompt brings up the main Geometric Tolerance dialog box, as shown in the following image. This box contains all tolerance zone boxes and datum identifier areas.

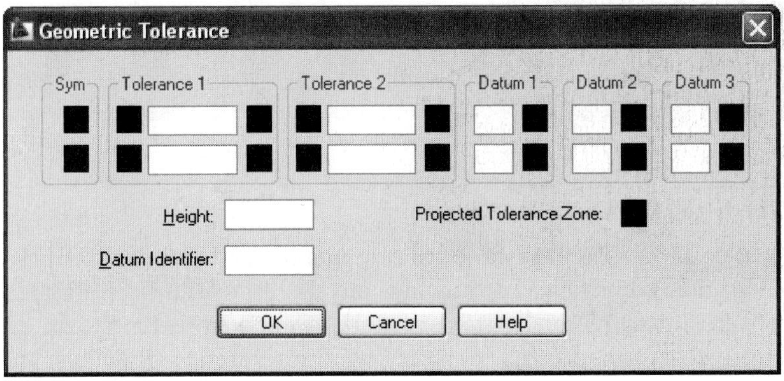

Figure 11.63

Clicking in one of the dark boxes under the Sym area displays the Symbol dialog box illustrated on the left in the following image. This dialog box contains all major geometric dimensioning and tolerancing symbols. Choose the desired symbol by clicking the specific symbol; this returns you to the main Geometric Tolerance dialog box.

The illustration on the right in the following image shows a chart outlining all geometric tolerancing symbols supported in the Symbol dialog box. Alongside each symbol is the characteristic controlled by the symbol, in addition to a tolerance type. Tolerances of form such as Flatness and Straightness can be applied to surfaces without a datum being referenced. On the other hand, tolerances of orientation such as Angularity and Perpendicularity require datums as reference.

Symbol	Purpose	Symbol	Purpose
⏥	Flatness	⊥	Perpendicularity
—	Straightness	//	Parallelism
○	Roundness	⌖	Position
⌭	Cylindricity	◎	Concentricity
⌒	Profile of a Line	⌯	Symmetry
⌓	Profile of a Surface	↗	Circular Runout
∠	Angularity	⌰	Total Runout

Figure 11.64

With the symbol placed inside this dialog box, you now assign such items as tolerance values, maximum material condition modifiers, and datums. In the following image, the tolerance of Parallelism is to be applied at a tolerance value of 0.005 units to Datum "A."

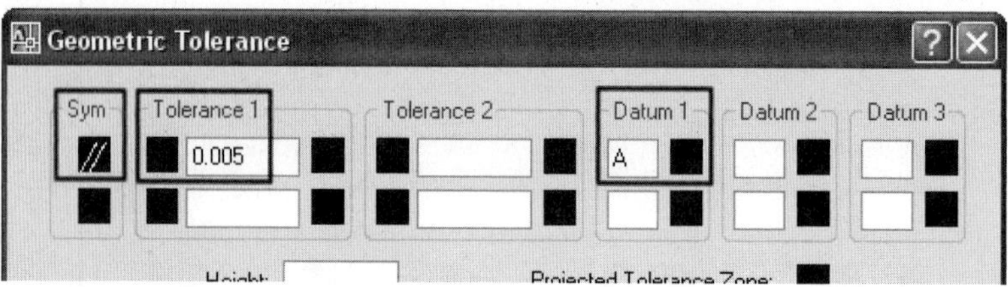

Figure 11.65

DIMENSION SYMBOLS

In today's global economy, the transfer of documents in the form of drawings is becoming a standard way of doing business. As drawings are shared with a subsidiary of an overseas company, two different forms of language may be needed to interpret the dimensions of the drawing. In the past, this has led to confusion in interpreting drawings, and as a result, a system of dimensioning symbols has been developed. It is hoped that a symbol will be easier to recognize and interpret than a note in a different language about the particular feature being dimensioned.

The following image shows some of the more popular dimensioning symbols in use today on drawings. Notice how the symbols are designed to make as clear and consistent an interpretation of the dimension as possible. As an example, the Deep or Depth symbol displays an arrow pointing down. This symbol is used to identify how far into a part a drill hole goes. The Arc Length symbol identifies the length of an arc, and so on.

Symbol	Description	Symbol	Description
⌒	Arc Length	2X	Number of Times
X.XX (boxed)	Basic Dimension	R	Radius
⌲	Conical Taper	(X.XX)	Reference Dimension
⌴	Counterbore/Spotface	SØ	Spherical Diameter
⌵	Countersink	SR	Spherical Radius
↧	Deep or Depth	⌳	Slope
Ø	Diameter	□	Square
X.XX (underlined)	Not to Scale		

Figure 11.66

CHARACTER MAPPING FOR DIMENSION SYMBOLS

Illustrated in the following image is a typical counterbore operation and the correct dimension layout. The QLEADER command was used to begin the diameter dimension. The counterbore and depth symbols were generated with the MTEXT command along with the Unicode Character Mapping dialog box. In any case, the first step in using the dimension symbols is to first create a new text style, for which you can use any name.

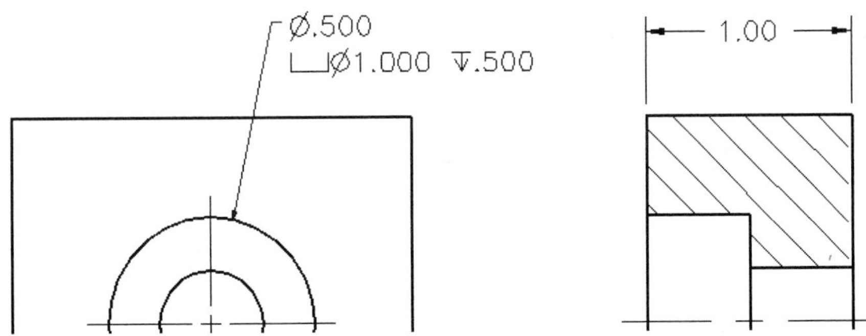

Figure 11.67

These extra symbol characters can be found by right-clicking inside the Text Formatting field. From the menu that appears, click Symbol followed by Other..., as shown in the following image.

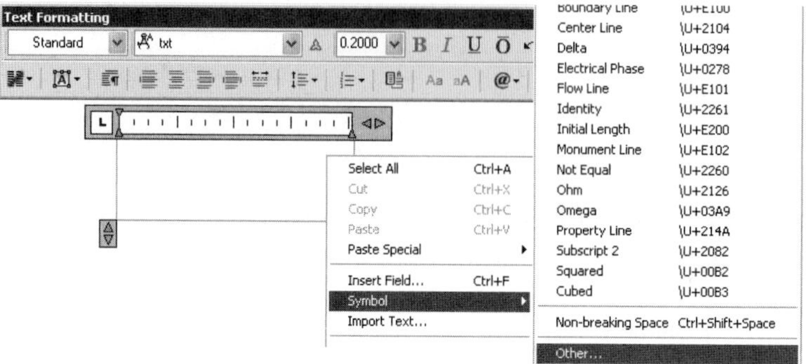

Figure 11.68

Clicking Other... displays the Character Map dialog box, as shown on the left in the following image. Notice, in the upper-left corner, that the current font is GDT, which holds all geometric tolerancing symbols along with the special dimensioning symbols such as counterbore, deep, and countersink. Be sure your current font is set to GDT. Once you identify a symbol, double-click it. A box appears around the symbol. Also, the symbol appears in the Characters to Copy area in the lower-left corner of the dialog box. Click the Copy button to copy this symbol to the Windows Clipboard. Then close the Character Map. Return to the Multiline Text Editor dialog box and press CTRL+V, which performs a paste operation. You may also paste the symbol by right-clicking in the Multiline Text Editor dialog box and clicking Paste. Because the counterbore symbol was copied to the Windows Clipboard, it pastes into the dialog box, as shown on the right in the following image.

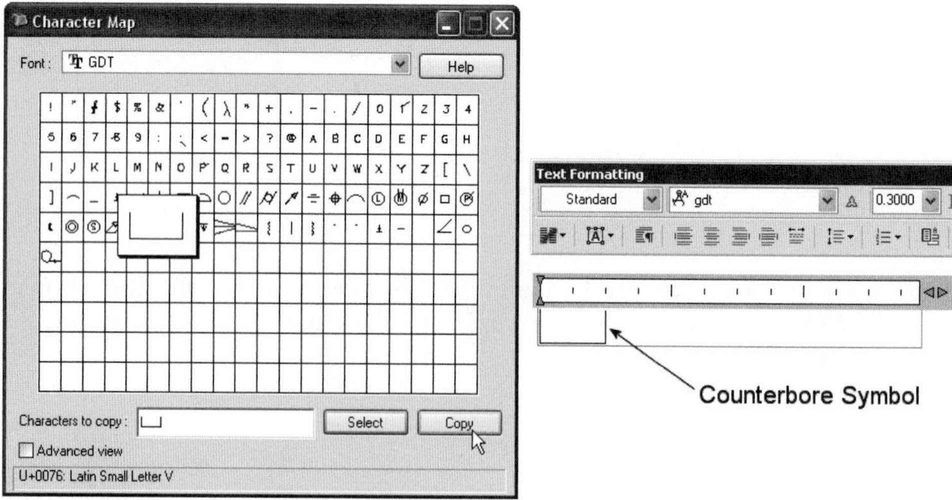

Figure 11.69

GRIPS AND DIMENSIONS

Grips have a tremendous amount of influence on dimensions. Grips allow dimensions to be moved to better locations; grips also allow the dimension text to be located at a better position along the dimension line.

Try It!: Open the drawing file 11_Dimgrip. In the example shown in the following image, notice the various unacceptable dimension placements. The 2.88 horizontal dimension lies almost on top of another extension line. To relocate this dimension to a better position, click the dimension. Notice that the grips appear and the entire dimension highlights. Now click the grip near "A," as illustrated on the left in the following image. (This grip is located at the left end of the dimension line.) When the grip turns red, the Stretch mode is currently active. Stretch the dimension above the extension line but below the 3.50 dimension. Press the ESC key to turn off the grips. The same results can be accomplished with the 4.50 horizontal dimension as it is stretched closer to the 3.50 dimension, as shown on the right in the following image. The DIMSPACE command could also be used to apply a standard space between dimensions, as shown on the right in the following image.

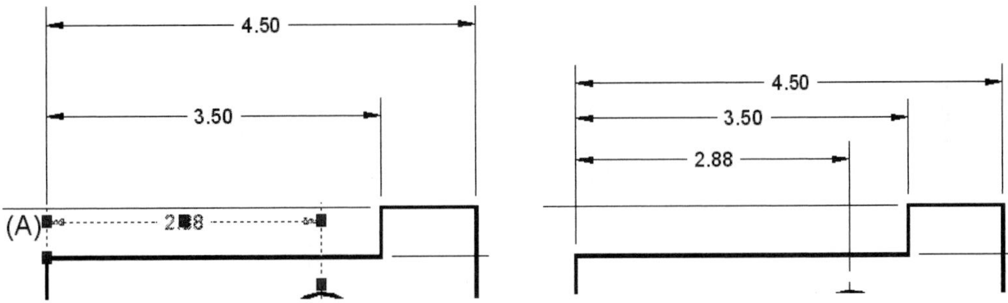

Figure 11.70

Notice that the two vertical dimensions shown on the left in the following image do not line up with each other; this would be poor practice. Pick both dimensions and notice the appearance of the grips, in addition to both dimensions being highlighted. Click the lower grip at "A" of the 1.25 dimension. When this grip turns red and places you in Stretch mode, select the grip at "B" of the opposite dimension. The result will be that both dimensions now line up with each other, as shown on the right in the following image.

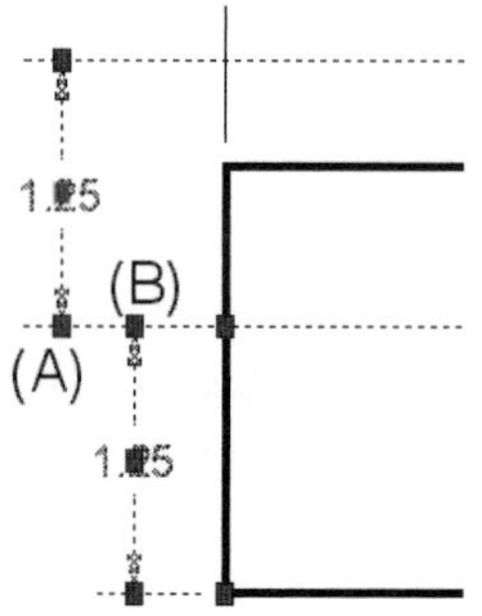

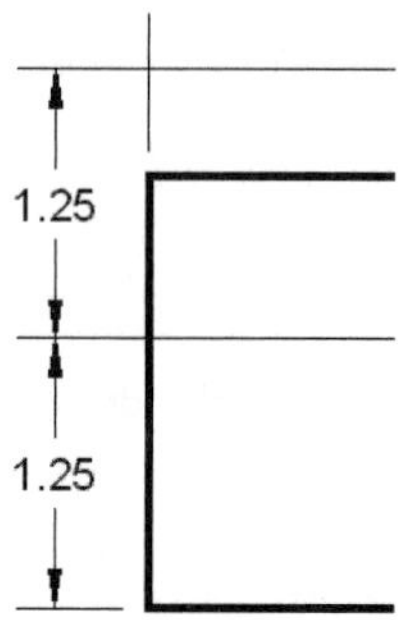

Figure 11.71

Illustrated on the left in the following image, the 2.50 dimension text is too close to the 2.00 vertical dimension on the right side of the object. Click the 2.50 dimension; the grips appear and the dimension highlights. Click in the grip representing the text location at "A." When this grip turns red, stretch the dimension text to a better location, as shown on the right in the following image.

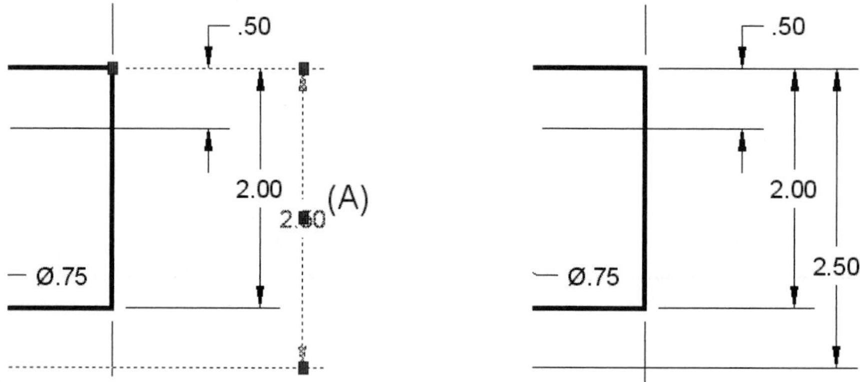

Figure 11.72

It is very easy to use grips to control the placement of diameter and radius dimensions. As shown on the left in the following image, click the diameter dimension; the grips appear in addition to the diameter dimension being highlighted. Click the grip that locates the dimension text at "A." When this grip turns red, relocate the diameter dimension text to a better location using the Stretch mode, as shown in the middle in the following image. The completed object, with dimensions edited through grips, is displayed on the right in the following image.

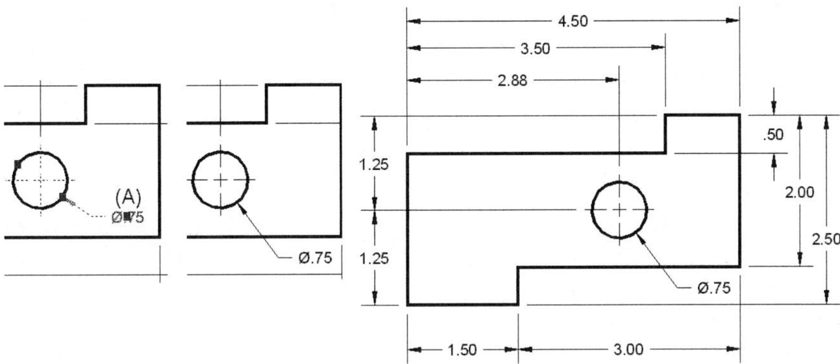

Figure 11.73

TUTORIAL EXERCISE: 11_FIXTURE.DWG

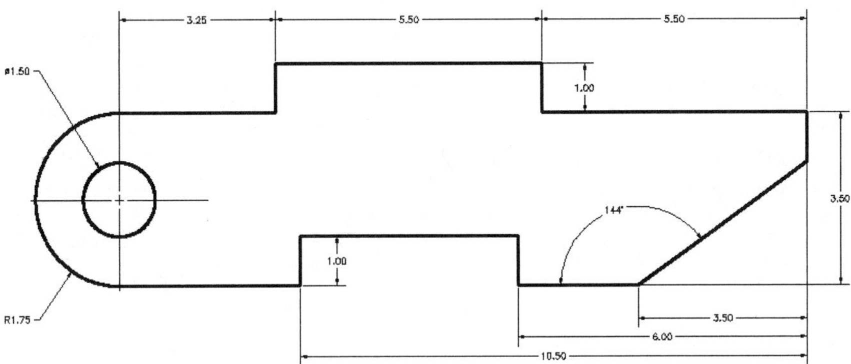

Figure 11.74

PURPOSE

The purpose of this tutorial is to add dimensions to the drawing of 11_Fixture.

SYSTEM SETTINGS

The drawing in the previous image is already constructed. Follow the steps in this tutorial for adding dimensions. Be sure the Endpoint and Intersection Object Snap modes are set.

LAYERS

Layers have already been created for this tutorial.

SUGGESTED COMMANDS

Use the DIMCENTER, DIMLINEAR, DIMCONTINUE, DIMBASELINE, DIMDIAMETER, and DIMRADIUS commands for placing dimensions throughout this tutorial.

Step 1

Open the drawing file 11_Fixture, which can be found on the CD. A series of linear, baseline, continue, radius, and diameter dimensions will be added to this view. Before continuing, verify that running OSNAP is set to Endpoint and Intersection modes and that OSNAP is turned on.

A number of dimension styles have been created to control various dimension properties. You will learn more about dimension styles in chapter 12. For now, make the Center Mark dimension style current by clicking its name in the Dimension Styles Control box, as shown in the following image.

Note: The Dimension Styles control box can be found in the Styles or Dimension toolbar. The Styles toolbar automatically displays if you make AutoCAD Classic the current workspace.

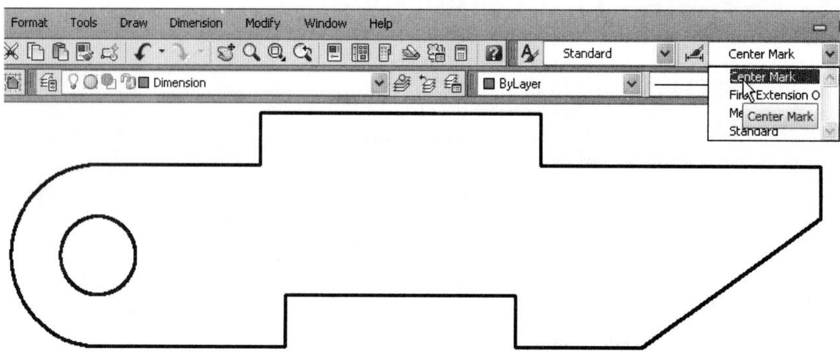

Figure 11.75

Step 2

Using the DIMCENTER command (DCE), add a center mark to identify the center of the circular features. Touch the edge of the arc to place the center mark. The Center Mark dimension style already allows a small and long dash to be constructed. See the following image.

```
Command: DCE (For DIMCENTER)
Select arc or circle: (Pick the edge of the arc in the following
    image)
```

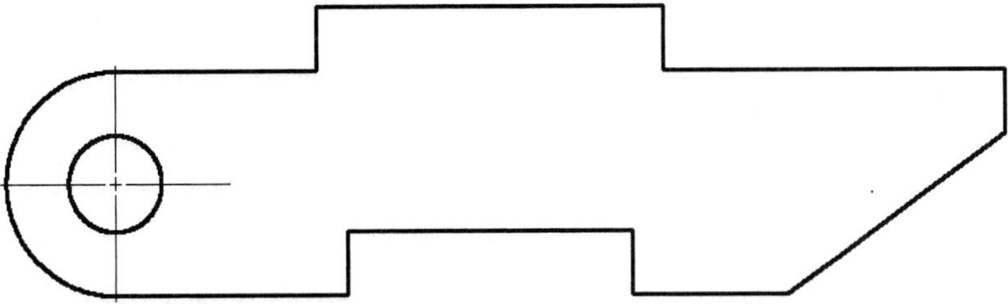

Figure 11.76

Step 3

Now that the center mark is properly placed, it is time to begin adding the horizontal and vertical dimensions. Before performing these tasks, first make the Mechanical dimension style current by clicking its name in the Dimension Styles Control box shown in the following image.

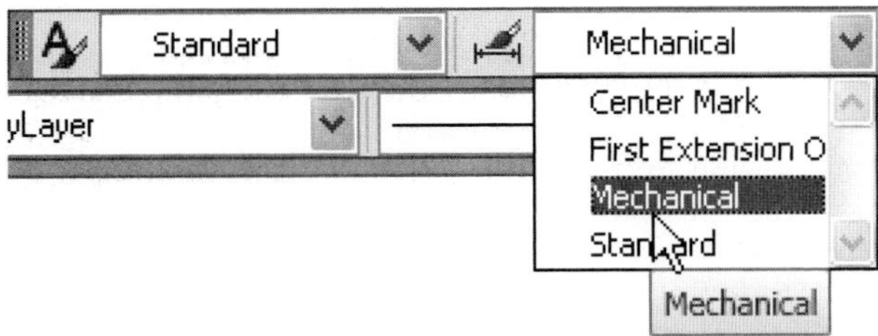

Figure 11.77

Step 4

Verify in the Status bar that OSNAP is turned on. Then place a linear dimension of 5.50 units from the intersection at "A" to the intersection at "B," as shown in the following image.

```
Command: DLI (For DIMLINEAR)
Specify first extension line origin or <select object>: (Pick the
    intersection at "A")
Specify second extension line origin: (Pick the intersection
    at "B")
Specify dimension line location or
[Mtext/Text/Angle/Horizontal/Vertical/Rotated]: (Locate the
    dimension in the following image)
Dimension text = 5.50
```

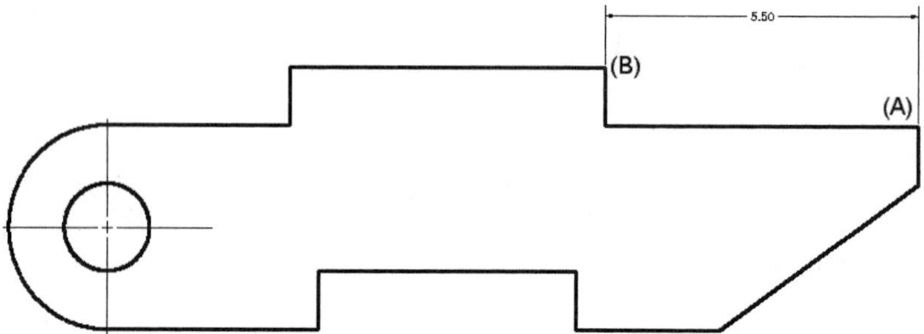

Figure 11.78

Step 5

Place two Continue dimensions at intersection "A" and endpoint "B" in the following image. These Continue dimensions know to calculate the new dimension from the second extension line of the previous dimension.

```
Command: DCO (For DIMCONTINUE)
Specify a second extension line origin or [Undo/Select] <Select>:
    (Pick the intersection at "A")
Dimension text = 5.50
Specify a second extension line origin or [Undo/Select] <Select>:
    (Pick the endpoint at "B")
Dimension text = 3.25
Specify a second extension line origin or [Undo/Select] <Select>:
    (Press ENTER)
Select continued dimension: (Press ENTER)
```

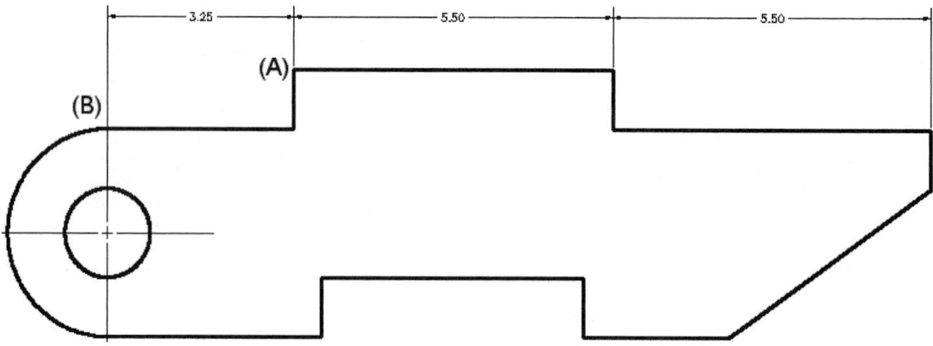

Figure 11.79

Step 6

Place a linear dimension of 3.50 units from the intersection at "A" to the intersection at "B," as shown in the following image.

```
Command: DLI (For DIMLINEAR)
Specify first extension line origin or <select object>: (Pick the
    intersection at "A")
Specify second extension line origin: (Pick the intersection at
    "B")
Specify dimension line location or
[Mtext/Text/Angle/Horizontal/Vertical/Rotated]: (Locate the
    dimension in the following image)
Dimension text = 3.50
```

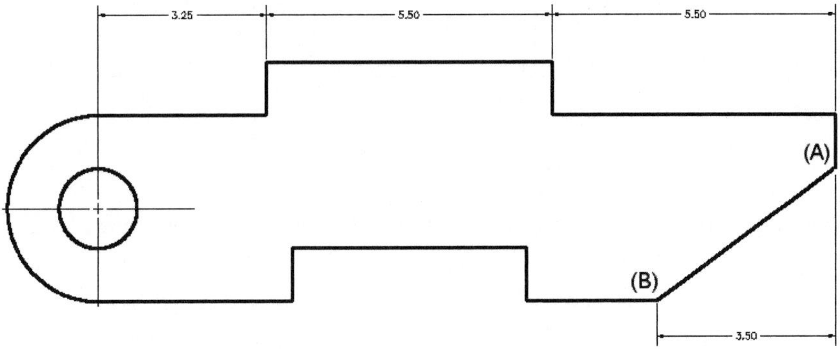

Figure 11.80

Step 7

Add a series of baseline dimensions, as shown in the following image. All baseline dimensions are calculated from the first extension line of the previous dimension (the linear dimension measuring 3.50 units).

```
Command: DBA (For DIMBASELINE)
Specify a second extension line origin or [Undo/Select] <Select>:
    (Pick the intersection at "A")
Dimension text = 6.00
Specify a second extension line origin or [Undo/Select] <Select>:
    (Pick the intersection at "B")
Dimension text = 10.50
Specify a second extension line origin or [Undo/Select] <Select>:
    (Press ENTER)
Select base dimension: (Press ENTER)
```

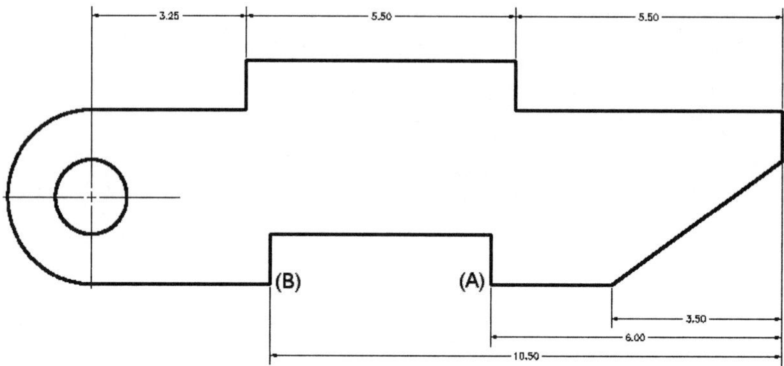

Figure 11.81

Step 8

Add a diameter dimension to the circle using the DIMDIAMETER command (DDI). Then add a radius dimension to the arc using the DIMRADIUS command (DRA), as shown in the following image.

```
Command: DDI (For DIMDIAMETER)
Select arc or circle: (Pick the edge of the circle at "A")
Dimension text = 1.50
Specify dimension line location or [Mtext/Text/Angle]: (Locate
    the diameter dimension in the following image)
```

```
Command: DRA (For DIMRADIUS)
Select arc or circle: (Pick the edge of the arc at "B")
Dimension text = 1.75
Specify dimension line location or [Mtext/Text/Angle]: (Locate
    the diameter dimension in the following image)
```

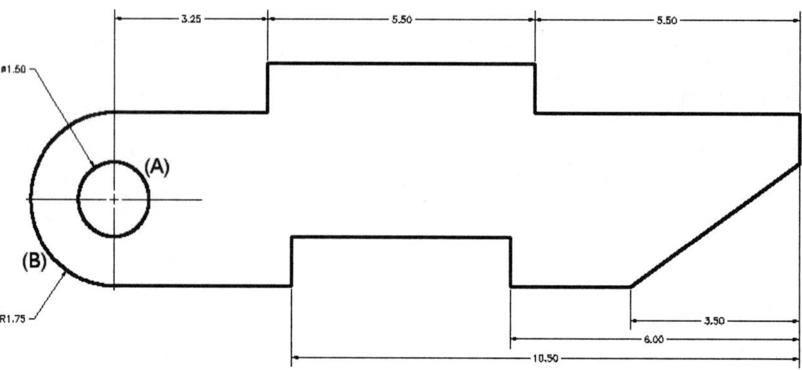

Figure 11.82

Step 9

Place a vertical dimension of 3.50 units using the DIMLINEAR command (DLI). Pick the first extension line origin at the intersection at "A" and the second extension line origin at the intersection at "B," as shown in the following image.

```
Command: DLI (For DIMLINEAR)
Specify first extension line origin or <select object>: (Pick the
    intersection at "A")
Specify second extension line origin: (Pick the intersection at
    "B")
Specify dimension line location or
[Mtext/Text/Angle/Horizontal/Vertical/Rotated]: (Locate the
    dimension in the following image)
Dimension text = 3.50
```

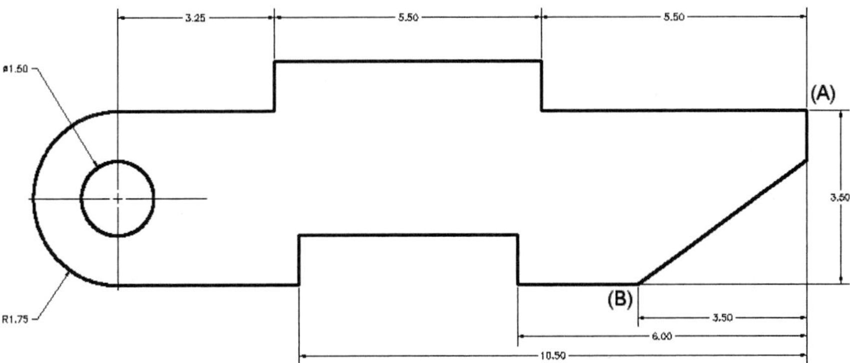

Figure 11.83

Step 10

Add an angular dimension of 144° using the DIMANGULAR command (DAN), as shown in the following image.

```
Command: DAN (For DIMANGULAR)
Select arc, circle, line, or <specify vertex>: (Pick the line at
    "A")
Select second line: (Pick the line at "B")
Specify dimension arc line location or [Mtext/Text/Angle/
    Quadrant]: (Locate the dimension in the following image)
Dimension text = 144
```

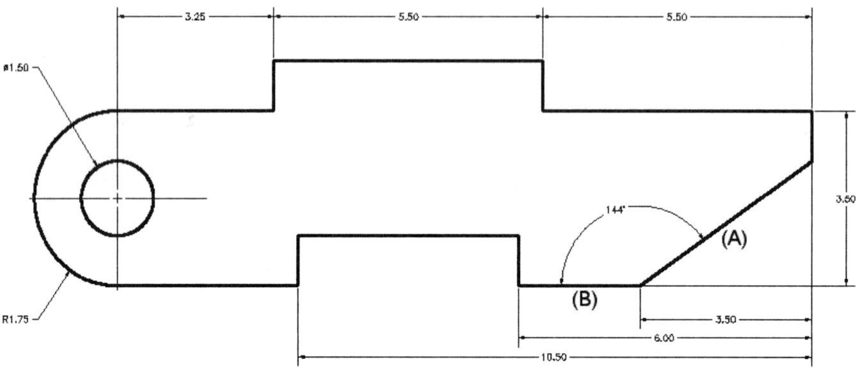

Figure 11.84

Step 11

The next series of dimensions involves placing linear dimensions so that one extension line is constructed by the dimension and the other is actually an object line that already exists. You need to turn off one extension line while leaving the other extension line turned on. This is accomplished by using an existing dimension style. Make the First Extension Off dimension style current by clicking its name in the Dimension Styles Control box, as shown in the following image. This dimension style will suppress or turn off the first extension line while leaving the second extension line visible. This prevents the extension line from being drawn on top of the object line.

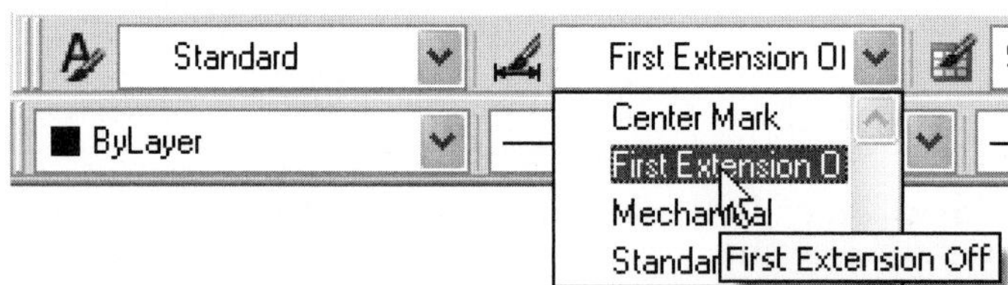

Figure 11.85

Step 12

Add the two linear dimensions to the slots by using the following command sequences and image as a guide.

```
Command: DLI (For DIMLINEAR)
Specify first extension line origin or <select object>:
(Pick the intersection at "A")
Specify second extension line origin:
(Pick the intersection at "B")
Specify dimension line location or
[Mtext/Text/Angle/Horizontal/Vertical/Rotated]: (Locate the
    dimension in the following image)
Dimension text = 1.00
```

```
Command: DLI (For DIMLINEAR)
Specify first extension line origin or <select object>:
(Pick the intersection at "C")
Specify second extension line origin:
(Pick the intersection at "D")
Specify dimension line location or
[Mtext/Text/Angle/Horizontal/Vertical/Rotated]: (Locate the
    dimension in the following image)
Dimension text = 1.00
```

The completed fixture drawing with all dimensions is shown in the following image.

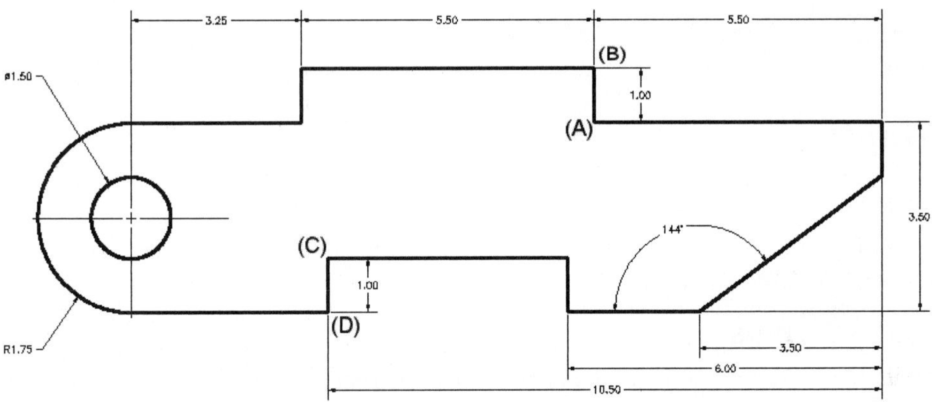

Figure 11.86

TUTORIAL EXERCISE: 11_DIMENSION VIEWS.DWG

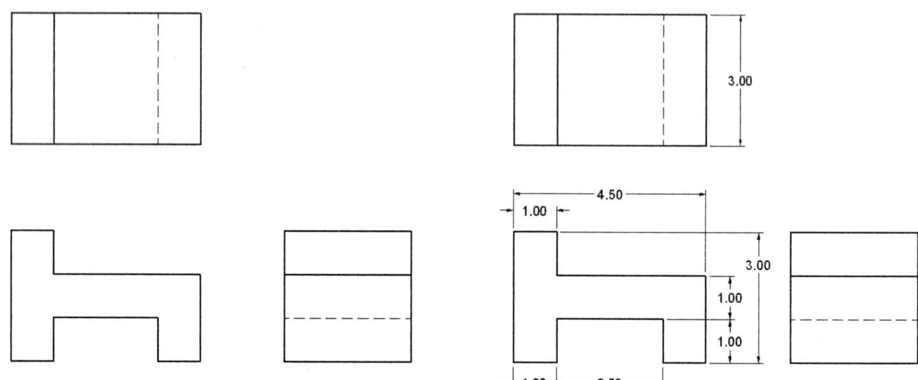

Figure 11.87

PURPOSE

The purpose of this tutorial is to add dimensions to the three-view drawing named 11_Dimension Views.dwg.

SYSTEM SETTINGS

The drawing in the previous image is already constructed. Dimensions must be added to various views to call out overall distances, in addition to features such as cuts and slots. Be sure the Object Snap modes Endpoint, Center, Intersect, and Extension are set.

LAYERS

Layers have already been created for this tutorial.

SUGGESTED COMMANDS

Use the DIMLINEAR command for horizontal and vertical dimensions on different views in this drawing.

Step 1

Open the drawing file 11_Dimension Views.dwg. Linear dimensions are placed identifying the overall length, width, and depth dimensions. The DIMLINEAR command is used to perform this task, as shown on the left in the following image.

Step 2

Detail dimensions that identify cuts and slots are placed. Because these cuts are visible in the Front view, the dimensions are placed there, as shown in the middle in the following image. The DIMCONTINUE command (DCO) could be used for the second 1.00 dimension.

Step 3

Once the spaces between the Front, Top, and Right Side views are used up by dimensions, the outer areas are used for placing additional dimensions such as the two horizontal dimensions on the right in the following image. Again, use DIMCONTINUE (DCO) for the second dimension. Use grips to adjust dimension text locations if necessary.

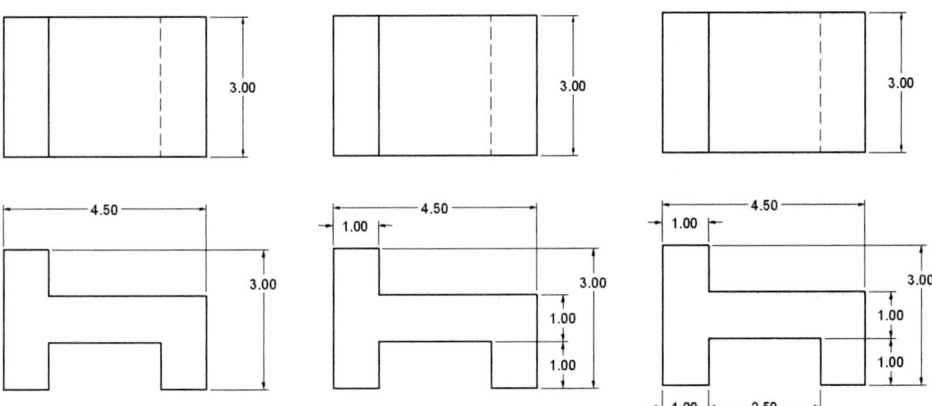

Figure 11.88

PROBLEMS FOR CHAPTER 11

DIRECTIONS FOR PROBLEMS 11–1 THROUGH 11–18

1. *Open each drawing located on the CD.*
2. *Use proper techniques to dimension each drawing.*
3. *Follow the steps involved in completing 11_Dimension Views.dwg.*

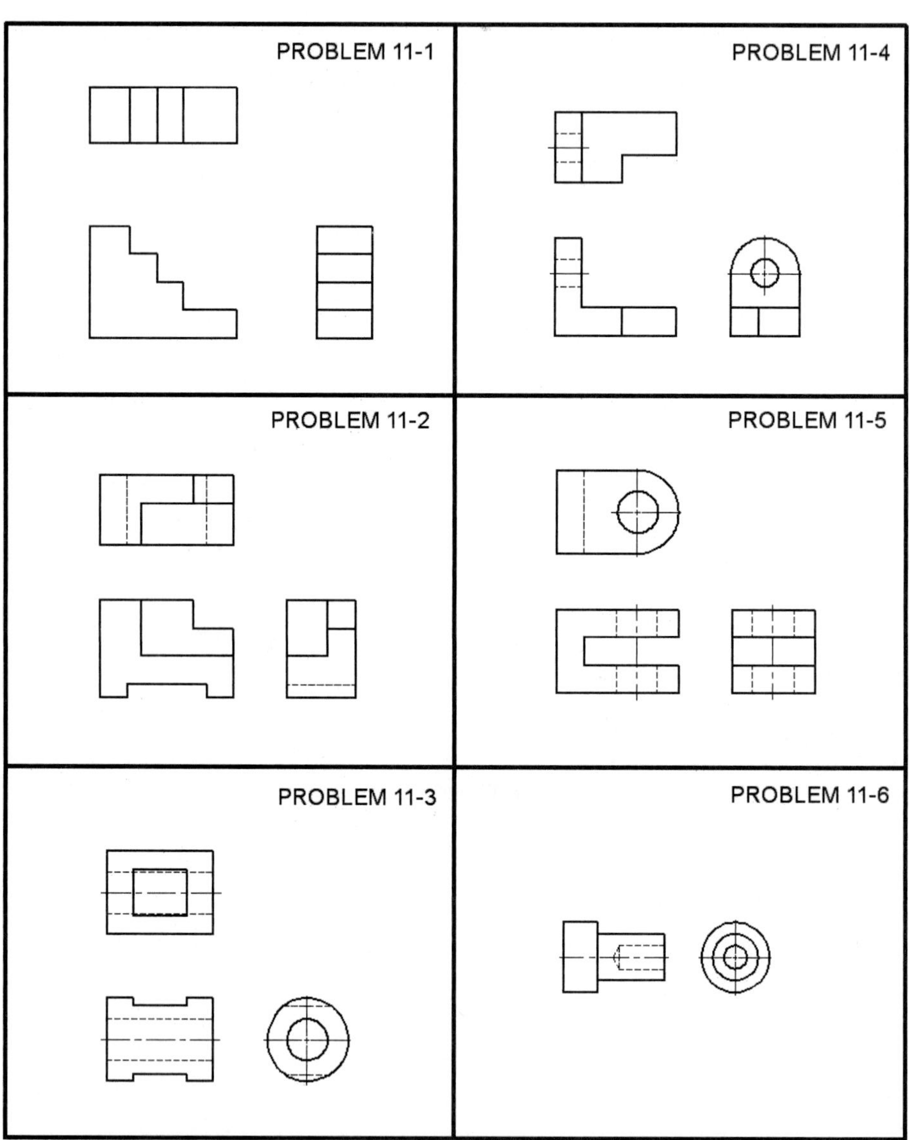
PROBLEM 11-1
PROBLEM 11-4
PROBLEM 11-2
PROBLEM 11-5
PROBLEM 11-3
PROBLEM 11-6

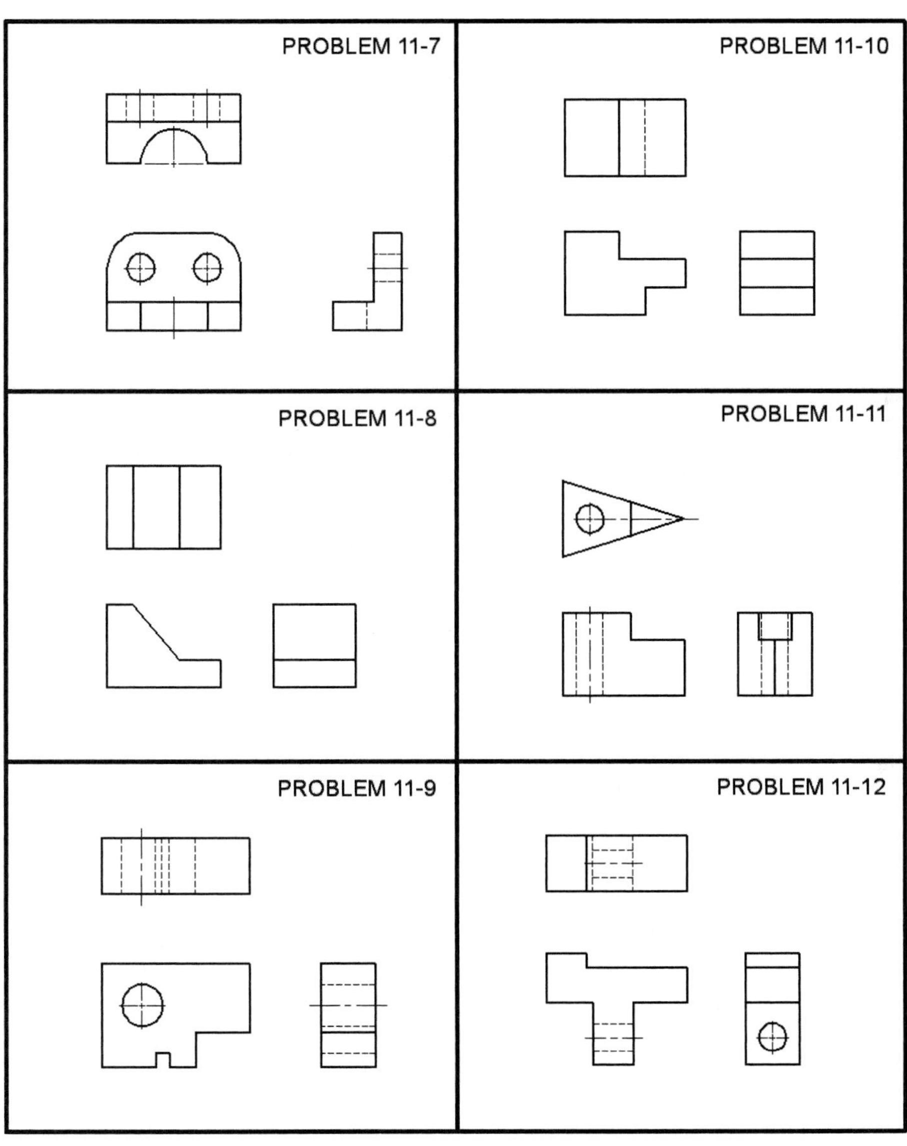
PROBLEM 11-7
PROBLEM 11-10
PROBLEM 11-8
PROBLEM 11-11
PROBLEM 11-9
PROBLEM 11-12

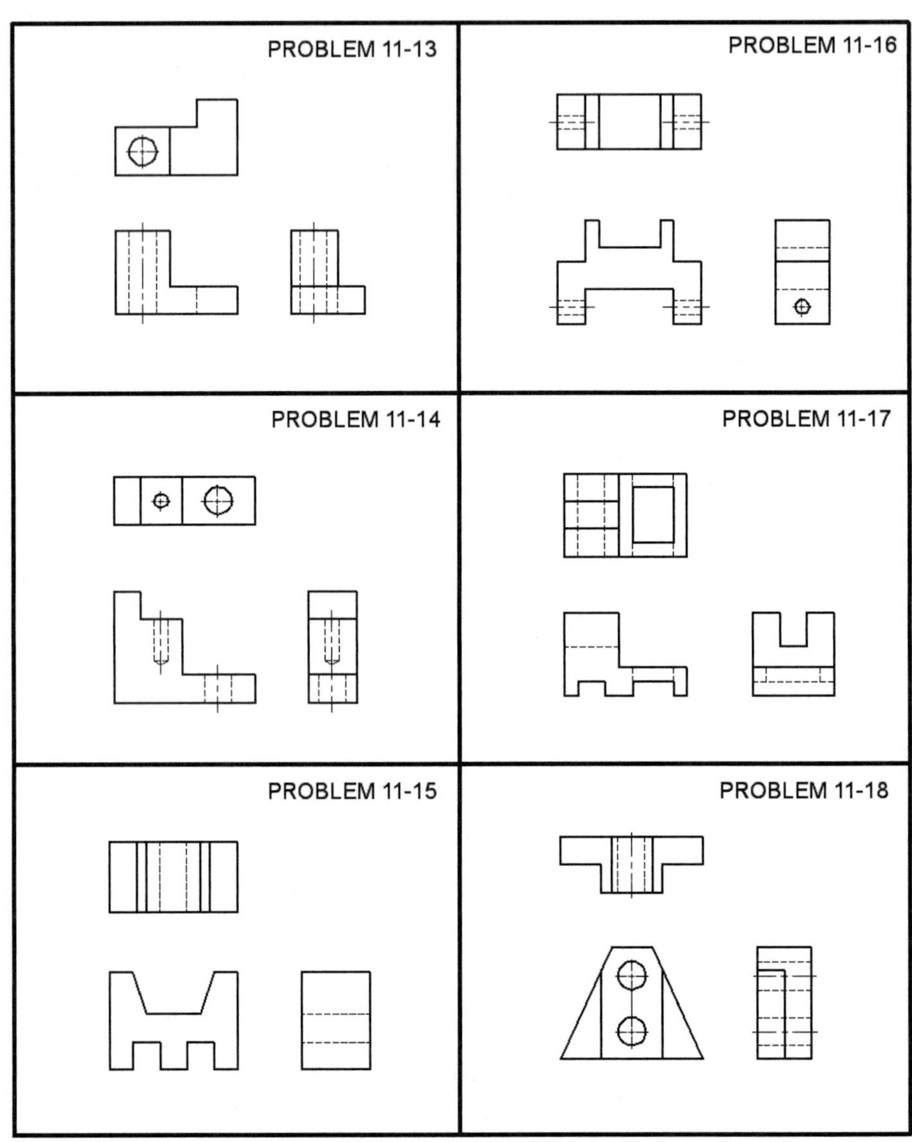
PROBLEM 11-13
PROBLEM 11-16
PROBLEM 11-14
PROBLEM 11-17
PROBLEM 11-15
PROBLEM 11-18

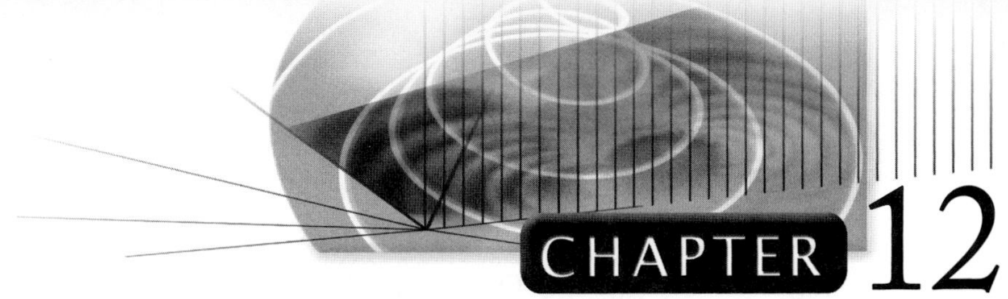

Managing Dimension Styles

Dimensions have different settings that could affect the group of dimensions. These settings include the control of the dimension text height, the size and type of arrowhead used, and whether the dimension text is centered in the dimension line or placed above the dimension line. These are but a few of the numerous settings available to you. In fact, some settings are used mainly for architectural applications, while other settings are only for mechanical uses. As a means of managing these settings, dimension styles are used to group a series of dimension settings under a unique name to determine the appearance of dimensions. This chapter will cover in detail the Dimension Style Manager dialog box and the following tabs associated with it: Lines, Symbols and Arrows; Text; Fit; Primary Units; Alternate Units; and Tolerances. Additional topics include overriding a dimension style and modifying the dimension style of an object.

THE DIMENSION STYLE MANAGER DIALOG BOX

Begin the process of creating a dimension style by choosing Dimension Style... from the Dimension pull-down menu, as shown on the left in the following image. You can also pick Dimension Style from the Dimension toolbar, Menu Browser, and Ribbon as shown in the following image. Entering the keyboard command DDIM or D is another way of accessing a Dimension Style.

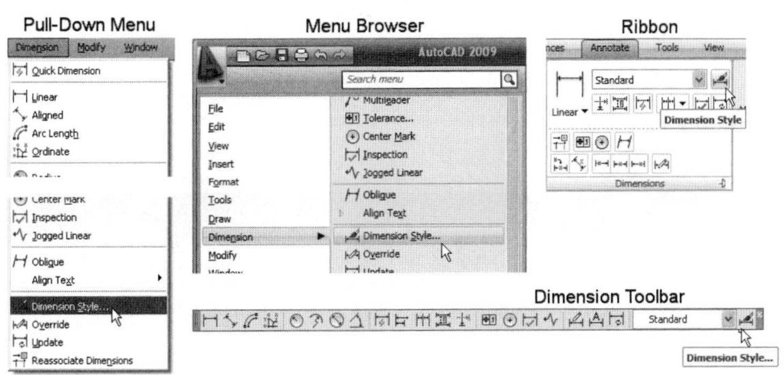

Figure 12.1

Performing any of the actions in the previous image will launch the Dimension Style Manager dialog box, shown in the following image. The current dimension style is listed as Standard. There is also an Annotative dimensions style present; this will be discussed in chapter 19. The Standard dimension style is automatically available when you create any new drawing. In the middle of the dialog box is an image icon that displays how some dimensions will appear based on the current value of the dimension settings. Various buttons are also available to set a dimension style to current, create a new dimension style, make modifications to an existing dimension style, display dimension overrides, and compare the differences and similarities of two dimension styles.

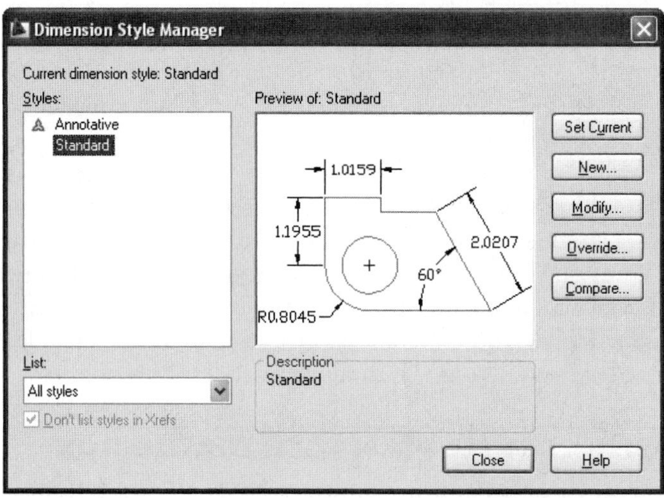

Figure 12.2

To create a new dimension style, click the New... button. This activates the Create New Dimension Style dialog box, as shown at the top of the following image. Enter a new name such as Mechanical in the New Style Name area. Then click the Continue button. This takes you to the New Dimension Style: Mechanical dialog box, as shown at the bottom of the following image, where a number of tabs hold all the settings needed in dimensioning.

Refer to the following table for a brief description of each tab located in the Dimension Style Manager dialog box:

Tab	Description
Lines	The Lines tab deals with settings that control dimension and extension lines.
Symbols and Arrows	The Lines and Arrows tab controls arrowheads and center marks.
Text	The Text tab contains the settings that control the appearance, placement, and alignment of dimension text.

Tab	Description
Fit	The Fit tab contains various fit options for placing dimension text and arrows, especially in narrow places.
Primary Units	You control the units used in dimensioning through the Primary Units tab.
Alternate Units	If you need to display primary and secondary units in the same drawing, the Alternate Units tab is used.
Tolerances	Finally, for mechanical applications, various ways to show tolerances are controlled in the Tolerances tab.

When you make changes to any of the settings under the tabs, the preview image updates to show these changes. This provides a quick way of previewing how your dimensions will appear in your drawing. All these areas will be explained in greater detail throughout this chapter.

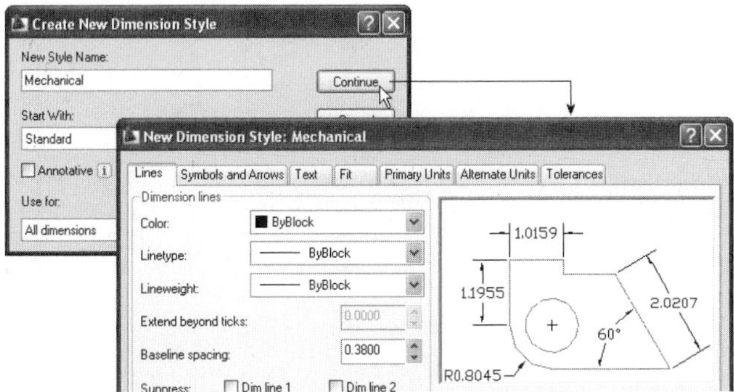

Figure 12.3

When you are finished making changes, click the OK button. This returns you to the Dimension Style Manager dialog box shown in the following image. Notice that the Mechanical dimension style has been added to the list of styles. Also, any changes made in the tabs are automatically saved to the dimension style you are creating or modifying. Click Mechanical and then pick the Set Current button to make this style the current dimension style. Clicking the Close button returns you to your drawing. Let's get a closer look at all the tabs and their settings. Type the letter D to reopen the Dimension Style Manager. Make sure Mechanical is highlighted and click the Modify... button.

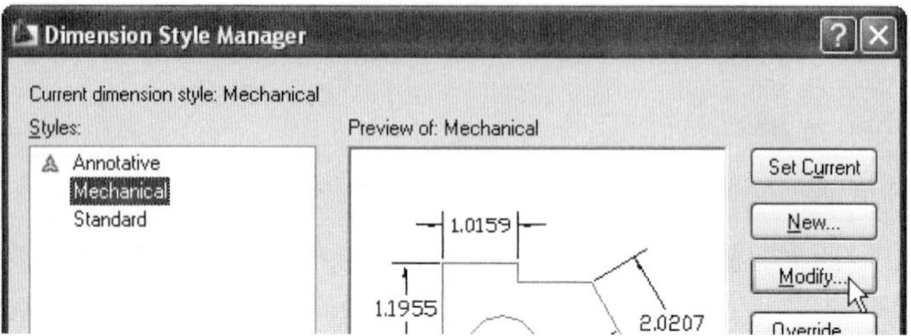

Figure 12.4

THE LINES TAB

The Modify Dimension Style: Mechanical dialog box in the following image displays the Lines tab, which will now be discussed in greater detail. This tab consists of four main areas dealing with dimension lines and extension lines.

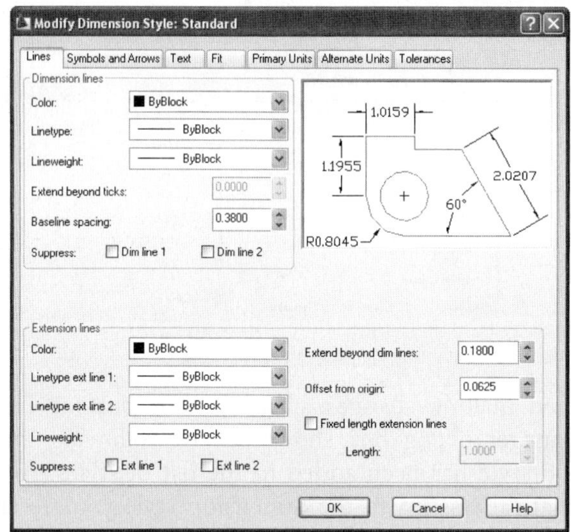

Figure 12.5

DIMENSION LINE SETTINGS

Use the Dimension lines area, shown in the following image, to control the color, lineweight, visibility, and spacing of the dimension line.

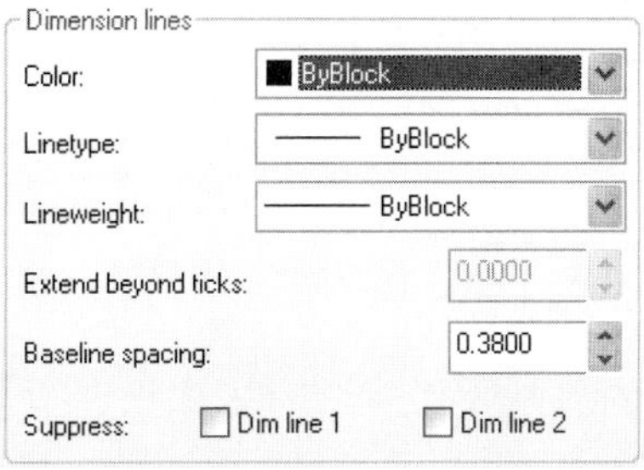

Dimension lines
Color: ByBlock
Linetype: ByBlock
Lineweight: ByBlock
Extend beyond ticks: 0.0000
Baseline spacing: 0.3800
Suppress: Dim line 1 Dim line 2

Figure 12.6

The Baseline spacing setting in the Dimension lines area controls the spacing of baseline dimensions, because they are placed at a distance from each other similar to the illustration on the left in the following image. This value affects the DIMBASELINE command (or DBA for short).

By default, dimension line suppression is turned off. To turn on suppression of dimension lines, place a check in the Dim line 1 or Dim line 2 box next to Suppress. This operation turns off the display of dimension lines for all dimensions placed under this dimension style. See the illustration on the right in the following image. This may be beneficial where tight spaces require that only the dimension text be placed.

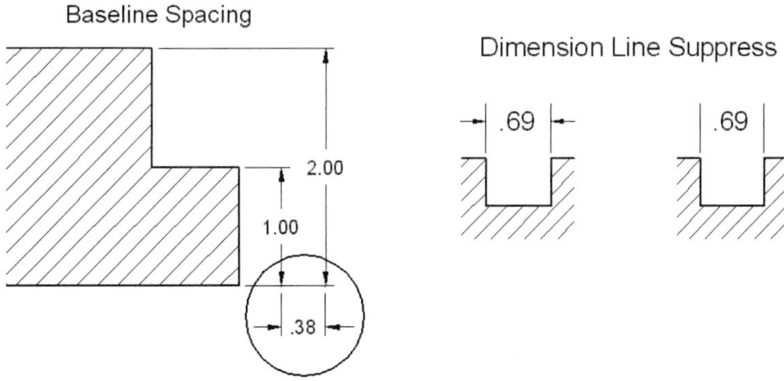

Figure 12.7

EXTENSION LINES

The Extension lines area controls the color, the distance the extension extends past the arrowhead, the distance from the object to the end of the extension line, and the visibility of extension lines, as shown in the following image. Additional controls allow you to set a different linetype for extension lines and provide the ability to use a fixed length value for all extension lines.

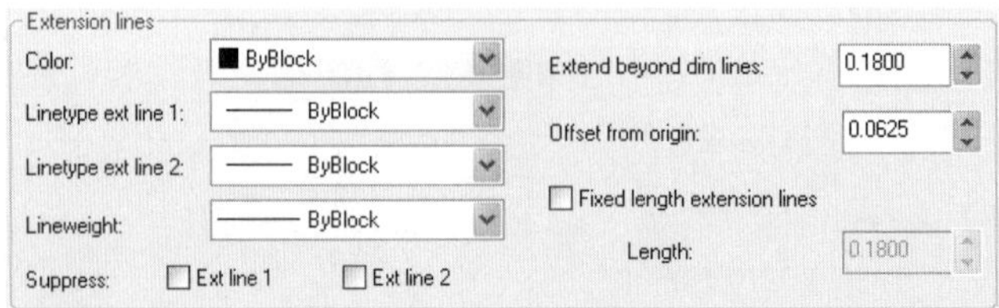

Figure 12.8

The Extend beyond dim lines setting controls how far the extension extends past the arrowhead or dimension line, as shown on the left in the following image. By default, a value of 0.18 is assigned to this setting.

The Offset from origin setting controls how far away from the object the extension line will start, as shown on the right in the following image. By default, a value of 0.06 is assigned to this setting.

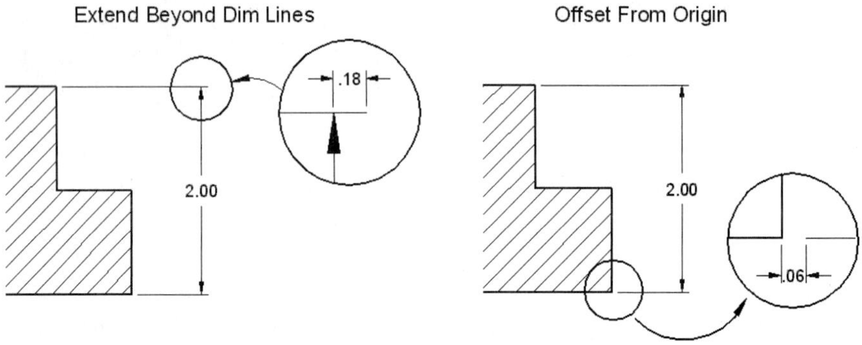

Figure 12.9

The Suppress Ext line 1 and Ext line 2 checkboxes control the visibility of extension lines. They are useful when you dimension to an object line and for avoiding placing the extension line on top of the object line. Placing a check in the Ext line 1 box of Suppress turns off the first extension line. Similarly, when you place a check in the Ext line 2 box of Suppress, the second extension line is turned off. Suppressing extension lines by checking these boxes suppresses all extension lines for dimensions placed under this dimension style. Study the examples in the following image to get a better idea about how suppression of extension lines operates.

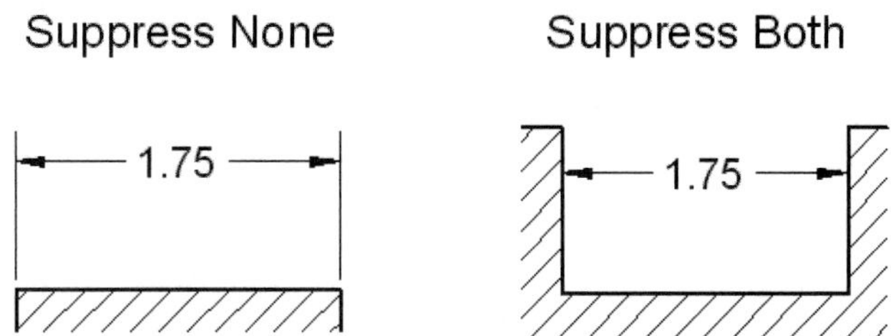

Figure 12.10

EXTENSION LINETYPES

You can assign linetypes presently loaded in your drawing, as shown in the following image. In this example, the Center linetype is assigned to extension lines. This provides an efficient means of applying custom linetypes to dimensions where appropriate, as in the dimensioning of the circular features in the plate.

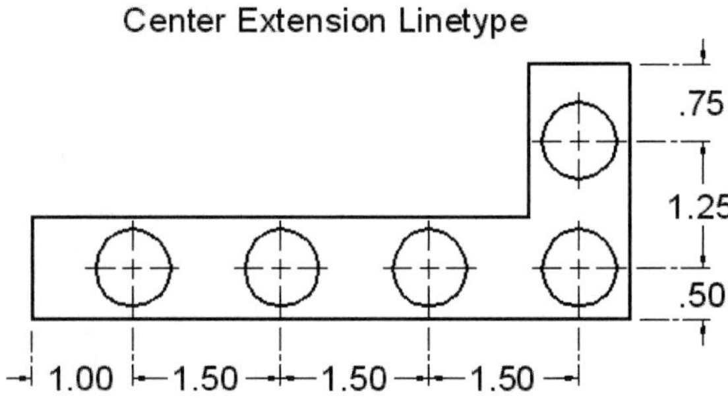

Figure 12.11

FIXED LENGTH EXTENSION LINES

Fixed length extension lines is a setting found under the Extension lines area of the Lines tab. It sets a user-defined length for all extension lines. In the illustration on the left in the following image, extension lines of varying lengths are displayed. All extension lines begin by default .0625 units away from the object. Illustrated on the right is another example of how extension lines are displayed at a fixed length of .25 units. Notice that all extension lines are the same length in this example. This value is calculated from where the endpoint of the arrow intersects with the extension to the bottom of the extension line. As you can see, this results in very short extension lines that are all the same length.

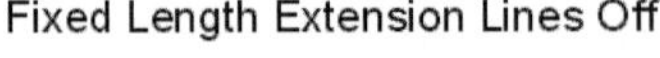

Fixed Length Extension Lines .25

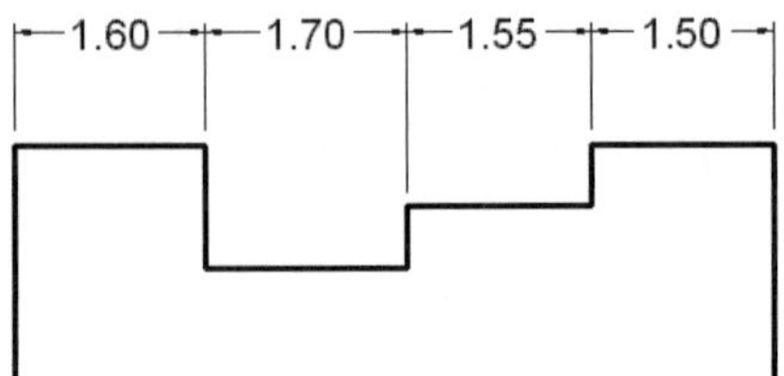

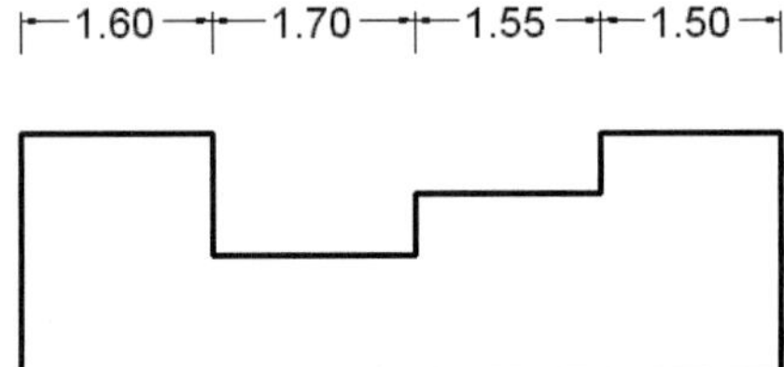

Figure 12.12

THE SYMBOLS AND ARROWS TAB

The Symbols and Arrows tab controls four main areas, as shown in the following image, namely, the size and type of arrowheads, the size and type of center marks, the style of the arc length symbol, and the angle formed when placing a radius jog dimension.

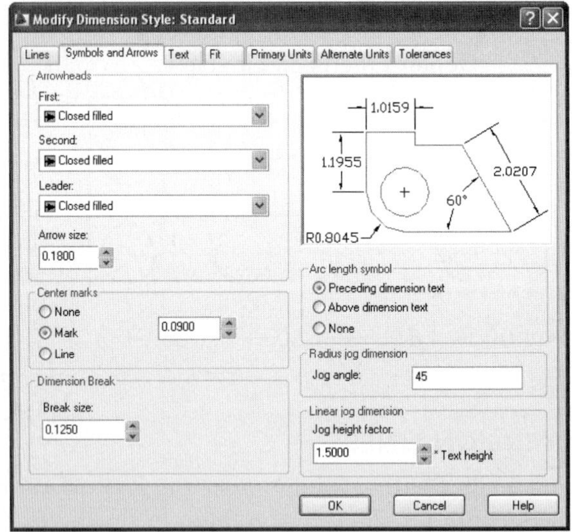

Figure 12.13

ARROWHEAD SETTINGS

Use the Arrowheads area to control the type of arrowhead terminator used for dimension lines and leaders, as shown in the following image. This dialog box also controls the size of the arrowhead.

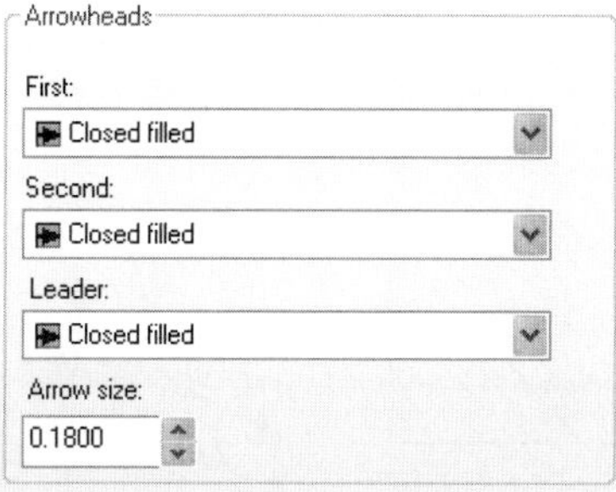

Figure 12.14

Clicking in the First box displays a number of arrowhead terminators. Choose the desired terminator from the list, as shown on the left in the following image. When you choose the desired arrowhead from the First box, the Second box automatically updates to the selection made in the First box. If you choose an arrowhead from the Second field, the first and second arrowheads may be different at opposite ends of the dimension line; this is desired in some applications. Choosing a terminator in the Leader box displays the arrowhead that will be used whenever you place a leader.

Illustrated on the right in the following image is the complete set of arrowheads, along with their names. The last arrow type in the image on the left is a User Arrow, which allows you to create your own custom arrowhead.

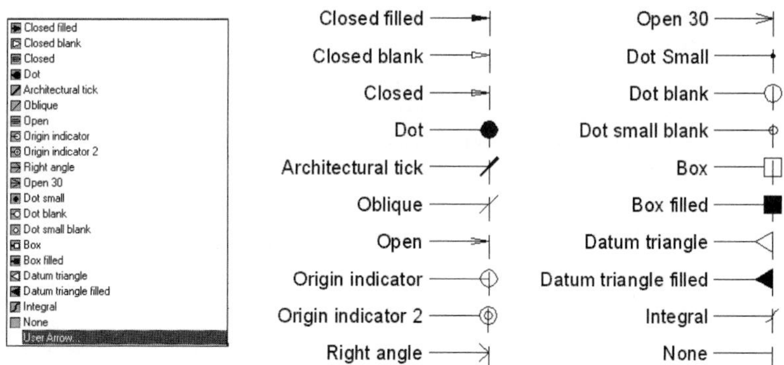

Figure 12.15

Use the Arrow size setting to control the size of the arrowhead terminator. By default, the arrow size is set to .18 units, as shown in the following image.

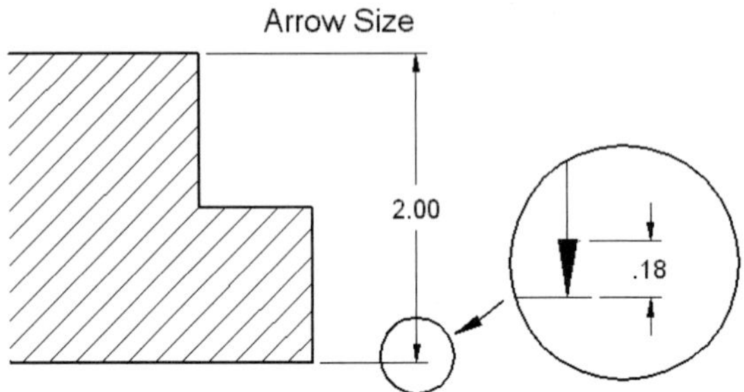

Figure 12.16

CENTER MARK SETTINGS

The Center marks for circles area allows you to control the type of center marker used when identifying the centers of circles and arcs. You can make changes to these settings by clicking the three center mark modes: None, Mark, or Line, as shown on the left in the following image. The Size box controls the size of the small plus mark (+). If Mark or Line is chosen, these lines will show up when placing radius and diameter dimensions.

The three types of center marks are illustrated on the right in the following image. The Mark option places a plus mark in the center of the circle. The Line option places the plus mark and extends a line past the edge of the circle. The None option displays a circle with no center mark.

Tip: It is good practice to add center marks to all circles and arcs before adding radius and diameter dimensions. To do this, first pick the Center Line type and use the DCE (Dimension Center) command to place the center marks. When finished, pick the None type in the center mark type area. Now all diameter and radius dimensions can be added without duplicating the center marks.

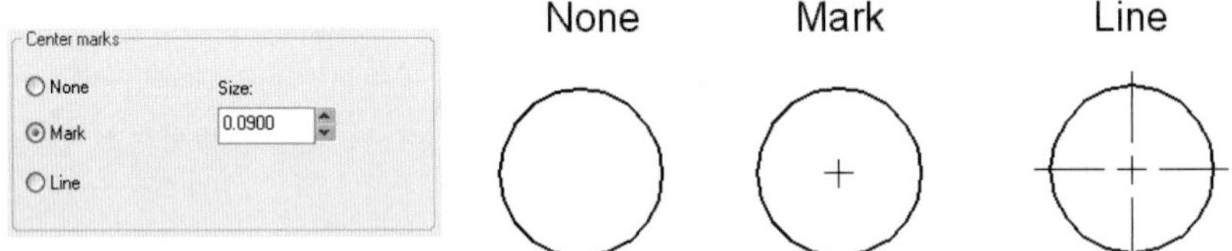

Figure 12.17

DIMENSION BREAK SETTINGS

Use this area to set the distance when breaking extension or dimension lines with the DIMBREAK command. The following image compares the default break distance of .125 with a new break distance of .25 and how the break is affected.

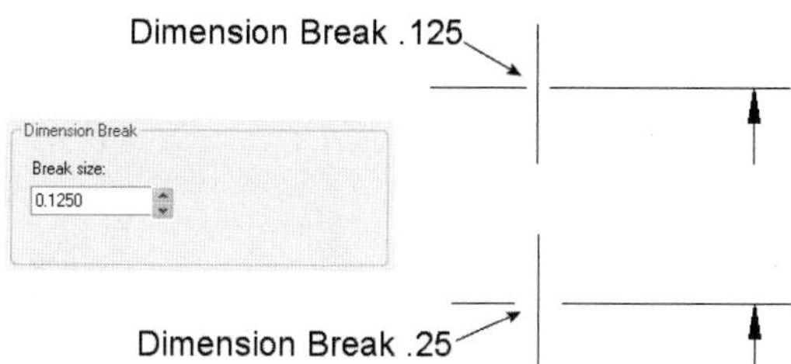

Figure 12.18

CONTROLLING DIMENSION ARC LENGTH SYMBOL

When placing an arc length dimension, you can control where the arc symbol is placed. In the following image, notice the three settings, namely, the arc symbol preceding the dimension text, the arc symbol placed above the dimension text, or no arc symbol displayed.

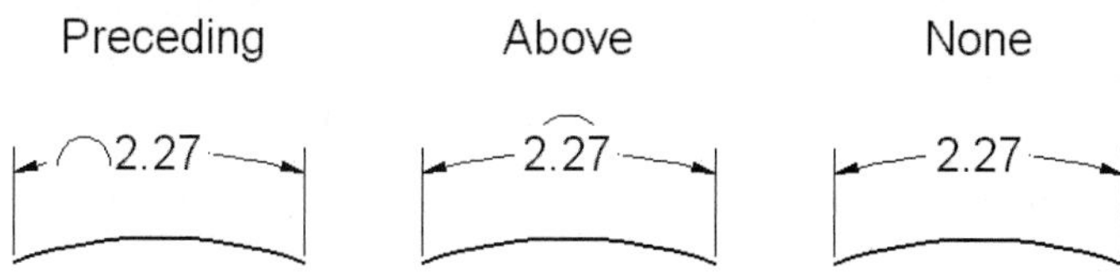

Figure 12.19

CONTROLLING THE RADIUS DIMENSION JOG ANGLE

You can control the angle of a jog dimension, as shown in the following image. By default, a 45° angle defines the jog, as shown on the left. Notice the effects of entering a 30° or 70° angle in the other examples in this image.

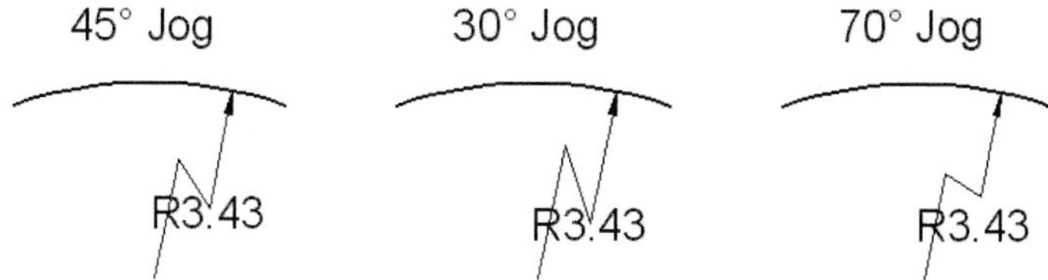

Figure 12.20

CONTROLLING THE LINEAR JOG HEIGHT FACTOR

This area is used to set the jog based on a multiplication factor. All jogs take this factor and multiply it by the current dimension text height. The following image compares the default 1.50 jog height factor with that of a 3.00 jog height factor. Notice that the lower jog is larger than the upper jog.

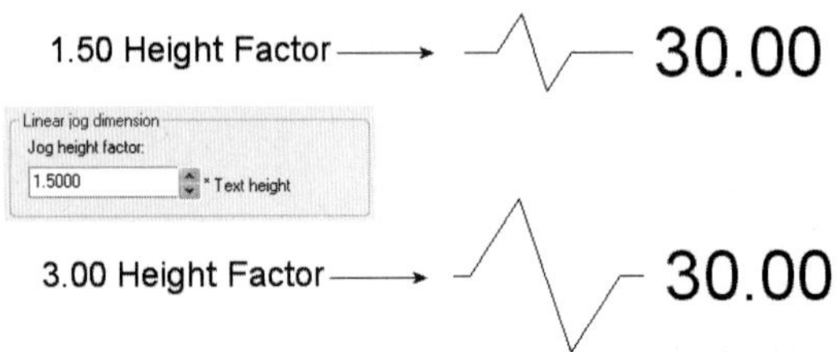

Figure 12.21

THE TEXT TAB

Use the Text tab shown in the following image to change the text appearance, for example, height; the text placement, for example, centered vertically and horizontally; and the text alignment, for example, always horizontal or parallel with the dimension line.

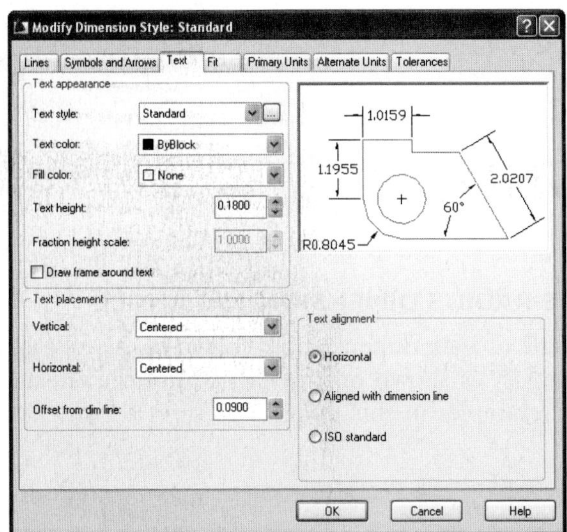

Figure 12.22

TEXT APPEARANCE

The Text height setting controls the size of the dimension text, as shown on the left in the following image. By default, a value of 0.18 is assigned to this setting.

Placing a check in the Draw frame around text box draws a rectangular box around all dimensions, as shown on the right in the following image. This can be done to emphasize the dimension text. It also is a way of calling out a basic dimension.

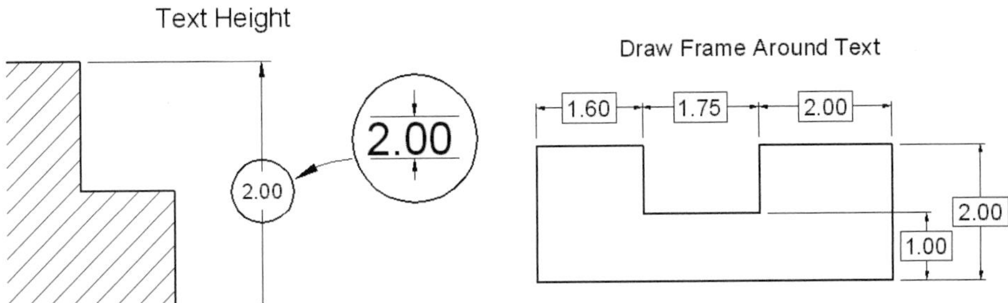

Figure 12.23

If the primary dimension units are set to architectural or fractional, the Fraction height scale activates. Changing this value affects the height of fractions that appear in the dimension. In the following image, a fractional height scale of 0.5000 will make the fractions as tall as the primary dimension number in the preview window.

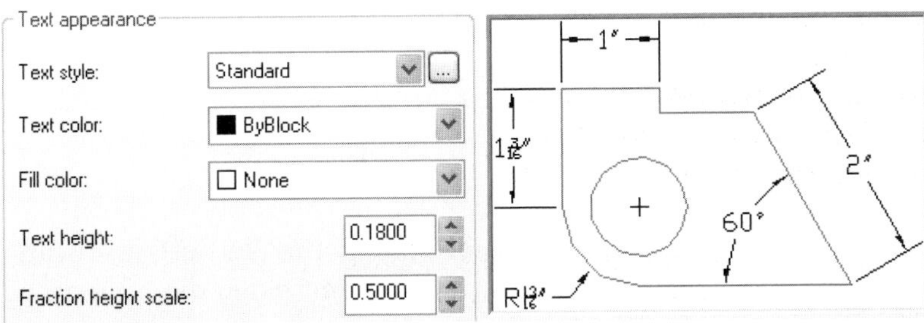

Figure 12.24

TEXT PLACEMENT

The Text placement area, shown in the following image, allows you to control the vertical and horizontal placement of dimension text. You can also set an offset distance from the dimension line for the dimension text.

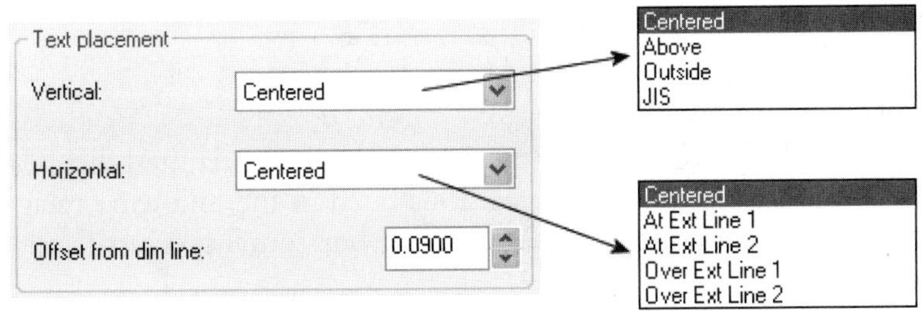

Figure 12.25

VERTICAL TEXT PLACEMENT

The Vertical area of Text placement controls the vertical justification of dimension text. Clicking in the drop-down field, as shown on the left in the following image, allows you to set vertical justification modes. By default, dimension text is centered vertically in the dimension line. Other modes include justifying vertically above the dimension line, justifying vertically outside the dimension line, and using the JIS (Japan International Standard) for placing text vertically.

Illustrated in the middle in the following image is the result of setting the Vertical justification to Centered. The dimension line will automatically be broken to accept the dimension text.

Illustrated on the right in the following image is the result of setting the Vertical justification to Above. Here the text is placed directly above a continuous dimension line. This mode is very popular for architectural applications.

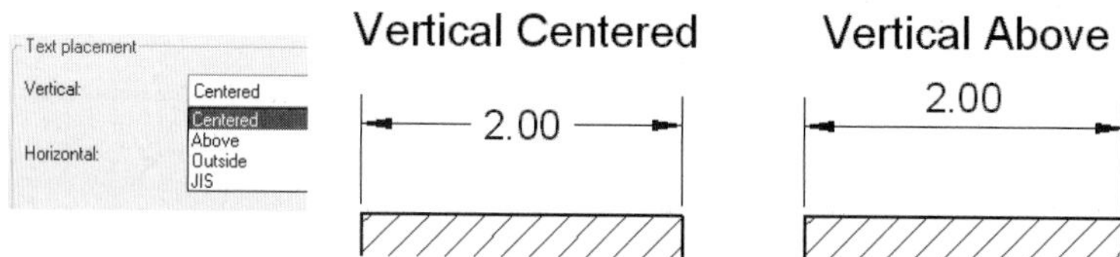

Figure 12.26

The following image illustrates the result of setting the Vertical justification mode to outside. All text, including that contained in angular and radial dimensions, will be placed outside the dimension lines and leaders.

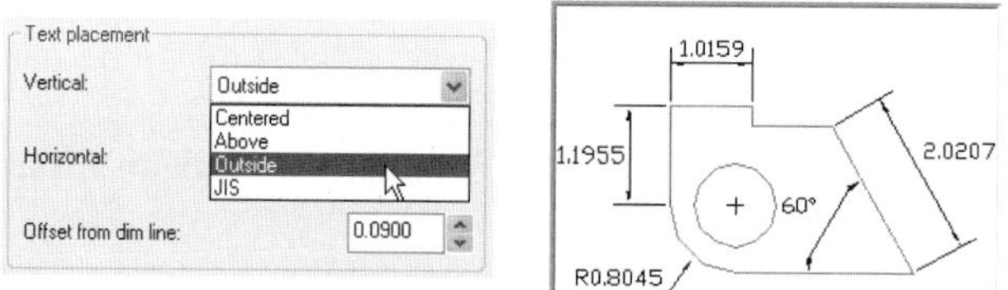

Figure 12.27

HORIZONTAL TEXT PLACEMENT

At times, dimension text needs to be better located in the horizontal direction; this is the purpose of the Horizontal justification area. Illustrated in the following image are the five modes of justifying text horizontally. By default, the horizontal text justification is centered in the dimension.

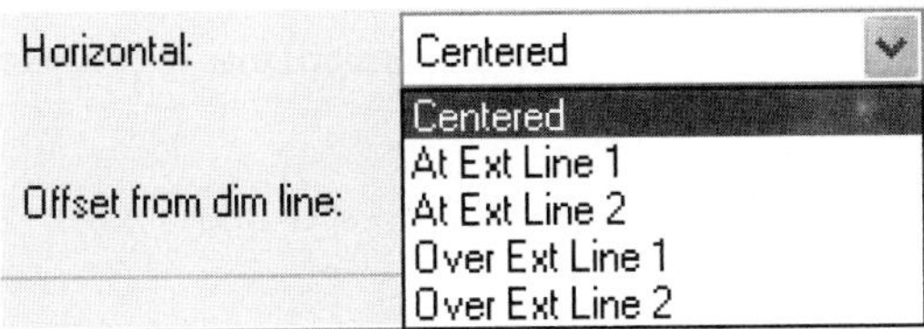

Figure 12.28

Clicking the Centered option of the Horizontal justification area displays the dimension text, as shown on the left in the following image. This is the default setting because it is the most commonly used text justification mode in dimensioning.

Clicking the At Ext Line 1 option displays the dimension text as shown in the middle in the following image, where the dimension text slides close to the first extension line. Use this option to position the text out of the way of other dimensions. Notice the corresponding option to have the text positioned nearer to the second extension line.

Clicking the Over Ext Line 1 option displays the dimension text parallel to and over the first extension line, as shown on the right in the following image. Notice the corresponding option to position dimension text over the second extension line.

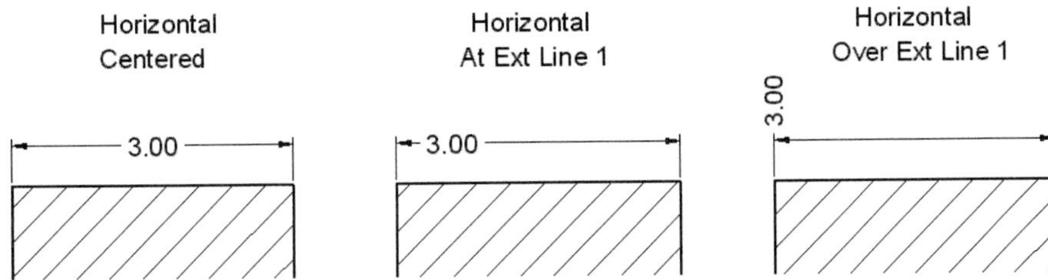

Figure 12.29

The last item of the Text placement area deals with setting an offset distance from the dimension line. When you place dimensions, a gap is established between the inside ends of the dimension lines and the dimension text. Entering different values depending on the desired results can control this gap. Study the examples in the middle and on the right in the following image, which have different text offset settings. Entering a value of zero (0) forces the dimension lines to touch the edge of the dimension text. Negative values are not supported.

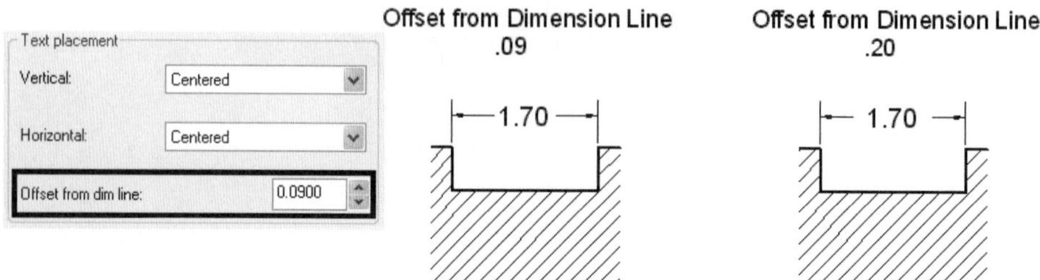

Figure 12.30

TEXT ALIGNMENT

Use the Text alignment area shown on the left in the following image to control the alignment of text. Dimension text can be placed either horizontally or parallel (aligned) to the edge of the object being dimensioned. An ISO (International Standards Organization) standard is also available for metric drawings. Click in the appropriate radio button to turn the desired text alignment mode on.

If the Horizontal radio button is clicked, all text will be read horizontally, as shown in the middle in the following image. This includes text located inside and outside the extension lines.

Clicking the Aligned with dimension line radio button displays the alignment results shown on the right in the following image. Here all text is read parallel to the edge being dimensioned. Not only will vertical dimensions align the text vertically, but the 2.06 dimension used to dimension the incline is also parallel to the edge.

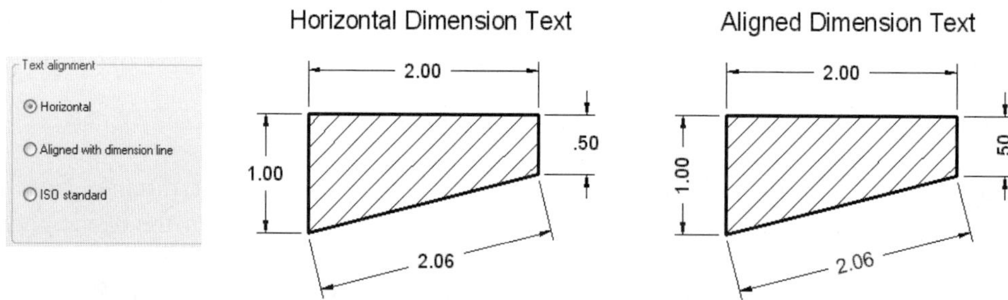

Figure 12.31

THE FIT TAB

Use the Fit tab in the following image to control how text and/or arrows are displayed if there isn't enough room to place both inside the extension lines. You could place the text over the dimension with or without a leader line. The Scale for dimension features area is very important when you dimension in Model Space or when you scale dimensions to Paper Space units. You can even fine-tune the placement of text manually and force the dimension line to be drawn between extension lines.

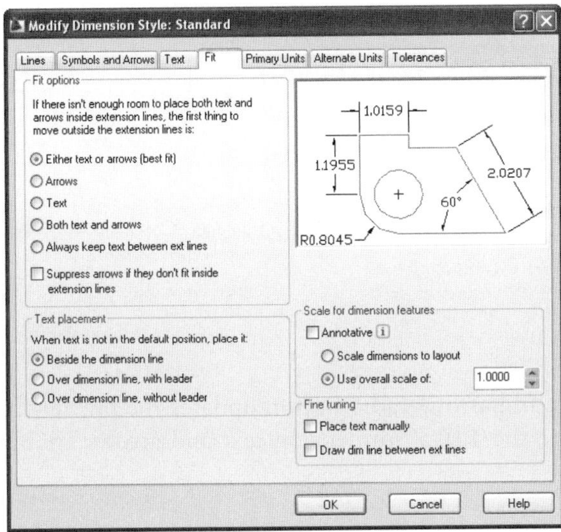

Figure 12.32

FIT OPTIONS

The Fit options area, shown on the left in the following image, has the radio button set for Either text or arrows (best fit). AutoCAD will decide to move either text or arrows outside extension lines. It will determine the item that fits the best. This setting is illustrated in the preview area on the right in the following image. It so happens that the preview image is identical when you click the radio button for Arrows. This tells AutoCAD to move the arrowheads outside the extension lines if there isn't enough room to fit both text and arrows.

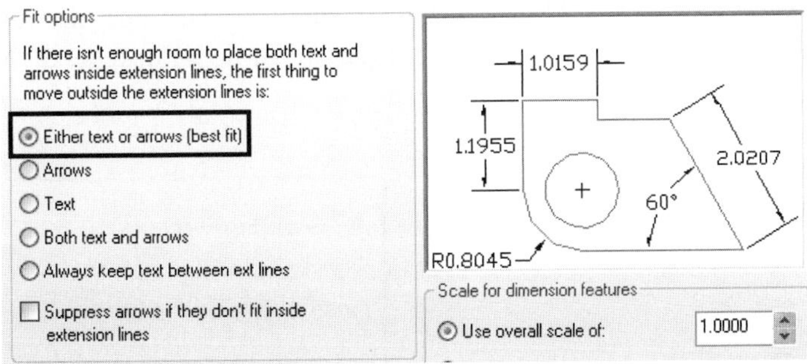

Figure 12.33

Clicking the Text radio button of the Fit options area updates the preview image, as shown in the following image. Here you are moving the dimension text outside the extension lines if the text and arrows do not fit. Since the value 1.0159 is the only dimension that does not fit, it is placed outside the extension lines, but the arrows are drawn inside the extension lines.

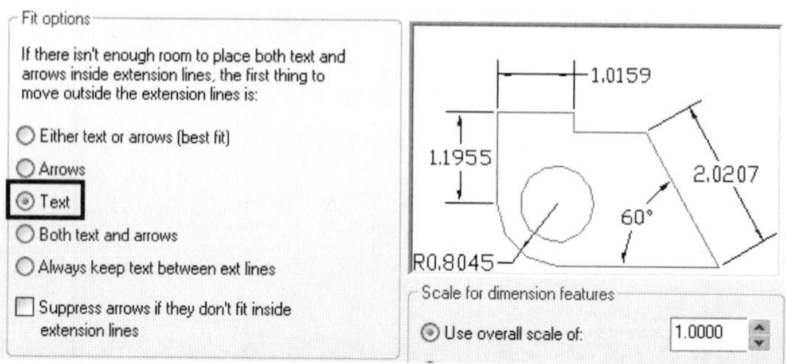

Figure 12.34

Clicking the Both text and arrows radio button updates the preview image, as shown in the following image, where the 1.0159 dimension text and arrows are both placed outside the extension lines.

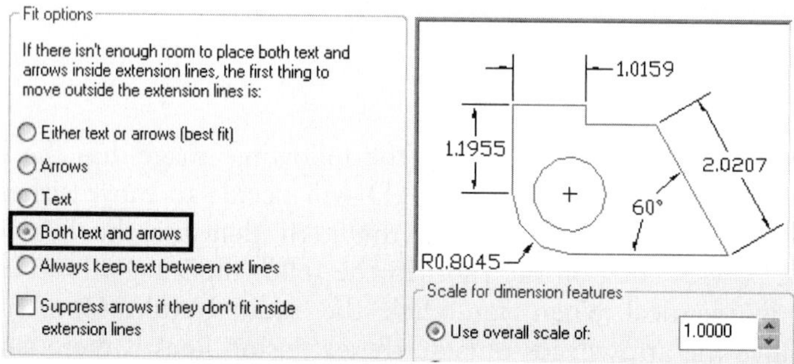

Figure 12.35

If you click the radio button for Always keep text between ext lines, the result is illustrated in the preview image in the following image. Here all dimension text, including the radius dimension, is placed between the extension lines.

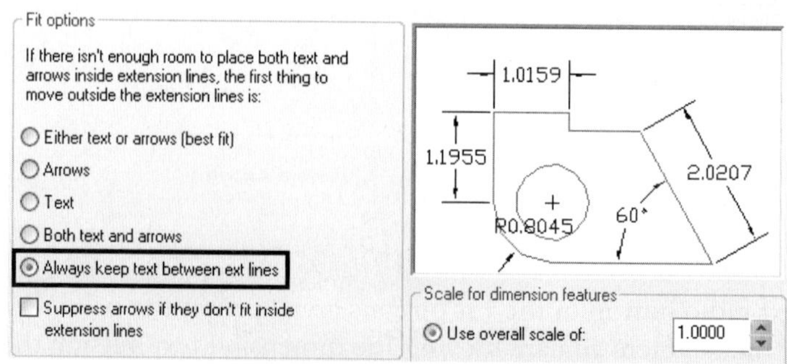

Figure 12.36

If you click the radio button for Either the text or the arrows (best fit) and you also place a check in the box for Suppress arrows if they don't fit inside the extension lines, the dimension line is turned off only for dimensions that cannot fit the dimension text and arrows. This is illustrated in the following image.

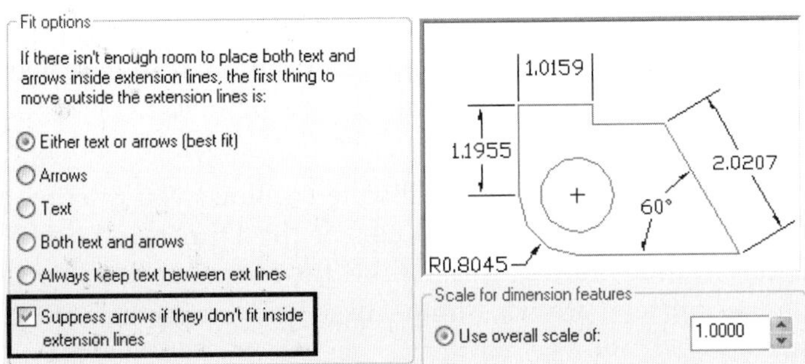

Figure 12.37

TEXT PLACEMENT

You control the placement of the text if it is not in the default position. Your choices, shown in the following image, are Beside the dimension line, which is the default, Over dimension line, with leader, or Over dimension line, without leader.

If you click the radio button for Text in the Fit options area and you click the radio button for Over dimension line, with leader, you get the result is illustrated in the middle in the following image. For the 1.0159 dimension that does not fit, the text is placed outside the dimension line with the text connected to the dimension line with a leader.

If you click the radio button for Text in the Fit options area and you click the radio button for Over dimension line, without leader, the result is illustrated on the right in the following image. For the 1.0159 dimension that does not fit, the text is placed outside the dimension. No leader is used.

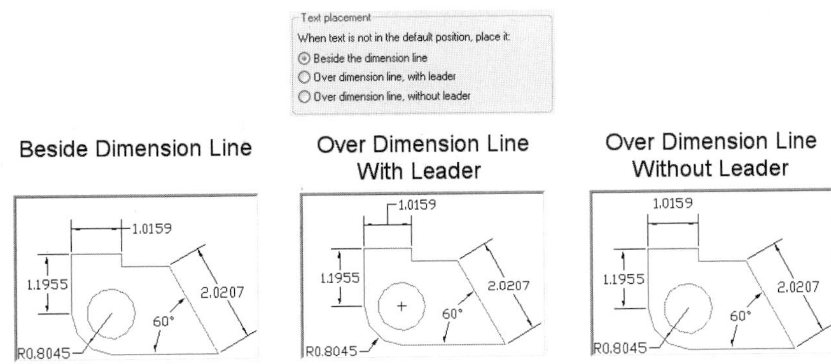

Figure 12.38

SCALE FOR DIMENSION FEATURES

This area allows you to set values that globally affect all current dimension settings that are specified by sizes or distances. The first setting, Annotative, can either be turned on or off. Annotative dimension styles create dimensions in which all the elements of the dimension, such as text, spacing, and arrows, scale uniformly by the annotation scale. This concept will be discussed in greater detail in chapter 19. Clicking the radio button next to Scale dimensions to layout allows you to have dimensions automatically scaled to Paper Space units inside a layout, as shown on the left in the following image. The scale of the viewport will control the scale of the dimensions.

Clicking the radio button next to Use overall scale of allows you to enter a multiplier that will affect all other values set in the various tabs of the Dimension Style Manager dialog box. The illustrations in the following image show the effects of overall scale factors of 1.00 and 2.00. The dimension text, arrows, origin offset, and extension beyond the arrow have all doubled in size.

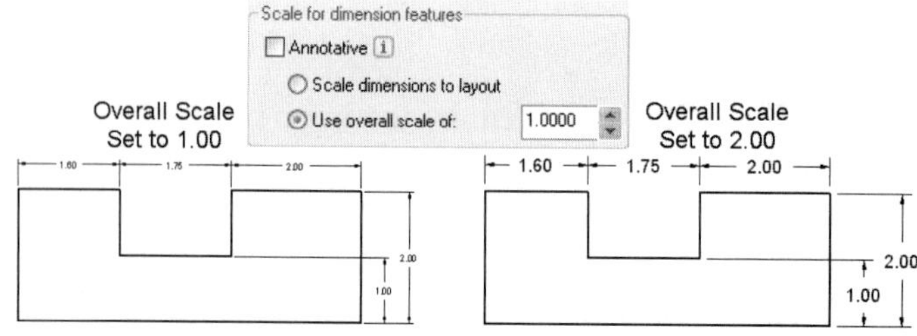

Figure 12.39

Try It!: Open the drawing file 12_Dimscale. A simple floor plan is displayed on the left in the following image. This floor plan is designed to be plotted at a scale of 1/2" = 1'0". Also displayed in this floor plan are dimensions in the magenta color. However, the dimensions are too small to be viewed. Set the Scale for dimension features found under the Fit tab of the Dimension Style dialog box to 24 (1' (12") divided by 1/2 is 24). The dimension values and tick marks should now be visible, as shown on the right in the following image.

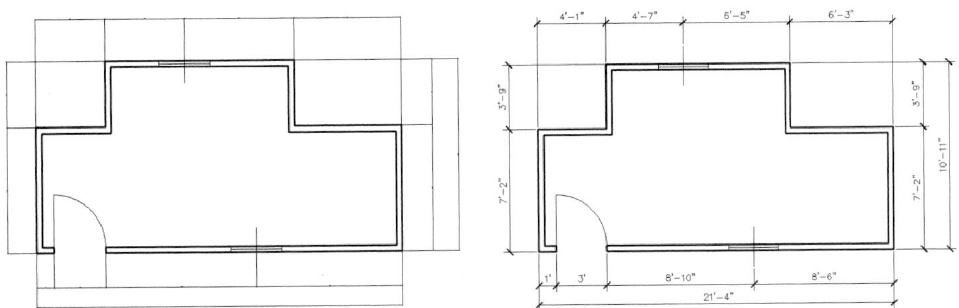

Figure 12.40

FINE-TUNING

You have two options to add further control of the fitting of dimension text in the Fine tuning area, as shown on the left in the following image. You can have total control for horizontally justifying dimension text if you place a check in the box for Place text manually when dimensioning. You can also force the dimension line to be drawn between extension lines by placing a check in this box. The results are displayed on the right in the following image.

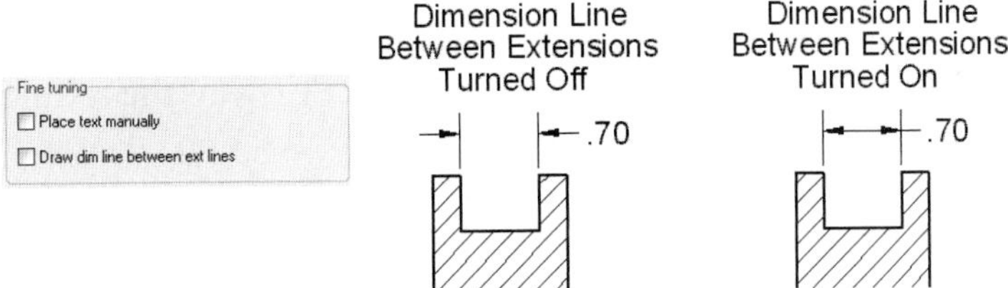

Figure 12.41

THE PRIMARY UNITS TAB

Use the Primary Units tab, shown in the following image, to control settings affecting the primary units. This includes the type of units the dimensions will be constructed in (decimal, engineering, architectural, and so on) and whether the dimension text requires a prefix or suffix.

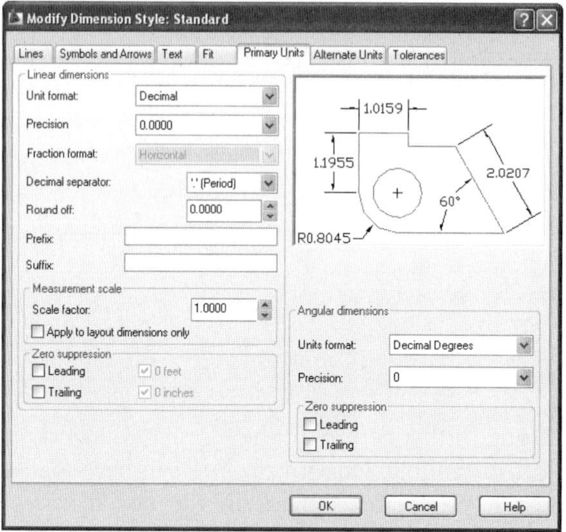

Figure 12.42

LINEAR AND ANGULAR DIMENSION UNITS

The Linear dimensions area of the Primary Units tab, shown in the previous image, has various settings that deal with primary dimension units. A few of the settings deal with the format when working with fractions. This area activates only if you are working in architectural or fractional units. Even though you may be drawing in architectural units, the dimension units are set by default to decimal. You can also designate the decimal separator as a Period, Comma, or Space.

Clicking in the box for Unit format, as shown on the left in the following image, displays the types of units you can apply to dimensions. You also control the precision of the primary units by clicking in the Precision box.

In a similar way, clicking in the box for Units format in the Angular dimensions category, as shown on the right in the following image, displays the various formats angles can be displayed in. The precision of these angle units can also be controlled by clicking in the Precision box.

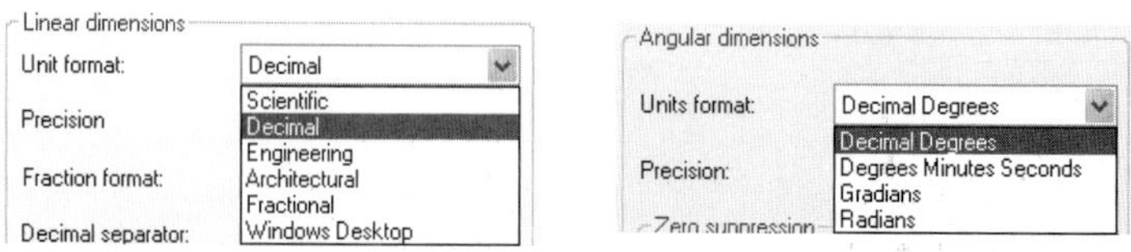

Figure 12.43

ROUNDING OFF DIMENSION VALUES

Use a Round off value to round off all dimension distances to the nearest unit based on the round off value. With a round off value of 0, the dimension text reflects the actual distance being dimensioned, as shown on the left in the following image. With a round off value set to .25, the dimension text reflects the next .25 increment, namely 2.50, as shown on the right in the following image.

Figure 12.44

APPLYING A DIMENSION PREFIX OR SUFFIX

A prefix allows you to specify text that will be placed in front of the dimension text whenever you place a dimension. Use the Suffix box to control the placement of a character string immediately after the dimension value. Both Prefix and Suffix control boxes are illustrated on the left in the following image. In the illustration on the right, the suffix "mm" was added to the dimension, signifying millimeters. Other common suffix letters include "M" for meters, "MI" for miles, and "in" for inches, to name just a few.

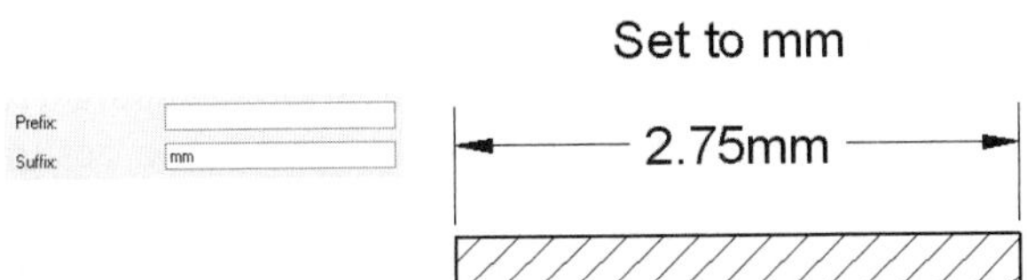

Figure 12.45

MEASUREMENT SCALE

The Measurement scale area, shown in the following image, acts as a multiplier for all linear dimension distances, including radius and diameter dimensions. When a dimension distance is calculated, the current value set in the Scale factor field is multiplied by the dimension to arrive at a new dimension value.

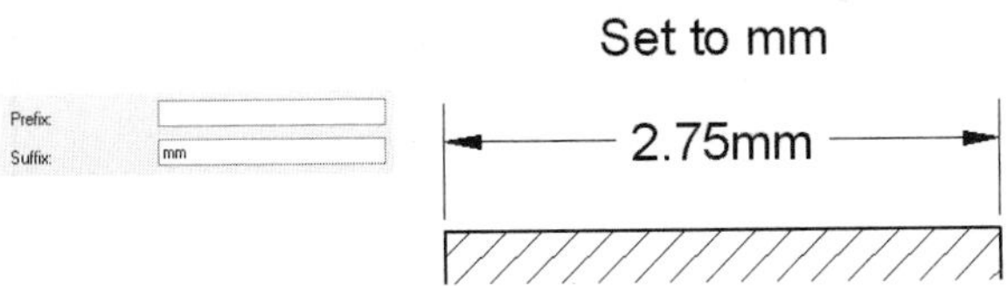

Figure 12.46

Illustrated on the left in the following image, and with a linear scale value of 1.00, the dimension distances are taken at their default values.

Illustrated on the right in the following image, the linear scale value has been changed to 2.00 units. This means 2.00 will multiply every dimension distance; the result is that the previous 3.00 and 2.00 dimensions are changed to 6.00 and 4.00, respectively. In a similar fashion, having a linear scale value set to 0.50 will reduce all dimension values to half their original values.

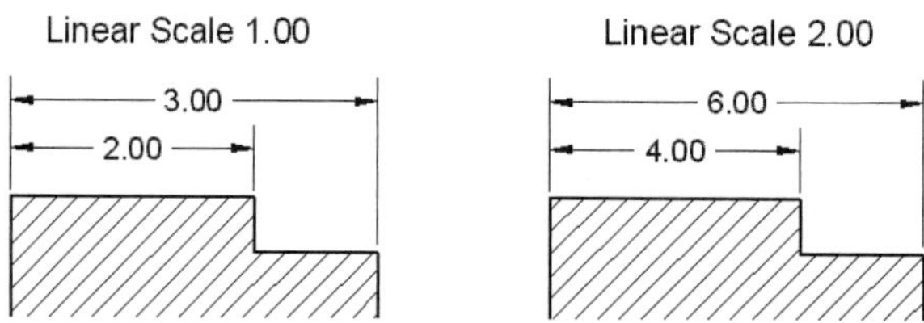

Figure 12.47

THE ALTERNATE UNITS TAB

Use the Alternate Units tab, shown in the following image, to enable alternate units, set the units and precision of the alternate units, set a multiplier for all units, use a round-off distance, and set a prefix and suffix for these units. You also have two placement modes for displaying these units. By default, alternate units are placed beside primary units. The alternate units are enclosed in square brackets. With two sets of units being displayed, your drawing could tend to become very busy. An application of using Alternate Units would be to display English and metric dimension values, since some design firms require both.

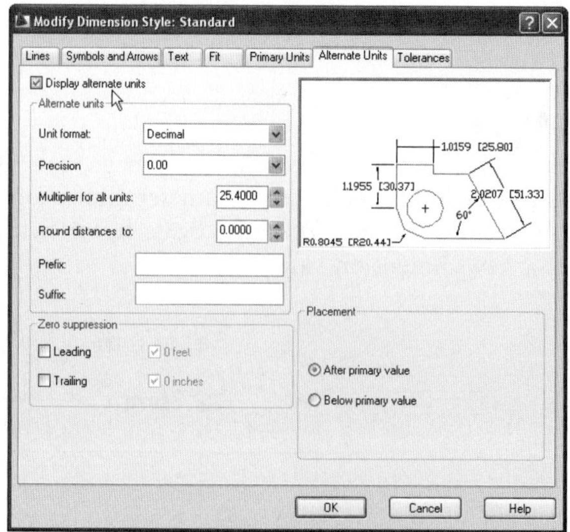

Figure 12.48

ALTERNATE UNITS

Once alternate units are enabled, all items in the Alternate Units area become active, as shown on the left in the following image. By default, the alternate unit value is placed in brackets alongside the calculated dimension value, as shown on the right in the following image. This value depends on the current setting in the Multiplier for all units field. This factor, set to 25.40 in the following image, is used as a multiplier for all calculated alternate dimension values.

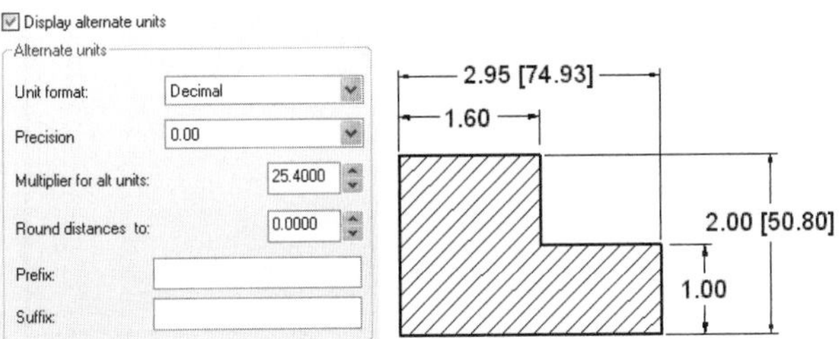

Figure 12.49

You could also click the radio button next to Below primary value, as shown on the left in the following image. This places the primary dimension above the dimension line and the alternate dimension below it, as shown on the right in the following image.

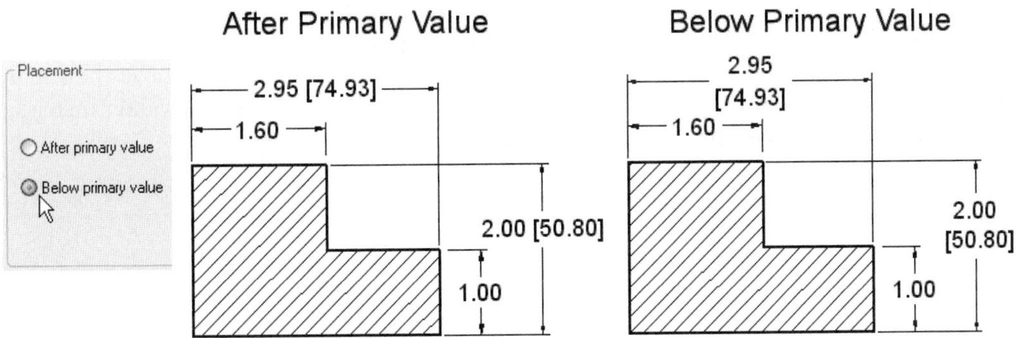

Figure 12.50

THE TOLERANCES TAB

The Tolerances tab shown in the following image consists of various fields used to control the five types of tolerance settings: None, Symmetrical, Deviation, Limits, and Basic. Depending on the type of tolerance being constructed, an Upper value and Lower value may be set to call out the current tolerance variance. The Vertical position setting allows you to determine where the tolerance will be drawn in relation to the location of the body text. The Scaling for height setting controls the text size of the tolerance. Usually this value is smaller than the body text size.

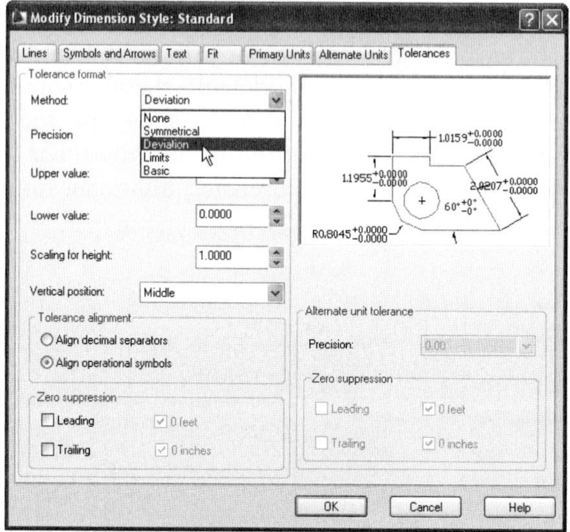

Figure 12.51

TOLERANCE FORMAT

The five tolerance types available in the drop-down list are illustrated in the following image. A tolerance setting of None uses the calculated dimension value without applying any tolerances. The Symmetrical tolerance uses the same value set in the Upper and Lower value. The Deviation tolerance setting will have a value set in the Upper value and an entirely different value set in the Lower value. The Limits tolerance will use the Upper and Lower values and place the results with the larger limit dimension placed above the smaller limit dimension. The Basic tolerance setting does not add any tolerance value; instead, a box is drawn around the dimension value.

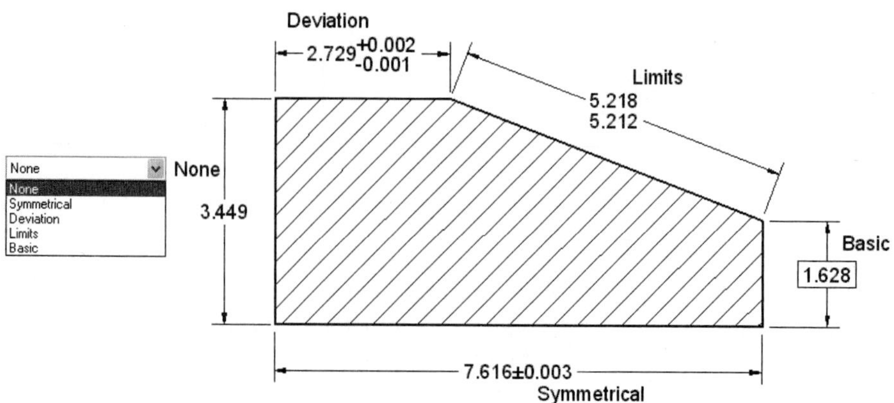

Figure 12.52

CONTROLLING THE ASSOCIATIVITY OF DIMENSIONS

THE DIMASSOC SYSTEM VARIABLE

The associativity of dimensions is controlled by the DIMASSOC variable. By default, this value is set to 2 for new drawings. This means that the dimension is associated with the object being dimensioned. Associativity means that if the object changes size or if some element of an object changes location, the dimension associated with the object will change as well.

When this variable is set to 1, the dimension is called nonassociative. Dimensions of this type exist in older AutoCAD drawings. This dimension is not associated with the object being dimensioned; however, it is possible to have the dimension value automatically updated using conventional AutoCAD editing commands.

The DIMASSOC variable can also be set to 0, which will create exploded dimensions. There is no association between the various elements of the dimension. All dimension lines, extension lines, arrowheads, and dimension text are drawn as separate objects. This means that grips will not have any effect on this dimension. You cannot stretch an object and have this dimension update to the object's new length. You cannot make a change in the Dimension Styles Manager dialog box and have this dimension affected. In other words, never set this variable to 0. The same effects can be achieved by using the EXPLODE command on a dimension; this is not recommended and is poor practice.

Try It!: Open the drawing 12_Dimassoc2. The object on the left in the following image has dimensions placed with the DIMASSOC variable set to 2. Another polyline shape on the right has various dimensions placed on its outside. Click the polyline on the left to activate its grips. Click the grip at "A" and stretch the polyline vertex up and to the left. Notice that all dimension locations, orientations, and values are automatically updated. Click the polyline vertex at "B" and stretch this vertex up and to the right. The same results occur, with the dimension elements being updated based on the new location of the object. Use grips to realign any dimension that is not updated to a new location.

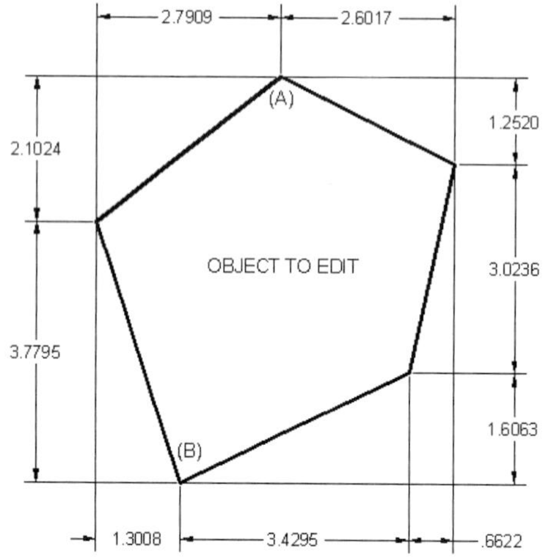

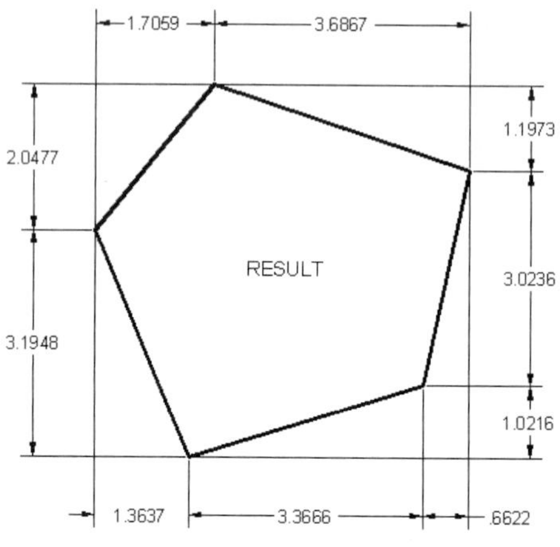

Figure 12.53

Tip: When you make changes to the DIMASSOC variable, these changes are stored in the drawing file.

REASSOCIATING DIMENSIONS

It is possible to change a nonassociative dimension into an associative dimension through the process called Reassociation. This command allows you to pick the nonassociative dimension and reestablish its endpoints with new endpoints located on the object being dimensioned. The following Try It! will demonstrate this capability.

Try It!: Open the drawing 12_Dim Reassoc. When the corner of the object at "B" stretches, the vertical 1.60 dimension does not readjust to the stretch operation. This is because the dimension may have been placed in a previous version of AutoCAD with DIMASSOC set to 1. Activate the DIMREASSOCIATE command from the Dimension pull-down menu, as shown on the left in the following image. Pick the 1.60 dimension as the object to reassociate. When the blue X appears, pick endpoints on the object to perform the reassociation, as shown in the middle of the following image.

```
Command: DIMREASSOCIATE
Select dimensions to reassociate ...
Select objects: (Pick the vertical 1.60 dimension)
Select objects: (Press ENTER to continue)
Specify first extension line origin or [Select object] <next>:
    (When the blue X appears at the end of the proper extension
    line, pick the endpoint of the object at "B")
Specify second extension line origin <next>: (The blue X will
    move to the opposite endpoint of the dimension. Pick the
    endpoint of the object at "C" in order to complete the
    dimension reassociation)
```

You can test to see whether this dimension is associative by selecting the object, picking either grip at "B" or "C," and stretching the object. The dimension should now update to the changes in the object's shape.

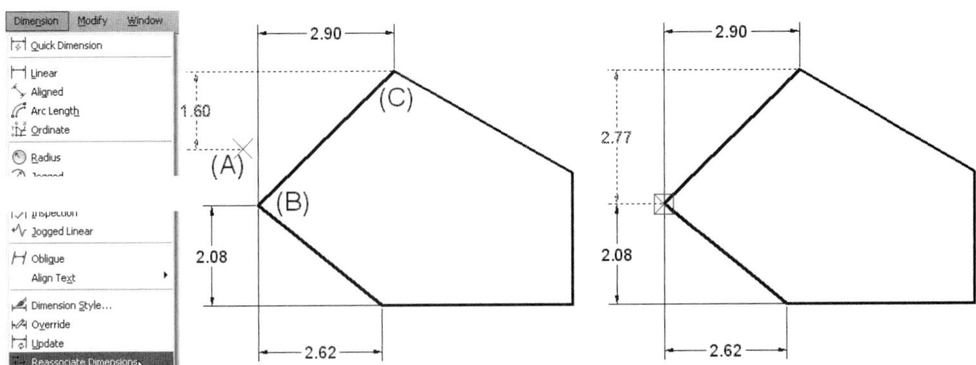

Figure 12.54

You will be able to identify associated dimensions by clicking the dimension while in the REASSOCIATE command. The familiar X is surrounded by a blue box, as shown on the right in the previous image. If you do not see the blue box, the dimension is not associative.

Reassociating works only on nonassociative dimensions that have their DIMASSOC value set to 1. The process of Reassociation does not have any effect on exploded dimensions (DIMASSOC set to 0).

While AutoCAD gives you this ability to reassociate dimensions, it would be very time-consuming to perform this task on hundreds of dimensions in a drawing. Reassociation is ideal when dealing with only a few dimensions.

USING DIMENSION TYPES IN DIMENSION STYLES

In addition to creating dimension styles, you can assign dimension types to dimension styles. The purpose of using dimension types is to reduce the number of dimension styles defined in a drawing. For example, the object in the following image consists of linear dimensions with three-decimal-place accuracy, a radius dimension with one-decimal-place accuracy, and an angle dimension with a box surrounding the number. Normally you would have to create three separate dimension styles to create this effect. However, the linear, radius, and angular dimensions consist of what are called dimension types.

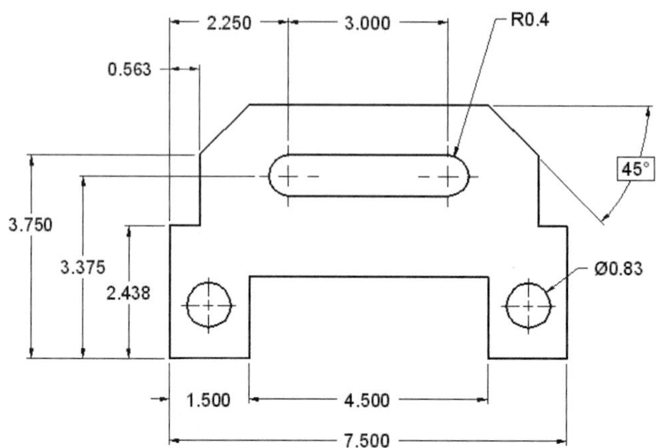

Figure 12.55

Dimension types appear similar to the following image when assigned to the Mechanical dimension style. Before creating an angular dimension, you create the dimension type and make changes in various tabs located in the Modify Dimension Style dialog box. These changes will apply only to the angular dimension type. To expose the dimension types, click the New button in the main Dimension Style Manager dialog box.

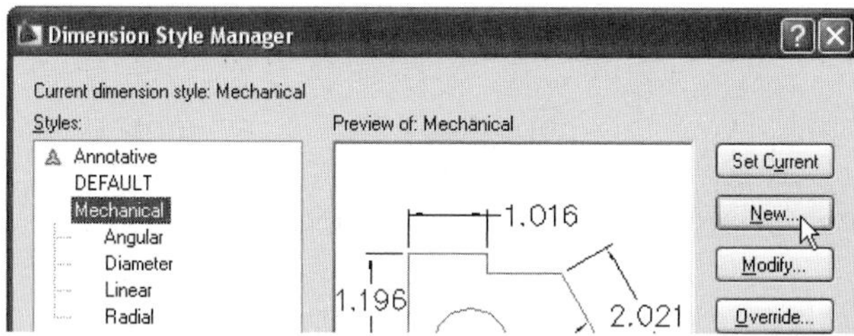

Figure 12.56

When the Create New Dimension Style dialog box appears, as shown in the following image, click in the Use for box. This box usually displays All dimensions. Notice all dimension types appearing. You can make changes to dimension settings that will apply only for linear, angular, radius, diameter, and ordinate dimensions. Also, a dimension type for Leaders and Tolerances is available. Highlight this option and then click the Continue button.

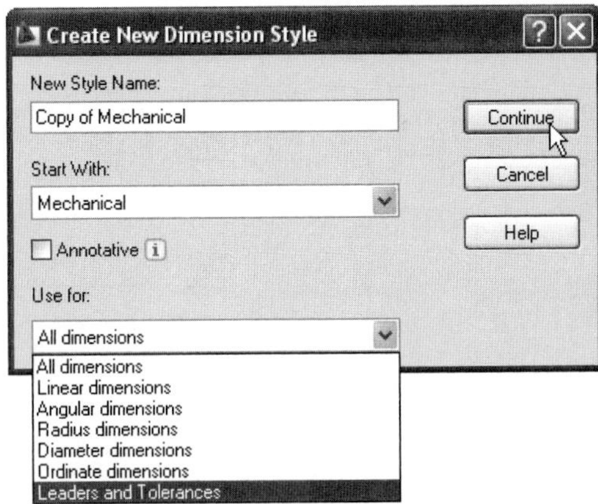

Figure 12.57

This takes you to the New Dimension Style: Mechanical: Leader dialog box, shown in the following image. Any changes you make in the tabs will apply only whenever you place a leader dimension. The use of dimension types is an efficient means of keeping the number of dimension styles down to a manageable level.

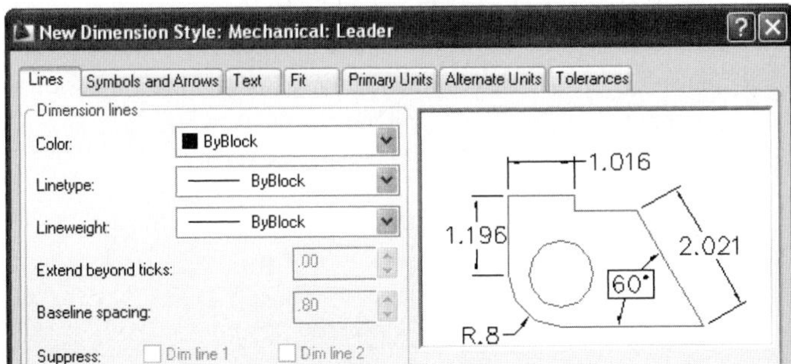

Figure 12.58

Try It!: Open the drawing file 12_Dimension Types. A series of Dimension Types have already been created. Use the illustration in the following image and add all dimensions to this object.

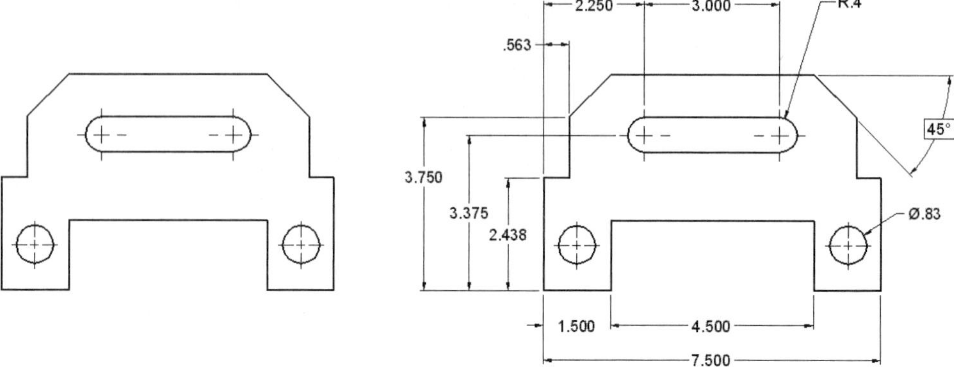

Figure 12.59

OVERRIDING A DIMENSION STYLE

Try It!: Open the drawing file 12_Dimension Override and activate the Dimension Style Manager dialog box. For special cases in which you need to change the settings of one dimension, a dimension override would be used. First launch the Dimension Style Manager dialog box. Clicking the Override button, as shown on the left in the following image, displays the Override Current Style: Mechanical dialog box. Under the Lines tab, check the Ext line 1 and Ext line 2 boxes in the Extension lines area for suppression, as shown on the right in the following image. One dimension needs to be constructed without displaying extension lines.

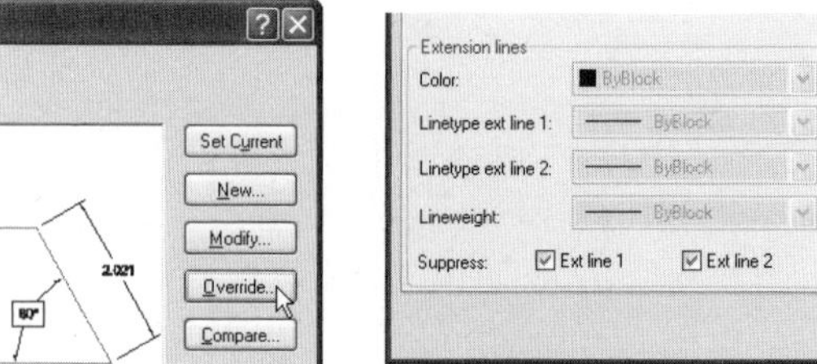

Figure 12.60

Clicking the OK button returns you to the main Dimension Style Manager dialog box, as shown in the following image. Notice that, under the Mechanical style, a new dimension type has been created called <style overrides> (Angular, Diameter, Linear, and Radial were already existing in this example). Also notice that the image in the Preview box shows sample dimensions displayed without extension lines. The <style overrides> dimension style is also the current style. Click the Close button to return to your drawing.

Note: If the Preview box does not show the correct sample image when you create the new <style overrides>, close and immediately reopen the Dimension Style Manager dialog box. The Preview box will show the correct sample image.

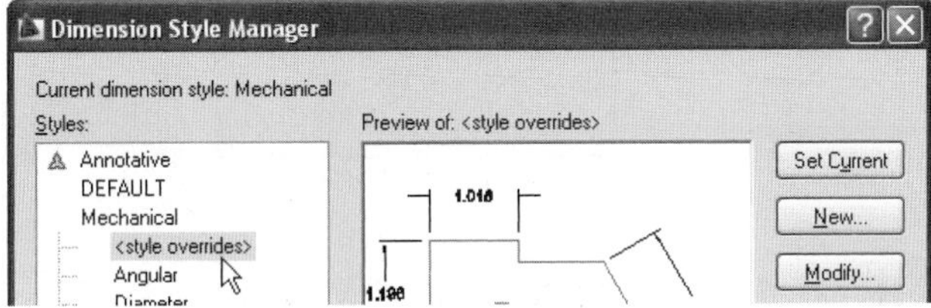

Figure 12.61

In the illustration on the right in the following image, a linear dimension that identifies the vertical distance of 1.250 is placed. To avoid the mistake of placing extension lines on top of existing object lines, the extension lines are not drawn due to the style override.

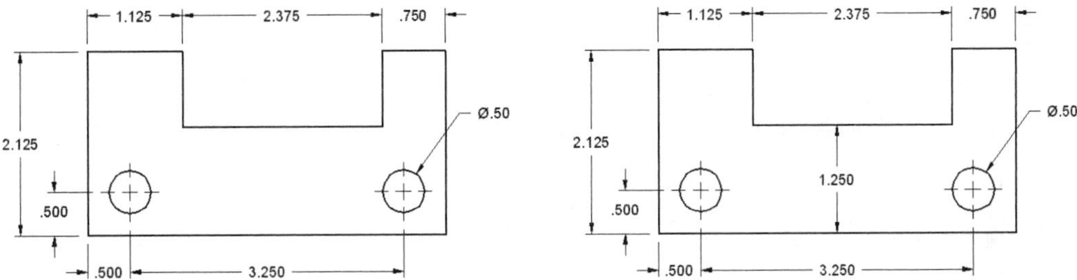

Figure 12.62

Unfortunately, if you place other linear dimensions, these will also lack extension lines. The Dimension Style Manager dialog box is once again activated. Clicking an existing dimension style such as Mechanical and then clicking the Set Current button displays the AutoCAD Alert box shown in the following image. If you click OK, the style override disappears from the listing of dimension types. If you would like to save the overrides under a name, click the Cancel button, right-click the <style overrides> listing, and rename this style to a new name. This preserves the settings under this new name.

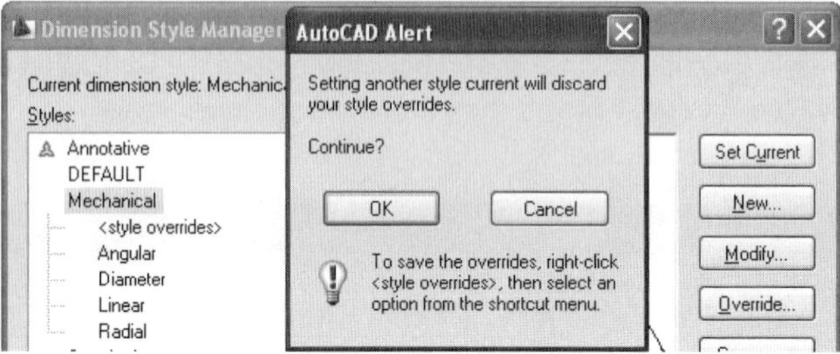

Figure 12.63

MODIFYING THE DIMENSION STYLE OF AN OBJECT

Existing dimensions in a drawing can easily be modified through a number of methods that will be outlined in this section. The following image illustrates the first of these methods: the use of the Dim Style Control box, which is part of the Dimension toolbar. Be sure to have this toolbar displayed on your screen before continuing.

Try It!: Open the drawing file 12_Dimension Edit. First click the dimension shown in the following image and notice that the dimension highlights and the grips appear. The current dimension style is also displayed in the Dim Style Control box. Click in the control box to display all other styles currently defined in the drawing. Click the style name TOLERANCE to change the highlighted dimension to that style. Press ESC to turn off the grips.

Tip: Be careful when docking this toolbar along the side of your screen. In lower screen resolutions, the Dim Style Control box will not display.

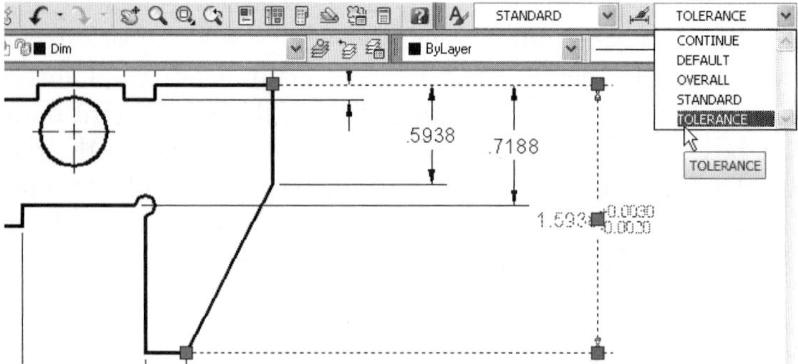

Figure 12.64

The second method of modifying dimensions is illustrated in the following image. In this example, select the dimension shown in this image; it highlights and grips appear. Right-clicking displays the cursor menu. Choosing the Dim Style heading displays the cascading menu of all dimension styles defined in the drawing. Click the TOLERANCE style to change the highlighted dimension to the new style.

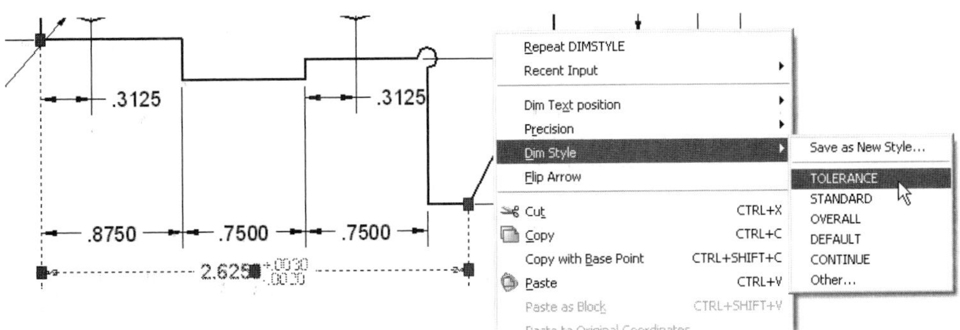

Figure 12.65

The third method of modifying dimensions begins with selecting the overall length dimension shown in the following image. Again, the dimension highlights and grips appear. Clicking the Properties button displays the Properties Window, shown in the following image. Click the Alphabetic tab to display all dimension settings and current values. If you click in the name box next to Dim style, all dimension styles will appear in the field. Click the OVERALL style to change the highlighted dimension to the new style. Close the Properties Palette and press the ESC key to turn off the grips.

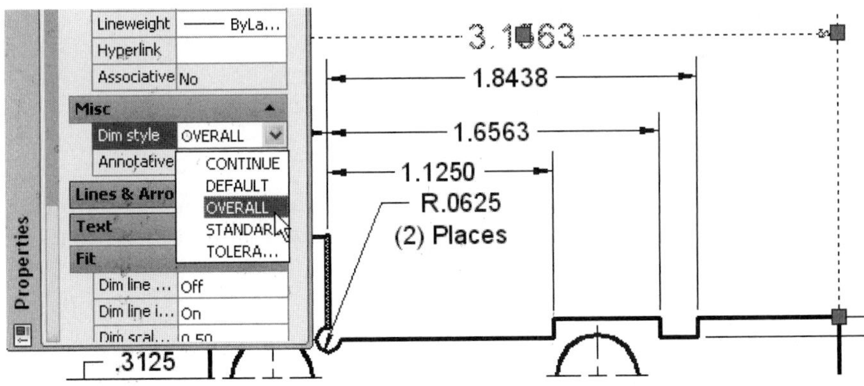

Figure 12.66

CREATING MULTILEADER STYLES

As with the Dimension Style Manager, you can create different styles of multileaders using the Multileader Style Manager. Choosing Multileader Style either from the toolbar or from the dashboard launches the Multileader Style Manager dialog box shown in the following image.

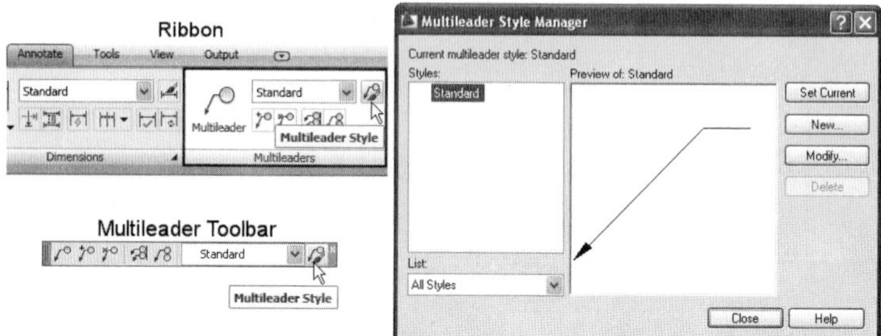

Figure 12.67

Clicking the New button launches the Create New Multileader Style dialog box, where you give a name for the new multiline style. In the following example, a new style called Circle Balloons will be created. This style will consist of a circle with text inside.

Figure 12.68

Clicking the Continue button takes you to three tabs that control the appearance of the multiline style. The first tab, Leader Format, controls the leader and includes type of leader (Straight, Spline, or None), the size of the arrowhead, and the leader break distance. The second tab, Leader Structure, controls the number of points that make up the leader, the length of the leader landing, and the scale factor of the leader. The last tab, Content, deals with the contents of the leader. By default, the Multileader type is made up of Mtext objects. Other options include Block or None. The other content includes basic text options and the type of leader connection.

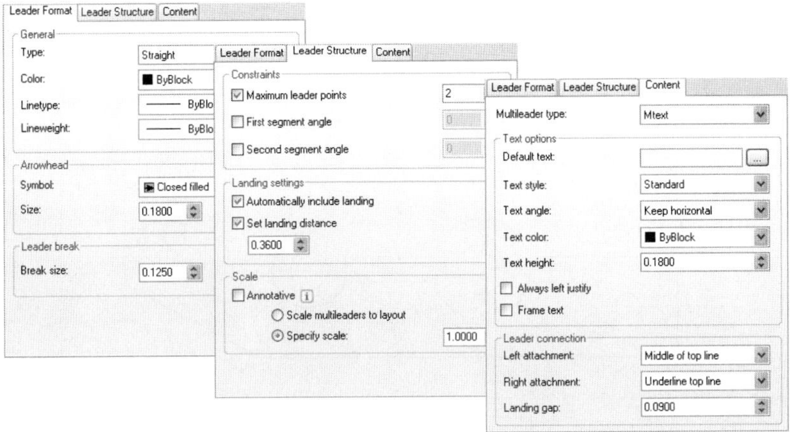

Figure 12.69

While in the Content tab, changing Mtext to Block activates block options, as shown in the following image. Changing the source block to circle updates the preview image of the multileader.

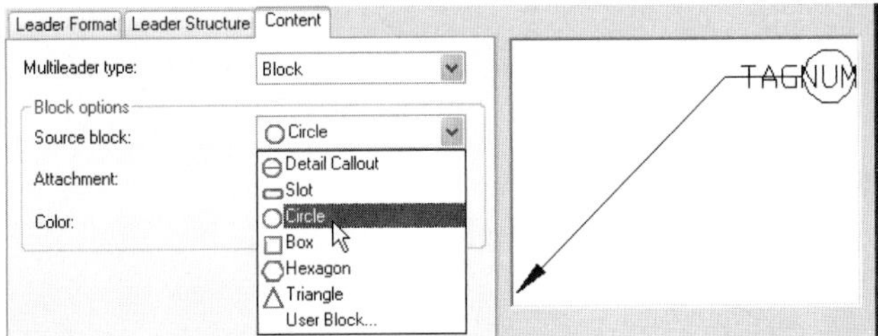

Figure 12.70

Returning back to the Multileader Style Manager dialog box allows you to make the new style, namely Circle Balloons, the current style, as shown on the left in the following image. The results of using this style in a drawing are shown on the right in the following image.

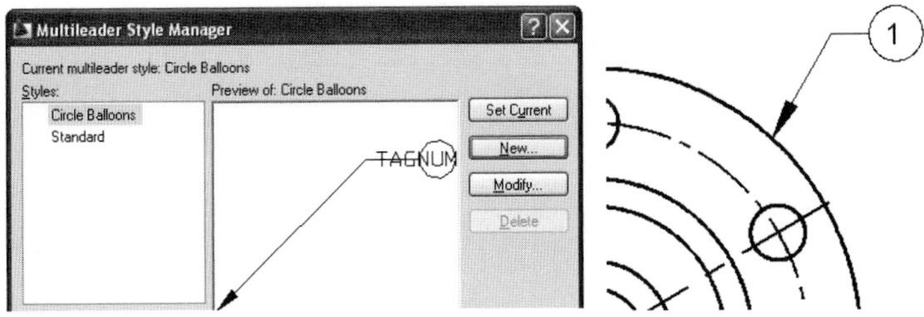

Figure 12.71

TUTORIAL EXERCISE: 12_DIMEX.DWG

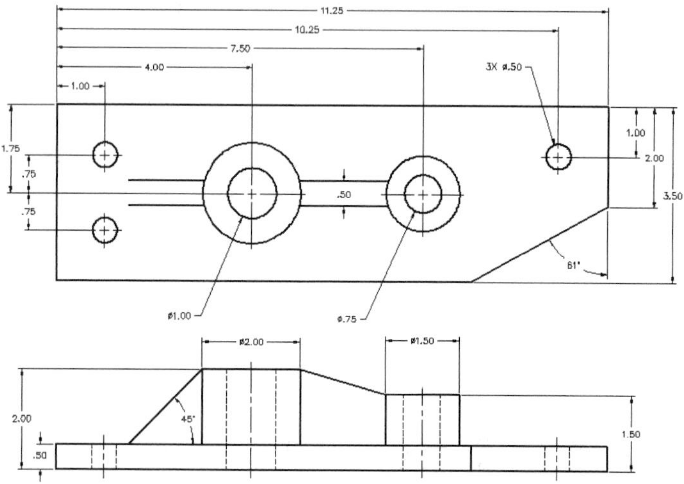

Figure 12.72

PURPOSE

The purpose of this tutorial is to place dimensions on the drawing of the two-view object illustrated in the previous image.

SYSTEM SETTINGS

No special system settings need to be made for this drawing file.

LAYERS

The drawing file 12_Dimex.Dwg has the following layers already created for this tutorial.

Name	Color	Linetype
Object	Magenta	Continuous
Hidden	Red	Hidden
Center	Yellow	Center
Dim	Yellow	Continuous

SUGGESTED COMMANDS

Open the drawing called 12_Dimex. The following dimension commands will be used: DIMLINEAR, DIMCONTINUE, DIMBASELINE, DIMCENTER, DIMRADIUS, DIMDIAMETER, DIMANGULAR, and LEADER. All dimension commands may be chosen from the Dimension toolbar or the Dimension pull-down menu, or entered from the keyboard. Use the ZOOM command to get a closer look at details and features that are being dimensioned.

Step 1

To prepare for the dimensioning of the drawing, type D to activate the Dimension Style Manager dialog box. Click the New button, which activates the Create New Dimension Style dialog box shown in the following image. In the New Style Name area, enter MECHANICAL. Click the Continue button to create the style, as shown in the following image.

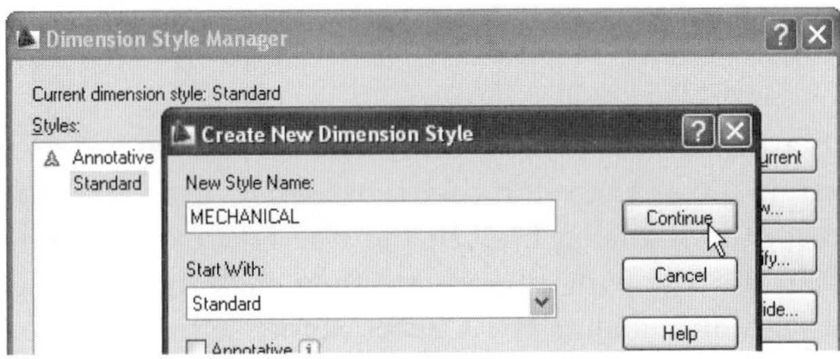

Figure 12.73

Step 2

When the New Dimension Style: MECHANICAL dialog box appears, make the following changes in the Lines tab, as shown on the left in the following image: Change the size of the Extend beyond dim lines from a value of 0.18 to a new value of 0.07. Switch to the Symbols and Arrows tab and change the Arrow size to from 0.18 to 0.12. Also change the Center marks to Line, as shown on the right in the following image.

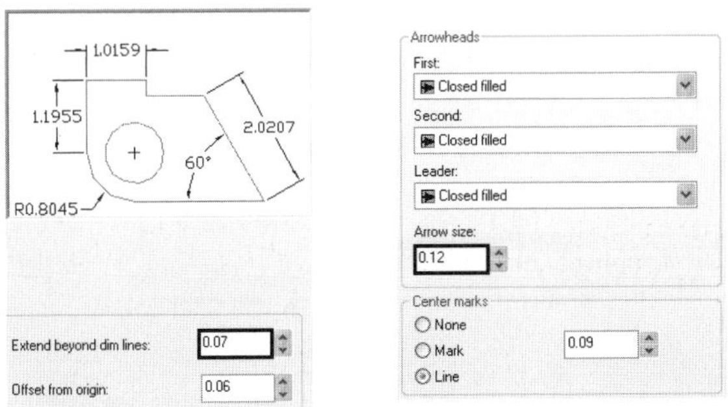

Figure 12.74

Step 3

Next, click the Text tab and change the Text height from a value of 0.18 to a new value of 0.12, as shown on the left in the following image. Then click the Primary Units tab and change the number of decimal places from 4 to 2. Also place a check in the box next to Leading in the Zero suppression area, as shown on the right in the following image. This turns off the leading zero for dimension values under 1 unit. When finished, click the OK button to return to the main Dimension Style Manager dialog box.

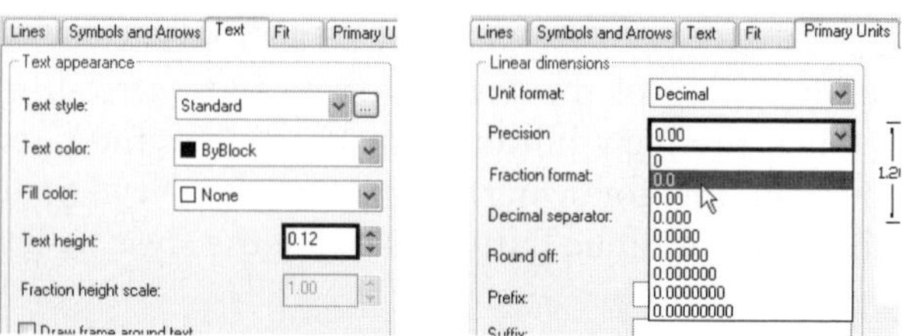

Figure 12.75

Step 4

You will now create a dimension type dealing with all diameter dimensions and make a change to the center mark settings for this type. Click the New button to display the Create New Dimension Style dialog box. In the Use for box, click Diameter dimensions, as shown on the left in the following image, and then click the Continue button. Click the Symbols and Arrows tab and change the Center marks setting to None, as shown on the right in the following image. This prevents center marks from being displayed when placing diameter dimensions. When finished, click the OK button to return to the main Dimension Styles Manager dialog box.

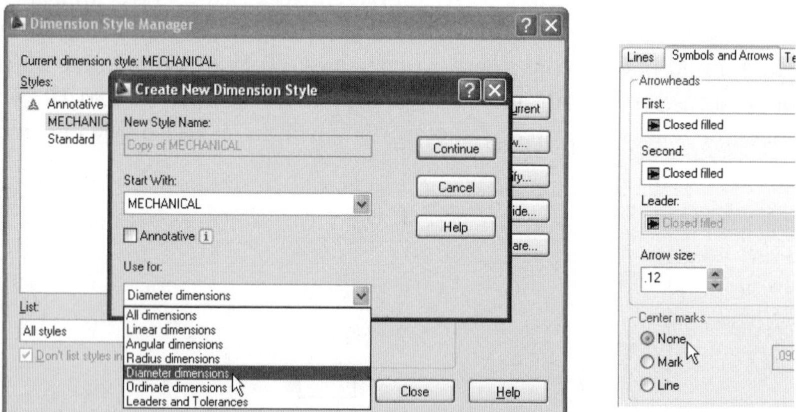

Figure 12.76

Step 5

In the Dimension Style Manager dialog box, click MECHANICAL in the Styles area. Then click the Set Current button to make MECHANICAL the current dimension style. Your display should appear similar to the following image. Click the Close button to save all changes and return to the drawing.

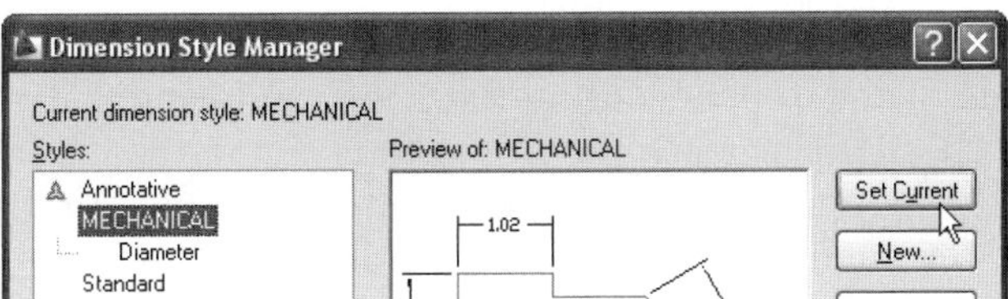

Figure 12.77

Step 6

Begin placing center markers to identify the centers of all circular features in the Top view. Use the DIMCENTER command (or the shortcut DCE) to perform this operation on circles "A" through "E," as shown in the following image.

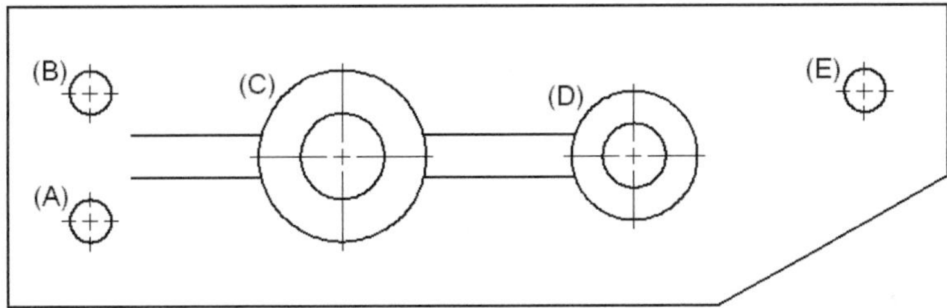

Figure 12.78

Step 7

Use the QDIM command to place a string of baseline dimensions. First select the individual lines labeled "A" through "F," as shown in the following image. When the group of dimensions previews as continued dimensions, change this grouping to baseline and click a location to place the baseline group of dimensions, as shown in the following image. Use grips to place the text as shown.

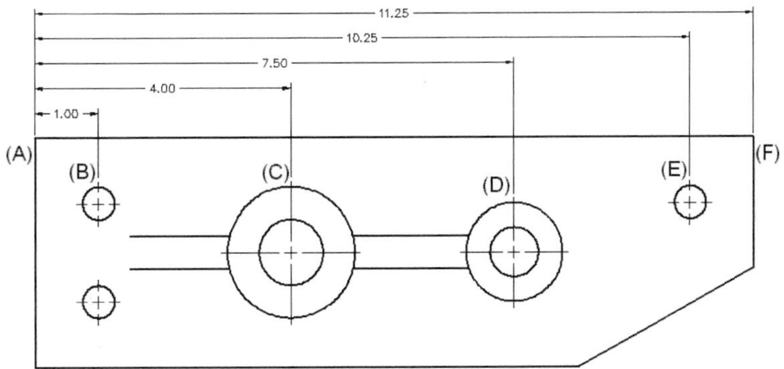

Figure 12.79

Step 8

Magnify the left side of the Top view using the ZOOM command. Then use the DIMLINEAR command (or DLI shortcut) to place the .75 vertical dimension, as shown on the left in the following image. Next, use the DIMCONTINUE command (or DCO shortcut) to place the next dimension in line with the previous dimension, as shown in the middle in the following image. Then use the DIMLINEAR command (or DLI shortcut) to place the 1.75 vertical dimension, as shown on the right in the following image. Use grips to place the text as shown.

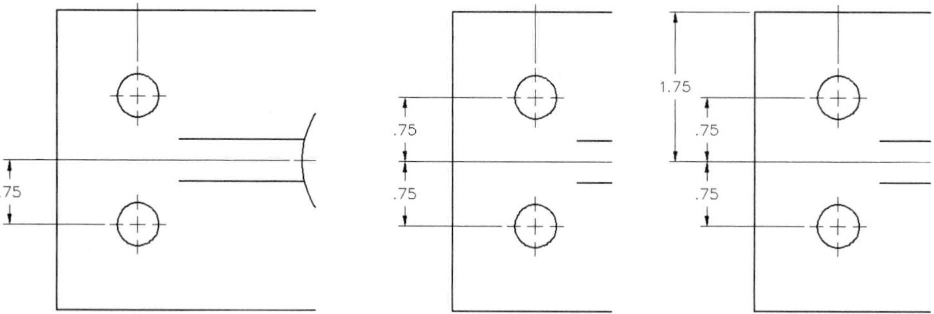

Figure 12.80

Step 9

Use the PAN command to slide over to the right side of the Top view while keeping the same zoom percentage. Then use the QDIM command and select the lines "A" through "C" in the following image. When the group of dimensions previews as continued dimensions, change this grouping to baseline and click a location to place the baseline group of dimensions, as shown in the following image. You may have to identify a new base point (datumPoint) in order for your dimension to match the illustration. Then use the DIMANGULAR command (or DAN shortcut) to place the 61° dimension in the following image.

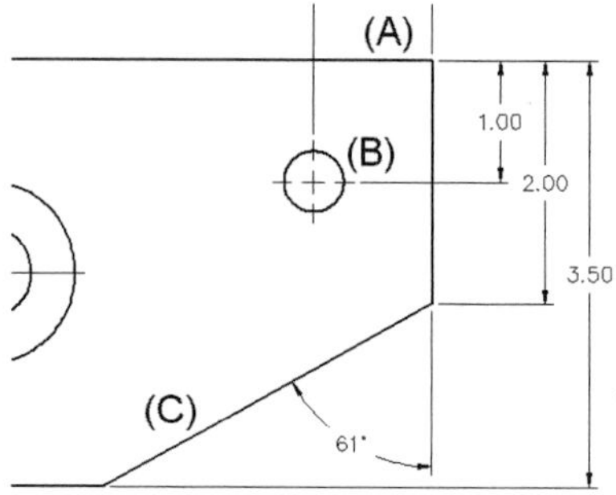

Figure 12.81

Step 10

Use the ZOOM command and the Extents option to display both the Front and Top views. Then use the DIMDIAMETER command (or DDI shortcut) to place two diameter dimensions, as shown in the following image.

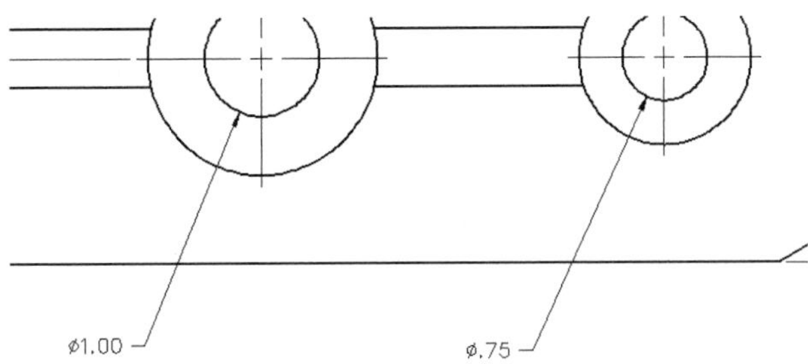

Figure 12.82

Step 11

Place a diameter dimension using the DIMDIAMETER (DDI) command on the circle, as shown in the following image.

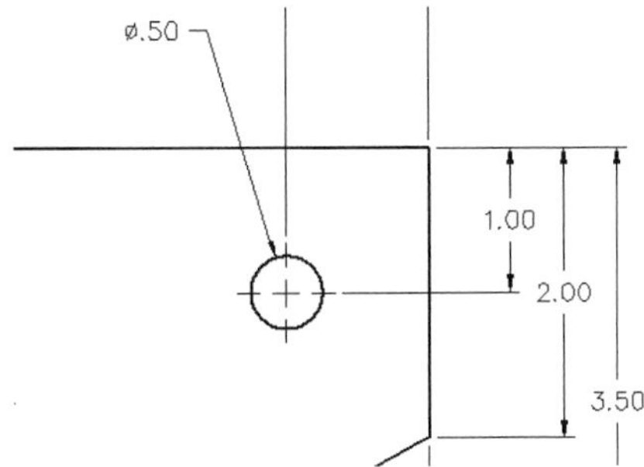

Figure 12.83

Step 12

Since the two other smaller holes share the same diameter value, use the DDEDIT command (or ED shortcut) to edit this dimension value. Clicking the diameter value activates the Text Formatting dialog box. Begin by typing 3X to signify three holes of the same diameter, as shown on the left in the following image. Click the OK button to return to your drawing. The results are illustrated on the right in the following image.

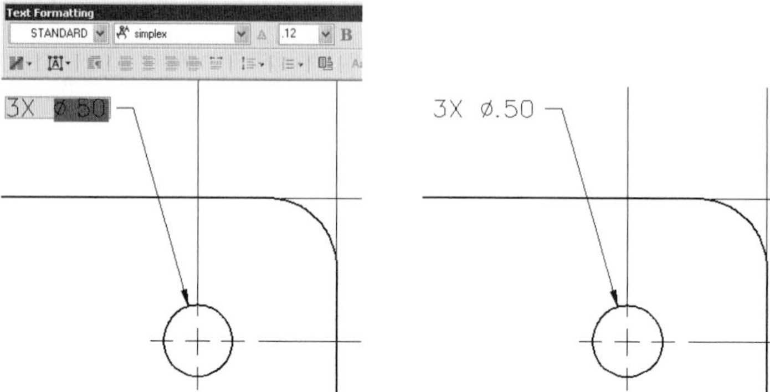

Figure 12.84

Step 13

One more dimension needs to be placed in the Top view: the .50 width of the rib. Unfortunately, because of the placement of this dimension, extension lines will be drawn on top of the object's lines. This is poor practice. To remedy this, a dimension override will be created. To do this, activate the Dimension Style Manager dialog box and click the Override button, as shown on the left in the following image. This displays the Override Current Style: MECHANICAL dialog box. In the Lines tab, place checks in the boxes to Suppress (turn off) Ext line 1 and Ext line 2 in the Extension lines area, as shown in the middle in the following image.

Click the OK button and notice the new <style overrides> listing under MECHANICAL, as shown on the right in the following image. Notice also that the preview image lacks extension lines. Click the Close button to return to the drawing.

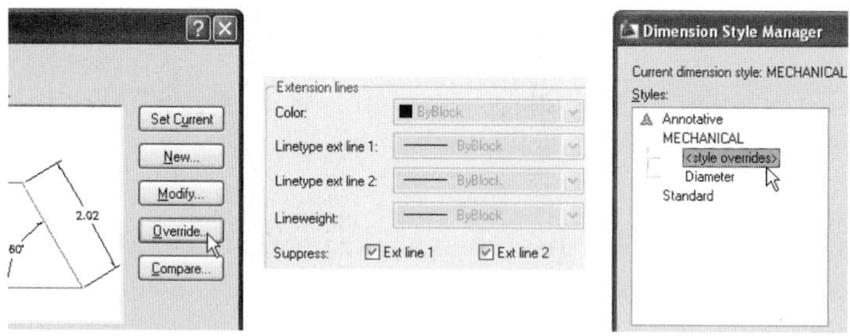

Figure 12.85

Step 14

Pan to the middle area of the Top view and place the .50 rib-width dimension while in the dimension style override, as shown in the following image. This dimension should be placed without having any extension lines visible. Also, use grips to drag the dimension text inside the visible object lines (temporarily turn off OSNAP to better accomplish this task).

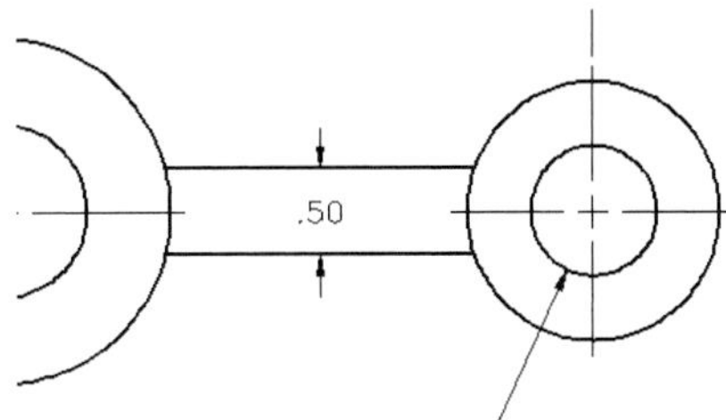

Figure 12.86

Step 15

The completed Top view, including dimensions, is illustrated in the following image. Use this figure to check that all dimensions have been placed and all features such as holes and fillets have been properly identified.

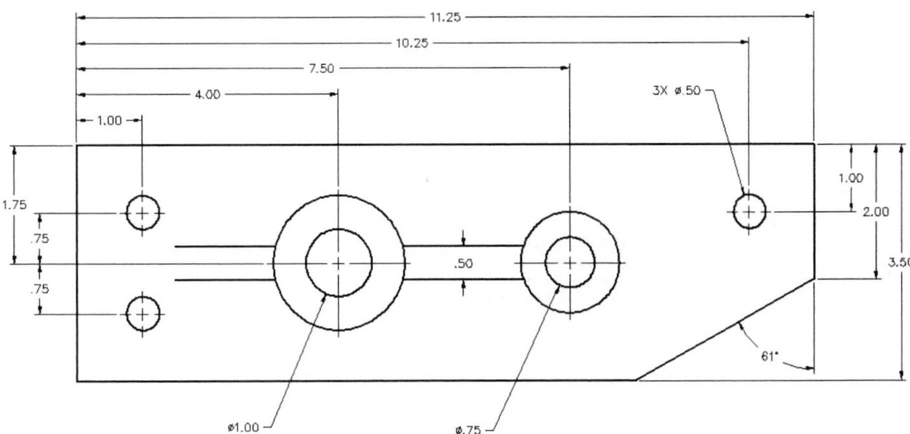

Figure 12.87

Step 16

The Front view in the following image will be the focus for the next series of dimensioning steps. Again use the ZOOM and PAN commands whenever you need to magnify or slide to a better drawing view position.

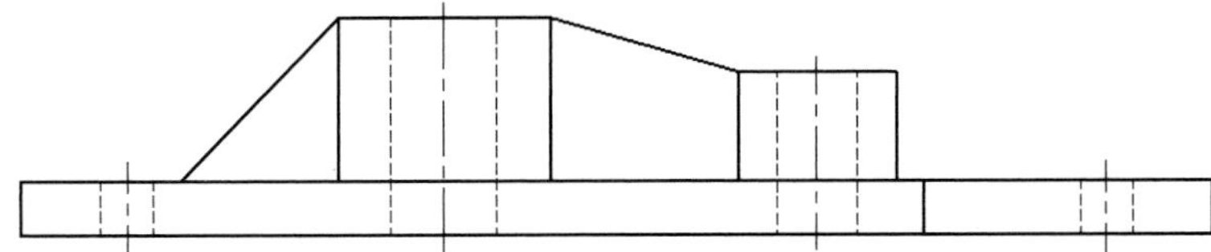

Figure 12.88

Step 17

Before continuing, activate the Dimension Style Manager dialog box, click the MECHANICAL style, and then click the Set Current button. An AutoCAD Alert box displays, as shown in the following image. Making MECHANICAL current discards any style overrides. Click the OK button to discard the changes because you need to return to having extension lines visible in linear dimensions.

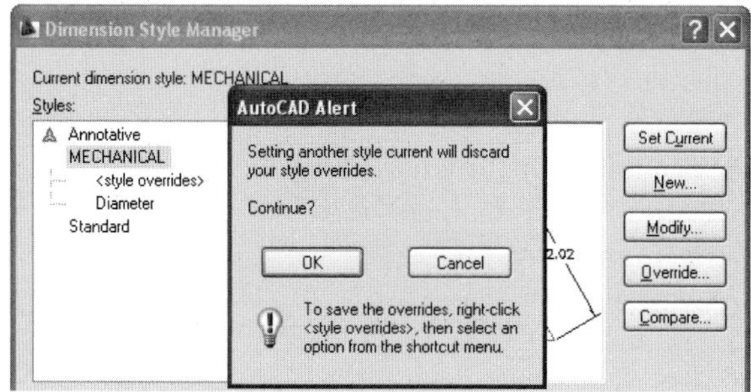

Figure 12.89

Step 18

Place two horizontal linear dimensions that signify the diameters of the two cylinders using the DIMLINEAR command, as shown in the following image. Also add the 1.50 vertical dimension to the side of the Front view, as shown in the following image.

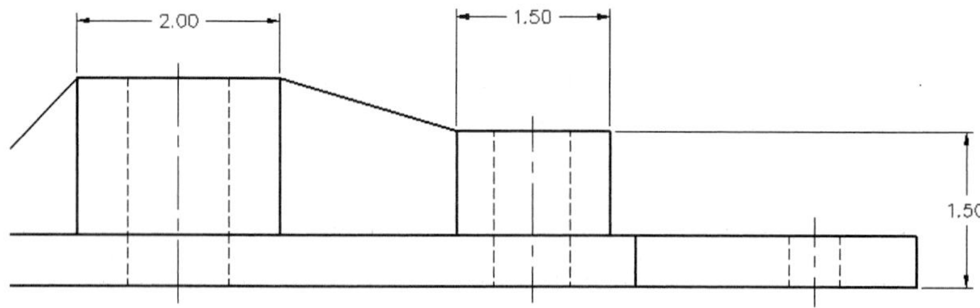

Figure 12.90

Step 19

Use the QDIM command to place the vertical baseline dimensions, as shown in the following image. Also place the 45° angular dimension in this image.

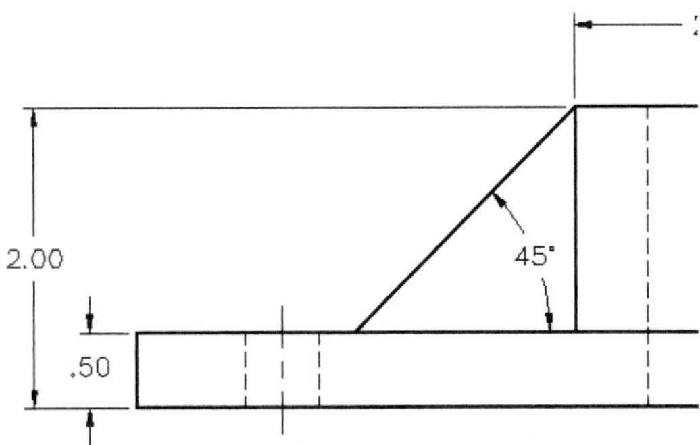

Figure 12.91

Step 20

The remaining step is to add the diameter symbol to the 2.00 and 1.50 dimensions, since the dimension is placed on a cylindrical object. Use the Text Formatting dialog box shown in the following image to accomplish this task.

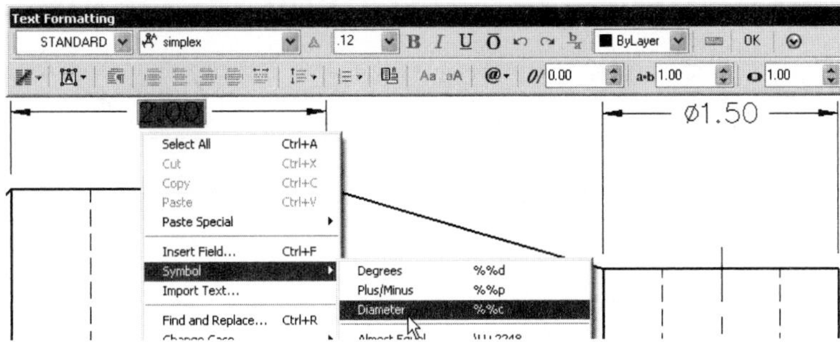

Figure 12.92

Once you have completed all dimensioning steps, your drawing should appear similar to the following image. Use grips to close the gaps between the extension and centerlines.

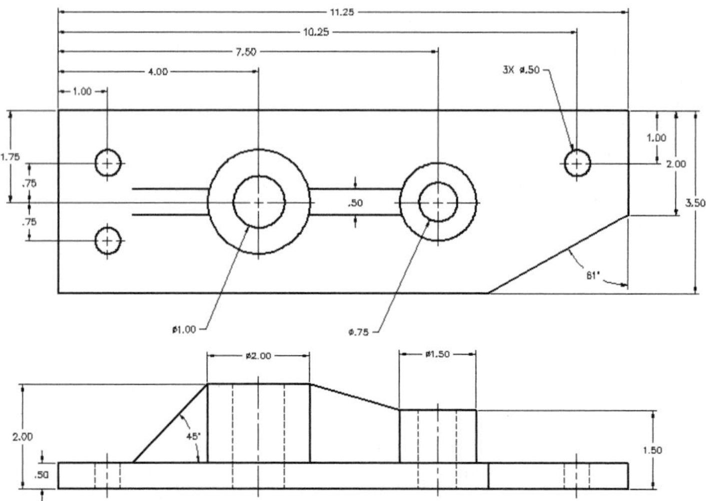

Figure 12.93

TUTORIAL EXERCISE: 12_ARCHITECTURAL DIMENSION.DWG

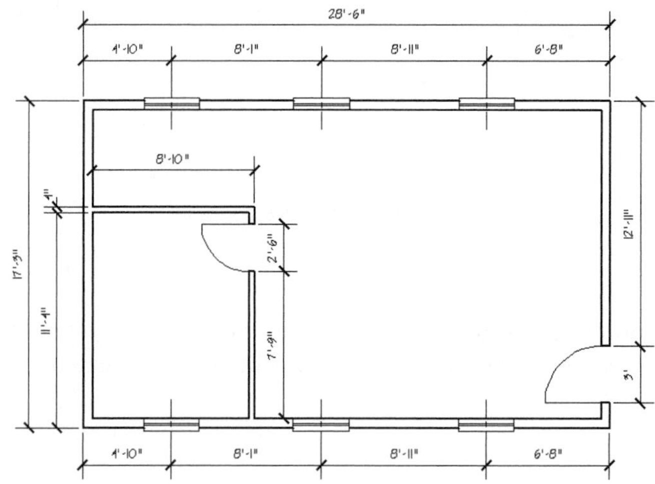

Figure 12.94

PURPOSE

This tutorial is designed to develop and apply dimension styles to an architectural drawing project shown in the previous image. The dimensions will be applied to a Model Space environment.

SYSTEM SETTINGS

No special settings are needed for this project.

LAYERS

Layers have already been created for this drawing.

SUGGESTED COMMANDS

An architectural dimension style will be created to reflect the settings required for an architectural drawing. Various horizontal and vertical dimensions will be placed around the floor plan to call out distances between windows, doors, and walls.

Step 1

Open the drawing file 12_Architectural Dimension, shown in the following image. Check to make sure that the dimension layer is the current layer.

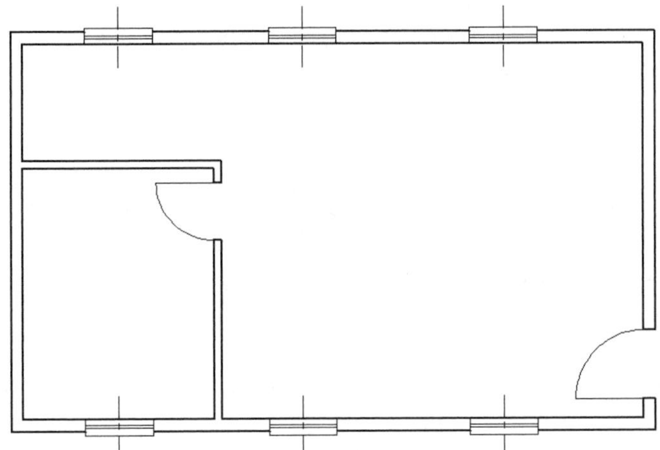

Figure 12.95

Step 2

The Dimension Style Manager will be used to set up characteristics of the dimensions used in architectural applications. Activate the Dimension Style Manager dialog box by clicking Dimension Style, located in the Format pull-down menu. Clicking the New... button launches the Create New Dimension Style dialog box, as shown in the following image. Replace Copy of Standard in the New Style Name field with Architectural. This dimension style will initially be based on the current standard dimension style and applied to all dimensions that are placed in the drawing. Click the Continue button to make various changes to the dimension settings.

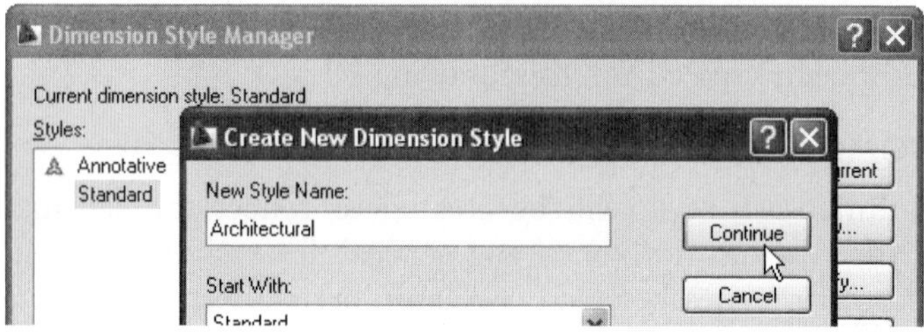

Figure 12.96

Step 3

Note that the name Architectural is now displayed in the title bar of the New Dimension Style dialog box, as shown in the following image. In the Symbols and Arrows section, change the arrowhead style to Architectural tick. When you change the First arrowhead, it also changes the Second arrowhead. Also change the arrow size to 1/8", or .125, as shown in the following image.

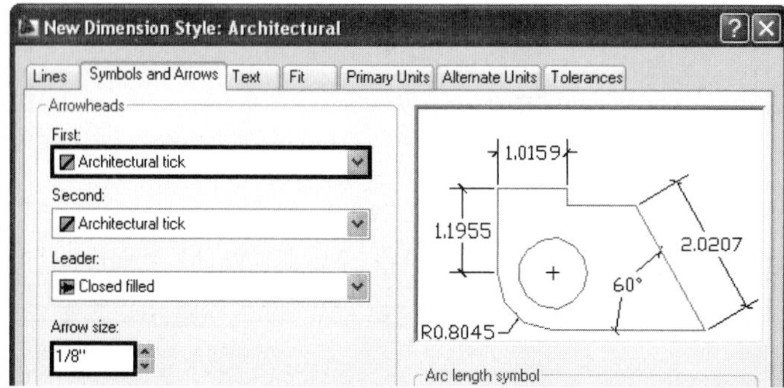

Figure 12.97

Step 4

Select the Text tab and make changes to the following items, as shown in the following image. Change the Text style from Standard to Architectural (this text style already exists in the drawing). As with the arrowheads, change the Text height to 1/8". Next, change the Vertical Text placement from Centered to Above. This will place dimension text above the dimension line, which is the architectural standard. Finally, in the Text alignment area, click the radio button next to Aligned with dimension line. This forces the dimension text to display parallel to the dimension line. This also means that for vertical dimensions the dimension must be read from the side, which represents another architectural standard.

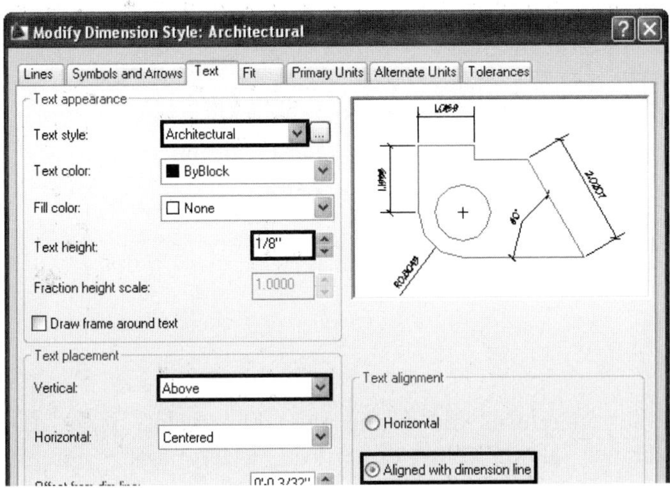

Figure 12.98

Step 5

Next, select the Fit tab and change the overall scale of the dimension. The overall scale is based on the plotting scale when plotting is done from the model tab. The anticipated plotting scale for this drawing is 1/4" = 1'. Because 1/4" divides 48 times into 1', type 48 in the Use overall scale of field, as shown in the following image.

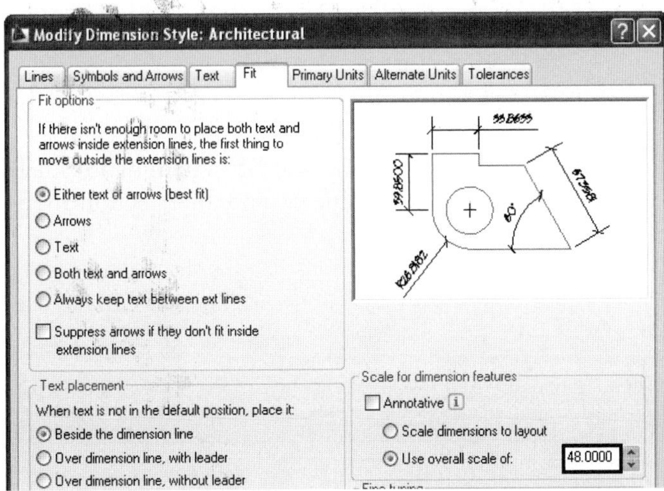

Figure 12.99

Step 6

Next, choose the Primary Units tab. In the Linear dimensions section, change the Unit format option to Architectural. Also, change the precision to 0'-0". Your display should appear similar to the following image.

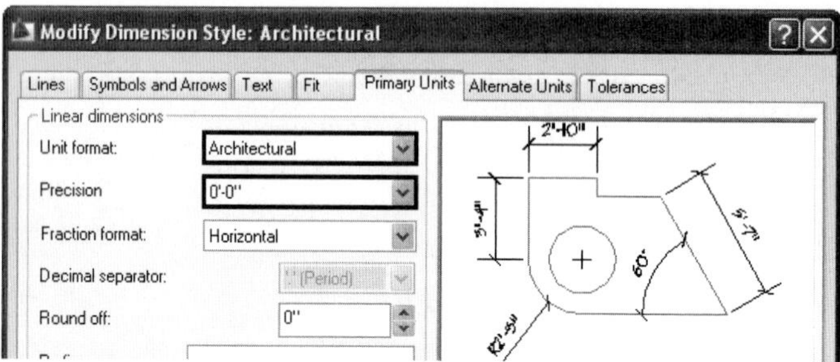

Figure 12.100

Step 7

Clicking the OK button returns you to the main Dimension Style Manager dialog box. Notice the appearance of the new dimension style, namely Architectural. As this dimension style is created, it is not current. Double-clicking Architectural or clicking the Set Current button makes this the current dimension style, as shown in the following image. When finished, click the Close button to return to the drawing editor.

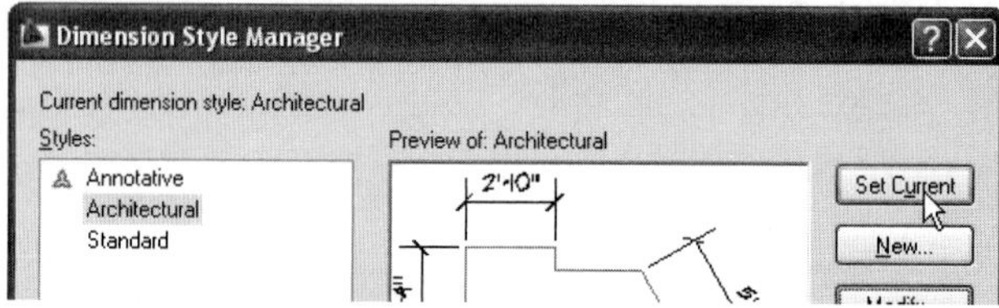

Figure 12.101

Step 8

Place a horizontal dimension using the LINEAR DIMENSION command (DLI), as shown in the following image. Be sure that OSNAP Endpoint and Intersection modes are active.

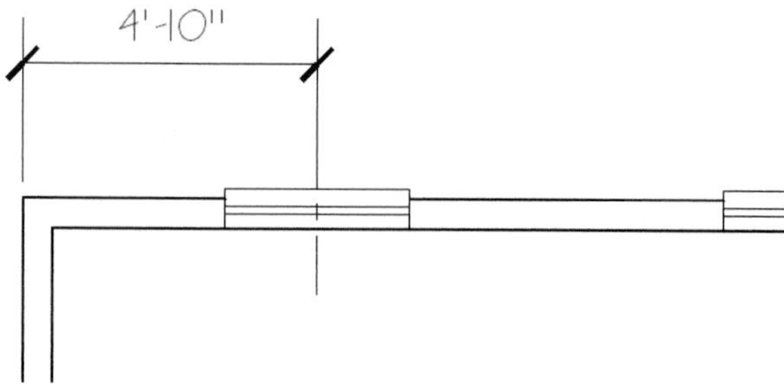

Figure 12.102

Step 9

Use the CONTINUE DIMENSION command (DCO) to apply dimensions in succession to dimension the window locations, as shown in the following image.

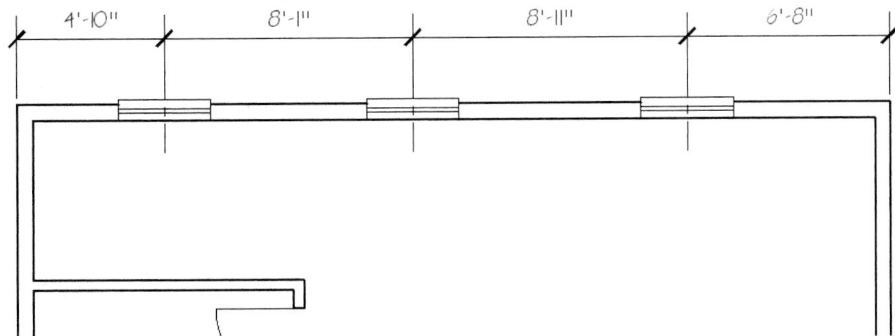

Figure 12.103

Step 10

Next add an overall length dimension using the LINEAR DIMENSION command (DLI), as shown in the following image.

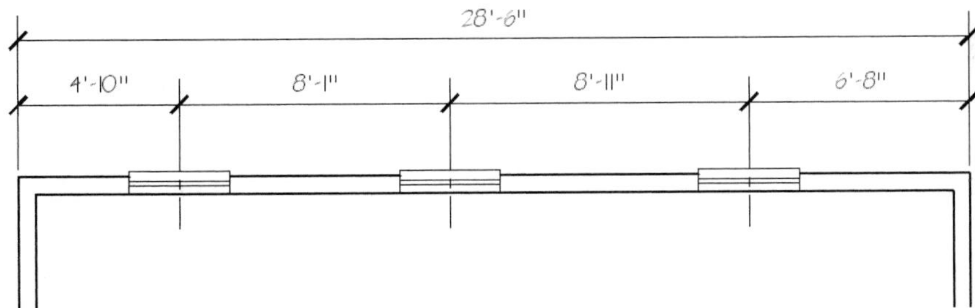

Figure 12.104

Step 11

Continue adding other dimensions to the floor plan, including those defining vertical and interior dimensions of the building, as shown in the following image.

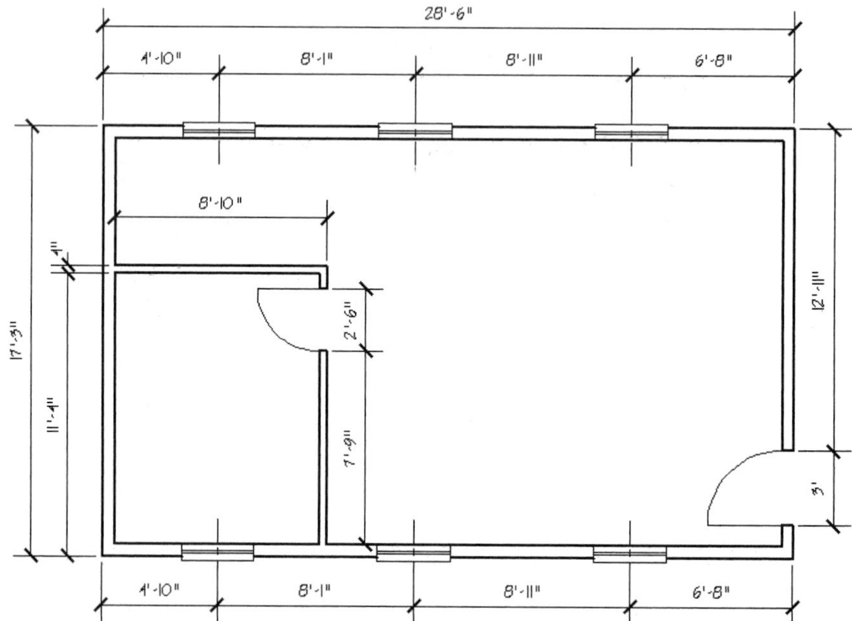

Figure 12.105

PROBLEMS FOR CHAPTER 12

DIRECTIONS FOR PROBLEMS 12–1 THROUGH 12–18

1. *Open each drawing, which is located on the CD that is supplied with this book.*
2. *Use proper techniques to dimension each drawing.*

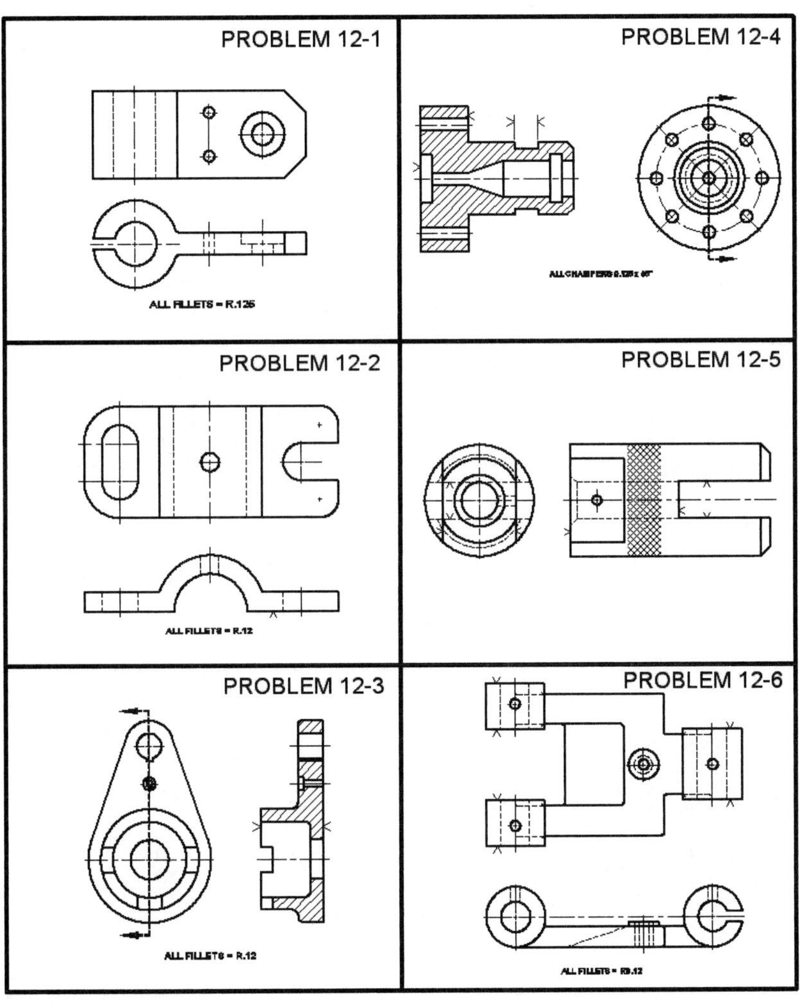
PROBLEM 12-1
ALL FILLETS = R.125
PROBLEM 12-4
PROBLEM 12-2
ALL FILLETS = R.12
PROBLEM 12-5
PROBLEM 12-3
ALL FILLETS = R.12
PROBLEM 12-6

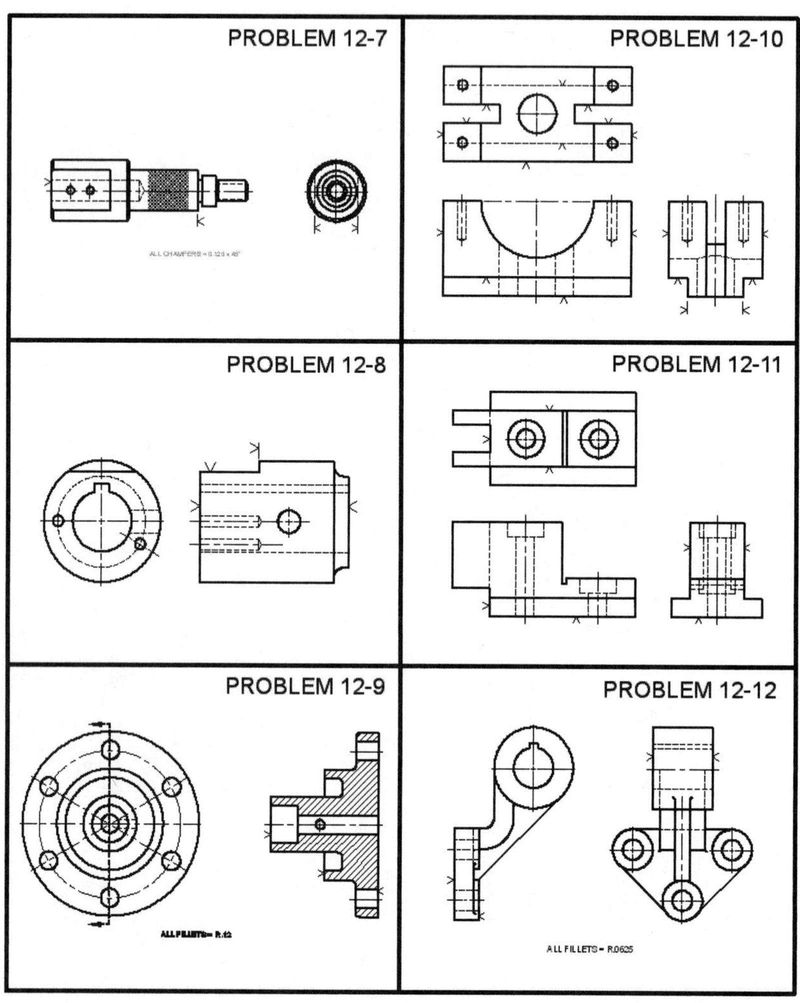
PROBLEM 12-7
ALL CHAMFERS = 0.125 x 45°
PROBLEM 12-10
PROBLEM 12-8
PROBLEM 12-11
PROBLEM 12-9
PROBLEM 12-12
ALL FILLETS = R.0625

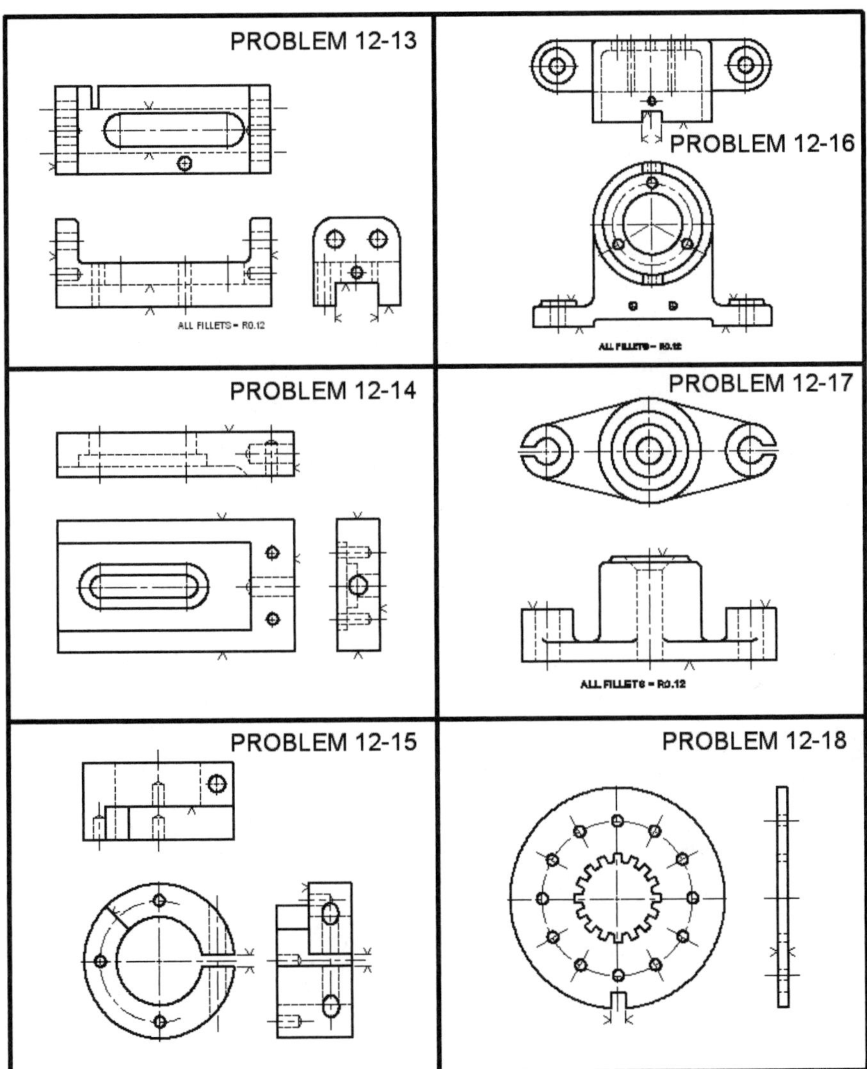

DIRECTIONS FOR PROBLEMS 12–19 THROUGH 12–24

1. *Use the isometric drawings provided to develop orthographic view drawings, showing as many views as necessary to communicate the design.*
2. *Fully dimension your drawings.*

PROBLEM 12–19

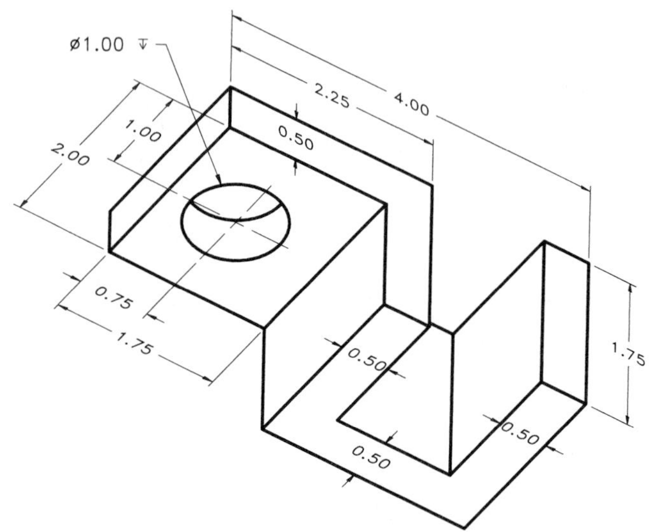

PROBLEM 12–20

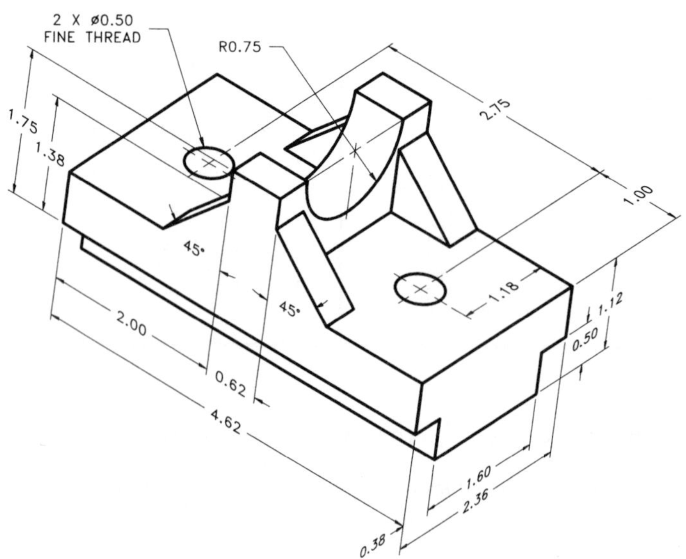

PROBLEM 12–21

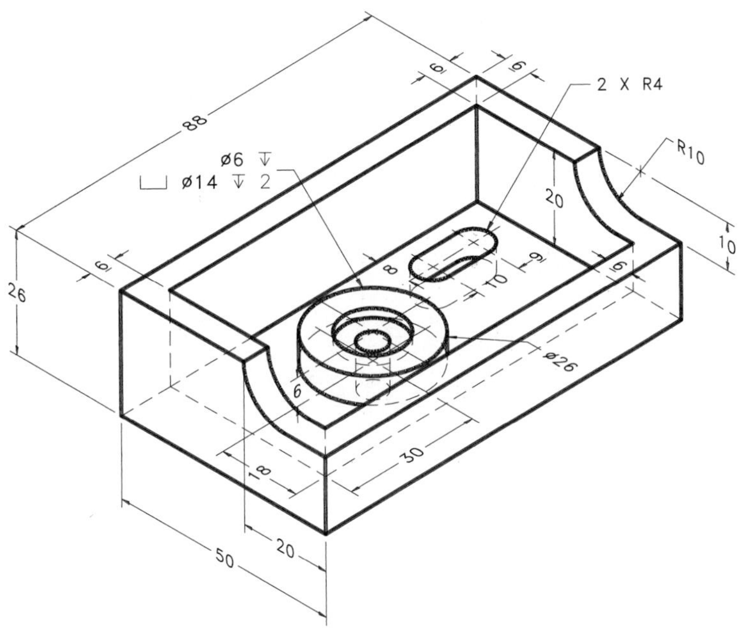

PROBLEM 12–22

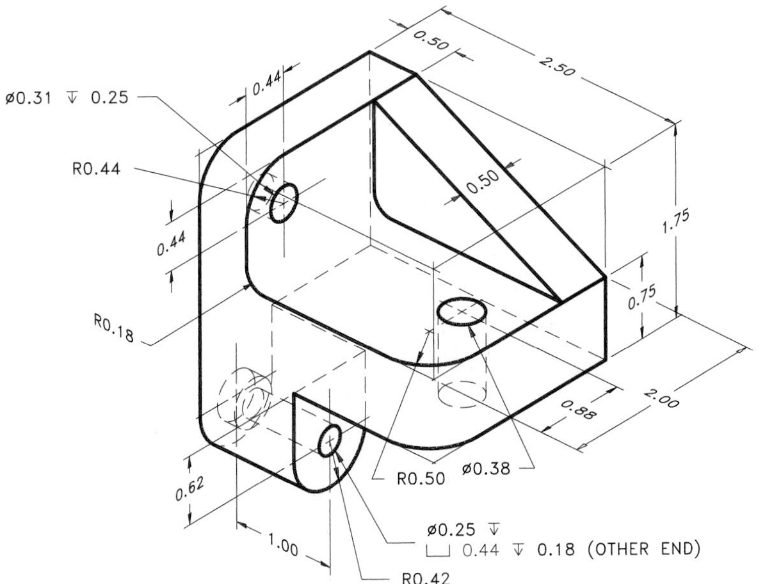

PROBLEM 12–23

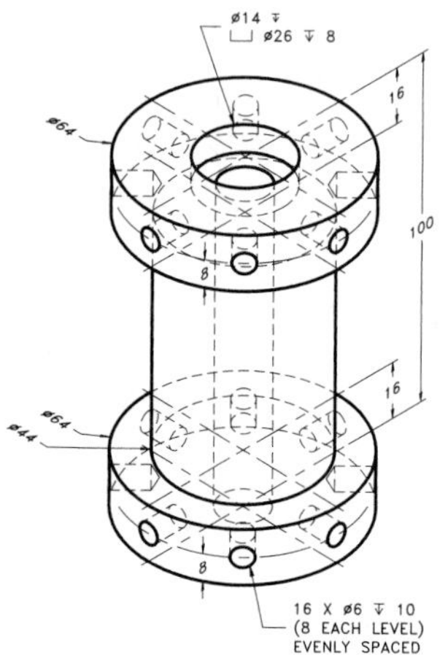

PROBLEM 12–24

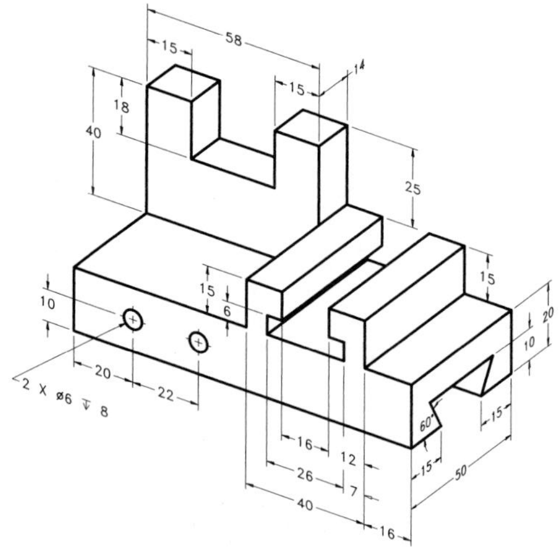

CHAPTER 13

Analyzing 2D Drawings

This chapter will show how a series of commands may be used to calculate distances and angles of selected objects. It will also show how surface areas may be calculated on complex geometric shapes. The following pages highlight all of AutoCAD's inquiry commands and show how you can use them to display useful information on an object or group of objects.

USING INQUIRY COMMANDS

You can choose AutoCAD's inquiry commands by using the Inquiry toolbar, the Menu Browser, the pull-down menu, the Ribbon, or by entering the commands at the keyboard, as illustrated in the following image.

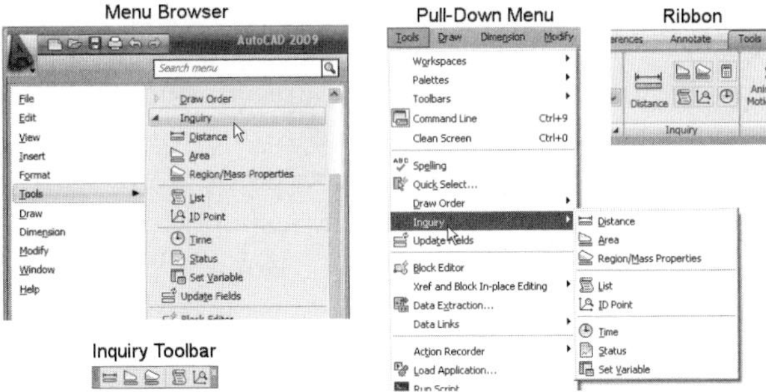

Figure 13.1

The following is a list of the inquiry commands with a short description of each:

Button	Tool	Shortcut	Description
	Distance	DI	Calculates the distance between two points. Also provides the delta XYZ coordinate values, the angle in the XY plane, and the angle from the XY plane.

Button	Tool	Shortcut	Description
	Area	AA	Calculates the surface area after you specify a series of points or select a polyline or circle. You can add or subtract multiple objects to calculate the area with holes and cutouts.
	Mass Properties	MASSPROP	Used for determining the mass properties of a region or 3D solid model.
	List	LI	Used to list important information on an object.
	Locate	ID	Used to determine the X, Y, and Z coordinates of a point.
	Time	TIME	Displays the time spent in the drawing editor.
	QuickCalc	QC	Displays a calculator for performing mathematical functions.
	Status	STATUS	Displays important information on the current drawing.
	Set Variable	SETVAR	Used for making changes to one of the many system variables internal to AutoCAD.

The following pages give a detailed description of how these commands are used in interaction with different AutoCAD objects.

FINDING THE AREA OF AN ENCLOSED SHAPE

The AREA command is used to calculate the area through the selection of a series of points. Select the endpoints of all vertices in the following image with the OSNAP-Endpoint option. Once you have selected the first point along with the remaining points in either a clockwise or counterclockwise pattern, respond to the prompt "Next point:" by pressing ENTER to calculate the area of the shape. Along with the area is a calculation for the perimeter.

Try It!: Open the drawing file 13_Area1. Use the following image and the prompt sequence below for finding the area by identifying a series of points.

```
Command: AA (For AREA)
Specify first corner point or [Object/Add/Subtract]: (Select the
    endpoint at "A")
Specify next corner point or press ENTER for total: (Select the
    endpoint at "B")
Specify next corner point or press ENTER for total: (Select the
    endpoint at "C")
Specify next corner point or press ENTER for total: (Select the
    endpoint at "D")
Specify next corner point or press ENTER for total: (Select the
    endpoint at "E")
Specify next corner point or press ENTER for total: (Select the
    endpoint at "A")
```

```
Specify next corner point or press ENTER for total: (Press ENTER to
    calculate the area)
Area = 25.25, Perimeter = 20.35
```

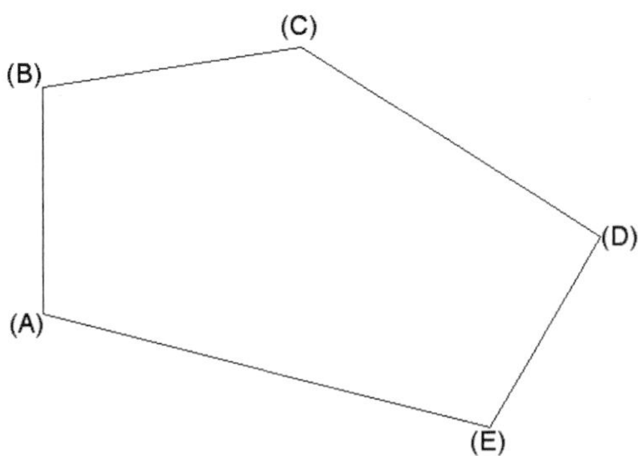

Figure 13.2

FINDING THE AREA OF AN ENCLOSED POLYLINE OR CIRCLE

The previous example showed how to find the area of an enclosed shape using the AREA command and identifying the corners and intersections of the enclosed area by a series of points. For a complex area, this could be a very tedious operation. As a result, the AREA command has a built-in Object option that calculates the area and perimeter of a polyline and the area and circumference of a circle. Finding the area of a polyline can only be accomplished if one of the following conditions is satisfied:

- The shape must have already been constructed through the PLINE command.
- The shape must have already been converted to a polyline through the PEDIT command if originally constructed from individual objects.

Try It!: Open the drawing file 13_Area2. Use the following image and the prompt sequence below for finding the area of both shapes.

```
Command: AA (For AREA)
Specify first corner point or [Object/Add/Subtract]: O (For
    Object)
Select objects: (Select the polyline at "A")
Area = 24.88, Perimeter = 19.51

Command: AA (For AREA)
Specify first corner point or [Object/Add/Subtract]: O (For
    Object)
Select objects: (Select the circle at "B")
Area = 7.07, Circumference = 9.42
```

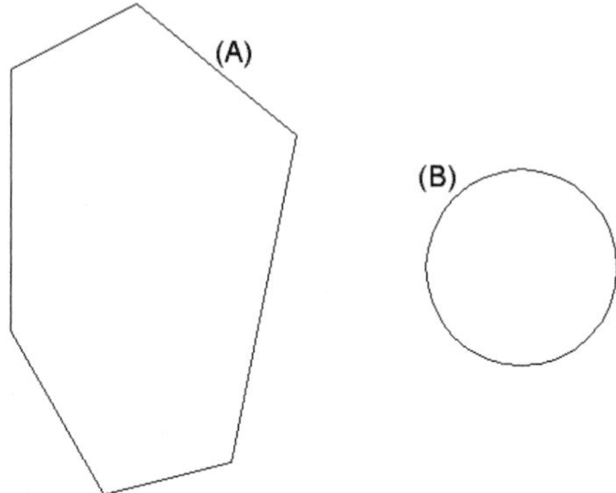

Figure 13.3

Try It!: Open the drawing file 13_Extrude1. In the following image, either convert all line segments into a single polyline object or use the BOUNDARY command to trace a polyline on the top of all the line segments. In either case, answer Question 1 regarding 13_Extrude1.

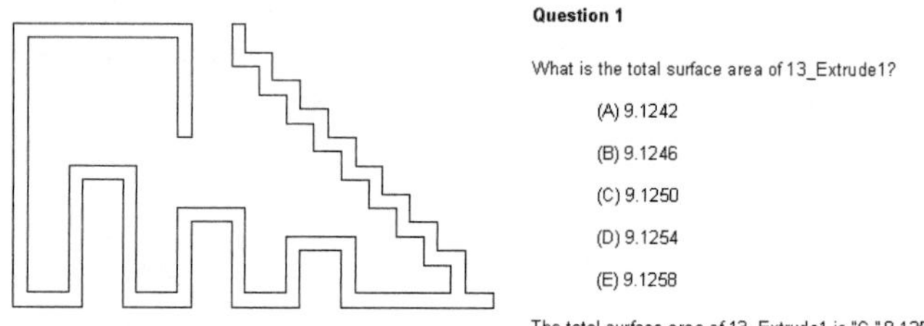

Question 1
What is the total surface area of 13_Extrude1?
(A) 9.1242
(B) 9.1246
(C) 9.1250
(D) 9.1254
(E) 9.1258
The total surface area of 13_Extrude1 is "C," 9.1250.

Figure 13.4

Try It!: Open the drawing file 13_Extrude2. In the following image, either convert all line segments into a single polyline object or use the BOUNDARY command to trace a polyline on the top of all the line segments. In either case, answer Question 1 regarding 13_Extrude2.

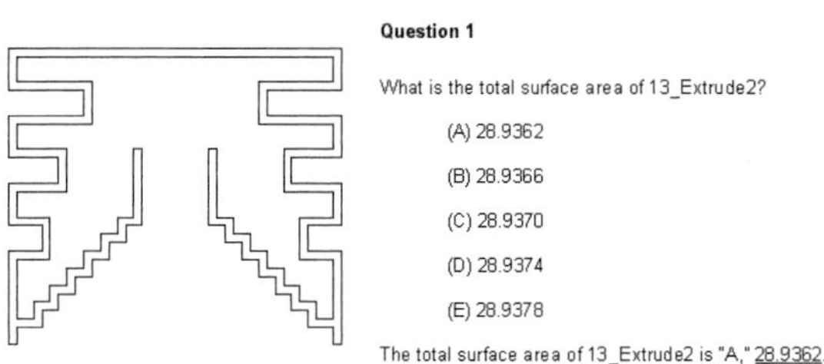

Question 1
What is the total surface area of 13_Extrude2?
(A) 28.9362
(B) 28.9366
(C) 28.9370
(D) 28.9374
(E) 28.9378
The total surface area of 13_Extrude2 is "A," 28.9362.

Figure 13.5

FINDING THE AREA OF A SHAPE BY SUBTRACTION

You may wish to calculate the area of a shape with holes cut through it. The steps you use to calculate the total surface area are: (1) calculate the area of the outline and (2) subtract the objects inside the outline. All individual objects, except circles, must first be converted to polylines through the PEDIT command. Next, find the overall area and add it to the database using the Add mode of the AREA command. Exit the Add mode and remove the inner objects using the Subtract mode of the AREA command. Remember that all objects must be in the form of a circle or polyline. This means that the inner shape at "B" in the following image must also be converted to a polyline through the PEDIT command before the area is calculated. Care must be taken when selecting the objects to subtract. If an object is selected twice, it is subtracted twice and may yield an inaccurate area in the final calculation.

In the following image, the total area with the circle and rectangle removed is 30.4314.

Try It!: Open the drawing file 13_Area3. Use the following image and the prompt sequence to verify this area calculation.

```
Command: AA (For AREA)
Specify first corner point or [Object/Add/Subtract]: A (For Add)
Specify first corner point or [Object/Subtract]: O (For Object)
(ADD mode) Select objects: (Select the polyline at "A")
Area = 47.5000, Perimeter = 32.0000
Total area = 47.5000
(ADD mode) Select objects: (Press ENTER to exit ADD mode)
Specify first corner point or [Object/Subtract]: S (For Subtract)
Specify first corner point or [Object/Add]: O (For Object)
(SUBTRACT mode) Select objects: (Select the polyline at "B")
Area = 10.0000, Perimeter = 13.0000
Total area = 37.5000
(SUBTRACT mode) Select objects: (Select the circle at "C")
Area = 7.0686, Circumference = 9.4248
Total area = 30.4314
(SUBTRACT mode) Select objects: (Press ENTER to exit SUBTRACT mode)
Specify first corner point or [Object/Add]: (Press ENTER to exit
    this command)
```

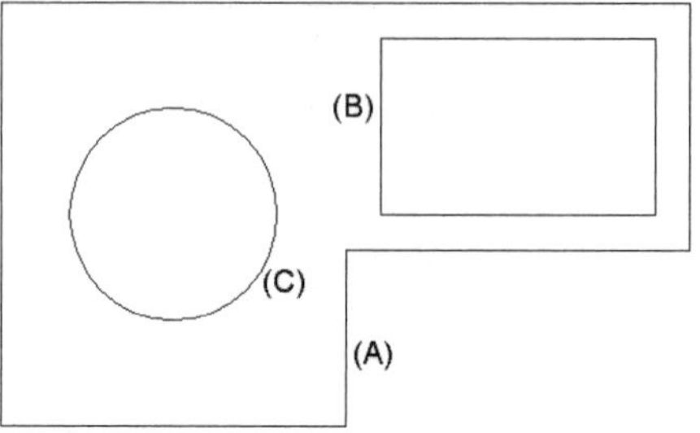

Figure 13.6

Try It!: Open the drawing file 13_Shield. In the following image, use the BOUNDARY command to trace a polyline on the top of all the line segments. Subtract all four-sided shapes from the main shape using the AREA command. Answer Question 1 regarding 13_Shield.

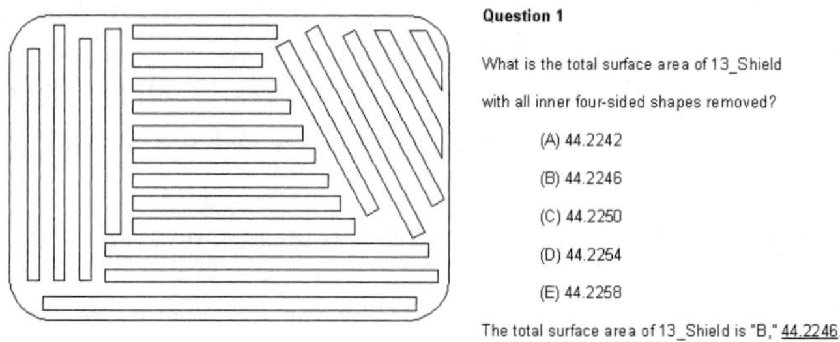

Question 1
What is the total surface area of 13_Shield
with all inner four-sided shapes removed?
(A) 44.2242
(B) 44.2246
(C) 44.2250
(D) 44.2254
(E) 44.2258
The total surface area of 13_Shield is "B," 44.2246.

Figure 13.7

USING FIELDS IN CALCULATIONS

Fields can be of great use when performing area calculations on geometric shapes. In the following image, the area of the room of the partial office plan will be calculated using fields. When the geometric shape changes, the field can be updated to display the new area calculation. Begin this process by first adding the following multiline text object, (AREA=). Then click the Insert Field button in Text Formatting toolbar.

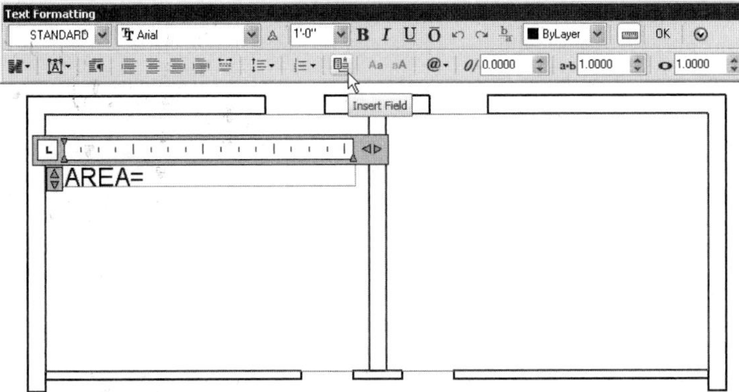

Figure 13.8

This action activates the Field dialog box, as shown in the following image. Locate Object under the Field Names area. Under the Object Type heading, click the Select object button.

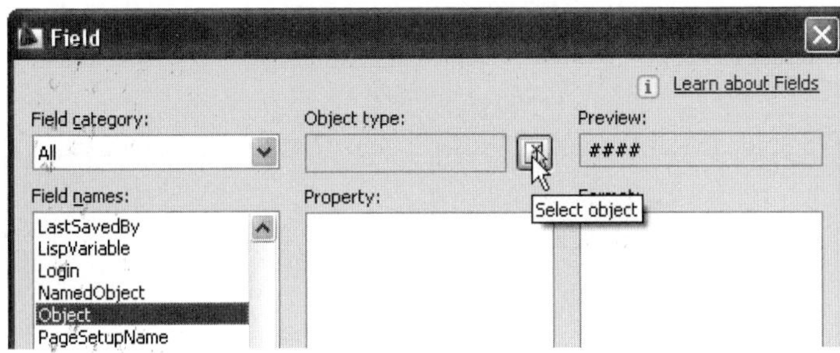

Figure 13.9

Clicking the Select object button in the previous image returns you to the drawing editor. Once you are there, select the rectangular polyline shape, as shown in the following image.

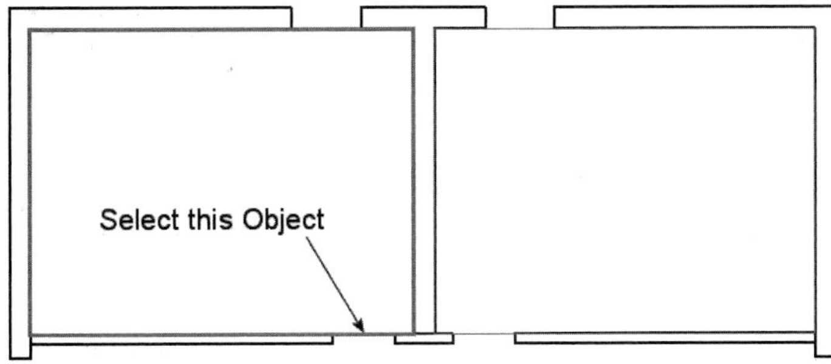

Figure 13.10

This returns you back to the Field dialog box, as shown in the following image. Notice that the property changed to Area and the Format changed to Architectural. The Preview area shows the current area calculation. Click the OK button to exit the Field dialog box.

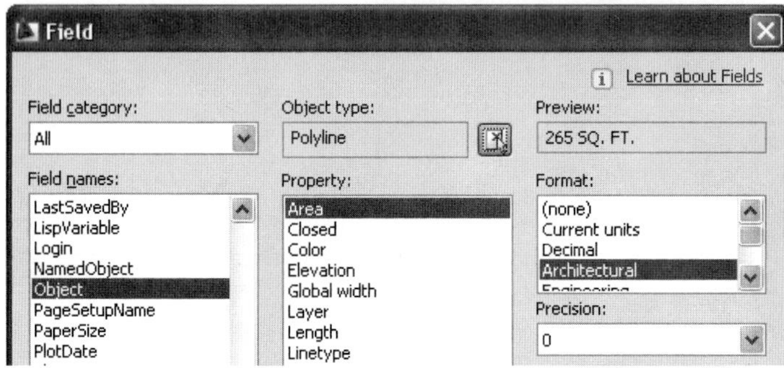

Figure 13.11

The results are illustrated in the following image. A field consisting of the square footage information is added to the end of the multiline text. Notice also that the field is identified by the text with the gray background.

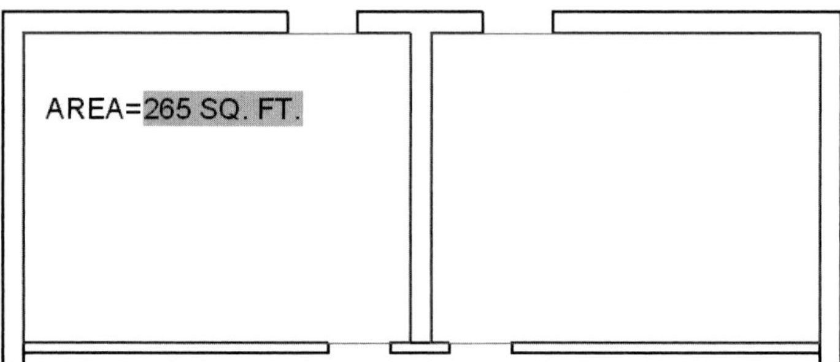

Figure 13.12

In the following image, the room was stretched 6' to the left. Regenerating the drawing screen updates the field to reflect the change in the square footage calculation, as shown in the following image.

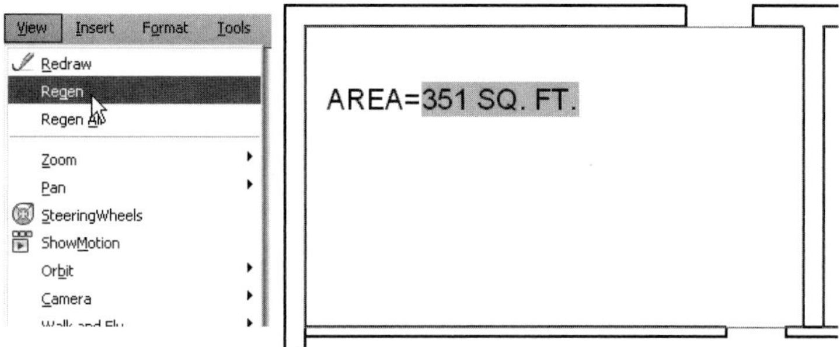

Figure 13.13

THE DIST (DISTANCE) COMMAND

The DIST command calculates the linear distance between two points on an object, whether it be the distance of a line, the distance between two points, or the distance from the quadrant of one circle to the quadrant of another circle. The following information is also supplied when you use the DIST command: the angle in the XY plane, the angle from the XY plane, and the delta XYZ coordinate values. The angle in the XY plane is given in the current angular mode set by the Drawing Units dialog box. The delta XYZ coordinate is a relative coordinate value taken from the first point identified by the DIST command to the second point.

Try It!: Open the drawing file 13_Distance. Use the following image and the prompt sequences below for the DIST command.

```
Command: DI (For DIST)
Specify first point: (Select the endpoint at "A")
Specify second point: (Select the endpoint at "B")
Distance = 6.36, Angle in XY Plane = 45.00, Angle from XY Plane
    = 0.00
Delta X = 4.50, Delta Y = 4.50, Delta Z = 0.00

Command: DI (For DIST)
Specify first point: (Select the endpoint at "C")
Specify second point: (Select the endpoint at "D")
Distance = 9.14, Angle in XY Plane = 192.75, Angle from XY Plane
    = 0.00
Delta X = -8.91, Delta Y = -2.02, Delta Z = 0.00
```

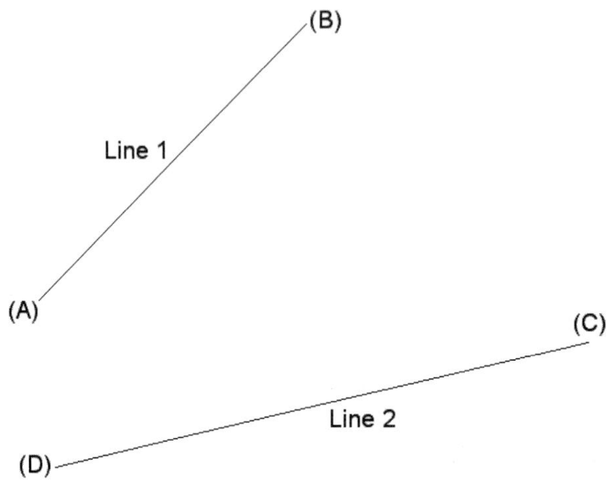

Figure 13.14

INTERPRETATION OF ANGLES USING THE DIST COMMAND

Previously, it was noted that the DIST command yields information regarding distance, delta X,Y coordinate values, and angle information. Of particular interest is the angle in the XY plane formed between two points. Picking the endpoint of the line segment at "A" as the first point followed by the endpoint of the line segment at "B" as the second point displays an angle of 42°, as shown on the left in the following image. This angle is formed from an imaginary horizontal line drawn from the endpoint of the line segment at "A" in the zero direction.

Take care when using the DIST command to find an angle on an identical line segment, illustrated on the right in the following image, compared to the example on the left. However, notice that the two points for identifying the angle are selected differently. With the DIST command, you select the endpoint of the line segment at "B" as the first point, followed by the endpoint of the segment at "A" for the second point. A new angle in the XY plane of 222° is formed. On the right in the following image, the angle is calculated by the construction of a horizontal line from the endpoint at "B," the new first point of the DIST command. This horizontal line is also drawn in the zero direction. Notice the relationship of the line segment to the horizontal baseline. In other words, be careful when identifying the order of line segment endpoints for extracting angular information.

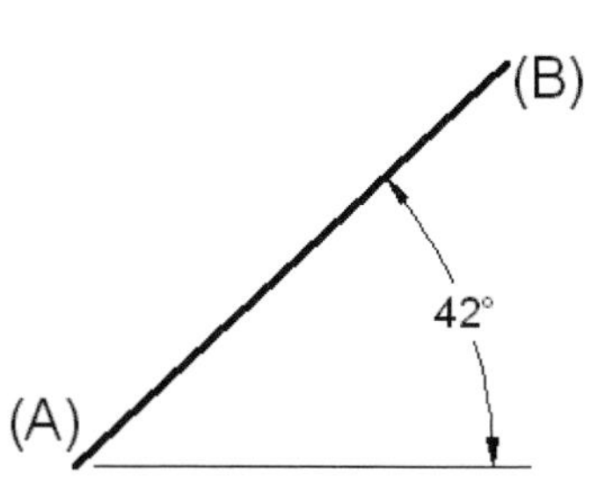

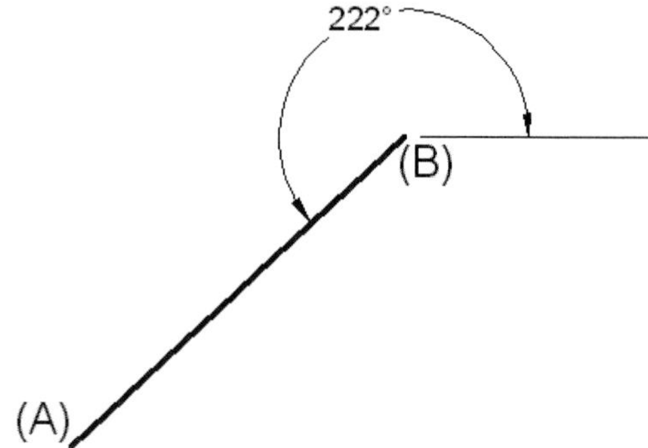

Figure 13.15

THE ID (IDENTIFY) COMMAND

The ID command is probably one of the more straightforward of the inquiry commands. ID can stand for "Identify" or "Locate Point" and allows you to obtain the current absolute coordinate listing of a point along or near an object.

In the following image, the coordinate value of the center of the circle at "A" was found through the use of ID and the OSNAP-Center mode. The coordinate value of the starting point of text string "B" was found with ID and the OSNAP-Insert mode. The coordinate value of the midpoint of the line segment at "CD" was found with ID and the OSNAP-Midpoint mode. Finally, the coordinate value of the current position of point "E" was found with ID and the OSNAP-Node mode.

Try It!: Open the drawing file 13_ID. Follow the prompt sequences below for calculating the XYZ coordinate point of the objects in the following image.

Command: **ID** Specify point: **Cen** of *(Select the edge of circle "A")* X = 2.00 Y = 7.00 Z = 0.00	Command: **ID** Specify point: **Mid** of *(Select line "CD")* X = 5.13 Y = 3.08 Z = 0.00
Command: **ID** Specify point: **Ins** of *(Select the text at "B")* X = 5.54 Y = 7.67 Z = 0.00	Command: **ID** Specify point: **Nod** of *(Select the point at "E")* X = 9.98 Y = 1.98 Z = 0.00

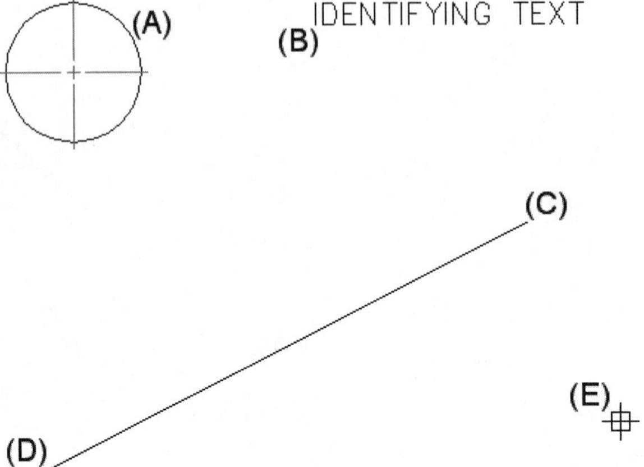

Figure 13.16

THE LIST COMMAND

Use the LIST command to obtain information about an object or group of objects. In the following image, two rectangles are displayed along with a circle, but are the rectangles made up of individual line segments or a polyline object? Using the LIST command on each object informs you that the first rectangle at "A" is a polyline. In addition to the object type, you also can obtain key information such as the layer that the object resides on, area and perimeter information for polylines, and circumference information for circles.

Try It!: Open the drawing file 13_List. Study the prompt sequence below for using the LIST command.

```
Command: LI (For LIST)
Select objects: (Select the object at "A")
Select objects: (Press ENTER to list the information on this
    object)
LWPOLYLINE Layer: "Object"
Space: Model space
Handle = 2B
Closed
Constant width 0.0000
area 3.3835
perimeter 7.6235
at point X= 4.7002 Y= 4.1846 Z= 0.0000
at point X= 7.1049 Y= 4.1846 Z= 0.0000
at point X= 7.1049 Y= 5.5916 Z= 0.0000
at point X= 4.7002 Y= 5.5916 Z= 0.0000
```

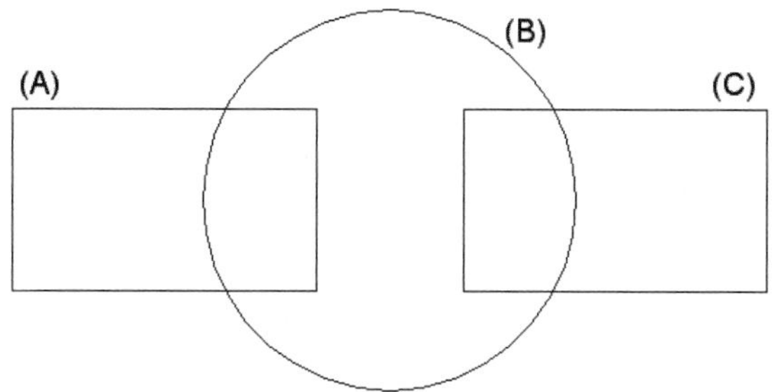

Figure 13.17

TUTORIAL EXERCISE: C-LEVER.DWG

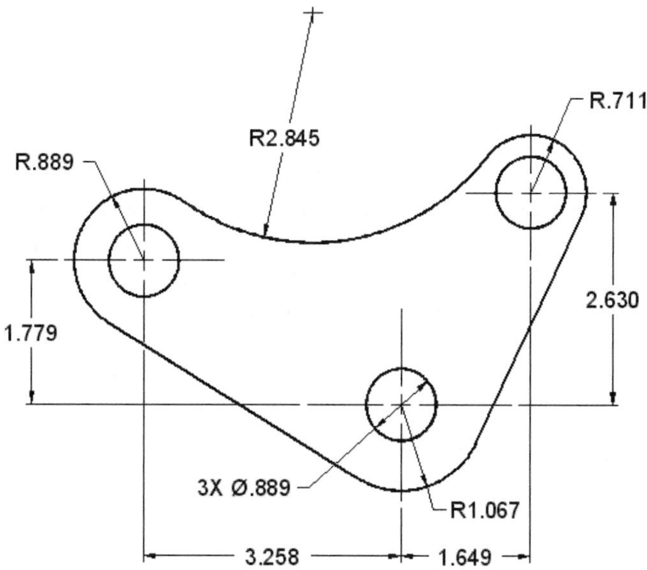

Figure 13.18

PURPOSE

This tutorial is designed to show you various methods of constructing the C-Lever object in the previous image. Numerous questions will be asked about the object, requiring the use of the AREA, DIST, ID, and LIST commands.

SYSTEM SETTINGS

Use the Drawing Units dialog box and change the number of decimal places past the zero from four units to three units. Keep the default drawing limits at (0.000,0.000) for the lower-left corner and (12.000,9.000) for the upper-right corner. Check to see that the following Object Snap modes are already set: Endpoint, Extension, Intersection, and Center.

LAYERS

Create the following layers with the format:

Name	Color	Linetype
Boundary	Magenta	Continuous
Object	Yellow	Continuous

SUGGESTED COMMANDS

Begin drawing the C-Lever with point "A" illustrated in the previous image at absolute coordinate 7.000,3.375. Begin laying out all circles. Then draw tangent lines and arcs. Use the TRIM command to clean up unnecessary objects. To prepare to answer the AREA command question, convert the profile of the C-Lever to a polyline using the BOUNDARY command. Other questions pertaining to distances, angles, and point identifications follow. Do not dimension this drawing.

Step 1

Make the Object layer current. Then construct one circle of 0.889 diameter with the center of the circle at absolute coordinate 7.000, 3.375, as shown in the following image. Construct the remaining circles of the same diameter by using the COPY command. Be sure Dynamic Input is turned on. Turn on Osnap Center and pick the center of the first circle drawn at coordinate 7.000, 3.375. Then enter coordinates to identify the second point of the remaining circles; namely 1.649, 2.630 and -3.258, 1.779. The negative value in one of the coordinates will copy to the left from the previous point.

```
Command: C (For CIRCLE)
Specify center point for circle or [3P/2P/Ttr (tan tan radius)]:
    7.000,3.375
Specify radius of circle or [Diameter]: D (For Diameter)
Specify diameter of circle: 0.889

Command: CP (For COPY)
Select objects: L (For Last)
Select objects: (Press ENTER to continue)
Specify base point or [Displacement] <Displacement>: (Be sure
    Osnap Center is turned on; then select the center of the
    existing circle)
Specify second point or <use first point as displacement>:
    1.649,2.630
Specify second point or [Exit/Undo] <Exit>: -3.258,1.779
Specify second point or [Exit/Undo] <Exit>: (Press ENTER to exit
    this command)
```

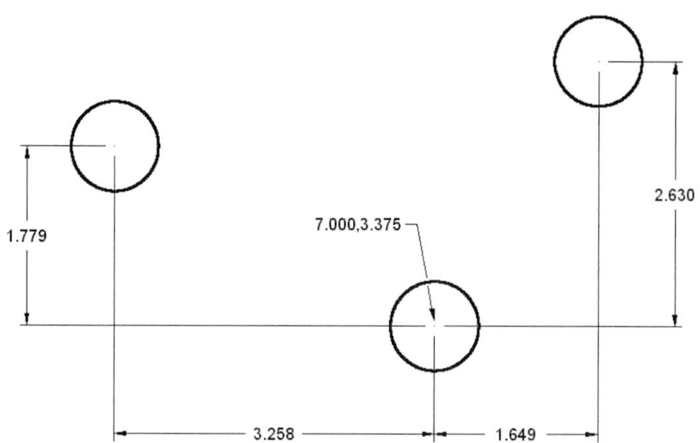

Figure 13.19

Step 2

Construct three more circles, as shown in the following image. Even though these objects actually represent arcs, circles will be drawn now and trimmed later to form the arcs.

```
Command: C (For CIRCLE)
Specify center point for circle or [3P/2P/Ttr (tan tan radius)]:
    (Select the edge of the circle at "A" to snap to its center)
Specify radius of circle or [Diameter] <0.445>: 1.067

Command: C (For CIRCLE)
Specify center point for circle or [3P/2P/Ttr (tan tan radius)]:
    (Select the edge of the circle at "B" to snap to its center)
Specify radius of circle or [Diameter] <1.067>: 0.889

Command: C (For CIRCLE)
Specify center point for circle or [3P/2P/Ttr (tan tan radius)]:
    (Select the edge of the circle at "C" to snap to its center)
Specify radius of circle or [Diameter] <0.889>: 0.711
```

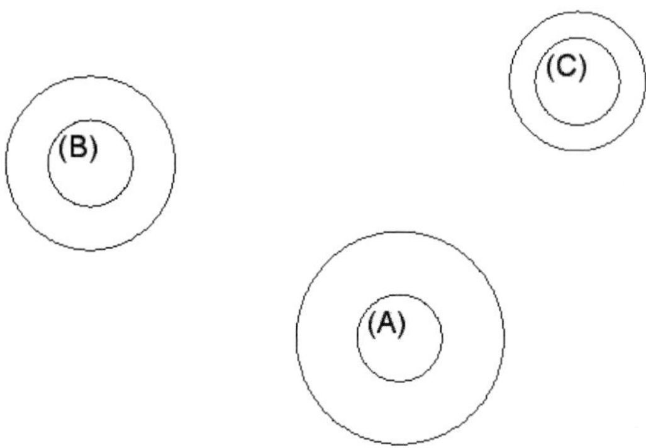

Figure 13.20

Step 3

Construct lines tangent to the three outer circles, as shown in the following image.

```
Command: L (For LINE)
Specify first point: Tan
to (Select the outer circle near "A")
Specify next point or [Undo]: Tan
to (Select the outer circle near "B")
Specify next point or [Undo]: (Press ENTER to exit this command)
```

```
Command: L (For LINE)
Specify first point: Tan
to (Select the outer circle near "C")
Specify next point or [Undo]: Tan
to (Select the outer circle near "D")
Specify next point or [Undo]: (Press ENTER to exit this command)
```

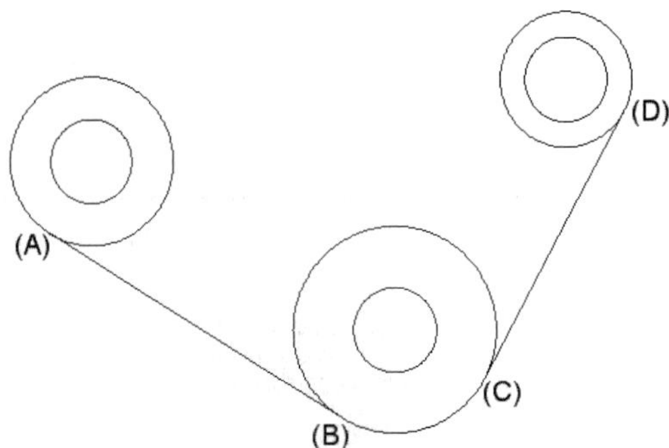

Figure 13.21

Step 4

Construct a circle tangent to the two circles in the following image, using the CIRCLE command with the Tangent-Tangent-Radius option (TTR).

```
Command: C (For CIRCLE)
Specify center point for circle or [3P/2P/Ttr (tan tan radius)]:
    TTR
Specify point on object for first tangent of circle: (Select the
    outer circle near "A")
Specify point on object for second tangent of circle: (Select
    the outer circle near "B")
Specify radius of circle <0.711>: 2.845
```

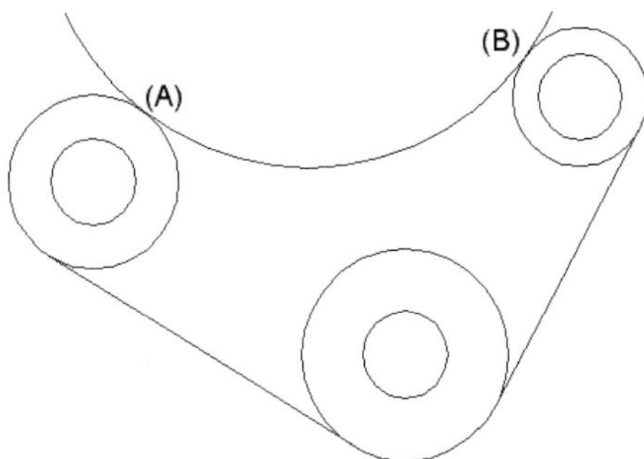

Figure 13.22

Step 5

Use the TRIM command to clean up and form the finished drawing. Pressing ENTER at the "Select objects:" prompt selects all objects as cutting edges although the objects do not highlight, as shown in the following image. Study the following prompts for selecting the objects to trim.

```
Command: TR (For TRIM)
Current settings: Projection=UCS Edge=None
Select cutting edges ...
Select objects or <select all>: (Press ENTER)
Select object to trim or shift-select to extend or
[Fence/Crossing/Project/Edge/eRase/Undo]: (Select the circle at
    "A")
Select object to trim or shift-select to extend or
[Fence/Crossing/Project/Edge/eRase/Undo]: (Select the circle at
    "B")
Select object to trim or shift-select to extend or
[Fence/Crossing/Project/Edge/eRase/Undo]: (Select the circle at
    "C")
Select object to trim or shift-select to extend or
[Fence/Crossing/Project/Edge/eRase/Undo]: (Select the circle at
    "D")
Select object to trim or shift-select to extend or
[Fence/Crossing/Project/Edge/eRase/Undo]: (Press ENTER to exit this
    command)
```

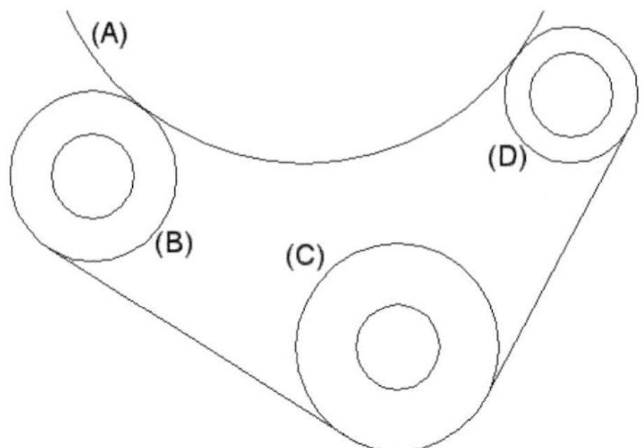

Figure 13.23

CHECKING THE ACCURACY OF C-LEVER.DWG

Once the C-Lever has been constructed, answer the following questions to determine the accuracy of this drawing. Use the following image to assist in answering the questions.

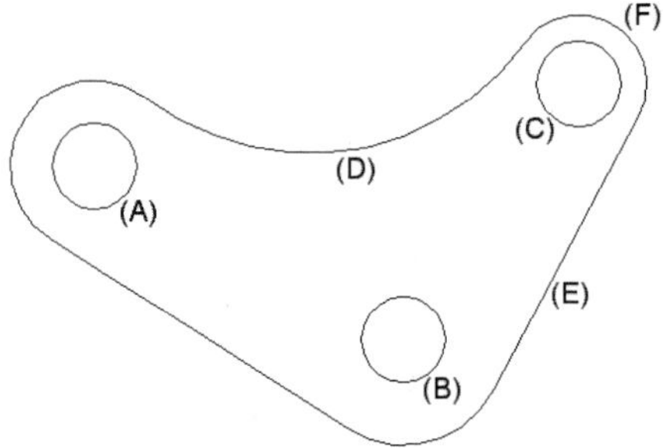

Figure 13.24

1. What is the total area of the C-Lever with all three holes removed?

   a. 13.744

   b. 13.749

   c. 13.754

   d. 13.759

   e. 13.764

2. What is the total distance from the center of circle "A" to the center of circle "B"?

   a. 3.692

   b. 3.697

   c. 3.702

   d. 3.707

   e. 3.712

3. What is the angle formed in the XY plane from the center of circle "C" to the center of circle "B"?

   a. 223°

   b. 228°

   c. 233°

   d. 238°

   e. 243°

4. What is the delta X,Y distance from the center of circle "C" to the center of circle "A"?

   a. -4.907,-0.851

   b. -4.907,-0.856

   c. -4.907,-0.861

   d. -4.907,-0.866

   e. -4.907,-0.871

5. What is the absolute coordinate value of the center of arc "D"?

   a. 5.869,8.218

   b. 5.869,8.223

   c. 5.869,8.228

   d. 5.869,8.233

   e. 5.869,8.238

6. What is the total length of line "E"?

   a. 3.074

   b. 3.079

   c. 3.084

   d. 3.089

   e. 3.094

7. What is the total length of arc "F"?
   a. 2.051
   b. 2.056
   c. 2.061
   d. 2.066
   e. 2.071

A solution for each question follows, complete with the method used to arrive at the answer. Apply these methods to any type of drawing that requires the use of inquiry commands.

Tip: Before performing an area calculation that requires you to remove numerous objects, turn the BLIPMODE variable on. This places a small mark on the object you select. In this way, you can keep better track of the objects you select. When finished, turn BLIPMODE off.

SOLUTIONS TO THE QUESTIONS ON C-LEVER

QUESTION 1

What is the total area of the C-Lever with all three holes removed?

a. 13.744
b. 13.749
c. 13.754
d. 13.759
e. 13.764

First make the Boundary layer current. Then use the BOUNDARY command and pick a point inside the object at "A" in the following image. This traces a polyline around all closed objects on the Boundary layer.

```
Command: BO (For BOUNDARY)
(When the Boundary Creation dialog box appears, click the Pick
    points button.)
Select internal point: (Pick a point inside of the object at "Y")
Selecting everything...
Selecting everything visible...
Analyzing the selected data...
Analyzing internal islands...
Select internal point: (Press ENTER to create the boundaries)
BOUNDARY created 4 polylines
```

Next, turn off the Object layer. All objects on the Boundary layer should be visible. Then use the AREA command to add and subtract objects to arrive at the final area of the object.

```
Command: AA (For AREA)
Specify first corner point or [Object/Add/Subtract]: A (For Add)
```

```
Specify first corner point or [Object/Subtract]: O (For Object)
(ADD mode) Select objects: (Select the edge of the shape near "X")
Area = 15.611, Perimeter = 17.771
Total area = 15.611
(ADD mode) Select objects: (Press ENTER to continue)
Specify first corner point or [Object/Subtract]: S (For Subtract)
Specify first corner point or [Object/Add]: O (For Object)
(SUBTRACT mode) Select objects: (Select circle "A")
Area = 0.621, Perimeter = 2.793
Total area = 14.991
(SUBTRACT mode) Select objects: (Select circle "B")
Area = 0.621, Perimeter = 2.793
Total area = 14.370
(SUBTRACT mode) Select objects: (Select circle "C")
Area = 0.621, Perimeter = 2.793
Total area = 13.749
(SUBTRACT mode) Select objects: (Press ENTER to continue)
Specify first corner point or [Object/Add]: (Press ENTER to exit
    this command)
```

The total area of the C-Lever with all three holes removed is (B), 13.749.

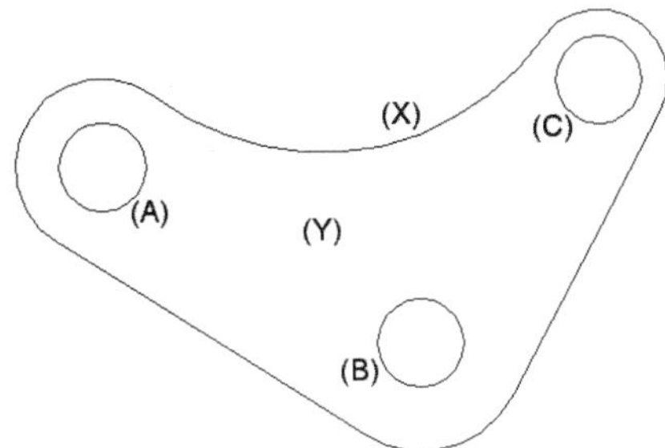

Figure 13.25

QUESTION 2

What is the total distance from the center of circle "A" to the center of circle "B"?

a. 3.692

b. 3.697

c. 3.702

d. 3.707

e. 3.712

Use the DIST (Distance) command to calculate the distance from the center of circle "A" to the center of circle "B" in the following image. Be sure to use the OSNAP-Center mode for locating the centers of all circles.

```
Command: DI (For DIST)
Specify first point: (Select the edge of the circle at "A")
```

```
Specify second point: (Select the edge of the circle at "B")
Distance = 3.712, Angle in XY Plane = 331, Angle from XY Plane = 0
Delta X = 3.258, Delta Y = -1.779, Delta Z = 0.000
```

The total distance from the center of circle "A" to the center of circle "B" is (E), 3.712.

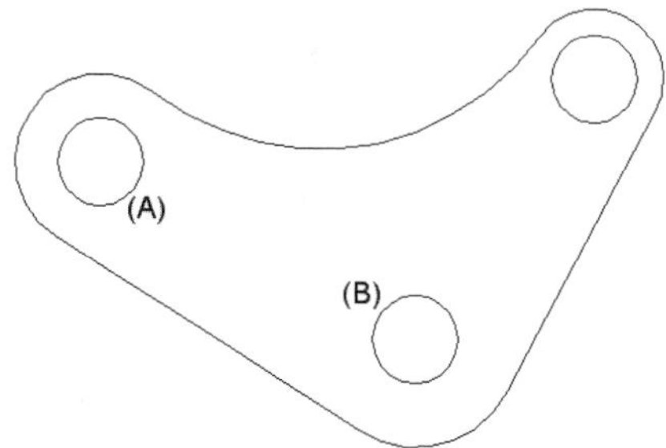

Figure 13.26

QUESTION 3

What is the angle formed in the XY plane from the center of circle "C" to the center of circle "B"?

a. 223°

b. 228°

c. 233°

d. 238°

e. 243°

Use the DIST (Distance) command to calculate the angle from the center of circle "C" to the center of circle "B" in the following image. Be sure to use the OSNAP-Center mode for locating the centers of all circles.

```
Command: DI (For DIST)
Specify first point: (Select the edge of the circle at "C")
Specify second point: (Select the edge of the circle at "B")
Distance = 3.104, Angle in XY Plane = 238, Angle from XY Plane = 0
Delta X = -1.649, Delta Y = -2.630, Delta Z = 0.000
```

The angle formed in the XY plane from the center of circle "C" to the center of circle "B" is (D), 238°.

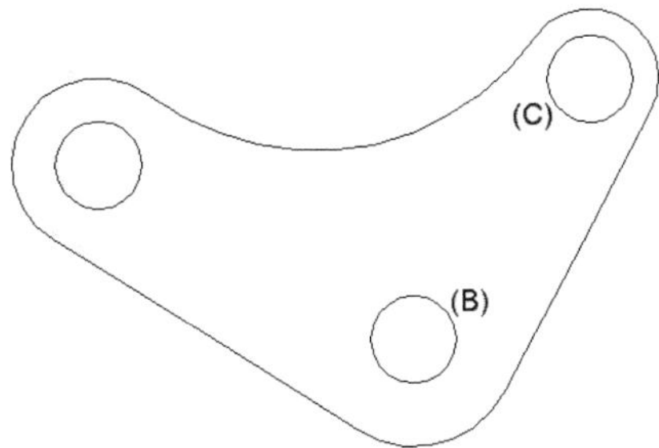

Figure 13.27

QUESTION 4

What is the delta X,Y distance from the center of circle "C" to the center of circle "A"?

a. -4.907,-0.851

b. -4.907,-0.856

c. -4.907,-0.861

d. -4.907,-0.866

e. -4.907,-0.871

Use the DIST (Distance) command to calculate the delta X,Y distance from the center of circle "C" to the center of circle "A" in the following image. Be sure to use the OSNAP-Center mode. Notice that additional information is given when you use the DIST command. For the purpose of this question, we will be looking only for the delta X,Y distance. The DIST command displays the relative XYZ distances. Since this is a 2D problem, only the X and Y values will be used.

```
Command: DI (For DIST)
Specify first point: (Select the edge of the circle at "C")
Specify second point: (Select the edge of the circle at "A")
Distance = 4.980, Angle in XY Plane = 190, Angle from XY Plane = 0
Delta X = -4.907, Delta Y = -0.851, Delta Z = 0.000
```

The delta X,Y distance from the center of circle "C" to the center of circle "A" is (A), -4.907,-0.851.

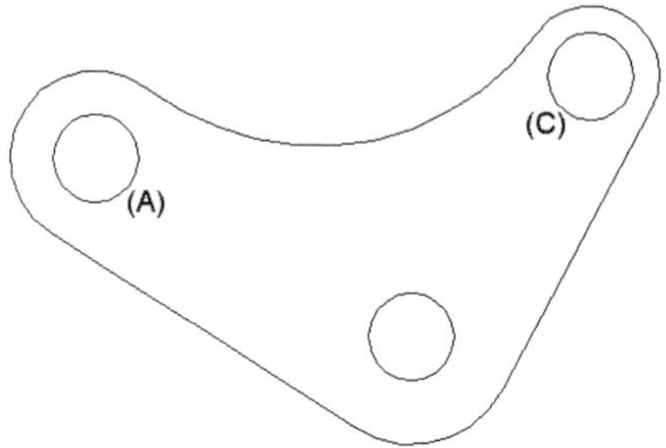

Figure 13.28

QUESTION 5

What is the absolute coordinate value of the center of arc "D"?

a. 5.869,8.218

b. 5.869,8.223

c. 5.869,8.228

d. 5.869,8.233

e. 5.869,8.238

The ID command is used to get the current absolute coordinate information on a desired point, as shown in the following image. This command displays the XYZ coordinate values. Since this is a 2D problem, only the X and Y values will be used.

```
Command: ID
Specify point: Cen
of (Select the edge of the arc at "D")
X = 5.869 Y = 8.223 Z = 0.000
```

The absolute coordinate value of the center of arc "D" is (B), 5.869,8.223.

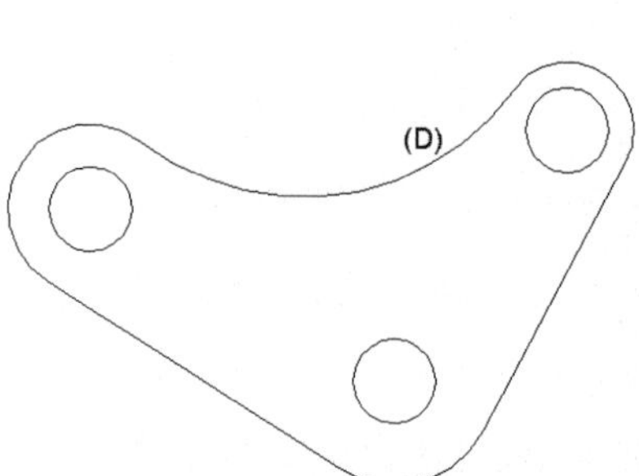

Figure 13.29

QUESTION 6

What is the total length of line “E”?

a. 3.074

b. 3.079

c. 3.084

d. 3.089

e. 3.094

Use the DIST (Distance) command to find the total length of line “E” in the following image. Be sure to use the OSNAP-Endpoint mode. Notice that additional information is given when you use the DIST command. For the purpose of this question, we will be looking only for the distance.

```
Command: DI (For DIST)
Specify first point: (Select the endpoint of the line at “X”)
Specify second point: (Select the endpoint of the line at “Y”)
Distance = 3.084, Angle in XY Plane = 64, Angle from XY Plane = 0
Delta X = 1.328, Delta Y = 2.783, Delta Z = 0.000
```

The total length of line “E” is (C), 3.084.

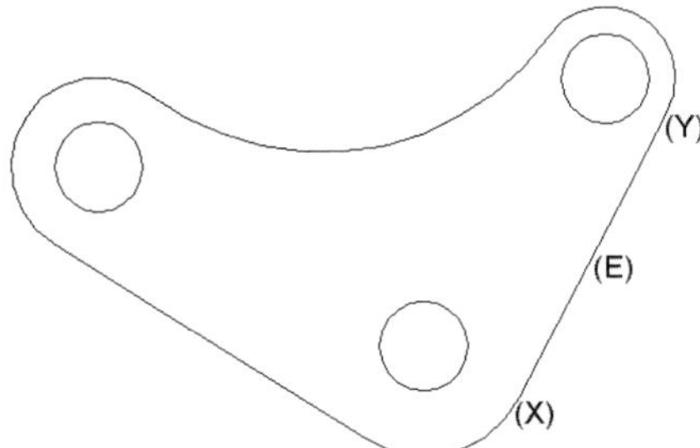

Figure 13.30

QUESTION 7

What is the total length of arc “F”?

a. 2.051

b. 2.056

c. 2.061

d. 2.066

e. 2.071

The LIST command is used to calculate the lengths of arcs. However, a little preparation is needed before you perform this operation. If arc “F” is selected, as shown on the left in the following image, notice that the entire outline is selected because it is a polyline. Use the EXPLODE command to break the outline into individual objects. Use the LIST command to get a listing of the arc length, as shown on the right in the following image.

```
Command: X (For EXPLODE)
Select objects: (Select the edge of the dashed polyline in the
    following image)
Select objects: (Press ENTER to perform the explode operation)

Command: LI (For LIST)
Select objects: (Select the edge of the arc at "F" in the
    following image)
Select objects: (Press ENTER to continue)
ARC Layer: "Boundary"
Space: Model space
Handle = 94
center point, X= 8.649 Y= 6.005 Z= 0.000
radius 0.711
start angle 334
end angle 141
length 2.071
```

The total length of arc "F" is (E), 2.071.

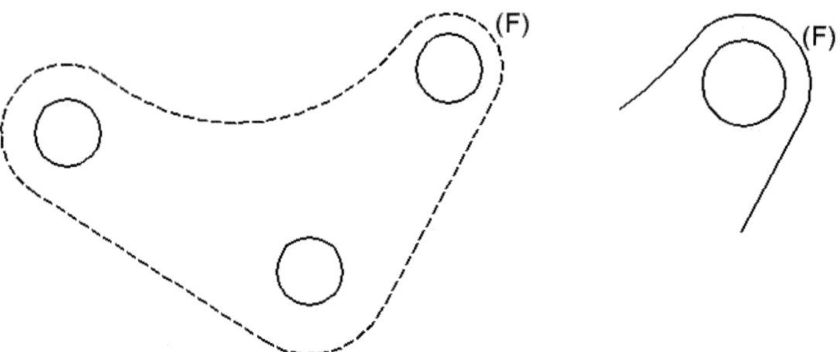

Figure 13.31

TUTORIAL EXERCISE: 13_FIELD_CALC.DWG

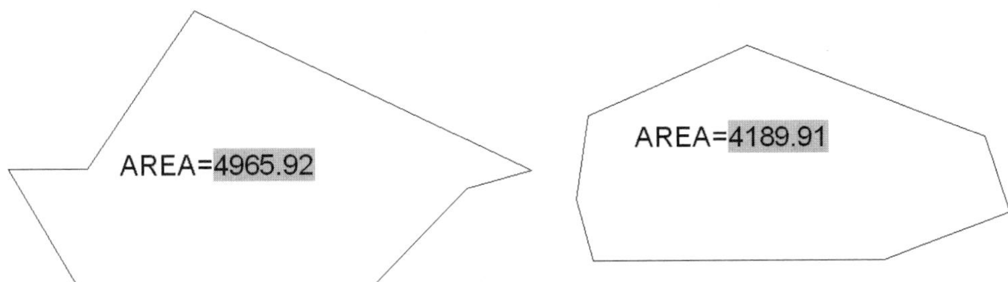

Figure 13.32

PURPOSE

This tutorial is designed to utilize the Field function of AutoCAD to calculate the area of an object.

SYSTEM SETTINGS

This drawing has drawing settings already created

LAYERS

Layers already exist in this drawing.

SUGGESTED COMMANDS

Open the drawing 13_Field_Calc.Dwg and activate the multiline text command. Set the text height to 12 and add a text header dealing with the Area. Then activate the Field dialog box and select a polyline object in the drawing. This retrieves the area information and adds this text calculation to the Text Formatting dialog box. Once the field is added, stretching the original polyline shape and regenerating the drawing recalculates the area.

Step 1

Open the drawing 13_Field_Calc.Dwg. Activate the Text Formatting dialog box, change the text height to 5, and add the text (AREA=), as shown in the following image. Continue by clicking on the Insert Field button.

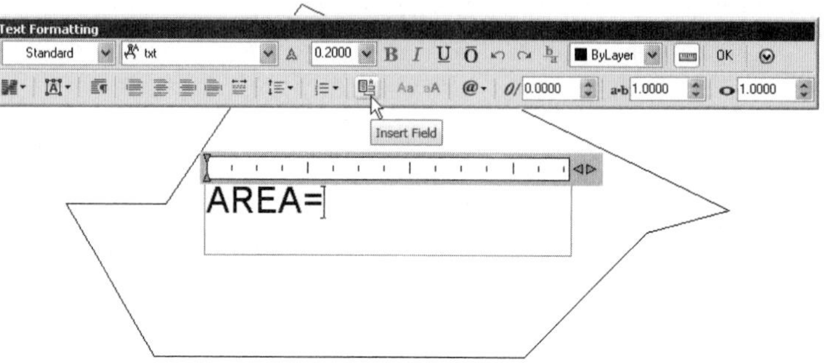

Figure 13.33

Step 2

When the Field dialog box appears, select the Object Field name, as shown in the following image. This prepares the dialog box to display field information based on the object that you select. Continue by clicking the Select object button as shown in the following image.

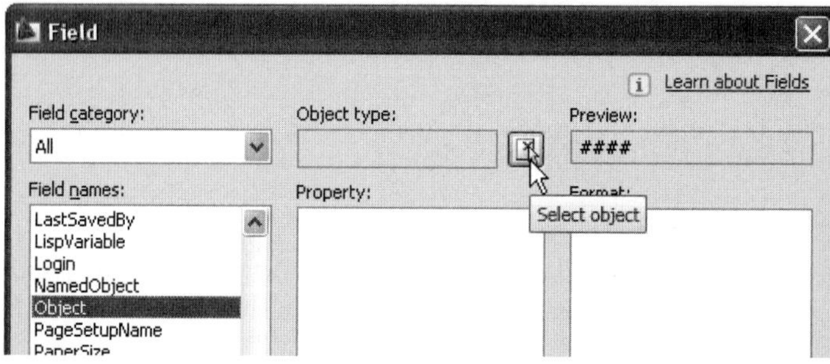

Figure 13.34

Step 3

Clicking the Select object button returns you to the drawing editor, where you can select the polyline shape of the object. This action again returns you to the Field dialog box. Verify the following changes in this dialog box: under the Field Names category, pick Object; under the Property category, pick Area; under the Format category, pick Decimal. Then click the OK button to exit the Field dialog box.

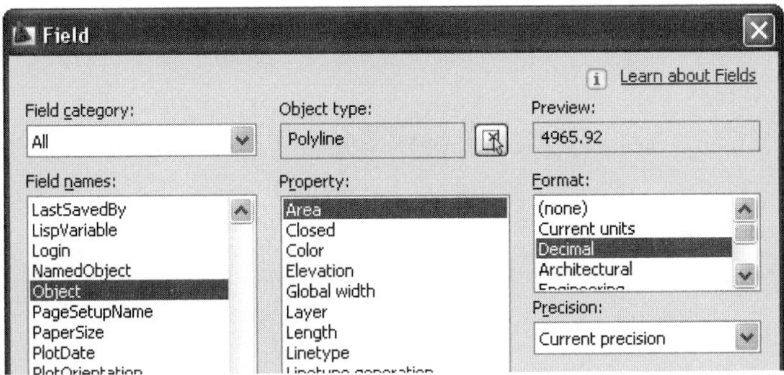

Figure 13.35

Step 4

The results are illustrated in the following image. A field consisting of the square footage information is added to the end of the multiline text.

Figure 13.36

Step 5

Now click the polyline shape to display grips, choose a number of grips, and stretch the vertices of the polyline to new locations as shown in the following image. Your display may not match the following image exactly. You will notice that the square footage does not automatically update itself.

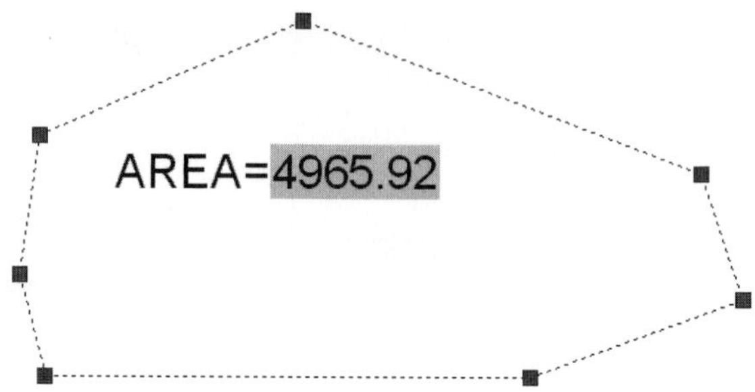

Figure 13.37

Step 6

To update the field information and change the area for the new polyline shape, click the REGEN command found under the View pull-down menu, as shown in the following image.

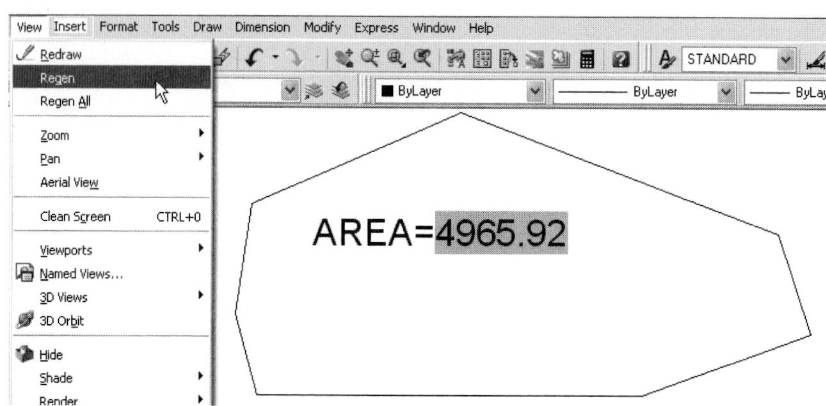

Figure 13.38

Step 7

The results are illustrated in the following image. Notice that the area field has changed to reflect the new size of the closed polyline shape.

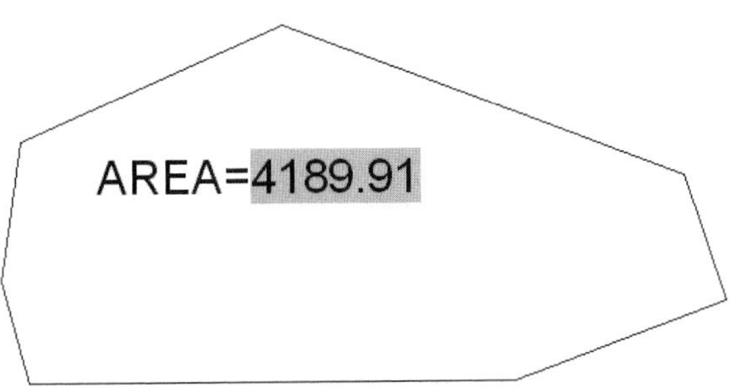

Figure 13.39

PROBLEMS FOR CHAPTER 13

PROBLEM 13–1 ANGLEBLK.DWG

DIRECTIONS FOR ANGLEBLK.DWG

Use the Drawing Units dialog box to set the units to decimal. Set the precision to two. Be sure the system of angle measure is set to decimal degrees and the number of decimal places for the display of angles is zero.

Keep all remaining default unit values. Keep the default settings for the drawing limits.

Begin the drawing shown in the following image by locating the lower-left corner of Angleblk, identified by "X" at coordinate (2.35,3.17).

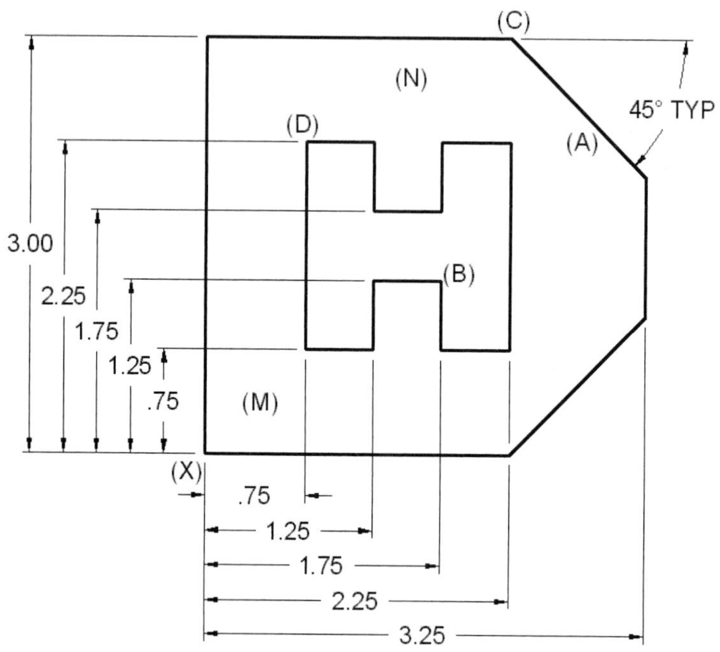

Refer to the drawing of the Angleblk and answer the following questions.

1. What is the total surface area of the Angleblk with the inner "H" shape removed?
   a. 6.66
   b. 6.77
   c. 6.89
   d. 7.00
   e. 7.11
2. What is the total area of the inner "H" shape?
   a. 1.75
   b. 1.81
   c. 1.87
   d. 1.93
   e. 1.99

3. What is the total length of line "A"?

   a. 1.29

   b. 1.35

   c. 1.41

   d. 1.47

   e. 1.53

4. What is the absolute coordinate value of the endpoint of the line at "B"?

   a. 4.04,4.42

   b. 4.10,4.42

   c. 4.16,4.42

   d. 4.22,4.42

   e. 4.28,4.42

5. What is the absolute coordinate value of the endpoint of the line at "C"?

   a. 4.60,6.11

   b. 4.60,6.17

   c. 4.60,6.23

   d. 4.60,6.29

   e. 4.60,6.35

6. Use the stretch command and extend the inner "H" shape a distance of 0.37 units in the 180° direction. Use "N" as the first corner of the crossing window and "M" as the other corner. Use the endpoint of "D" as the base point of the stretching operation. What is the new surface area of Angleblk with the inner "H" shape removed?

   a. 6.63

   b. 6.69

   c. 6.75

   d. 6.82

   e. 6.87

7. Use the scale command with the endpoint of the line at "D" as the base point. Reduce the size of just the inner "H" using a scale factor of 0.77. What is the new surface area of Angleblk with the inner "H" removed?

   a. 7.48

   b. 7.54

   c. 7.60

   d. 7.66

   e. 7.72

PROBLEM 13–2 LEVER2.DWG

DIRECTIONS FOR LEVER2.DWG

Use the Drawing Units dialog box to set the units to decimal. Keep the precision at four places.

Be sure the system of angle measure is set to decimal degrees and the number of decimal places for the display of angles is zero.

Keep the remaining default unit values.

Begin the drawing shown in the following image by locating the center of the 1.0000-diameter circle at absolute coordinate (2.2500,4.0000).

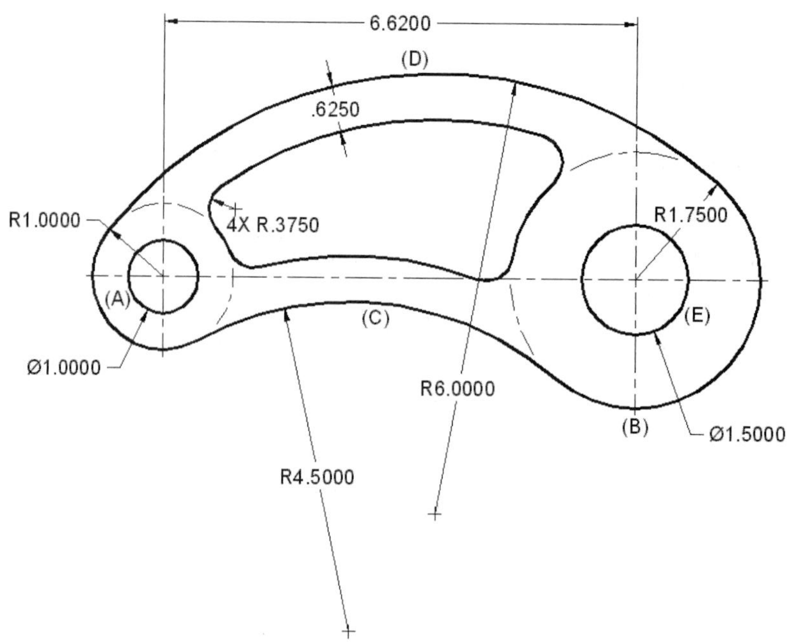

Refer to the drawing of Lever2 and answer the following questions.

1. What is the total area of Lever2 with the inner irregular shape and both holes removed?
   a. 17.6813
   b. 17.6819
   c. 17.6825
   d. 17.6831
   e. 17.6837
2. What is the absolute coordinate value of the center of the 4.5000-radius arc "C"?
   a. 4.8944,-0.0822
   b. 4.8944,-0.8226
   c. 4.8950,-0.8232
   d. 4.8956,-0.8238
   e. 4.8962,-0.8244
3. What is the absolute coordinate value of the center of the 6.0000-radius arc "D"?
   a. 6.0828,0.7893
   b. 6.0834,0.7899
   c. 6.0840,0.7905
   d. 6.0846,0.7911
   e. 6.0852,0.7917
4. What is the total length of arc "C"?
   a. 5.3583
   b. 5.3589
   c. 5.3595
   d. 5.3601
   e. 5.3607

5. What is the distance from the center of the 1.0000-diameter circle "A" to the intersection of the circle and centerline at "B"?
   a. 6.8456
   b. 6.8462
   c. 6.8474
   d. 6.8480
   e. 6.8486
6. What is the angle formed in the XY plane from the upper quadrant of arc "D" to the center of the 1.5000-circle "E"?
   a. 313°
   b. 315°
   c. 317°
   d. 319°
   e. 321°
7. What is the delta X,Y distance from the upper quadrant of arc "C" to the center of the 1.0000-hole "A"?
   a. -2.6444,0.3220
   b. -2.6444,0.3226
   c. -2.6444,0.3232
   d. -2.6444,0.3238
   e. -2.6444,0.3244

PROBLEM 13–3 MAIN ROCKER.DWG

DIRECTIONS FOR MAIN ROCKER.DWG

Begin the construction of Main Rocker by keeping the default units set to decimal but changing the number of decimal places past the zero from four to two. Be sure the system of angle measure is set to decimal degrees and the number of decimal places for the display of angles is zero. Keep the remaining default unit values. Use the LIMITS command to change the drawing limits to (0.00,0.00) for the lower-left corner and (15.00,12.00) for the upper-right corner.

Begin the drawing by placing the center of the regular hexagon and 2.25-radius arc at coordinate (6.25,6.50).

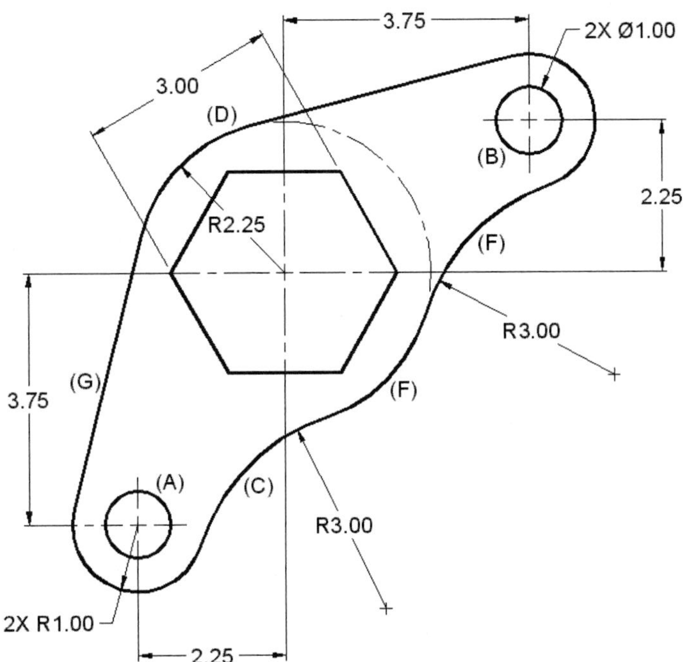

Refer to the drawing of Main Rocker and answer the following questions.

1. What is the distance from the center of the 1.00-diameter hole "A" to the center of the other 1.00-diameter hole "B"?

   a. 8.39

   b. 8.42

   c. 8.46

   d. 8.49

   e. 8.52

2. What is the absolute coordinate value of the center of the 3.00-radius arc "C"?

   a. 7.79,1.48

   b. 7.79,1.51

   c. 7.79,1.54

   d. 7.76,1.51

   e. 7.76,1.48

3. What is the total length of line "G"?

   a. 4.14

   b. 4.19

   c. 4.24

   d. 4.29

   e. 4.34

4. What is the length of the 2.25-radius arc segment "E"?

   a. 2.10

   b. 2.13

   c. 2.16

   d. 2.19

   e. 2.22

5. What is the angle formed in the XY plane from the center of the 1.00-diameter circle "B" to the center of the 2.25-radius arc "D"?

   a. 203°

   b. 205°

   c. 207°

   d. 209°

   e. 211°

6. What is the total surface area of Main Rocker with the hexagon and both 1.00-diameter holes removed?

   a. 20.97

   b. 21.00

   c. 21.03

   d. 21.06

   e. 21.09

7. What is the total length of arc "D"?
   a. 2.31
   b. 2.36
   c. 2.41
   d. 2.46
   e. 2.51

PROBLEM 13–4 PATTERN4.DWG

DIRECTIONS FOR PATTERN4.DWG

Start a new drawing called Pattern4. Even though this is a metric drawing, no special limits need be set. Keep the default setting of decimal units precision to zero. Be sure the system of angle measure is set to decimal degrees and the number of decimal places for the display of angles is zero. Keep all remaining default unit values.

Begin the drawing, as shown in the following image, by constructing Pattern4 with vertex "A" at absolute coordinate (50,30).

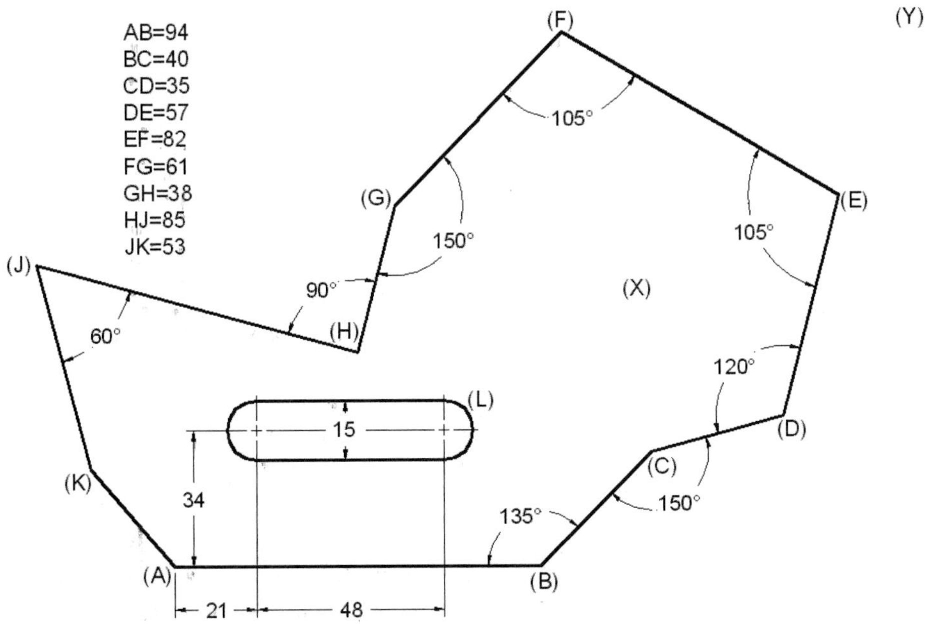

Refer to the drawing of Pattern4 and answer the following questions.

1. What is the total distance from the intersection of vertex "K" to the intersection of vertex "A"?
   a. 33
   b. 34
   c. 35
   d. 36
   e. 37
2. What is the total area of Pattern4 with the slot removed?
   a. 14493
   b. 14500
   c. 14529
   d. 14539
   e. 14620
3. What is the perimeter of the outline of Pattern4?
   a. 570
   b. 578
   c. 586
   d. 594
   e. 602
4. What is the distance from the intersection of vertex "A" to the intersection of vertex "E"?
   a. 186
   b. 190
   c. 194
   d. 198
   e. 202

5. What is the absolute coordinate value of the intersection at vertex "G"?
   a. 104,117
   b. 105,118
   c. 106,119
   d. 107,120
   e. 108,121
6. What is the total length of arc "L"?
   a. 20
   b. 21
   c. 22
   d. 23
   e. 24
7. Stretch the portion of Pattern4 around the vicinity of angle "E." Use "Y" as the first corner of the crossing box. Use "X" as the other corner. Use the endpoint of "E" as the base point of the stretching operation. For the new point, enter a polar coordinate value of 26 units in the 40° direction. What is the new degree value of the angle formed at vertex "E"?
   a. 78°
   b. 79°
   c. 80°
   d. 81°
   e. 82°

PROBLEM 13–5 ARCHITECTURAL APPLICATION—APARTMENT.DWG

DIRECTIONS FOR APARTMENT.DWG

Change from decimal units to architectural units. Keep all remaining default values. The thickness of all walls measures 4".

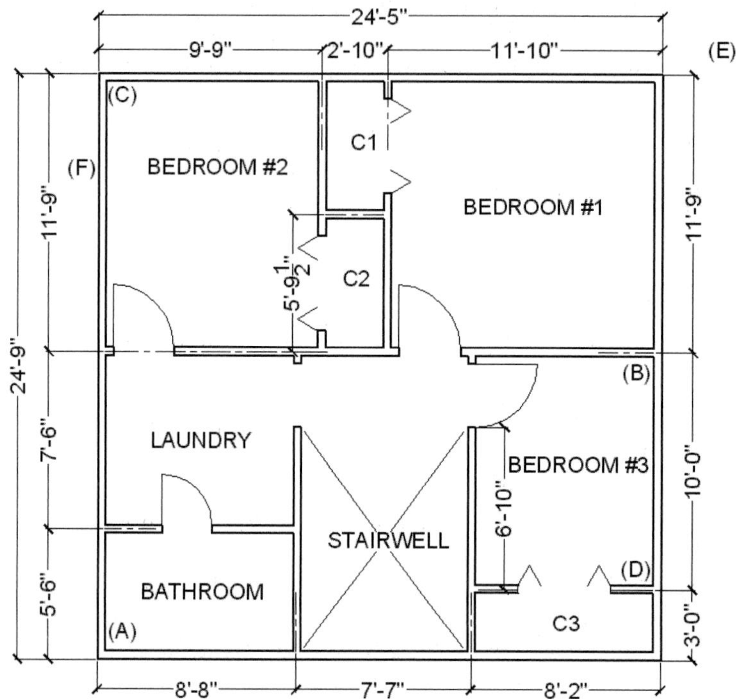

Refer to the drawing of Apartment and answer the following questions.

1. What is the total area in square feet of all bedrooms? (Pick the closest value.)
   a. 296 sq. ft.
   b. 306 sq. ft.
   c. 316 sq. ft.
   d. 326 sq. ft.
   e. 336 sq. ft.
2. What is the distance from the intersection of the corner at "A" to the intersection of the corner at "B"? (Pick the closest value.)
   a. 25'-2"
   b. 25'-7"
   c. 26'-0"
   d. 26'-5"
   e. 26'-10"

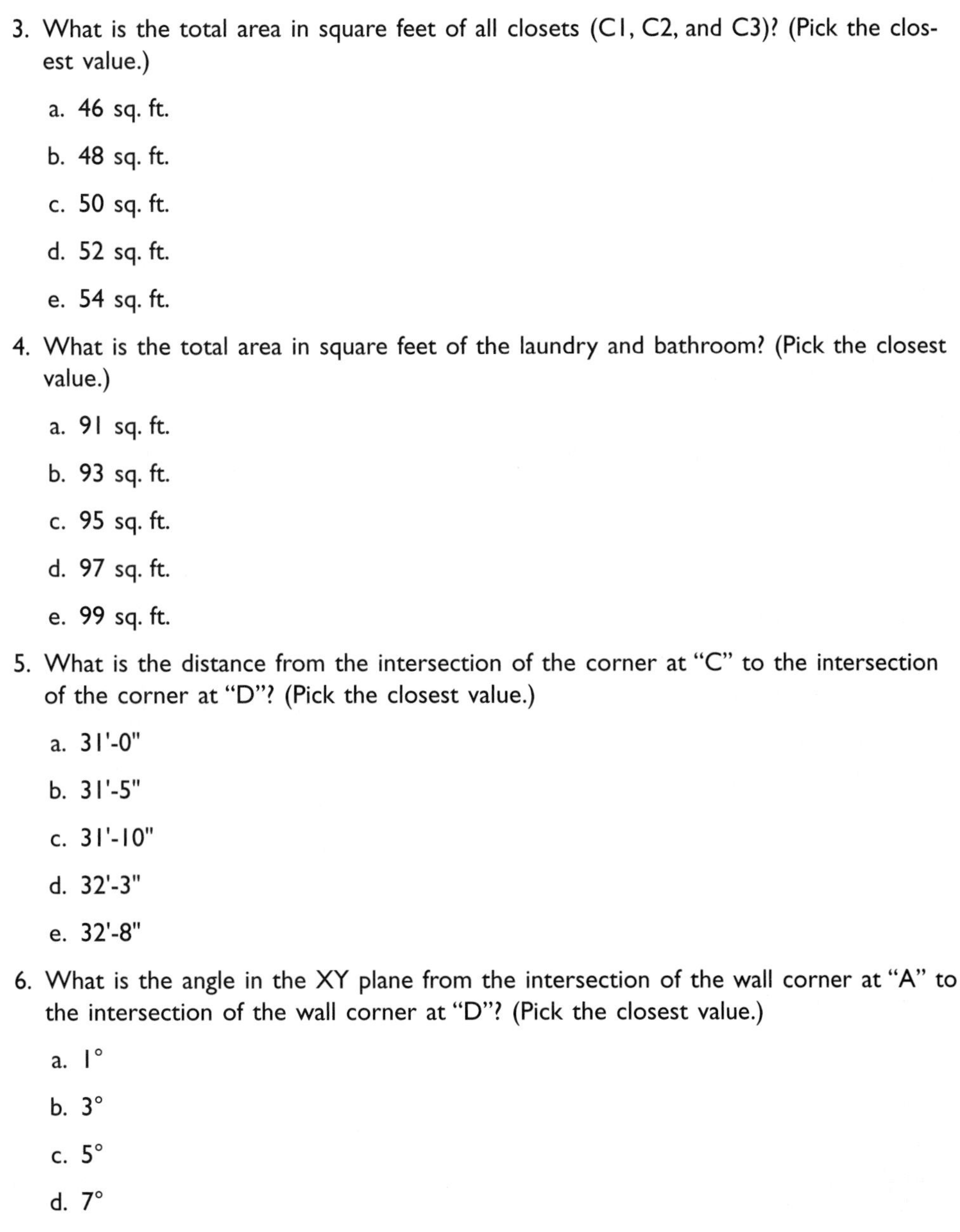

3. What is the total area in square feet of all closets (C1, C2, and C3)? (Pick the closest value.)
   a. 46 sq. ft.
   b. 48 sq. ft.
   c. 50 sq. ft.
   d. 52 sq. ft.
   e. 54 sq. ft.
4. What is the total area in square feet of the laundry and bathroom? (Pick the closest value.)
   a. 91 sq. ft.
   b. 93 sq. ft.
   c. 95 sq. ft.
   d. 97 sq. ft.
   e. 99 sq. ft.
5. What is the distance from the intersection of the corner at "C" to the intersection of the corner at "D"? (Pick the closest value.)
   a. 31'-0"
   b. 31'-5"
   c. 31'-10"
   d. 32'-3"
   e. 32'-8"
6. What is the angle in the XY plane from the intersection of the wall corner at "A" to the intersection of the wall corner at "D"? (Pick the closest value.)
   a. 1°
   b. 3°
   c. 5°
   d. 7°
   e. 9°

7. Stretch the floor plan up at a distance of 2'-5". Use "E" and "F" as the locations of the crossing box. What is the new total area of Bedroom #1, Bedroom #2, C1, and C2? (Pick the closest value.)
   a. 315 sq. ft.
   b. 319 sq. ft.
   c. 323 sq. ft.
   d. 327 sq. ft.
   e. 331 sq. ft.

INTERMEDIATE-LEVEL DRAWINGS

PROBLEM 13–6 GASKET2.DWG

DIRECTIONS FOR GASKET2.DWG

Use the Drawing Units dialog box to set the units to decimal. Set the precision to two. Be sure the system of angle measure is set to decimal degrees and the number of decimal places for the display of angles is zero.

Keep the remaining default unit values.

Begin the drawing by locating the center of the 6.00 × 3.00 rectangle at absolute coordinate (6.00,4.75).

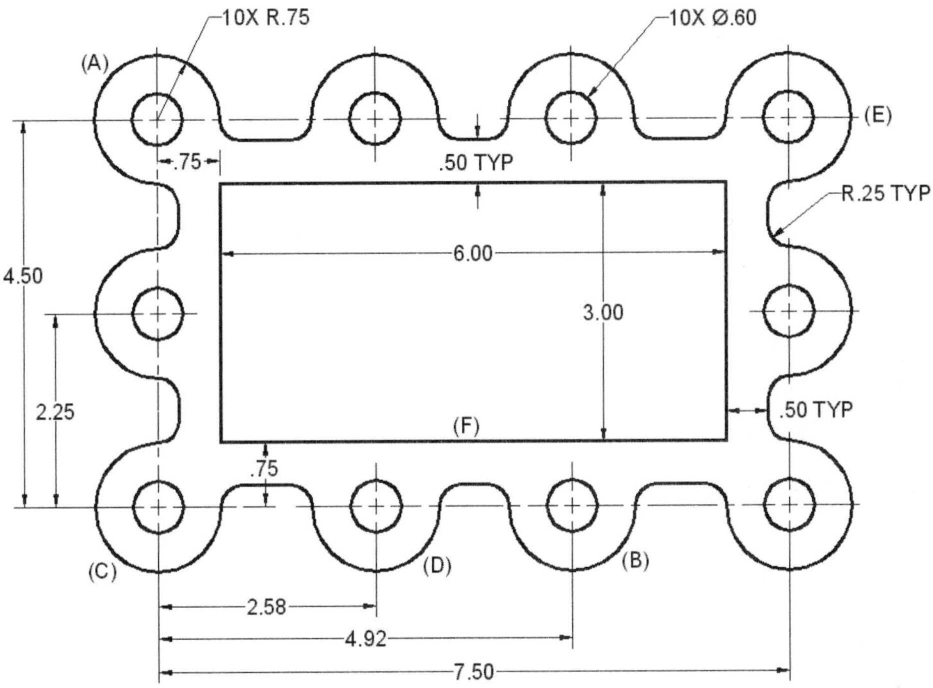

Refer to the drawing of Gasket2 and answer the following questions.

1. What is the total surface area of Gasket2 with the rectangle and all ten holes removed?

   a. 21.46

   b. 21.48

   c. 21.50

   d. 21.52

   e. 21.54

2. What is the distance from the center of arc "A" to the center of arc "B"?

   a. 6.63

   b. 6.67

   c. 6.71

   d. 6.75

   e. 6.79

3. What is the length of arc segment "C"?
   a. 3.44
   b. 3.47
   c. 3.50
   d. 3.53
   e. 3.56
4. What is the absolute coordinate value of the center of the 0.75-radius arc "D"?
   a. 4.83,2.50
   b. 4.83,2.47
   c. 4.83,2.53
   d. 4.80,2.50
   e. 4.83,2.56
5. What is the angle formed in the XY plane from the center of the 0.75-radius arc "D" to the center of the 0.75-radius arc "A"?
   a. 116°
   b. 118°
   c. 120°
   d. 122°
   e. 124°
6. What is the delta X,Y distance from the intersection at "E" to the midpoint of the line at "F"?
   a. -4.50,3.65
   b. -4.50,-3.65
   c. -4.50,-3.70
   d. -4.50,3.75
   e. -4.50,-3.75

7. Use the SCALE command to reduce the size of the inner rectangle. Use the midpoint of the line at "F" as the base point. Use a scale factor of 0.83 units. What is the new total surface area with the rectangle and all ten holes removed?

   a. 26.99

   b. 27.04

   c. 27.09

   d. 27.14

   e. 27.19

PROBLEM 13–7 HANGER.DWG

DIRECTIONS FOR HANGER.DWG

Use the Drawing Units dialog box to set units to decimal. Set precision to three. Be sure the system of angle measure is set to decimal degrees and the number of decimal places for the display of angles is zero. Keep all remaining default unit values. Use the LIMITS command and set the upper-right corner of the screen area to a value of (250.000,150.000). Perform a Zoom All operation.

Begin drawing the hanger by locating the center of the 40.000-radius arc at absolute coordinate (55.000,85.000).

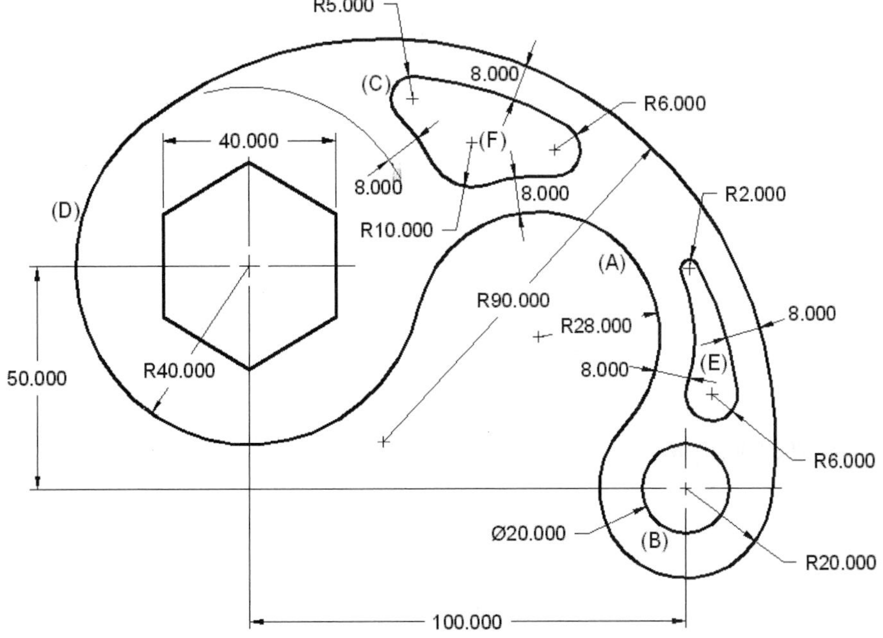

Refer to the drawing of the hanger and answer the following questions.

1. What is the area of the outer profile of the hanger?
   a. 9970.567
   b. 9965.567
   c. 9975.567
   d. 9975.005
   e. 9980.347
2. What is the area of the hanger with the polygon, circle, and irregular shapes removed?
   a. 7304.089
   b. 7305.000
   c. 7303.890
   d. 7304.000
   e. 7306.098
3. What is the absolute coordinate value of the center of the 28.000-radius arc "A"?
   a. 120.000,69.000
   b. 121.082,68.964
   c. 121.520,68.237
   d. 121.082,69.000
   e. 121.082,66.937
4. What is the absolute coordinate value of the center of the 20.000-diameter circle "B"?
   a. 154.000,35.000
   b. 156.000,36.000
   c. 156.147,35.256
   d. 155.000,35.000
   e. 156.000,37.000

5. What is the angle formed in the XY plane from the center of the 5.000-radius arc "C" to the center of the 40.000-radius arc "D"?

   a. 219°

   b. 221°

   c. 223°

   d. 225°

   e. 227°

6. What is the total area of irregular shape "E"?

   a. 271.613

   b. 271.723

   c. 271.784

   d. 271.801

   e. 271.822

7. What is the total area of irregular shape "F"?

   a. 698.511

   b. 698.621

   c. 699.817

   d. 699.856

   e. 699.891

PROBLEM 13–8 HOUSING1.DWG

DIRECTIONS FOR HOUSING1.DWG

Begin constructing Housing1 by keeping the default units set to decimal but changing the precision to three. Be sure the system of angle measure is set to decimal degrees and the number of decimal places for the display of angles is zero. Place the center of the 1.500-radius circular centerline at absolute coordinate (6.500,5.250).

Before constructing the outer ellipse, set the PELLIPSE command to a value of 1. This draws the ellipse as a polyline object.

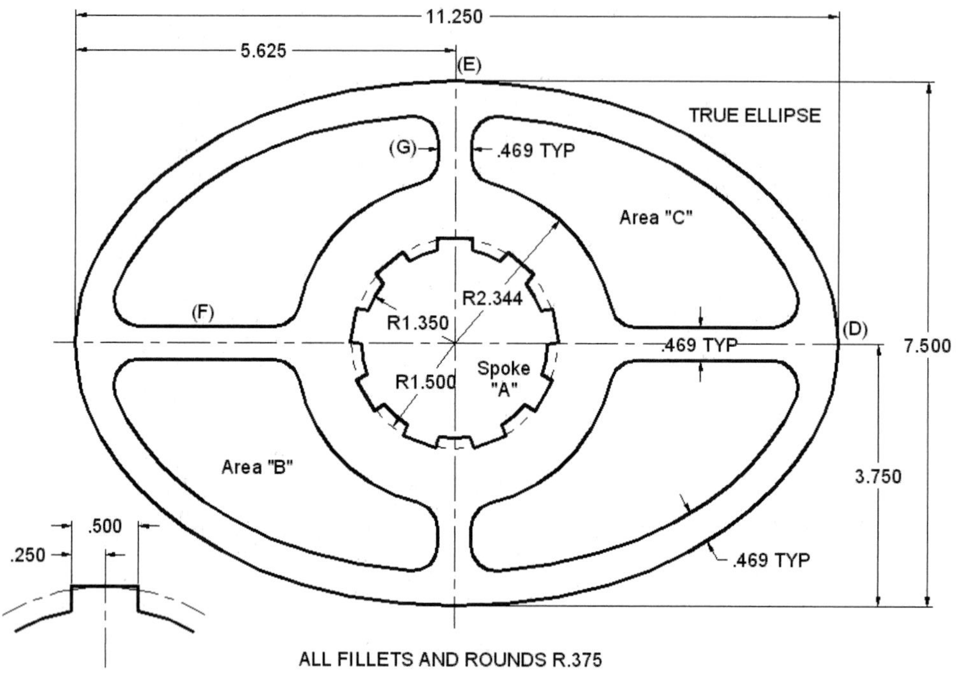

Refer to the drawing of Housing1 and answer the following questions.

1. What is the perimeter of Spoke "A"?
   a. 11.564
   b. 11.570
   c. 11.576
   d. 11.582
   e. 11.588
2. What is the perimeter of Area "B"?
   a. 12.513
   b. 12.519
   c. 12.525
   d. 12.531
   e. 12.537

3. What is the total area of Area "C"?
   a. 7.901
   b. 7.907
   c. 7.913
   d. 7.919
   e. 7.928
4. What is the absolute coordinate value of the intersection of the ellipse and centerline at "D"?
   a. 12.125,5.244
   b. 12.125,5.250
   c. 12.125,5.256
   d. 12.125,5.262
   e. 12.125,5.268
5. What is the total surface area of Housing1 with the spoke and all slots removed?
   a. 27.095
   b. 28.101
   c. 28.107
   d. 28.113
   e. 28.119
6. What is the distance from the midpoint of the horizontal line segment at "F" to the midpoint of the vertical line segment at "G"?
   a. 4.235
   b. 4.241
   c. 4.247
   d. 4.253
   e. 4.259

7. Increase Spoke "A" in size using the SCALE command. Use the center of the 1.500-radius arc as the base point. Use a scale factor of 1.115 units. What is the new total area of Housing1 with the spoke and all slots removed?

   a. 26.535

   b. 26.541

   c. 26.547

   d. 26.553

   e. 26.559

PROBLEM 13–9 ROTOR.DWG

DIRECTIONS FOR ROTOR.DWG

Start a new drawing called Rotor, as shown in the following image. Keep the default setting of decimal units precision to three.

Be sure the system of angle measure is set to decimal degrees and the number of decimal places for the display of angles is zero. Keep all remaining default unit values. Begin the drawing by constructing the center of the 6.250-unit-diameter circle at "A" at absolute coordinate (5.500,5.000).

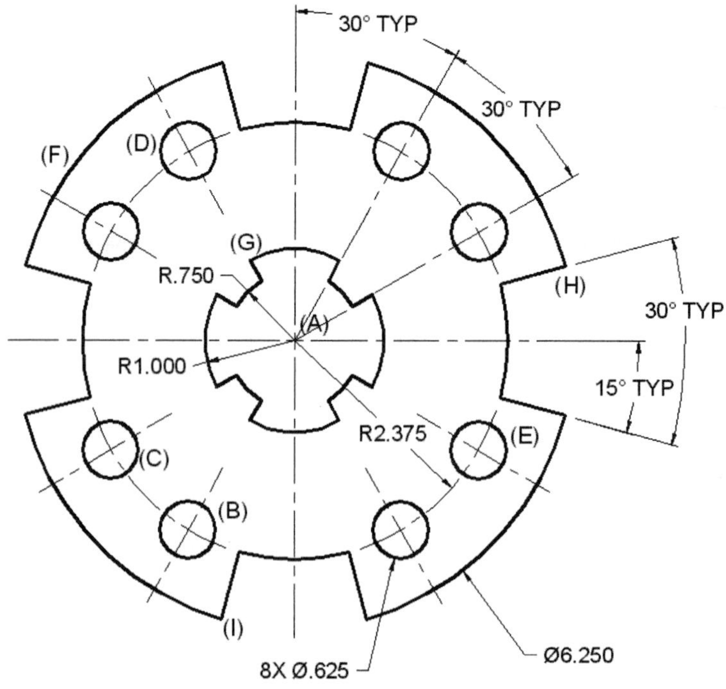

Refer to the drawing of the rotor and answer the following questions.

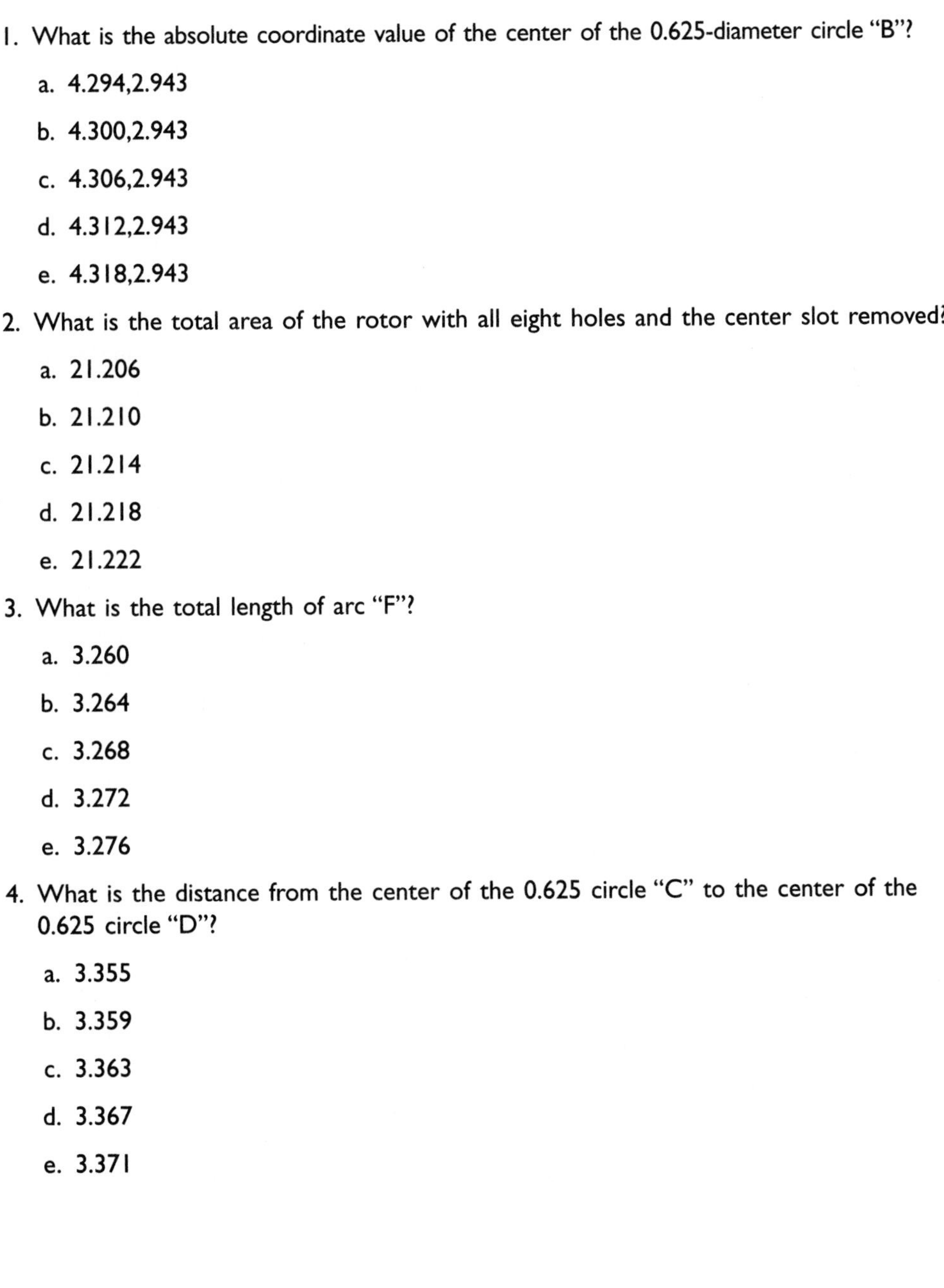

1. What is the absolute coordinate value of the center of the 0.625-diameter circle "B"?
   a. 4.294,2.943
   b. 4.300,2.943
   c. 4.306,2.943
   d. 4.312,2.943
   e. 4.318,2.943
2. What is the total area of the rotor with all eight holes and the center slot removed?
   a. 21.206
   b. 21.210
   c. 21.214
   d. 21.218
   e. 21.222
3. What is the total length of arc "F"?
   a. 3.260
   b. 3.264
   c. 3.268
   d. 3.272
   e. 3.276
4. What is the distance from the center of the 0.625 circle "C" to the center of the 0.625 circle "D"?
   a. 3.355
   b. 3.359
   c. 3.363
   d. 3.367
   e. 3.371

5. What is the angle formed in the XY plane from the center of the 0.625 circle "B" to the center of the 0.625 circle "E"?

   a. 11°

   b. 13°

   c. 15°

   d. 17°

   e. 19°

6. What is the delta X,Y distance from the intersection at "H" to the intersection at "I"?

   a. -3.827,-3.827

   b. -3.827,3.827

   c. -3.834,3.820

   d. -3.841,3.813

   e. -3.848,-3.806

7. Use the SCALE command to increase the size of just the center slot "G." Use the center of arc "F" as the base point. Use a scale factor of 1.500 units. What is the new surface area of the rotor with the center slot and all eight holes removed?

   a. 17.861

   b. 17.868(E) 17.889

   c. 17.875

   d. 17.882

PROBLEM 13–10 PATTERN5.DWG

DIRECTIONS FOR PATTERN5.DWG

Begin the construction of Pattern5 by keeping the default units set to decimal but changing the precision to zero. Be sure the system of angle measure is set to decimal degrees and the number of decimal places for the display of angles is zero. Begin the drawing by placing Vertex "A" at absolute coordinate (190,30).

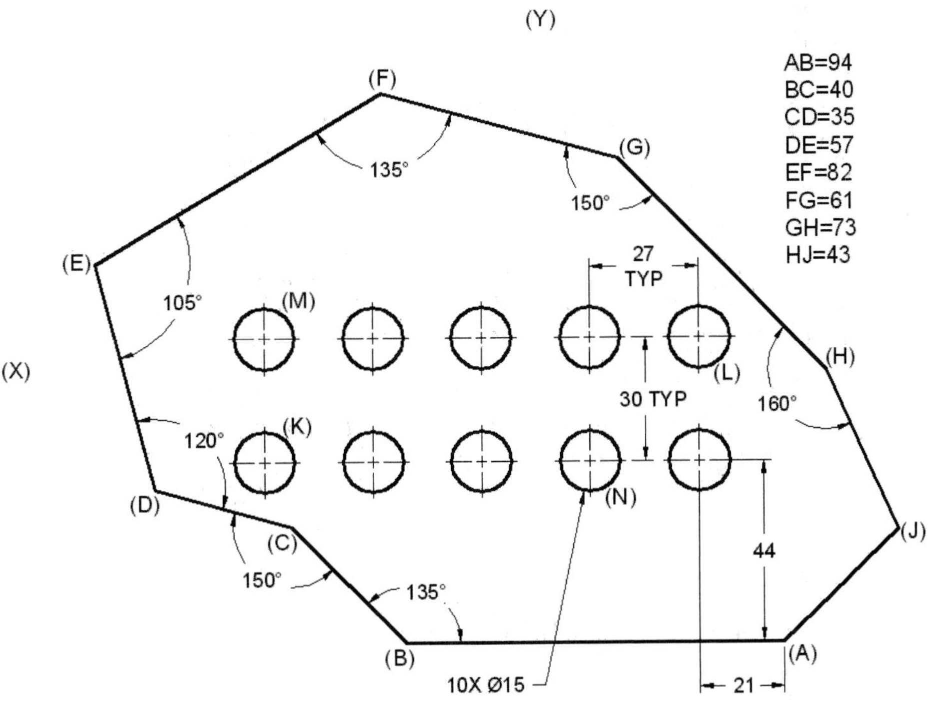

Refer to the drawing of Pattern5 and answer the following questions.

1. What is the distance from the intersection of vertex "J" to the intersection of vertex "A"?

   a. 38

   b. 39

   c. 40

   d. 41

   e. 42

2. What is the perimeter of Pattern5?

   a. 523

   b. 524

   c. 525

   d. 526

   e. 527

3. What is the total area of Pattern5 with all 10 holes removed?
   a. 16369
   b. 16370
   c. 16371
   d. 16372
   e. 16373
4. What is the distance from the center of the 15-diameter hole "K" to the center of the 15-diameter hole "L"?
   a. 109
   b. 110
   c. 111
   d. 112
   e. 113
5. What is the angle formed in the XY plane from the center of the 15-diameter hole "M" to the center of the 15-diameter hole "N"?
   a. 340°
   b. 342°
   c. 344°
   d. 346°
   e. 348°
6. What is the absolute coordinate value of the intersection at "F"?
   a. 90,163
   b. 92,163
   c. 94,165
   d. 96,167
   e. 98,169

7. Use the STRETCH command to lengthen Pattern5. Use "Y" as the first point of the stretch crossing box. Use "X" as the other corner. Pick the intersection at "F" as the base point and stretch Pattern5 a total of 23 units in the 180° direction. What is the new total area of Pattern5 with all ten holes removed?

   a. 17746

   b. 17753

   c. 17760

   d. 17767

   e. 17774

PROBLEM 13–11 RATCHET.DWG

DIRECTIONS FOR RATCHET.DWG

Use the Drawing Units dialog box to change the precision to two. Be sure the system of angle measure is set to decimal degrees and the number of decimal places for the display of angles is zero. Keep the remaining default unit values. Begin by drawing the center of the 1.00-radius arc of the ratchet at absolute coordinate (6.00,4.50), as shown in the following image.

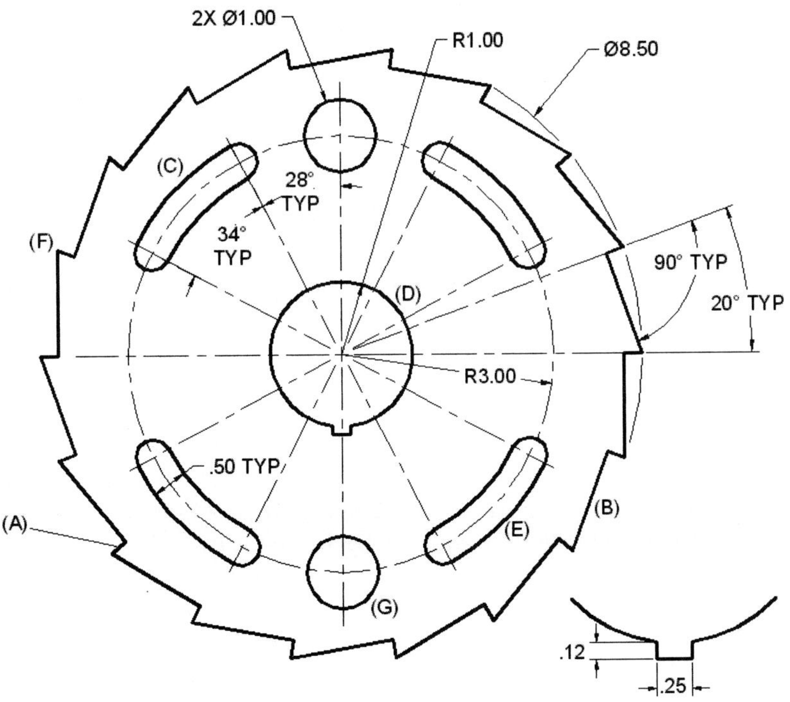

Refer to the drawing of the ratchet and answer the following questions.

1. What is the total length of the short line segment "A"?
   a. 0.22
   b. 0.24
   c. 0.26
   d. 0.28
   e. 0.30
2. What is the total length of line "B"?
   a. 1.06
   b. 1.19
   c. 1.33
   d. 1.45
   e. 1.57
3. What is the total length of arc "C"?
   a. 1.81
   b. 1.93
   c. 2.08
   d. 2.19
   e. 2.31
4. What is the perimeter of the 1.00-radius arc "D" with the 0.25 × 0.12 keyway?
   a. 6.52
   b. 6.63
   c. 6.77
   d. 6.89
   e. 7.02
5. What is the total surface area of the ratchet with all 4 slots, the two 1.00-diameter holes, and the 1.00-radius arc with the keyway removed?
   a. 42.98
   b. 43.04
   c. 43.10
   d. 43.16
   e. 43.22

6. What is the absolute coordinate value of the endpoint at "F"?

   a. 2.01,5.10

   b. 2.01,5.63

   c. 2.01,5.95

   d. 2.20,5.95

   e. 2.37,5.95

7. What is the angle formed in the XY plane from the endpoint of the line at "F" to the center of the 1.00-diameter hole "G"?

   a. 304°

   b. 306°

   c. 308°

   d. 310°

   e. 312°

PROBLEM 13–12 GENEVA.DWG

DIRECTIONS FOR GENEVA.DWG

Start a new drawing called Geneva. Keep the default settings of decimal units, and set precision to two. Be sure the system of angle measure is set to decimal degrees and the number of decimal places for the display of angles is zero. Begin the drawing by constructing the 1.50-diameter arc at absolute coordinate (7.50,5.50).

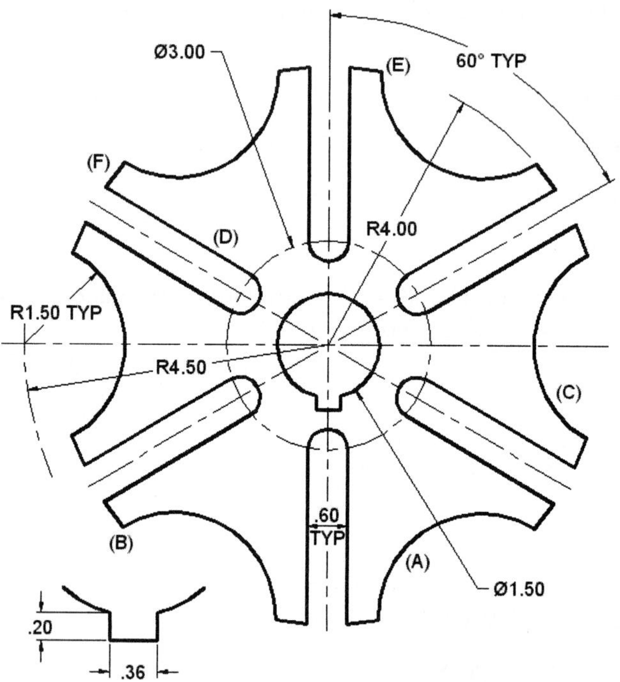

Refer to the drawing of the Geneva and answer the following questions.

1. What is the total length of arc "A"?
   a. 3.00
   b. 3.10
   c. 3.21
   d. 3.30
   e. 3.40
2. What is the angle formed in the XY plane from the intersection at "B" to the center of arc "C"?
   a. 11°
   b. 13°
   c. 15°
   d. 17°
   e. 19°

3. What is the absolute coordinate value of the midpoint of line "D"?
   a. 5.27,7.13
   b. 5.27,7.17
   c. 5.23,7.13
   d. 5.31,7.13
   e. 5.31,7.09
4. What is the total area of the Geneva with the 1.50-diameter hole and keyway removed?
   a. 27.40
   b. 27.44
   c. 27.48
   d. 27.52
   e. 27.66
5. What is the total distance from the midpoint of arc "F" to the center of arc "A"?
   a. 8.24
   b. 8.29
   c. 8.34
   d. 8.39
   e. 8.44
6. What is the delta X,Y distance from the intersection at "E" to the center of arc "C"?
   a. 3.93,-3.75
   b. 3.93,3.75
   c. 3.75,3.80
   d. 3.75,3.93
   e. 3.75,-3.93

7. Use the SCALE command to reduce the Geneva in size. Use 7.50,5.50 as the base point; use a scale factor of 0.83 units. What is the absolute coordinate value of the intersection at "E"?

   a. 8.12,8.71

   b. 8.12,8.76

   c. 8.12,8.81

   d. 8.12,8.86

   e. 8.12,8.91

PROBLEM 13–13 STRUCTURAL APPLICATION—GUSSET.DWG

DIRECTIONS FOR GUSSET.DWG

Begin the construction of the Gusset by keeping the default units set to decimal, but changing the precision to two.

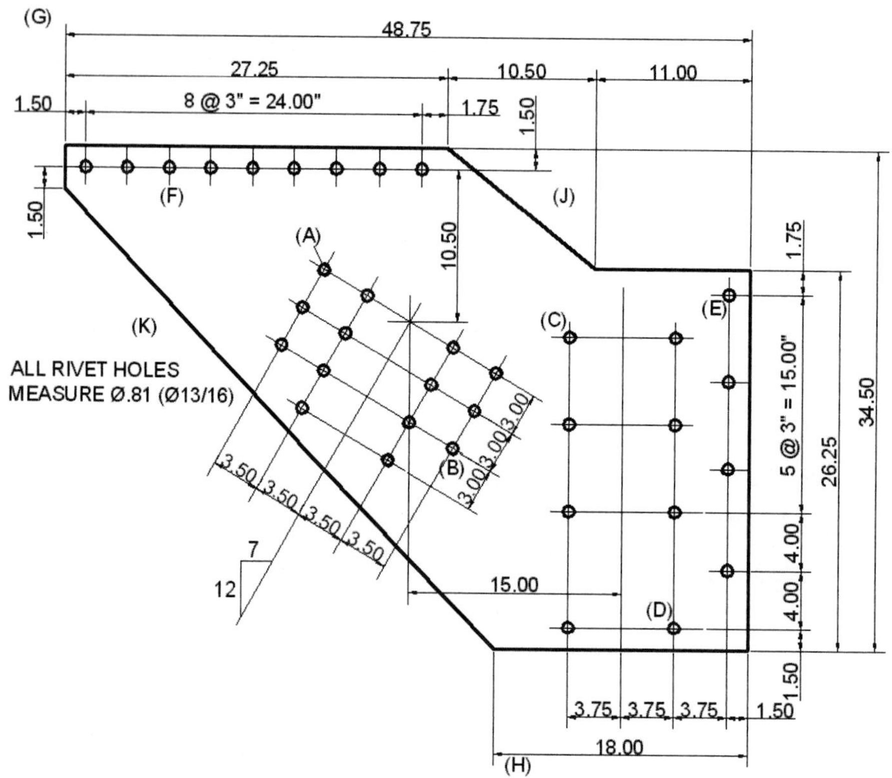

Refer to the drawing of the Gusset and answer the following questions.

1. What is the total area of the gusset plate with all thirty-five rivet holes removed?

   a. 1045.46

   b. 1050.67

   c. 1055.21

   d. 1060.40

   e. 1065.88

2. What is the delta X,Y distance from the center of hole "A" to the center of hole "B"?

   a. 9.07,-12.19

   b. 9.07,-12.24

   c. 9.07,-12.29

   d. 9.07,-12.34

   e. 9.07,-12.39

3. What is the angle formed in the XY plane from the center of hole "C" to the center of hole "D"?

   a. 285°

   b. 288°

   c. 291°

   d. 294°

   e. 297°

4. What is the angle formed in the XY plane from the center of hole "E" to the center of hole "F"?

   a. 165°

   b. 168°

   c. 171°

   d. 174°

   e. 177°

5. What is the total length of line "J"?
   a. 13.15
   b. 13.20
   c. 13.25
   d. 13.30
   e. 13.35
6. What is the total length of line "K"?
   a. 44.02
   b. 44.07
   c. 44.12
   d. 44.17
   e. 44.22
7. Stretch the gusset plate directly to the left using a crossing box from "H" to "G" and at a distance of 3.00 units. What is the new area of the gusset plate with all 35 rivet holes removed?
   a. 1121.98
   b. 1126.72
   c. 1131.11
   d. 1136.59
   e. 1141.34

CHAPTER 14

Working with Drawing Layouts

INTRODUCING DRAWING LAYOUTS

Up to this point in the book, all drawings have been constructed in Model Space. This consisted of drawings composed of geometry such as arcs, circles, lines, and even dimensions. You can even plot your drawing from Model Space; however, this gets a little tricky, especially when you attempt to plot the drawing at a known scale, or even trickier if multiple details need to be arranged on a single sheet at different scales.

You can elect to work in two separate spaces on your drawing, namely Model Space and Layout mode, sometimes referred to as Paper Space. Typically, part geometry and dimensions are drawn in Model Space. However, items such as notes, annotations, and title blocks are laid out separately in a drawing layout, which is designed to simulate an actual sheet of paper. To arrange a single view of a drawing or multiple views of different drawings, you arrange a series of viewports in the drawing layout to view the images. These viewports are mainly rectangular in shape and can be made any size. You can even create circular and polygonal viewports. A viewports tool allows you to scale the image inside the viewport. In this way, a series of images may be scaled differently even though the drawing layout will be plotted out at a scale of 1:1.

This chapter introduces you to the controls used in a drawing layout to manage information contained in a viewport. A tutorial exercise is provided to help you practice creating drawing layouts.

MODEL SPACE

Before starting on the topic of Paper Space, you must first understand the environment in which all drawings are originally constructed, namely Model Space. It is here that the drawing is drawn full size or in real-world units. Model Space is easily identified by the appearance of the User Coordinate System icon located in the lower-left corner of the active drawing area in the following image. This icon is associated with the Model Space environment. Some past and present users of AutoCAD have felt that it clutters the display screen and gets in the way of geometry, and they have used the UCSICON command to turn the icon off. While these points may seem true, the icon should be present at all times to identify the space you are working in. If you are utilizing the AutoCAD Classic workspace, another indicator that you are in Model Space is the presence of the Model tab, located just below the User Coordinate System icon, also in the following image.

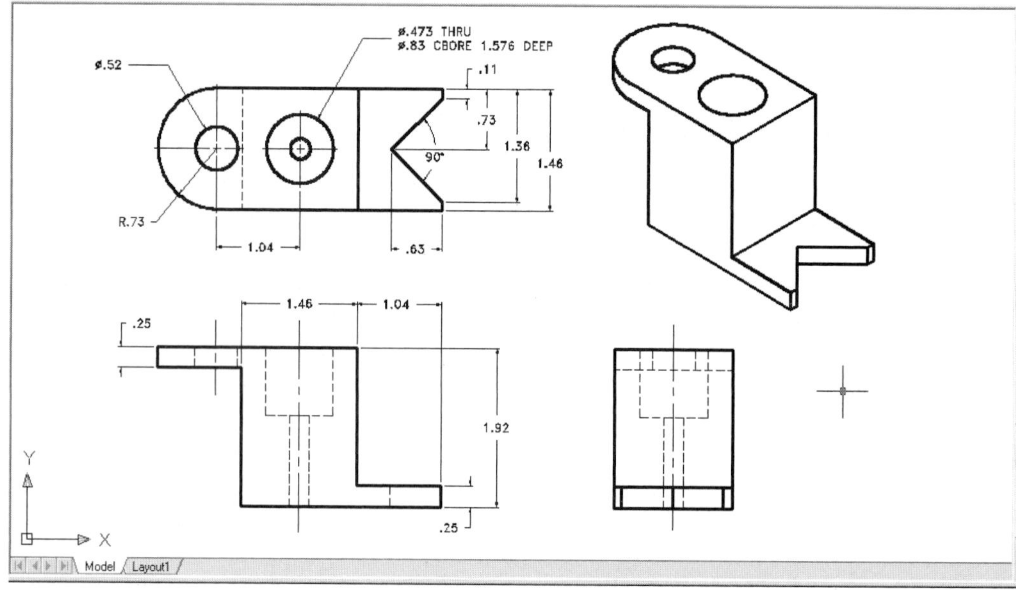

Figure 14.1

MODEL SPACE AND LAYOUTS

You have already seen that Model Space is the environment set aside for constructing your drawing in real-world units or full size, as shown on the left in the following image. Model space can be easily recognized by the familiar User Coordinate System icon found in the lower-left corner of every model space screen.

In the AutoCAD Classic workspace, clicking the Layout1 tab found in the lower-left corner of the display screen activates the layout environment. Paper Space is considered an area used to lay out your drawing before it is plotted. Layouts are also used to place title block information and notes associated with the drawing. The drawing illustrated on the right in the following image has been laid out in Paper Space and shows a sheet of paper with the drawing surrounded by a viewport. The dashed lines along the outer perimeter of the sheet are referred to as margins. Anything inside the margins will plot; for this reason, this is called the printable area. Notice also at the bottom of the screen that the Layout1 tab is activated. Both tabs at the bottom of the screen can be used to easily display your drawing in either Model Space or Paper Space. Also notice the icon in the lower-left corner of the illustration; this icon is in the form of a triangle and is used to quickly identify the Paper Space environment.

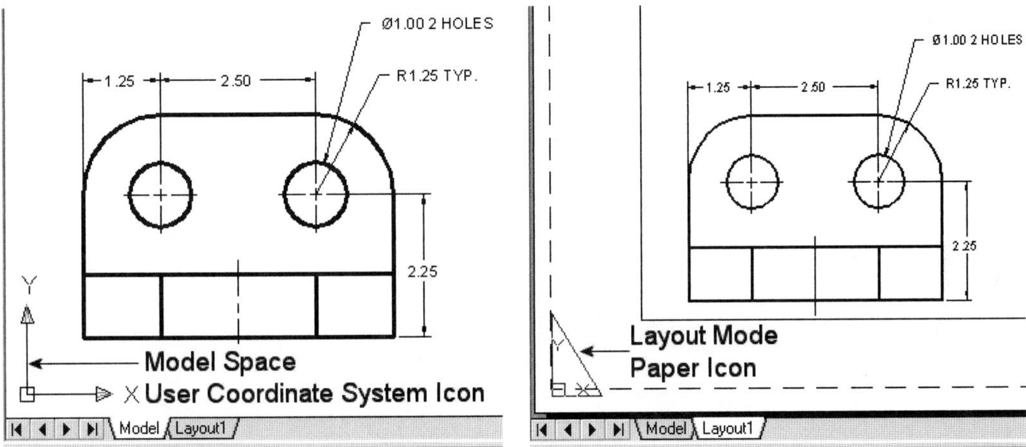

Figure 14.2

LAYOUT FEATURES

When you draw and plot from Model Space, the limits of the drawing have to be properly set depending on the scale of the drawing and the sheet of paper the drawing is planned for. Also, title block information has to be scaled depending on the scale of the drawing. The text height that made up the notes in a drawing also has to be scaled depending on the scale of the drawing. Finally, the scale of the drawing is applied to the Plot Settings tab of the Plot dialog box, which scales the drawing for placement on a certain sheet of paper. There can be many problems when you use this method. For instance, a certain design office assigned an individual to plot all drawings produced by the engineers and designers. It was this individual's responsibility to plot the drawings at the proper scale. However, with numerous drawings to plot, it was easy to get the plot scales confused. As a result, drawings were plotted at the wrong scale. This not only cost paper but valuable time. This scenario is the reason for using layouts; all drawings, no matter how many details are arranged, are plotted at 1 = 1 with little or no error encountered, through the use of layouts. Plotting becomes the fundamental reason for using the layout environment. Here are a few other reasons for arranging a drawing in a layout:

- Layouts are based on the actual sheet size. If you are plotting a drawing on a D-size sheet of paper, you use the actual size (36 × 24) in Paper Space.
- Title blocks and the text used for notes do not have to be scaled when placed in a layout.
- Viewports created in a layout are user-defined. Viewports created in Model Space are dependent on a configuration set by AutoCAD. In a layout, the viewports can be different sizes and shapes, depending on the information contained in the viewport.

- Multiple viewports can be created in a single layout, as in Model space. However, the images assigned to different layout viewports can be assigned different scales. Also, the control of layers in a layout is viewport-dependent. In other words, layers turned on in one viewport can be turned off or frozen in another viewport.
- All drawings, no matter how many viewports are created or details arranged, are plotted out at a scale of 1 = 1.

SETTING UP A PAGE

Before creating viewports in a layout, it is customary to first set up a page based on the sheet size, plotted scale, and plot style. To begin the process of setting up a page, right-click Layout1 or Quick View Layouts, depending on your workspace, and pick Page Setup Manager, as shown in the following image.

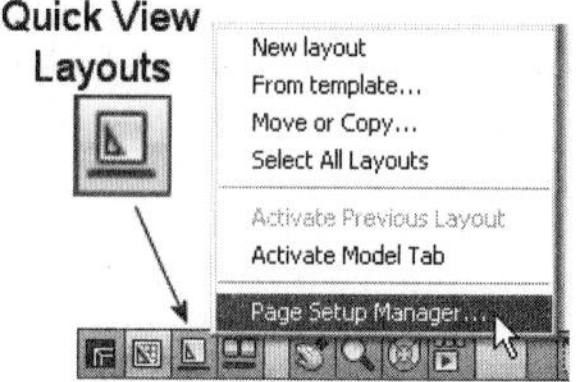

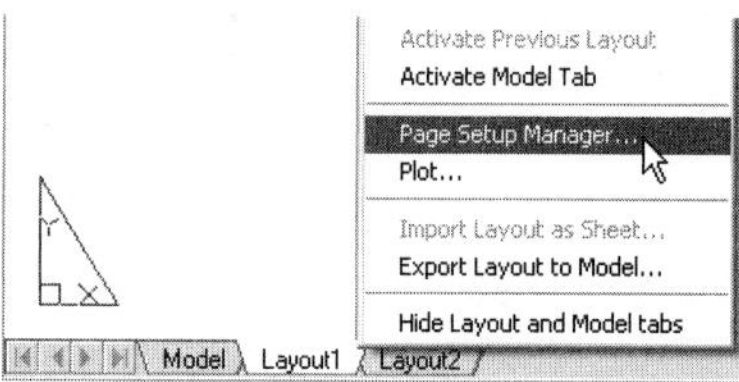

Figure 14.3

Performing the previous task displays the Page Setup Manager dialog box shown on the left in the following image. The purpose of this initial dialog box is to create a series of page setups under unique names. These page setup names can hold information such as plotter name, page sizes, and plot style tables used for pen assignments when plotting. To create a page setup, click the New button to display the New Page Setup dialog box shown on the right in the following image. When creating the new page setup name, try to give the page a name that will give you a hint for its intended purpose. In this dialog box, a new page setup name was entered: C-Size (DWF6). The "C" refers to the size of the drawing sheet. The DWF6 refers to the type of plot device. DWF stands for Drawing Web Format and is used to plot a drawing out to a file. Once in this DWF form, the file can be sent to others literally around the globe for review.

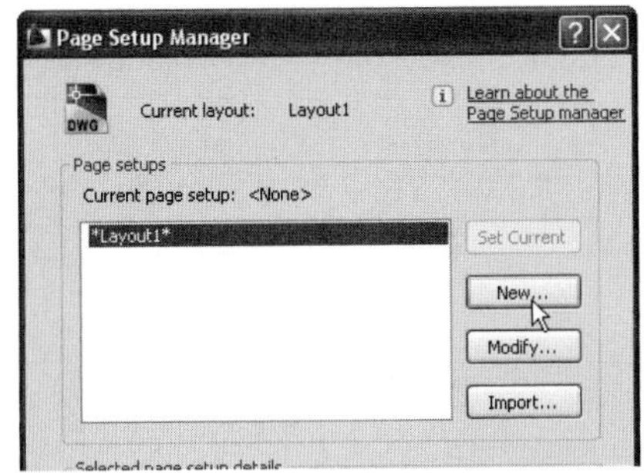

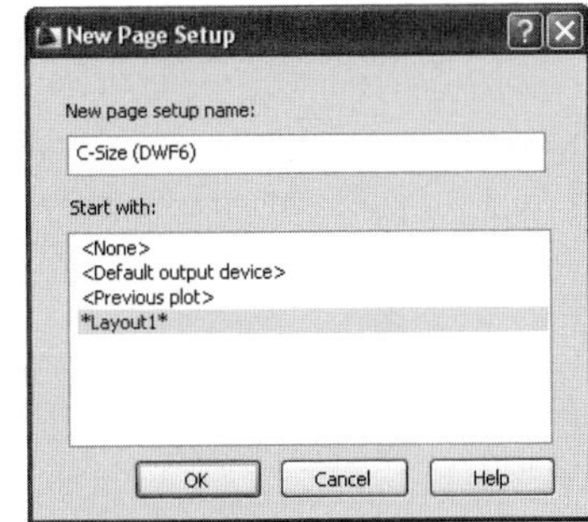

Figure 14.4

Clicking the OK button in the New Page Setup dialog box displays a larger, more comprehensive, Page Setup dialog box, as shown in the following image. The following areas will now be explained:

> **Printer/plotter—Area A:** Use this area to choose the plotter you will be outputting to. Even though you may not be ready to plot, you can still designate a plotter. In this image the DWF6 ePlot.pc3 file will be used.
>
> **Paper size—Area B:** This area holds all paper sizes. It is very important to know that these paper sizes are based on the plotter you selected back in Area A.
>
> **Plot style table (pen assignments)—Area C:** This area is used to communicate with AutoCAD. Plot style will be assigned to this page for plotting purposes.

Another important area to take note of is the Plot scale. This is the plotted scale that the layout will be based on. Since the main purpose of creating a layout is to plot at a scale of 1 to 1, the scale 1:1 reflects this practice. This also means that the information contained in the layout has already been scaled. This will be explained in greater detail later in this chapter.

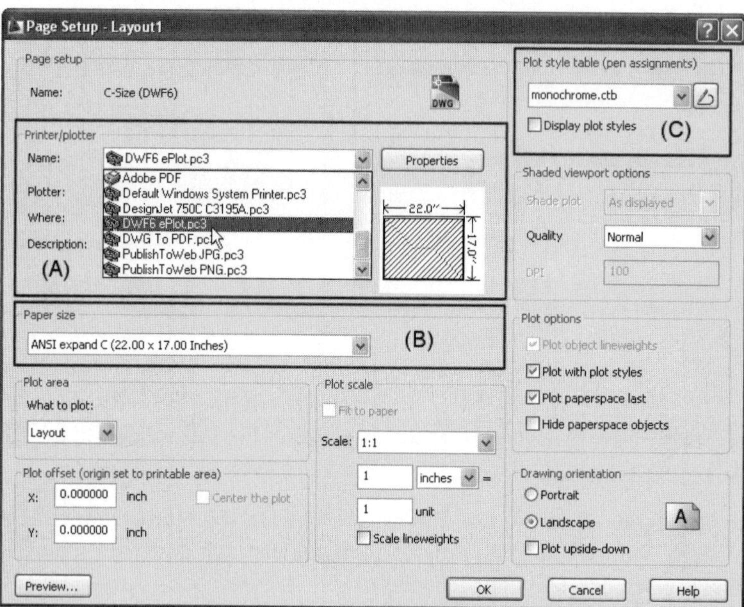

Figure 14.5

A more detailed look at the available paper sizes is illustrated on the left in the following image. Again, these paper sizes are a direct result of the printer/plotter you are using. Notice in this image a number of paper sizes that have the word "expand" associated with them. These paper sizes have a larger printable area as opposed to the paper sizes that are not identified with "expand."

Illustrated on the right in the following image are the default Plot style tables that are available when AutoCAD is loaded. Plot style tables control the appearance of a printed drawing, which can include color, linetype, and even lineweight. For example, the acad.ctb table is designed for a color plot. Whatever objects are red will plot out in the color red (provided you are using a color printer). The monochrome.ctb plot style takes all colors and plots them out as black objects on a white sheet of paper. Notice other color tables that plot out objects in different shades of gray (Grayscale.ctb) and a number of screening plot styles. These are useful when you want to fade away a group of objects. A screening factor of 25% would have the objects faded more than a screen factor of 50%, and so on.

ARCH D (24.00 x 36.00 Inches)
ARCH expand C (24.00 x 18.00 Inches)
ARCH expand C (18.00 x 24.00 Inches)
ARCH C (24.00 x 18.00 Inches)
ARCH C (18.00 x 24.00 Inches)
ANSI expand E (34.00 x 44.00 Inches)
ANSI E (34.00 x 44.00 Inches)
ANSI expand D (34.00 x 22.00 Inches)
ANSI expand D (22.00 x 34.00 Inches)
ANSI D (34.00 x 22.00 Inches)
ANSI D (22.00 x 34.00 Inches)
ANSI expand C (22.00 x 17.00 Inches)
ANSI expand C (17.00 x 22.00 Inches)
ANSI C (22.00 x 17.00 Inches)
ANSI C (17.00 x 22.00 Inches)

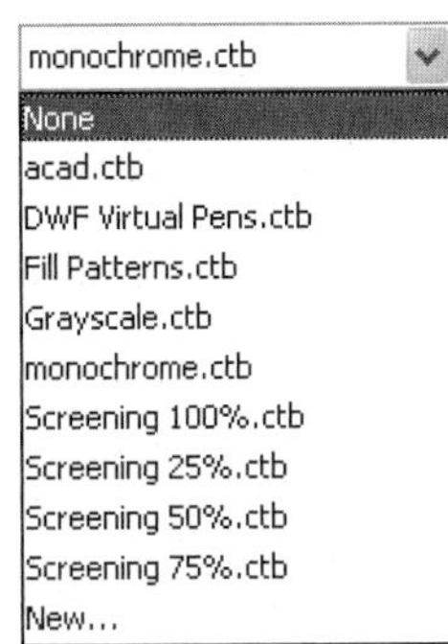

Figure 14.6

Once the proper page settings are made, clicking the OK button in the previous dialog box returns you to the Page Setup Manager. Notice that the new page setup is listed. Clicking the Set Current button assigns the new settings to your layout. Clicking the Close button returns you to the drawing layout tab.

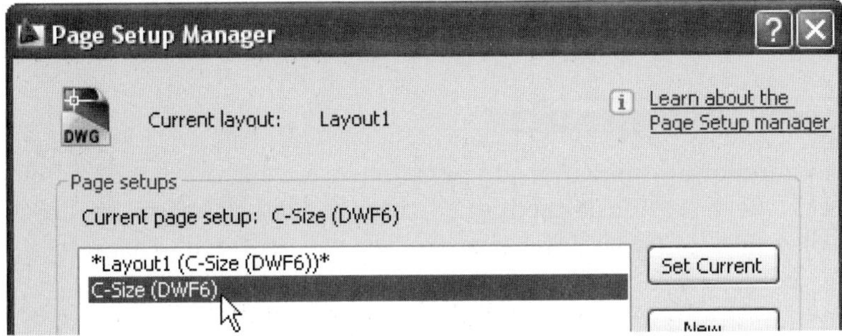

Figure 14.7

Note: When you make a number of changes in the Page Setup dialog box and then save your drawing, these settings are automatically saved inside the drawing.

FLOATING MODEL SPACE

Notice the appearance of the drafting triangle icon in the lower-left corner of the display screen. This signifies that you are presently in the Paper Space environment. Another indicator in the AutoCAD Classic workspace is that the Paper button will activate in the status bar located at the bottom of the display screen. While in Paper Space, operations such as adding a border, title block, and general notes are usually performed. Paper Space is also the area in which you plot your drawing.

When a viewport is created in a layout, the drawing image automatically fills up the viewport. Some operations need to be performed in Model Space without leaving the Layout environment; this is referred to as working in floating Model Space. To activate floating Model Space, double-click inside a viewport. The UCS icon will normally display, as shown on the right in the following image. Notice that the icon appears inside the viewport. In the AutoCAD Classic workspace, you can also activate floating model space by clicking the Paper button in the status bar and changing it to Model.

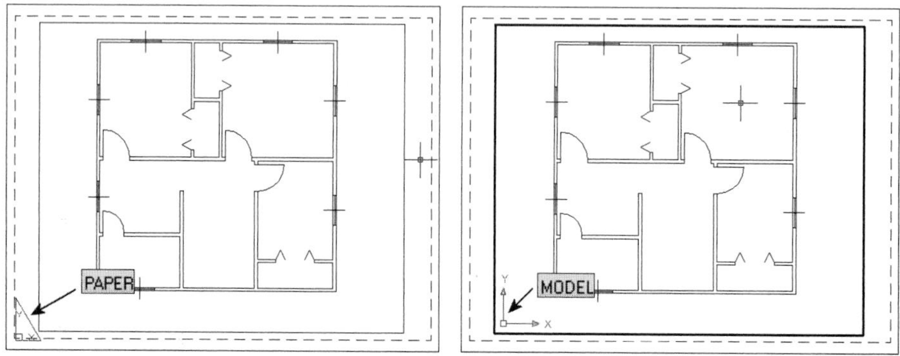

Figure 14.8

Note: If you need to switch back to the Layout mode, double-click outside the viewport. The Paper Space drafting triangle reappears.

SCALING VIEWPORT IMAGES

One of the more important operations to perform on a viewport is the scaling of the image to Paper Space. It was previously mentioned that by default, images that are brought into viewports display in a ZOOM-Extents appearance. To scale an image to a viewport, use the next series of steps and refer to the following image.

- Have the Viewports toolbar present somewhere on your screen.
- While in Paper Space, click the edge of the viewport that will be scaled.
- In the Viewports toolbar, click in the scale box to display other scales.
- Select various scales until the image fits in the viewport.

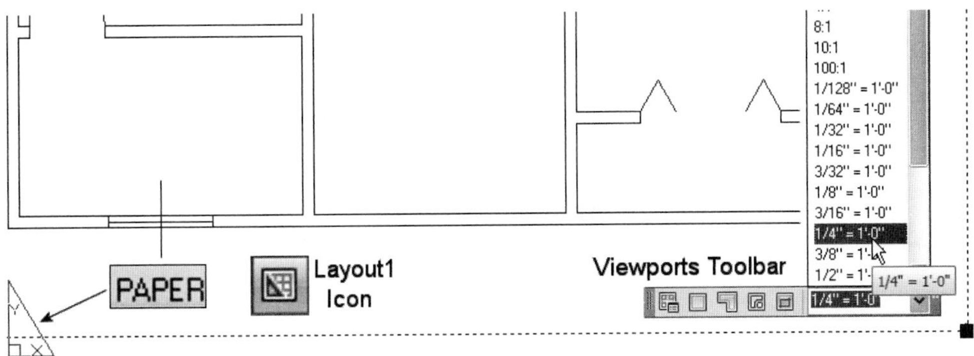

Figure 14.9

Note: You can also apply a scale to a viewport by double-clicking inside the viewport to activate floating Model Space and setting the proper scale.

If you find that the image is very close to fitting but is cut off by one edge of a viewport, there are two options to resolve the problem:

- Stretch the viewport (using grips or the STRETCH command).
- Pan the image into position.

A combination of both methods usually proves to be the most successful.

To pan the image, double-click inside the viewport to start floating Model Space. You will notice the thick borders on the viewport. Using the PAN command or holding down the wheel on a wheel mouse, move the image to fit the viewport. Once the image is panned into position, double-click outside the viewport on the drawing surface (usually the area surrounding the drawing sheet) to return from the floating Model Space back to the layout space.

CONTROLLING THE LIST OF SCALES

You will notice that when setting the scale of the image inside the Paper Space viewport, the list of scales is very long. You may find yourself using only certain scales and ignoring others. You could also be confronted with assigning a scale that is not present on the list. All these situations can be easily handled through the use of the Edit Scale List dialog box, which is activated by picking Scale List from the Format pull-down menu, as shown in the following image.

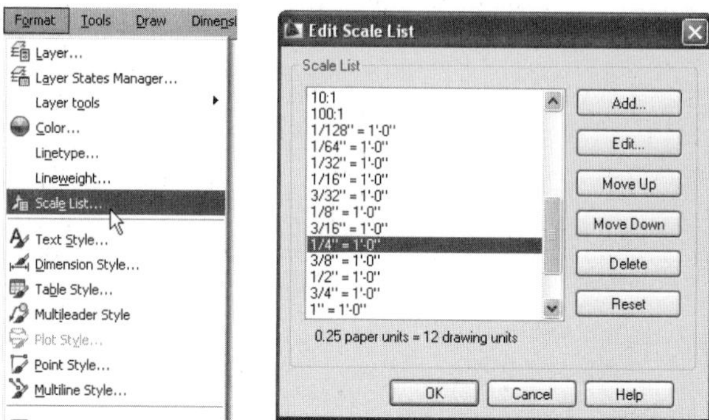

Figure 14.10

Various buttons are available that allow you to add, edit, or delete scales from the list. Clicking the Add button displays the Add Scale dialog box, as shown on the left in the following image. In this dialog box, a new scale of 1:1000 is being created since it is not available in the default scale list. Clicking the Edit button displays the Edit Scale dialog box, as shown on the right in the following image. Here an existing scale is being edited to reflect the new scale of 1:500.

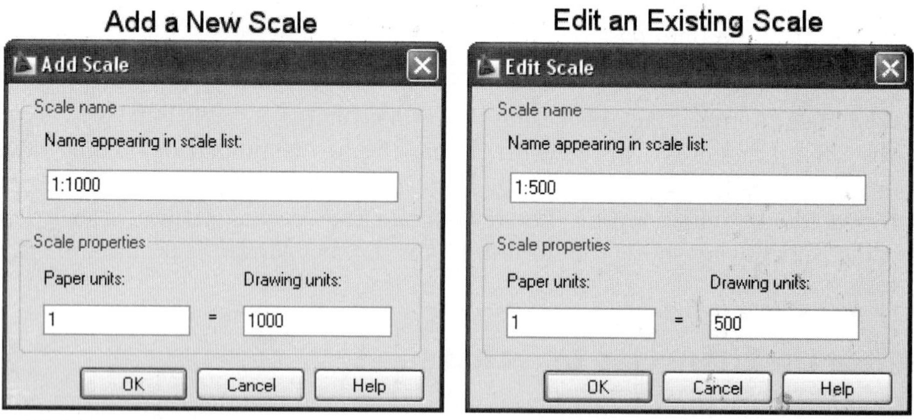

Figure 14.11

Clicking the Delete button removes a scale from this list. You can also rearrange scales so that your most popular scales always appear at the top of the list. Simply click the Move Up button to move the selected scale up the list or the Move Down button to arrange the selected scale at the bottom of the list.

LOCKING VIEWPORTS

When you have scaled the contents of a floating Model Space viewport, it is very easy to accidentally change this scale by rolling the wheel of the mouse or performing another ZOOM operation. To prevent accidental panning and zooming of a viewport image once it has been scaled, a popular technique is to lock the viewport. To perform this task, use these steps and refer to following image:

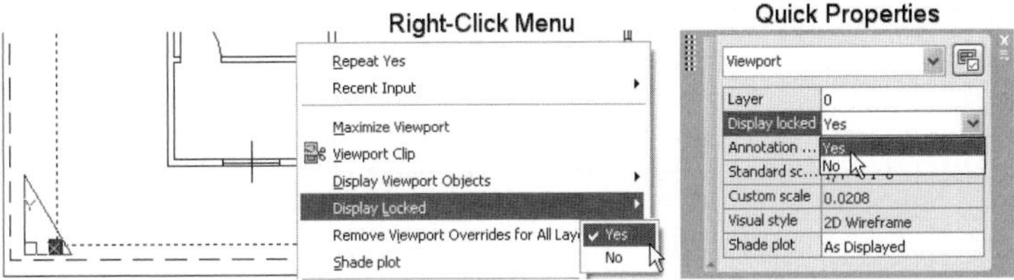

Figure 14.12

- Be sure you are in Layout mode (look for the drafting triangle in the lower-left corner).
- Select the viewport to lock. It will highlight and grips will appear. If Quick Properties are activated, as shown in the image on the right, you can lock the display by selecting "Yes."
- You can also use a right-click menu to lock the viewport. Right-click your screen. A menu will appear.
- Select Display Locked followed by "Yes."

Yet another method that can be used for locking viewports is shown in the following image. After the edge of a viewport is selected, you can click the Lock/Unlock Viewport button that is available at the right end of the status bar. This provides a more convenient area for locking or unlocking viewports.

Figure 14.13

Note: When you lock a viewport, your scale will be grayed out inside the Viewports toolbar. If you need to change the scale of an image inside a viewport, you must first unlock the viewport.

MAXIMIZING A VIEWPORT

The VPMAX command is designed to maximize the size of a viewport in a layout. This is especially helpful when editing drawings with small viewports. It also eliminates the need to constantly switch between Model Space and a Layout. The VPMAX command can be entered at the keyboard. You could also click the Maximize Viewport button located in the status bar at the bottom of the display screen, as shown in the following image.

Note: This button is visible only when you are in a Layout; it is not displayed in Model Space. Also, if numerous viewports are created, arrow buttons appear, allowing you to move between viewports. Click the Maximize Viewport button a second time to minimize the viewport and return to the Layout environment.

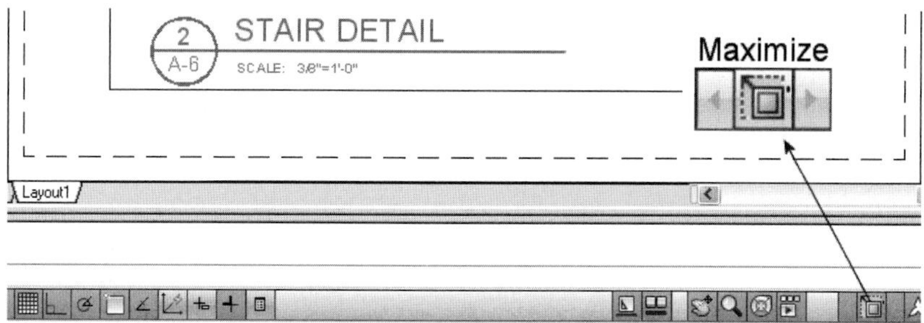

Figure 14.14

The following image displays a drawing that has its viewport maximized. To return to Paper Space or Layout mode, double-click the red outline or click the Minimize Viewport button.

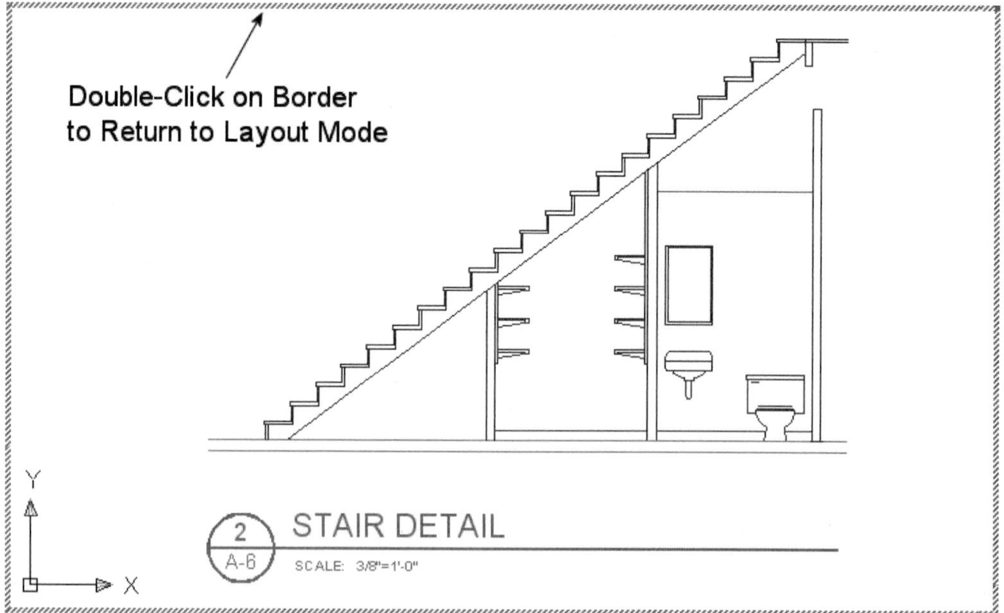

Figure 14.15

HIDING LAYOUT AND MODEL TABS

If you are utilizing the AutoCAD Classic workspace, the Layout and Model tabs will normally be displayed as shown in the following image. As a means of regaining valuable screen space, you can hide the layout and model tabs. To do this, right-click one of the tabs and choose Hide Layout and Model tabs from the menu as shown in the following image. If you are utilizing the 2D Drafting & Annotation workspace, the tabs are already hidden.

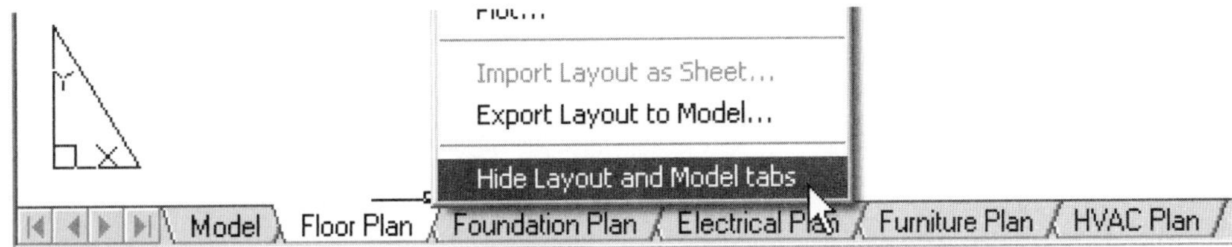

Figure 14.16

To redisplay the tabs on the screen, right-click the Model or Layout buttons, right-click, and pick Display Layout and Model Tabs, as shown in the following image.

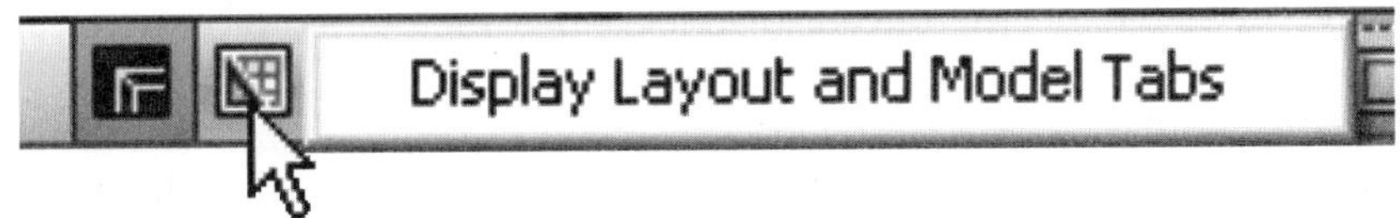

Figure 14.17

CREATING A LAYOUT

Try It!: Open the drawing file 14_Gasket1. The illustration on the left in the following image shows the drawing originally created in Model Space. This drawing is required to be laid out in Paper Space.

Click the Quick View Layout button to display all model and layout views. Pick the Layout1 icon; a layout is automatically created, as shown in the following image.

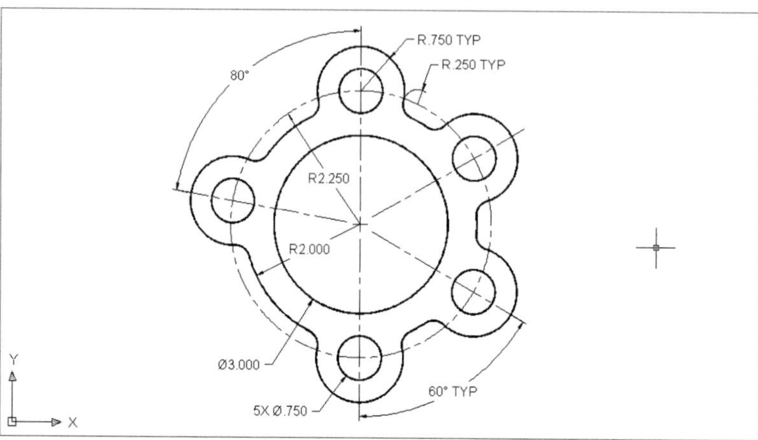

Figure 14.18

It is good practice to rename the Layout1 tab located at the bottom of your screen. Do this by first picking the Quick View Layouts button and double-clicking the Layout1 text. Rename Layout1 to One View Drawing, as shown on the right in the following image. Quick View Layouts will be discussed in greater detail later on in this chapter along with Quick View Drawings.

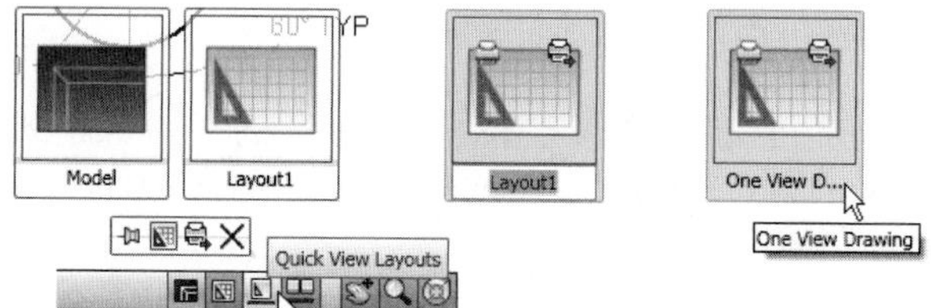

Figure 14.19

Next, right-click the One View Drawing icon and pick Page Setup Manager from the menu, as shown on the left in the following image. This launches the Page Setup Manager dialog box, illustrated in the middle in the following image. Click the New button. When the New Page Setup dialog box appears, enter the name B-Sized (DWF6), as shown on the right in the following image. When finished, click the OK button.

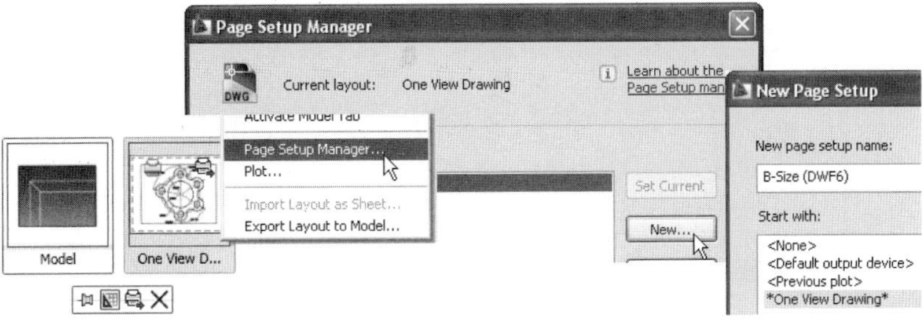

Figure 14.20

When the Page Setup dialog box appears, as shown in the following image, click in the Printer/plotter field and change the name of the plot device to DWF6 ePlot. Next, click in the Paper size field and select ANSI expand B (17.00 × 11.00 Inches) from the available list. Finally, click in the field under Plot style table and change the plot style to Monochrome. When you are finished making these changes, click the OK button in the Page Setup dialog box.

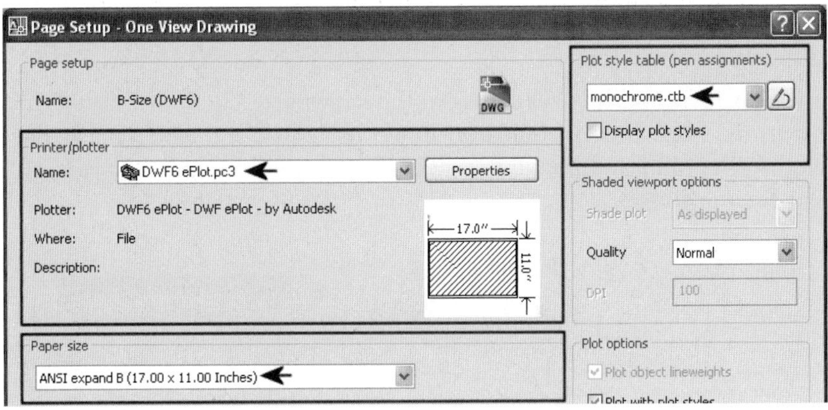

Figure 14.21

When you return to the Page Setup Manager dialog box; double-click the B-Size (DWF6) layout to make it current, as shown in the following image. Click the Close button to continue with laying out this drawing.

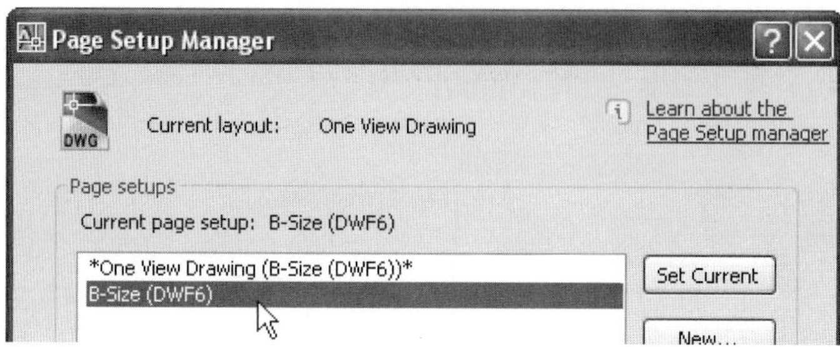

Figure 14.22

Because you increased the size of the sheet, the image of the gasket in the viewport did not grow in size. The size of the viewport depends on the scale applied in this next step.

Rather than activating floating Model Space, simply click the edge of the viewport, as shown in the following image. Locate the Viewports button on the right side of the status bar. Clicking this button displays the Viewports toolbar, as shown on the right in the following image. The number inside the field in this image is the current scale of the image inside the viewport. Whenever you create a drawing layout, the image is automatically zoomed to the extents of the viewport. For this reason, the desired scale will never be correctly displayed in this area.

To have the image properly scaled inside the viewport, click the down arrow in the Viewports button. This displays the drop-down list of all standard scales. These include standard engineering and architectural scales. The image is assigned a scale of 1:1, which is applied to the viewport.

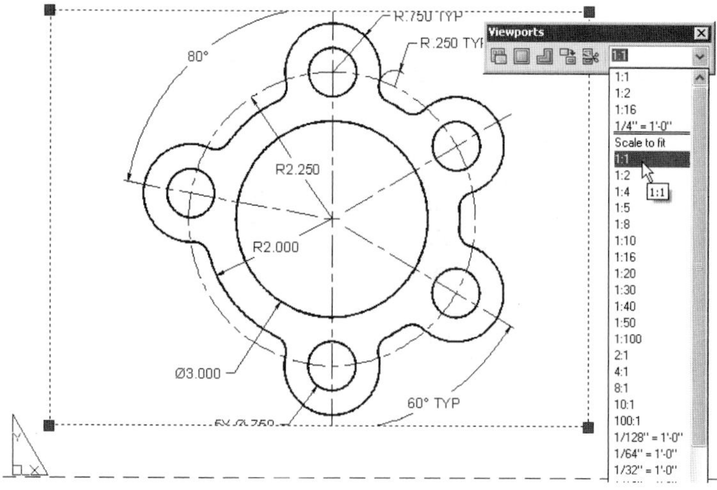

Figure 14.23

You can see on the left in the following image that the gasket has grown inside the viewport. However, it is properly scaled. The viewport needs to be increased in size to accommodate the scaled image of the gasket. Clicking the viewport once displays grips located at the four corners, as shown on the left in the following image. Use these grips to size the viewport to the image. Accomplish this by clicking one of the grips in the corner and stretching the viewport to a new location, as shown on the right in the following image.

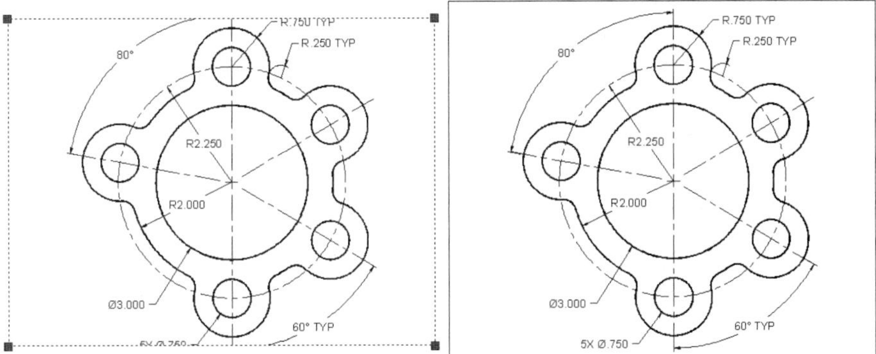

Figure 14.24

The results of this operation are displayed in the following image. Press ESC to remove the object highlight and the grips.

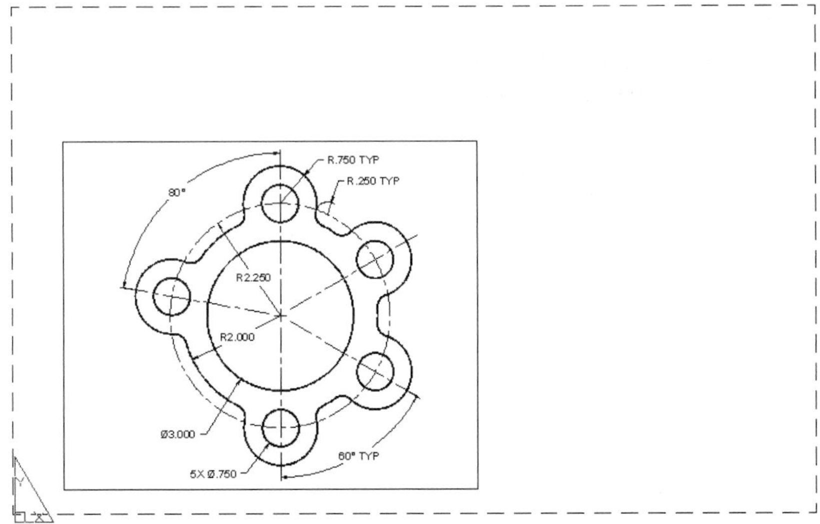

Figure 14.25

Before the drawing is plotted, a title block containing information such as drawing scale, date, title, company name, and designer is inserted in the Paper Space environment. Accomplish this by first choosing Block... from the Insert pull-down menu, as shown on the left in the following image. This activates the Insert dialog box, shown on the right in the following image. This feature of AutoCAD will be covered in greater detail in chapter 16. Be sure the block ANSI B title block is listed in the Name field.

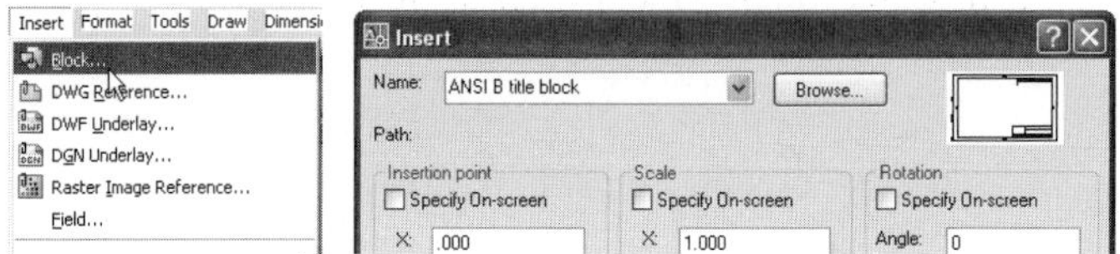

Figure 14.26

Click the OK button to place the title block in the following image. Be sure the title block displays completely inside the paper margins; otherwise, part of it may not plot.

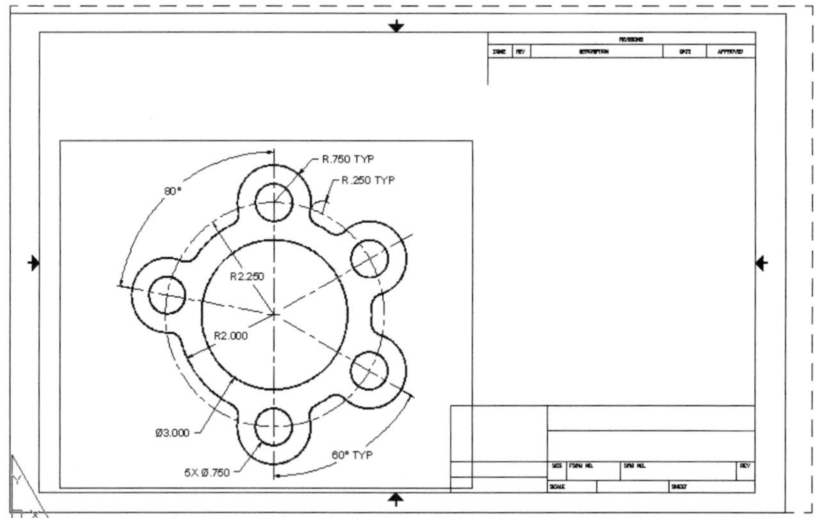

Figure 14.27

At this point, the MOVE command is used to move the viewport containing the drawing image to a better location based on the title block. Unfortunately, if the drawing were to be plotted out at this point, the viewport would also plot. It is considered good practice to assign a layer to the viewport and then turn that layer off before plotting. You could also set the viewport layer to No Plot. This would allow the viewport to be visible in drawing mode yet not plot out. In the following image, the viewport is first selected and grips appear. Click in the Layer Control box and click on the Viewports layer. The viewport is now on the Viewports layer. Then click the lightbulb icon in the Viewports layer row to turn it off.

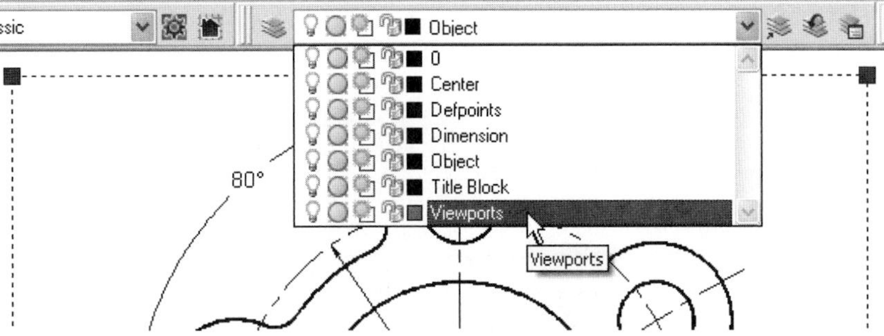

Figure 14.28

The drawing display will appear similar to the following image, with the drawing laid out and properly scaled in Paper Space.

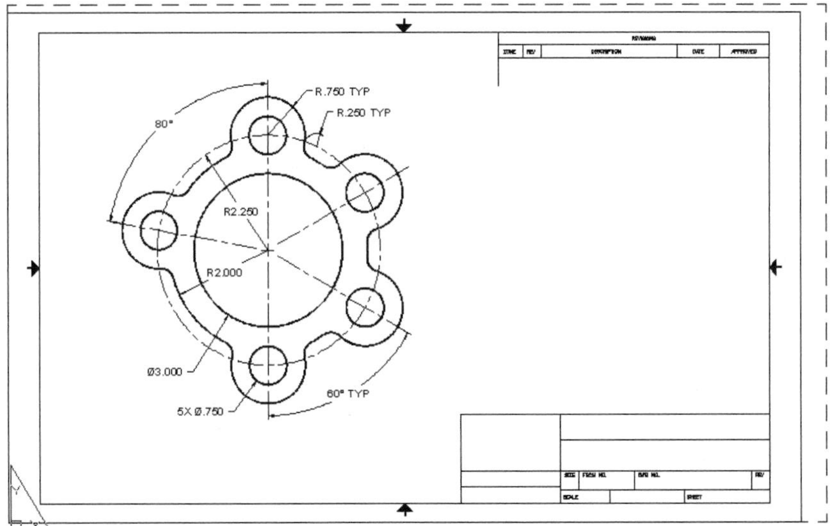

Figure 14.29

USING A WIZARD TO CREATE A LAYOUT

The AutoCAD Create Layout wizard can be especially helpful in laying out a drawing in the Paper Space environment.

Clicking Create Layout... displays the Create Layout-Begin dialog box, as shown on the right in the following image. You cycle through the different categories, and when you are finished, the drawing layout is displayed, complete with title block and viewport.

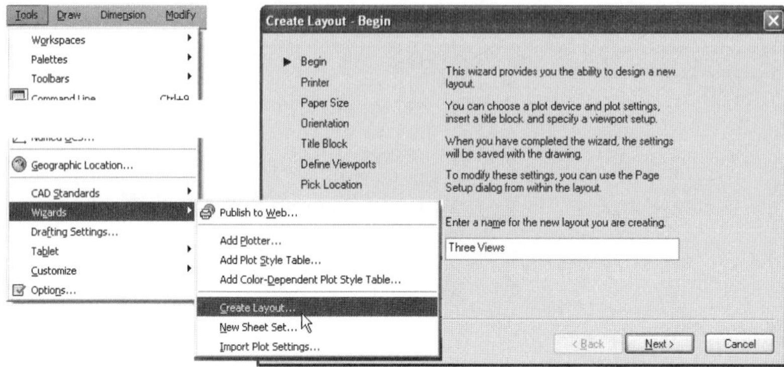

Figure 14.30

Functions of the Create Layout wizard dialog box are briefly described in the following table:

Create Layout Wizard

Category	Purpose
Begin	Change the name of the layout here.
Printer	Use the Printer category to select a configured plotter from the list provided.
Paper Size	The Paper Size category displays all available paper sizes supported by the currently configured plotter.
Orientation	Use the Orientation category to designate whether to plot the drawing in Landscape or Portrait mode.
Title Block	Depending on the paper size, choose a corresponding title block to automatically be inserted in the layout sheet.
Define View-ports	The Define Viewports category is used to either create a viewport or leave the drawing layout empty of viewports.
Pick Location	The Pick Location category creates the viewport to hold the image in Paper Space. If you click the Next > button, the viewport will be constructed to match the margins of the paper size. If you click the Select location < button, you return to the drawing and pick two diagonal points to define the viewport.
Finish	The Finish category alerts you to a new layout name that will be created. Once it is created, modifications can be made through the Page Setup dialog box.

ARRANGING ARCHITECTURAL DRAWINGS IN A LAYOUT

Architectural drawings pose certain challenges when you lay them out in Paper Space based on the scale of the drawing. For example, the following image shows a drawing of a stair detail originally created in Model Space. The scale of this drawing is 3/8" = 1'-0". This scale will be referred to later on in this segment on laying out architectural drawings in Paper Space.

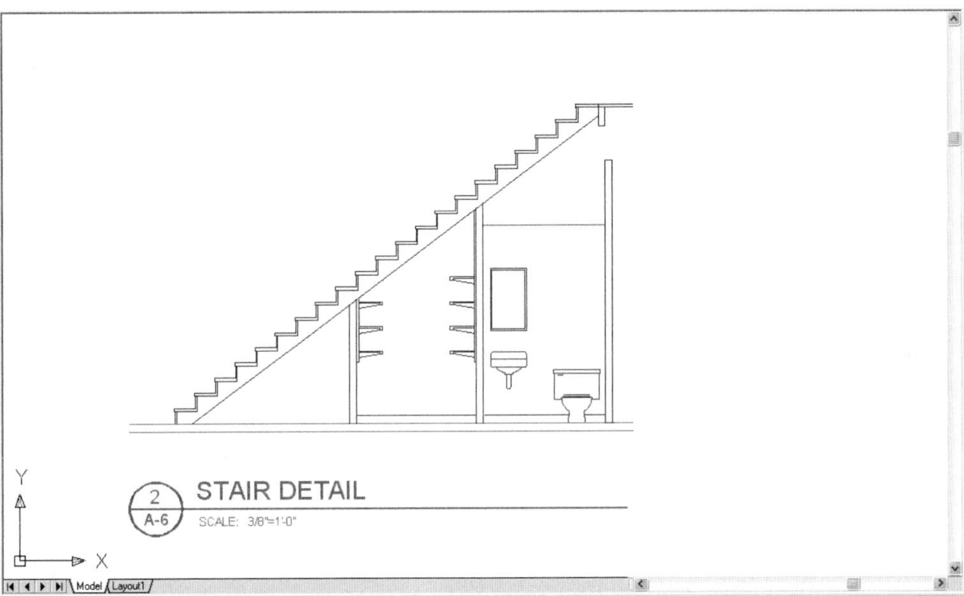

Figure 14.31

Try It!: Open the drawing file 14_Stair Detail. Click the Layout1 tab (if the tabs are hidden, click the Layout1 button on the status bar) or use the Create Layout wizard to arrange the stair detail in Paper Space so that your drawing appears similar to the following image. Use the Page Setup dialog box to create a new page setup that uses the DWF6 ePlot plot device, the ANSI expand C sheet, and the Monochrome plot style table. In the figure, we see the drawing sheet complete with a viewport that holds the image and an ANSI C title block that has been inserted. Because the illustration in the following image is now visible in Paper Space, it is not necessarily scaled to 3/8" = 1'-0". Click the edge of the viewport. Activate the Viewports toolbar and scale the image in the viewport to the 3/8" = 1'-0"scale. Notice, in the figure, a complete listing of the more commonly used architectural scales.

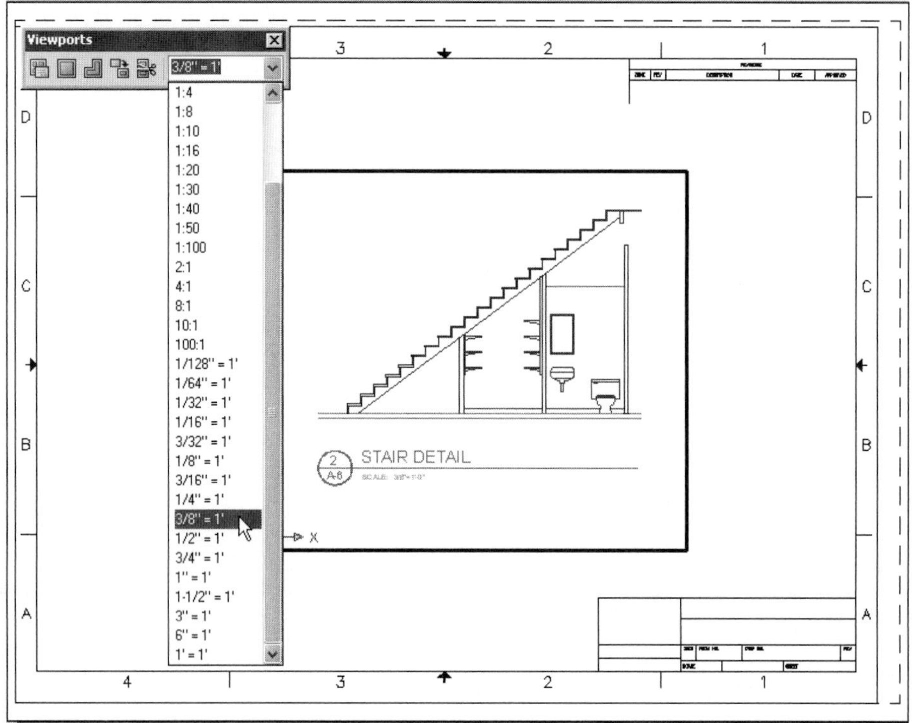

Figure 14.32

Changing the viewport to a different layer and then turning that layer off displays the completed layout of the stair detail in the following image. Since the purpose of Paper Space is to lay out a drawing and scale the image inside the viewport, this drawing will be plotted at a scale of 1:1.

Note: Another common practice is to set the layer that contains all viewports to a No Plot state in the Layer Properties Manager palette. The viewport will always remain visible in your layout. However, when previewing your image before plotting, the viewport will disappear. Exiting Plot Preview will make the viewport reappear.

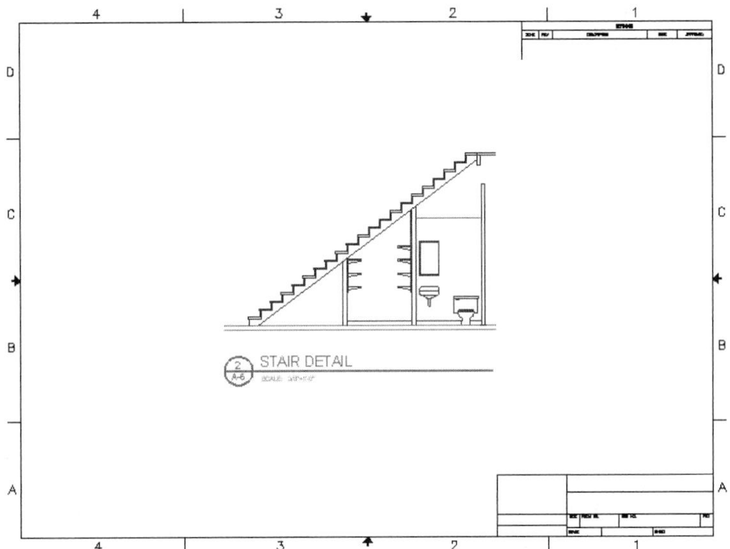

Figure 14.33

CREATING MULTIPLE DRAWING LAYOUTS

The methods explained so far have dealt with the arrangement of a single layout in Paper Space. AutoCAD provides for greater flexibility when working in Paper Space by enabling you to create multiple layouts of the same drawing. This will be explained through the following image, which shows a floor plan complete with electrical plan. Individual layouts will be created to display separate images of the floor and electrical plans.

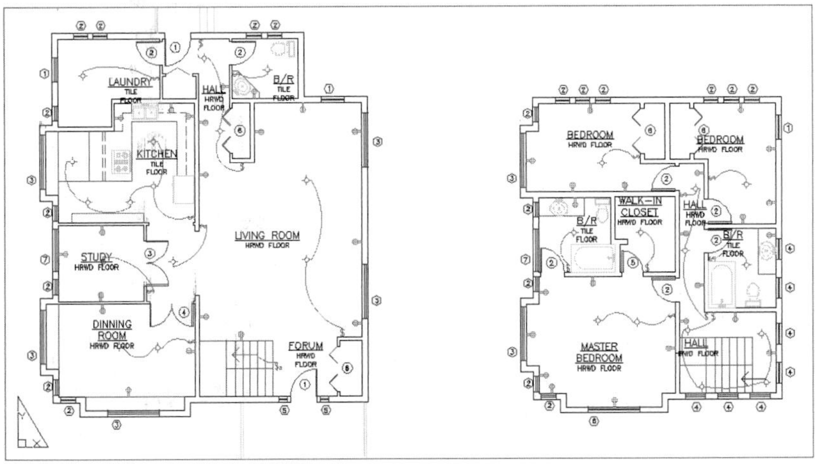

Figure 14.34

First verify that the Model and Layout tabs are displayed on the screen. If they are not, right-click the Model or Layout1 button on the status bar and select "Display Model and Layout tabs." To create a new layout, click on the Floor Plan layout tab as shown on the left in the following image. Then press the Ctrl key while dragging your cursor. You will notice a small page icon with a "plus" sign indicating the Floor Plan layout will be copied. The results are displayed on the right in the following image. Notice the addition of the layout Floor Plan (2) in the list of layouts.

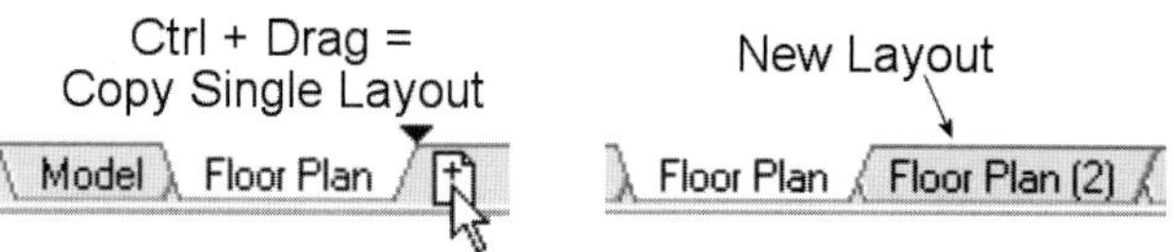

Figure 14.35

You can also create multiple layouts. In this example, first click on the Floor Plan and hold down the Shift key while you select the Electrical Plan layout tab as shown on the left in the following image. Then press the CTRL key while dragging your cursor. You will notice a small multiple page icon with a "plus" sign indicating the Floor Plan and Electrical Plan layouts will be copied. The results are displayed on the right in the following image. Notice the addition of the layout Floor Plan (2) and Electrical Plan (2) in the list of layouts.

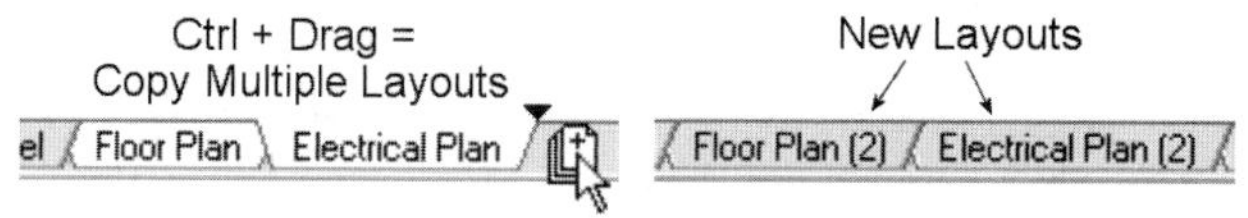

Figure 14.36

When you want to move a layout to a new location and, in effect, change the order of the layouts, click on the layout and drag it to its new location as shown in the following image. In this example, the CTRL key is not utilized since you are moving the layout and not copying it.

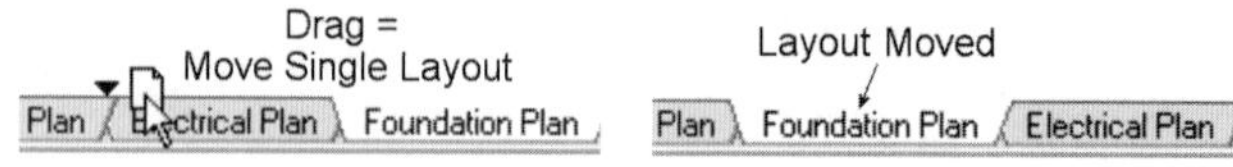

Figure 14.37

When you copy a layout, all information inside the viewport, such as title block, viewport, and viewport scale/layer state information, is copied. Renaming layouts more appropriately is customary at this stage. In the following image, the Foundation Plan layout was renamed to Furniture Plan. This was accomplished by double-clicking on the Foundation Plan layout tab. This will highlight all text in the tab. Entering new text renames the layout.

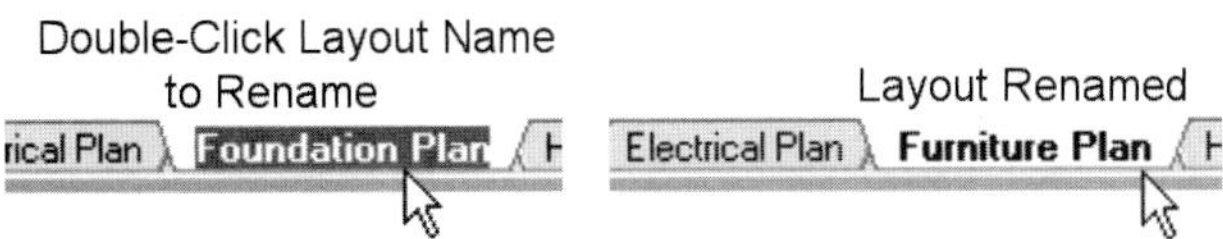

Figure 14.38

At times, layouts need to be deleted or removed from the drawing entirely. To delete the HVAC Plan layout in the next example, right-click on the HVAC Plan layout tab and select Delete from the menu as shown on the left in the following image. An AutoCAD alert box will prompt you to delete the layout by clicking on the OK button. Notice from the alert box that the Model tab cannot be deleted.

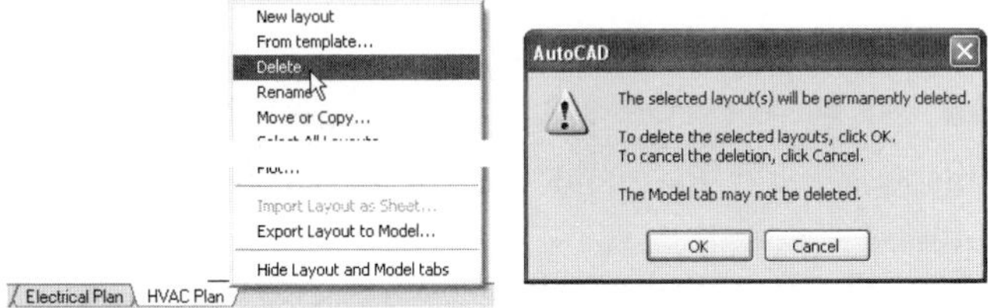

Figure 14.39

Try It!: Open the drawing 14_Facilities_Plan. Notice that a number of layouts were created that correspond to the various room numbers, as shown in the following image. However, the room numbers are out of order and need to be rearranged starting with the lowest room number and going to the highest room number. The Overall layout should be reordered directly after the Model tab. Use the drag and drop technique illustrated in Figure 14.37 to rearrange all the layout tabs in order.

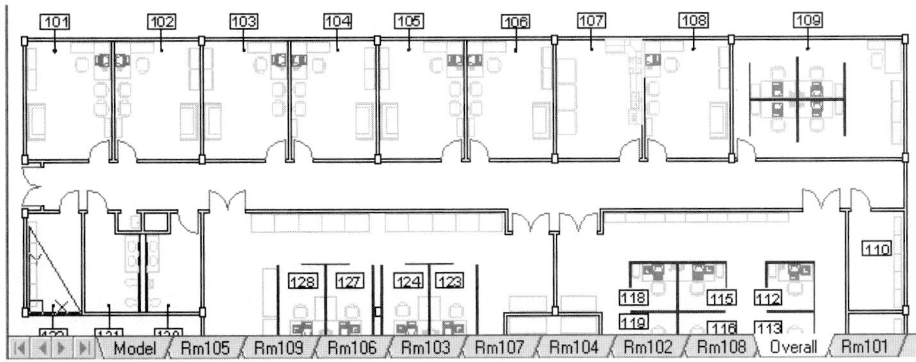

Figure 14.40

After performing the reordering operation, your screen should appear similar to the following image.

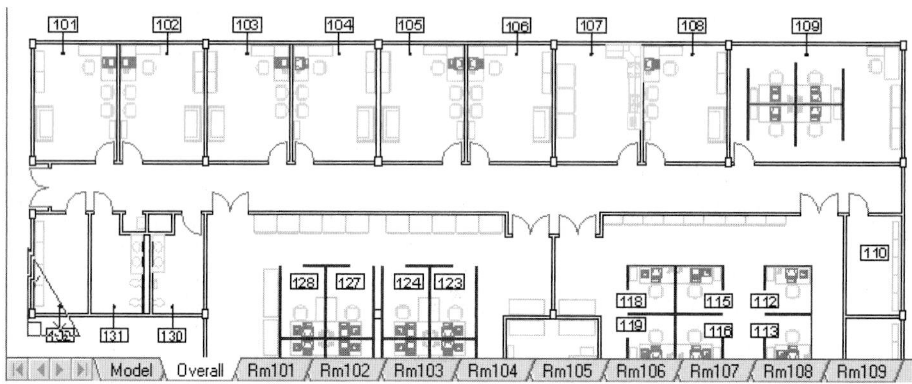

Figure 14.41

USING LAYERS TO MANAGE MULTIPLE LAYOUTS

In the following image, a Floor Plan and Electrical Plan layout are already created. Also, the Floor Plan layout is currently active. You want to see only the floor plan inside this viewport and none of the objects on electrical layers.

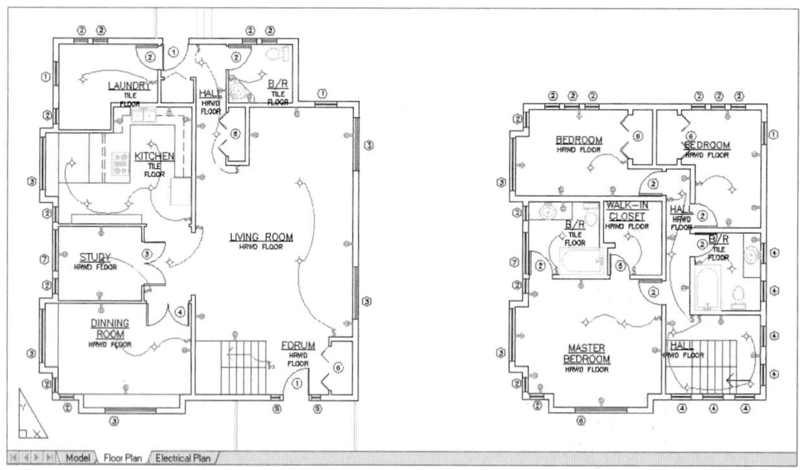

Figure 14.42

First, double-click inside the viewport in order to switch to floating Model Space. Activating the Layer Properties Manager palette displays the layer information, as shown in the following image. When you are in floating Model Space and use this palette, a number of additional layer modes are added. Two of these modes provide for the ability to freeze layers only in the active viewport and the ability to freeze layers in new viewports. Freezing layers in all viewports is not an effective means of controlling layers when you create multiple layouts. You need to be very familiar with the layers created in order to perform this task. To display only the floor plan information, notice that two layers

(Lighting and Power) have been frozen under the Viewport Freeze heading, as shown in the following image.

All other layers remain visible in this viewport.

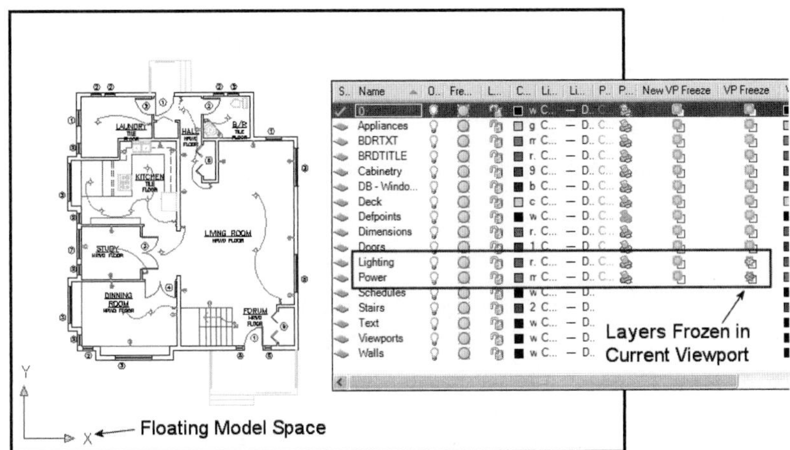

Figure 14.43

The resulting image is displayed in the following image with only the floor plan information visible in the viewport. The electrical layers have been frozen only in this viewport and do not display.

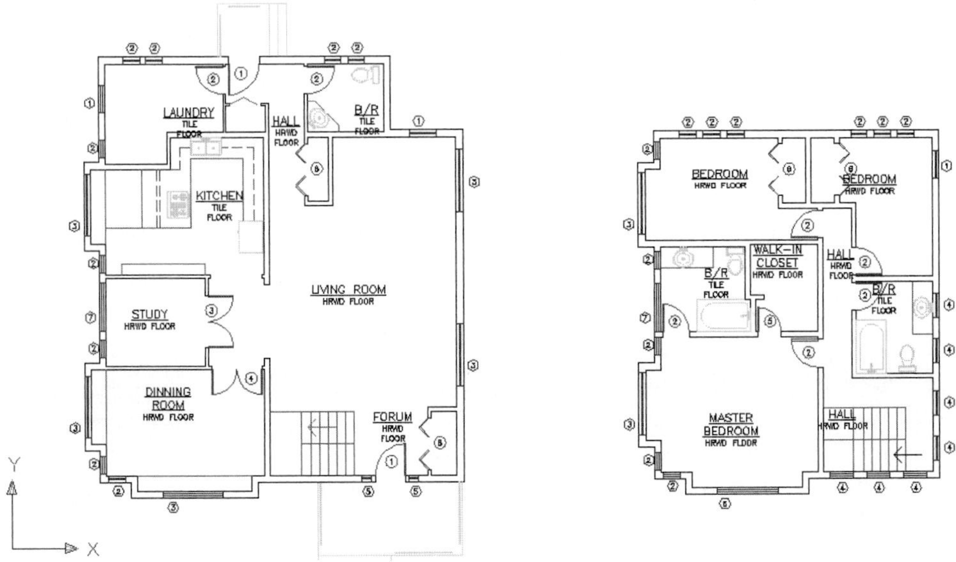

Figure 14.44

Note: You can also freeze layers in the current viewport through the Layer control box, as shown in the following image.

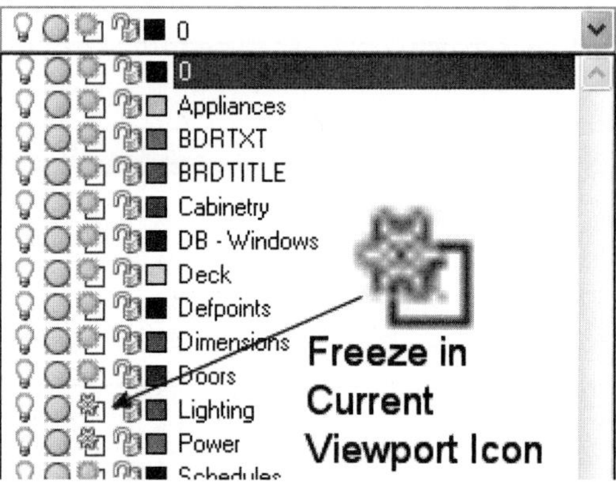

Figure 14.45

ADDITIONAL LAYER TOOLS THAT AFFECT VIEWPORTS

Additional controls on layers in viewports are available through the Layer Properties Manager palette. The following image illustrates the Electrical Plan layout that was created in the previous segment. Sometimes, you want to add special effects to your layouts such as changing the color, linetype, or even the lineweight just in the current viewport.

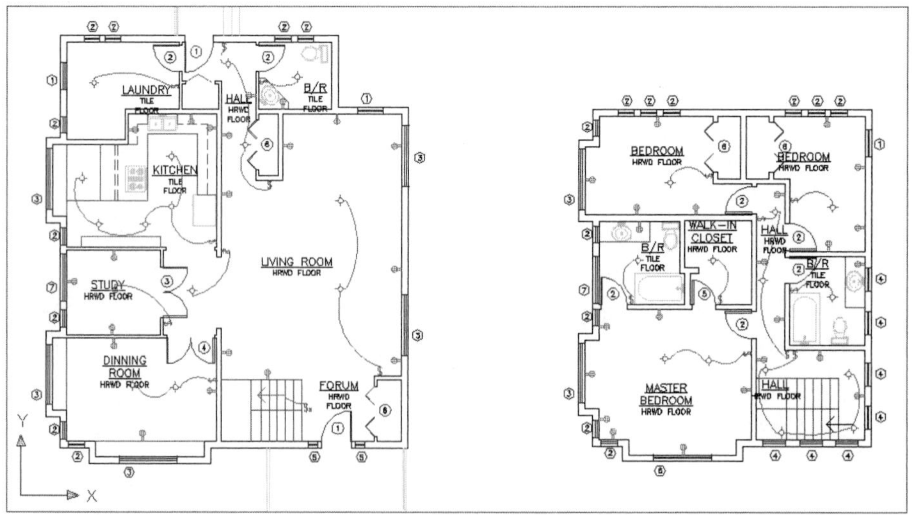

Figure 14.46

After double-clicking inside the Electrical Plan layout, launch the Layer Properties Manager palette to display all layers in the drawing. For the purposes of this example, all layers except for those dealing with electrical components will have the color changed to a light gray. The reason is to dim or fade the floor plan out in order to give emphasis on the electrical layers. In the following image, a number of layers are selected and then their colors changed to 9 or light gray through the VP Color column. Notice also in the following image the presence of the VP Linetype and VP Lineweight columns. You can also change the linetype and lineweight of selected layers. These changes are only present in the current viewport.

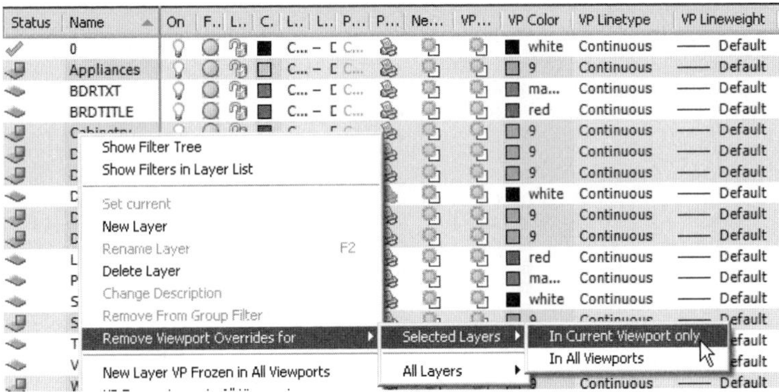

Figure 14.47

The results of performing this operation are illustrated in the following image where the floor plan components display faintly while the electrical components stand out.

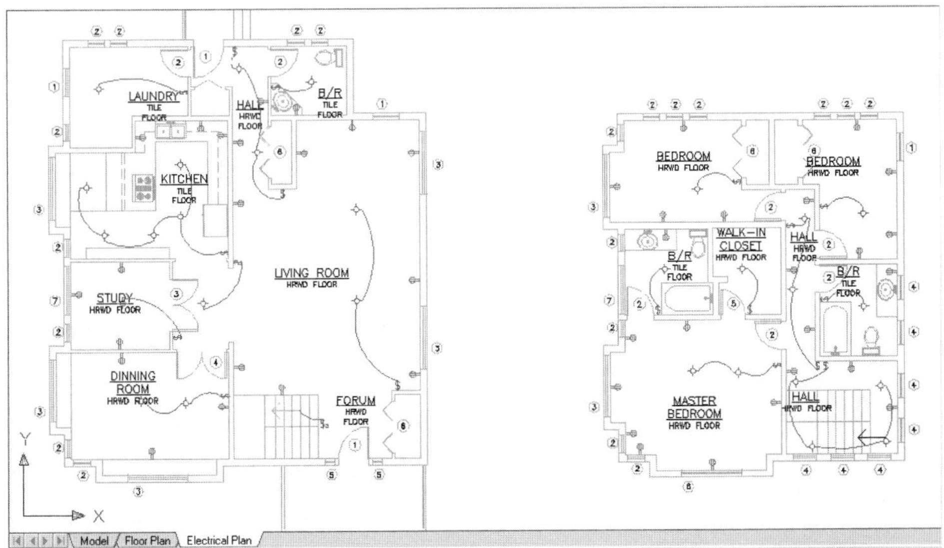

Figure 14.48

The following image identifies those layers that have had either the color, linetype, or lineweight overridden. These layers are highlighted with a light blue background. If you need to return the control of colors, linetypes, or lineweights back to their original state, highlight the layers you want to affect and right-click. The menu illustrated in the following image will appear. Click "Remove Viewport Overrides for" followed by "Selected Layers" and then "In Current Viewport only."

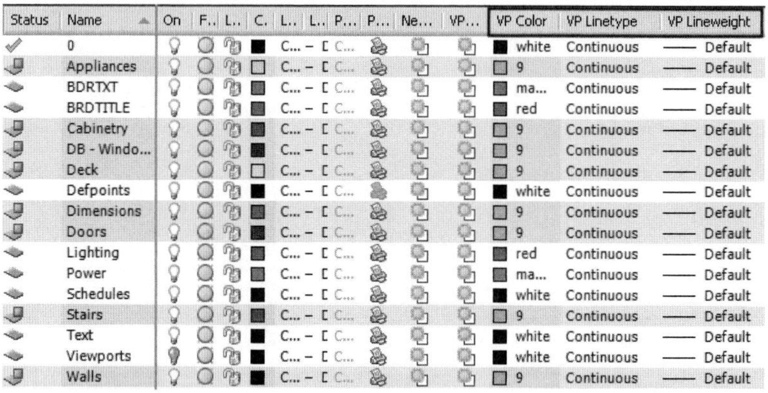

Figure 14.49

Whenever performing viewport color, linetype, or lineweight overrides, a special layer group is automatically created to help you manage these items. In the following image of the Layer Properties Manager palette, notice the new layer group called Viewport Overrides. Clicking this name displays the layers that had their color overridden in the previous step.

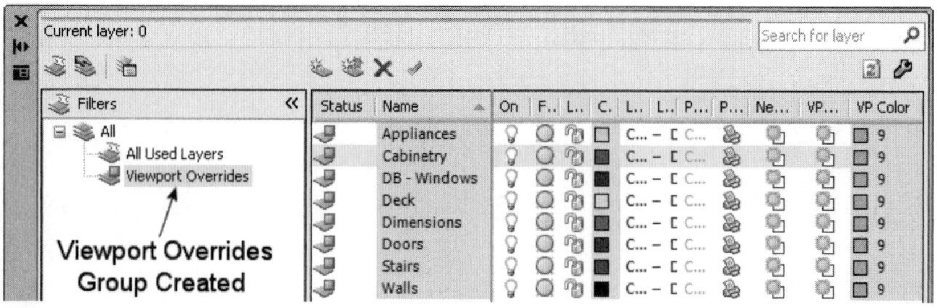

Figure 14.50

Note: The information contained in the Layer Properties Manager palette can get quite numerous and overwhelming. To better manage this information, a special menu exists that allows you to turn off those layer states that you do not use on a regular basis. To activate this menu, move your cursor into one of the layer state headings and right-click to display the menu shown in the following image. Additional controls allow you to maximize all columns or a single column. You can even reset all columns back to their default widths.

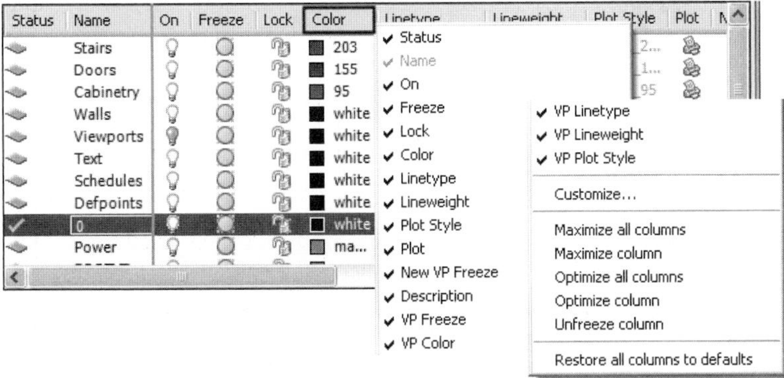

Figure 14.51

ASSOCIATIVE DIMENSIONS AND LAYOUTS

Dimensioning all objects in Model Space was once considered the only reasonable and reliable method to dimension multiple objects. The main reason was because there was no associativity with the Model Space objects and dimensions placed in a Layout or Paper Space. This has changed. Once you scale the Model Space objects inside a layout, the scale of the layout has a direct bearing on the dimensions being placed. This is accomplished only if the DIMASSOC variable is set to 2.

Try It!: Open the drawing file 14_Inlay. Two viewports are arranged in a single layout called Inlay Floor Tile, as shown in the following image. The images inside these viewports are scaled differently; the main inlay pattern in the left viewport is scaled to 3/4" = 1'-0". The detail image in the right viewport has been scaled to 3" = 1'-0". The DIMASSOC variable is currently set to 2. This allows you to dimension the objects in Paper Space at two different scales and still have the correct dimensions appear.

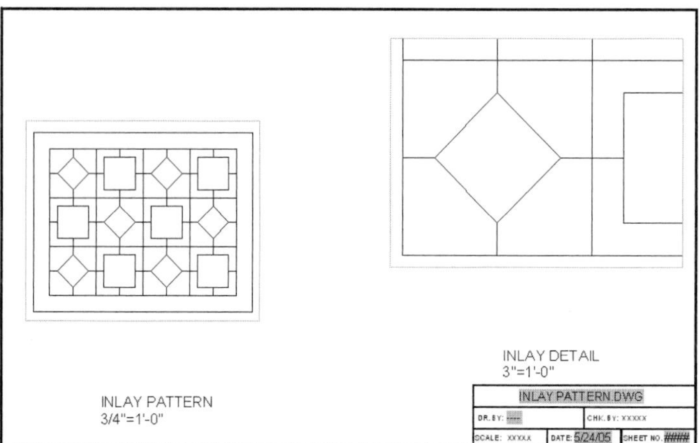

Figure 14.52

While in Paper Space, use the DIMLINEAR command and place a few linear dimensions on the main inlay plan in the left viewport. Now switch and add a few linear dimensions in the right viewport. Zoom in to a few of the dimensions and see that they reflect the Model Space distances. Place more dimensions on the main inlay pattern and detail using a combination of dimension commands. Observe the correct values being placed.

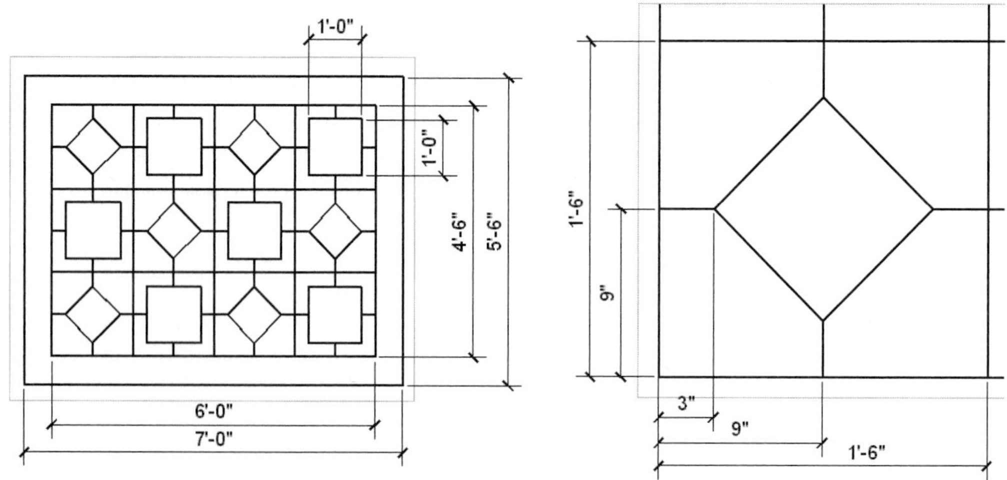

Figure 14.53

Next, click and drag each viewport to a more convenient location on the drawing screen. Notice that the dimensions move along with the viewports.

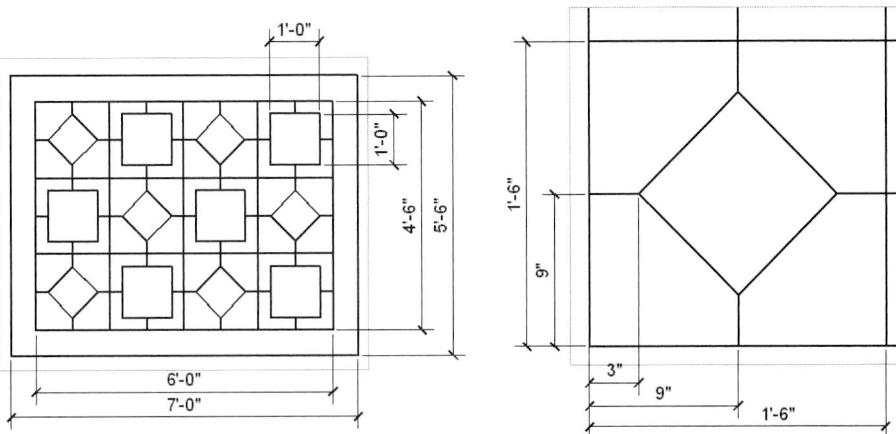

Figure 14.54

Note: If you drag viewports around and the dimensions do not keep up with the viewports, try entering the DIMREGEN command from the keyboard. This should update all dimensions to the new viewport positions.

HATCH SCALING RELATIVE TO PAPER SPACE

While in a layout, you have the opportunity to scale a hatch pattern relative to the current Paper Space scale in a viewport. In this way, you can easily have AutoCAD calculate the hatch pattern scale, since this is determined by the scale of the image inside a viewport. Open the file 14_Valve Gasket. This drawing consists of a small gasket that does not have crosshatching applied to the thin inner border. If you hatched this shape using the default hatch settings while in Model Space, the gasket would appear similar to the illustration in the following image. Notice that the hatch pattern spacing is too large. You would need to experiment with various hatch scales in order to achieve the desired hatch results.

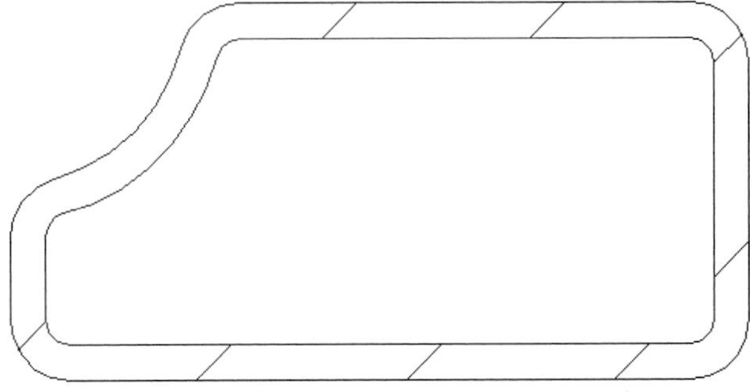

Figure 14.55

A better way to approach this problem would be to enter Paper Space, activate floating Model Space, and produce the hatch pattern from there. While inside a floating Model

Space viewport, activate the Hatch and Gradient dialog box. Notice the Relative to paper space checkbox. Placing a check in this box activates this feature, as shown on the left in the following image. This feature is grayed out if you try to hatch in Model Space.

The finished gasket with hatching applied is illustrated on the right in the following image. Notice how the hatch scale is based on the viewport scale.

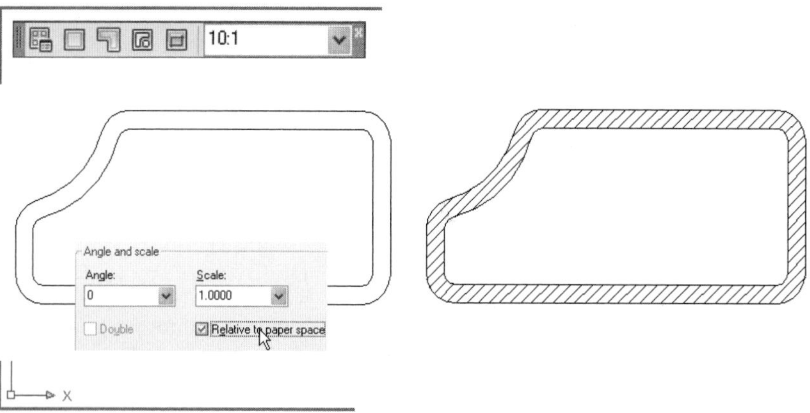

Figure 14.56

Try It!: Open the drawing file 14_Hatch Partial Plan. The inner walls of the object in the following image need to be crosshatched. Apply the technique of making the hatch scale relative through the Hatch and Gradient dialog box. The results are illustrated on the right in the following image.

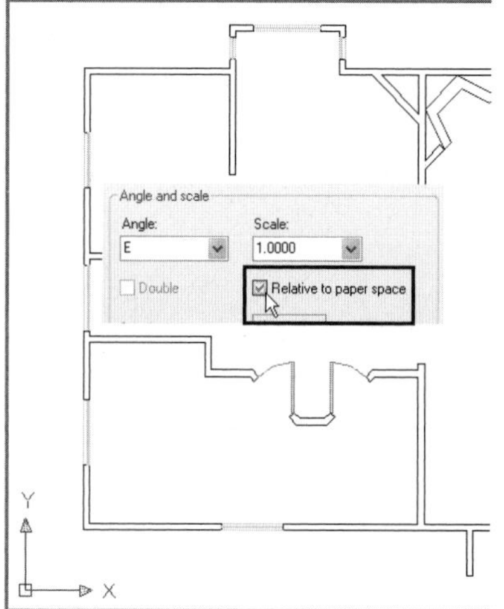

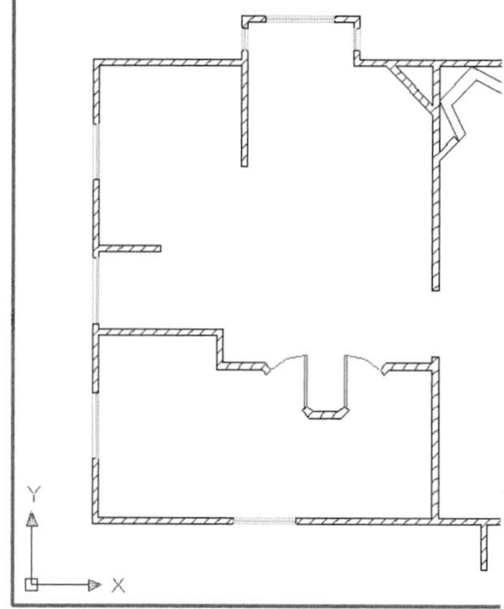

Figure 14.57

QUICK VIEW LAYOUTS

The Quick View tool allows you to preview and switch between open drawings and layouts associated with drawings. Two modes are available through the Quick View tool; namely Quick View Layouts and Quick View Drawings. Each tool is activated from the status bar located at the bottom of the display screen as shown in the following image.

Clicking on the Quick View Layouts button as shown on the left in the following image will display the model space and layouts of the current drawing in a row.

Clicking on the Quick View Drawings button as shown on the right in the following image will display all drawings currently opened.

Figure 14.58

The following image illustrates various tabs that represent drawing layouts. Rather than click on the layout tab to launch the layout drawing, moving your cursor over a layout, such as HVAC Plan, will preview this layout as shown in the following image.

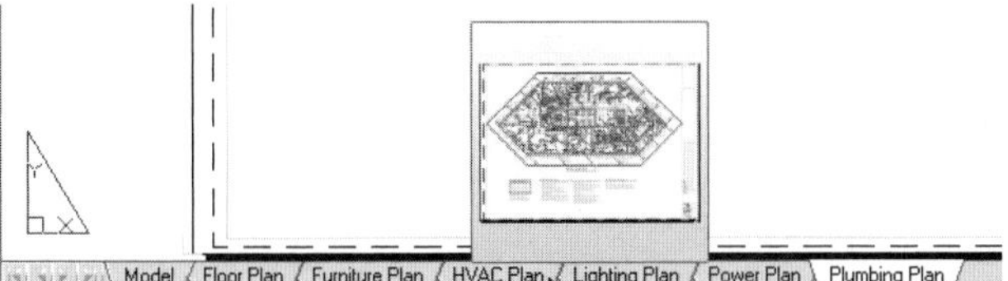

Figure 14.59

You can even preview images of multiple layouts of a drawing. This is accomplished by clicking on the Quick View Layouts button at the bottom of the display screen as shown in the following image. Once a number of layouts are displayed, you can click on an image to make this layout current. Additional buttons are displayed on each image that allow you to plot and publish a drawing from this image. These images can even be resized dynamically by holding down the CTRL key as you roll the wheel on a mouse.

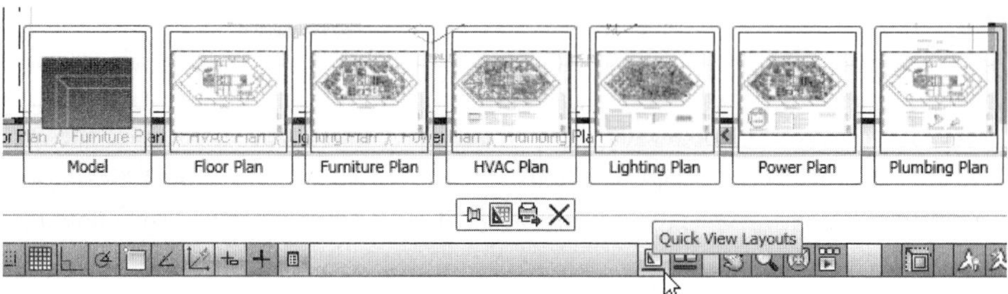

Figure 14.60

The following image illustrates a special toolbar that is displayed below all Quick View Layout images. These buttons are explained from left to right. The Pin icon allows you to pin images of layouts so they will always be visible even while you are working on a drawing; the New Layout icon creates a layout and displays as a Quick View image at the end of the row; the Publish icon launches the Publish dialog box for the purpose of publishing layouts; and the Close icon closes all Quick View layout images.

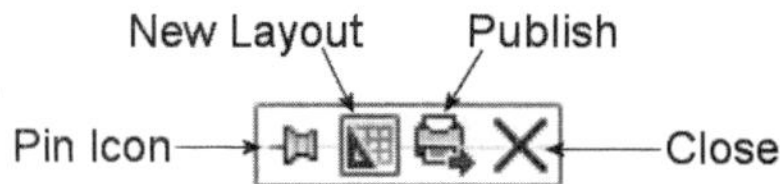

Figure 14.61

QUICK VIEW DRAWINGS

As with Quick View layouts, Quick View drawings allow you to display every drawing currently open as shown in the following image. The image of the current drawing will appear highlighted.

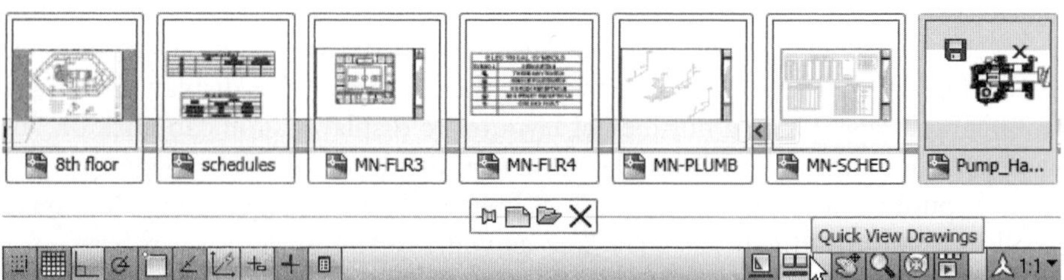

Figure 14.62

If you move your cursor over an image of a drawing that contains layouts, all layouts for that drawing are displayed above the Quick View drawing as shown in the following image. From there, you can make drawings or layouts current by double clicking on the image.

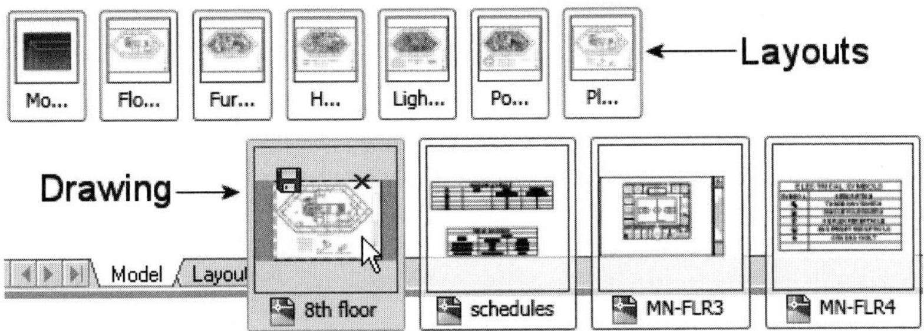

Figure 14.63

The following image illustrates a special toolbar that is displayed below all Quick View Drawing images. These buttons are explained from left to right. The Pin icon allows you to pin images of drawings so they will always be visible even while you are working on a drawing; the New icon creates a new drawing file and displays the file as a Quick View image at the end of the row; the Open icon launches the Open dialog box for the purpose of opening drawing files; and the Close icon closes all Quick View Drawing images.

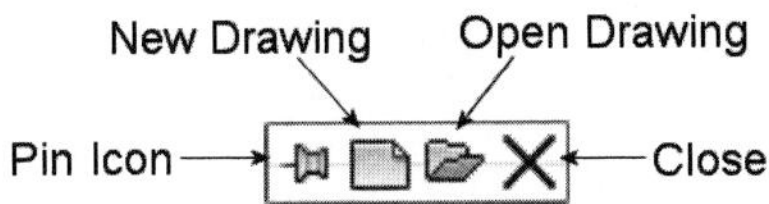

Figure 14.64

TUTORIAL EXERCISE: 14_HVAC.DWG

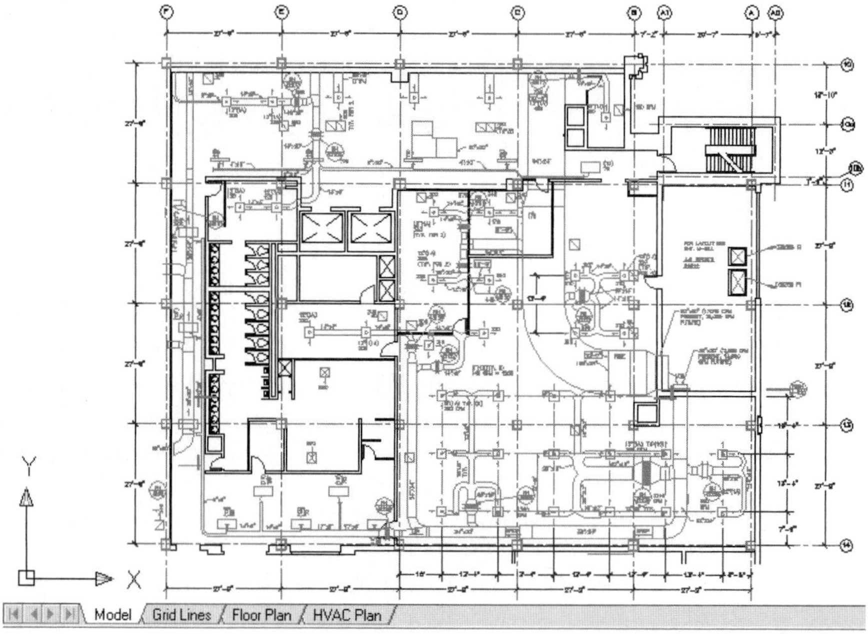

Figure 14.65

PURPOSE

This tutorial is designed to create multiple layouts of the HVAC drawing in the previous image in Paper Space.

SYSTEM SETTINGS

All unit, limit, and plotter settings have already been made in this drawing.

LAYERS

Layers have already been created for this exercise.

SUGGESTED COMMANDS

You will begin by opening the drawing file 14_Hvac.Dwg. Layout1 has already been created for you, An architectural title block has been inserted into Paper Space and the image has already been scaled to 1/8" = 1'-0". This layout, which will be renamed Grid Lines, will be modified by freezing the layers that pertain to the floor and HVAC. This is accomplished while in floating Model Space such that the layers are only frozen in that viewport. From this layout, create another layout called Floor Plan. While in floating Model Space, freeze the layers that pertain to the grid lines and HVAC plans. From this layout, create another layout called Hvac Plan. While in floating Model Space, freeze the layers that pertain to the grid lines. Turn off all viewports and edit drawing titles for each layout.

Step 1

Illustrated in the following image is a layout of an HVAC plan (heating, venting, and air conditioning). A viewport already exists and the image inside the viewport has been scaled to a value of 1/8" = 1'-0". A border also exists in this layout. The goal is to create two extra layouts that show different aspects of the HVAC plan. One layout will display only the grid pattern used to lay out the plan. Another layout will show just the floor plan information. The third layout will show the HVAC and floor plans together.

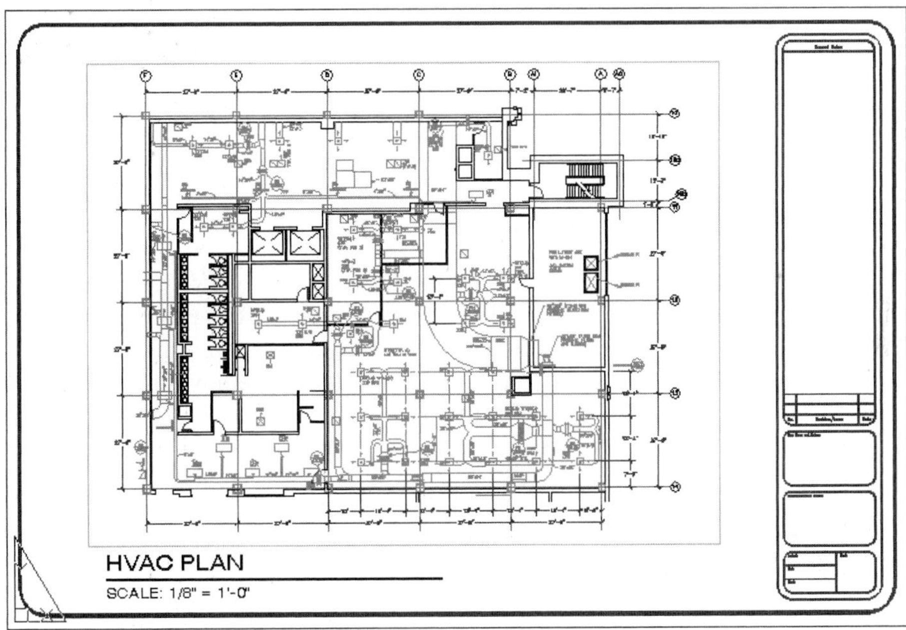

Figure 14.66

Step 2

Before you start to create the new layouts, the viewport present in the existing layout needs to be locked. This will prevent any accidental zooming in and out while inside floating Model Space. To lock a viewport, click the edge of the viewport and pick the Lock/Unlock Viewport icon shown in the following image. When the viewport is locked, you cannot change the Viewport Scale. Another method used for locking a viewport is to click the edge of the viewport, right-click, and choose Display Locked followed by Yes from the menu, Locking viewports is good practice and should be performed before creating any extra layouts. Now that this viewport is locked, the copied viewports created in other layouts will also be locked.

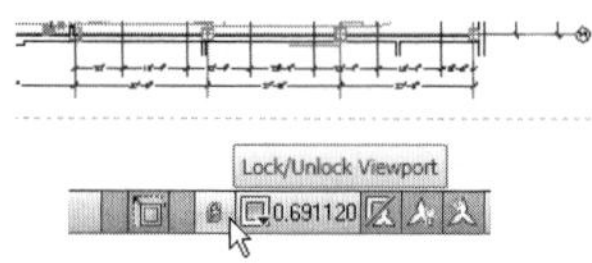

Figure 14.67

Step 3

Next, double-click on the Layout1 tab, located in the lower-left corner of the display screen. Rename this layout to Grid Lines, as shown in the following image. It is always good practice to give your layouts meaningful names.

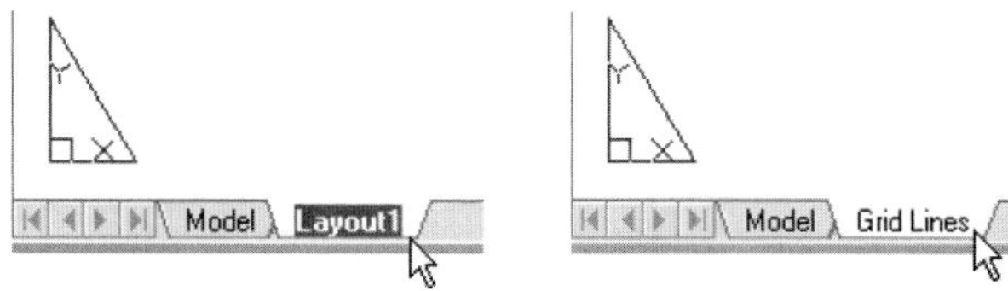

Figure 14.68

Step 4

Activate the Quick Properties tool. Click the drawing title (HVAC PLAN); the title should highlight and grips will appear. When the Quick Properties palette displays, change HVAC PLAN to GRID LINES PLAN, as shown in the following image. Pressing ENTER automatically updates the text to the new value. Press ESC to remove the grip and dismiss the Quick Properties palette.

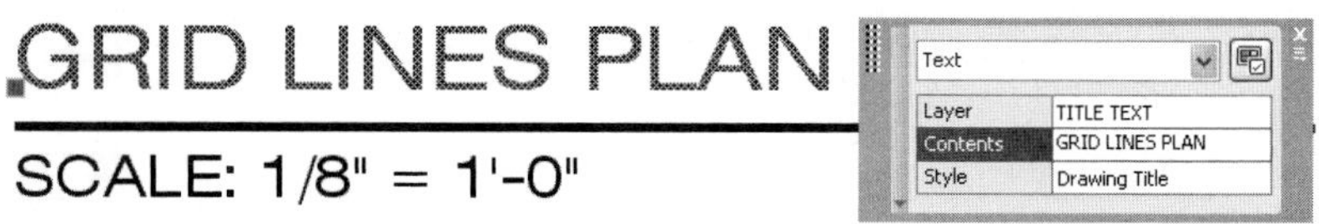

Figure 14.69

Step 5

Prepare to create the second layout by first holding down the CTRL key while pressing and dragging the Grid Lines tab, as shown on the left in the following image. The new layout, Grid Lines (2), is created, as shown on the right in the following image.

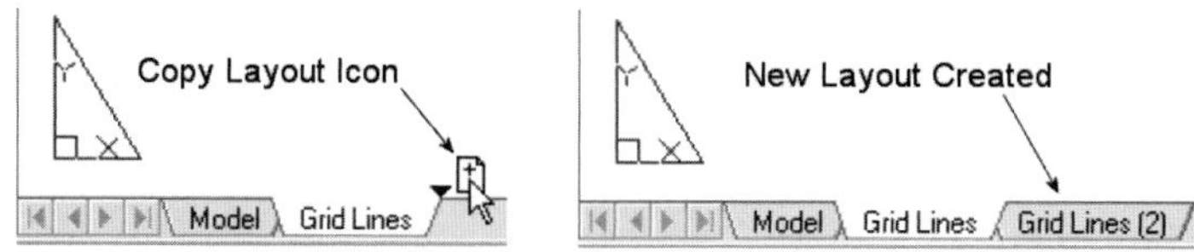

Figure 14.70

Step 6

Before continuing, double-click the new layout Grid Lines (2), as shown on the left in the following image. Change this current layout name from Grid Lines (2) to the new name Floor Plan, as shown on the right in the following image.

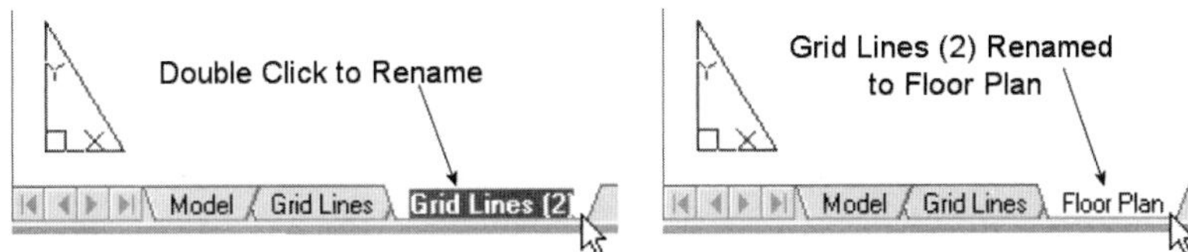

Figure 14.71

Step 7

Use the previous steps to help create one final layout. Create a copy of the Floor Plan layout and change its name to HVAC Plan. When you are finished creating and renaming all the layouts, the lower-left corner of your display should appear similar to the following image.

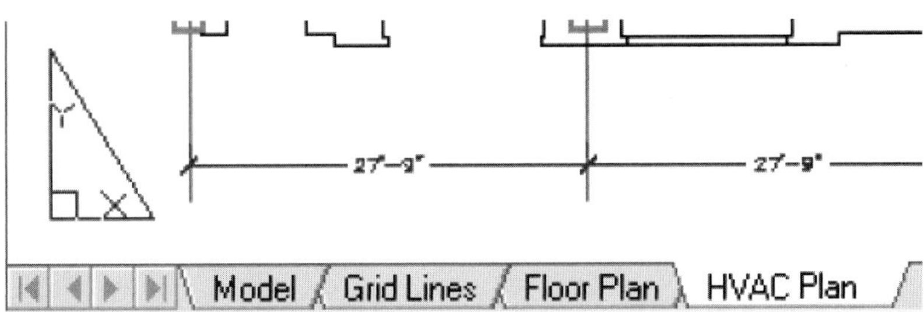

Figure 14.72

Step 8

Click the Grid Lines tab. Double-click anywhere inside the viewport, as shown in the following image; this places you in floating Model Space. You will need to turn off all layers that deal with the floor plan and HVAC, which leave only the layers with the grid lines visible.

Activate the Layer control box, as shown in the following image, and freeze the layers DOORS, FLOOR, HVAC DIM, and HVAC SUP in the current viewport by clicking the appropriate icon. The frozen layers pertain to the floor plan and HVAC plan.

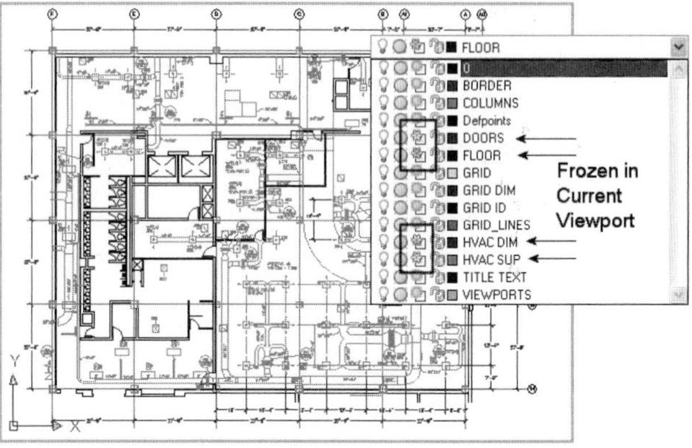

Figure 14.73

Double-click anywhere outside the viewport to return to Paper Space. Your drawing should appear similar to the following image. Notice that only the columns and grid lines appear visible in this layout.

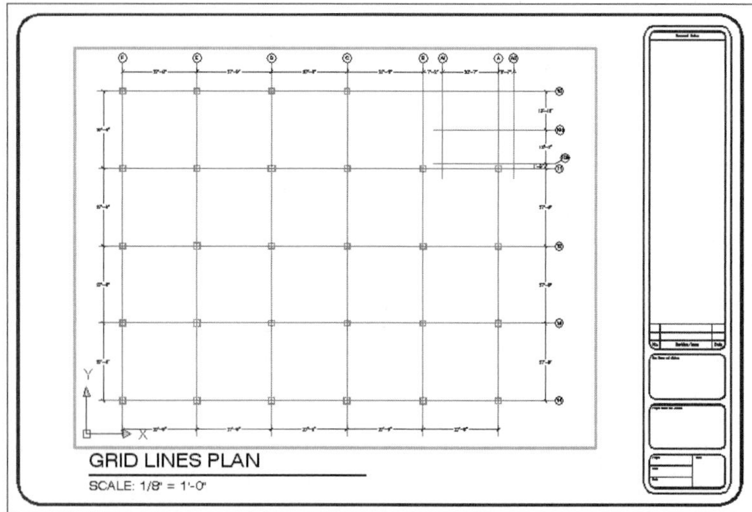

Figure 14.74

Step 9

Click the Floor Plan tab. Double-click anywhere inside the viewport, as shown in the following image; this places you in floating Model Space. You will need to turn off all layers that deal with the grid lines and HVAC.

Activate the Layer control box, shown in the following image, and freeze the layers in the current viewport that pertain to the grid lines plan and HVAC plan (GRID, GRID DIM, GRID ID, GRID_LINES, HVAC DIM, and HVAC SUP).

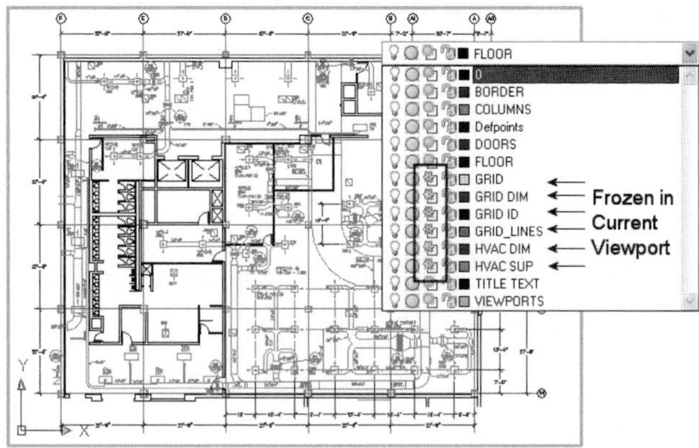

Figure 14.75

Double-click anywhere outside the viewport to return to Paper Space. Use the Quick Properties palette to change the drawing title from GRID LINES PLAN, as shown in the following image, to FLOOR PLAN. Do this by first selecting GRID LINES PLAN, which will launch the Quick Properties palette allowing you to change the content of the text. When finished, press ESC to remove the palette and grip.

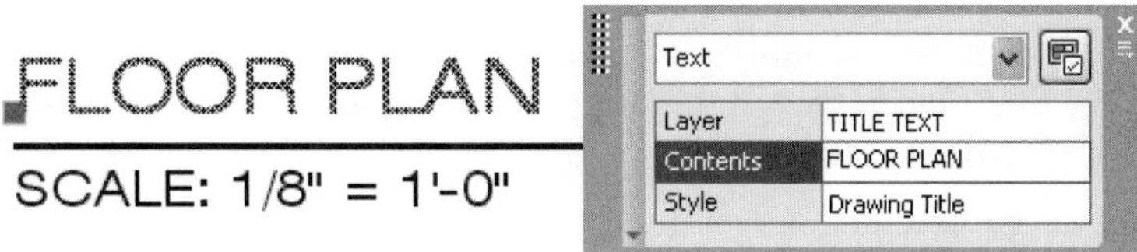

Figure 14.76

Your drawing now displays only floor plan information, as shown in the following image, while the HVAC and GRID layers are frozen only in this viewport.

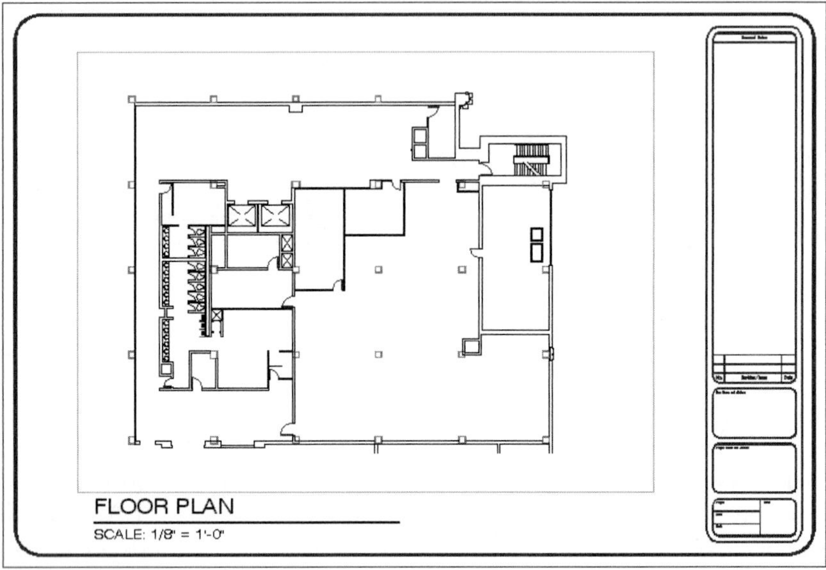

Figure 14.77

Step 10

Click the HVAC Plan tab. Double-click anywhere inside the viewport, as shown in the following image; this places you in floating Model Space. You will need to turn off all layers that deal with the grid lines plan.

Activate the Layer control box, as shown in the following image, and freeze the layers in the current viewport that pertain to the grid lines plan (COLUMNS, GRID, GRID DIM, GRID ID, and GRID_LINES).

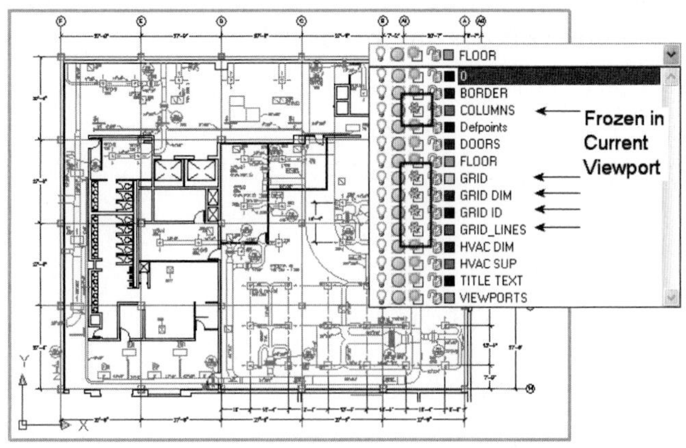

Figure 14.78

Double-click anywhere outside the viewport to return to Paper Space. Use the Quick Properties palette to change the drawing title from GRID LINES PLAN, as shown in the following image, to HVAC PLAN. When finished, press ESC to remove the palette and grip.

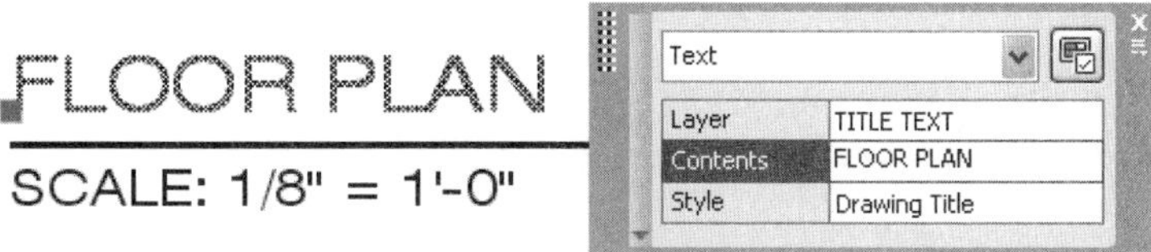

Figure 14.79

Your drawing should appear similar to the following image. In this viewport, you see the floor plan and HVAC ductwork but no grid lines or columns.

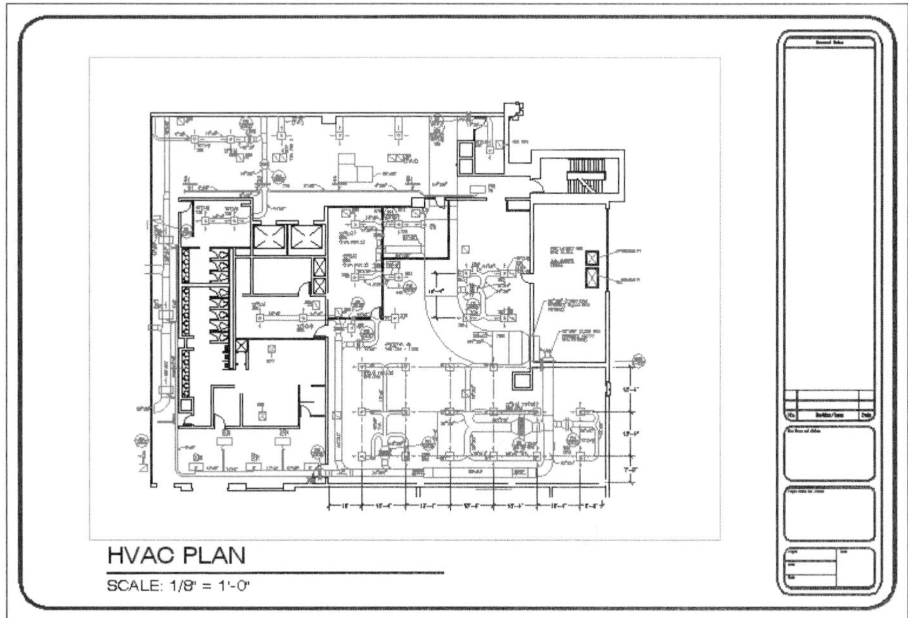

Figure 14.80

Step 11

One additional layer control technique needs to be performed. To better distinguish the objects that represent the HVAC Plan, the floor plan layer needs to be changed to a different color. This change in color must only occur in the current viewport and not affect any other viewports or even objects in Model Space. Double-click inside of this viewport and launch the Layer Properties Manager palette. Click the Floor layer and change the color under the VP Color column to 9 as shown in the following image. This color represents light gray and affects only the Floor Plan layer. Click the OK button to exit this dialog box.

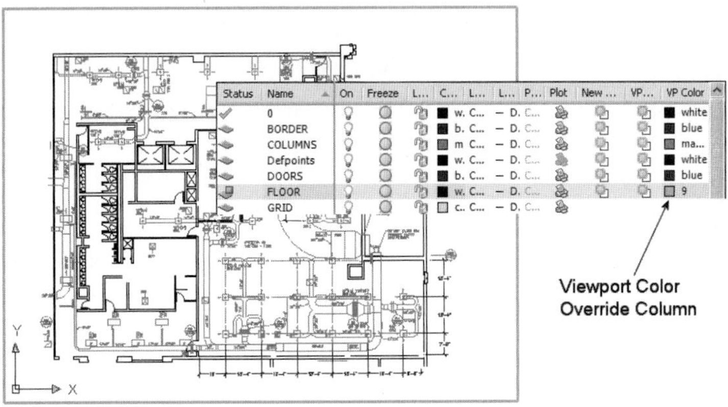

Figure 14.81

The results are illustrated in the following image. Notice how the HVAC objects stand out compared to the floor plan objects.

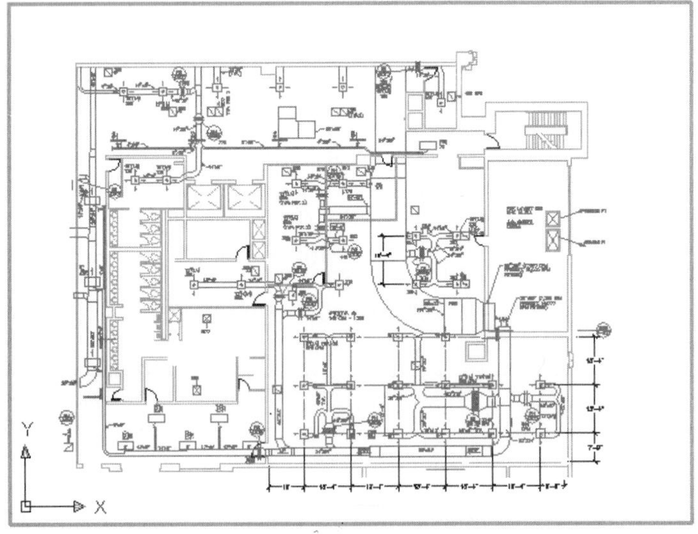

Figure 14.82

Step 12

In the event you need to change the color of the Floor layer back to its original color assignment, activate the Layer Properties Manager palette and right-click Color 9 under the VP Color column to display the menu shown in the following image. From this menu, click Remove Viewport Overrides for followed by Color and then In Current Viewport Only. This changes the color of the Floor layer back to its original color assignment.

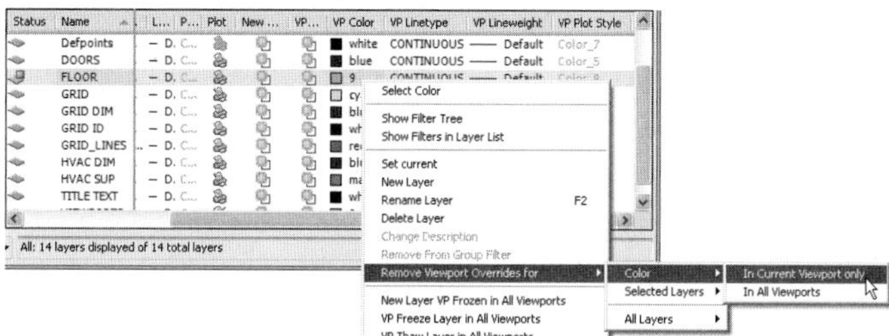

Figure 14.83

Step 13

Turn off the Viewports layer. The completed drawing is displayed in the following image. This exercise illustrated how to create multiple layouts. It also illustrated how to freeze layers in one viewport and have the same layers visible in other viewports. If the Viewports layer is set to No Plot through the Layer Properties Manager palette, performing a plot preview will not display the Viewports layer.

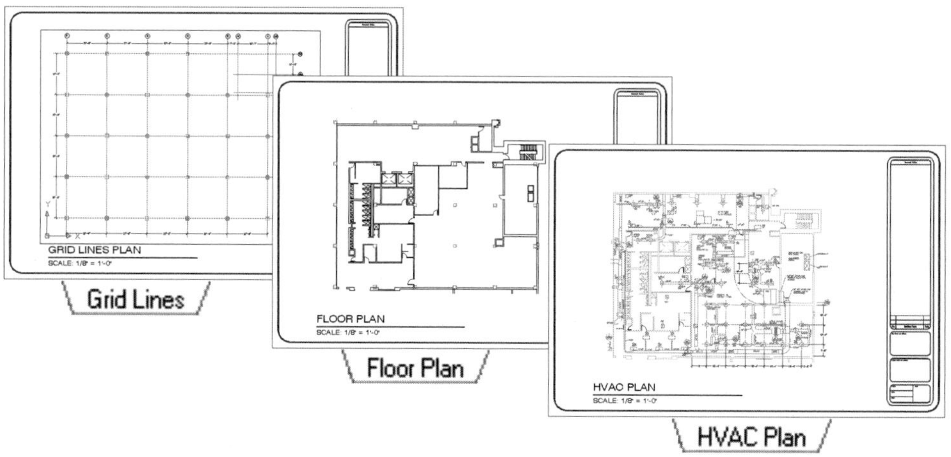

Figure 14.84

CHAPTER 15

Plotting Your Drawings

This chapter discusses plotting through a series of tutorial exercises designed to perform the following tasks:

- Configure a new plotter
- Plot from a drawing layout (Paper Space)
- Control lineweights
- Create a Color Dependent plot style table
- Publish multiple drawing sheets
- Create a web page consisting of various drawing layouts for viewing over the Internet

CONFIGURING A PLOTTER

Before plotting, you must first establish communication between AutoCAD and the plotter. This is called configuring. From a list of supported plotting devices, you choose the device that matches the model of plotter you own. This plotter becomes part of the software database, which allows you to choose this plotter many times. If you have more than one output device, each device must be configured before being used. This section discusses the configuration process used in AutoCAD.

Step 1

Begin the plotter configuration process by choosing Plotter Manager from the File pull-down menu, as shown on the left in the following image (the File heading in the Menu Browser can also be used to choose Plotter Manager). This activates the Plotters program group, as shown on the right in the following image, which lists all valid plotters that are currently configured. The listing in this image displays the default plotters configured after the software is loaded. Except for the DWF devices, which allow you to publish a drawing for viewing over the Internet, or a popular DWG to PDF device that allows you to create a PDF (Adobe) document directly from an AutoCAD DWG file, a plotter has not yet been configured. Double-click the Add-A-Plotter Wizard icon to continue with the configuration process.

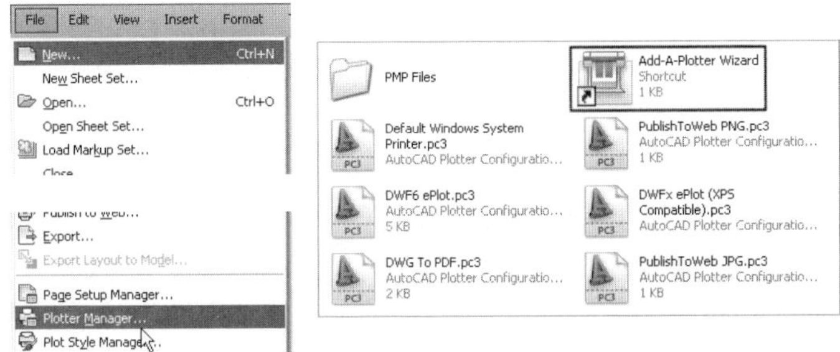

Figure 15.1

Step 2

Double-clicking the Add-A-Plotter Wizard icon displays the Add Plotter - Introduction Page dialog box, as shown in the following image. This dialog box states that you are about to configure a Windows or non-Windows system plotter. This configuration information will be saved in a file with the extension .PC3. Click the Next > button to continue on to the next dialog box.

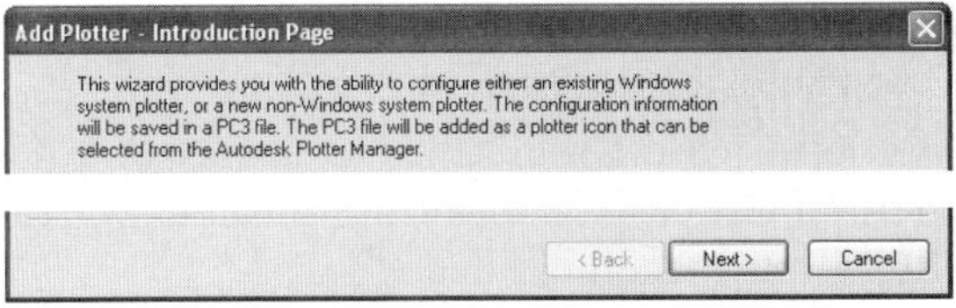

Figure 15.2

Step 3

In the Add Plotter-Begin dialog box shown on the left in the following image, decide how the plotter will be controlled by the computer you are currently using, by a network plot server, or by an existing system printer where changes can be made specifically for AutoCAD. Click the radio button next to My Computer. Then click the Next > button to continue on to the next dialog box.

Use the Add Plotter-Plotter Model dialog box, as shown on the right in the following image, to associate your plotter model with AutoCAD. You would first choose the appropriate plotter manufacturer from the list provided. Once this is done, all models supported by the manufacturer appear to the right. If your plotter model is not listed, you are told to consult the plotter documentation for a compatible plotter. For the purposes of this tutorial, click Hewlett-Packard in the list of Manufacturers. Click the DesignJet 750C Plus C3195A for the plotter model. A Driver Info dialog box may appear, giving you more directions regarding the type of HP DeskJet plotter selected. Click the Continue button to move on to the next dialog box used in the plotter configuration process.

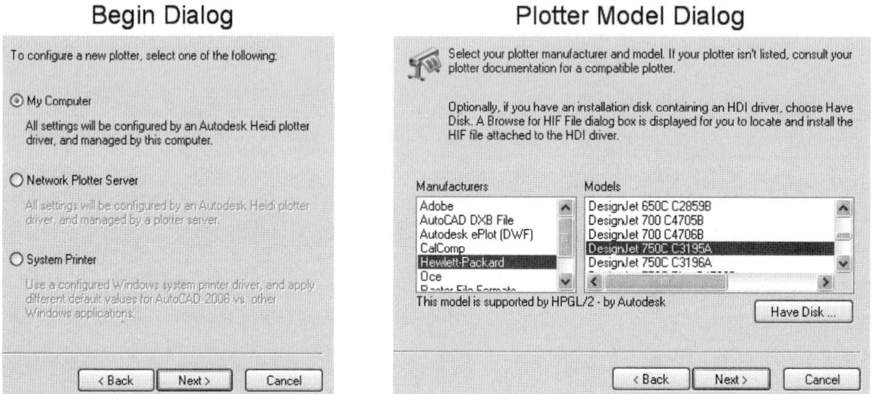

Figure 15.3

Step 4

PCP and PC2 files have been in existence for many years. They were designed to hold plotting information such as pen assignments. In this way, you used the PCP or PC2 files to control pen settings instead of constantly making pen assignments every time you perform a plot; at least this is how pen assignments were performed in past versions of AutoCAD. The Add Plotter-Import PCP or PC2 dialog box, shown on the left in the following image, allows you to import those files for use in AutoCAD in a PC3 format. If you will not be using any PCP or PC2 files from previous versions of AutoCAD, click the Next > button to move on to the next dialog box.

In the Add Plotter-Ports dialog box, shown on the right in the following image, click the port used for communication between your computer and the plotter. The LPT1 port will be used for the purposes of this tutorial. Place a check in its box and then click the Next > button to continue on to the next dialog box.

Import Pcp or Pc2 Dialog

To import plotter specific information from a previously saved PCP or PC2 file, choose Import File. Paper size, plot optimization level, network share names, and port names can be imported into the new PC3 file.

Import File ...

NOTE: Use the Add Plot Style Table wizard to import pen assignment information. Use the Import PCP/PC2 Settings wizard to import PCP or PC2 page setup information.

< Back | Next > | Cancel

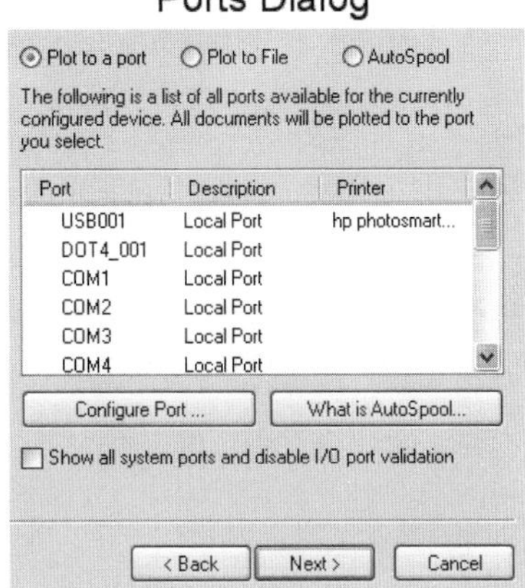

Figure 15.4

Step 5

In the Add Plotter-Plotter Name dialog box, shown on the left in the following image, you have the option of giving the plotter a name other than the name displayed in the dialog box. For the purpose of this tutorial, accept the name that is given. This name will be displayed whenever you use the Page Setup and Plot dialog boxes.

The last dialog box is displayed on the right in the following image. In the Add Plotter-Finish dialog box, you can modify the default settings of the plotter you just configured. You can also test and calibrate the plotter if desired. Click the Finish button to dismiss the Add Plotter-Finish dialog box.

Plotter Name Dialog

The model name you selected is the default plotter configuration name. You can accept the default name, or enter a new name to identify the new PC3 file you have created. The name you apply will be displayed in the Page Setup and Plot dialog boxes.

Plotter Name:

DesignJet 750C C3195A

Note: If you enter a name that is exactly the same as a System Printer's name, you will not see the System Printer listed in the AutoCAD 2006 Plot or Page Setup dialog boxes.

< Back | Next > | Cancel

Finish Dialog

The plotter DesignJet 750C C3195A has been installed with its default configuration settings. To modify the default settings, choose Edit Plotter Configuration.

Edit Plotter Configuration ...

Optionally, to perform a plot calibration test on the newly configured plotter, and verify that your drawing measurements plot accurately, choose Calibrate Plotter.

Calibrate Plotter ...

< Back | Finish | Cancel

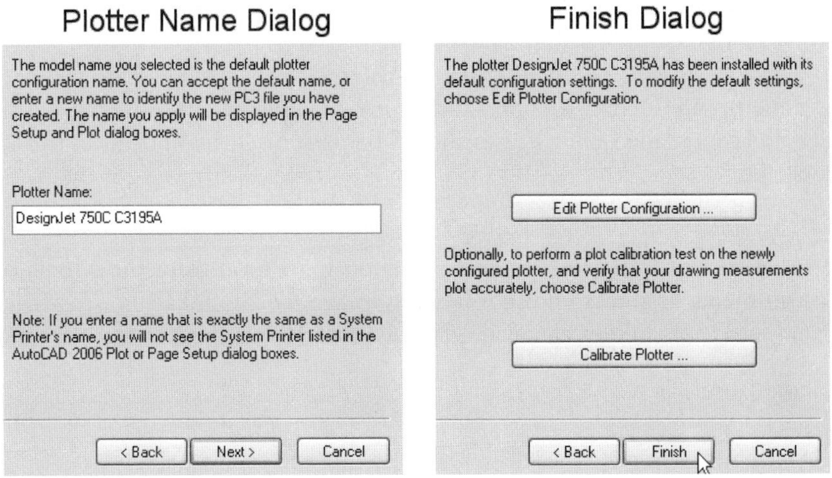

Figure 15.5

Step 6

Exiting the Add Plotter dialog box returns you to the Plotters program group, shown in the following image. Notice that the icon for the DesignJet 750C Plus plotter has been added to this list. This completes the steps used to configure the DesignJet 750C Plus plotter. Follow these same steps if you need to configure another plotter.

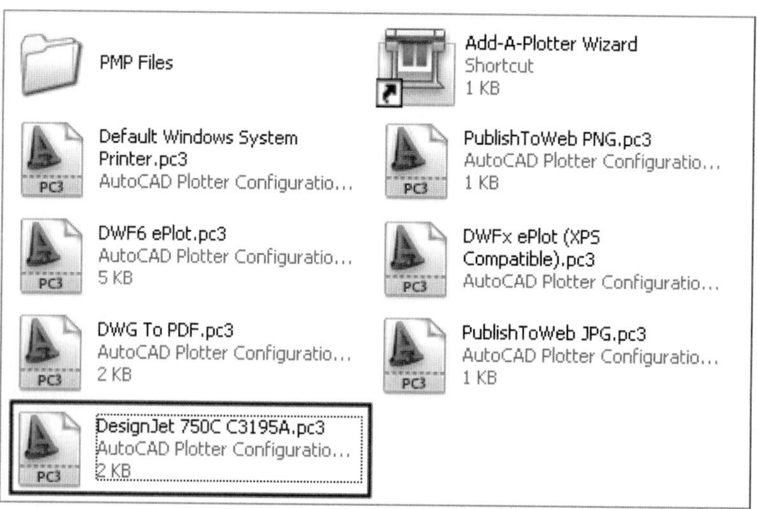

Figure 15.6

PLOTTING FROM A LAYOUT

Open the drawing 15-Center Guide to step through the process of plotting the drawing from Layout mode or Paper Space.

Step 1

Open the drawing 15_Center_Guide.Dwg. This drawing should already be laid out in Paper Space. A layout called Four Views should be present at the bottom of the screen next to the Model tab.

Step 2

Begin the process of plotting this drawing by choosing Plot... from the Menu Browser or File pull-down menu, as shown in the following image. You could also choose the plot icon from the Standard toolbar. You can also type the word Plot or hold down the CTRL key and type the letter P.

Figure 15.7

This activates the Plot dialog box, shown in the following image. For the purposes of this tutorial, the DWF6 ePlot is being used as the output device. You can opt to create additional copies of your plot; by default this value is set to 1. The DWF6 ePlot can plot only one copy, so this value is grayed out. Next, make sure the paper size is currently set to ANSI expand C (22.00 × 17.00 inches). In the Drawing orientation area, make sure that the radio button adjacent to Landscape is selected. You could also plot the drawing out in Portrait mode, where the short edge of the paper is the top of the page. For special plots, you could plot the drawing upside down.

In the Plot area, the Layout radio button is selected. Since you created a layout, this is the obvious choice. The Extents mode allows you to plot the drawing based on all objects that make up the drawing. Plotting the Display plots your current drawing view, but be careful: If you are currently zoomed in to your drawing, plotting the Display will plot only this view. In this case, it would be more practical to use Layout or Extents to plot. When you plot a layout in Paper Space, the Plot scale will be set to 1:1. Since you pre-scaled the drawing to the Paper Space viewport using the Viewports toolbar, all drawings in Paper Space are designed to be plotted at this scale. The Plot offset is designed to move or shift the location of your plot on the paper if it appears off center. In the Plot options area, you have more control over plots by applying lineweights, using existing plot styles, plotting Paper Space last (Model Space first), or even hiding objects, on a 3D solid model.

One other area to change is in the Plot style table area, located in the upper-right corner of the dialog box. This area controls the appearance of the plot, for instance, a colored plot versus a monochromatic plot (black lines on white paper). You can even create your own plot style, which will be discussed later in this chapter. For the purposes of this tutorial, monochrome will be used, as shown in the following image.

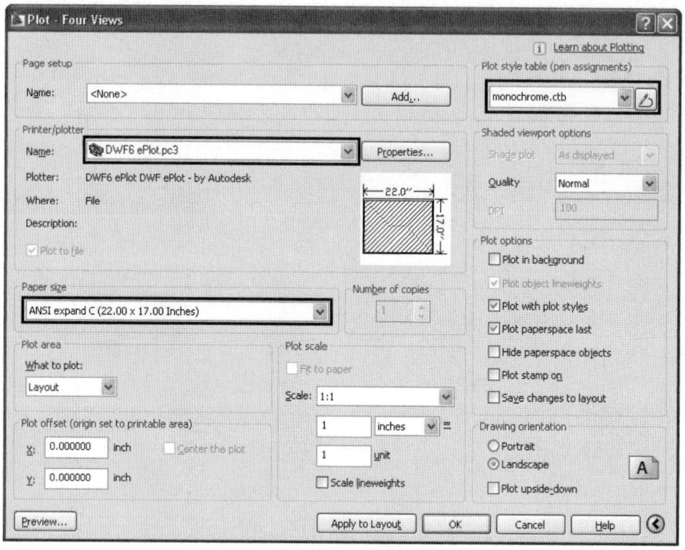

Figure 15.8

Step 3

One of the more efficient features of plotting is that you can preview your plot before sending the plot information to the plotter. In this way, you can determine whether the entire drawing will plot based on the sheet size (this includes the border and title block). Clicking the Preview... button activates the image shown in the following image. The sheet size is shown along with the border and four-view drawing. Right-clicking anywhere on this preview image displays the cursor menu, allowing you to perform various display functions such as ZOOM and PAN to assist with the verification process. If everything appears satisfactory, click the Plot option to send the drawing information to the plotter. Clicking the Exit option returns you to the Plot dialog box, where you can make changes in the Plot Device or Plot Settings tabs.

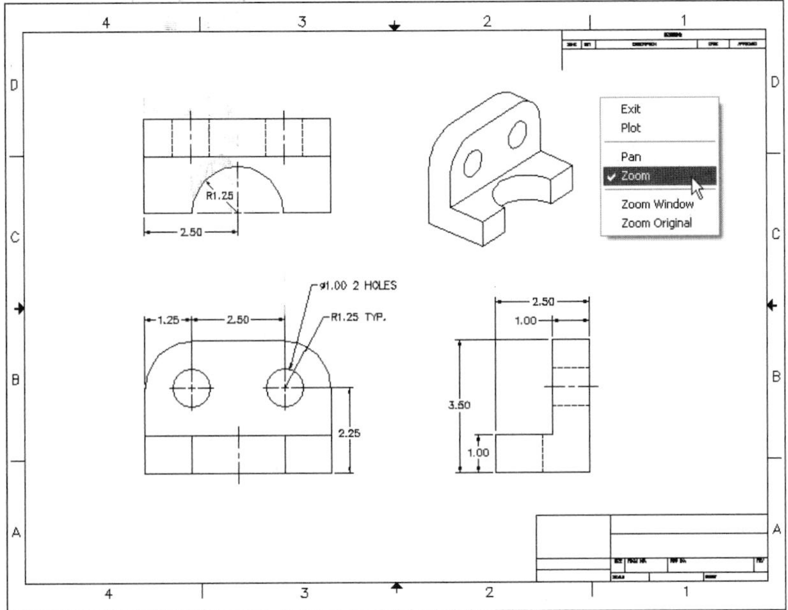

Figure 15.9

Note: You can also plot from Model Space. However, this is more involved because you must bring borders, title blocks, and notes into Model Space. Also, the sizes of notes and dimensions in Model Space must be properly sized to plot correctly.

ENHANCING YOUR PLOTS WITH LINEWEIGHTS

This section on plotting describes the process of assigning lineweights to objects and then having the lineweights appear in the finished plot. Open the drawing called 15_V_Step. Dwg, shown in the following image, and notice that you are currently in Model Space (the Model tab is current at the bottom of the screen). Follow the next series of steps to assign lineweights to a drawing before it is plotted.

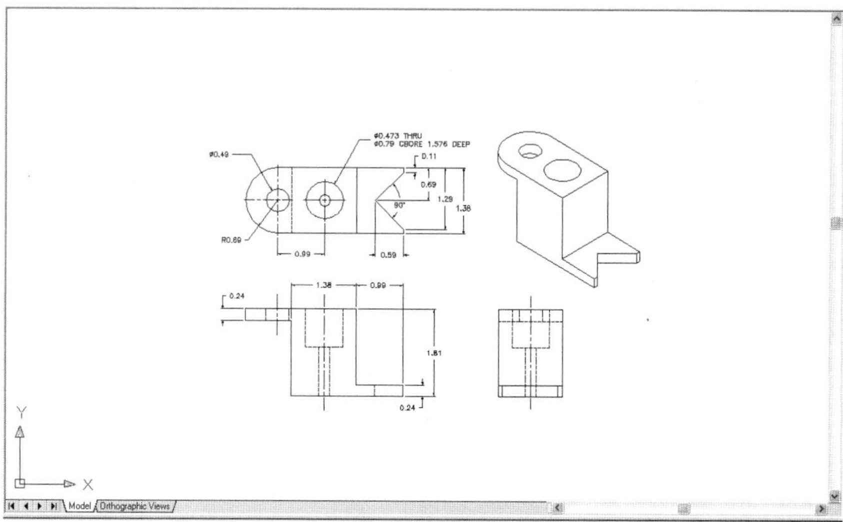

Figure 15.10

Step 1

From the illustration of the drawing in the previous image, all lines on the Object layer need to be assigned a lineweight of 0.60 mm. All objects on the Hidden layer need to be assigned a lineweight of 0.30 mm. There is also a title block that will be used with this drawing. The Title Block layer needs a lineweight assignment of 0.80 mm. Click in the Layer Properties Manager palette, shown in the following image, and make these lineweight assignments. When you are finished, click OK to save the lineweight assignments and return to Model Space.

Status	Name	On	Freeze	Lock	Color	Linetype	Lineweight	Plot Style	Plot
	0				white	CONTINUOUS	Default	Color_7	
	Center				green	CENTER2	Default	Color_3	
	Defpoints				white	CONTINUOUS	Default	Color_7	
	Dimension				green	CONTINUOUS	Default	Color_3	
	Hidden				red	HIDDEN2	0.30 ...	Color_1	
	Object				white	CONTINUOUS	0.60 ...	Color_7	
	Text				green	CONTINUOUS	Default	Color_3	
	Title Block				white	CONTINUOUS	0.80 ...	Color_7	
	Viewport				green	CONTINUOUS	Default	Color_3	

Figure 15.11

Step 2

Click the LWT button in the Status bar to display the lineweights, as shown in the following image. Either of the buttons, icon or text, will do.

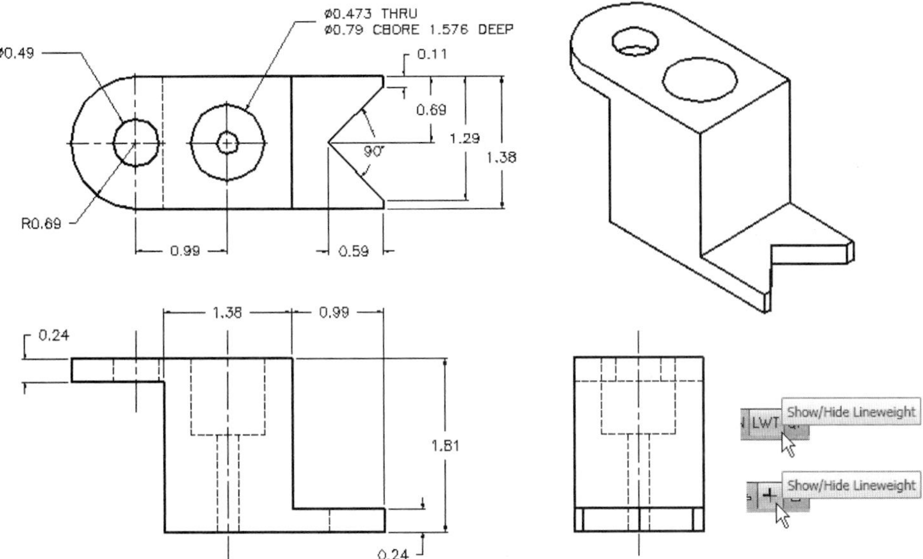

Figure 15.12

You can control the lineweight scale to have the linetypes give off a more pleasing appearance in your drawing. Clicking Lineweight in the Format pull-down menu displays the Lineweight Settings dialog box, as shown in the following image. Notice the slider bar in the area called Adjust Display Scale. This controls only the way lineweights display. Use this slider bar to reduce or increase the scale of the lineweights when they are viewed in your drawing. In the following image, the slider bar has been adjusted toward the minimum side of the scale. This should make the lineweights better to read in the drawing. Click the OK button to return to the drawing and observe the results.

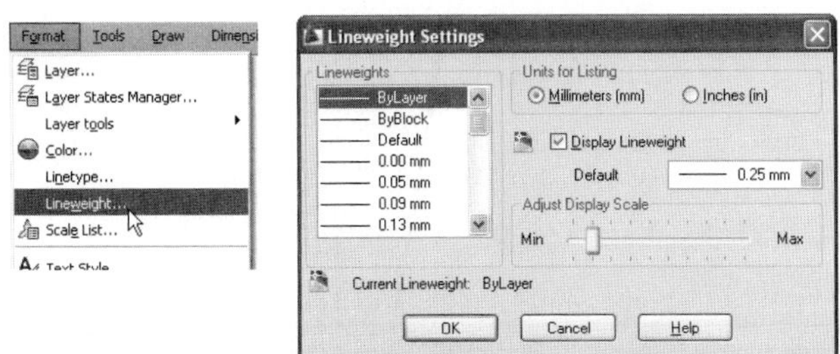

Figure 15.13

All hidden and object lines have been reduced in scale in the following image. This action adds to the clarity and appearance of the drawing. If after making changes to the lineweights you decide to increase the scale, return to the Lineweight Settings dialog box and readjust the slider bar until you achieve the desired results.

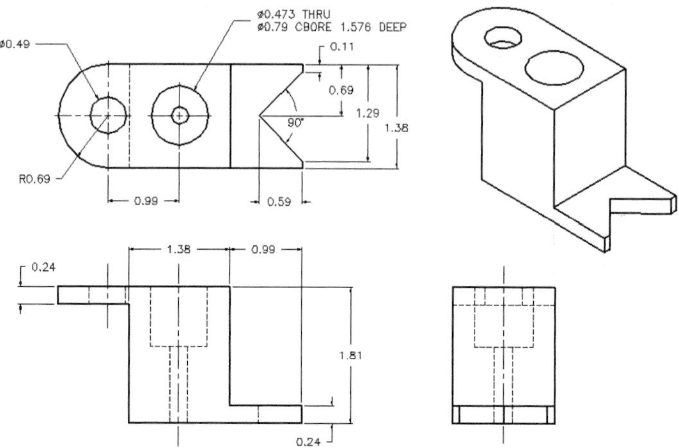

Figure 15.14

Step 3

Clicking the Orthographic Views tab switches you to Paper Space, as shown in the following image. The lineweights do not appear in Paper Space at first glance. Zooming in to the drawing displays all lineweights at their proper widths.

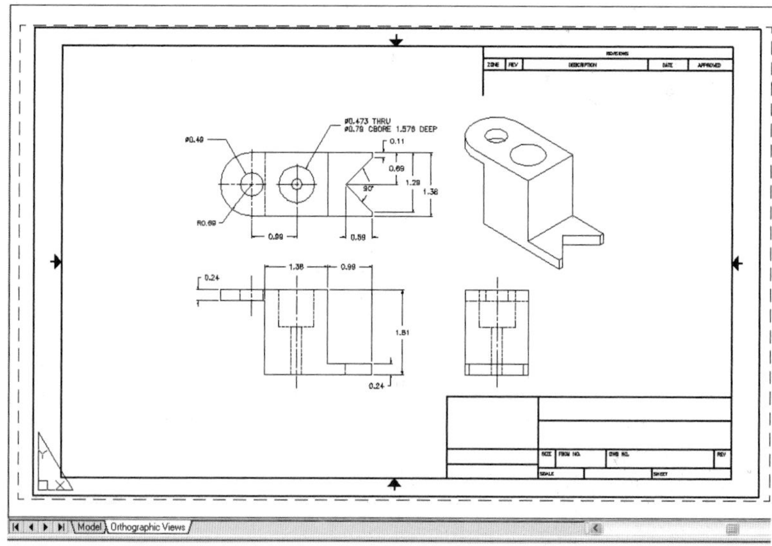

Figure 15.15

Step 4

Activate the Plot dialog box, as shown in the following image. Verify that the current plot device is the DWF6 ePlot. If this is not the device, click the Plot Device tab to activate this plotter.

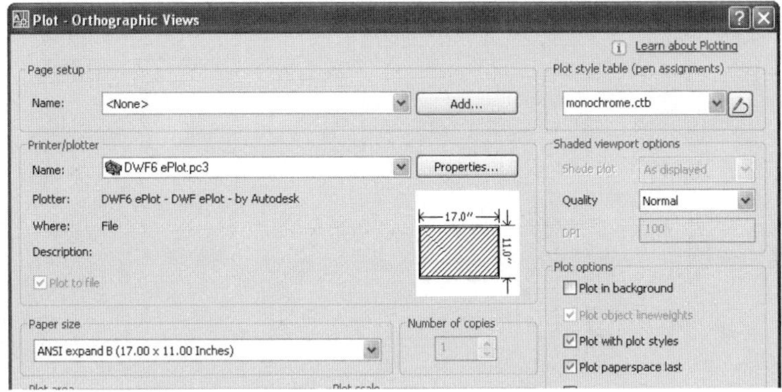

Figure 15.16

Step 5

Click the Preview... button; your display should appear similar to the following image. Notice that the viewport is not present in the plot preview. The Viewports layer was either turned off or set to a non-plot state inside the Layer Properties Manager palette. To view the lineweights in Preview mode, zoom in to segments of your drawing.

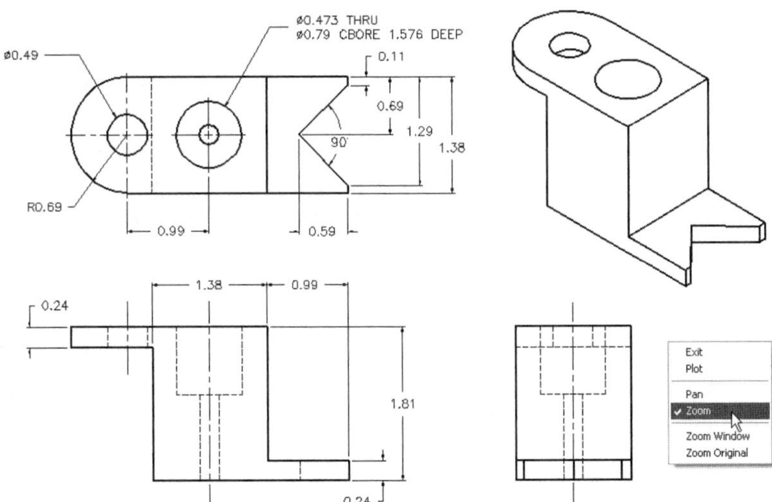

Figure 15.17

CREATING A COLOR-DEPENDENT PLOT STYLE TABLE

This section of the chapter is devoted to the creation of a Color Dependent plot style table. Once the table is created, it will be applied to a drawing. From there, the drawing will be previewed to see how this type of plot style table affects the final plot. Open the drawing 15_Color_R-Guide.Dwg. Your display should appear similar to the following image. A two-view drawing together with an isometric view is arranged in a layout called Orthographic Views.

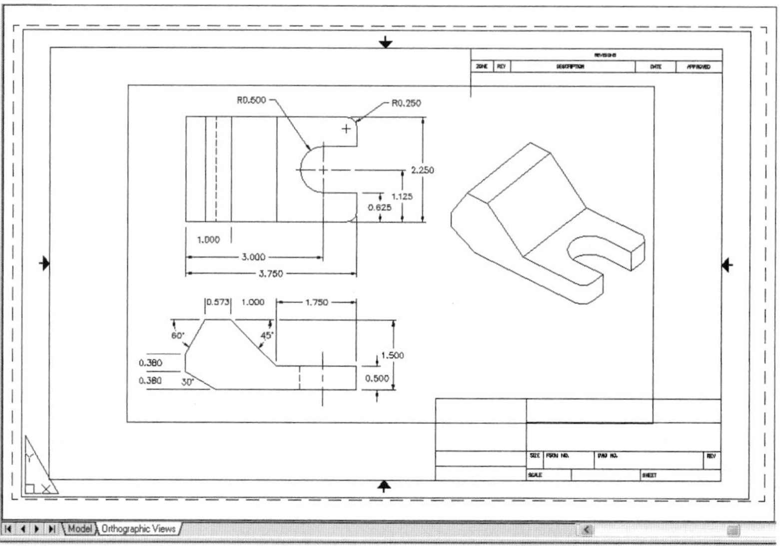

Figure 15.18

The drawing is also organized by layer names and color assignments. The object is to create a Color Dependent plot style table where all layers will plot out black. Also, through the Color Dependent plot style table, the hidden lines will be assigned a lineweight of 0.30 mm, object lines 0.70 mm, and the title block 0.80 mm, as shown in the following table. Follow the next series of steps to perform this task.

Color	Layer	Lineweight
Red	Hidden	0.30
Yellow	Center	Default
Green	Viewports	Default
Cyan	Text	Default
Blue	Title Block	0.80
Magenta	Dimensions	Default
Black	Object	0.70

Step 1

Begin the process of creating a Color Dependent plot style table by choosing Plot Style Manager from the File pull-down menu, as shown on the left in the following image. This activates the Plot Styles dialog box, as shown on the right in the following image. Various Color Dependent and Named plot styles already exist in this dialog box. To create a new plot style, double-click the Add-A-Plot Style Table Wizard.

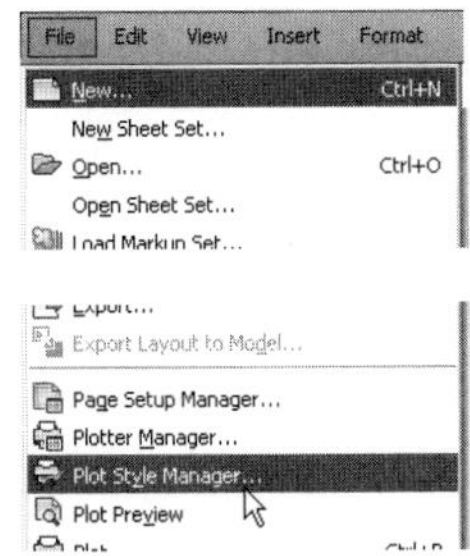

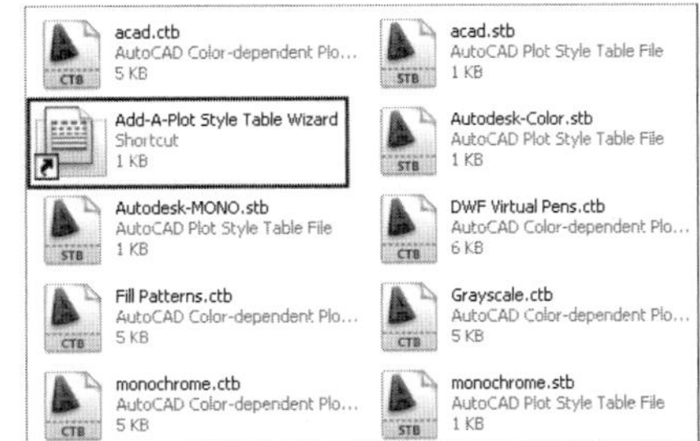

Figure 15.19

Step 2

A partial illustration of the Add Plot Style Table dialog box appears, as shown in the following image, and introduces you to the process of creating plot style tables. Plot styles contain plot definitions for color, lineweight, linetype, end capping, fill patterns, and screening. You are presented with various choices in creating a plot style from scratch, using the parameters in an existing plot style, or importing pen assignment information from a PCP, PC2, or CFG file. You also have the choice of saving this plot style information in a CTB (Color Dependent) or STB (Named) plot style. This chapter will discuss only the creation of a Color Dependent plot style. A Next > button is displayed at the bottom of all Plot Style Wizard dialog boxes. Even though they are not displayed in these images, click it on your display to move on to the next dialog box.

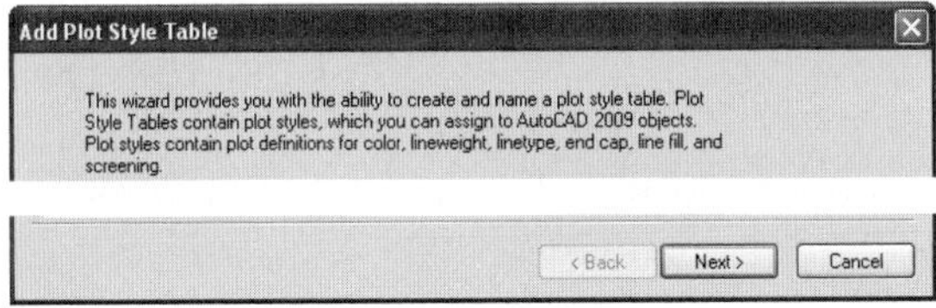

Figure 15.20

Step 3

In the partial illustration of the Add Plot Style Table-Begin dialog box, shown on the left in the following image, four options are available for you to choose, depending on how you want to create the plot style table. Click the Start from scratch radio button to create this plot style from scratch. If you have made pen assignments from previous releases of AutoCAD, you can import them through this dialog box. Click the Next > button to display the next dialog box. This Plot Style Table will be started from scratch.

In the partial illustration of the Add Plot Style Table-Pick Plot Style Table dialog box, shown on the right in the following image, click the Color-Dependent Plot Style Table radio button to make this the type of plot style you will create. Click the Next > button to display the next dialog box.

Begin Dialog

Start from scratch
Create a new plot style table from scratch.
Use an existing plot style table
Create a new plot style table based on an existing plot style table.
Use My R14 Plotter Configuration (CFG)
Import the pen table properties from a R14 CFG file.
Use a PCP or PC2 file
Import the pen table properties from an existing PCP or PC2 file.

Pick Plot Style Table Dialog

To name the style properties, select Named Plot Style Table. To create a plot style table that references each object's AutoCAD 2006 color, select Color-Dependent Plot Style Table.
Color-Dependent Plot Style Table
255 plot styles will be created. The information will be saved in a plot style table (CTB) file.
Named Plot Style Table
A plot style table will be created that contains one plot style named Normal. New plot styles can be added in the Plot Style Table Editor.

Figure 15.21

Step 4

Use the Add Plot Style Table-File Name dialog box, shown as a partial image on the left in the following image, to assign a name to the plot style table. Enter the name Ortho_Drawings in the File name area. The extension CTB is automatically added to this file name. Click the Next > button to display the next dialog box.

The Finish dialog box, shown on the right in the following image, alerts you that a plot style called Ortho_Drawings.ctb has been created. However, you want to have all colors plot out black and you need to assign different lineweights to a few of the layers. To accomplish this, click the Plot Style Table Editor... button as shown in this image.

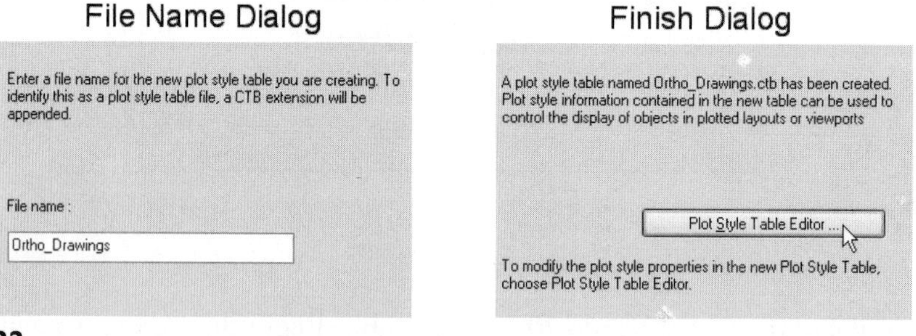

File Name Dialog
Enter a file name for the new plot style table you are creating. To identify this as a plot style table file, a CTB extension will be appended.
File name :
Ortho_Drawings
Finish Dialog
A plot style table named Ortho_Drawings.ctb has been created. Plot style information contained in the new table can be used to control the display of objects in plotted layouts or viewports
Plot Style Table Editor ...
To modify the plot style properties in the new Plot Style Table, choose Plot Style Table Editor.

Figure 15.22

Step 5

Clicking the Plot Style Table Editor... button displays the Plot Style Table Editor dialog box, as shown in the following image. Notice the name of the plot style table present at the top of the dialog box. Also, three tabs are available for making changes to the current plot style table (Ortho-Drawings.ctb). The first tab is General and displays file information about the current plot style table being edited. It is considered good practice to add a description to further document the purpose of this plot style table. It must be pointed out at this time that this plot style table will be used on other drawings besides the current one. Rather, if layers are standard across projects, the same plot style dialog box can be used. This is typical information that can be entered in the Description area.

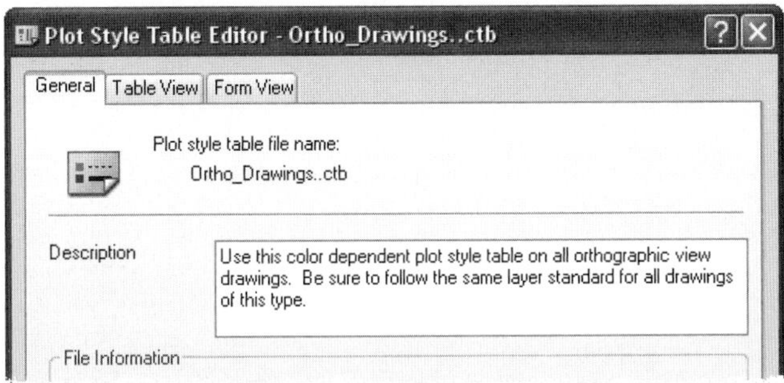

Figure 15.23

Step 6

Clicking the Table View tab activates the dialog box, as shown on the left in the following image. You can use the horizontal scroll bar to get a listing of all 255 colors along with special properties that can be changed. This information is presented in a spreadsheet format. You can click in any of the categories under a specific color and make changes, which will be applied to the current plot style table. The color and lineweight changes will be made through the next tab.

Clicking the Form View tab displays the dialog box shown on the right in the following image. Here the colors are arranged vertically with the properties displayed on the right.

Click Color 1 (Red) and change the Color property to Black. Whatever is red in your drawing will plot out in the color black. Since red is used to identify hidden lines, click in the Lineweight area and set the lineweight for all red lines to 0.30 mm.

For Color 2 (Yellow), Color 3 (Green), Color 4 (Cyan), and Color 6 (Magenta), change the color to Black in the Properties area. All colors can be selected at one time by holding down CTRL while picking each of them. Changes can then be made to all colors simultaneously. Whatever is yellow, green, cyan, or magenta will plot out in the color black. No other changes need to be made in the dialog box for these colors.

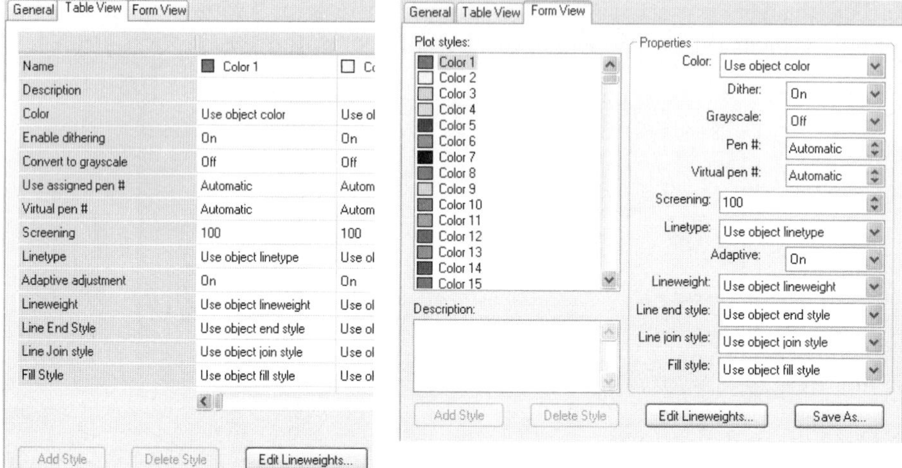

Figure 15.24

Step 7

Click Color 5 (Blue), as shown on the left in the following image, and change the Color property to Black. Whatever is blue in your drawing will plot out in the color black. Since blue is used to identify the title block lines, click in the Lineweight area and set the lineweight for all blue lines to 0.80 mm.

Click Color 7 (Black), as shown on the right in the following image, and change the Color property to Black. Since black is used to identify the object lines, click in the Lineweight area and set the lineweight for all object lines to 0.70 mm.

This completes the editing process of the current plot style table. Click the Save & Close button; this returns you to the Finish dialog box. Clicking the Finish button returns you to the Plot Styles dialog box. Close this box to display your drawing.

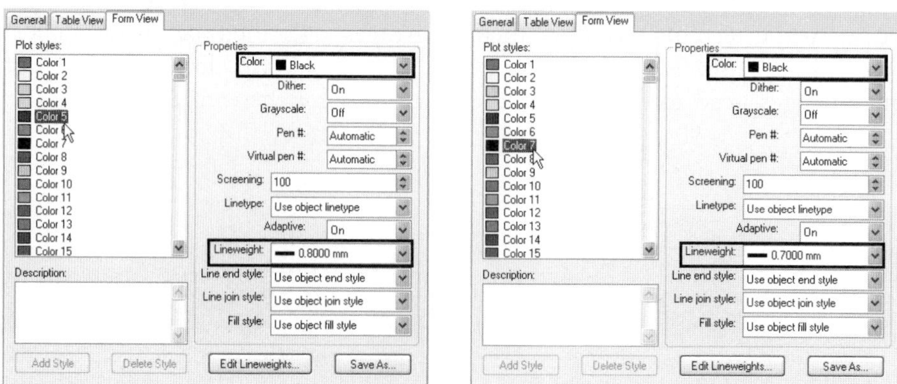

Figure 15.25

Step 8

Activate the Plot dialog box in the Plot device tab, verify that the plotter is the DWF6 ePlot plotter (this plotter has been used in all plotting tutorials throughout this chapter). In the Plot style table (pen assignments) area, make the current plot style Ortho_Drawings.ctb, as shown on the left in the following image. Notice at the top of the dialog box that the plot style table will be saved to this layout. This means that if you need to plot this drawing again in the future, you will not have to look for the desired plot style table.

Click the Plot Settings tab in the Plot dialog box and verify in the lower-right corner of the dialog box under Plot options that you will be plotting with plot styles, as shown on the right in the following image. Click the Preview button to display the results. Unless a plotter is attached to your computer, it is not necessary to click the OK button.

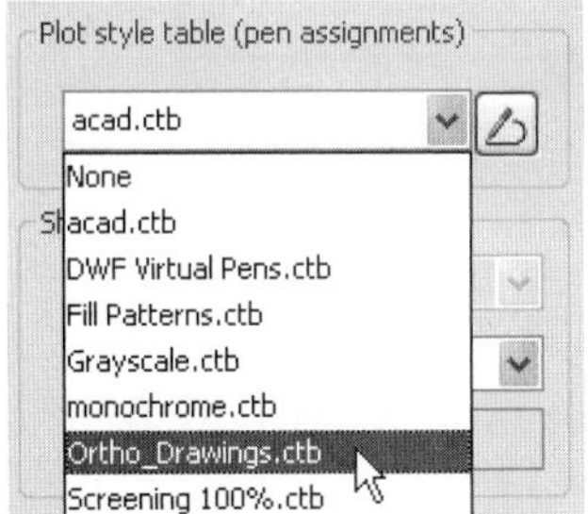

Figure 15.26

Step 9

The results of performing a plot preview are illustrated in the following image. Notice that all lines are black even though they appear in color in the drawing file.

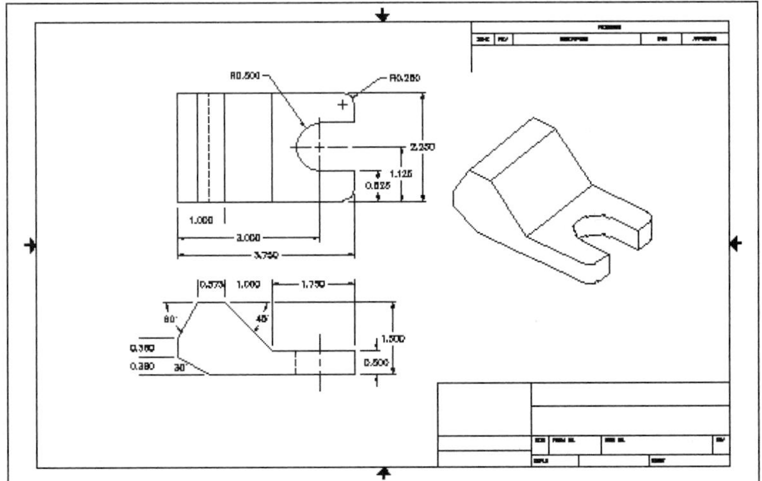

Figure 15.27

Notice that when you zoom in on a part of the preview, different lineweights appear, as in the following image, even though they all appear the same in the drawing file. This is the result of using a Color Dependent plot style table on this drawing. This file can also be attached to other drawings that share the same layer names and colors.

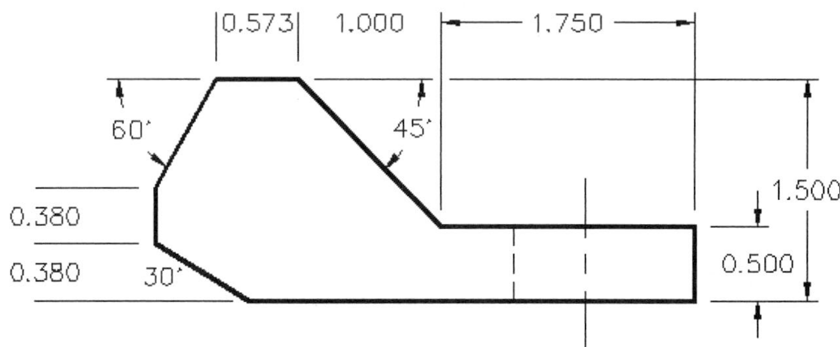

Figure 15.28

PUBLISHING MULTIPLE DRAWING SHEETS

You also have the ability to arrange a number of drawing layouts from other drawings under a single dialog box and perform the plot in this manner. This is accomplished through the Publish dialog box.

Step 1

Clicking on Publish from the File pull-down menu as shown on the left in the following image (or from the Menu Browser) will activate the Publish dialog box as shown on the right in the following image. A number of the controls will be explained through the various steps that follow. The first step is to populate the list area of the dialog box with drawings. Clicking on the Add Sheets icon begins this process.

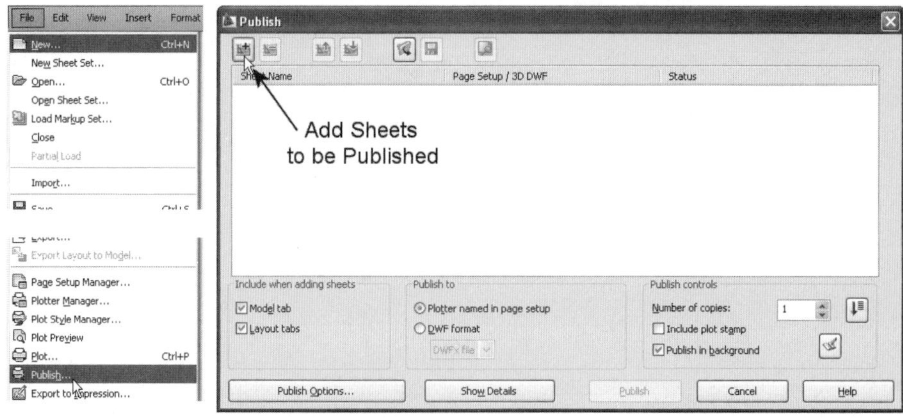

Figure 15.29

Step 2

Clicking on the Add Sheets icon in the previous image activates the Select drawings dialog box as shown on the left in the following image. It is here where you select multiple drawings to publish (or plot.) Once the drawings are selected, clicking the Select button at the bottom of the Select Drawings dialog box (not shown in this illustration) will return you to the Publish dialog box. Notice all of the drawing information that is now visible in the list area of this dialog box. Listed are drawings to be plotted from Model and Layouts as shown on the right in the following image by the different icons. Notice also that the three Model icons are struck with a red line signifying that a page setup has not been created for these. The main reason for this is that most drawings are plotted from layout mode.

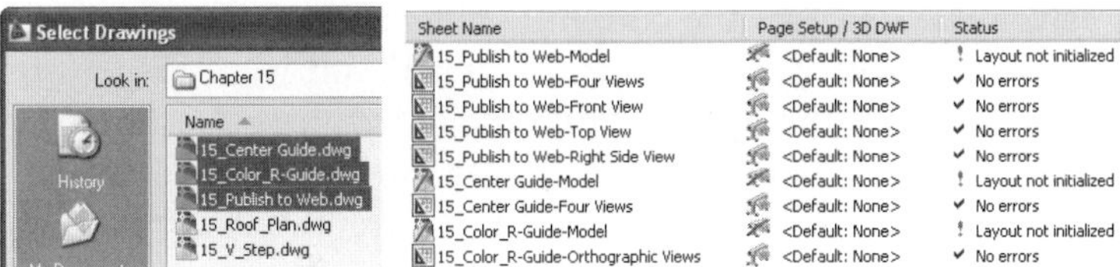

Figure 15.30

Step 3

A closer inspection of this issue in the past step is illustrated below. Notice on the left in the following image that all model and layout drawings are displayed since both Model and Layout tabs are checked in the Publish dialog box. If you will not be plotting from the Model tab, you can remove the check from the box before you begin adding drawings as shown on the right in the following image. The results show only the layout drawings being prepared to be plotted.

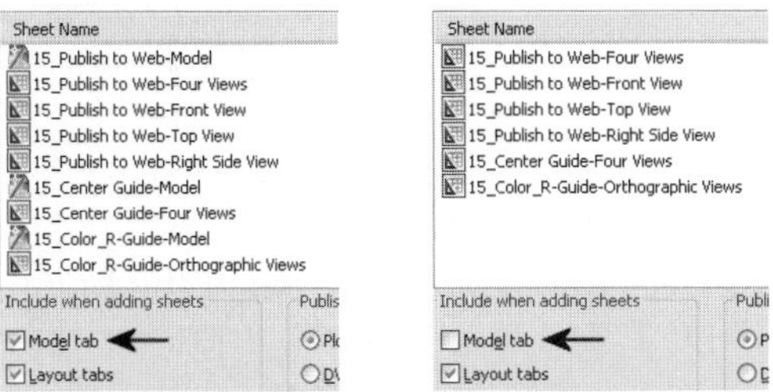

Figure 15.31

Step 4

Another way of controlling the contents of what will be plotted is through a series of icons that allow you to add or remove drawings and even change their order. Any drawing such as the 15-Publish to Web-Model as shown on the left in the following image can be flagged and removed using these buttons. The four buttons illustrated on the right include Add Sheets, Remove Sheets, Move Sheet Up, and Move Sheet Down. This provides more control as you build the list of drawings to be published.

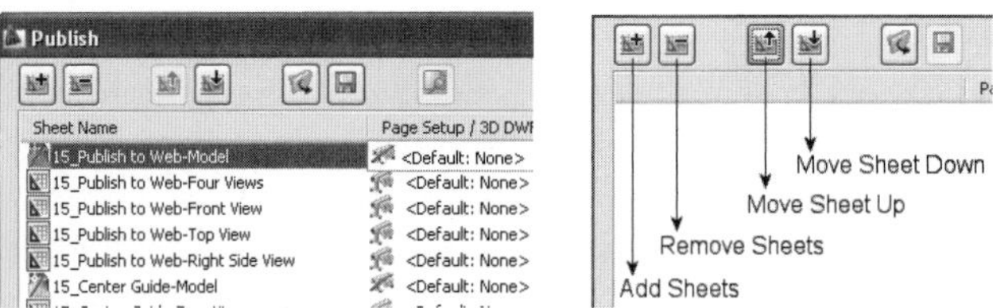

Figure 15.32

Step 5

When it is time to plot the drawings, click on the Publish button located at the bottom of the Publish dialog box. Before the publish operation is executed, an alert dialog box appears asking if you want to save the current list of sheets under a name. The purpose of creating a name is to retrieve this information later if you want to perform plots on the same drawings and eliminate the need to build the list of drawings from scratch. Whether you create a name or not, clicking on the Yes or No buttons will begin publishing all drawings in the background while you still have the current drawing present on the screen. There are many more features to the Publish dialog box and this series of steps will get you started in arranging numerous drawings sheets to be plotted.

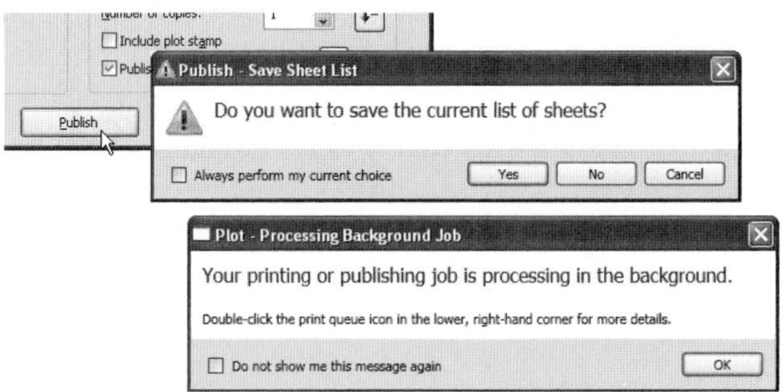

Figure 15.33

PUBLISHING TO THE WEB

Yet another way of applying electronic plots is through the Publish to the Web utility that is provided with AutoCAD. This feature creates a project that consists of a formatted HTML page and your drawing content in either DWF, JPG, or PNG image formats. Through the Publish to Web wizard, you can select how your layout will look from a number of preformatted designs. Once the HTML page is created, you can post the page to an Internet location through the wizard. Follow the next series of steps, which demonstrate how easy it is to publish drawings to the web.

Step 1

Open the drawing file 15_Publish to Web. Notice that four layouts exist, namely Four Views, Front View, Top View, and Right Side View. All four of these layouts will be arranged in a single web page to demonstrate how easy it is to perform this task. This example demonstrates how you can publish various layouts of the same drawing to the web; you can also publish different drawings as well.

Begin the process of publishing to the web by clicking the File pull-down menu followed by Publish to Web..., as shown on the left in the following image. This launches the Publish to Web wizard as shown on the right in the following image. In this first dialog box (called Begin), be sure the radio button next to Create New Web Page is selected. If you already have an existing web page that needs updating or editing, you can click the radio button next to Edit Existing Web Page. When finished, click the Next > button and continue on to the next step of this wizard.

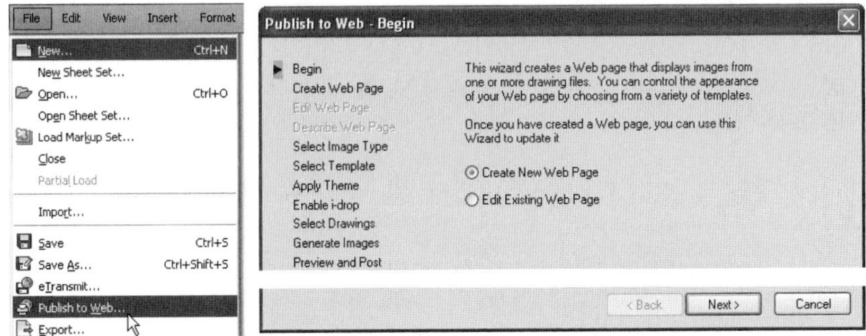

Figure 15.34

Step 2

In the Create Web Page dialog box shown in the following image, add the name of the web page as Four Views with Details; then specify the location of this web page. It is also good practice to add a description of the web page in the event that others will be manipulating your web page. After completing the description, click the Next > button to go to the next step.

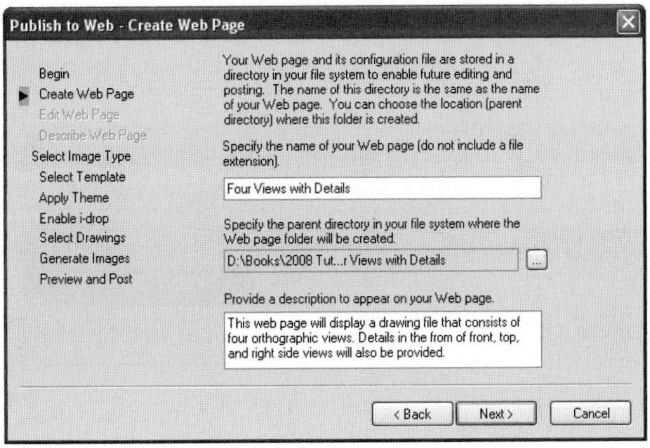

Figure 15.35

Step 3

When the Select Image Type dialog box appears, you have the opportunity to pick the type of image that the drawings will be generated in. By default, your web page will be created in DWFx format, as shown on the left in the following image. These are vector-based images of your drawing that are designed to be viewed using Internet Explorer 7 or with the free Autodesk Design Review viewer. A second image type is DWF. These are vector-based images of your drawing that are also viewed with the free Autodesk Design Review viewer. Another image type option consists of JPG images. JPG files consist of raster images that do not perform very well if your drawing contains a lot of text. However, for this tutorial exercise, JPG will be the image format used. Yet another image format is displayed on the right in the following image. The PNG formats are also raster-based representations of your AutoCAD drawing. They produce high-quality images when compared with JPG files. Be sure that the image type reads JPG and click the Next > button when finished.

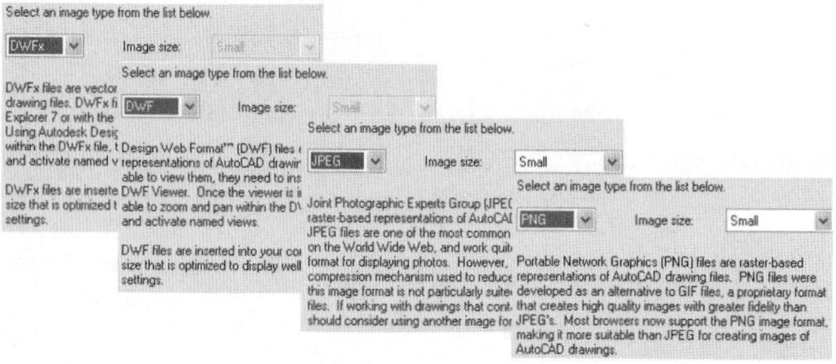

Figure 15.36

Step 4

The next step in the process of publishing to the web is to select a web template. The Select Template dialog box in the following image displays four different templates for you to choose from. Click each one to preview each template at the right of the dialog box. For this tutorial, click Array of Thumbnails and click the Next > button when finished.

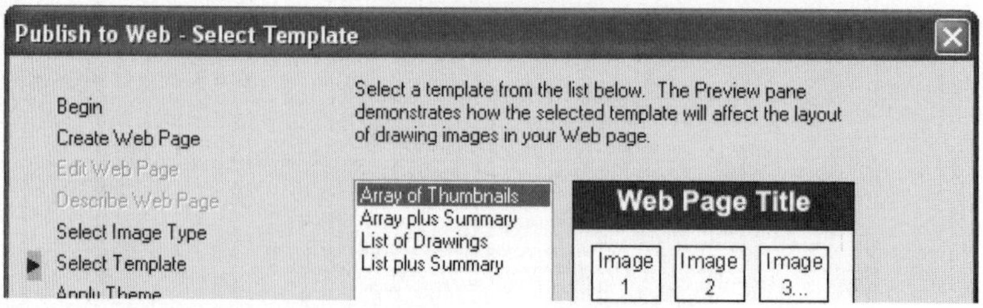

Figure 15.37

Step 5

The next step when publishing to the web is the selection of a theme. The illustration on the left in the following image shows seven possible themes that you can apply. Click each one to preview the contents. For this tutorial, the Classic theme will be used, which is illustrated on the right in the following image. When finished, click the Next > button to continue.

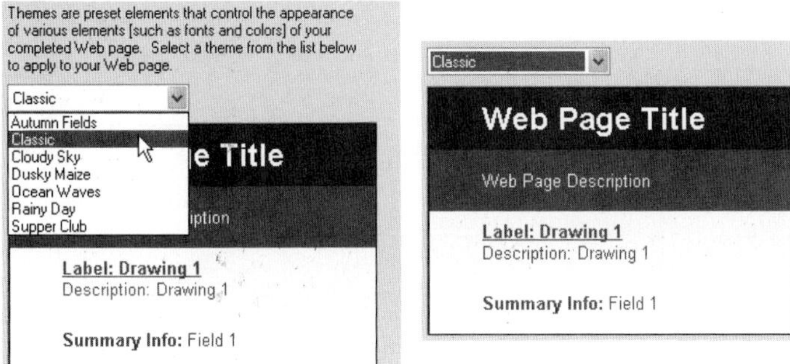

Figure 15.38

Step 6

The next step displays the Enable i-drop dialog box, shown in the following image. Placing a check in the Enable i-drop box allows those who visit your web page to drag and drop drawing files into their session of AutoCAD. While this is a very powerful feature, we will not be demonstrating it during this tutorial. Click the Next > button to continue.

Figure 15.39

Step 7

The next dialog box allows you to add the drawings that will make up the web page. As illustrated on the left in the following image, the drawing name is listed along with layout and label information. The label can be any name or series of names that you wish to appear in your web page. After filling in a description of this drawing, click the Add -> button to add the label to the image list. Under the Layout heading shown on the right in the following image, click Front View as the new layout, add a description, and click the Add -> button to add this label to the image list.

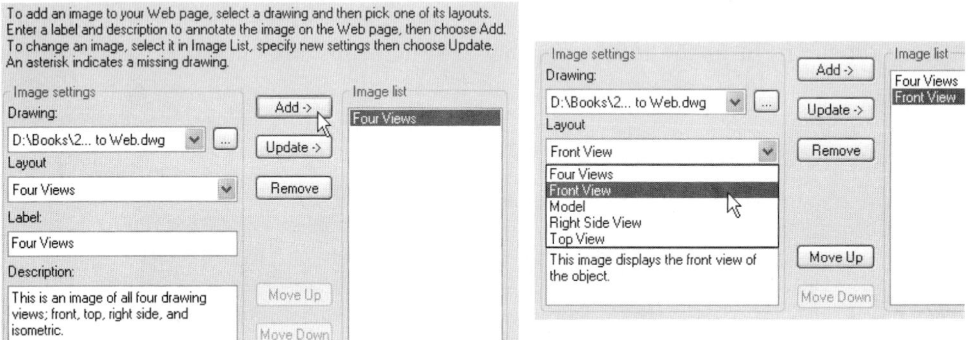

Figure 15.40

Do the same for the Right Side View and Top View. After adding the four layouts to the image list, your dialog should appear similar to the following image. Click the Next > button to continue.

Figure 15.41

Step 8

The Generate Images dialog box, shown on the left in the following image, creates the web page in the folder that you specified earlier. Clicking the Next > button of this dialog box pauses your system while all the layouts are plotted. This may take time depending on the number and complexity of drawings being published. Be sure the radio button next to "Regenerate images for drawings that have changed" is selected.

After the regeneration of all images, the Preview and Post dialog box appears, as shown on the right in the following image. You can either preview your results or post the web page to a web site at a later time. Click the Preview button.

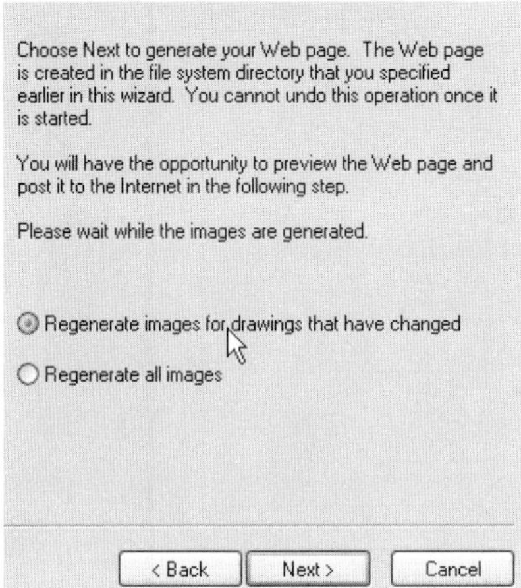
Generate Images
Choose Next to generate your Web page. The Web page is created in the file system directory that you specified earlier in this wizard. You cannot undo this operation once it is started.
You will have the opportunity to preview the Web page and post it to the Internet in the following step.
Please wait while the images are generated.
Regenerate images for drawings that have changed
Regenerate all images
< Back
Next >
Cancel

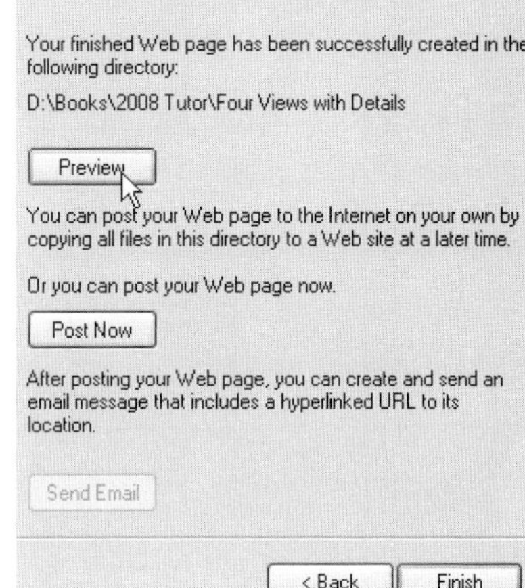
Preview and Post
Your finished Web page has been successfully created in the following directory:
D:\Books\2008 Tutor\Four Views with Details
Preview
You can post your Web page to the Internet on your own by copying all files in this directory to a Web site at a later time.
Or you can post your Web page now.
Post Now
After posting your Web page, you can create and send an email message that includes a hyperlinked URL to its location.
Send Email
< Back
Finish

Figure 15.42

Step 9

Clicking the Preview button in the previous step launches Microsoft Internet Explorer and displays your web page, as shown in the following image.

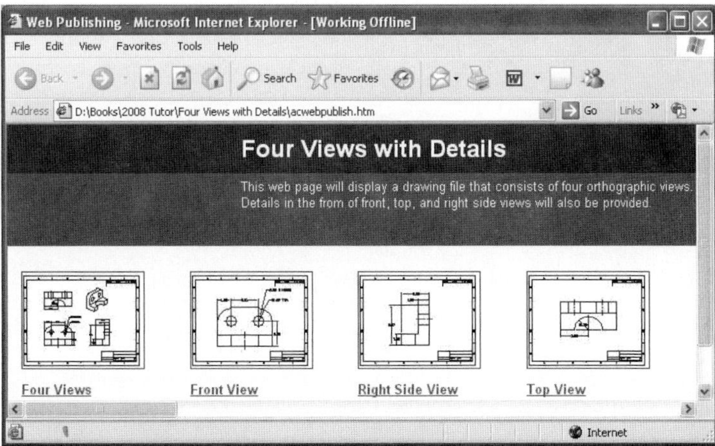
Web Publishing - Microsoft Internet Explorer - [Working Offline]
File Edit View Favorites Tools Help
Back Search Favorites
Address D:\Books\2008 Tutor\Four Views with Details\acwebpublish.htm
Go Links
Four Views with Details
This web page will display a drawing file that consists of four orthographic views. Details in the from of front, top, and right side views will also be provided.
Four Views
Front View
Right Side View
Top View
Internet

Figure 15.43

To view each image separately, click the thumbnail to enlarge the image and show more detail (see the image of the four views in the following image). Dismissing Internet Explorer takes you back to the Preview and Post dialog box. Click the Finish button to end this task.

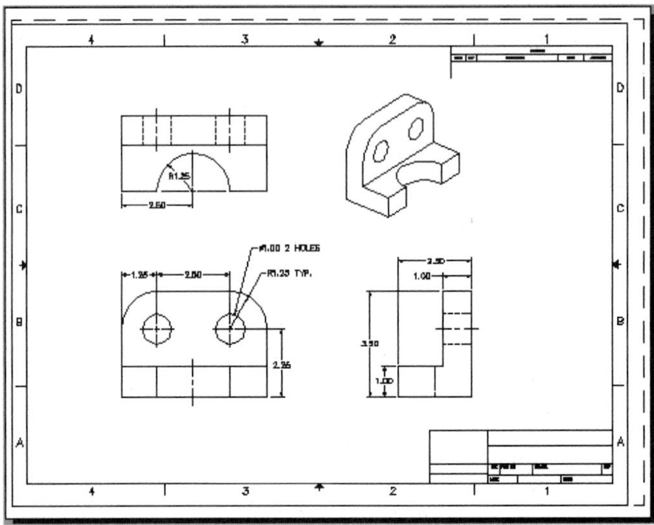
4
3
2
1
D
C
B
A
R1.25
2.50
Ø1.00 2 HOLES
R1.25 TYP.
1.25
2.50
2.25
2.50
1.00
3.50
1.00

Figure 15.44

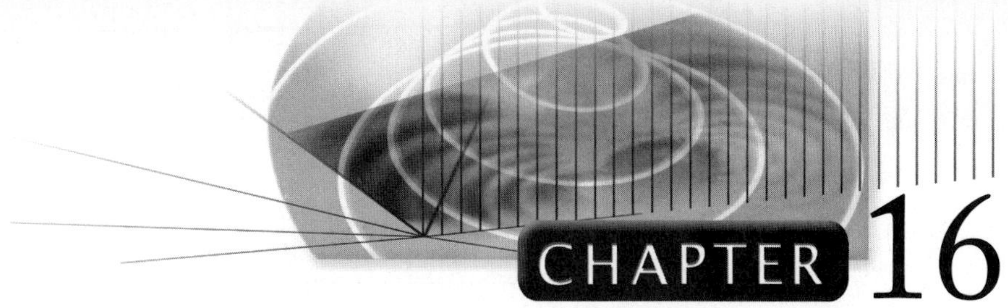

Working with Blocks

This chapter begins the study of how blocks are created and merged into drawing files. This is a major productivity enhancement and is considered an electronic template similar to those methods used with block templates in manual drafting practices. The first segment of this chapter will discuss what blocks are and how they are created. Blocks are typically inserted in the current drawing but can be inserted in any drawing by utilizing the proper commands and techniques. The chapter continues by discussing other topics such as redefining blocks and the effects blocks have on table objects. Dynamic blocks provide even greater control of block objects. A number of dynamic block techniques will be utilized. Next, this chapter continues with a discussion about using the Insert dialog box and the DesignCenter to bring blocks into drawings. The DesignCenter is a special feature that allows blocks to be inserted in drawings with drag and drop techniques. In addition, blocks found in other drawings can easily be inserted in the current drawing from the DesignCenter. Yet another feature, the Tool Palette, allows you to organize blocks and hatch patterns in one convenient area. These object types can then be shared with the current drawing through drag and drop techniques. Finally, the chapter will discuss the use of MDE (Multiple Design Environment). This feature allows the opening of multiple drawings within a single AutoCAD session and provides a convenient method of exchanging data, such as blocks, between one drawing and another. As an added bonus to AutoCAD users, a series of block libraries is supplied with the package. These block libraries include such application areas as mechanical, architectural, electrical, piping, and welding, to name just a few.

WHAT ARE BLOCKS?

Blocks usually consist of smaller components of a larger drawing. Typical examples include doors and windows for floor plans, nuts and bolts for mechanical assemblies, and resistors and transistors for electrical schematics. In the following image, which shows an electrical schematic, all resistors, capacitors, tetrodes, and diodes are considered blocks that make up the total drawing of the electrical schematic. The capacitor is highlighted as one of these components.

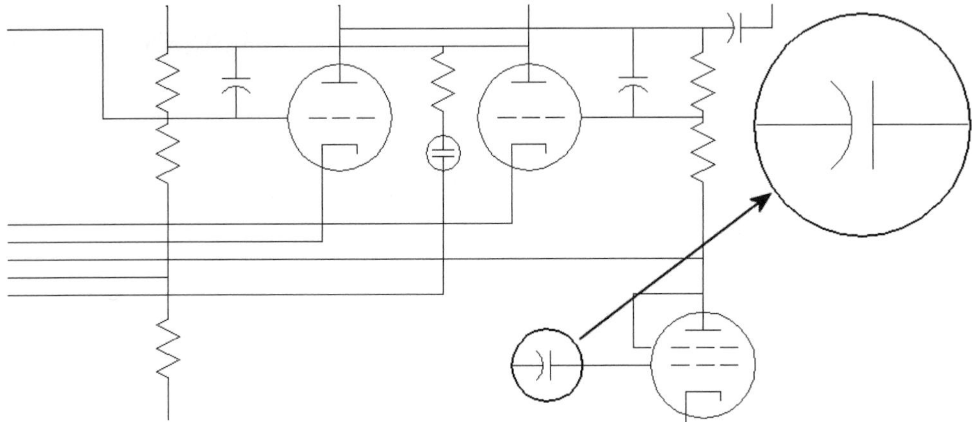

Figure 16.1

Blocks are created and then inserted in a drawing. When creating the block, you must first provide a name for the block. The capacitor illustrated on the left in the following image was assigned the name CAPACITOR. Also, when you create a block, an insertion point is required. This acts as a reference point from which the block will be inserted. In the illustration on the left in the following image, the insertion point of the block is the left end of the line.

At times, blocks have to be rotated into position. In the illustration on the right in the following image, notice that one capacitor is rotated at a 45° angle while the other capacitor is rotated 270°. In this way the same block can be used numerous times even though it is positioned differently in the drawing.

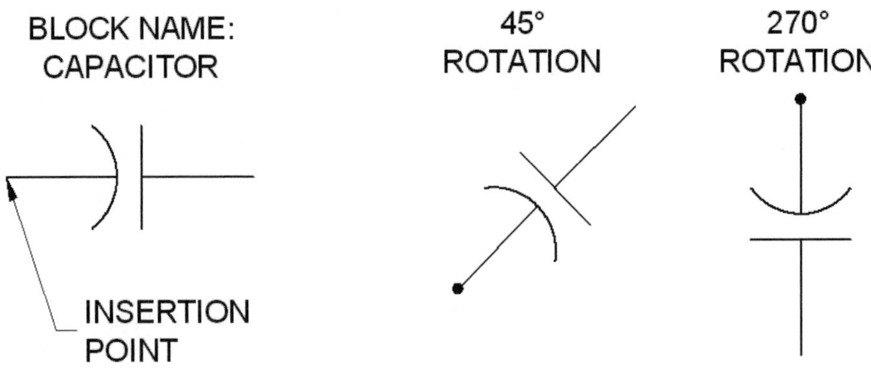

Figure 16.2

METHODS OF SELECTING BLOCK COMMANDS

Commands that deal with blocks can be selected from one of four menu areas; Menu Browser, Pull-Down Menu, Ribbon, and Draw Toolbar as shown in the following image.

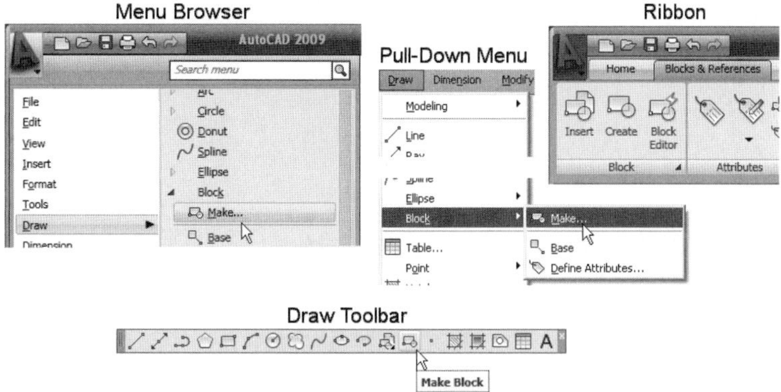

Figure 16.3

You can also enter block-related commands from the keyboard, either using their entire name or through command aliases, as in the following examples. These commands will be discussed in greater detail in the following pages.

> Enter B to activate the Block Definition dialog box (BLOCK).
>
> Enter I to activate the Insert dialog box (INSERT).
>
> Enter W to activate the Write Block dialog box (WBLOCK).
>
> Enter ADC to activate the DesignCenter (ADCENTER).

CREATING A LOCAL BLOCK USING THE BLOCK COMMAND

The illustration on the left in the following image is a drawing of a hex head bolt. This drawing consists of one polygon, representing the hexagon, a circle, indicating that the hexagon is circumscribed about the circle, and two centerlines. Rather than copy these individual objects numerous times throughout the drawing, you can create a block using the BLOCK command. Selecting Block from the Draw pull-down menu and then selecting Make activates the Block Definition dialog box, illustrated on the right in the following image. Entering the letter B from the keyboard also activates the Block Definition dialog box.

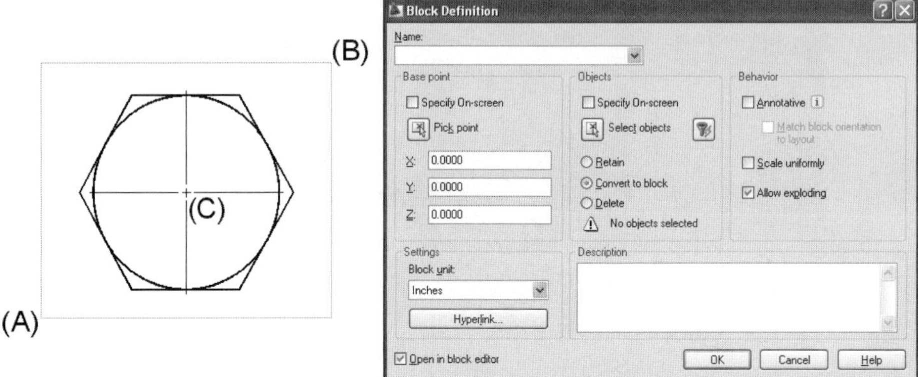

Figure 16.4

When you create a block, the numerous objects that make up the block are all considered a single object. This dialog box allows for the newly created block to be merged into the current drawing file. This means that a block is available only in the drawing it was created in; it cannot be shared directly with other drawings. (The WBLOCK command is used to create global blocks that can be inserted in any drawing file. This command will be discussed later on in this chapter.)

To create a block through the Block Definition dialog box, enter the name of the block, such as "Hexbolt," in the Name field, as shown in the following image. You can use up to 255 alphanumeric characters when naming a block.

Next, click the Select objects button; this returns you to the drawing editor. Create a window from "A" to "B" around the entire hex bolt to select all objects, as shown in the previous image. When finished, press ENTER. This returns you to the Block Definition dialog box, where a previewed image of the block you are about to create is displayed in the upper-right corner of the dialog box, as shown in the following image. Whenever you create a block, you can elect to allow the original objects that made up the hex bolt to remain on the screen (click Retain button), to be replaced with an instance of the block (click Convert to Block button), or to be removed after the block is created (click Delete button). Click the Delete radio button to erase the original objects after the block is created. If the original objects that made up the hex bolt are unintentionally removed from the screen during this creation process, the OOPS command can be used to retrieve all original objects to the screen.

The next step is to create a base point or insertion point for the block. This is considered a point of reference and should be identified at a key location along the block. By default, the Base point in the Block Definition dialog box is located at 0,0,0. To enter a more appropriate base-point location, click the Pick point button; this returns you to the drawing editor and allows you to pick the center of the hex bolt at "C," as shown in the previous image. Once this point is selected, you are returned to the Block Definition dialog box; notice how the Base point information now reflects the key location along the block, as shown in the following image.

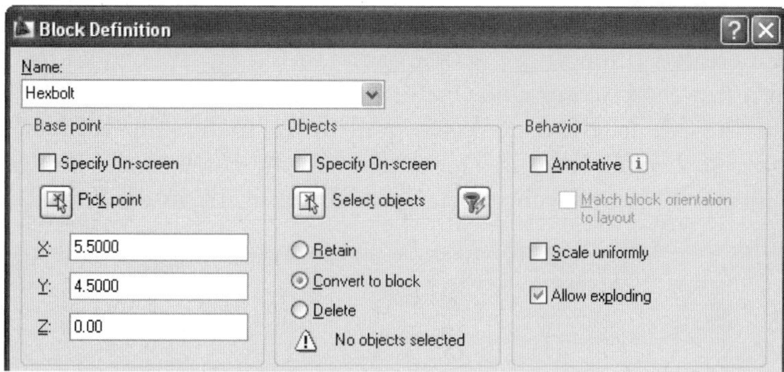

Figure 16.5

Yet another feature of the Block Definition dialog box allows you to add a description for the block. Many times the name of the block hides the real meaning of the block, especially if the block name consists of only a few letters. Click in the Description field and add the following statement: "This is a hexagonal head bolt." This allows you to refer to the description in case the block name does not indicate the true intended purpose for the block, as shown in the following image. Specifying the Block unit determines the type of units utilized for scaling the block when it is inserted from the DesignCenter.

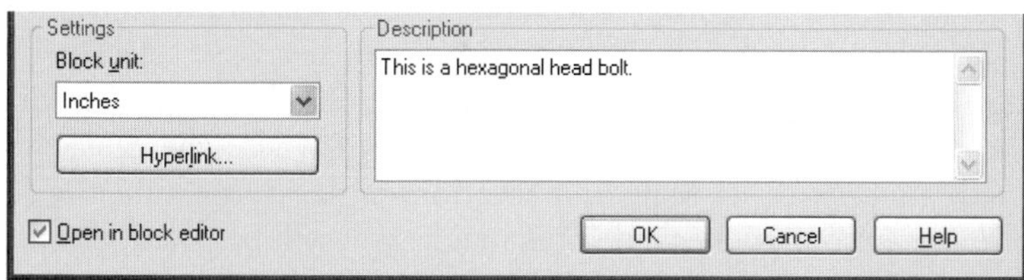

Figure 16.6

As the block is written to the database of the current drawing, the individual objects used to create the block automatically disappear from the screen (due to our earlier selection of the Delete radio button). When the block is later inserted in a drawing, it will be placed in relation to the insertion point.

Note: In addition to creating blocks by grouping several objects into one, you can also write an entire drawing out to a file using the WBLOCK command. Entering W at the command prompt will display the Write Block dialog box. As with the Block Definition dialog box, you will designate a base point and select the objects. You will also be required to enter a file name in order for the objects to be written out to a drawing file.

INSERTING BLOCKS

Once blocks are created through the BLOCK command, they are merged or inserted in the drawing through the INSERT command. Enter INSERT or I from the keyboard or choose the command from the Menu Browser, Insert pull-down menu, the Ribbon, or the Draw toolbar, as shown in the following image, to activate the Insert dialog box, also shown in the following image. This dialog box is used to dynamically insert blocks.

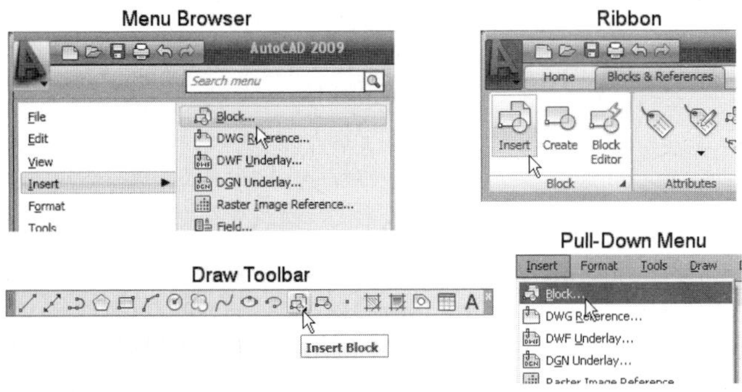

Figure 16.7

Clicking on Insert from either of the menus in the previous illustration will activate the Insert dialog box, as shown in the following image. This dialog box is used to dynamically insert blocks. First, by clicking the Name drop-down list box, select a block from the current drawing (clicking the Browse button locates global blocks or drawing files). After you identify the name of the block to insert, the point where the block will be inserted must be specified, along with its scale and rotation angle. By default, the Insertion point area's Specify On-screen box is checked. This means you will be prompted for the insertion point at the command prompt area of the drawing editor. The default values for the scale and rotation insert the block at the original size and orientation. It was already mentioned that a block consists of several objects combined into a single object. The Explode box determines whether the block is inserted as one object or the block is inserted and then exploded back to its individual objects. Once the name of the block, such as Hexbolt, is selected, click the OK button.

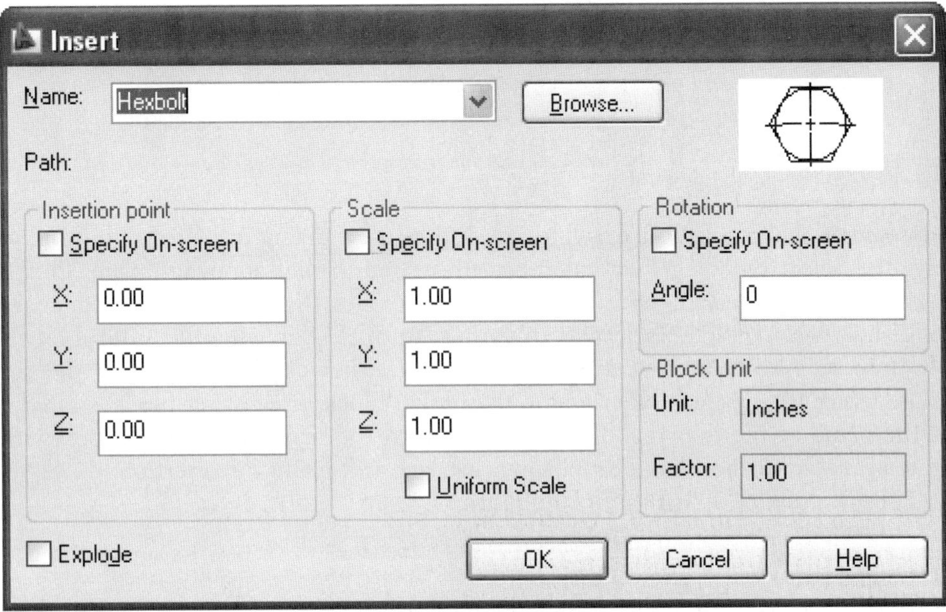

Figure 16.8

The following prompts complete the block insertion operation:

```
Specify insertion point or [Basepoint/Scale/X/Y/Z/Rotate/PScale/
    PX/PY/PZ/PRotate]: (Mark a point at "A" in the following
    image to insert the block)
```

If the Specify On-screen boxes are also checked for Scale and Rotation, the following prompts will complete the block insertion operation.

```
Specify insertion point or [Basepoint/Scale/X/Y/Z/Rotate/PScale/
    PX/PY/PZ/PRotate]: (Mark a point at "A," as shown in the
    following image, to insert the block)
Enter X scale factor, specify opposite corner, or [Corner/XYZ]
    <1>: (Press ENTER to accept default X scale factor)
Enter Y scale factor <use X scale factor>: (Press ENTER to accept
    default)
Specify rotation angle <0>: (Press ENTER to accept the default
    rotation angle and insert the block, as shown in the
    following image)
```

If blocks are already defined as part of the database of the current drawing, they may be selected from the Name drop-down list box, as shown on the right in the following image.

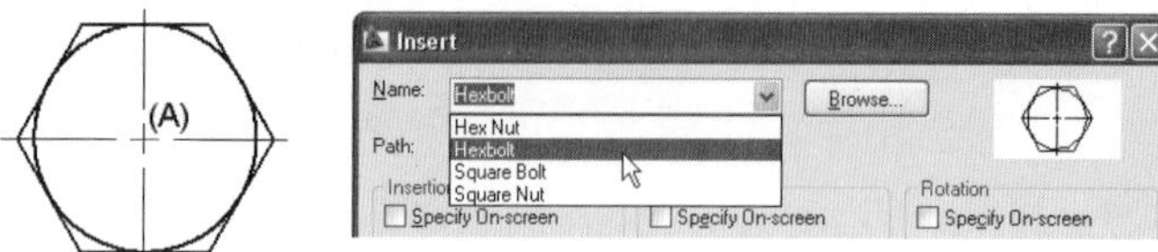

Figure 16.9

For inserting global blocks in a drawing, select the Insert dialog box Browse button, which displays the Select Drawing File dialog box illustrated in the following image. This is the same dialog box associated with opening drawing files. In fact, you will be unable to distinguish between global blocks and any other AutoCAD drawing file. Global blocks are simply drawing files created in a unique way, and either can be inserted in the current drawing. Select the desired folder where the global block or drawing file is located; then select the name of the drawing. This returns you to the main Insert dialog box with the file now available in the Name drop-down list box and ready for insertion.

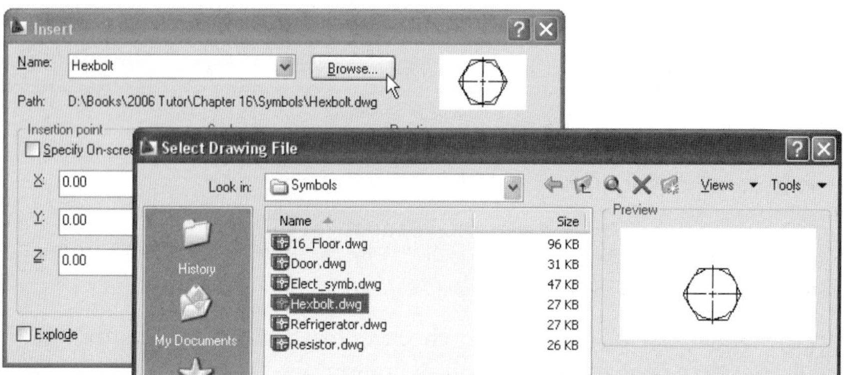

Figure 16.10

APPLICATIONS OF BLOCK INSERTIONS

The following image shows the results of entering different scale factors and rotation angles when blocks are inserted in a drawing file. The image at "A" shows the block inserted in a drawing with its default scale and rotation angle values. The image at "B" shows the result of inserting the block with a scale factor of 0.50 and a rotation angle of 0°. The image appears half its normal size. The image at "C" shows the result of inserting the block with a scale factor of 1.75 and a rotation angle of 0°. In this image, the scale factor increases the block in size while keeping the same proportions. The image at "D" shows the result of inserting the block with different X and Y scale factors. In this image, the X scale factor is 0.50 while the Y scale factor is 2.00 units. Notice how out of proportion the block appears. There are certain applications where different scale factors are required to produce the desired effect. The image at "E" shows the result of inserting the block with the default scale factor and a rotation angle of 30°. As with all rotations, a positive angle rotates the block in the counterclockwise direction; negative angles rotate in the clockwise direction.

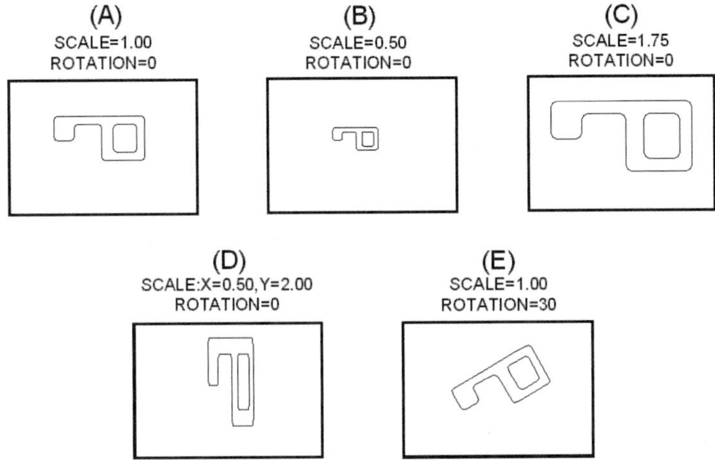

Figure 16.11

TRIMMING TO BLOCK OBJECTS

The ability to trim to a block object is possible through the TRIM command. As you select cutting edges on the block to trim to, the command isolates the cutting edges from the remainder of the objects that make up the block. In the following image, the outside edges of the bolt act as cutting edges.

Try It!: To test this feature, open the drawing 16_Trim Plates and use the following image as a guide. The bolt is a block. Activate the TRIM command, pick the edges of the bolt at "A" and "B" as cutting edges, and pick the lines at "C" through "E" as the objects to trim.

The result of using the TRIM command is illustrated on the right in the following image.

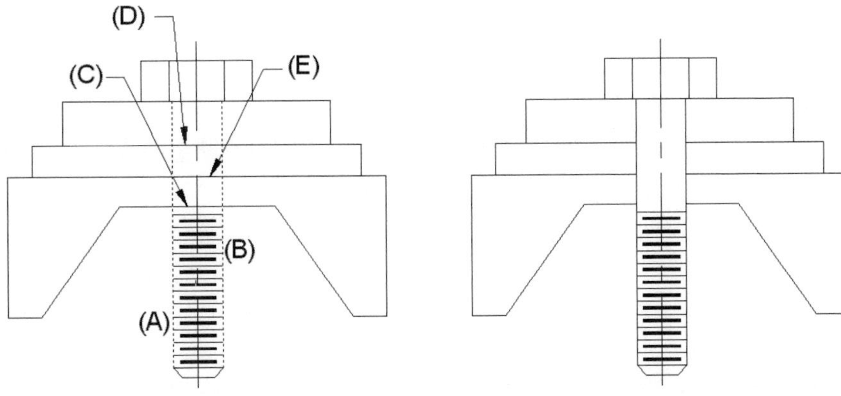

Figure 16.12

EXTENDING TO BLOCK OBJECTS

As with the TRIM command, objects can be extended to boundary edges that are part of a block with the EXTEND command. Selecting boundary edges that consist of block components isolates these objects. This enables you to extend objects to these boundary edges.

Try It!: To test this feature, open the drawing 16_Extend Plates and use the following image as a guide. Again the bolt is a block. Activate the EXTEND command, pick the edges of the bolt at "A" and "B" as boundary edges, and pick the lines at "C" through "H" as the objects to extend.

The result of using the EXTEND command on block objects as cutting edges is illustrated on the right in the following image.

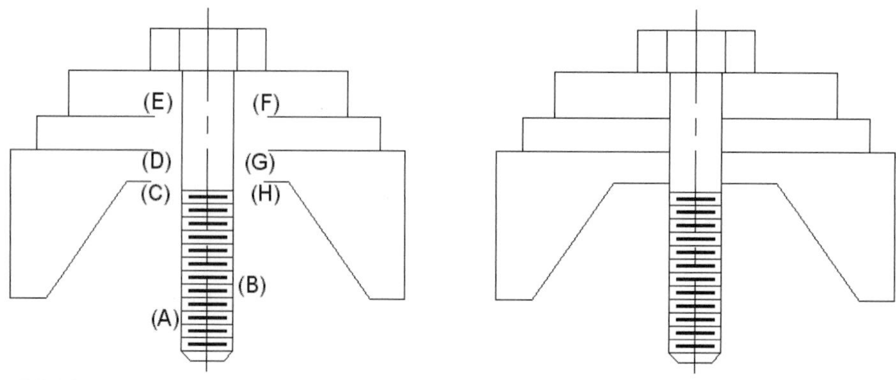

Figure 16.13

EXPLODING BLOCKS

It has already been mentioned that as blocks are inserted in a drawing file, they are considered one object even though they consist of numerous individual objects. At times it is necessary to break a block up into its individual parts. The EXPLODE command is used for this.

Illustrated on the left in the following image is a block that has been selected. Notice that the entire block highlights and the insertion point of the block is identified by the presence of the grip. Using the EXPLODE command on a block results in the illustration on the right in the following image. Here, when one of the lines is selected, only that object highlights. Exploding a block breaks the block back up into its individual objects. As a result, you must determine when it is appropriate to explode a block.

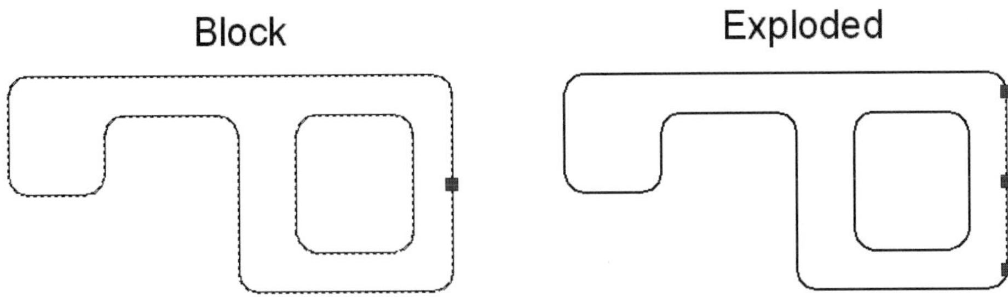

Figure 16.14

MANAGING UNUSED BLOCK DATA WITH THE PURGE DIALOG BOX

AutoCAD stores named objects (blocks, dimension styles, layers, linetypes, multiline styles, plot styles, shapes, table styles, and text styles) with the drawing. When the drawing is opened, AutoCAD determines whether other objects in the drawing reference each named object. If a named object is unused and not referenced, you can remove the definition of the named object from the drawing by using the PURGE command. This is a very important productivity technique used for compressing or cleaning up the database of the drawing. Picking File from the pull-down menu, followed by Drawing Utilities, displays Purge..., as shown on the left in the following image. Clicking Purge... displays the Purge dialog box, as shown on the right in the following image.

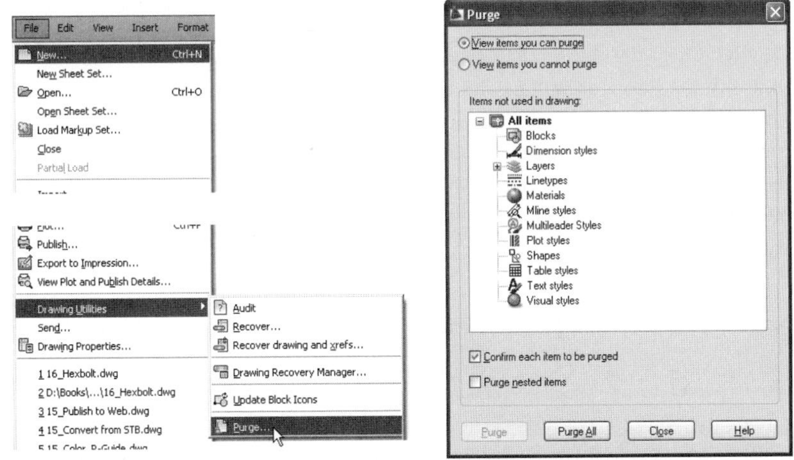

Figure 16.15

The Blocks category is expanded by clicking the "+." This produces the list of all the items that are currently unused in the drawing that can be removed, as shown on the left in the following image. Clicking the item SOFA2, right-clicking, and then picking Purge from the shortcut menu (or simply clicking the button at the bottom of the dialog box) displays the Confirm Purge dialog box, as shown in the middle in the following image. Click the Yes button and the block is removed from the listed items, as shown on the right in the following image.

Figure 16.16

Other controls are available in the Purge dialog box, such as the ability to view items that cannot be purged and the capability of purging nested items. An example of purging a nested item would be purging a block definition that lies inside another block definition. Entering PU at the Command prompt also activates the Purge dialog box.

Tip: The layer "0," Standard text and dimension styles, and the Continuous linetype cannot be purged from a drawing.

Try It!: Open the drawing file 16_Purge, as shown on the left in the following image. Activate the Purge dialog box and click the Purge All button. You will be prompted to remove all items individually through the Confirm Purge dialog box, as shown on the right in the following image. Answer Yes to each item you want purged or Yes to All, until all items, including blocks, layers, linetypes, text styles, dimension styles, and multiline styles, have been removed from the database of the drawing.

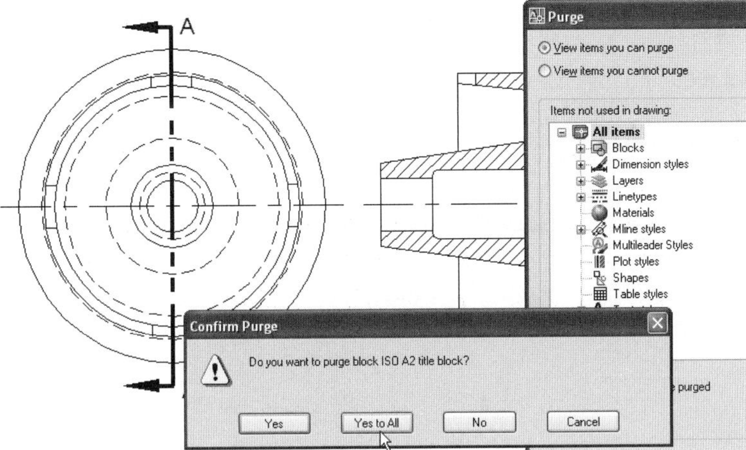

Figure 16.17

REDEFINING BLOCKS

At times, a block needs to be edited. Rather than erase all occurrences of the block in a drawing, you can redefine it. This is considered a major productivity technique because all blocks that share the same name as the block being redefined will automatically update to the latest changes. The illustration on the left in the following image shows various blocks inserted in a drawing. The block name is Step Guide and the insertion point is at the lower-left corner of the object.

Try It!: The drawing file 16_Block Redefine has been provided on CD for you to experiment with redefining blocks. When the drawing appears on your screen, as shown on the left in the following image, double-click any block in the drawing. This launches the Edit Block Definition dialog box, as shown on the right in the following image. A list of all blocks defined in the drawing appear. Click Step Guide from the list and click the OK button.

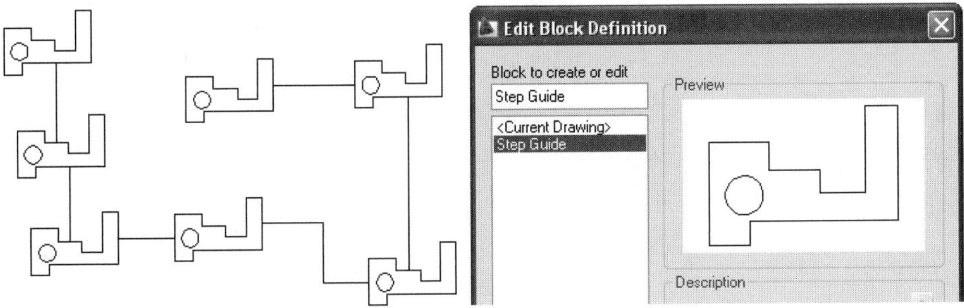

Figure 16.18

After you click the OK button, the Block Editor area appears, as shown on the left in the following image. A majority of the tools displayed in the palette on the left are designed for creating dynamic blocks, which will be explained in detail later in this chapter. For now, use this environment to make changes to the geometry that makes up the Step Guide. You can add new objects, change the properties of objects, or make modifications to this object. For this example, erase the circle and stretch the upper-right corner of the object down by 2 units, as shown on the right in the following image.

To return to your drawing, you must first save the changes made to the object by clicking the Save Block Definition button, also shown on the right in the following image. To exit the Block Editor area, click the Close Block Editor button.

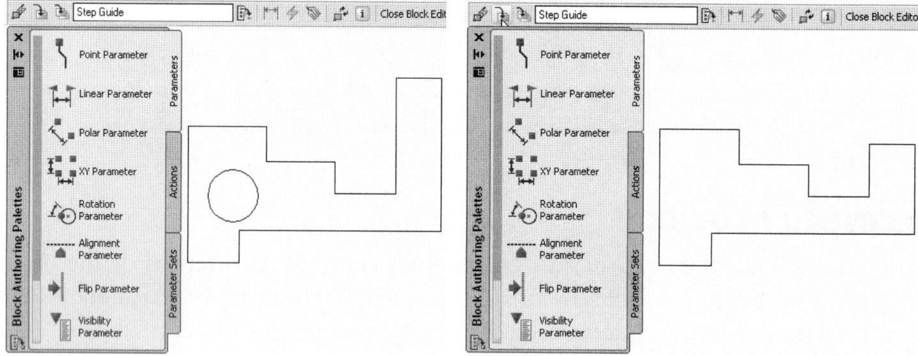

Figure 16.19

When you return to the drawing editor, notice that all Step Guide blocks have been updated to the changes made in the Block Editor. This provides a very productive method of making changes to all blocks in a drawing. This method works only if blocks have not been exploded in a drawing. It is for this reason that you must exercise caution when exploding blocks in a drawing.

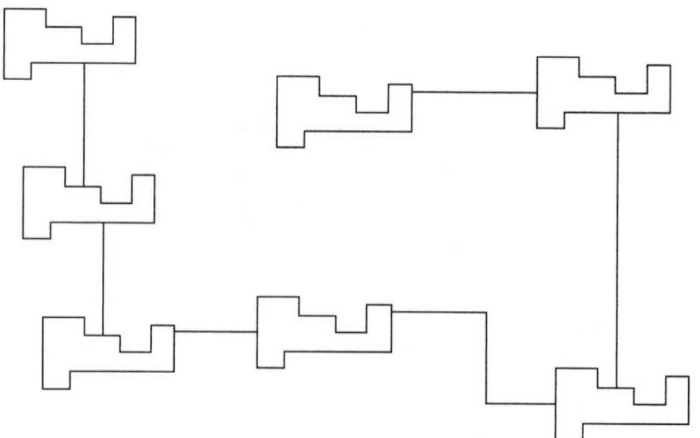

Figure 16.20

BLOCKS AND THE DIVIDE COMMAND

The DIVIDE command was already covered in chapter 5. It allows you to select an object, give the number of segments, and place point objects at equally spaced distances depending on the number of segments. The DIVIDE command has a Block option that allows you to place blocks at equally spaced distances.

Try It!: Open the drawing file 16_Speaker. In the following image, a counter-bore hole and centerline need to be copied 12 times around the elliptical centerline so that each hole is equally spaced from others. (The illustration of the block CBORE is displayed at twice its normal size.) Because the ARRAY command is used to copy objects in a rectangular or circular pattern, that command cannot be used for an ellipse. The DIVIDE command's Block option allows you to specify the name of the block and the number of segments. In the following image, the elliptical centerline is identified as the object to divide. Follow the command sequence to place the block CBORE in the elliptical pattern:

```
Command: DIV (For DIVIDE)
Select object to divide: (Select the elliptical centerline)
Enter the number of segments or [Block]: B (For Block)
Enter name or block to insert: CBORE
Align block with object? [Yes/No] <Y>: (Press ENTER to accept)
Enter the number of segments: 12
```

The results are illustrated on the right in the following image. Notice how the elliptical centerline is divided into 12 equal segments by 12 blocks called CBORE. In this way, any object may be divided through the use of blocks.

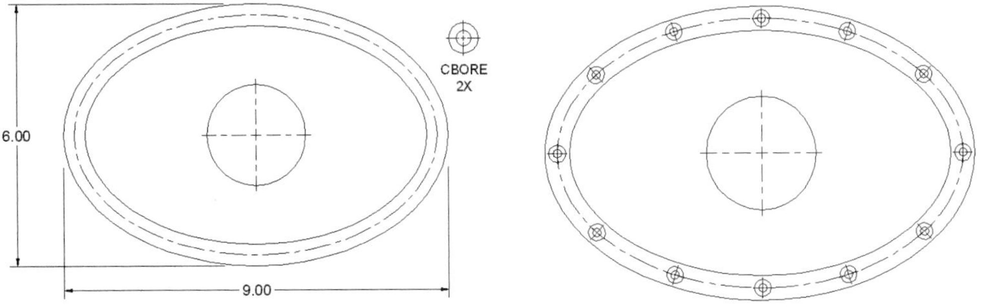

Figure 16.21

BLOCKS AND THE MEASURE COMMAND

As with the DIVIDE command, the MEASURE command offers increased productivity when you measure an object and insert blocks at the same time.

Try It!: Open the drawing file 16_Chain. In the following image, a polyline path will be divided into 0.50-length segments using the block CHAIN2. The perimeter of the polyline was calculated by the LIST command to be 22.00 units, which is evenly divisible by 0.50 and will allow the insertion of 44 blocks. Follow the command prompt sequence below for placing a series of blocks called CHAIN2 around the polyline path to create a linked chain.

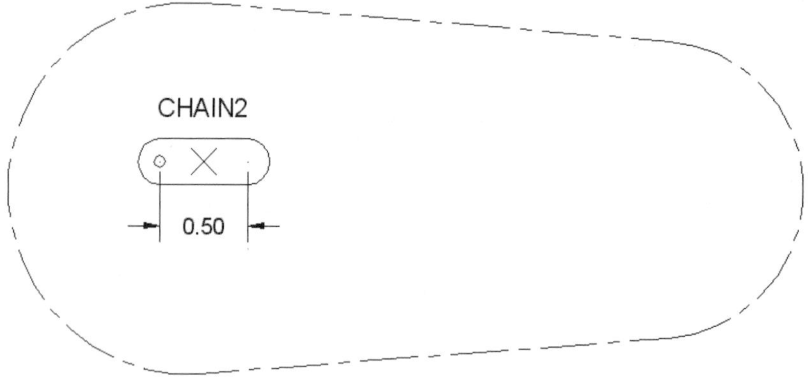

Figure 16.22

```
Command: ME (For MEASURE)
Select object to measure: (Pick the polyline path)
Specify length of segment or [Block]: B (For Block)
Enter name of block to insert: CHAIN2
Align block with object? [Yes/No] <Y>: (Press ENTER to accept)
Specify length of segment: 0.50
```

The result is illustrated on the left in the following image, with all chain links being measured along the polyline path at increments of 0.50 units.

Answering No to the prompt "Align block with object? [Yes/No] <Y>" displays the results, as shown on the right in the following image. Here all blocks are inserted horizontally and travel in 0.50 increments. While the polyline path has been successfully measured, the results are not acceptable for creating the chain.

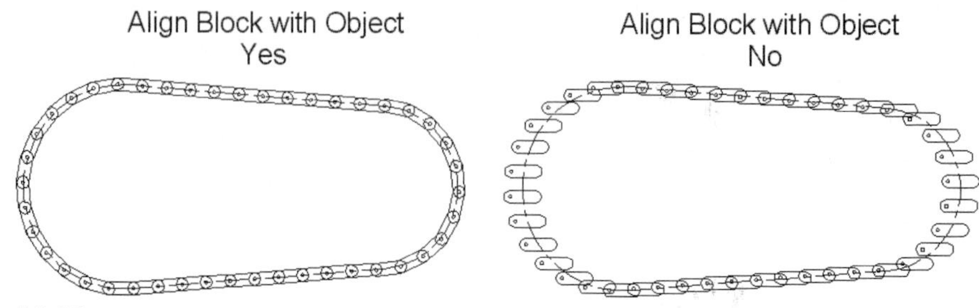

Figure 16.23

RENAMING BLOCKS

Blocks can be renamed to make their meanings more clear through the RENAME command. This command can be found in the Format pull-down menu, as shown on the left in the following image, and when selected displays a dialog box similar to the one illustrated on the right in the following image. Clicking Blocks in the dialog box lists all blocks defined in the current drawing. One block with the name REF1 was abbreviated, and we wish to give it a full name. Clicking the name REF1 pastes it in the Old Name field. Type the desired full name REFRIGERATOR in the Rename To field and click the Rename To button to rename the block.

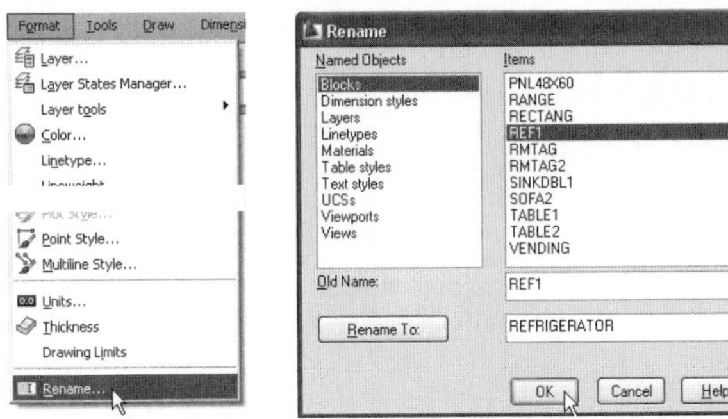

Figure 16.24

TABLES AND BLOCKS

In the following image, electrical symbols and their descriptions are arranged in the legend to call out the symbols in a table. To add blocks to a table, click inside the cell that will hold the block and right-click to display the menu; then click Insert, followed by Block....

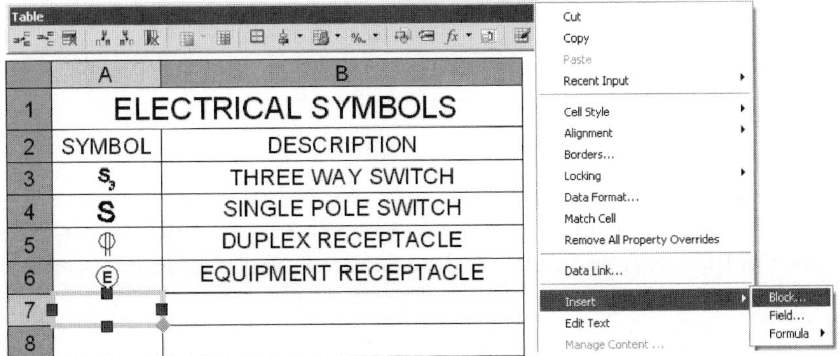

Figure 16.25

When the Insert a Block in a Table Cell dialog appears, locate the name of the block to insert, as shown in the following image.

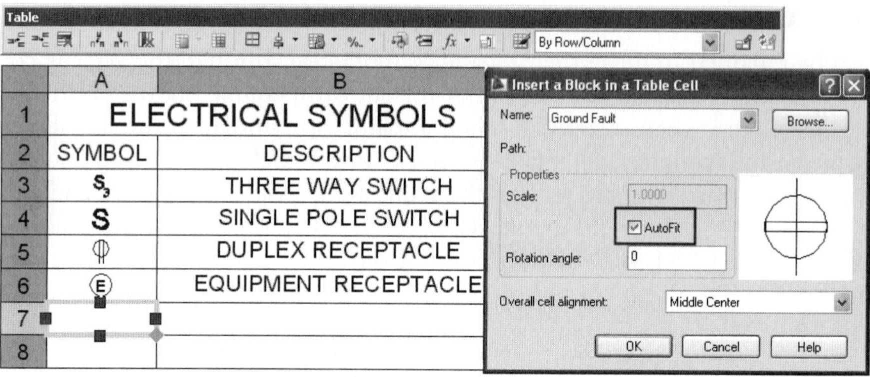

Figure 16.26

The results are illustrated in the following image, where the Ground Fault block is inserted into the highlighted cell of the table. Because the AutoFit function is checked in the previous figure, the block will be scaled to fit inside the cell, no matter how large or small the block is. The title of the symbol was also added to the table by double-clicking inside the cell to launch the Text Formatting toolbar.

ELECTRICAL SYMBOLS	
SYMBOL	DESCRIPTION
S_3	THREE WAY SWITCH
S	SINGLE POLE SWITCH
	DUPLEX RECEPTACLE
Ⓔ	EQUIPMENT RECEPTACLE
	GROUND FAULT

Figure 16.27

Note: You can type the TINSERT command in at the command prompt to insert a block into a table cell. After selecting the cell in which to insert the block, the Insert Block in Table dialog box appears. Select a block from the list to be inserted into the selected table cell.

ADDITIONAL TIPS FOR WORKING WITH BLOCKS

CREATE BLOCKS ON LAYER 0

Blocks are best controlled, when dealing with layer colors, linetypes, and lineweights, by being drawn on Layer 0, because it is considered a neutral layer. By default, Layer 0 is assigned the color White and the Continuous linetype. Objects drawn on Layer 0 and then converted to blocks take on the properties of the current layer when inserted in the drawing. The current layer controls color, linetype, and lineweight.

CREATE BLOCKS FULL SIZE IF APPLICABLE

The illustration on the left in the following image shows a drawing of a refrigerator, complete with dimensions. In keeping with the concept of drawing in realworld units in CAD or at full size, individual blocks must also be drawn at full size in order for them to be inserted in the drawing at the correct proportions. For this block, construct a rectangle 28 units in the X direction and 24 units in the Y direction. Create a block called Refrigerator by picking the rectangle. When testing out this block, be sure to be in a drawing that is set to the proper units based on a scale such as ½" = 1'-0" or ¼" = 1'-0". These are typical architectural scales, and the block of the refrigerator will be in the correct proportions on these types of sheet sizes.

One exception to the full-size rule is illustrated on the right in the following image. Rather than create each door block separately to account for different door sizes, create the door so as to fit into a 1-unit by 1-unit square. The purpose of drawing the door block inside a 1-unit square is to create only one block of the door and insert it at a scale factor matching the required door size. For example, for a 2'-8" door, enter 2'8 or 32 when prompted for the X and Y scale factors. For a 3'-0" door, enter 3' or 36 when prompted for the scale factors. Numerous doors of different types can be inserted in a drawing using only one block of the door. Also try using a negative scale factor to mirror the door as it is inserted.

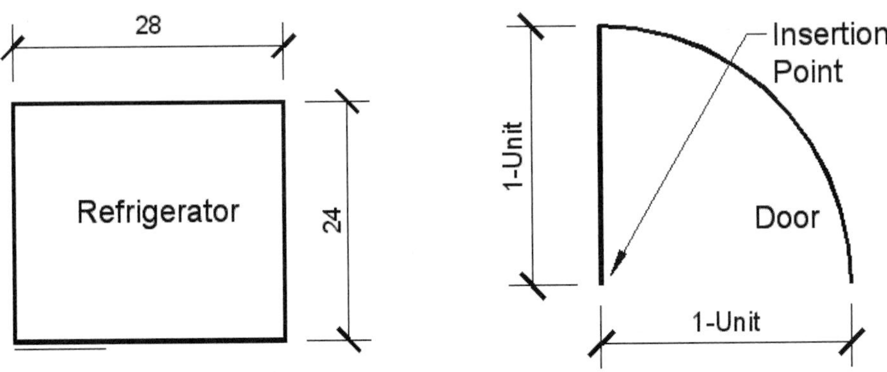

Figure 16.28

USE GRID WHEN PROPORTIONALITY, BUT NOT SCALE, IS IMPORTANT

Sometimes blocks represent drawings in which the scale of each block is not important. In the previous example of the refrigerator, scale was very important in order for the refrigerator to be drawn according to its full-size dimensions. This is not the case, as shown on the left in the following image, of the drawing of the resistor block. Electrical schematic blocks are generally not drawn to any specific scale; however, it is important that all blocks are proportional to one another. Setting up a grid is good practice in keeping all blocks at the same proportions. Whatever the size of the grid, all blocks are designed around the same grid size. The result is shown on the right in the following image; with four blocks being drawn with the same grid, their proportions look acceptable.

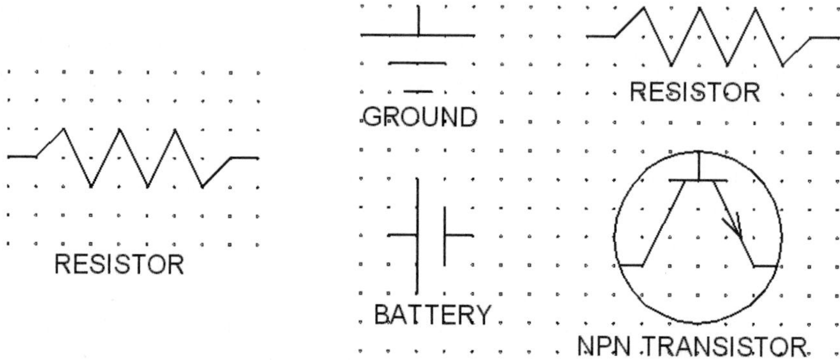

Figure 16.29

INSERTING BLOCKS WITH DESIGNCENTER

DesignCenter provides an additional means of inserting blocks and drawings even more efficiently than through the Insert dialog box. This feature has the distinct advantage of inserting specific blocks internal to one drawing into another drawing. If the DesignCenter is not present, you can load it by choosing Palettes and then DesignCenter from the Menu Browser as shown on the left in the following image or the Tools pull-down menu, as shown on the right in the following image.

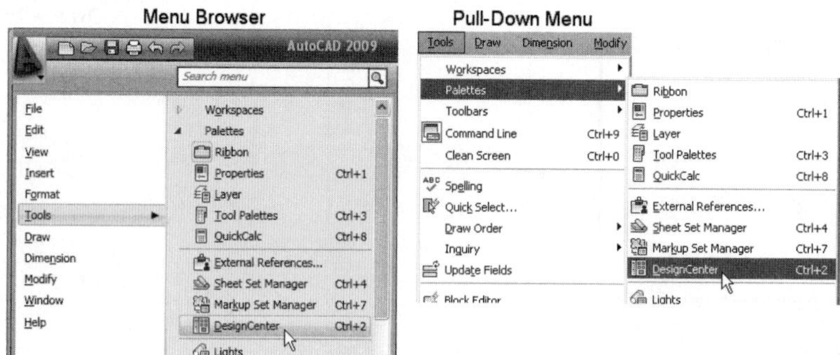

Figure 16.30

When used for the first time, the DesignCenter loads in the middle of the screen. An Auto-hide feature for the DesignCenter allows it to hide (collapse) once you move the cursor outside the DesignCenter window. This helps clear the drawing area when the DesignCenter is not being used. To expand it again, simply move the cursor over the DesignCenter title bar. A shortcut menu allows you to turn the Auto-hide and docking features on or off as desired, as shown in the following image. If you prefer that the DesignCenter remain on the screen, you can still dock it to the left or right side of the AutoCAD drawing screen. The DesignCenter can be resized as the user requires. You can unload it by clicking the X in the title bar, selecting DesignCenter from the Tools pull-down menu, clicking the DesignCenter button on the Standard toolbar, or entering the ADCCLOSE command at the keyboard.

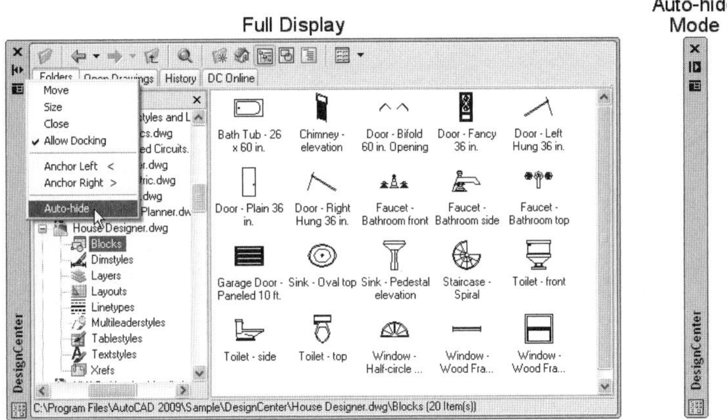

Figure 16.31

Tip: The DesignCenter can also be launched by pressing and holding down the CTRL key and typing 2 (CTRL+2).

Block libraries may be prepared in different formats in order for them to be used through the DesignCenter. One method is to place all global blocks in one folder. The DesignCenter identifies this folder and graphically lists all drawing files to be inserted. Another method of organizing blocks is to create one drawing containing all local blocks. When this drawing is identified through the DesignCenter, all blocks internal to this drawing display in the DesignCenter palette area.

DESIGNCENTER COMPONENTS

The DesignCenter is isolated in the following image. The following components of the dialog box are identified below and in the figure: Control buttons, Tree View or Navigation Pane, Palette or Content Pane, and Shortcut Menu.

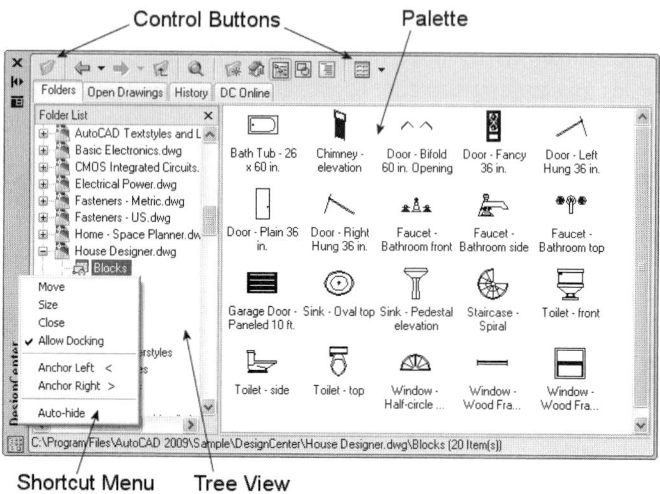

Figure 16.32

A more detailed illustration of the DesignCenter Control buttons is found in the following image. These buttons are identified as Load, Back, Forward, Up, Search, Favorites, Home, Tree View Toggle, Preview, Descriptions, and Views. It may be necessary to resize the DesignCenter to see all the buttons.

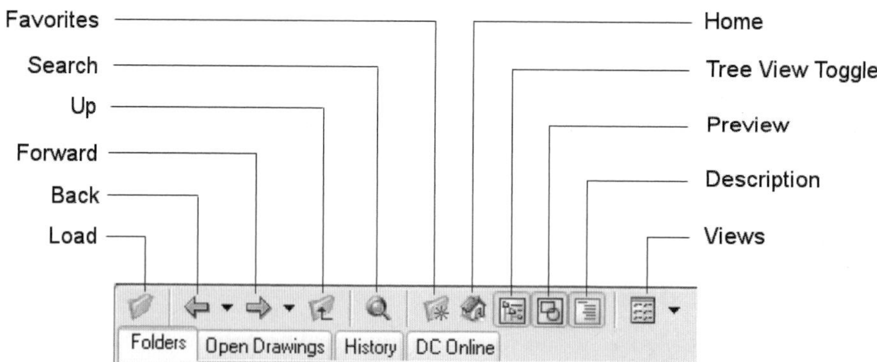

Figure 16.33

USING THE TREE VIEW

As mentioned earlier, clicking the Tree View button expands the DesignCenter to look similar to the illustration in the following image. The DesignCenter divides into two major areas: on the right is the familiar Palette area; on the left is the Tree View (a layout of network and local drive folders). Clicking a folder displays its contents in the Palette area, as shown in the following image.

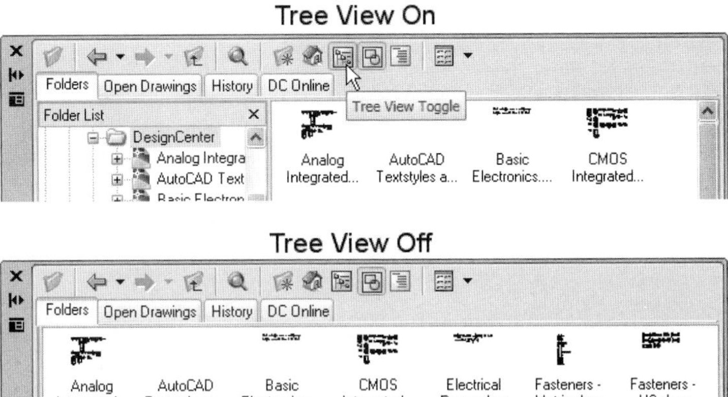

Figure 16.34

Clicking a drawing in the Tree View displays the drawing objects, as shown in the following image, that can be shared through the DesignCenter (Blocks, Dimstyles, Layers, Layouts, Linetypes, Tablestyles, Textstyles, and Xrefs).

Figure 16.35

Double-click the Blocks icon (or click the "+" symbol next to the drawing name and choose Blocks in the Tree View) to display the blocks available in the selected drawing, as shown in the following image.

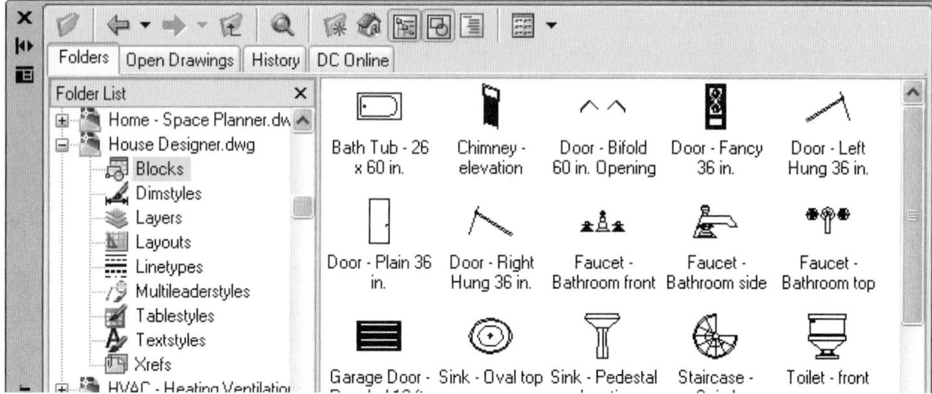

Figure 16.36

INSERTING BLOCKS THROUGH THE DESIGNCENTER

The following image displays a typical floor plan drawing along with the DesignCenter (docked to the left side of the screen), showing the blocks identified in the current drawing. If no blocks are found internal to the drawing, the Palette area will be empty.

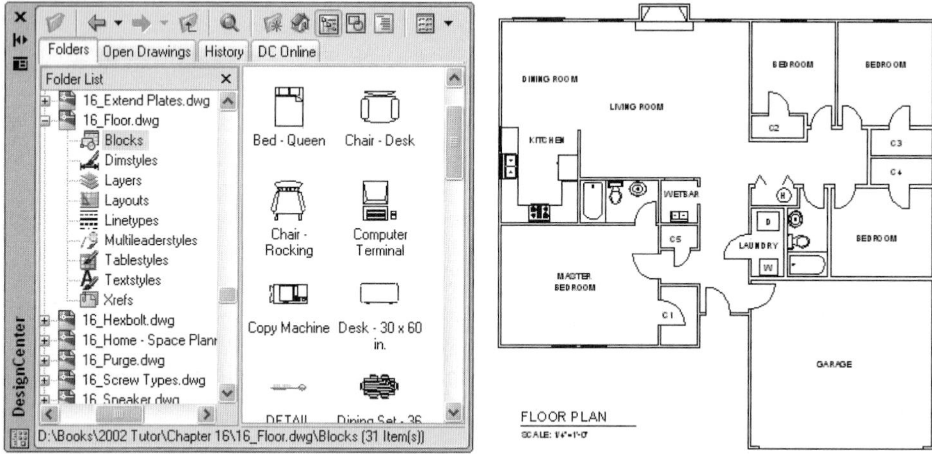

Figure 16.37

The DesignCenter operates on the "drag and drop" principle. Select the desired block located in the Palette area of the DesignCenter, drag it out, and drop it into the desired location of the drawing. The following image shows a queen-size bed dragged and dropped into the bedroom area of the floor plan.

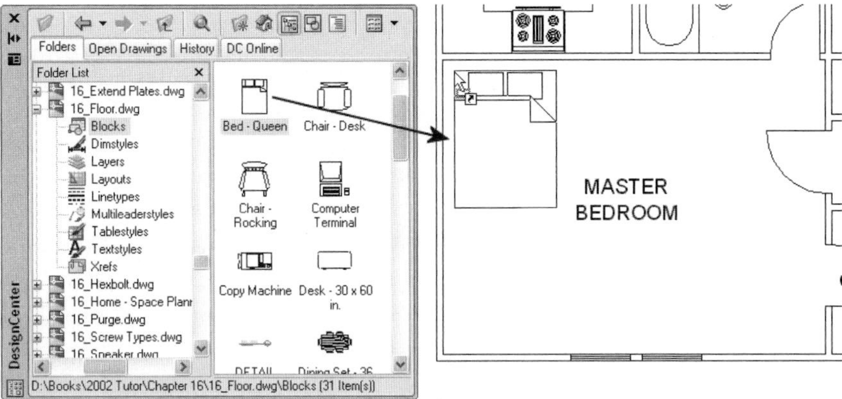

Figure 16.38

When performing a basic drag and drop operation with the left mouse button, all you have to do is identify where the block is located and drop it into that location (the use of running object snaps can ensure that the blocks are dropped in a specific location). What if the block needs to be scaled or rotated? If the drag and drop method is performed with the right mouse button, a shortcut menu is provided, as shown in the following image. Selecting Insert Block from the shortcut menu displays the Insert dialog box, covered earlier in this chapter, which allows you to specify different scale and rotation values. The following image shows a rocking chair inserted and rotated into position in the corner of the room. Instead of dragging the block with the right mouse button, double-click the block in the Palette area and the Insert dialog box is provided immediately.

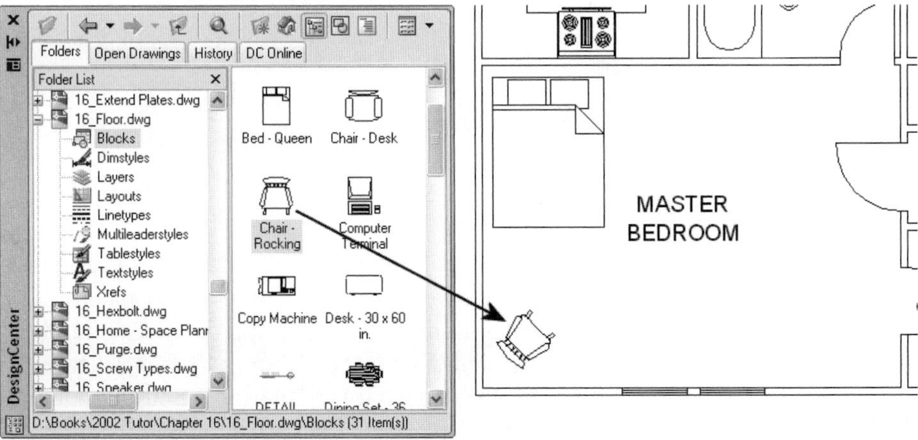

Figure 16.39

INSERTING BLOCKS USING THE TOOL PALETTE

Generally, tool palettes allow you to organize blocks and hatch patterns for insertion into your drawing. This feature is somewhat similar to the DesignCenter in its ability to drag and drop blocks, layers, dimension styles, text styles, and hatch patterns into a drawing. The Tool Palette, however, is specific to blocks, commands, and hatch patterns. You can organize and customize the Tool Palette to meet your individual drawing needs.

By default when AutoCAD is first loaded, the Tool Palette is positioned in the upper-right corner of your display screen. If the Tool Palette is not visible, it can be activated by selecting it from the Tools pull-down menu or from the Standard toolbar, as shown on the left and in the middle of the following image.

Tip: The Tool Palette can also be activated by pressing down the CTRL key and typing 3 (CTRL+3).

Once displayed, the Tool Palette provides a number of sample tabs for you to experiment with, as shown on the right in the following image. Notice in this figure how block icons are displayed with a lightning bolt graphic. This means that the block is considered dynamic. This feature of creating and using dynamic blocks will be covered later in this chapter.

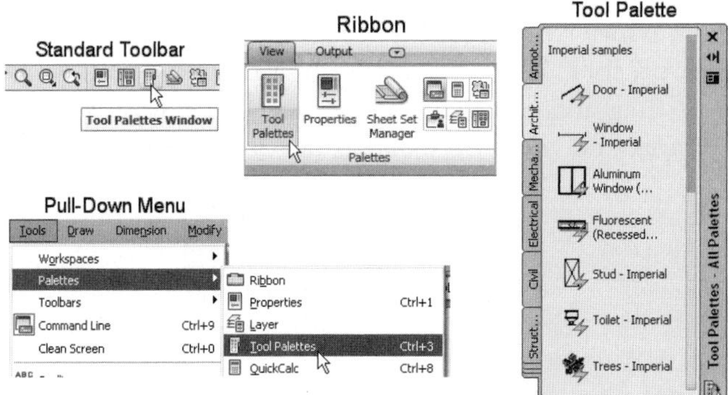

Figure 16.40

Additional tool palettes are available to assist in your design capabilities. To access these extra palettes, click the area located in the lower-left corner of any tool palette, as shown in the following image. A long list activates alongside the existing palette. A number of the palettes consist of blocks that can be dragged and dropped into a drawing. Other palettes such as Lighting and Cameras deal with the 3D rendering module contained in AutoCAD. Notice also a number of material palettes. These also deal with 3D rendering.

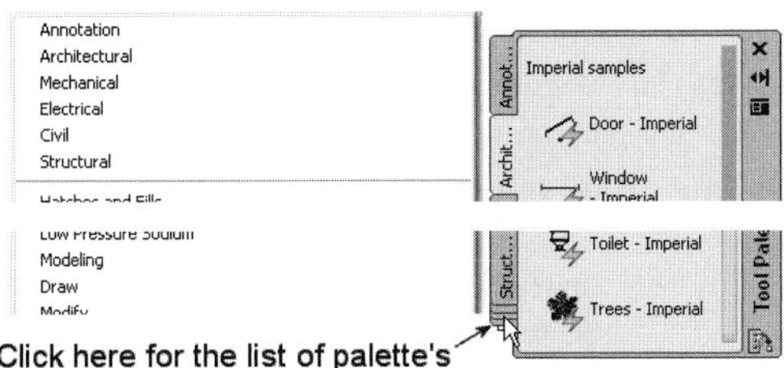

Figure 16.41

Illustrated in the following image is an example of placing hatch patterns. In this example, a concrete pattern is dragged from the Tool Palette, as shown on the right in the following image, and dropped into a closed area of your drawing, as in the fireplace hearth illustrated in the figure. The results are shown in the following image, with the hatching pattern applied to the closed area.

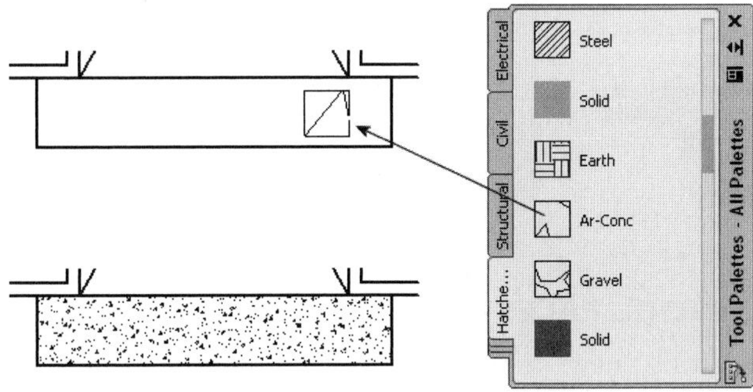

Figure 16.42

The Tool Palette has a number of very powerful features to automate its operation. For example, in the following image, a Vehicles symbol is selected. Right-clicking this block activates a shortcut menu. The following options are available:

Cut—Cutting the block from the Tool Palette to the Windows clipboard

Copy—Copying the block to the Windows clipboard

Delete—Deleting this block from the Tool Palette

Rename—Changing the name of this block in the Tool Palette

Properties—Changing the properties of this block in the Tool Palette

The Properties option allows you to modify the object's properties to suit your specific needs. Select the Properties option, as shown on the left in the following image. This activates the Tool Properties dialog box, as shown in the middle of the following image. All the information in the fields can be changed and applied to this symbol. For example, suppose you need to change the insertion scale of this block for a number of drawings. Changing the scale in the Tool Properties dialog box changes the block's scale as it is inserted into the drawing. This feature is also available for hatch patterns when using the Tool Palette. Scrolling down this dialog box displays a heading for Custom Properties. In the illustration on the right in the following image, numerous versions of the block are displayed. This is because the item selected was constructed as a dynamic block.

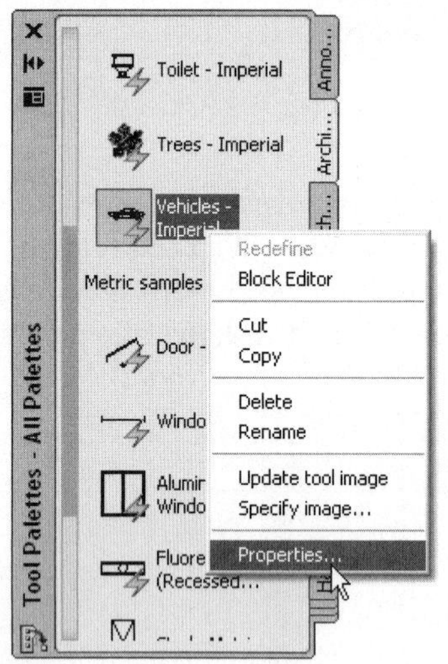

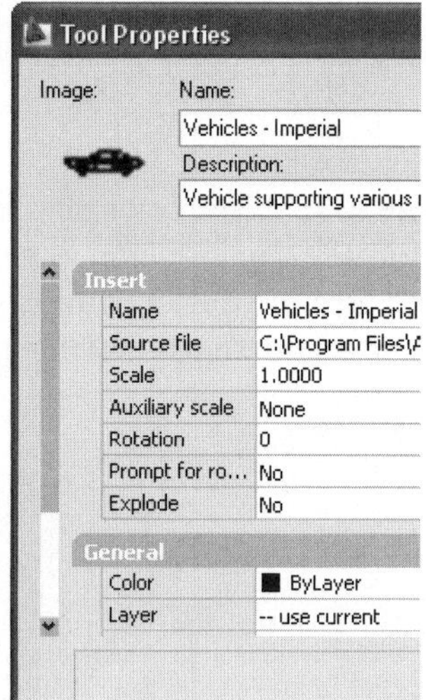

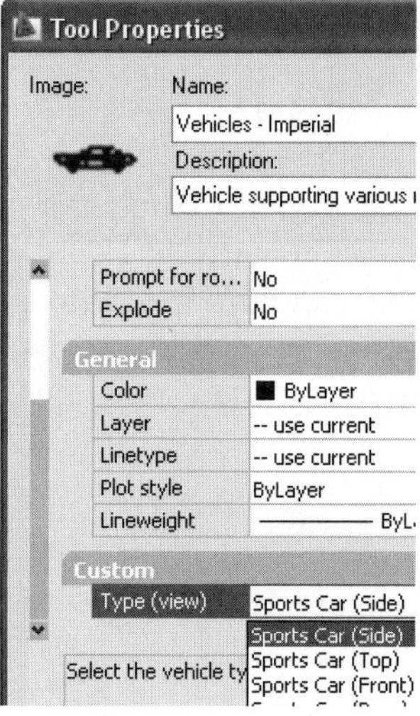

Figure 16.43

CREATING A NEW TOOL PALETTE

The process for creating new tool palettes is very easy and straightforward. First, right-click anywhere inside the Tool Palette to display the menu illustrated on the left in the following image. Options of this Tool Palette menu include:

Allow Docking—Allows the Tool Palette to be docked to the sides of your display screen. Removing the check disables this feature.

Auto-Hide—When checked, this feature collapses the Tool Palette so only the thin blue strip is displayed. When you move your cursor over the blue strip, the Tool Palette redisplays.

Transparency—Activates a Transparency dialog box, which controls the opaqueness of the Tool Palette.

View Options—Activates the View Options dialog box, which controls the size of the hatch and block icons and whether the icon is labeled or not.

Delete Tool Palette—Deletes the Tool Palette. A warning dialog box appears asking whether you really want to perform this operation.

Rename Tool Palette—Renames the Tool Palette.

Customize—Activates the Customize dialog box, which allows you to create a new Tool Palette.

New Palette—Creates a blank Tool Palette.

Click the New Palette option, as shown on the left in the following image. This automatically creates a blank Tool Palette. As illustrated on the right in the following image, a new Tool Palette name, Electrical, has been entered in the field.

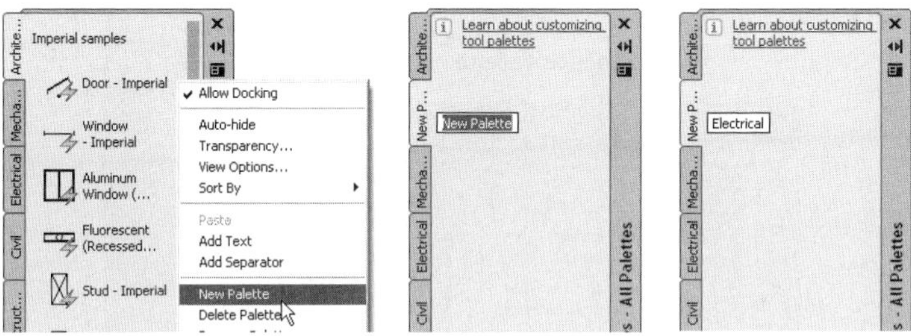

Figure 16.44

Once the Tool Palette name is entered, a tab is created for this palette, as shown on the right in the following image. To add blocks and hatch patterns to this new palette, activate the DesignCenter, search for the folder that contains the symbols you wish to place in the Tool Palette, and drag and drop these blocks or hatch patterns from the DesignCenter into the Tool Palette, as shown on the left in the following image.

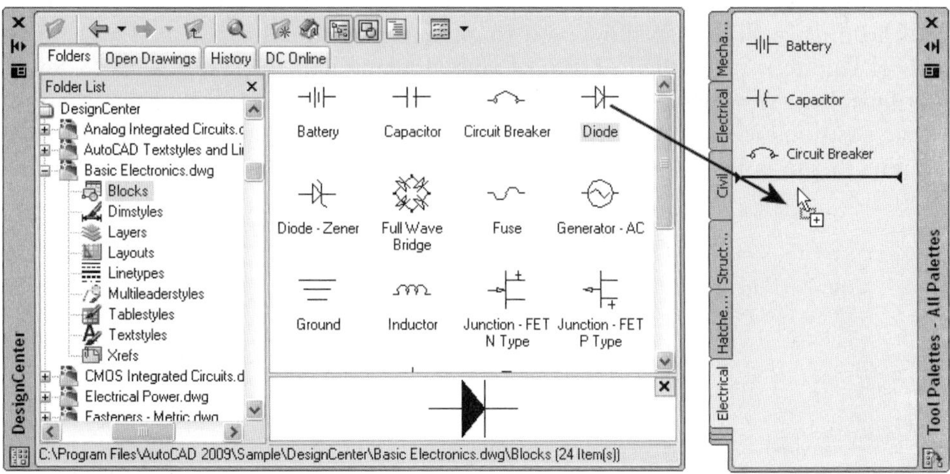

Figure 16.45

If you want to create a new Tool Palette from one whole drawing in the DesignCenter, activate the DesignCenter and go to the drawing that contains the blocks. Right-clicking this drawing displays the menu, as shown on the left in the following image. Clicking the Create Tool Palette option creates the Tool Palette using the same name as the DesignCenter drawing. The new Tool Palette contains all blocks from this drawing, as shown on the right in the following image.

Tip: If you right-click a folder and select Create Tool Palette of Blocks, a new Tool Palette will be created with the name of the folder and will contain the drawings from that folder.

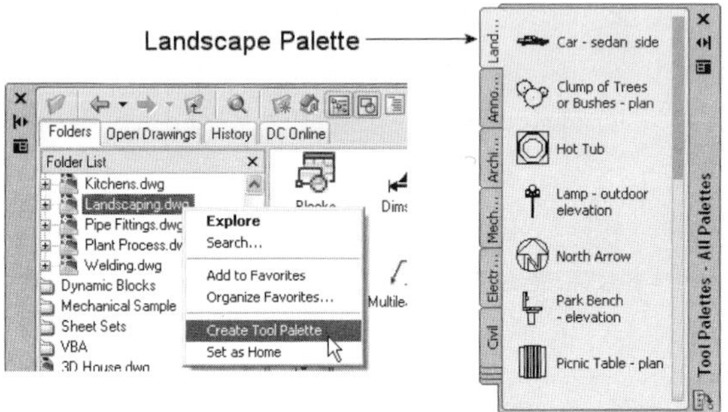

Figure 16.46

TRANSPARENCY CONTROLS FOR PALETTES

The Transparency feature of the menu, illustrated on the left in the following image, allows you to control the opaqueness of the Tool Palette. Picking Transparency activates the Transparency dialog box, illustrated in the middle in the following image. Under the General heading, a slider bar allows you to change from Solid to Clear. This means that if the Tool Palette is positioned on top of your drawing and the slider is set to Solid, you will not be able to view what is underneath it. Changing the slider bar to the location near the Clear position, as shown on the right in the following image, makes the Tool Palette transparent to the point that you will be able to see objects under the Tool Palette.

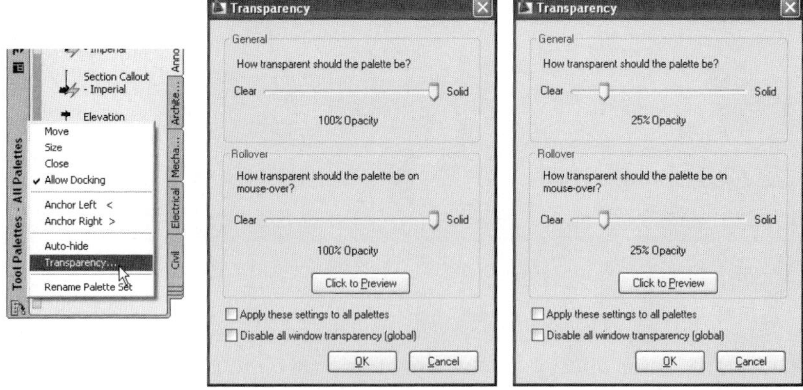

Figure 16.47

The results of setting the Transparency dialog box to near the Clear position are illustrated in the following image. You can see both the Tool Palette and the drawing at the same time. This is a very powerful option for displaying more information in your drawing.

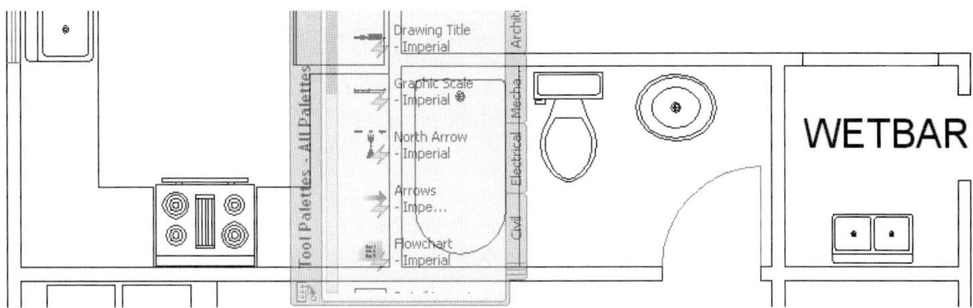

Figure 16.48

WORKING WITH MULTIPLE DRAWINGS

The Multiple Design Environment allows users to open multiple drawings within a single session of AutoCAD, as shown in the following image. This feature, like DesignCenter, allows the sharing of data between drawings. You can easily copy and move objects, such as blocks, from one drawing to another.

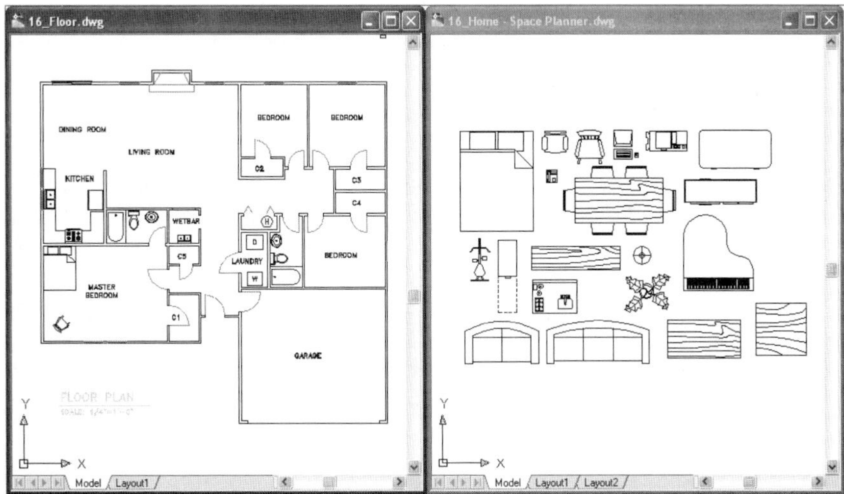

Figure 16.49

OPENING MULTIPLE DRAWINGS

Repeat the OPEN command as many times as necessary to open all drawings you will need. In fact, you can select multiple drawings in the Select File dialog box by holding down CTRL or SHIFT as you select the files, as shown in the following image.

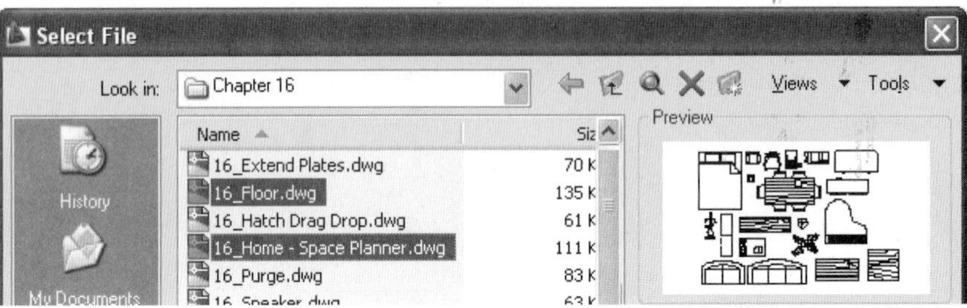

Figure 16.50

Once the drawings are open, use CTRL+F6 or CTRL+TAB to switch back and forth between the drawings. To efficiently work between drawings, you may wish to tile or cascade the drawing windows utilizing the Window pull-down menu, as shown in the following image. A list of all the open drawings is displayed at the bottom of the Window pull-down menu. Selecting one of the file names is another convenient way to switch between drawings. Remember to use the CLOSE command (Window or File pull-down menu) to individually close any drawings that are not being used. To close all drawings in a single operation, the CLOSEALL command (Window pull-down menu) can be used. If changes were made to any of the drawings, you will be prompted to save those changes before the drawing closes.

Figure 16.51

WORKING BETWEEN DRAWINGS

Once your drawings are opened and arranged on the screen, you are ready to cut and paste, copy and paste, or drag and drop objects between drawings. The first step is to cut or copy objects from a drawing. The object information is stored on the Windows clipboard until you are ready for the second step, which is to paste the objects into that same drawing or any other open drawing. These operations are not limited to AutoCAD. In fact, you can cut, copy, and paste between different Windows applications.

Use one of the following commands to cut and copy your objects:

CUTCLIP—To remove selected objects from a drawing and store them on the clipboard

COPYCLIP—To copy selected objects from a drawing and store them on the clipboard

COPYBASE—Similar to the COPYCLIP command, but allows the selection of a base point for locating your objects when they are pasted

Use one of the following commands to paste your objects:

PASTECLIP—Pastes the objects at the location selected

PASTEBLOCK—Similar to PASTECLIP command but objects are inserted as a block and an arbitrary block name is assigned

The commands listed can be typed at the keyboard, selected from the Edit pull-down menu as shown on the left in the following image, or selected by right-clicking the display screen when a command is not in progress, as shown in the middle in the following image.

Objects may also be copied between drawings with drag and drop operations. After selecting the objects, place the cursor over the objects (without selecting a grip) and then drag and drop the objects in the new location. Dragging with the right mouse button depressed provides a shortcut menu allowing additional control over pasting operations, as shown on the right in the following image.

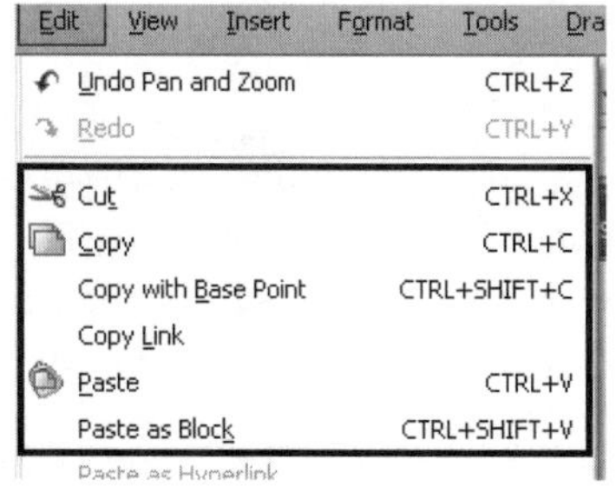

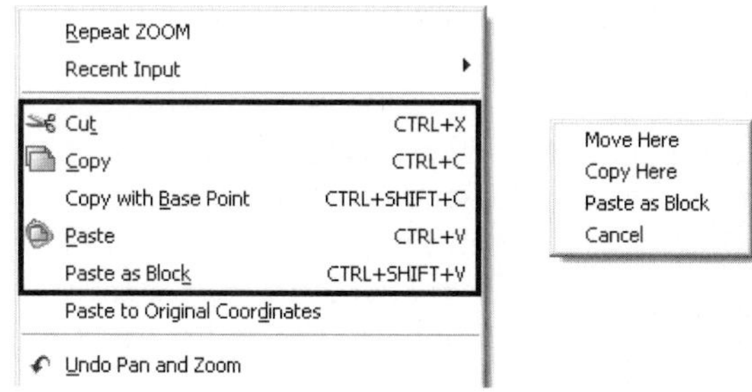

Figure 16.52

ADVANCED BLOCK TECHNIQUES—CREATING DYNAMIC BLOCKS

Dynamic blocks are blocks that change in appearance whenever they are edited through grips or through custom tables embedded inside the block. In the following image, a dynamic block called Drawing Title was located under the Architectural tab of the Tool Palettes, as shown on the right in the following image. This block was then dragged and dropped into the drawing. Clicking this block displays the normal insertion point grip. However, notice a second grip, which appears as an arrow pointing in the right direction. This grip identifies a stretching action associated with the dynamic block. When you click this arrow and move your cursor, that portion of the dynamic block changes, giving the block a different appearance without redefining or exploding the block. In this example, the arrow grip is designed to stretch the line. Notice also that dynamic blocks are identified in the Tool Palette by the appearance of a lightning bolt icon.

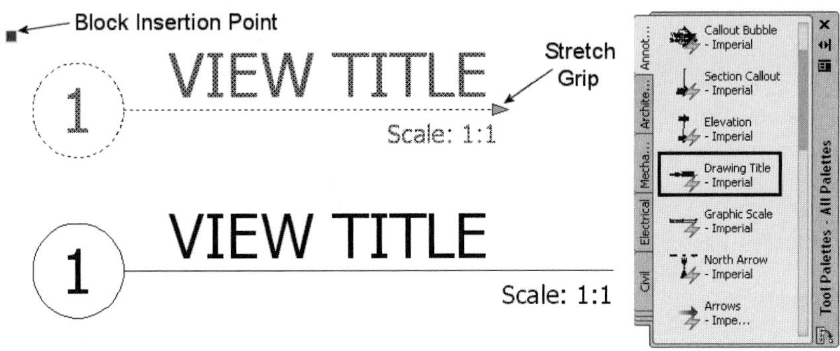

Figure 16.53

The creation process begins with the Block Editor. This can be accessed by double-clicking the block, which displays the Edit Block Definitions dialog box. This is the same dialog box used when redefining a block. You select the block name from the list and click the OK button. At this point, you enter the Block Editor environment, as shown in the following image. It is here that you make assignments to the dynamic block. The first step is to assign a parameter to the block. In the following image, a Linear Parameter was selected from the list of parameters in the toolbar and assigned to the geometry, in this case a line segment. This parameter usually is named Distance by default. In this example, the default name was renamed to a term with more meaning, Title Line Length. The Properties Palette is used for this renaming task. Once a parameter is present in the Block Editor, an action item such as Stretch is linked to the parameter. It is the action item that allows the block to change when edited back in the drawing editor. In this example, two action items are present, namely, Stretch and Move. The Stretch action allows the line to be stretched to different lengths. The Move action is required in order to move the VPSCALE attribute along with the line as it is being stretched.

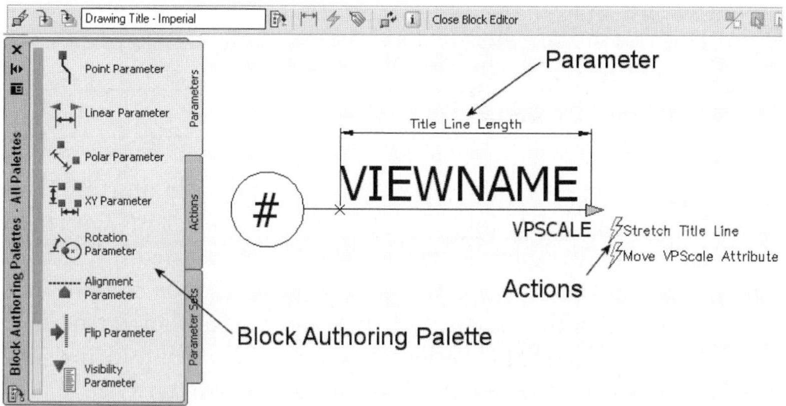

Figure 16.54

When you are satisfied with the parameter and action assignments and wish to test the features out on the block, you first click the Save button to save the changes to the block name and then click the Close Block Editor button, as shown in the following image, to return to the drawing editor.

Figure 16.55

The following table gives a brief description of all buttons found in the toolbar of the Block Editor:

Button	Description
	Switches back to the Edit Block Definition dialog box
	Saves any changes to the current block name
	Used to save any changes under a different block name
	Toggles the Block Authoring Palettes on or off
	Used for assigning parameters to the block
	Used for assigning action items to the block
	Launches the Attribute Definition dialog box for assigning attributes. Attributes are covered in chapter 17

Button	Description
	Used for updating the parameter and action item text size
	Used for learning more about dynamic blocks

The following image illustrates another, more powerful example of using dynamic blocks. Normally you would have to create four different blocks in order to show the various door swings. Through the use of Visibility States, all four doors pictured in the following image belong to a single block name. You simply pick the desired door opening from a list that displays with the block. Another feature of dynamic blocks illustrated in the following image is the ability to flip the door to different locations. Two Flip Grips allow you to flip the door along horizontal or vertical hinges. Notice also a Stretch Grip. This is present to stretch the door based on different wall thicknesses. As you can see, dynamic blocks can easily become a major productivity tool used to reduce the overall number of differently named blocks in your drawing.

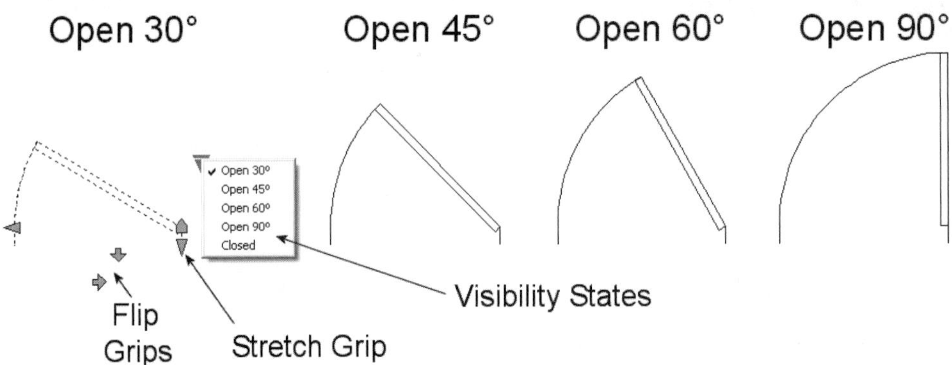

Figure 16.56

Yet another example of dynamic blocks is illustrated in the following image. Through the use of Visibility States, you have the ability to consolidate a number of blocks under a single name. In this image, notice the block name Trees – Imperial, as shown on the right in the Tool Palette. This single block name actually contains 12 different tree blocks. When you insert one of these dynamic tree blocks, the Visibility States grip appear. Clicking the grip activates the list of trees. You pick the tree from the list and the previous tree block changes based on what you select from the list.

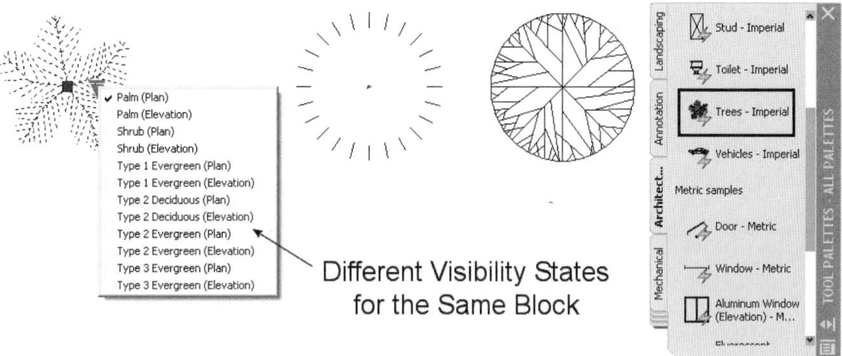

Figure 16.57

The following table gives a brief description of each dynamic block grip.

Dynamic Block Grip	Name	Description
■	Insertion Grip	The original grip that displays at the insertion point of a block; can also be used for moving the block to a new location
► ▼	Lengthen	Allows the block to be lengthened or shortened by stretching, scaling, or arraying
⬟	Alignment	Positions or aligns the block based on an object; the positioning occurs when you move the block near the object
⇦ ⇩	Flip	Allows the entire block or items inside the block to be flipped
●	Rotate	Allows the entire block or items inside the block to be rotated
▼	List	Allows you to choose from a list of items

Before you begin the process of designing dynamic blocks, here are a few items to consider:

- For what intended purpose are you designing this dynamic block?
- How do you want this block to change when it is being edited?
- What parameters are needed in order to create changes in the block?
- What actions need to be assigned to the parameters?
- Do you need the block to contain various size values in order to make incremental changes?
- Do you want to organize various blocks under a single name and control what is displayed through Visibility States?
- Do you want to create a table consisting of different values and change the size of a block through the table?

The next series of exercises will allow you to experiment with various capabilities of dynamic blocks.

WORKING WITH PARAMETERS AND ACTIONS

This exercise is designed to familiarize you with the basics of assigning parameters and actions to a block and testing its dynamic nature. You will assign a Parameter and Action item to a block and then test the results by stretching the block to new sizes without exploding the block.5 position="fixed" number="nonumber" differentiation="TryIt" identifier="chap16_FL013">

Try It!: Open the drawing file 16_Dynamic_Basics. An existing block called Table01 is already created. The size of this table is 6' by 4'. Double-click the existing block Table01 to launch the Edit Block Definition dialog box, as shown on the right in the following image.

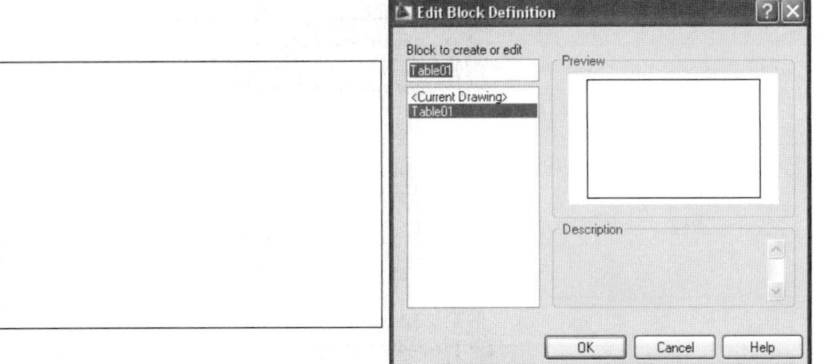

Figure 16.58

Clicking the OK button in the Edit Block Definition dialog box launches the Dynamic Block Editor, as shown in the following image. You can make changes to the geometry of the existing block or you can assign parameters and actions, making the block dynamic. These assignments are made through the Block Authoring Palettes, as shown in the following image.

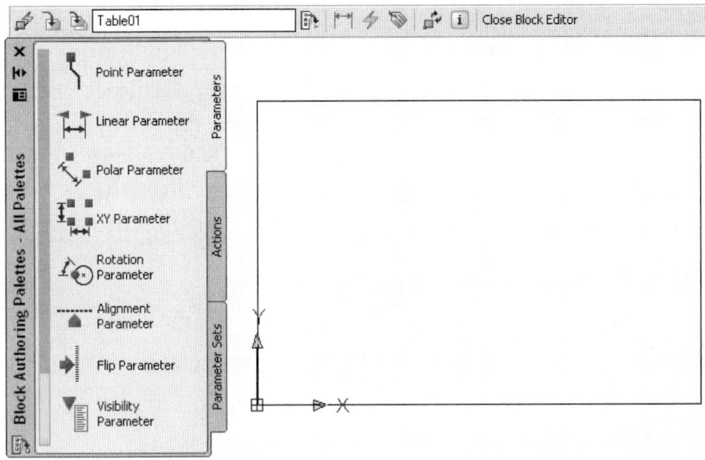

Figure 16.59

We want the ability to change the overall width of this table block. To do this, first make sure the Parameters tab is active and click the Linear Parameter tool, as shown in the following image. You will be directed through a series of prompts, which can be found in the command prompt area.

Just as in dimensioning, you pick a starting point and endpoint for the parameter, namely, the endpoints of the bottom line of the rectangle. You also specify the label location, as shown in the following image.

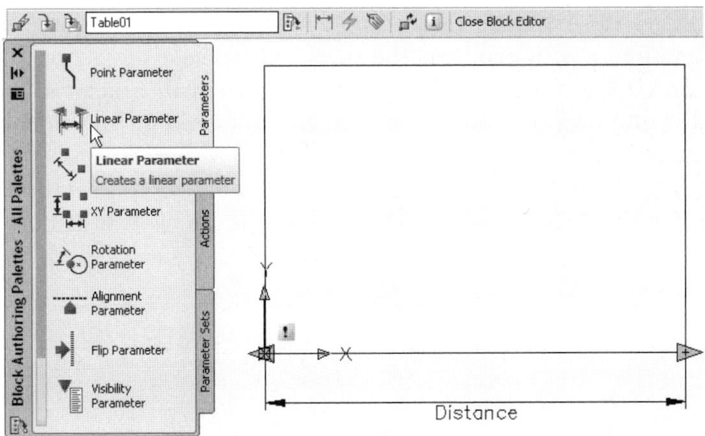

Figure 16.60

Adding a parameter is the first step in creating a dynamic block. An action must now be associated with this parameter. Select the Actions tab on the palette. Since you want the ability to change the width of the table, select the Stretch action, which will be used to accomplish this task. As with adding the parameter, the Stretch action comes with a lengthy series of command prompts.

When initiating the Stretch action, first pick the existing Distance parameter. Since the parameter was created with two endpoints, specify the endpoint to associate with the Stretch action. Select the rightmost parameter endpoint. Next, create a crossing box around the area to Stretch, as shown in the following image. Then select the rectangle as the object to stretch. Finally, specify a location for the icon that signifies the Stretch action.

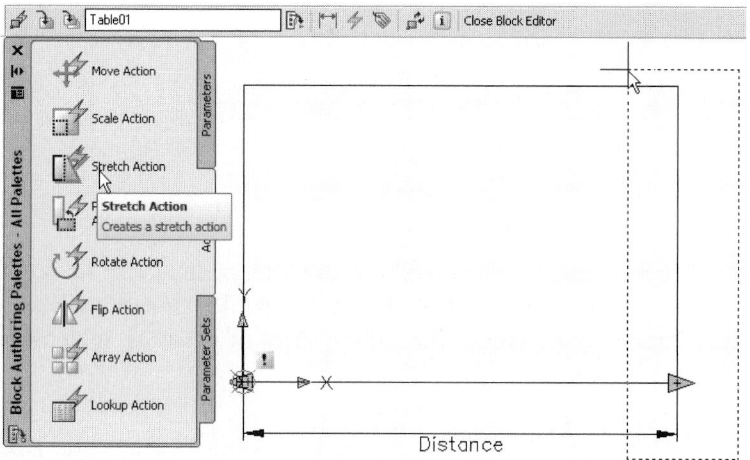

Figure 16.61

The following image illustrates all the components that make up the dynamic block. The Stretch icon should be positioned near the side of the object the action occurs at. Notice also the Alert icon. One more action needs to be created in order for the Alert icon to disappear. For the purposes of this exercise, we will not need an additional action because we will be stretching this rectangle only to the right. When you are finished assigning parameters and actions, click the Save Block Definition button to save these changes to the block. To return to the drawing editor, click the Close Block Editor button.

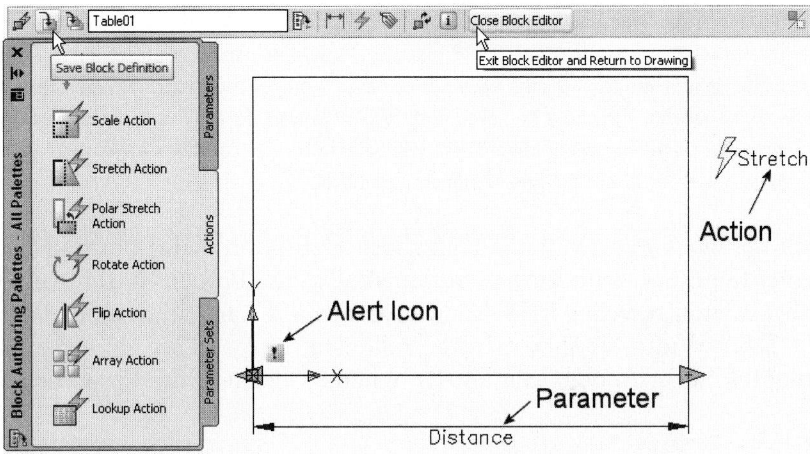

Figure 16.62

While back in the drawing, add a dimension to the existing block. Then click the rectangle and notice that two arrows appear. These arrows represent the parameter. Clicking the right arrow stretches the block to the right or left (be sure OSNAP is turned off). Notice also that the dimension changes to the new value. If you attempt to stretch the other side of the rectangle, the arrow will move but the rectangle will not. This is because the Stretch action was not assigned to this side.

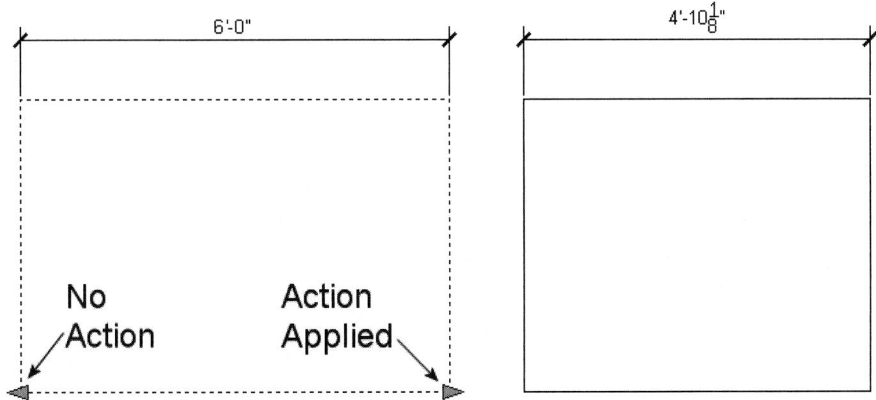

Figure 16.63

WORKING WITH VALUE SET PROPERTIES FOR PARAMETERS

Value sets are associated with parameters and are used to define custom property values for block references. Three types of value sets can be defined in a block reference; they are List, Increment, and None. This exercise will concentrate on using the Increment value set.

Try It!: Open the drawing file 16_Dynamic_Value_Sets. We will be using the same table block from the previous Try It! exercise. However, instead of stretching the table to the right at random lengths, we want to better control the stretching operation. We want to stretch this table in increments of 6". We also do not want the table to get narrower than 4' or wider than 8'. We will be able to control these items through the Value Set associated with the Distance parameter.

Double-click the block to activate the Edit Block Definition dialog box and click the OK button to enter the Block Authoring environment. Use the Properties command to activate the Properties Palette, and click the Distance parameter. Change the parameter name from Distance to Table Width, as shown in the following image. This gives the parameter a more meaningful name. Actions can also be renamed through the Properties Palette.

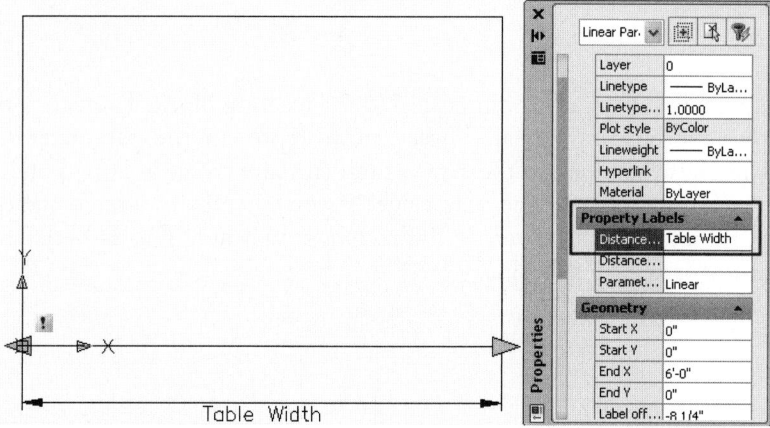

Figure 16.64

With the Properties Palette active and the Table Width parameter selected, scroll down to the bottom of the Properties Palette and notice the Value Set category. It is here that you activate one of the three Value Set types. For this example, make the Increment type active. This also displays other information related to this type. Change the Distance Increment field to 6", change the Distance Minimum field to 4', and change the Distance Maximum field to 8', as shown in the following image. As you make these changes, notice the appearance of tick marks signifying these three values.

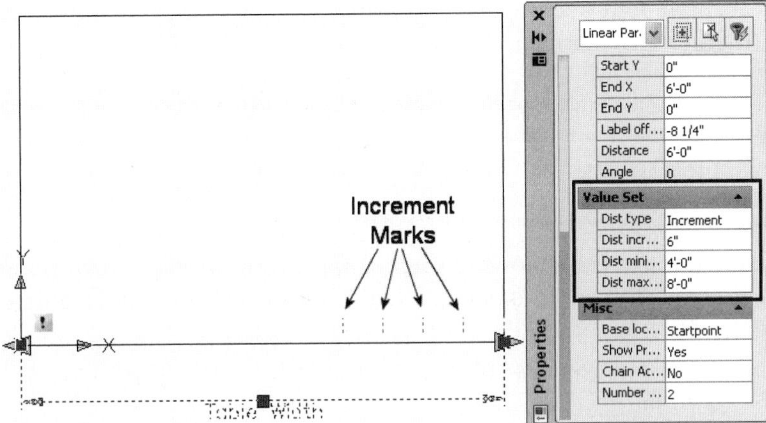

Figure 16.65

When you have finished creating these Value Sets for the Table Width parameter, save these changes to the current block name and then close the Block Editor. This returns you to your drawing. Click the block and then the rightmost parameter arrow. Notice the appearance of the two parameters and tick marks, as shown on the left in the following image. The tick marks are in increments of 6". The tick mark on the far left begins at the 4' distance of the table. The tick mark on the far right ends at the 8' distance. Slide your cursor to the right or left and notice your cursor snapping to the increment tick marks. Clicking one of the tick marks adjusts the table to that distance, as shown on the right in the following image.

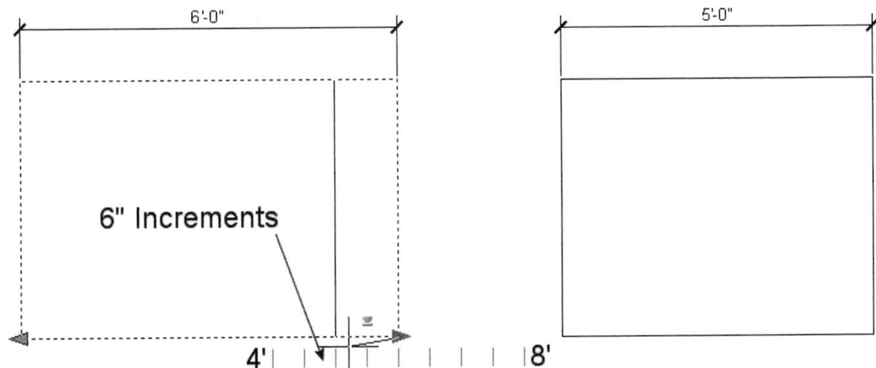

Figure 16.66

ALIGNMENT PARAMETER APPLICATIONS

Alignment Parameters allow you to drag a block object over an object and have the block automatically rotate and orientate itself relative to the geometry your cursor is over. This makes the Alignment Parameter one of the easiest modes of adding to a dynamic block.

Two types of Alignment Parameters are available: Perpendicular and Tangent. When creating an Alignment Parameter, you specify an alignment point and alignment angle.

Try It!: Open the drawing file 16_Dynamic_Alignment. A block of a sink is already defined in this drawing. Double-click the block to activate the Edit Block Definition dialog box, and click the OK button to enter the Block Authoring environment, as shown in the following image.

Before continuing, make the following preparations: turn on Polar, turn on and set OSNAP mode to Quadrant, and turn on OTRACK. Then click the Alignment Parameter tool in the Block Authoring Palette.

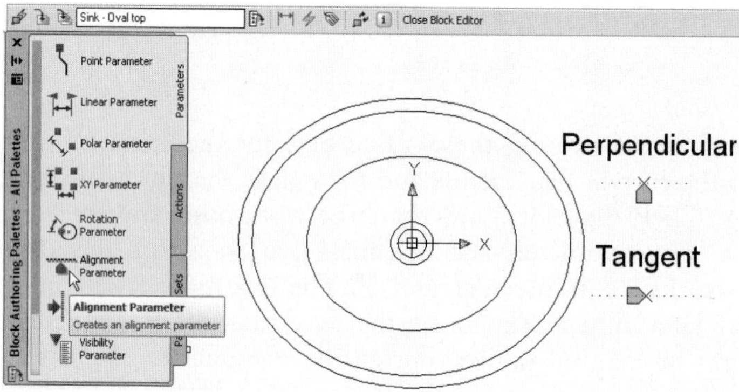

Figure 16.67

For the base point of the alignment, move your cursor over the top of the ellipse, track straight up, and enter a value of 2, as shown on the left in the following image. For the alignment direction, move your cursor to the right of the last known point and pick a blank part of your screen, as shown in the middle of the following image. The Alignment Parameter appears, as shown on the right in the following image. The Alignment Parameter being used for this application is Perpendicular.

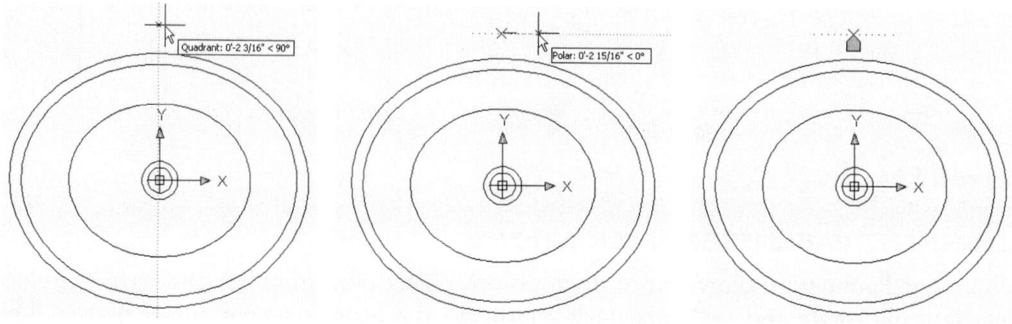

Figure 16.68

Saving this sink block and exiting the Block Editor returns you to the drawing editor. Clicking the sink block, as shown on the left in the following image, exposes the Perpendicular Alignment Parameter. Click this Alignment Parameter, move your cursor to any of the inner walls of the bathroom plan, and notice the sink automatically rotating at an angle perpendicular to the wall. When you are satisfied with the desired orientation, click to place the block, as shown on the right in the following image.

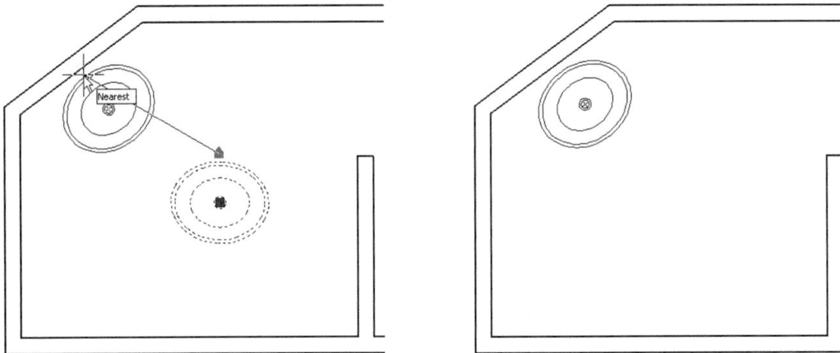

Figure 16.69

MULTIPLE ALIGNMENT POINT APPLICATIONS

You can assign multiple alignment points to a single block. Then, when you insert the block, you can cycle through the grip points by pressing the CTRL key while dragging the block into position until the desired insertion point is found. The next Try It! exercise will illustrate this capability of dynamic blocks.

Try It!: Open the drawing file 16_Dynamic_Insertion_Cycling. A block of a window is already defined in this drawing. Double-click the block to activate the Edit Block Definition dialog box and click the OK button to enter the Block Authoring environment, as shown in the following image. Activate the Alignment Parameter and create the three parameters shown in the following image. Use OSNAP endpoint for the alignment location. For the alignment angle, pick to the right or left of the endpoint, as shown in the following image. The three icons identify the Perpendicular alignment tool being used.

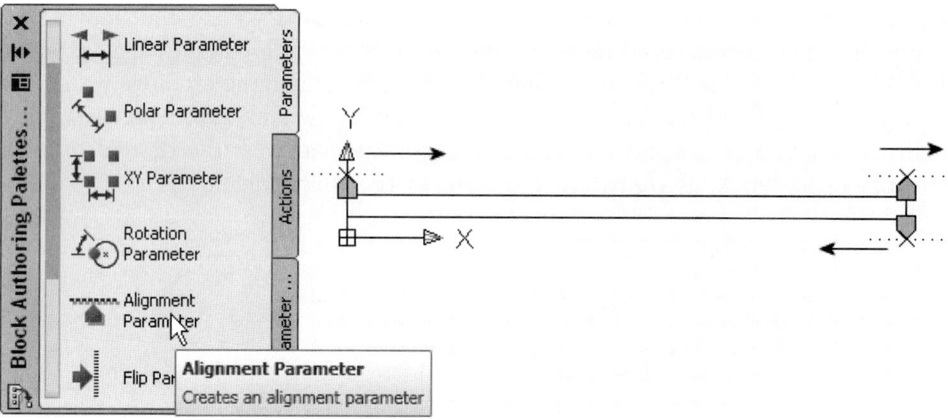

Figure 16.70

Save this window block and exit the Block Editor to return to the drawing editor. Clicking the existing block activates the three alignment points just created. The grip located in the lower-left corner of the block is the original insertion point. Now experiment by inserting the window and pressing the CTRL key to cycle through all insertion points. You should be able to place all windows in the following image without the need to move, copy, or rotate the block into position. Since the Perpendicular Alignment Parameter was used, the window aligns itself perpendicular (at a 90° angle) with each wall.

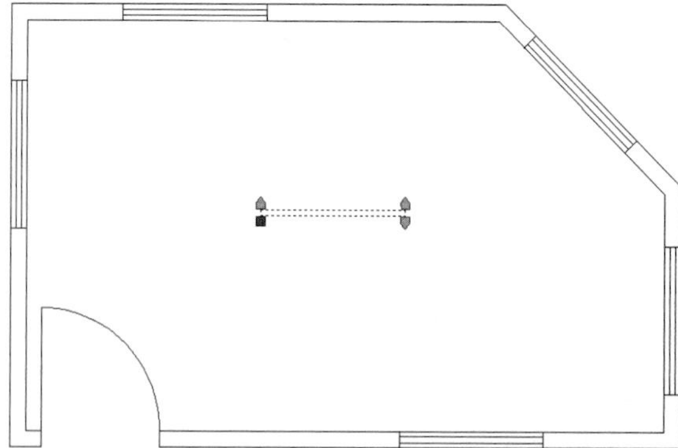

Figure 16.71

SCALE AND ANGLE OFFSET APPLICATIONS

This application of dynamic blocks allows you to increase or decrease the angle direction of a part of the block relative to a parameter.

Try It!: Open the drawing file 16_Dynamic_Offsets. Two door blocks are already defined in this drawing. Double-click the left door block to activate the Edit Block Definition dialog box and click the OK button to enter the Block Authoring environment. Activate the Linear Parameter. For the Start Point, click the bottom endpoint of the line, and for the End Point, click the bottom endpoint of the arc. Specify the label location below the door block; the label should read Distance.

Click the Distance Parameter you just created and activate the Properties Palette. In the palette, change the Property Label from Distance to Door Width. Under the Value Set area, change the Distance Type to Increment, change the Distance Increment to 2", change the Distance Minimum to 1', and change the Distance Maximum to 3'. Refer to the illustration on the right in the following image for these changes.

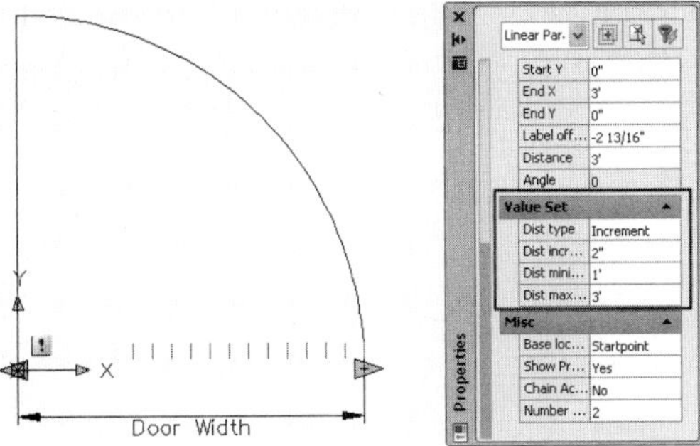

Figure 16.72

Next, activate the Scale Action, select Door Width as the parameter, and select the line and arc as the objects to scale. Press ENTER to continue. Specify the action location, as shown on the right in the following image.

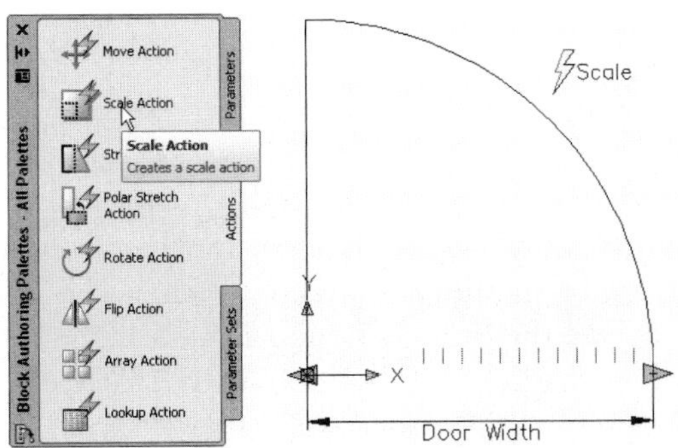

Figure 16.73

Click the Save Block Definition button to save these changes, and then click the Close Block Editor button to return to the drawing editor.

Click the door symbol, select the right arrow grip, and slide your cursor to one of the increment hash marks. Notice how the door symbol reduces in size in 2" increments. Notice also how the arc and line scale to accommodate these new distances.

Double-click the right door block to activate the Edit Block Definition dialog box, and click the OK button to enter the Block Authoring environment. This door symbol actually shows the thickness of the door. The same Linear Parameter used for the previous door symbol has already been applied to this door. Also, the Scale Action already exists.

However, in this example, the Scale Action has been applied only to the arc. As the door increases or decreases in size, the arc will be affected by these actions. However, you do not want the door thickness to increase or decrease. You want the door thickness to remain unchanged but the door height to increase or decrease along with the arc.

To have the door height be affected by changes in the Door Width parameter, first click the Stretch Action. Select the Door Width parameter. In order for the parameter point to be associated with the action, type S for second point. This should select the point at the bottom of the arc. Next, erect a crossing box, as shown on the left in the following image. Select the two vertical edges of the door along with the short horizontal top of the door.

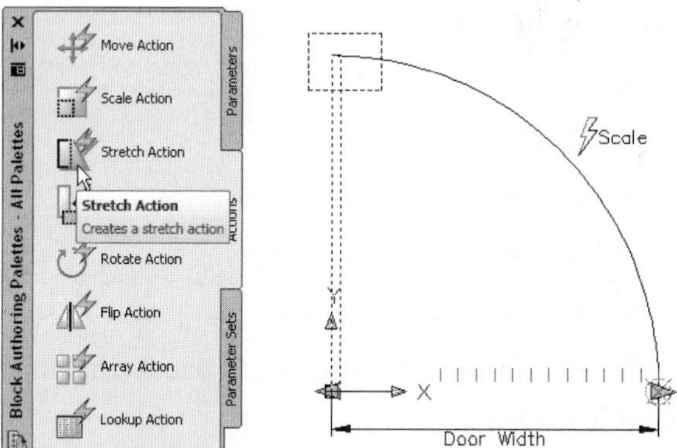

Figure 16.74

Press ENTER. The following prompt will display:

```
Specify action location or [Multiplier/Offset]:
```

Type O for Offset, and at the next prompt, type 90 for the offset angle. This stretches the top of the door straight up, or in the 90° direction, when the door symbol increases or decreases in size. Finally, locate the Stretch Action icon near the top of the door, as shown on the left in the following image.

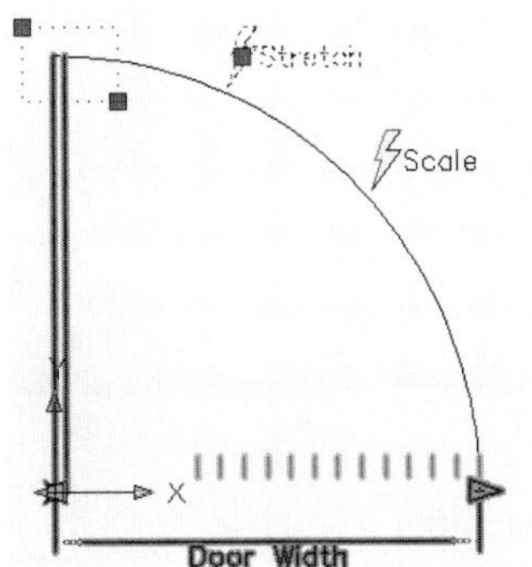

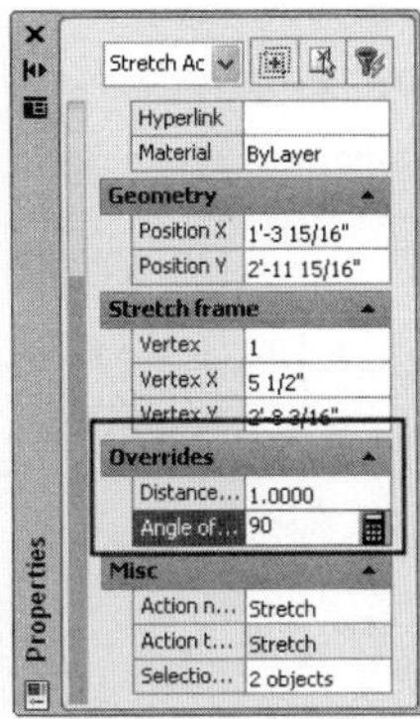

Figure 16.75

Click the Save Block Definition button to save these changes, and then click the Close Block Editor button to return to the drawing editor.

Click the door symbol, select the right arrow grip, and slide your cursor to one of the increment hash marks. Notice how the door symbol reduces in size in 2" increments. Notice also how the arc scales and door thickness stretch to accommodate these new distances.

ARRAY APPLICATIONS

Linking an Array Action to a dynamic block can have some interesting effects. As the dynamic block increases in width, items associated with the block can be added depending on the distance of the width. This application will utilize a table block consisting of four chairs positioned around a 4' table. As the table increases in width, extra chairs need to be added to the block.

Try It!: Open the drawing file 16_Dynamic_Array. A block consisting of four chairs arranged around a 4' table is displayed in the drawing editor. This exercise will illustrate how to stretch the table in 4' increments and have extra columns of chairs created. Double-click the table block to activate the Edit Block Definition dialog box and click the OK button to enter the Block Authoring environment.

Click the Table Width parameter and launch the Properties Palette. Make changes under the Value Set area, as shown in the following image. Set the Distance type to Increment, the Distance increment to 4', the Distance minimum to 4', and the Distance maximum to 16'.

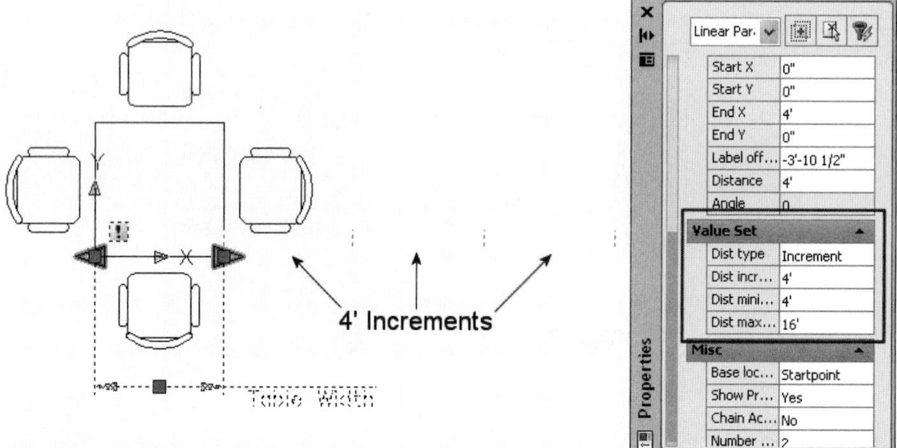

Figure 16.76

Next, assign a Stretch Action to the right end of the table and rightmost positioned chair, as shown in the following image. After activating the Stretch Action, pick the Table Width parameter. In order for the parameter point to associate with the action, type S for second point. Create a crossing box to surround the rightmost chair and right end of the table, and select the table and rightmost chair as the objects to stretch. Locate the Stretch Action icon in a convenient location of your screen.

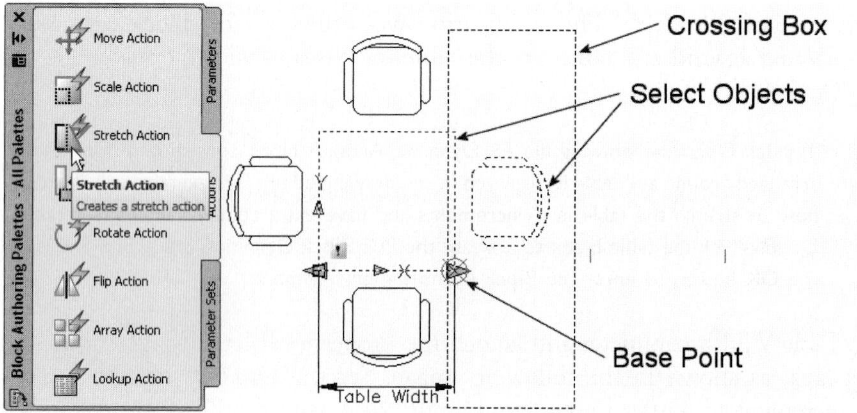

Figure 16.77

If you experiment by stretching this new block in the drawing editor, you will find that the table and rightmost chair stretch; however, the two chairs located in the 90° and 270° positions remain in their original positions. We want these two chairs to be copied or arrayed at a specified distance depending on how long the table stretches. To perform this operation, click the Array Action button back in the Block Authoring editor. Select the Table Width parameter, select both chairs, as shown in the following image, and enter a

Column Width distance of 3'11 as the distance from one chair to another. Locate the Array Action icon in a convenient screen location.

Click the Save Block Definition button to save these changes, and then click the Close Block Editor button to return to the drawing editor.

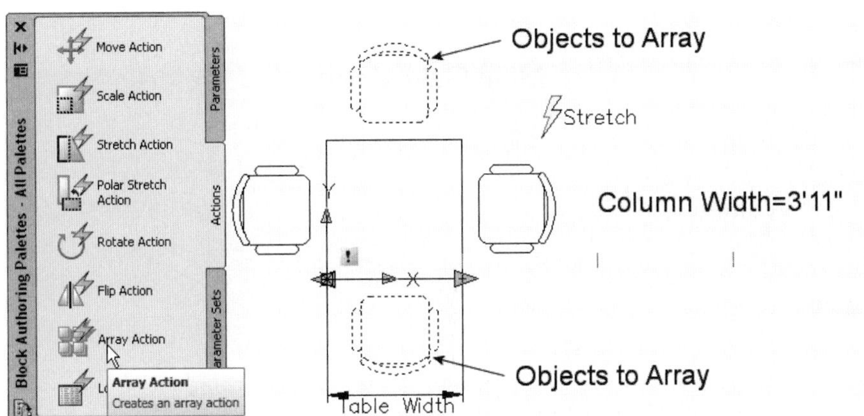

Figure 16.78

When you return to the drawing editor, click the block of the table and chairs, pick the right arrow grip as shown on the left in the following image, and begin stretching the table. When you stretch the table to a new length of 8', two new columns of chairs are added. Stretch to a new length of 16' and your screen should appear similar to the illustration on the right in the following image.

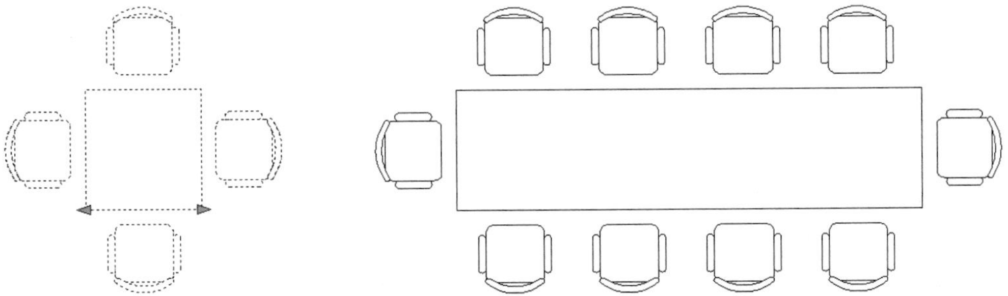

Figure 16.79

WORKING WITH VISIBILITY STATES

Using a visibility state on a block allows you to define various versions of the block under one unique name. For example, rather than create four separate blocks to represent the fittings of a gate valve, as shown in the following image, one master block consisting of all geometry native to all versions will be created. You create visibility states and turn off the geometry you do not want to see. This is an extremely productive way to keep the number of block names down in a drawing file without sacrificing the various types of blocks normally created individually.

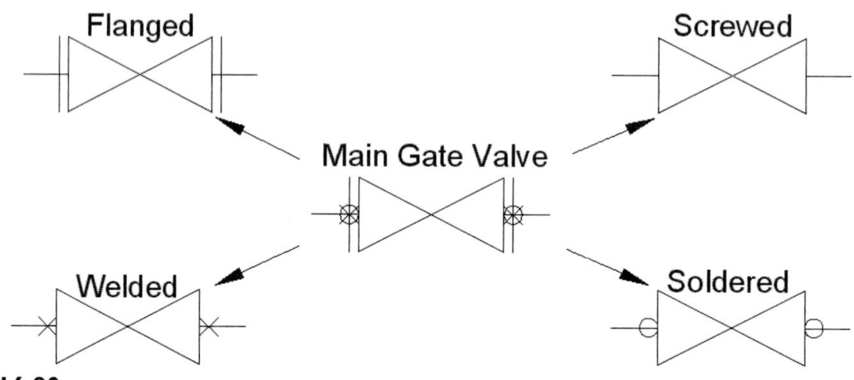

Figure 16.80

Try It!: Open the drawing file 16_Dynamic_Visibility. A block consisting of a gate valve is displayed in the drawing editor. This exercise will illustrate how to create various visibility states using the gate valve block provided. Double-click the gate valve block to activate the Edit Block Definition dialog box and click the OK button to enter the Block Authoring environment.

First, create a Visibility Parameter and locate the icon, as shown in the following image.

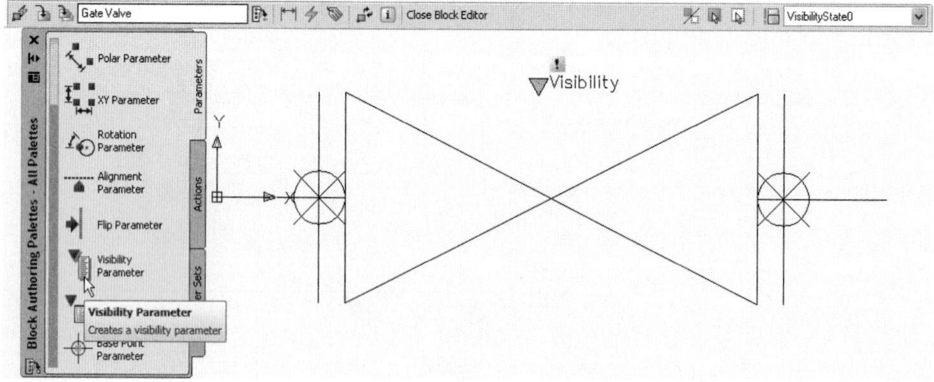

Figure 16.81

Next, click the Manage Visibility States button to launch the Visibility States dialog box, as shown in the following image. Immediately rename the default state (VisibilityState0) to Main Gate Valve. This is the main block that contains all geometry associated with all the similar but different gate valve blocks.

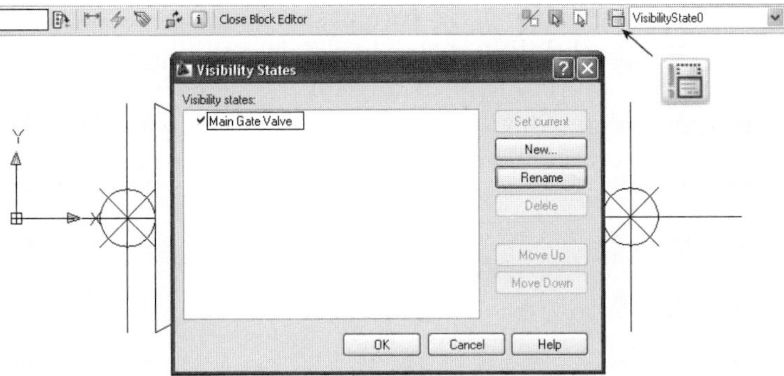

Figure 16.82

The following table gives a brief description of the visibility tools present in the Block Editor:

Button	Description
	Toggles between all visibility modes, visible or invisible
	Used for making objects in a visibility state visible
	Used for making objects in a visibility state invisible
	Launches the Visibility States dialog box used for managing visibility states

Next, click the New button and begin creating additional visibility states. When the New Visibility State dialog box appears, enter Soldered, as shown in the following image. Keep all remaining default options for new states unchanged. Click the OK button when finished. Use the same method to create three other visibility states: Flanged, Screwed, and Welded.

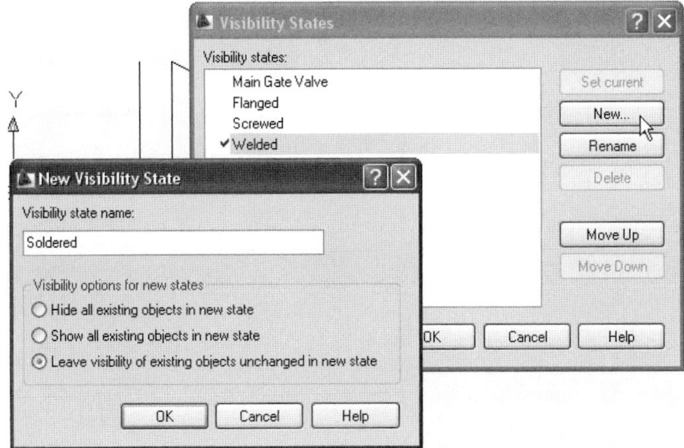

Figure 16.83

Next, make the Flanged visibility state current by clicking its name, located in the upper-right corner of the Block Editor dialog box. Select all the geometry that you do not want to be displayed when in the Flanged state. In the following image, both circles and four diagonal lines are highlighted. Click the Make Invisible button to turn this geometry off. The result is the gate valve symbol displaying only the geometry necessary to define a Flanged gate valve.

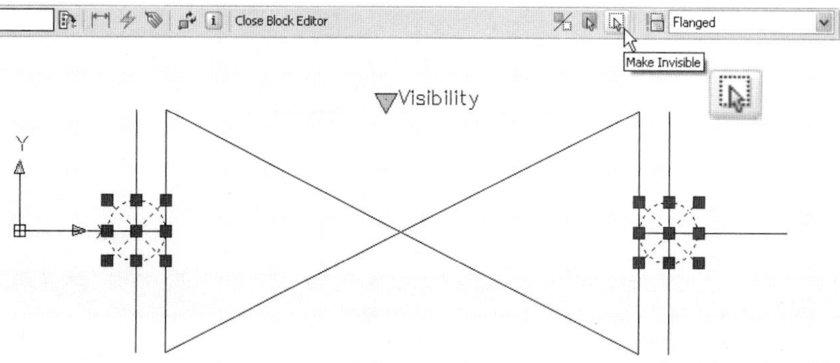

Figure 16.84

Continue by making the Screwed Visibility state current, noticing that all the geometry reappears in the block. For a screwed gate valve, select both circles, four diagonal lines, and two vertical line segments, as shown in the following image. Click the Make Invisible button, and the remaining geometry defines a screwed gate valve. Repeat this procedure for the Welded and Soldered visibility states. The Welded state should consist of the gate valve and the four diagonal lines. The Soldered state should consist of the gate valve and both circles.

Click the Save Block Definition button to save these changes, and then click the Close Block Editor button to return to the drawing editor.

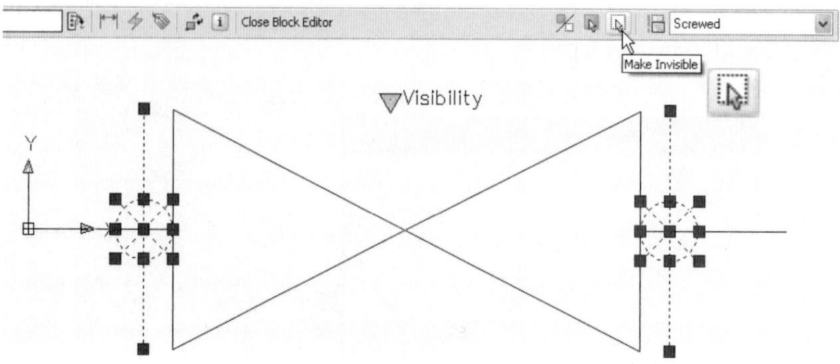

Figure 16.85

When you return to the drawing editor, create various copies of the gate valve. Then click an individual block and notice the visibility state grip. Click the grip and pick the desired gate valve fitting, as shown in the following image. This method provides a quick and efficient means of organizing blocks that are similar but different in appearance.

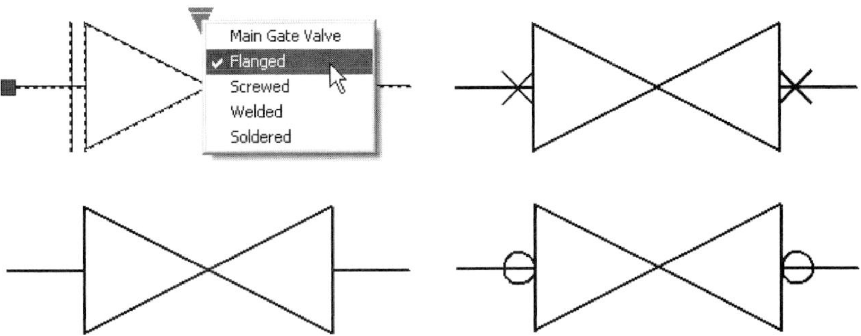

Figure 16.86

Note: Another technique of creating blocks through visibility states is to open an existing block in the Block Editor and rename the default visibility state. You then create a new visibility state, click the Hide All Objects button, and then close the Visibility States dialog box. Since the Block Editor remains empty, insert a block and then explode it at the origin of the Block Editor. Display the Visibility States dialog box and repeat this procedure for as many blocks as desired.

USING LOOKUP PARAMETERS AND TABLES

A Lookup Parameter allows you to create a lookup property whose value is controlled by a table of custom properties located in the block. While inside this table, you add a series of different input properties that are tied directly to a single parameter or multiple parameters assigned to the block. You also create lookup properties that are tied to the input properties. These lookup properties appear as a list inside the block once you return to the drawing editor. You then pick different lookup properties from the list in order to make changes to the block. This concept of creating lookup properties will be demonstrated in the next Try It! exercise.

Try It!: Open the drawing file 16_Dynamic_Lookup. A rectangular block with diagonal lines is already defined in this drawing. The name of this block is Basic Stud and represents a 2 × 4 construction object. This exercise will illustrate how to create a lookup table. After this is done, you will be able to choose various sizes of this basic stud ranging in size from the standard 2 × 4 to a 2 × 12. Begin by double-clicking the 2 × 4 Basic Stud block to activate the Edit Block Definition dialog box, and click the OK button to enter the Block Authoring environment.

Notice in the following image that a Linear Parameter (Stud Width) and Stretch Action (Stretch Right) are already defined in your drawing. The grips at both ends of the parameter name, normally visible, have been turned off through the Properties Palette. You will be controlling the sizes of the stud through the lookup list.

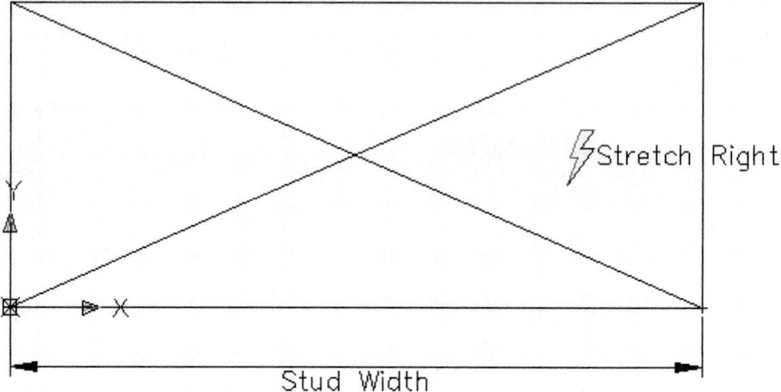

Figure 16.87

Begin by clicking Lookup Parameter located under the Parameters tab, as shown in the following image. Locate this parameter in a convenient location of the screen and use the Properties Palette to rename the parameter to Basic Stud Sizes.

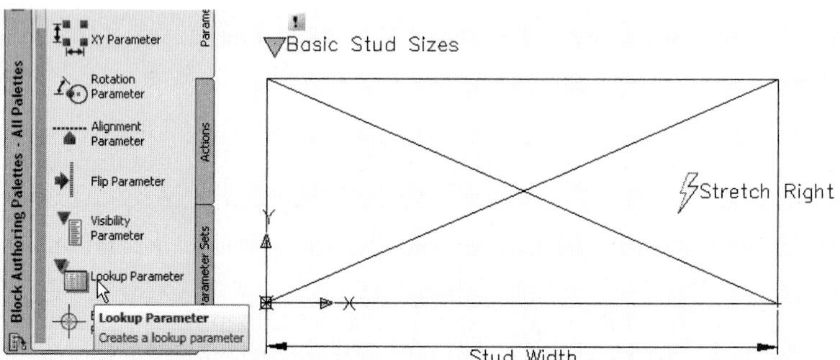

Figure 16.88

Create a Lookup Action item by selecting this item from the Block Authoring Palette, as shown on the left in the following image. Associate this action with a parameter by picking the Basic Stud Sizes Lookup Parameter that you just created in the previous step. Locate the Lookup Action next to the Basic Stud Sizes Lookup Parameter. After you select a screen position for the action location, the Property Lookup Table dialog box appears, as shown on the right in the following image. Click the Add Properties button.

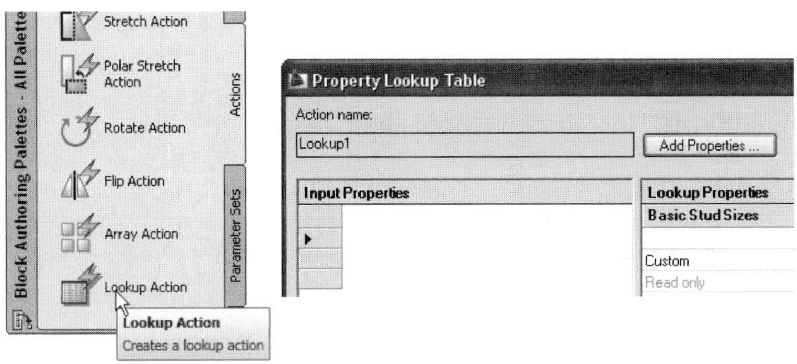

Figure 16.89

When the Add Parameter Properties dialog box appears, as shown on the left in the following image, notice the linear parameter listed. For this example, this is the only parameter that will be used. In more complicated Lookup Actions, you may have numerous parameter names to choose from. For now, click the OK button to accept this parameter name.

This returns you to the Property Lookup Table dialog box. Notice the column for Stud Width. Start entering the values for various stud widths, such as 3.5, 5.5, 7.5, 9.5, and 11.5, as shown on the left side of this dialog box.

You also need to fill in the right column of the dialog box under the heading Lookup Properties Basic Stud Sizes. These are the listings that will be visible when you return to the drawing editor and test the Basic Stud block for its various sizes. Under this column, enter the following sizes; 2 × 4, 2 × 6, 2 × 8, 2 × 10, and 2 × 12. Be certain these correspond with the Stud Width values on the left side of the dialog box. When you are finished, your dialog box should appear similar to the illustration on the right in the following image.

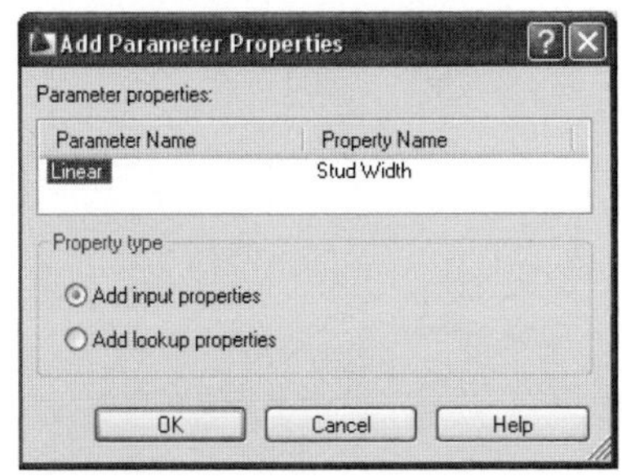

Property Lookup Table

Action name: Lookup1 | Add Properties ...

Input Properties	Lookup Properties
Stud Width	Basic Stud Sizes
3.5000	2 x 4
5.5000	2 x 6
7.5000	2 x 8
9.5000	2 x 10
11.5000	2 x 12
<Unmatched>	Custom
	Read only

Figure 16.90

One other very important operation needs to be performed inside this dialog box. On the right side of the dialog box, notice Custom. Click this item if it appears grayed out. This activates the Read only mode listed just below it. Click Read only, pick the down arrow, and choose Allow reverse lookup, as shown in the following image. This activates the lookup table when you click the block once back in the drawing editor. If this is not selected, the table will not display in the block when you return to the drawing editor.

Input Properties	Lookup Properties
Stud Width	Basic Stud Sizes
3.5000	2 x 4
5.5000	2 x 6
7.5000	2 x 8
9.5000	2 x 10
11.5000	2 x 12
<Unmatched>	Custom
	Allow reverse lookup
	Allow reverse lookup
	Read only

Figure 16.91

When you return to the Block Editor screen, change the name of the Lookup Action item to Basic Stud Table. Save the block and close the Block Editor to return to the drawing editor. Click the Block and notice the appearance of the Lookup Parameter. Click the down arrow in the Lookup grip and test the various Basic Stud Sizes by picking them from the list, as shown in the following image.

Figure 16.92

TUTORIAL EXERCISE: ELECTRICAL SCHEMATIC

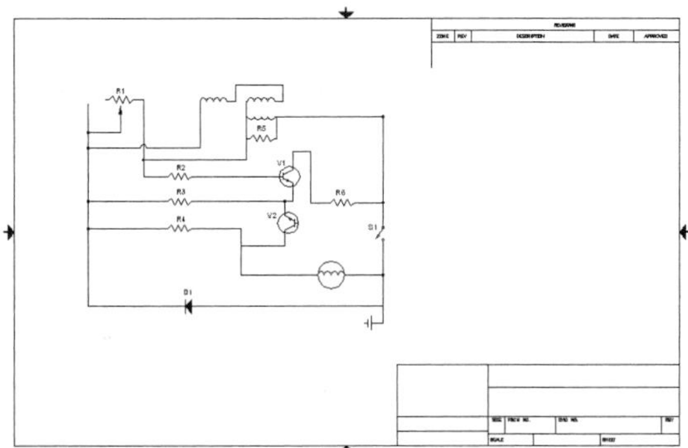

Figure 16.93

PURPOSE

This tutorial is designed to lay out electrical blocks such as resistors, transistors, diodes, and switches to form an electrical schematic, as shown in the previous image.

SYSTEM SETTINGS

Create a new drawing called ELECT_SYMB to hold the electrical blocks. Keep all default units and limits settings. Set the GRID and SNAP commands to 0.0750 units. This will be used to assist in the layout of the blocks. Once this drawing is finished, create another new drawing called ELECTRICAL_SCHEM1. This drawing will show the layout of an electrical schematic. Keep the default units but use the LIMITS command to set the upper-right corner of the display screen to (17.0000,11.0000). Grid and snap do not have to be set for this drawing.

LAYERS

Create the following layers for ELECTRICAL_SCHEM1 with the format:

Name	Color	Linetype
Border	White	Continuous
Blocks	Red	Continuous
Wires	Blue	Continuous

SUGGESTED COMMANDS

This tutorial begins by creating a new drawing called ELECT_SYMB, which will hold all electrical blocks. You will use the image below as a guide in drawing all electrical blocks. A grid with spacing 0.0750 would provide further assistance with the drawing of the blocks. Once all blocks are drawn, the BLOCK command is used to create blocks out of the individual blocks. You will save this drawing and create a new drawing file called ELECTRICAL_SCHEM1. The DesignCenter will be used to drag and drop the internal blocks from the drawing ELECT_SYMB into the new drawing, ELECTRICAL_SCHEM1. Finally, you will connect all blocks with lines that represent wires and electrical connections, add block identifiers, and save the drawing.

Step 1

Begin a new drawing and call it ELECT_SYMB. Be sure to save this drawing to a convenient location. It will be used along with the DesignCenter to bring the internal blocks into another drawing file.

Step 2

Using a grid/snap of 0.075 units and the following image as a guide, construct each block. The grid/snap will help keep all blocks proportional to each other. Create all the blocks on the neutral layer 0. Also, the X located on each block signifies its insertion point (do not draw the X).

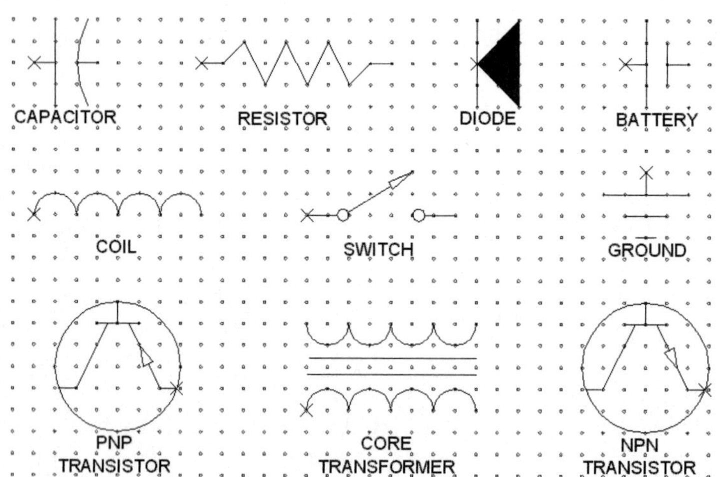

Figure 16.94

Step 3

For a more detailed approach to creating the blocks, the resistor block will be created. Using the illustration on the left in the following image as a guide, first construct the resistor with the LINE command while using the grid/snap to connect points until the resistor is created. Be sure Snap mode is active for this procedure. Then issue the BLOCK command, which will display the dialog box illustrated on the right in the following image. This will be an internal block to the drawing ELECT_SYMB; name the block RESISTOR. In the Select objects area of the dialog box, select all lines that represent the resistor (if you labeled the symbols, do not select the text as part of the block). Select the Retain radio button so that the symbol will remain visible in your drawing. Returning to the dialog box will show the objects selected in a small image icon on the right of the dialog box. Next, identify the base point of the resistor at the (A). It is very important to use Snap or an Object Snap mode when picking the base point. Under Settings, change the Block unit to Inches. When finished with this dialog box, click the OK button to create the block. Repeat this procedure for the remainder of the electrical symbols. When complete, save the drawing as ELECT_SYMB.

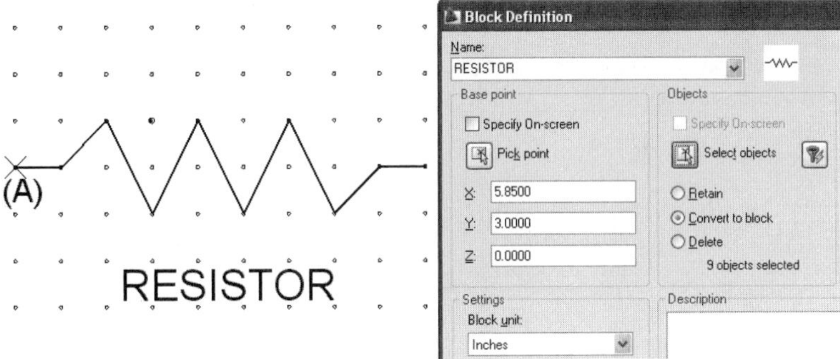

Figure 16.95

Step 4

Begin another new drawing called ELECTRICAL_SCHEM1. Activate the DesignCenter. Click the Tree View icon and expand the DesignCenter to display the folders on your hard drive. Locate the correct folder in which the drawing ELECT_SYMB was saved and click it. Double-click ELECT_SYMB and select Blocks to display the internal blocks, as illustrated in the following image. These internal blocks can now be inserted in any AutoCAD drawing file.

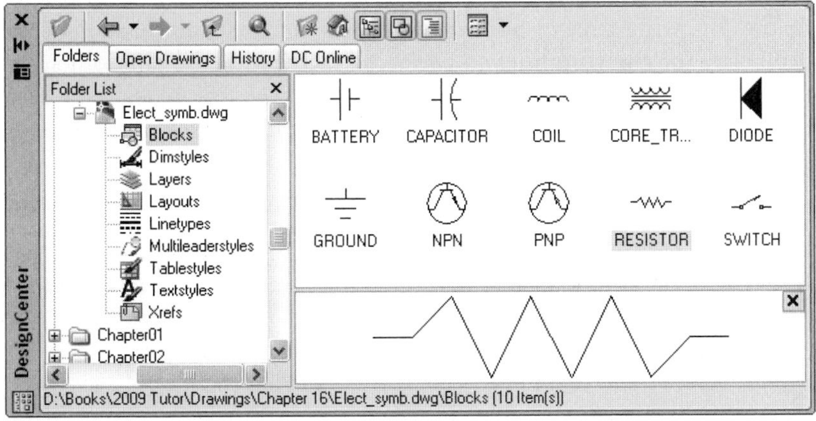

Figure 16.96

Step 5

Make the Blocks layer current and begin dragging and dropping the blocks into the drawing, as shown in the following image. Double-clicking an image of a block located in the DesignCenter launches the Insert dialog box should you need to change the scale or rotation angle of the block. In the case of the electrical schematic, it is not critical to place the blocks exactly, because they are moved to better locations when connected to lines that represent wires.

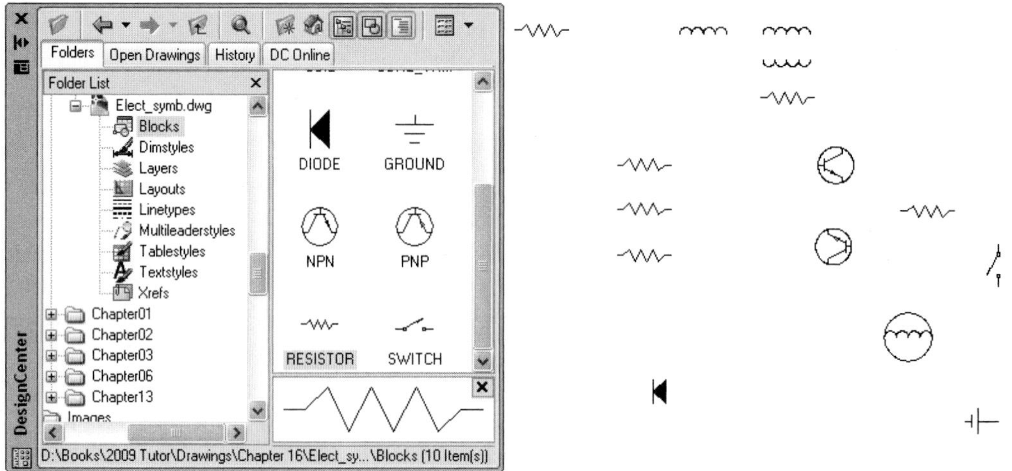

Figure 16.97

Step 6

Make the Wires layer current and begin connecting all the blocks to form the schematic, as shown in the following image. Some type of Object Snap mode such as Endpoint or Insert must be used for the wire lines to connect exactly with the endpoints of the blocks. If spaces get tight or if the blocks look too crowded, move the block to a better location and continue drawing lines to form the schematic. Hint: The STRETCH command is a fast and easy way to reposition the blocks without having to reconnect the wires.

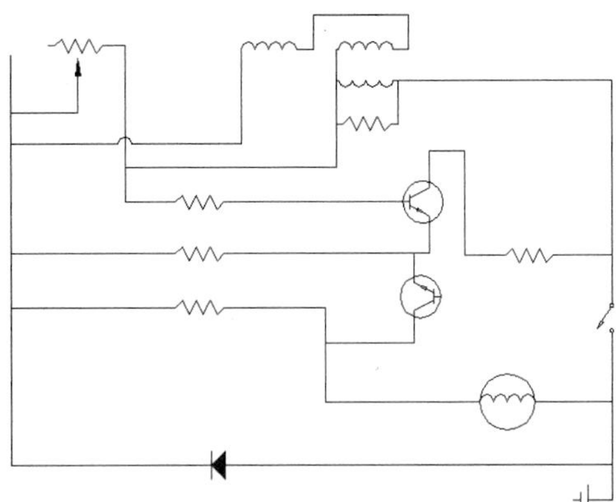

Figure 16.98

Step 7

Add text (place it on the Blocks layer) identifying all resistors, transistors, diodes, and switches by number. The text height used was 0.12; however, this could vary depending on the overall size of your schematic. Also, certain areas of the schematic show connections with the presence of a dot at the intersection of the connection. Use the DONUT command with an inside diameter of 0.00 and an outside diameter of 0.075. Place donuts on the Wires layer in all locations, as shown in the following image.

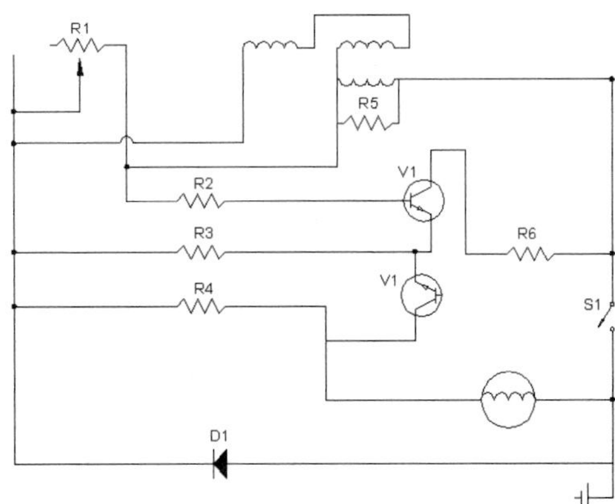

Figure 16.99

TUTORIAL EXERCISE: SWITCH PLATE.DWG

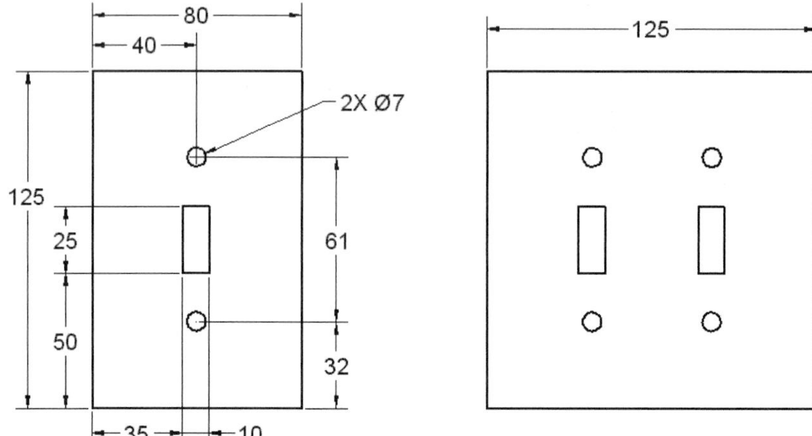

Figure 16.100

PURPOSE

This tutorial is designed to link an Array action to a dynamic block. As the dynamic block increases in width, items associated with the block can be added, depending on the width.

SYSTEM SETTINGS

No special system settings need to be made for this exercise.

LAYERS

No special layers need to be created for this exercise.

SUGGESTED COMMANDS

This application will utilize a switch plate consisting of a basic rectangular shape and three openings: two circular openings that represent screw holes and a rectangular opening used to accept the switch. As the switch plate increases in width, extra openings will automatically be added to the block.

Step 1

Open the drawing file 16_Dynamic_Switch_Plate. A block consisting of a single-opening switch plate is displayed in the drawing editor. Dimensions have been added to the switch plate to calculate the incremental distances. This exercise will illustrate how to stretch the switch plate in 45 mm increments and have additional switch place openings created. Double-click the switch-plate block to activate the Edit Block Definition dialog box and click the OK button to enter the Block Authoring environment.

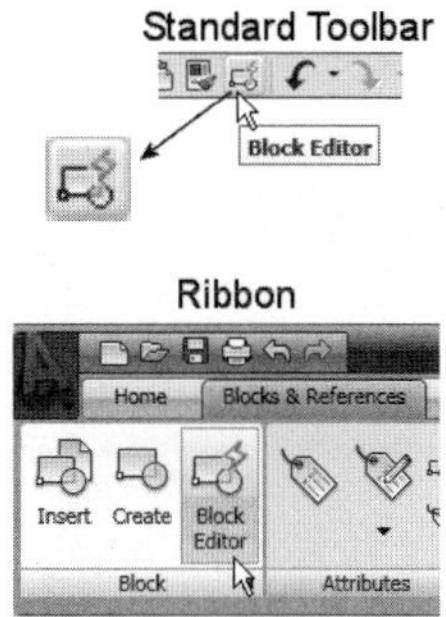

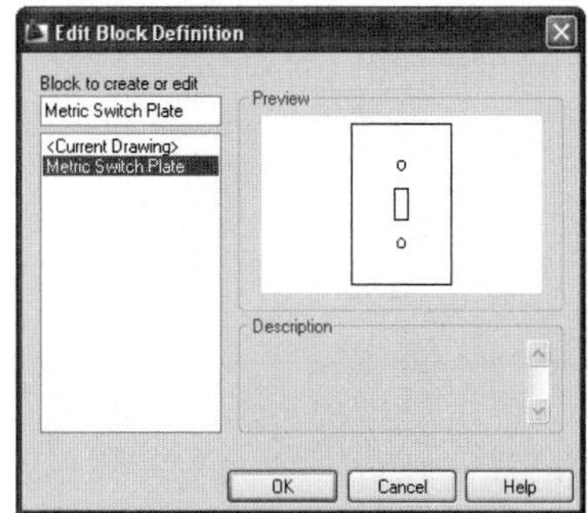

Figure 16.101

Step 2

Add a Linear Parameter to the top of the switch plate. Change the name of the Distance parameter to Plate Width, as shown in the following image.

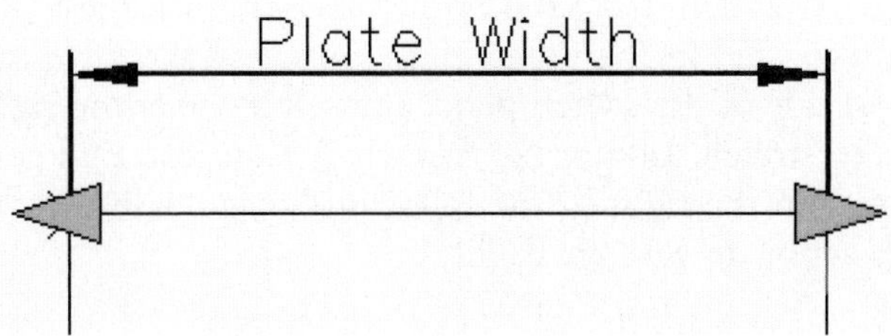

Figure 16.102

Step 3

Next, click the Plate Width parameter and launch the Properties Palette. Make changes under the Value Set area, as shown in the following image; set the Distance type to Increment, the Distance increment to 45, the Distance minimum to 80, and the Distance maximum to 305, as shown in the following image. When the switch plate is at its minimum distance of 80 mm, a single opening is displayed. When the switch plate is at its maximum distance of 305 mm, six openings are displayed.

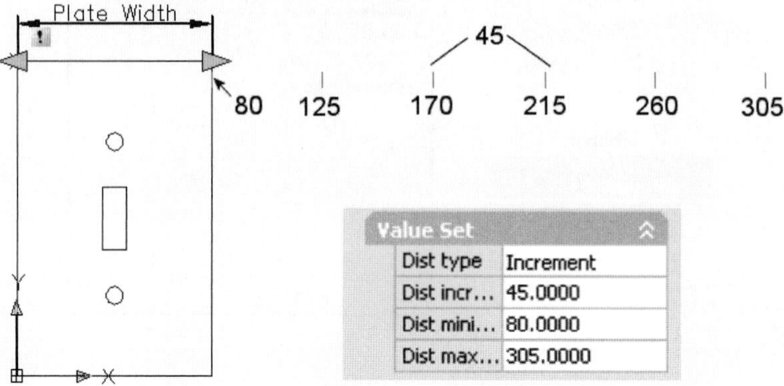

Figure 16.103

Step 4

Next, assign a Stretch Action to the right side of the switch plate, as shown on the left in the following image. After activating the Stretch Action, pick the Plate Width parameter, select the base point located in the upper-right corner of the table, create a crossing box to surround the right side of the switch plate, and select the entire right side of the switch plate as the objects to stretch. Locate the Stretch Action icon in a convenient location of your screen, as shown on the right in the following image. Use the Properties window to rename the Stretch action to Stretch Right.

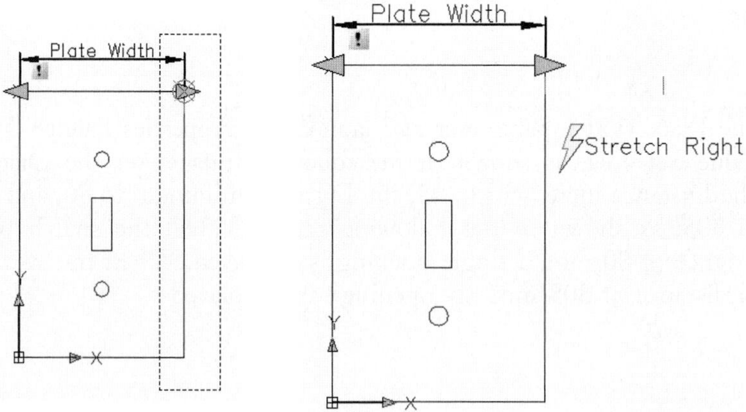

Figure 16.104

Step 5

Now assign a Stretch Action to the left side of the switch plate using the same steps performed in the previous step. After activating the Stretch Action, pick the Plate Width parameter, select the base point located in the upper-left corner of the switch plate, create a crossing box to surround the left side of the switch plate, and select the entire left side of the switch plate as the objects to stretch. Locate the Stretch Action icon in a convenient location of your screen, as shown in the following image. Use the Properties window to rename the Stretch action to Stretch Left.

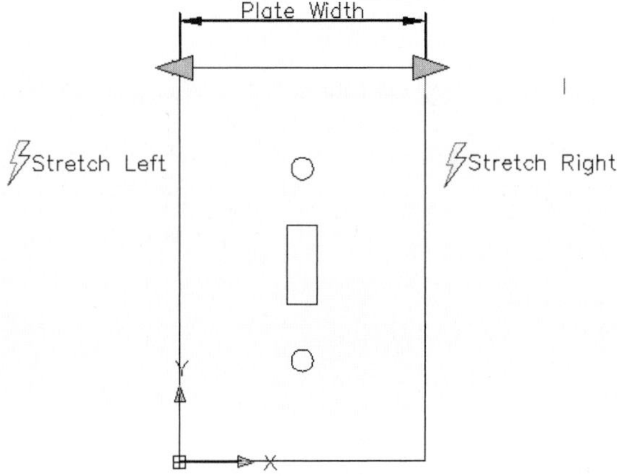

Figure 16.105

Step 6

If you experiment by stretching this new block back in the drawing editor, you will find that the switch plate will stretch; however, the screw holes and switch opening will remain in their original positions. We want these openings to be copied or arrayed at a specified distance, depending on how much the switch plate stretches. To perform this operation, click the Array Action button back in the Block Authoring editor. Select the Plate Width parameter, select all three openings, as shown in the following image, and enter a Column Width distance of 45 mm as the distance from one set of openings to another. Locate the Array Action icon in a convenient screen location.

Click the Save Block Definition button to save these changes, and then click the Close Block Editor button to return to the drawing editor.

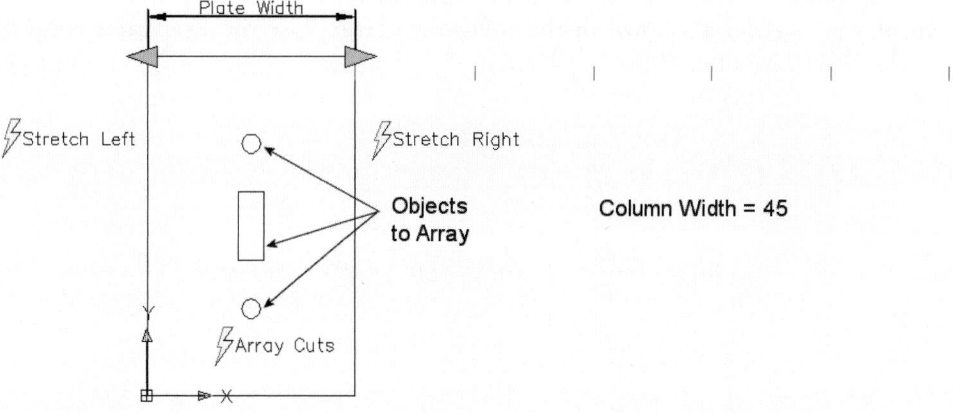

Figure 16.106

Step 7

When you return to the drawing editor, click the block of the switch plate, pick the right arrow grip, and begin stretching the switch plate. When you stretch the switch plate to a new length of 305 mm, which represents the maximum distance, a series of six switch openings are created, as shown on the right in the following image. This completes this tutorial exercise on creating a dynamic block of a switch plate.

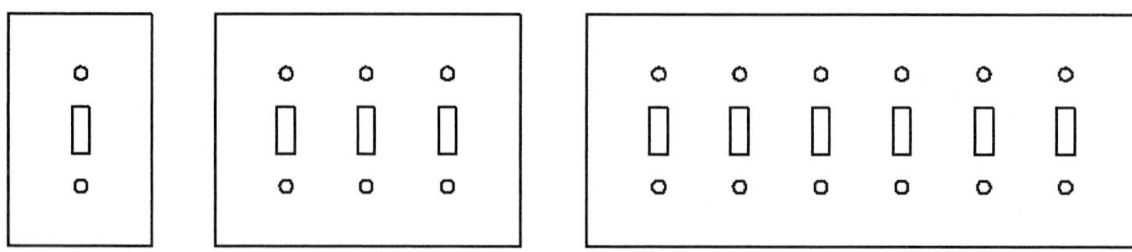

Figure 16.107

TUTORIAL EXERCISE: ELECTRICAL SYMBOLS.DWG

PURPOSE

This tutorial is designed to create visibility states to have numerous electrical symbols contained in a single block name.

SYSTEM SETTINGS

No special system settings need to be made for this tutorial.

LAYERS

No special layers need to be created for this tutorial.

SUGGESTED COMMANDS

The concept of creating visibility states in dynamic blocks begins with the creation of a block using the Edit Block Definition dialog box. You then create a visibility state while inside the Block Authoring environment. During this process, you insert an electrical symbol and immediately explode it into individual objects. You then enter the Visibility States dialog box and change the name of the current state to match that of the electrical symbol. You then create a new visibility state, name this state the next electrical symbol, and hide all existing objects in this new state. When you return to the Block Authoring environment, you insert the next electrical symbol, explode it, and follow the same process for arranging a number of electrical symbols in the same visibility state, as shown in the following image.

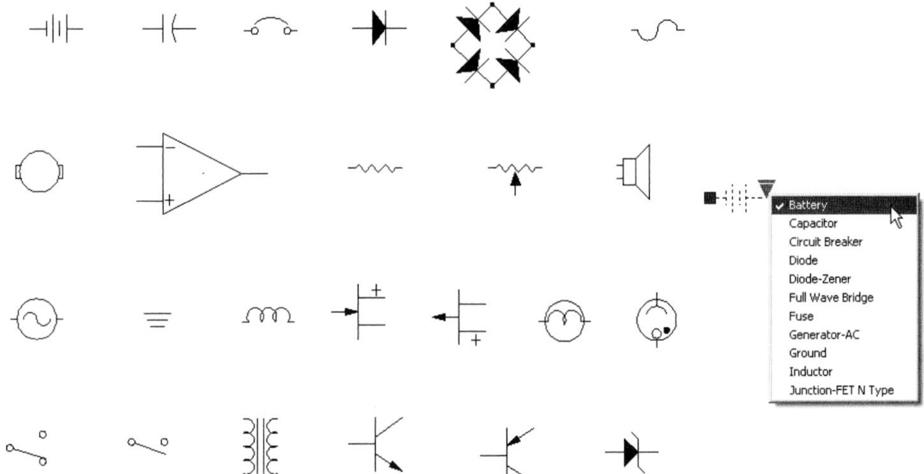

Figure 16.108

Step 1

Before creating visibility states, you must create a new general block name that will hold all individual electrical symbols. You must also enter the block editor to create the visibility states. First, click the Block Editor button located in the Standard toolbar or the Ribbon, as shown on the left in the following image. When the Edit Block Definition dialog box appears, as shown on the right in the following image, enter Electrical Symbols as the block name and click the OK button. This action opens the new block name in the Block Editor.

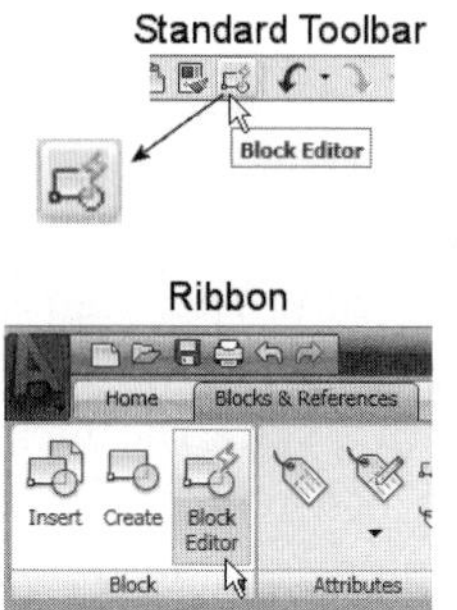

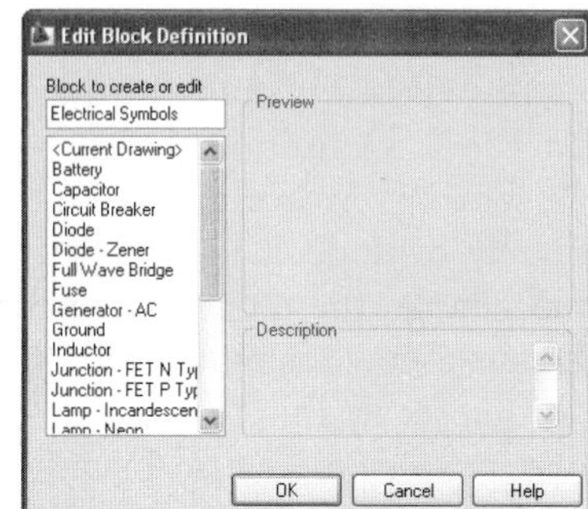

Figure 16.109

Step 2

Insert the first block, called Battery, in the Block Editor, as shown in the following image. Use an insertion point of 0,0. In fact, all individual electrical blocks will be inserted at 0,0 for consistency. Once the block appears in the Block Editor, use the EXPLODE command to break up the block into individual objects. Then, locate the Visibility Parameter in the Parameters tab of the Block Authoring Palette and place the marker to the right and below the battery symbol, as shown in the following image.

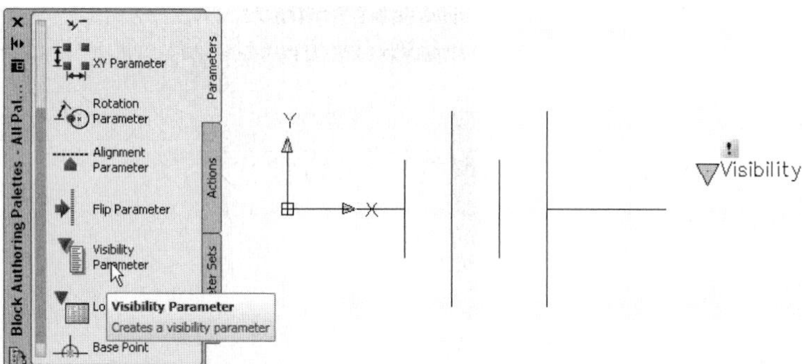

Figure 16.110

Step 3

Notice in the upper-right corner of the Block Authoring Palette the appearance of a toolbar. Use this toolbar for managing visibility states. First, click the Manage Visibility States button, as shown in the following image.

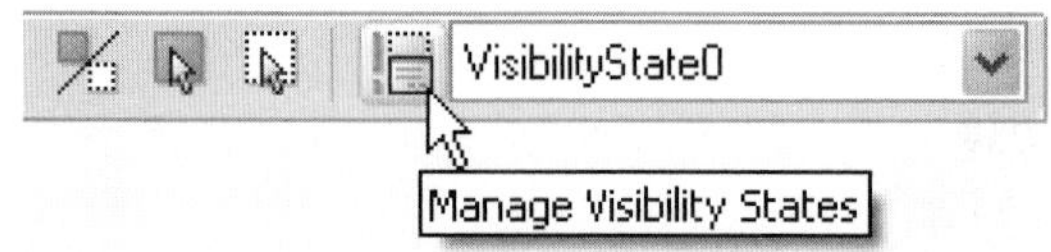

Figure 16.111

Step 4

This launches the Visibility States dialog box. When you use this dialog box for the first time, a default state called VisibilityState0 is already created. Use the Rename button to change the name of this visibility state to Battery. This state coincides with the object currently displayed in the Block Authoring screen. Next, a new visibility state needs to be created by clicking the New button in the Visibility States dialog box.

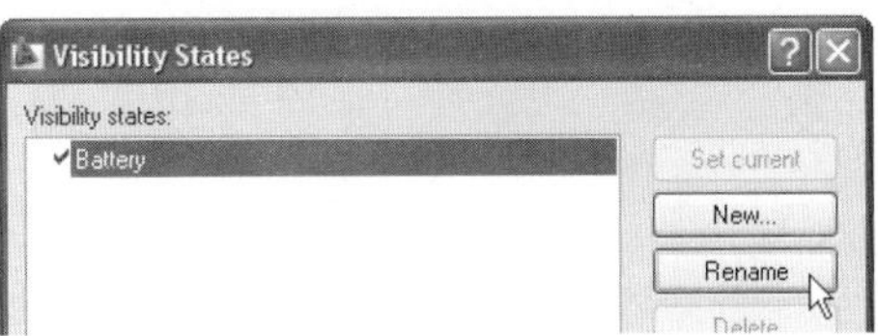

Figure 16.112

Step 5

This launches the New Visibility State dialog box, as shown on the left in the following image. In the Visibility state name field, enter the name Capacitor. Also, be sure to check the radio button next to Hide all existing objects in new state. This will hide all objects when you return to the Block Editor. Click the OK button to continue.

Step 6

When you return to the Block Editor, insert the block Capacitor at an insertion point of 0,0 and explode the block back into its individual objects. The objects that remain in the Block Editor form the Capacitor visibility state.

Launch the Manage Visibility States dialog box again. Create another new visibility state called Circuit Breaker, and make sure Hide all existing objects in new state is selected. Insert the Circuit Breaker block at an insertion point of 0,0 and explode the block back into its individual objects.

Repeat the procedure for creating more visibility states based on the other electrical block symbols. Close the Block Editor and save the changes to the block.

Insert a Battery block in the drawing and test the visibility states by selecting them from the list, as shown on the right in the following image. If you created visibility states of all electrical symbols, your list will be longer.

Figure 16.113

PROBLEMS FOR CHAPTER 16

PROBLEM 16–1

Construct this battery pack using the block library supplied with AutoCAD: Basic Electronics. dwg. This drawing is not to scale.

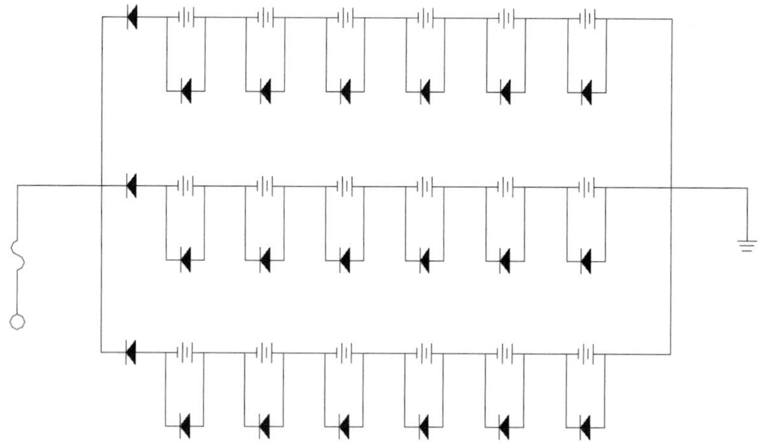

PROBLEM 16–2

Construct this electrical schematic using the block library supplied with AutoCAD: Basic Electronics.dwg. This drawing is not to scale.

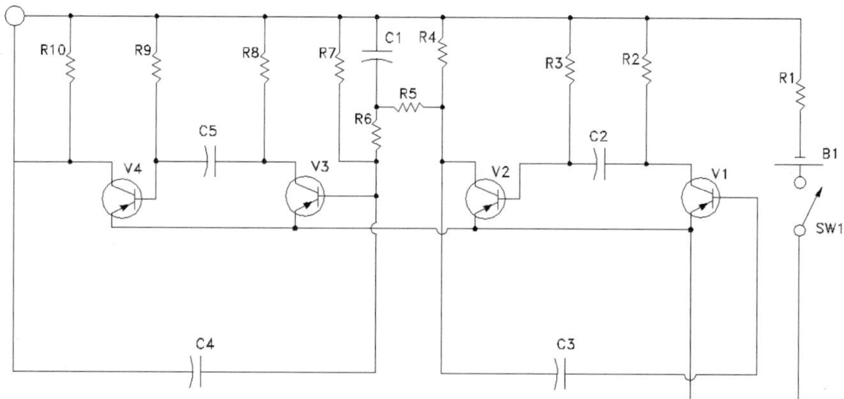

PROBLEM 16–3

Construct this electrical schematic using the block library supplied with AutoCAD: Basic Electronics.dwg. This drawing is not to scale.

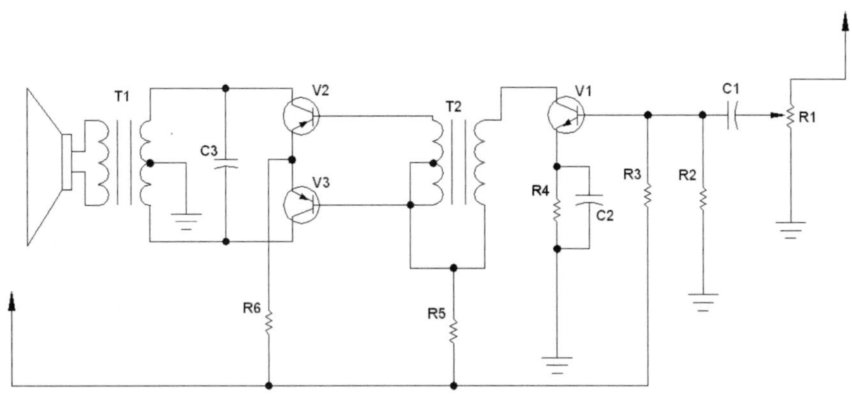

PROBLEM 16–4

Construct this logic gate schematic using the following block libraries supplied with AutoCAD: CMOS Integrated Circuits.dwg and Basic Electronics.dwg. This drawing is not to scale.

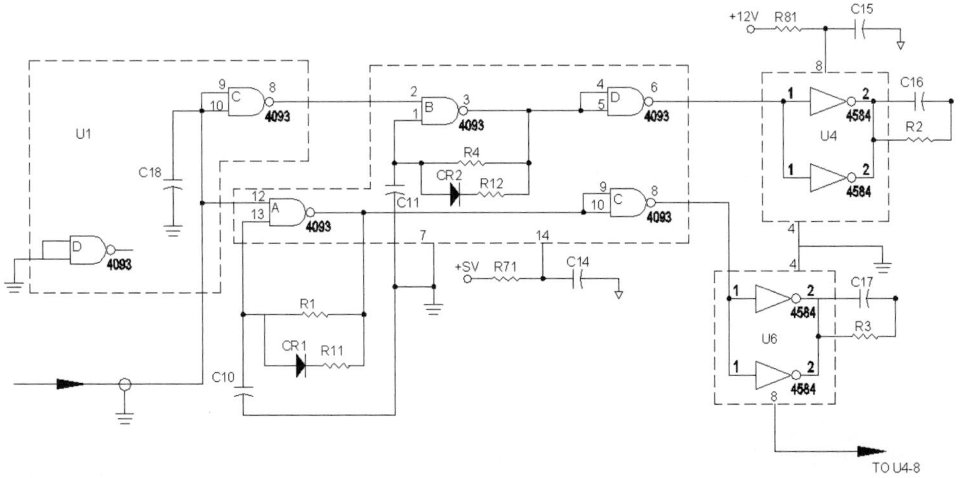

INTERMEDIATE LEVEL DRAWINGS

PROBLEM 16–5

Construct this instrumentation pipe diagram using the following block library supplied with AutoCAD: Pipe Fittings.dwg. This drawing is not to scale.

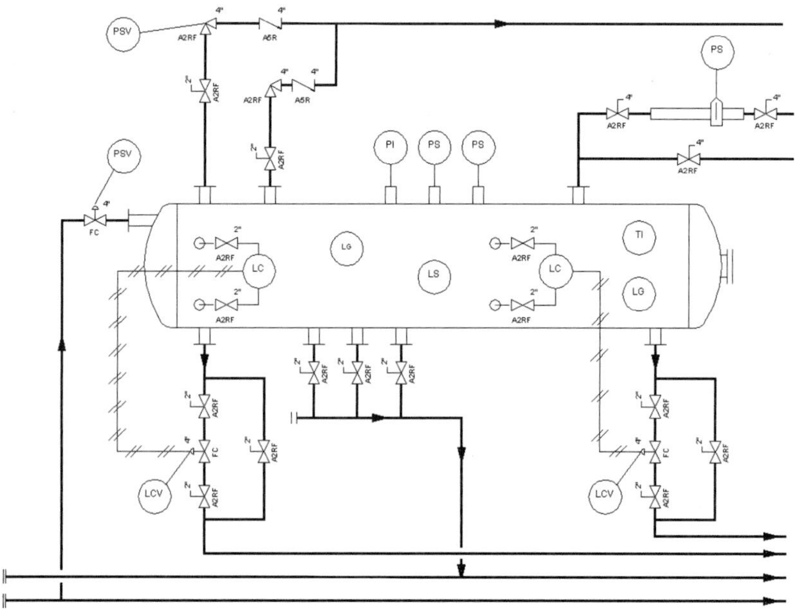

PROBLEM 16–6

Construct the floor plan of the house using the dimensions provided. Create the appropriate layers to separate the walls from the dimensions, and so on. Once the floor plan is completed, create an interior plan consisting of furniture and appliances. Place the blocks in their appropriate spaces using the designated room titles. Use the following block libraries supplied with AutoCAD: House Designer.dwg, Home – Space Planner.dwg, and Kitchens.dwg.

Use the following suggestions for the creation of doors and windows:

All windows measure 2'-8" wide.

All bedroom doors measure 2'-6" wide.

The bathroom door measures 2'-0" wide.

The main entrance door measures 3'-0" wide.

The kitchen and laundry doors each measure 2'-8" wide.

All closet openings measure 4'-0" wide.

All interior walls measure 4".

The exterior walls with brick veneer measures 5".

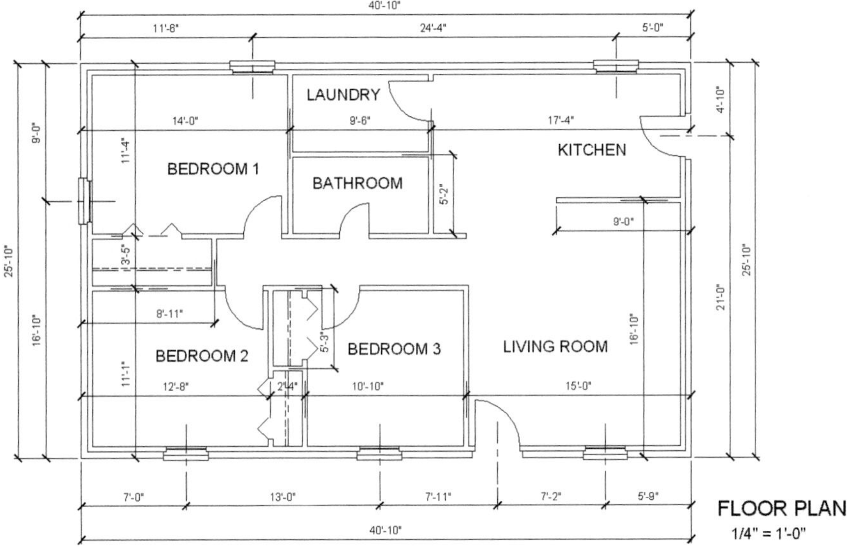

CHAPTER 17

Working with Attributes

This chapter begins the study of how to create, display, edit, and extract attributes. Attributes consist of intelligent text data that is attached to a block. The data could consist of part description, part number, catalog number, and price. Whenever the block is inserted in a drawing, you are prompted for information that, once entered, becomes attribute data. Attributes could also be associated with a title block for entering such items as drawing name, who created, checked, revised, and approved the drawing, drawing date, and drawing scale. Once included in the drawing, the attributes can be extracted and shared with other programs.

WHAT ARE ATTRIBUTES?

An attribute may be considered a label that is attached to a block. This label is called a tag and can contain any type of information that you desire. Examples of attribute tags are illustrated on the left in the following image. The tags are RESISTANCE, PART_NAME, WATTAGE, and TOLERANCE. They relate to the particular symbol they are attached to, in this case an electrical symbol of a resistor. Attribute tags are placed in the drawing with the symbol. When the block, which includes the tags and symbol, is inserted, AutoCAD requests the values for the attributes. The same resistor with attribute values is illustrated on the right in the following image. When you create the attribute tags, you determine what information is requested, the actual prompts, and the default values for the information requested. Once the values are provided and inserted in the drawing, the information contained in the attributes can be displayed, extracted, and even shared with spreadsheet or database programs.

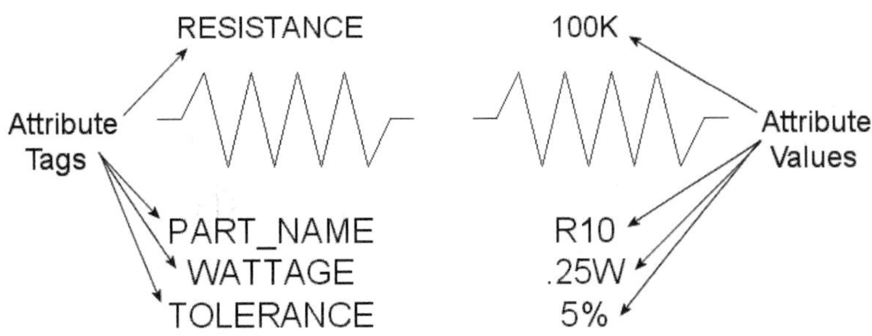

Figure 17.1

Once attributes are inserted in the drawing, the following additional tasks can be performed on attributes:

1. Attributes can be turned on, turned off, or displayed normally through the use of the ATTDISP command.
2. The individual attributes can be changed through the Enhanced Attribute Editor dialog box, which is activated through the EATTEDIT command.
3. Attributes can be globally edited using the -ATTEDIT and FIND commands.
4. Once edited, attributes can be extracted into tables or text files through the Data Extraction wizard, which is activated by the DATAEXTRACTION command.
5. The text file created by the attribute extraction process can be exported to a spreadsheet or database program.

The following commands, which will be used throughout this chapter, assist in the creation, editing, and manipulation of attributes:

ATTDEF—Activates the Attribute Definition dialog box used for the creation of attributes

ATTDISP—Used to control the visibility of attributes in a drawing

-ATTEDIT—Used to edit attributes singly or globally; the hyphen (-) in front of the command activates prompts from the command line

FIND—Used to find and replace text including attribute values

BATTMAN—Activates the Block Attribute Manager dialog box, used to edit attribute properties of a block definition

EATTEDIT—Activates the Enhanced Attribute Editor dialog box, used to edit the attributes of a block

EATTEXT—Activates the Attribute Extraction wizard, used to extract attributes

CREATING ATTRIBUTES THROUGH THE ATTRIBUTE DEFINITION DIALOG BOX

Selecting Define Attributes from the Block cascading menu on the Draw pull-down menu (or the Menu Browser), shown on the left in the following image, activates the Attribute Definition dialog box, illustrated on the right in the following image. The following components of this dialog box will be explained in this section: Attribute Tag, Attribute Prompt, Attribute Default, and Attribute Mode.

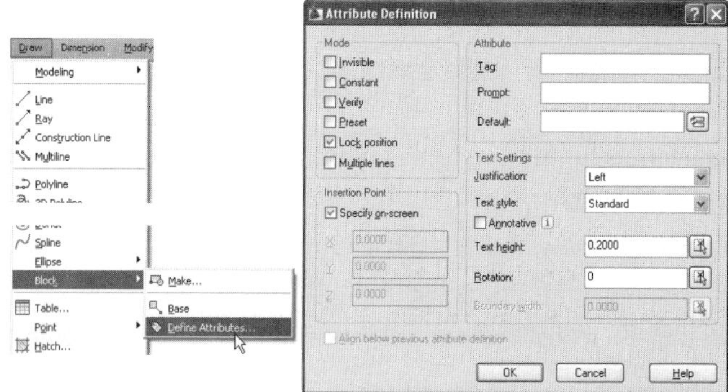

Figure 17.2

ATTRIBUTE TAG

A tag is the name given to an attribute. Typical attribute tags could include PART_NAME, CATALOG_NUMBER, PRICE, DRAWING_NAME, and SCALE. The underscore is used to separate words because spaces are not allowed in tag names.

ATTRIBUTE PROMPT

The attribute prompt is the text that appears on the text line when the block containing the attribute is inserted in the drawing. If you want the prompt to be the same as the tag name, enter a null response by leaving the Prompt field blank. If the Constant mode is specified for the attribute, the prompt area is not available.

ATTRIBUTE DEFAULT

The attribute default is the default value displayed when the attribute is inserted in the drawing. This value can be accepted or a new value entered as desired. The attribute value is handled differently if the attribute mode selected is Constant or Preset.

ATTRIBUTE MODES

Invisible—This mode is used to determine whether the label is invisible when the block containing the attribute is inserted in the drawing. If you later want to make the attribute visible, you can use the ATTDISP command.

Constant—Use this mode to give every attribute the same value. This might be very useful when the attribute value is not subject to change. However, if you designate an attribute to contain a constant value, it is difficult to change it later in the design process.

Verify—Use this mode to verify that every value is correct. This is accomplished by prompting the user twice for the attribute value.

Preset—This allows for the presetting of values that can be changed. However, you are not prompted to enter the attribute value when inserting a block. The attribute values are automatically set to their default values.

Lock Position—This mode locks the location of the attribute located inside of the block reference. When unlocked, the attribute can be moved relative to the rest of the block using grips.

Multiple lines—Specifies that the attribute value can contain multiple lines of text.

Note: The effects caused by invoking the Verify and Preset modes are apparent only when the ATTDIA system variable is off (the Enter Attributes dialog box is not displayed). When entering data at the command prompt, you will be asked twice for data that is to be verified and you will not be asked at all to supply data for attributes that are preset.

TUTORIAL EXERCISE: B TITLE BLOCK.DWG

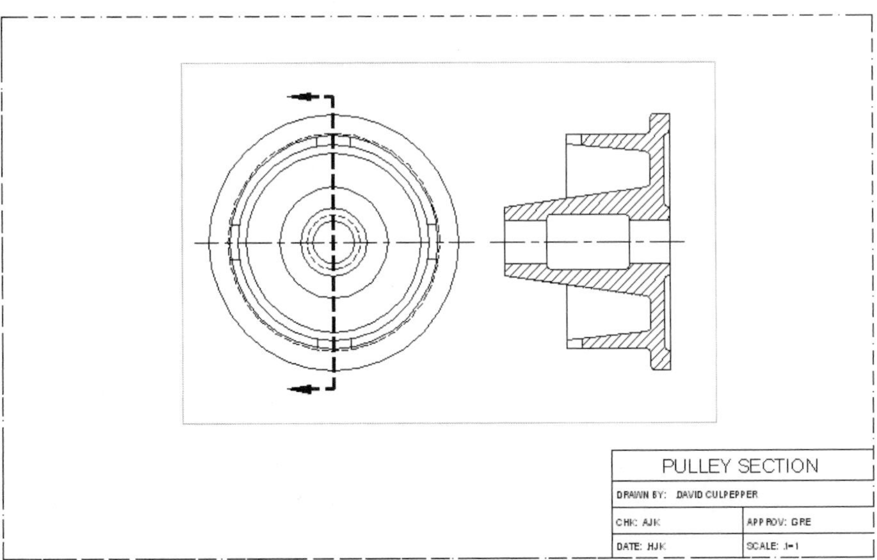

Figure 17.3

PURPOSE

This tutorial is designed to assign attributes to a title block. The completed drawing is illustrated in the previous image.

SYSTEM SETTINGS

Because the drawings B Title Block and 17_Pulley Section are provided on the CD, all units and drawing limits have already been set.

LAYERS

All layers have already been created for this drawing.

SUGGESTED COMMANDS

Open the drawing B Title Block. Using the Attribute Definition dialog box (activated through the ATTDEF command), assign attributes consisting of the following tag names: DRAWING_NAME, DRAWN_BY, CHECKED_BY, APPROVED_BY, DATE, and SCALE. Save this drawing file with its default name. Open the drawing 17_Pulley Section, insert the title block in this drawing, and answer the attribute prompts designed to complete the title block information.

Step 1

Open the drawing B Title Block and observe the title block area, shown in the following image. Various point objects are present to guide you in the placement of the attribute information. All points are located on the layer Points.

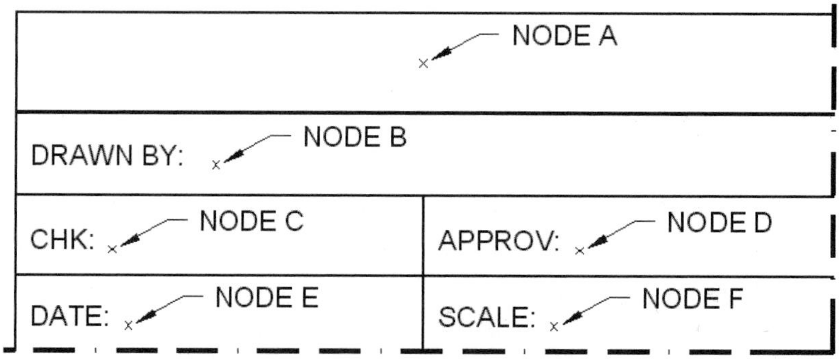

Figure 17.4

Step 2

Activate the Attribute Definition dialog box, as shown on the left in the following image. Leave all items unchecked in the Mode area of this dialog box with the exception of Lock position. In the Attribute area, make the following changes: Enter DRAWING_NAME in the Tag field. In the Prompt field, enter: What is the name of the drawing? In the Default field, enter UNNAMED. In the Text Settings area, change the Justification to Middle and the Height to 0.25 units. When finished, click the OK button. This returns you to your drawing. Using OSNAP-Node, pick the point at "A," as shown on the upper right in the following image. The DRAWING_NAME tag is added to the title block area, as shown on the lower right in the following image.

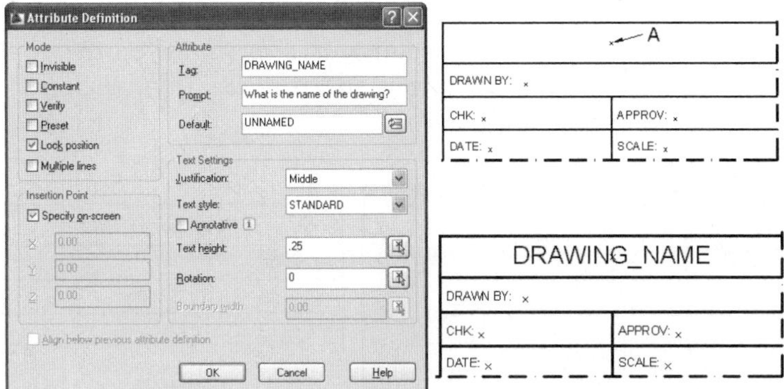

Figure 17.5

Step 3

To define the next attribute, activate the Attribute Definition dialog box, as shown on the left in the following image. Leave all items unchecked in the Mode area of this dialog box except for Lock position. In the Attribute area, make the following changes: Enter DRAWN_BY in the Tag field. In the Prompt field, enter: Who created this drawing? In the Default field, enter UNNAMED. In the Text Settings area, verify that Justification is set to Left, and change the Height to 0.12 units. When finished, click the OK button. This returns you to your drawing. Using OSNAP-Node, pick the point at "B," as shown on the upper right in the following image. The DRAWN_BY tag is added to the title block area, as shown on the lower right in the following image.

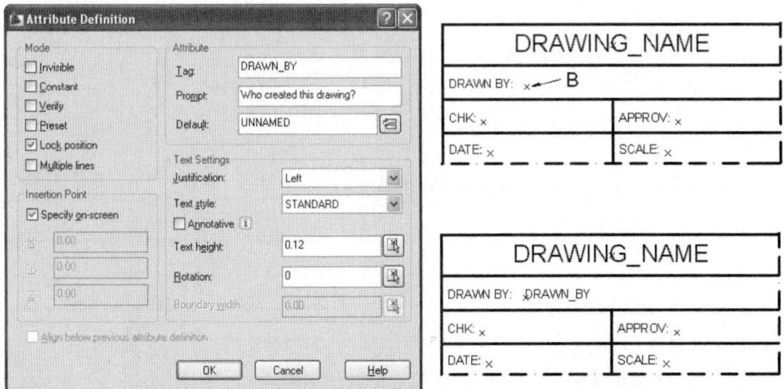

Figure 17.6

Step 4

Activate the Attribute Definition dialog box, as shown on the left in the following image. Leave all items unchecked in the Mode area of this dialog box except for Lock position. In the Attribute area, make the following changes: Enter CHECKED_BY in the Tag field. In the Prompt field, enter: Who will be checking this drawing? In the Default field, enter CHIEF DESIGNER. In the Text Settings area, verify that Justification is set to Left and the Height is 0.12 units. When finished, click the OK button. This returns you to your drawing. Using OSNAP-Node, pick the point at "C," as shown on the upper right in the following image. The CHECKED_BY tag is added to the title block area, as shown on the lower right in the following image.

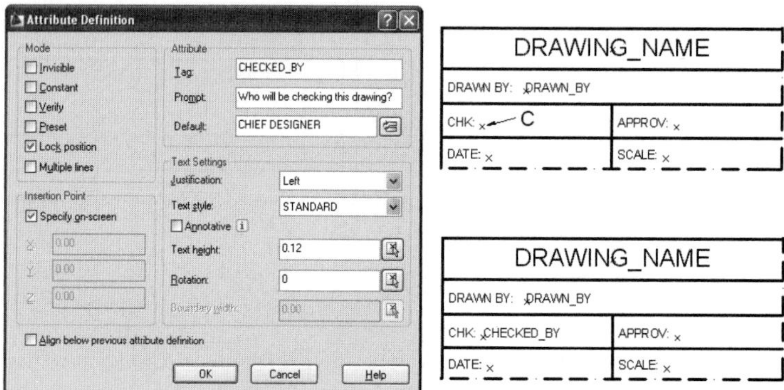

Figure 17.7

Step 5

Activate the Attribute Definition dialog box, as shown on the left in the following image. Leave all items unchecked in the Mode area of this dialog box except for Lock position. In the Attribute area, make the following changes: Enter APPROVED_BY in the Tag field. In the Prompt field, enter: Who will be approving this drawing? In the Default field, enter CHIEF ENGINEER. In the Text Settings area, verify that Justification is set to Left and the Height is 0.12 units. When finished, click the OK button. This returns you to your drawing. Using OSNAP-Node, pick the point at "D," as shown on the upper right in the following image. The APPROVED_BY tag is added to the title block area, as shown on the lower right in the following image.

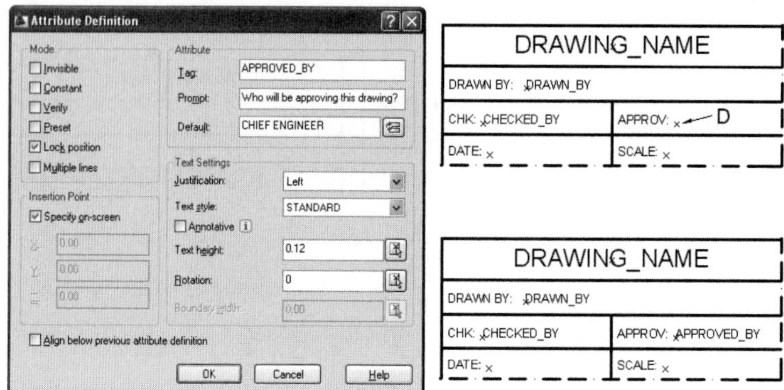

Figure 17.8

Step 6

Activate the Attribute Definition dialog box, as shown on the left in the following image. Leave all items unchecked in the Mode area of this dialog box except for Lock position. In the Attribute area, make the following changes: Enter DATE in the Tag field. In the Prompt field, enter: When was this drawing completed? In the Default field, enter UNDATED. In the Text Settings area, verify that Justification is set to Left and the Height is 0.12 units. When finished, click the OK button. This returns you to your drawing. Using OSNAP-Node, pick the point at "E," as shown on the upper right in the following image. The DATE tag is added to the title block area, as shown on the lower right in the following image.

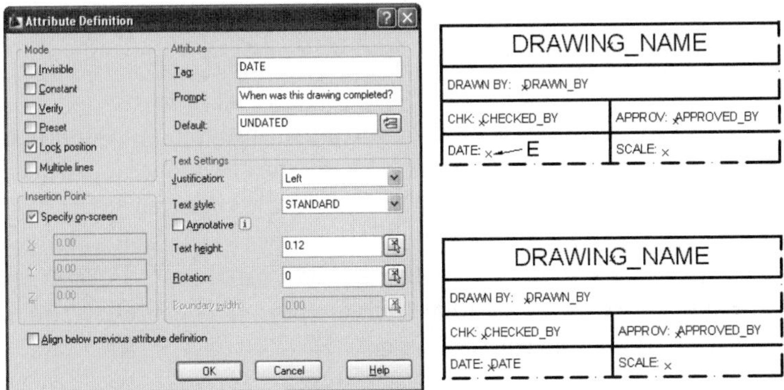

Figure 17.9

Step 7

Activate the Attribute Definition dialog box, as shown on the left in the following image. Leave all items unchecked in the Mode area of this dialog box except for Lock position. In the Attribute area, make the following changes: Enter SCALE in the Tag field. In the Prompt field, enter: What is the scale of this drawing? In the Default field, enter 1 = 1. In the Text Settings area, verify that Justification is set to Left and the Height is 0.12 units. When finished, click the OK button. This returns you to your drawing. Using OSNAP-Node, pick the point at "F," as shown on the upper right in the following image. The SCALE tag is added to the title block area, as shown on the lower right in the following image.

You have now completed creating the attributes for the title block; turn off the Points layer. Close and save the changes to this drawing with its default name.

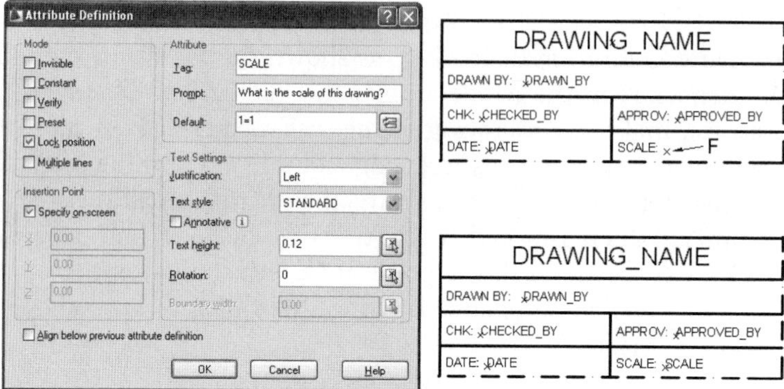

Figure 17.10

Step 8

To test the attributes and see how they function in a drawing, first open the drawing file 17_Pulley Section. Notice that this image is viewed from inside a Layout (or Paper Space). The current page setup is based on the DWF6-ePlot.pc3 file, with the current sheet size as ANSI Expand B (17.00 × 11.00 inches). The viewport that holds the two views of the Pulley has the No Plot state assigned in the Layer Properties Manager dialog box. You could turn the Viewports layer off to hide the viewport if you prefer. You will now insert the title block with attributes in this drawing.

Verify that the system variable ATTDIA is set to 1 (This can be typed in from the keyboard). This allows you to enter your attribute values through the use of a dialog box. Make the Title Block layer current and activate the Insert Block dialog box. Browse for the Title Block drawing just created. Clear the Specify On-screen checkbox in the Insertion point area. This automatically places the title block at the 0,0,0 location of the layout, in the lower-left corner of the printable area indicators. Click OK. Before the title block can be placed, you must first fill in the boxes in the Edit Attributes dialog box, illustrated in the following image (if this dialog box does not appear and you are prompted for values at the command line, it is because ATTDIA is set to 0). Complete all boxes; enter appropriate names and initials as directed.

Note: Pressing the TAB key is a quick way of moving from one box to another while inside any dialog box.

Edit Attributes

Block name: B Title Block

What is the name of the drawing?	PULLEY SECTION
Who created this drawing?	Enter your Name Here
Who will be checking this drawing?	Enter the Initials of a Co-Worker
Who will be approving this drawing?	Enter Your Instructors/Supervisors Name
When was this drawing completed?	Enter Todays Date
What is the scale of this drawing?	1=1

Figure 17.11

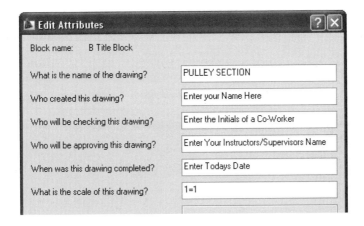

The completed drawing with title block and attributes inserted is illustrated in the following image.

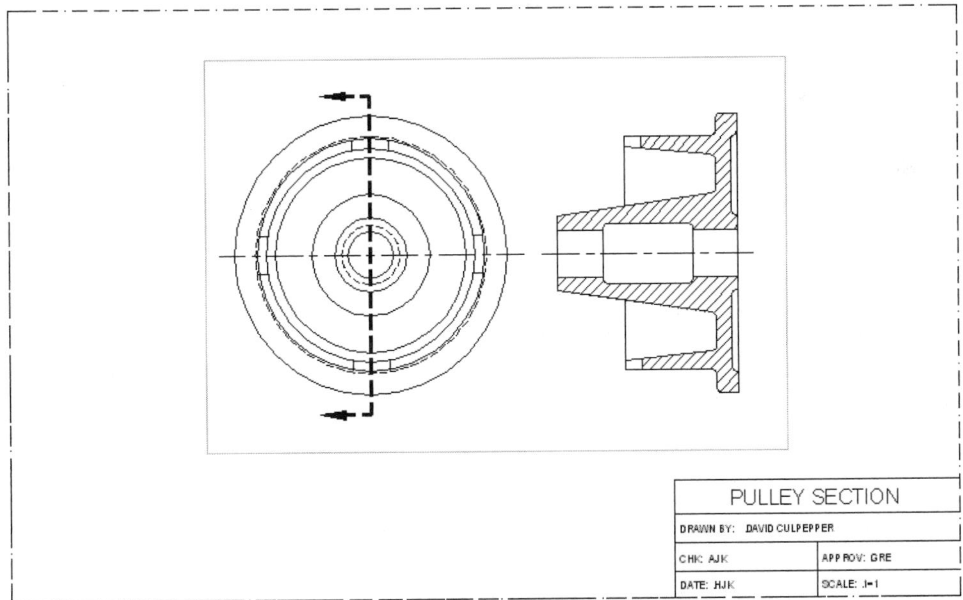

Figure 17.12

CREATING MULTIPLE LINES OF ATTRIBUTES

The previous example illustrated the creation of individual attributes for the title block. Attributes can also be created with multiple lines of text. Located in the Attribute Definition dialog box is a Multiple lines mode. When checked, the default attribute value is grayed out and an Open Multiline Editor button is present. After creating the tag CLIENT_INFORMATION, as shown in the following image, click the Open Multiline Editor button.

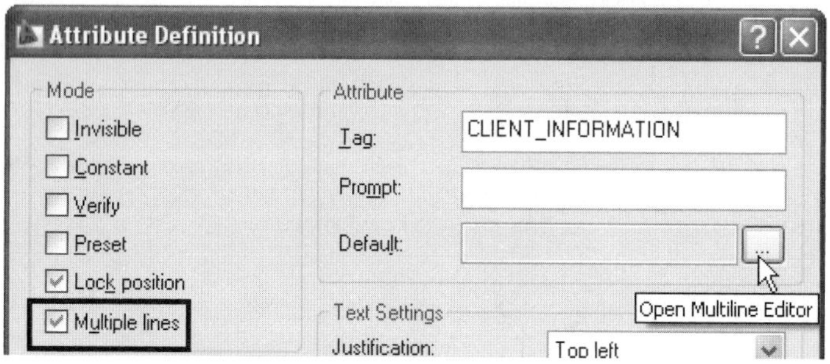

Figure 17.13

A simplified version of the Text Formatting toolbar allows multiple lines of attributes to be created. Illustrated on the left in the following image are a number of information and address attributes that are created. Clicking the OK button returns you to the Attribute Definition dialog box, as shown on the right in the following image. Notice that all of the information created through the Text Formatting is grouped in the Default field.

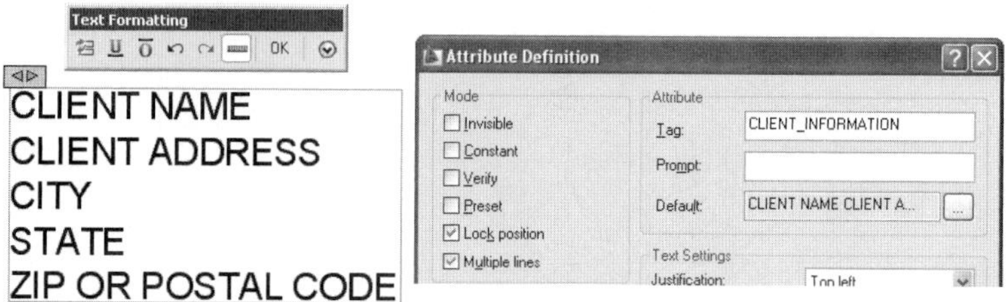

Figure 17.14

After creating the multiple line attribute, the attribute tag is grouped with geometry and a new block definition is created, as shown in the following image that includes the multiple line attribute.

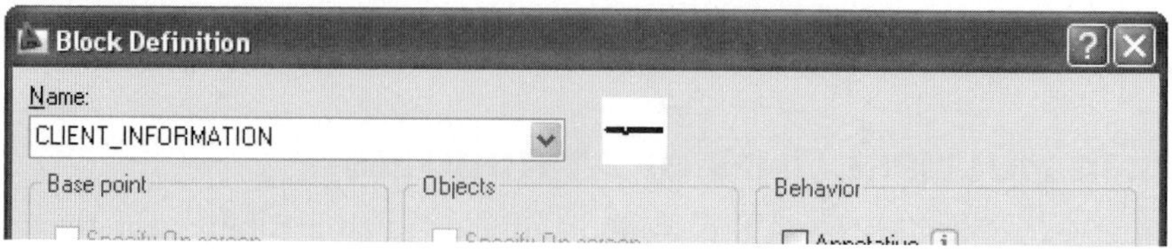

Figure 17.15

When inserting the block containing the multiple line attribute, all lines of the attribute display as shown on the left in the following image. When an insertion point is specified for the block, the Edit Attributes dialog box displays as shown on the right in the following image. To change the various attribute fields, click the Open Multiline Editor button.

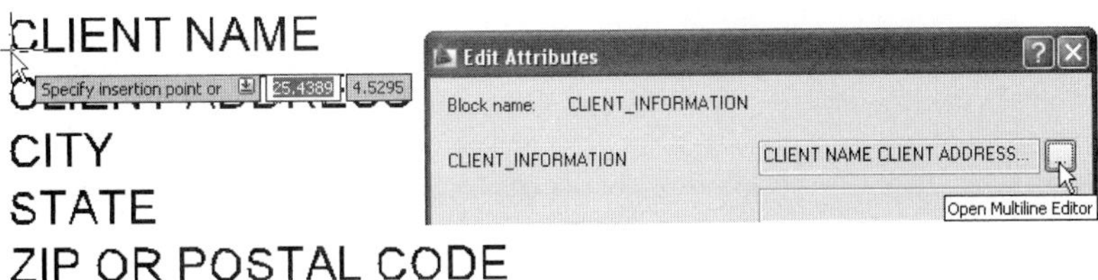

Figure 17.16

When the Text Formatting toolbar displays on the left in the following image, replace each field such as CLIENT NAME and CLIENT ADDRESS, with actual names and locations, as shown on the right in the following image.

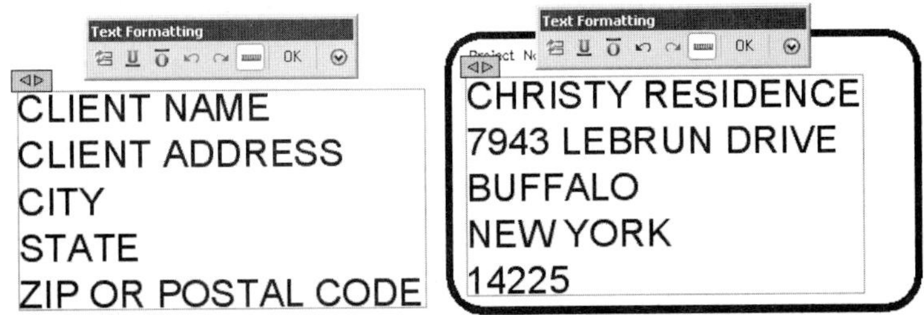

Figure 17.17

FIELDS AND ATTRIBUTES

When defining attributes for a specific drawing object such as a title block, the attribute value can be converted into a field, as shown in the following image. In this figure, the attribute tag (PROJECT_NAME) and prompt (What is the project name?) are created using conventional methods. Before entering a default value, click the Insert field button. This displays the Field dialog box. In this figure, the SheetSet field category is being used to assign a custom property called Project Name.

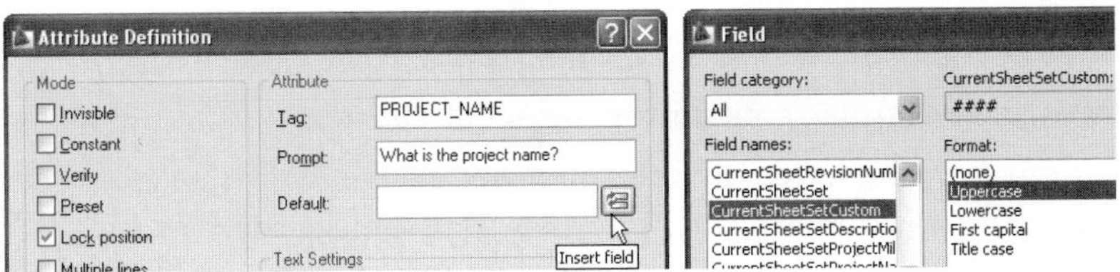

Figure 17.18

When you work with attribute values as fields, the actual default entry is illustrated, as shown in the following image. In this example, #### was inserted into this cell from the Field dialog box in the previous figure. When a title block that contains this attribute is inserted into a drawing, you will be prompted to enter a new value for the PROJECT_NAME tag.

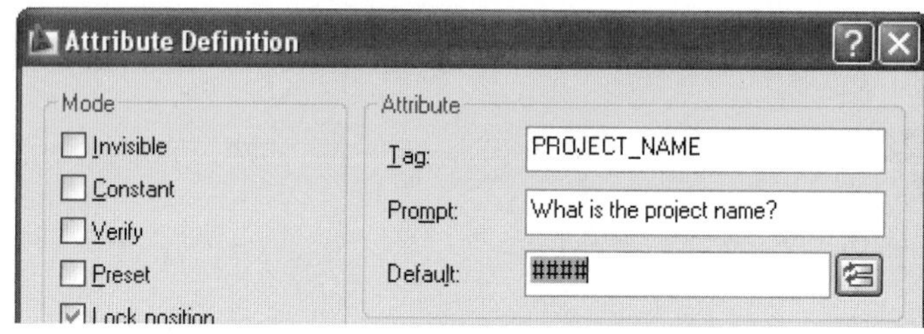

Figure 17.19

CONTROLLING THE DISPLAY OF ATTRIBUTES

You don't always want attribute values to be visible in the entire drawing. The ATTDISP command is used to determine the visibility of the attribute values. This command can be entered from the keyboard or can be selected from the View pull-down menu under Display > Attribute Display, as shown in the following image.

```
Command: ATTDISP
Enter attribute visibility setting [Normal/ON/OFF] <Normal>:
(Enter the desired option)
```

The following three modes are used to control the display of attributes:

VISIBILITY NORMAL

This setting displays attributes based on the mode set through the Attribute Definition dialog box. If some attributes were created with the Invisible mode turned on, these attributes will not be displayed with this setting, which makes this setting popular for displaying certain attributes and hiding others.

VISIBILITY ON

Use this setting to force all attribute values to be displayed on the screen. This affects even attribute values with the Invisible mode turned on.

VISIBILITY OFF

Use this setting to force all attribute values to be turned off on the screen. This is especially helpful in busy drawings that contain lots of detail and text.

All three settings are illustrated in the following image. With ATTDISP set to Normal, attribute values defined as Invisible are not displayed. This is the case for the WATTAGE and TOLERANCE tags. The Invisible mode was turned on inside the Attribute Definition dialog box when they were created. As a result, these values are not visible. With ATTDISP set to on, all resistor values are forced to be visible. When set to off, this command makes all resistor attribute values invisible.

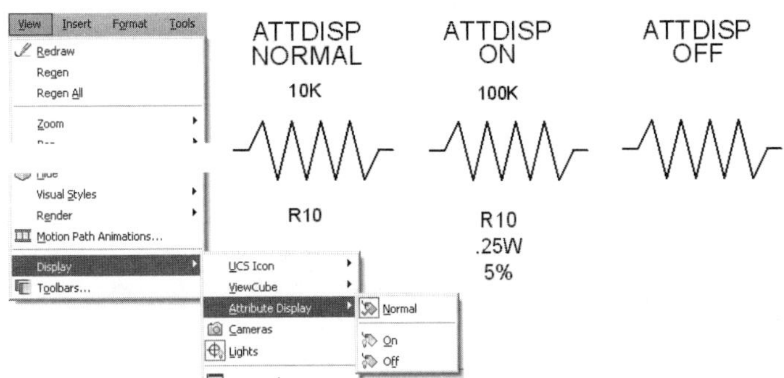

Figure 17.20

EDITING ATTRIBUTES

A number of attribute tools in the form of dialog boxes are available to edit or make modifications on attributes. Two commonly used tools are the Enhanced Attribute Editor dialog box and the Block Attribute Manager dialog box. The Enhanced Attribute Editor dialog box can be displayed by selecting Object from the Menu Browser or Modify pull-down menu, followed by Attribute, and then Single, as shown in the following image. The Block Attribute Manager dialog box can also be selected from this area of the Object > Attribute cascading menu. These tools can also be selected from the Modify II toolbar or from the Ribbon, as shown in the following image.

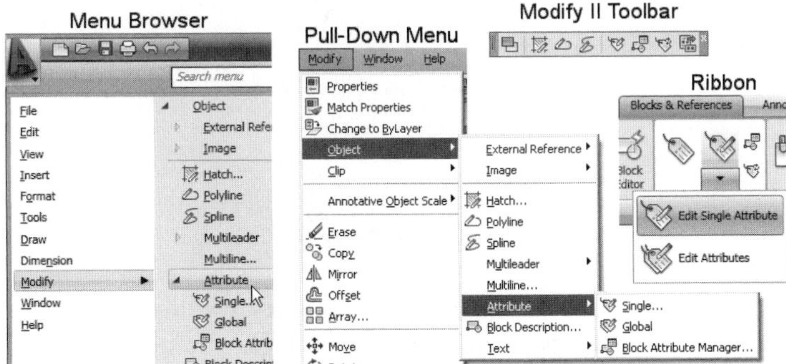

Figure 17.21

The following table gives a brief description of the Attribute modification tools available.

Button	Tool	Command	Function
	Edit Attribute	EATTEDIT	Launches a dialog box designed to edit the attributes in a block reference
	Block Attribute Manager	BATTMAN	Launches a dialog box used for managing attributes for blocks in the current drawing

Button	Tool	Command	Function
	Synchronize Attributes	ATTSYNC	Updates all instances of the selected block with the currently defined attribute properties
	Data Extraction	EATTEXT	Exports the attributes found in a block reference out to a table or external file

THE ENHANCED ATTRIBUTE EDITOR DIALOG BOX

Click the Edit Attribute toolbar button to display the Enhanced Attribute Editor dialog box, as shown in the following image. This dialog box can also be activated by entering EATTEDIT at the command prompt. In the image, the Dryer block was selected. Notice that the attribute value is based on the selected tag. The value can then be modified in the Value field. To change to a different tag, select it with your cursor. Notice that this dialog box has three tabs. The Attribute tab allows you to select the attribute tag that will be edited.

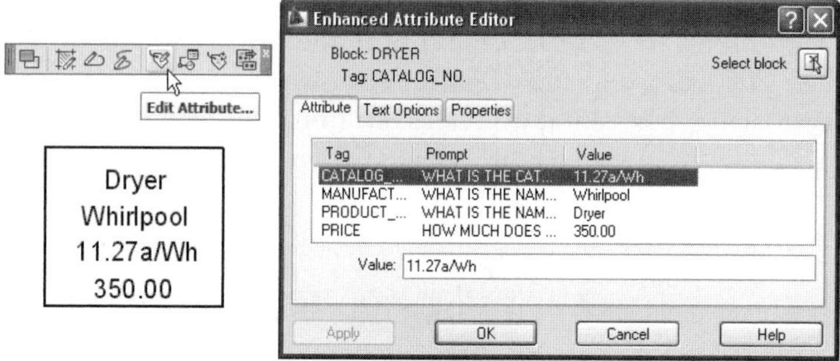

Figure 17.22

Note: Double-clicking a block containing attribute values also launches the Enhanced Attribute Editor dialog box, illustrated in the previous image.

Clicking the Text Options tab displays the dialog box illustrated on the left in the following image. Use this area to change the properties of the text associated with the attribute tag selected, such as text style, justification, height, rotation, and so on. Once the changes are made, click the Apply button.

Clicking the Properties tab displays the dialog box illustrated on the right in the following image. Use this area to change properties such as layer, color, lineweight, and so on. When you have completed the changes to the attribute, click the Apply button.

All three tabs of the Enhanced Attribute Editor dialog box have a Select block button visible in the upper-right corner of each dialog box. Once you have edited a block, use this button to select a different block for editing, if desired.

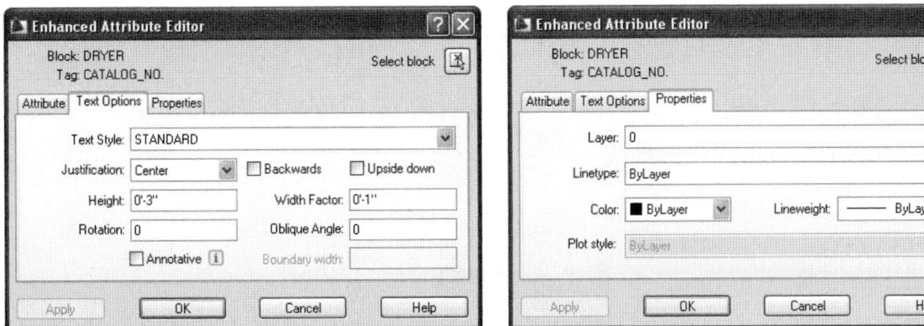

Figure 17.23

THE BLOCK ATTRIBUTE MANAGER

Clicking the Block Attribute Manager toolbar button or entering BATTMAN at the command prompt displays the Block Attribute Manager dialog box, as shown in the following image. This dialog box is displayed as long as attributes are defined in your drawing. It does not require you to pick any blocks because it searches the database of the drawing and automatically lists all blocks with attributes. These are illustrated in the drop-down list, also shown in the following image. As will be shown, the Block Attribute Manager dialog box allows you to edit attributes globally.

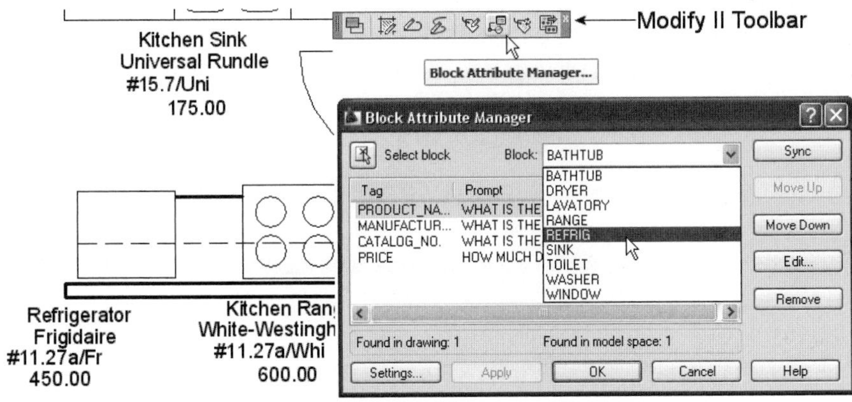

Figure 17.24

Clicking the Edit button displays the Edit Attribute dialog box, as shown in the following image. Three tabs similar to those in the Enhanced Attribute Editor dialog box are displayed. Use these tabs to edit the attribute characteristics, the text options (style, justification, etc.), or the properties (layer, etc.) of the attribute. Notice at the bottom of the dialog box that, as changes are made, you automatically see these changes because the Auto preview changes checkbox is selected.

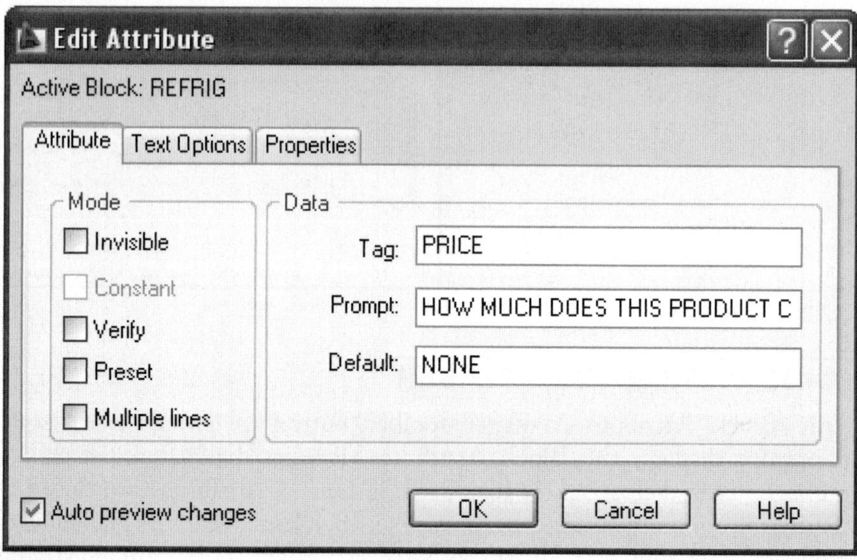

Figure 17.25

Clicking the Settings button in the main Block Attribute Manager dialog box displays the Settings dialog box, as shown on the left in the following image. If you click one of the properties such as Height and click the OK button, this property will be displayed in the main Block Attribute Manager dialog box, as shown on the right in the following image.

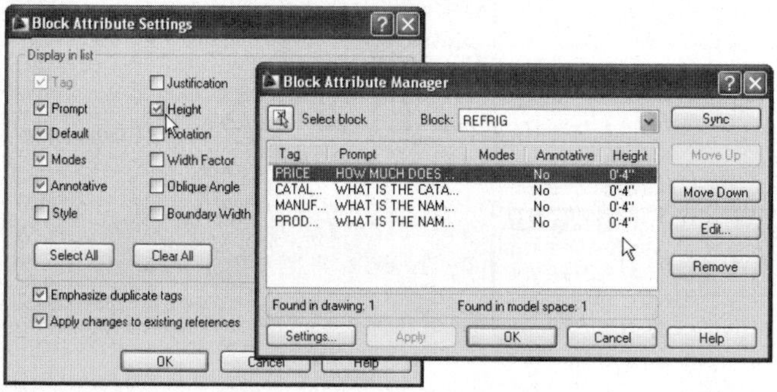

Figure 17.26

TUTORIAL EXERCISE: 17_BLOCK ATTRIB MGR.DWG

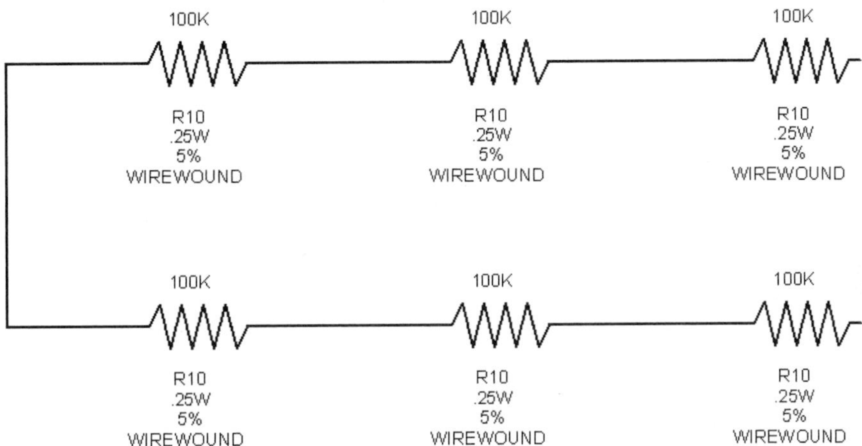

Figure 17.27

PURPOSE

This tutorial is designed to modify the properties of all attributes with the Block Attribute Manager dialog box, using the drawing illustrated in the previous image.

SYSTEM SETTINGS

All units and drawing limits have already been set. Attributes have also been assigned to the resistor block symbols.

LAYERS

All layers have already been created for this drawing.

SUGGESTED COMMANDS

The Block Attribute Manager dialog box will be used to turn the Invisible mode on for a number of attribute values. These changes will be automatically seen as they are made.

Step 1

Open the drawing file 17_Block Attrib Mgr, as shown in the previous image. The following attribute tags need to have their Invisible modifier turned on: TYPE, TOLERANCE, and WATTAGE. Activate the Block Attribute Manager dialog box (BATTMAN command), as shown in the following image, and select the TYPE tag. Then click the Edit button.

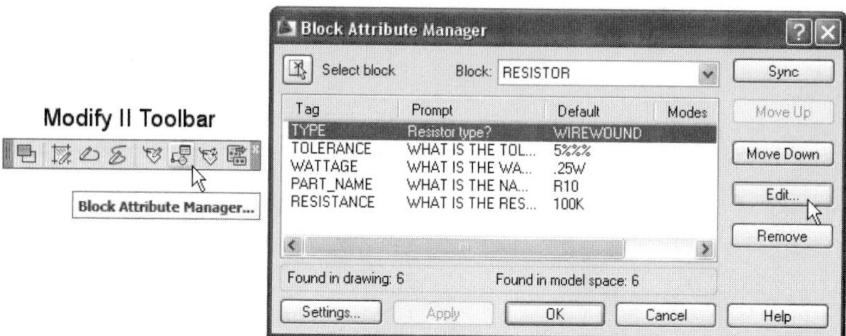

Figure 17.28

Step 2

This takes you to the Edit Attribute dialog box, as shown in the following image. In the Attribute tab, place a check next to the Invisible mode and click the OK button. Notice that the changes are global and take place automatically. The attribute value WIREWOUND is no longer visible for any of the resistor blocks.

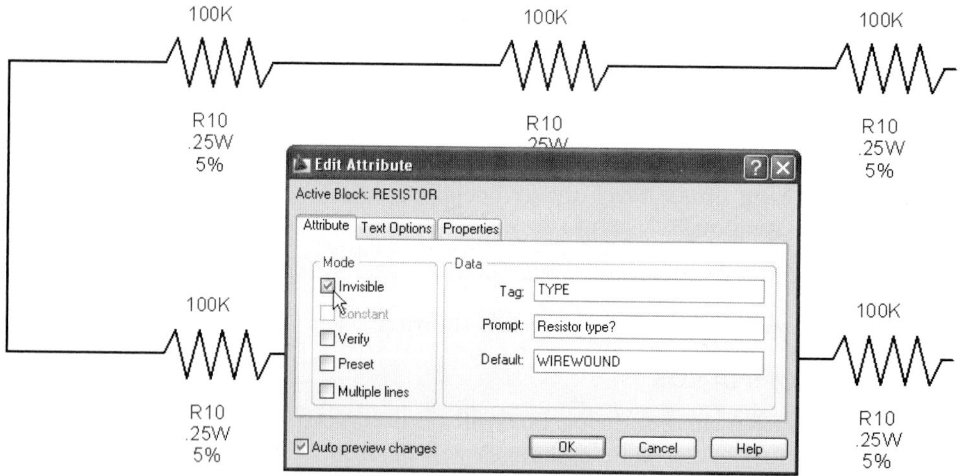

Figure 17.29

Step 3

Turn the Invisible mode on for the TOLERANCE and WATTAGE tags. Your display should appear similar to the following image, with all attributed tags invisible except for PART_NAME, and RESISTANCE.

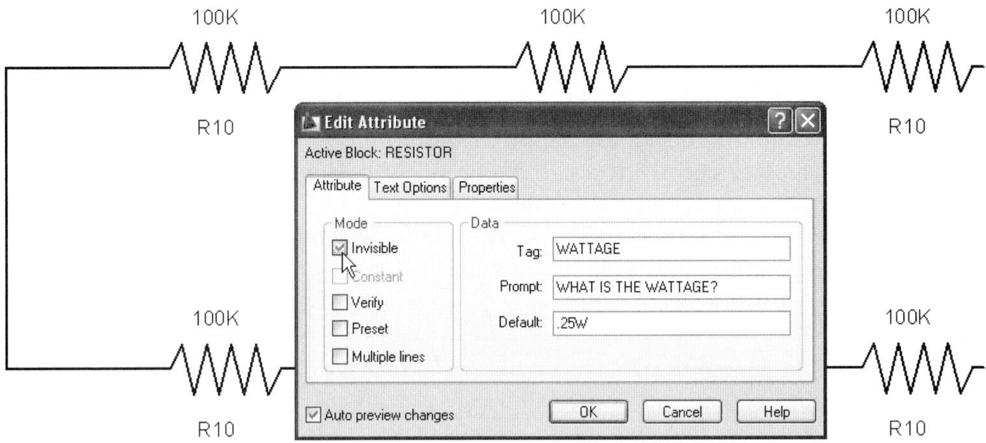

Figure 17.30

EDITING ATTRIBUTE VALUES GLOBALLY

Global editing is used to edit multiple attributes at one time. The criteria you specify will limit the set of attributes selected for editing. While the BATTMAN command provided a method to edit attribute properties, the FIND command can help you globally edit specific attribute values.

Choose this command from the Edit pull-down menu, as shown on the left in the following image. Clicking Find will display the Find and Replace dialog box as shown on the right in the following image.

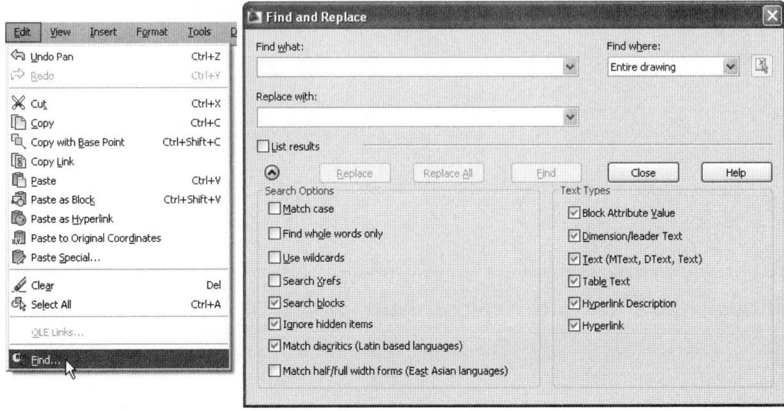

Figure 17.31

Note: To ensure the FIND command will edit block attribute values, expand the Find and Replace dialog box and verify that a check is placed next to the block attribute value.

Let us examine how this process of globally editing attributes works through the Find and Replace dialog box. In the following image, suppose the tolerance value of 5% needs to be changed to 10% on all blocks containing this attribute value. You could accomplish this by editing the attributes one at a time. However, this could be time consuming especially if the edit affects a large number of attributes. A more productive method would be to edit the group of attributes globally through the Find and Replace dialog box. Follow the examples below for accomplishing this task.

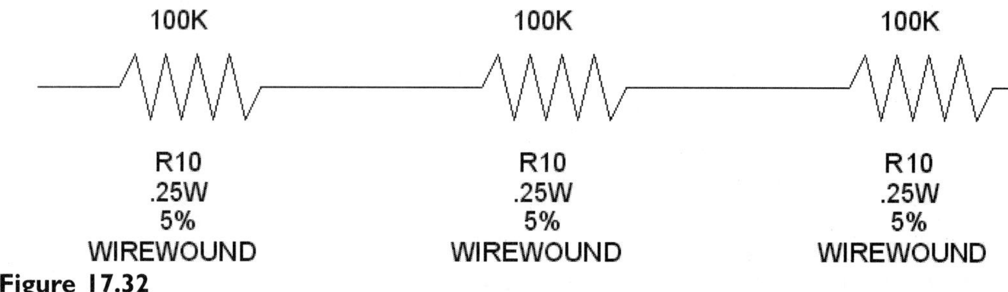

Figure 17.32

While in the Find and Replace dialog box, enter 5% in the Find What edit box. Then enter 10% in the Replace With edit box as shown on the left in the following image. Clicking the Replace All button will display a second dialog box stating how many matches were found and how many objects were changed as shown on the right in the following image.

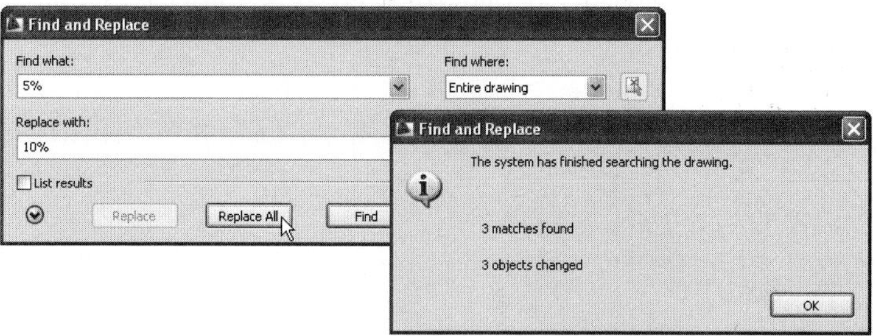

Figure 17.33

The result is shown in the following image, where all attributes were changed in a single command through the Find and Replace dialog box.

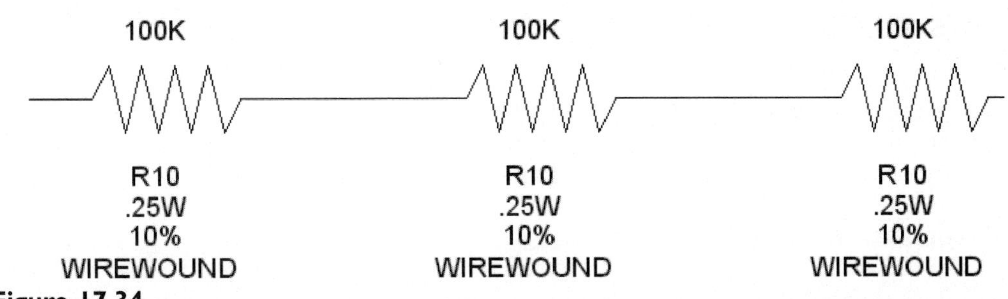

Figure 17.34

Note: In order for the FIND command to identify block attribute values, they must not be assigned the Invisible mode. You can use the BATTMAN command to temporarily remove the Invisible mode if needed.

TUTORIAL EXERCISE: 17_COMPUTERS.DWG

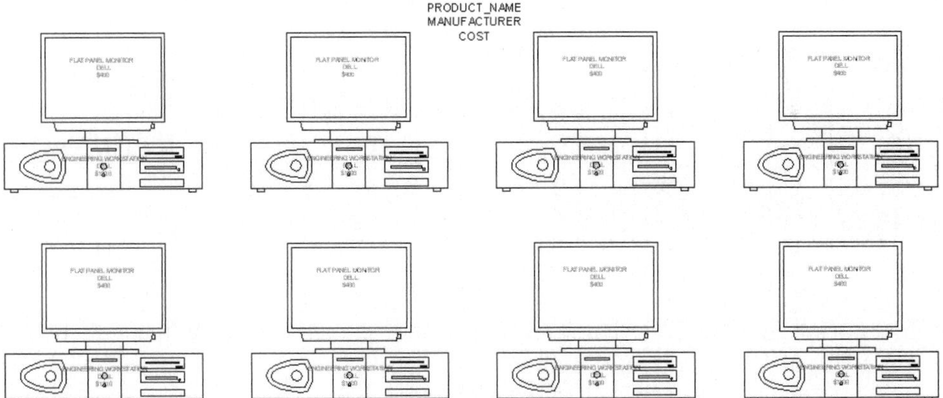

Figure 17.35

PURPOSE

This tutorial is designed to globally edit attribute values using the FIND command on the group of computers in the previous image.

SYSTEM SETTINGS

Open the drawing 17_Computers. All units and drawing limits have already been set. Attributes have also been assigned to various computer components.

LAYERS

All layers have already been created for this drawing.

SUGGESTED COMMANDS

Utilize the FIND command to remove all dollar signs from the COST attribute values. This editing operation would be necessary if you needed to change the data in the COST box from character to numeric values for an extraction operation (extracting attribute data will be discussed in more detail later in this chapter).

Step 1

Open the drawing 17_Computers. The following image shows an enlarged view of two of the computer workstations. For each workstation, $400 needs to be changed to 400, and $1300 needs to be changed to 1300. In other words, the dollar sign needs to be removed from all values for each workstation.

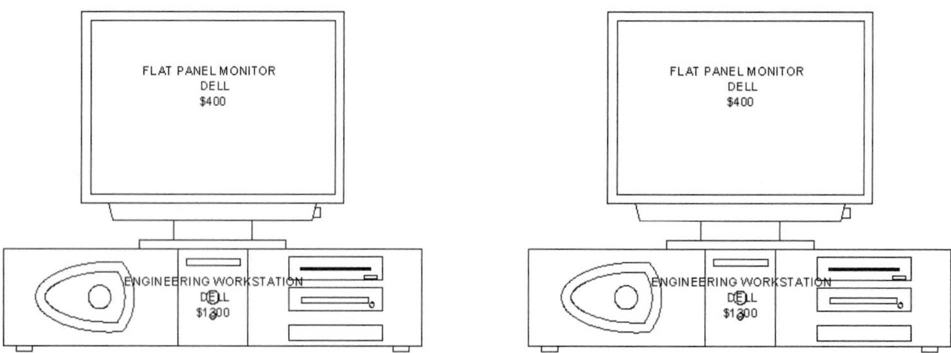

Figure 17.36

Step 2

Launch the Find and Replace dialog box and enter $ in the Find What edit box. Then leave the Replace With edit box empty as shown on the left in the following image. Clicking the Replace All button will display a second dialog box stating how many matches were found (16) and how many objects were changed (16) as shown on the right in the following image.

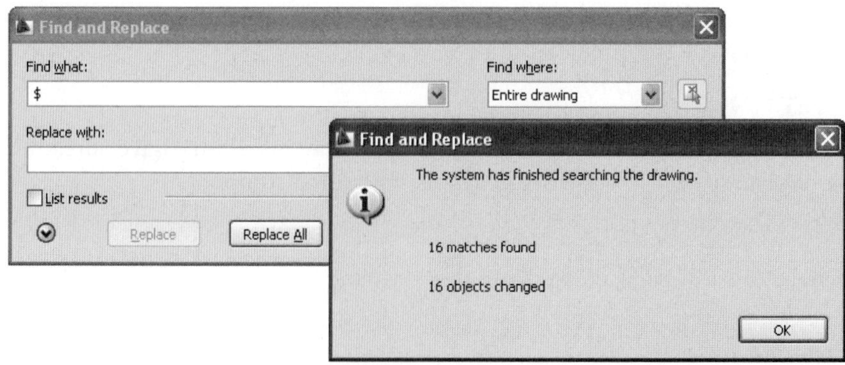

Figure 17.37

The results are illustrated in the following image, with all dollar signs removed from all attribute values through the Find and Replace dialog box.

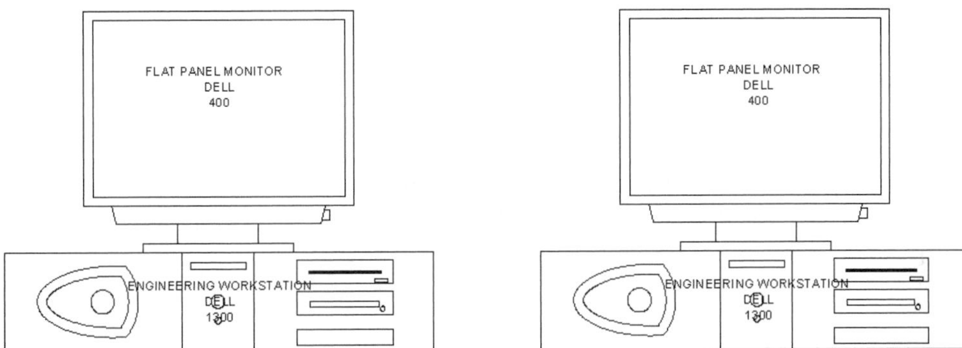

Figure 17.38

Note: The -ATTEDIT command is also available at the command line to globally edit all attribute values and remove all dollar signs from all blocks. Pick this command from the Modify pull-down menu by clicking on Object followed by Attribute and then Global. The command sequence is listed as follows:

```
Command: -ATTEDIT
Edit attributes one at a time? [Yes/No] <Y>: N (For No)
Performing global editing of attribute values.
Edit only attributes visible on screen? [Yes/No] <Y>: (Press ENTER
    to accept this default)
Enter block name specification <*>: (Press ENTER to accept this
    default)
Enter attribute tag specification <*>: COST
Enter attribute value specification <*>: (Press ENTER to accept
    this default)
Select Attributes: (Select all attributes using a window
    selection box)
Select Attributes: (Press ENTER)
16 attributes selected.
Enter string to change: $
Enter new string: (Press ENTER; this removes the dollar sign from
    all attribute values)
```

REDEFINING ATTRIBUTES

If more sweeping changes need to be made to attributes, a mechanism exists that allows you to redefine the attribute tag information globally. You follow a process similar to redefining a block, and the attribute values are also affected. A few examples of why you would want to redefine attributes might be to change their mode status (Invisible, Constant, etc.), change the name of a tag, reword a prompt, change a value to something completely different, add a new attribute tag, or delete an existing tag entirely.

Any new attributes assigned to existing block references will use their default values. Old attributes in the new block definition retain their old values. If you delete an attribute tag, AutoCAD deletes any old attributes that are not included in the new block definition.

EXPLODING A BLOCK WITH ATTRIBUTES

Before redefining an attribute, you should first copy an existing block with attributes and explode it. This returns the block to its individual objects and return the attribute values to their original tag information. The following image shows a kitchen sink with attribute values on the left and, on the right, the same sink but this time with attribute tags. The EXPLODE command was used on the right block to return the attribute values to their tags.

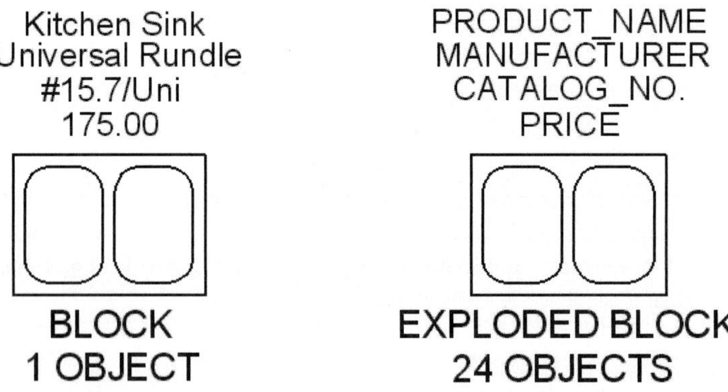

Figure 17.39

USING THE PROPERTIES PALETTE TO EDIT TAGS

A useful tool in making changes to attribute information is the Properties Palette. For example, clicking the CATALOG_NO. tag and then activating the Properties Palette, shown in the following image, allows you to make changes to the attribute prompt and values. You could even replace the attribute tag name with something completely different without having to use the Attribute Definition dialog box, as shown in the middle in the following image. Scrolling down the Properties Palette, as shown on the right in the following image, exposes the four attribute modes. If an attribute was originally created to be visible, you can make it invisible by changing the Invisible modifier here from No to Yes.

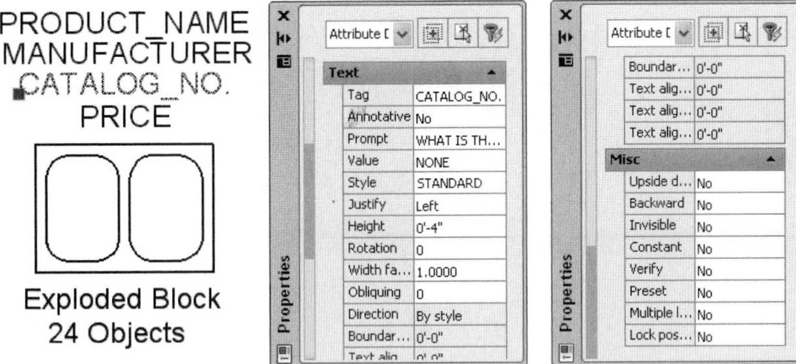

Figure 17.40

You can also double-click the attribute tag in order to make any changes. Double-clicking the CATALOG_NO. attribute tag, illustrated in the following image, displays the Edit Attribute Definition dialog box (DDEDIT command). Use this dialog box to make changes to the attribute tag, prompt, and default.

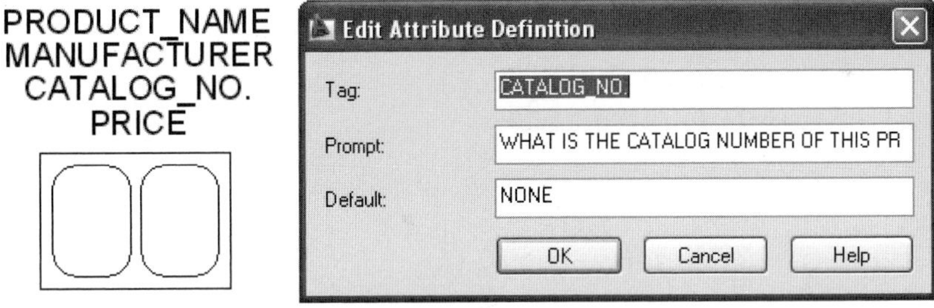

Figure 17.41

REDEFINING THE BLOCK WITH ATTRIBUTES

Once you have made changes to the attribute tags, you are ready to redefine the block along with the attributes with the ATTREDEF command. Use the following prompt sequence and illustration for accomplishing this task.

```
Command: ATTREDEF
Enter name of the block you wish to redefine: SINK
Select objects for new Block...
Select objects: (Pick a point at "A" in the following image)
Specify opposite corner: (Pick a point at "B")
Select objects: (Press ENTER to continue)
Specify insertion base point of new Block: MID
of (Pick the midpoint of the sink at "C")
```

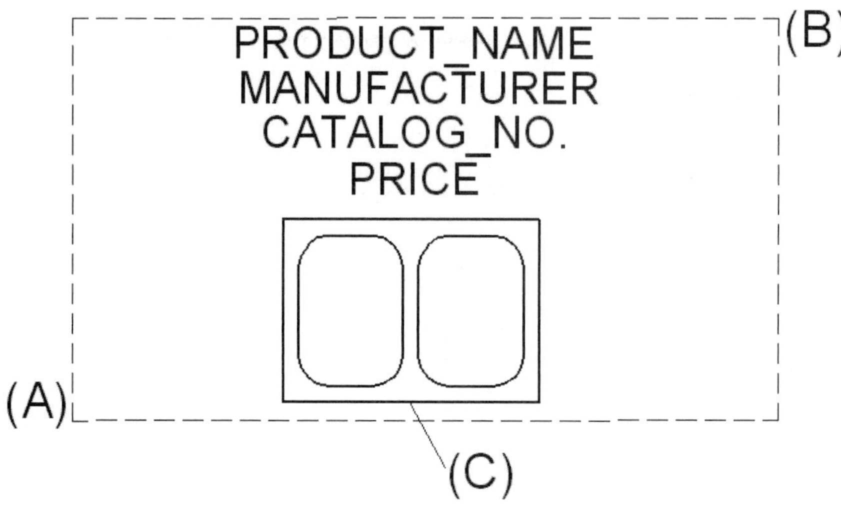

Figure 17.42

TUTORIAL EXERCISE: 17_RESISTORS.DWG

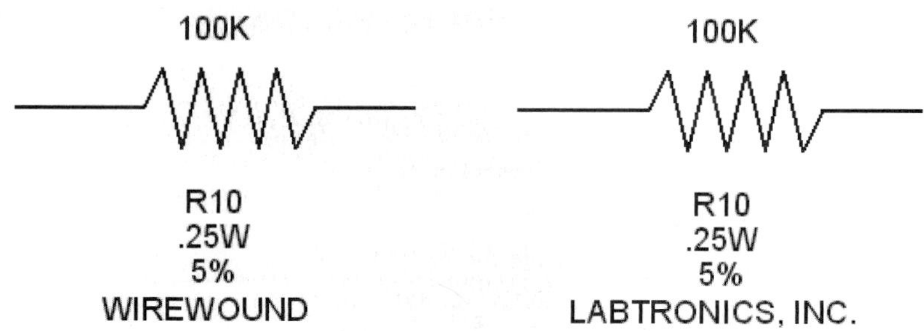

Figure 17.43

PURPOSE

This tutorial is designed to redefine attribute tags and update existing attribute values, as shown in the previous image.

SYSTEM SETTINGS

All units and drawing limits have already been set. Attributes have also been assigned to computer components.

LAYERS

All layers have already been created for this drawing.

SUGGESTED COMMANDS

The PROPERTIES and DDEDIT commands will be used to edit various characteristics of attributes.

Step 1

Open the drawing file 17_Resistors. In the following image a series of resistors is arranged in a partial circuit design. A number of changes need to be made to the original attribute definitions. Once the changes are made, the block and attributes will be redefined and all the blocks will automatically be updated.

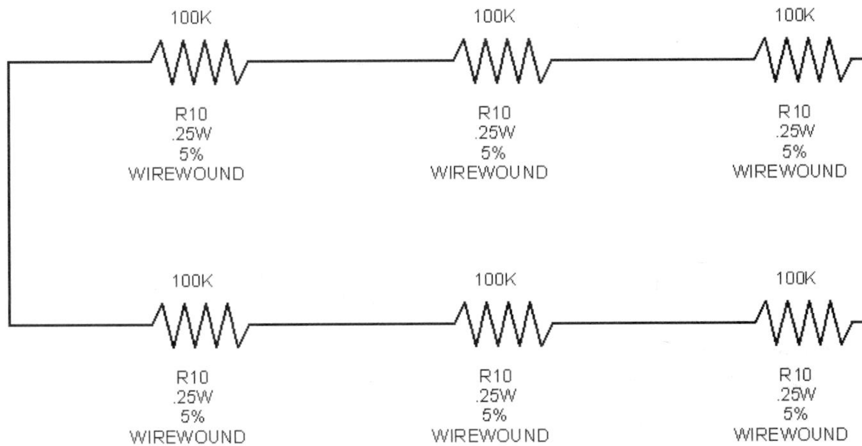

Figure 17.44

Step 2

Copy one of the resistor symbols to a blank part of your screen. Then explode this block. This should break the block down into individual objects and return the attribute values to their tags, as shown on the left in the following image.

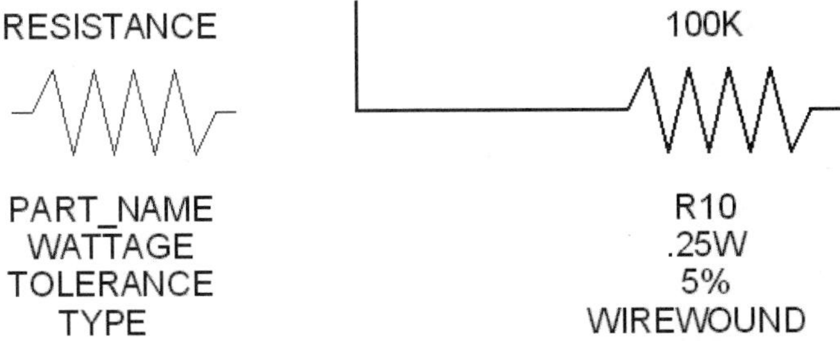

Figure 17.45

Step 3

Let us examine what needs to be accomplished. The ATTDISP command is set to Normal; however, all the attribute values are displayed. Some of this information will be made invisible to simplify the drawing. Once invisible, this data can still be viewed by setting the ATTDISP command to on. In this step, the tags WATTAGE, TOLERANCE, and TYPE will have their attribute mode changed to Invisible.

First, select the WATTAGE, TOLERANCE, and TYPE tags. These items will be highlighted, along with grips being displayed. Activate the Properties Palette and observe at the top that three Attributes are currently selected. Scroll down to the bottom of this window and under the Misc section, take note of the Invisible attribute mode. Click No, pick the down arrow in this field, and set the Invisible mode to Yes, as shown in the following image. Move the cursor into the drawing area and press ESC to remove the object highlight and the grips. These three tags have now been changed to Invisible.

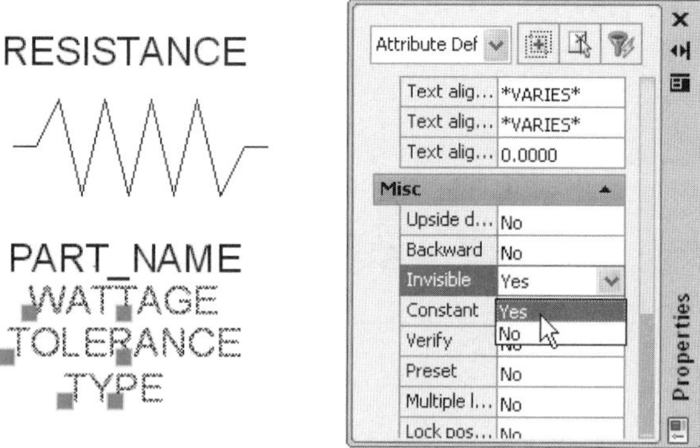

Figure 17.46

Step 4

In this step, the tag TYPE needs to be replaced with a new tag. This also means creating a new prompt and default value. Double-click the TYPE tag to display the Edit Attribute Definition dialog box, as shown in the following image, and change the Tag from TYPE to SUPPLIER. In the Prompt field, change the existing prompt to Supplier Name? In the Default field, change WIREWOUND to LABTRONICS, INC, as shown in the following image. When finished, click the OK button to accept the changes and dismiss the dialog box.

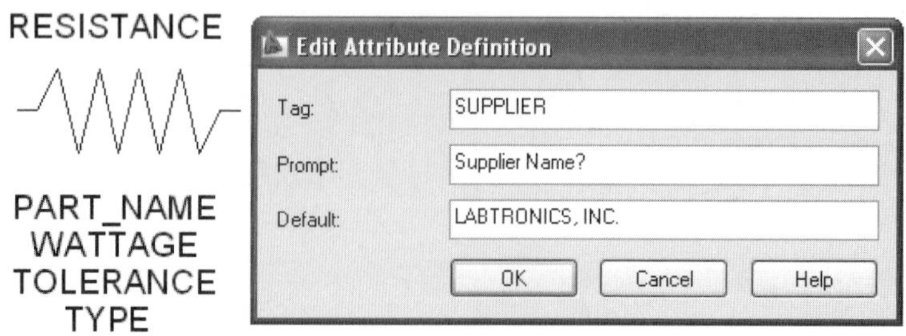

Figure 17.47

Step 5

After these changes are made to the exploded block, the ATTREDEF command will be used to redefine the block and automatically update all resistor blocks to their new values and states. Activate the ATTREDEF command. Enter RESISTOR as the name of the block to redefine. Select the block and attributes, as shown on the left in the following image. Pick the new insertion point at "C"; as the drawing regenerates, all blocks are updated to the new attribute values, as shown on the right in the following image.

```
Command: ATTREDEF
Enter name of the block you wish to redefine: RESISTOR
Select objects for new Block...
Select objects: (Pick a point at "A" in the following image)
Specify opposite corner: (Pick a point at "B")
16 found
Select objects: (Press ENTER to continue)
Specify insertion base point of new Block: End (For Object Snap
     Endpoint mode)
of (Pick the endpoint of the resistor at "C")
```

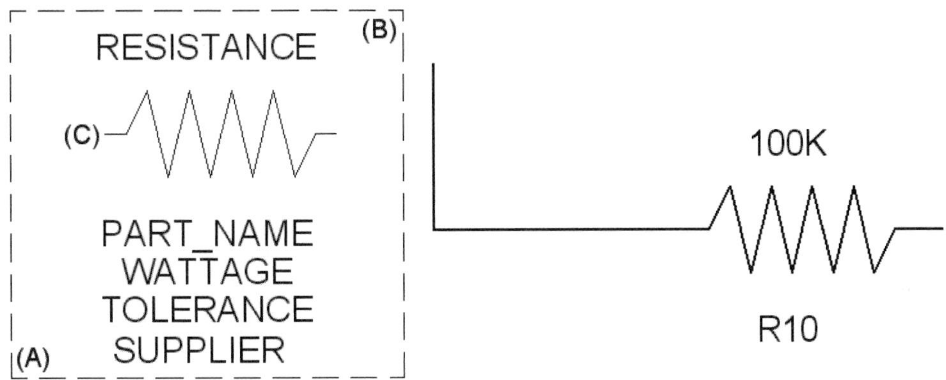

Figure 17.48

Step 6

To see if the SUPPLIER tag replaced the TYPE tag, activate the ATTDISP command and turn all attribute values on. The results are displayed in the following image. This action forces all attributes to be turned on.

```
Command: ATTDISP
Enter attribute visibility setting [Normal/ON/OFF] <Normal>: ON
```

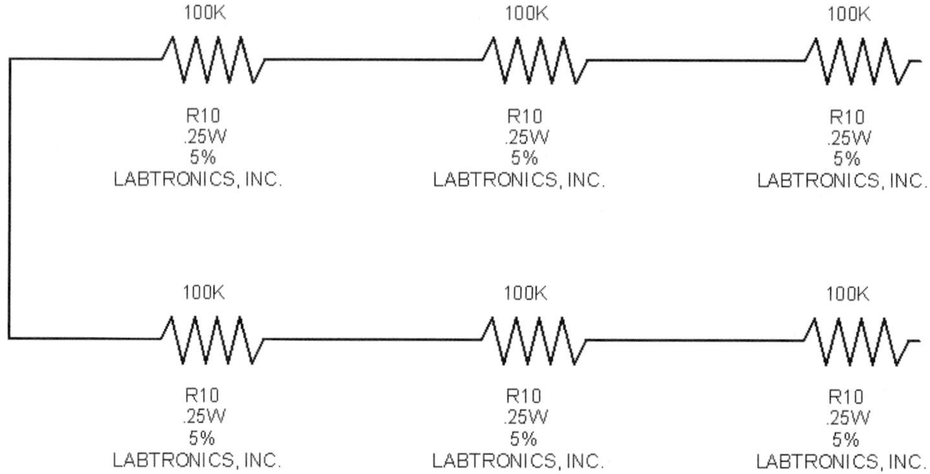

Figure 17.49

EXTRACTING ATTRIBUTES

Once attributes have been created, displayed, and edited, one additional step would be to extract attributes out to a file. The attributes can then be imported into Microsoft Excel or even brought into an existing AutoCAD drawing as a table object. This process is handled through the Attribute Extraction wizard, which is activated by selecting Attribute Extraction from the Tools pull-down menu, Menu Browser, or Modify II Toolbar, as shown in the following image.

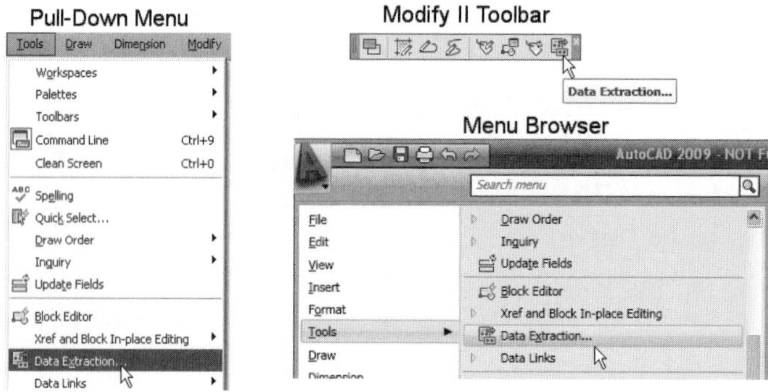

Figure 17.50

Clicking the Data Extraction toolbar button or entering EATTEXT at the command prompt activates the Data Extraction wizard, shown in the following image. Use this wizard to step you through the attribute extraction process. You can start with a template or an extraction file, if you have previously created one. The templates are saved with a BLK extension and the extraction files with an EXT extension. Each extraction you create will be saved and can be reused at any time to extract the latest information. You will be able to individually select the attributes as well as specific block information to extract. You can preview the extraction file, save the template, and export the results in either TXT (tab-separated file), CSV (comma separated), XLS (Microsoft Excel), or MDB (Microsoft Access) format or extract the results into an AutoCAD table object.

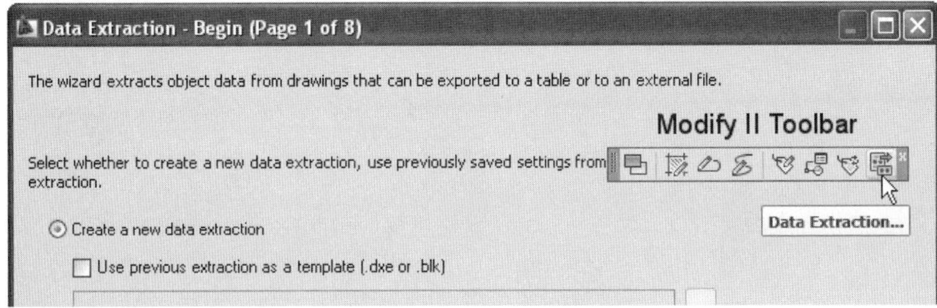

Figure 17.51

The following table gives a brief description of each page used for extracting attributes using the Attribute Extraction wizard.

Attribute Extraction Wizard

Page and Title	Description
Page 1—Begin	Create a new data extraction or edit an existing data extraction
Page 2—Define Data Source	Select the current drawing or browse for other drawings
Page 3—Select Objects	Select the blocks and attributes to include as row and column headers
Page 4—Select Properties	Select the properties you want to extract
Page 5—Refine Data	Either keep the existing order of information or reorder the rows and columns
Page 6—Choose Output	Select to output to a table or to an external file
Page 7—Table Style	Select a table style
Page 8—Finish	Click this button to finish the extraction

TUTORIAL EXERCISE: 17_CAD DEPT MODULES.DWG

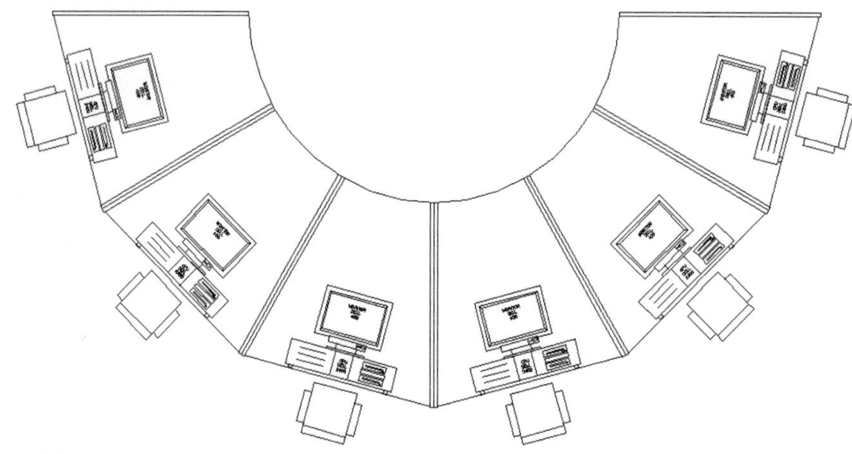

Figure 17.52

PURPOSE

This tutorial is designed to extract attribute data through the use of the Data Extraction wizard for the drawing 17_CAD Dept Modules, illustrated in the previous image.

SYSTEM SETTINGS

All units and drawing limits have already been set. Attributes have also been assigned to computer components.

LAYERS

All layers have already been created for this drawing.

SUGGESTED COMMANDS

This drawing appears inside a layout that has already been created. A table consisting of the extracted attributes will be inserted into this layout. Follow the steps provided by the Data Extraction wizard to extract the attributes out to an AutoCAD table.

Step 1

Open the drawing file 17_CAD Dept Modules.dwg. Activate the Data Extraction wizard, as shown in the following image, by clicking Attribute Extraction from the Tools pull-down menu, as shown on the left. You will create a new data extraction. Click the Next button to continue to the next step.

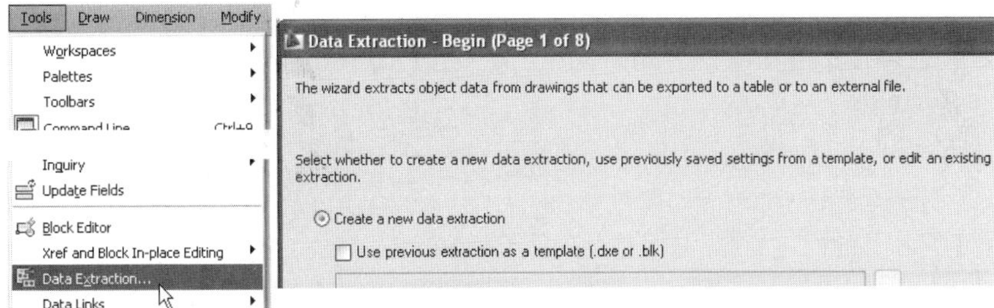

Figure 17.53

When the Save Data Extraction As dialog box appears as in the following image, enter the name of the data extraction file as CAD Dept Modules and select an appropriate folder in the Save in drop down list. When finished, click the Save button.

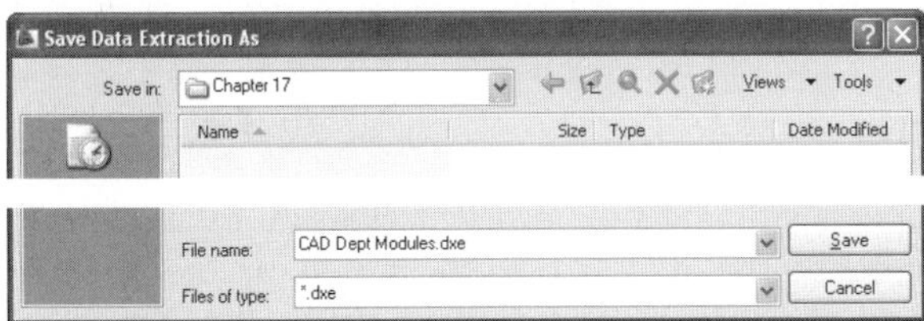

Figure 17.54

Step 2

In this next step, you pick the data source from which the attributes will be extracted. When the Define Data Source dialog box appears, verify that Drawings/Sheet set is picked and that the Include current drawing box is checked as shown in the following image. Click the Next > button to continue.

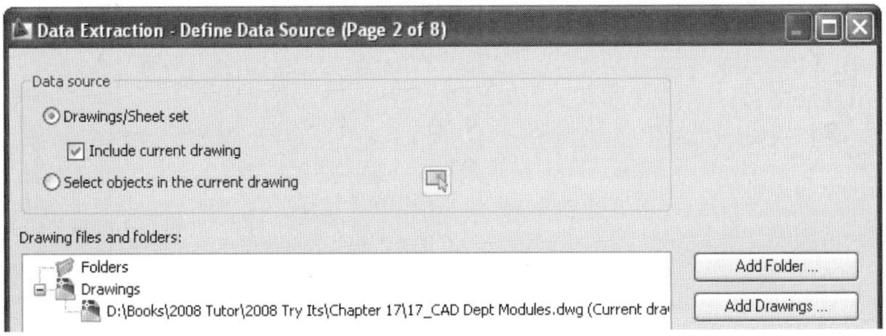

Figure 17.55

Step 3

When the Select Objects dialog box displays, verify that the Display all object types box is not checked and that the Display blocks with attributes only box is checked at the bottom of the dialog box as shown in the following image. When the three items display at the top of this dialog box, remove the check from the box next to Architectural Title. The remaining two items, CPU and Monitor, should remain checked. Click the Next > button to continue.

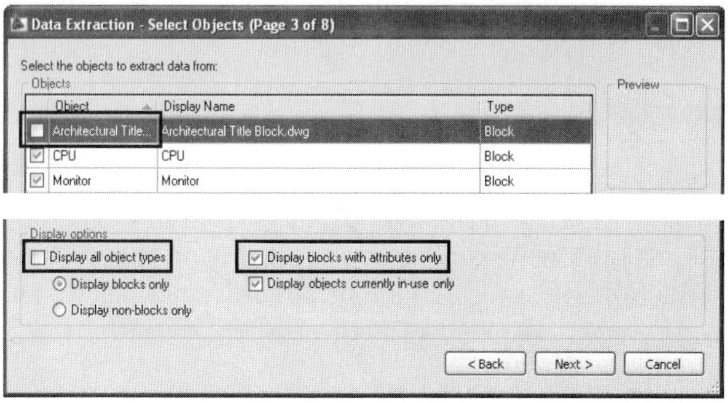

Figure 17.56

Step 4

When the Select Properties dialog box appears, clear all of the checks from the following boxes on the right side of the dialog box as shown in the following image: Drawing, General, Geometry, and Misc. The only Category filter checked should be the box next to Attribute. Also, verify that all three properties on the left side of the dialog box, COST, MANUFACTURER, and PRODUCT_NAME, are checked. Click the Next > button to continue.

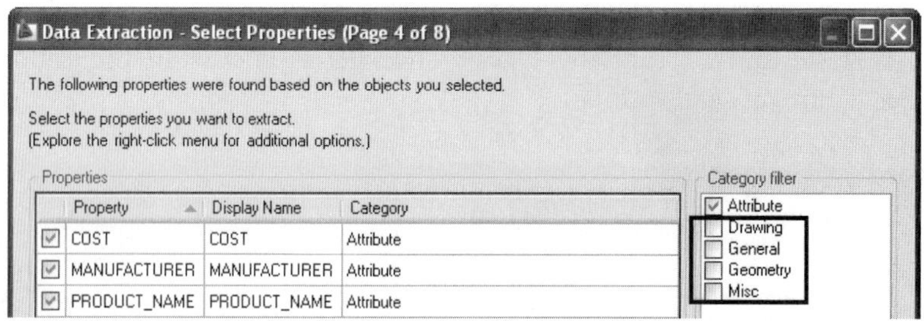

Figure 17.57

Step 5

When the Refine Data dialog box appears, keep the display of all information as is. This dialog box allows you to reorder, hide, and sort the columns in order to display the information in a different format. A button located in the lower-right corner of this dialog box allows you to create an external data link. Click the Next > button to continue.

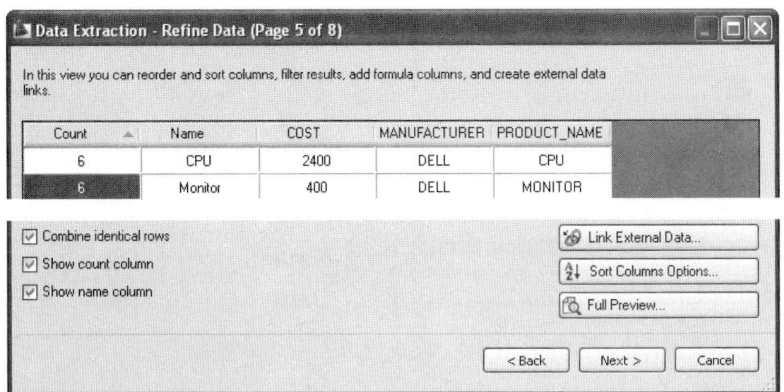

Figure 17.58

Step 6

When the Choose Output dialog box displays, place a check in the box next to Insert data extraction table into drawing, as shown in the following image. This creates an AutoCAD table that displays in the current drawing. Click the Next > button to continue.

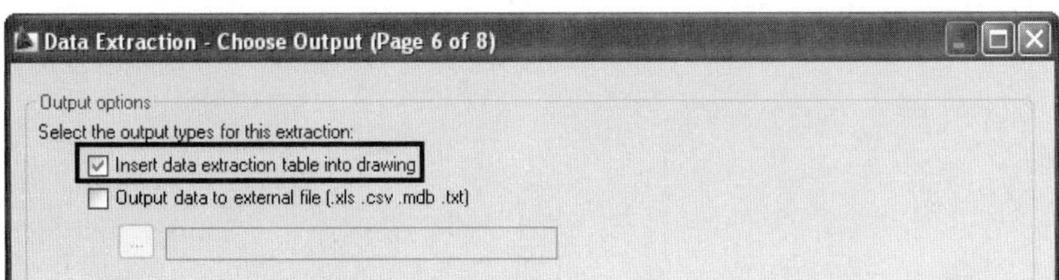

Figure 17.59

Step 7

This next dialog box deals with the components of the table that the attributes will be extracted to. In the following image, you need to enter a title for the table. In this exercise, the title of this table will be ATTRIBUTE EXTRACTION RESULTS. You can also select a table style. In this exercise, a table style called Attributes was already created. This table also displays in the dialog box, as shown on the right in the following image.

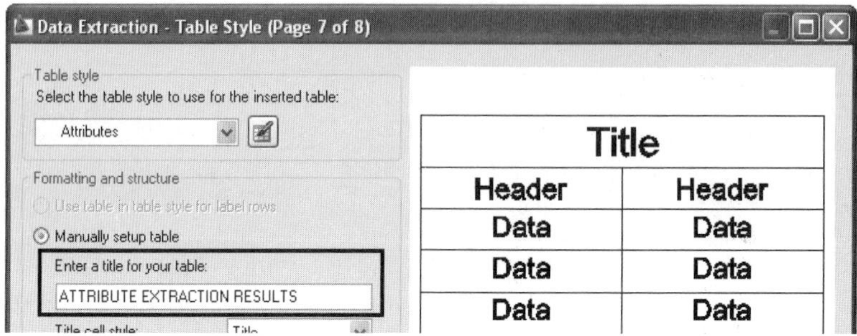

Figure 17.60

Step 8

The final dialog box states that if you extract the attributes to a table, you will be prompted for an insertion point after you click the Finish button. If you are creating external files, these will be created when clicking the Finish button.

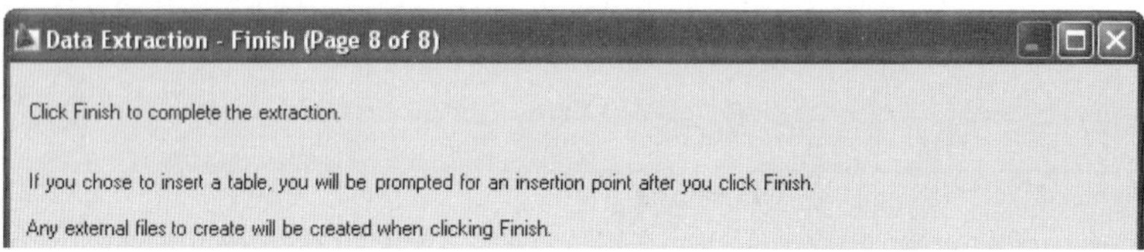

Figure 17.61

Step 9

Clicking the Finish button in the previous step brings you back into the drawing editor. You will be prompted to insert the table; do so in a convenient location of the drawing, as shown in the following image. In this image, notice that the Count, Block Name, COST, MANUFACTURER, and PRODUCT_NAME were all extracted.

ATTRIBUTE EXTRACTION RESULTS				
Count	Name	COST	MANUFACTURER	PRODUCT_NAME
6	CPU	2400	DELL	CPU
6	Monitor	400	DELL	MONITOR

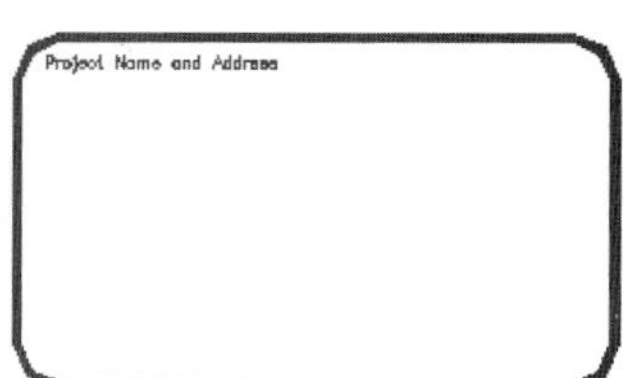

Figure 17.62

SYSTEM VARIABLES THAT CONTROL ATTRIBUTES

ATTREQ

Determines whether the INSERT command uses default attribute settings during insertion of blocks. The following settings can be used:

0	No attribute values are requested; all attributes are set to their default values.
1	Turns on prompts or a dialog box for attribute values, as specified by ATTDIA.

```
Command: ATTREQ
Enter new value for ATTREQ <1>:
```

ATTMODE

Controls the display of attributes. The following settings can be used:

0	Off: Makes all attributes invisible.
1	Normal: Retains current visibility of each attribute; visible attributes are displayed, invisible attributes are not.
On:	Makes all attributes visible.

```
Command: ATTMODE
Enter new value for ATTMODE <1>:
```

ATTDIA

Controls whether the INSERT command uses a dialog box for attribute value entry. The following settings can be used:

0	Issues prompts on the command line.
1	Initiates a dialog box for attribute value entry.

```
Command: ATTDIA
Enter new value for ATTDIA <1>:
```

CHAPTER 18

Working with External References and Raster Image and DWF Files

This chapter begins the study of External References and how they differ from blocks. When changes are made to a drawing that has been externally referenced into another drawing, these changes are updated automatically. There is no need to redefine the external reference in the same way blocks are redefined. When the sharing of data is important, the ETRANSMIT command is used to collect all support files that make up a drawing. This guarantees to the individual reviewing the drawing that all support files come with the drawing. Working with raster images is also discussed in this chapter. Raster images can be attached to a drawing file for presentation purposes. The brightness, contrast, and fade factor of the images can also be adjusted. To help you control the order in which images display in your drawing, the DRAWORDER command will be explained.

COMPARING EXTERNAL REFERENCES AND BLOCKS

You have already seen in chapter 16 how easy and popular it is to insert blocks and drawing files into other drawing files. One advantage in performing these insertions is a smaller file size, which results because all objects connected with each block are grouped into a single object. It is only when blocks are exploded that the drawing size once again increases because the objects that made up the block have been separated into their individual objects. Another advantage of blocks is their ability to be redefined in a drawing.

External references are similar to blocks in that they are considered one object. However, external references are attached to the drawing file, whereas blocks are inserted in the drawing. This attachment actually sets up a relationship between the current drawing file and the external referenced drawing. For example, take a floor plan file and externally reference it into a current drawing file. All objects associated with the floor plan are brought in, including their current layer and linetype qualities. With the floor plan acting as a guide, such items as electrical symbols, furniture, and dimensions can be added to the floor plan and saved to the current drawing file. Now a design change needs to be made to the floor plan.

Open the floor plan, stretch a few doors and walls to new locations, and save the file. The next time the drawing holding the electrical symbols and furniture arrangement is opened, the changes to the floor plan are made automatically to the current drawing file. This is one of the primary advantages of external references over blocks. Also, even though blocks are known to reduce file size, using external references reduces the file size even more than using blocks. One disadvantage of an external reference is that items such as layers and other blocks that belong to the external reference can be viewed but have limited capabilities for manipulation.

CHOOSING EXTERNAL REFERENCE COMMANDS

The XREF (XR is the alias) command activates a palette that allows you to attach and control external references in a drawing. This command can be accessed through the Menu Browser, Pull-Down menu, Ribbon, or Reference toolbar as shown in the following image.

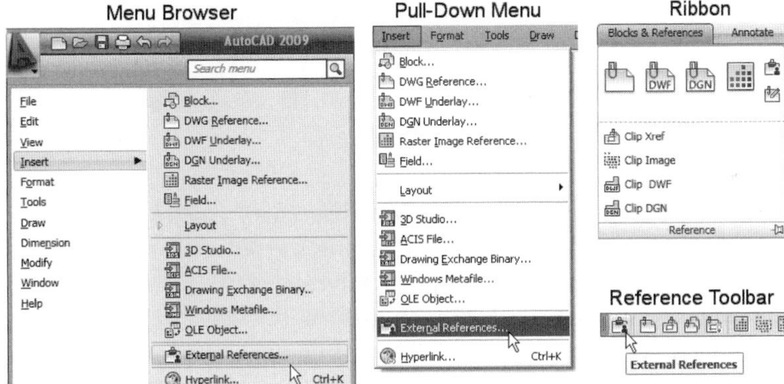

Figure 18.1

Whichever method you use, the External References palette displays, as shown in the following image. In this image, the drawing file 8th floor is listed at the top of this palette. Underneath this file are a number of additional files dealing with the furniture, hvac, lighting, plan, plumbing, and power; these files represent external references that were attached to the 8th floor drawing file.

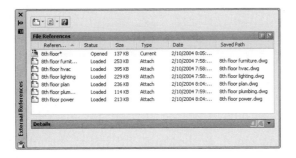

File References

Referen...	Status	Size	Type	Date	Saved Path
8th floor*	Opened	137 KB	Current	2/10/2004 8:05:...	
8th floor furnit...	Loaded	253 KB	Attach	2/10/2004 7:58:...	8th floor furniture.dwg
8th floor hvac	Loaded	395 KB	Attach	2/10/2004 7:58:...	8th floor hvac.dwg
8th floor lighting	Loaded	229 KB	Attach	2/10/2004 7:58:...	8th floor lighting.dwg
8th floor plan	Loaded	236 KB	Attach	2/10/2004 8:04:...	8th floor plan.dwg
8th floor plum...	Loaded	114 KB	Attach	2/10/2004 7:59:...	8th floor plumbing.dwg
8th floor power	Loaded	213 KB	Attach	2/10/2004 8:04:...	8th floor power.dwg

Details

External References

Figure 18.2

The lower portion of the External References palette illustrated in the previous image has a Details heading. Double-clicking Details displays the information about the selected external reference, as shown on the left in the following image. In this example, the 8th floor furniture external reference is selected and has various items such as status, file size, type, and date, to name a few listed. Clicking the Preview button displays a preview of the selected external reference, as shown on the right in the following image.

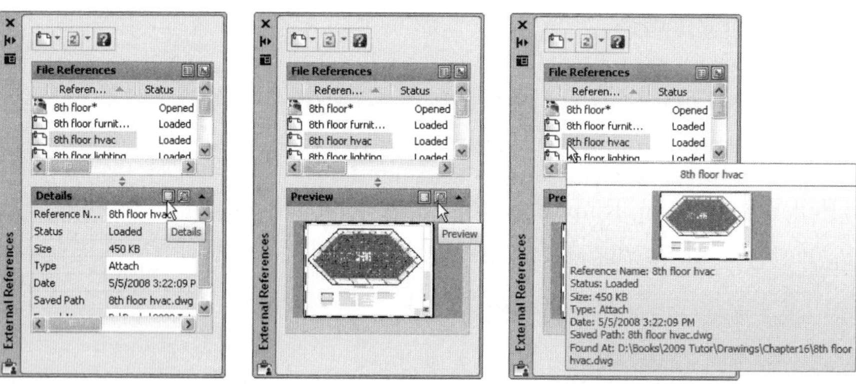

Figure 18.3

Located in the upper-left corner of the External References palette are buttons that are used for attaching or reloading external references. Clicking the button located on the left, as shown in the following image, displays three different attach modes. Attach DWG activates the Select Reference File dialog box, used for selecting an existing drawing file to merge into the current drawing. Attach Image activates the Select Image File dialog box, used for selecting a raster image for merging into the current drawing. Attach DWF activates the Select DWF File dialog box, used for selecting a DWF (Drawing Web Format) file for merging into the current drawing. Attach DGN activates the Select DGN File dialog box, used for importing MicroStation® V8 DGN drawing files into DWG files.

When changes to original drawing files that are referenced into the current drawing occur, click Refresh to update the information located in the External References palette. You can also choose Reload All References as a means of updating your current drawing to any changes that have occurred to the external references in the drawing.

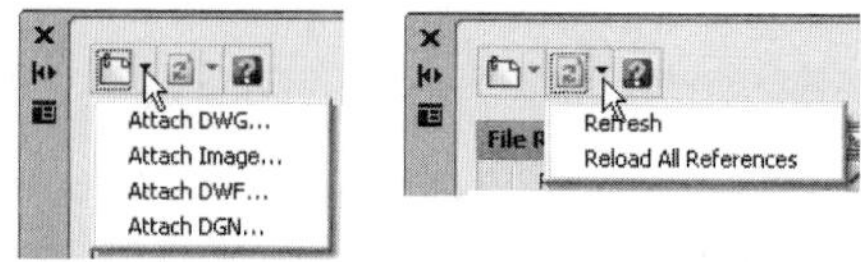

Figure 18.4

External references can also be attached and reloaded by right-clicking in the palette area and activating a shortcut menu, as shown on the left in the following image. Once the External References palette is populated with various references, select one of the references and right-click to display the shortcut menu shown on the right in the following image. With this menu you can elect to open the original drawing file being referenced or click Attach to merge another drawing file. The Unload, Reload, Detach, and Bind options are discussed in detail later in this chapter.

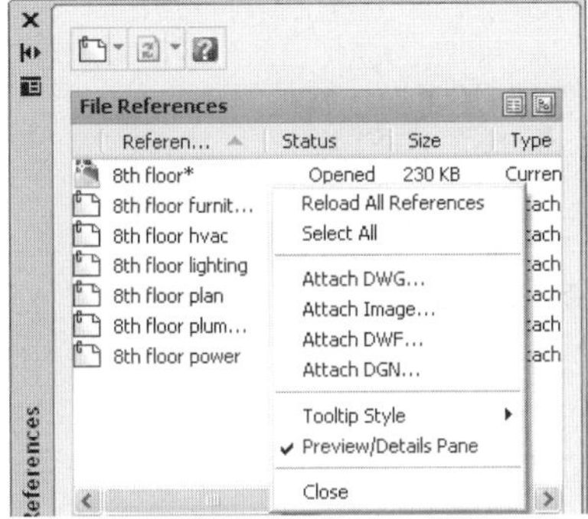

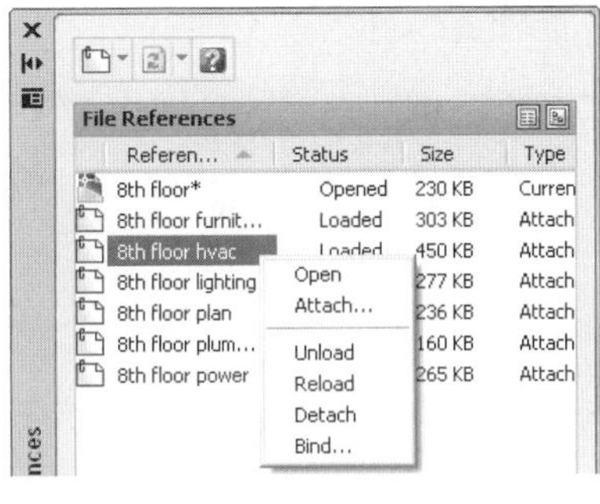

Figure 18.5

One of the more popular ways to access all commands that deal with External References is through the Reference toolbar previously shown. The following table outlines the main commands found in this toolbar.

Button	Tool	Function
	External References	Launches the External References Palette, used for managing external references in a drawing
	Attach Xref	Launches the External Reference dialog box, used for attaching an external reference into a drawing
	Clip Xref	Used to isolate a portion of an external reference by clipping away unnecessary objects
	Xbind	Used to convert named objects, such as blocks and layers that belong to an external reference, to usable items
	Xref Frame	Controls the frame used to clip an external reference

ATTACHING AN EXTERNAL REFERENCE

Use the ATTACH XREF command to attach an external drawing file into the current drawing file. This can be compared with the INSERT command for inserting blocks into drawing files. The Attach option sets up a path that looks for the external reference every time the drawing containing it is loaded.

Note: To accomplish the Try It! exercises in this chapter, copy the chapter 18 files to a convenient location/folder on your hard drive. This way you will be able to modify and save the drawings as required.

Try It!: Start a new drawing file from scratch. Then click the Attach Xref button in the Reference toolbar to activate the Select Reference File dialog box, as shown in the following image. This dialog box is very similar to the one used for selecting drawing files to initially load into AutoCAD. Click the file 18_Asesmp1 to attach and click the Open button.

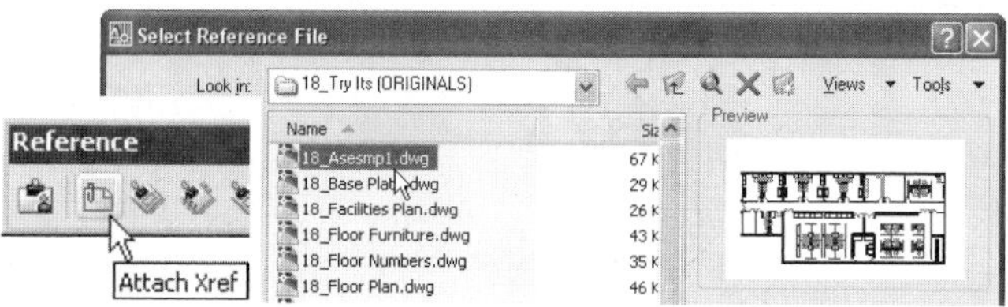

Figure 18.6

Once the file 18_Asesmp1 is selected, the External Reference dialog box appears, as shown in the following image. Some of the information contained in this dialog box is similar to the information in the Insert dialog box, such as the insertion point, scale, and rotation angle of the external reference. Of importance, in the upper part of the dialog box is the path information associated with the external reference. If, during file management, an externally referenced file is moved to a new location, the new path of the external reference must be reestablished; otherwise, it does not load into the drawing it was attached to. Remove the check from the Specify On-screen box under Insertion point. This inserts this external reference at absolute coordinate 0,0,0. Clicking the OK button returns you to the drawing editor and attaches the file to the current drawing.

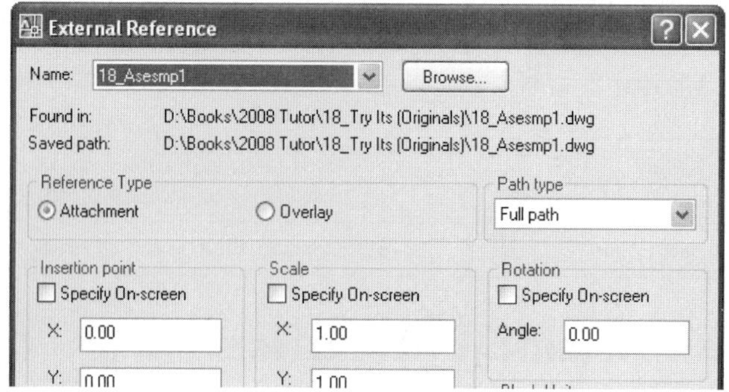

Figure 18.7

Since you will not see the floor plan on the small drawing sheet, perform a ZOOM-All. Your results should appear similar to the illustration in the following image.

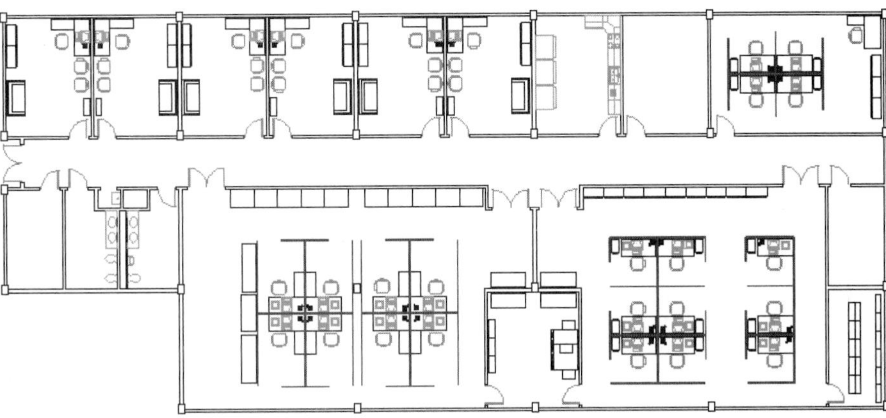

Figure 18.8

With the external reference attached to the drawing file, layers and blocks that belong to the external reference are displayed in a unique way. Illustrated in the following image are the current layers that are part of the drawing. However, notice how a number of layers begin with the same name (18_Asesmp1); also, what appears to be a vertical bar separates 18_Asesmp1 from the names of the layers. Actually, the vertical bar represents the Pipe symbol on the keyboard, and it designates that the layers belong to the external reference, namely 18_Asesmp1. These layers can be turned off, can be frozen, or can even have the color changed. However, you cannot make these layers current for drawing because they belong to the external reference.

Status	Name	On	Freeze	Lock	Color	Linetype	Lineweight	Plot Style	Plot
	0				white	Continuous	—— Default	Color_7	
	18_Asesmp1|BORDER				gre...	Continuous	—— Default	Color_3	
	18_Asesmp1|CHAIRS				red	Continuous	—— Default	Color_1	
	18_Asesmp1|CPU				ma...	Continuous	—— Default	Color_6	
	18_Asesmp1|DOOR				red	Continuous	—— Default	Color_1	
	18_Asesmp1|FIXTURES				ma...	Continuous	—— Default	Color_6	
	18_Asesmp1|FURNITURE				white	Continuous	—— Default	Color_7	
	18_Asesmp1|KITCHEN				ma...	Continuous	—— Default	Color_6	
	18_Asesmp1|PARTITIONS				blue	Continuous	—— Default	Color_5	
	18_Asesmp1|ROOM NUMBERS				8	Continuous	—— Default	Color_8	
	18_Asesmp1|WALLS				white	Continuous	—— Default	Color_7	

Figure 18.9

One of the real advantages of using external references is the way they are affected by drawing changes. To demonstrate this, first save your drawing file under the name 18_Facilities. Close this drawing and open the original floor plan called 18_Asesmp1. Use the Insert dialog box to insert a block called ROOM NUMBERS, as shown in the following image, into the 18_Asesmp1 file. Use an insertion point of 0,0,0 for placing this block and click OK in the Insert dialog box. Save this drawing file and open the drawing 18_Facilities.

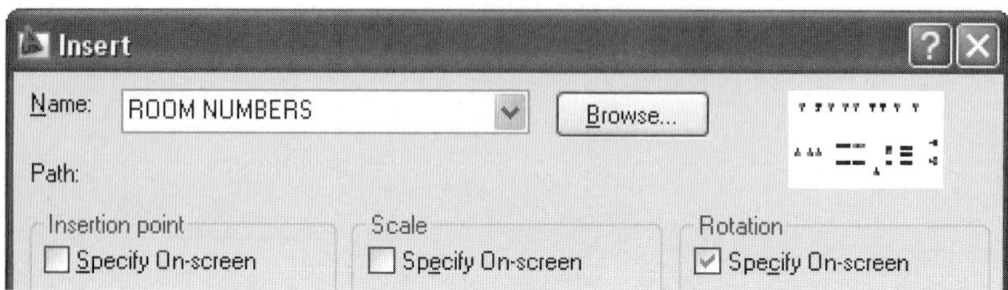

Figure 18.10

In the following image, the drawing file 18_Asesmp1 is opened again. Notice how room tags have been added to the floor plan. Once the drawing holding the external reference is loaded, all room tags, which belong to the external reference, are automatically displayed.

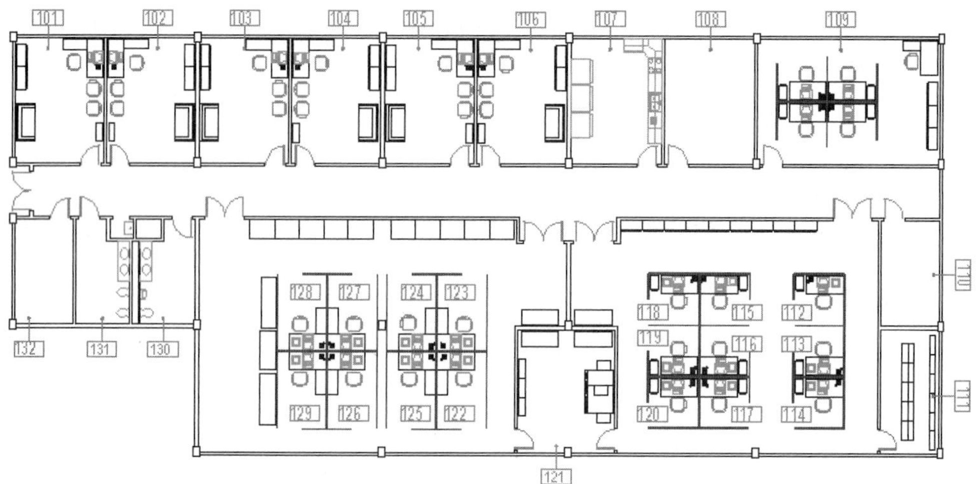

Figure 18.11

OVERLAYING AN EXTERNAL REFERENCE

Suppose in the last example of the facilities floor plan that some design groups need to see the room number labels while other design groups do not. This is the purpose of overlaying an external reference instead of attaching it. All design groups will see the information if the external reference is attached. If information is overlaid and the entire drawing is externally referenced, the overlaid information does not display; it's as if it is invisible. This option is illustrated in the next Try It! exercise.

Try It!: Open the drawing 18_Floor Plan. Then use the External Reference dialog box, shown in the following image, to attach the file 18_Floor Furniture (the steps for accomplishing this were demonstrated in the previous Try It! exercise). Use an insertion point of 0,0,0 and be sure the reference type is Attachment.

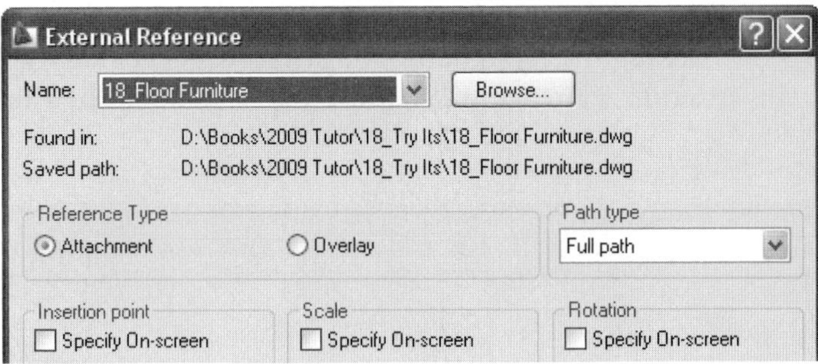

Figure 18.12

Your display should be similar to the illustration in the following image. You want all design groups to view this information.

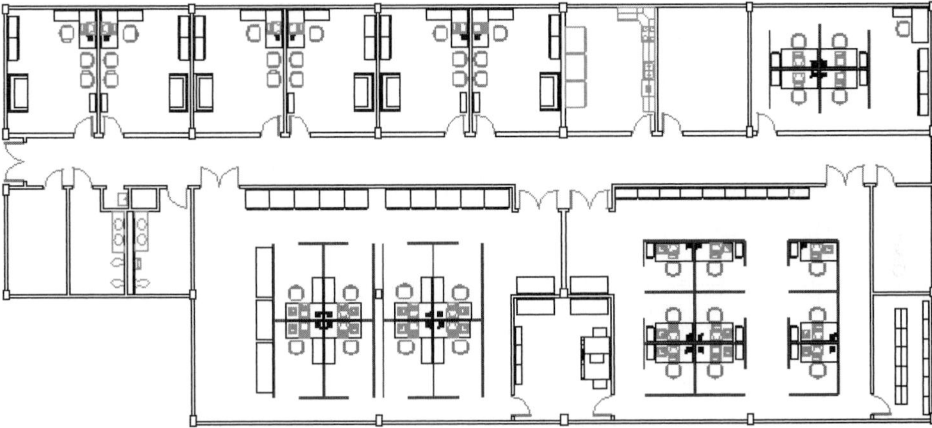

Figure 18.13

Now we will overlay an external reference. In the External Reference palette click the Attach Dwg button. When the Select Reference File dialog box appears, as shown in the following image, pick the file 18_Floor Numbers and click the Open button.

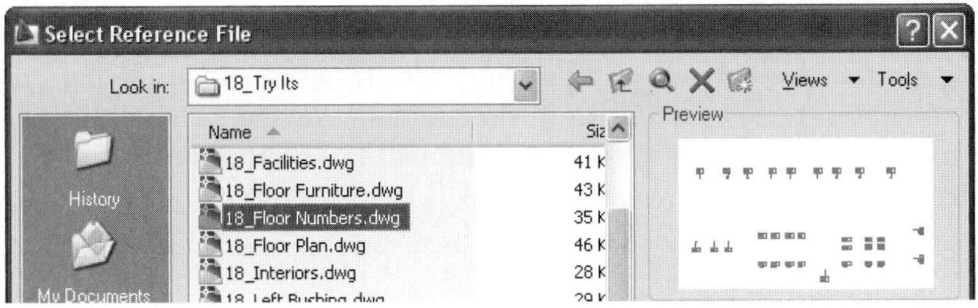

Figure 18.14

When the External Reference dialog box appears, as shown in the following image, change the Reference Type by clicking the radio button next to Overlay. Under the Insertion Point heading, be sure this external reference will be inserted at 0,0,0. When finished, click the OK button.

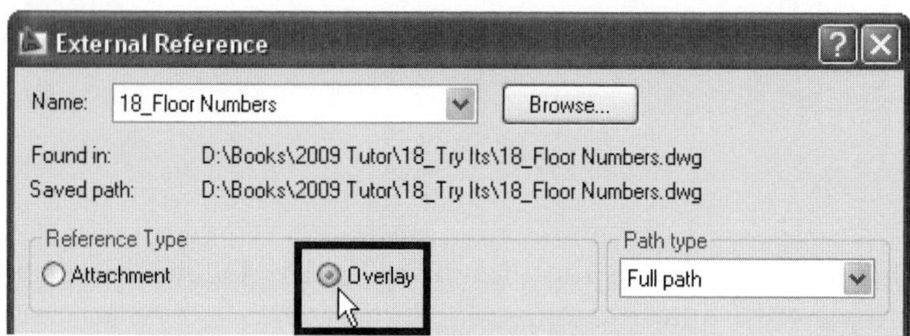

Figure 18.15

Your display should appear similar to the following image. Save this drawing file under its original name of 18_Floor Plan and then close the drawing file.

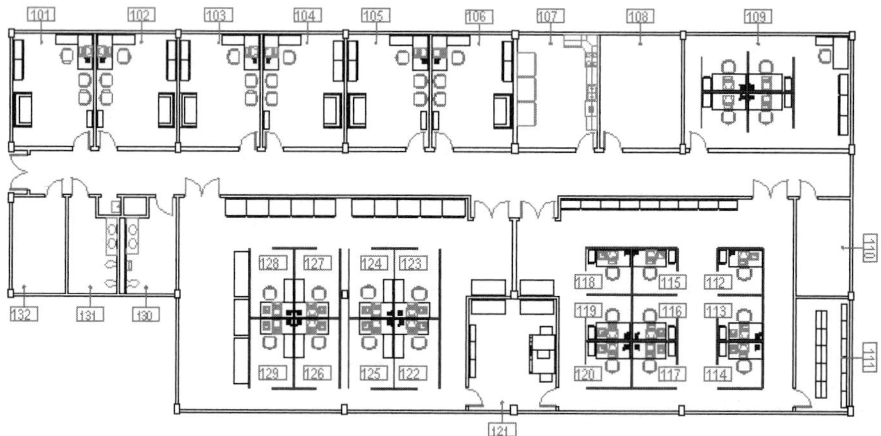

Figure 18.16

Now start a new drawing file from scratch. In the External Reference palette click the Attach Dwg button and attach the drawing file 18_Floor Plan. When the External Reference dialog box appears, as shown in the following image, be sure that the Reference Type is set to Attachment and that the drawing will be inserted at 0,0,0. When finished, click the OK button.

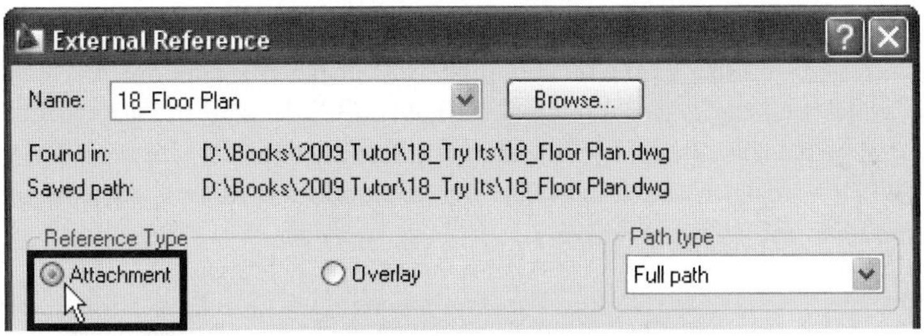

External Reference

Name: 18_Floor Plan Browse...

Found in: D:\Books\2009 Tutor\18_Try Its\18_Floor Plan.dwg

Saved path: D:\Books\2009 Tutor\18_Try Its\18_Floor Plan.dwg

Reference Type

Attachment Overlay

Path type

Full path

Figure 18.17

Your drawing is not displayed because the default limits are too small, so perform a ZOOM-All operation and observe the results, shown in the following image. Notice that the room numbers do not display because they were originally overlaid in the file 18_Floor Plan. This completes the Try It! exercise.

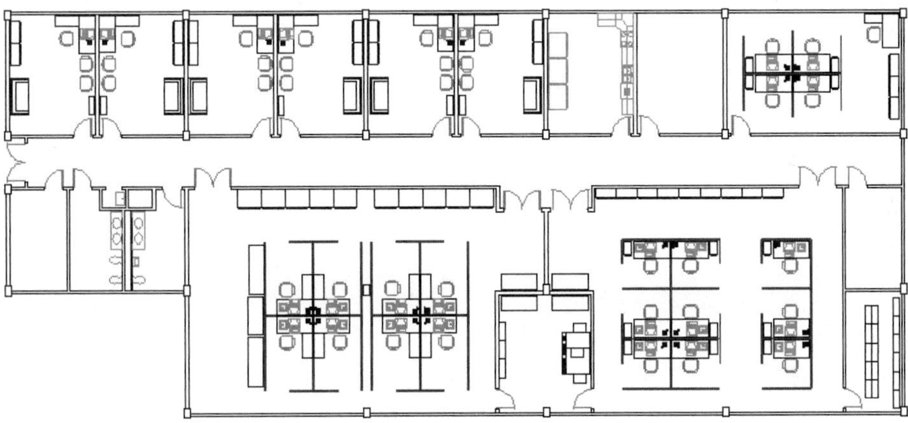

Figure 18.18

THE XBIND COMMAND

Earlier, it was mentioned that layers belonging to external references cannot be used in the drawing they were externally referenced into. The same is true with blocks that were inserted into the drawing. When blocks are listed, the name of the external reference is given first, with the Pipe symbol following, and finally the actual name of the block, as in the example 18_Asesmp1|DESK2 where DESK2 is the block and 10_Asesmp1 is the external reference. As with layers, these blocks cannot be used in the drawing because of the presence of the Pipesymbol, which is not supported in the name of the block. There is a way, however, to convert a block or layer into a usable object through the XBIND command.

Try It!: Open the drawing file 18_Facilities Plan. Then select the XBIND command from the Reference toolbar, as shown in the following image. This activates the Xbind dialog box, which lists the external reference 18_ASESMP1. Expand the listing of all named objects, such as blocks and layers associated with the external reference. Then expand the Block heading to list all individual blocks associated with the external reference, as shown in the following image.

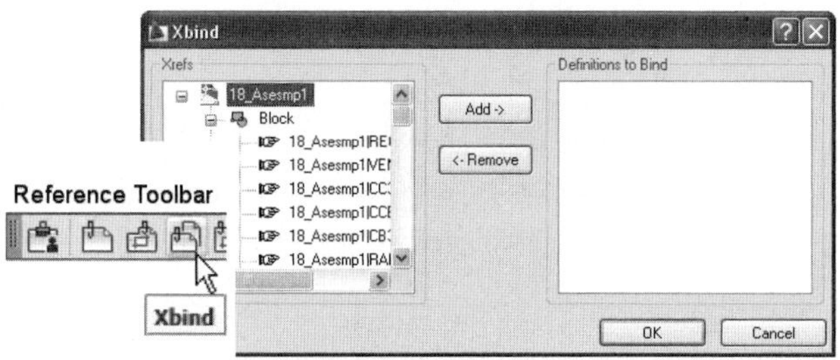

Figure 18.19

Click the block 18_Asesmp1|DESK2 in the listing on the left; then click the Add -> button. This moves the block name over to the right under the listing of Definitions to Bind, as shown in the following image. Do the same for 18_Asesmp1|DESK3 and 18_Asesmp1|DESK4. When finished adding these items, click the OK button to bind the blocks to the current drawing file.

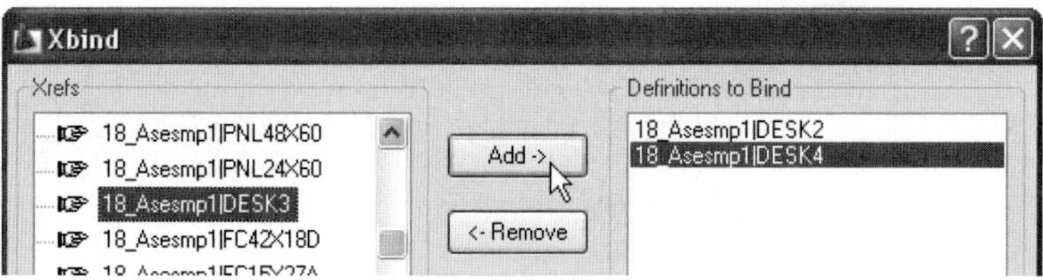

Figure 18.20

Test to see that new blocks have in fact been bound to the current drawing file. Activate the Insert dialog box, click the Name drop-down list box, and notice the display of the blocks, as shown in the following image. The three symbols just bound from the external reference still have the name of the external reference, namely, 18_Asesmp1. However, instead of the Pipe symbol separating the name of the external reference and block names, the characters 0 are now used. This is what makes the blocks valid in the drawing. Now these three blocks can be inserted in the drawing file even though they used to belong only to the external reference 18_Asesmp1.

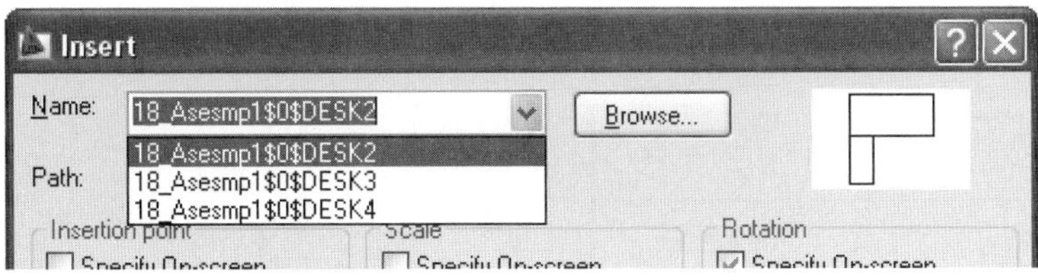

Figure 18.21

IN-PLACE REFERENCE EDITING

In-Place Reference Editing allows you to edit a reference drawing from the current drawing, which is externally referencing it. You then save the changes back to the original drawing file. This becomes an efficient way of making a change to a drawing file from an externally referenced file.

Try It!: Open the drawing file 18_Pulley Assembly, shown in the following image. Both holes located in the Base Plate, Left Support, and Right Support need to be stretched 0.20 units to the left and right. This will center the holes along the Left and Right Supports. Since all images that make up the Pulley Assembly are external references, the In-Place Reference Editing feature will be illustrated.

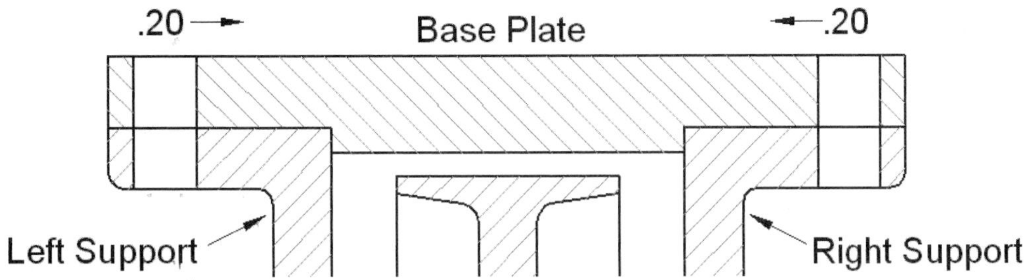

Figure 18.22

Begin by accessing the Edit Reference In-Place (REFEDIT) command from the Refedit toolbar, as shown in the following image.

Once in the REFEDIT command, you are prompted to select the reference you wish to modify. Pick the top line of the Base Plate, as shown in the following image. This displays the Reference Edit dialog box, as shown on the right in the following image. The reference to be edited, which is 18_Base Plate in our case, should be selected. Nested references may also be displayed; one of these could be selected for editing instead, if desired. Clicking the OK button returns you to the screen.

Note: Double-clicking an external reference also launches the Reference Edit dialog box.

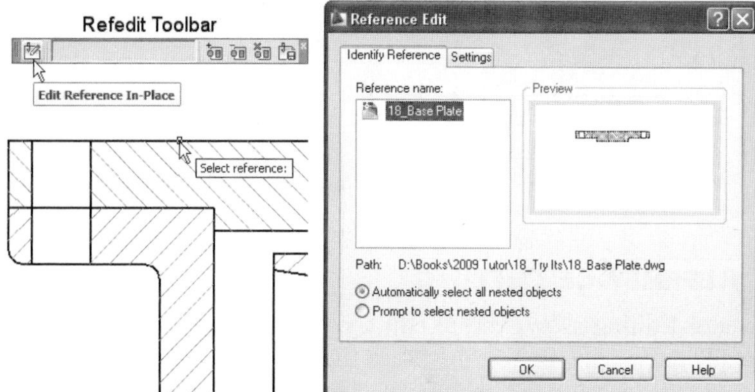

Figure 18.23

If it is not already displayed, the Refedit toolbar will be automatically provided, as shown in the following image, and you will be returned to the command prompt. To better identify the external reference being edited, the other external references in the drawing take on a slightly faded appearance. This faded effect returns to normal when you exit the Refedit process. You can now make modifications to the Base Plate by using the STRETCH command to stretch both holes a distance of 0.20 units to the inside.

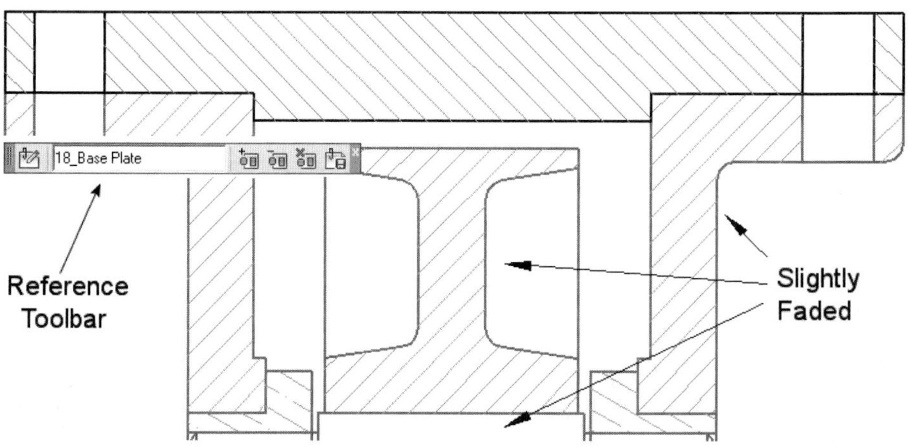

Figure 18.24

When performing the STRETCH operation, notice that even if the crossing window were to extend across both parts, only the holes in the Base Plate will be modified, as shown in the following image. When you are satisfied with the changes, select the Save Reference Edits button on the toolbar. An AutoCAD alert box will ask you to confirm the saving of reference changes, as shown in the following image.

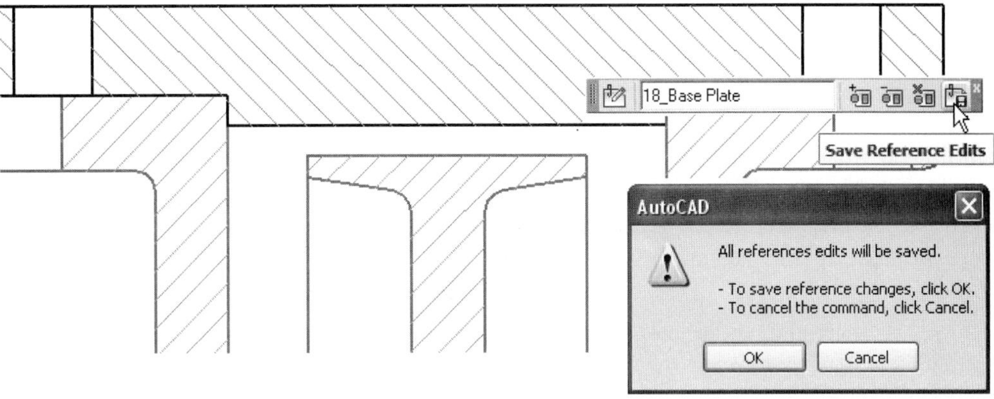

Figure 18.25

After you click OK, the results can be seen, as shown in the following image. Notice that only the objects that belong to the external reference (in this case, the holes that are part of the Base Plate) are affected. Also, the Refedit toolbar automatically closes.

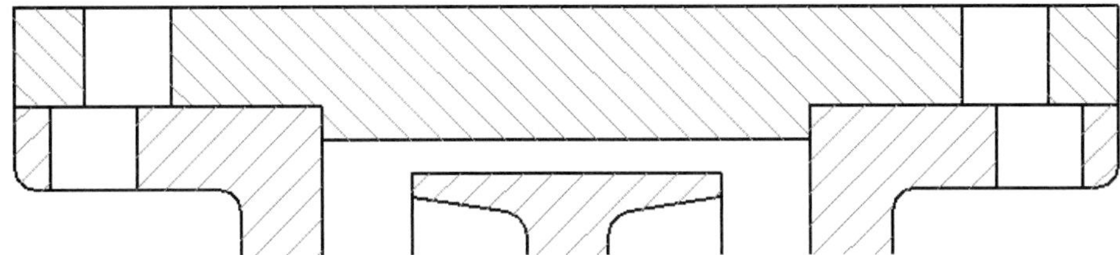

Figure 18.26

Perform the same series of steps using In-Place Reference Editing separately for the Left and Right supports. Stretch the hole located on the Left Support a distance of 0.20 units to the right. Stretch the hole located on the Right Support a distance of 0.20 units to the left. The final results of this In-Place Reference Editing Try It! exercise are displayed in the following image.

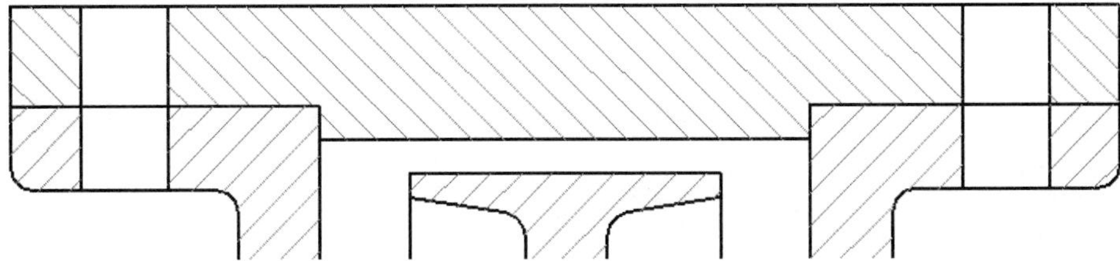

Figure 18.27

BINDING AN EXTERNAL REFERENCE

Binding an external reference to a drawing makes the external reference a permanent part of the drawing and no longer an externally referenced file. In actuality, the external reference is converted into a block object. To bind the entire database of the Xref drawing, including all its xref-dependent named objects such as blocks, dimension styles, layers, linetypes, and text styles, use the Bind option of the External Reference palette. Since binding an external reference breaks the link with the original drawing file, this becomes a very popular technique at the end of a project when archiving final drawing files is important.

To bind an external reference, click the External References button located in the Reference toolbar to activate the External References palette. Select the external reference that you want to bind to the current drawing, right-click, and pick Bind from the menu, as shown on the left in the following image. This option activates the Bind Xrefs dialog box, illustrated on the right in the following image. Two options are available inside this dialog box: Bind and Insert.

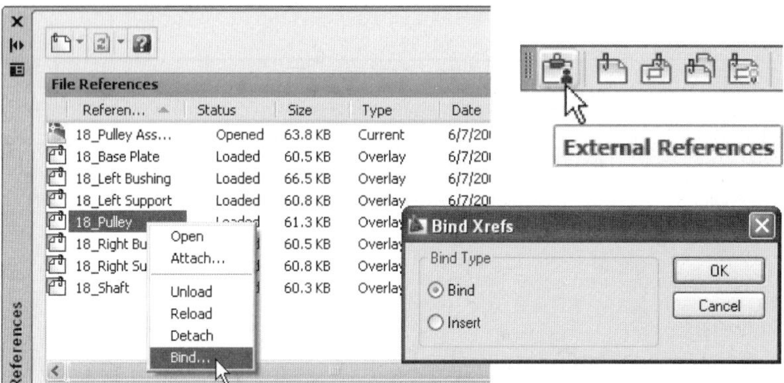

Figure 18.28

The Bind option binds to the current drawing file all blocks, layers, dimension styles, and so on that belonged to an external reference. After you perform this operation, layers can be made current and blocks inserted in the drawing. For example, a typical block definition belonging to an external reference is listed in the symbol table as XREFname|-BLOCKname. Once the external reference is bound, all block definitions are converted to XREFname0BLOCKname. In the layer display in the following image, the referenced layers were converted to usable layers with the Bind option.

Referenced Layers	Converted Layers
18_Pulley|Hatch	18_Pulley0Hatch
18_Pulley|Object	18_Pulley0Object
18_Pulley|Text	18_Pulley0Text

Note: The same naming convention is true for blocks, dimension styles, and other named items. The result of binding an external reference is similar to a drawing that was inserted into another drawing.

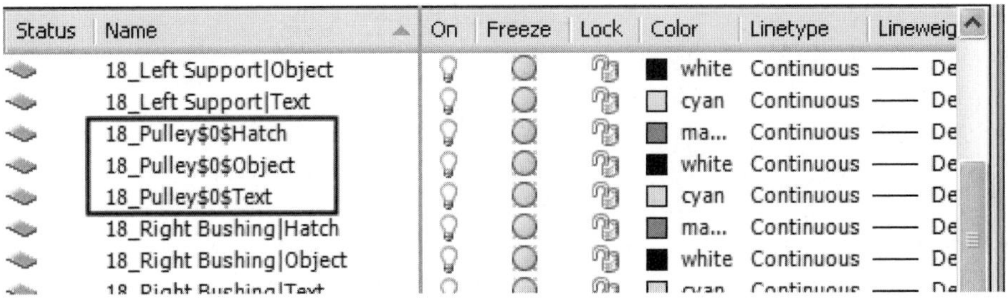

Status	Name	On	Freeze	Lock	Color	Linetype	Lineweig
	18_Left Support|Object				white	Continuous	De
	18_Left Support|Text				cyan	Continuous	De
	18_Pulley0Hatch				ma...	Continuous	De
	18_Pulley0Object				white	Continuous	De
	18_Pulley0Text				cyan	Continuous	De
	18_Right Bushing|Hatch				ma...	Continuous	De
	18_Right Bushing|Object				white	Continuous	De
	18_Right Bushing|Text				cyan	Continuous	De

Figure 18.29

The Insert option of the Bind Xrefs dialog box is similar to the Bind option. However, instead of named items such as blocks and layers being converted to the format XREFname0-BLOCKname, the name of the external reference is stripped, leaving just the name of the block, layer, or other named item (BLOCKname, LAYERname, etc.). In the following image, the following referenced layers were converted to usable layers with the Insert option:

Referenced Layers	Converted Layers
18_Pulley|Hatch	Hatch
18_Pulley|Object	Object
18_Pulley|Text	Text

Note: It is considered an advantage to use the Bind option over the Insert option. This way, you can identify the layer that was tied to the previously used external reference.

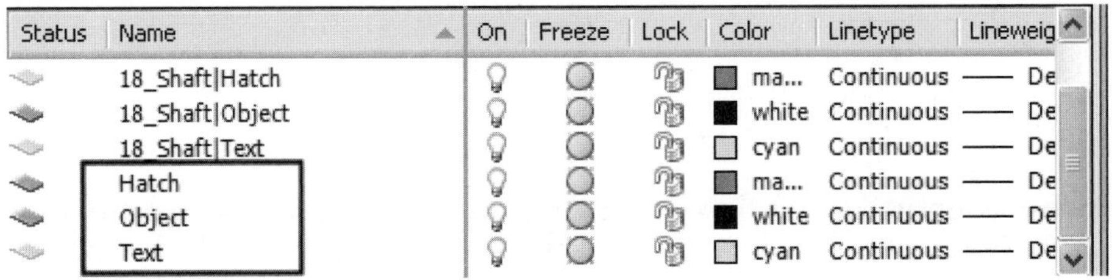

Figure 18.30

CLIPPING AN EXTERNAL REFERENCE

Since attaching an external reference displays the entire reference file, you have the option of displaying a portion of the file. This is accomplished by clipping the external reference with the XCLIP command. Choose this command from either the Reference toolbar or the Modify > Clip > Xref pull-down menu, as shown in the following image.

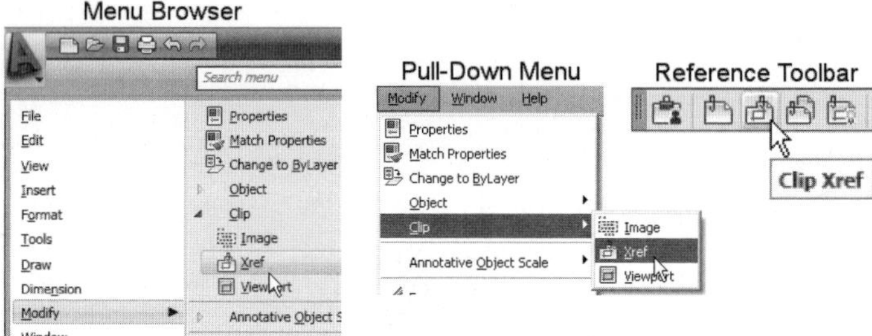

Figure 18.31

This operation is very useful when you want to emphasize a particular portion of your external reference file. Clipping boundaries include polylines in the form of rectangles, regular polygonal shapes, or even irregular polyline shapes. All polylines must form closed shapes. Also, clipping can take two forms, namely Outside and Inside modes. With the Outside mode, objects outside of the clipping boundary are hidden. With the Inside mode, objects inside of the clipping boundary are hidden.

Try It!: Open the drawing file 18_Facilities Plan. Follow the illustration on the left in the following image and the command prompt sequence below for performing an outside clipping operation.

```
Command: XCLIP
Select objects: (Pick the external reference)
Select objects: (Press ENTER to continue)
Enter clipping option
[ON/OFF/Clipdepth/Delete/generate Polyline/New boundary] <New>: N
    (For New)
Outside mode - Objects outside boundary will be hidden.
Specify clipping boundary or select invert option:
[Select polyline/Polygonal/Rectangular/Invert clip]
    <Rectangular>: (Press ENTER)
Specify first corner: (Pick a point at "A," as shown on the left
    in the following image)
Specify opposite corner: (Pick a point at "B")
```

The results are displayed on the right in the following image. If you want to return the clipped image to the full external reference, use the XCLIP command and use the OFF clipping option. This temporarily turns off the clipping frame. To permanently remove the clipping frame, use the Delete clipping option.

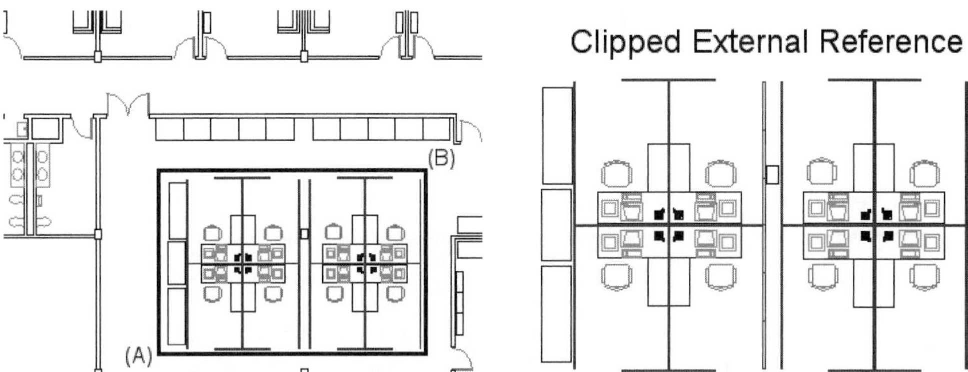

Figure 18.32

The following image illustrates the results of clipping an external reference based on Inside mode. In this example, the objects defined inside of the clipping boundary are removed.

```
Command: XCLIP
Select objects: (Pick the external reference in the
    previous image)
Select objects: (Press ENTER to continue)
Enter clipping option
[ON/OFF/Clipdepth/Delete/generate Polyline/New boundary] <New>: N
    (For New)
Outside mode - Objects outside boundary will be hidden.
Specify clipping boundary or select invert option:
[Select polyline/Polygonal/Rectangular/Invert clip]
    <Rectangular>: I (For Invert clip)
Inside mode - Objects inside boundary will be hidden.
Specify clipping boundary or select invert option:
[Select polyline/Polygonal/Rectangular/Invert clip]
    <Rectangular>: (Press ENTER)
Specify first corner: (Pick a point at "A," as shown in the
    previous image)
Specify opposite corner: (Pick a point at "B")
```

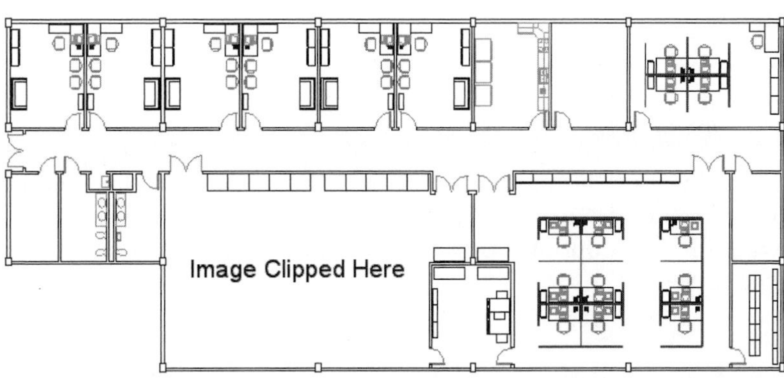

Figure 18.33

OTHER OPTIONS OF THE EXTERNAL REFERENCES PALETTE

The following additional options of the External References palette will now be discussed, using the following image as a guide.

File References

Reference...	Status	Size	Type	Date	Saved
18_Pulley Assembly	Opened	63.8 KB	Current	6/7/2007 8:21:3...	
18_Base Plate		60.5 KB	Overlay	6/7/2007 8:17:0...	D:\Tut
18_Left Bushing		66.5 KB	Overlay	6/7/2007 8:20:0...	D:\Tut
18_Left Support		60.8 KB	Overlay	6/7/2007 8:18:2...	D:\Tut
18_Pulley		61.3 KB	Overlay	6/7/2007 8:19:4...	D:\Tut
18_Right Bushing		60.5 KB	Overlay	6/7/2007 8:20:3...	D:\Tut
18_Right Support		60.8 KB	Overlay	6/7/2007 8:20:5...	D:\Tut
18_Shaft		60.3 KB	Overlay	6/7/2007 8:21:1...	D:\Tut

Open
Attach...
Unload
Reload
Detach
Bind...

Figure 18.34

UNLOAD

Unload is similar to the Detach option, with the exception that the external reference is not permanently removed from the database of the drawing file. When an external reference is unloaded, it is still listed in the External References palette, as shown in the following image. Notice that an arrow (facing down) is displayed to signify the unloaded status. Since this option suppresses the external reference from any drawing regenerations, it is used as a productivity technique. Reload the external reference when you want it returned to the screen.

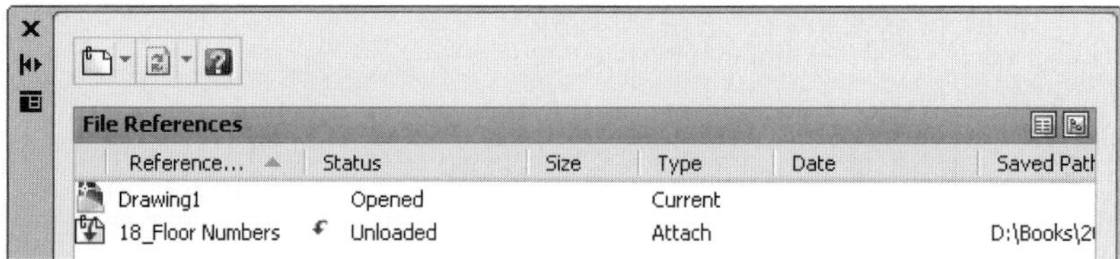

Figure 18.35

RELOAD

This option loads the most current version of an external reference. It is used when changes to the external reference are made while the external reference is currently being used in another drawing file. This option works well in a networked environment, where all files reside on a file server.

DETACH

Use this option to permanently detach or remove an external reference from the database of a drawing.

EXTERNAL REFERENCE NOTIFICATION TOOLS

A series of tools and icons are available to assist with managing external references. When a drawing consisting of external references opens, an icon appears in the extreme lower-right corner of your screen, as shown on the left in the following image. Clicking this icon launches the External References palette. In the event that the source file was changed or modified, the next time you return to the external reference drawing, a yellow caution icon appears, as shown on the right in the following image, informing you that the external reference was changed. This example refers to the Pulley Assembly, in which the Base Plate was modified and saved. When you return to the Pulley Assembly, the yellow caution icon signifies the change to the file 18_Base Plate.

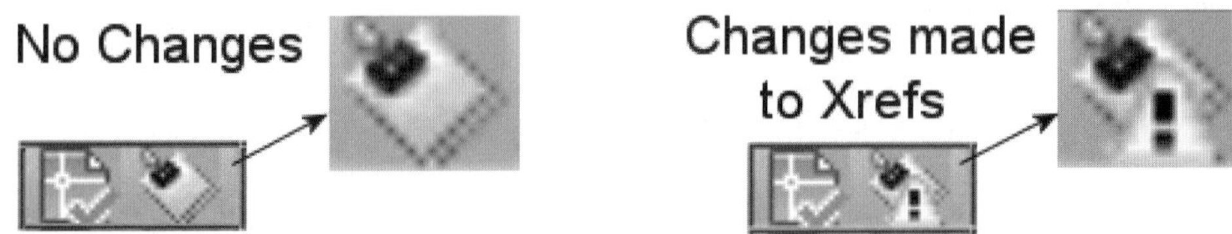

Figure 18.36

Clicking the yellow caution icon launches the External References palette. When you click the Refresh button, you will notice the appearance of a yellow caution triangle, as shown in the following image. The status of this file alerts you that 18_Base Plate needs to be reloaded in order for the change to be reflected in the Pulley Assembly drawing.

File Re[ferences] — Refresh

Reference...	Status	Size	Type	Date
18_Pulley Assembly	Opened	63.8 KB	Current	6/7/2007 8:21:3...
18_Base Plate	Needs reloading	58.5 KB	Overlay	5/9/2008 4:58:0...
18_Left Bushing	Loaded	66.5 KB	Overlay	6/7/2007 8:20:0...
18_Left Support	Loaded	60.8 KB	Overlay	6/7/2007 8:18:2...
18_Pulley	Loaded	61.3 KB	Overlay	6/7/2007 8:19:4...

Figure 18.37

Right-clicking 18_Base Plate displays a menu, as shown in the following image. Clicking Reload while the external reference is highlighted reloads the external reference.

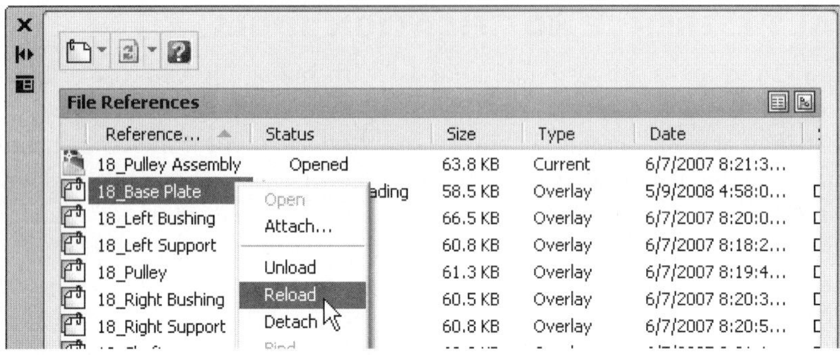

Figure 18.38

Illustrated in the following image is another feature of external references. The left support of the pulley assembly was selected. When you right-click, the shortcut menu appears, as shown on the right in the following image. Use this menu to perform the following tasks:

Edit Xref In-place—This option activates the Reference Edit dialog box for the purpose of editing the external reference in-place.

Open Xref—This option opens the selected external reference in a separate window.

Clip Xref—This option launches the XCLIP command for the purpose of clipping a portion of the external reference

External References—This option launches the External References palette.

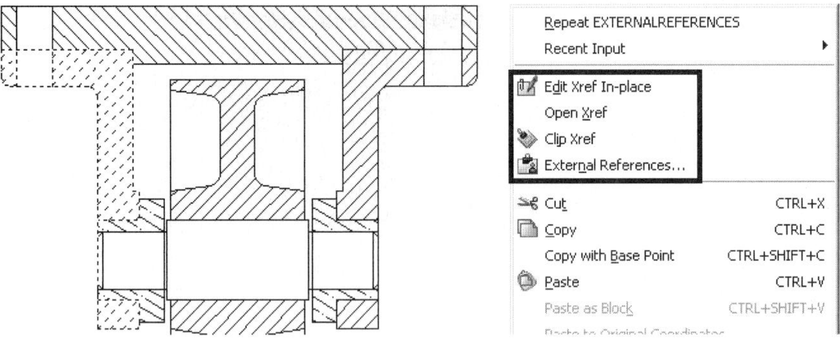

Figure 18.39

USING ETRANSMIT

ETransmit is an AutoCAD utility that is helpful in reading the database of your drawing and listing all support files needed. Once these files are identified, you can have this utility gather all files into one zip file. You can then copy these files to a disk or CD, or transmit the files over the Internet. Clicking eTransmit..., located in the File pull-down menu shown on the left in the following image, displays the Create Transmittal dialog box, as shown in the middle in the following image. Notice in the Files Tree tab a listing of all support files grouped by their specific category. For instance, an External References category exists. Clicking the "+" sign lists all external references set to be transmitted.

Clicking the Files Table tab, as shown on the right in the following image, displays a list of all files that will be included in the transmittal set.

When you are finished examining all the support files, clicking the OK button takes you to the Specify Zip File dialog box. It is here that you enter a file name. All support files listed under the Files Tree tab will be grouped into a single zip file for easy sharing with other individuals or companies.

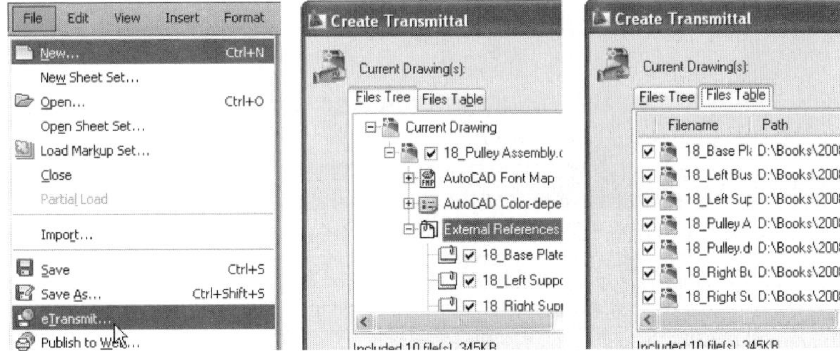

Figure 18.40

WORKING WITH RASTER IMAGES

Raster images in the form of JPG, GIF, TIF, and so on can easily be merged with your vector-based AutoCAD drawings. The addition of raster images can give a new dimension to your drawings. Typical examples of raster images are digital photographs of an elevation of a house or an isometric view of a machine part. Whatever the application, working with raster images is very similar to what was just covered with external references. You attach the raster image to your drawing file. Once it is part of your drawing, additional tools are available to manipulate and fine-tune the image for better results. To attach a raster image to a drawing, you can use the External References palette (Attach Image button), the Insert pull-down menu (click Raster Image Reference, as shown on the left in the following image), or the Reference toolbar (click the Attach Image button). The Reference toolbar also contains a number of tools for manipulating raster images, as shown on the right in the following image.

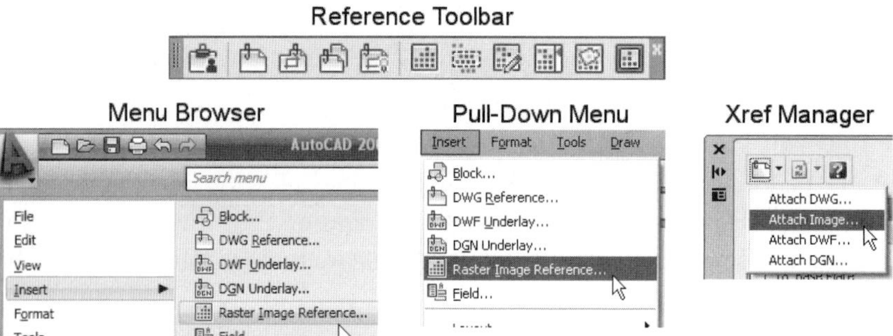

Figure 18.41

The following table outlines the main raster image commands found in the Reference toolbar.

Button	Tool	Function
	Attach Image	Used for attaching a raster image to a drawing
	Clip Image	Used for cropping or clipping the raster image
	Adjust Image	Used for adjusting the brightness, contrast, and fade factor of a raster image
	Image Quality	Uses two settings, high or draft, to control the quality of raster images
	Image Transparency	Controls whether the background of a raster image is transparent or opaque
	Image Frame	Used for turning the rectangular frame on or off for a raster image

To attach a raster image utilizing the Reference toolbar, click the Attach Image button. This launches the Select Image File dialog box, as shown in the following image. Choose the desired raster image to attach from the list and then click the Open button.

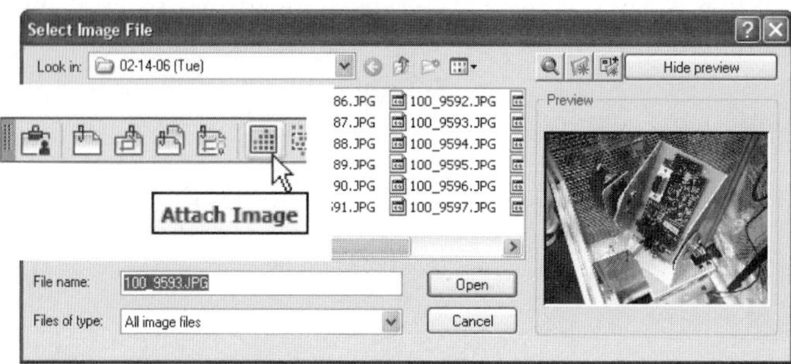

Figure 18.42

After choosing the correct raster image, the Image dialog box appears, as shown on the left in the following image. From this dialog box, change the insertion point, scale, or rotation parameters if desired. Clicking the OK button returns you to your drawing, where you pick an insertion point and change the scale factor of the image. A typical raster image is illustrated on the right in the following image.

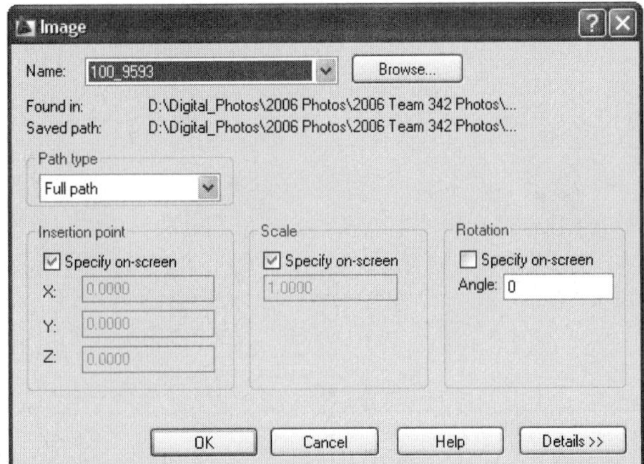

Figure 18.43

To manage all images in a drawing, click the External References button located in the Reference toolbar. The same dialog box used for managing external references is used for managing image files, as shown in the following image.

File References

Reference...	Status	Size	Type	Date	Saved Pat
Drawing1	Opened		Current		
100_2871	Loaded	758 KB	JPG	1/9/2007 6:55:5...	D:\Digital_
100_2878	Loaded	649 KB	JPG	1/9/2007 6:57:4...	D:\Digital_
100_2879	Loaded	677 KB	JPG	1/9/2007 6:58:0...	D:\Digital_

Figure 18.44

Try It!: Begin this exercise on working with raster images by opening the drawing file 18_Linkage, illustrated in the following image. In this exercise a raster image will be attached and placed in the blank area to the right of the two-view drawing.

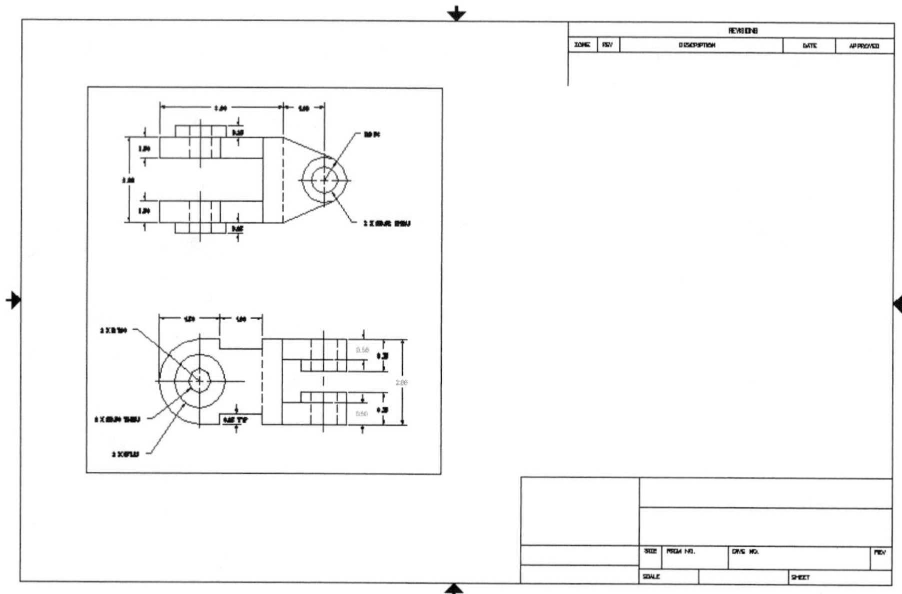

Figure 18.45

Click the Attach Image button located in the Reference toolbar. When the Select Image File dialog box appears, click the file Linkage1.jpg. Then click the Open button.

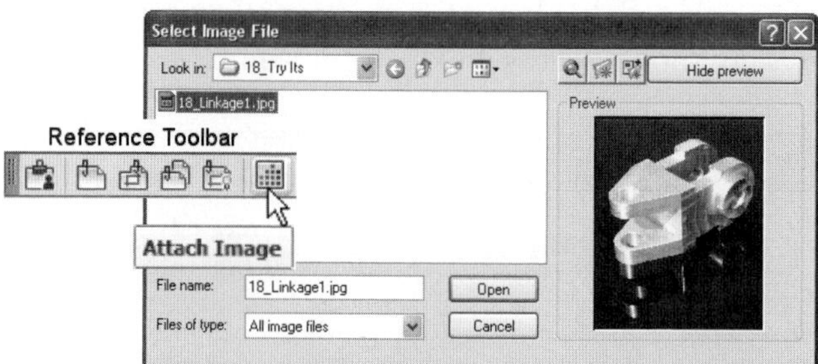

Figure 18.46

When the Image dialog box appears, as shown in the following image, leave all default settings. You will be specifying the insertion point on the drawing screen. Click the OK button to continue.

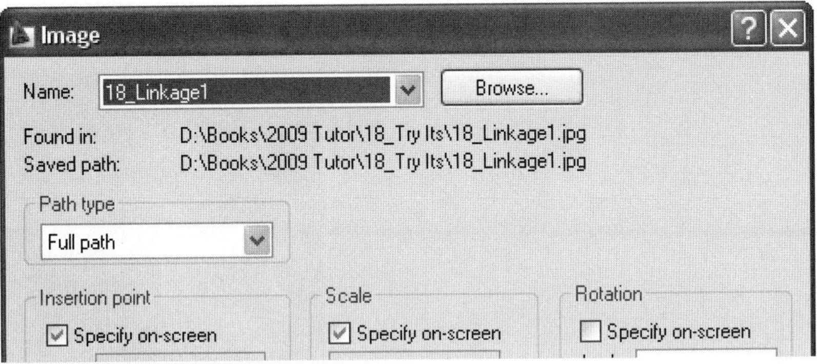

Figure 18.47

Once back in the drawing file, pick a point anchoring the lower-left corner of the graphic, as shown in the following image. Now move your cursor in an upward right direction to scale the image as necessary. Pick a second point at a convenient location to display the image.

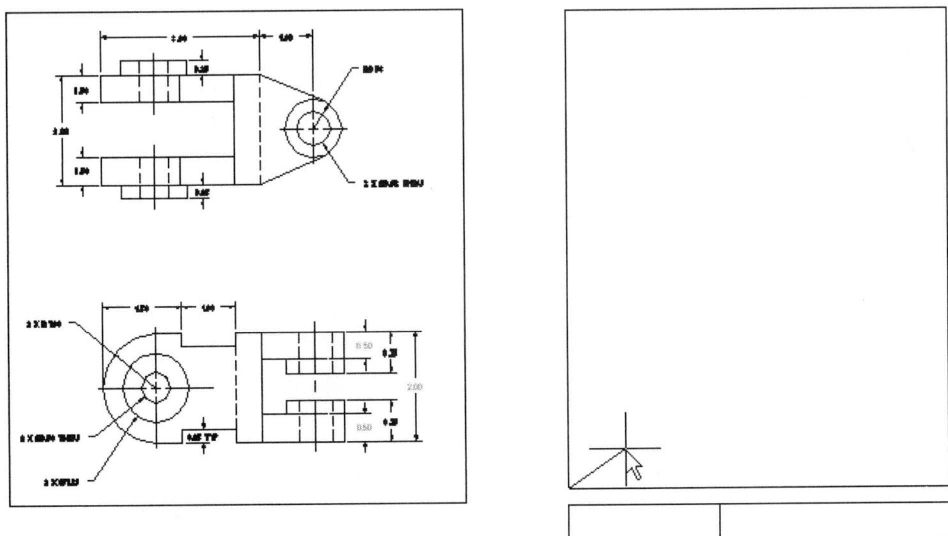

Figure 18.48

As the image displays on your screen, you can adjust its size very easily. Click the edge of the image (the image frame) and notice the grips appearing at the four corners of the image. Clicking a grip and then moving your cursor increases or decreases the image's size, as shown in the following image.

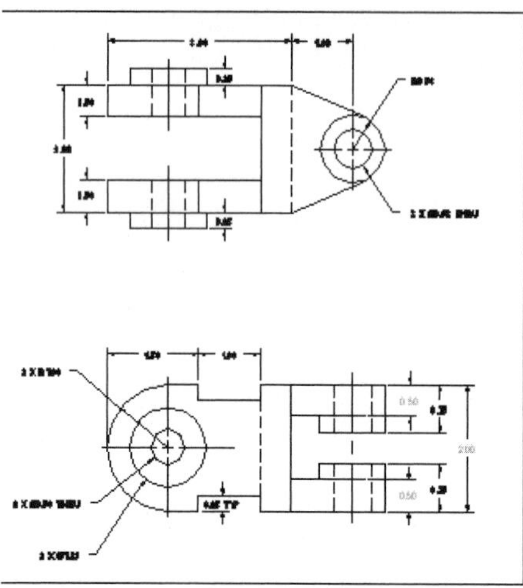

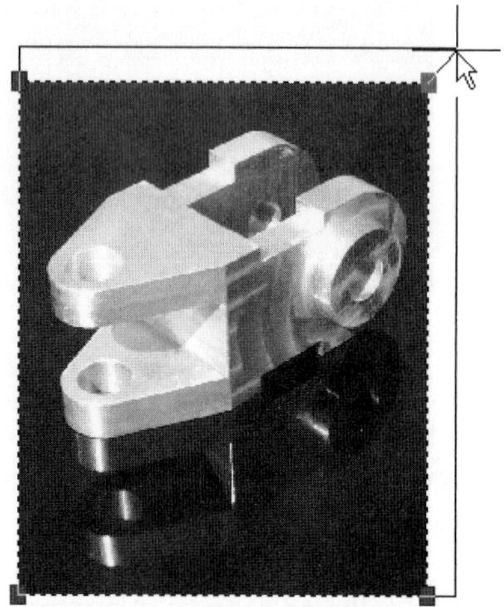

Figure 18.49

From the Reference toolbar shown in the following image, identify the Adjust Image button and click it. Clicking the edge of the raster image frame displays the Image Adjust dialog box, as shown in the following image. Adjust the Brightness, Contrast, and Fade settings and notice that each of these affects the image previewed to the right of this dialog box. You can easily return to the original image by clicking the Reset button in the lower-left corner of this dialog box.

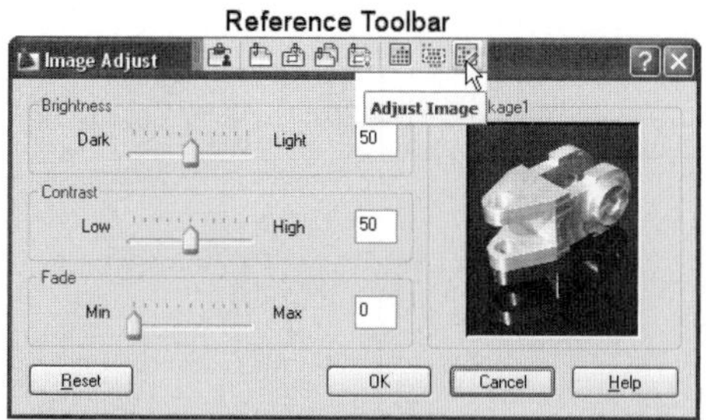

Figure 18.50

Clicking the OK button in the Image Adjust dialog box returns you to the drawing. The final results are illustrated in the following image, with the vector drawing and raster image sharing the same layout.

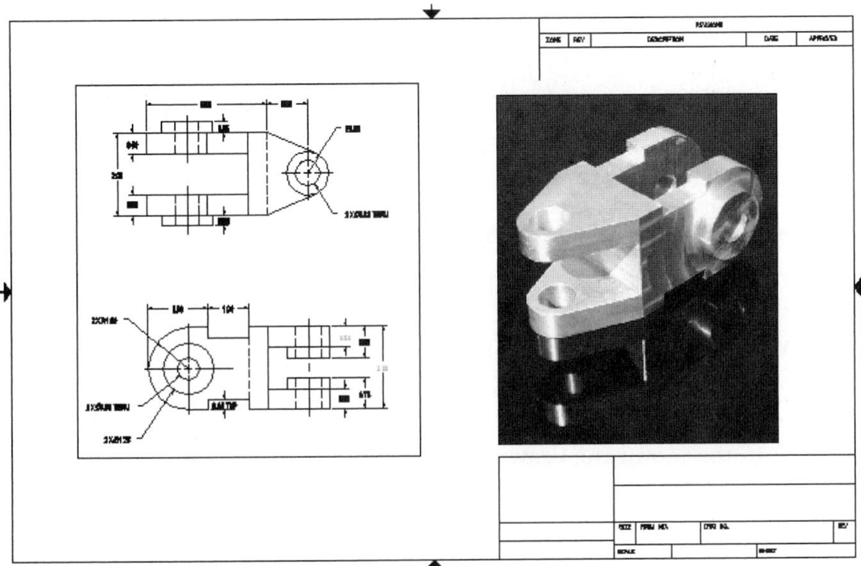

Figure 18.51

USING THE DRAWORDER COMMAND

With the enhancements made to raster images and the ability to merge raster images with vector graphics, it is important to control the order in which these images are displayed. The DRAWORDER command is used to provide this level of control over raster images. The DRAWORDER command can be selected from the Modify II toolbar or from the Tools pull-down menu, as shown in the following image. A dedicated Draw Order toolbar is also available.

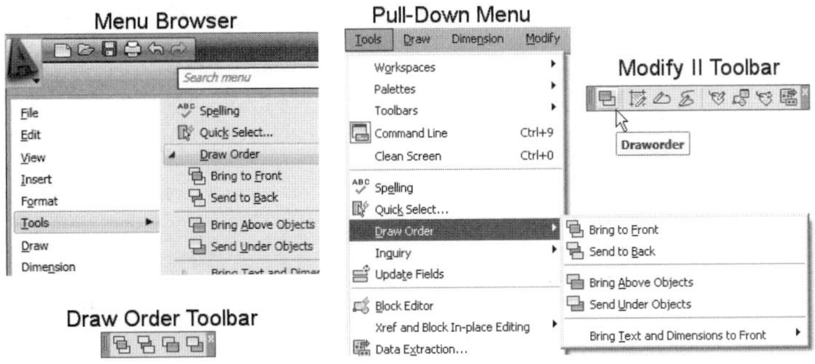

Figure 18.52

The four modes of the Draw Order toolbar are described as follows:

Button	Tool	Function
	Bring to Front	The selected object is brought to the top of the drawing order
	Send to Back	The selected object is sent to the bottom of the drawing order
	Bring Above Objects	The selected object is brought above a specified reference object
	Send Under Objects	The selected object is sent below the specified reference object

In the illustration on the left in the following image, the image of camera unit has been merged with two text objects, BASE PLATE and SWIVEL. Unfortunately, because of the drawing order of operations, the image is masking part of both text objects.

To correct the problem, select the Bring to Front tool. Both text objects are selected as the objects to bring to the top of the drawing order, and the results are illustrated on the right in the following image. Both text objects are now at the top of the drawing order and can be read in full.

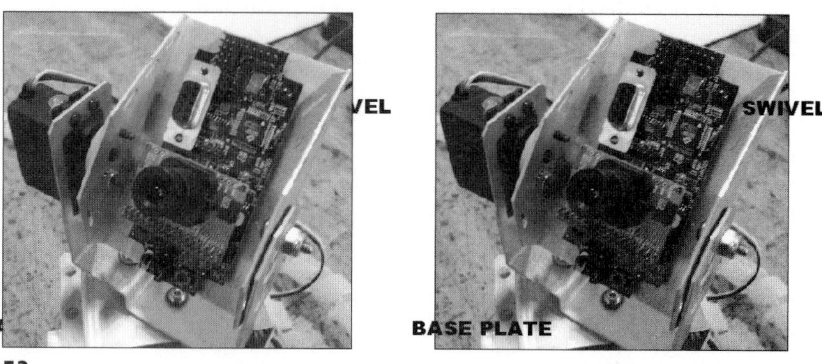

Figure 18.53

TUTORIAL EXERCISE: EXTERNAL REFERENCES

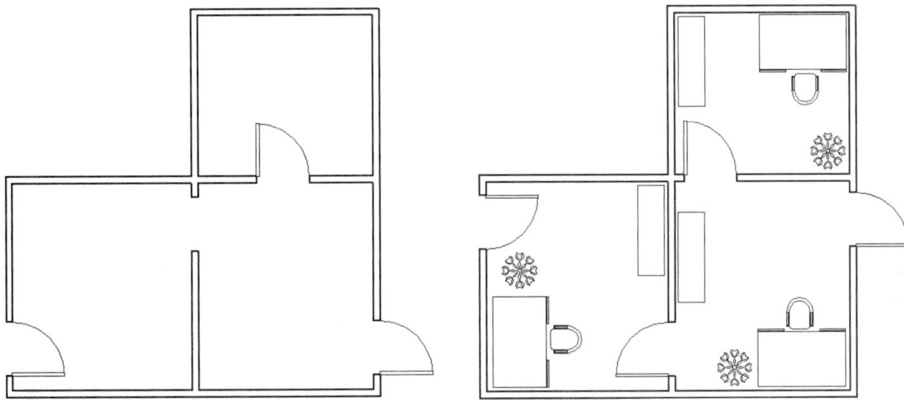

Figure 18.54

PURPOSE

This tutorial is designed to use the office floor plan to create an interior plan consisting of various interior symbols such as desks, chairs, shelves, and plants. The office floor plan will be attached to another drawing file through the XREF command.

SYSTEM SETTINGS

Since these drawings are provided on the CD, all system settings have been made.

LAYERS

The creation of layers is not necessary because layers already exist for both drawing files you will be working on.

SUGGESTED COMMANDS

Begin this tutorial by opening the drawing 18_Office.Dwg, which is located on the CD, and viewing its layers and internal block definitions. Then open the drawing 18_Interiors.Dwg, also located on the CD, and view its layers and internal blocks. The file 18_Office.Dwg will then be attached to 18_Interiors.Dwg. Once this is accomplished, chairs, desks, shelves, and plants will be inserted in the office floor plan for laying out the office furniture. Once 18_Interiors.Dwg is saved, a design change needs to be made to the original office plan; open 18_Office.Dwg and stretch a few doors to new locations. Save this file and open 18_Interiors.Dwg; notice how the changes to the doors are automatically made. The Xbind dialog box will also be shown as a means for making a block that had previously belonged to an external reference usable in the file 18_Interiors.Dwg.

Step 1

Open 18_Office.Dwg, which can be found on the CD, and observe a simple floor plan consisting of three rooms, as shown on the left in the following image. Furniture will be laid out using the floor plan as a template.

Step 2

While in 18_Office.Dwg, use the Layer Properties Manager palette and observe the layers that exist in the drawing for such items as doors, walls, and floor, as shown on the right in the following image. These layers will appear differently once the office plan is attached to another drawing through the XREF command.

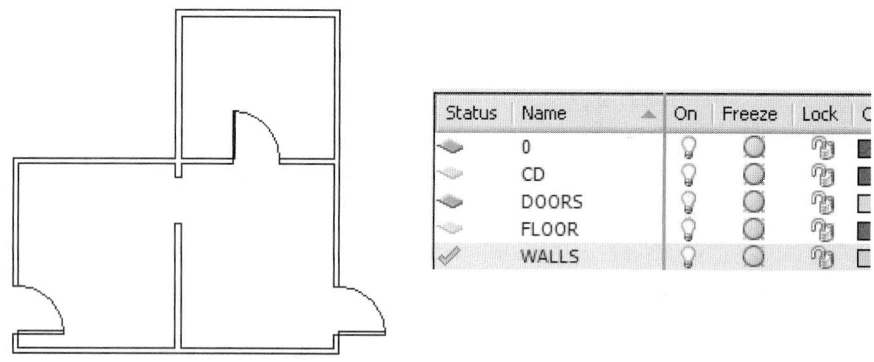

Figure 18.55

Step 3

While in the office plan, activate the Insert dialog box through the INSERT command. At times, this dialog box is useful for displaying all valid blocks in a drawing. Clicking the Name drop-down list displays the results shown in the following image. Two blocks are currently defined in this drawing; as with the layers, once the office plan is merged into another drawing through the XREF command, these block names will change. When you have finished viewing the defined blocks, close 18_Office.Dwg.

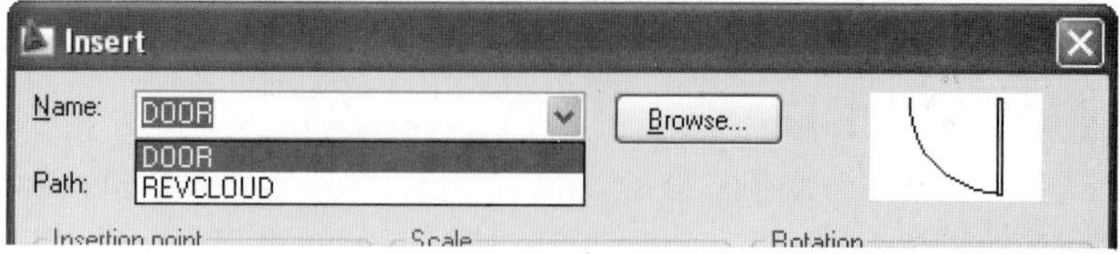

Figure 18.56

Step 4

This next step involves opening 18_Interiors.Dwg, also found on the CD, and looking at the current layers found in this drawing. Once this drawing is open, use the Layer

Properties Manager palette to observe that layers exist in this drawing for such items as floor and furniture, as shown in the following image.

Status	Name	On	Freeze	Lock	Color	Linetype	Lineweight	Plot Style	Plot
	0				white	CONTINU...	—— Default	Color_7	
	FLOOR				white	CONTINU...	—— Default	Color_7	
✓	FURNITURE				red	CONTINU...	—— Default	Color_1	

Figure 18.57

Step 5

As with the office plan, activate the Insert dialog box through the INSERT command to view the blocks internal to the drawing, as shown in the following image. The four blocks listed consist of various furniture items and will be used to lay out the interior plan.

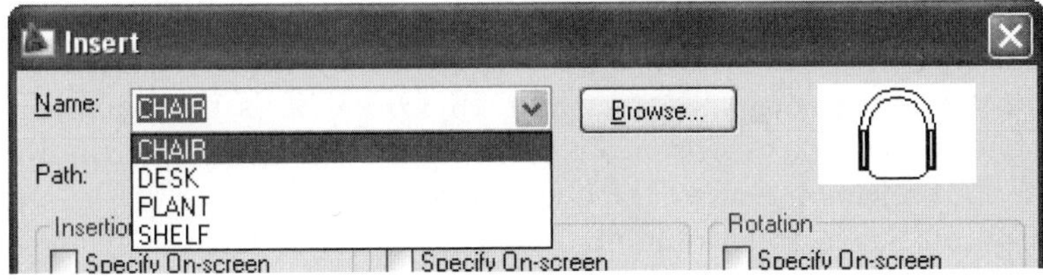

Figure 18.58

Step 6

Close 18_Office.Dwg and verify that 18_Interiors.Dwg is still open and active. The office floor plan will now be attached to the interior plan in order for the furniture to be properly laid out. Make the Floor layer current. Rather than insert the office plan as a block, use the External Reference palette to attach the drawing. This palette will activate when you enter the XREF command from the keyboard or choose External References from the Insert pull-down menu, as shown on the left in the following image. After the palette displays, click the Attach DWG button (or the drop-down list, if the button is not displayed), as shown on the right in the following image.

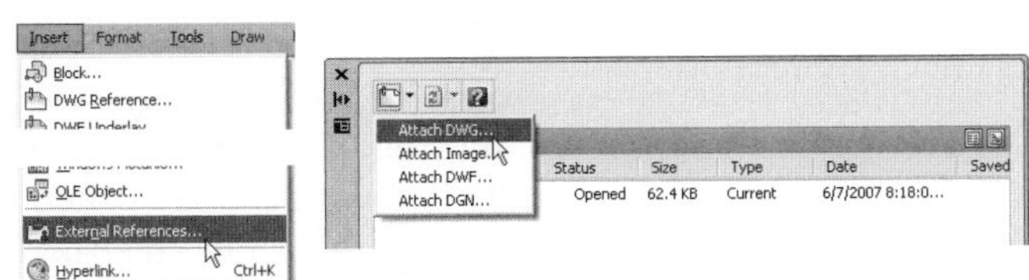

Figure 18.59

Step 7

Clicking Attach DWG, shown in the previous image, displays the Select Reference File dialog box, as shown in the following image. Find the appropriate folder and click the drawing file 18_Office.Dwg.

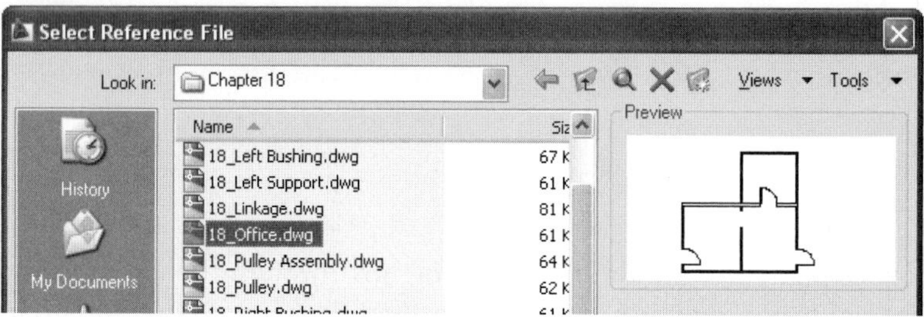

Figure 18.60

Step 8

Selecting the file 18_Office.Dwg displays the External Reference dialog box, as shown on the left in the following image. Notice that 18_OFFICE is the name of the external reference file chosen for attachment in the current drawing. Verify that the Reference Type is an attachment. If selected, remove the checkmarks from the Specify On-screen boxes for the Insertion point, Scale, and Rotation. In the External Reference dialog box, click the OK button to attach 18_Office.Dwg to 18_Interiors.Dwg.

The floor plan is now attached to the Interiors drawing at the insertion point 0,0, as shown on the right in the following image.

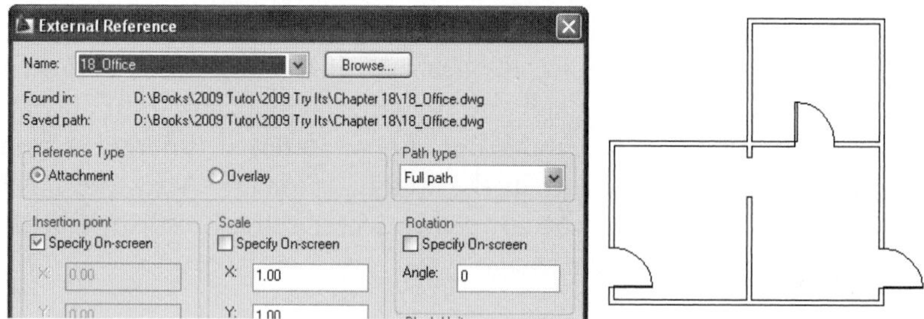

Figure 18.61

With the floor plan attached to the interiors drawing, notice how both drawings appear in the External References palette, as shown in the following image. When you have finished studying the information located in this palette, you can dismiss the palette from the screen by clicking the "X" located in the upper-left corner.

File References

Reference...	Status	Size	Type	Date	Sav
18_Interiors*	Opened	62.4 KB	Current	6/7/2007 8:18:0...	
18_Office	Loaded	60.6 KB	Attach	6/7/2007 8:18:4...	D:\B

Figure 18.62

Step 9

Once again, activate the Layer Properties Manager palette, paying close attention to the display of the layers. Using the following image as a guide, you can see the familiar layers of Floor and Furniture. However, notice the group of layers beginning with 18_Office; the layers actually belonging to the external reference file have the designation of XREF|LAYER. For example, the layer 18_Office|DOORS represents a layer located in the file 18_Office.Dwg that holds all door symbols. The "|," or Pipe symbol, is used to separate the name of the external reference from the layer. The layers belonging to the external reference file may be turned on or off, locked, or even frozen. However, these layers cannot be made current for drawing on.

Status	Name	On	Freeze	Lock	Color	Linetype	Lineweight	Plot Style	Plo
	0				white	CONTINU...	—— Default	Color_7	
	18_Office|CD				red	CONTINU...	—— Default	Color_1	
	18_Office|DOORS				cyan	CONTINU...	—— Default	Color_4	
	18_Office|FLOOR				red	CONTINU...	—— Default	Color_1	
	18_Office|WALLS				gre...	CONTINU...	—— Default	Color_3	
	FLOOR				white	CONTINU...	—— Default	Color_7	
✓	FURNITURE				red	CONTINU...	—— Default	Color_1	

Figure 18.63

Step 10

Make the Furniture layer current and begin inserting the desk, chair, shelf, and plant symbols in the drawing using the INSERT command or through the DesignCenter. The external reference file 18_OFFICE is to be used as a guide throughout this layout. It is not important that your drawing match exactly the image shown on the left in the following image. After positioning all symbols in the floor plan, save your drawing under its original name of 18_Interiors.Dwg but do not close the file.

Even with the interior drawing file still open, make the original drawing file, 18_Office. Dwg, current by opening this file, and make the following modifications: stretch all three doors as indicated to new locations and mirror one of the doors so it is positioned closer to the wall; stretch the wall opening over to the other end of the room, as shown on the right in the following image. Finally, save these changes under the original name of 18_Office.Dwg and close the file.

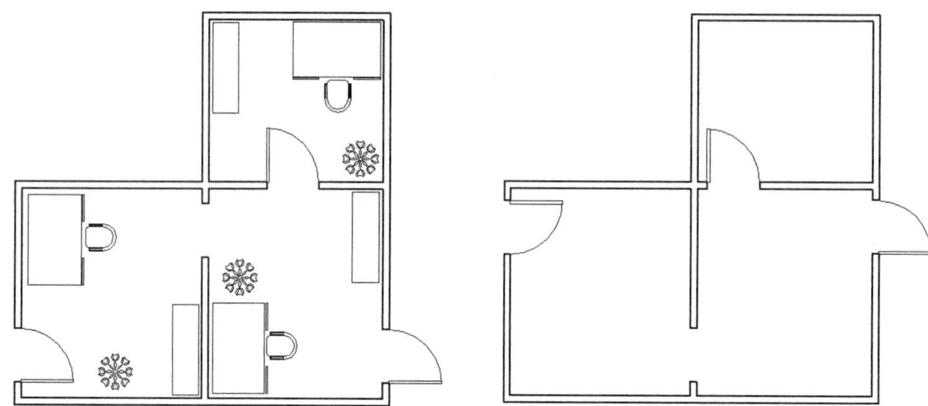

Figure 18.64

Step 11

After you close the office plan, your display returns to the interiors plan. At this point, nothing in the drawing will appear to have changed. However, notice in the lower-right corner of your display screen a tray of buttons. One of these buttons (Manage Xrefs) deals specifically with external reference files. Once the office plan is changed and saved, a Manage Xrefs button appears in this portion of your screen. It is also displayed on the left in the following image. Right-clicking this button displays two options. One of the options allows you to reload all external reference files. Click this option and notice the effects on the floor plan, illustrated on the left of the following image, where changes to the office plan are now reflected in the interiors drawing.

In this case, observe how some of the furniture is now in the way of the doorways. If the office plan had been inserted in the interiors drawing as a block, these changes would not have occurred this easily.

Because of the changes in the door openings, edit the drawing by moving the office furniture to better locations. The illustration on the right in the following image can be used as a guide, although your drawing may appear slightly different.

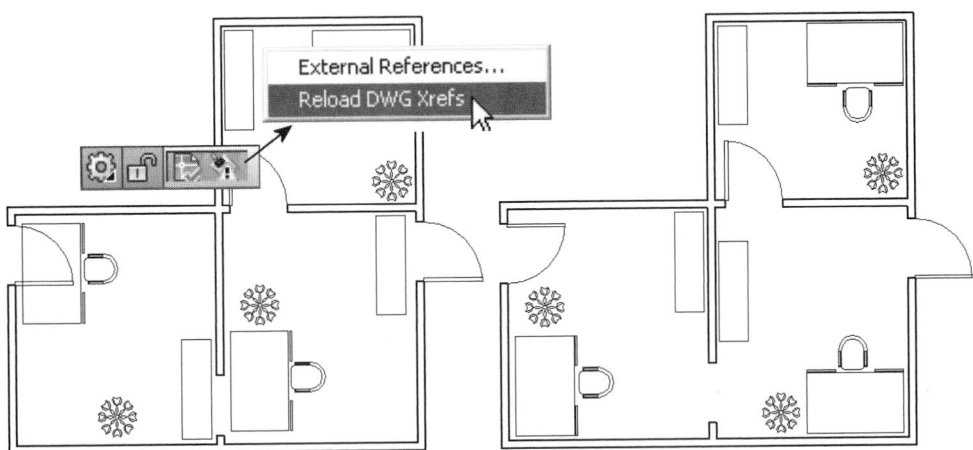

Figure 18.65

Step 12

A door needs to be added to an opening in one of the walls of the office plan. However, the door symbol belongs to the externally referenced file 18_Office.Dwg. The door block is defined as 18_Office|DOOR; the "|" character is not valid in the naming of the block and, therefore, cannot be used in the current drawing. The block must first be bound to the current drawing before it can be used. Use the XBIND command to make the external reference's Door block available in the 18_Interiors.Dwg. This command can be selected from the Reference toolbar, as shown in the following image, which displays the Xbind dialog box. While in this dialog box, click the "+" symbol next to the file 18_Office and then click the "+" symbol next to Block. This displays all blocks that belong to the external reference.

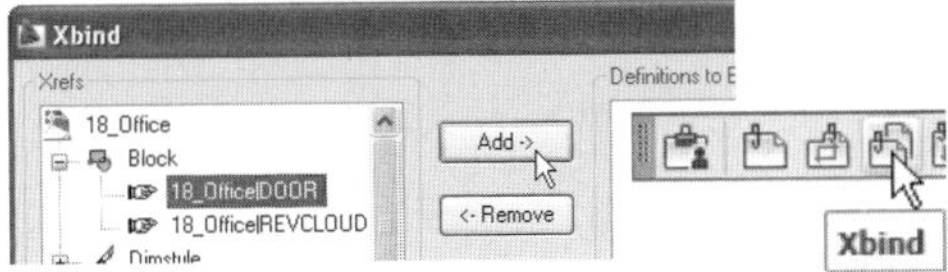

Figure 18.66

Step 13

Clicking 18_OFFICE|DOOR followed by the Add -> button, as shown in the previous image, moves the block of the door to the Definitions to Bind area in the following image. Click OK to dismiss the dialog box; the door symbol is now a valid block that can be inserted in the drawing.

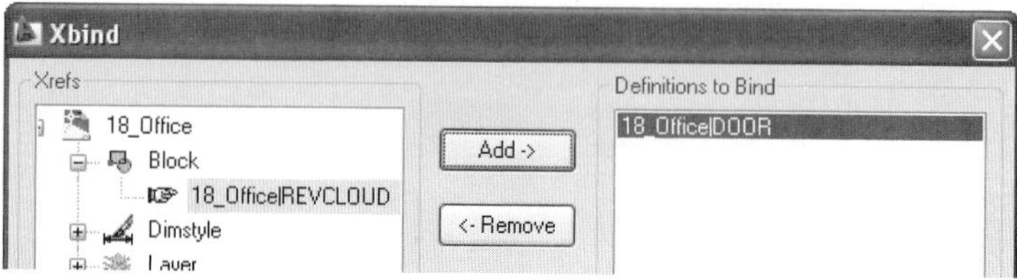

Figure 18.67

Step 14

Activate the Insert dialog box through the INSERT command, click the Name drop-down list, and notice the name of the door, as shown on the left in the following image. It is now listed as 18_Office0DOOR; the "|" character was replaced by the "0," making the block valid in the current drawing. This is AutoCAD's standard way of converting blocks that belong to external references to blocks that can be used in the current drawing file. This same procedure works on layers belonging to external references as well.

Step 15

Make the Floor layer current and insert the door symbol into the open gap, as shown on the right in the following image, to complete the drawing.

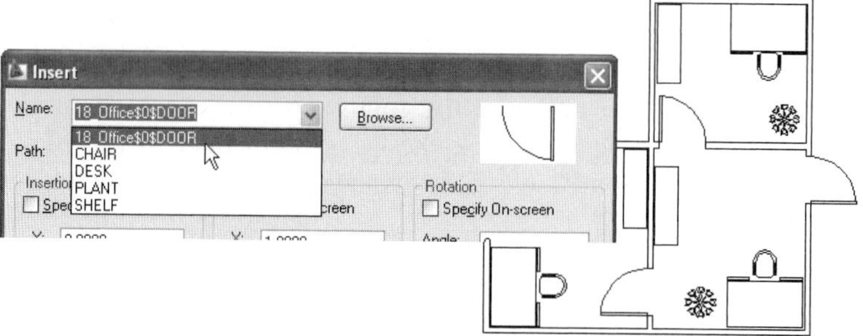

Figure 18.68

Advanced Layout Techniques

This chapter is a continuation of chapter 14, "Working with Drawing Layouts." First, we will demonstrate how multiple images of the same drawing file can be laid out at different scales. This demonstration will include a number of layering and dimensioning techniques to achieve success. This chapter continues by demonstrating how to create and manage annotation scales. Arranging multiple details of different drawing files in the same layout will also be demonstrated.

ARRANGING DIFFERENT VIEWS OF THE SAME DRAWING

This chapter begins immediately with a tutorial exercise designed to lay out two views of the same drawing in the same layout. Both images will be scaled at different values inside their respective veiwports. Follow the next series of steps for creating this type of layout.

TUTORIAL EXERCISE: 19_ROOF PLAN.DWG

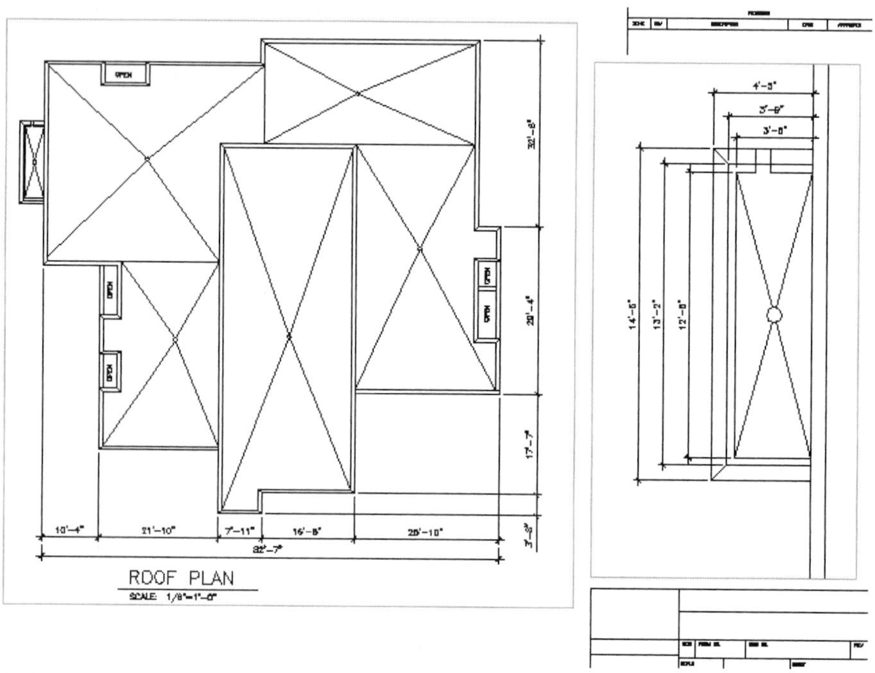

Figure 19.1

PURPOSE

This tutorial exercise is designed to lay out the two architectural views displayed in the previous image in the Layout mode (Paper Space). The two views consist of an overall and a detail view of a roof plan.

SYSTEM SETTINGS

Open an existing drawing called 19_Roof Plan. Keep all default settings for the units and limits of the drawing.

LAYERS

Layers have already been created for this drawing.

SUGGESTED COMMANDS

A new viewport will be created to house a detail view of the roof plan. Dimensions will also be added to this viewport. However, the Dimension scale setting of the drawing needs to be set to "Scale dimension to layout." This will enable the scale of the dimensions in the new viewport to match those dimensions in the main roof plan viewport.

Step 1

Open the drawing file 19_Roof Plan. Create a new viewport, as shown in the following image. This viewport should be long and narrow to accommodate the detail. Notice that when you are creating the viewport, the entire image zooms to the extents of the viewport. This is normal and you will arrange the detail in the next series of steps.

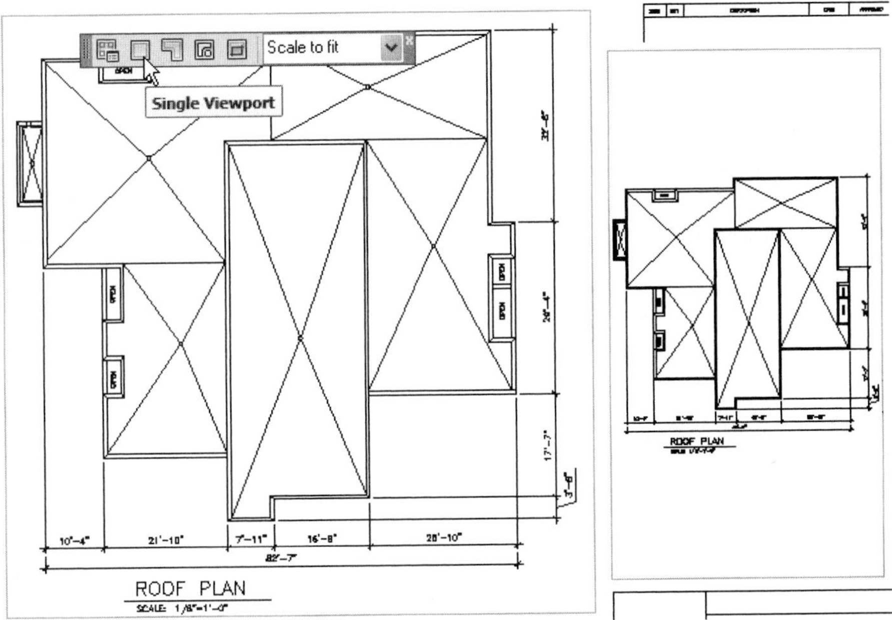

Figure 19.2

Step 2

Double-click inside the new viewport to make it active. Then pan the area of the roof plan so it appears similar to the following image. When applying a scale to this viewport, it is easier to see the results if the image you are detailing is approximately in the middle of the viewport. If it is not, it may be difficult to locate the area of the detail.

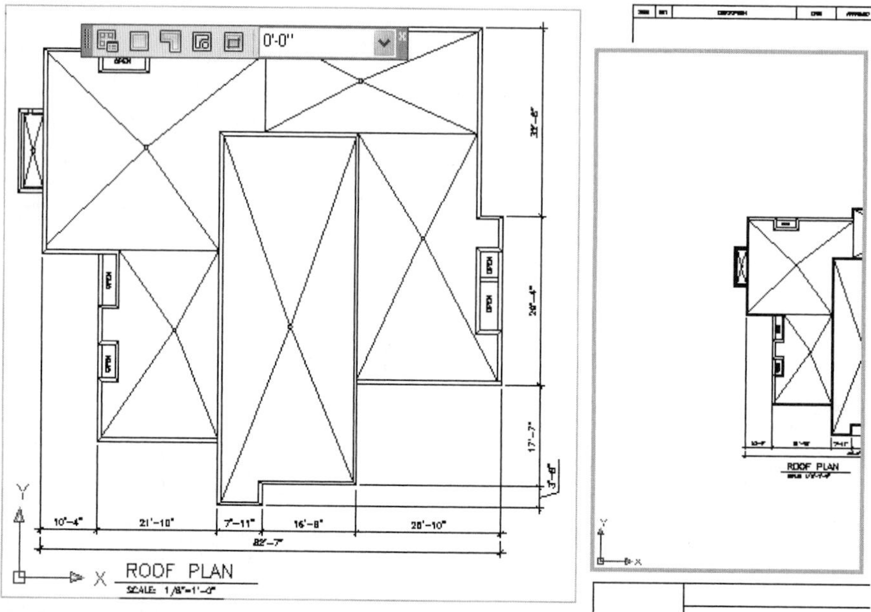

Figure 19.3

Step 3

Activate the various scales from the Viewports toolbar and pick 1/2" = 1'-0". This should increase the size of the detail in this viewport. Use the PAN command to center the long rectangular portion of the roof in the viewport, as shown in the following image.

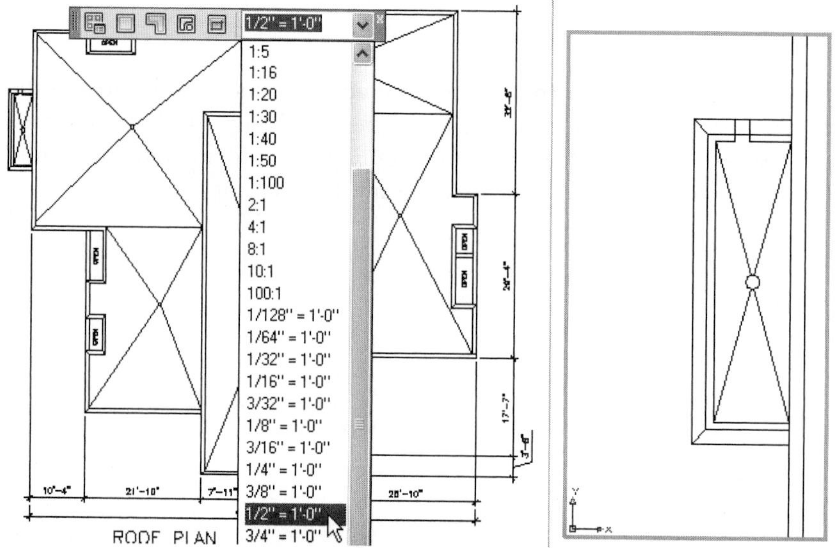

Figure 19.4

Step 4

Make the Detail Dim layer current. This layer will be used to hold all dimensions that will be added to the detail view. Before adding dimensions, one more item needs to be taken care of. First activate the main floor plan viewport by clicking inside it to make it active. Then launch the Dimension Styles Manager dialog box, click the Modify button, and pick the Fit tab. Under the Scale for dimension features area, click the radio button next to Scale dimensions to layout, as shown in the following image. When finished, click the OK button to dismiss this dialog box and close the main Dimension Style Manager dialog box. If the dimension scale does not automatically set for you (the dimensions are unreadable), you will have to update all dimensions in this viewport to reflect the changes to this dimension style. This can be accomplished by picking Update from the Dimension pull-down menu and selecting all the dimensions in the viewport.

Note: You can verify that the scale is correct by using the Properties Window or the LIST command on a dimension and noting that the Dimscale system variable has been overridden and set to 96.

Step 5

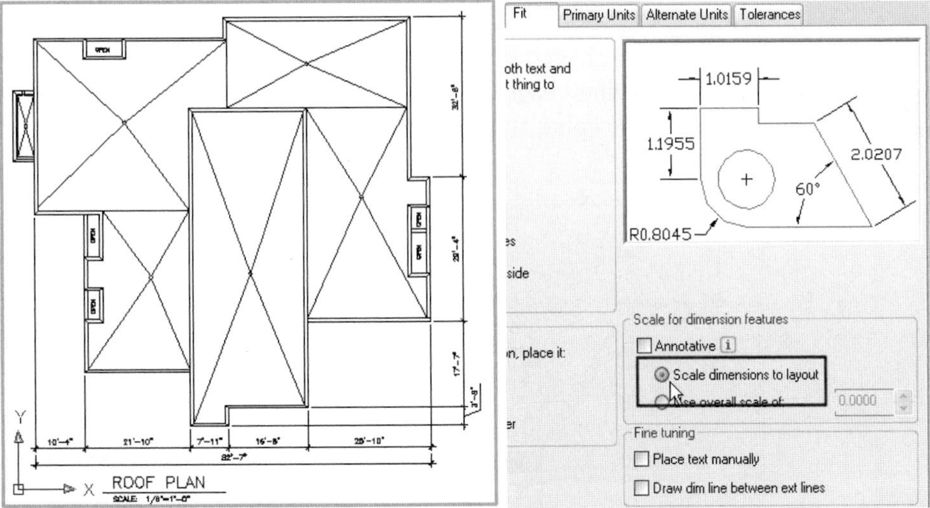

Figure 19.5

Activate the viewport containing the roof detail and place the linear dimensions as shown in the following image. Notice that the sizes of these dimensions exactly match those in the main floor plan. The scale of the viewport, which we set earlier to 1/2" = 1'-0", sets the scale of the dimensions. This occurs automatically because the "Scale dimensions to layout" button was selected, as shown in the previous step. You should also notice that as these dimensions are placed in the detail, they also appear in the main floor plan viewport, as shown in the following image.

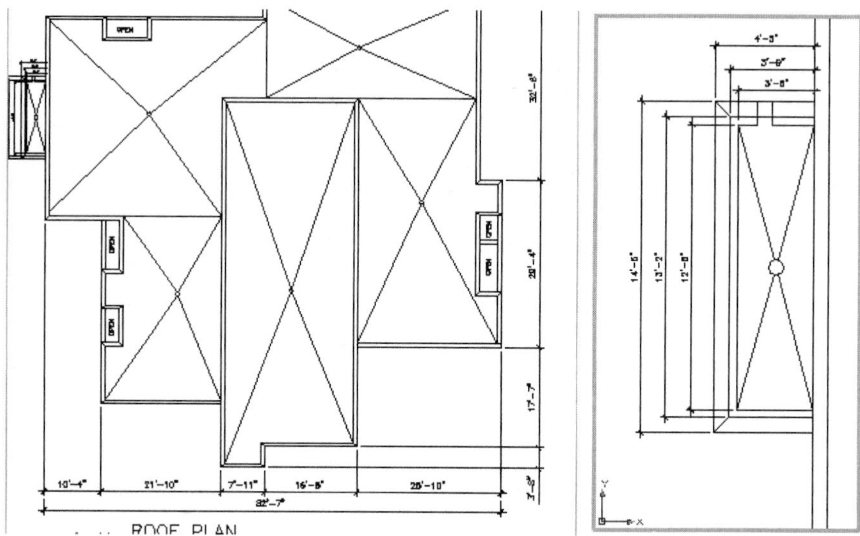

Figure 19.6

Step 6

Activate the main floor plan viewport by clicking inside it. Then launch the Layer Properties Manager palette. Identify the Dim Detail layer and freeze it in the current viewport by clicking the button as shown in the following image. This action will freeze the Dim Detail layer in the main floor plan viewport while keeping the layer visible in the detail viewport.

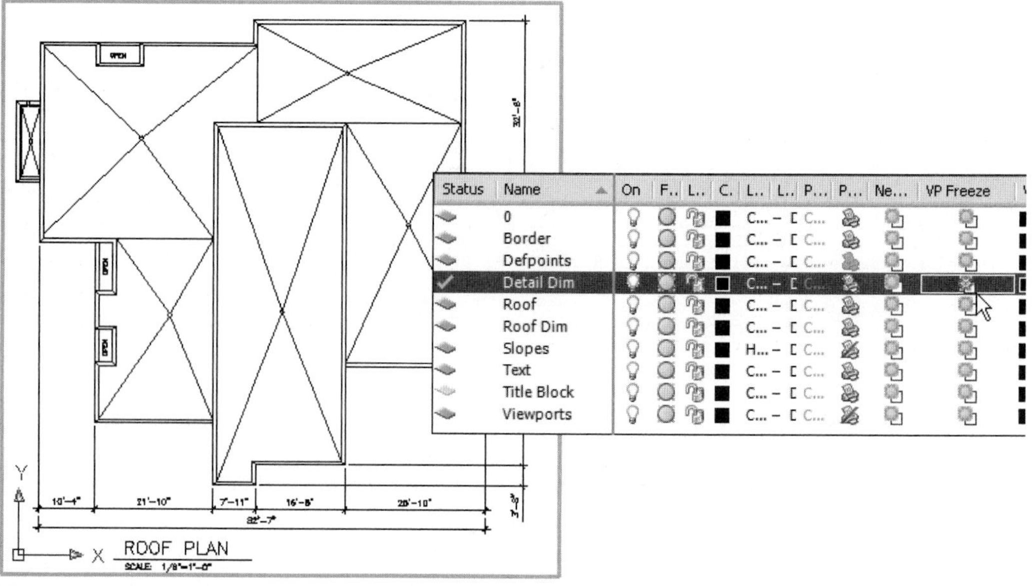

Figure 19.7

Step 7

The completed layout is shown in the following image.

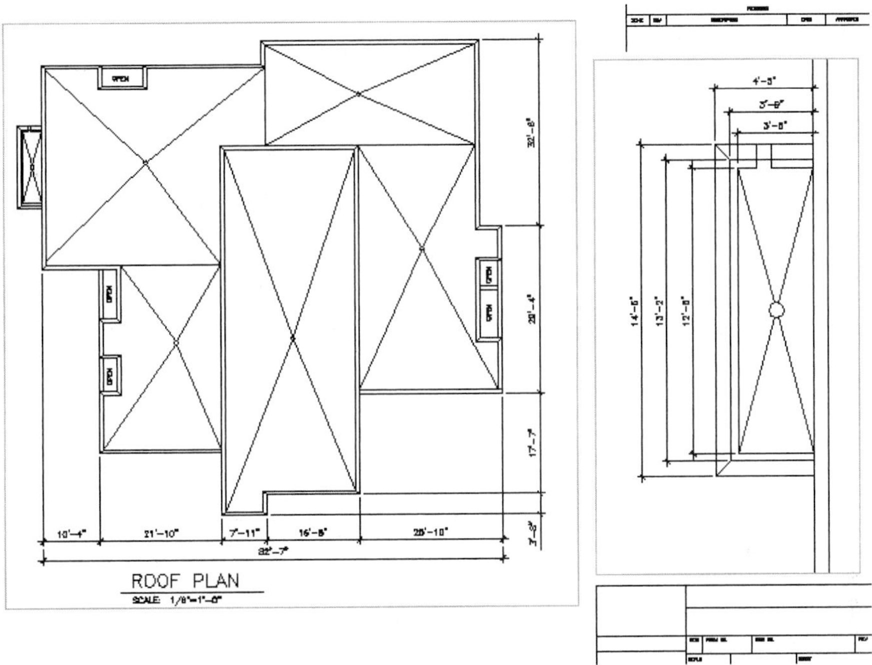

Figure 19.8

Step 8

Switch back to Model Space and notice the appearance of dashed lines representing the slope lines of the roof, as shown on the left in the following image. Switch back to Paper Space and notice that the dashed lines disappear. To display the dashed lines in the Paper Space viewports, change the LTSCALE value to 1.00; the dashed lines will appear as shown on the right in the following image. The ability to have linetypes display in Paper Space is controlled by the PSLTSCALE (Paper Space Scale) system variable, which is also set to a default value of 1.00. In general, when utilizing layouts, you should leave the LTSCALE set to 1.00 and linetypes will automatically be displayed in the viewport at the designated viewport scale.

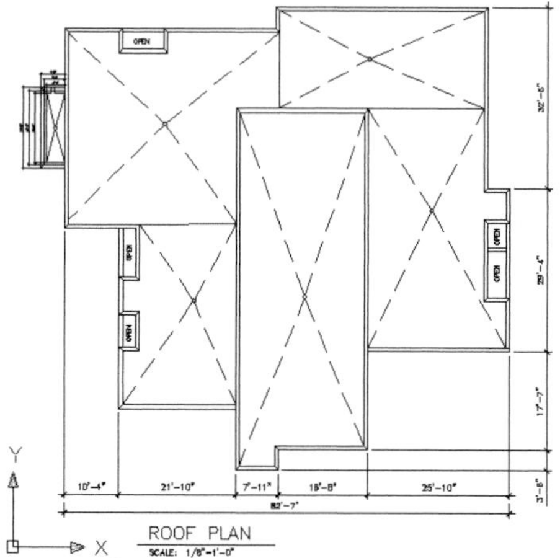

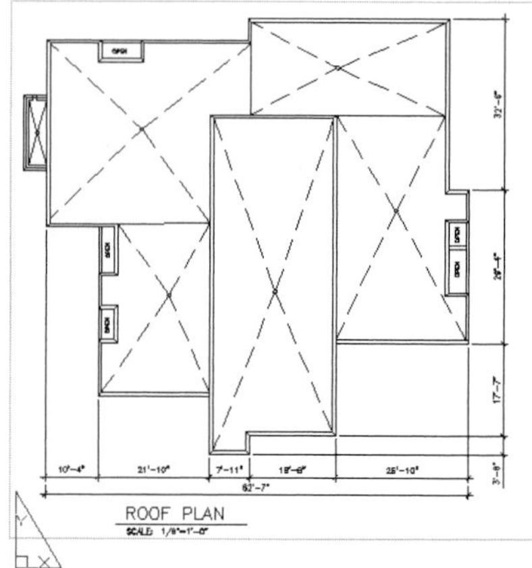

Figure 19.9

CREATING A DETAIL PAGE IN LAYOUT MODE

This next discussion focuses on laying out on the same sheet a series of details composed of different drawings, which can be at different scales. As the viewports are laid out in Paper Space and images of the drawings are displayed in floating Model Space, all images will appear in all viewports. This is not a major problem, because Paper Space allows for layers to be frozen in certain viewports and not in others.

Try It!: Open the drawing file 19_Bearing Details. The following image shows three objects: a body, a bushing, and a bearing all arranged in Model Space. The body and bearing will be laid out at a scale of 1 = 1. The bushing will be laid out at a scale of 2 = 1 (enlarged to twice its normal size). The dimension scales have been set for all objects in order for all dimension text to be displayed at the same size. Also, layers have been created for each object.

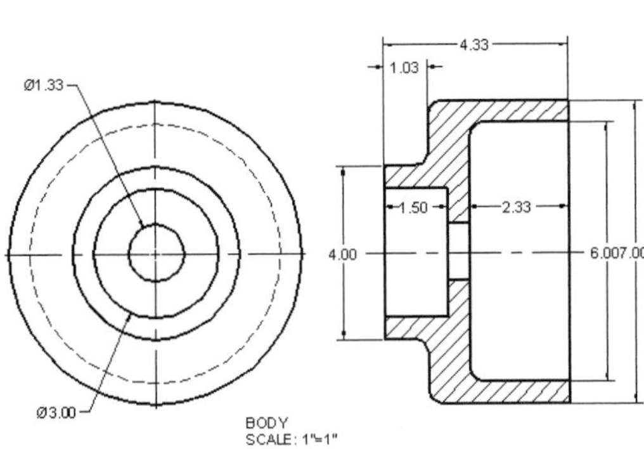

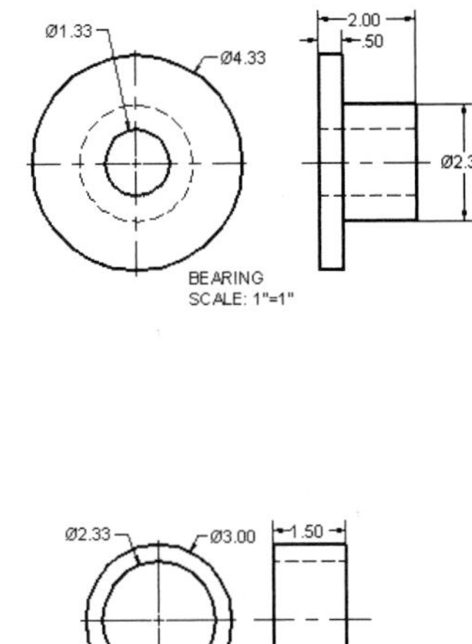

Figure 19.10

The following items have already been added to this drawing:

- A Vports layer was created. This layer holds all viewport information, and when the drawing is completed can be frozen or turned off.
- In Layout mode, an ANSI D-size sheet was selected for Page Setup.
- A D-size title block was inserted onto the Title Block layer.

Click the layout name ANSI-D. Except for the title block, the drawing sheet is empty. Using the Single Viewport tool located in the Viewports toolbar, create three viewports, as shown in the following image. The exact size of these viewports is not important at this time since you will have to adjust them at the end of the exercise. When creating these viewports, notice that all three objects (body, bearing, and bushing) appear in all three viewports. You will now freeze layers inside these viewports in order to show a different part in each viewport.

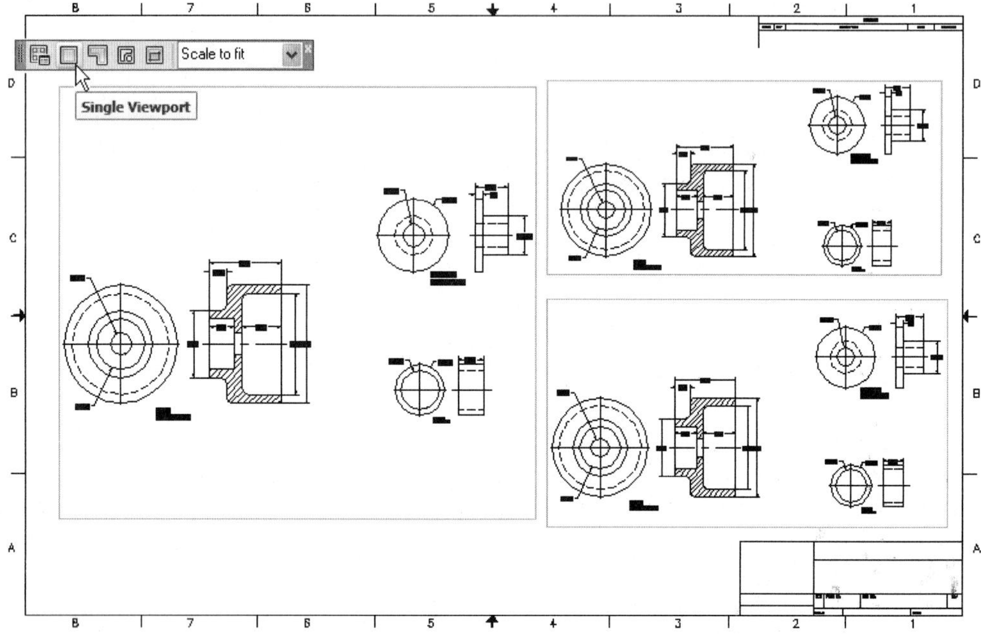

Figure 19.11

Begin the process of isolating one part per viewport by making the large viewport on the left active by double-clicking inside it. You will notice the viewport taking on the familiar thick border appearance, and the UCS icon is present in the lower-left corner of the viewport, as shown on the left in the following image. Then activate the Layer Properties Manager palette. All layers that begin with Body need to be visible in this viewport. Pick all layers that begin with Bearing and Bushing and freeze these layers only in this viewport, as shown on the right in the following image.

Note: To select multiple layers at one time, hold down the CTRL key as you select each layer name.

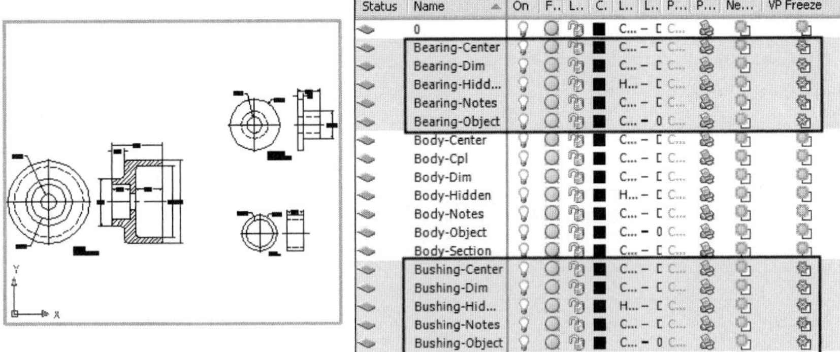

Figure 19.12

Next, click inside the viewport located in the upper-right corner to make it active, as shown on the left in the following image. Activate the Layer Properties Manager palette. All layers that begin with Bearing need to be visible in this viewport. Pick all layers that begin with Body and Bushing and freeze these layers only in this viewport, as shown on the right in the following image.

Note: To select all the layers at one time, select the first layer (Body-Center) then hold down the SHIFT key as you select the last layer name (Bushing-Object).

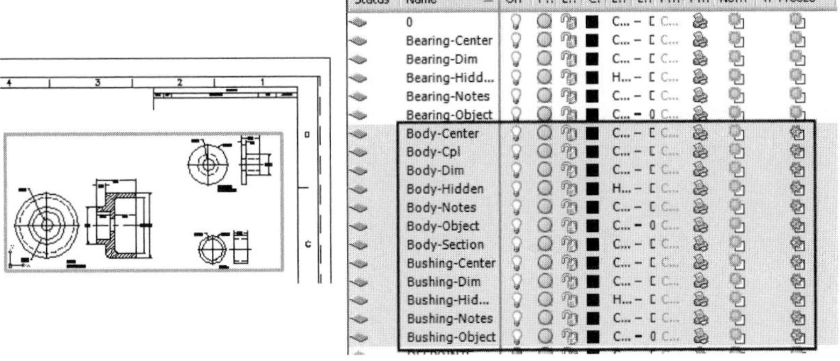

Figure 19.13

Finally, click inside the viewport located in the lower-right corner to make it active, as shown on the left in the following image. Activate the Layer Properties Manager palette. All layers that begin with Bushing need to be visible in this viewport. Pick all layers that begin with Body and Bearing and freeze these layers only in this viewport, as shown on the right in the following image.

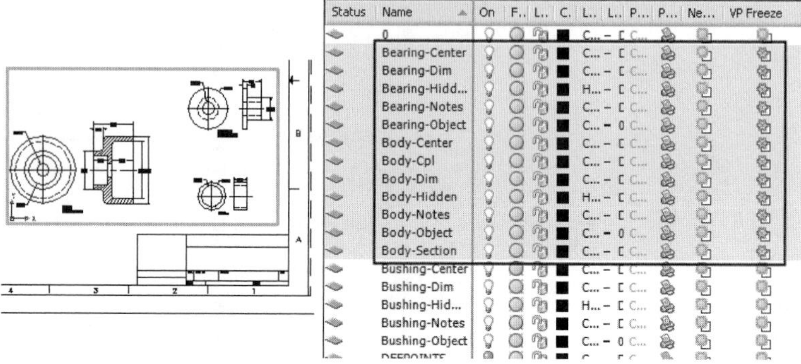

Figure 19.14

When finished, each viewport should contain a different part file. Items dealing with the part body are visible in the large viewport on the left. The bearing views are visible in the upper-right viewport and the bushing views are visible in the lower-right viewport. One other step is needed to better organize your work. Click in each port and pan each image so it appears centered in each viewport, as shown in the following image. Return to Paper Space by double-clicking outside the viewports.

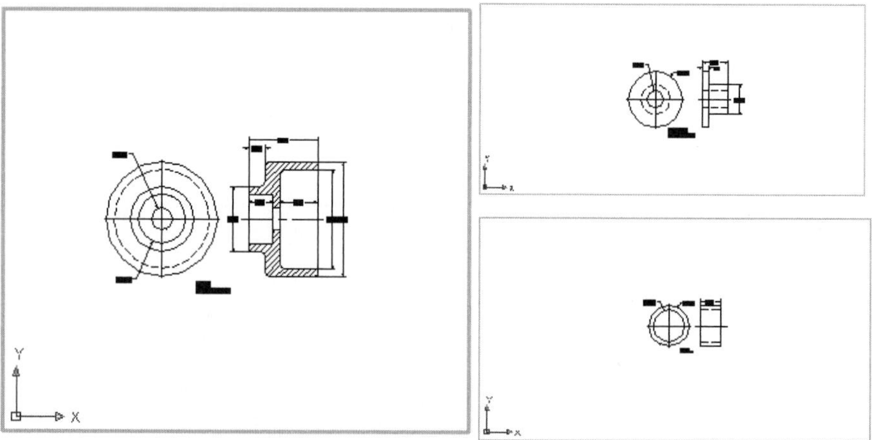

Figure 19.15

Each image will now be scaled to each viewport. Be sure the Viewports toolbar is visible somewhere on the display screen. Two of the viewports share the same scale factor. Click the edge of the two viewports to highlight them as shown in the following image. Then select the scale 1:1 from the Viewports toolbar to scale the body and bearing views to their respective viewports, as shown in the following image. Press ESC to remove the grips.

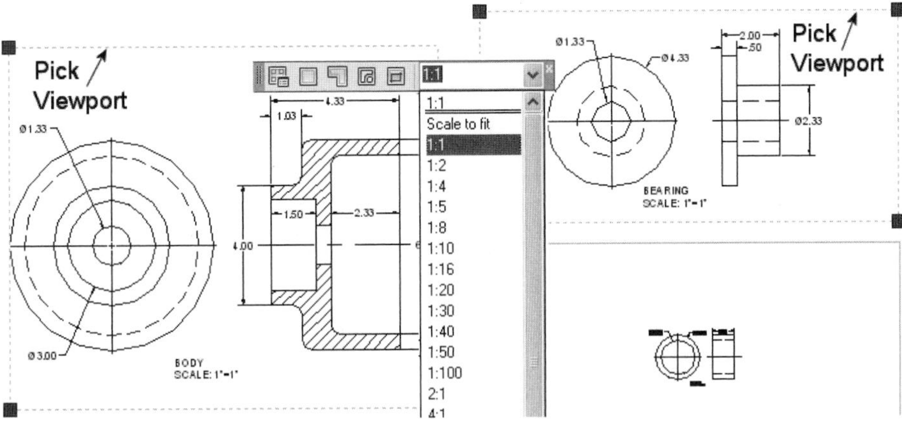

Figure 19.16

Next, click the edge of the viewport located in the lower-right corner to highlight it. Then select the scale 2:1 from the Viewports toolbar to scale the bushing views to this viewport, as shown in the following image. The scale of 2:1 will double the size of this view since it is smaller than the others in Model Space.

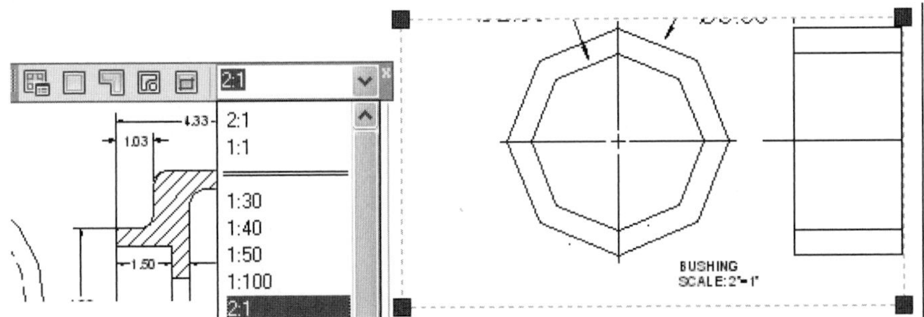

Figure 19.17

Adjust the viewports by clicking the edges, picking corner grips, and stretching the viewports in order to see all the drawings and dimensions. The dimensions in the lower-right corner viewport may appear larger than the others due to the larger scale. Activate this viewport by double-clicking inside it. Then launch the Dimension Style Manager dialog box, click the Modify button, select the Fit tab, and click the button next to Scale dimensions to layout (Paper Space). All dimensions inside this viewport will be automatically scaled to the viewport. You will have to update the dimension scales in other viewports by selecting the Update option in the Dimension pull-down or the Dimension toolbar. The final layout, consisting of three different details, some at different scales, is shown in the following image.

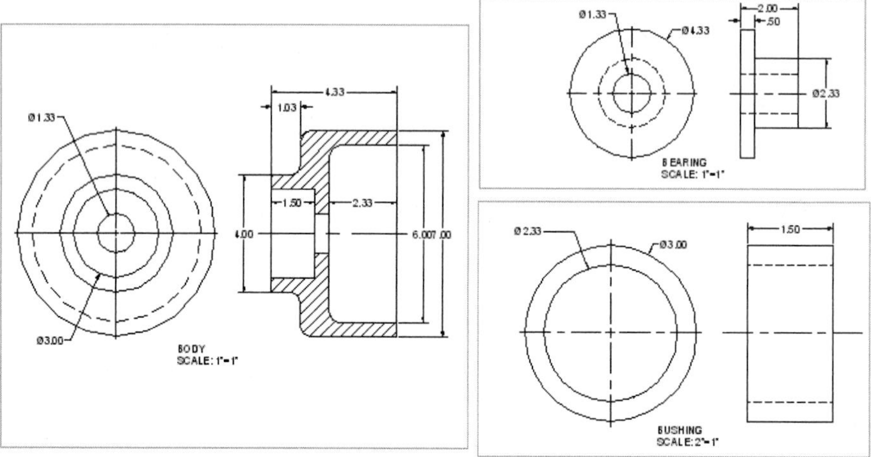

Figure 19.18

A good test is to perform a Plot Preview on this drawing. Notice in the following image the absence of any viewports. This is due to the No Plot setting being applied to the VPORTS layer.

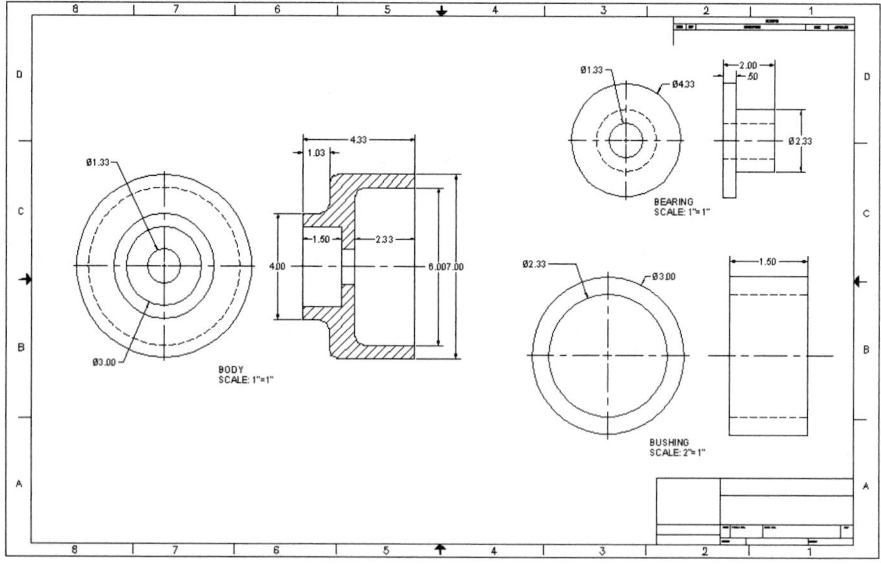

Figure 19.19

ADDITIONAL VIEWPORT CREATION METHODS

When constructing viewports, you are not limited to rectangular or square shapes. Although you have been using the Scale drop-down list box in the Viewports toolbar, as shown in the following image, to scale the image inside the viewport, other buttons are available and act on the shape of the viewport. You can clip an existing viewport to reflect a different shape, convert an existing closed object into a viewport, construct a multisided closed or polygonal viewport, or display the Viewports dialog box (you have already constructed a single viewport in a previous exercise).

Figure 19.20

Refer to the following table for a brief description of the extra commands available in the Viewports toolbar.

Button	Command	Description
	VPORTS	Displays the Viewports dialog box
	VPORTS Single	Creates a single viewport
	VPORTS Polygonal	Used for creating a polygonal viewport
	VPORTS Object	Converts existing object into a viewport
	VCLIP	Used for clipping an existing viewport

Try It!: Open the drawing file 19_Floor Viewports. Notice that Vports is the current layer. You will convert the large rectangular viewport into a polygonal viewport by a clipping operation. First click the Clip Existing Viewport button. Pick the rectangular viewport and begin picking points to construct a polygonal viewport around the perimeter of the floor plan dimensions, as shown in the following image. You can turn ORTHO on to assist with this operation. It is not critical that all lines are orthogonal (horizontal or vertical). The following command sequence will also aid with this operation.

```
Command: VPCLIP
Select viewport to clip: (Select the rectangular viewport)
Select clipping object or [Polygonal] <Polygonal>: (Press ENTER)
Specify start point: (Pick at "A")
Specify next point or [Arc/Length/Undo]: (Pick at "B")
Specify next point or [Arc/Close/Length/Undo]: (Pick at "C")
Specify next point or [Arc/Close/Length/Undo]: (Pick at "D")
Specify next point or [Arc/Close/Length/Undo]: (Pick at "E")
Specify next point or [Arc/Close/Length/Undo]: (Pick at "F")
Specify next point or [Arc/Close/Length/Undo]: C (To close the
    shape)
```

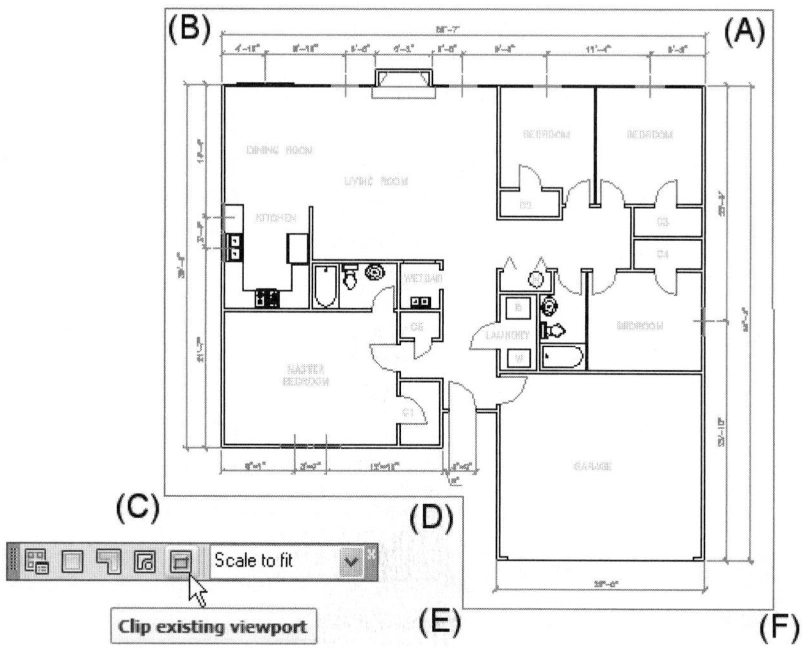

Figure 19.21

When you are finished, move the viewport with the image of the floor plan to the right of the screen. Then construct a circle in the upper-left corner of the title block. Pick the Convert Object to Viewport button and select the circle you just constructed. Notice that the circle converts to a viewport with the entire floor plan displayed inside its border. Double-click inside this new viewport to make it current and change the scale of the image to the 1/2" = 1'-0" scale using the Viewports toolbar.

```
Command: -VPORTS
Specify corner of viewport or
[ON/OFF/Fit/Shadeplot/Lock/Object/Polygonal/Restore/LAyer/2/3/4]
    <Fit>: O (For Object)
Specify object to clip viewport: (Pick the circle)
```

Pan inside the circular viewport until the laundry and bathroom appear, as shown in the following image. When finished, double-click outside the edge of the viewport to switch to Paper Space. Adjust the size of the circular viewport with grips if the image is too large or too small.

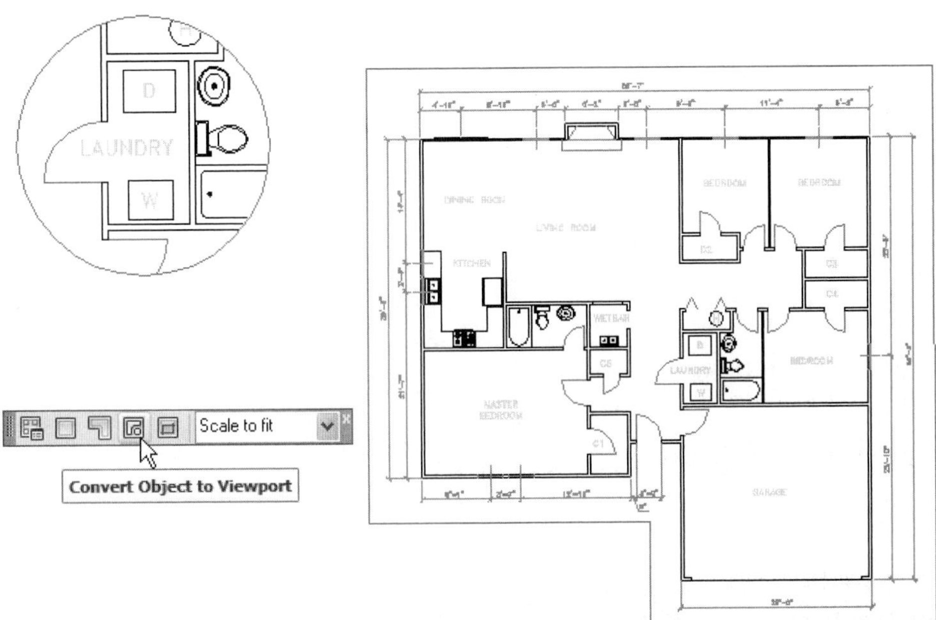

Figure 19.22

Click the Polygonal Viewport button and construct a multisided viewport similar to the one located in the following image. Close the shape and do not be concerned that a few lines may not be orthogonal.

```
Command: -VPORTS
Specify corner of viewport or
[ON/OFF/Fit/Shadeplot/Lock/Object/Polygonal/Restore/LAyer/2/3/4]
    <Fit>: P (For Polygonal)
Specify start point: (Pick at "A")
Specify next point or [Arc/Length/Undo]: (Pick at "B")
Specify next point or [Arc/Close/Length/Undo]: (Pick at "C")
Specify next point or [Arc/Close/Length/Undo]: (Pick at "D")
Specify next point or [Arc/Close/Length/Undo]: (Pick at "E")
Specify next point or [Arc/Close/Length/Undo]: (Pick at "F")
Specify next point or [Arc/Close/Length/Undo]: C (To close the
    shape)
Regenerating model.
```

As the image of the floor plan appears in this new viewport, double-click inside the new viewport to make it current. Scale the image inside the viewport to the scale 3/8" = 1'0". Pan until the kitchen area and master bathroom are visible. Your display should appear similar to the following image.

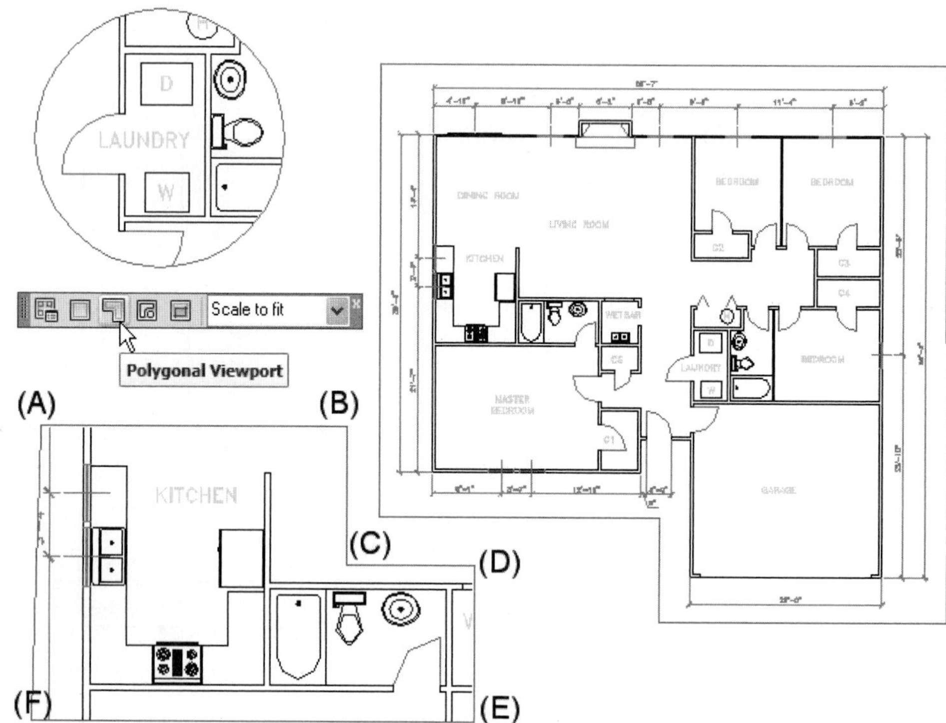

Figure 19.23

Double-click outside this viewport to return to Paper Space. Make any final adjustments to viewports using grips. When finished, turn off the Vports layer. Your display should appear similar to the following image.

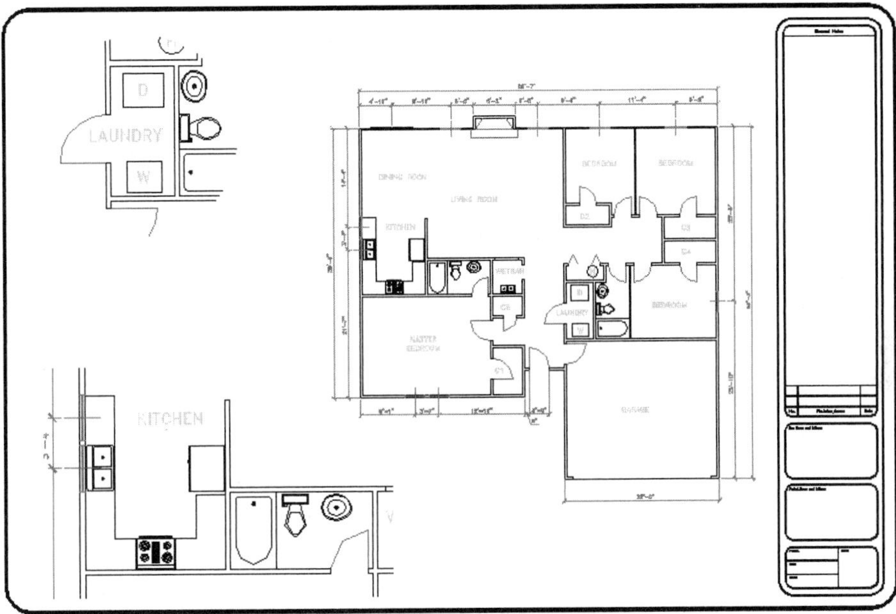

Figure 19.24

MATCHING THE PROPERTIES OF VIEWPORTS

In chapter 7, the MATCHPROP (Match Properties) command was introduced as a means of transferring all or selected properties from a source object to a series of destination objects. In addition to transferring layer information, dimension styles, hatch properties, and text styles from one object to another, you can also transfer viewport information from a source viewport to other viewports. Information such as viewport layer and the viewport scale are a few of the properties to transfer to other viewports. When you enter the MATCHPROP command and select the Settings option, the dialog box in the following image will appear. When transferring viewport properties, be sure the Viewport option of this dialog box is checked.

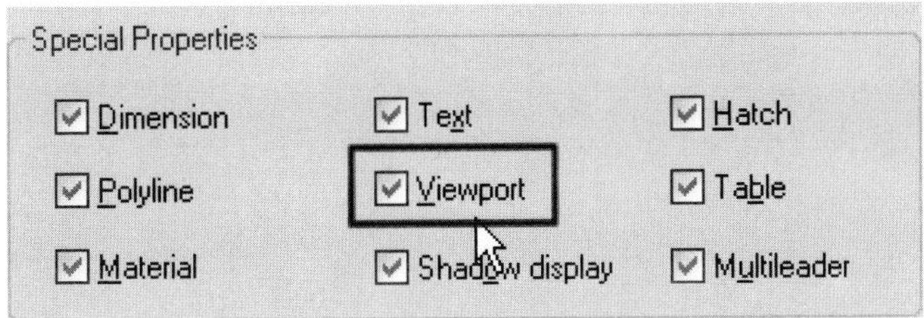

Figure 19.25

Try It!: Open the drawing file 19_Matchprop Viewports, as shown in the following image. This drawing consists of four viewports holding different object types. The object in the first viewport (labeled "A" in the following image) is already scaled to 1 = 1. All other viewports do not belong to the Viewports layer. They are all scaled differently. These three viewports need to have the same properties as the first; namely, all viewports need to belong to the Viewports layer and all images inside all viewports need to be scaled to 1 = 1. Rather than perform these operations on each individual viewport, the MATCHPROP command will be used to accomplish this task.

```
Command: MA (For MATCHPROP)
Select source object: (Pick the edge of Viewport "A")
Current active settings: Color Layer Ltype Ltscale Lineweight
    Thickness
PlotStyle Dim Text Hatch Polyline Viewport Table Material Shadow
    display
Multileader
Select destination object(s) or [Settings]: (Pick the edge of
    Viewport "B")
Select destination object(s) or [Settings]: (Pick the edge of
    Viewport "C")
Select destination object(s) or [Settings]: (Pick the edge of
    Viewport "D")
Select destination object(s) or [Settings]: (Press ENTER to exit
    this command)
```

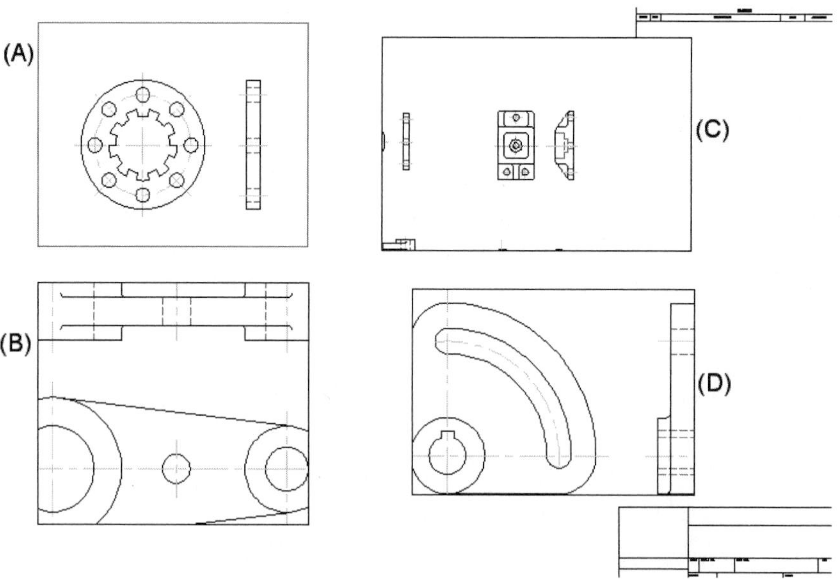

Figure 19.26

The results are illustrated in the following image. All viewports share the same layer and are scaled to 1 = 1 after using the MATCHPROP command.

Figure 19.27

ANNOTATION SCALE CONCEPTS

When annotating a drawing, which involves adding text, dimensions, and even hatch objects, problems would always occur when working with multiple viewports and the images in these viewports were at different scales. In the following image, the layout on the left is scaled to 1:1 while the layout on the right is scaled to 2:1, or double the original size. Notice how the dimension text appears much smaller on the left than the right. This was a typical problem encountered by individuals heavily involved in drawing layouts. One fix for this problem is to create extra layers for the dimensions. Then the dimensions were duplicated on top of each other and assigned to the extra layers. Depending on the viewport scale, certain dimensions were frozen while others were kept visible. While this process still works, it tends to be very time consuming and confusing. This is the reason for using Annotative Scales.

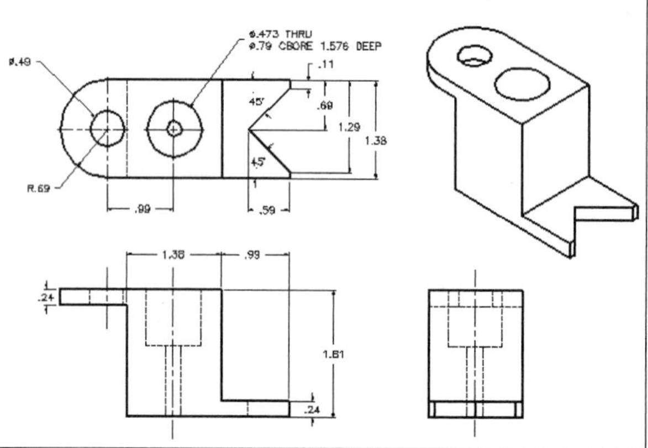

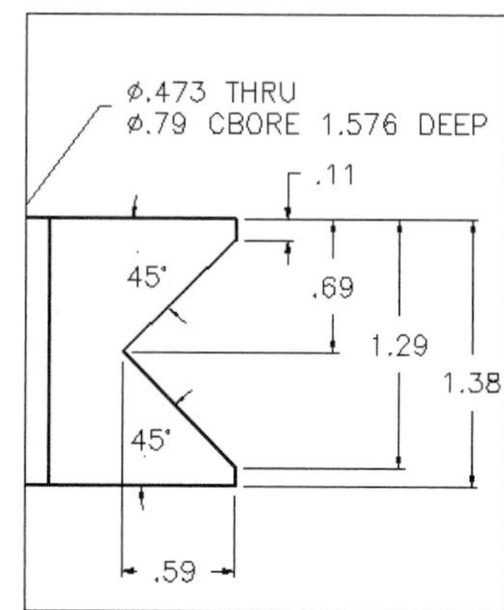

Figure 19.28

The following image is almost identical to the previous one with the exception that all of the dimension text is at the same height even though the scale of the viewports is different. Instead of using the cumbersome layer assignments and duplicate dimensions, an Annotative property was assigned to the dimensions. This property allows you to automate the process of scaling annotations. This means that when the viewports are scaled, the dimension text is scaled to the Annotation scale, allowing the text to be the same size in your layouts no matter what the scale.

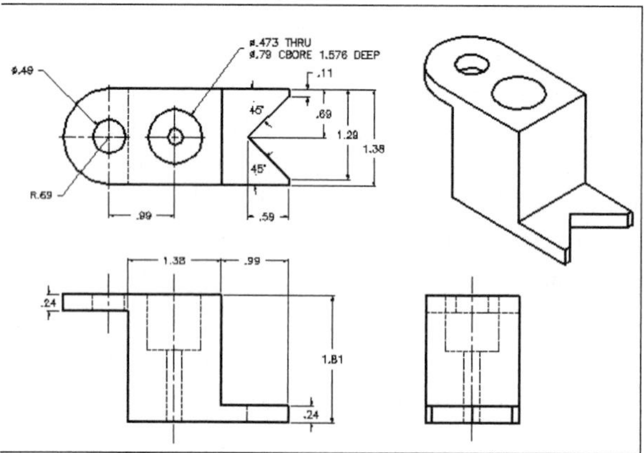

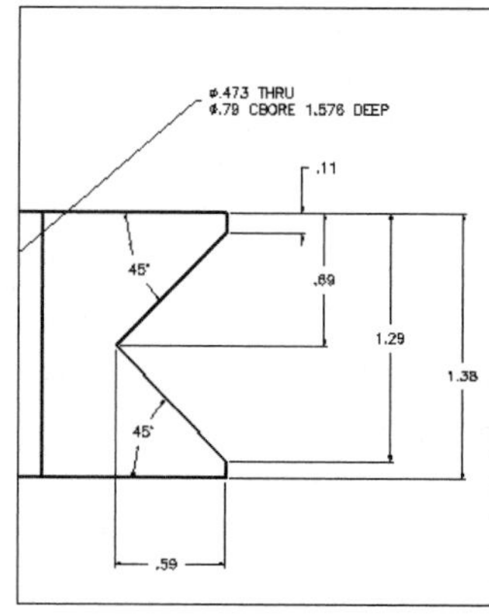

Figure 19.29

The following image illustrates a number of dialog boxes that come equipped with an Annotative property setting. Checking the appropriate box makes that style and objects associated with the style annotative. The Annotative property can be set when creating attributes (A), block definitions (B), multileader styles (C), hatch and gradient patterns (D), text styles (E), and dimension styles (F).

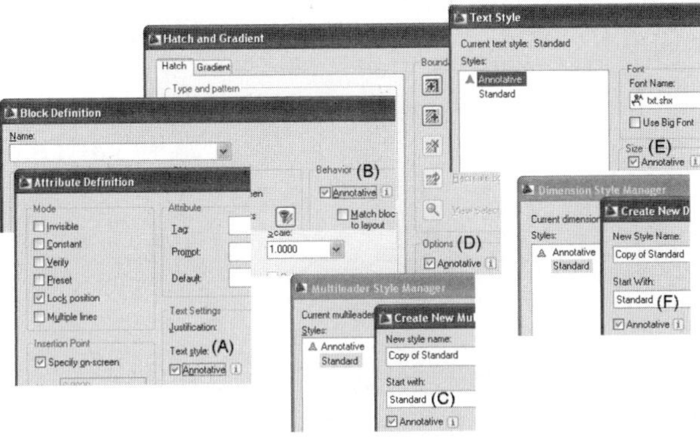

Figure 19.30

CREATING AN ANNOTATIVE STYLE

The following image illustrates the typical Text Style dialog box with a number of text styles already created. When creating a new text style or modifying an existing text style, place a check in the box next to Annotative, as shown in the following image.

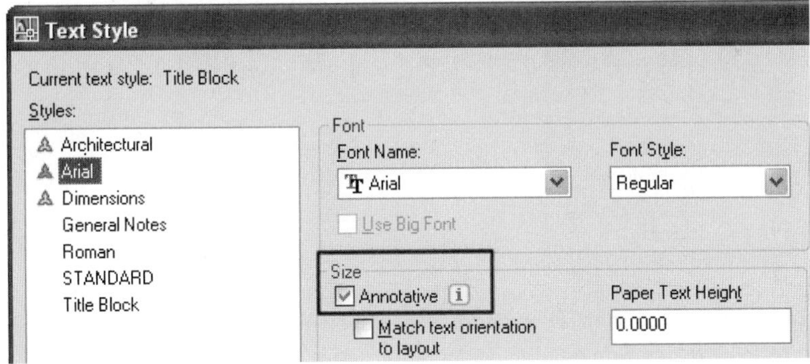

Figure 19.31

Annotative styles can be distinguished from traditional styles by a scale icon that appears next to the styles name, as shown in the following image. In this image, notice the scale icon present in Dimension Style Manager and the Multileader Style Manager.

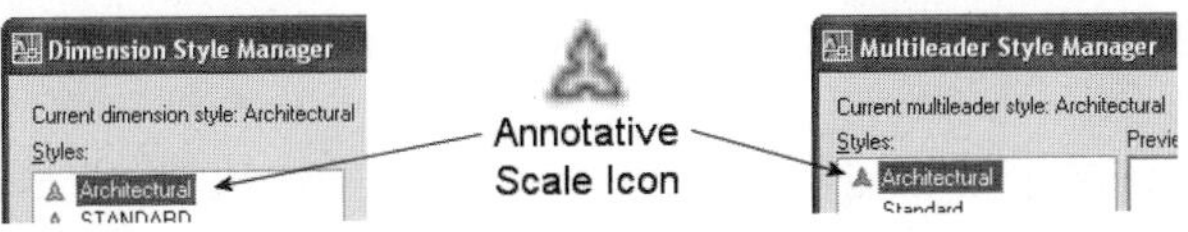

Figure 19.32

The following image illustrates an Mtext object that was created under the control of an Annotative text style. When you hover your cursor over this text object, the Annotative icon appears as shown in the following image. This provides a quick way of identifying annotative text compared with regular, traditional text.

Figure 19.33

ANNOTATIVE SCALING TECHNIQUES

While in model space, set the Annotation scale to the appropriate scale for your drawing (1 1/2"=1'0") as shown on the left in the following image. Any annotative text or dimensions placed in the drawing will be automatically sized per the scale specified. While in a drawing layout, you select the edge or activate a viewport and the status bar displays VP (Viewport) Scale. Selecting the VP Scale, as shown in the following image on the right, displays a list of scales that are identical to those used in the Viewports toolbar. Once the scale is selected, the image in the viewport zooms to the size required by the scale and any annotative dimensions or text will appear correctly sized in the layout.

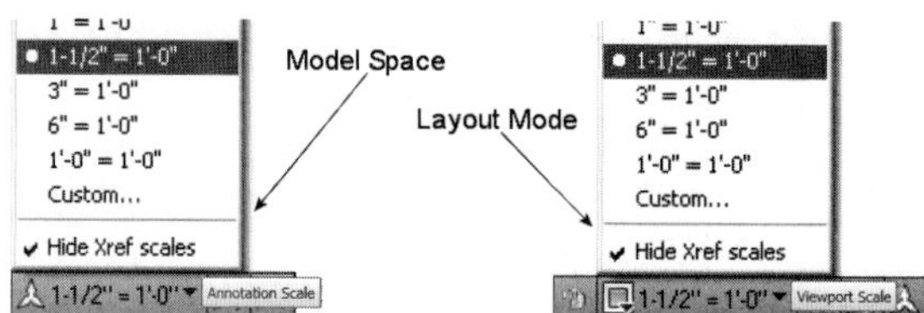

Figure 19.34

Note: Do not use the Viewports toolbar, as shown on the left in the following image, to scale viewports when utilizing annotation scales. Use the status bar Viewport Scale button only, as shown on the right in the following image, to ensure proper automatic scaling and viewing of annotative objects.

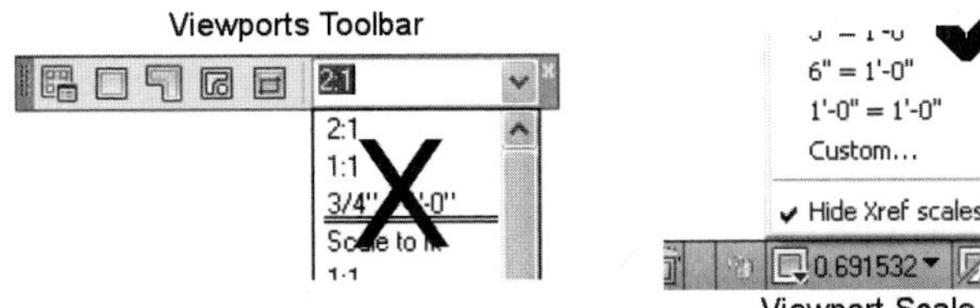

Figure 19.35

VIEWING CONTROLS FOR ANNOTATIVE SCALES

Located in the Status bar are two additional buttons used for controlling how annotative scales are viewed in the current viewport. The two buttons are "Annotation Visibility" and "Automatically add scales to annotative objects when the annotation scale changes."

The Annotation Visibility button, as shown in the following image, is either on or off. When Annotation Visibility is turned off in a layout, only annotative objects that use or match the current scale display. This means that if you change the VP Scale to a different value, the annotation objects may disappear from the screen. This setting prevents you from having to freeze dimensions, text, and hatching in viewports where they would appear at the wrong size.

When Annotation Visibility is turned on in a layout, all annotative objects that use all scales will display.

Figure 19.36

Another button controls the automatic adding of scales to annotative objects, as shown in the following image. When this button is turned off, annotative scales are not automatically added to objects inside a viewport. It is interesting to note that objects can have more than one annotative scale assigned to them. This allows the same annotative object to appear at the correct size in multiple viewports.

When this button is turned on, all annotative objects are automatically updated to match the new annotative scale.

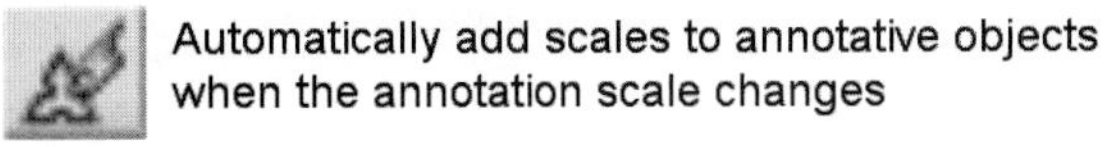

Automatically add scales to annotative objects
when the annotation scale changes

Figure 19.37

A representation of the annotative object is created for each scale. As you hover your cursor over the annotative object, you can see the various scale representations shown in the following image. Notice also that a multiple scales icon is present that signifies that the annotative object has multiple scales associated with it.

Figure 19.38

While there is no limit to the number of scales that can be used to represent annotative objects, too many scales can be difficult to interpret, as shown in the following image.

Figure 19.39

To add or delete annotation scales from selected objects, first enter floating Model Space, click the annotation object, right-click to display the menu as shown on the left in the following image, and select the Annotation Object Scale item. Clicking Add/Delete Scales... displays the Annotation Object Scale dialog box shown on the right in the following image. Click the Add button and you will be provided a list of scales to choose from. To delete a scale, select it from the list shown, such as 4:1, and click the now activated Delete button. Note that this operation adds or deletes the annotation scale for only the selected object.

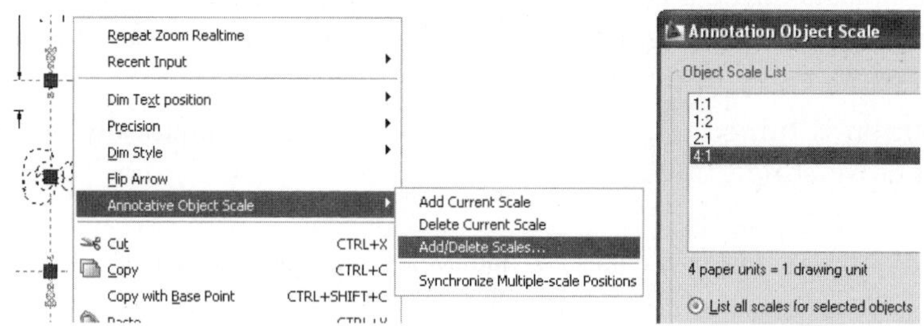

Figure 19.40

Note: The Annotation Object Scale dialog box can also be accessed by entering OBJECTSCALE when you are prompted to select the annotative objects.

Yet another method of activating the Annotation Object Scale dialog box is illustrated in the following image. In this example, an annotative dimension was first selected; then the

Properties Palette was displayed as shown on the left in the following image. Notice under the Misc heading the Annotative category and the 2:1 scale field. Clicking in this field displays three dots or ellipses. Clicking the ellipses button launches the Annotation Object Scale dialog box. Clicking the Add... button displays the list of scales to add to the selected annotation object.

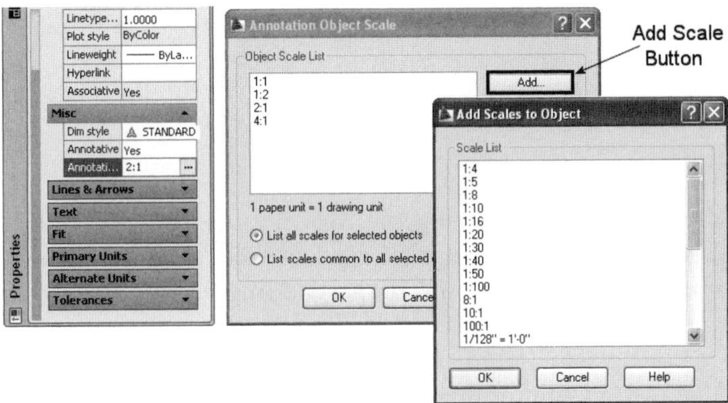

Figure 19.41

Note: To delete an annotation scale from the list for all annotation objects, enter the OBJECTSCALE command. At the Select Annotation Objects prompt, enter All to select all annotation objects. When the list of scales appears in the Annotation Object Scale dialog box, pick the scale from this list to delete.

ANNOTATIVE LINETYPE SCALING

The following Try It! exercise demonstrates how Annotative scale controls the scale of linetypes displayed in both Model and Paper Space. The system variable MSLTSCALE should be set to a value of 1 in order for the linetypes to display properly in Model Space. The default value is 1 but it may be set to 0 in older drawings. The system variable PSLTSCALE should also be set to a value of 1 in order for the linetypes to display properly in Paper Space.

Try It!: Open the drawing 19_Land Plat. The drawing opens up in Model Space as shown on the left in the following image. Also, the current value of LTSCALE (Linetype Scale) is 1.00. Since this drawing will be plotted at a scale of 1:30, the land plat outline is too large to view the linetypes. You could zoom in to a segment of the plat; however, even in this magnified view the number of short and long dashes is numerous. This problem, however, is fixed when switching to the Paper Space layout. Here, as shown on the right in the following image, the linetypes are properly scaled thanks to the Paper Space linetype scaling function being turn on (PSLTSCALE = 1) by default.

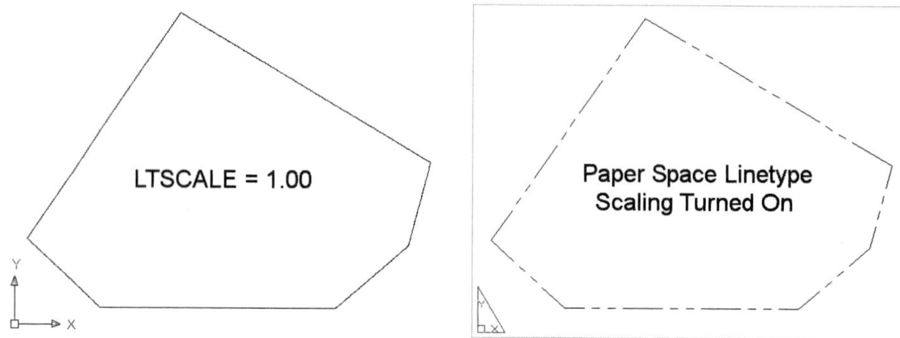

Figure 19.42

Switch back to Model Space by clicking the Model tab and change the LTSCALE value from 1.00 to 30.00. The results are displayed on the left in the following image with the linetypes representing the outline of the property being visible. Then click the layout tab and observe the results shown on the right in the following image. Even with Paper Space linetype scaling turned on, the LTSCALE value affects the Paper Space image where the linetype scale value is too large to display the linetypes. In other words, you either display the linetypes in the Model tab or the Layout tab but not both.

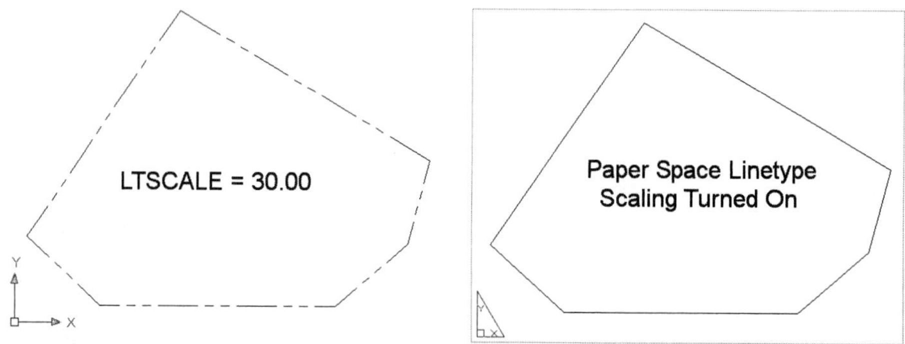

Figure 19.43

The solution to this dilemma of the scaling of linetypes is illustrated in the following image. Return to Model Space by selecting the Model tab and set the LTSCALE back to 1.00. In the status bar set the Annotation Scale to 1:30 and perform a REGEN, as illustrated on the left in the following image. The linetype scale appears correct thanks to the Model Space linetype scale being turned on (MSLTSCALE = 1). When you are switching to the Layout tab, the linetypes are still visible in this environment thanks to the Paper Space linetype scale being turned on (PSLTSCALE = 1), as illustrated on the right in the following image.

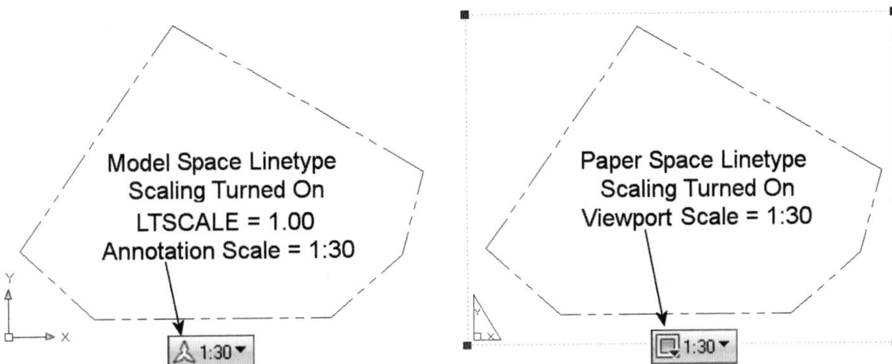

Figure 19.44

CREATING AN ANNOTATIVE TEXT STYLE

The following Try It! exercise demonstrates how annotative scales affect text added to a drawing.

Try It!: Open the drawing 19_Anno_Duplex. Use the following steps and images for creating an annotative text style and applying it to this drawing.

1. Create a new text style called Room Names. Assign the Arial font and check the box next to Annotative.

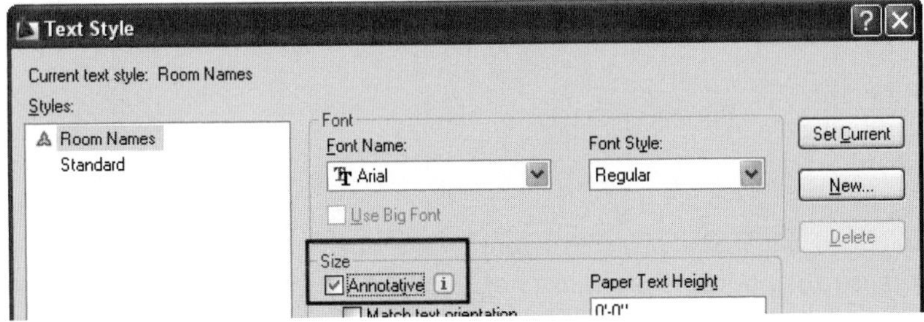

Figure 19.45

2. Use MTEXT to create a room name. Notice the mtext height reads 3/16, as shown in the following image of the command prompt. In a full size drawing, this text will not be readable. Press the ESC key to exit the command.

```
MTEXT Current text style:  "Room Names"  Text height:  3/16"  Annotative:  Yes
Specify first corner:
```

Figure 19.46

3. While still in Model Space, change the current Annotation Scale to ¼" = 1'-0" as shown on the left in the following image.

Reenter the MTEXT command and notice that the height of the text has changed to 9 5/8" based on the current annotation scale. Add the text BEDROOM 1 to the room and exit the Text Formatting toolbar. Hover your cursor over the text and notice the appearance of the scale icon signifying that this text is considered an annotative object, as shown on the right in the following image.

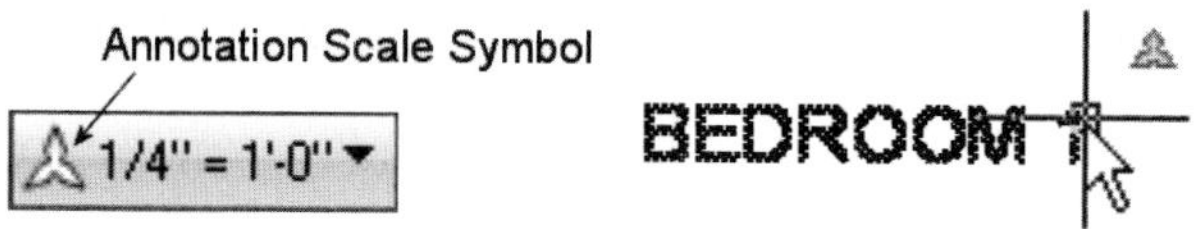

Figure 19.47

4. Continue using MTEXT to add names to all of the remaining rooms of the duplex, as shown in the following image.

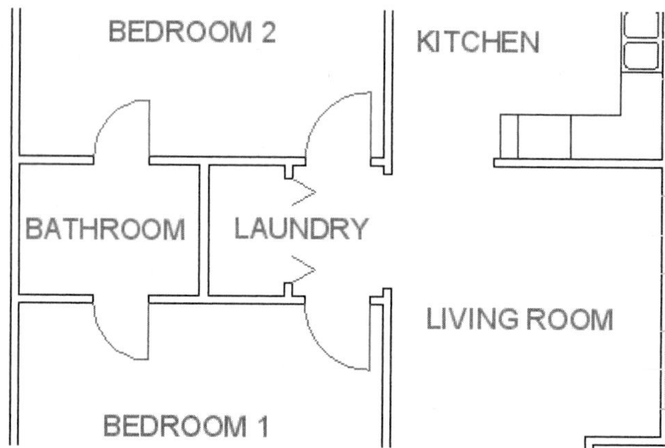

Figure 19.48

5. Switch to the B-Size (DWF6) layout. None of the room names display. This is because an annotative scale has not been set for this viewport. Click the edge of the viewport and change the VP Scale to ¼" = 1'-0", as shown on the left in the following image.

   Notice that the image inside of the viewport changes to reflect the current scale. Also, the text reappears since it matches the original annotative scale of ¼" = 1'-0", as shown on the right in the following image.

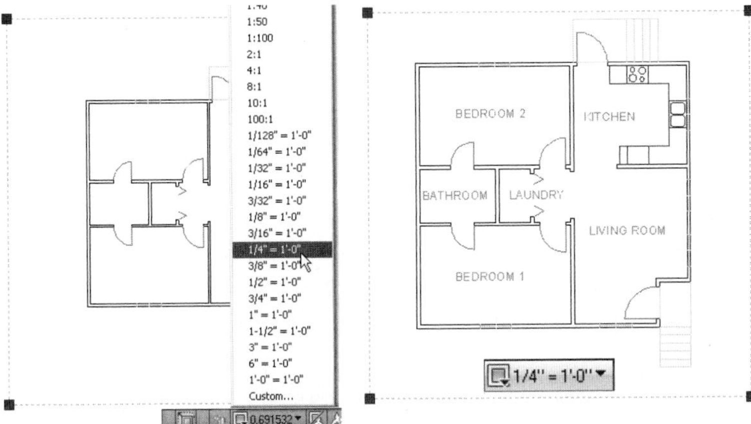

Figure 19.49

6. Set the "Automatically add scales to annotative objects when the annotation scale changes" button to on, as shown on the left in the following image.

   Click the edge of the viewport and change the scale of the viewport to 3/16" = 1'-0", as shown on the right in the following image. Even though the floor plan inside of the viewport is smaller, the size of the text is the same as with the previous scale.

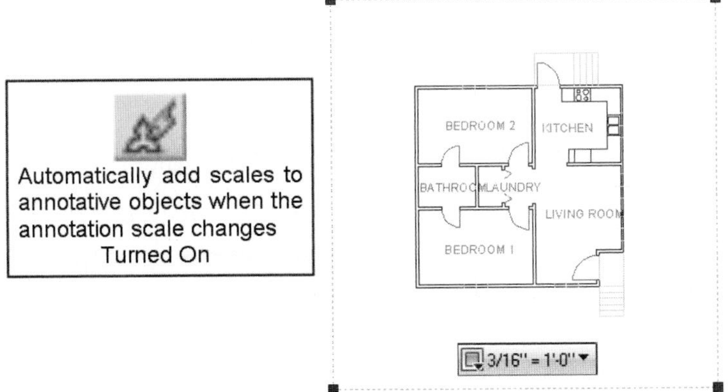

Figure 19.50

7. Switch back to Model Space by clicking the Model tab. Click one of the text objects and hover your cursor over this object. Looking carefully at the selected text; two text objects actually appear, one smaller than the other. The two text objects reflect the larger and smaller annotative scales that were used on the viewport back in the layout.

Figure 19.51

8. With the text object still selected, activate the Properties Palette and observe the information contained under the Text heading, as shown on the left in the following image.

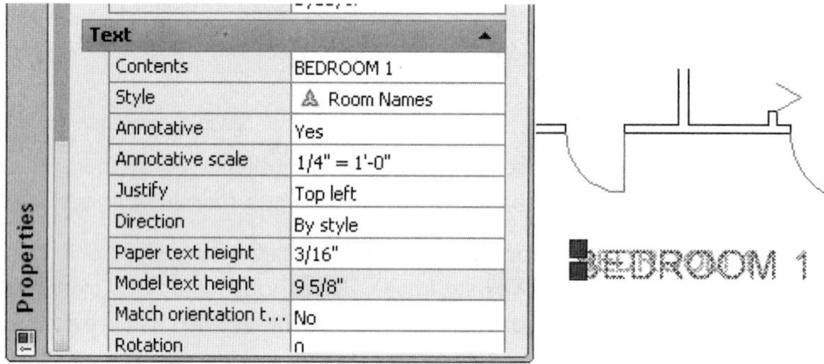

Figure 19.52

9. Click inside the Annotation Scale field. When the three dots (ellipses) appear, click these, as shown on the left in the following image, and select the text object again; then press the ENTER key to launch the Annotation Object Scale dialog box. The two annotation scales should be listed as shown on the right in the following image. Click the Cancel button when you are finished.

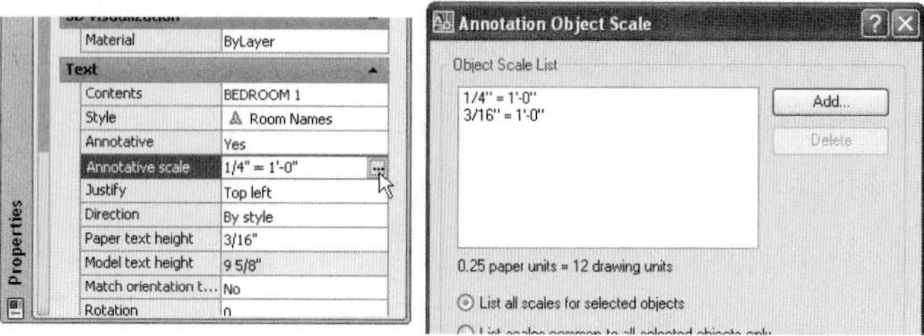

Figure 19.53

10. Switch back to the B-Size (DWF6) layout and turn off the "Automatically add scales to annotative objects when the annotation scale changes" button and verify that the Annotation Visibility button is turned off, as shown in the following image. Typically, these

buttons should remain off in a layout when working with annotation scales so that annotative objects will only appear in the viewports they are correctly sized for and you will not be continuously creating scales every time you change a viewport scale.

Figure 19.54

11. Change the scale of the viewport to ½" = 1'-0". The image will increase in size; however, the room names disappear, as shown on the left in the following image, because the Annotation Visibility button was turned off. Also, this scale was not automatically added to the Annotation Object Scale dialog box because the "Automatically add scales to annotative objects when the annotation scale changes" button was turned off.

    Change the viewport scale back to ¼" = 1'-0", as shown on the right in the following image.

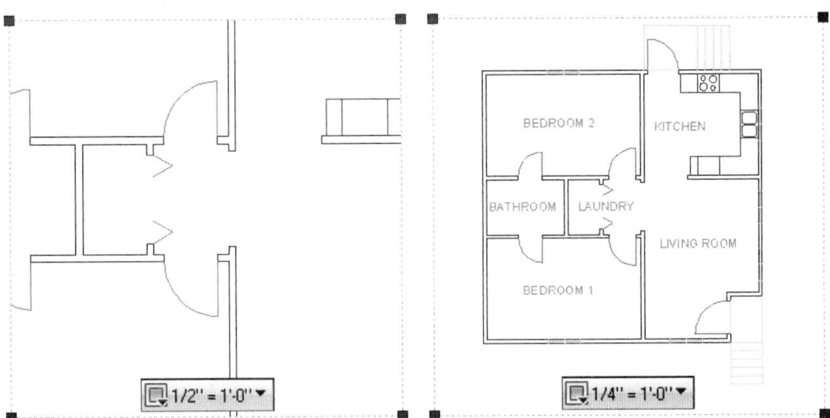

Figure 19.55

CREATING AN ANNOTATIVE DIMENSION STYLE

The following Try It! exercise demonstrates the affects that annotative scales have on adding dimensions to a drawing.

Try It!: Open the drawing 19_Anno_Dimension. This file picks up from the previous exercise. Two annotative scales are already present in this drawing; namely the ¼" = 1'-0" and 3/16" = 1'-0" scales. Use the following steps and images for creating an annotative dimension style and applying it to this drawing.

1. Create a new dimension style called Arch_Anno, as shown in the following image. Place a check in the Annotative box, and then click the Continue button.

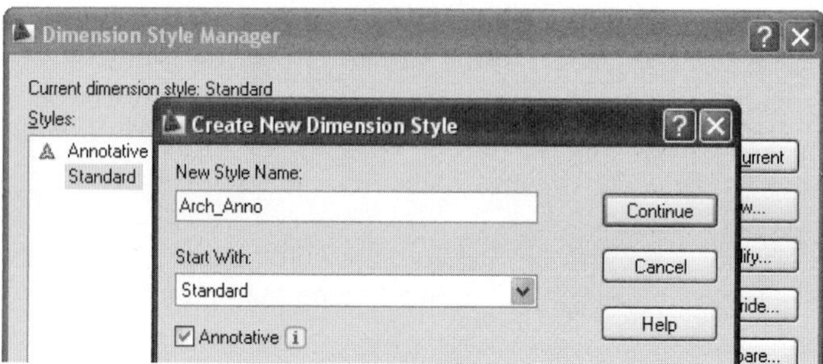

Figure 19.56

2. Use the table below for making changes while inside of the Dimension Styles Manager dialog box.

Dimension Styles Dialog Box		
Tab	Setting	Change To
Symbols and Arrows	Arrowheads	Architectural Tick
Symbols and Arrows	Arrow Size	1/8"
Text	Text Height	1/8"
Text	Text Placement – Vertical	Above
Text	Text Alignment	Align with dimension line
Primary Units	Unit Format	Architectural
Primary Units	Precision	0'-0"

Verify in the Fit tab that the Annotative box is checked in the Scale for dimension area, as shown in the following image. After making changes to the dimension settings, click the OK button.

Figure 19.57

3. When you return back to the main Dimension Styles Manager dialog box, notice the Arch_Anno dimension style present in the list. Notice also the appearance of the scale icon next to this dimension style name, signifying that all dimensions in this style will be controlled by the annotative scale feature. Click the Close button to exit the Dimension Styles Manager dialog box and continue with this exercise.

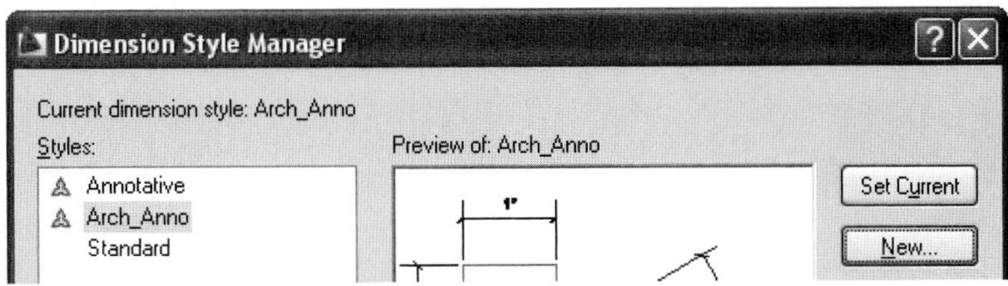

Figure 19.58

4. With the Annotation Scale already set to ¼" = 1'-0" in Model Space, begin placing linear and continue dimensions in the various locations of the floor plan, as shown on the left in the following image.

   Switch to the B-Size (DWF6) layout. Observe that the all dimensions are visible based on the current annotation scale of ¼" = 1'-0" matching the VP Scale, as shown on the right in the following image.

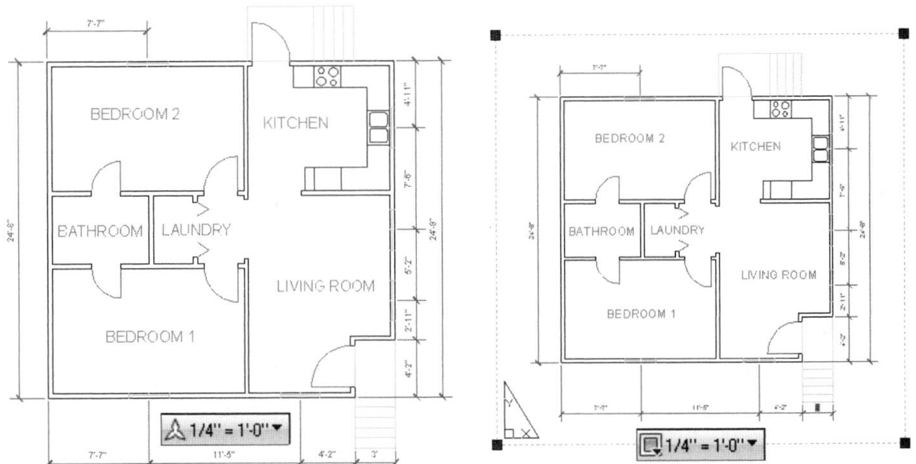

Figure 19.59

5. Turn on the Automatically add scales button, as shown in the following image.

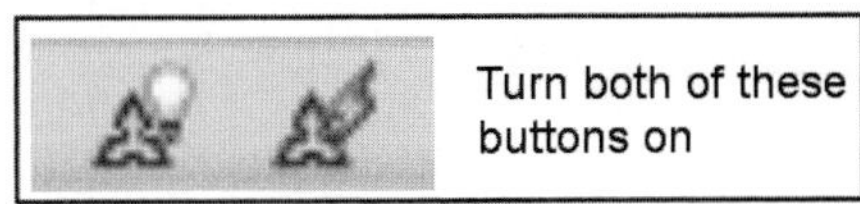

Figure 19.60

6. Change the VP Scale to 3/16" = 1'-0", as shown on the left in the following image. Notice in this image that as the image of the floor plan gets smaller, the dimensions remain their original plotting height of 1/8".

Turn the Automatically add scales button back off and switch the VP Scale back to ¼" = 1'-0", as shown on the right in the following image.

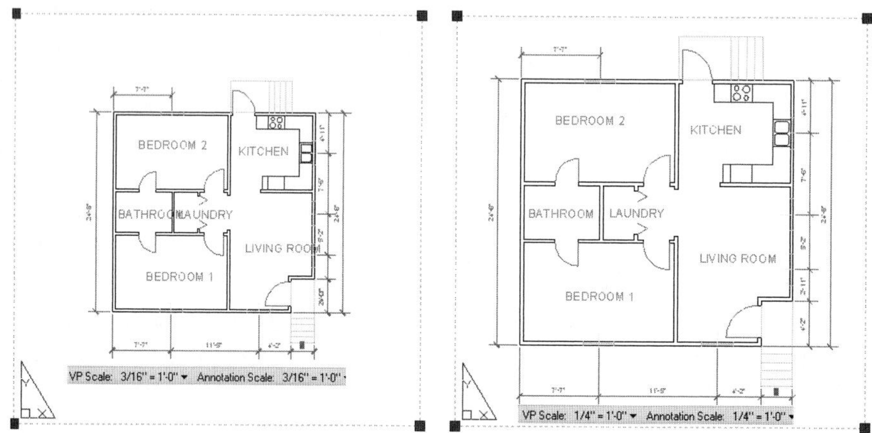

Figure 19.61

WORKING WITH ANNOTATIVE HATCHING

The following Try It! exercise demonstrates how annotative scales affect how an object is crosshatched when using the Hatch and Gradient dialog box.

Try It!: Open the drawing 19_Anno_Hatch. This file picks up from the previous exercise. Two annotative scales are already present in this drawing, namely the ¼" = 1'-0" and 3/16" = 1'-0" scales. Dimensions are also present in this drawing. Use the following steps and images for working with annotative hatching in this drawing.

1. Activate the Boundary Hatch dialog box. Change the hatch pattern scale to 48, as shown on the left in the following image. Also place a check next to Annotative to turn on Annotative scaling of the hatch pattern, as shown on the right in the following image.

Figure 19.62

2. Click the Pick Points button and select all internal areas that represent the floor plan, as shown on the left in the following image. When you return back to the

boundary hatch dialog box, click the OK button to place the hatch pattern as shown on the right in the following image.

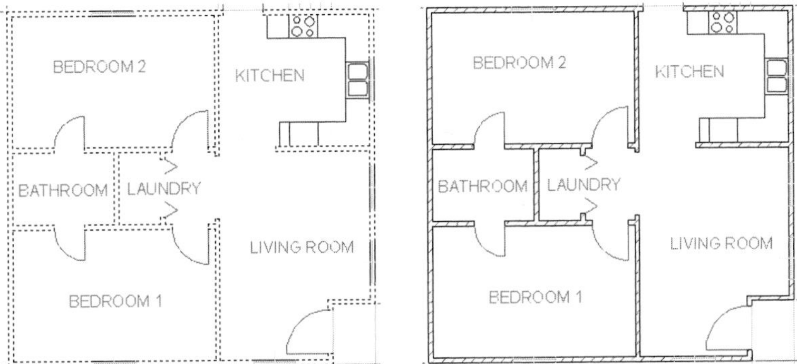

Figure 19.63

3. Switch to the B-Size (DWF6) layout and notice that the hatch pattern is visible based on the current Viewport and Annotation scales of ¼" = 1'-0", as shown in the following image.

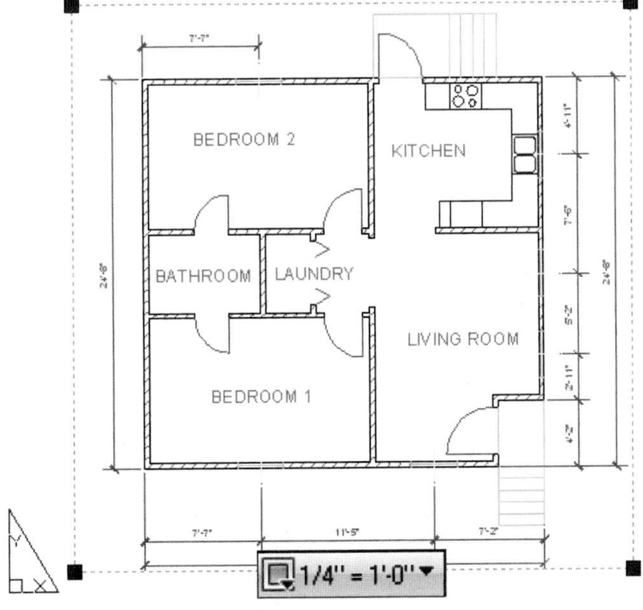

Figure 19.64

4. Change the VP Scale to 3/16" = 1'-0". Notice that as the image of the floor plan gets smaller, the hatch pattern disappears because the hatch scale does not match the

viewport scale. Turn the Annotation visibility button on to see the hatching but notice that the hatch is incorrectly sized for this viewport scale, as shown in the following image.

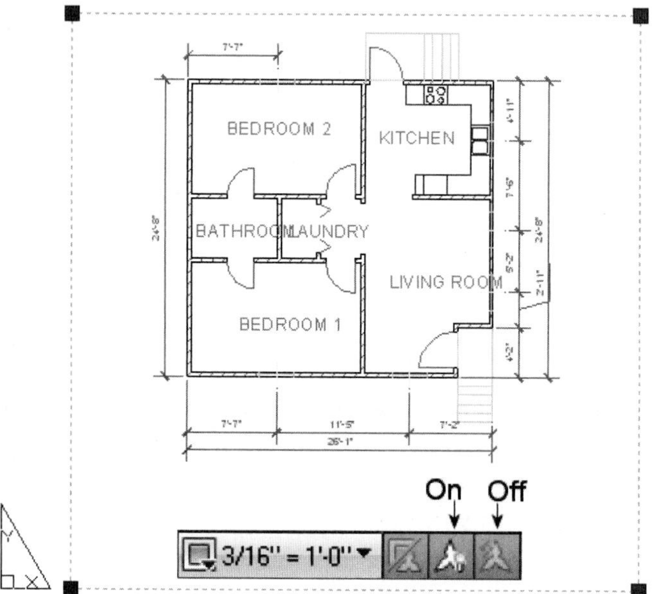

Figure 19.65

TUTORIAL EXERCISE: 19_ARCHITECTURAL DETAILS.DWG

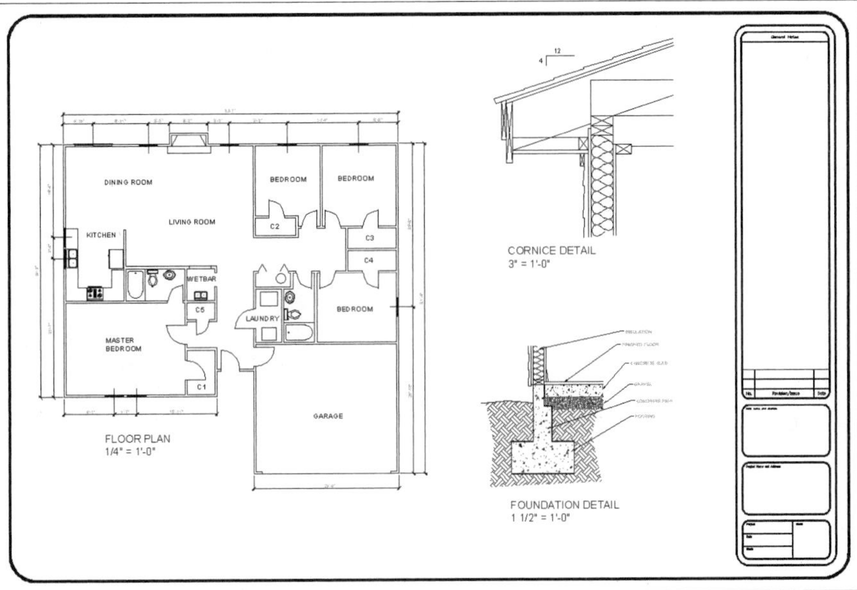

Figure 19.66

PURPOSE

This tutorial exercise is designed to lay out three architectural details, displayed in the previous image, in the Layout mode (Paper Space). The three details consist of a floor plan, cornice detail, and foundation detail.

SYSTEM SETTINGS

Open an existing drawing called 19_Architectural Details. Keep all default settings for the units and limits of the drawing.

LAYERS

Layers have already been created for this drawing.

SUGGESTED COMMANDS

An existing drawing will be used to create a new layout. This layout will hold three viewports displaying three different architectural details at different scales. The Layout wizard will be used to create the initial layout in Paper Space. Additional viewports will be added to hold the remaining details. Annotation scales will be used to scale the details relative to their individual viewports. Grips will be used to adjust the viewports. A plot preview will be performed to observe the images floating on the drawing sheet.

Step 1

Open the drawing 19_Architectural Details.Dwg. Notice the floor plan and the two small details. The three architectural images will need to be arranged in separate viewports with separate scales. The floor plan will need to be laid out at a scale of 1/4" = 1'-0"; the cornice detail at 3" = 1'-0" and the foundation detail at 1 ½" = 1'-0".

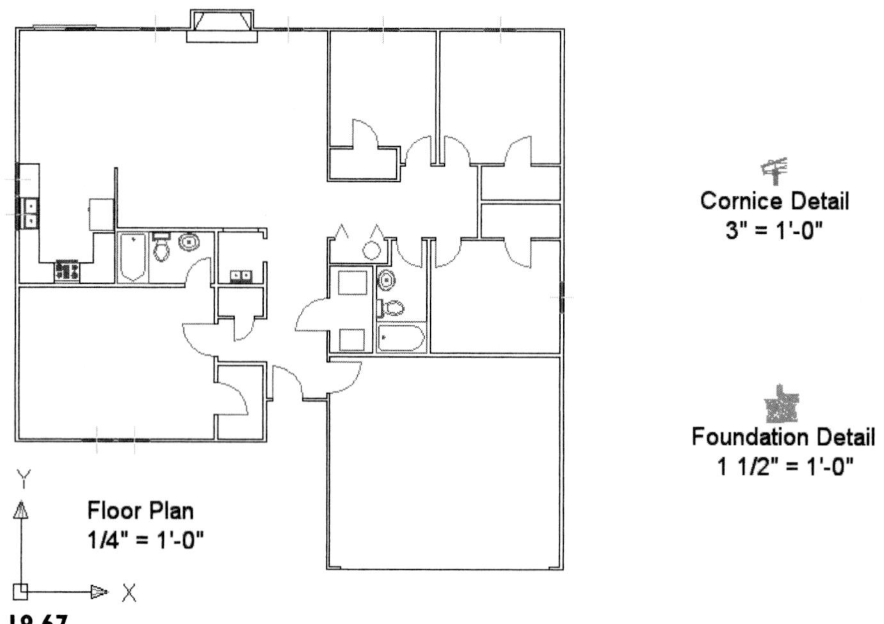

Figure 19.67

Step 2

While in Model Space, change the Annotation Scale to ¼" = 1'-0", as shown in the following image.

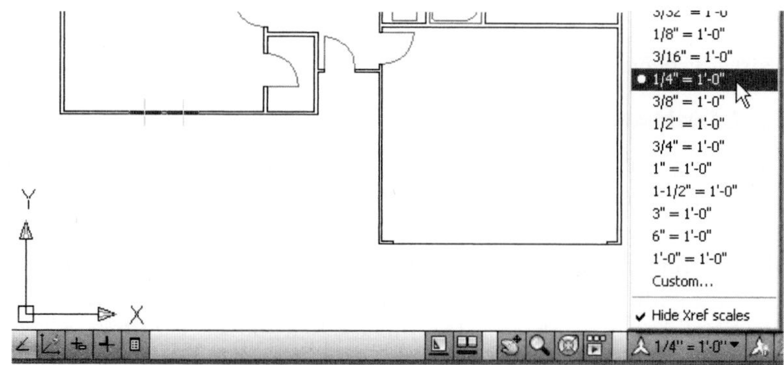

Figure 19.68

Step 3

Change to the Text layer and add the room names to the floor plan using the following image as a guide. Use the MTEXT command and a height of .20 for all room names. The Text Style "Titles" is already created and is set to Annotative, as shown in the following image.

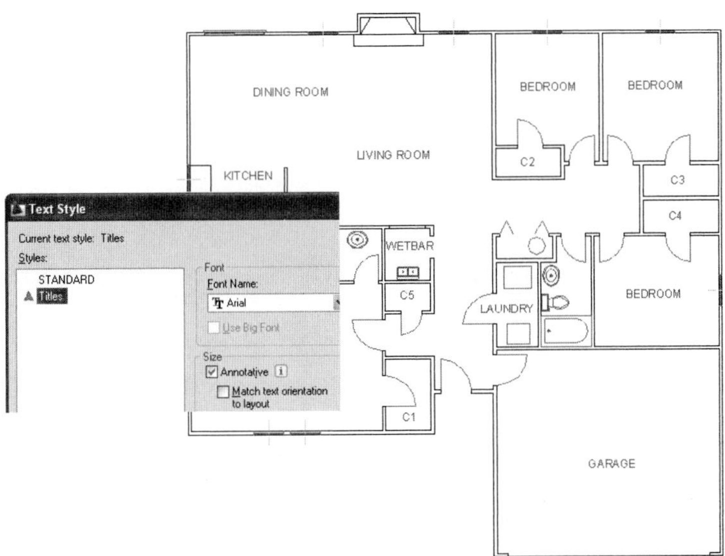

Figure 19.69

Step 4

Change the layer to Dimensions. Add all dimensions to the floor plan using the following image as a guide. The Dimension Style "Architectural" is already created and set to Annotative, as shown in the following image.

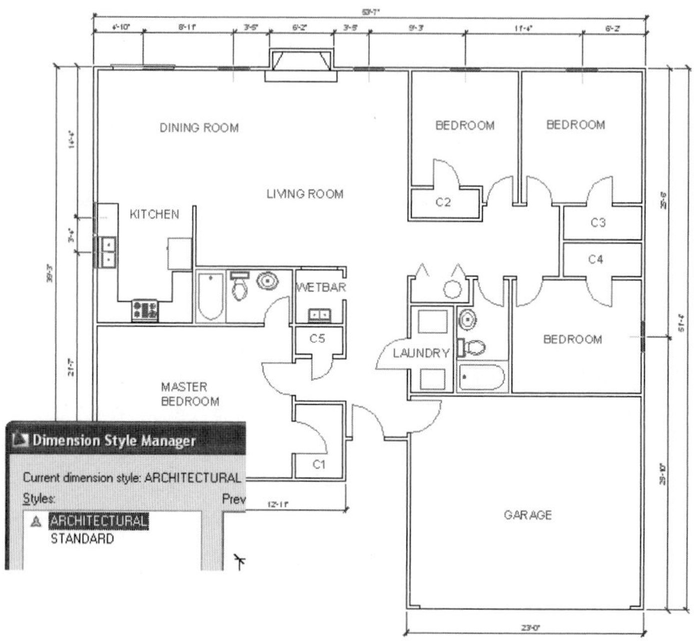

Figure 19.70

Step 5

Activate the Layout wizard by clicking in the pull-down menu, as shown on the left in the following image. When the Create Layout dialog box appears, change the name of the layout to Detail Sheet, as shown on the right in the following image. Continue inside this wizard by making the following changes in the additional dialog boxes that appear:

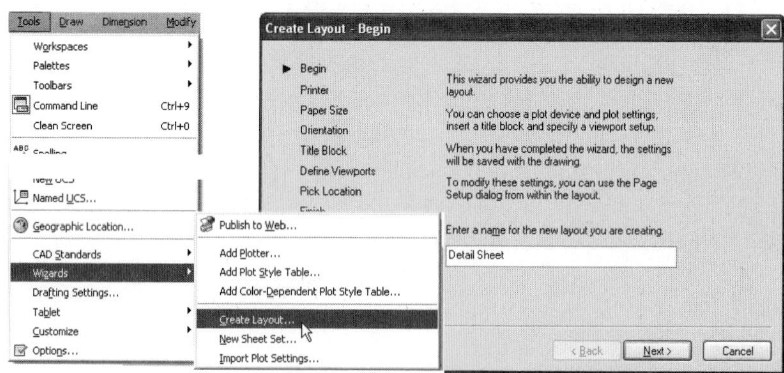

Figure 19.71

- Printer—Select DWF6 ePlot.pc3 device from the list of available printers.
- Paper Size—Select ARCH expanded D (36.00 × 24.00 Inches).

- Orientation—Verify that Landscape is the current setting in this dialog box.
- Title Block—From the available list, pick Architectural Title Block.dwg.
- Define Viewports—Verify that Single is set for the Viewport Setup. Use the default Scale to Fit listing in the Viewport scale field.
- Pick Location—Click the Select Location button. This returns you momentarily to the drawing editor. Pick two points at "A" and "B" as the corners of the viewport, as shown in the following image.
- Finish—Click the Finish button to exit the wizard and return to the drawing editor.

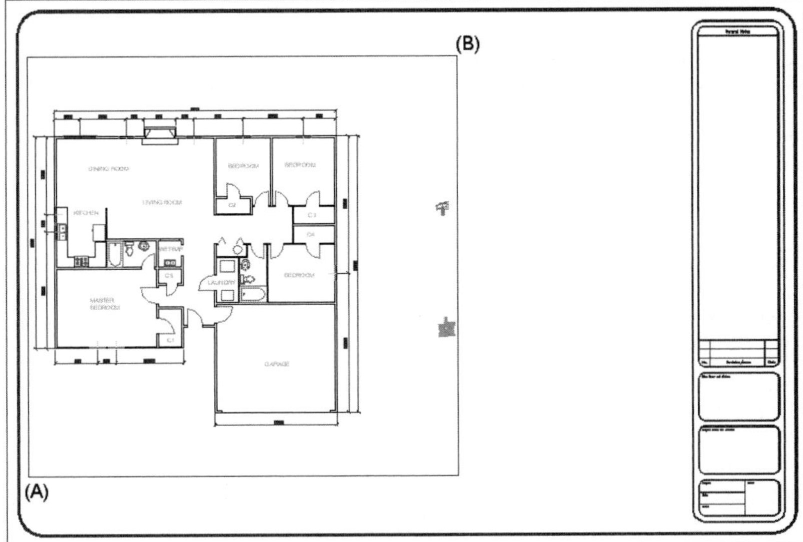

Figure 19.72

Step 6

Create two additional viewports using the Single Viewport button, located in the Viewports toolbar as shown in the following image. Construct the first viewport from "A" to "B" and the second viewport from "C" to "D." These locations are approximate. Notice that the images of all the details appear in both viewports.

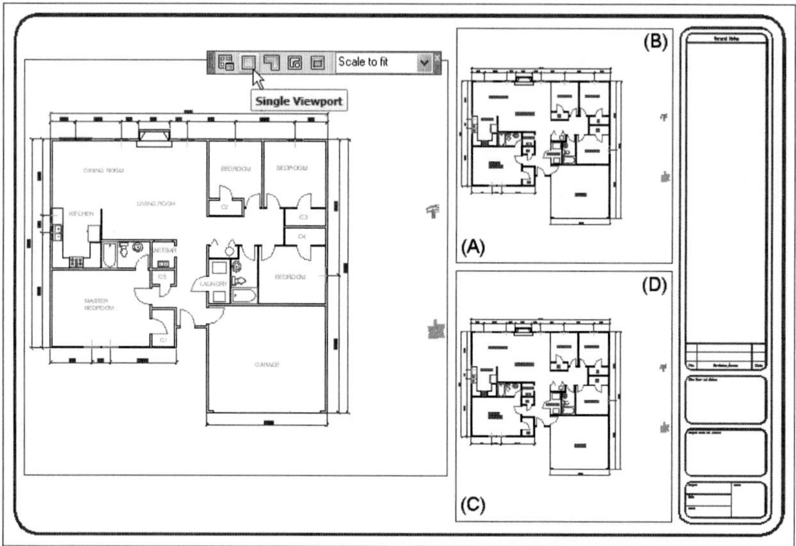

Figure 19.73

Step 7

Click the edge of all three viewports and notice in the Layer Control box the name of the layer. Clicking the down arrow and selecting the Viewports layer changes all three viewports to this layer, as shown in the following image.

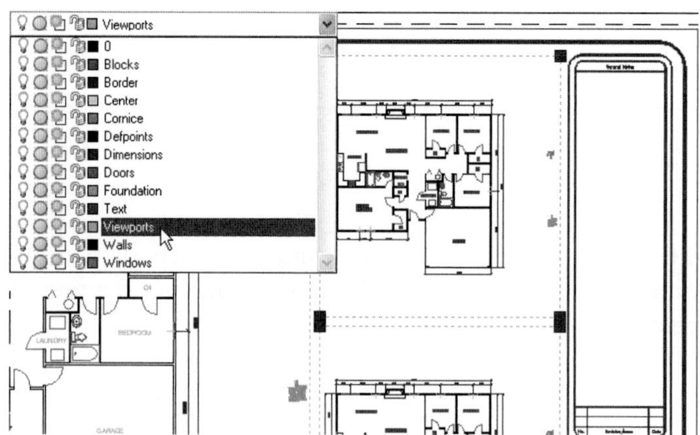

Figure 19.74

Step 8

Click the edge of the large viewport and change the VP Scale to ¼" = 1'-0", as shown in the following image.

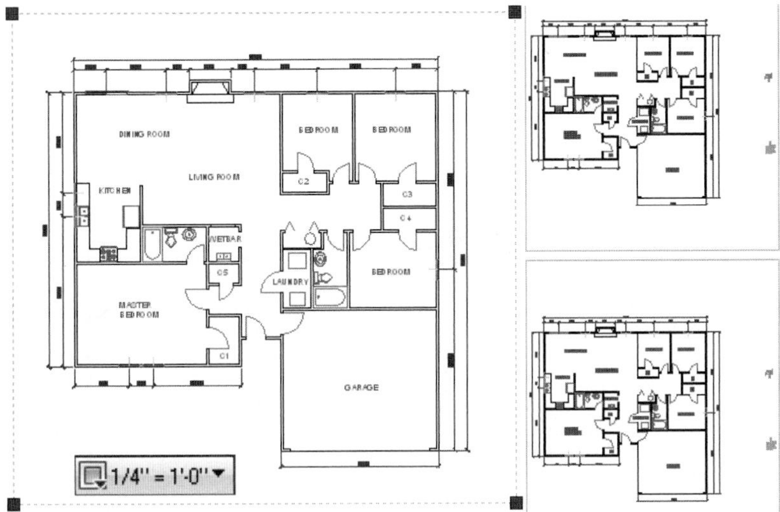

Figure 19.75

Step 9

Double-click inside the upper-right viewport to make it current. Pan the image of the Cornice approximately to the middle of the top viewport. It is helpful to perform this step on small details that are about to be scaled. Sometimes images get lost in the viewport if this step is not performed.

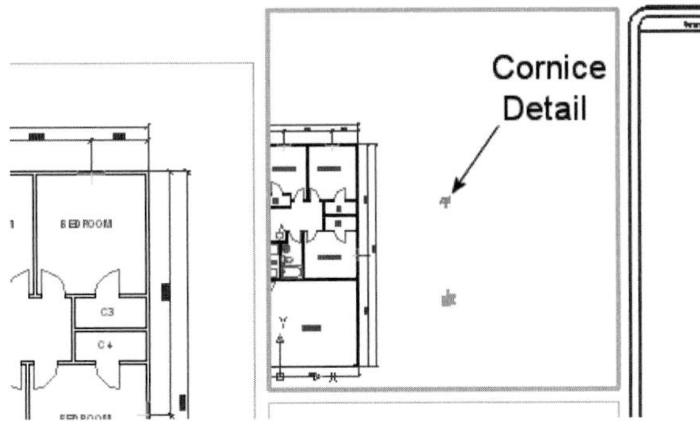

Figure 19.76

Step 10

Change the VP Scale of this image to 3" = 1'-0", as shown in the following image. The results of this operation are illustrated in the following image. You may have to pan the image in order to center it better inside the viewport.

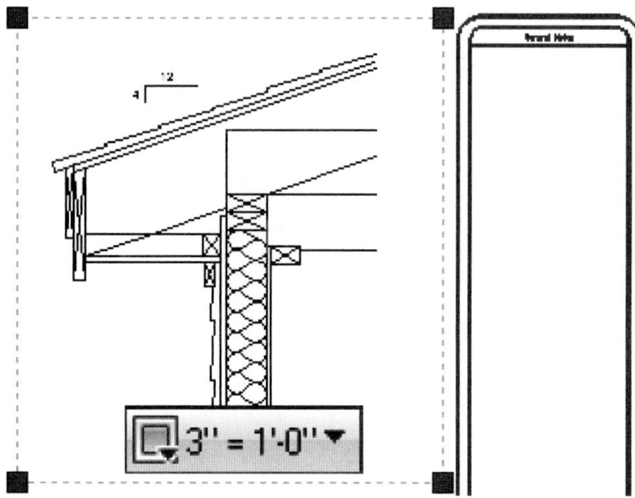

Figure 19.77

Step 11

Perform the same operations on the bottom-right viewport in order to scale the foundation detail inside the viewport. Click inside this viewport to make it current. Pan the image of the Foundation detail approximately to the middle of the bottom viewport.

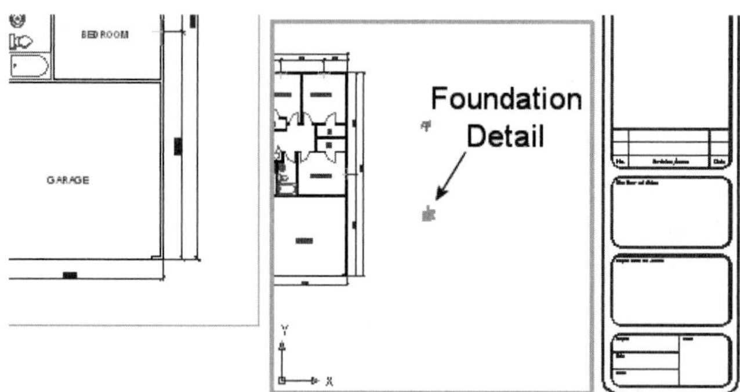

Figure 19.78

Step 12

Change the VP Scale of this image to 1-1/2" = 1'-0", as shown in the following image. The results of this operation are illustrated in the following image. You may have to pan the image in order to better position it inside the viewport. Note that this detail favors the left side of the viewport; you will need this room to label various parts of the detail using multileaders in a later step.

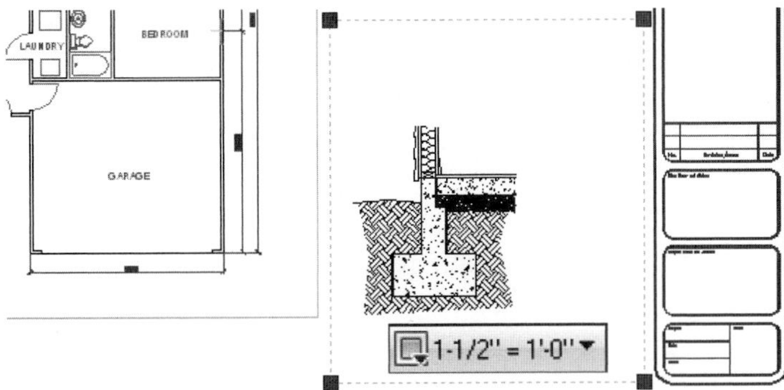

Figure 19.79

Step 13

Switch back to Paper Space. Click the edges of all three viewports and click the Lock button, as shown in the following image. This prevents any accidental scaling of the images if you have to enter floating Model Space. Notice also that the VP Scale reads VARIES. This is because all three viewports were selected and the scales of each viewport are different.

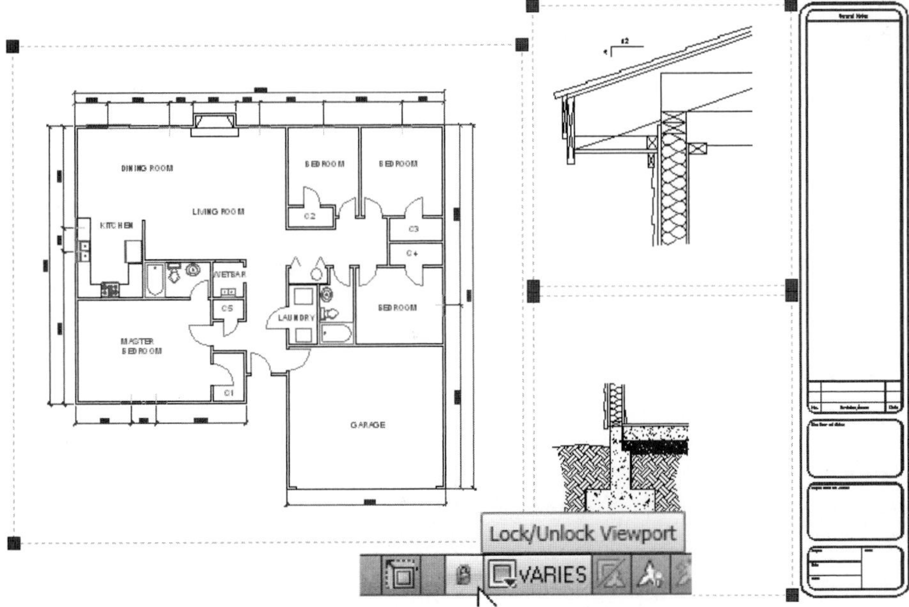

Figure 19.80

Step 14

Change to the Text layer. Double-click inside the large viewport to activate floating Model Space. Use the MTEXT command and a text height of .30 to place the title of this image

(FLOOR PLAN) and the scale (¼" = 1'-0") as shown in the following image. The size of the text is scaled and visible since the text style was set to Annotative.

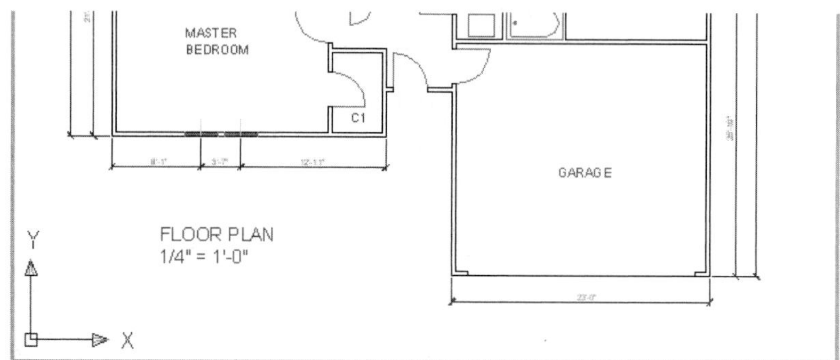

Figure 19.81

Step 15

Click inside of the viewport containing the cornice detail to make it active and use the MTEXT command and a text height of .30 to place the title of this image and the scale, as shown on the left in the following image. Click inside the viewport containing the foundation detail to make it active and use the MTEXT command to place the title of this image and the scale as shown on the right in the following image.

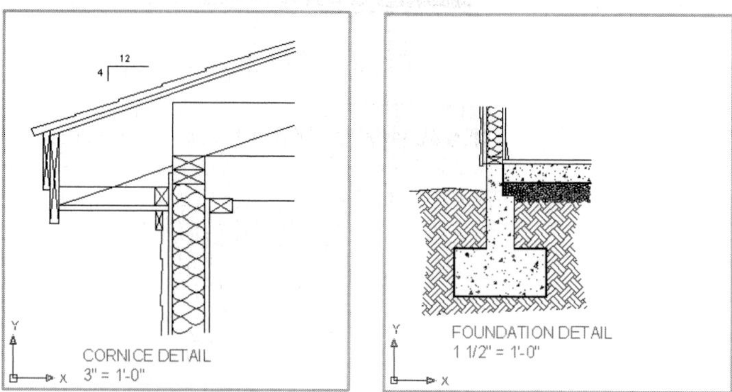

Figure 19.82

Step 16

Verify that the foundation plan is still the active viewport; the User Coordinate System icon should be visible. You will label various elements that make up the foundation detail with multileaders. Notice on the left in the following image that a multileader style is already created and set to Annotative. Then choose the MLEADER command from the Dimension pull-down menu or the Ribbon, as shown in the following image. Place the following multileader notes: INSULATION, FINISHED FLOOR, CONCRETE SLAB, GRAVEL, CONCRETE PIER, and FOOTING as shown on the right in the following image.

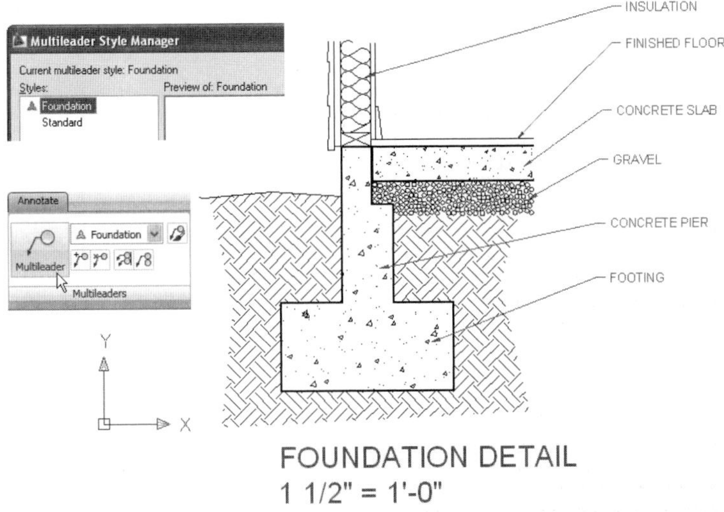

Figure 19.83

Step 17

Switch back to Paper Space and perform a plot preview. Your display should appear similar to the following image.

Note: You could also add crosshatching to the walls of the floor plan. To do this, click the Model tab to return to Model Space. Activate the Hatch and Gradient dialog box. Use the ANSI31 hatch pattern and change the scale of the pattern to 48. Also place a check in the box next to Annotative.

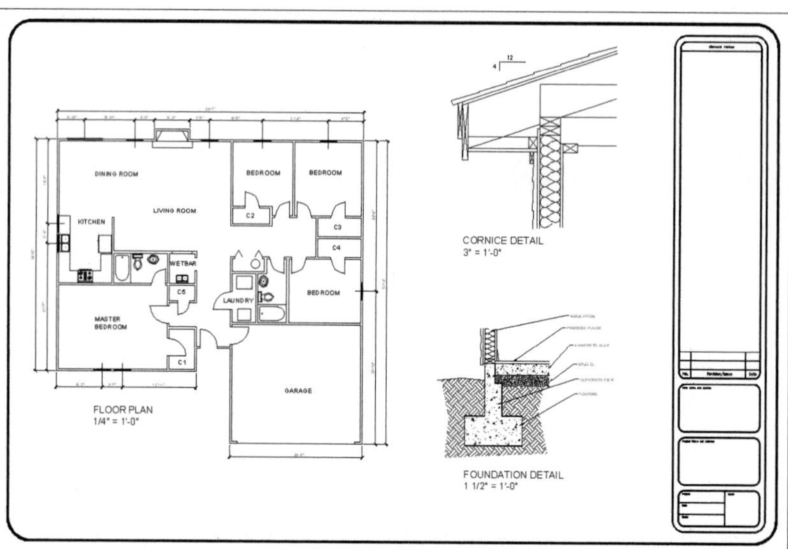
12
4
DINING ROOM
LIVING ROOM
KITCHEN
BEDROOM
BEDROOM
C2
C3
C4
WETBAR
C5
LAUNDRY
BEDROOM
MASTER
BEDROOM
C1
GARAGE
FLOOR PLAN
1/4" = 1'-0"
CORNICE DETAIL
3" = 1'-0"
FOUNDATION DETAIL
1 1/2" = 1'-0"

Figure 19.84

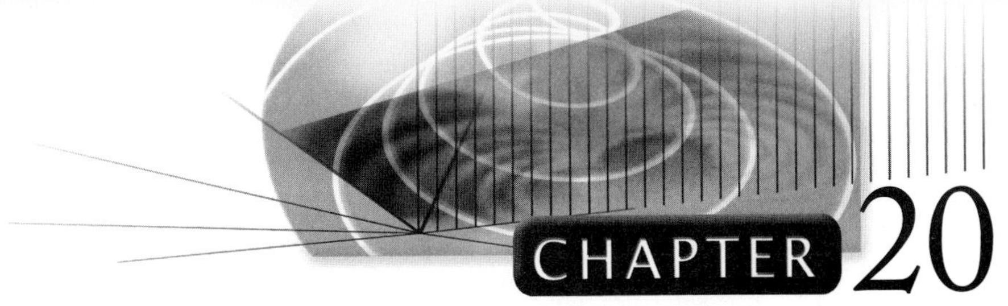

Solid Modeling Fundamentals

Solid models are mathematical models of actual objects that can be analyzed through the calculation of such items as mass properties, center of gravity, surface area, moments of inertia, and much more. Before solid models are built, various options of the UCS command need to be understood. The use and creation of visual styles are also discussed in this chapter. The solid model starts the true design process by defining objects as a series of primitives. Boxes, cones, cylinders, spheres, and wedges are all examples of primitives. These building blocks are then joined together or subtracted from each other through modify commands. Intermediate solid modeling commands such as EXTRUDE, REVOLVE, SWEEP, and LOFT are discussed in this chapter, as well as the ability to create a spiral or helix object. Fillets and chamfers can be created to give the solid model a more realistic appearance and functionality. Two-dimensional views can be extracted from the solid model along with a cross-section of the model.

3D MODEL TYPES

A 3D model can be broken down into various types. The more popular types of 3D models include wireframe, surface, and solid. Each of these are described in greater detail later in the chapter.

WIREFRAME MODELS

A 3D wireframe model is shown on the left in the following image. This is the simplest type of 3D model and, unlike the 2D isometric drawing, it can be viewed from any direction by simply changing the viewpoint. A wireframe model, however, as lines intersect from the front and back of the object, is sometimes difficult to visualize and interpret.

In 2D drawings the Z value of an XYZ coordinate remains set to zero. In 3D modeling, height is shown by providing values for Z. Values for height can be entered at the command prompt, selected using object snaps, or selected through the use of aids such as XYZ point filters. In this method, values are temporarily saved for later use inside a command. The values can be retrieved and a coordinate value entered to complete the command. Manipulation of a user-defined coordinate system (UCS) is another especially useful method for providing height when creating a 3D object. A new coordinate system can be defined along any plane on which to place objects.

SURFACE MODELS

Surfacing picks up where the wireframe model leaves off. Because of the complexity of the wireframe and the number of intersecting lines, surfaces are applied to the wireframe. A major advantage of a surfaced model over a wireframe model is the capability of performing a hidden line removal operation. A HIDE is performed to view the model without interference of other faces. The surfaces created on the right in the following image are in the form of opaque objects, called 3D faces. Three-dimensional faces are created by selecting either three or four points in space, as required, to create the surface needed. Surfacing an existing wireframe can be easily accomplished by utilizing OSNAP options to assist in point selection. The 3DFORBIT (3D Free Orbit) command may be used to view the model in different 3D positions to make sure all sides of the model have been surfaced.

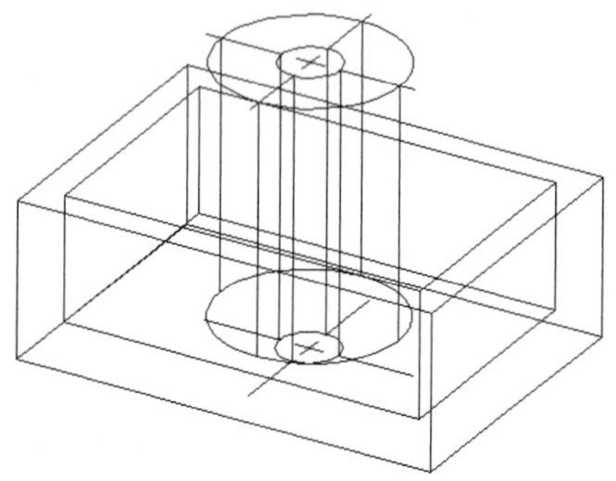

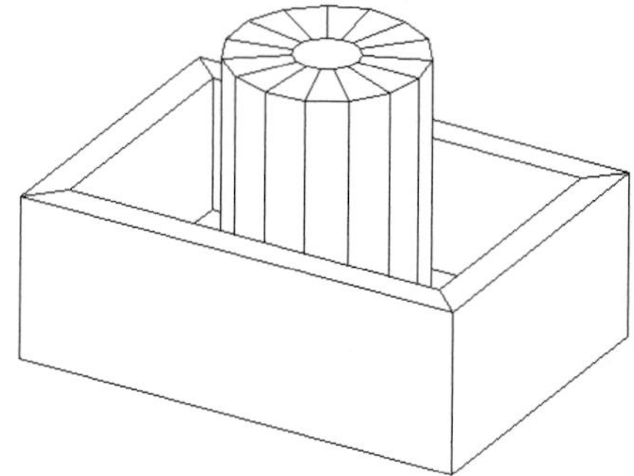

Figure 20.1

SOLID MODELS

The creation of a solid model remains the most complete way to represent a model in 3D, as shown in the following image. It is also the most versatile 3D representation of an object. Wireframe models and surfaced models can be analyzed by taking distance readings and the identification of key points of the model. Key orthographic views such as Front, Top, and Right Side views can be extracted from wireframe models. Surfaced models can be imported into shading packages for increased visualization. Solid models do all these operations and more. This is because the solid model, as it is called, is a solid representation of the actual object. From cylinders to spheres, wedges to boxes, all objects that go into the creation of a solid model have volume. This allows a model to be constructed of what are referred to as primitives. These primitives are then merged into one using addition, subtraction, and intersection operations. What remains is the most versatile of 3D drawings. This method of creating models will be the main focus of this chapter.

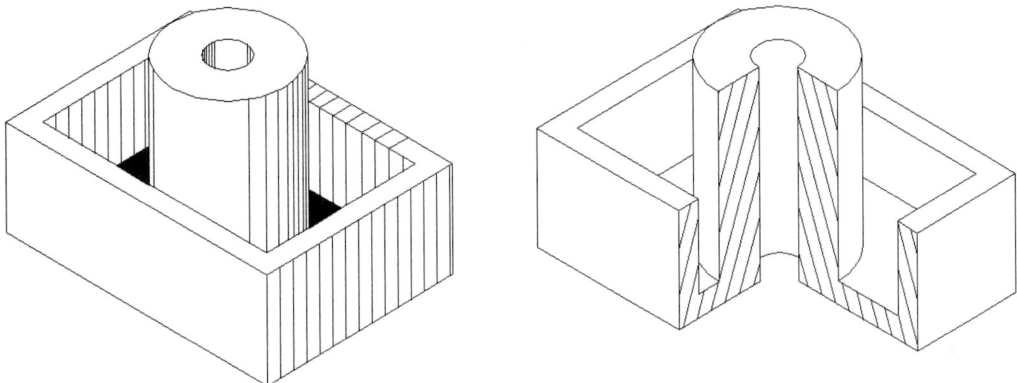

Figure 20.2

SOLID MODELING BASICS

All objects, no matter how simple or complex, are composed of simple geometric shapes, or primitives. The shapes include boxes, wedges, cylinders, cones, pyramids, spheres, and tori. Solid modeling allows for the creation of these primitives. Once created, the shapes are either merged or subtracted to form the final object. Follow the next series of steps to form the object shown on the left in the following image.

Begin the process of solid modeling by constructing a solid slab that represents the base of the object. You can do this by using the BOX command. Supply the length, width, and height of the box, and the result is a solid slab, as shown in the middle of the following image.

Next, construct a cylinder using the CYLINDER command. This cylinder will eventually form the curved end of the object. One of the advantages of constructing a solid model is the ability to merge primitives together to form composite solids. Using the UNION command, combine the cylinder and box to form the complete base of the object, as shown on the right in the following image.

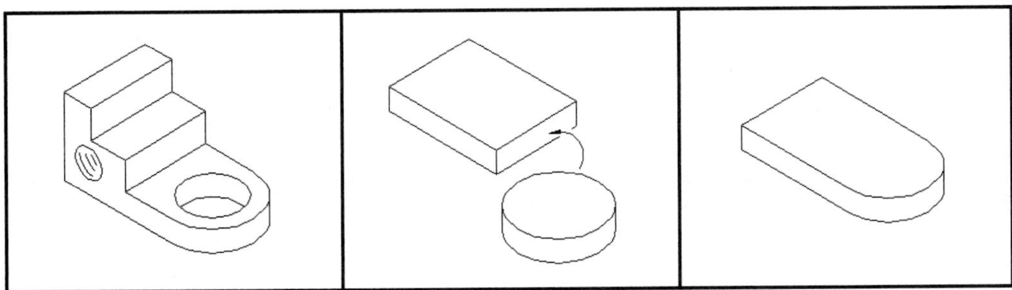

Figure 20.3

As the object progresses, create another solid box and move it into position on top of the base. Then use the UNION command to join this new block with the base. As you add new shapes during this process, they all become part of the same solid, as shown on the left in the following image.

Create yet another box and move it into position. However, instead of combining the new box with the block, it is removed from the solid. Use the SUBTRACT command to perform this subtraction operation and create the step, as shown on the right in the following image.

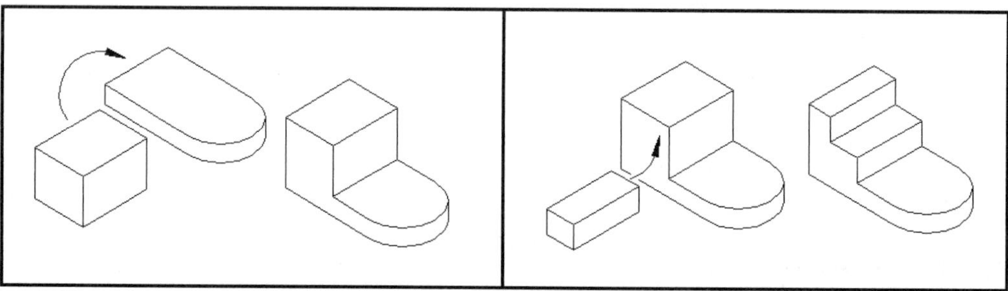

Figure 20.4

You can form holes in a similar fashion. First, use the CYLINDER command to create a cylinder with the diameter and depth of the desired hole. Next, move the cylinder into the solid and then subtract it using the SUBTRACT command. Again, the object shown on the left in the following image represents a composite solid object.

Using existing AutoCAD tools such as User Coordinate Systems (discussed in detail later in this chapter), create another cylinder with the CYLINDER command. It too is moved into position, where it is subtracted through the SUBTRACT command, as shown on the right in the following image.

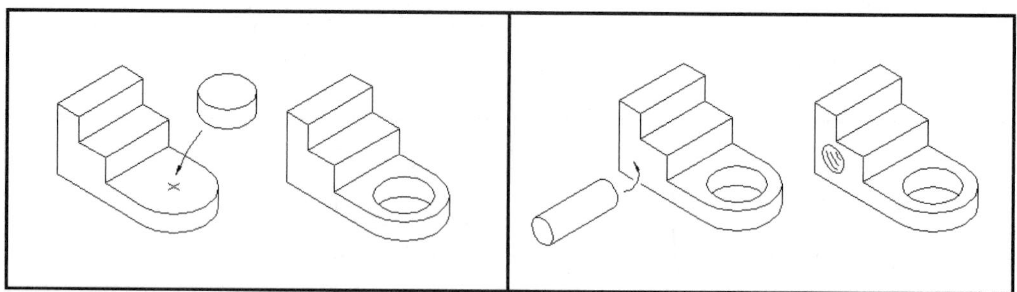

Figure 20.5

The complete solid model of the object is shown on the left in the following image. Advantages of constructing a solid model out of an object come in many forms. Two-dimensional profiles of different surfaces of the solid model can be taken to create orthographic views. Section views of solids can automatically be formed and crosshatched, as shown on the right in the following image.

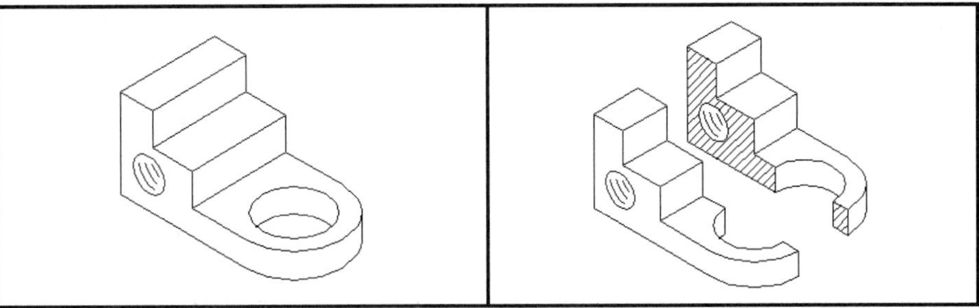

Figure 20.6

A very important analysis tool is the MASSPROP command, which is short for "mass property." Information such as the mass and volume of the solid object is automatically calculated, along with centroids and moments of inertia, components that are used in computer-aided engineering (CAE) as shown in the following image.

Mass:	223.2 gm	
Volume:	28.39 cu cm (Err: 3.397)	
Bounding box:		X:–1.501—6.001 cm
		Y: –0.0009944—2.75 cm
		Z: –3.251—3.251 cm
Centroid:		X: 2.36 cm
		Y: 0.7067 cm
		Z: 0.07384 cm
Moments of inertia:		X: 612.2 gm sq cm (Err: 76.16)
		Y: 2667 gm sq cm (Err: 442.6)
		Z: 2444 gm sq cm (Err: 383.2)

Figure 20.7

WORKING IN THE 3D MODELING WORKSPACE

As a means of allowing you to work in a dedicated custom, task-orientated environment, predefined workspaces are already created in AutoCAD. These workspaces consist of menus, toolbars, and palettes that are organized around a specific task. When you select a workspace, only those menus, toolbars, and palettes that relate directly to the task are displayed.

The workspaces can be found by picking the Workspaces toolbar as shown on the left in the following image or from the status bar located at the bottom of the display screen, as shown on the right in the following image. Of these workspaces, you will find it productive to use 3D Modeling throughout this chapter in the book.

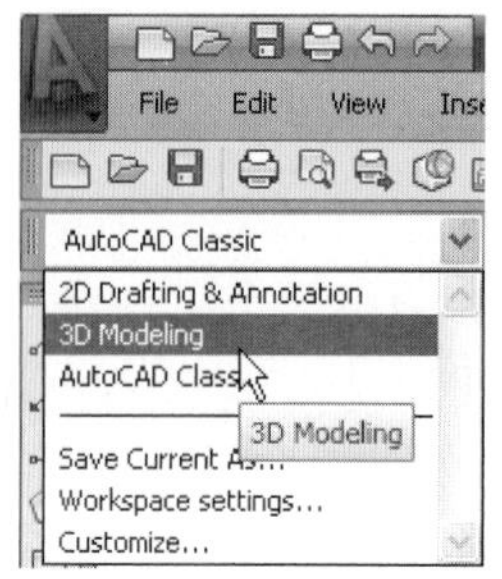

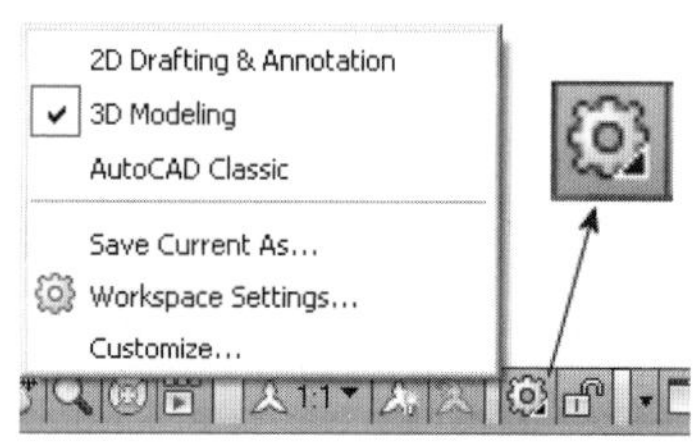

Figure 20.8

THE 3D MODELING WORKSPACE

Clicking on the 3D Modeling workspace in the previous image changes your AutoCAD screen to appear similar to the following image. In this image, your screen contains only 3D-related menus and palettes.

Note: In addition to the Ribbon, the tool palette also displays when you activate the 3D Modeling workspace. The tool palette has been turned off in the following image.

When you make changes to your drawing display (such as moving, hiding, or displaying a toolbar or a tool palette group) and you want to preserve the display settings for future use, you can save the current settings to a workspace.

The Ribbon of the 3D Modeling workspace consists of buttons and controls used primarily for 3D modeling, 3D navigation, controlling lights, controlling visual styles, creating and applying materials, and producing renderings as shown in the following image. The use of the Ribbon eliminates the need to display numerous toolbars, which tend to clutter up your screen. This enables you to have more screen real estate for constructing your 3D models. When you activate the 3D Modeling workspace, the Ribbon automatically displays. The Ribbon can also be displayed in any workspace by clicking the Tools pull-down menu, followed by Palettes, and then Ribbon.

When the Ribbon first displays, the Home tab is active. At this point, the following panels are available based on the 3D Modeling workspace; 3D Modeling, Draw, Solid Editing, Modify, View, Layers, Properties, and Utilities, as shown in the following image.

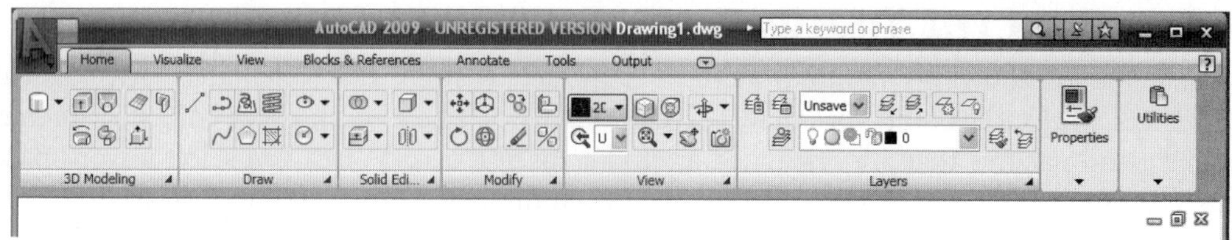

Figure 20.9

THE VISUALIZE TAB OF THE RIBBON

Another helpful tab used for working in a 3D environment is the Visualize tab. Clicking on the Visualize tab of the Ribbon will expose the following panels; Visual Styles, Edge Effects, Lights, Sun, Time & Location, and Materials, as shown in the following image.

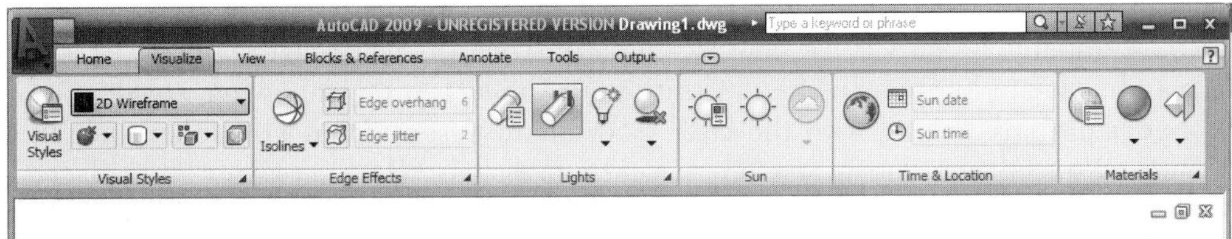

Figure 20.10

THE VIEW TAB OF THE RIBBON

The View tab of the Ribbon is another helpful area when working in 3D. Clicking on the View tab of the Ribbon will expose the following panels; UCS, Viewports, Palettes, 3D Palettes, Window, and Window Elements, as shown in the following image.

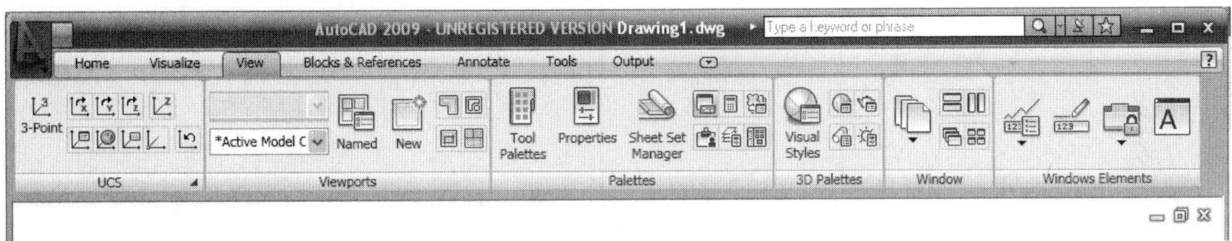

Figure 20.11

THE AUTOCAD CLASSIC WORKSPACE AND 3D

3D modeling can also be performed while in the AutoCAD Classic workspace, as shown in the following image. The only drawback is that you will need to pre-load toolbars that are commonly used for constructing 3D models. A few of these toolbars are illustrated in the following image.

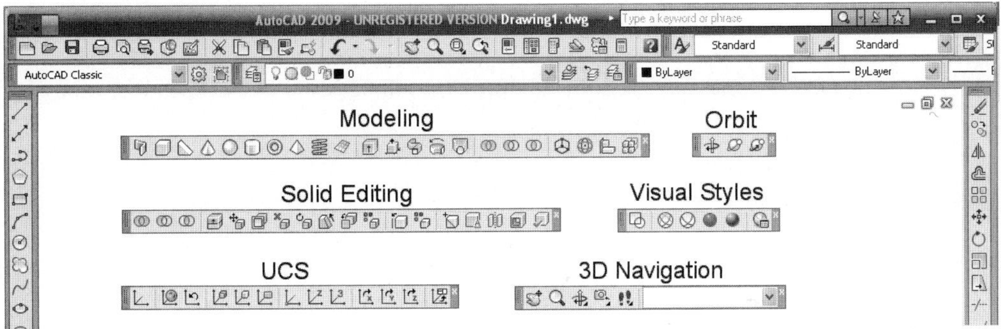

Figure 20.12

CREATING USER COORDINATE SYSTEMS

Two-dimensional computer-aided design still remains the most popular form of representing drawings for most applications. However, in applications such as architecture and manufacturing, 3D models are becoming increasingly popular for creating rapid prototype models or for creating tool paths from the 3D model. To assist in this creation process, User Coordinate Systems are used to create construction planes where features such as holes and slots are located. In the illustration on the left in the following image, a model of a box is displayed along with the User Coordinate System icon. The appearance of the UCS icon can change depending on the current visual style the model is displayed in. Visual styles are discussed later in this chapter. For now, the UCS icons illustrated on the right can take on a 2D or 3D appearance. Both UCS icon examples show the directions of the three User Coordinate System axes. The icon arrows indicate the current positive coordinate direction.

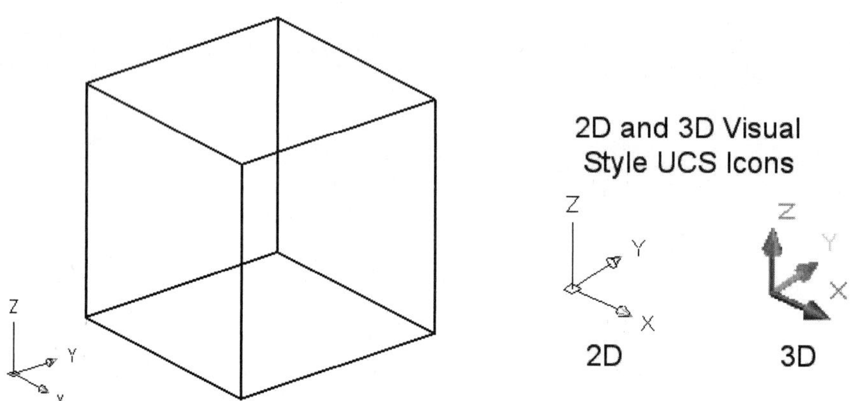

Figure 20.13

Phase 1 of creating a User Coordinate system begins with an understanding of how the UCS command operates. The command line sequence follows, along with all options; the command and options can also be selected from the Menu Browser, Tools pull-down menu, View tab of the Ribbon, or the UCS toolbar, as illustrated in the following image.

```
Command: UCS
Current ucs name: *WORLD*
Specify origin of UCS or [Face/NAmed/OBject/Previous/View/World/
    X/Y/Z/ZAxis]
<World>: N (For New)
Specify origin of new UCS or [ZAxis/3point/OBject/Face/View/X/
    Y/Z] <0,0,0>:
```

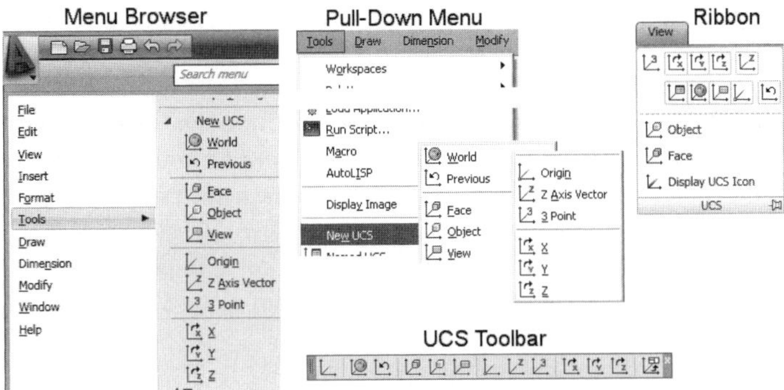

Figure 20.14

The following table gives a brief description of each User Coordinate System mode.

Button	Tool	Function
	UCS	Activates the UCS command located in the Command Prompt area
	World	Switches to the World Coordinate System from any previously defined User Coordinate System
	UCS Previous	Sets the UCS icon to the previously defined User Coordinate System
	Face UCS	Creates a UCS based on the selected face of a solid object
	Object	Creates a UCS based on an object selected
	View	Creates a UCS parallel to the current screen display
	Origin	Used to specify a new origin point for the current UCS
	Z Axis Vector	Creates a new UCS based on two points that define the Z axis
	3 Point	Creates a new UCS by picking three points
	X	Used for rotating the current UCS along the X-axis
	Y	Used for rotating the current UCS along the Y-axis
	Z	Used for rotating the current UCS along the Z-axis
	Apply	Sets the current User Coordinate System setting to a specific viewport(s)

THE UCS–SPECIFY ORIGIN OF UCS AND ORIGIN OPTION

The default sequence of the UCS command defines a new User Coordinate System by first defining a new origin (translating – picking a new 0,0,0 position) and second, if needed, changing the direction of the X-axis (rotating the coordinate system).

The illustration on the left in the following image is a sample model with the current coordinate system being the World Coordinate System. Follow the illustrations in the middle and on the right in the following image to define a new User Coordinate System using the default command sequence.

Try It!: Open the drawing file 20_Ucs Origin. Activate the UCS command. Identify a new origin point for 0,0,0 at "A," as shown in the middle in the following image. This should move the User Coordinate System icon to the point that you specify. If the icon remains in its previous location and does not move, use the UCSICON command with the Origin option to display the icon at its new origin point. To prove that the corner of the box is now 0,0,0, construct a circle at the bottom of the 5" cube, as shown on the right in the following image.

```
Command: UCS
Current ucs name: *WORLD*
Specify origin of UCS or [Face/NAmed/OBject/Previous/View/World/
    X/Y/Z/ZAxis]
<World>: End
of (Select the endpoint of the line at "A")
Specify point on X-axis or <Accept>: (Press ENTER to accept since
    you do not want to rotate the coordinate system)

Command: C (For CIRCLE)
Specify center point for circle or [3P/2P/Ttr (tan tan radius)]:
    2.5,2.5,0
Specify radius of circle or [Diameter]: 2
```

Notice a small "plus" at the corner of the User Coordinate System icon (see the illustration on the right in the following image). This indicates that the UCS icon is displayed at the origin (0,0,0). A small box, as shown on the left in the following image, indicates that the icon is displaced at the World Coordinate System (WCS).

It is also important to note that the circle shown on the right in the following image is a 2D object and can only be drawn in or parallel to the XY plane (for example, the bottom or top of the box). To draw a circle in the side of the box, we will need to not only move (translate) our coordinate system but rotate it as well.

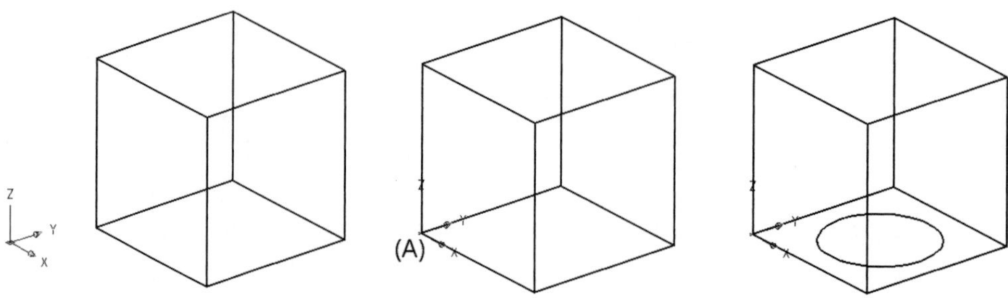

Figure 20.15

Another popular method of changing the UCS origin is provided through the Origin option. This option is available through the UCS toolbar, Ribbon, Menu Browser, and

Tools pull-down menu. It allows you select a new 0,0,0 position but does not provide any prompts for rotating the coordinate system. The Origin option is not listed in the command prompts but can be entered as shown below.

```
Command: UCS
Current ucs name: *WORLD*
Specify origin of UCS or [Face/NAmed/OBject/Previous/View/World/
    X/Y/Z/ZAxis]
<World>: O (For Origin)
Specify new origin point <0,0,0>: (Pick desired point)
```

THE UCS–3POINT OPTION

Use the 3point option of the UCS command to specify a new User Coordinate System by identifying an origin and new directions of its positive X- and Y-axes (translate and rotate). This option, like the Origin option, is not listed in the command sequence but can be entered anyway. The option is displayed in the UCS toolbar, Ribbon, Menu Browser, and Tools pull-down menu (New UCS).

Try It!: Open the drawing file 20_Ucs 3p. The illustration on the left in the following image shows a 3D cube in the World Coordinate System. To construct objects on the front panel, first define a new User Coordinate System parallel to the front. Use the following command sequence to accomplish this task.

```
Command: UCS
Current ucs name: *WORLD*
Specify origin of UCS or [Face/NAmed/OBject/Previous/View/World/
    X/Y/Z/ZAxis]
<World>: 3 (For 3point)
Specify new origin point <0,0,0>: End
of (Select the endpoint of the model at "A" as shown in the
    middle of the following image)
Specify point on positive portion of X-axis <>: End
of (Select the endpoint of the model at "B")
Specify point on positive-Y portion of the UCS XY plane <>: End
of (Select the endpoint of the model at "C")
```

With the Y-axis in the vertical position and the X-axis in the horizontal position, any type of object can be constructed along this plane, such as the lines shown on the right in the following image.

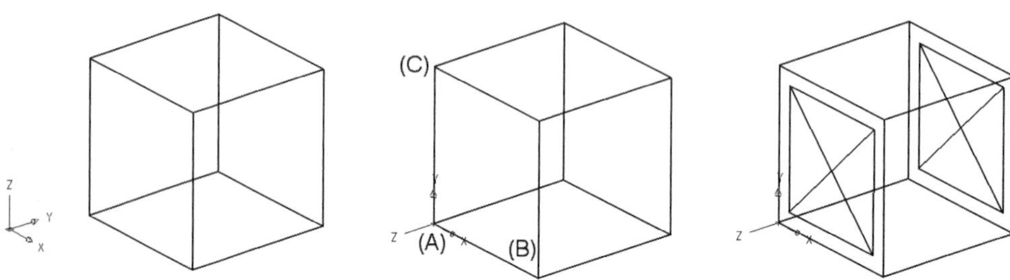

Figure 20.16

Try It!: Open the drawing file 20_Incline. The 3point method of defining a new User Coordinate System is quite useful in the example shown in the following image, where a UCS needs to be aligned with the inclined plane. Use the intersection at "A" as the origin of the new UCS, the intersection at "B" as the direction of the positive X-axis, and the intersection at "C" as the direction of the positive Y-axis.

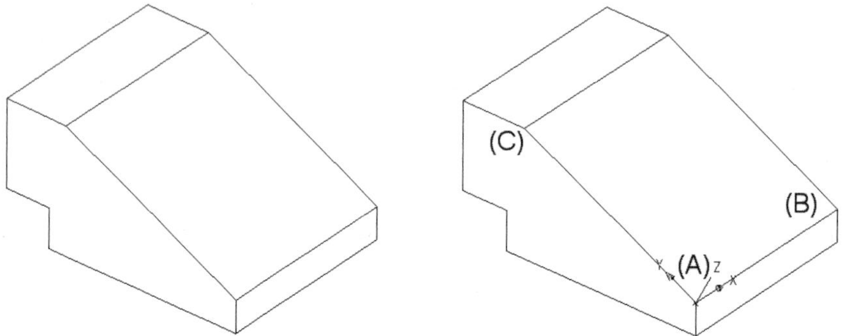

Figure 20.17

THE UCS–X/Y/Z ROTATION OPTIONS

Using the X/Y/Z rotation options rotates the current user coordinate around the specific axis. Select the X, Y, or Z option to establish the axis that will act as the pivot; a prompt appears asking for the rotation angle about the pivot axis. The right-hand rule is used to determine the positive direction of rotation around an axis. Think of the right hand gripping the pivot axis with the thumb pointing in the positive X, Y, or Z direction. The curling of the fingers on the right hand determines the positive direction of rotation. The following image illustrates positive rotation about each axis. By viewing down the selected axis, the positive rotation is seen to be counterclockwise.

Axis Rotations

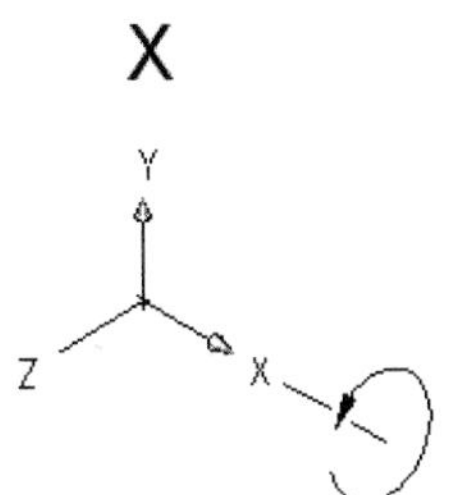

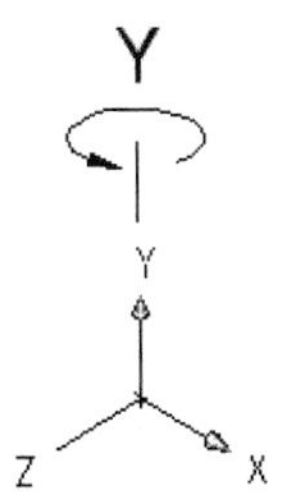

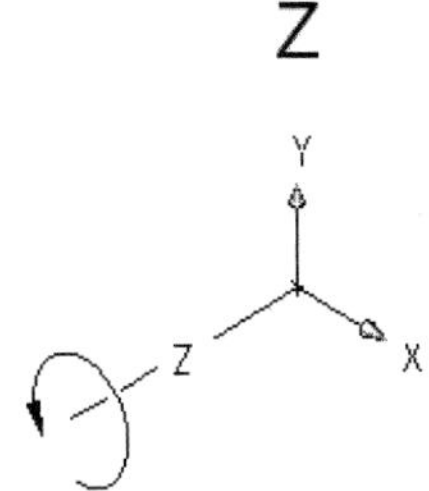

Figure 20.18

Try It!: Open the drawing file 20_Ucs Rotate. Given the cube shown on the left in the following image in the World Coordinate System, the X option of the UCS command will be used to stand the icon straight up by entering a 90° rotation value, as in the following prompt sequence.

```
Command: UCS
Current ucs name: *NO NAME*
Specify origin of UCS or [Face/NAmed/OBject/Previous/View/World/
    X/Y/Z/ZAxis]
<World>: X
Specify rotation angle about X-axis <90>: (Press ENTER to accept
    90° of rotation)
```

The X-axis is used as the pivot of rotation; entering a value of 90° rotates the icon the desired degrees in the counterclockwise direction, as shown on the right in the following image.

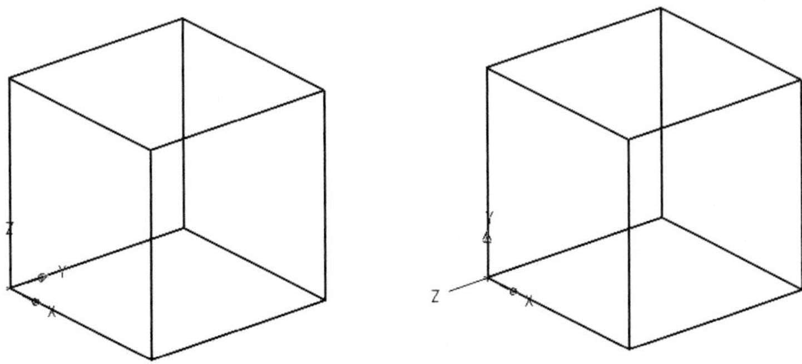

Figure 20.19

THE UCS–OBJECT OPTION

Another option for defining a new User Coordinate System is to select an object and have the User Coordinate System align to that object (translate and rotate).

Try It!: Open the drawing file 20_Ucs Object. Given the 3D cube shown on the left in the following image, use the following command sequence and the illustration on the right to construct the circle on the proper plane.

```
Command: UCS
Current ucs name: *WORLD*
Specify origin of UCS or [Face/NAmed/OBject/Previous/View/World/
    X/Y/Z/ZAxis]
<World>: OB (For Object)
Select object to align UCS: (Select the circle shown on the
    right in the following image)
```

The type of object selected determines the alignment (translation and rotation) of the User Coordinate System. In the case of the circle, the center of the circle becomes the origin of the User Coordinate System. The point where the circle was selected becomes the point through which the positive X-axis aligns. Other types of objects that can be selected include arcs, dimensions, lines, points, plines, solids, traces, 3dfaces, text, and blocks.

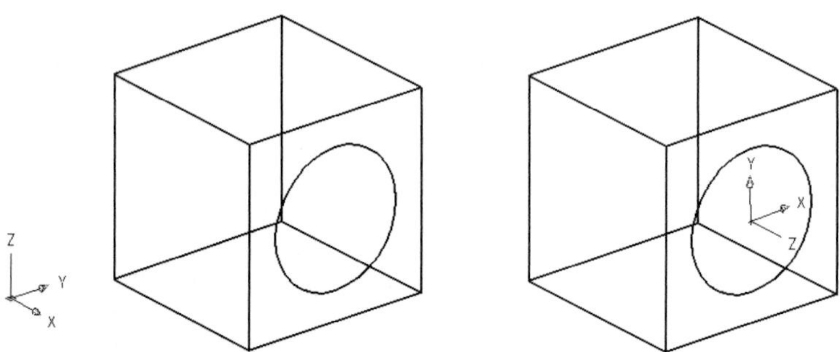

Figure 20.20

THE UCS–FACE OPTION

The Face option of the UCS command allows you to establish a new coordinate system aligned to the selected face of a solid object.

Try It!: Open the drawing file 20_Ucs Face. Given the current User Coordinate System, as shown on the left in the following image, and a solid box, follow the command sequence below to align the User Coordinate System with the Face option.

```
Command: UCS
Current ucs name: *WORLD*
Specify origin of UCS or [Face/NAmed/OBject/Previous/View/World/
    X/Y/Z/ZAxis]
<World>: F (For Face)
Select face of solid object: (Select the edge of the model at
    point "A," as shown on the right in the following
    image—select the edge near the corner that you want as the
    origin)
```

```
Enter an option [Next/Xflip/Yflip] <accept>: (Press ENTER to
    accept the UCS position)
```

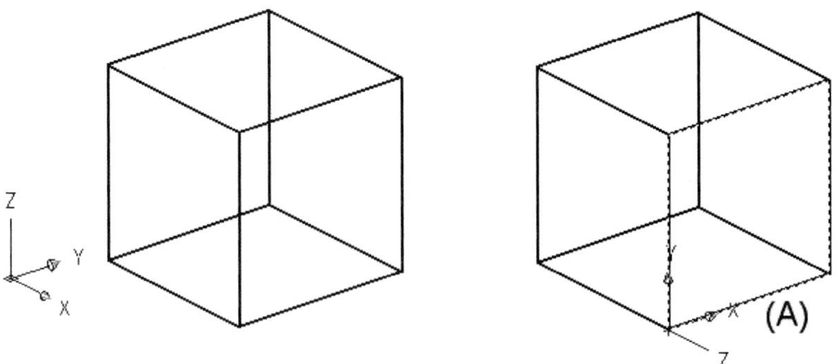

Figure 20.21

THE UCS–VIEW OPTION

The View option of the UCS command allows you to establish a new coordinate system where the XY plane is perpendicular to the current screen viewing direction; in other words, it is parallel to the display screen.

Try It!: Open the drawing file 20_Ucs View. Given the current User Coordinate System, as shown on the left in the following image, follow the prompts below along with the illustration on the right to align the User Coordinate System with the View option.

```
Command: UCS
Current ucs name: *WORLD*
Specify origin of UCS or [Face/NAmed/OBject/Previous/View/World/
    X/Y/Z/ZAxis]
<World>: V (For View)
```

The results are displayed on the right in the following image, with the User Coordinate System aligned parallel to the display screen.

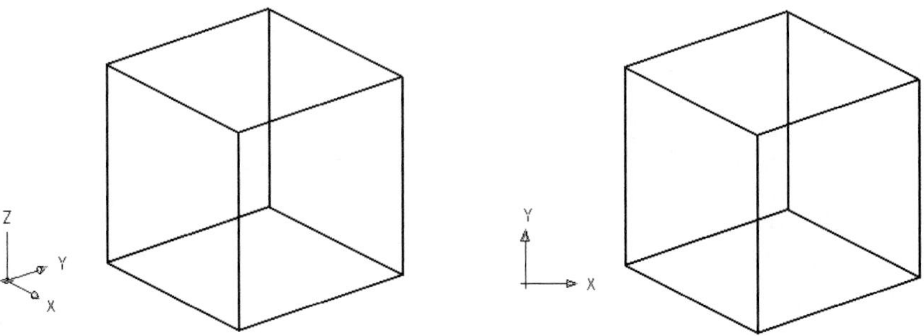

Figure 20.22

USING DYNAMIC UCS MODE

While inside the UCS command, you can automatically switch the plane of the UCS by simply hovering your cursor over the face of a 3D solid object. This special function is available when the DUCS (Dynamic UCS) button is turned on in the status bar, as shown in the following image. The next Try It! exercise illustrates how this method of manipulating the UCS dynamically is accomplished.

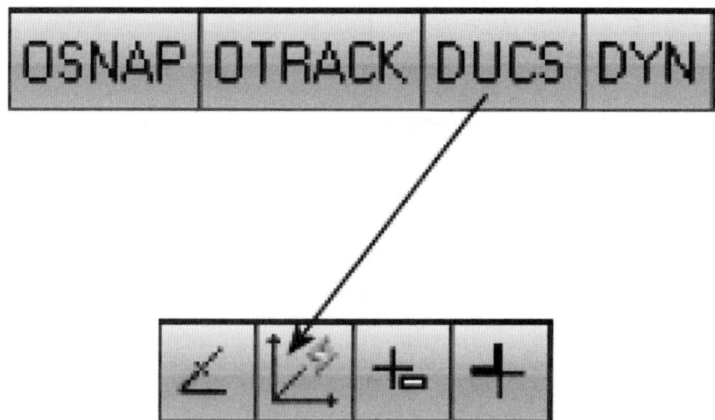

Figure 20.23

Try It!: Open the drawing file 20_Dynamic_Ucs. Given the current User Coordinate System, as shown on the left in the following image, follow the prompts below, along with the illustrations, to dynamically align the User Coordinate System to a certain face and location.

In this first example of dynamically setting the UCS, hover your cursor along the front face of the object until it highlights, as illustrated on the left in the following image, and pick the endpoint at "A" to locate the UCS as shown on the right in the following image.

```
Command: UCS
Current ucs name: *WORLD*
Specify origin of UCS or [Face/NAmed/OBject/Previous/View/World/
    X/Y/Z/ZAxis]
<World>: (Move the Dynamic UCS icon over the front face, as
    shown on the left in the following image. Then pick the
    endpoint at "A")
Specify point on X-axis or <Accept>: (Press ENTER to accept)
```

With the new UCS defined, it is good practice to save the position of the UCS under a unique name. These named User Coordinate Systems can then be easily retrieved for later use.

```
Command: UCS
Current ucs name: *NO NAME*
Specify origin of UCS or [Face/NAmed/OBject/Previous/View/World/
    X/Y/Z/ZAxis]
<World>: NA (For NAmed)
```

```
Enter an option [Restore/Save/Delete/?]: S (For Save)
Enter name to save current UCS or [?]: Front
```

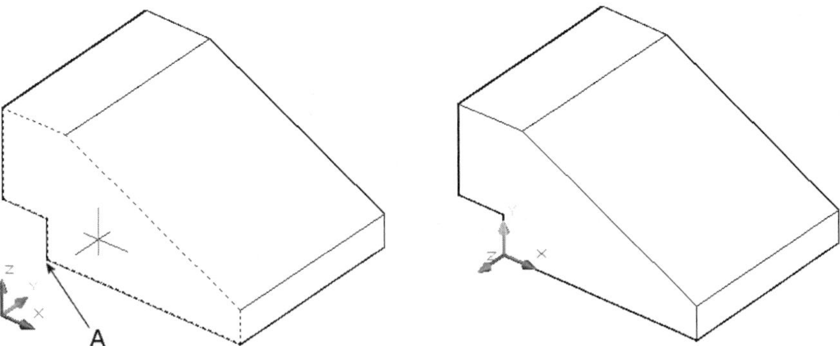

Figure 20.24

This next example requires you to pick the endpoint to better define the X-axis while dynamically locating the UCS. Hover your cursor along the top face of the object illustrated on the left in the following image and pick the endpoint at "A" to locate origin of the UCS. Continue by picking the endpoint at "B" as the X-axis. Save this UCS as "Top."

```
Command: UCS
Current ucs name: Front
Specify origin of UCS or [Face/NAmed/OBject/Previous/View/World/
    X/Y/Z/ZAxis]
<World>: (Move the Dynamic UCS icon over the top face as shown
    on the left in the following image. Then pick the endpoint
    at "A")
Specify point on the X-axis or <Accept>: (Pick the endpoint at
    "B" to align the X axis)
Specify point on the XY plane or <Accept>: (Press ENTER to accept)

Command: UCS
Current ucs name: *NO NAME*
Specify origin of UCS or [Face/NAmed/OBject/Previous/View/World/
    X/Y/Z/ZAxis]
<World>: NA (For NAmed)
Enter an option [Restore/Save/Delete/?]: S (For Save)
Enter name to save current UCS or [?]: Top
```

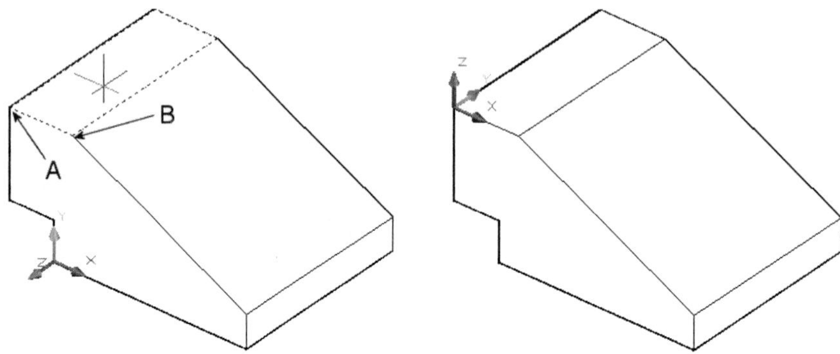

Figure 20.25

Next, hover your cursor along the side face of the object illustrated on the left in the following image and pick the endpoint at "A" to locate the origin of the UCS. Continue by picking the endpoint at "B" as the X-axis and the endpoint at "C" to define the XY plane. Save this UCS as "Side."

```
Command: UCS
Current ucs name: Top
Specify origin of UCS or [Face/NAmed/OBject/Previous/View/World/
    X/Y/Z/ZAxis]
<World>: (Move the Dynamic UCS icon over the side face as shown
    on the left in the following image. Then pick the endpoint
    at "A")
Specify point on X-axis or <Accept>: (Pick the endpoint at "B"
    to align the X axis)
Specify point on the XY plane or <Accept>: (Pick the endpoint at
    "C" to align the XY plane)

Command: UCS
Current ucs name: *NO NAME*
Specify origin of UCS or [Face/NAmed/OBject/Previous/View/World/
    X/Y/Z/ZAxis]
<World>: NA (For NAmed)
Enter an option [Restore/Save/Delete/?]: S (For Save)
Enter name to save current UCS or [?]: Side
```

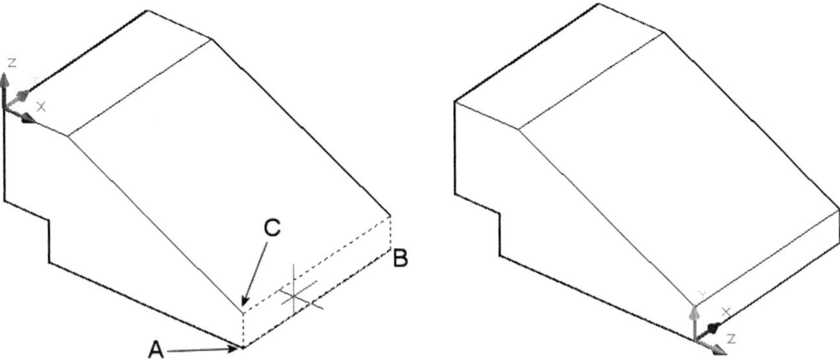

Figure 20.26

Finally, hover your cursor along the inclined face of the object illustrated on the left in the following image and pick the endpoint at "A" to locate the origin of the UCS. Continue by picking the endpoint at "B" as the X axis and the endpoint at "C" to define the XY plane. Save this UCS as "Auxiliary."

```
Command: UCS
Current ucs name: Side
Specify origin of UCS or [Face/NAmed/OBject/Previous/View/World/
    X/Y/Z/ZAxis]
<World>: (Move the Dynamic UCS icon over the inclined face as
    shown on the left in the following image. Then pick the
    endpoint at "A")
Specify point on X-axis or <Accept>: (Pick the endpoint at "B"
    to align the X axis)
Specify point on the XY plane or <Accept>: (Pick the endpoint at
    "C" to align the XY plane)

Command: UCS
Current ucs name: *NO NAME*
Specify origin of UCS or [Face/NAmed/OBject/Previous/View/World/
    X/Y/Z/ZAxis]
<World>: NA (For NAmed)
Enter an option [Restore/Save/Delete/?]: S (For Save)
Enter name to save current UCS or [?]: Auxiliary
```

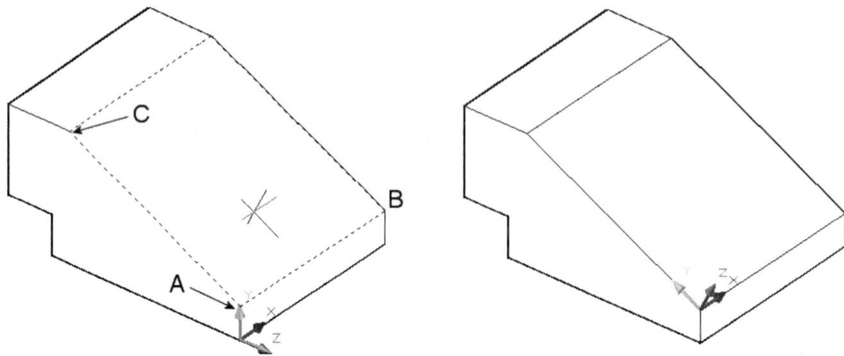

Figure 20.27

USING THE UCS DIALOG BOX

As stated earlier, considerable drawing time can be saved by assigning a name to a User Coordinate System once it has been created. Once numerous User Coordinate Systems have been defined in a drawing, using their names instead of re-creating each coordinate system easily restores them. You can accomplish this on the command line by using the Save and Restore options of the UCS command. A method of retrieving previously saved User Coordinate Systems is to choose Named UCS from the Tools pull-down menu illustrated on the left in the following image (the UCSMAN command). This displays the UCS dialog box, as shown on the right in the following image, with the Named UCSs tab selected. All User Coordinate Systems previously defined in the drawing are listed here. To make one of these coordinate systems current, highlight the desired UCS name and select the Set Current button, as shown on the right in the following image. A named UCS can also be made current by simply double-clicking it. To define (save) a coordinate system you must have first translated and rotated the UCS into a desired new position; then use the dialog box to select the "Unnamed" UCS and rename it. The UCS dialog box provides a quick method of restoring previously defined coordinate systems without entering them at the keyboard.

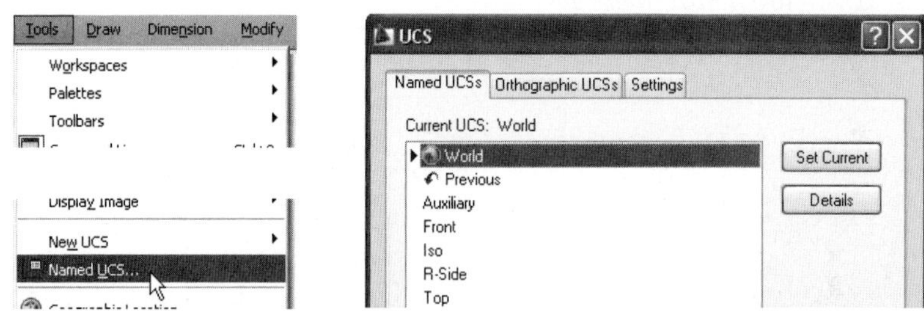

Figure 20.28

Clicking on the Orthographic UCSs tab of the UCS dialog box allows you to rotate the User Coordinate System so that the XY plane is parallel and oriented to one of the six orthographic views: Front, Top, Back, Right Side, Left Side, and Bottom.

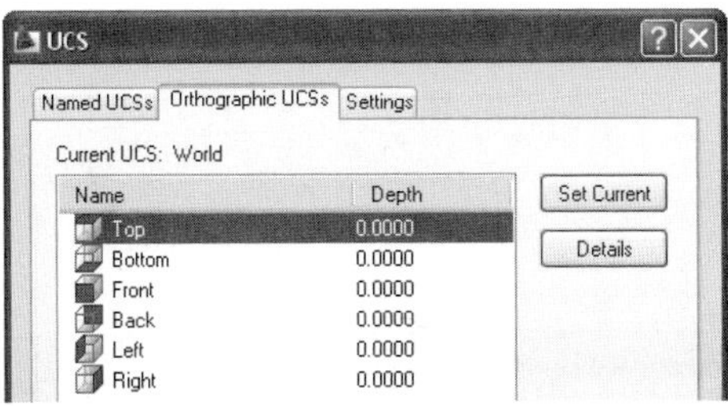

Figure 20.29

Another technique for quickly changing from one named UCS to another is illustrated in the following image. The ViewCube, which will be discussed later in this chapter, has a down arrow next to WCS (World Coordinate System). Clicking on this button displays all named user coordinate systems defined in the model. Clicking on one of these named coordinate systems will switch you to that coordinate system.

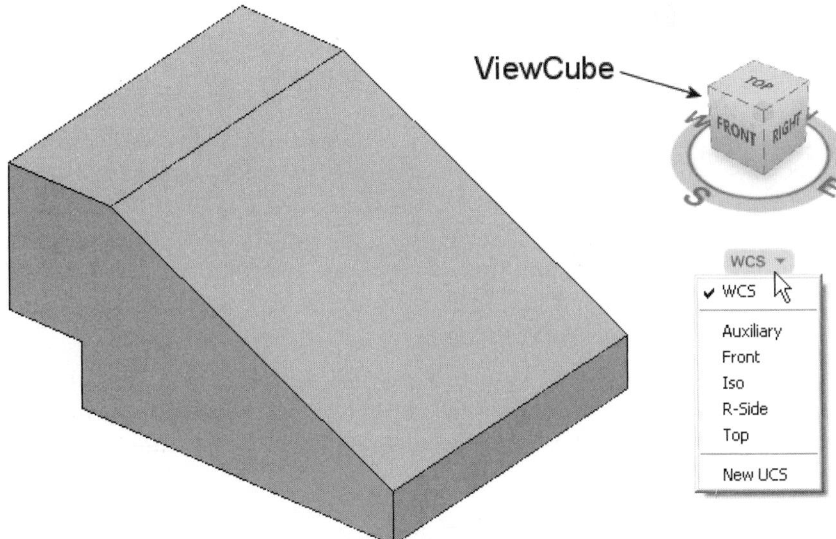

Figure 20.30

CONTROLLING THE DISPLAY OF THE UCS ICON

The display of the UCS icon is controlled by clicking on View in the pull-down menu (or in the Menu Browser) followed by Display and UCS Icon. By default, the icon is turned on. If you want to turn it off, click the On icon. The square box around the icon in the menu will disappear signifying the UCS icon is turned off. Repeat this procedure to turn the icon back on.

Notice also in the following image that this menu allows you to control the visibility of the ViewCube.

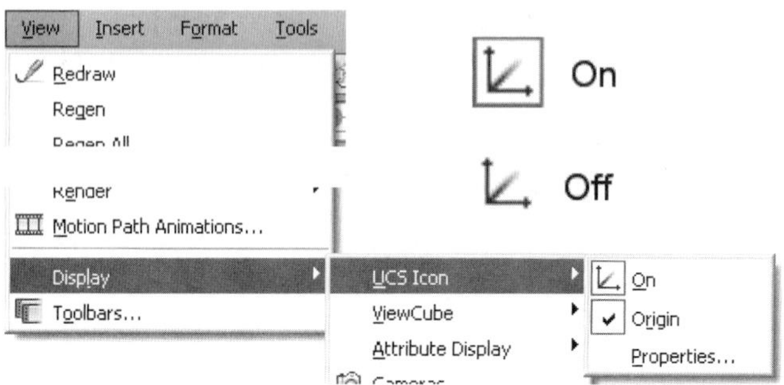

Figure 20.31

THE APPEARANCE OF THE 3D COORDINATE SYSTEM ICON

The 3D Coordinate System icon can take on different forms during the drawing process. Viewing the icon in plan view displays the following image at "A." The box at the intersection of the X- and Y-axes indicates the World Coordinate System (WCS). In the following image at "B," a "+" inside the box indicates that the WCS icon is displayed at 0,0,0. The icon must be totally displayed on the screen; if the icon cannot be totally shown at its origin, the icon locates itself in the lower-left corner of the display screen and no "+" is shown.

Note: The UCSICON command can be used to turn the coordinate system icon on or off, show it in 3D or 2D style, and determine whether it will be displayed at the origin or always in the lower-left corner (the Noorigin option).

The icon displayed at "C" represents the Paper Space environment or layout mode. This environment is strictly two-dimensional and is normally used as a layout tool for arranging views and displaying title block information.

A User Coordinate System (UCS) icon (notice that the box is missing) is shown in the following image at "D." This coordinate system has been translated and/or rotated from the World Coordinate System. In the following image at "E," the "+" indicates that the UCS icon is displayed at 0,0,0.

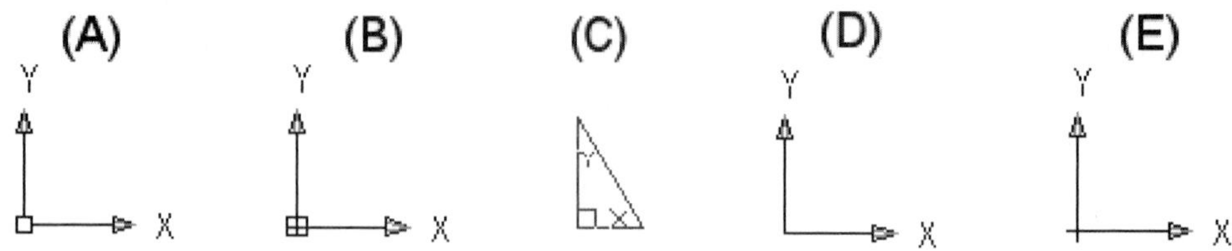

Figure 20.32

THE PLAN COMMAND

When you design in 3D space, it is sometimes helpful after you change the User Coordinate System, to view your model in plan view or parallel to the XY plane. This is the purpose of the PLAN command. For example, illustrated on the left in the following image is a 3D model in which the X- and Y-axes are positioned along the front face of the object. Activating the PLAN command and accepting the default value of <Current> displays the model as shown on the right in the following image. This gives you a 2D view of the solid model, which can be used to better see how certain features such as holes and slots are located along a 2D plane. To switch back to the 3D view, perform a ZOOM-Previous operation.

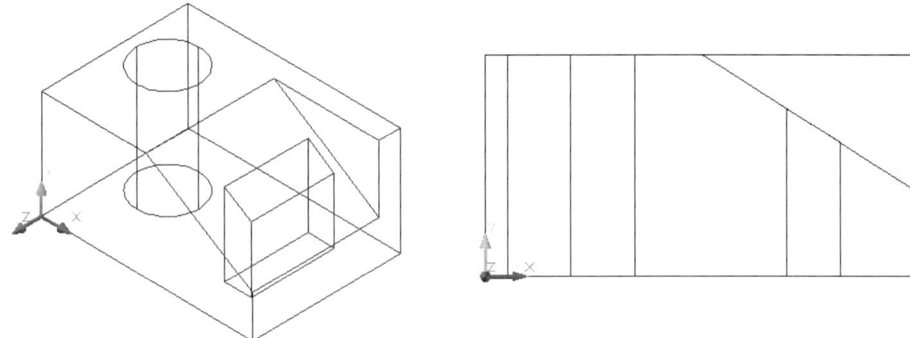

Figure 20.33

Note: A system variable called UCSFOLLOW is available to automate the PLAN command. Here is how it works. When UCSFOLLOW is set to a value of 1 (or turned on), a plan view is automatically generated in the designated viewport whenever you change to a new User Coordinate System. Setting UCSFOLLOW to 0 (zero) turns this mode off.

VIEWING 3D MODELS WITH THE VIEWCUBE

Various methods are available for viewing a model in 3D. One of the more efficient ways is through the 3DFORBIT (3D Free Orbit) command, which can be selected from the View menu, as shown on the left in the following image, or from the Ribbon, as shown in the middle in the following image. Command options as well as other viewing tools are located in the 3D Navigation toolbar, shown on the right in the following image.

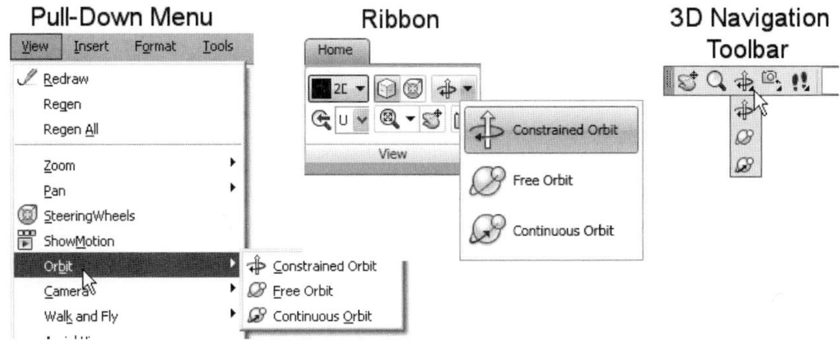

Figure 20.34

The following table gives a brief description of each command found in the 3D Navigation toolbar.

Button	Tool	Function
	3D Pan	Used to pan a 3D model around the display screen
	3D Zoom	Used to perform real-time zooming operations in 3D
	Constrained Orbit	Constrains the 3D orbit along the XY plane or Z axis
	Free Orbit	Used to dynamically rotate a 3D model around the display screen
	Continuous Orbit	Allows you to view an object in a continuous orbit motion
	Swivel	Allows you to view an object with a motion that is similar to looking through a camera viewfinder
	Adjust Distance	Allows you to view an object closer or farther away
	Walk	Changes the view of a 3D display so that you appear to be walking through the model
	Fly	Allows you to fly through a 3D model
	Walk and Fly Settings	Allows you to change settings used for producing an animation of a 3D Walk or 3D Fly

VIEWING WITH FREE ORBIT

Choosing the Free Orbit button (3DFORBIT) from the 3D Navigation toolbar displays your model, similar to the following image. Use the large circle to guide your model through a series of dynamic rotation maneuvers. For instance, moving your cursor outside the large circle at "A" allows you to dynamically rotate (drag) your model only in a circular direction. Moving your cursor to either of the circle quadrant identifiers at "B" and "C" allows you to rotate your model horizontally. Moving your cursor to either of the circle quadrant identifiers at "D" or "E" allows you to rotate your model vertically. Moving your cursor inside the large circle at "F" allows you to dynamically rotate your model to any viewing position.

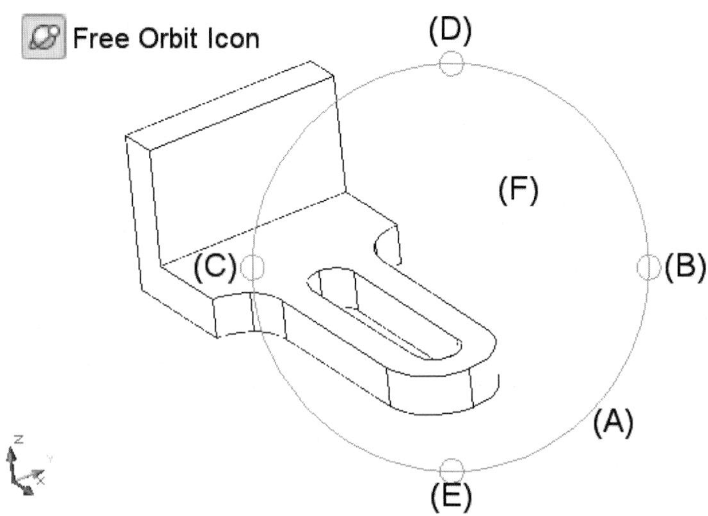

Figure 20.35

VIEWING WITH CONSTRAINED ORBIT

Another way to rotate a 3D model is through the Constrained Orbit button (3DORBIT), which can be selected from the 3D Navigation toolbar. However unlike the FREE ORBIT command, performing a constrained orbit prevents you from rolling the 3D model completely over, adding to confusion in interpreting the model. It is easy to orbit around the geometry; however, as you begin to attempt to orbit above or below the 3D model, the orbiting stops when you reach the top or bottom.

Note: A quick and efficient way of activating the Constrained Orbit tool is to press and hold down the SHIFT key while pressing on the middle button or wheel of the mouse.

VIEWING WITH CONTINUOUS ORBIT

Performing a continuous orbit, (the 3DCORBIT command) rotates your 3D model continuously. After entering the command, press and drag your cursor in the direction you want the continuous orbit to move. Then, when you release the mouse button, the 3D model continues to rotate in that direction.

VIEWING 3D MODELS WITH THE VIEWCUBE

To further assist in rotating and viewing models in 3D, a ViewCube is available whenever the 3D graphics system is enabled (or your visual style is something other than 2D Wireframe). The basic function of this tool is to view your model in either standard or isometric views. By default, the ViewCube is displayed in the upper right corner of the graphics screen. Also by default, the ViewCube takes on a transparent appearance signifying it is currently inactive, as shown on the left in the following image. Moving your cursor over the ViewCube activates this tool and takes on an opaque appearance. Right clicking on the ViewCube will display a menu for controlling how the model can be viewed, as shown on the right in the following image.

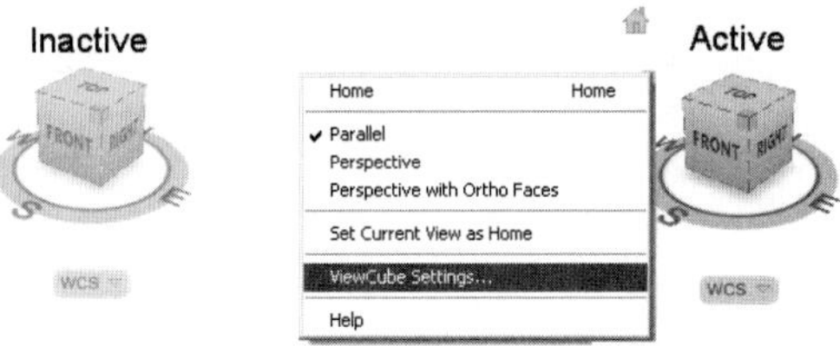

Figure 20.36

Clicking on ViewCube Settings, as shown in the previous image, displays the ViewCube Settings dialog box as shown in the following image. You can control the opacity level of the ViewCube in addition to its size and on-screen position.

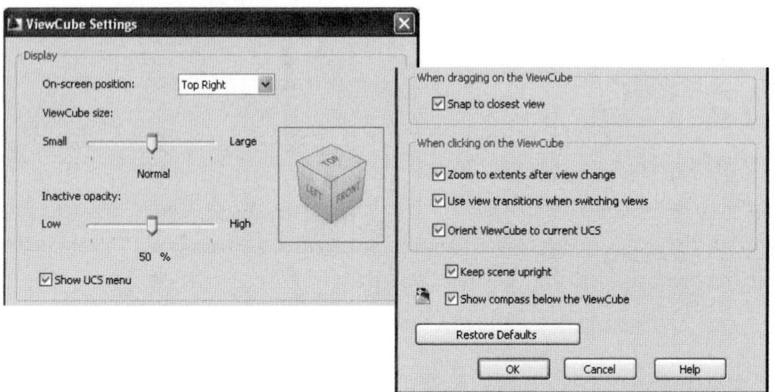

Figure 20.37

Various viewing modes are available through the ViewCube depending on what part of the ViewCube is picked. These modes are all displayed in the following image. For example, clicking on one of the corners of the ViewCube will display a model in an isometric mode. If you want to view a model orthogonally, then one of the six face-viewing modes would work. The edge pivot mode pivots a model along a selected edge. Once in an orthogonal view, such as a Top View, rotation arrows display that allow you to rotate or swivel a model. To return back to the default view of a model, click on the house icon signifying the home position.

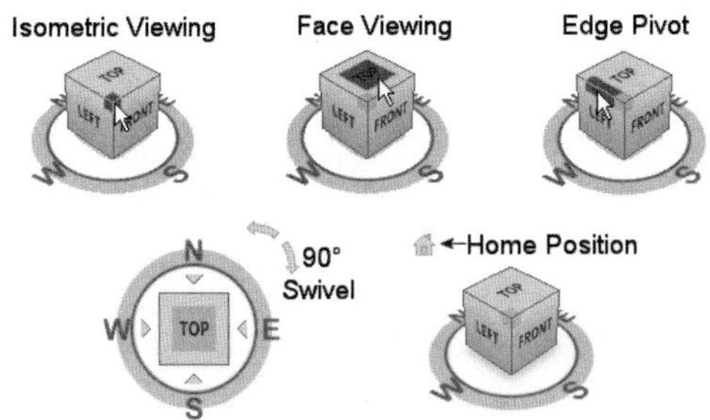

Figure 20.38

An example of how the ViewCube operates is illustrated in the following image. You can either click on one of the box corners to get a standard isometric view, or press and hold down the mouse on a corner. In this case, moving the mouse will rotate the model dynamically as shown in the following image.

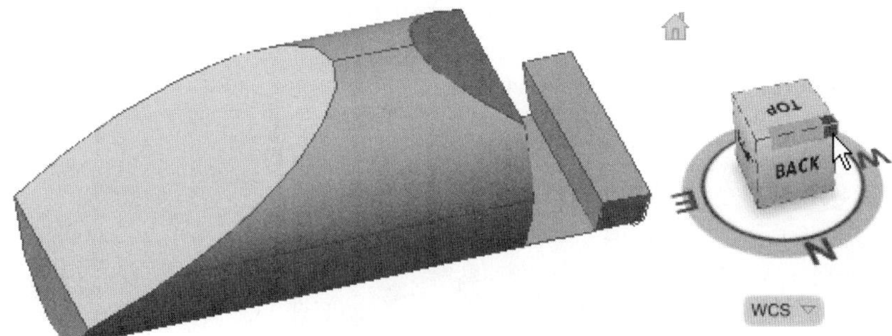

Figure 20.39

An example of displaying a model in its default location is illustrated in the following image. No matter how your model is currently displayed, you can always return the model to its default location by clicking on the Home icon whenever the ViewCube is displayed.

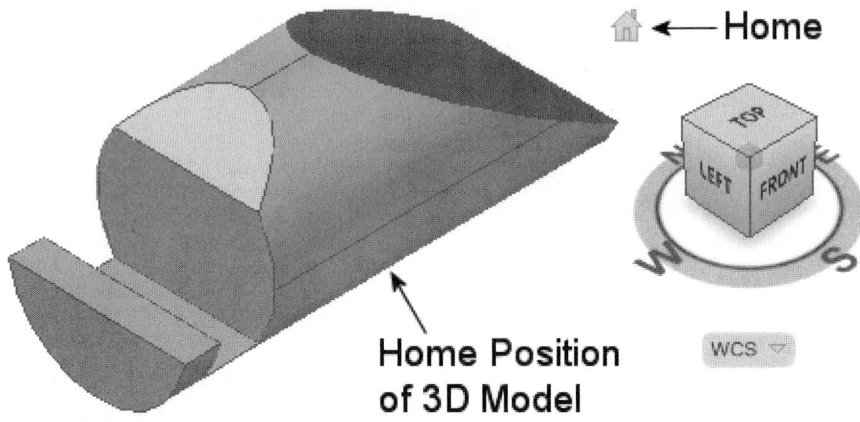

Figure 20.40

Based on how a model is created, it is very easy to view the model orthographically, as shown in the following image. This is accomplished by clicking on any one of the six standard face-viewing modes located in the ViewCube. The following image is displayed from its top viewing mode of the ViewCube.

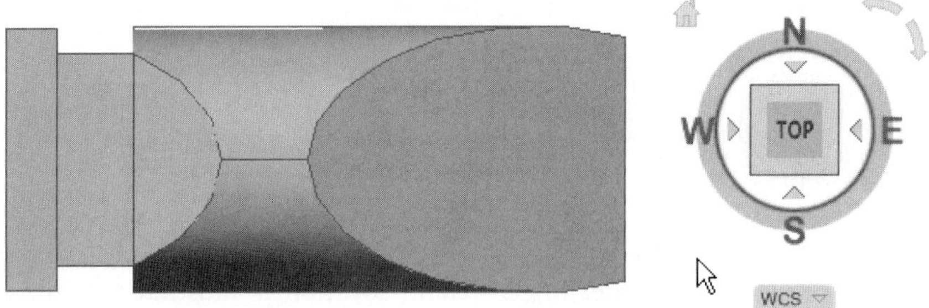

Figure 20.41

Note: The ViewCube can be turned on or off through the View pull-down menu or Menu Browser and the following menu picks; View > Display > ViewCube > On.

USING THE STEERING WHEEL

Another tool called the steering wheel is available to assist in viewing models in 3D. The steering wheel is activated from a special icon located in the status bar at the bottom of the display screen, as shown on the left in the following image. When the steering wheel displays, eight modes are available that allow you to perform the following operations on a 3D model; Zoom, Rewind, Pan, Orbit, Center, Walk, Look, and Up/Down. Zoom mode allows you to zoom in or out. Rewind mode allows you to use a series of images created to zoom to previous views. Pan mode lets you move the view to a new location. Orbit mode allows you to rotate the view. Center mode centers the view based on the position of your cursor on the model. Walk mode allows you to swivel the viewpoint. Look mode moves the view without rotating the viewpoint. Up/Down mode changes the viewpoint of the model vertically. To assist with the selection of these modes, a tooltip is available to illustrate the purpose of each mode, as shown in the following image. You can even click on the arrow in the lower-right corner of the steering wheel to display the menu, as shown on the right in the following image. Use this menu to change settings associated with the steering wheel or launch the Steering Wheel Settings dialog box in order to make further changes or restore the steering wheel back to its original settings.

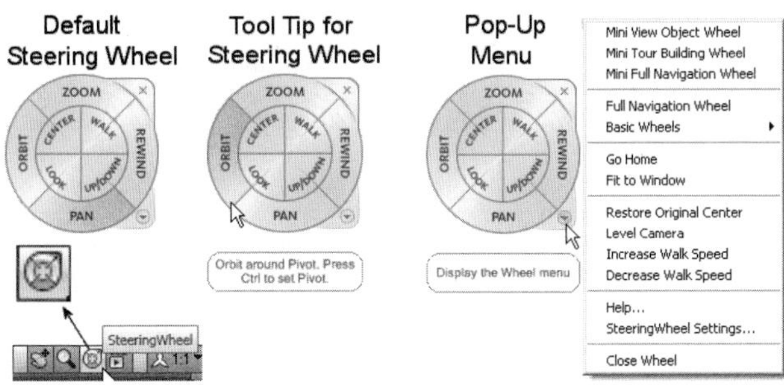

Figure 20.42

THE VIEW TOOLBAR

A View toolbar is available, as shown in the following image. The View toolbar has the extra advantage of displaying icons that guide you in picking the desired viewpoint. You can also access most 3D viewing modes through the View pull-down menu by clicking on 3D Views, also shown in the following image.

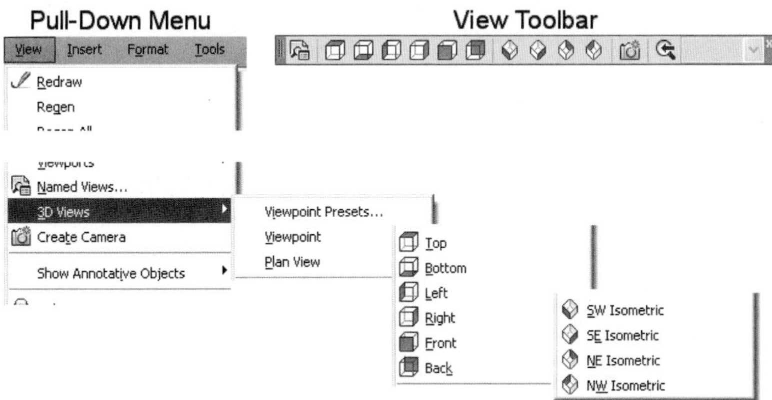

Figure 20.43

The following table gives a brief description of each View mode.

Button	Tool	Function
	Named Views	Launches the View Manager dialog box used for creating named views
	Top View	Orientates a 3D model to display the top view
	Bottom View	Orientates a 3D model to display the bottom view
	Left View	Orientates a 3D model to display the left view
	Right View	Orientates a 3D model to display the right view
	Front View	Orientates a 3D model to display the front view
	Back View	Orientates a 3D model to display the back view
	SW Isometric	Orientates a 3D model to display the southwest isometric view
	SE Isometric	Orientates a 3D model to display the southeast isometric view
	NE Isometric	Orientates a 3D model to display the northeast isometric view
	NW Isometric	Orientates a 3D model to display the northwest isometric view
	Create Camera	Used to set up a point from which to view a 3D model and the point that you are viewing
	View Previous	Displays the previous view

SHADING SOLID MODELS

Various shading modes are available to help you better visualize the solid model you are constructing. Access the five shading modes by choosing Visual Styles from the Menu Browser or View pull-down menu, as shown in the following image, or by clicking a button on the Visual Styles toolbar. An area is also available in the Ribbon for working with visual styles, as shown on the right in the following image.

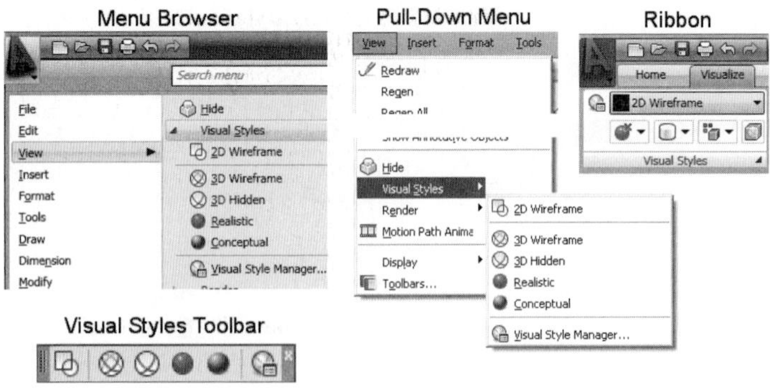

Figure 20.44

The following table gives a brief description of each visual style mode.

Options of the SHADEMODE Command

Button	Visual Style	Description
	2D Wireframe	Displays the 3D model as a series of lines and arcs that represent boundaries
	3D Wireframe	Displays the 3D model that is similar in appearance to the 2D option; the UCS icon appears as a color shaded image
	3D Hidden	Displays the 3D model with hidden edges removed
	Realistic	Shades the objects and smooths the edges between polygon faces; if materials are attached to the model, they will display when this visual style is chosen
	Conceptual	This mode also shades the objects and smooths the edges between polygon faces; however, the shading transitions between cool and warm colors
	Manage Visual Styles	Launches the Visual Styles Manager palette, used for creating new visual styles and applying existing visual styles to a 3D model

The following image illustrates all five visual styles and how they affect the appearance of a 3D solid model.

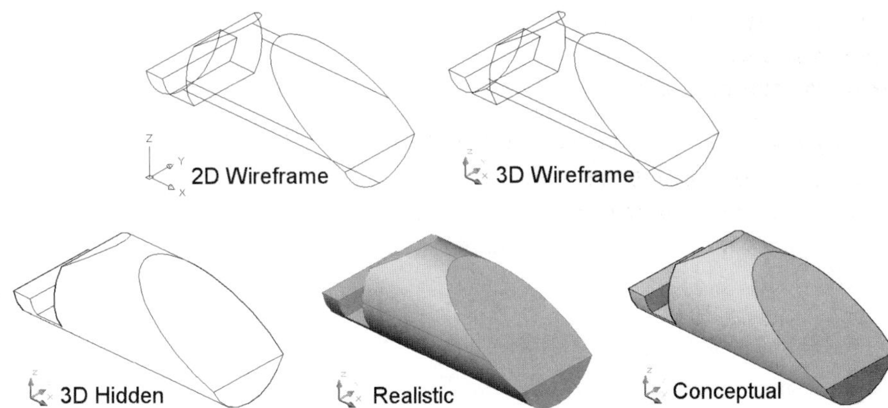

Figure 20.45

Try It!: Open the drawing file 20_Visual Styles. Experiment with the series of shading modes. Your results should be similar to the images provided in the previous illustrations.

Note: When you perform such operations as Free, Constrained, and Continuous Orbit, the current visual style mode remains persistent. This means that if you are in a Realistic visual style and you rotate your model using one of the previously mentioned operations, the model remains shaded throughout the rotation operation.

CREATING A VISUAL STYLE

Custom visual styles can be created to better define how a 3D model will appear when shaded. Click on the Manage Visual Styles button located in the Visualize tab of the Ribbon, as shown on the left in the following image, to launch the Visual Styles Manager palette, as shown in the middle of the following illustration. The five visual styles that are located in the Visual Styles toolbar can also be found in this palette. Use the palette for creating new visual styles and applying them to the current viewport. A visual style area is also located in the Ribbon under the Edge Effects panel, as shown on the right in the following image. This area of the Ribbon consists of various buttons that can be turned on or off and that affect the visual style. Slider bars are also available to experiment with changing the values of certain settings that will automatically be reflected in the 3D solid model.

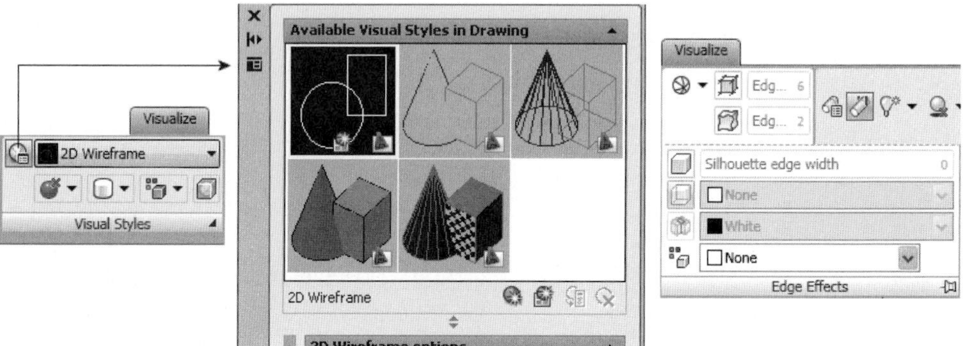

Figure 20.46

SOLID MODELING COMMANDS

The following image illustrates the Menu Browser, pull-down menu, Ribbon, and the Modeling toolbar for accessing solid modeling commands. Choosing Modeling from the Draw pull-down menu displays six groupings of commands. The first grouping displays POLYSOLID, BOX, SPHERE, CYLINDER, CONE, WEDGE, TORUS, and PYRAMID, which are considered the building blocks of the solid model and are used to construct basic "primitives." The second grouping displays the PLANESURF command, designed to create a planar surface between two closed shapes or through two diagonal points. The third grouping displays the "sweep" commands—EXTRUDE, REVOLVE, SWEEP, and LOFT—which provide an additional way to construct solid models. Polyline outlines or circles can be extruded (swept) to a thickness that you designate. You can also revolve (sweep) other polyline outlines about an axis. The fourth grouping of solid modeling commands enables you to create a section plane in order to look at the interior of a solid model. The fifth grouping displays a set of commands that create or manipulate mesh surfaces. Commands include SOLID, 3DFACE, 3DMESH, EDGE, REVSURF, TABSURF, RULESURF, and EDGESURF. The last grouping, Setup, displays three commands designed to extract orthographic views from a solid model. The three commands are SOLDRAW, SOLVIEW, and SOLPROF. Many of these same solid modeling commands can be accessed from the Modeling toolbar and Ribbon, as shown in the following image.

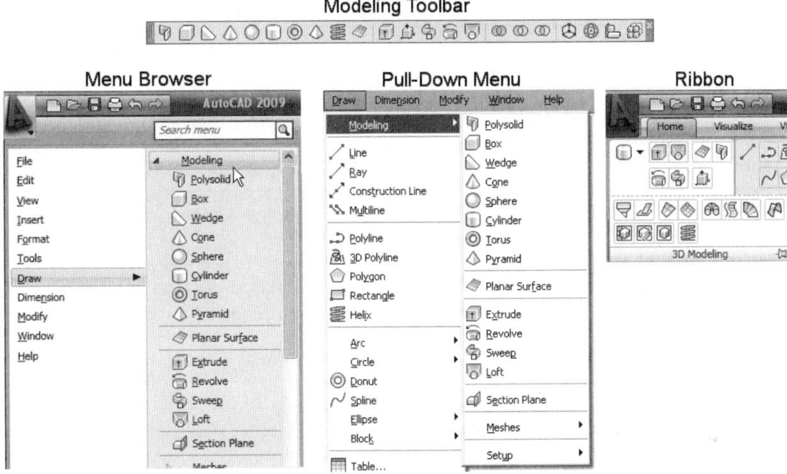

Figure 20.47

The following table gives a brief description of each command located in the Modeling toolbar.

Button	Tool	Shortcut	Function
	Polysolid	POLYSOLID	Creates a solid shape based on a direction, width and height of the solid
	Box	BOX	Creates a solid box
	Wedge	WE	Creates a solid wedge
	Cone	CONE	Creates a solid cone
	Sphere	SPHERE	Creates a solid sphere
	Cylinder	CYL	Creates a solid cylinder
	Torus	TOR	Creates a solid torus
	Pyramid	PYR	Creates a solid pyramid
	Helix	HELIX	Creates a 2D or 3D helix
	Planar Surface	PLANESURF	Creates a planar surface
	Extrude	EXT	Creates a solid by extruding a 2D profile

Button	Tool	Shortcut	Function
	Presspull	PRESSPULL	Presses or pulls closed areas resulting in a solid shape or a void in a solid
	Sweep	SWEEP	Creates a solid based on a profile and a path
	Revolve	REV	Creates a solid by revolving a 2D profile about an axis of rotation
	Loft	LOFT	Creates a lofted solid based on a series of cross-section shapes
	Union	UNI	Joins two or more solids together
	Subtraction	SU	Removes one or more solids from a source solid shape
	Intersect	INT	Extracts the common volume shared by two or more solid shapes
	3D Move	3DMOVE	Moves objects a specified distance in a specified direction based on the position of a move grip tool
	3D Rotate	3DROTATE	Revolves objects around a base point based on the rotate grip tool
	3D Align	3DALIGN	Aligns objects with other objects in 2D and 3D
	3D Array	3DARRAY	Creates patterns of objects along the X, Y, and Z axes

CREATING SOLID PRIMITIVES

Seven different commands are available for creating 3D solids in basic geometric shapes; namely boxes, wedges, cylinders, cones, spheres, tori (donut shaped objects), and pyramids as shown in the following image. These solid shapes are often called primitives because they are used as building blocks for more complex solid models. They are seldom useful by themselves, but these primitives can be combined and modified into a wide variety of geometric shapes. The following command sequences allow you to practice creating one example of each primitive.

Try It!: Open the drawing file 20_Primitive_Examples which represents a drawing void of any objects. Follow the next series of seven prompts to create each primitive illustrated in the following image.

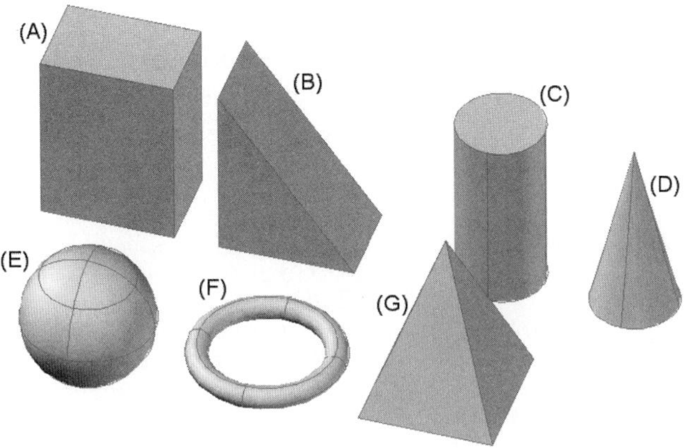

Figure 20.48

CREATING BOX PRIMITIVES

Box primitives consist of brick-shaped solid objects. They have six rectangular sides, which are either perpendicular or parallel to one another. Boxes are probably the most often used primitive, as many of the objects that are modeled are made up of rectangles and squares. There are many ways of creating boxes; the following prompt sequence illustrates the length option for creating a box (see the previous image at "A").

```
Command: BOX
Specify first corner or [Center]: 4.00,9.00
Specify other corner or [Cube/Length]: L (For Length)
Specify length <5.0000>: 3.00 (point your cursor in the positive
    X direction)
Specify width <2.0000>: 2.00 (point your cursor in the positive
    Y direction)
Specify height or [2Point] <1.0000>: 4.00 (point your cursor in
    the positive Z direction)
```

CREATING WEDGE PRIMITIVES

Wedges are like boxes that have been sliced diagonally edge to edge. They have a total of five sides, three of which are rectangular, two that are triangular. The top rectangular side slopes down in the X direction. The two sides opposite this sloping side are perpendicular to each other. The bottom rectangular surface is on the XY plane. The following prompt sequence illustrates the length option for creating a wedge (see the previous image at "B").

```
Command: WE (For WEDGE)
Specify first corner or [Center]: 8.00,9.00
Specify other corner or [Cube/Length]: L (For Length)
Specify length <5.0000>: 3.00 (point your cursor in the positive
    X direction)
Specify width <2.0000>: 2.00 (point your cursor in the positive
    Y direction)
Specify height or [2Point] <3.0000>: 4.00 (point your cursor in
    the positive Z direction)
```

CREATING CYLINDER PRIMITIVES

Cylinders are probably the second most often used primitive. Cylinders can be created as either circular or elliptical. The following prompt sequence illustrates the diameter option for creating a cylinder (see the previous image at "C").

```
Command: CYL (For CYLINDER)
Specify center point of base or [3P/2P/Ttr/Elliptical]:
    14.00,10.00
Specify base radius or [Diameter] <1.5000>: D (For Diameter)
Specify diameter <3.0000>: 2.00
Specify height or [2Point/Axis endpoint] <4.0000>: 4.00 (point
    your cursor in the positive Z direction)
```

CREATING CONE PRIMITIVES

Cone primitives are closely related to cylinders. They have the same round or elliptical cross section; but they taper either to a point or a specified height with different radius forming a truncated cone. The following prompt sequence illustrates the radius option for creating a cone (see the previous image at "D").

```
Command: CONE
Specify center point of base or [3P/2P/Ttr/Elliptical]:
    17.00,10.00
Specify base radius or [Diameter] <1.0000>: 1.00
Specify height or [2Point/Axis endpoint/Top radius] <4.0000>:
    4.00 (point your cursor in the positive Z direction)
```

CREATING SPHERE PRIMITIVES

Creating spheres is the most straight-forward process in creating primitives. Specify the sphere's center point and then specify either the radius or diameter of the sphere. The following prompt sequence illustrates the diameter option for creating a sphere (see the previous image at "E").

```
Command: SPHERE
Specify center point or [3P/2P/Ttr]: 6.00,6.00
Specify radius or [Diameter] <2.0000>: D (For Diameter)
Specify diameter <4.0000>: 3.00
```

CREATING A TORUS PRIMITIVE

Although the torus is not often needed, it is the most interesting and flexible of the seven primitive solids. The basic shape of a torus is that of a donut, but it can also take on a football shape. To be made properly, this primitive requires a torus radius value and a tube radius value, as shown in the following image. The following prompt sequence illustrates creating a torus (see the previous image at "F").

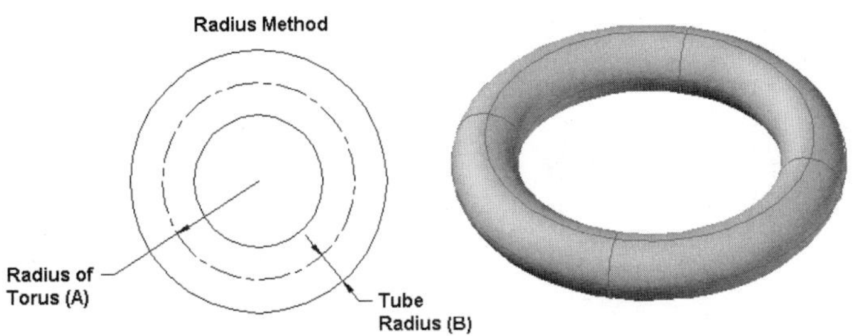

Figure 20.49

```
Command: TOR (For TORUS)
Specify center point or [3P/2P/Ttr]: 10.00,6.00
Specify radius or [Diameter] <1.5000>: 1.50
Specify tube radius or [2Point/Diameter]: .25
```

CREATING PYRAMID PRIMITIVES

Solid pyramids actually have similar prompts used for creating cylinders with the exception that the base of the pyramid consists of edges and are noncircular in shape. The following prompt sequence illustrates the creation of a pyramid (see Figure 20-48 at "G").

```
Command: PYR (For PYRAMID)
4 sides Circumscribed
Specify center point of base or [Edge/Sides]: 14.00,6.00
Specify base radius or [Inscribed] <1.5000>: 1.50 (point your
    cursor in the positive X direction)
Specify height or [2Point/Axis endpoint/Top radius] <4.0000>:
    4.00 (point your cursor in the positive Z direction)
```

USING BOOLEAN OPERATIONS ON SOLID PRIMITIVES

To combine one or more primitives to form a composite solid, a Boolean operation is performed. Boolean operations must act on at least a pair of primitives, regions, or solids. These operations in the form of commands are located in the Menu Browser or Modify pull-down menu under Solids Editing. They can also be selected from the Solids Editing panel of the Ribbon, the Modeling toolbar, or Solid Editing toolbar. Boolean operations allow you to add two or more objects together, subtract a single object or group of objects from another, or find the overlapping volume—in other words, form the solid common to both primitives. Displayed in the following image are the UNION, SUBTRACT, and INTERSECT commands that you use to perform these Boolean operations.

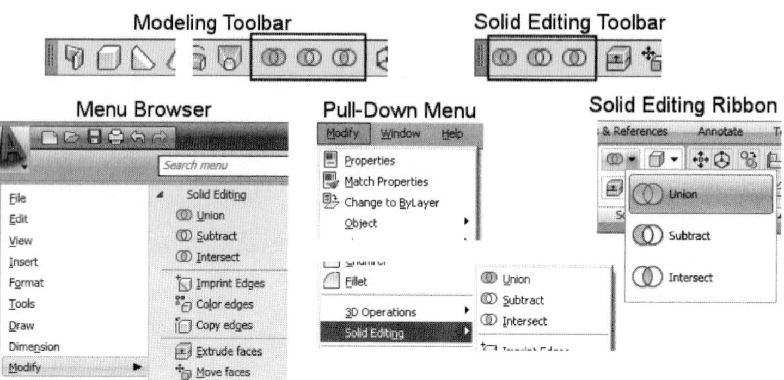

Figure 20.50

In the following image, a cylinder has been constructed along with a box. Depending on which Boolean operation you use, the results could be quite different. In the following image at "Union," both the box and cylinder are considered one solid object. This is the purpose of the UNION command: to join or unite two solid primitives into one. The image at "Subtract" goes on to show the result of removing or subtracting the cylinder from the box—a hole is formed inside the box as a result of using the SUBTRACT command. The image at "Intersect" illustrates the intersection of the two solid primitives or the area that both solids have in common. This solid is obtained through the INTERSECT command. All Boolean operation commands can work on numerous solid primitives; that is, if you want to subtract numerous cylinders from a box, you can subtract all of the cylinders at one time.

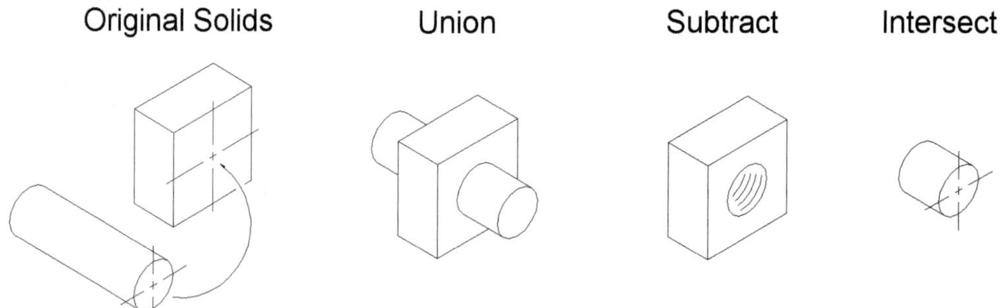

Figure 20.51

CREATING SOLID UNIONS

This Boolean operation, the UNION command, joins two or more selected solid objects together into a single solid object.

Try It!: Open the drawing file 20_Union. Use the following command sequence and image for performing this task.

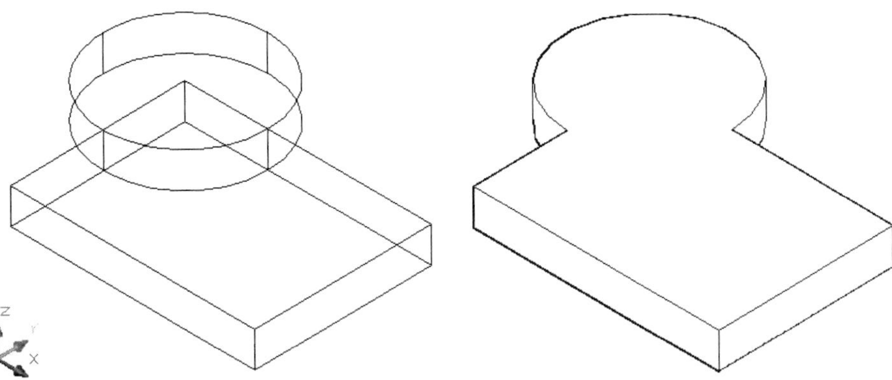

Figure 20.52

```
Command: UNI (For UNION)
Select objects: (Pick the box and cylinder)
Select objects: (Press ENTER to perform the union operation)
```

SUBTRACTING SOLIDS

Use the SUBTRACT command to subtract one or more solid objects from a source object, as shown in the following image. Choose this command in one of the following ways:

- From the Solids Editing or Modeling toolbars
- From the Ribbon of the 3D Modeling workspace
- From the pull-down menu or Menu Browser (Modify > Solids Editing > Subtract)
- From the keyboard (SU or SUBTRACT)

Try It!: Open the drawing file 20_Subtract. Use the following command sequence and image for performing this task.

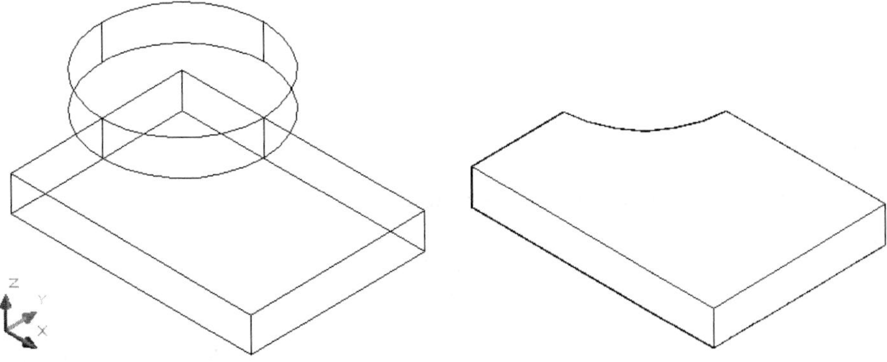

Figure 20.53

```
Command: SU (For SUBTRACT)
Select solids and regions to subtract from...
```

```
Select objects: (Pick the box)
Select objects: (Press ENTER to continue with this command)
Select solids and regions to subtract...
Select objects: (Pick the cylinder)
Select objects: (Press ENTER to perform the subtraction operation)
```

CREATING INTERSECTIONS

Use the INTERSECT command to find the solid common to a group of selected solid objects, as shown in the following image. Choose this command in one of the following ways:

- From the Solids Editing or Modeling toolbars
- From the Ribbon of the 3D Modeling workspace
- From the pull-down menu or Menu Browser (Modify > Solids Editing > Intersect)
- From the keyboard (IN or INTERSECT)

Try It!: Open the drawing file 20_Intersection. Use the following command sequence and image for performing this task.

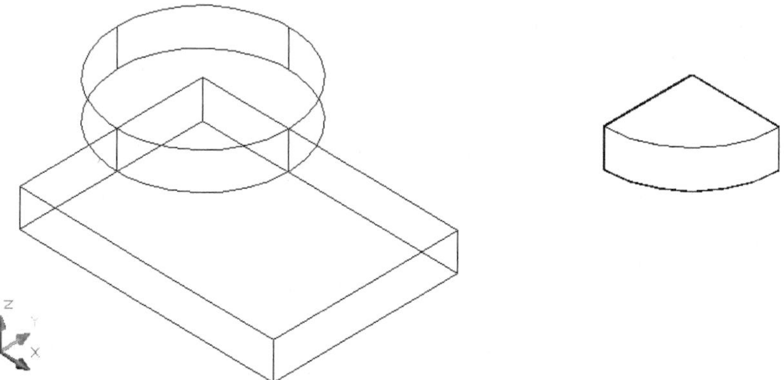

Figure 20.54

```
Command: IN (For INTERSECT)
Select objects: (Pick the box and cylinder)
Select objects: (Press ENTER to perform the intersection
    operation)
```

3D APPLICATIONS OF UNIONING SOLIDS

The following image shows an object consisting of one horizontal solid box, two vertical solid boxes, and two extruded semicircular shapes. All primitives have been positioned with the MOVE command. To join all solid primitives into one solid object, use the UNION command. The order of selection of these solids for this command is not important.

Try It!: Open the drawing file 20_3D App Union. Use the following prompts and image for performing these tasks.

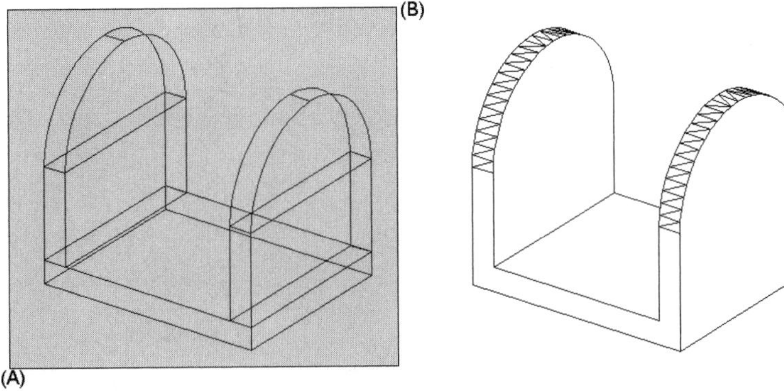

Figure 20.55

```
Command: UNI (For UNION)
Select objects: (Pick your screen at "A")
Select objects: (Pick your screen at "B")
Select objects: (Press ENTER to perform the union operation)
```

3D APPLICATIONS OF MOVING SOLIDS

Using the same problem from the previous example, let us now add a hole in the center of the base. The cylinder will be created through the CYLINDER command. It will then be moved to the exact center of the base. You can use the MOVE command along with the OSNAP-Tracking mode to accomplish this. Tracking mode automatically activates the ORTHO mode when it is in use. See the following image.

```
Command: CYL (For CYLINDER)
Specify center point of base or [3P/2P/Ttr/Elliptical]: 3.00,3.00
Specify base radius or [Diameter]: 0.75
Specify height or [2Point/Axis endpoint]: 0.25
```

```
Command: M (For MOVE)
Select objects: (Select the cylinder at "A")
Select objects: (Press ENTER to continue with this command)
Specify base point or [Displacement] <Displacement>: Cen
of (Select the bottom of the cylinder at "A")
Specify second point or <use first point as displacement>: M2P
    (To activate Midpoint between two points)
First tracking point: Mid
of (Select the midpoint of the bottom of the base at "B")
Next point (Press ENTER to end tracking): Mid
of (Select the midpoint of the bottom of the base at "C")
Next point (Press ENTER to end tracking): (Press ENTER to end
    tracking and perform the move operation)
```

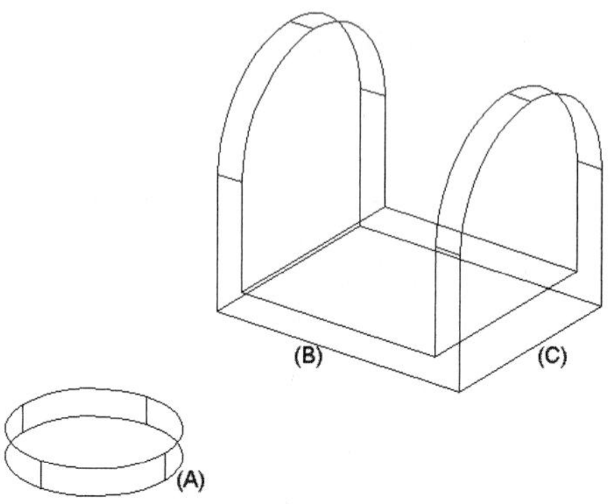

Figure 20.56

3D APPLICATIONS OF SUBTRACTING SOLIDS

Now that the solid cylinder is in position, use the SUBTRACT command to remove the cylinder from the base of the main solid and create a hole in the base, as shown in the following image.

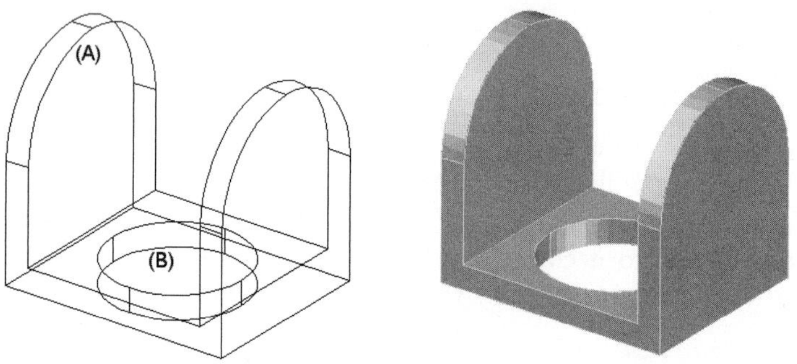

Figure 20.57

```
Command: SU (For SUBTRACT)
Select solids and regions to subtract from...
Select objects: (Select the main solid as source at "A")
Select objects: (Press ENTER to continue with this command)
Select solids and regions to subtract...
Select objects: (Select the cylinder at "B")
Select objects: (Press ENTER to perform the subtraction operation)
Command: HIDE (For hidden line removal view)
Command: RE (For REGEN to return to wireframe view)
```

CREATING INTERSECTIONS

Intersections remain one of the most misunderstood concepts to grasp in solid modeling. However, once you become comfortable with this concept, it can be one of the more powerful solid modeling tools available. Creating an intersection involves creating a solid model from the common volumes of two or more overlapping solids. The objects labeled

"A" and "B" in the following image represent existing solid models. These models are then moved to overlap each other as shown at "C" in the following image. After creating the intersection, the nonoverlapping portions of the model are removed leaving the solid model as shown at "D" in the following image.

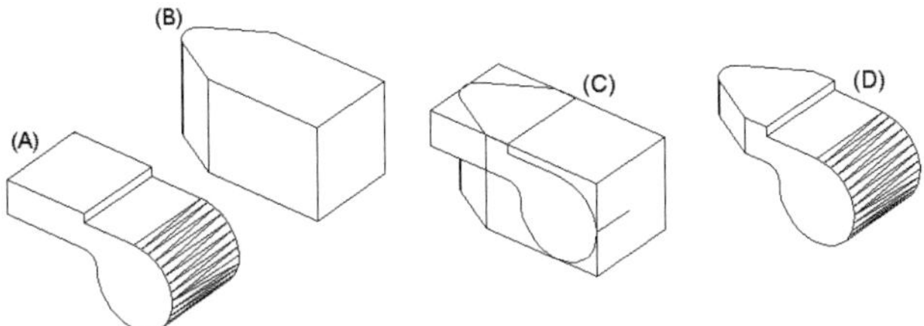

Figure 20.58

Try It!: Open the drawing file 20_Int1. In the following image, the object at "C" represents the finished model. Look at the sequence beginning at "A" to see how to prepare the solid primitives for an Intersection operation. Two separate 3D Solid objects are created at "A." One object represents a block that has been filleted along with the placement of a hole drilled through. The other object represents the U-shaped extrusion. With both objects modeled, they are moved on top of each other at "B." The OSNAP-Midpoint was used to accomplish this. Finally, the INTERSECT command is used to find the common volume shared by both objects; namely the illustration at "C."

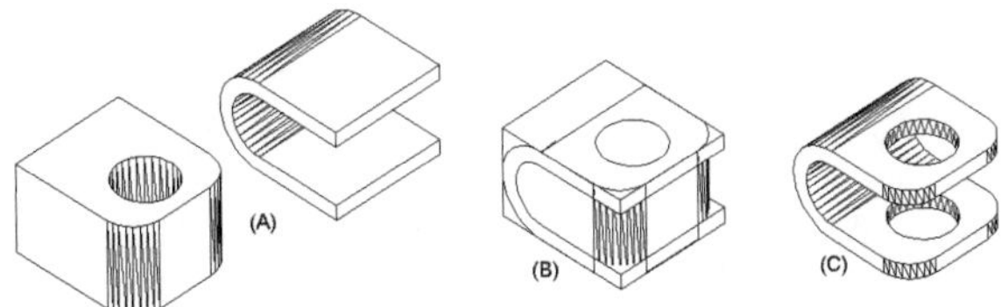

Figure 20.59

Try It!: Open the drawing file 20_Int2. The object in the following image is another example of how the INTERSECT command may be applied to a solid model. For the results at "B" to be obtained from a cylinder that has numerous cuts, the cylinder is first created as a separate model. Then the cuts are made in another model at "A." Again, both models are moved together (use the Quadrant and Midpoint OSNAP modes), and then the INTERSECT command is used to achieve the results at "B." Before undertaking any solid model, first analyze how the model is to be constructed. Using intersections can create dramatic results, which would normally require numerous union and subtraction operations.

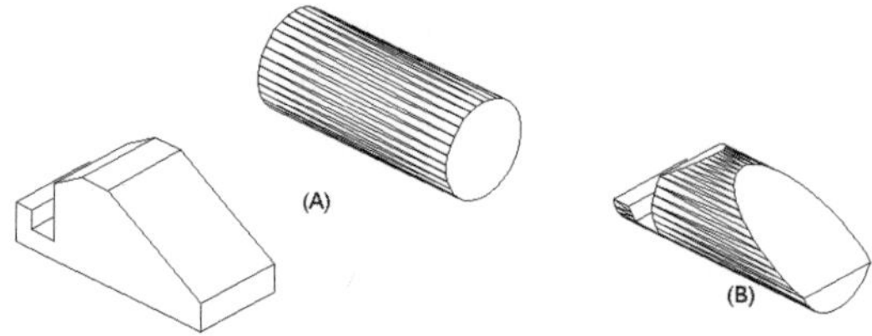

Figure 20.60

CREATING SOLID EXTRUSIONS

The EXTRUDE command creates a solid by extrusion. Choose this command in one of the following ways:

- From the Modeling Toolbar
- From the Ribbon of the 3D Modeling workspace
- From the pull-down menu or Menu Browser (Draw > Modeling > Extrude)
- From the keyboard (EXT or EXTRUDE)

Only regions or closed, single-entity objects such as circles and closed polylines can be extruded into a solid. Other options of the EXTRUDE command include the following: Extrude by Direction, in which you specify two points that determine the length and direction of the extrusion; Extrude by Path, in which the extrusion is created based on a path that consists of a predefined object; and Extrude by Taper Angle, in which a tapered extrusion is created based on an angle value between –90 and +90.

Use the following prompts to construct a solid extrusion of the closed polyline object in the following image. For the height of the extrusion, you can enter a numeric value or you can specify the distance by picking two points on the display screen.

Try It!: Open the drawing file 20_Extrude1. Use the following command sequence and image for performing this task.

```
Command: EXT (For EXTRUDE)
Current wire frame density: ISOLINES=4
Select objects to extrude: (Select the polyline object at "A")
Select objects to extrude: (Press ENTER to continue with this
    command)
Specify height of extrusion or [Direction/Path/Taper angle]: 1.00
```

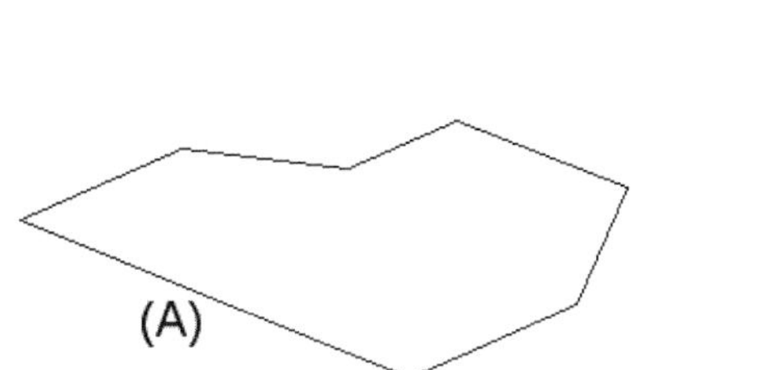

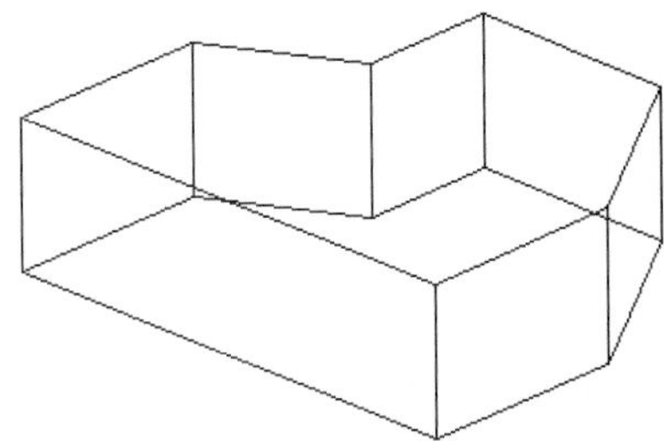

Figure 20.61

You can create an optional taper along with the extrusion by utilizing the Taper angle option provided.

Try It!: Open the drawing file 20_Extrude2. Use the following command sequence and image for performing this task.

```
Command: EXT (For EXTRUDE)
Current wire frame density: ISOLINES=4
Select objects to extrude: (Select the polyline object at "B" in
    the following image)
Select objects to extrude: (Press ENTER to continue with this
    command)
Specify height of extrusion or [Direction/Path/Taper angle]: T
    (For Taper angle)
Specify angle of taper for extrusion <0>: 15
Specify height of extrusion or [Direction/Path/Taper angle]: 1.00
```

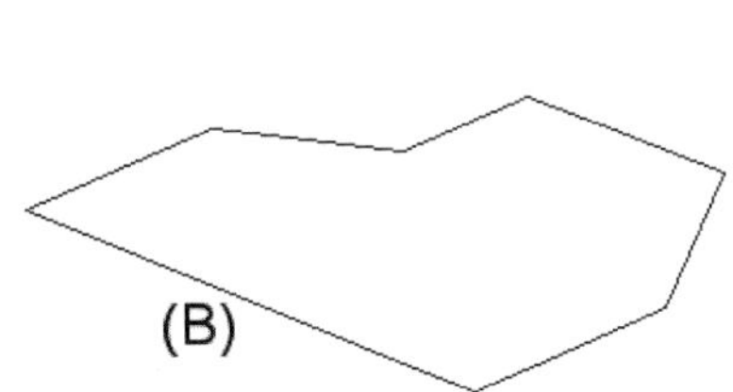

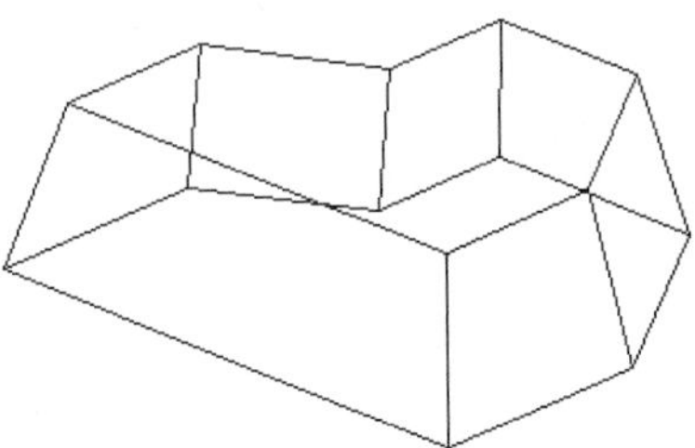

Figure 20.62

You can also create a solid extrusion by selecting a path to be followed by the object being extruded. Typical paths include regular and elliptical arcs, 2D and 3D polylines, or splines. The extruded pipe in the following image was created using the following steps: First the polyline path was created. Then, a new User Coordinate System was established through the UCS command along with the Z-axis option; the new UCS was positioned at the end of the polyline with the Z-axis extending along the polyline. A circle was constructed with its center point at the end of the polyline. Finally, the circle was extruded along the polyline path.

Try It!: Open the drawing file 20_Extrude Pipe. Use the following command sequence and image for performing this task.

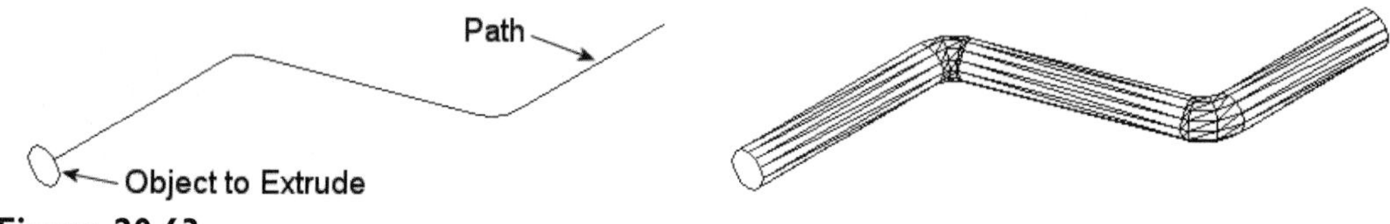

Figure 20.63

```
Command: EXT (For EXTRUDE)
Current wire frame density: ISOLINES=4
Select objects to extrude: (Select the small circle as the
    object to extrude)
Select objects to extrude: (Press ENTER to continue)
Specify height of extrusion or [Direction/Path/Taper angle]: P
    (For Path)
Select extrusion path or [Taper angle]: (Select the polyline
    object representing the path)
```

EXTRUDING AN EXISTING FACE

Once a solid object is created, existing faces of the model can be used as profiles to further extrude shapes. Illustrated on the left in the following image is a wedge-shaped 3D model. After entering the EXTRUDE command, press and hold down the CTRL key and select the inclined face. You can either enter a value or drag your cursor to define the height of the extrusion, as shown in the middle in the following image. The results are shown on the right in the following image. The new extruded shape created from the inclined face is considered a separate solid object. If both shapes need to be considered one, use the UNION command and select both extruded boxes to join them as one solid.

Try It!: Open the drawing file 20_Extrude Face. Use the following command sequence and image for performing an extrusion on an existing face.

```
Command: EXT (For EXTRUDE)
Current wire frame density: ISOLINES=4
Select objects to extrude: (Press and hold down the CTRL key and
    select the inclined face as shown on the left in the
    following image)
Select objects to extrude: (Press ENTER to continue)
Specify height of extrusion or [Direction/Path/Taper angle]: 5.00
```

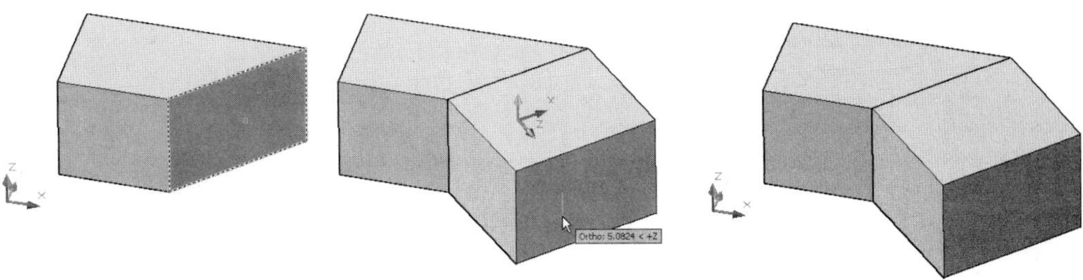

Figure 20.64

CREATING REVOLVED SOLIDS

The REVOLVE command creates a solid by revolving an object about an axis of revolution. Choose this command in one of the following ways:

- From the Modeling Toolbar
- From the Ribbon of the 3D Modeling workspace
- From the pull-down menu or Menu Browser (Draw > Modeling > Revolve)
- From the keyboard (REV or REVOLVE)

Only regions or closed, single-entity objects such as polylines, polygons, circles, ellipses, and 3D polylines can be revolved to create a solid. If a group of objects is not in the form of a single entity, group them together using the PEDIT command or create a closed polyline/region with the BOUNDARY command. The following image represents a revolved 3D Solid object.

Try It!: Open the drawing file 20_Revolve1. Many practical applications require creating a composite solid from primitives and extrusions. The CYLINDER command was used to construct the cylindrical primitive and the REVOLVE command was used to create the revolved solid. Use the following command sequence on the objects as illustrated on the left in the following image for creating a revolved solid.

```
Command: REV (For REVOLVE)
Current wire frame density: ISOLINES=4
Select objects to revolve: (Select profile "A" as the object to
     revolve)
Select objects to revolve: (Press ENTER to continue with this
     command)
Specify axis start point or define axis by [Object/X/Y/Z]
     <Object>: O (For Object)
Select an object: (Select line "B")
Specify angle of revolution or [STart angle] <360>: (Press ENTER
     to accept the default and perform the revolving operation)
```

Use the Center option of OSNAP along with the MOVE command to position the revolved solid inside the cylinder as shown on the right in the following image.

```
Command: M (For MOVE)
Select objects: (Select the revolved solid in the following
    image)
Select objects: (Press ENTER to continue with this command)
Specify base point or [Displacement] <Displacement>: Cen
of (Select the center of the revolved solid at "C")
Specify second point or <use first point as displacement>: Cen
of (Select the center of the cylinder at "D")
```

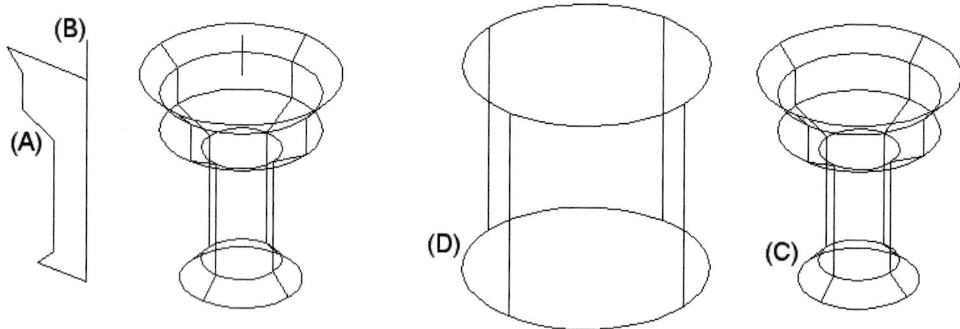

Figure 20.65

Once the revolved solid is positioned inside the cylinder, use the SUBTRACT command to subtract the revolved solid from the cylinder, as shown in the following image. Use the HIDE command to perform a hidden line removal at "B" to check that the solid is correct (this would be difficult to interpret in wireframe mode).

```
Command: SU (For SUBTRACT)
Select solids and regions to subtract from...
Select objects: (Select the cylinder as source)
Select objects: (Press ENTER to continue with this command)
Select solids and regions to subtract...
Select objects: (Select the revolved solid)
Select objects: (Press ENTER to perform the subtraction operation)
Command: HI (For HIDE)
```

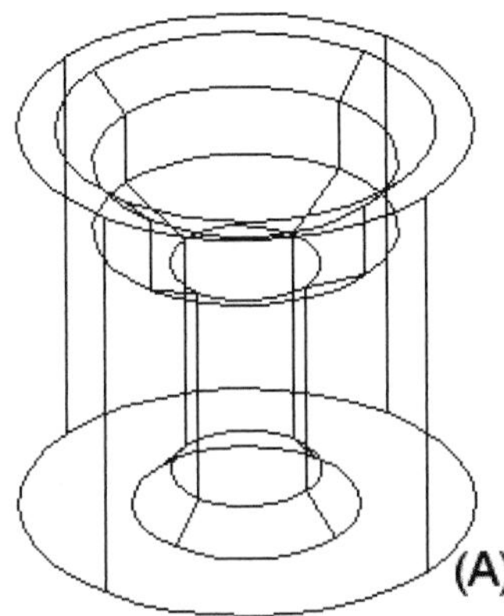

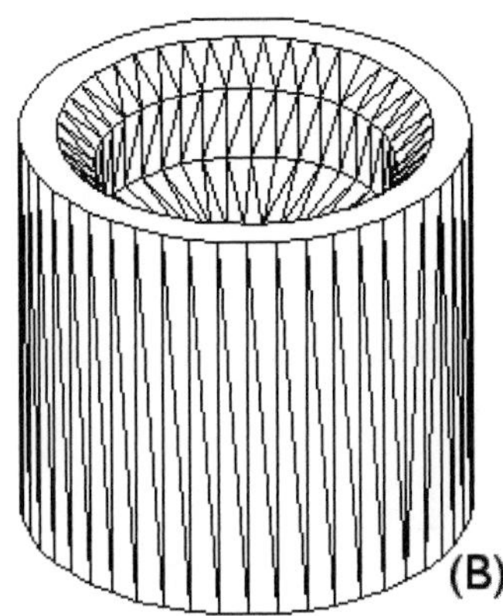

Figure 20.66

Note: As with the EXTRUDE command, you can also revolve an existing face of a 3D model to create a revolved feature of a 3D model.

CREATING A SOLID BY SWEEPING

The SWEEP command creates a solid by sweeping a profile along an open or closed 2D or 3D path. The result is a solid in the shape of the specified profile along the specified path. If the sweep profile is closed, a solid is created. If the sweep profile is open, a swept surface is created. Choose this command in one of the following ways:

- From the Modeling toolbar
- From the Ribbon of the 3D Modeling workspace
- From the pull-down menu or Menu Browser (Draw > Modeling > Sweep)
- From the keyboard (SWEEP)

Try It!: Open the drawing file 20_Sweep. Use the following command sequence and image for performing this task.

Illustrated on the left in the following image is an example of the geometry required to create a swept solid. Circles "A" and "B" represent profiles, while arc "C" represents the path of the sweep. Notice in this illustration that the circles do not have to be connected to the path; however, both circles must be constructed in the same plane in order for both to be included in the sweep operation. Use the following command sequence for creating a swept solid. The results of this operation are illustrated on the right in the following image.

```
Command: SWEEP
Current wire frame density: ISOLINES=4
Select objects to sweep: (Pick circles "A" and "B")
Select objects to sweep: (Press ENTER to continue)
Select sweep path or [Alignment/Base point/Scale/Twist]: (Pick
    arc "C")
```

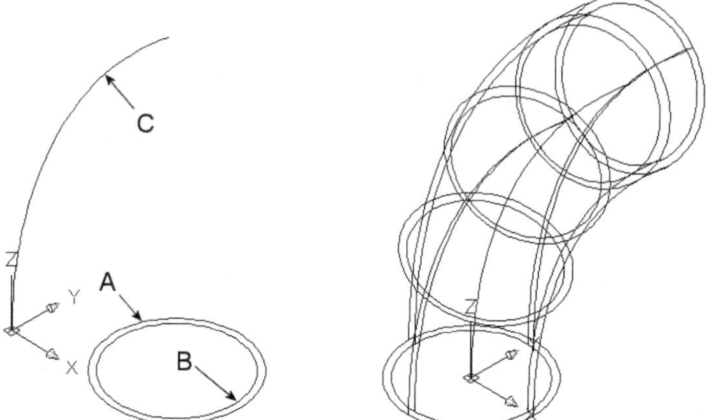

Figure 20.67

Illustrated on the left is the shaded solution to sweeping two circles along a path consisting of an arc. Notice, however, that an opening is not created in the shape; instead, the inner swept shape is surrounded by the outer swept shape. Both swept shapes are considered individual objects. To create the opening, subtract the inner shape from the outer shape using the SUBTRACT command. The results are illustrated on the right in the following image.

```
Command: SU (For SUBTRACT)
Select solids and regions to subtract from:
Select objects: (Select the outer sweep shape)
Select objects: (Press ENTER)
Select solids and regions to subtract ..
Select objects: (Select the inner sweep shape)
Select objects: (Press ENTER to perform the subtraction)
```

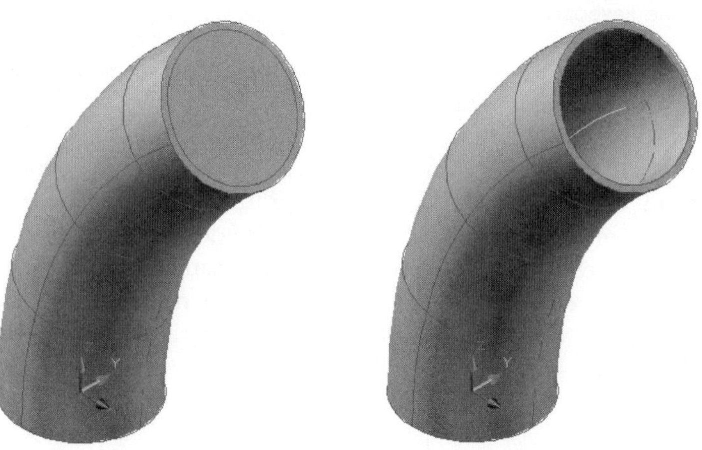

Figure 20.68

CREATING A SOLID BY LOFTING

The LOFT command creates a solid based on a series of cross-sections. These cross-sections define the shape of the solid. If the cross-sections are open, a surface loft is created. If the cross-sections are closed, a solid loft is created. When you are performing lofting operations, at least two cross-sections must be created. Choose this command in one of the following ways:

- From the Modeling toolbar
- From the Ribbon of the 3D Modeling workspace
- From the pull-down menu or Menu Browser (Draw > Modeling > Loft)
- From the keyboard (LOFT)

The following image illustrates a lofting operation based on open spline shaped objects. In the illustration on the left, the cross-sections of the plastic bottle are selected individually and in order starting with the left of the bottle and ending with the profile on the right of the bottle. The results are illustrated on the right in the following image, with a surface that is generated from the open profiles.

```
Command: LOFT
Select cross-sections in lofting order: (Select all cross-
    sections from the rear to the front)
Select cross-sections in lofting order: (Press ENTER to continue)
Enter an option [Guides/Path/Cross-sections only] <Cross-sections
    only>: G (For Guides)
Select guide curves: (Select both guide curves)
Select guide curves: (Press ENTER to create the loft)
```

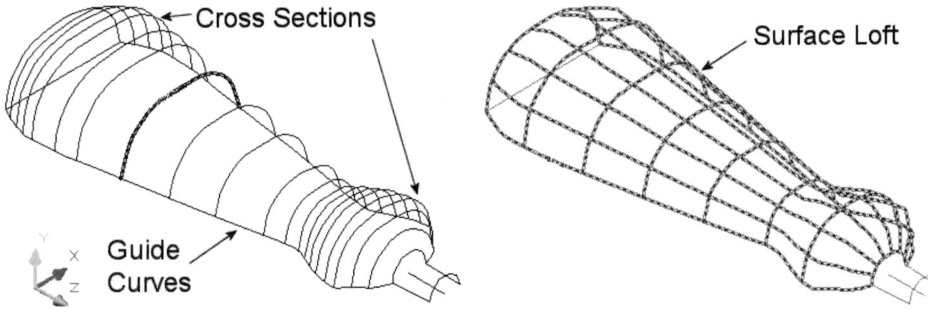

Figure 20.69

In the following image, instead of open profiles, all of the cross-sections consist of closed profile shapes. The same rules apply when creating lofts; all profiles must be selected in the proper order. The results of lofting closed profiles are illustrated on the right in the following image with the creation of a solid shape.

```
Command: LOFT
Select cross-sections in lofting order: (Select all cross-
    sections from the rear to the front)
Select cross-sections in lofting order: (Press ENTER to continue)
Enter an option [Guides/Path/Cross-sections only] <Cross-sections
    only>: G (For Guides)
Select guide curves: (Select both guide curves)
Select guide curves: (Press ENTER to create the loft)
```

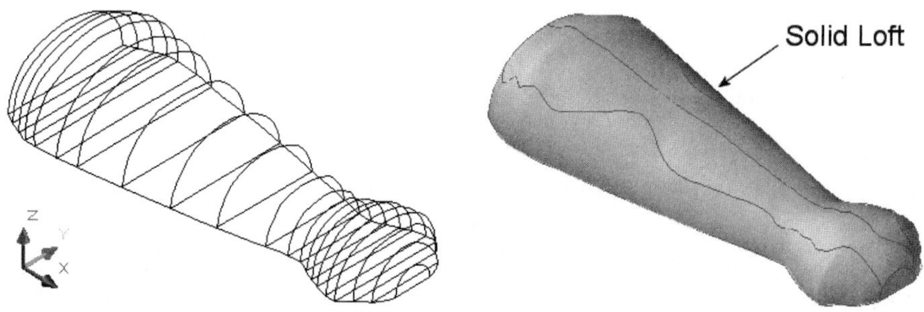

Figure 20.70

Try It!: Open the drawing file 20_Bowling Pin. You will first create a number of cross-section profiles in the form of circles that represent the diameters at different stations of the bowling pin. The LOFT command is then used to create the solid shape.

Begin by turning Dynamic Input (DYN) off. Then, create all the circles that make up the cross-sections of the bowling pin. Use the table shown on the left to construct the circles shown on the right in the following image.

Circle Coordinates	Circle Diameters
0,0,9.5	Ø0.75
0,0,8.5	Ø1.50
0,0,6.5	Ø1.00
0,0,3.5	Ø3.00
0,0,0.5	Ø2.00
0,0,0	Ø1.75

Figure 20.71

With all profiles created, activate the LOFT command and pick the cross-sections of the bowling pin beginning with the bottom circle and working your way up to the top circle. When the Loft Settings dialog box appears, as shown on the left in the following image, verify that Smooth Fit is selected and click the OK button to produce the loft that is illustrated in wireframe mode in the middle of the following image.

```
Command: LOFT
Select cross-sections in lofting order: (Select the six cross-
    sections of the bowling pin in order)
Select cross-sections in lofting order: (Press ENTER to continue)
Enter an option [Guides/Path/Cross-sections only] <Cross-sections
    only>: (Press ENTER to accept this default value)
Enter an option [Guides/Path/Cross-sections only] <Cross-sections
    only>: (Press ENTER to perform the loft)
```

To complete the bowling pin, use the FILLET command to round off the topmost circle of the bowling pin. Then view the results by clicking on the Realistic or Conceptual visual style. Your display should appear similar to the illustration as shown on the right in the following image.

```
Command: F (For FILLET)
Current settings: Mode = TRIM, Radius = 0.00
Select first object or [Undo/Polyline/Radius/Trim/Multiple]: R
    (For Radius)
Specify fillet radius <0.00>: 0.50
Select first object or [Undo/Polyline/Radius/Trim/Multiple]:
    (Pick the edge of the upper circle)
Enter fillet radius <0.50>: (Press ENTER to accept this value)
Select an edge or [Chain/Radius]: (Press ENTER to perform the
    fillet operation)
1 edge(s) selected for fillet.
```

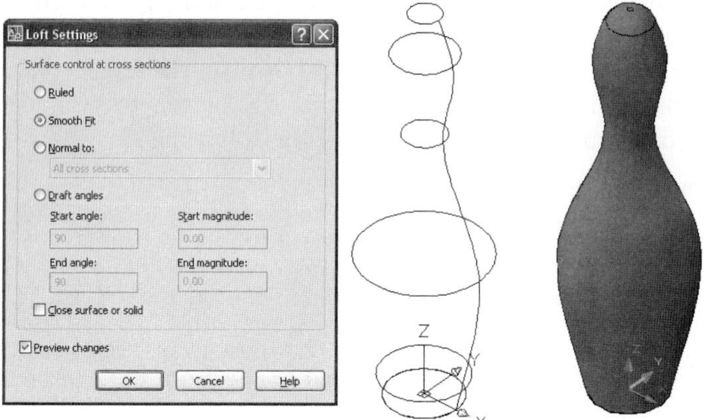

Figure 20.72

CREATING A HELIX

The HELIX command creates a 2D or 3D spiral object. Choose this command in one of the following ways:

- From the Modeling toolbar
- From the Ribbon of the 3D Modeling workspace (expand the 3D Modeling panel)
- From the pull-down menu or Menu Browser (Draw > Modeling > Helix)
- From the keyboard (HELIX)

Try It!: Open the drawing file 20_Helix. Use the following command sequences and images for creating a 2D, 3D, and spiral helix.

Using the table below, experiment with the following command prompts for constructing a 2D helix, 3D helix, and 3D spiral, as shown on the left in the following images.

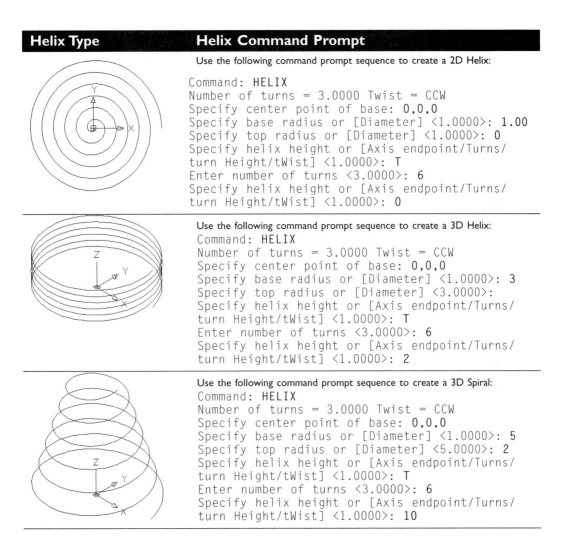

Helix Type	Helix Command Prompt
	Use the following command prompt sequence to create a 2D Helix: Command: **HELIX** Number of turns = 3.0000 Twist = CCW Specify center point of base: **0,0,0** Specify base radius or [Diameter] <1.0000>: **1.00** Specify top radius or [Diameter] <1.0000>: **0** Specify helix height or [Axis endpoint/Turns/ turn Height/tWist] <1.0000>: **T** Enter number of turns <3.0000>: **6** Specify helix height or [Axis endpoint/Turns/ turn Height/tWist] <1.0000>: **0**
	Use the following command prompt sequence to create a 3D Helix: Command: **HELIX** Number of turns = 3.0000 Twist = CCW Specify center point of base: **0,0,0** Specify base radius or [Diameter] <1.0000>: **3** Specify top radius or [Diameter] <3.0000>: Specify helix height or [Axis endpoint/Turns/ turn Height/tWist] <1.0000>: **T** Enter number of turns <3.0000>: **6** Specify helix height or [Axis endpoint/Turns/ turn Height/tWist] <1.0000>: **2**
	Use the following command prompt sequence to create a 3D Spiral: Command: **HELIX** Number of turns = 3.0000 Twist = CCW Specify center point of base: **0,0,0** Specify base radius or [Diameter] <1.0000>: **5** Specify top radius or [Diameter] <5.0000>: **2** Specify helix height or [Axis endpoint/Turns/ turn Height/tWist] <1.0000>: **T** Enter number of turns <3.0000>: **6** Specify helix height or [Axis endpoint/Turns/ turn Height/tWist] <1.0000>: **10**

Figure 20.73

HELIX APPLICATIONS

Typically, a wireframe model of a helix does not fully define how an object like a spring should look. It would be beneficial to show the spring as a thin wire wrapping around a cylinder to form the helical shape. To produce this type of object, use the wireframe of a circle as the object to sweep around the helix to produce the spring.

Try It!: Open the drawing file 20_Helix Spring. A helix is already created, along with small circular profile. With the helix as a path and the circle as the object to sweep, use the following command prompt and images to create a spring.

```
Command: SWEEP
Current wire frame density: ISOLINES=4
Select objects to sweep: (Select the small circle)
Select objects to sweep: (Press ENTER to continue)
Select sweep path or [Alignment/Base point/Scale/Twist]: (Select
    the helix)
```

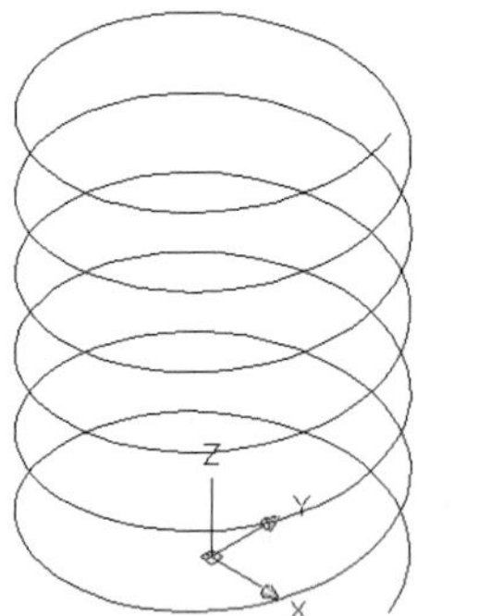

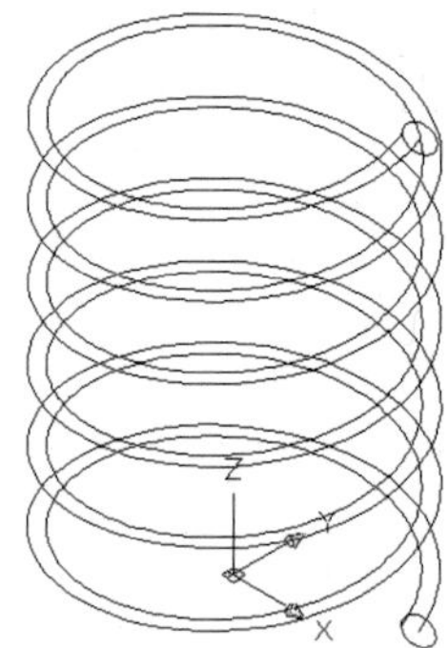

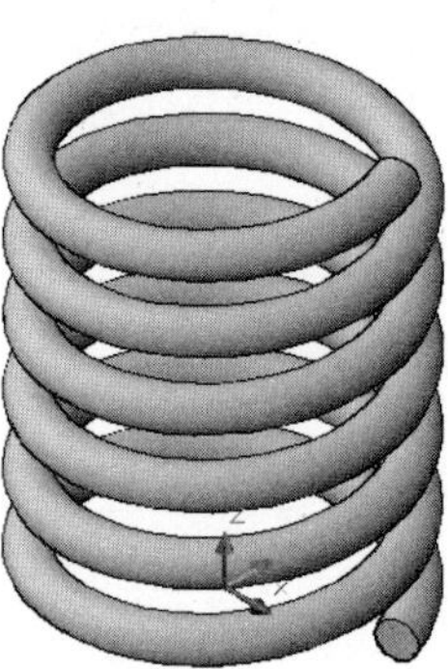

Figure 20.74

CREATING POLYSOLIDS

A polysolid is created in a fashion similar to the one used to create a polyline. The POLYSOLID command is used to create this type of object. Choose this command in one of the following ways:

- From the Modeling toolbar
- From the Ribbon of the 3D Modeling workspace
- From the pull-down menu or Menu Browser (Draw > Modeling > Polysolid)
- From the keyboard (POLYSOLID)

You pick points and can use the direct distance mode of entry to designate the distances of the polysolid, as shown on the left in the following image, where a width has been entered. The main difference between polysolids and polylines is that you can designate the height of a polysolid. In this way, polysolids are ideal for creating such items as walls in your model, as shown on the right in the following image.

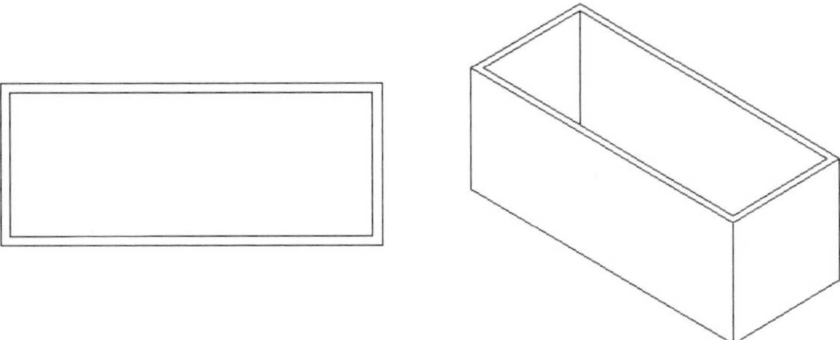

Figure 20.75

Polysolids can also be created from existing 2D geometry such as lines, polylines, arcs, and even circles. Illustrated on the left in the following image is a 2D polyline that has had fillets applied to a number of corners. Activating the POLYSOLID command and selecting the polyline changes the appearance of the object to match the illustration, as shown in the middle in the following image. Here a width has been automatically applied to the polysolid. When the polysolid is viewed in 3D using the SE Isometric, the polysolid is displayed with a height, as shown on the right in the following image.

```
Command: POLYSOLID
Specify start point or [Object/Height/Width/Justify] <Object>:
    (Press ENTER to accept Object)
Select object: (Pick the polyline object as shown on the left in
    the following image)
```

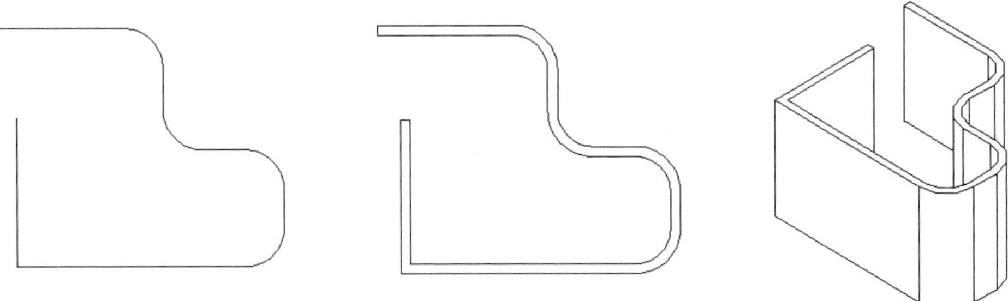

Figure 20.76

Try It!: Open the drawing file 20_Polysolid Walls. Use the following command sequence and image to construct the 3D walls using the POLYSOLID command.

```
Command: POLYSOLID
Specify start point or [Object/Height/Width/Justify] <Object>: H
    (For Height)
Specify height <0'-4">: 8'
Specify start point or [Object/Height/Width/Justify] <Object>: W
    (For Width)
Specify width <0'-0 1/4">: 4
```

```
Specify start point or [Object/Height/Width/Justify] <Object>:
    (Pick a point in the lower-left corner of the display
    screen)
Specify next point or [Arc/Undo]: (Move your cursor to the right
    and enter 30')
Specify next point or [Arc/Undo]: (Move your cursor up and enter
    10')
Specify next point or [Arc/Close/Undo]: (Move your cursor to the
    left and enter 5')
Specify next point or [Arc/Close/Undo]: (Move your cursor up and
    enter 5')
Specify next point or [Arc/Close/Undo]: (Move your cursor to the
    left and enter 25')
Specify next point or [Arc/Close/Undo]: C (For Close)
```

When finished, rotate your model using the SE Isometric view or the 3DFORBIT (3D Free Orbit) command to view the results.

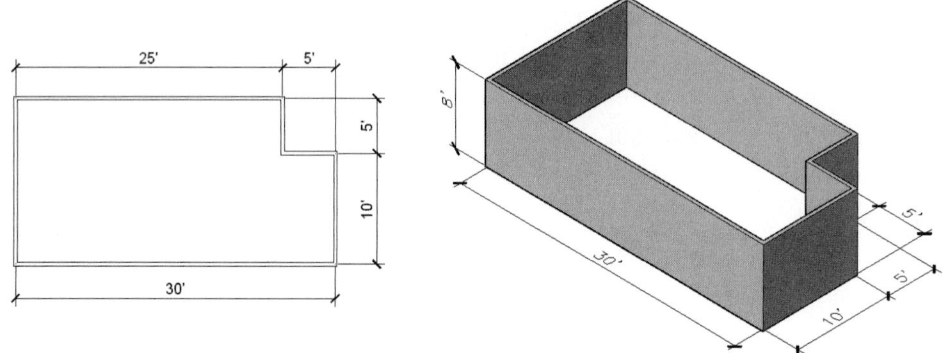

Figure 20.77

FILLETING SOLID MODELS

Filleting of simple or complex objects is easily handled with the FILLET command. This is the same command as the one used to create a 2D fillet.

Try It!: Open the drawing file 20_Tee Fillet. Use the following prompt sequence and image for performing this task.

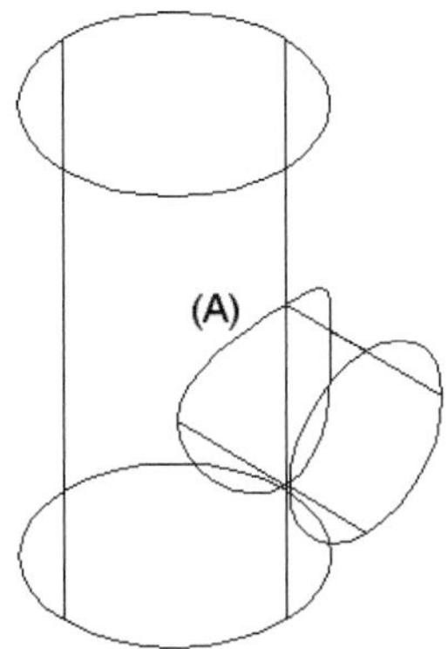

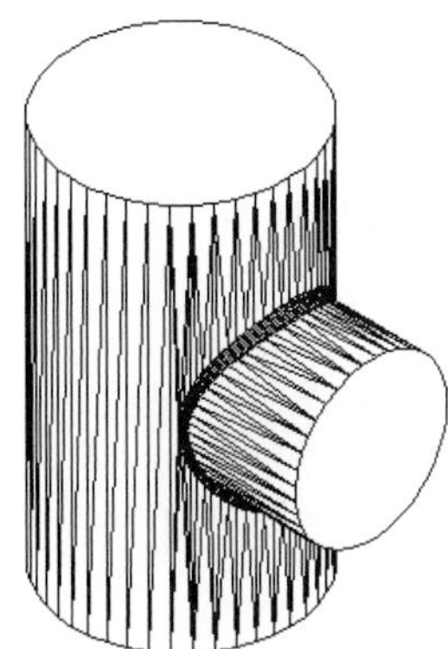

Figure 20.78

```
Command: F (For FILLET)
Current settings: Mode = TRIM, Radius = 0.5000
Select first object or [Undo/Polyline/Radius/Trim/Multiple]:
    (Select the edge at "A," which represents the intersection
    of both cylinders)
Enter fillet radius <0.5000>: 0.25
Select an edge or [Chain/Radius]: (Press ENTER to perform the
    fillet operation)
1 edge(s) selected for fillet.
```

Try It!: Open the drawing file 20_Slab Fillet. A group of objects with a series of edges can be filleted with the Chain option of the FILLET command. Use the following prompt sequence and image for performing this task.

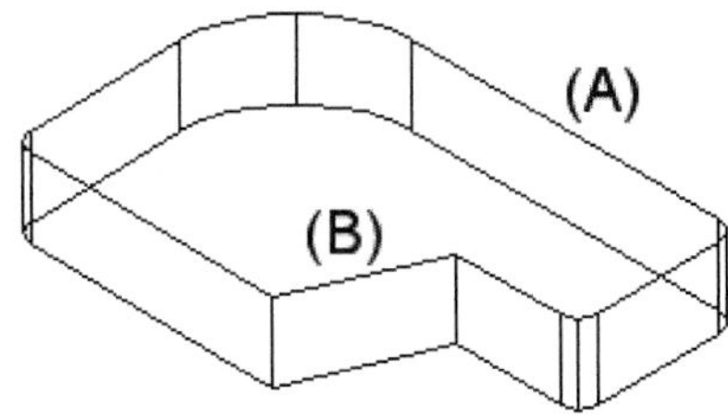

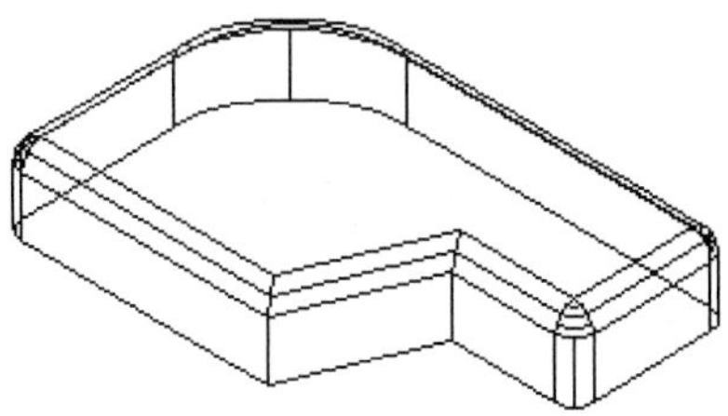

Figure 20.79

```
Command: F (For FILLET)
Current settings: Mode = TRIM, Radius = 0.5000
Select first object or [Undo/Polyline/Radius/Trim/Multiple]:
    (Select the edge at "A")
Enter fillet radius <0.5000>: (Press ENTER to accept the default
    value)
Select an edge or [Chain/Radius]: C (For chain mode)
Select an edge chain or [Edge/Radius]: (Select edge "A" again;
    notice how the selection is chained until it reaches an
    abrupt corner)
Select an edge chain or [Edge/Radius]: (Select the edge at "B")
```

```
Select an edge chain or [Edge/Radius]: (Press ENTER when finished
    selecting all edges to perform the fillet operation)
10 edge(s) selected for fillet.
```

CHAMFERING SOLID MODELS

Just as the FILLET command uses the Chain mode to group a series of edges together, the CHAMFER command uses the Loop option to perform the same type of operation.

Try It!: Open the drawing file 20_Slab Chamfer. Use the following prompt sequence and image for performing this task.

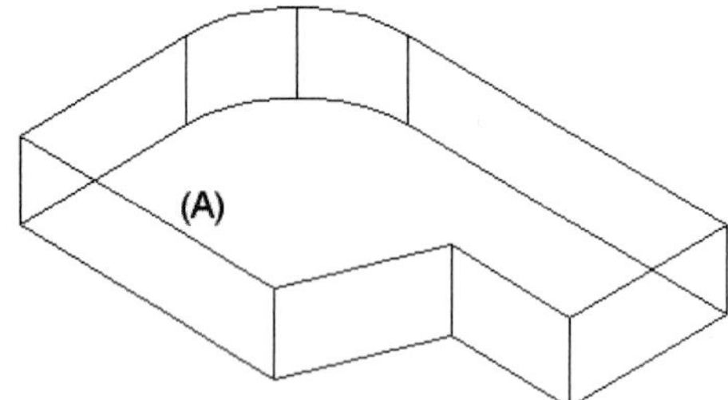

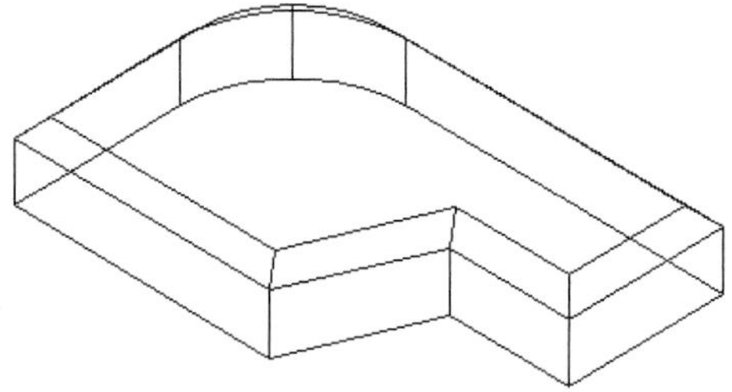

Figure 20.80

```
Command: CHA (For CHAMFER)
(TRIM mode) Current chamfer Dist1 = 0.5000, Dist2 = 0.5000
Select first line or [Undo/Polyline/Distance/Angle/Trim/mEthod/
    Multiple]: (Pick the edge at "A")
Base surface selection...
Enter surface selection option [Next/OK (current)] <OK>: N (This
    option selects the other surface shared by edge "A"; the top
    surface should highlight. If not, use this option again)
Enter surface selection option [Next/OK (current)] <OK>: (Press
    ENTER to accept the top surface)
Specify base surface chamfer distance <0.5000>: (Press ENTER to
    accept the default)
Specify other surface chamfer distance <0.5000>: (Press ENTER to
    accept the default)
Select an edge or [Loop]: L (To loop all edges together into
    one)
Select an edge loop or [Edge]: (Pick any top edge; notice that
    the loop option does not stop at abrupt corners)
Select an edge loop or [Edge]: (Press ENTER to perform the
    chamfering operation)
```

OBTAINING MASS PROPERTIES OF A SOLID MODEL

The MASSPROP command calculates the mass properties of a solid model. Choose this command in one of the following ways:

- From the Inquiry toolbar
- From the Ribbon of the 3D Modeling workspace (Tools tab – Inquiry panel)
- From the pull-down menu or Menu Browser (Tools > Inquiry > Region/Mass Properties)
- From the keyboard (MASSPROP)

Try It!: Open the drawing file 20_Tee Massprop. Use the MASSPROP command to calculate the mass properties of a selected solid, as shown in the following image. All calculations are based on the current position of the User Coordinate System. You will be given the option of writing this information to a file if desired.

```
Command: MASSPROP
Select objects: (Select the model)
Select objects: (Press ENTER to continue with this command)
```

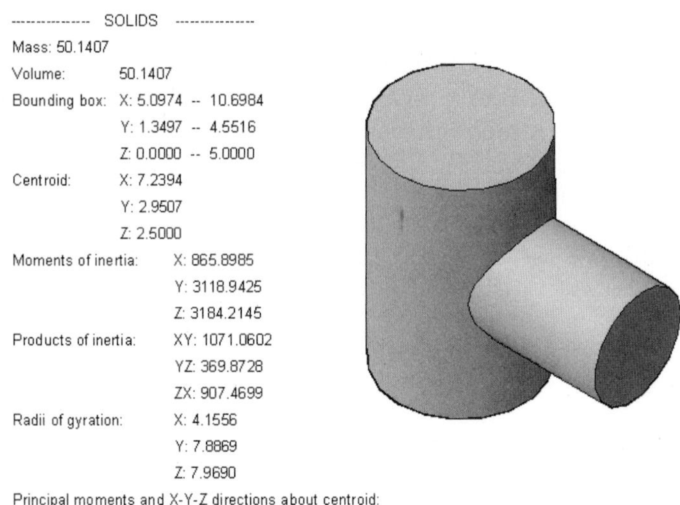

```
---------------- SOLIDS ----------------
Mass: 50.1407
Volume:            50.1407
Bounding box:      X: 5.0974 -- 10.6984
                   Y: 1.3497 -- 4.5516
                   Z: 0.0000 -- 5.0000
Centroid:          X: 7.2394
                   Y: 2.9507
                   Z: 2.5000
Moments of inertia:     X: 865.8985
                        Y: 3118.9425
                        Z: 3184.2145
Products of inertia:    XY: 1071.0602
                        YZ: 369.8728
                        ZX: 907.4699
Radii of gyration:      X: 4.1556
                        Y: 7.8869
                        Z: 7.9690
Principal moments and X-Y-Z directions about centroid:
     I: 115.9688 along [1.0000 0.0000 0.0000]
     J: 177.7542 along [0.0000 1.0000 0.0000]
     K: 119.8558 along [0.0000 0.0000 1.0000]
```

Figure 20.81

SYSTEM VARIABLES THAT AFFECT SOLID MODELS

System variables are available to control the appearance of your solid model whenever you perform shading or hidden line removal operations. Three in particular are useful: ISOLINES, FACETRES, and DISPSILH. The following text describes each system variable in detail.

THE ISOLINES (ISOMETRIC LINES) SYSTEM VARIABLE

Tessellation refers to the lines that are displayed on any curved surface, such as those shown in the following image, to help you visualize the surface. Tessellation lines are automatically formed when you construct solid primitives such as cylinders and cones. These lines are also calculated when you perform solid modeling operations such as SUBTRACT and UNION.

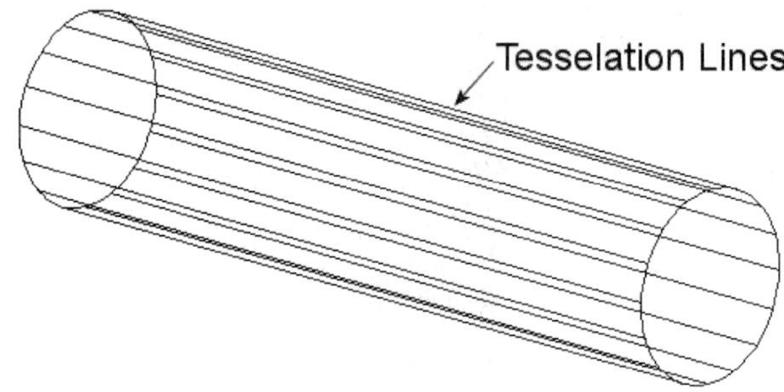

Figure 20.82

The number of tessellation lines per curved object is controlled by the system variable called ISOLINES. By default, this variable is set to a value of 4. The following image shows the results of setting this variable to other values, such as 9 and 20. After the isolines have been changed, regenerate the screen to view the results. The more lines used to describe a curved surface, the more accurate the surface will look in wireframe mode; however, it will take longer to process screen regenerations.

Try It!: Open the drawing file 20_Tee Isolines. Experiment by changing the ISOLINES system variable to numerous values. Perform a drawing regeneration using the REGEN command each time that you change the number of isolines.

THE FACETRES (FACET RESOLUTION) SYSTEM VARIABLE

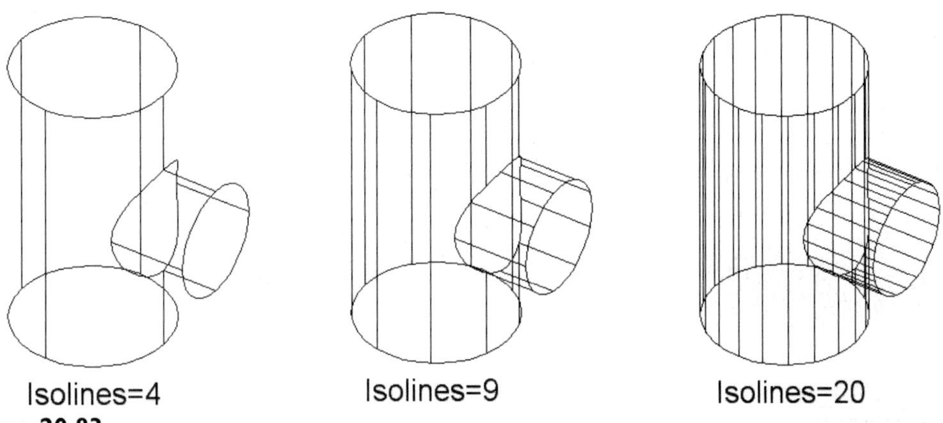

Figure 20.83

When you perform hidden line removals on solid objects, the results are similar to those displayed in the following image. The curved surfaces are now displayed as flat triangular surfaces (faces), and the display of these faces is controlled by the FACETRES system variable. The cylinder with FACETRES set to 0.50 processes much more quickly than the cylinder with FACETRES set to 2, because there are fewer surfaces to process in such operations as hidden line removals. However, the image with FACETRES set to a large value (as high as 10) shows a more defined circle. The default value for FACETRES is 0.50, which seems adequate for most applications.

Try It!: Open the drawing file 20_Tee Facetres. Experiment by changing the FACETRES system variable to numerous values (valid values are .01 to 10). Perform a HIDE each time you change the number of facet lines to view the results.

THE DISPSILH (DISPLAY SILHOUETTE) SYSTEM VARIABLE

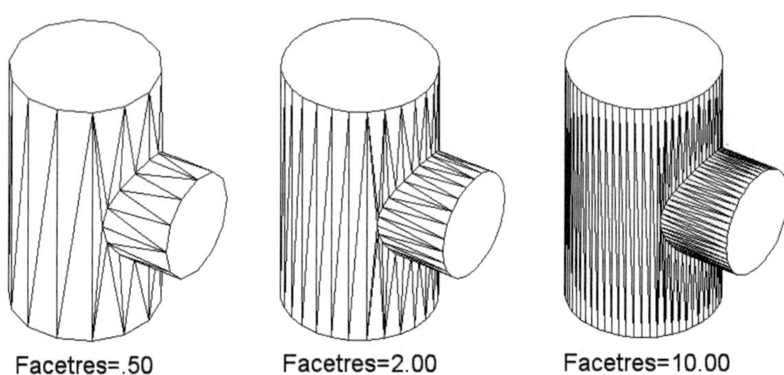

Figure 20.84

To have the edges of your solid model take on the appearance of an isometric drawing when displayed as a wireframe, use the DISPSILH system variable. This system variable means "display silhouette" and is used to control the display of silhouette curves of solid objects while in either the wireframe or hide mode. Silhouette edges are turned either on or off, with the results displayed in the following image; by default they are turned off. When a HIDE is performed, this system variable controls whether faces are drawn or suppressed on a solid model.

Try It!: Open the drawing file 20_Tee Dispsilh. Experiment by turning the display of silhouette edges on and off. Regenerate your display each time you change this mode to view the results. Also use the HIDE command to see how the hidden line removal image is changed.

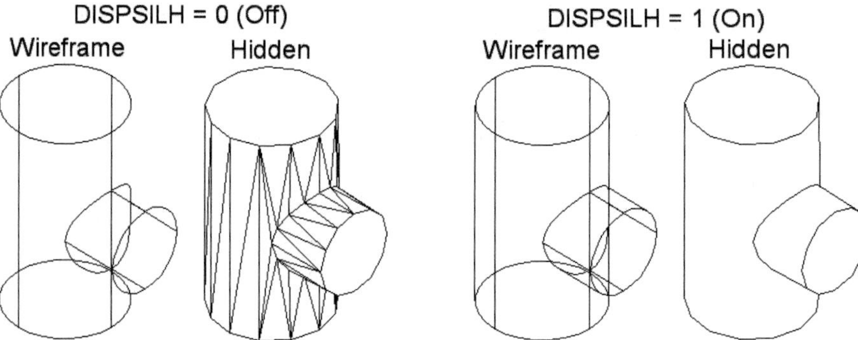

Figure 20.85

TUTORIAL EXERCISE: COLLAR.DWG

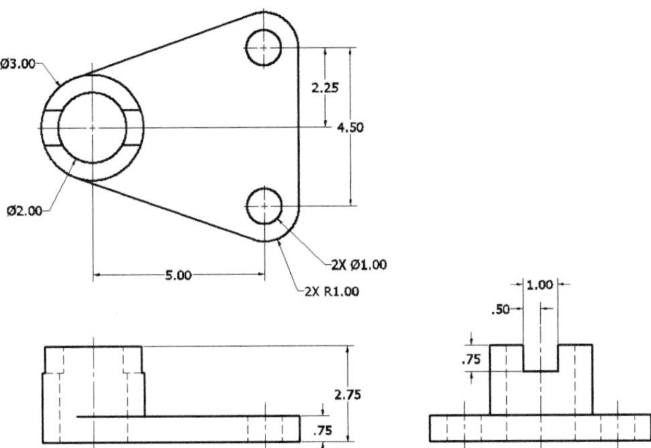

Figure 20.86

PURPOSE

This tutorial is designed to construct a solid model of the Collar using the dimensions in the previous image.

SYSTEM SETTINGS

Use the current limits set to 0,0 for the lower-left corner and (12,9) for the upper-right corner. Change the number of decimal places from four to two using the Drawing Units dialog box.

LAYERS

Create the following layer:

Name	Color	Linetype
Model	Cyan	Continuous

SUGGESTED COMMANDS

Begin this tutorial by laying out the Collar in plan view and drawing the basic shape outlined in the Top view. Convert the objects to a polyline and extrude the objects to form a solid. Draw a cylinder and combine this object with the base. Add another cylinder and then subtract it to form the large hole through the model. Add two small cylinders and subtract them from the base to form the smaller holes. Construct a solid box and subtract it to form the slot across the large cylinder.

Step 1

Begin the Collar by setting the Model layer current. Then draw the three circles shown on the left in the following image using the CIRCLE command. Perform a ZOOM-All after all three circles have been constructed.

Step 2

Draw lines tangent to the three arcs using the LINE command and the OSNAP-Tangent mode, as shown on the right in the following image. Notice that the User Coordinate System icon has been turned off in this image.

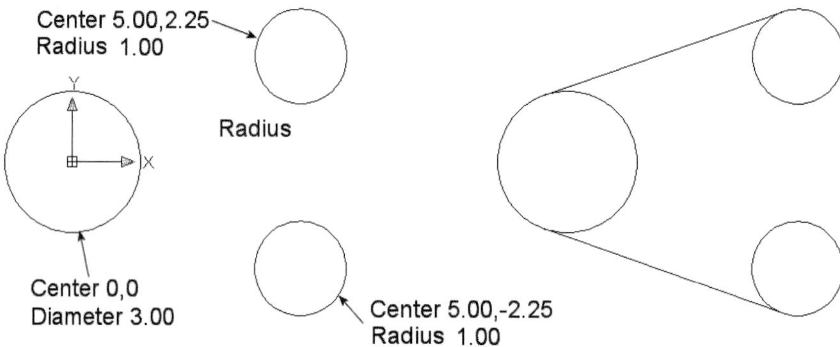

Figure 20.87

Step 3

Use the TRIM command to trim the circles. When prompted to select the cutting edge object, press ENTER; this makes cutting edges out of all objects in the drawing. Perform this trimming operation so your display appears similar to the illustration on the left in the following image.

Step 4

Prepare to construct the base by viewing the object in 3D. Select a viewpoint by choosing 3D Views from the View pull-down menu and then choosing SE Isometric. Your display should appear similar to the illustration on the right in the following image.

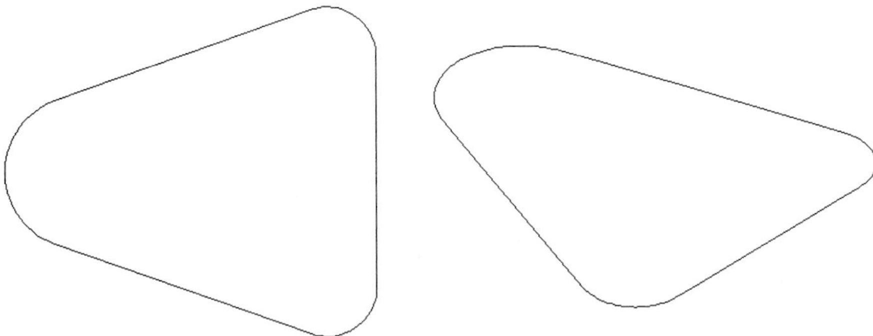

Figure 20.88

Step 5

Convert all objects to a polyline using the Join option of the PEDIT command, as shown on the left in the following image.

Step 6

Use the EXTRUDE command to extrude the base to a thickness of 0.75 units, as shown on the right in the following image.

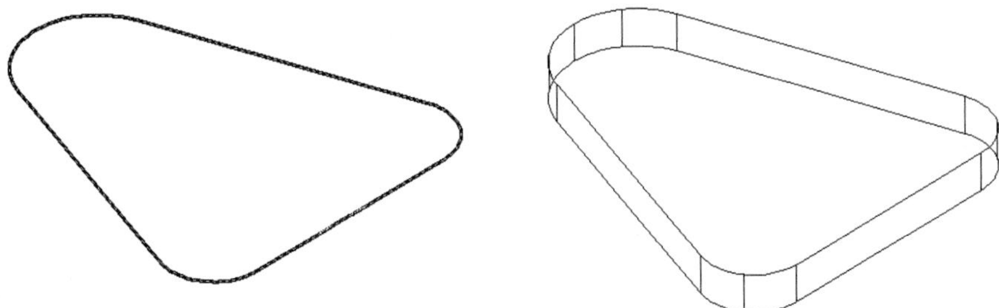

Figure 20.89

Step 7

Turn off Dynamic UCS on the status bar. Create a cylinder using the CYLINDER command. Begin the center point of the cylinder at 0,0,0, with a diameter of 3.00 units and a height of 2.75 units. You may have to perform a ZOOM-All to display the entire model, as shown on the left in the following image.

Step 8

Merge the cylinder just created with the extruded base using the UNION command, as shown on the right in the following image.

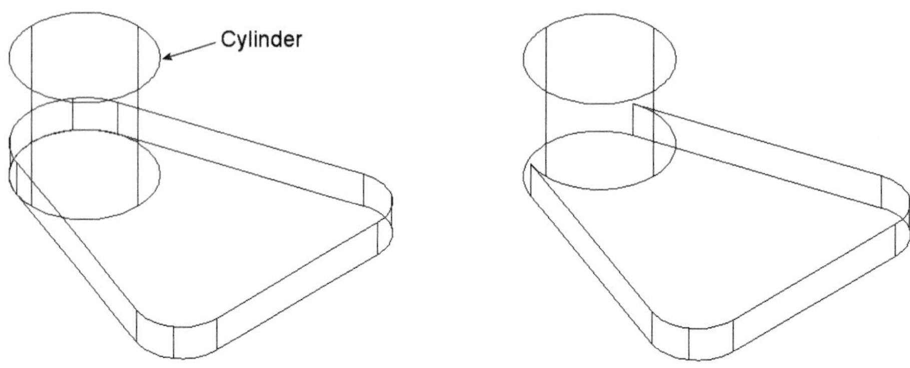

Figure 20.90

Step 9

Use the CYLINDER command to create a 2.00-unit-diameter cylinder representing a through hole, as shown on the left in the following image. The height of the cylinder is 2.75 units, with the center point at 0,0,0.

Step 10

To cut the hole through the outer cylinder, use the SUBTRACT command. Select the base as the source object; select the inner cylinder as the object to subtract. Use the HIDE command to view the results, as shown on the right in the following image.

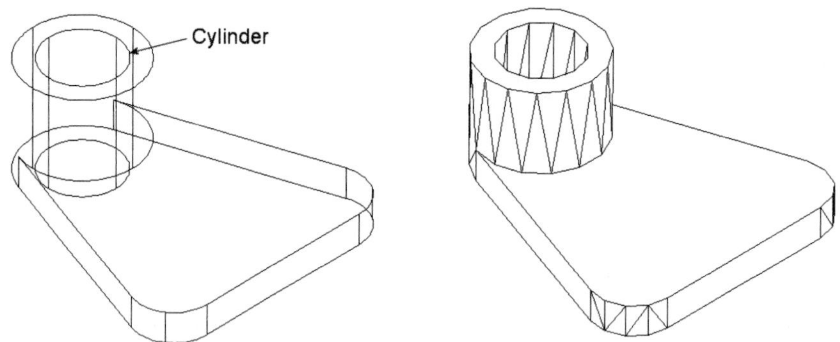

Figure 20.91

Step 11

Use the REGEN command to regenerate your screen and return to Wireframe mode. Begin placing the two small drill holes (1.00 diameter and .75 high) in the base using the CYLINDER command. Use the OSNAP-Center mode to place each cylinder at the center of arcs "A" and "B," as shown on the left in the following image.

Step 12

Subtract both 1.00-diameter cylinders from the base of the model using the SUBTRACT command. Use the HIDE command to view the results, as shown on the right in the following image.

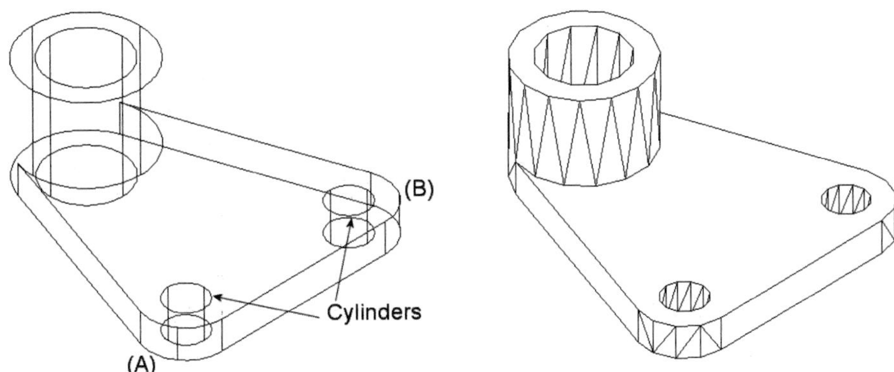

Figure 20.92

Step 13

Begin constructing the rectangular slot that will pass through the two cylinders. Use the BOX command and Center option to accomplish this. Locate the center of the box at 0,0,2.75 and make the box 4 units long, 1 unit wide, and 1.50 units high (using the Center option) as shown in the following image.

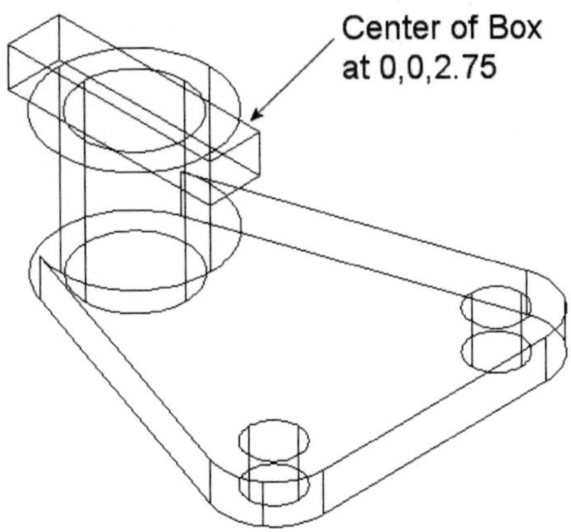

Figure 20.93

Step 14

Use the SUBTRACT command to subtract the rectangular box from the solid model, as shown on the left in the following image.

Step 15

Change the facet resolution to a higher value using the FACETRES system variable and a value of 5. Then perform a hidden line removal to see the appearance of the completed model, as shown on the right in the following image.

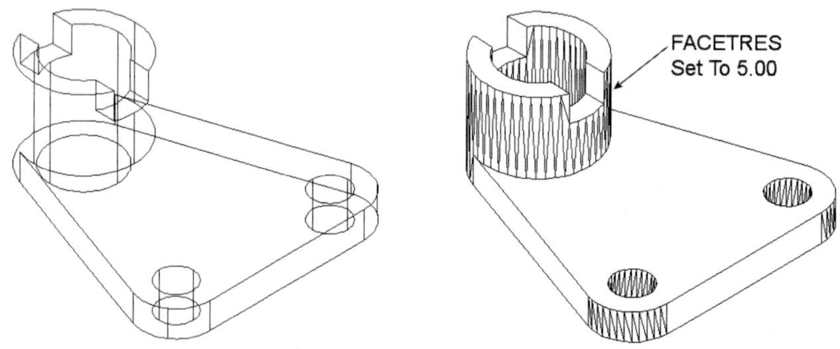

Figure 20.94

TUTORIAL EXERCISE: VACUUM ATTACHMEN.DWG

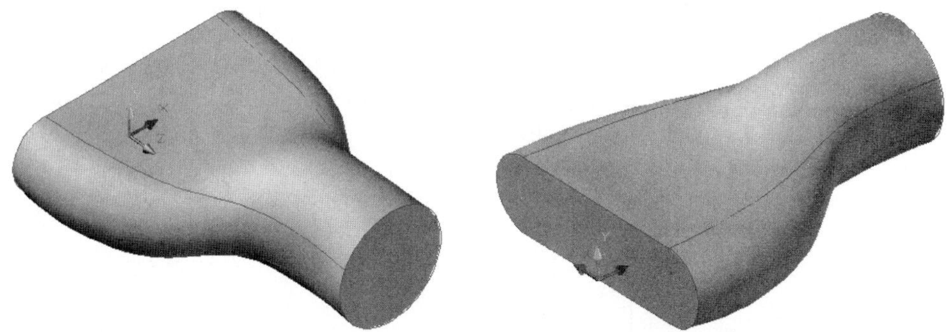

Figure 20.95

PURPOSE

This tutorial is designed to construct a solid model of the vacuum cleaner attachment using the Loft tool.

SYSTEM SETTINGS

Use the current limits, which are set to 0,0 for the lower-left corner and (12,9) for the upper-right corner. Change the number of decimal places from four to two using the Drawing Units dialog box.

LAYERS

Create the following layer:

Name	Color	Linetype
Model	Cyan	Continuous

SUGGESTED COMMANDS

Begin this tutorial by laying out a slot shape and circle. Next move the midpoint of the bottom of the slot to 0,0,0 and then copy it to 0,0,2. Move the bottom quadrant of the circle to 0,0,3.5 and then copy it to 0,0,5. These steps form the four cross-sections used for creating the loft.

Step 1

First create a new drawing file. Then, begin the construction of this 3D model by constructing 2D geometry that will define the final shape of the vacuum cleaner attachment. Use the PLINE command to construct the slot shape as a closed polyline using the dimensions shown on the left in the following image. Then construct the circle using the diameter dimension shown on the right in the following image.

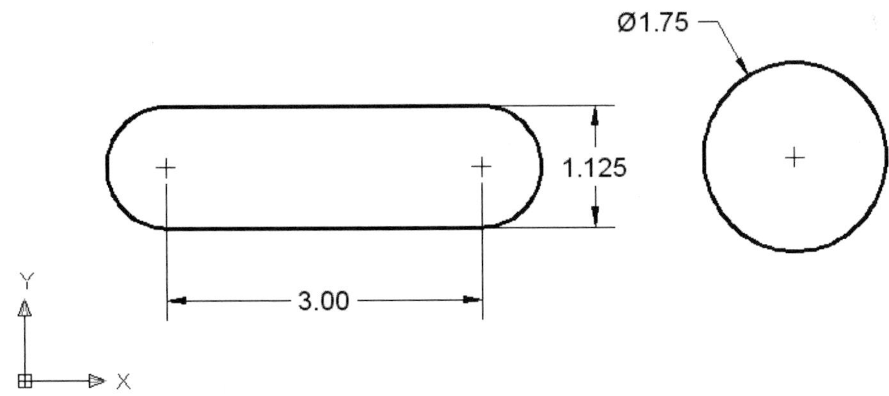

Figure 20.96

Step 2

Switch to SE Isometric viewing, as shown in the following image. This viewing position can be found by selecting View from the pull-down menu, then selecting 3D Views. Before continuing, be sure that Dynamic Input is turned off for this segment. Next, move the slot shape from the midpoint of the bottom line to point 0,0,0.

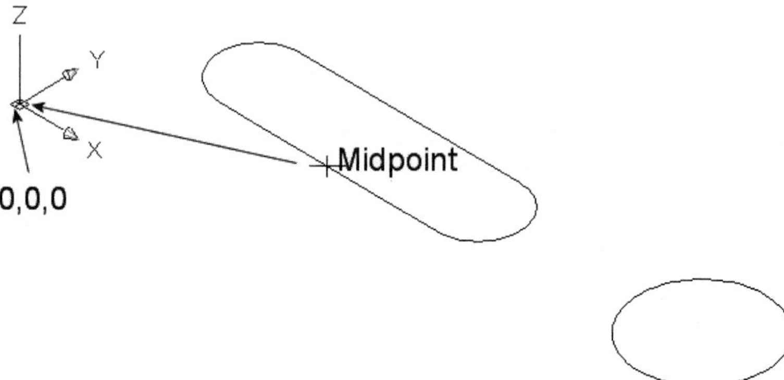

Figure 20.97

Step 3

With the slot moved to the correct position, use the COPY command to create a duplicate shape of the slot using a base point of 0,0,0 and a second point at 0,0,2. When finished, your display should appear similar to the following image.

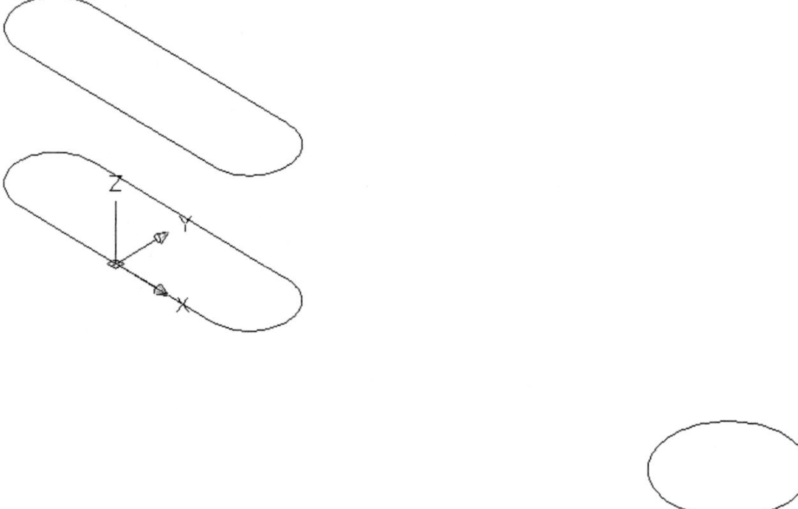

Figure 20.98

Step 4

You will next move the circle. Enter the MOVE command and pick the base point of the move using the bottom quadrant of the circle, as shown on the right in the following image. For the second point of displacement, enter the coordinate value of 0,0,3.5 from the keyboard.

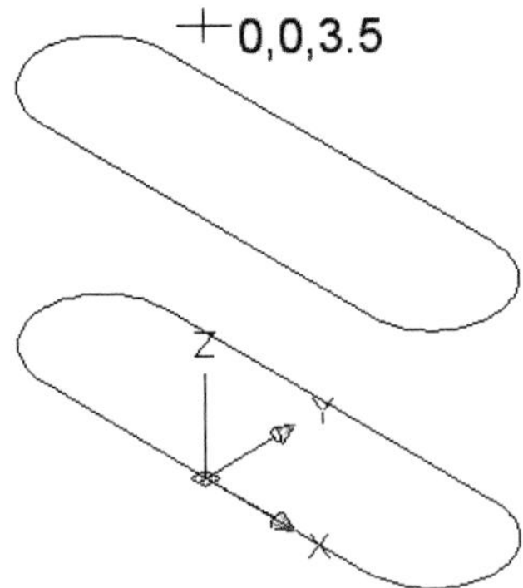

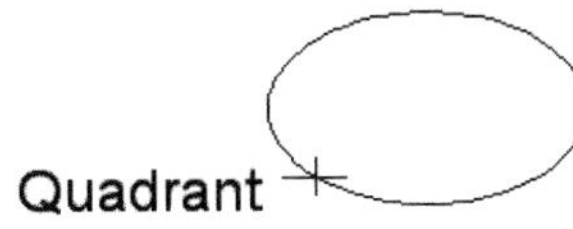

Figure 20.99

Step 5

With the circle moved to the correct location, use the COPY command to duplicate this circle. Copy this circle from the bottom quadrant to a second point located at 0,0,5, as shown on the left in the following image. When finished, your display should appear similar to the illustration on the right in the following image. These form the four cross-sectional shapes that make up the vacuum cleaner attachment.

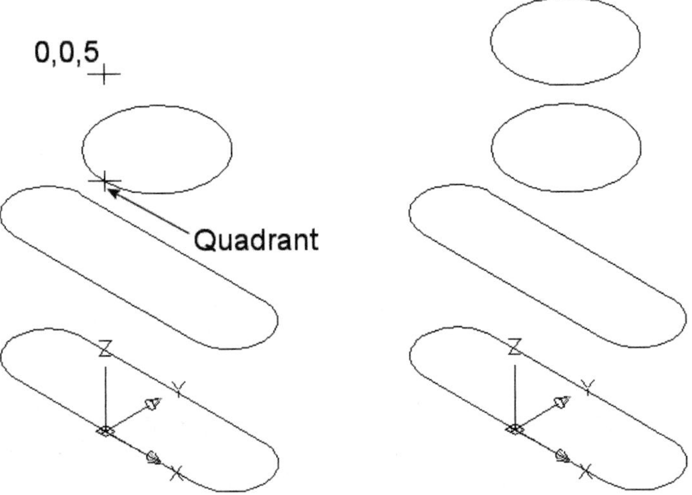

Figure 20.100

Step 6

Issue the PLAN command through the keyboard and observe that the bottom midpoints of both slots and the bottom quadrants of both circles are all aligned to 0,0,0 as shown in the following image. When finished, perform a ZOOM-Previous operation to return to the SE Isometric view.

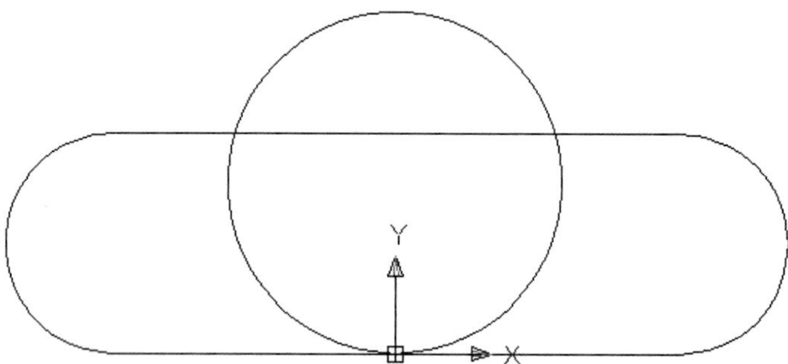

Figure 20.101

Step 7

Use the 3DFORBIT (Free Orbit) command and rotate your model so it appears similar to the following image.

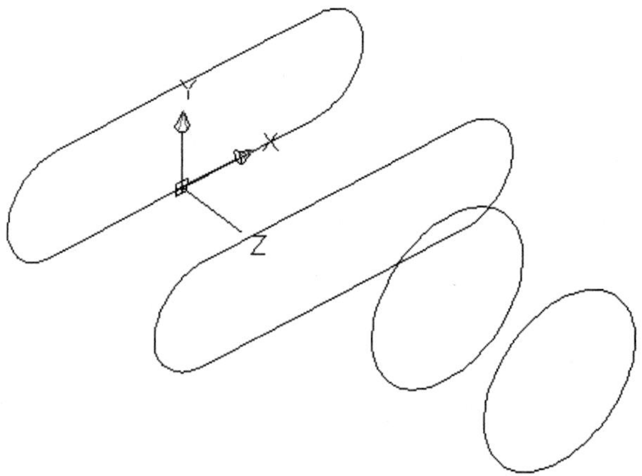

Figure 20.102

Step 8

Activate the LOFT command and pick the four cross-sections in the order labeled 1 through 4, as shown on the left in the following image. When the Loft Settings dialog box appears, keep the default Surface control at cross-sections setting at Smooth Fit and click the OK button, as shown on the right in the following image.

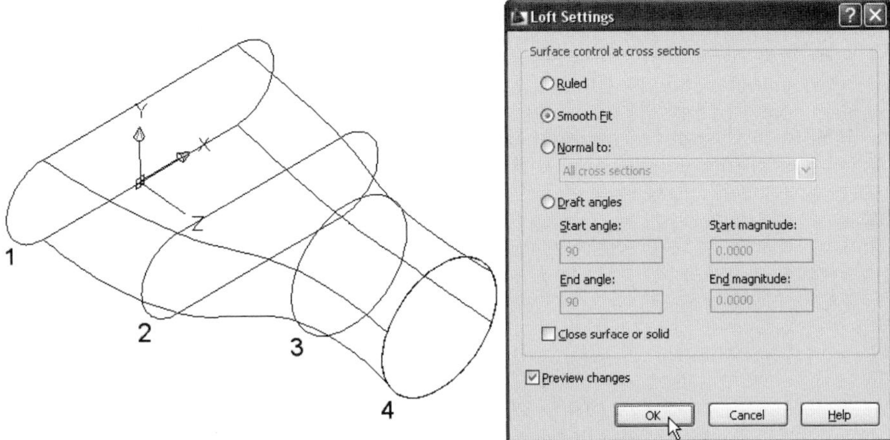

Figure 20.103

Step 9

The results are illustrated in the following image.

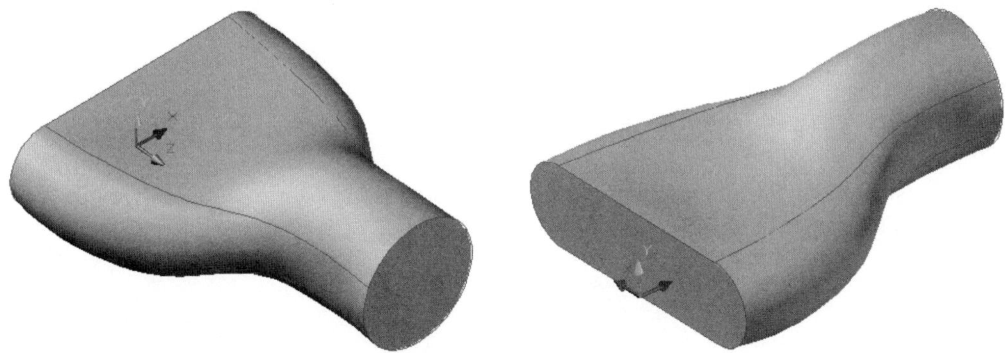

Figure 20.104

Step 10

An alternate step would be to activate the SOLIDEDIT command and use the Shell option to produce a thin wall of .10 units on the inside of the vacuum attachment. This command and option will be discussed in greater detail in the next chapter. Use the following command prompt sequence and images to perform this operation.

```
Command: SOLIDEDIT
Solids editing automatic checking: SOLIDCHECK=1
Enter a solids editing option [Face/Edge/Body/Undo/eXit] <eXit>:
    B (For Body)
Enter a body editing option
[Imprint/seParate solids/Shell/cLean/Check/Undo/eXit] <eXit>: S
    (For Shell)
Select a 3D solid: (Select the vacuum attachment)
```

```
Remove faces or [Undo/Add/ALL]: (Pick the face at "A")
Remove faces or [Undo/Add/ALL]: (Pick the face at "B")
Remove faces or [Undo/Add/ALL]: (Press ENTER to continue)
Enter the shell offset distance: .10
Solid validation started.
Solid validation completed.
Enter a body editing option
[Imprint/seParate solids/Shell/cLean/Check/Undo/eXit] <eXit>:
    (Press ENTER)
Solids editing automatic checking: SOLIDCHECK=1
Enter a solids editing option [Face/Edge/Body/Undo/eXit] <eXit>:
    (Press ENTER)
```

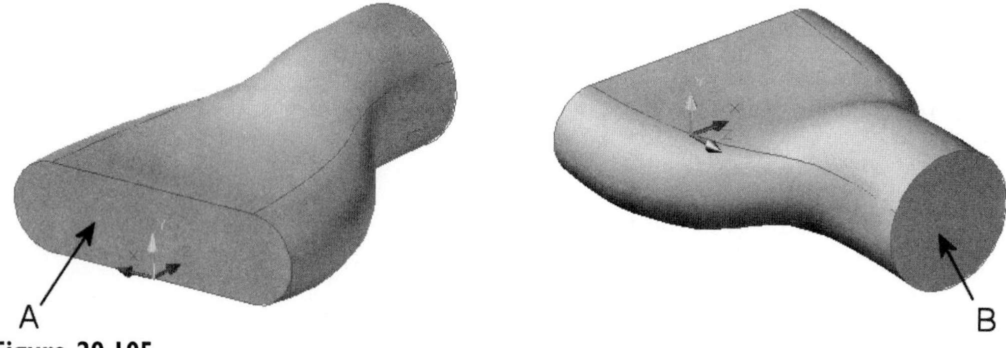

Figure 20.105

Step 11

When finished completing the Shell option of the SOLIDEDIT command, your model should appear similar to the following image.

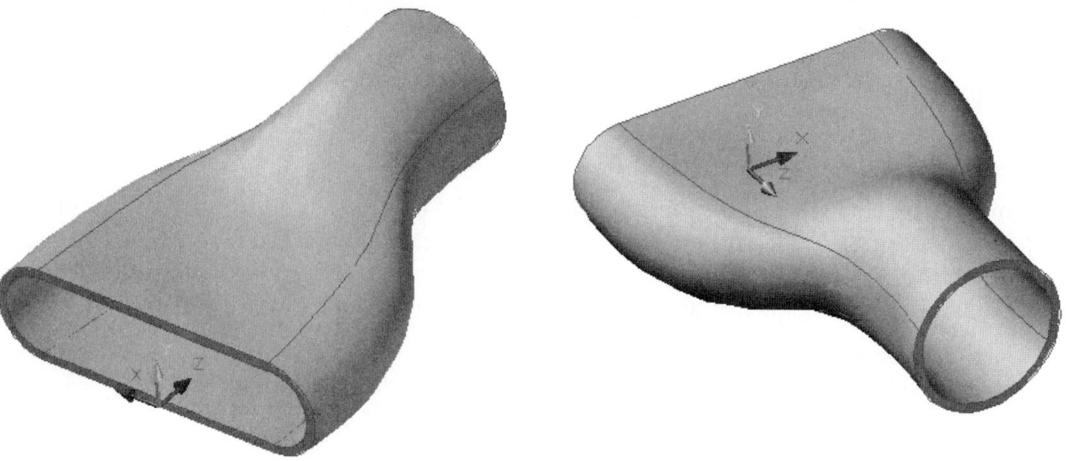

Figure 20.106

PROBLEMS FOR CHAPTER 20

DIRECTIONS FOR PROBLEMS 20–1 THROUGH 20–20

1. *Create a 3D solid model of each object on Layer "Model."*
2. *When finished, calculate the volume of the solid model using the* MASSPROP *command.*

PROBLEM 20–1

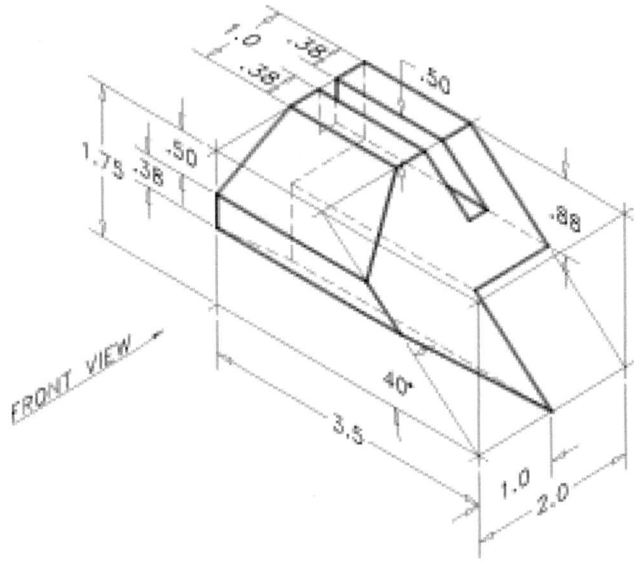

PROBLEM 20–2

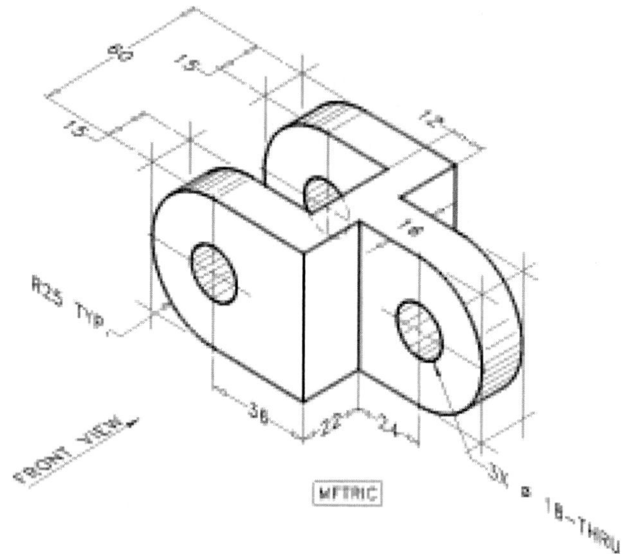

PROBLEM 20–3

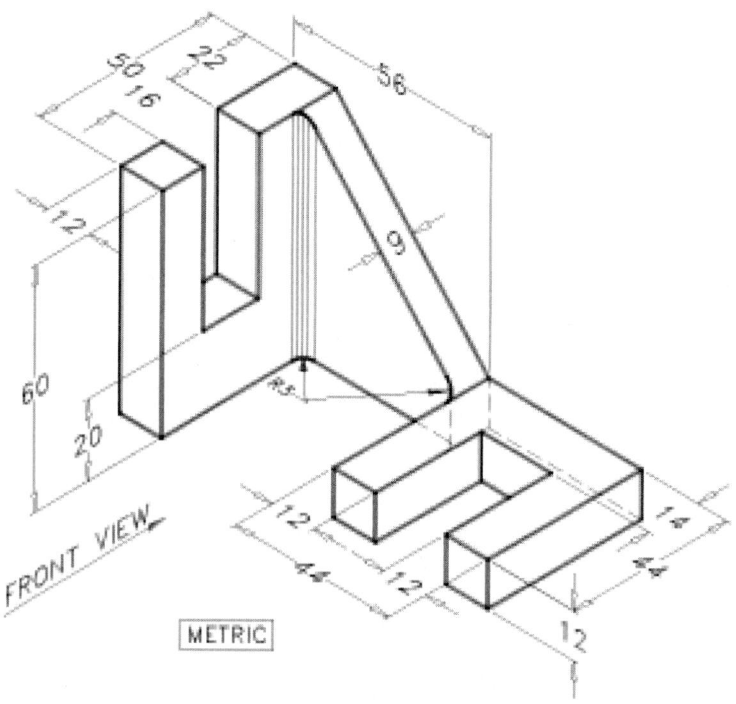

PROBLEM 20–4

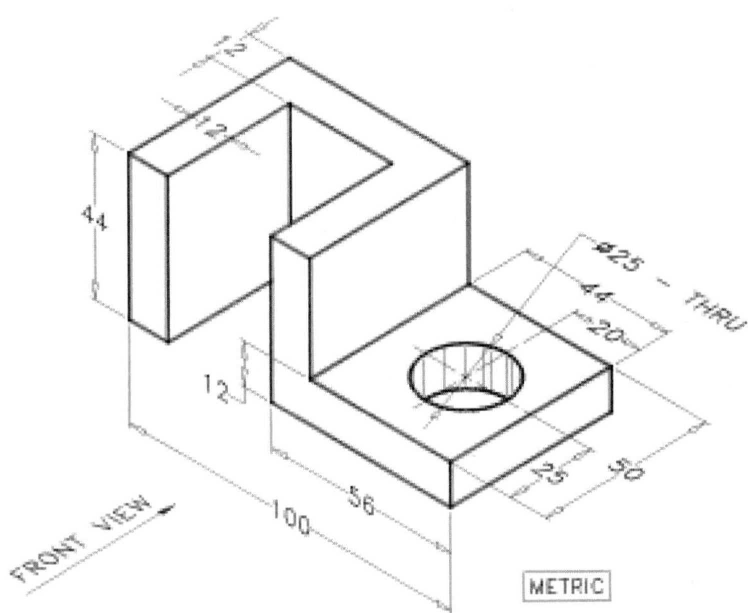

PROBLEM 20–5

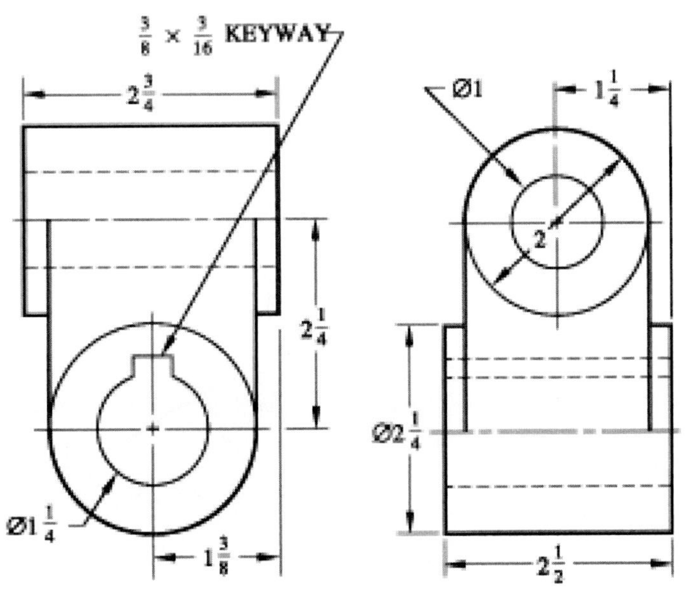

PROBLEM 20–6

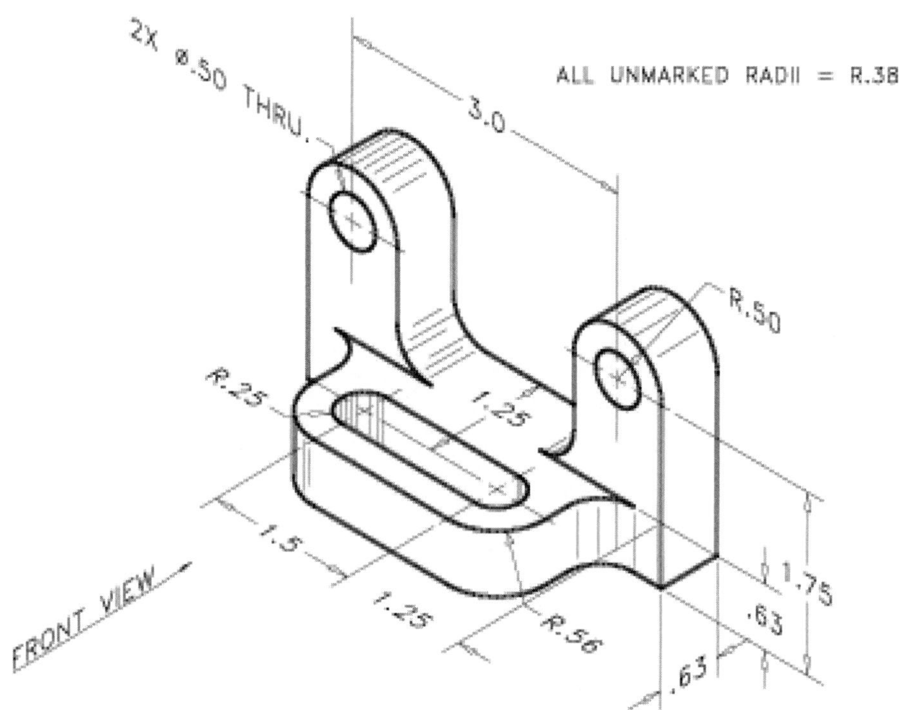

PROBLEM 20–7

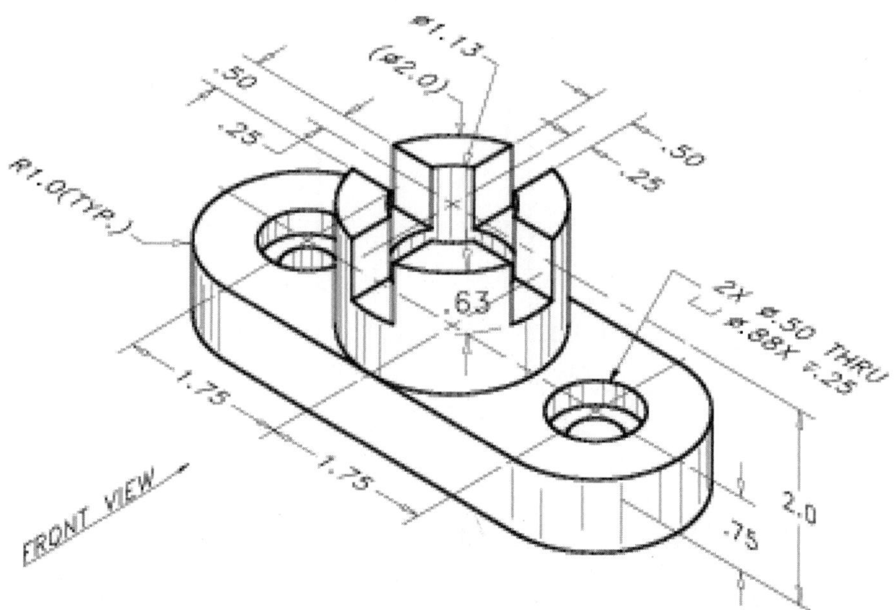

PROBLEM 20–8

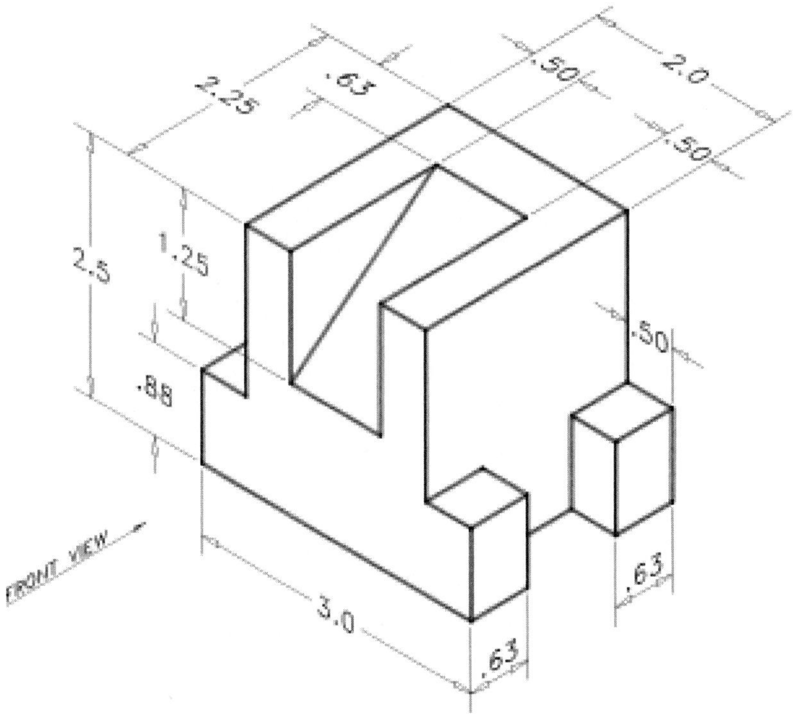

PROBLEM 20–9

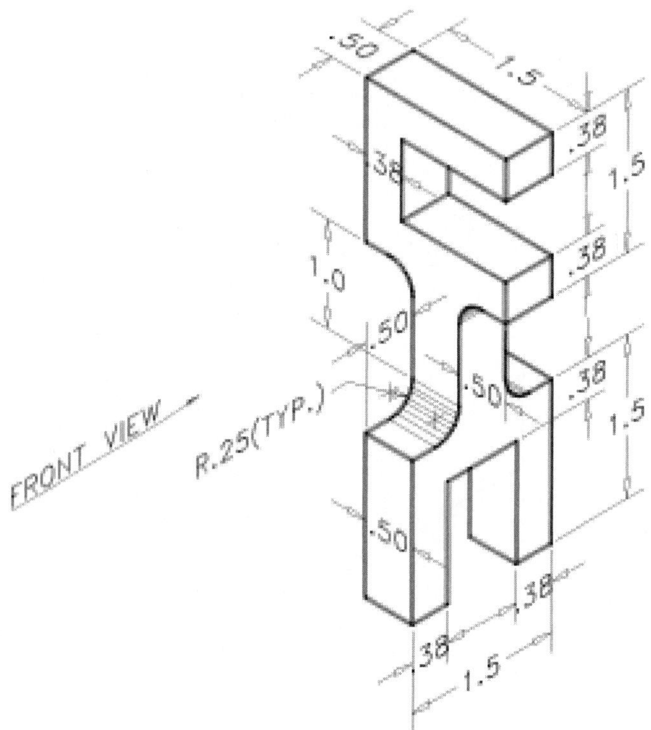

PROBLEM 20–10

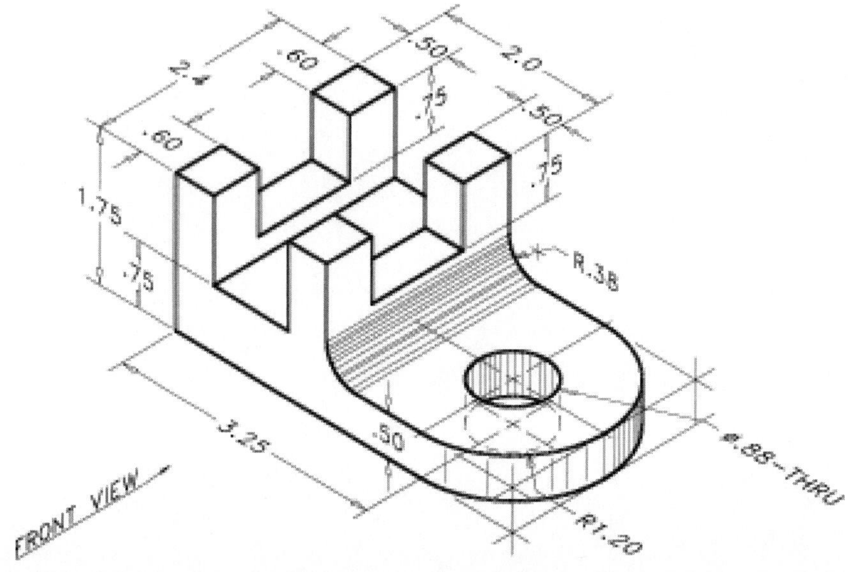

PROBLEM 20–11

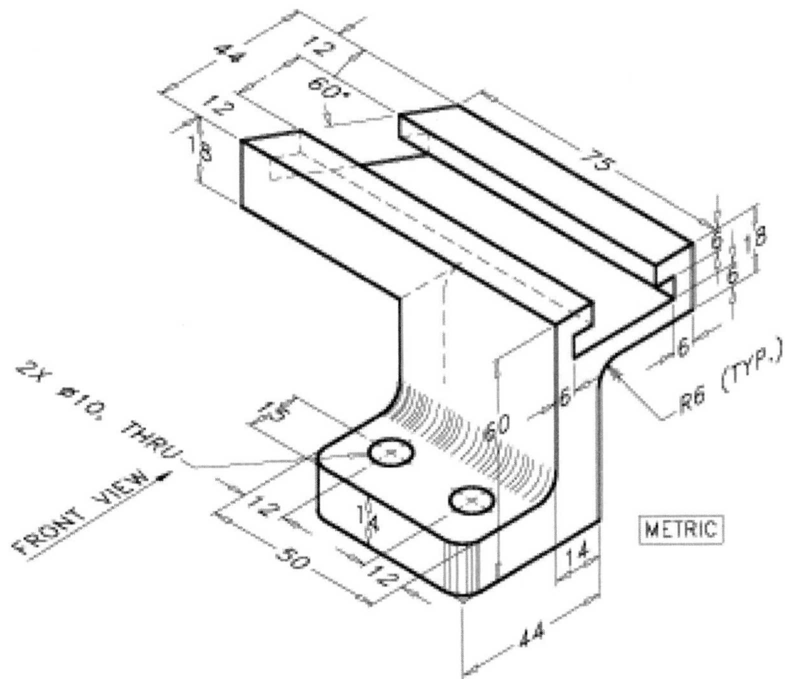

PROBLEM 20–12

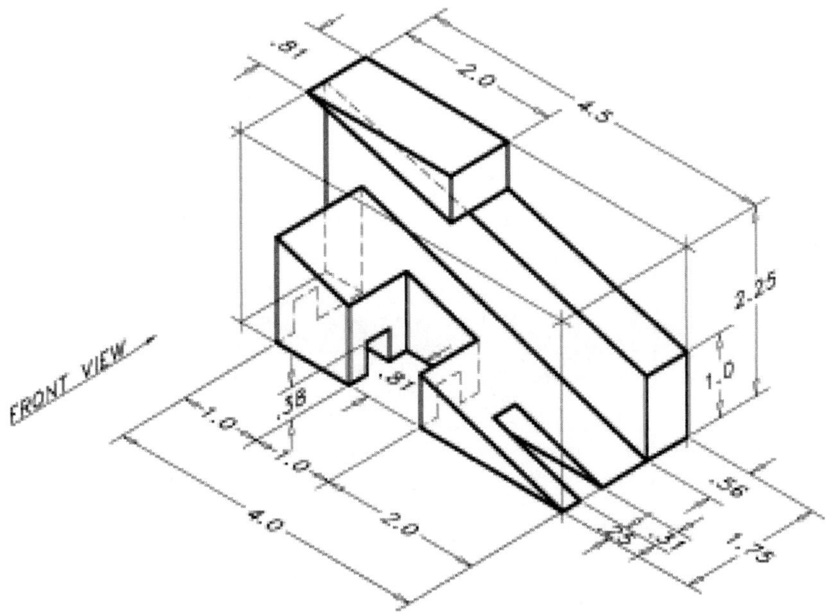

PROBLEM 20–13

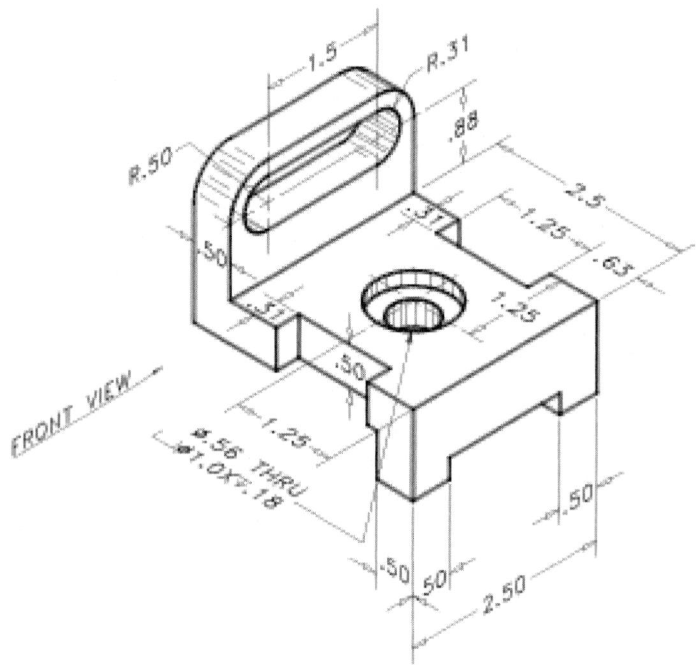

PROBLEM 20–14

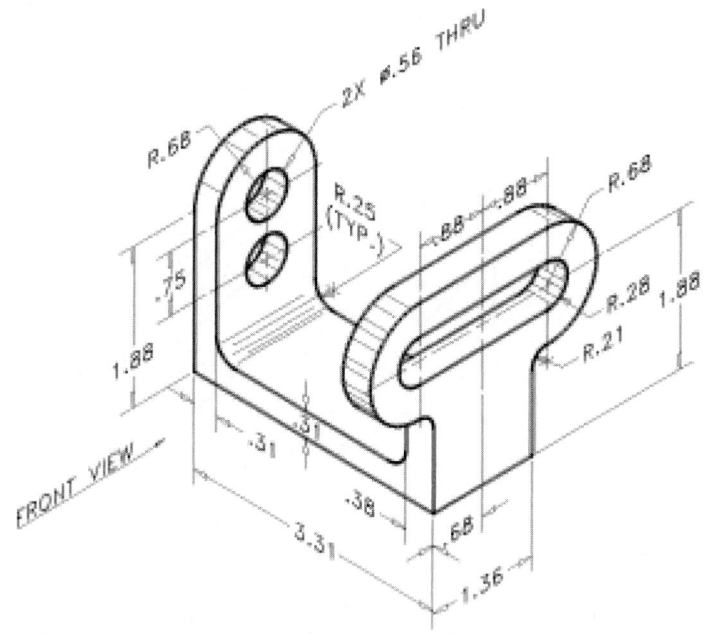

PROBLEM 20–15

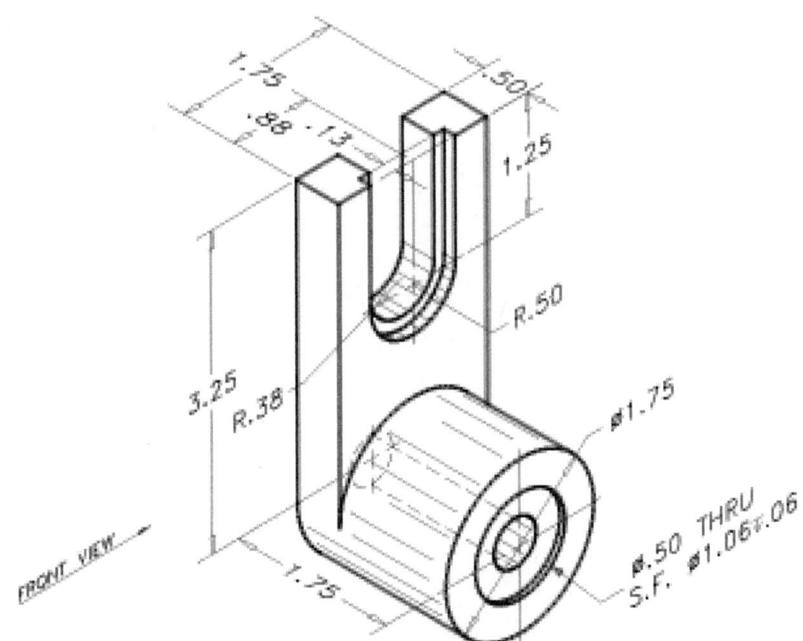

PROBLEM 20–16

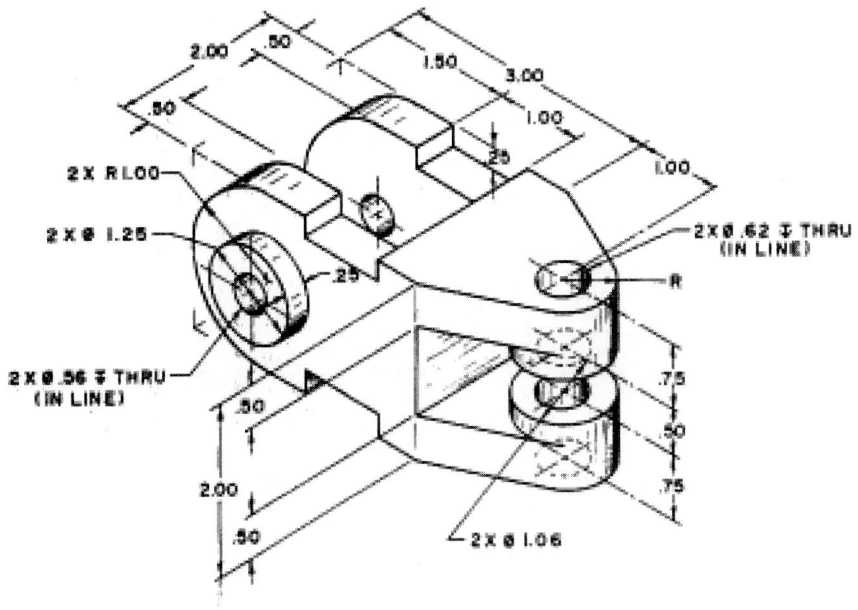

PROBLEM 20–17

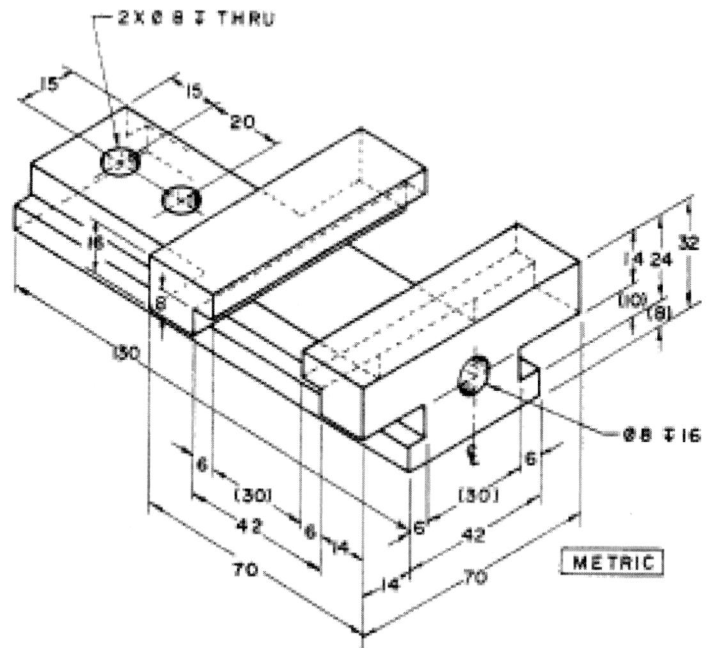

PROBLEM 20–18

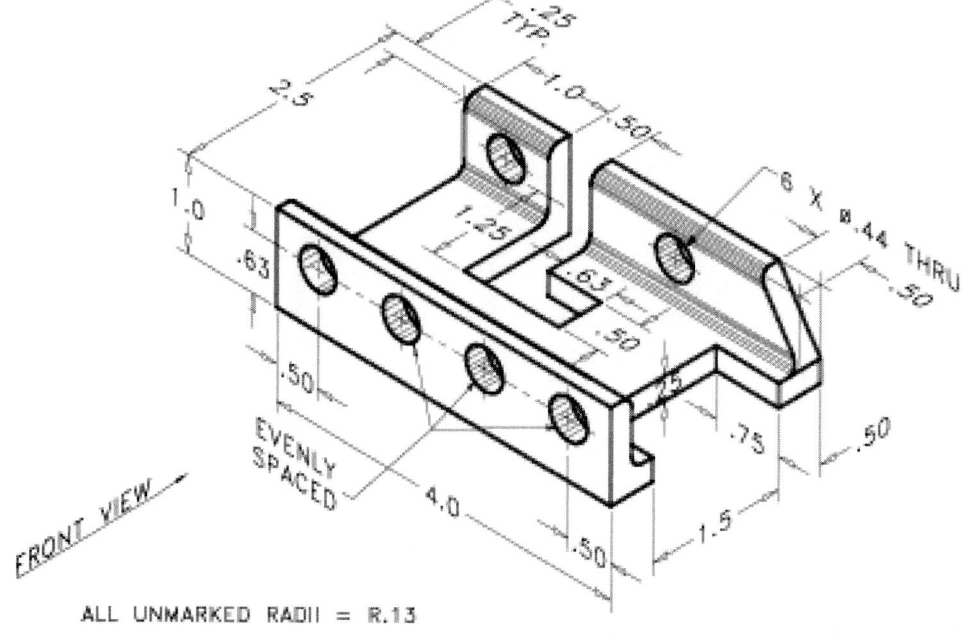

PROBLEM 20–19

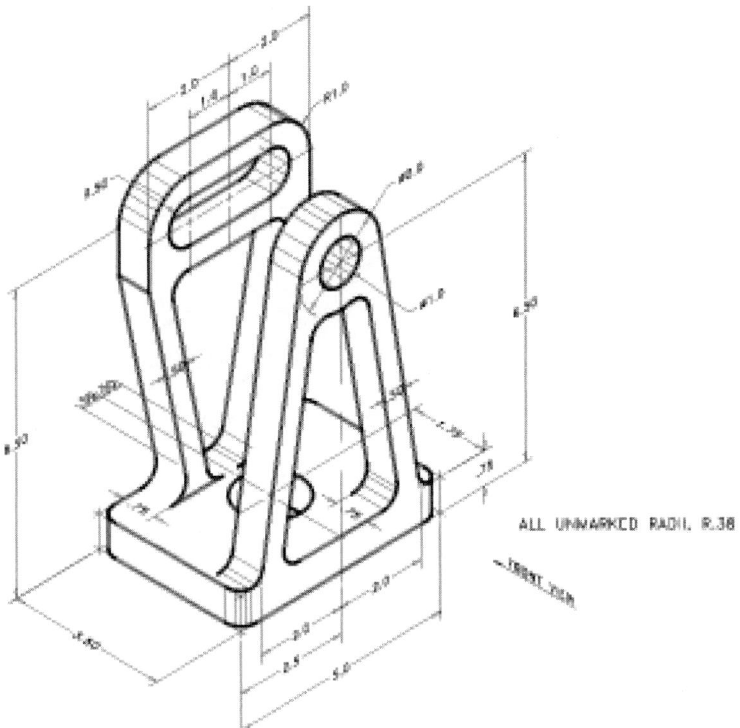

PROBLEM 20–20

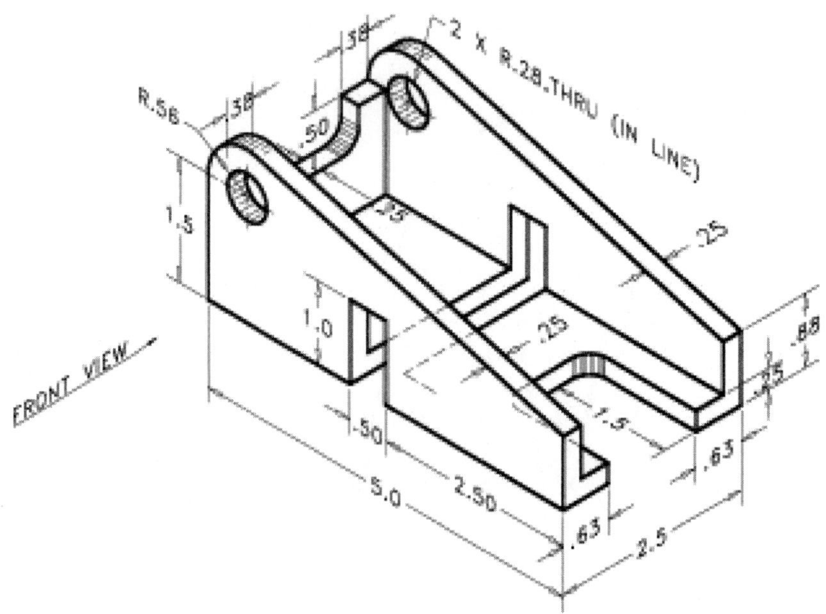

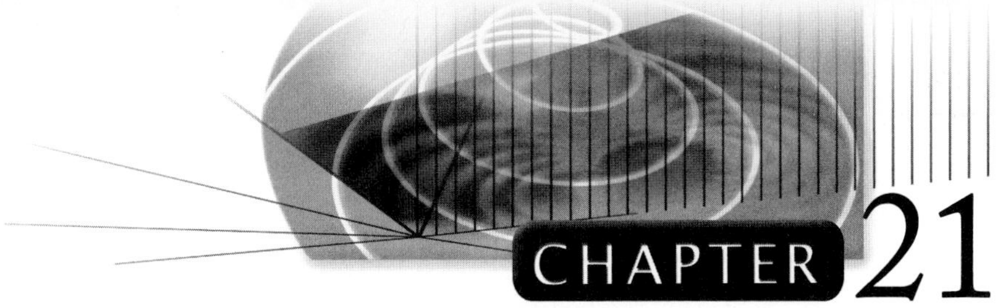

Concept Modeling and Editing Solids

This chapter begins with the study of creating concept models using a press-and-pull drag method. The use of subshapes for editing solid models is discussed, as is the use of grips for editing solids. Once a solid model is created, various methods are available to edit the model. The methods discussed in this chapter include the following 3D operation commands: (3DMOVE, 3DROTATE, 3DALIGN, MIRROR3D, and ARRAY3D). Also included in this chapter is a segment on using the many options of the SOLIDEDIT command. Various 3D models will be opened and used to illustrate these commands and options.

CONCEPTUAL MODELING

Chapter 20 dealt with the basics of creating solid primitives and how to join, subtract, or intersect these shapes to form complex solid models. Sometimes you do not need the regimented procedures outlined in the previous chapter to get your point across about a specific product you have in mind. You can simply create a concept solid model by dragging shapes together to form your idea. Various concept-modeling techniques will be explained in the next segment of this chapter.

DRAGGING BASIC SHAPES

Constructing a 3D model with a concept in mind is easier than ever. Where exact distances are not important, the following image illustrates the construction of a solid block using the BOX command. First view your model in one of the many 3D viewing positions, such as SE (Southeast) Isometric. Next enter the BOX command and pick first and second corner points for the box on the screen, as shown on the left in the following image. When prompted for the height of the box, move your cursor up and notice the box increasing in height, as shown on the right in the following image. Click to locate the height, BOX command. All solid primitives can be constructed using this technique.

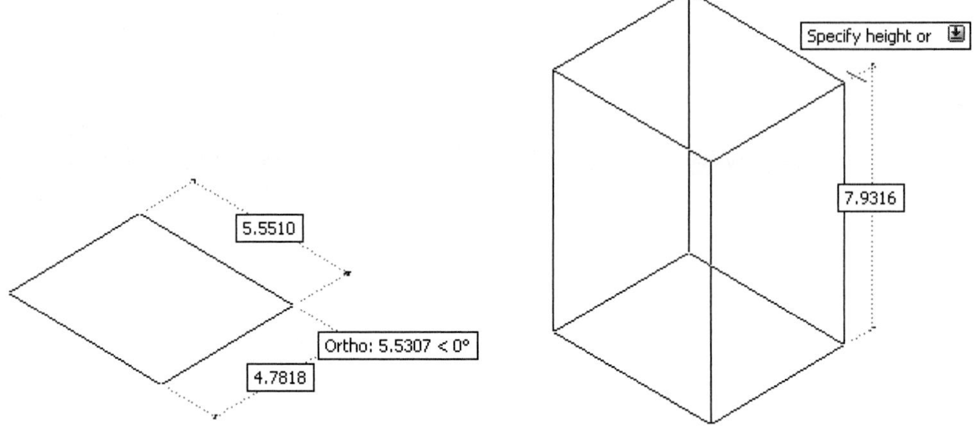

Figure 21.1

Try It!: Open the drawing file 21_Concept_Box. This drawing file does not contain any objects. Also, it is already set up to be viewed in the Southeast Isometric position (SE Iso). Activate the BOX command, pick two points to define the first and second corner points of the rectangular base, and move your cursor up and pick to define the height of the box. You could also experiment using this technique for creating cylinders, pyramids, spheres, cones, wedges, and a torus. When finished, exit this drawing without saving any changes.

USING DYNAMIC UCS TO CONSTRUCT ON FACES

Once a basic shape is created, it is very easy to create a second shape on an existing face with Dynamic UCS (DUCS) mode. Here is how it works. In the following image, a cylinder will be constructed on one of the faces. Turn on DUCS (located in the status bar), activate the CYLINDER command, and when prompted for the base or center point, hover your cursor over the face, as shown on the left in the following image. The face highlights (the edge appears dashed) to indicate that it has been acquired. The Dynamic UCS cursor adjusts itself to this face by aligning the XY plane parallel to the face. When you click a point on this face for the start of the cylinder, the UCS icon changes to reflect this change and the base of the cylinder can be seen, as shown in the middle in the following image. Pick a point to specify the radius. To specify the height of the cylinder, simply drag your cursor away from the face and you will notice the cylinder taking shape, as shown on the right in the following image. Clicking a point defines the height and exit the CYLINDER command.

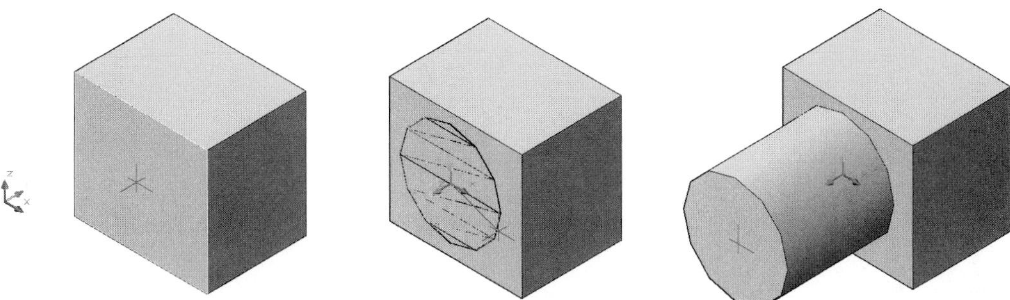

Figure 21.2

Try It!: Open the drawing file 21_Concept_Cylinder. Issue the CYLINDER command and, for the base or center point, hover your cursor over the front face in the previous image and pick a point. The UCS changes to reflect this new position. Move your cursor until you see the circle forming on the face and pick to define the radius of the cylinder. When prompted for the height, move your cursor forward and notice the cylinder taking shape. Pick to define the cylinder height and to exit the command.

When the cylinder is created, it is considered an individual object separate from the solid block, as shown on the left in the following image. After the primitives are constructed, the UNION command is used to join all primitives together as a single solid object, as shown on the right in the following image.

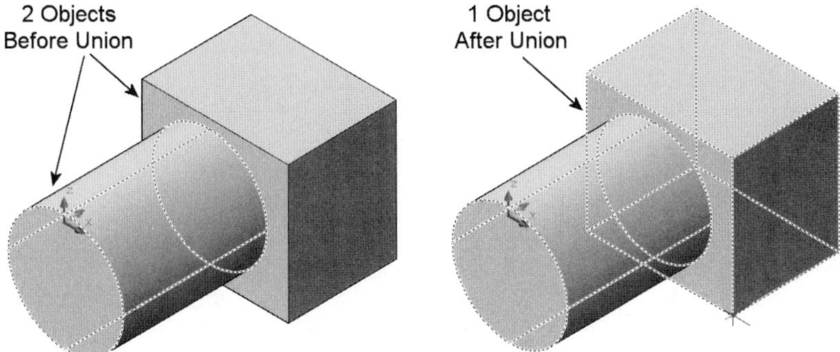

Figure 21.3

USING GRIPS TO MODIFY SOLID MODELS

Whenever a solid primitive is selected when no command is active, grips are displayed, as with all types of objects that make up a drawing. The grips that appear on solid primitives, as shown in the following image, range in shape from squares to arrows. You can perform an edit operation by selecting either the square or arrow shapes. The type of editing that occurs depends on the type of grip selected.

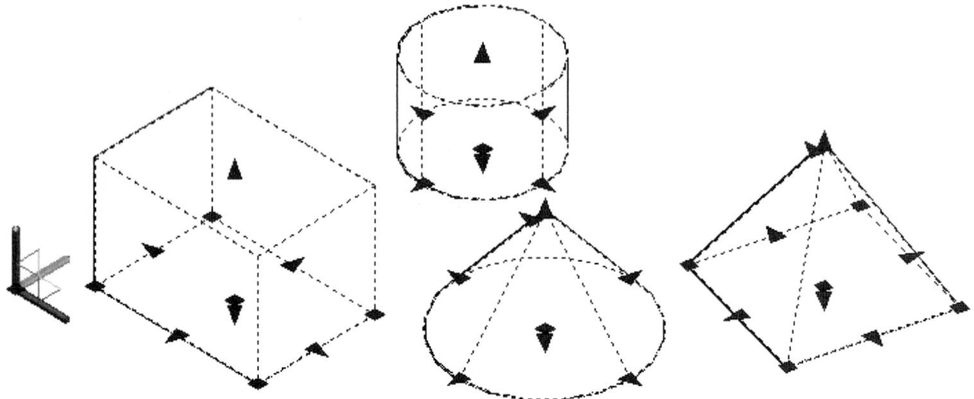

Figure 21.4

KEY GRIP LOCATIONS

The difference between the square and arrow grips is illustrated in the following image of a solid box primitive. The square grip located in the center of a primitive allows you to change the location of the solid. Square grips displayed at the corner (vertex) locations of a primitive allow you to resize the base shape. The arrows located along the edges of the rectangular base allow each individual side to be modified. Arrow grips that point vertically also appear in the middle of the top and bottom faces of the box primitive. These grips allow you to change the height of the primitive.

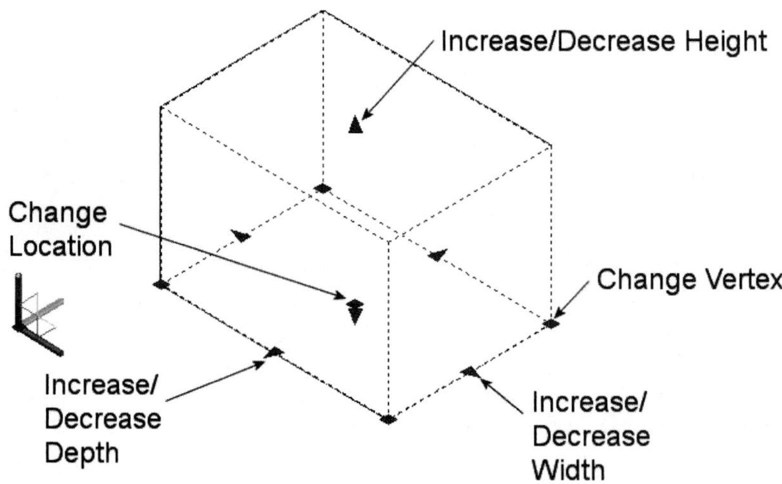

Figure 21.5

GRIP EDITING A CONE

The previous image outlined the various types of grips that display on a box primitive. The grips that appear on cylinders, pyramids, cones, and spheres have similar editing capabilities. The following Try It! exercise illustrates the effects of editing certain arrow grips on a cone.

Try It!: Open the drawing file 21_GripEdit_Cone. Click the cone and the grips appear, as shown on the left in the following image. Click the arrow grip at "A" and stretch the base of the cone in to match the object illustrated on the right in the following image. Next, click the arrow grip at "B" (not the grip that points up) and stretch this grip away from the cone to create a top surface similar to the object illustrated on the right in the following image. Experiment further with grips on the cone by clicking the top grip, which points up, and stretch the cone (now referred to as a frustrum) up.

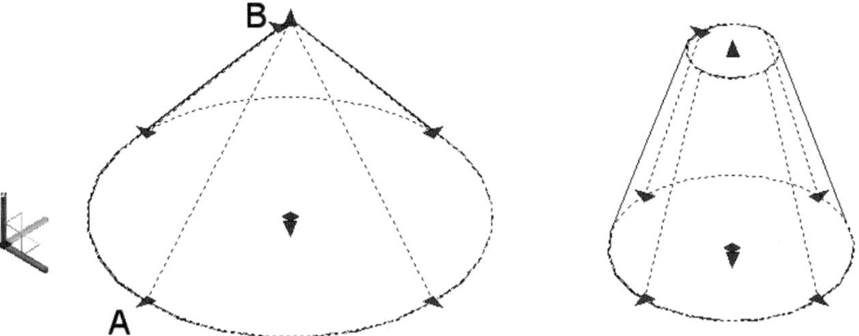

Figure 21.6

EDITING WITH GRIPS AND DYNAMIC INPUT

You have already seen how easy it is to select grips that belong to a solid primitive and stretch the grips to change the shape of the primitive. This next Try It! exercise deals with grip editing of primitives that make up a solid model in an accurate manner. To accomplish this, Dynamic Input must be turned on.

Try It!: Open the drawing file 21_GripEdit_Shape01. This 3D model consists of three separate primitives, namely, two boxes and one cylinder. The height of the cylinder needs to be lowered. To accomplish this, click the cylinder and observe the positions of the grips and numeric values, as shown on the left in the following image. The total height of 6.0000 needs to be changed to 3.50 units. Click the arrow grip at the top of the cylinder that is pointing up, press the Tab key to highlight the overall height value of 6.0000, and change this value to 3.50. Pressing ENTER to accept this new value lowers the cylinder, as shown on the right in the following image. Feel free to experiment with both boxes by either increasing or decreasing their heights using this grip-editing method.

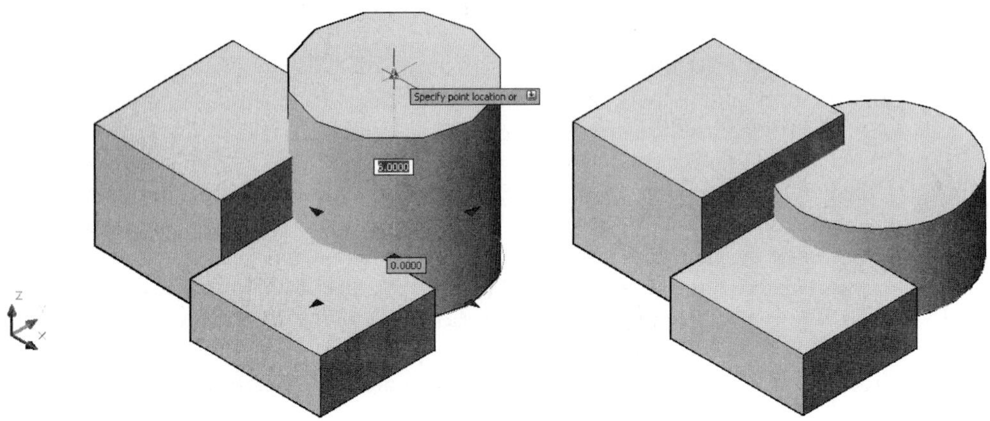

Figure 21.7

MANIPULATING SUBOBJECTS

A subobject is any part of a solid. It could be a face, an edge, or a vertex (corner). The following image illustrates a 3D box and pyramid. Pressing the CTRL key allows you to select a subobject. Notice in this illustration that each subobject selected has a grip associated with it. If you accidentally select a subobject, press CTRL + SHIFT and pick it again to deselect it.

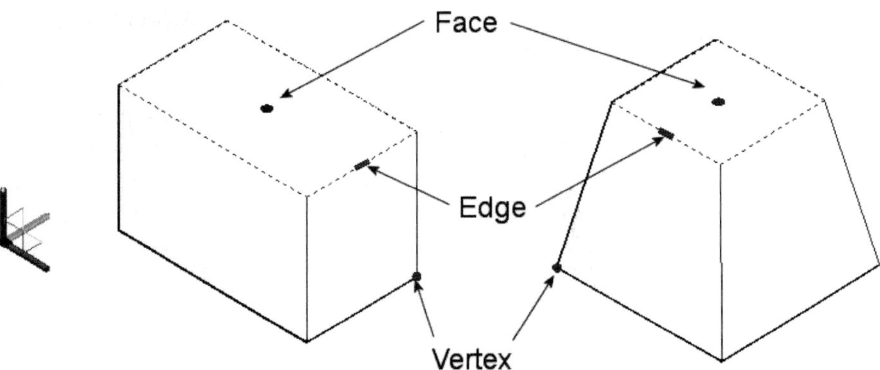

Figure 21.8

Once a subobject is selected, click the grip to activate the grip Stretch, Move, Rotate, Scale, and Mirror modes. You can drag your cursor to a new location or enter a direct distance value from the keyboard. In the following image, the edge of each solid object was selected as the subobject and dragged to a new location.

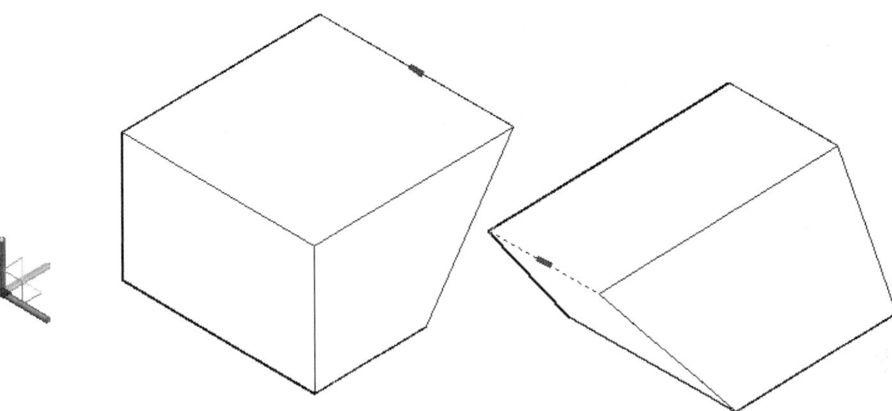

Figure 21.9

EDITING SUBOBJECTS OF A SOLID MODEL

You have seen how easy it is to isolate a subobject of a solid model by pressing the CTRL key while selecting the subobject. The same technique can be used to isolate a primitive that is already consumed or made part of a solid model. This would allow you to edit a specific primitive while leaving other primitives of the solid model unselected. The next Try It! exercise illustrates this technique.

Try It!: Open the drawing file 21_GripEdit_Shape02. This drawing represents three separate primitives joined into one solid model using the UNION command. The height of the front block needs to be increased. First, verify that ORTHO is turned on to assist in this operation. Then, press and hold down the CTRL key while clicking the front block. Notice that only this block highlights and displays various grips, as shown on the left in the following image. The height of this block needs to be increased by 3.00 units. Click the arrow grip at the top of the block that is pointing up, and drag the geometry up until the tool tip reads approximately 3.00 units, as shown in the middle in the following image. Type a value of 3.00 and press ENTER to increase the block, as shown on the right in the following image. Feel free to experiment with the other block and cylinder by either increasing or decreasing their heights using this grip-editing method.

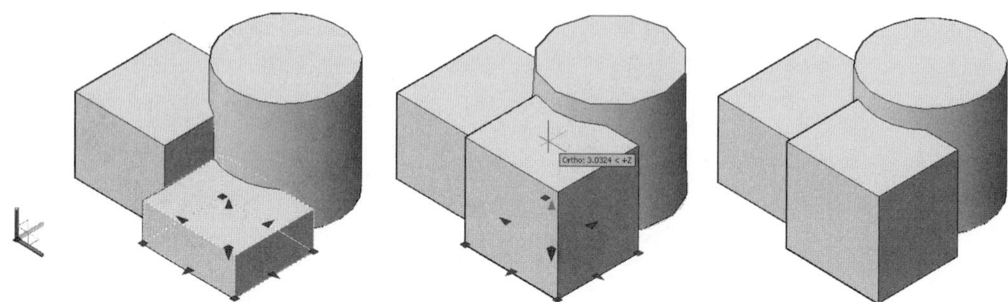

Figure 21.10

ADDING EDGES AND FACES TO A SOLID MODEL

When building solid models, use the UNION, SUBTRACT, and INTERSECT commands as the primary means of joining, removing, or creating a solid that is common to two intersecting primitives. You can also imprint regular objects such as lines, circles, or polylines—for example, the line segment shown on the left in the following image—directly onto the face of a solid model. This line, once imprinted, becomes part of the solid and in our case divides the top face into two faces. Then, using the subobject technique, we can change the shape of the solid model. The command used to perform this operation is IMPRINT, which can be found in the Ribbon or the Solids Editing toolbar, as shown on the right in the following image.

```
Command: IMPRINT
Select a 3D solid: (Select the 3D solid model)
Select an object to imprint: (Pick the line segment constructed
    across the top surface of the solid model)
Delete the source object [Yes/No] <N>: Y (For Yes)
Select an object to imprint: (Press ENTER to perform this
    operation and exit the command)
```

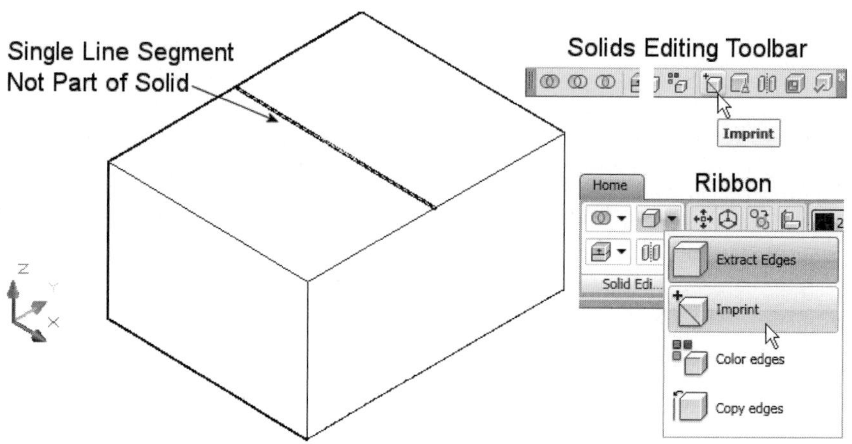

Figure 21.11

After the imprint operation is performed on the line segment, this object becomes part of the solid model. When the imprinted line is selected as a subobject, the grip located on the line can be selected and dragged up or down, as shown in the following image. The results can dramatically change the shape of the solid model.

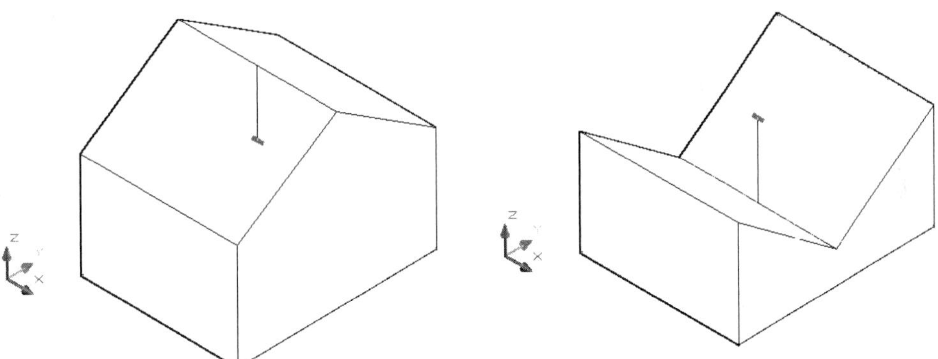

Figure 21.12

Try It!: Open the drawing file 21_Imprint_Roof. First, verify that ORTHO is turned on to assist in this operation. Issue the IMPRINT command, select the solid block, as shown on the left in the following image, and pick the single line segment as the object to imprint. Next, press and hold down the CTRL key while clicking the imprinted line. Notice that only this line highlights and displays a grip at its midpoint, as shown in the middle of the following image. Pick this grip, slowly move your cursor up, and notice the creation of the roof peak. You could also move your cursor down to create a V-shaped object.

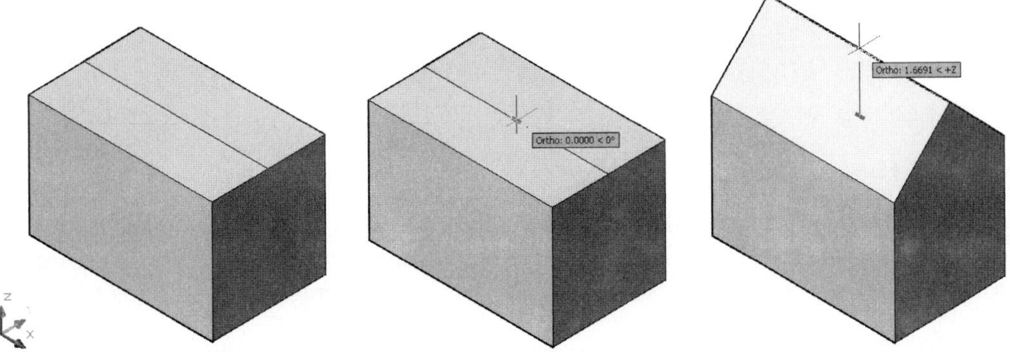

Figure 21.13

PRESSING AND PULLING BOUNDING AREAS

An additional technique used for constructing solid models is available to speed up the construction and modification processes. The technique is called pressing and pulling, which can be accessed by picking the PRESSPULL command from the Ribbon or the Modeling toolbar, as shown in the following image. Any closed area that can be hatched can be manipulated using the PRESSPULL command.

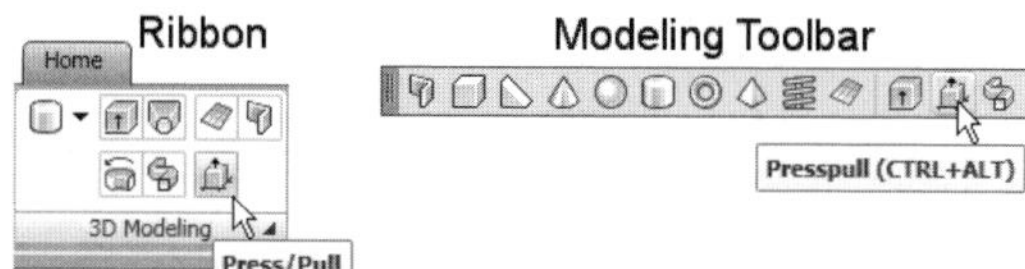

Figure 21.14

Try It!: Open the drawing file 21_PressPull_Shapes. Activate the PRESSPULL command from the Dashboard, pick inside of the circular shape, as shown on the left in the following image, move your cursor up, and enter a value of 6 units. Activate the PRESSPULL command again, pick inside of the closed block shape, as shown in the middle in the following image, move your cursor up, and enter a value of 4 units. Use the PRESSPULL command one more time on the remaining closed block shape, as shown on the right in the following image, move your cursor up, and enter a value of 2 units. Performing this task results in the creation of three separate solid shapes. Use the UNION command to join all shapes into one 3D solid model.

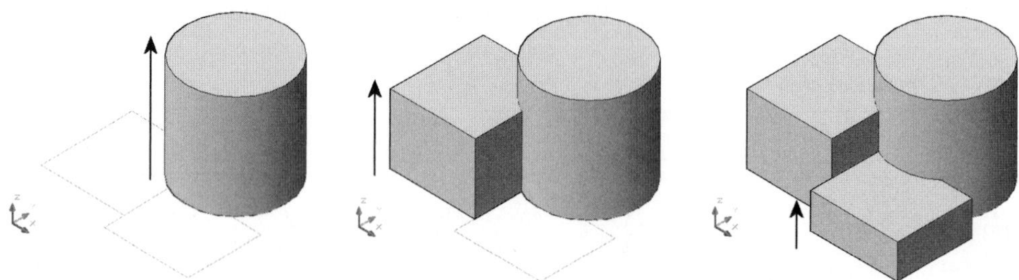

Figure 21.15

Another technique to activate the press-and-pull feature is to press and hold the CTRL + ALT keys, and then pick the area. You can perform the operation the same way as using the PRESSPULL command from the dashboard. The next Try It! exercise illustrates this technique.

Try It!: Open the drawing file 21_PressPull_Hole. Press and hold down the CTRL+ALT keys and pick inside of the circular shape, as shown on the left in the following image. Moving your cursor up creates the cylinder, as shown on the right in the following image.

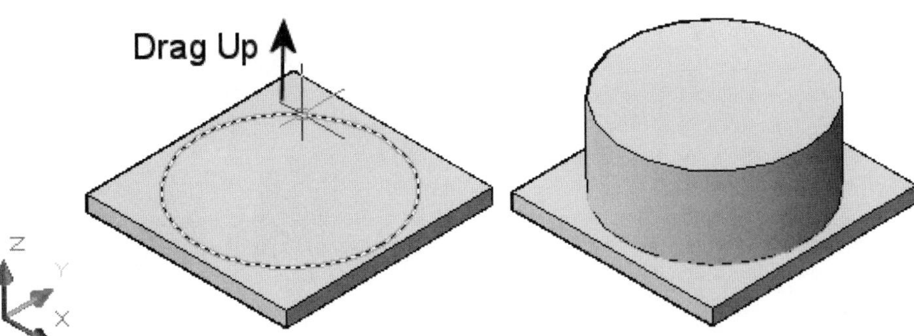

Figure 21.16

Undo the previous operation, activate press and pull by pressing and holding down the CTRL + ALT keys, and pick inside of the circular shape, as shown on the left in the following image. However, instead of moving your cursor up to form a cylinder, drag your cursor down into the thin block to perform a subtraction operation and create a hole in the block.

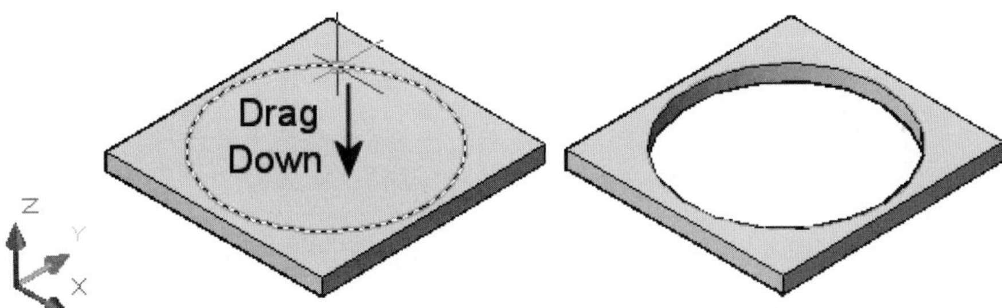

Figure 21.17

Try It!: Open the drawing file 21_PressPull Plan. Press and hold down the CTRL + ALT keys while picking inside of the bounding area created by the inner and outer lines, as shown on the left in the following image. Move your cursor up and enter a value of 8' to construct a solid model of the walls, as shown on the right in the following image.

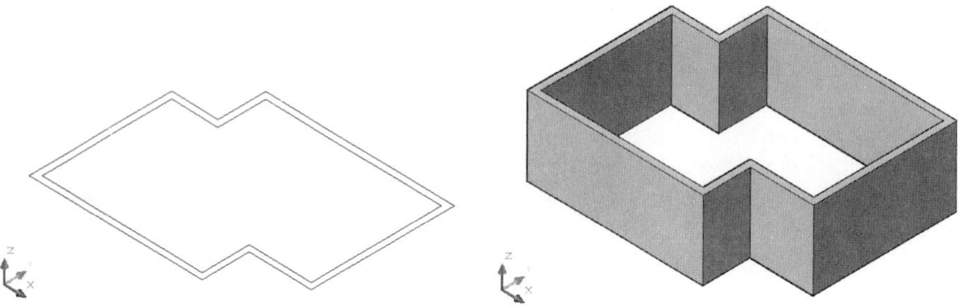

Figure 21.18

Try It!: Open the drawing file 21_PressPull Openings. Press and hold down the CTRL + ALT keys while picking inside of one of the rectangles that signify a door or window opening. Move your cursor into the wall and pick inside of the model to create the opening. Use this technique for the other openings as well.

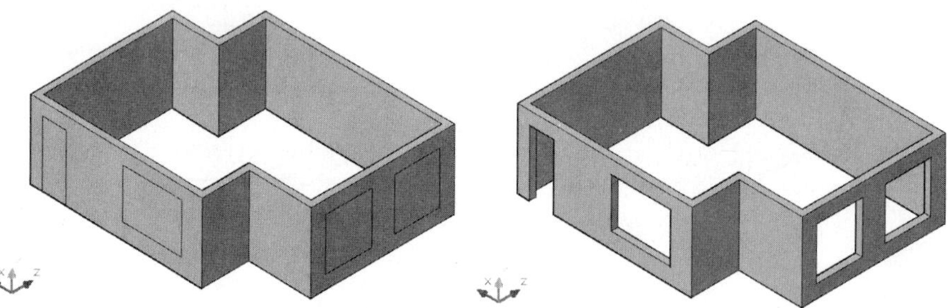

Figure 21.19

USING PRESS AND PULL ON BLOCKS

Pressing and pulling to create solid shapes is not limited to just closed shapes such as rectangles, circles, or polylines. The press-and-pull feature can also be used on block objects. In some cases, depending on how you use the press-and-pull feature, it is possible to convert the 2D block into a 3D solid model. The next Try It! exercise illustrates this.

Try It!: Open the drawing file 21_PressPull_Bed. All objects that describe this 2D bed are made up of a single block. You will use the PRESSPULL command to highlight certain closed boundary areas and pull the area to a new height. In this way, PRESSPULL is used to convert a 2D object into a 3D model.

Activate the PRESSPULL command, move your cursor inside the area, as shown on the left in the following image, and press and hold down the left mouse button as you move your cursor up. When you let go of the mouse button, type a value of 12 to extrude this area a distance of 12 units up, as shown on the right in the following image.

```
Command: PRESSPULL
Click inside bounded areas to press or pull. (Move your cursor
    over the area, as shown on the left in the following image.
    Press and hold down the mouse button and pull in an upward
    direction. Let go of the mouse button, enter a value of 12
    in the designated field, and press ENTER when finished)
1 loop extracted.
1 Region created.
12 (Value entered)
```

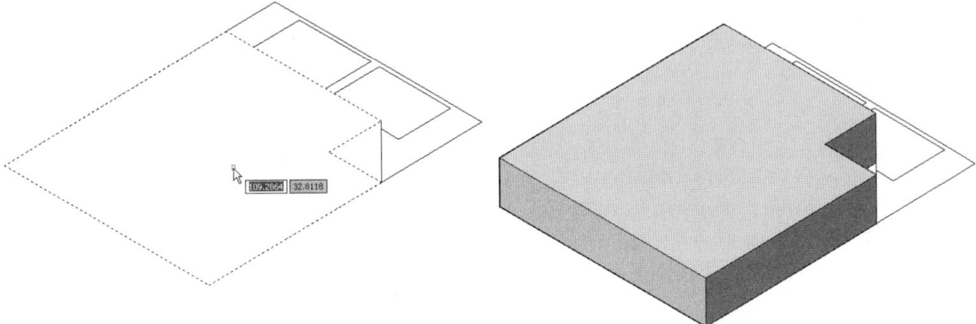

Figure 21.20

Activate the PRESSPULL command, move your cursor inside the small triangular area, as shown on the left in the following image, and press and hold down the left mouse button as you move your cursor up. When you let go of the mouse button, type a value of 12.5 to extrude this area a distance of 12.5 units up, as shown on the right in the following image.

```
Command: PRESSPULL
Click inside bounded areas to press or pull. (Move your cursor
    over the triangular, area, as shown on the left in the
    following image. Press and hold down the mouse button and
    pull in an upward direction. Let go of the mouse button,
    enter a value of 12.5 in the designated field, and press ENTER
    when finished)
1 loop extracted.
1 Region created.
12.5 (Value entered)
```

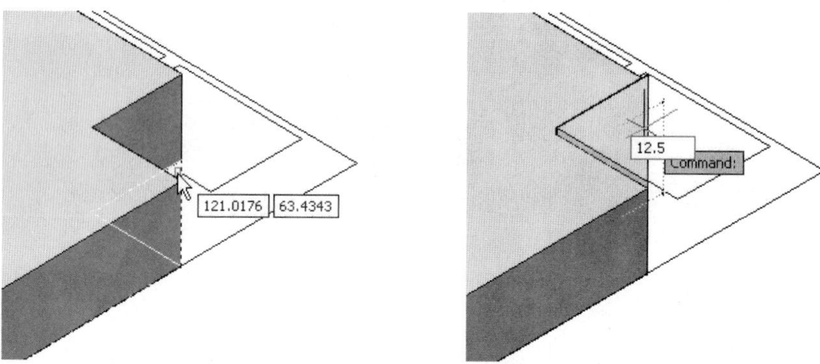

Figure 21.21

Activate the PRESSPULL command, move your cursor inside the area, as shown on the left in the following image, and press and hold down the left mouse button as you move your cursor up. When you let go of the mouse button, type a value of 12 to extrude this area a distance of 12 units up, as shown on the right in the following image.

```
Command: PRESSPULL
Click inside bounded areas to press or pull. (Move your cursor
    over the back area of the bed, as shown on the left in the
    following image. Press and hold down the mouse button and
    pull in an upward direction. Let go of the mouse button,
    enter a value of 12 in the designated field, and press ENTER
    when finished)
1 loop extracted.
1 Region created.
12 (Value entered)
```

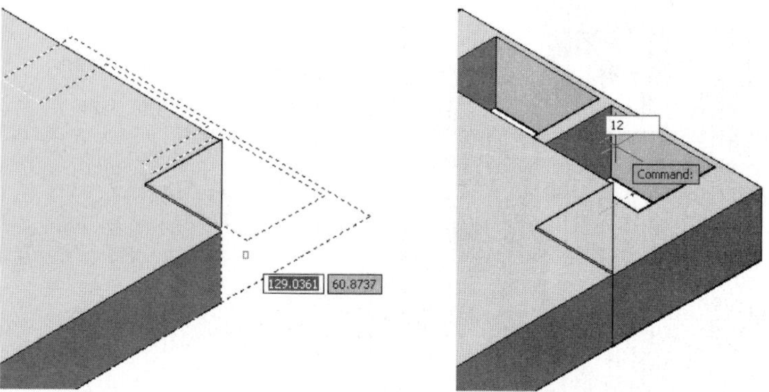

Figure 21.22

Activate the PRESSPULL command, move your cursor inside the rectangular area represented as a pillow, as shown on the left in the following image, and press and hold down the left mouse button as you move your cursor up. When you let go of the mouse button, type a value of 15 to extrude this area a distance of 15 units up, as shown on the right in the following image.

```
Command: PRESSPULL
Click inside bounded areas to press or pull. (Move your cursor
    over the area representing the pillow, as shown on the left
    in the following image. Press and hold down the mouse button
    and pull in an upward direction. Let go of the mouse button,
    enter a value of 15 in the designated field, and press ENTER
    when finished)
1 loop extracted.
1 Region created.
15 (Value entered)
```

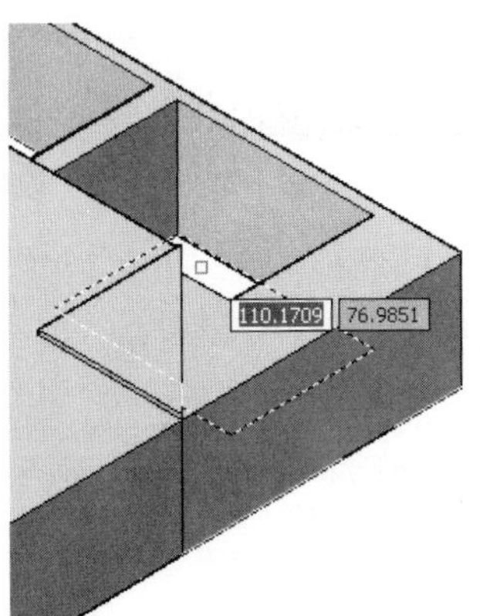

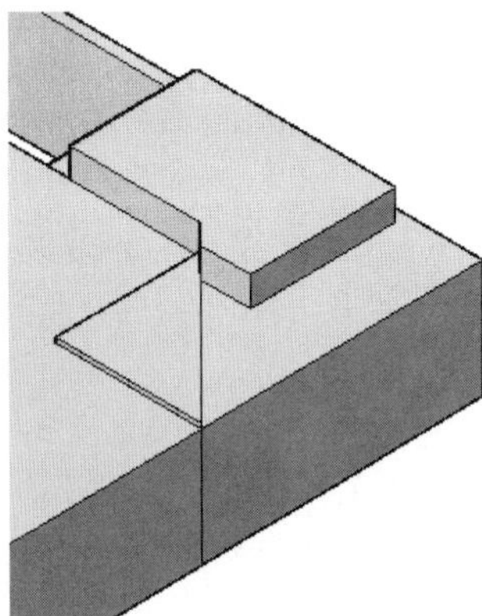

Figure 21.23

Perform the same press-and-pull operation on the second pillow. Extrude this shape a value of 15 units, as shown on the left in the following image. Since the 3D objects created by the press-and-pull operations are all considered individual primitives, use the UNION command to join all primitives into a single 3D solid model. An alternate step would be to use the FILLET command to round off all corners and edges of the bed and pillows, as shown on the right in the following image.

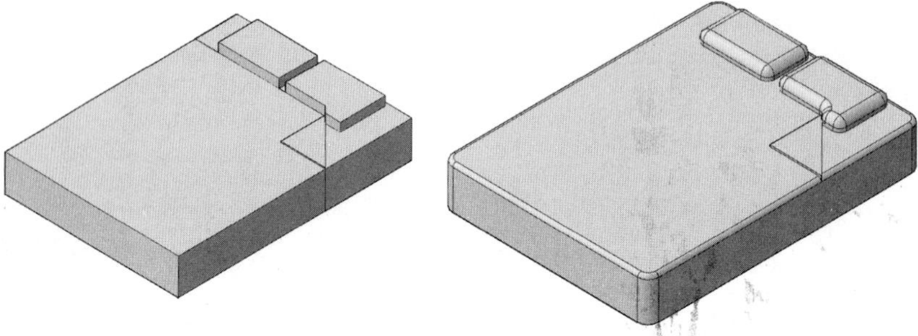

Figure 21.24

ADDITIONAL METHODS FOR EDITING SOLID MODELS

Additional methods are available to edit the solid model. These methods include the 3D Operation commands (3DMOVE, 3DALIGN, 3DROTATE, MIRROR3D, and 3DARRAY) and the SOLIDEDIT command. Various 3D drawings will be opened and used to illustrate these commands.

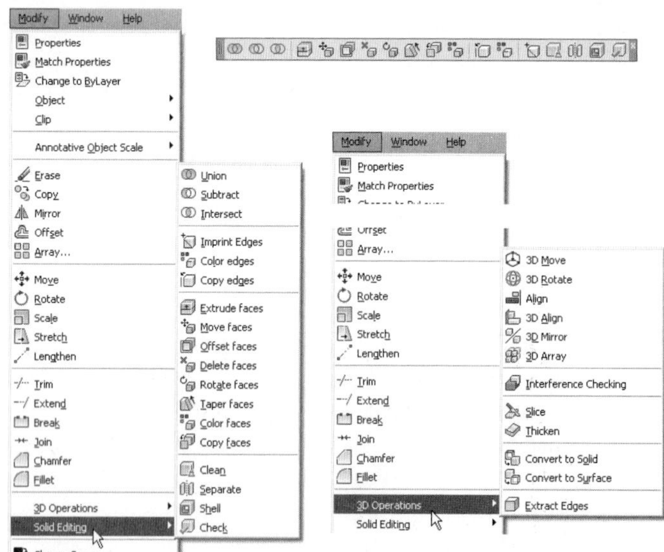

Figure 21.25

USING THE 3DMOVE COMMAND

To assist in the positioning of objects, the 3DMOVE tool is available. This tool displays the move grip tool, which displays an axis for the purpose of moving objects a specified direction and distance. Select this command either from the Ribbon, as shown on the left in the following image, or from the Modify pull-down menu, as shown on the right in the following image.

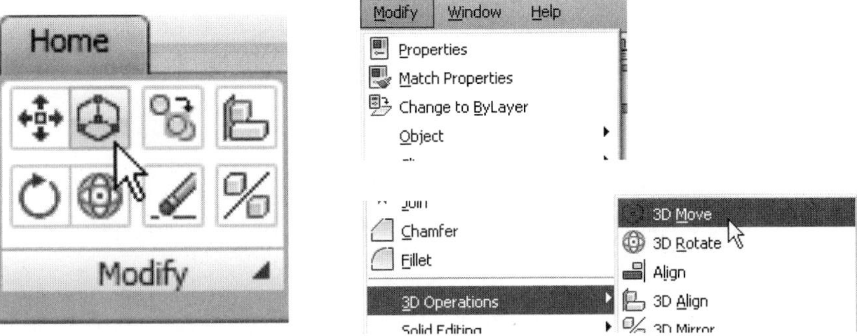

Figure 21.26

When using this command, you can select either objects or subobjects to move. To select subobjects, press and hold down the CTRL key as you select. After you press ENTER to signify that you are done with the selection process, the move grip tool displays attached to your cursor, as shown in the following image.

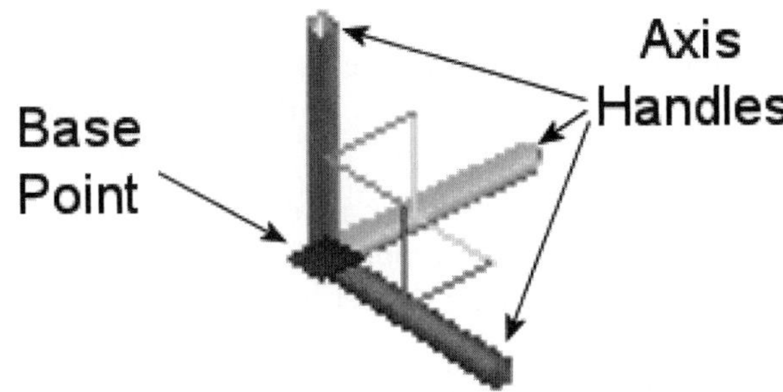

Figure 21.27

Pick any convenient point to utilize as the base point for the move. Object snap modes can be used to accomplish this task if desired. You then move your cursor over one of the three axis handles to define the direction of the move. As you hover over one of the handles, it turns yellow and a direction vector displays. Click this axis handle to lock in the direction vector. Then enter a value to move the solid model or move your cursor along the direction vector, as shown in the following image, and pick a location to move.

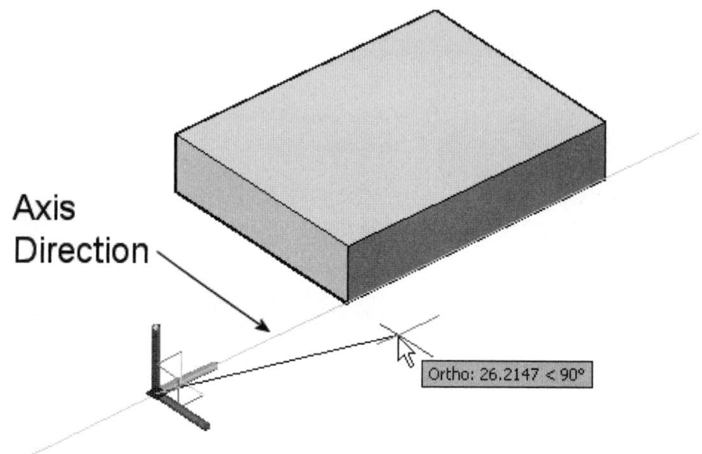

Figure 21.28

THE 3DALIGN COMMAND

Use the 3DALIGN command to specify up to three points to define the source plane of one 3D solid model, followed by up to three points to define the destination plane where the first solid model will be moved or aligned to.

When you specify points, the first source point is referred to as the base point. This point is always moved to the first destination point. Selecting second and third source or destination points results in the 3D solid model being rotated into position.

Try It!: Open the drawing file 21_3DAlign. The objects on the right in the following image need to be positioned or aligned to form the assembled object shown in the small isometric view in this image. At this point, it is unclear at what angle the objects are currently rotated. When you use the 3DALIGN command, it is not necessary to know this information. Rather, you line up source points with destination points. When the three sets of points are identified, the object moves and rotates into position. The first source point acts as a base point for a move operation. The first destination point acts as a base point for rotation operations. The second and third sets of source and destination points establish the direction and amount of rotation required to align the objects.

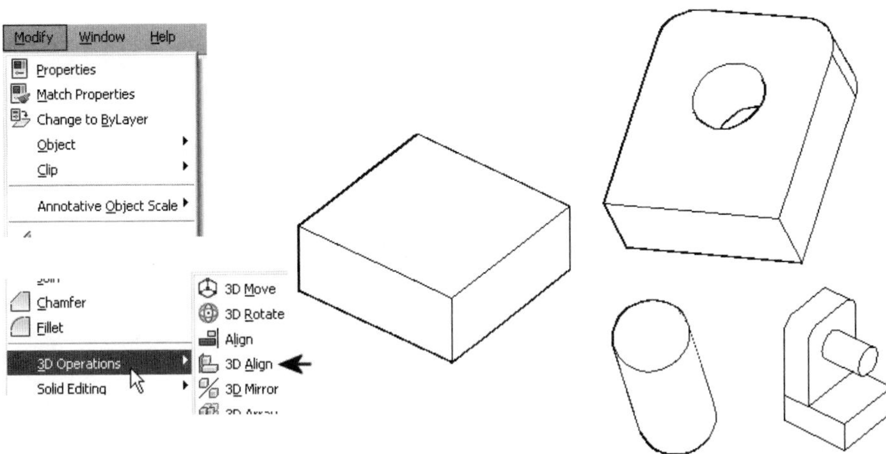

Figure 21.29

Follow the prompt sequence below and the illustration in the following image for aligning the hole plate with the bottom base. The first destination point acts as a base point to which the cylinder locates.

```
Command: 3DALIGN
Select objects: (Select the object with the hole)
Select objects: (Press ENTER to continue)
Specify source plane and orientation ...
Specify base point or [Copy]: (Select the endpoint at "A")
Specify second point or [Continue] <C>: (Select the endpoint
    at "B")
Specify third point or [Continue] <C>: (Select the endpoint
    at "C")
Specify destination plane and orientation ...
Specify first destination point: (Select the endpoint at "D")
Specify second destination point or [eXit] <X>: (Select the
    endpoint at "E")
Specify third destination point or [eXit] <X>: (Select the
    endpoint at "F")
```

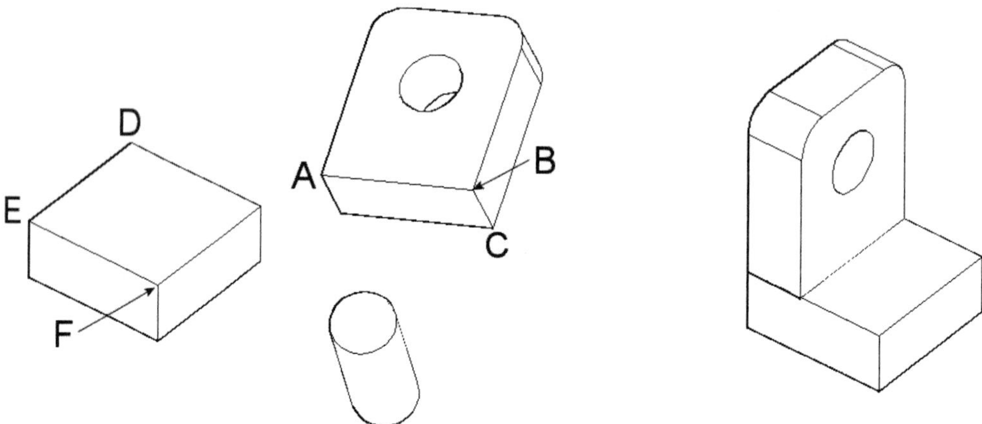

Figure 21.30

The results of the previous step are illustrated on the left in the following image. Next, align the cylinder with the hole. Circular shapes often need only two sets of source and destination points for the shapes to be properly aligned.

```
Command: 3DALIGN
Select objects: 1 found
Select objects:
Specify source plane and orientation ...
Specify base point or [Copy]: (Select the center of circle "A")
Specify second point or [Continue] <C>: (Select the center of
    circle "B")
Specify third point or [Continue] <C>: (Press ENTER to continue)
Specify destination plane and orientation ...
Specify first destination point: (Select the center of
    circle "C")
Specify second destination point or [eXit] <X>: (Select the
    center of circle "D")
Specify third destination point or [eXit] <X>: (Press ENTER
    to exit)
```

The completed 3D model is illustrated on the right in the following image. The ALIGN command provides an easy means of putting solid objects together to form assembly models.

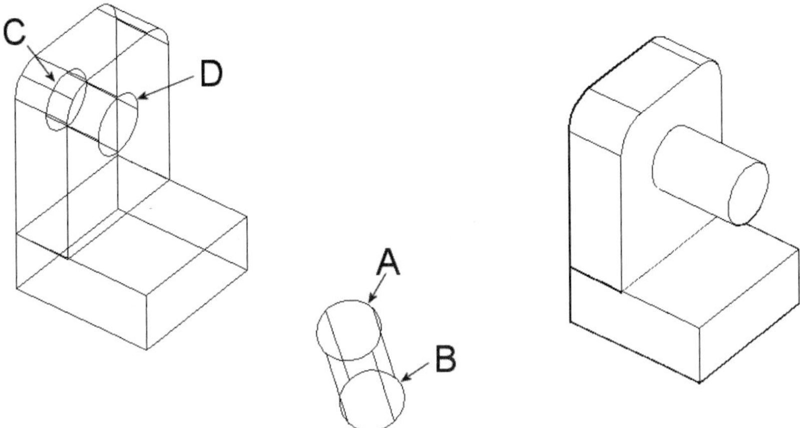

Figure 21.31

THE 3DROTATE COMMAND

The 3DROTATE command uses a special rotate grip tool to rotate objects around a base point. Select this command from either the Ribbon, as shown on the left in the following image, or the Modify pull-down menu, as shown on the left in the following image.

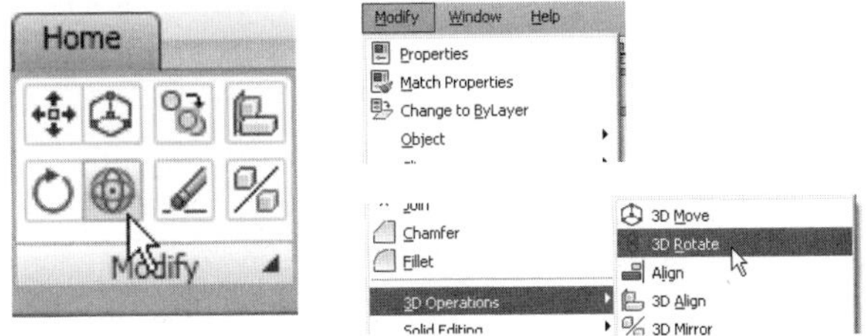

Figure 21.32

After you select the object or objects to rotate, you are prompted to pick a base point, which will act as the pivot point of the rotation. After you pick this base point, the rotate grip tool appears, as shown in the following image. You then hover your cursor over an axis handle until it turns yellow and an axis vector appears. If this axis is correct, click it to establish the axis of rotation. Then enter the start and end angles to perform the rotation.

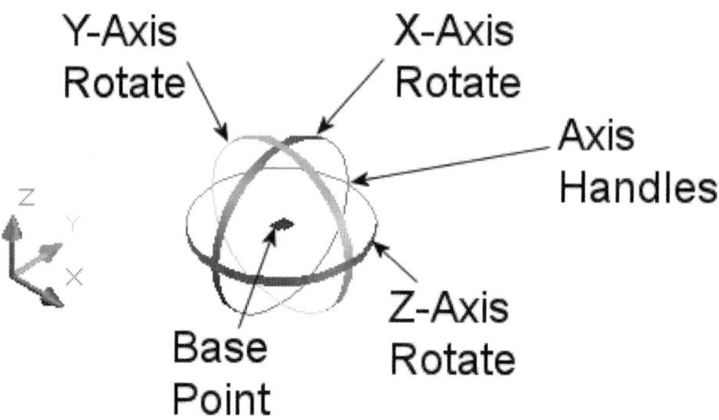

Figure 21.33

Try It!: Open the drawing 21_3DRot. In the following image, a base containing a slot needs to be joined with the two rectangular boxes to form a back and side. First select box "A" as the object to rotate in 3D. You will be prompted to define the base point of rotation. Next you will be prompted to choose a rotation axis by picking the appropriate axis handle on the grip tool. This axis will serve as a pivot point where the rotation occurs. Entering a negative angle of 90° rotates the box in the counterclockwise direction (as you would look down the Y-axis). Negative angles rotate in the clockwise direction. An easy way to remember the rotation direction is to use the right-hand rule. Point your thumb in the axis direction and your fingers will curl in the positive rotation direction.

```
Command: 3DROTATE
Current positive angle in UCS: ANGDIR=counterclockwise ANGBASE=0
Select objects: (Select box "A")
Select objects: (Press ENTER to continue)
Specify base point: (Pick the endpoint at "A")
Pick a rotation axis: (When the rotate grip tool appears, hover
    on the axis handle until the proper axis appears, as shown
    in the following image; then pick with the cursor)
Specify angle start point: (Pick the endpoint at "A")
Specify angle end point: -90
```

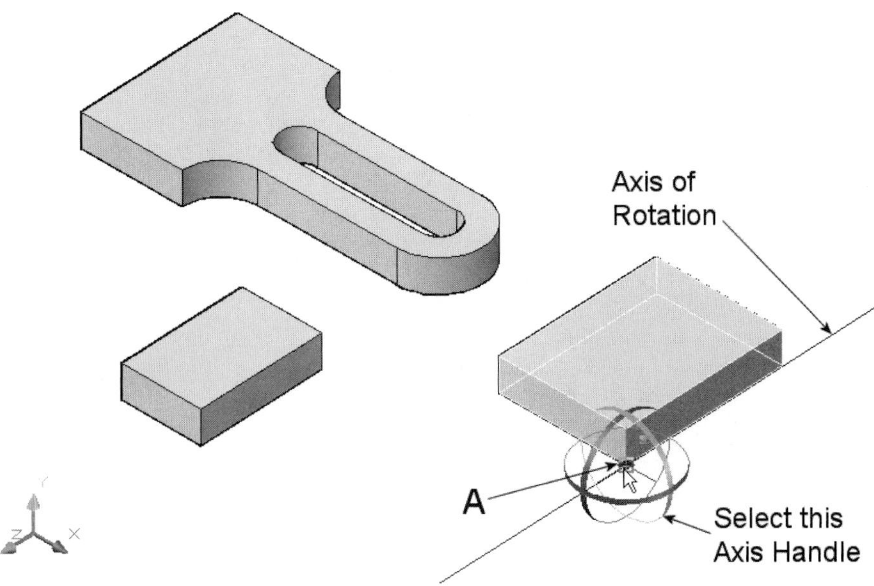

Figure 21.34

Next, the second box, as shown in the following image, is rotated 90° after the proper rotation axis is selected.

```
Command: 3DROTATE
Current positive angle in UCS: ANGDIR=counterclockwise ANGBASE=0
Select objects: (Select box "A")
Select objects: (Press ENTER to continue)
Specify base point: (Pick the endpoint at "A")
Pick a rotation axis: (When the rotate grip tool appears, click
    the axis handle until the proper axis appears, as shown in
    the following image; then pick this axis with the cursor)
Specify angle start point: (Pick the endpoint at "A")
Specify angle end point: 90
```

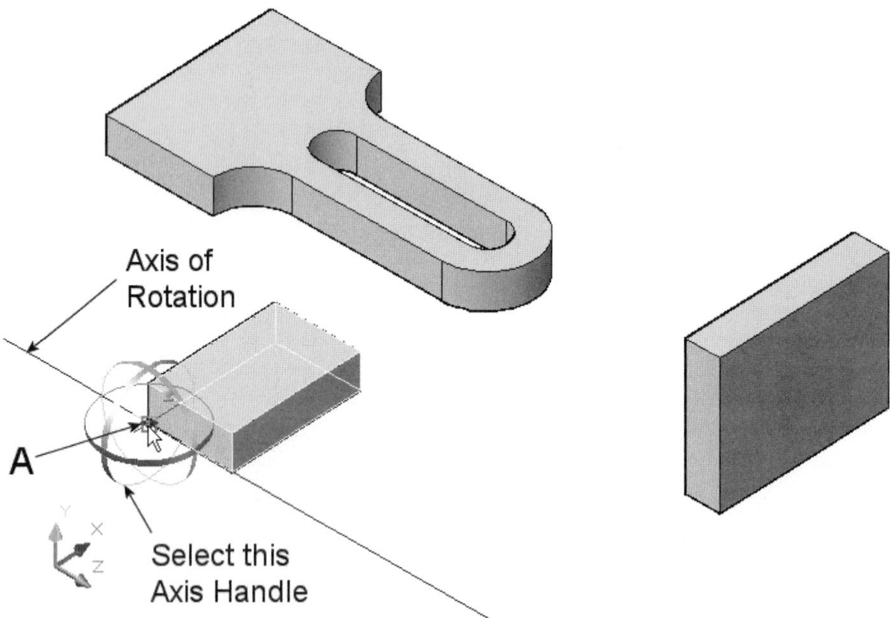

Figure 21.35

The results of these operations are illustrated on the left in the following image. Once the boxes are rotated to the correct angles, they are moved into position using the MOVE command and the appropriate Object Snap modes. Box "A" is moved from the endpoint of the corner at "A" to the endpoint of the corner at "C." Box "B" is moved from the endpoint of the corner at "B" to the endpoint of the corner at "C." Once moved, they are then joined to the model through the UNION command, as shown on the right in the following image.

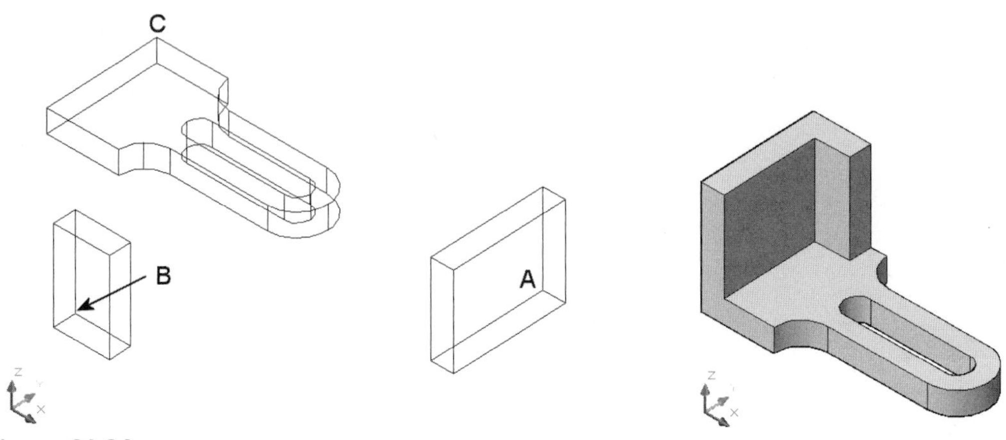

Figure 21.36

THE MIRROR3D COMMAND

The MIRROR3D command is a 3D version of the MIRROR command. In this command, however, instead of flipping over an axis, you mirror over a plane. A thorough understanding of the User Coordinate System is a must in order to properly operate this command.

Try It!: Open the drawing 21_Mirror3D. As illustrated on the left in the following image, only half of the object is created. The symmetrical object on the right in the following image is needed and can easily be created by using the MIRROR3D and UNION commands.

```
Command: MIRROR3D
Select objects: (Select the part)
Select objects: (Press ENTER to continue)
Specify first point of mirror plane (3 points) or
[Object/Last/Zaxis/View/XY/YZ/ZX/3points] <3points>: YZ (For
    YZ plane)
Specify point on YZ plane <0,0,0>: (Select the endpoint at "A")
Delete source objects? [Yes/No] <N>: N (Keep both objects)
Command: UNI (For UNION)
Select objects: (Select both solid objects)
Select objects: (Press ENTER to perform the union operation)
Command: HI (For HIDE; the solid should appear as illustrated on
    the right in the following image)
```

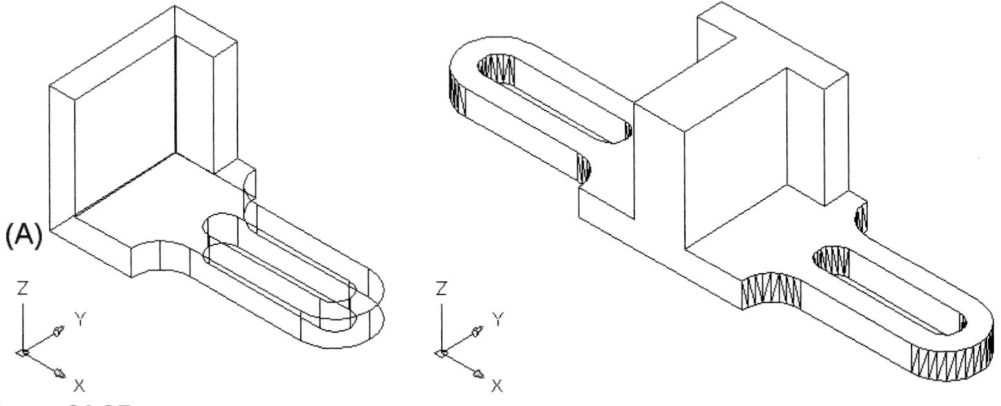

Figure 21.37

THE 3DARRAY COMMAND

The 3DARRAY command, like the ARRAY command, allows you to create both polar and rectangular arrays. For the 3D polar array, however, you select any 3D axis to rotate about, and in the 3D version of the rectangular array, you have not only rows and columns but levels as well.

Try It!: Open the drawing 21_3DArray. As illustrated on the left in the following image, six arms need to be arrayed around the hub in the center. To accomplish this we will perform a polar array about an axis running through the hub. We will rotate throughout a full 360°.

Note: If you perform a polar array at an angle other than 360°, take note of the direction in which you select the axis because this will affect the direction of rotation (right-hand rule).

```
Command: 3A (For 3DARRAY)
Select objects: (Select the Arm "A")
Select objects: (Press ENTER to continue)
Enter the type of array [Rectangular/Polar] <R>: P (For Polar)
Enter the number of items in the array: 6
Specify the angle to fill (+=ccw, -=cw) <360>: (Press ENTER to
    accept the default)
Rotate arrayed objects? [Yes/No] <Y>: Y
Specify center point of array: Cen
of (Specify the center of the back hub circle at "B")
Specify second point on axis of rotation: Cen
of (Specify the center of the front hub circle at "C")
Command: UNI (For UNION)
Select objects: (Select the six arms and the hub)
Select objects: (Press ENTER to perform the union operation)
Command: HI (For HIDE, the solid should appear as illustrated on
    the right in the following image)
Command: RE (For REGEN; this will convert the image back to
    wireframe mode)
```

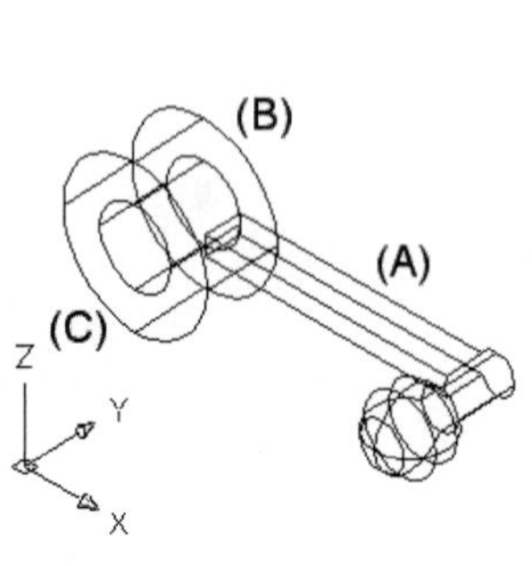

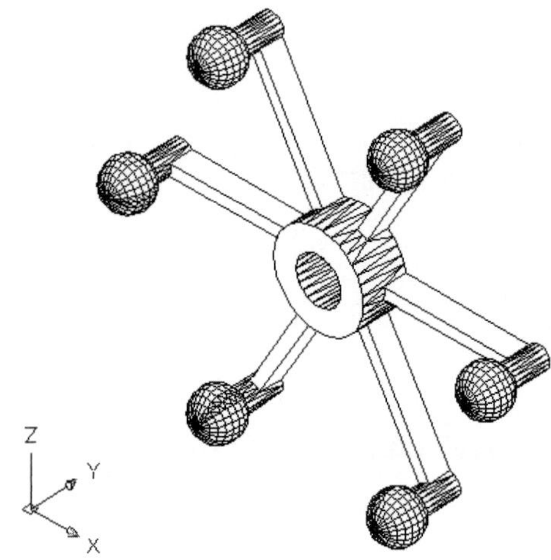

Figure 21.38

DETECTING INTERFERENCES OF SOLID MODELS

 The INTERFERE command identifies any interference and highlights the solid models that overlap. Choose this command in one of the following ways:

- From the 3D Make control panel located in the dashboard
- From the pull-down menu (Modify > 3D Operations > Interference Checking)
- From the keyboard (INTERFERE)

Try It!: Open the drawing file 21_Pipe Interference. Use this command to find any interference shared by a series of solid pipe objects, as shown on the left in the following image. Click the Interference checking button, located in the dashboard, as shown on the right in the following image, to begin this command.

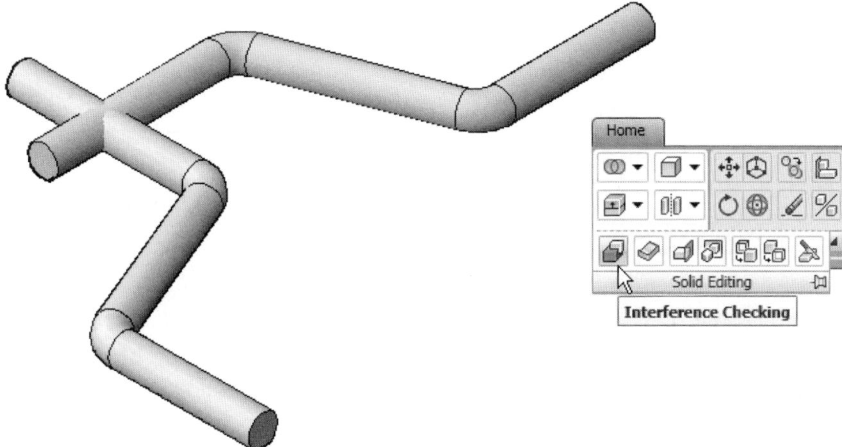

Figure 21.39

After launching this command, you will be prompted to select first and second sets of solids. In each case, pick both pipes separately. Pressing ENTER at the end of the command sequence changes the solid objects to a wireframe display to expose the areas of the objects considered to be interfering with one another, as shown on the left in the following image. An Interference Checking dialog box also appears, as shown on the right in the following image. The information in the dialog box allows you to verify the first and second sets of interfering objects. If they exist, you can also cycle through additional interferences using the Previous and Next buttons. In addition, a checkbox is provided, which allows you to create the interference solid, if desired.

```
Command: INTERFERE
Select first set of objects or [Nested selection/Settings]: (Pick
    one of the pipe objects)
Select first set of objects or [Nested selection/Settings]:
    (Press ENTER to continue)
```

```
Select second set of objects or [Nested selection/checK first
    set] <checK>: (Pick the second pipe object)
Select second set of objects or [Nested selection/checK first
    set] <checK>: (Press ENTER to perform the interference check)
```

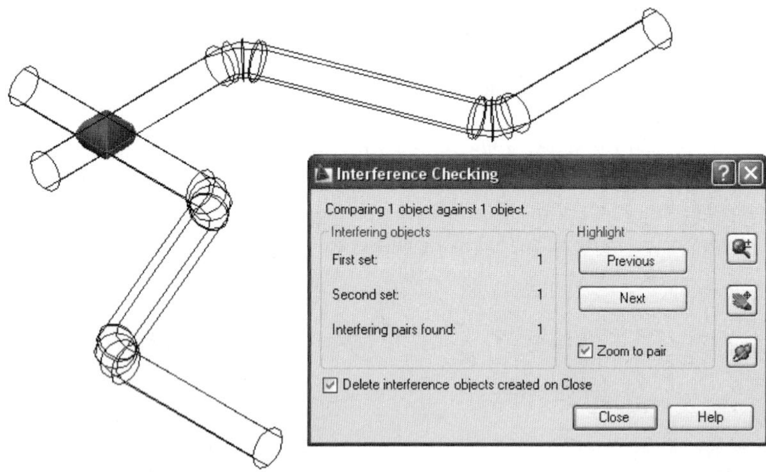

Figure 21.40

SLICING SOLID MODELS

Yet another way of cutting sections in 3D solid models is through the SLICE command. This command creates new solids from the existing ones that are sliced. You can also retain one or both halves of the sliced solid. Slicing a solid requires some type of cutting plane. The default method of creating this plane is by picking two points. You can also define the cutting plane or surface by specifying three points, by picking a surface, by using another object, or by basing the cutting plane line on the current positions of the XY, YZ, or ZX planes.

Choose the SLICE command in one of the following ways:

- From the 3D Make control panel located in the dashboard
- From the pull-down menu (Modify > 3D Operations > Slice)
- From the keyboard (SL or SLICE)

Try It!: Open the drawing file 21_Tee Slice. This command is similar to the SECTION command except that the solid model is actually cut or sliced at a plane that you define. In the example on the left in the following image, this plane is defined by the User Coordinate System. Before the slice is made, you also have the option of keeping either one or both halves of the object. The MOVE command is used to separate both halves, as shown on the right in the following image.

```
Command: SL (For SLICE)
Select objects: (Select the solid object)
Select objects: (Press ENTER to continue with this command)
```

```
Specify first point on slicing plane by [Object/Zaxis/View/XY/YZ/
    ZX/3points] <3points>: XY
Specify a point on XY-plane <0,0,0>: (Press ENTER to accept this
    default value)
Specify a point on desired side of the plane or [keep Both
    sides]: B (To keep both sides)
```

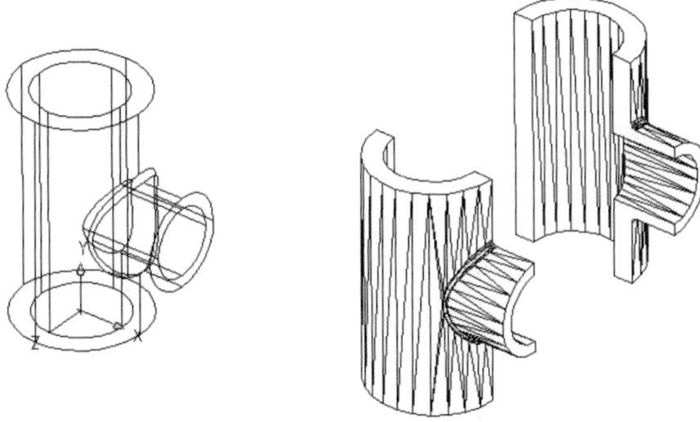

Figure 21.41

SLICING A SOLID WITH A SURFACE

A solid object can also be sliced by a surface. A surface is created by performing a 3D operation such as extrusion or revolution on an open object. Once the surface is created, it is positioned inside of the 3D solid model, where a slicing operation is performed. The next Try It! exercise illustrates the use of this technique.

Try It!: Open the drawing file 21_Surface Flow. You will first extrude a spline to create a surface. The surface will then be used to slice a solid block. You will keep the bottom portion of the solid.

Before slicing the solid block, first extrude the spline object, as shown on the left in the following image, a distance equal to the depth of the block (from “A” to “B”). Since the spline represents an open shape, the result of performing this operation is the creation of a surface instead of a solid, as shown on the right in the following image.

```
Command: EXT (For EXTRUDE)
Current wire frame density: ISOLINES=4
Select objects to extrude: (Pick the spline object)
Select objects to extrude: (Press ENTER to continue)
Specify height of extrusion or [Direction/Path/Taper angle]: D
    (For Direction)
Specify start point of direction: (Pick the endpoint at "A")
Specify end point of direction: (Pick the endpoint at "B")
```

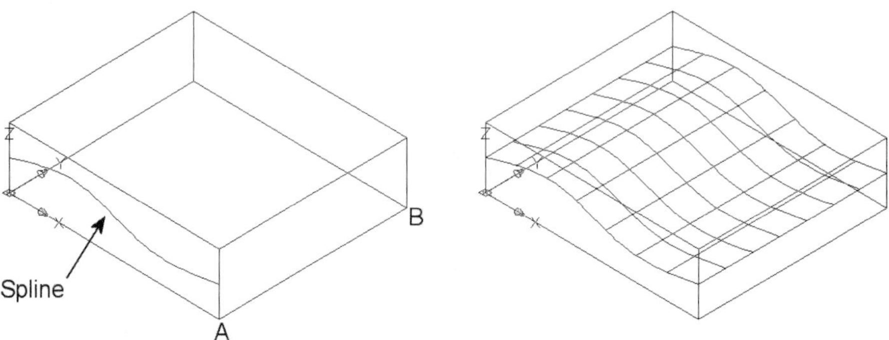

Figure 21.42

With the newly created surface positioned inside of the solid block, issue the SLICE command. Pick the solid block as the object to slice and select the surface as the slicing plane, as shown on the left in the following image. You will also be prompted to select the portion of the solid to keep. Here is where you pick the bottom of the solid, as shown on the left in the following image.

```
Command: SL (For SLICE)
Select objects to slice: (Pick the solid block)
Select objects to slice: (Press ENTER to continue)
Specify start point of slicing plane or [planar
Object/Surface/Zaxis/View/XY/YZ/ZX/3points] <3points>: S (For
    Surface)
Select a surface: (Pick the surface, as shown on the left in the
    following image)
Select solid to keep or [keep Both sides] <Both>: (Pick the
    bottom of the solid)
```

The results are displayed on the right in the following image, with the solid block being cut by the surface.

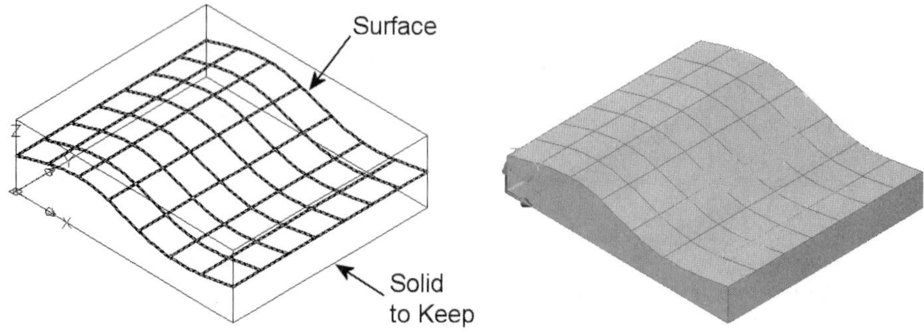

Figure 21.43

TERRAIN MODELING

A unique type of solid model can be created when a model is sliced by a surface created using a lofting operation. In the following image, four different splines have been applied to the edge faces of a solid block. Using the LOFT command, two splines are selected as cross sections and the other two splines as guides or rails. Once this specialized surface is created, slice the solid block using the surface and keep the lower portion of the model.

Try It!: Open the drawing file 21_Terrain. You will create a lofted surface by selecting the two cross sections and the two guide curves, as shown on the left in the following image. The results are displayed on the right in the following image, with a complex surface being created from the loft operation.

```
Command: LOFT
Select cross-sections in lofting order: (Select cross section #1)
Select cross-sections in lofting order: (Select cross section #2)
Select cross-sections in lofting order: (Press ENTER to continue)
Enter an option [Guides/Path/Cross-sections only] <Cross-sections
    only>: G (For Guides)
Select guide curves: (Select guide curve #1)
Select guide curves: (Select guide curve #2)
Select guide curves: (Press ENTER to create the surface)
```

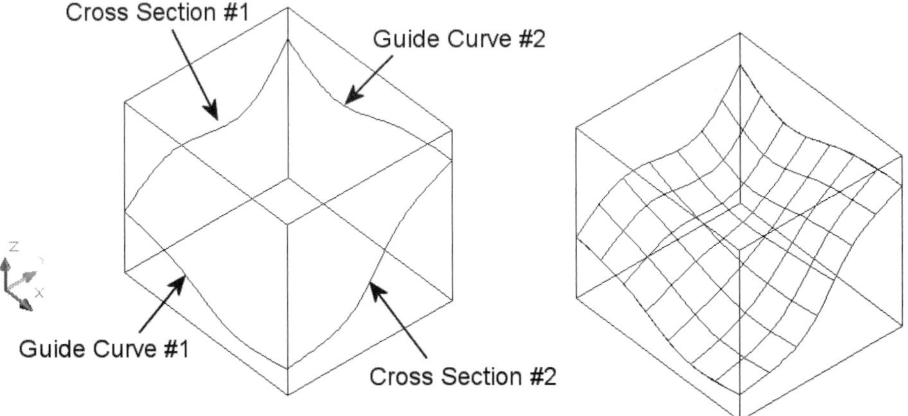

Figure 21.44

With the surface created, activate the SLICE command, pick the solid block as the object to slice, select the surface as the slicing plane, and, finally, pick the lower portion of the solid as the portion to keep, as shown on the left in the following image. The results are illustrated on the right in the following image.

```
Command: SL (For SLICE)
Select objects to slice: (Select the solid block)
Select objects to slice: (Press ENTER to continue)
Specify start point of slicing plane or [planar
```

```
Object/Surface/Zaxis/View/XY/YZ/ZX/3points] <3points>: S (For
    Surface)
Select a surface: (Select the surface)
Select solid to keep or [keep Both sides] <Both>: (Pick the
    lower portion of the solid block)
```

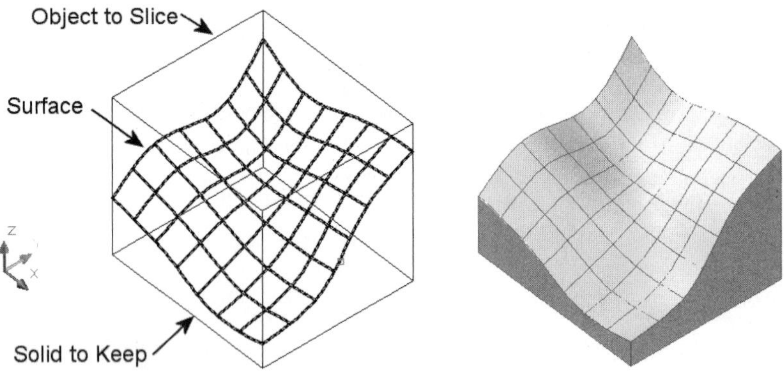

Figure 21.45

THE SOLIDEDIT COMMAND

Once features such as holes, slots, and extrusions are constructed in a solid model, the time may come to make changes to these features. This is the function of the SOLIDEDIT command. This command contains numerous options, which can be selected from the Solids Editing toolbar, as shown in the following image. You could also choose Solids Editing from the Modify pull-down menu. The menus are arranged in three groupings, namely, Face, Edge, and Body editing. These groupings are discussed in the pages that follow. Also, it is recommended that instead of entering the SOLIDEDIT command at the command prompt, you use the Solids Editing toolbar or pull-down menu to perform editing operations. This eliminates a number of steps and make it easier to locate the appropriate option under the correct grouping.

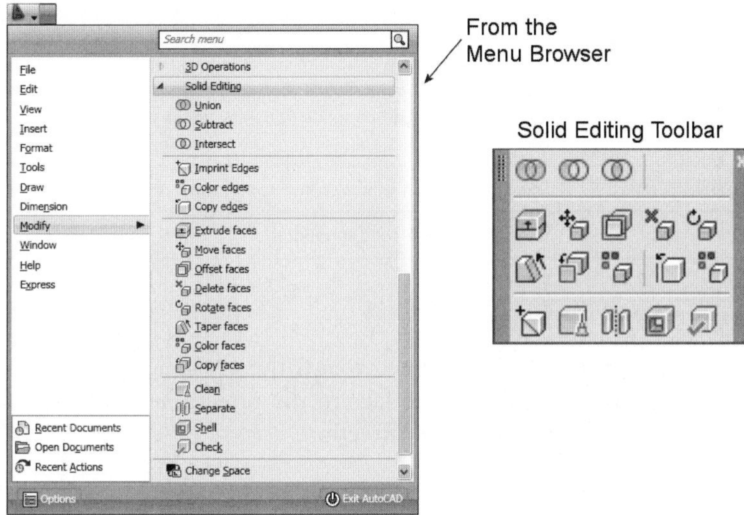

Figure 21.46

The following table gives a brief description of each mode for performing solid editing operations.

Button	Tool	Function
	Extrude Faces	Used for lengthening or shortening faces on a solid
	Move Faces	Used for moving a solid shape to a new location
	Offset Faces	Used for offsetting faces at a specified distance on a solid
	Delete Faces	Used for removing faces and fillets from a solid
	Rotate Faces	Used for rotating faces on a selected solid
	Taper Faces	Used for tapering selected faces of a solid at a draft angle along a vector direction
	Copy Faces	Used for copying selected faces of a solid. These copied faces can take the form of regions or bodies.
	Color Faces	Used for assigning unique colors to individual faces
	Copy Edges	Used to copy edges from a solid. These new edges are often used to create new solids.
	Color Edges	Used for assigning unique colors to individual solid edges
	Imprint	Used for adding construction geometry to a solid model
	Clean	Used to remove imprints from a solid

Button	Tool	Function
	Separate	Used to separate a solid into multiple parts as long as those parts do not intersect at any point. Solids sometimes act as a single entity even though they appear to be separate solids (unioning solids together that do not touch or removing part of a solid so that the remaining pieces do not touch).
	Shell	Used to create a thin wall in a solid model
	Check	Used to prove a solid is valid

EXTRUDING (FACE EDITING)

Faces may be lengthened or shortened through the Extrude option of the SOLIDEDIT command. A positive distance extrudes the face in the direction of its normal. A negative distance extrudes the face in the opposite direction.

Try It!: Open the drawing file 21_Extrude. In the following image, the highlighted face at "A" needs to be decreased in height.

Note: You will achieve better results when selecting a face for the SOLIDEDIT command if you pick on the inside of the face rather than on the edge of the face.

```
Command: SOLIDEDIT
Solids editing automatic checking: SOLIDCHECK=1
Enter a solids editing option [Face/Edge/Body/Undo/eXit] <eXit>:
    F (For Face)
Enter a face editing option
[Extrude/Move/Rotate/Offset/Taper/Delete/Copy/coLor/mAterial/
    Undo/eXit] <eXit>: E (For Extrude)
Select faces or [Undo/Remove]: (Select the face inside the area
    represented by "A")
Select faces or [Undo/Remove/ALL]: (Press ENTER to continue)
Specify height of extrusion or [Path]: -10.00
Specify angle of taper for extrusion <0>: (Press ENTER)
Solid validation started.
Solid validation completed.
Enter a face editing option
[Extrude/Move/Rotate/Offset/Taper/Delete/Copy/coLor/mAterial/
    Undo/eXit] <eXit>: (Press ENTER)
Solids editing automatic checking: SOLIDCHECK=1
Enter a solids editing option [Face/Edge/Body/Undo/eXit] <eXit>:
    (Press ENTER)
```

The result is illustrated in the following image at "B."

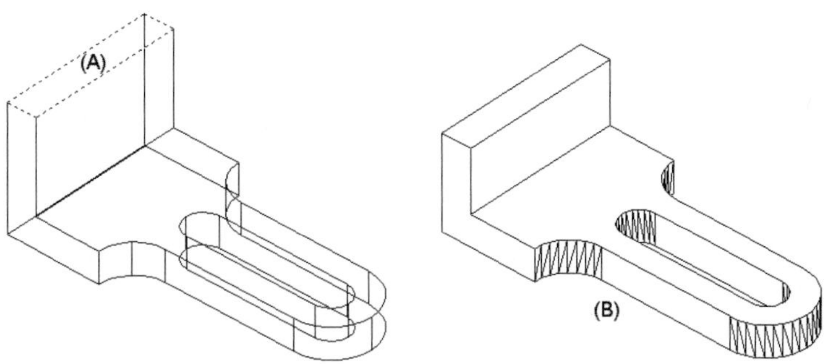

Figure 21.47

MOVING (FACE EDITING)

Try It!: Open the drawing 21_Move. The object in the following image illustrates two intersecting cylinders. The two horizontal cylinders need to be moved 1 unit up from their current location. The cylinders are first selected at "A" and "B" through the SOLIDEDIT command along with the Move option.

```
Command: SOLIDEDIT
Solids editing automatic checking: SOLIDCHECK=1
Enter a solids editing option [Face/Edge/Body/Undo/eXit] <eXit>:
    F (For Face)
Enter a face editing option
[Extrude/Move/Rotate/Offset/Taper/Delete/Copy/coLor/mAterial/
    Undo/eXit] <eXit>: M (For Move)
Select faces or [Undo/Remove]: (Select both highlighted faces at
    "A" and "B")
Select faces or [Undo/Remove/ALL]: (Press ENTER to continue)
Specify a base point or displacement: (Pick any point on the
    screen)
Specify a second point of displacement: @0,0,1
Solid validation started.
Solid validation completed.
Enter a face editing option
[Extrude/Move/Rotate/Offset/Taper/Delete/Copy/coLor/mAterial/
    Undo/eXit] <eXit>: (Press ENTER)
Solids editing automatic checking: SOLIDCHECK=1
Enter a solids editing option [Face/Edge/Body/Undo/eXit] <eXit>:
    (Press ENTER)
```

The results are illustrated on the right in the following image.

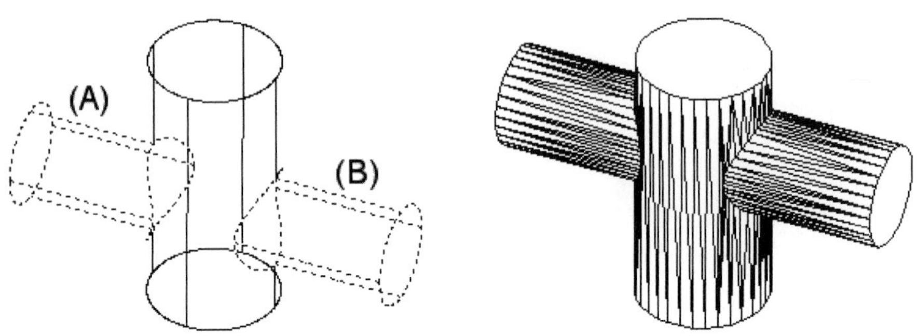

Figure 21.48

ROTATING (FACE EDITING)

Try It!: Open the drawing file 21_Rotate. In the following image, the triangular cutout needs to be rotated 45° in the clockwise direction. Use the Rotate Face option of the SOLIDEDIT command to accomplish this. You must select all faces of the triangular cutout at "A," "B," and "C." During the selection process you will have to remove the face that makes up the top of the rectangular base at "D" before proceeding.

```
Command: SOLIDEDIT
Solids editing automatic checking: SOLIDCHECK=1
Enter a solids editing option [Face/Edge/Body/Undo/eXit] <eXit>:
    F (For Face)
Enter a face editing option
[Extrude/Move/Rotate/Offset/Taper/Delete/Copy/coLor/mAterial/
    Undo/eXit] <eXit>: R (For Rotate)
Select faces or [Undo/Remove]: (Select all faces that make up
    the triangular extrusion. Pick inside areas at "A," "B," and
    "C"; select "C" twice to highlight it)
Select faces or [Undo/Remove/ALL]: R (For Remove)
Remove faces or [Undo/Add/ALL]: (Select the face at "D" to
    remove)
Remove faces or [Undo/Add/ALL]: (Press ENTER to continue)
Specify an axis point or [Axis by object/View/Xaxis/Yaxis/Zaxis]
    <2points>: Z (For Zaxis)
Specify the origin of the rotation <0,0,0>: (Select the endpoint
    at "E")
Specify a rotation angle or [Reference]: -45 (To rotate the
    triangular extrusion 45° in the clockwise direction)
Solid validation started.
Solid validation completed.
Enter a face editing option
[Extrude/Move/Rotate/Offset/Taper/Delete/Copy/coLor/mAterial/
    Undo/eXit] <eXit>: (Press ENTER)
Solids editing automatic checking: SOLIDCHECK=1
Enter a solids editing option [Face/Edge/Body/Undo/eXit] <eXit>:
    (Press ENTER)
```

The results are illustrated on the right in the following image.

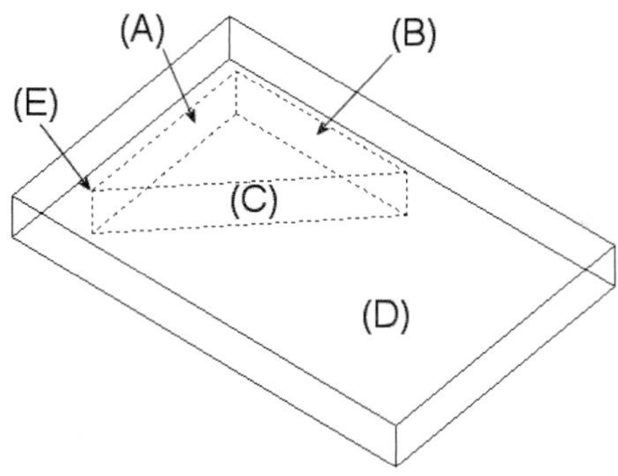

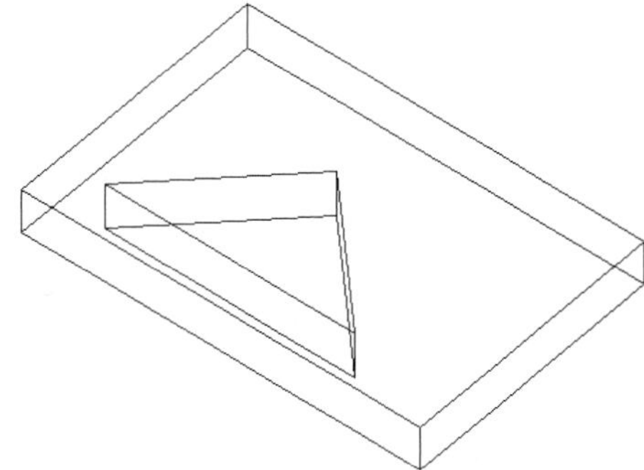

Figure 21.49

OFFSETTING (FACE EDITING)

Try It!: Open the drawing file 21_Offset. In the following image, the holes need to be resized. Use the Offset Face option of the SOLIDEDIT command to increase or decrease the size of selected faces. Using positive values increases the volume of the solid. Therefore, the feature being offset gets smaller, similar to the illustration at "B." Entering negative values reduces the volume of the solid; this means that the feature being offset gets larger, as in the figures at "C." Study the prompt sequence and the following image for the mechanics of this command option.

```
Command: SOLIDEDIT
Solids editing automatic checking: SOLIDCHECK=1
Enter a solids editing option [Face/Edge/Body/Undo/eXit] <eXit>:
     F (For Face)
Enter a face editing option
[Extrude/Move/Rotate/Offset/Taper/Delete/Copy/coLor/mAterial/
     Undo/eXit] <eXit>: O (For Offset)
Select faces or [Undo/Remove]: (Select inside the edges of the
     two holes at "D" and "E")
Select faces or [Undo/Remove/ALL]: (Press ENTER to continue)
Specify the offset distance: .50
Solid validation started.
Solid validation completed.
Enter a face editing option
[Extrude/Move/Rotate/Offset/Taper/Delete/Copy/coLor/mAterial/
     Undo/eXit] <eXit>: (Press ENTER)
Solids editing automatic checking: SOLIDCHECK=1
Enter a solids editing option [Face/Edge/Body/Undo/eXit] <eXit>:
     (Press ENTER)
```

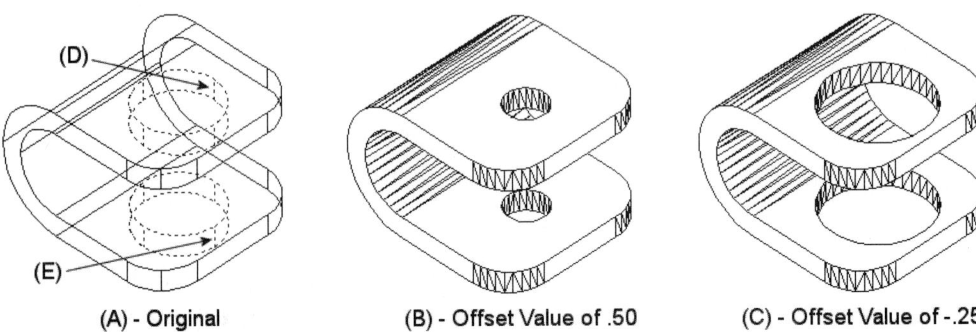

Figure 21.50

TAPERING (FACE EDITING)

Try It!: Open the drawing 21_Taper. The object at "A" in the following image represents a solid box that needs to have tapers applied to its sides. Using the Taper Face option of the SOLIDEDIT command allows you to accomplish this task. Entering a positive angle moves the location of the second point into the part, as shown in the middle in the following image. Entering a negative angle moves the location of the second point away from the part, as shown on the right in the following image.

```
Command: SOLIDEDIT
Solids editing automatic checking: SOLIDCHECK=1
Enter a solids editing option [Face/Edge/Body/Undo/eXit] <eXit>:
    F (For Face)
Enter a face editing option
[Extrude/Move/Rotate/Offset/Taper/Delete/Copy/coLor/mAterial/
    Undo/eXit] <eXit>: T (For Taper)
Select faces or [Undo/Remove]: (Select faces "A" through "D")
Select faces or [Undo/Remove/ALL]: (Press ENTER to continue)
Specify the base point: (Select the endpoint at "E")
Specify another point along the axis of tapering: (Select the
    endpoint at "B")
Specify the taper angle: 10 (For the angle of the taper)
Solid validation started.
Solid validation completed.
Enter a face editing option
[Extrude/Move/Rotate/Offset/Taper/Delete/Copy/coLor/mAterial/
    Undo/eXit] <eXit>: (Press ENTER)
Solids editing automatic checking: SOLIDCHECK=1
Enter a solids editing option [Face/Edge/Body/Undo/eXit] <eXit>:
    (Press ENTER)
```

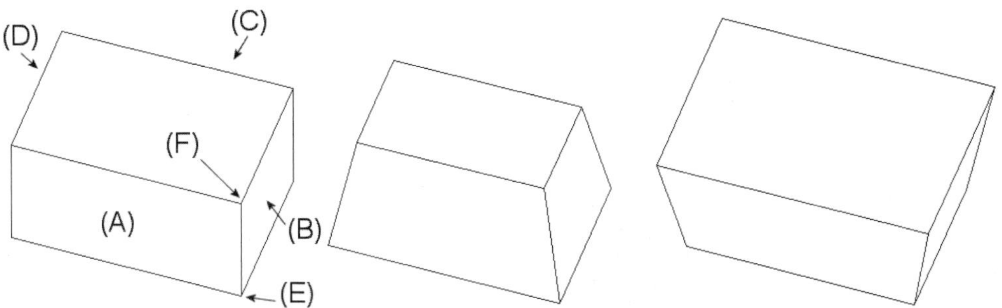

Figure 21.51

DELETING (FACE EDITING)

Faces can be erased through the Delete Face option of the SOLIDEDIT command.

Try It!: Open the drawing file 21_Delete. In the following image, select the hole at "A" as the face to erase.

```
Command: SOLIDEDIT
Solids editing automatic checking: SOLIDCHECK=1
Enter a solids editing option [Face/Edge/Body/Undo/eXit] <eXit>:
    F (For Face)
Enter a face editing option
[Extrude/Move/Rotate/Offset/Taper/Delete/Copy/coLor/mAterial/
    Undo/eXit] <eXit>: D (For Delete)
Select faces or [Undo/Remove]: (Select inside the hole at "A")
Select faces or [Undo/Remove/ALL]: (Press ENTER to continue)
Solid validation started.
Solid validation completed.
Enter a face editing option
[Extrude/Move/Rotate/Offset/Taper/Delete/Copy/coLor/mAterial/
    Undo/eXit] <eXit>: (Press ENTER)
Solids editing automatic checking: SOLIDCHECK=1
Enter a solids editing option [Face/Edge/Body/Undo/eXit] <eXit>:
    (Press ENTER)
```

The results are illustrated on the right in the following image.

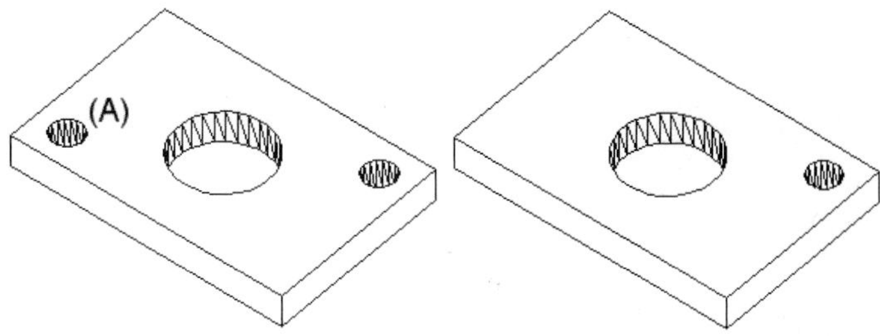

Figure 21.52

COPYING (FACE EDITING)

You can copy a face for use in the creation of another solid model using the Copy Face option of the SOLIDEDIT command.

Try It!: Open the drawing file 21_Copy. In the following image, the solid model at "A" will be used to create a region from the face at "B." While in the command, select the face by picking in the area at "B." Notice that all objects making up the face, such as the rectangle and circles, are highlighted. Picking a base point and second point copies the face at "C." The resulting object at "C" is actually a region. The region could be exploded back into individual lines and circles, which could then be used to create a new object. A region can also be extruded by using the EXTRUDE command to create another solid model such as the one illustrated on the right in the following image.

```
Command: SOLIDEDIT
Solids editing automatic checking: SOLIDCHECK=1
Enter a solids editing option [Face/Edge/Body/Undo/eXit] <eXit>:
    F (For Face)
Enter a face editing option
[Extrude/Move/Rotate/Offset/Taper/Delete/Copy/coLor/mAterial/
    Undo/eXit] <eXit>: C (For Copy)
Select faces or [Undo/Remove]: (Select the top face in area "B")
Select faces or [Undo/Remove/ALL]: (Press ENTER to continue)
Specify a base point or displacement: (Pick a point to
    copy from)
Specify a second point of displacement: (Pick a point to
    copy to)
Enter a face editing option
[Extrude/Move/Rotate/Offset/Taper/Delete/Copy/coLor/mAterial/
    Undo/eXit] <eXit>: (Press ENTER)
Solids editing automatic checking: SOLIDCHECK=1
Enter a solids editing option [Face/Edge/Body/Undo/eXit] <eXit>:
    (Press ENTER)
```

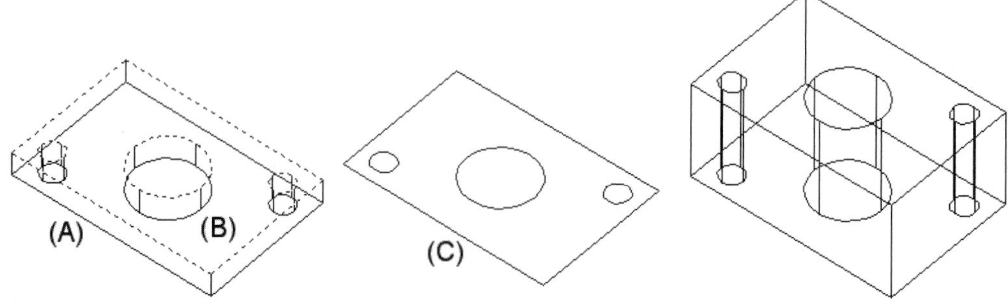

Figure 21.53

IMPRINTING (BODY EDITING)

An interesting and powerful method of adding construction geometry to a solid model is though the process of imprinting.

Try It!: Open the drawing file 21_Imprint. In the following image at "A," a box along with slot is already modeled in 3D. A line was constructed from the midpoints of the top surface of the solid model. The SOLIDEDIT command will be used to imprint this line to the model, which results in dividing the top surface into two faces.

```
Command: SOLIDEDIT
Solids editing automatic checking: SOLIDCHECK=1
Enter a solids editing option [Face/Edge/Body/Undo/eXit] <eXit>:
    B (For Body)
Enter a body editing option
[Imprint/seParate solids/Shell/cLean/Check/Undo/eXit] <eXit>: I
    (For Imprint)
Select a 3D solid: (Select the solid model)
Select an object to imprint: (Select line "B")
Delete the source object [Yes/No] <N>: Y (For Yes; this erases
    the line)
Select an object to imprint: (Press ENTER to complete the imprint
    operation)
Enter a body editing option
[Imprint/seParate solids/Shell/cLean/Check/Undo/eXit] <eXit>:
    (Press ENTER)
Solids editing automatic checking: SOLIDCHECK=1
Enter a solids editing option [Face/Edge/Body/Undo/eXit] <eXit>:
    (Press ENTER)
```

The segments of the lines that come in contact with the 3D solid remain on the part's surface. However, these are no longer line segments; rather, these lines now belong to the part. The lines actually separate the top surface into two faces.

Use the Extrude Face option of the SOLIDEDIT command on one of the newly created faces at "C" and increase the face in height by an extra 1.50 units. The results are illustrated on the right in the following image.

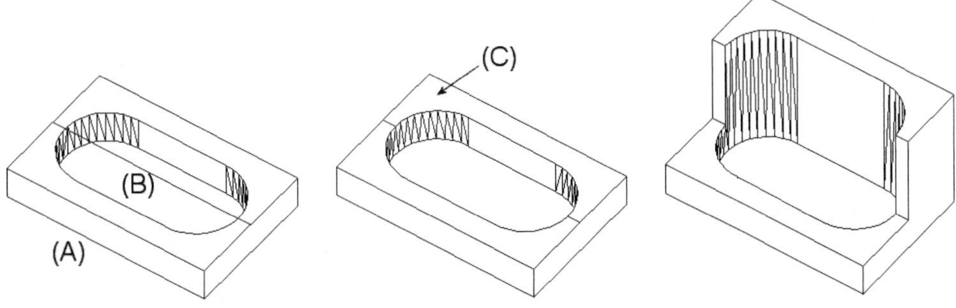

Figure 21.54

SEPARATING SOLIDS (BODY EDITING)

Sometimes when performing union and subtraction operations on solid models, you can end up with models that do not actually touch (intersect) but act as a single object. The Separating Solids option of the SOLIDEDIT command is used to correct this condition.

Try It!: Open the drawing file 21_Separate. In the following image, the model at "A" is about to be sliced in half with a thin box created at the center of the circle and spanning the depth of the rectangular shelf.

Use the SUBTRACT command and subtract the rectangular box from the solid object. After you subtract the box, pick the solid at "B" and notice that both halves of the object highlight even though they appear separate. To convert the single solid model into two separate models, use the SOLIDEDIT command followed by the Body and Separate options.

```
Command: SOLIDEDIT
Solids editing automatic checking: SOLIDCHECK=1
Enter a solids editing option [Face/Edge/Body/Undo/eXit] <eXit>:
    B (For Body)
Enter a body editing option
[Imprint/seParate solids/Shell/cLean/Check/Undo/eXit] <eXit>: P
    (For Separate)
Select a 3D solid: (Pick the solid model at "B")
Enter a body editing option
[Imprint/seParate solids/Shell/cLean/Check/Undo/eXit] <eXit>:
    (Press ENTER)
Solids editing automatic checking: SOLIDCHECK=1
Enter a solids editing option [Face/Edge/Body/Undo/eXit] <eXit>:
    (Press ENTER)
```

This action separates the single model into two. When you select the model in the following image at "C," only one half highlights.

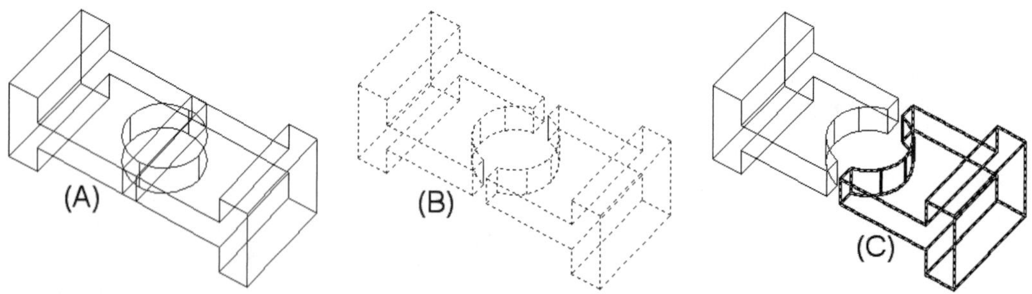

Figure 21.55

SHELLING (BODY EDITING)

Shelling is the process of constructing a thin wall inside or outside of a solid model. Positive thickness produces the thin wall inside; negative values for thickness produce the thin wall outside. This wall thickness remains constant throughout the entire model. Faces may be removed during the shelling operation to create an opening.

Try It!: Open the drawing file 21_Shell. For clarity, it is important to first rotate your model or your viewpoint such that any faces to be removed are visible. Use the 3DFORBIT (Free Orbit) command and the following image to change the model view from the one shown at "A" to the one shown at "B." Use the 3D Hidden Visual Style option in the 3DFORBIT command to ensure that the bottom surface at "C" is visible. Now the Shell option of the SOLIDEDIT command can be used to "hollow out" the part and remove the bottom face. An additional note: only one shell is permitted in a model.

```
Command: SOLIDEDIT
Solids editing automatic checking: SOLIDCHECK=1
Enter a solids editing option [Face/Edge/Body/Undo/eXit] <eXit>:
    B (For Body)
Enter a body editing option
[Imprint/seParate solids/Shell/cLean/Check/Undo/eXit] <eXit>: S
    (For Shell)
Select a 3D solid: (Select the solid model at "B")
Remove faces or [Undo/Add/ALL]: (Pick a point at "C"; because
    the model is already highlighted, it is not obvious that the
    face is selected but "1 face found, 1 removed" will be
    indicated)
Remove faces or [Undo/Add/ALL]: (Press ENTER to continue)
Enter the shell offset distance: 0.20
Solid validation started.
Solid validation completed.
Enter a body editing option
[Imprint/seParate solids/Shell/cLean/Check/Undo/eXit] <eXit>:
    (Press ENTER)
Solids editing automatic checking: SOLIDCHECK=1
Enter a solids editing option [Face/Edge/Body/Undo/eXit] <eXit>:
    (Press ENTER)
```

The results are illustrated on the right in the following image.

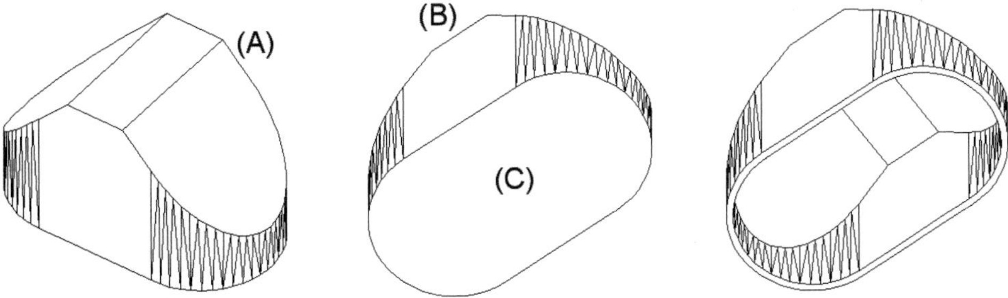

Figure 21.56

CLEANING (BODY EDITING)

When imprinted lines that form faces are not used, they can be deleted from a model by the Clean (Body Editing) option of the SOLIDEDIT command.

Try It!: Open the drawing file 21_Clean. The object at "A" in the following image illustrates lines originally constructed on the top of the solid model. These lines were then imprinted at "B." Since these lines now belong to the model, the Clean option is used to remove them. The results are illustrated on the right in the following image.

```
Command: SOLIDEDIT
Solids editing automatic checking: SOLIDCHECK=1
Enter a solids editing option [Face/Edge/Body/Undo/eXit] <eXit>:
    B (For Body)
Enter a body editing option
[Imprint/seParate solids/Shell/cLean/Check/Undo/eXit] <eXit>: L
    (For Clean)
Select a 3D solid: (Select the solid model at "B")
Enter a body editing option
[Imprint/seParate solids/Shell/cLean/Check/Undo/eXit] <eXit>:
    (Press ENTER)
Solids editing automatic checking: SOLIDCHECK=1
Enter a solids editing option [Face/Edge/Body/Undo/eXit] <eXit>:
    (Press ENTER)
```

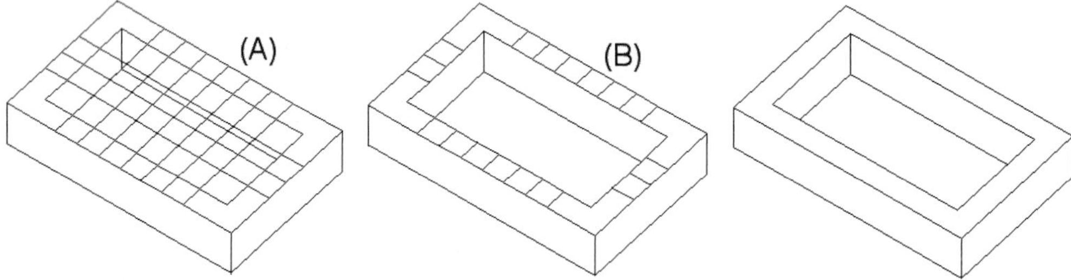

Figure 21.57

Additional options of the SOLIDEDIT command include the ability to apply a color or material to a selected face. You can also apply color to an edge or copy edges from a model so that they can be used for construction purposes on other models.

TUTORIAL EXERCISE: 21_3DROTATE.DWG

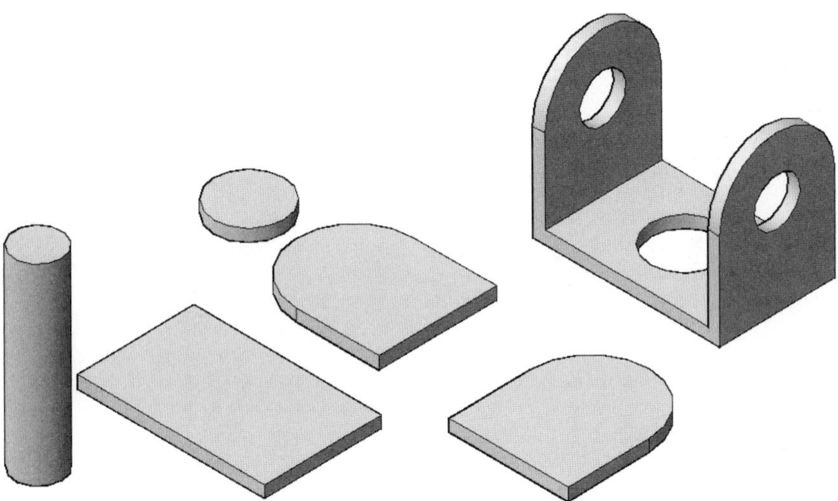

Figure 21.58

PURPOSE

This tutorial exercise is designed to use the 3DROTATE command to rotate the objects into position before creating unions or performing subtractions, as shown in the following image.

SYSTEM SETTINGS

The drawing units, limits, grid, and snap values are already set for this drawing. Check to see that the following Object Snap modes are already set: Endpoint, Center, Intersection, and Extension. Also check to see that OSNAP is turned on. These are located in the bottom status bar.

LAYERS

Make sure the current layer is Model.

SUGGESTED COMMANDS

Open the drawing file 21_3DRotate.dwg.

Use the 3DROTATE command to rotate the sides to the proper angle before they are moved into place and joined. The cylinders are also rotated and moved into place before being subtracted from the base and sides to create holes.

Step 1

Use the MOVE command to position the center of the large cylinder "A" in the middle of the base, between "B" and "C," as identified in the following image. Use the OSNAP-m2p option (midpoint between two points) to accomplish this.

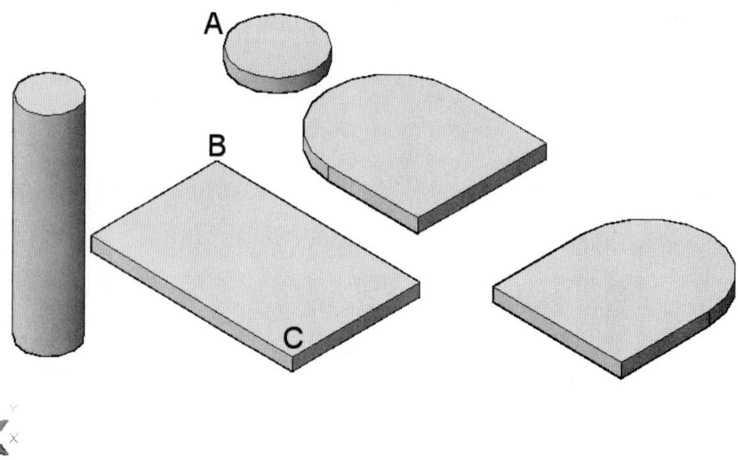

Figure 21.59

Step 2

Begin rotating the shapes into place using the 3DROTATE command. Select solid "A" as the object to rotate. Pick the endpoint at "B" as the base point of the rotation. When the rotate grip tool appears, click the axis handle to display the axis, as shown in the following image. Enter a value of -90 degrees to perform the rotate operation.

```
Command: 3DROTATE
Current positive angle in UCS: ANGDIR=counterclockwise ANGBASE=0
Select objects: (Select solid "A")
Select objects: (Press ENTER to continue)
Specify base point: (Pick the endpoint at "B")
Pick a rotation axis: (When the rotate grip tool appears, hover
     on the axis handle until the proper axis appears, as shown
     in the following image; then pick with the mouse)
Specify angle start point: (Pick the endpoint at "B")
Specify angle end point: -90
```

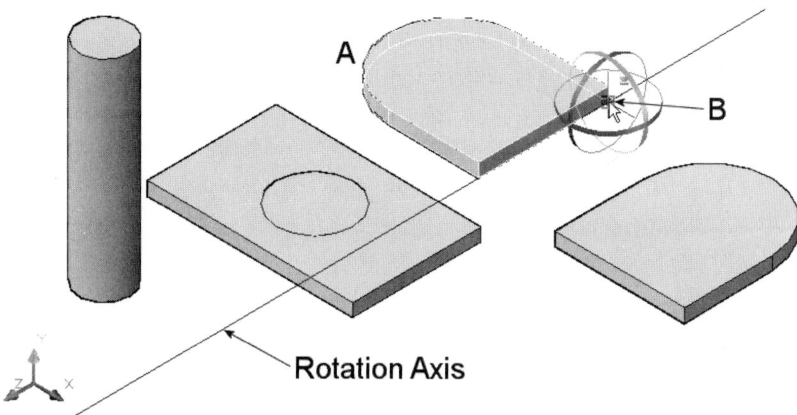

Figure 21.60

Step 3

Position the endpoint of the side panel at "A" that you just rotated to the endpoint of the top of the base at "B" using the MOVE command, as shown in the following image.

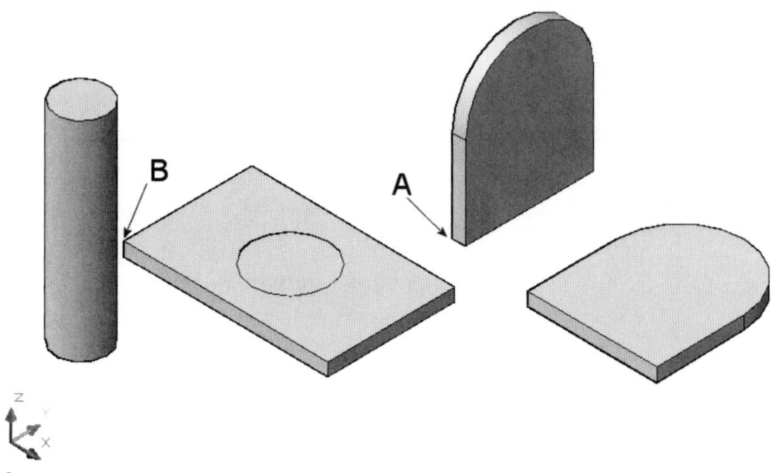

Figure 21.61

Step 4

Now rotate the second side panel into position. This will take two steps to accomplish. First, rotate the panel 90° using the rotation axis, as shown in the following image.

```
Command: 3DROTATE
Current positive angle in UCS: ANGDIR=counterclockwise ANGBASE=0
Select objects: (Select the other side panel)
Select objects: (Press ENTER to continue)
Specify base point: (Pick the endpoint at "A")
```

```
Pick a rotation axis: (When the rotate grip tool appears, hover
    on the axis handle until the proper axis appears, as shown
    in the following image; then pick with the cursor)
Specify angle start point: (Pick the endpoint at "A")
Specify angle end point: 90
```

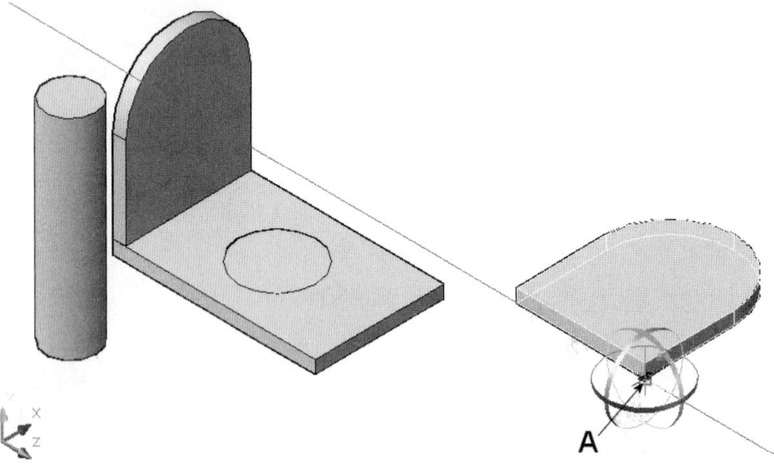

Figure 21.62

Step 5

Now rotate the same panel -90° using the rotation axis, as shown in the following image.

```
Command: 3DROTATE
Current positive angle in UCS: ANGDIR=counterclockwise ANGBASE=0
Select objects: (Select the same side panel)
Select objects: (Press ENTER to continue)
Specify base point: (Pick the endpoint at "A")
Pick a rotation axis: (When the rotate grip tool appears, hover
    on the axis handle until the proper axis appears, as shown
    in the following image; then pick with the cursor)
Specify angle start point: (Pick the endpoint at "A")
Specify angle end point: -90
```

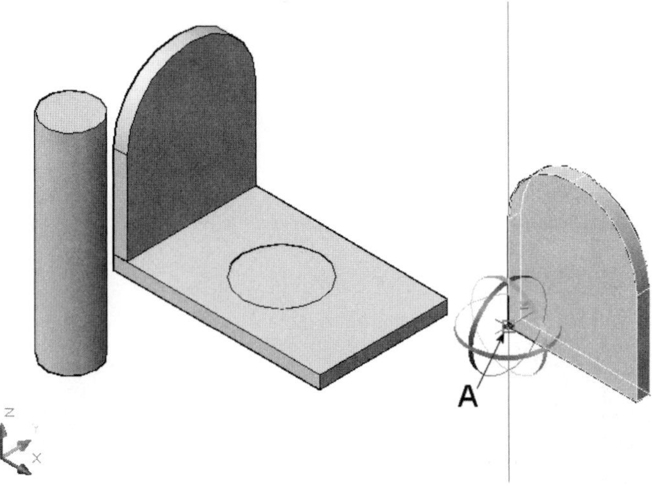

Figure 21.63

Step 6

Now position the endpoint of the panel at "A" along the top of the endpoint of the base at "B" using the MOVE command, as shown on the left in the following image. When finished, the display should appear similar to the illustration on the right in the following image.

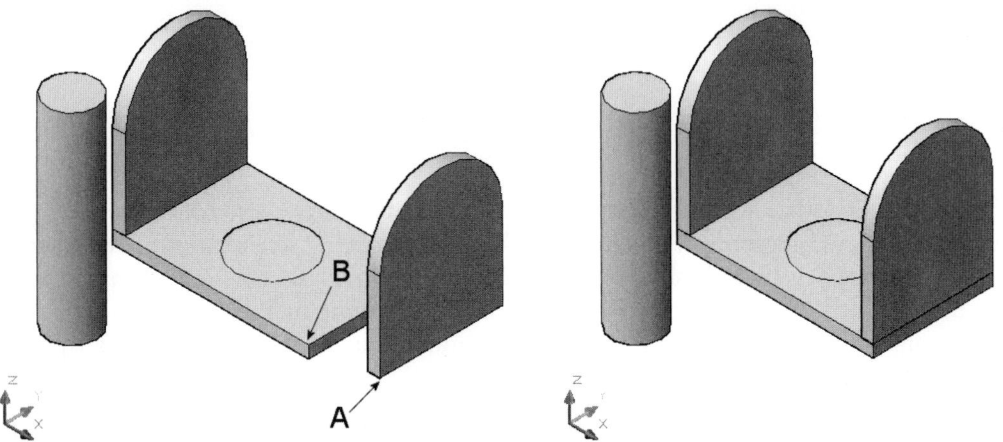

Figure 21.64

Step 7

Next, rotate the long cylinder -90° along the rotation axis, as shown on the left in the following image. Then, move the cylinder from its center at "B" to the center of the arc on the side panel at "C," as shown on the right in the following image.

```
Command: 3DROTATE
Current positive angle in UCS: ANGDIR=counterclockwise ANGBASE=0
Select objects: (Select the tall cylinder)
Select objects: (Press ENTER to continue)
Specify base point: (Pick the center of the cylinder at "A")
Pick a rotation axis: (When the rotate grip tool appears, click
    the axis handle until the proper axis appears, as shown in
    the following image; then pick this axis with the cursor)
Specify angle start point: (Pick the center of the cylinder
    at "A")
Specify angle end point: -90
```

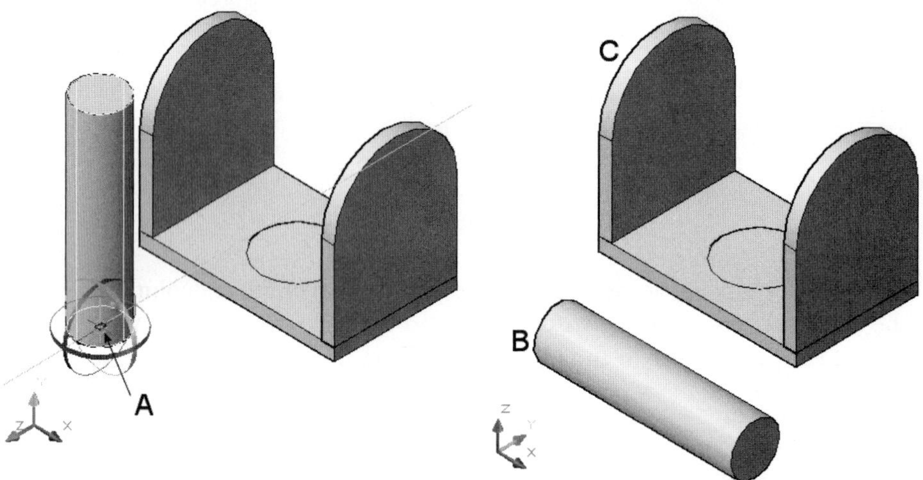

Figure 21.65

Step 8

Finally, use the SUBTRACT command to subtract the two cylinders from the three plates using the following prompt sequence and illustration on the left in the following image. The side panels will automatically be unioned to the base when you complete this command. The completed object is displayed on the right in the following image. The ALIGN command could also have been used to rotate and move the extruded shapes into their proper positions.

```
Command: SU (For SUBTRACT) (can be selected from the
    pull-down menu [Modify > Solids Editing > Subtract])
Select solids and regions to subtract from
Select objects: (Select the two side panels and the base)
Select objects: (Press ENTER to continue)
Select solids and regions to subtract.
Select objects: (Select the two cylinders to subtract)
Select objects: (Press ENTER to perform the union/subtraction
    operations)
```

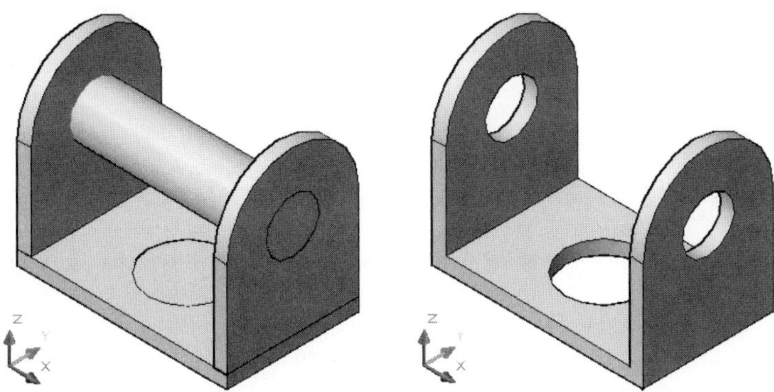

Figure 21.66

PROBLEMS FOR CHAPTER 21

Problem 21–1

Utilize the skills that are laid out in this chapter on problems that are found in chapters 9, 10, and 20. Many of the problems in these chapters apply the editing skills found in this chapter.

One option is to take a drawing that was completed as a two-dimensional drawing and re-create it as a 3D solid.

CHAPTER 22

Creating 2D Multiview Drawings from a Solid Model

One of the advantages of creating a 3D image in the form of a solid model is the ability to use the data of the solid model numerous times for other purposes. The purpose of this chapter is to generate 2D multiview drawings from the solid model. Two commands are used together to perform this operation: SOLVIEW and SOLDRAW. The SOLVIEW command is used to create a layout for the 2D views. This command automatically creates layers used to organize visible lines, hidden lines, and dimensions. The SOLDRAW command draws the requested views on the specific layers that were created by the SOLVIEW command (this includes the drawing of hidden lines to show hidden features, and even hatching if a section is requested).

Included in the layout of 2D multiview projections will be the creation of orthographic, auxiliary, section, and isometric views for a drawing. Also explained will be the FLATSHOT command, used to create 2D views that are saved as blocks. A tutorial is available at the end of this chapter to demonstrate how to dimension orthographic views that are laid out with the SOLVIEW and SOLDRAW commands.

THE SOLVIEW AND SOLDRAW COMMANDS

Once you create the solid model, you can lay out and draw the necessary 2D views using a number of Modeling Setup commands. Choosing Modeling from the Menu Browser or Draw pull-down menu and then Setup, as shown in the following image, exposes the Drawing (SOLDRAW), View (SOLVIEW), and Profile (SOLPROF) commands. Only SOLDRAW and SOLVIEW will be discussed in this chapter. These commands can also be accessed by clicking on the Home tab and then the 3D Modeling panel of the Ribbon, as shown on the right in the following image.

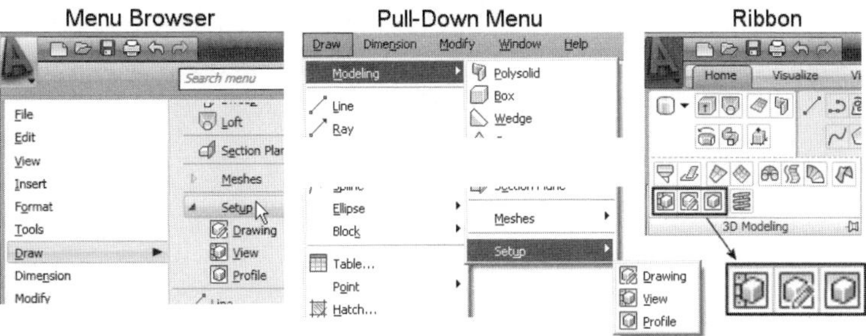

Figure 22.1

Once the SOLVIEW command is entered, the display screen automatically switches to the first layout or Paper Space environment. Using SOLVIEW lays out a view based on responses to a series of prompts, depending on the type of view you want to create. Usually, the initial view that you lay out serves as the starting point for other orthogonal views and is based on the current User Coordinate System. This needs to be determined before you begin this command. Once an initial view is created, it is very easy to create Orthographic, Auxiliary, Section, and even Isometric views.

As SOLVIEW is used as a layout tool, the images of the views created are still simply plan views of the original solid model. In other words, after you lay out a view, it does not contain any 2D features, such as hidden lines. As shown in the previous image, clicking Drawing activates the SOLDRAW command, which is used to actually create the 2D profiles once it has been laid out through the SOLVIEW command.

Try It!: Open the drawing file 22_Solview. Before using the SOLVIEW command, study the illustration of this solid model, as shown at "A" in the following image. In particular, pay close attention to the position of the User Coordinate System icon. The current position of the User Coordinate System will start the creation process of the base view of the 2D drawing.

Tip: Before you start using the SOLVIEW command, remember to load the Hidden linetype. This automatically assigns this linetype to any new layer that requires hidden lines for the drawing mode. If the linetype is not loaded at this point, it must be manually assigned later to each layer that contains hidden lines through the Layer Properties Manager palette.

Activating SOLVIEW automatically switches the display to the layout or Paper Space environment. Since this is the first view to be laid out, the UCS option will be used to create the view based on the current User Coordinate System. The view produced is similar to looking down the Z-axis of the UCS icon. A scale value may be entered for the view. For the View Center, click anywhere on the screen and notice the view being constructed. You can pick numerous times on the screen until the view is in a desired location. The placement of this first view is very important because other views will most likely be positioned from this one. When this step is completed, press ENTER to place the view. Next, you are prompted to construct a viewport around the view. Remember to make this viewport large enough for dimensions to fit inside. Once the view is given a name, it is laid out similar to the illustration at "B" in the following image.

```
Command: SOLVIEW
Enter an option [Ucs/Ortho/Auxiliary/Section]: U (For Ucs)
Enter an option [Named/World/?/Current] <Current>: (Press ENTER)
Enter view scale <1.0000>: (Press ENTER)
Specify view center: (Pick a point near the center of the screen
    to display the view; keep picking until the view is in the
    desired location)
Specify view center <specify viewport>: (Press ENTER to place
    the view)
Specify first corner of viewport: (Pick a point at "D")
Specify opposite corner of viewport: (Pick a point at "E")
Enter view name: FRONT
Enter an option [Ucs/Ortho/Auxiliary/Section]: (Press ENTER to
    exit this command)
```

Once the view has been laid out through the SOLVIEW command, use SOLDRAW to actually draw the view in two dimensions. The Hidden linetype was loaded for you in this drawing and since it was loaded prior to using the SOLVIEW command, hidden lines will automatically be assigned to layers that contain hidden line information. The result of using the SOLDRAW command is shown at "C" in the following image. You are no longer looking at a 3D solid model but at a 2D drawing created by this command.

```
Command: SOLDRAW
(If in Model Space, you are switched to Paper Space)
Select viewports to draw.
Select objects: (Pick anywhere on the viewport at "C" in the
    following image)
Select objects: (Press ENTER to perform the SOLDRAW operation)
One solid selected.
```

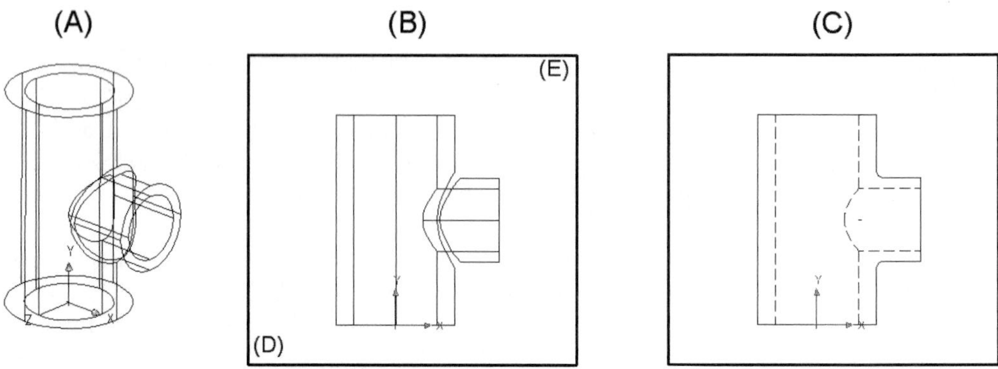

Figure 22.2

The use of layers in 2D-view layout is so important that when you run the SOLVIEW command, the layers shown in the following image are created automatically. With the exception of Model and 0, the layers that begin with "FRONT" and the VPORTS layer were all created by the SOLVIEW command. The FRONT-DIM layer is designed to hold dimension information for the Front view. FRONT-HID holds all hidden lines information for the Front view; FRONT-VIS holds all visible line information for the Front view. All Paper Space viewports are placed on the VPORTS layer. The Model layer has automatically been frozen in the current viewport to hide the 3D model and show the 2D visible and hidden lines.

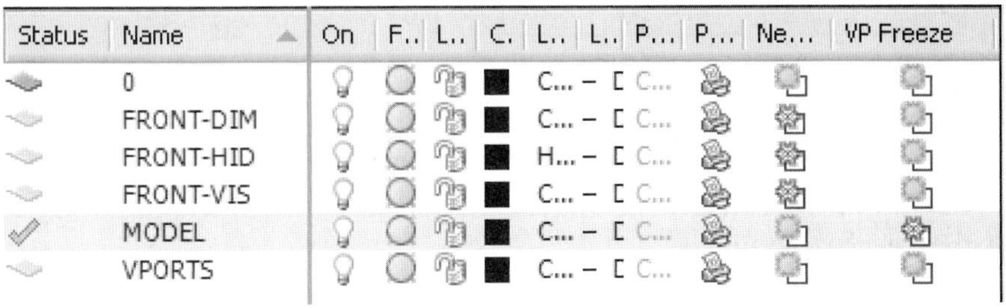

Status	Name	On	F..	L..	C.	L..	L..	P...	P...	Ne...	VP Freeze
	0					C...	– [	C...			
	FRONT-DIM					C...	– [	C...			
	FRONT-HID					H...	– [	C...			
	FRONT-VIS					C...	– [	C...			
	MODEL					C...	– [	C...			
	VPORTS					C...	– [	C...			

Figure 22.3

In order for the view shown on the left in the following image to be dimensioned in model space, three operations must be performed. First, double-click inside the Front view to be sure it is the current floating Model Space viewport. Next, make FRONT-DIM the current layer. Finally, if it is not already positioned correctly, set the User Coordinate System to the current view using the View option of the UCS command. The UCS icon should be similar to the illustration on the left in the following image (you should be looking straight down the Z-axis). Now add all dimensions to the view using conventional dimensioning commands with the aid of Object snap modes. When you work on adding dimensions to another view, the same three operations must be made in the new view: make the viewport active by double-clicking inside it, make the appropriate dimension layer current, and update the UCS, if necessary, to the current view with the View option.

When you draw the views using the SOLDRAW command and then add the dimensions, switching back to the solid model by clicking the Model tab displays the illustration shown on the right in the following image. In addition to the solid model of the object, the constructed 2D view and dimensions are also displayed. All drawn views from Paper Space display with the model. To view just the solid model you would have to use the Layer Properties Manager palette along with the Freeze option and freeze all drawing-related layers.

Tip: Any changes made to the solid model will not update the drawing views. If changes are made, you must erase the previous views and run SOLVIEW and SOLDRAW again to generate a new set of views.

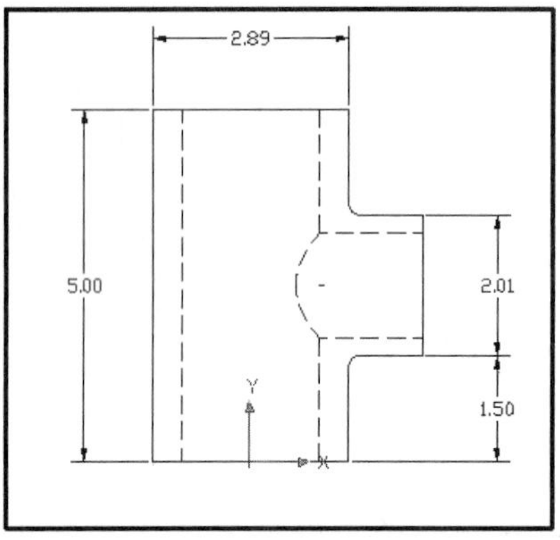

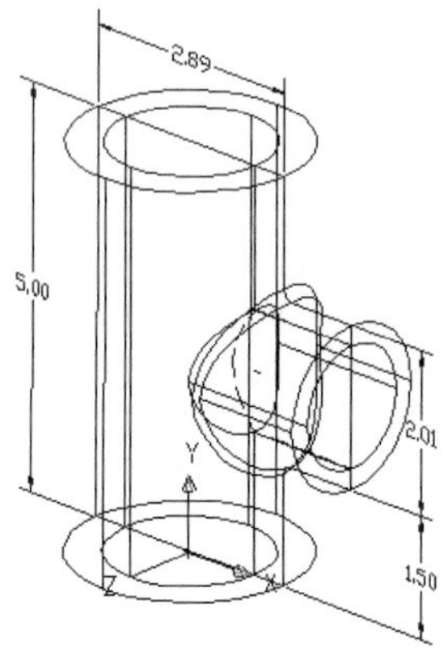

Figure 22.4

CREATING ORTHOGRAPHIC VIEWS

Once the first view is created, orthographic views can easily be created with the Ortho option of the SOLVIEW command.

Try It!: Open the drawing file 22_Ortho. Notice that you are in a layout and a Front view is already created. Follow the command prompt sequence below to create two orthographic views. When finished, your drawing should appear similar to the following image.

```
Command: SOLVIEW
Enter an option [Ucs/Ortho/Auxiliary/Section]: O (For Ortho)
Specify side of viewport to project: (Select the top of the
    viewport at "A"—a midpoint Osnap will be automatically
    provided)
Specify view center: (Pick a point above the front view to
    locate the top view)
Specify view center <specify viewport>: (Press ENTER to place
    the view)
Specify first corner of viewport: (Pick a point at "B")
Specify opposite corner of viewport: (Pick a point at "C")
Enter view name: TOP
Enter an option [Ucs/Ortho/Auxiliary/Section]: O (For Ortho)
Specify side of viewport to project: (Select the right side of
    the viewport at "D")
Specify view center: (Pick a point to the right of the front
    view to locate the right side view)
Specify view center <specify viewport>: (Press ENTER to place
    the view)
Specify first corner of viewport: (Pick a point at "E")
Specify opposite corner of viewport: (Pick a point at "F")
Enter view name: R_SIDE
Enter an option [Ucs/Ortho/Auxiliary/Section]: (Press ENTER to
    exit this command)
```

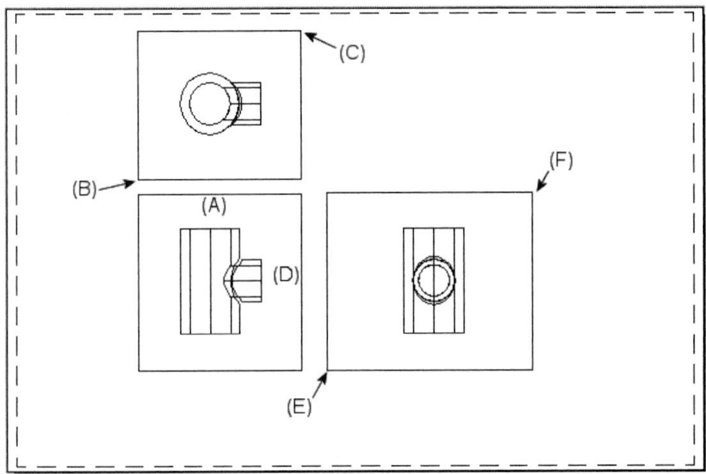

Figure 22.5

Running the SOLDRAW command on the three views displays the illustration shown in the following image. Notice the appearance of the hidden lines in all views. The VPORTS layer is turned off to display only the three views.

```
Command: SOLDRAW
Select viewports to draw.
Select objects: (Select the three viewports that contain the
    Front, Top, and Right Side view information)
Select objects: (Press ENTER to perform the SOLDRAW operation)
```

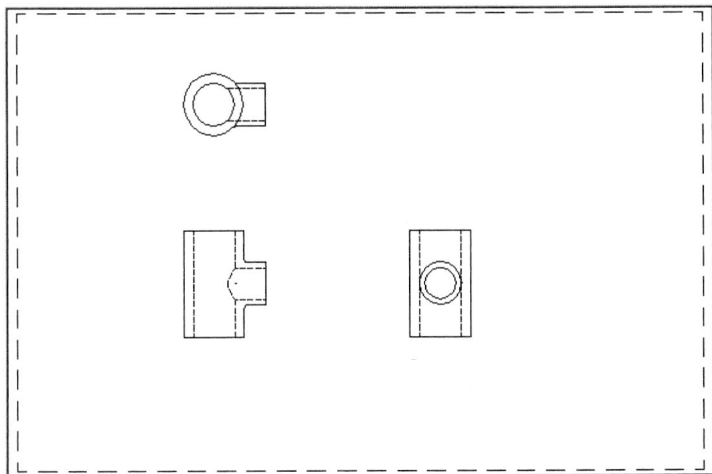

Figure 22.6

CREATING AN AUXILIARY VIEW

In the illustration on the left in the following image, the true size and shape of the inclined surface containing the large counterbore hole cannot be shown with the standard orthographic view. An auxiliary view must be used to properly show these features.

Try It!: Open the drawing file 22_Auxiliary. From the 3D model in the following image use the SOLVIEW command to create a Front view based on the current User Coordinate System. The results are shown on the right in the following image.

```
Command: SOLVIEW
Enter an option [Ucs/Ortho/Auxiliary/Section]: U (For Ucs)
Enter an option [Named/World/?/Current] <Current>: (Press ENTER)
Enter view scale <1.0000>: (Press ENTER)
Specify view center: (Pick a point to locate the view, as shown
    on the right in the following image)
Specify view center <specify viewport>: (Press ENTER to place
    the view)
Specify first corner of viewport: (Pick a point at "A")
Specify opposite corner of viewport: (Pick a point at "B")
Enter view name: FRONT
Enter an option [Ucs/Ortho/Auxiliary/Section]: (Press ENTER to
    exit this command)
```

Tip: Normally you do not end the SOLVIEW command after each view is laid out. Once you finish creating a view you simply enter the appropriate option (Ucs, Ortho, Auxiliary, or Section) and create the next one. This process can continue until all necessary views are provided.

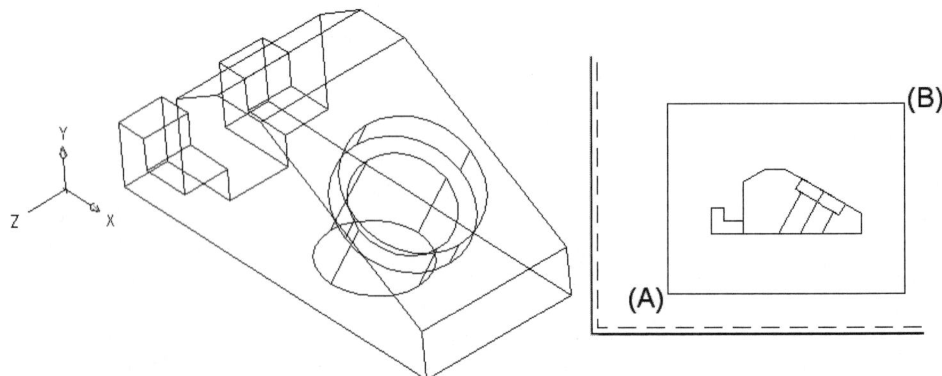

Figure 22.7

Now begin the process of constructing an auxiliary view, as shown on the left in the following image. After selecting the Auxiliary option of the SOLVIEW command, click the endpoints at "A" and "B" to establish the edge of the surface to view. Pick a point at "C" to indicate the side from which to view the auxiliary view. Notice how the Paper Space icon tilts perpendicular to the edge of the auxiliary view. Pick a location for the auxiliary view and establish a viewport. The result is illustrated in the middle in the following image.

```
Command: SOLVIEW
Enter an option [Ucs/Ortho/Auxiliary/Section]: A (For Auxiliary)
Specify first point of inclined plane: End
of (Pick the endpoint at "A")
Specify second point of inclined plane: End
of (Pick the endpoint at "B")
Specify side to view from: (Pick a point inside of the viewport
    at "C")
Specify view center: (Pick a point to locate the view, as shown
    in the middle in the following image)
Specify view center <specify viewport>: (Press ENTER to place
    the view)
Specify first corner of viewport: (Pick a point at "D")
Specify opposite corner of viewport: (Pick a point at "E")
Enter view name: AUXILIARY
Enter an option [Ucs/Ortho/Auxiliary/Section]: (Press ENTER to
    exit this command)
```

Run the SOLDRAW command and turn off the VPORTS layer. The finished result is illustrated on the right in the following image. Hidden lines display only because this linetype was previously loaded.

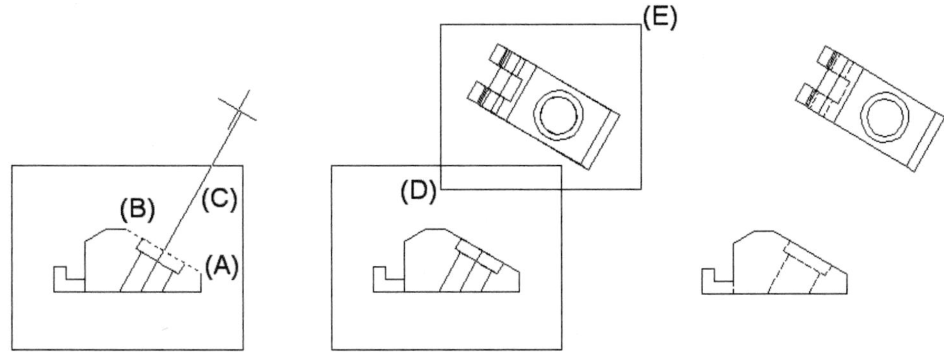

Figure 22.8

CREATING A SECTION VIEW

The SOLVIEW and SOLDRAW commands can also be used to create a full section view of an object. This process automatically creates section lines and place them on a layer (*-HAT) for you.

Try It!: Open the drawing file 22_Section. From the model illustrated on the left in the following image, create a Top view based on the current User Coordinate System, as shown on the right in the following image.

```
Command: SOLVIEW
Regenerating layout.
Enter an option [Ucs/Ortho/Auxiliary/Section]: U (For Ucs)
Enter an option [Named/World/?/Current] <Current>: (Press ENTER)
Enter view scale <1.0000>: (Press ENTER)
Specify view center: (Pick a point to locate the view, as shown
    on the right in the following image)
Specify view center <specify viewport>: (Press ENTER to place
    the view)
Specify first corner of viewport: (Pick a point at "A")
Specify opposite corner of viewport: (Pick a point at "B")
Enter view name: TOP
Enter an option [Ucs/Ortho/Auxiliary/Section]: (Press ENTER to
    exit this command)
```

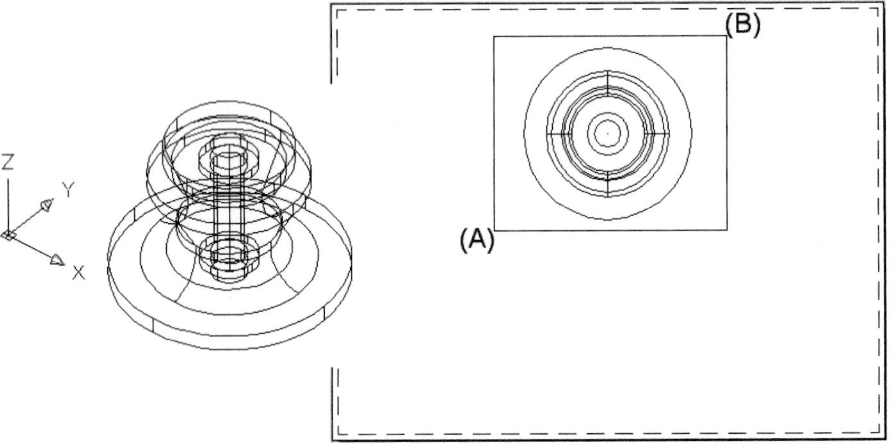

Figure 22.9

Begin the process of creating the section. You must first establish the cutting plane line in the Top view, as shown on the left in the following image. After the cutting plane line is drawn, select the side from which to view the section. Then locate the section view. This is similar to the process of placing an auxiliary view.

```
Command: SOLVIEW
Enter an option [Ucs/Ortho/Auxiliary/Section]: S (For Section)
Specify first point of cutting plane: Qua
of (Pick a point at "A")
Specify second point of cutting plane: (Turn Ortho on, pick a
    point at "B")
Specify side to view from: (Pick a point inside of the viewport
    at "C")
Enter view scale <1.0000>: (Press ENTER)
Specify view center: (Pick a point below the top view to locate
    the view, as shown on the right in the following image)
Specify view center <specify viewport>: (Press ENTER to place
    the view)
Specify first corner of viewport: (Pick a point at "D")
Specify opposite corner of viewport: (Pick a point at "E")
```

```
Enter view name: FRONT_SECTION
Enter an option [Ucs/Ortho/Auxiliary/Section]: (Press ENTER to
    continue)
```

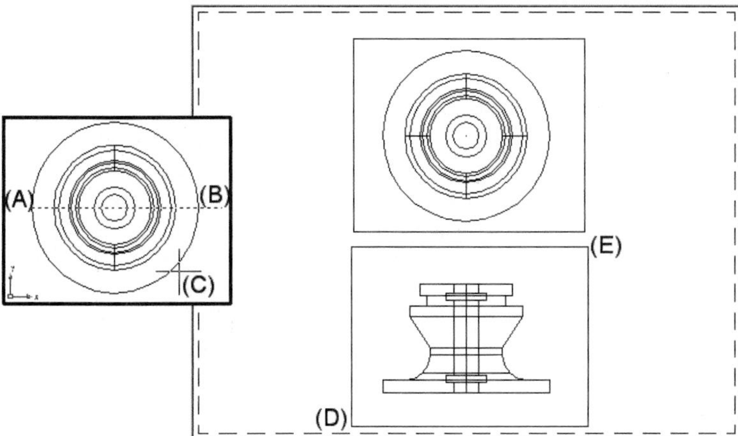

Figure 22.10

Running the SOLDRAW command on the viewports results in the illustration on the left in the following image. You can also activate the viewport displaying the section view and use the HATCHEDIT command to edit the hatch pattern. In the illustration on the right in the following image, the hatch pattern scale was increased to a value of 2.00 and the viewports were turned off.

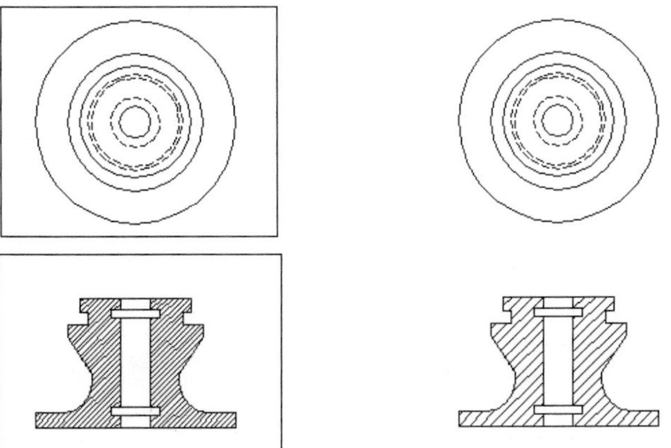

Figure 22.11

CREATING AN ISOMETRIC VIEW

Once orthographic, section, and auxiliary views are projected, you also have an opportunity to project an isometric view of the 3D model. This type of projection is accomplished using the UCS option of the SOLVIEW command and relies entirely on the viewpoint and User Coordinate System setting for your model.

Try It!: Open the drawing file 22_Iso. This 3D model should appear similar to the illustration on the left in the following image. To prepare this image to be projected as an isometric view, first define a new User Coordinate System based on the current view. See the prompt sequence below to accomplish this task. Your image and UCS icon should appear similar to the illustration on the right in the following image.

```
Command: UCS
Current UCS name: *WORLD*
Specify origin of new UCS or [Face/NAmed/OBject/Previous/View/
    World/X/Y/Z/ZAxis] <World>: V (For View)
```

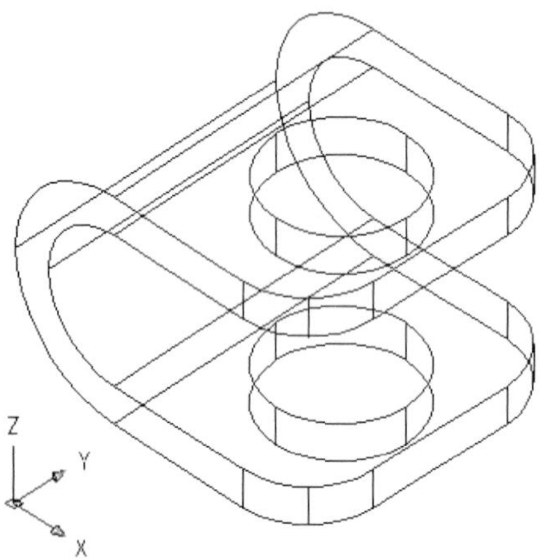

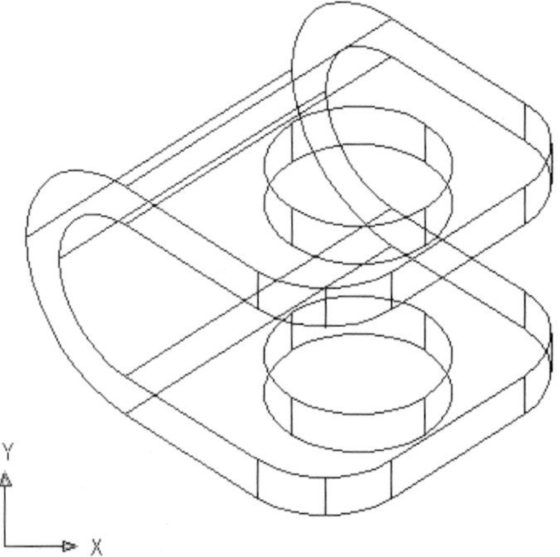

Figure 22.12

Next, run the SOLVIEW command based on the current UCS. Locate the view and construct a viewport around the isometric, as shown in the sample layout in the following image. Since dimensions are normally placed in the orthographic view drawings and not on an isometric, you can tighten up on the size of the viewport.

```
Command: SOLVIEW
Regenerating layout.
Enter an option [Ucs/Ortho/Auxiliary/Section]: U (For Ucs)
Enter an option [Named/World/?/Current] <Current>: (Press ENTER)
Enter view scale <1.0000>: (Press ENTER)
Specify view center: (Pick a point to locate the view in the
    following image)
Specify view center <specify viewport>: (Press ENTER to place
    the view)
Specify first corner of viewport: (Pick a point at "A")
```

```
Specify opposite corner of viewport: (Pick a point at "B")
Enter view name: ISO
Enter an option [Ucs/Ortho/Auxiliary/Section]: (Press ENTER to
    exit this command)
```

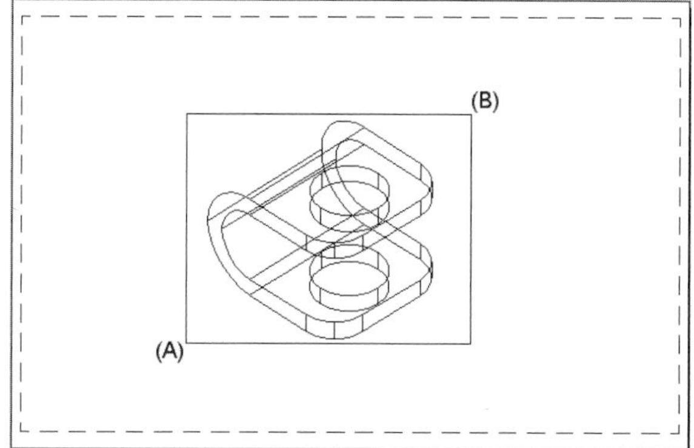

Figure 22.13

Running the SOLDRAW command on the isometric results in visible lines as well as hidden lines being displayed, as shown on the left in the following image. Generally, hidden lines are not displayed in an isometric view. The layer called ISO-HID, which was created by SOLVIEW, contains the hidden lines for the isometric drawing. Use the Layer Properties Manager palette to turn off this layer. The results of this operation are illustrated on the right in the following image.

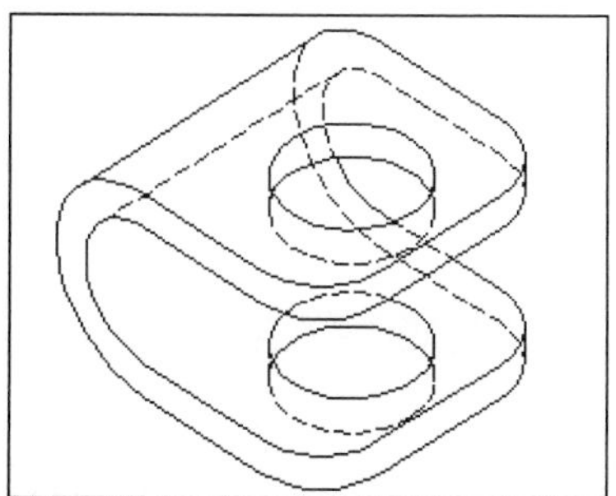

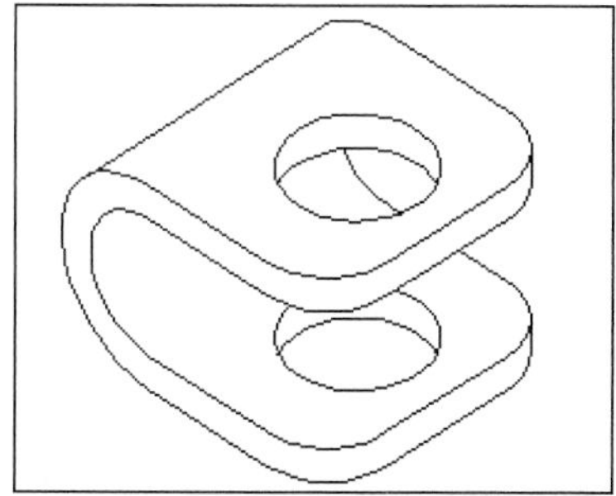

Figure 22.14

EXTRACTING 2D VIEWS WITH FLATSHOT

The FLATSHOT command is used to create a quick 2D view of a 3D solid model based on the current view. First align your view of the 3D solid and Flatshot projects an object onto the XY plane. The object created is in the form of a block and can be inserted and modified if necessary, since the block consists of 2D geometry. Choose Flatshot from the Solid Editing panel of the Ribbon, as shown in the following image.

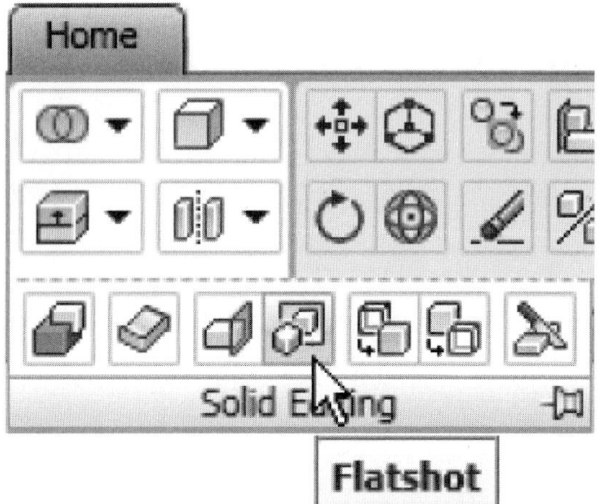

Figure 22.15

Begin the process of creating a flattened 2D view from a 3D solid model by first aligning the screen for the view that you want captured. Illustrated on the left in the following image is a 3D solid model that is currently being viewed in the Front direction. Notice also the alignment of the XY plane; Flatshot will project the geometry to this plane.

When you activate the FLATSHOT command, the dialog box illustrated on the right in the following image displays. The Destination area is used for inserting a new block or replacing an existing block. You can even export the geometry to a file with the familiar DWG extension that can be read directly by AutoCAD.

The Foreground lines area allows you to change the color and linetype of the lines considered visible.

In the Obscured lines area, you have the option of showing or not showing these lines. Obscured lines are considered invisible to the view and should be assigned the HIDDEN linetype if showing this geometry.

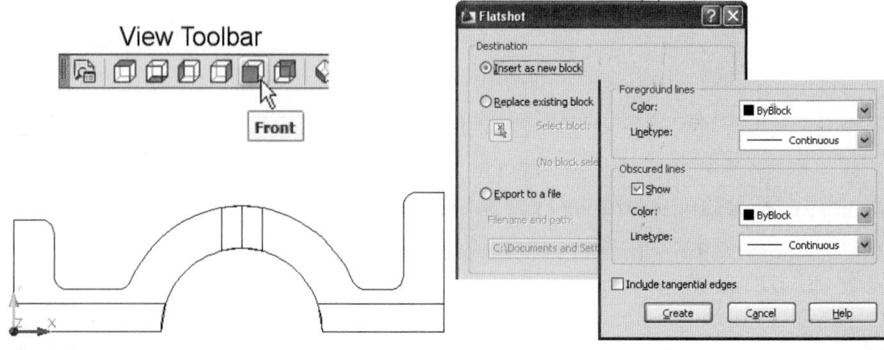

Figure 22.16

After clicking the Create button, you will be prompted to insert the block based on the view. In the following image, the object on the left is the original 3D solid model, and the object on the right is the 2D block generated by the Flatshot operation. An isometric view is used in this image to illustrate the results performed by Flatshot.

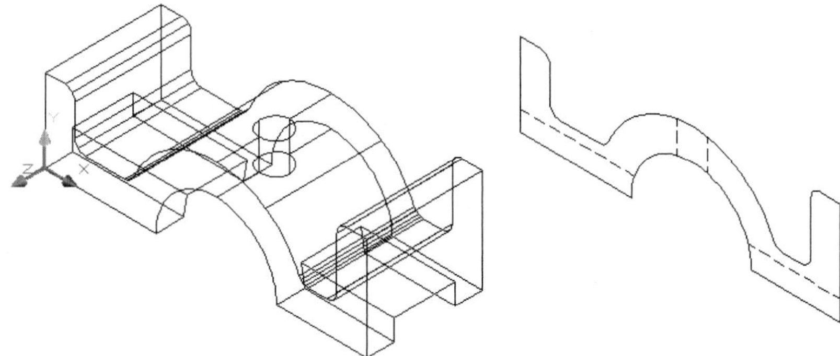

Figure 22.17

It was pointed out that Flatshot creates a block. However, during the creation process, you are never asked to input the name of the block. This is because Flatshot creates a block with a randomly generated name (sometimes referred to as anonymous). This name is illustrated in the following image, where the Rename dialog box (RENAME command) is used to change the name of the block to something more meaningful, such as Front View. In fact, it is considered good practice to immediately rename the block generated by Flatshot to something more recognizable.

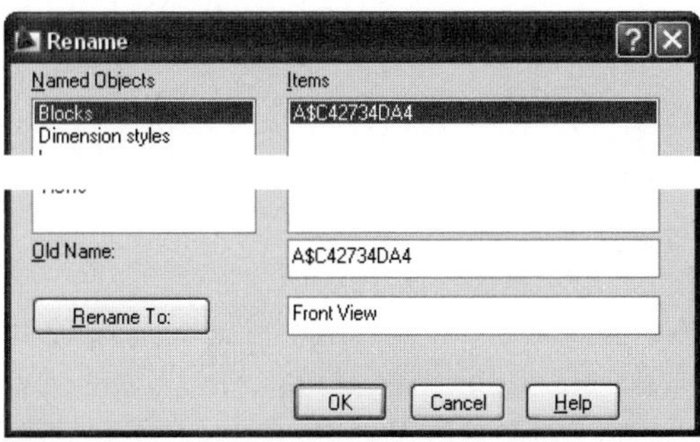

Figure 22.18

By default, whenever creating a block with Flatshot, you must immediately insert this block in Model space. It is also considered good practice to insert all blocks in a layout. In this way, Model space will hold the 3D solid model information and the layout will hold the 2D geometry, as shown in the following image.

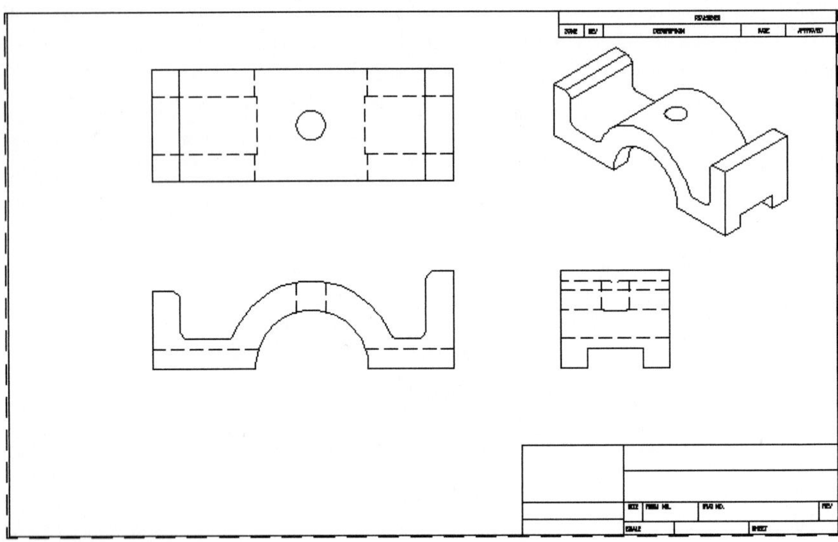

Figure 22.19

Note: While FLATSHOT can be used with layouts, it also provides a quick way to generate a 2D view of a 3D model while in Model Space.

TUTORIAL EXERCISE: 22_SOLID DIMENSION.DWG

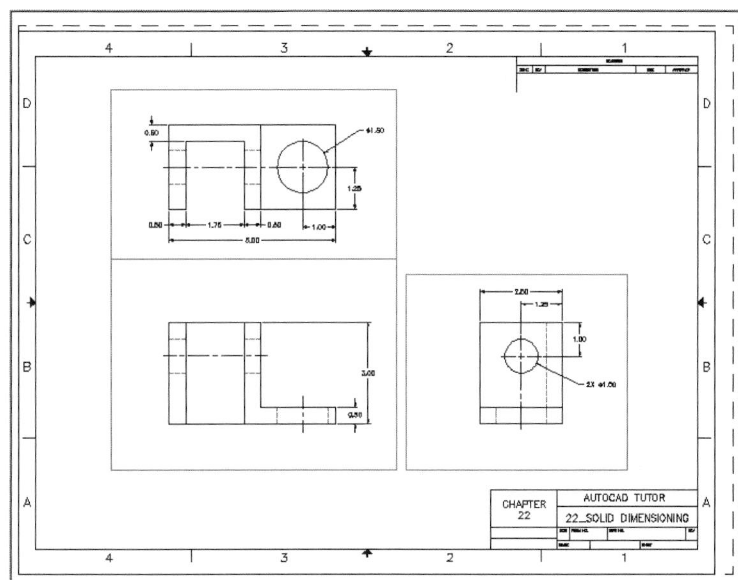

Figure 22.20

PURPOSE

This tutorial exercise is designed to add dimensions to a solid model that has had its views extracted using the SOLVIEW and SOLDRAW commands, as shown in the previous image.

SYSTEM SETTINGS

Drawing and dimension settings have already been changed for this drawing.

LAYERS

Layers have already been created for this tutorial exercise.

SUGGESTED COMMANDS

In this tutorial you will activate the front viewport of the 22_Solid Dimension drawing and make the Front-DIM layer current. Then add dimensions to the Front view. As these dimensions are placed in the Front view, the dimensions do not appear in the other views. This is because the SOLVIEW command automatically creates layers and then freezes those layers in the appropriate viewports. Next, activate the top viewport, make the Top-DIM layer current, and add dimensions to the Top view. Finally, activate the right side viewport, make the Right Side-DIM layer current, and add the remaining dimensions to the Right Side views.

Step 1

Begin this tutorial by opening the drawing 22_Solid Dimension. Your display should appear similar to the following image. Viewports have already been created and locked in this drawing. A locked viewport means that as you activate the viewport and accidentally zoom, the image inside of the viewport does not zoom; rather, the entire drawing is affected by the zoom operation. Centerline layers have also been created and correspond to the three viewports. Centerlines have already been placed in their respective views.

Be sure Osnap is turned on with Endpoint being the active mode.

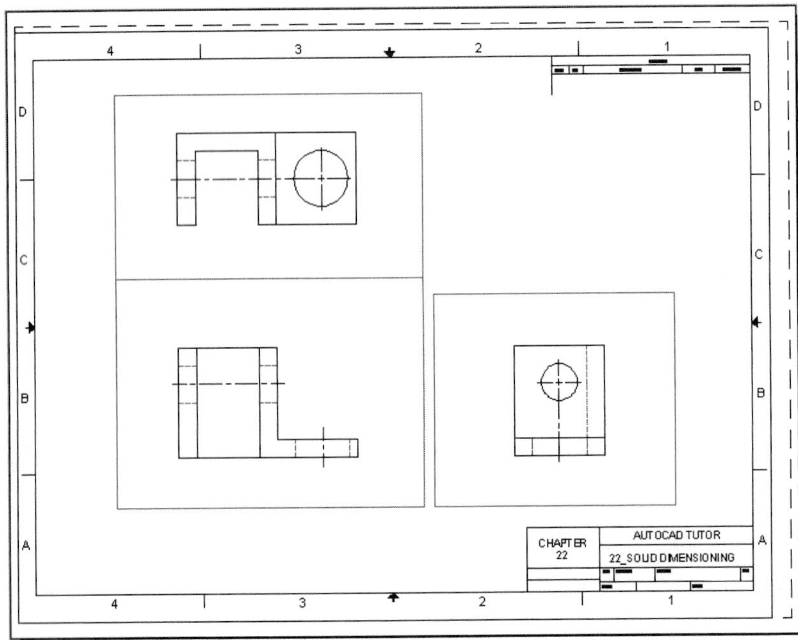

Figure 22.21

Step 2

Activate the viewport that contains the Front view by double-clicking inside of it. You know you have accomplished this if the floating model space icon appears. Then make the Front-DIM layer current, as shown in the following image.

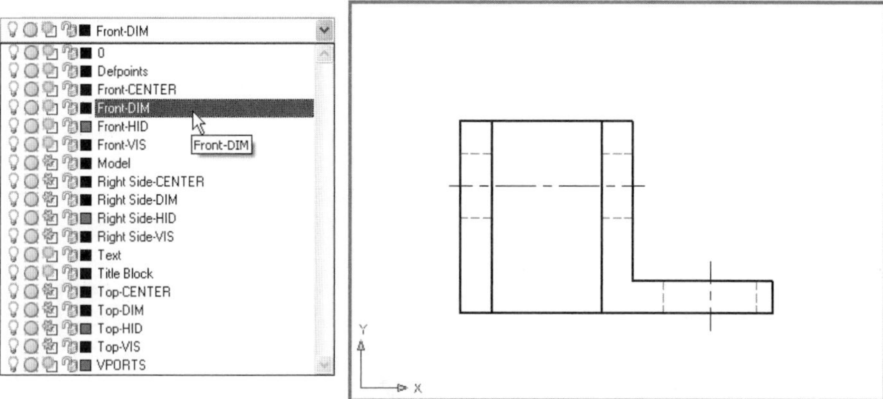

Figure 22.22

Step 3

Most of the layers displayed in the Layer Control box were created by the SOLVIEW command. Notice how they correspond to a particular viewport. For example, study the following table regarding the layer names dealing with the front viewport:

Layer Name	Purpose
Front-CENTER	Centerlines
Front-DIM	Dimension lines
Front-HID	Hidden lines
Front-VIS	Visual (Object) lines

Notice the Layer Front-DIM in the following image. Because the viewport holding the Front view is current, this layer is thawed, meaning the dimensions placed in this viewport will be visible. Notice that the other dimension layers (Right Side-DIM and Top-DIM) are frozen. The SOLVIEW command automatically set up the dimension layers to be visible in the current viewport and frozen in the other viewports.

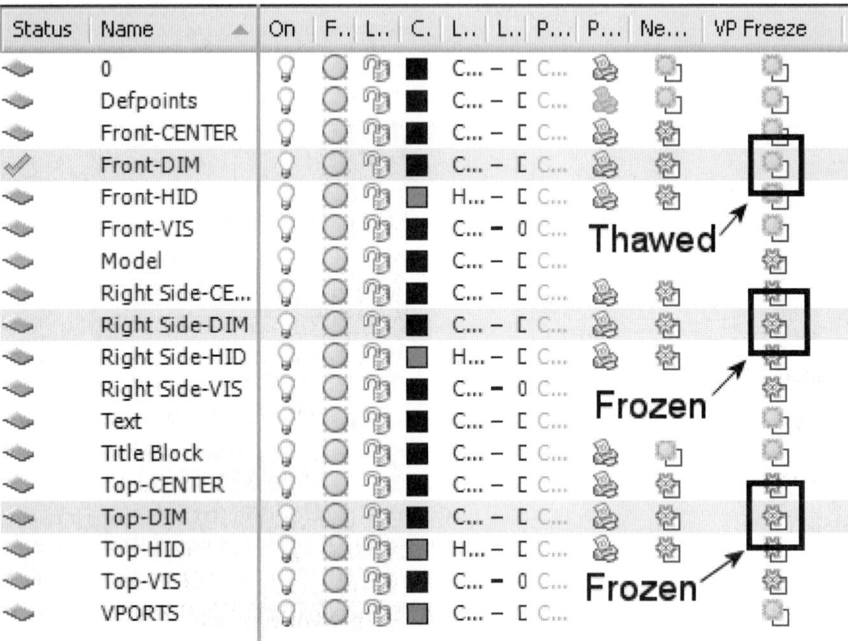

Figure 22.23

Step 4

Add the two dimensions to the Front view, as shown in the following image. Linear and baseline dimension commands could be used to create these dimensions. Grips can be used to stretch the text of the 0.50 dimension so that it is located in the middle of the extension lines.

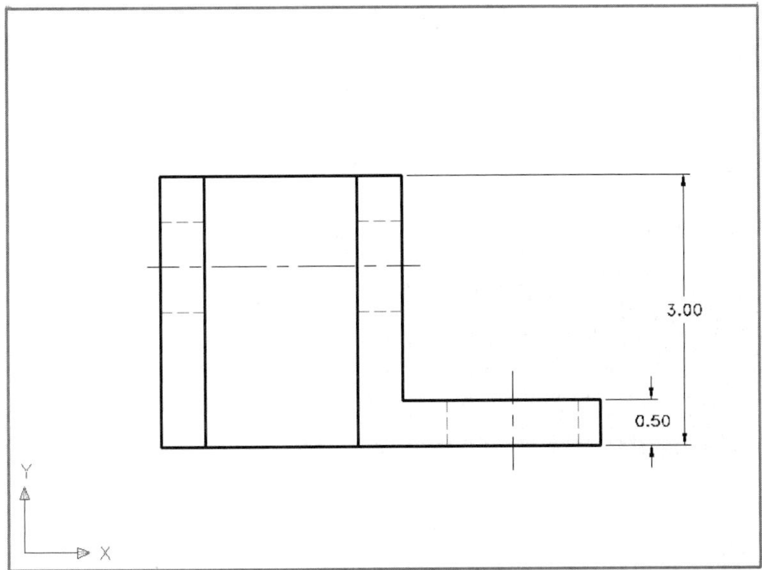

Figure 22.24

Step 5

Activate the viewport that contains the Top view by clicking inside it. Then make the Top-DIM layer current, as shown in the following image. This layer is designed to show dimensions visible in the top viewport and make dimensions in the front and right side viewports frozen (or invisible).

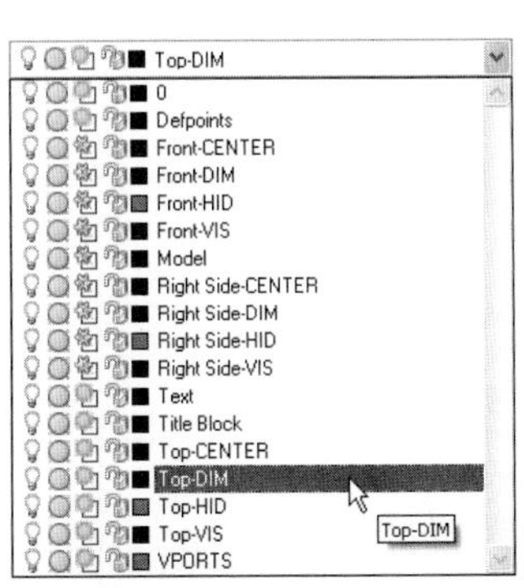

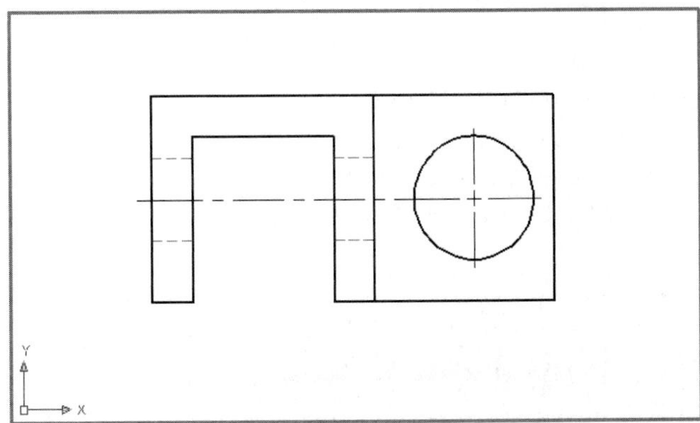

Figure 22.25

Step 6

Add dimensions to the Top view using the following image as a guide.

Tip: Because the viewports are locked, use the zoom and pan operations freely while dimensioning in floating model space.

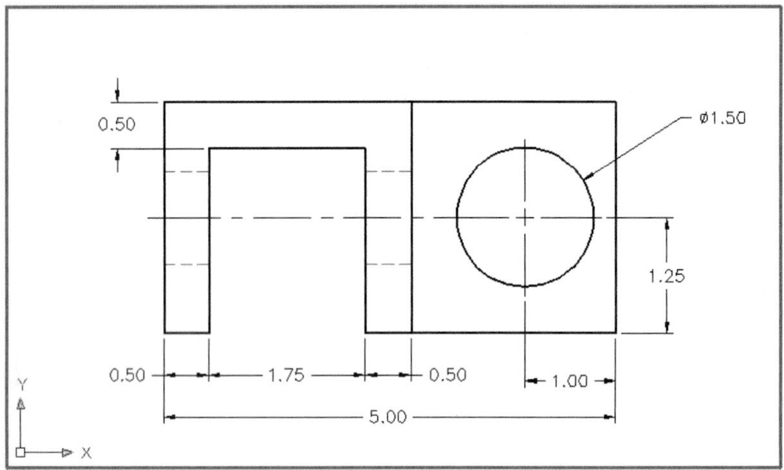

Figure 22.26

Step 7

Activate the viewport that contains the Right Side view by clicking inside the viewport. Then make the Right Side-DIM layer current, as shown in the following image. This layer is designed to make dimensions visible in the right side viewport and make dimensions in the front and top viewports frozen (or invisible).

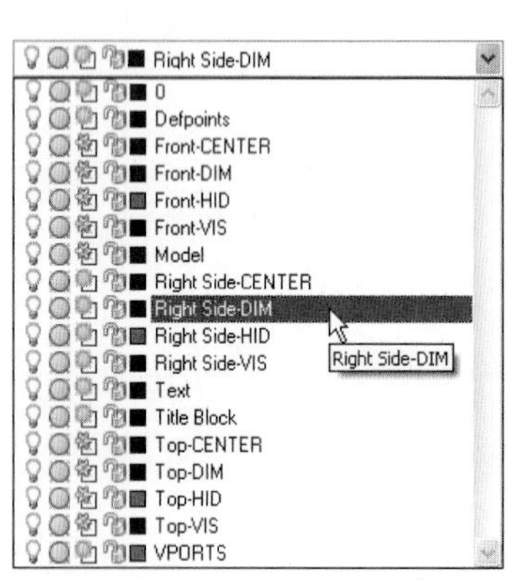

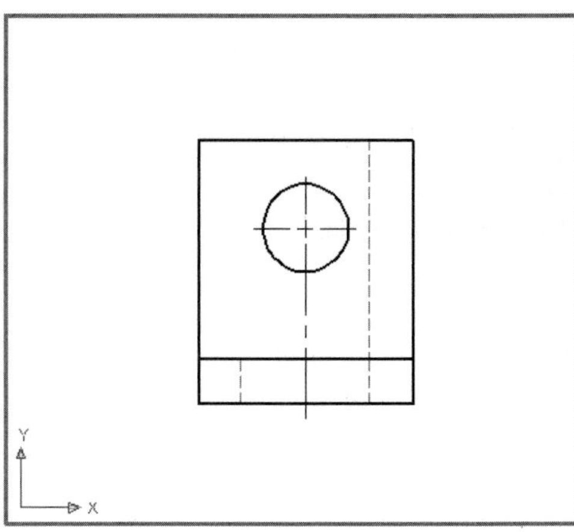

Figure 22.27

Step 8

Add dimensions to the Right Side view using the following image as a guide. Edit the diameter dimension text to reflect two holes. Use the DDEDIT command to accomplish this. When the Text Formatting dialog box appears, type "2X" and click the OK button.

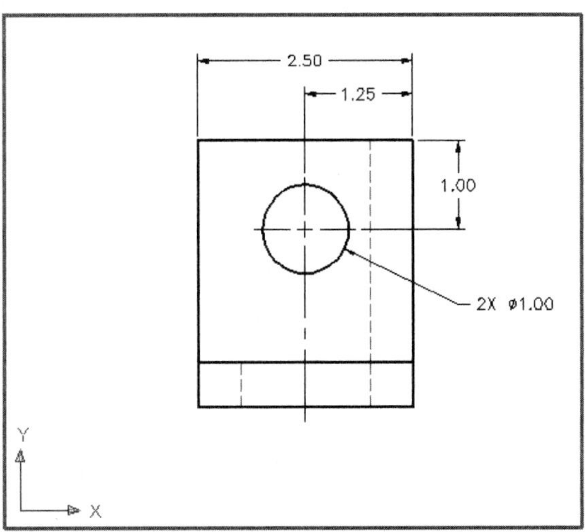

Figure 22.28

Step 9

The completed dimensioned solid model is illustrated in the following image.

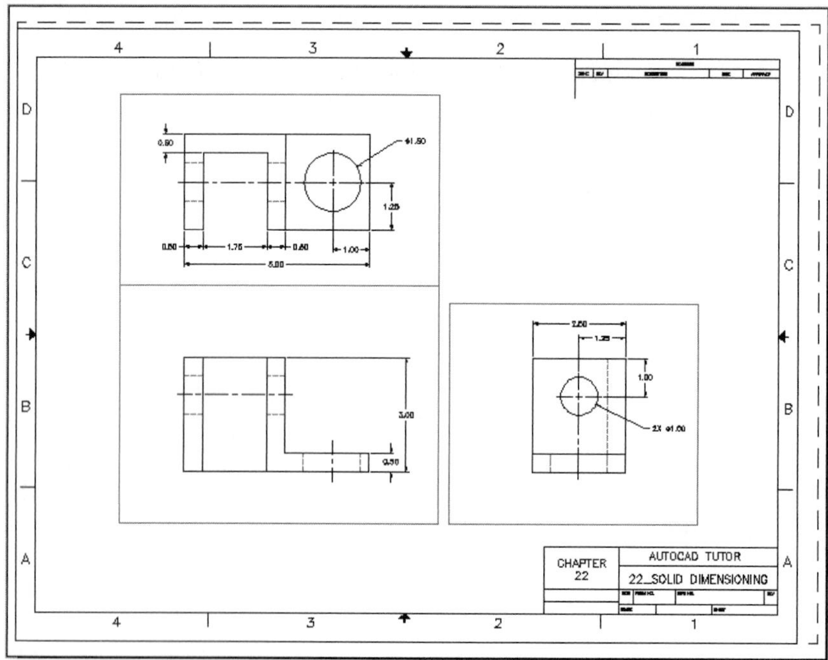

Figure 22.29

PROBLEMS FOR CHAPTER 22

1. Create an engineering drawing of Problems 20–1 through 20–20 consisting of Front, Top, Right Side, and Isometric views.
2. Add proper dimensions to each drawing.

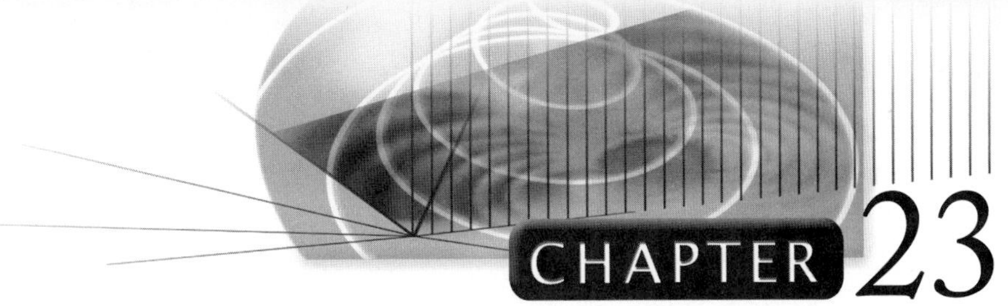

Producing Renderings and Motion Studies

This chapter introduces you to renderings in AutoCAD and how to produce realistic renderings of 3D models. The heart of the rendering operation is the controls found in the dashboard. Adding lights will be the next topic of discussion. With this important feature you produce unlimited special effects, depending on the location and intensity of the lights. The Point, Distant, and Spotlights features are discussed, as is the ability to simulate sunlight. You will be able to attach materials to your models to provide additional realism. Materials can be selected from a library available in AutoCAD, or you can even make your own materials. You will be shown how to walk and fly through a 3D model. You will also be given instruction on how to make a motion path animation of a 3D model.

AN INTRODUCTION TO RENDERINGS

Engineering and architectural drawings are able to pack a vast amount of information into a 2D outline drawing supplemented with dimensions, some symbols, and a few terse notes. However, training, experience, and sometimes imagination are required to interpret them, and many people would rather see a realistic picture of the object. Actually, realistic pictures of a 3D model are more than just a visual aid for the untrained. They can help everyone visualize and appreciate a design, and can sometimes even reveal design flaws and errors.

Shaded, realistic pictures of 3D models are called renderings. Until recently, they were made with colored pencils and pens or with paintbrushes and airbrushes. Now they are often made with computers, and AutoCAD comes with a rendering program that is automatically installed by the typical AutoCAD installation and is ready for your use. The following image shows, for comparison, the solid model of a bracket in its wireframe form, as it looks when the HIDE command has been invoked, and when it is rendered.

This chapter is designed to give you an overview on how to create pleasing, photorealistic renderings of your 3D models.

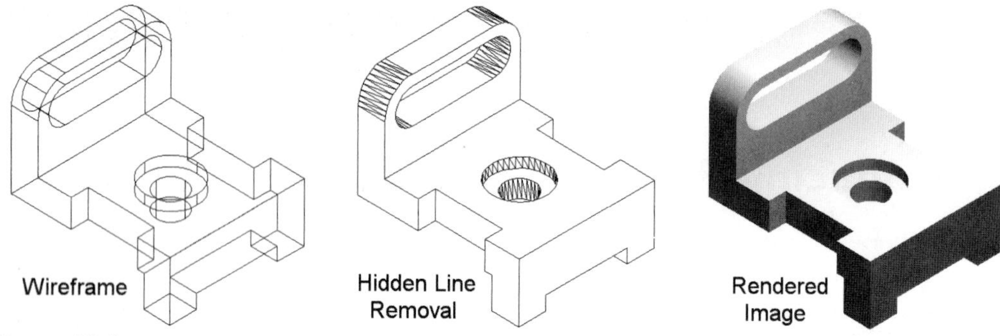

Figure 23.1

The following image displays numerous ways of accessing rendering commands. Clicking View in the pull-down menu, followed by Render, displays a number of tools used for rendering, as shown on the left in the following image. The Render toolbar, also shown in the following image, is another convenient way to access rendering commands. The Menu Browser, shown on the right in the following image, can also be used to access light, material, and rendering commands.

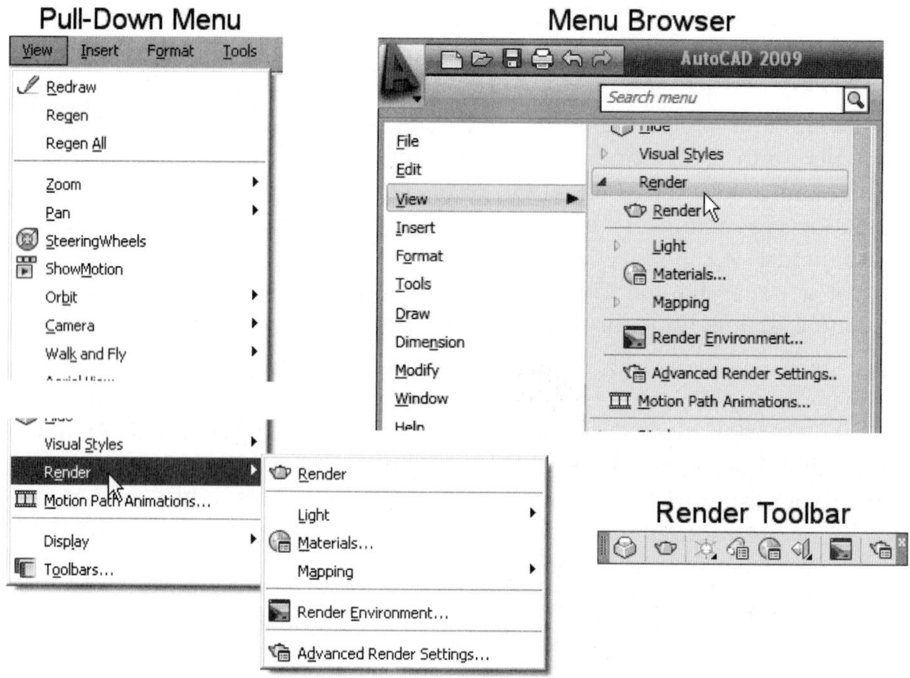

Figure 23.2

The following table gives a brief description of each rendering tool.

Button	Tool	Function
	Hide	Performs a hidden line removal on a 3D model
	Render	Switches to the Render window, where a true rendering of the 3D model is performed
	Lights	Contains six additional buttons used for controlling lights in a 3D model
	Light List	Displays the Lights in Model palette used for managing lights that already exist in a 3D model
	Materials	Displays the Materials palette, used for creating and applying materials to a 3D model
	Planar Mapping	Contains four additional buttons, used for mapping materials to planar, box, cylindrical, and spherical surfaces
	Render Environment	Displays the Rendering Environment dialog box that is used mainly for controlling the amount of fog applied to a 3D model
	Advanced Render Settings	Displays the Advanced Rendering Settings palette, used for making changes to various rendering settings

AN OVERVIEW OF PRODUCING RENDERINGS

The object illustrated in the following image consists of a 3D model that has a Realistic visual style applied. The color of the model comes from the color set through the Layer Properties Manager. After a series of lights are placed in a 3D model, and when materials have been applied, the next step in the rendering process is to decide how accurate a rendering to make.

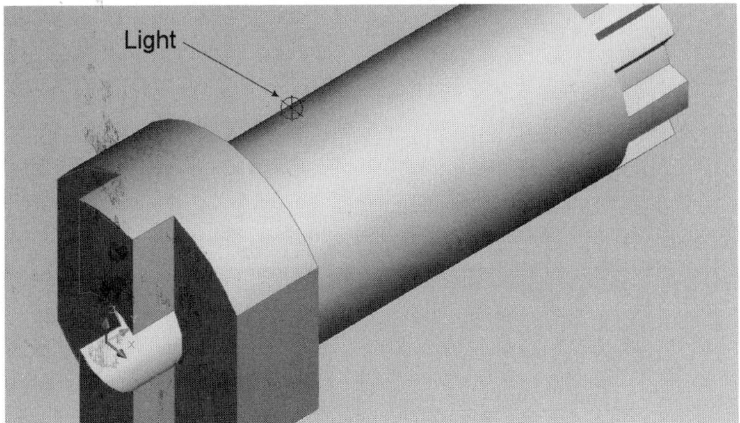

Figure 23.3

As shown in the following image of the Ribbon, the Render panel displays five different rendering modes, or presets, as they are called. These presets range from Draft to

Presentation and control the quality of the final rendered image. For example, when you perform a rendering in draft mode, the processing speed of the rendering is very fast; however, the quality of the rendering is very poor. This render preset is used, for example, to perform a quick rendering when you are unsure about the positioning of lights. When you are pleased with the lighting in the 3D model, you can switch to a higher render preset, such as High or Presentation. These modes process the rendering very slowly; however, the quality is photo-realistic. To perform a rendering, click the Render button, as shown in the following image.

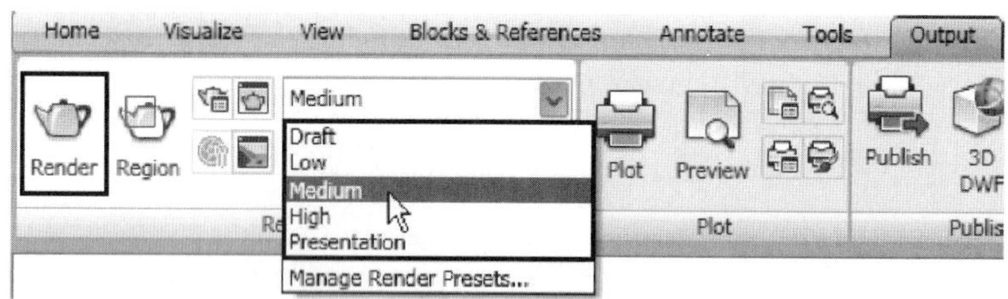

Figure 23.4

Clicking the Render button, as shown in the previous image, switches your screen to the render window, as shown in the following image. Notice the shadows that are cast to the base of the 3D model. Shadow effects are one of many special rendering tools used to make a 3D model appear more realistic. Illustrated on the right of the rendering window is an area used for viewing information regarding the rendered image. Also, when you produce a number of renderings, they are saved in a list at the bottom of the rendering window.

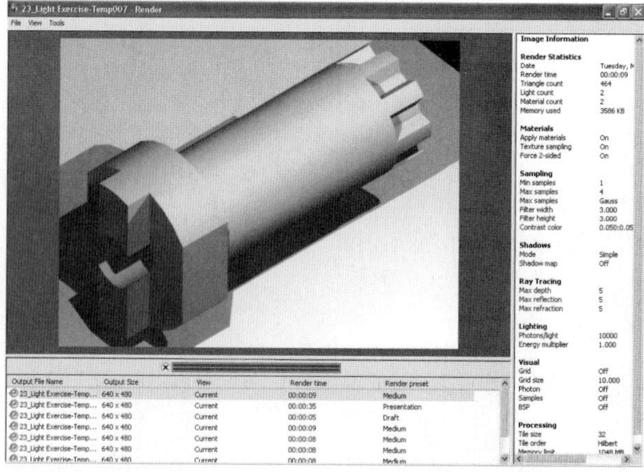

Figure 23.5

As mentioned earlier, the following image displays the results of performing a Draft versus Presentation rendering. In the draft image, notice that the edges of the model do not look as sharp as they do in the presentation model. Also, the draft image does not apply shadows when being rendered. All these factors speed up the rendering of the draft image; however, the quality of the image suffers.

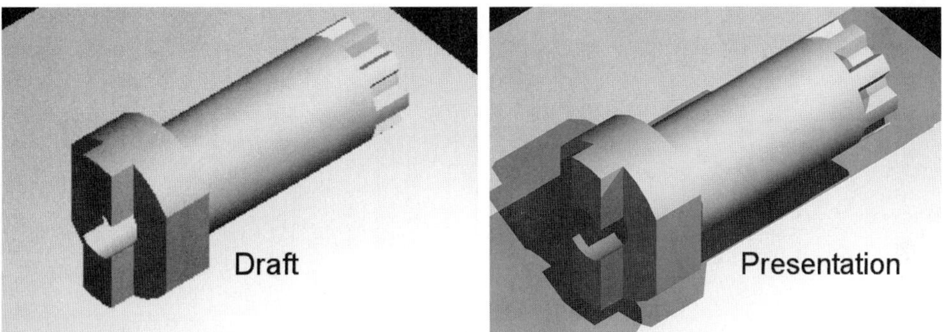

Figure 23.6

When you have produced a quality rendered image, you can save this image under one of the many file formats illustrated in the following image. Supported raster image formats include BMP, PCX, TGA, TIF, JPEG, and PNG.

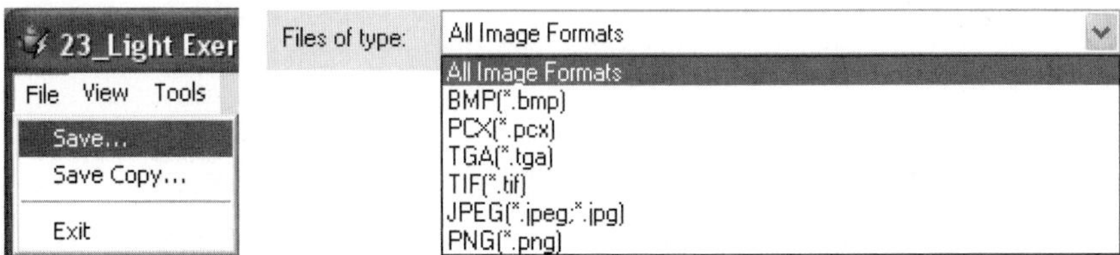

Figure 23.7

Try It!: This exercise consists of an overview of the rendering process. Open the drawing file 23_Render_Presets. You will notice a container and two glasses resting on top of a flat platform, as shown in the following image. Also shown in this image are two circular shapes with lines crossing through their centers. These shapes represent lights in the model. A third light representing a spotlight is also present in this model, although it is not visible in the following image.

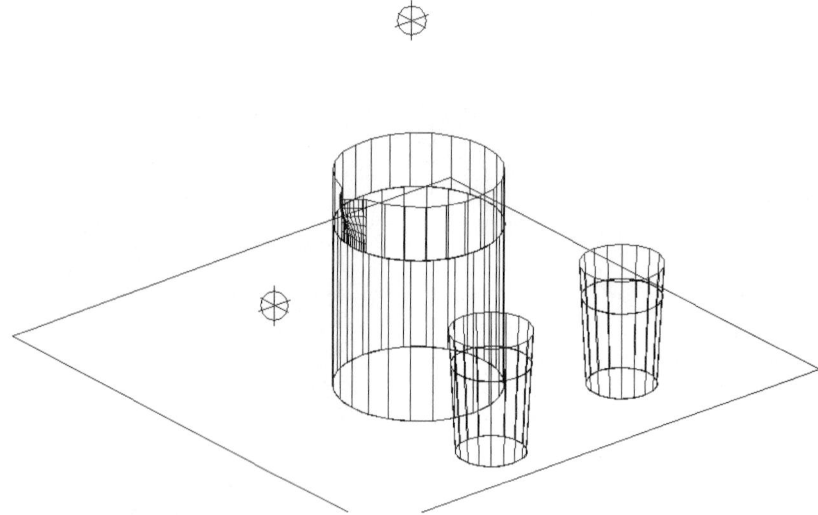

Figure 23.8

Activate the 3D Modeling workspace, click on the Output tab, and click the Render button. The results are illustrated on the following image, with materials, lights, and a background being part of the rendered scheme. Notice in the Render panel the presence of Medium. This represents one of the many render presets used to control the quality of the rendered image.

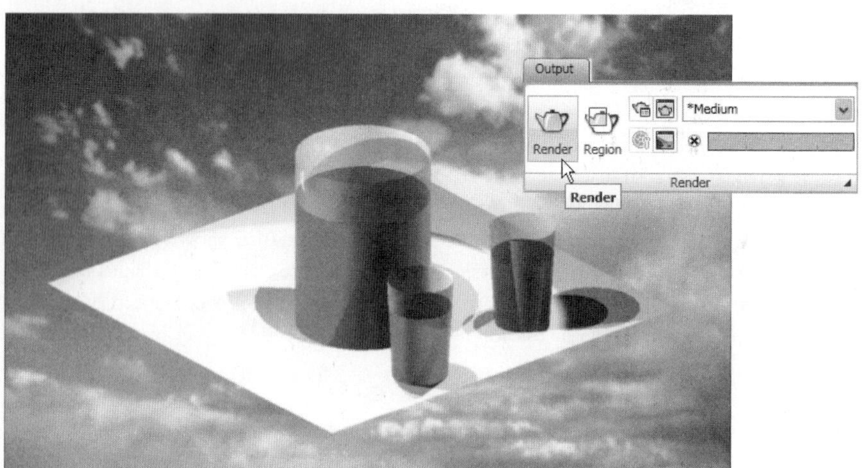

Figure 23.9

To see how the Render Preset value affects the rendered image, change the render preset in the panel from Medium to Draft and click on the Render button, as shown in the following image. Notice that the quality of the rendered image looks choppy; also, shadows are lost. However, the Draft Render Preset is always useful when testing out the lighting of the rendered scene.

The processing time of this preset is very fast compared to other render presets, although the quality of the Draft Render Preset does not look very appealing.

Figure 23.10

In the previous image, the lighting looks too bright and overpowering in the rendered image. To edit the intensity of existing lights, click the Lights List button, located in the Visualize panel of the Ribbon, as shown on the left in the following image. This launches the Lights in Model palette, as shown in the middle of the following image. Double-clicking Pointlight1 launches the Properties palette for that light, as shown on the right in the following image. Locate the Intensity factor under the General category and change the default intensity value of 1.00 to a new value of 0.50. This reduces the intensity of this light by half. Perform this same operation on Pointlight2 and Spotlight1 by changing their intensities from 1.00 to 0.50.

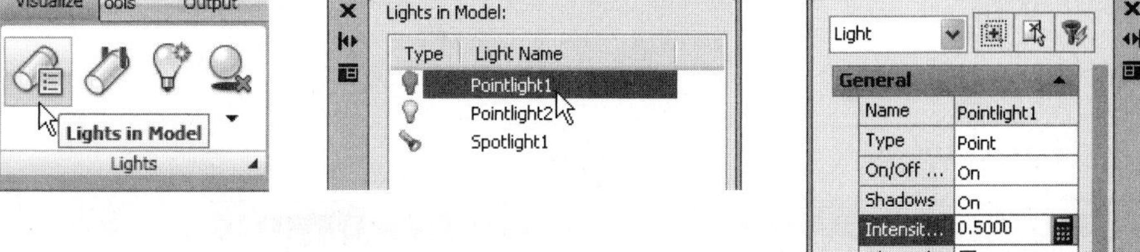

Figure 23.11

After changing the intensity of each light in the model, change the render preset from Draft to High in the Ribbon and then click the Render button. The results are displayed in the following image, with the lights being less intense and the shadows being more pronounced.

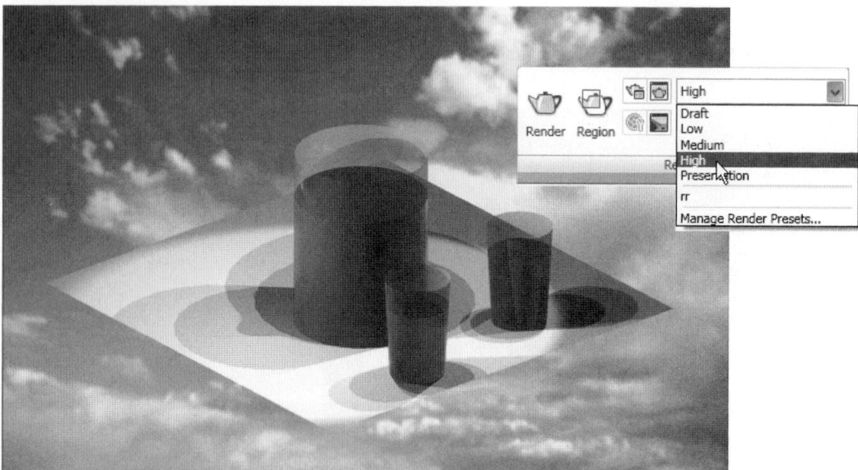

Figure 23.12

Next, click the Materials button, located under the Materials tab of the Ribbon, as shown on the left in the following image. The Materials palette displays all materials created in the model. Clicking the material displays the properties of this material, as shown on the left in the following image. We will experiment more with materials later in this chapter. This concludes the exercise.

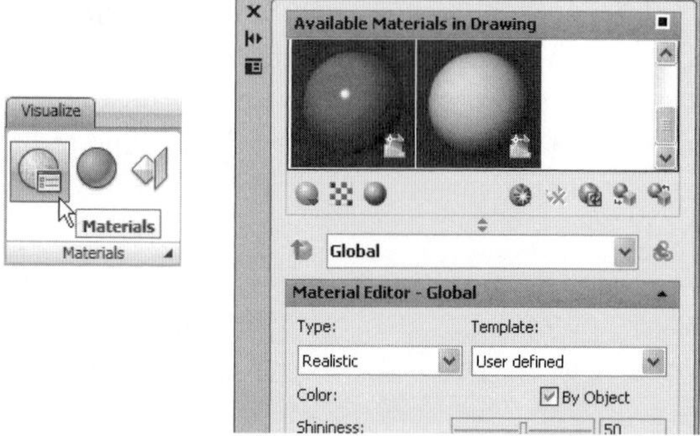

Figure 23.13

CREATING AND PLACING LIGHTS FOR RENDERING

The ability to produce realistic renderings is dependent on what type of lighting is used and how these lights are placed in the 3D model. Once the lights are placed, their properties can be edited through the Properties palette. Before placing the light, you need to decide on the type of light you wish to use: Point Light, Spotlight, or Distant Light. The following image displays three areas for obtaining light commands: the View pull-down menu, on the left; the Lights toolbar, in the middle; and the Lights panel, located in the Ribbon on the right.

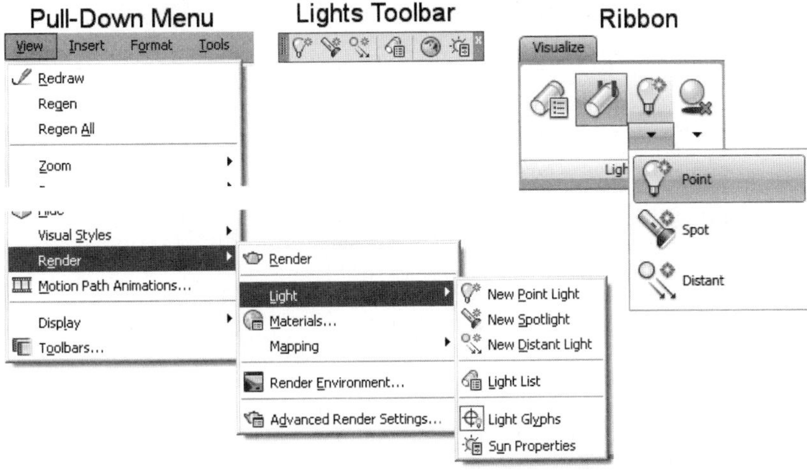

Figure 23.14

You can place numerous lights in a single 3D model and then adjust these lights depending on the desired affect. Each of the standard light types also allows for shadows to be cast, giving depth to your 3D model. The following table displays each light button along with the name of the light and its function.

Button	Tool	Function
	New Point Light	Creates a new point light, similar to a lightbulb
	New Spotlight	Creates a new spotlight given a source and target
	New Distant Light	Creates a new distant light, similar to the sun
	Light List	Displays a list of all lights defined in a model
	Geographic Location	Displays a dialog box that allows you to select a geographic location from which to calculate the sun angle
	Sun Properties	Displays the Sun Properties palette, which allows you to make changes to the properties of the sun

The Geographic Location button is unique in that it allows you to control the position of the sun depending on a location, as shown in the following image. Clicking the Use Map button, as shown on the left in the following image, displays regional maps of the world, which allow you to produce sun studies based on these locations.

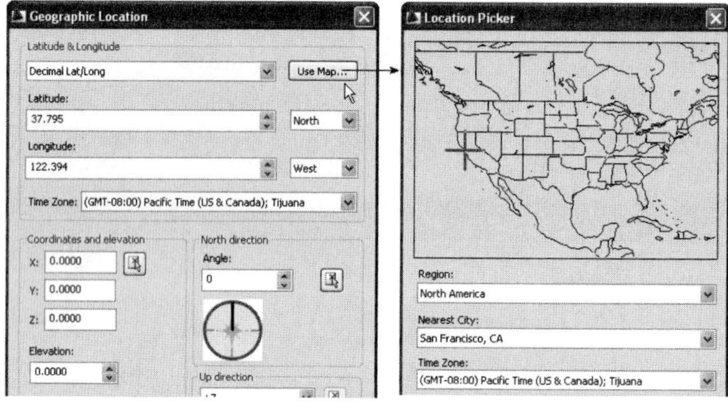

Figure 23.15

A number of lights are also available through the Tool Palette, as shown on the left in the following image. Five light categories are available to give your renderings a more realistic appearance. One of the categories deals with generic lights, as shown on the right in the following image. Through this category, you can drag and drop point, spot, and distant lights into your 3D model. The remaining categories of lights are classified as photometric.

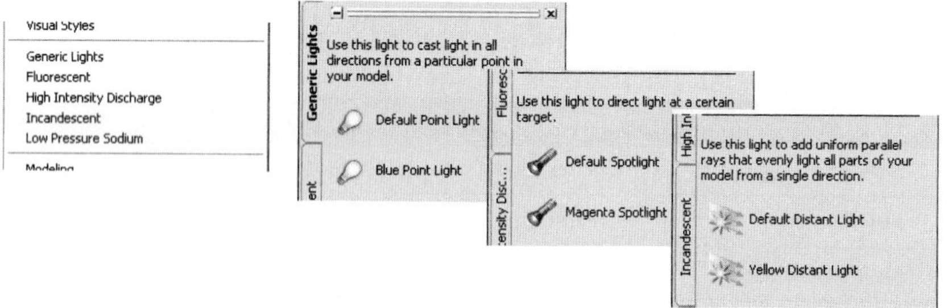

Figure 23.16

Photometric lights allow you to illuminate your scenes for an even more realistic rendering. The four categories of photometric lights are fluorescent, high intensity discharge, incandescent, and low pressure sodium, all shown in the following image. This means that if your model requires the use of a 100W halogen bulb, you select this light from the tool palette and drag and drop it into place.

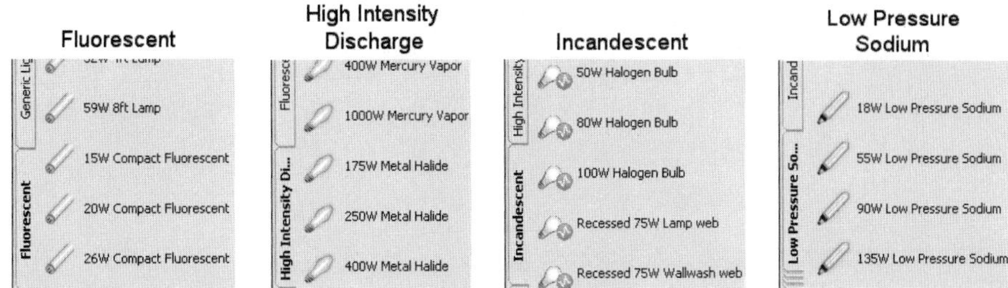

Figure 23.17

Try It!: Open the file 23_Valve Head Lights. This drawing file, shown in the following image, contains the solid model of a machine component that we will use for a rendering. You will place a number of point lights and one spotlight to illuminate this model.

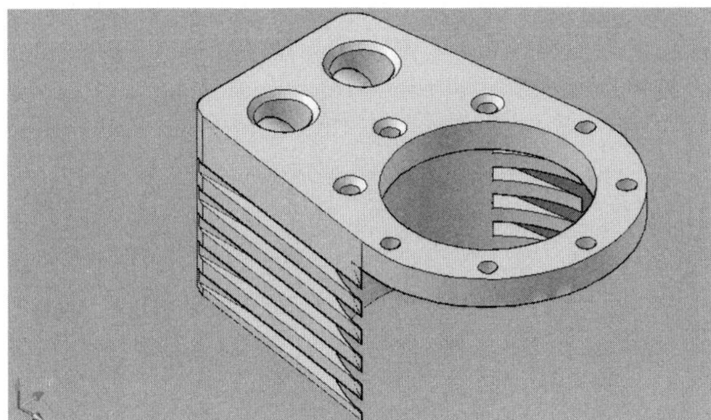

Figure 23.18

Switch to Plan view, as shown in the following image. It will be easier to place the lights while viewing the model from this position. Activate the Ribbon, and in the Light panel, click the Point button, as shown on the right in the following image.

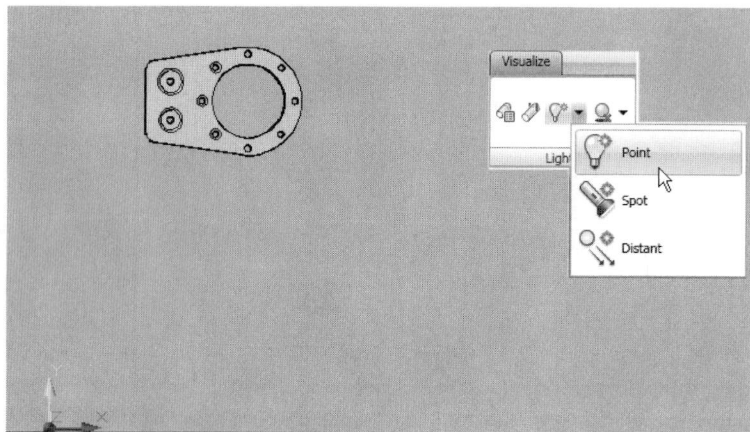

Figure 23.19

Place three point lights using the following prompts. You will place the lights at the approximate locations indicated in the following image. You will change the intensity and the default names of each light through the following command prompts.

For the first light, enter the following information:

```
Command: _pointlight (Pick from the Light control panel of the
    dashboard)
Specify source location <0,0,0>: (Pick at "A")
Enter an option to change [Name/Intensity/Status/shadoW/
    Attenuation/Color/eXit]
<eXit>: I (For Intensity)
Enter intensity (0.00 - max float) <1.0000>: .25
Enter an option to change [Name/Intensity/Status/shadoW/
    Attenuation/Color/eXit]
<eXit>: N (For Name)
Enter light name <Pointlight1>: Overhead Light
Enter an option to change [Name/Intensity/Status/shadoW/
    Attenuation/Color/eXit]
<eXit>: (Press ENTER to create the light)
```

For the second light, enter the following information:

```
Command: _pointlight (Pick from the Light control panel of the
    dashboard)
Specify source location <0,0,0>: (Pick at "B")
Enter an option to change [Name/Intensity/Status/shadoW/
    Attenuation/Color/eXit]
<eXit>: I (For Intensity)
Enter intensity (0.00 - max float) <1.0000>: .25
Enter an option to change [Name/Intensity/Status/shadoW/
    Attenuation/Color/eXit]
<eXit>: N (For Name)
Enter light name <Pointlight2>: Lower Left Light
```

```
Enter an option to change [Name/Intensity/Status/shadoW/
    Attenuation/Color/eXit]
<eXit>: (Press ENTER to create the light)
```

For the third light, enter the following information:

```
Command: _pointlight (Pick from the Light control panel of the
    dashboard)
Specify source location <0,0,0>: (Pick at "C")
Enter an option to change [Name/Intensity/Status/shadoW/
    Attenuation/Color/eXit]
<eXit>: I (For Intensity)
Enter intensity (0.00 - max float) <1.0000>: .25
Enter an option to change [Name/Intensity/Status/shadoW/
    Attenuation/Color/eXit]
<eXit>: N (For Name)
Enter light name <Pointlight3>: Upper Left Light
Enter an option to change [Name/Intensity/Status/shadoW/
    Attenuation/Color/eXit]
<eXit>: (Press ENTER to create the light)
```

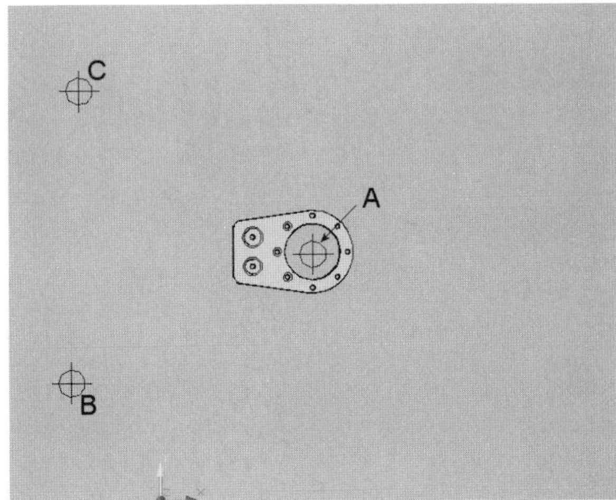

Figure 23.20

When you have finished placing all three point lights, activate the View Manager dialog box (the VIEW command), select the SE Zoomed view, and click the Set Current button, as shown in the following image. This changes your model to a zoomed-in version of the Southeast Isometric view.

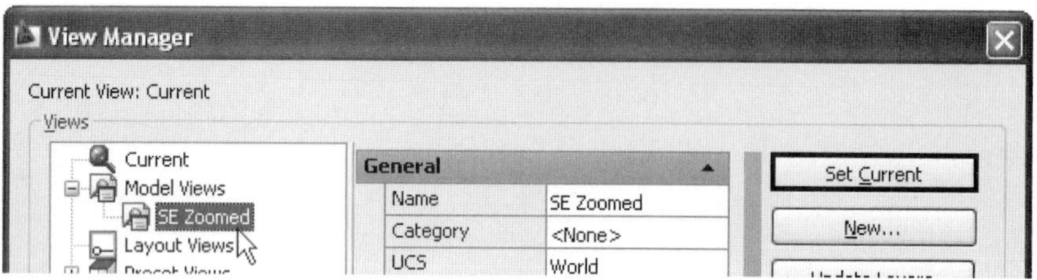

Figure 23.21

When all three point lights were placed, unfortunately all of these lights were located on the top of the base plate, as shown in the following image. The lights need to be assigned an elevation, or Z coordinate. To perform this task, click the Lights in Model button located in the Ribbon to display the Lights in Model palette, as shown on the right in the following image.

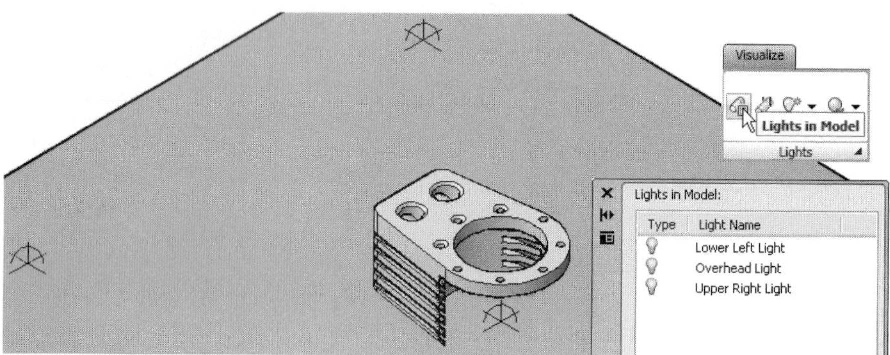

Figure 23.22

Double-click the Lower Left Light to display the Properties palette on this light. Locate the Position Z coordinate, located under the Geometry category of this palette, as shown in the following image, and change the value from 0 to 200. This elevates the Lower Left Light a distance of 200 mm, as shown in the following image.

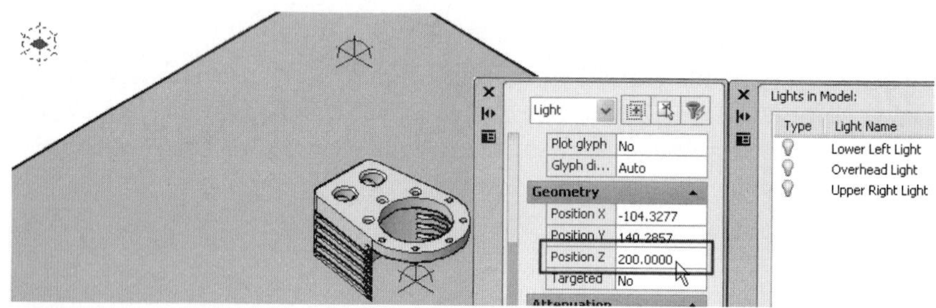

Figure 23.23

Continue changing the elevations of the remaining point lights. Change the Position Z coordinate value of the Overhead Light from 0 mm to 300 mm and the Position Z coordinate value of the Upper Left Light from 0 mm to 200 mm. Your display should appear similar to the following image.

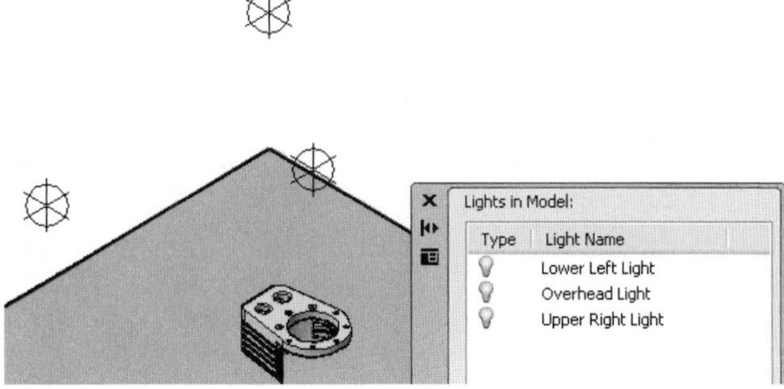

Figure 23.24

Next, place a spotlight into the 3D model by clicking the Spot button, as shown on the right in the following image. Place the source for the spotlight at the approximate location at "A" and the spotlight target at the center of the bottom of the valve head at "B," as shown in the following image.

```
Command: spotlight (Pick from the Light control panel of the
    dashboard)
Specify source location <0,0,0>: (Pick the approximate location
    for the spotlight at "A")
Specify target location <0,0,-10>: (Pick the bottom center of the
    valve head at "B")
Enter an option to change
[Name/Intensity/Status/Hotspot/Falloff/shadoW/Attenuation/Color/
    eXit] <eXit>: I (For Intensity)
Enter intensity (0.00 - max float) <1.0000>: .50
Enter an option to change
[Name/Intensity/Status/Hotspot/Falloff/shadoW/Attenuation/Color/
    eXit] <eXit>: N (For Name)
Enter light name <Spotlight5>: Spotlight
Enter an option to change
[Name/Intensity/Status/Hotspot/Falloff/shadoW/Attenuation/Color/
    eXit] <eXit>: (Press ENTER to create the light)
```

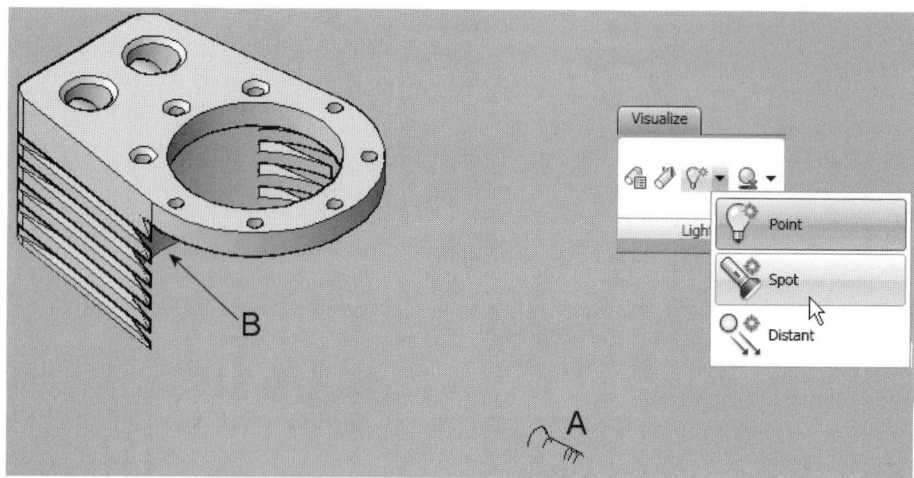

Figure 23.25

As with the point lights, the source of the spotlight is located at an elevation of 0 and needs to be changed to a different height. Double-click the spotlight icon to display the Properties palette, and change the Position Z coordinate value to 200 mm, as shown on the right in the following image.

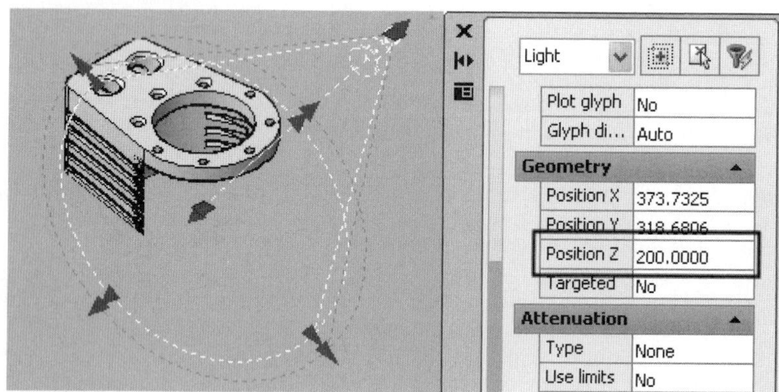

Figure 23.26

Finally, check to see that the render preset value is set to Medium and click the Render button to display the model, as shown in the following image. Shadows are automatically applied to the model from the lights. This concludes the exercise.

Figure 23.27

AN INTRODUCTION TO MATERIALS

Another way to make models more realistic and lifelike is to apply a material to the 3D model. A library of sample materials is available through a number of tabs located in the Tool palette, as shown in the following image. The materials can be dragged and dropped onto the model.

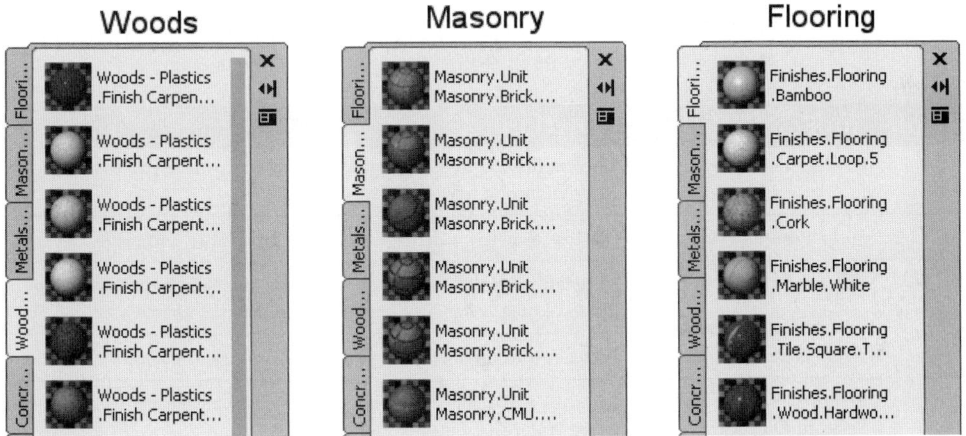

Figure 23.28

Materials can also be created through the Materials palette, as shown on the left in the following image. This palette can be activated from the Ribbon, as shown on the right, or from the View pull-down menu, as shown in the middle.

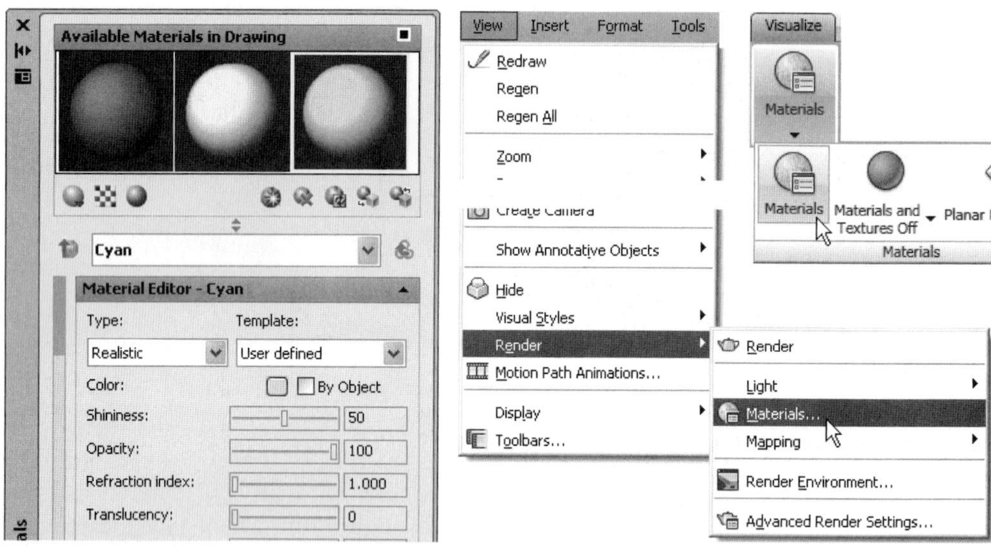

Figure 23.29

Located in the Materials palette are a number of buttons used for creating and applying materials to the 3D model. These buttons and their functions are explained in the following table.

Button	Tool	Function
	Create New Material	Displays the Create New Material dialog box, where you enter a name and description of a new material
	Purge from Drawing	Used to eliminate a material from the database of a drawing
	Indicate Materials in Use	Materials currently used in a drawing are identified by a drawing icon located in the lower-right corner of the material swatch
	Apply Material to Objects	Attaches a material to a single object or group of objects
	Remove Materials from Selected Objects	Removes a material from a selected object(s) but keeps the material loaded in the Materials palette

Before a material can be applied to a model, it must first be loaded. One method of accomplishing this is to drag the material from the Tool palette, as shown on the right in the following image, and drop it anywhere in the drawing. The material then appears in the Materials palette, as shown on the left in the following image.

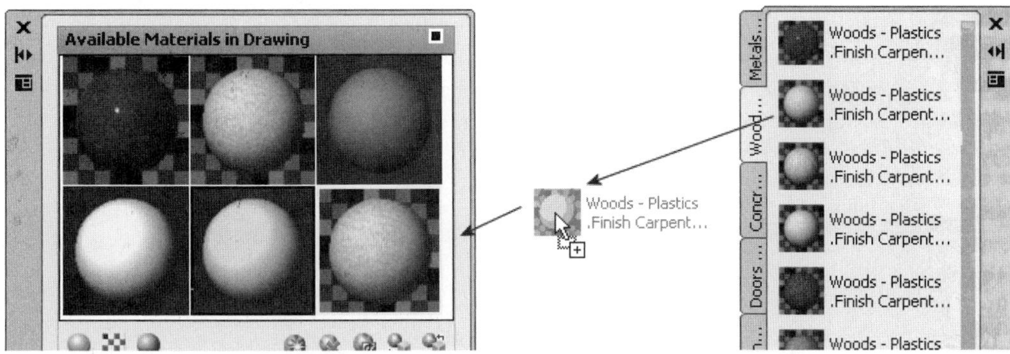

Figure 23.30

The Create New Material button in the Materials palette allows you to create your own custom materials. For creating materials, a number of templates are available, as shown on the left in the following image. These templates already have properties set, depending on the purpose of the material. For instance, for the Mirror template, a number of settings have already been made to ensure that the material reflects correctly. As you create materials, you can further modify these settings, as shown on the right in the following image. Clicking in the Diffuse area allows you to change the color of the material. You can also control such properties as shininess, translucency, and opacity in this palette.

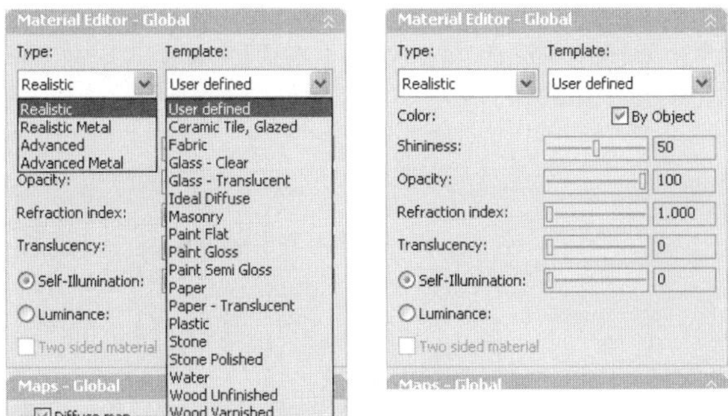

Figure 23.31

WORKING WITH MATERIALS

Once a material or group of materials has been loaded into a 3D model drawing, the next step is to attach these materials to drawing shapes and components. This can be accomplished a number of ways, as outlined in the following Try It! exercise.

Try It!: Open the drawing file 23_Connecting Rod. A single point light source has already been created and placed in this model. A realistic visual style is currently being applied to the model, as shown in the following image. You will attach a material from the Tool palette and observe the rendering results. After removing this material, you will create a new material, change a few settings, and observe these rendering results.

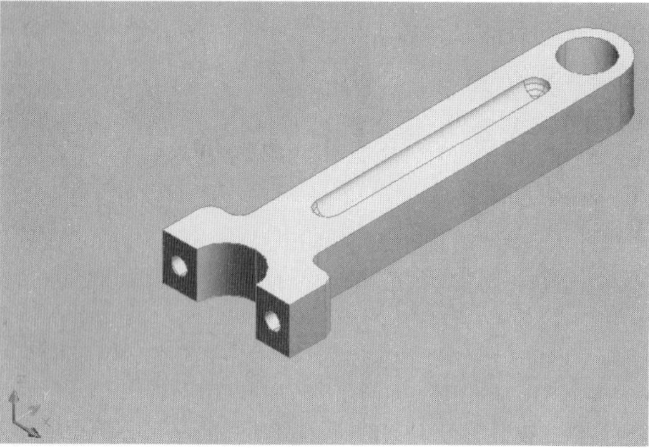

Figure 23.32

First, activate the Tool palette and locate the Metals – Materials Sample tab. Then, locate the Copper material in this tab, as shown on the right in the following image. Click this material once and, at the Select Objects prompt, pick the edge of the 3D model, as shown on the left in the following image.

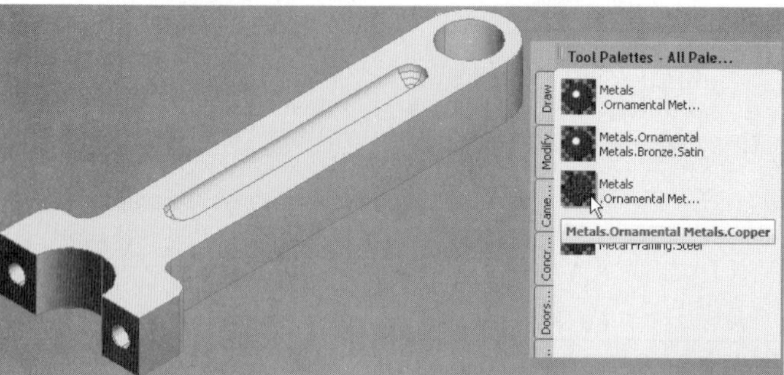

Figure 23.33

Since you are in a realistic visual style, materials and textures are automatically turned on through the Materials panel located under the Visualize tab of the Ribbon, as shown in the following image. The results of applying the copper material to the model are also shown in the following image.

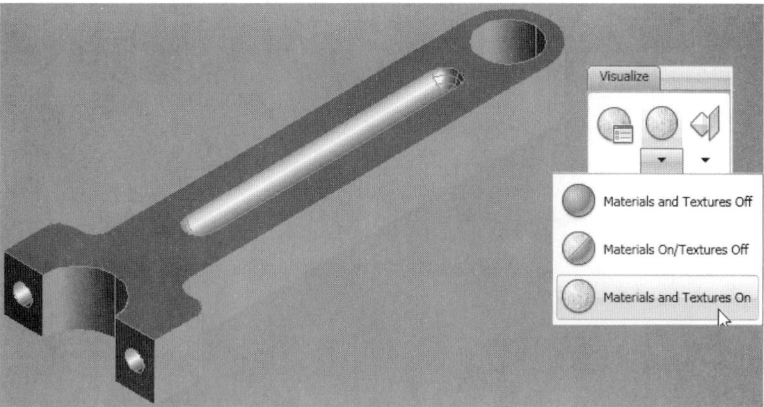

Figure 23.34

Before producing the rendering of this object, activate the Materials Editor by clicking the Materials button located in the Ribbon, as shown on the left in the following image. While inside the Materials Editor, change the Shininess to a value of 50. Click the radio button next to the Self-Illumination value to turn this mode on and change its value to 50, as shown on the right in the following image.

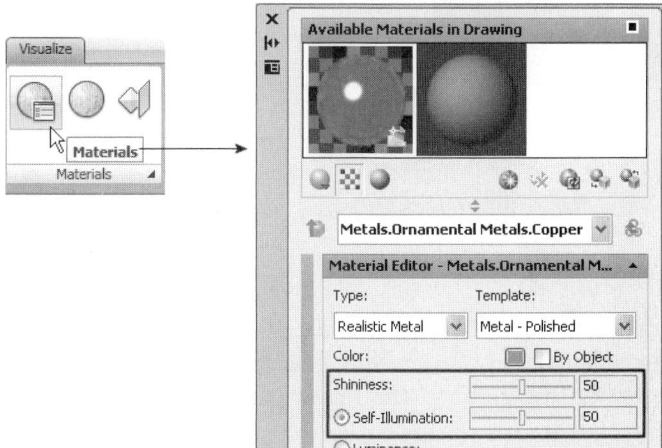

Figure 23.35

To display shadows, perform a rendering with the rendering style set to Medium. The results of performing the rendering are shown in the following image. In addition to the copper material being applied, shadows are cast along a flat surface from the existing light source.

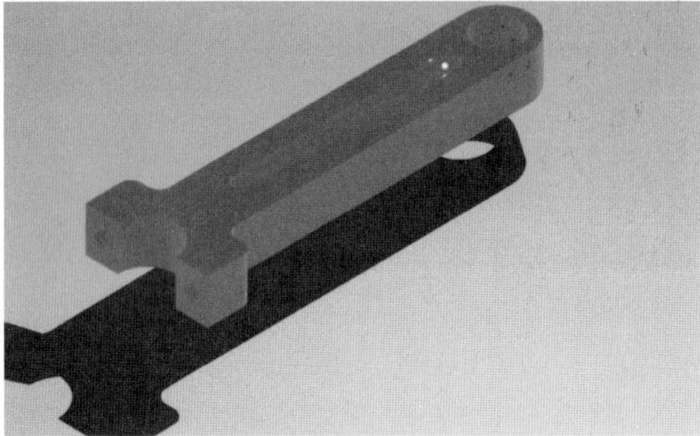

Figure 23.36

Before continuing to the next material, click the Remove Materials from Selected Objects button in the Materials palette and pick the edge of the model to remove this material from the model but not from the Materials palette.

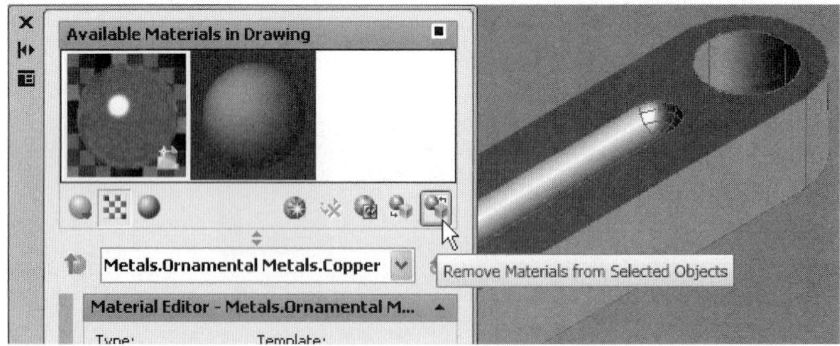

Figure 23.37

To create a new material, click the Create New Material button, as shown on the left in the following image. When the Create New Material dialog box appears, enter the name Red Metal, as shown on the right in the following image. Click OK.

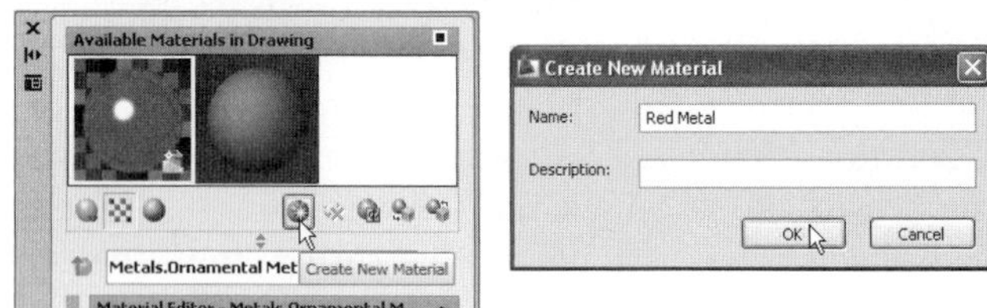

Figure 23.38

Keep the default template set to Realistic and click the Color box, as shown on the left in the following image, to change the color of the material. When the Select Color dialog box appears, change to the Index Color tab and change the color to the one shown on the right in the following image. Click the OK button to accept the material and return to the Materials palette.

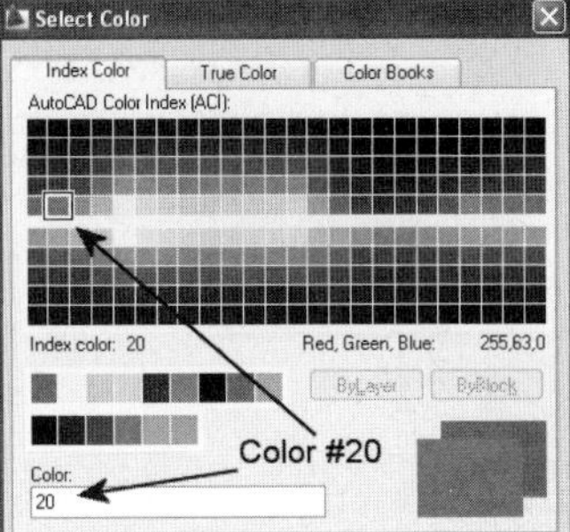

Figure 23.39

When you return to the Materials palette, click the Apply Material to Objects button, as shown on the left in the following image. At the Select objects prompt, select the 3D model shown on the right in the following image.

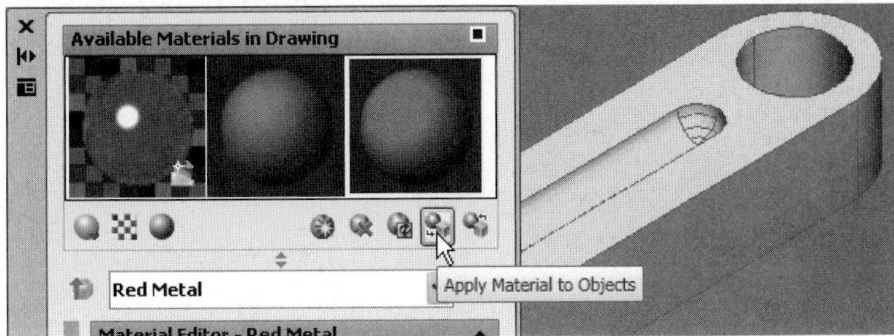

Figure 23.40

Producing a rendering of the 3D model with the new material should result in an image similar to the following image. When finished, exit the rendering mode and return to the 3D model.

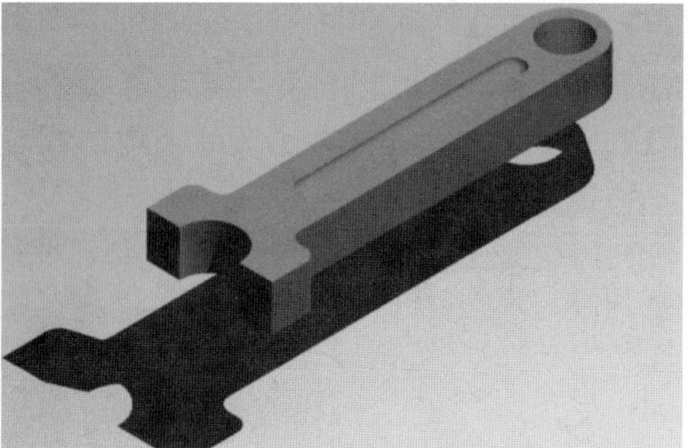

Figure 23.41

While still using the red metal material, expand the Material Editor to expose the Opacity heading, and change the value to 20 by sliding the bar, as shown on the left in the following image. The material will update to reflect this change in opacity, which will change the material to include a degree of transparency.

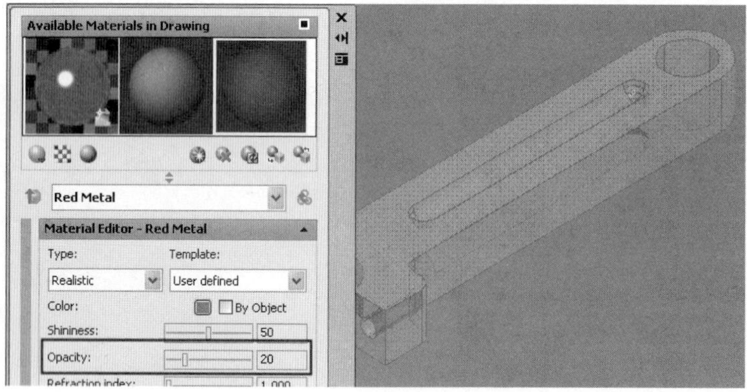

Figure 23.42

Producing a new rendering displays the model, as shown in the following image, complete with transparent material and shadows. This concludes the exercise.

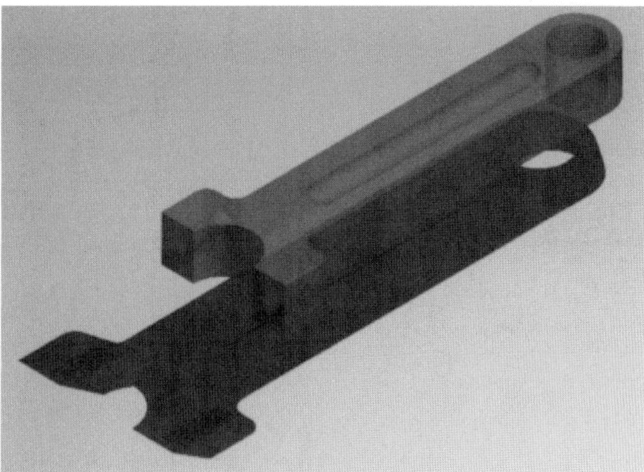

Figure 23.43

USING MATERIAL TEMPLATES

Using a materials template is one way to automate the creation process for new materials. The next Try It! exercise illustrates the use of materials templates.

Try It!: Open the file 23_Piston Mirror, as shown in the following image. A number of lights have already been placed in this model. Also, the model is being viewed through the Realistic visual style. In this exercise, you will create two new materials. One of the materials will contain mirror properties and be applied to the piston. When you perform a render operation, the reflection of one of the piston rings will be visible in the top of the piston.

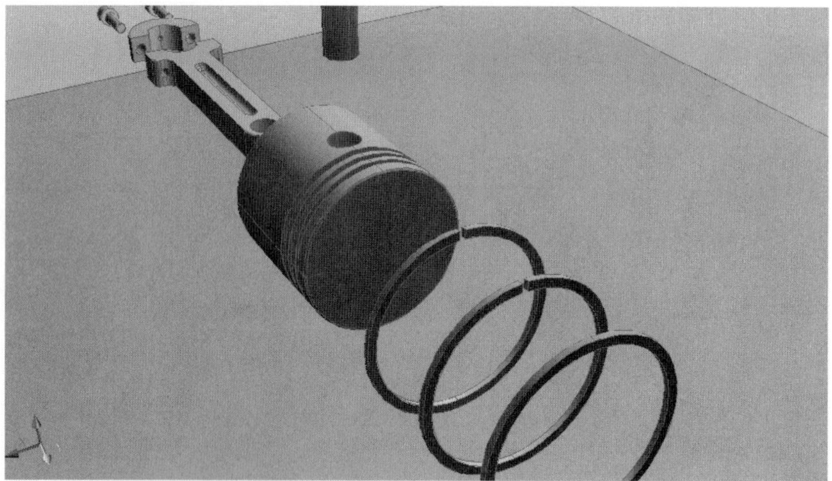

Figure 23.44

Begin by launching the Materials palette and clicking the Create New Material button, as shown on the left in the following image. Enter a new material name called Mirror in the Create New Material dialog box, as shown on the right in the following image. Click OK.

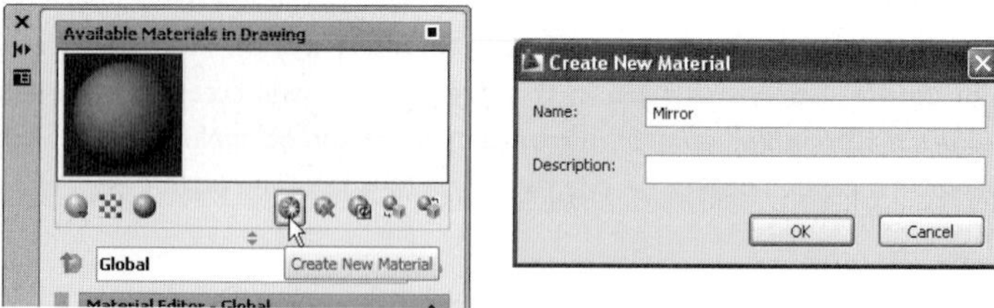

Figure 23.45

Change the Type to Realistic Metal, as shown on the left in the following image. A number of templates are available to assist with the creation of special material types such as a mirror property. Click the down arrow in the Template area and pick the Mirror template from this list, as shown on the left in the following image. A few other items need to be fine-tuned to produce the best mirror property. First, click the Color area; this launches the Color dialog box. Click the Index Color tab and choose one of the shades of yellow. Then, in the Shininess area, adjust the slider bar to read approximately 75%. Adjust the Self-Illumination slider bar to read approximately 50%. The Materials palette should appear similar to the illustration shown on the right in the following image.

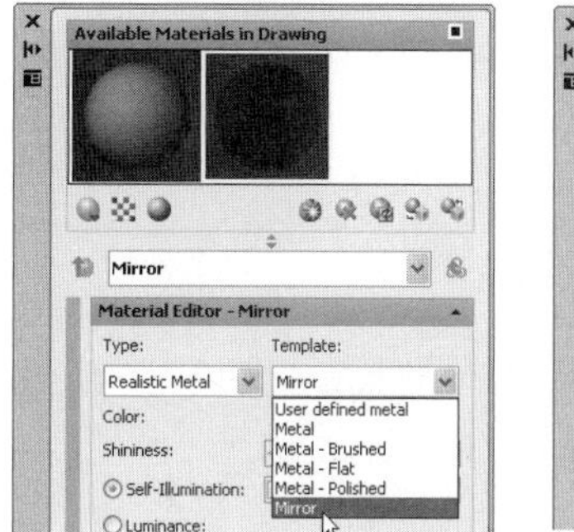

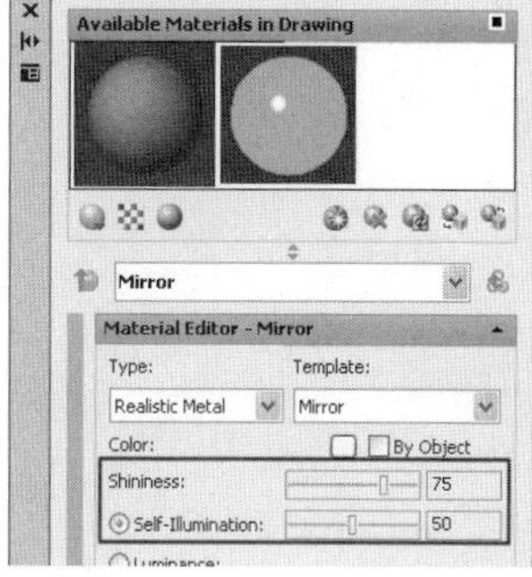

Figure 23.46

Next, click the Apply Material to Objects button and pick the edge of the piston to apply the material to this 3D model, as shown in the following image.

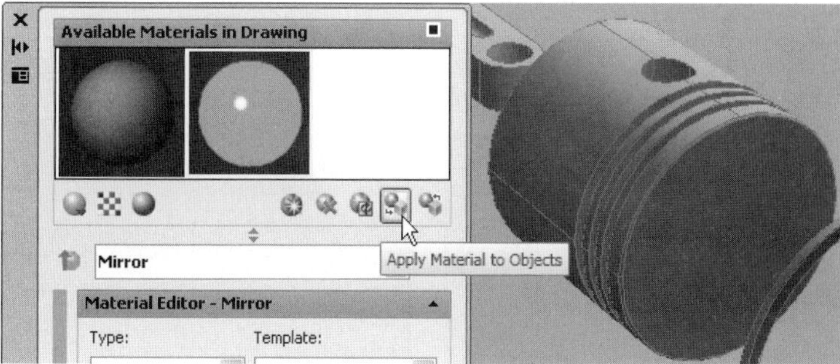

Figure 23.47

Test the mirror property by performing a render with the render style set to Medium. Notice in the following image that the piston ring and shadows are visible in the piston due to the mirror material.

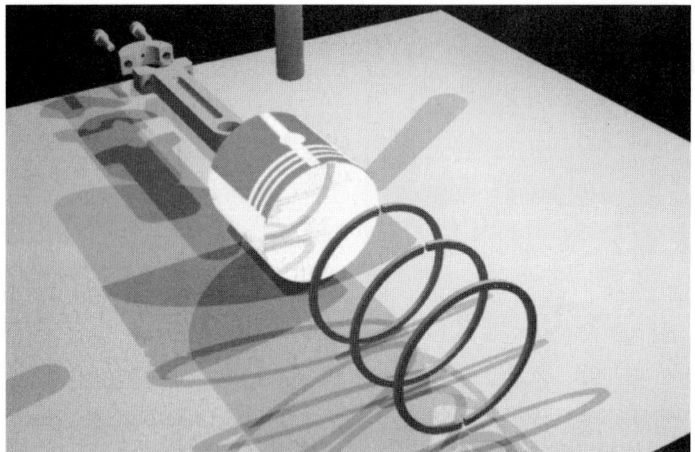

Figure 23.48

Create another material called Piston Support Parts, as shown in the following image. This material will be applied to the remainder of the parts that form the piston assembly.

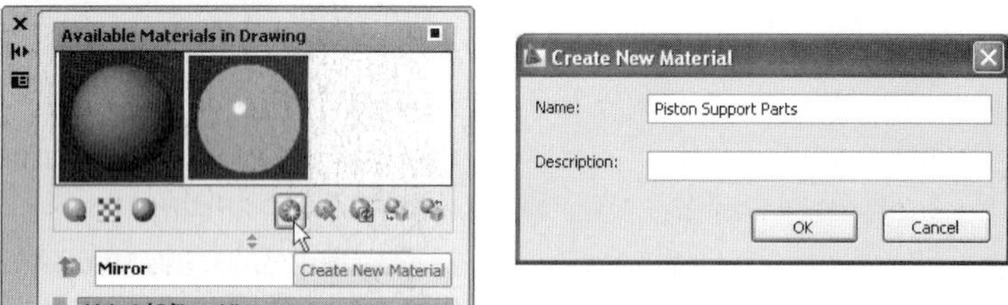

Figure 23.49

Keep the default material Type set to Realistic. Click in the Color area and change the color of this material to a different shade of green. When finished, click the Apply Material to Objects button and pick the remainder of the piston parts, as shown in the following image. (Leave the main piston set to the mirror material.)

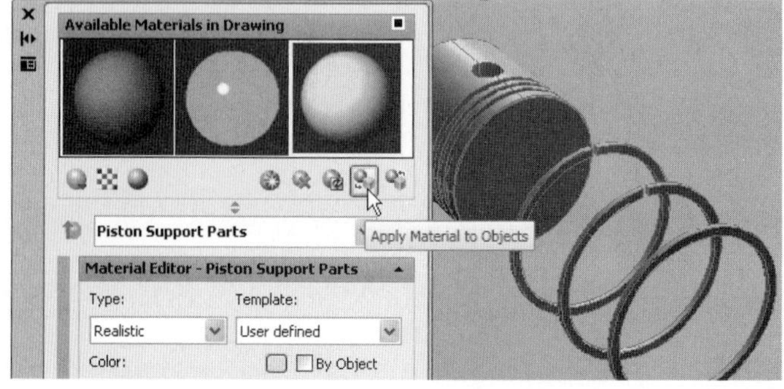

Figure 23.50

Perform another rendering test. Your image should appear similar to the following image. This concludes the exercise.

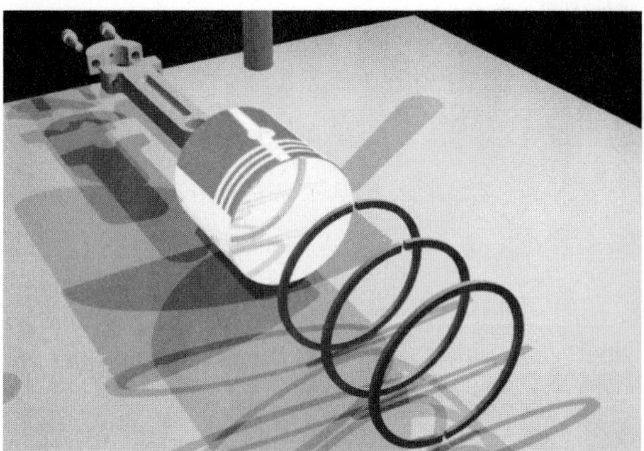

Figure 23.51

ASSIGNING MATERIALS BY LAYER

One way to render detailed reflections is to use ray tracing, which gives the most realistic reflections. We can create our own material with attributes that will take advantage of this type of rendering.

Try It!: Open the file 23_Interior Materials, as shown in the following image. You will drag and drop existing materials from the Tool palette into the drawing. You will then assign these materials to specific layers and perform the rendering. Lights have already been created for this exercise.

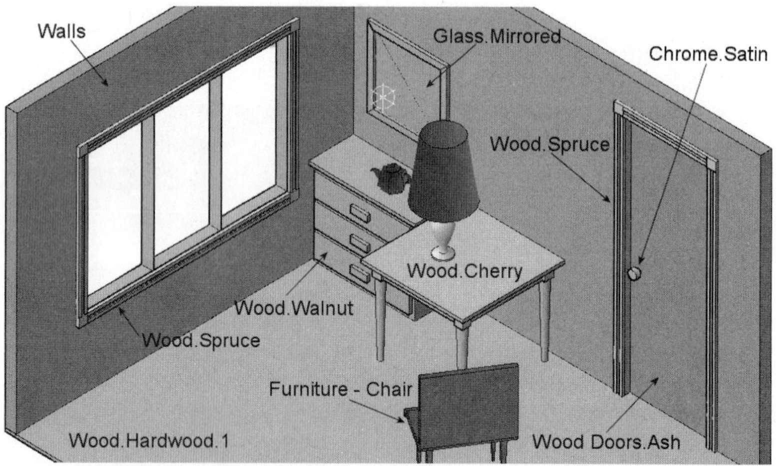

Figure 23.52

Activate the Materials palette and notice that two materials have already been created in this drawing. The first material, shown on the left in the following image, is a fabric color designed to be applied to the chair. The second material, shown on the right in the following image, is a paint color to be applied to the walls of the 3D model.

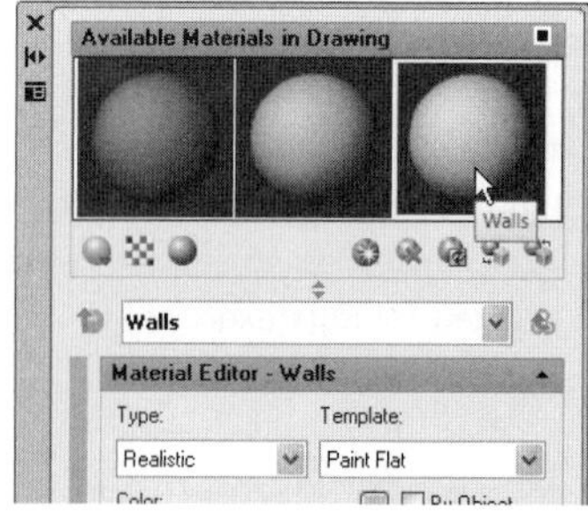

Figure 23.53

All other materials will be obtained from the Tool palette. You will be dragging and dropping a number of materials into a blank area of your drawing. This drag-and-drop action will load the materials into the drawing for your use. You will be selecting materials from three tabs located in this palette. The first set of materials will be loaded from the Doors and Windows Materials Library, as shown on the left in the following image. You will have to move your cursor over a material and leave it stationary in order for the whole material name to be displayed. Locate the first material, Doors – Windows.Door Hardware.Chrome.Satin, as shown on the right in the following image. Press and hold down your mouse button over this material, drag the material icon into your drawing, and drop it to load it.

Figure 23.54

Continue loading the following materials located in the appropriate tabs using the table below as a guide:

Materials Sample Tool Palette Tab	Material
Doors and Windows	Doors – Windows.Door Hardware.Chrome.Satin
Doors and Windows	Doors – Windows.Glazing.Glass.Mirrored
Doors and Windows	Doors – Windows.Wood Doors.Ash
Flooring	Finishes.Flooring.Wood.Hardwood.1
Woods and Plastics	Wood – Plastics.Finish Carpentry.Wood Cherry
Woods and Plastics	Wood – Plastics.Finish Carpentry.Wood Walnut
Woods and Plastics	Wood – Plastics.Finish Carpentry.Wood Spruce

When you have finished loading all materials, you can check the status of the load by clicking the Materials button, located in the Ribbon, as shown on the left in the following image. This launches the Materials palette and displays all materials that can be applied to 3D models in the drawing, as shown on the right in the following image.

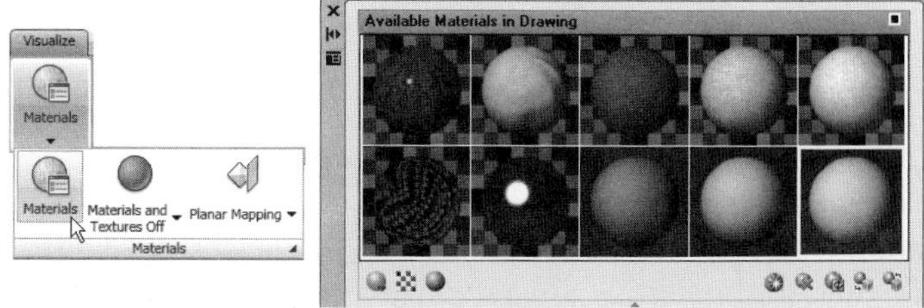

Figure 23.55

The next step is to assign a material to a layer. From the Ribbon, click the Attach By Layer button, as shown on the left in the following image. This launches the Material Attachment Options dialog box, as shown on the right in the following image. Now you will drag a material located in the left column of the dialog box and drop it onto a layer located in the right column of the dialog box. In this example, the Chrome.Satin material has been dropped onto the Door Knob layer.

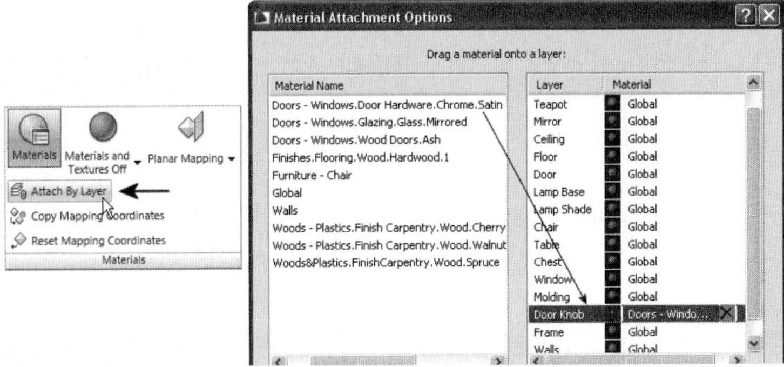

Figure 23.56

Using the following image as a guide, continue assigning materials to layers by dragging and dropping the materials onto the appropriate layers.

Layer	Material	
0	Global	
Teapot	Global	
Mirror	Doors - Windows.Glazing.Glass.Mirrored	X
Ceiling	Global	
Floor	Finishes.Flooring.Wood.Hardwood.1	X
Door	Doors - Windows.Wood Doors.Ash	X
Lamp Base	Global	
Lamp Shade	Global	
Chair	Furniture - Chair	X
Table	Woods - Plastics.Finish Carpentry.Wood.Cherry	X
Chest	Woods - Plastics.Finish Carpentry.Wood.Walnut	X
Window	Global	
Molding	Woods&Plastics.FinishCarpentry.Wood.Spruce	X
Door Knob	Doors - Windows.Door Hardware.Chrome.Satin	X
Frame	Global	
Walls	Walls	X

Figure 23.57

When all material assignments have been made to the layers, click OK. Render out the design and verify that the materials are properly assigned to the correct 3D objects, as shown in the following image.

Figure 23.58

APPLYING A BACKGROUND

Images can be placed behind a 3D model for the purpose of creating a background effect. Backgrounds can further enhance a rendering. For example, if you have designed a 3D house, you could place a landscape image behind the rendering. To place a background image, the image must be in a raster format such as BMP, TGA, or TIF. The process of assigning a background image begins with the View Manager dialog box, as shown in the following image. You create a new view and associate a background with the view. In the following image, the New button is picked, which launches the New View dialog box. Enter a new name for the view and place a check in the box next to Override default background. This launches the Background dialog box, where you can pick a background from three different types.

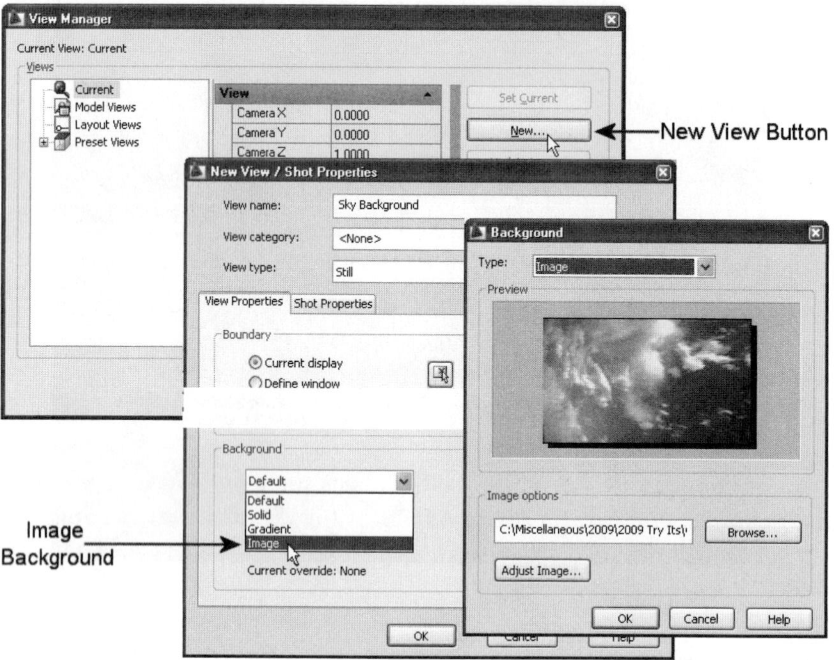

Figure 23.59

These background types are explained as follows.

Solid—A solid background means that AutoCAD replaces the default white (or black) background of the drawing screen with another color. You choose the color from the Colors section of the dialog box.

Gradient—A gradient means that the color changes from one end of the screen to the other, such as from red at the bottom to light blue at the top (to simulate a sunset). AutoCAD gives you three ways to control a linear gradient—look carefully: they are tucked into the lower-right corner. Horizon specifies where the lower color ends; a value closer to 0 moves the lower color lower down. When Height is set to 0, you get a two-color gradient; any other value gives you a three-color gradient. Rotation rotates the gradient.

Image—You select a raster image for the background. The image can be in BMP (Bitmap), GIF, PNG, TGA (Targa), TIF (Tagged Image File Format), JPG (JPEG File Interchange Format), or PCX (PC Paintbrush) format.

Try It!: Open the drawing file 23_Piston Background. The Realistic visual style is applied to this model, as shown in the following image. Also applied are four point lights. A special mirror material is attached to the sphere.

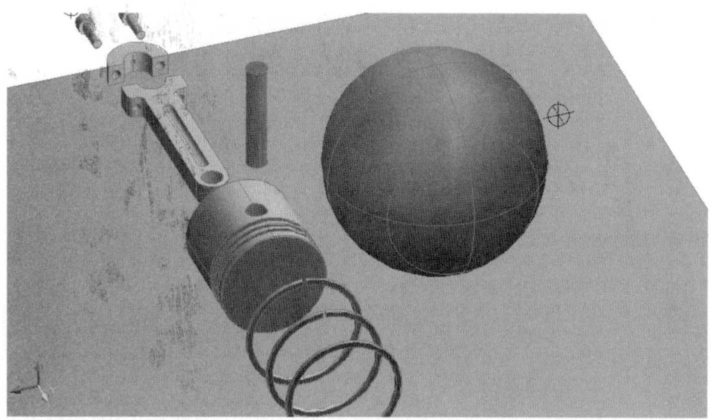

Figure 23.60

Begin by performing a render using the Medium rendering style. Your image should appear similar to the following image, in which the reflections of various piston parts appear in the sphere due to its mirror property. All that is missing is a background that will be applied to the 3D models.

Figure 23.61

All visual style and rendering backgrounds are controlled in the View Manager dialog box, as shown in the following image. Begin the selection of a background by clicking the New button.

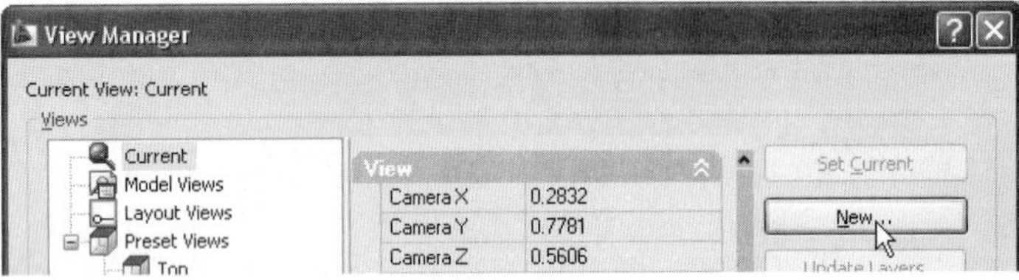

Figure 23.62

You must first create a new view and then assign a background to this view. When the New View/Shot Properties dialog box appears, as shown on the left in the following image, enter a name for the view, such as Sky Background. In the background area, click in the default field and pick Image, as shown on the left in the following image. This launches the Background dialog box shown on the right in the following image. Click the Browse button to search for valid image files.

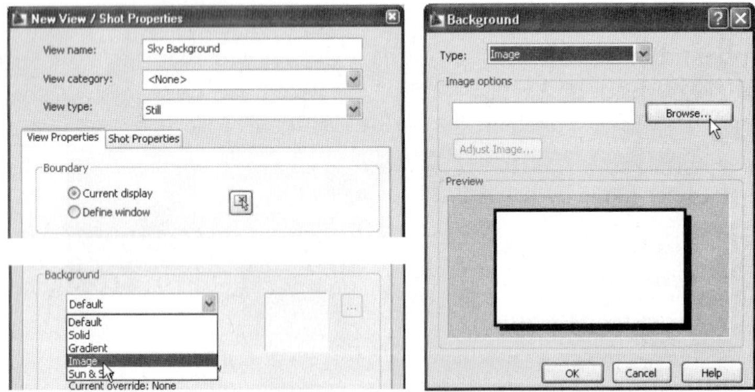

Figure 23.63

Once you click the Browse button in the Background dialog box, the Select File dialog box appears, as shown in the following image. Locate the folder in which the chapter 23 Try It! exercises are located. From this list, find sky.tga, select it, and click the Open button.

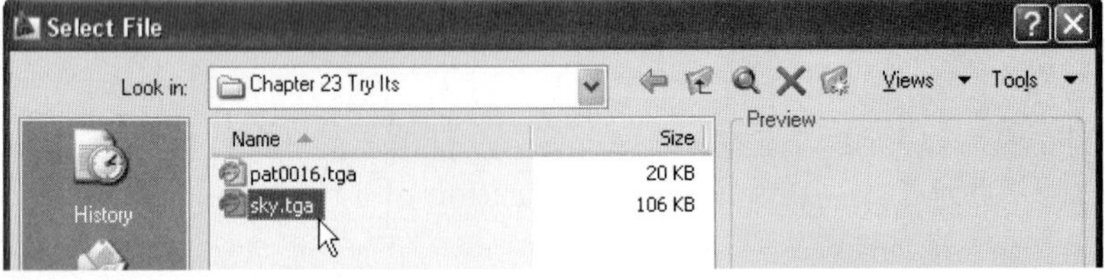

Figure 23.64

This takes you to the Background dialog box again. The sky graphic appears small but centered on the sheet, as shown on the left in the following image. To control the display of this file in the final rendering, click the Adjust Image button to launch the Adjust

Background Image dialog box and change the Image position to Stretch. This should make the sky graphic fill the entire screen, as shown on the right in the following image.

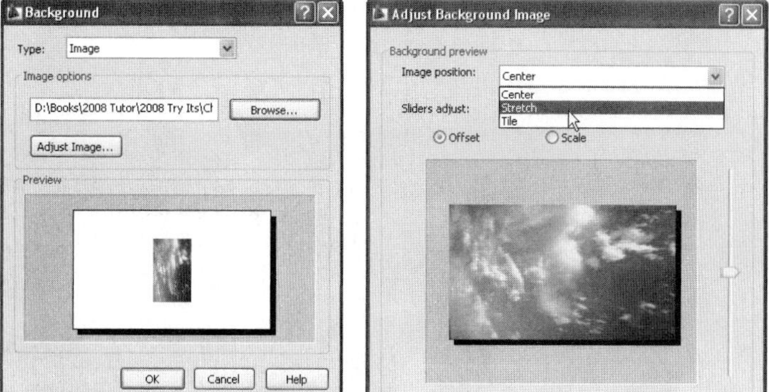

Figure 23.65

Click the OK buttons in the Adjust Background Image and Background dialog boxes to return to the View Manager dialog box, where the Sky Background is now part of the Views list. Click the Set Current button to make this view current in the drawing and click the OK button to dismiss this dialog box.

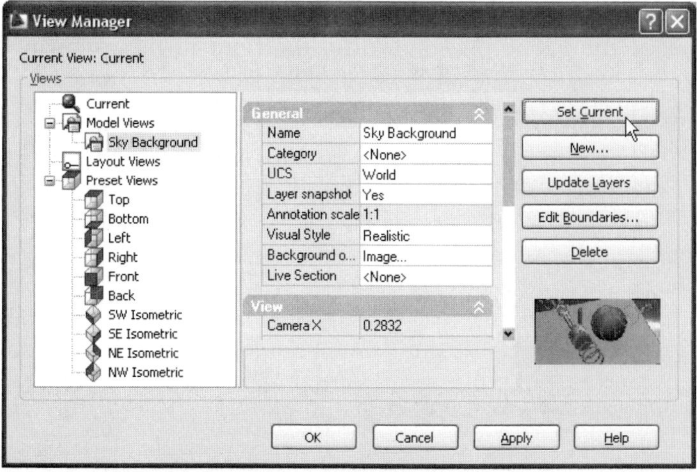

Figure 23.66

Perform a render using the Medium rendering style and notice the results, as shown in the following image. With the image applied, it appears that the flat base sheet is floating in air. Also, since the sphere still has the mirror material property, the sky is reflected here in addition to the piston parts. This concludes the exercise.

Figure 23.67

WALKING AND FLYING THROUGH A MODEL

To further aid with visualization of a 3D model, walking and flying actions can be simulated through the 3DWALK and 3DFLY commands. Both can be selected from the View pull-down menu, as shown on the left in the following image. When walking through a 3D model, you travel along the XY plane. When flying through the model, you move the cursor to look over the top of the model.

Additional controls for walking and flying through a model can be found under the Animations panel of the Ribbon, as shown on the right in the following image.

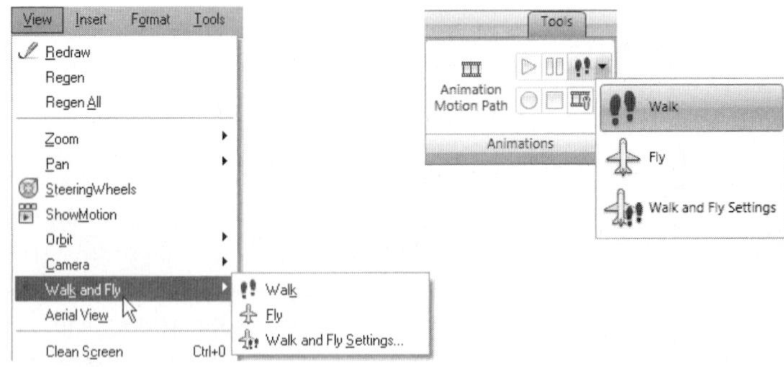

Figure 23.68

When you first activate the 3DWALK command, a warning dialog box appears stating that you must be in Perspective mode to walk or fly through your model. Click the Change button to enter Perspective mode.

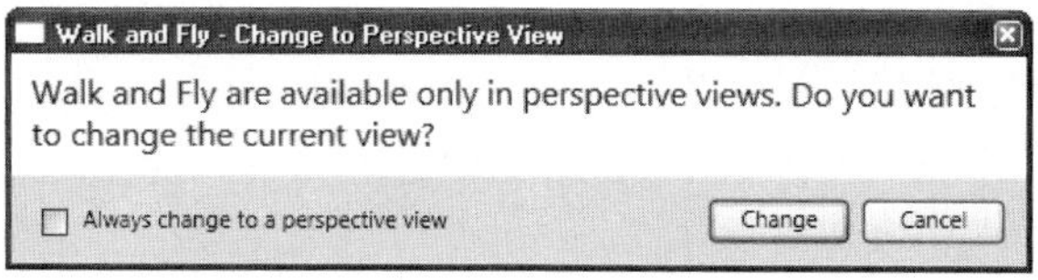

Walk and Fly - Change to Perspective View

Walk and Fly are available only in perspective views. Do you want to change the current view?

Always change to a perspective view

Change Cancel

Figure 23.69

When you first enter the walk or fly mode, a Position Locator palette appears, as shown on the left in the following image. It gives you an overall view of the position of the camera and target in relation to the 3D model. You can drag on the camera location inside the preview pane of the Position Locator to change its position. You can also change the target as you adjust the viewing points of the 3D model. Right-clicking displays the shortcut menu, as shown in the following image. Use this menu for changing to various modes that assist in the rotating of the model.

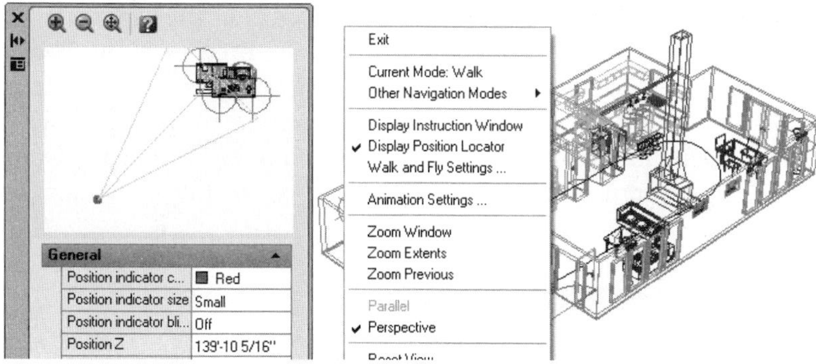

Figure 23.70

ANIMATING THE PATH OF A CAMERA

The ability to walk or fly through a model has just been discussed. This last segment will concentrate on creating a motion path animation by which a camera can follow a predefined polyline path to view the contents of a 3D model. Clicking Motion Path Animations, found under the View pull-down menu shown on the left in the following image, launches the Motion Path Animation dialog box shown on the right in the following image. You select the path for the camera and target in addition to changing the number of frames per second and the number of frames that will make up the animation.

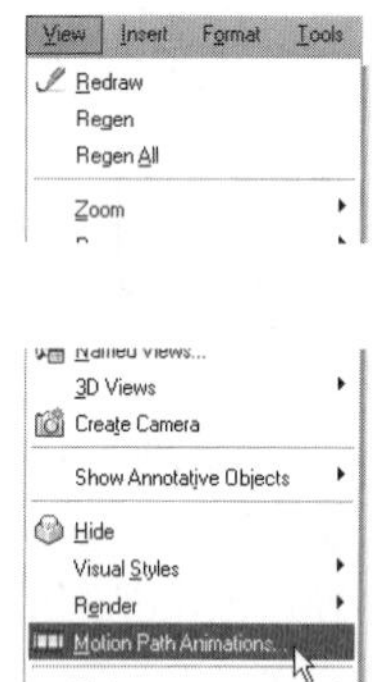

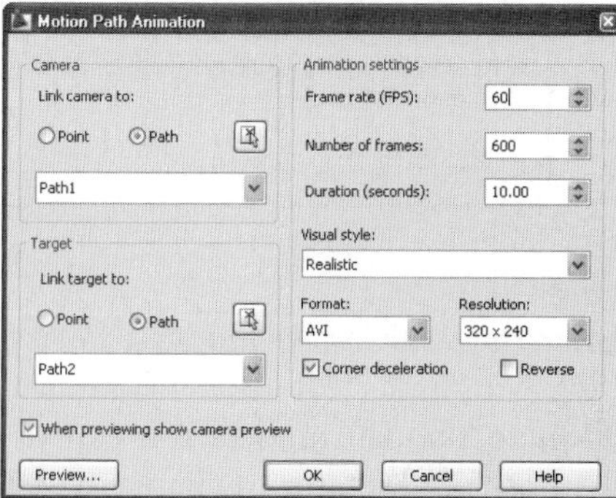

Figure 23.71

After making changes to the Motion Path Animation dialog box, you have the opportunity to preview the animation before actually creating it. A sample animation preview is shown in the following image. You can even see the relative position the camera is in as it passes through the 3D model along the polyline path. After the preview is finished, clicking the OK button in the main Motion Path Animation dialog box creates the animation and writes the results out to a dedicated file format. Supported formats include AVI, MOV, MPG, and WMV. Depending on the resolution and number of frames, this process could take a long time. However, it gives you the capability of creating an animation using any kind of 3D model.

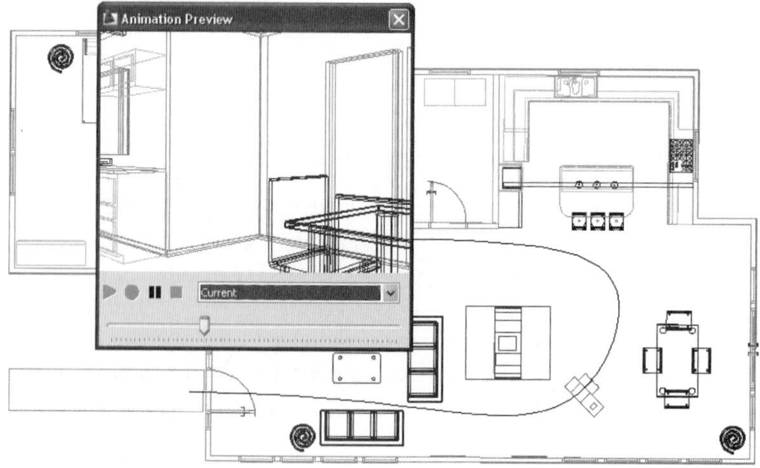

Figure 23.72

Try It!: Open the drawing file 3D_House Motion. A polyline path has already been created at an elevation of 4' to simulate an individual walking through this house. Once you have created the motion path animation, this polyline path will not be visible when the animation is played back.

Begin by clicking Motion Path Animations, which is found under the View pull-down menu. When the Motion Path Animation dialog box appears, as shown in the following image, make the following changes:

- Click the Select Path button in the Camera area and pick the polyline displayed in the floor plan. If necessary, change the path name to Path1. If an AutoCAD Alert box appears, click the Yes button to override the existing path name.
- Click the Select Path button in the Target area and pick the polyline displayed in the floor plan. If necessary, change the path name to Path2. If an AutoCAD Alert box appears, click the Yes button to override the existing path name. Both the Camera and Target will share the same polyline path.
- Change the Frame rate (frames per second) from 30 to 60.
- Change the Number of frames from 30 to 600. This updates the Duration from 1 to 10 seconds.
- Change the Visual style to Realistic.
- Change the Format to AVI.
- Keep the resolution set to 320 × 240.

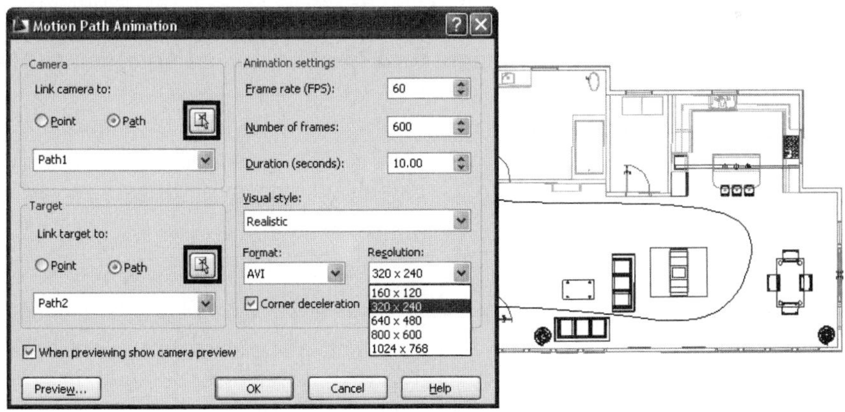

Figure 23.73

When you have finished making these changes, click the Preview button to preview the results of the motion animation. Play the animation preview as many times as you like. Once you close the preview, you will return to the Motion Path Animation dialog box. Click the OK button and enter the name of the AVI file as House Motion Study. Clicking the OK button in this dialog box begins the processing of the individual frames that will make up the animation. The total processing time to produce the animation should be between 5 and 10 minutes.

When finished, launch one of the many Windows Media Player applications and play the AVI file. This concludes the exercise.

TUTORIAL EXERCISE: SUNLIGHT STUDY

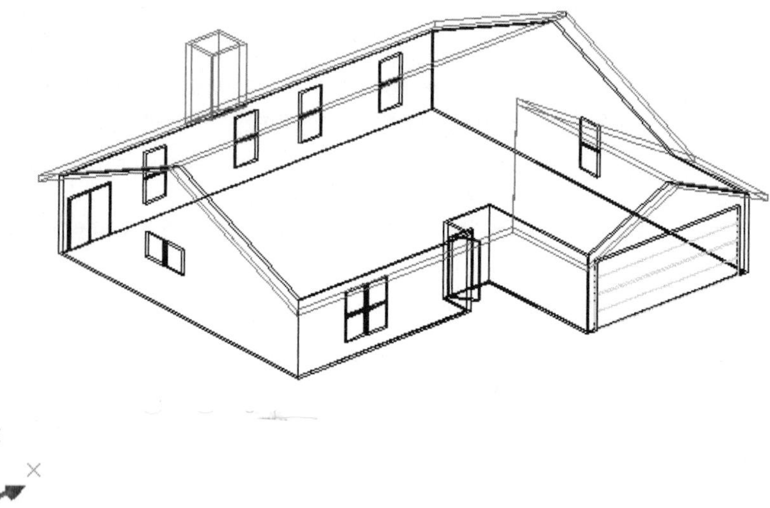

Figure 23.74

PURPOSE

This tutorial is designed to simulate sunlight on a specific day and time, and to observe the shadows that are cast by the house.

SYSTEM SETTINGS

Since this drawing is provided on the CD-ROM, all system settings have already been made.

LAYERS

The creation of layers is not necessary.

SUGGESTED COMMANDS

Begin this tutorial by opening up the drawing 23_House Plan Rendering, as shown in the previous image. You will be performing a study based on the current location of the house and the position of the sun on a certain date, time, and geographic location. Shadow casting will be utilized to create a more realistic study.

Step 1

Make the 3D Modeling workspace current. With the Ribbon active, locate the Visualize tab and click the Sun Status button, as shown on the left in the following image, to turn on the sun. A dialog box appears, informing you that you cannot display sunlight if the default lighting mode is turned on and asking you whether you want to turn off default lighting. Click the area as shown on the right in the following image to accept the recommended setting to turn off the default lighting.

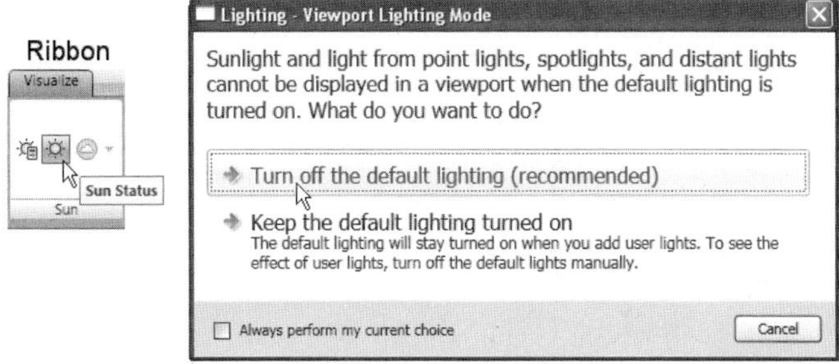

Figure 23.75

Step 2

Next, click the Edit the Sun button, located in the Ribbon shown on the left in the following image, to display the Sun Properties palette shown on the right in the following image. In this palette, you can change various properties that deal with the sun, such as shadows, date, time, azimuth, altitude, and source vector of the sun.

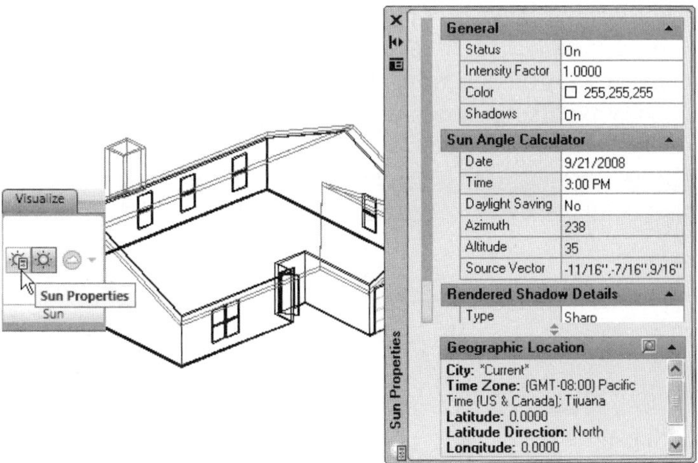

Figure 23.76

Step 3

Click the launch Geographic Location button, as shown on the left in the following image, to change the location of the 3D model. When the Define Geographic Location dialog box appears, as shown on the right in the following image, click on the area to enter the location values.

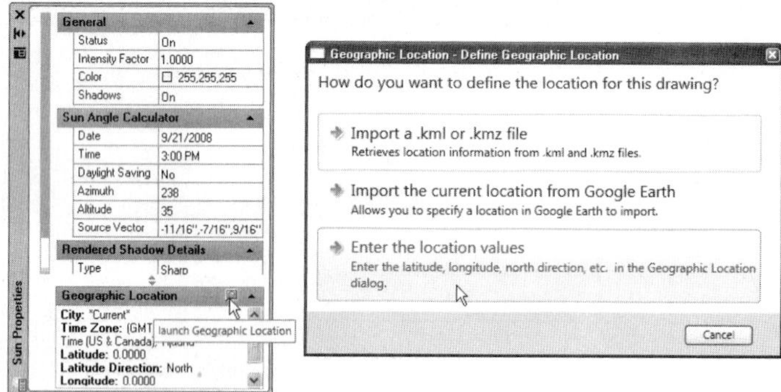

Figure 23.77

Step 4

When the Geographic Location dialog box appears, click on the Use Map button, as shown on the left in the following image. This will launch the Location Picker dialog box as shown on the right in the following image. Click the coast of South Carolina and check to see that Charleston, SC, appears. You could also use the Nearest City drop-down list to select the desired location. Other maps from throughout the world also are available. When you are satisfied with the location, click the OK button to leave the Location Picker and Geographic Location dialog boxes.

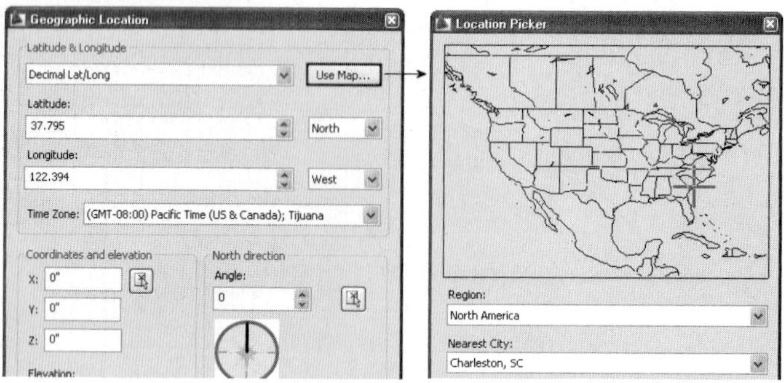

Figure 23.78

Step 5

A dialog box appears, informing you that the time zone has been automatically updated with the change in the geographic location. Click the area of this dialog box shown in the following image to accept updated time zone.

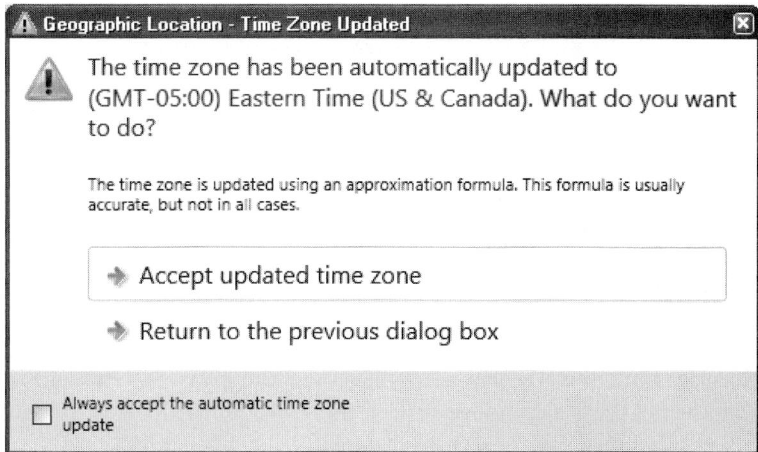

Figure 23.79

Step 6

Next you will change the date and time to perform a sun study when the sun is positioned on a fall day. Click the Date area in the Sun Properties palette, click the three dots (ellipses), and change the date to October 18, 2008, as shown in the following image.

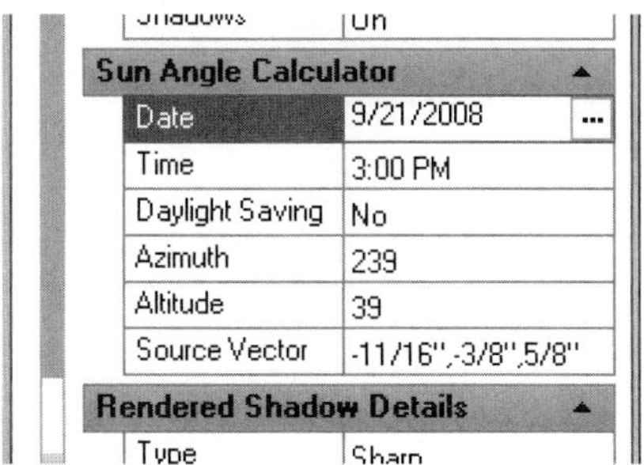

Figure 23.80

Step 7

Then, click in the Time area. When an arrow appears, click it to display a number of times of day in 15-minute increments, and click on 11:00 AM, as shown on the left in the following image. When you perform this change, the date and time information is also updated in the Ribbon, as shown on the right in the following image.

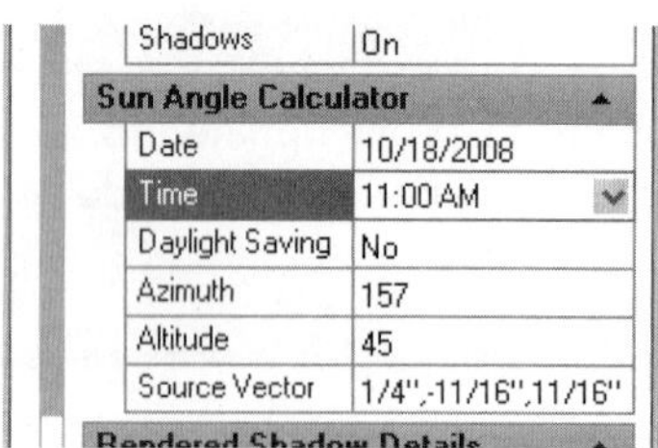

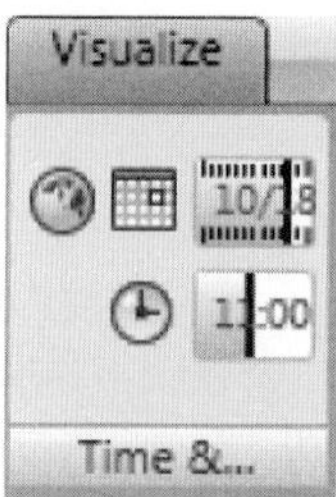

Figure 23.81

Step 8

Clicking the Render button in the Ribbon will render the house, as shown in the following image. The shadows cast by the house reflect the time of 11:00 AM in mid-October in Charleston, South Carolina, in the United States.

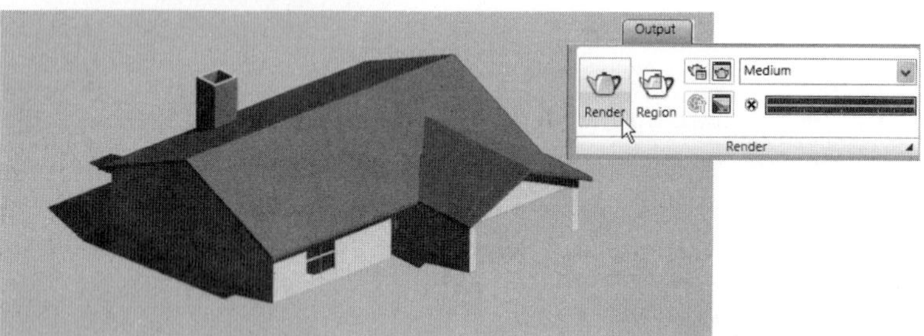

Figure 23.82

INDEX

SYMBOLS

A

B

D

E

F

G

H

I

J

L

M

P

Q

R

S

T

U

V

W

X

Z